NATIONAL ACADEMIES Sciences Engineering Medicine

NATIONAL ACADEMIES PRESS
Washington, DC

Oil in the Sea IV

Inputs, Fates, and Effects

Committee on Oil in the Sea IV

Ocean Studies Board

Division on Earth and Life Studies

Consensus Study Report

NATIONAL ACADEMIES PRESS 500 Fifth Street, NW Washington, DC 20001

This study was supported by the American Petroleum Institute under Award Number 2018-112630, the Bureau of Ocean Energy Management under Award Number 10004961, the Bureau of Safety and Environmental Enforcement under Award Number 140E0119P0018, Fisheries and Oceans Canada under Award Number 10005009, the Gulf of Mexico Research Initiative under Award Number SA-232014, and the National Academies of Sciences, Engineering, and Medicine's Presidents' Circle Fund. Any opinions, findings, conclusions, or recommendations expressed in this publication do not necessarily reflect the views of any organization or agency that provided support for the project.

International Standard Book Number-13: 978-0-309-27429-6
International Standard Book Number-10: 0-309-27429-X
Digital Object Identifier: https://doi.org/10.17226/26410
Library of Congress Catalog Number: 2022943781

This publication is available from the National Academies Press, 500 Fifth Street, NW, Keck 360, Washington, DC 20001; (800) 624-6242 or (202) 334-3313; http://www.nap.edu.

Printed in the United States of America.

Suggested citation: National Academies of Sciences, Engineering, and Medicine. 2022. *Oil in the Sea IV: Inputs, Fates, and Effects*. Washington, DC: The National Academies Press. https://doi.org/10.17226/26410.

The **National Academy of Sciences** was established in 1863 by an Act of Congress, signed by President Lincoln, as a private, nongovernmental institution to advise the nation on issues related to science and technology. Members are elected by their peers for outstanding contributions to research. Dr. Marcia McNutt is president.

The **National Academy of Engineering** was established in 1964 under the charter of the National Academy of Sciences to bring the practices of engineering to advising the nation. Members are elected by their peers for extraordinary contributions to engineering. Dr. John L. Anderson is president.

The **National Academy of Medicine** (formerly the Institute of Medicine) was established in 1970 under the charter of the National Academy of Sciences to advise the nation on medical and health issues. Members are elected by their peers for distinguished contributions to medicine and health. Dr. Victor J. Dzau is president.

The three Academies work together as the **National Academies of Sciences, Engineering, and Medicine** to provide independent, objective analysis and advice to the nation and conduct other activities to solve complex problems and inform public policy decisions. The National Academies also encourage education and research, recognize outstanding contributions to knowledge, and increase public understanding in matters of science, engineering, and medicine.

Learn more about the National Academies of Sciences, Engineering, and Medicine at **www.nationalacademies.org**.

Consensus Study Reports published by the National Academies of Sciences, Engineering, and Medicine document the evidence-based consensus on the study's statement of task by an authoring committee of experts. Reports typically include findings, conclusions, and recommendations based on information gathered by the committee and the committee's deliberations. Each report has been subjected to a rigorous and independent peer-review process and it represents the position of the National Academies on the statement of task.

Proceedings published by the National Academies of Sciences, Engineering, and Medicine chronicle the presentations and discussions at a workshop, symposium, or other event convened by the National Academies. The statements and opinions contained in proceedings are those of the participants and are not endorsed by other participants, the planning committee, or the National Academies.

Rapid Expert Consultations published by the National Academies of Sciences, Engineering, and Medicine are authored by subject-matter experts on narrowly focused topics that can be supported by a body of evidence. The discussions contained in rapid expert consultations are considered those of the authors and do not contain policy recommendations. Rapid expert consultations are reviewed by the institution before release.

For information about other products and activities of the National Academies, please visit **www.nationalacademies.org/about/whatwedo**.

COMMITTEE ON OIL IN THE SEA IV

KIRSI K. TIKKA (NAE) (*Chair*), Maritime Advisor and Board Member at Maritime Companies, London, England, American Bureau of Shipping (ret.)
EDWIN "ED" LEVINE (*Vice Chair*), Scientific Support & Coordination, LLC, Neptune, New Jersey, NOAA (ret.)
AKUA ASA-AWUKU, Department of Chemical and Biomolecular Engineering, University of Maryland, College Park
CYNTHIA BEEGLE-KRAUSE, SINTEF Trondheim, Norway (through October 2021)
VICTORIA BROJE, Shell Projects and Technology, Houston, Texas
STEVEN BUSCHANG, Texas General Land Office, Austin, Texas (through August 2021)
DAGMAR SCHMIDT ETKIN, Environmental Research Consulting, Cortland Manor, New York
JOHN FARRINGTON, Woods Hole Oceanographic Institution, Massachusetts (ret.)
JULIA FOGHT, Department of Biological Sciences, University of Alberta, Canada (ret.)
BERNARD GOLDSTEIN (NAM), Graduate School of Public Health, University of Pittsburgh, Pennsylvania (ret.)
CARYS L. MITCHELMORE, University of Maryland Center for Environmental Science, Chesapeake Biological Laboratory, Solomons
NANCY RABALAIS, Louisiana State University, College of the Coast and Environment, Baton Rouge
JEFFREY SHORT, JWS Consulting, Juneau, Alaska
SCOTT SOCOLOFSKY, Zachry Department of Civil and Environmental Engineering, Texas A&M University, College Station
BERRIN TANSEL, Department of Civil and Environmental Engineering, Florida International University, Miami
HELEN K. WHITE, Department of Chemistry, Haverford College, Pennsylvania
MICHAEL ZICCARDI, One Health Institute, University of California, Davis

Staff

KELLY OSKVIG, Study Director
MEGAN MAY, Associate Program Officer (through June 2021)
KENZA SIDI-ALI-CHERIF, Program Assistant (through December 2021)
GRACE CALLAHAN, Program Assistant

OCEAN STUDIES BOARD

Preface

Through the request and sponsorship of many agencies and industry, the National Academies conducted consensus studies leading to a series of published reports in 1975, 1985, and 2003 on inputs (including anthropogenic sources and natural seeps), fates, and effects of petroleum-based hydrocarbon mixtures in the sea. The reports have assessed the scope of the challenge and made recommendations for improvement in oil spill science, prevention, and mitigation of the impact of harmful discharges on the environment, and for reduction of inputs from operational and accidental discharges.

Almost 20 years have passed since the publication of *Oil in the Sea III: Inputs, Fates, and Effects* in 2003, and it has been more than a decade now since the *Deepwater Horizon* oil spill. Over this time, significant advances have been made in scientific methods to study detection, fates, and effects of oil in marine environments. In the fall of 2020, a committee of 17 members was convened for a consensus study titled Oil in the Sea IV: Inputs, Fates, and Effects, sponsored by the American Petroleum Institute, the Bureau of Ocean Energy Management, the Bureau of Safety and Environmental Enforcement, Fisheries and Oceans Canada, the Gulf of Mexico Research Initiative, and the National Academies of Sciences, Engineering and Medicine's Presidents' Circle Fund.

The committee was tasked to update the 2003 report with an emphasis on North American waters. Whereas earlier National Academies reports focused more on quantifying hydrocarbon inputs in the sea, the current report has extensive sections on fates and effects that incorporate increased knowledge in these fields, in addition to new estimates on the input inventories. The committee has also included the latest understanding on the effects of oil in the marine environment on human health, which was not covered by the previous report. Additionally, the report covers accidental spill mitigation, focusing on both prevention and response as it relates to reduction of inputs and minimizing effects. Across all topics in this report, an increased focus was also paid to the Arctic marine environment in this report.

The committee held several virtual public meetings where experts from federal and local governments, industry, indigenous communities, not-for-profit organizations, and academia were invited to talk to the committee on the current state and future needs of oil spill science and the impact of oil in the sea on coastal communities. These diverse voices helped inform and enrich the writing of this report, and we wish to thank each person for their service.

Committee members also reviewed an extensive amount of scientific literature and deliberated on the material in closed session. All meetings were held virtually due to the COVID-19 pandemic. Although the conditions were not ideal for a consensus study, the committee worked collaboratively and diligently for more than 1 year, putting in long hours to prepare a report covering large amounts of new information. The committee members worked constructively as a team to formulate findings, conclusions, and recommendations for further advancement of knowledge of fates and effects, and for improvement in oil spill prevention and response. The updated volume is a major leap forward in depth and breadth of material included in the report. We would like to thank the committee members for their dedication, insightfulness, expertise, and invaluable contributions to complete the work under challenging circumstances.

We would also like to thank the National Academies staff for their support to the committee. We would like to express our special gratitude to Kelly Oskvig, Senior Program Officer, for her tireless energy and patience to organize meetings, presentations, and endless lists of other tasks. Additional support was provided by Susan Roberts, Ocean Studies Board Director, Megan May, Kenza Sidi-Al-Cherif, Grace Callahan, Stacee Karras, Caroline Bell, Elizabeth Costa, Leighann Martin, Safah Wyne, and other staff.

Finally, we would like to extend our gratitude to the sponsors who provided resources to make this important work possible. We hope that the report will be a useful reference and a source of information to the sponsors as well as to others who can benefit in their work or in their daily lives from the increased knowledge contained in *Oil in the Sea IV*. To that end, the committee has made numerous recommendations for improving research and other activities in the future. Although the world is transitioning to non-hydrocarbon energy sources to reduce greenhouse gas emissions from human activities, hydrocarbons will remain in the energy and petrochemical streams until other sources are available at sufficient scale and competitive cost. It is nevertheless important to continue to advance the knowledge and understanding of the environmental concerns associated with fates and effects of oil in the sea. Preserving the natural ocean environment is essential to the well-being of our planet and we would like to acknowledge all the professionals and scientists who are dedicated to keeping our seas healthy.

Kirsi Tikka (NAE), *Chair*
Ed Levine, *Vice Chair*
Committee on Oil in the Sea IV

Reviewers

This Consensus Study Report was reviewed in draft form by individuals chosen for their diverse perspectives and technical expertise. The purpose of the independent review process is to provide candid and critical comments that will assist the National Academies of Sciences, Engineering, and Medicine in making each published report as sound as possible and to ensure that it meets the institutional standards for quality, objectivity, evidence, and responsiveness to the study charge. The review comments and draft manuscript remain confidential to protect the integrity of the deliberative process.

We thank the following individuals for their review of this report:

JENNIFER FIELD, Oregon State University, Corvallis
MERV FINGAS, Environment Canada (ret.)
MARIA HARTLEY, Chevron Services Company, Houston, Texas
IRA LEIFFER, University of California, Santa Barbara
MICHAEL MACRANDER, Integral Consulting, Seattle, Washington
NAJMEDIN MESHKATI, University of Southern California, Los Angeles
STEVEN MURAWSKI, University of South Florida, St. Petersburg
ED OVERTON, Louisiana State University, Baton Rouge
GARY SHIGENAKA, Office of Response and Restoration, National Oceanic and Atmospheric Administration (ret.)
ROGER SUMMONS, Massachusetts Institute of Technology, Cambridge
DAVID VALENTINE, University of California, Santa Barbara
JEFFREY WICKLIFFE, University of Alabama, Birmingham
PHOEBE ZITO, University of New Orleans, Louisiana

Although the reviewers listed above provided many constructive comments and suggestions, they were not asked to endorse the conclusions or recommendations of this report, nor did they see the final draft before its release. The review of this report was overseen by **RICHARD SEARS,** Stanford University, and **DAVID DZOMBAK,** Carnegie Mellon University. They were responsible for making certain that an independent examination of this report was carried out in accordance with the standards of the National Academies and that all review comments were carefully considered. Responsibility for the final content rests entirely with the authoring committee and the National Academies.

Acknowledgments

This report was greatly enhanced by input from outside consultants and from a number of public information-gathering meetings during the study process. The committee thanks the consultant teams for their contribution to the report, including Rabia Ahmed, Gretchen Greene, and Jeri Sawyer (Greene Economics); Alyssa Garvin, Dong-Joo Joung, John Kessler, and Thomas Weber (University of Rochester), Samira Danashagar, Ian McDonald, Carrie O'Reilly, and Mauricio Silva (Florida State University); and Bill Meurer (ExxonMobil). The committee would like to thank all of the experts who presented during these meetings: Rita Colwell (Gulf of Mexico Research Institute), Tim Steffek (American Petroleum Institute), Candi Hudson (Bureau of Safety and Environmental Enforcement [BSEE]), Walter Johnson (Bureau of Ocean Energy Management), Kenneth Lee (Fisheries and Oceans Canada), Carol Baillie (Interagency Coordinating Committee on Oil Pollution Research), Anabelle Nicolas-Kepec (International Tanker Owners Pollution Federation Limited), Rob Cox (IPIECA), Eddie Murphy (Pipeline and Hazardous Material Safety Administration), Vicki Cornish (Marine Mammal Conservation), Henry Huntington (Ocean Conservancy), Alan Mearns (National Oceanic and Atmospheric Administration [NOAA]), Bryan Domangue (BSEE), Jason Mathews (BSEE), Mary Kang (McGill University), Tara Yacovitch (Aerodyne), Wesley Williams (Oakridge National Laboratory), Jim Elliott (Teichman Group), Steve Hampton (California Department of Fish and Wildlife, ret.), Joost De Gouw (University of Colorado Boulder), Dana Tulis (U.S. Coast Guard [USCG]), Jeffrey Lantz (USCG), Odd Brakstad (SINTEF), Charles Greer (National Research Council Canada), Collin Ward (Woods Hole Oceanographic Institution), Roger Prince (ExxonMobil, ret.), Michel Boufadel (New Jersey Institute of Technology), Deborah French McCay (RPS Group), Sandro Galea (Boston University), Maureen Lichtveld (University of Pittsburgh), Mace Barron (U.S. Environmental Protection Agency [U.S. EPA]), Adriana Bejarano (Shell), Lori Schwacke (National Marine Mammal Foundation), Steve Murawski (University of Southern Florida), Ed Wirth (NOAA), John Incardona (NOAA), Jordi Dachs (Institute of Environmental Assessment and Water Research, CSIS), Greg Challenger (Polaris Applied Sciences), Porfiro Álvarez-Torres (Consortium of Marine Research Institutions of the Gulf of Mexico and the Caribbean), Casey Hubert (University of Calgary), Mindi Farber-DeAnda (Energy Information Administration), Dean Foreman (American Petroleum Institute), Tim Nedwed (ExxonMobil), Jodi Harney (CSA Ocean Sciences Inc.), Richard Camilli (Woods Hole Oceanographic Institution), Ellen Ramirez (NOAA), Jacqui Michel (Research Planning Inc.), Adam Davis (NOAA), Jeffrey Wickliffe (University of Alabama at Birmingham), Dan Villeneuve (U.S. EPA), Adam Biales (U.S. EPA), Amy Kukulya (Woods Hole Oceanographic Institution), Kelsey Leonard (University of Waterloo), Mây Nguyễn (Lowland Center), Simon Lambert (University of Saskatchewan), Vera Metcalf (Eskimo Walrus Commission), Ken Paul (Wolastoqey Nation), and Chief Shirell Parfair-Dardar (Grand Caillou/Dulac Band of Biloxi-Chitimacha-Choctaw). Their varied perspectives and expertise were invaluable to the creation of this report.

Contents

Summary

The growing energy demand in industrialized North America, as in the rest of the world, has been satisfied mostly by use of fossil fuel hydrocarbons. As consumption, exploration, transportation, and production of hydrocarbons offshore and in coastal areas have increased, scientists have recognized and studied impacts of oil in the sea. Regulators have implemented new requirements, spill responders have developed innovative techniques, and the industry has employed updated operational practices and safety measures to limit the impact on the marine environment. Although significant progress has been made to better understand and reduce quantities and effects of oil in the sea, the risks have not been eliminated. A serious reminder of this was the 2010 *Deepwater Horizon* (DWH) disaster, an explosion that resulted in the largest oil spill in North American waters.

With the support of many agencies and industry, the National Academies published reports in 1975, 1985, and 2003 on inputs, fates, and effects of petroleum-based hydrocarbon mixtures in the sea, both from natural sources and human activities. The reports assessed the scope of the challenge and made recommendations for improvement in oil spill science, prevention, and mitigation of the impact of harmful discharges on the environment, and for reduction of inputs from operational and accidental discharges. This study is the fourth report in a series. Nearly two decades have passed since the third report was released, over which time there have been significant advances in technology and science. There are almost 20 additional years of research on long-term effects of oil spills on the environment from incidents such as the *Exxon Valdez* oil spill and six large (more than 10,000 barrels) spills in North American waters, including the DWH explosion and oil spill.

The Committee on the Oil in the Sea IV was appointed to document the current state of knowledge of sources, volumes, fate, and effects of oil entering the marine environment; to identify important gaps in research and understanding in each of those areas; and to make recommendations on reducing the inputs into the sea and the effects of oil on the environment (the Statement of Task is included in Chapter 1, Box 1.2). The committee's deliberations and report writing were informed by review of scientific literature and by a series of public meetings and presentations, drawing in expertise from academic, governmental, and non-governmental communities.

SOURCES OF OIL IN THE SEA

The life cycle of "oil in the sea" starts with the input of oil to the marine environment. Oil enters the sea from a variety of sources, including:

- natural oil and gas seeps;
- extraction of petroleum (spills from production and drilling platforms, deposition from platform air emission, produced water, and gas condensate discharges disaggregated by platform type [e.g., deep water, ultra-deep, shallow waters, leaks associated with platform decommissioning]);
- transportation of petroleum (pipeline spills, tanker spills, operational discharges, coastal facility spills, deposition from tanker exhaust and volatile organic carbon [VOC] emissions); and
- consumption of petroleum (land-based runoff, recreational vehicle discharge, spills from commercial vessels, operational discharges from commercial vessels, atmospheric deposition from land-based sources, aircraft dumping during emergencies).

The past 20 years have brought many changes to the ways in which oil and natural gas are extracted, transported, and consumed. The North American energy landscape has changed with the production of shale oil and natural gas both in the United States and Canada. In the past, the United States was a net importer of hydrocarbons, but now, because of the increased production and changes in legislation allowing export of crude oil, the United States has become an exporter of crude oil and natural gas, which has had an impact on terminals and on shipping routes.

Increased urban populations in coastal areas, changes in consumer behavior, improved fuel efficiency of vehicles, and introduction of electric cars are all impacting land-based consumption and, although more difficult to quantify, inputs of oil into the sea.

Other major changes that have had an impact on inputs of oil entering the sea since the *Oil in the Sea III* report include new regulations on design and operation of ships carrying hydrocarbons as cargo or as fuel, and on engines used on recreational vessels. Following the DWH oil spill, major changes were introduced to the regulations governing offshore oil and gas operations, and major advances were made in blowout prevention and source control. New performance measures and enforcement mechanisms have been introduced to improve pipeline safety. In general, there is an increased focus on improving the safety culture in the oil and gas as well as in the shipping industry.

Although specific data are lacking for estimating many inputs of fossil fuel hydrocarbons into North American waters, enough data do exist to understand recent trends and, in many cases, to provide more precise estimations of annual volumes. Overall, looking at oil inputs to the sea from 2010 to 2019, as compared to previous reported estimates from 1990 to 1999, if the DWH oil spill event had not occurred, it could be concluded that regulatory changes, advances in science and technology, and, for the most part, attention to safety have helped to reduce the amount of oil pollution in North American waters.

However, there are also new potential sources for oil pollution in North American and global waters related to the aging infrastructure in the sea and along the coasts, deteriorating shipwrecks releasing oil, sea-level rise and increased intensity and frequency of storms, use and transport of new types of oil, changes in shipping routes including travel through arctic regions, shifts in workforce, expertise and asset ownership in the energy industry, warfare, and others.

The committee's best estimates of the mass of oil entering the sea through natural sources, land-based sources, operational discharge, and accidental spills are shown in Figure S.1 (refer to Chapter 3 for details on the estimates). The general trends are described below:

1. **The estimates of land-based sources by far outweigh other sources, even when including the DWH oil spill and worst-case projections for the Mississippi Canyon 20 oil spill in the estimates.** Land-based runoff was calculated using the same approach as *Oil in the Sea III* and is based on several assumptions; it is therefore not possible to determine the quantitative increase with confidence; however, these estimates are in line with global estimates (which are closer to 30 times higher than those reported two decades ago in *Oil in the Sea III*).
2. **The second highest input is from natural oil seeps.** The committee's estimates of natural seeps are roughly one-third lower than reported in *Oil in the Sea III*, and reflect updated estimates for the Gulf of Mexico; new data are not available in other North American regions.
3. **Spills represent the third highest input, even when including an annualized estimate for the DWH oil spill.** In North America, spills occurred more frequently in offshore waters than nearshore waters and predominantly in the Gulf of Mexico. Over 20 years, the volume of spills decreased significantly for pipelines, tank vessels, non-tank vessels, and coastal refineries.
4. **The other major source is operational discharges, which, in this report, include produced water from offshore oil wells and discharges from machinery operations on commercial vessels.** Assuming full compliance of regulations, all discharges from commercial vessel operations are small, less than 10 metric tonnes per year. Although it can be safely assumed that discharges from recreational boating greatly decreased with regulatory actions to ban sales of the two-stroke engine, there are no data available for estimating actual discharges of oil in the marine environment from this source. Produced water estimates are higher in total than they were 20 years ago, reflective of increased offshore hydrocarbon production.

It should be noted that not all sources of oil in the sea are equal in terms of impact on marine life; chronic or continuous inputs have very different effects on the environment than accidental spills. The volume and the rate of discharge, as well as other factors, are important for determining both the fate and the effects of the oil in the marine environment.

This report also recognizes "that there was a significant lack of systematic data collection concerning petroleum hydrocarbon discharges entering the oceans" (NRC, 1975), indicating little progress since the original *Oil in the Sea* report in 1975 (titled *Petroleum in the Marine Environment*) and more recently in 2003. While the Oil in the Sea IV committee found more specific data on some inputs (spill volumes, permitted discharges), values in other input categories remain vague (land-based runoff, sewage discharge, two-stroke engines, atmospheric deposition, and emergency aerial dumping) to nonexistent.

More precise estimates of fossil fuel inputs in relation to natural sources such as oil seeps may permit identification of particular anthropogenic inputs that are significant in comparison with local natural inputs, are likely to have adverse effects on marine organisms, and are amenable to efforts to reduce or otherwise mitigate these inputs.

Recommendation: In recognition of decades of inaction on past *Oil in the Sea* report recommendations to measure natural and anthropogenic oil inputs

to inform mitigation of the effects on the marine environment, the committee recommends that an independent group report on measures and responsibilities of North American agencies to acquire more comprehensive data to better achieve quantification of oil inputs to the sea. The following actions should be taken to improve quantification of oil inputs to the marine environment:

- **Federal agencies should work with industry and academia to use existing techniques, along with exploration of new technologies, for identification and quantification of inputs from natural seeps including less recently studied areas such as the North American Pacific margin, North American Arctic margin, and newly discovered seeps in the North American Atlantic margin. Priority should be given to areas with offshore energy exploration and production and along marine shipping routes to better assess background levels of oil in the sea and inform damage assessment.**
- **Federal agencies should work with state and local authorities to undertake regular monitoring of oil inputs from land-based sources of water (runoff, rivers, harbors, and direct ocean sewage discharge) to determine oil inputs into marine environments.**
- **Federal agencies should support research to refine estimates of land-based inputs of oil that are transported via the atmosphere to the sea (including fuel jettison) by expanding the geographic and temporal coverage of data collection, and by refining understanding of the source of hydrocarbons measured in marine atmospheres and surface seawater.**
- **Relevant agencies should work with industry to gain a better understanding of composition and concentrations in produced water (from offshore exploration and production activities) released into the marine environment and implement practices to reduce potential environmental impacts of these discharges.**
- **A study should be conducted on the level of the International Convention for the Prevention of Pollution from Ships (MARPOL) Annex I compliance to establish a baseline to monitor changes in compliance.**

PREVENTION OF OIL IN THE SEA

Because land-based runoff is a significant source of oil in coastal seas (resulting largely from urbanization and an increase in the number of vehicles in operation), preventative measures could include initiatives to improve capture and processing of stormwater and sewage discharge, reduce gasoline vehicle usage (e.g., carpooling, increased use of public transportation), and continue improvements in fuel efficiency and increase usage of electric vehicles, as well as improving car maintenance and the replacement of older vehicles.

Advances have been made by industry and federal and state agencies in developing and implementing technologies and best practices for preventing accidental spills from onshore and offshore pipelines and facilities and marine transportation.

Recommendation: During the transition to more renewable energy sources, the following steps should be taken by industry and by federal and state agencies to prevent future spills in North American waters:

- **Consistent with previous reports, government and industry should continue their efforts to develop and implement technologies and best practices to prevent and reduce the magnitude of accidental spills from onshore and offshore pipelines and facilities and marine transportation. The committee recognizes the risk of complacency following periods of reduced spillage and advises government and industry to maintain vigilance.**
- **Government should review whether the technical recommendations arising from the extensive investigations in the aftermath of the DWH oil spill regarding blowout preventers, operational issues, and safety culture have been implemented, and identify those that have not yet but could be implemented.**
- **Government should conduct a comprehensive review of the integrity of coastal onshore and offshore energy infrastructure to determine if it can withstand increased frequency and intensity of extreme weather events and other natural hazards. This should involve:**
    - **review and update of design criteria for extreme events in light of new data;**
    - **assessment of modifications to existing structures needed to prevent or mitigate damage or spillage of oil resulting from extreme events; and**
    - **development of response plans and corresponding response capabilities to reduce and mitigate spills in case of damage due to extreme weather events.**
- **To mitigate potential spills from aging infrastructure, appropriate agencies should take inventory of the existing remnants of oil storage, transport, or production activities that still need to be identified. Salvage or capping of these facilities should be prioritized based on the potential impact if the infrastructure fails.**
- **In order to maintain response readiness, government should assess the economic and environmental impacts of changes in marine vessel transportation on pollution risk, such as introduction of new fuel types, increased vessel size, new types of cargoes (liquefied natural gas, other gases, biofuels, and diluted bitumen), and new traffic patterns and offshore infrastructure.**

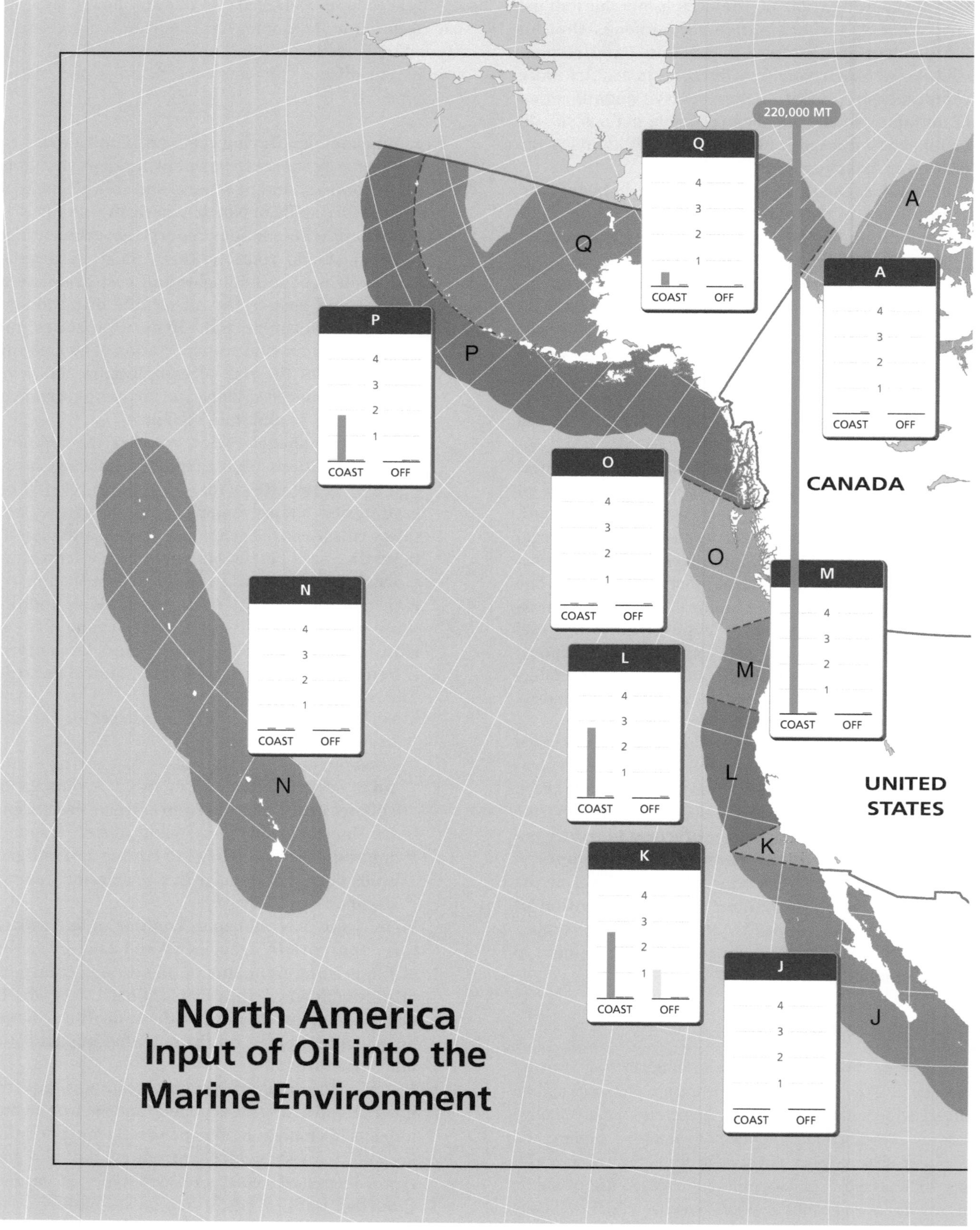

FIGURE S.1 Map of North America showing estimates of oil in the sea through natural sources, land-based sources, operational discharges, and accidental spills.

NOTES: Lettering identifies geographic regions in North American waters defined in Appendix A. Numerical values represent 10,000 MT/year.

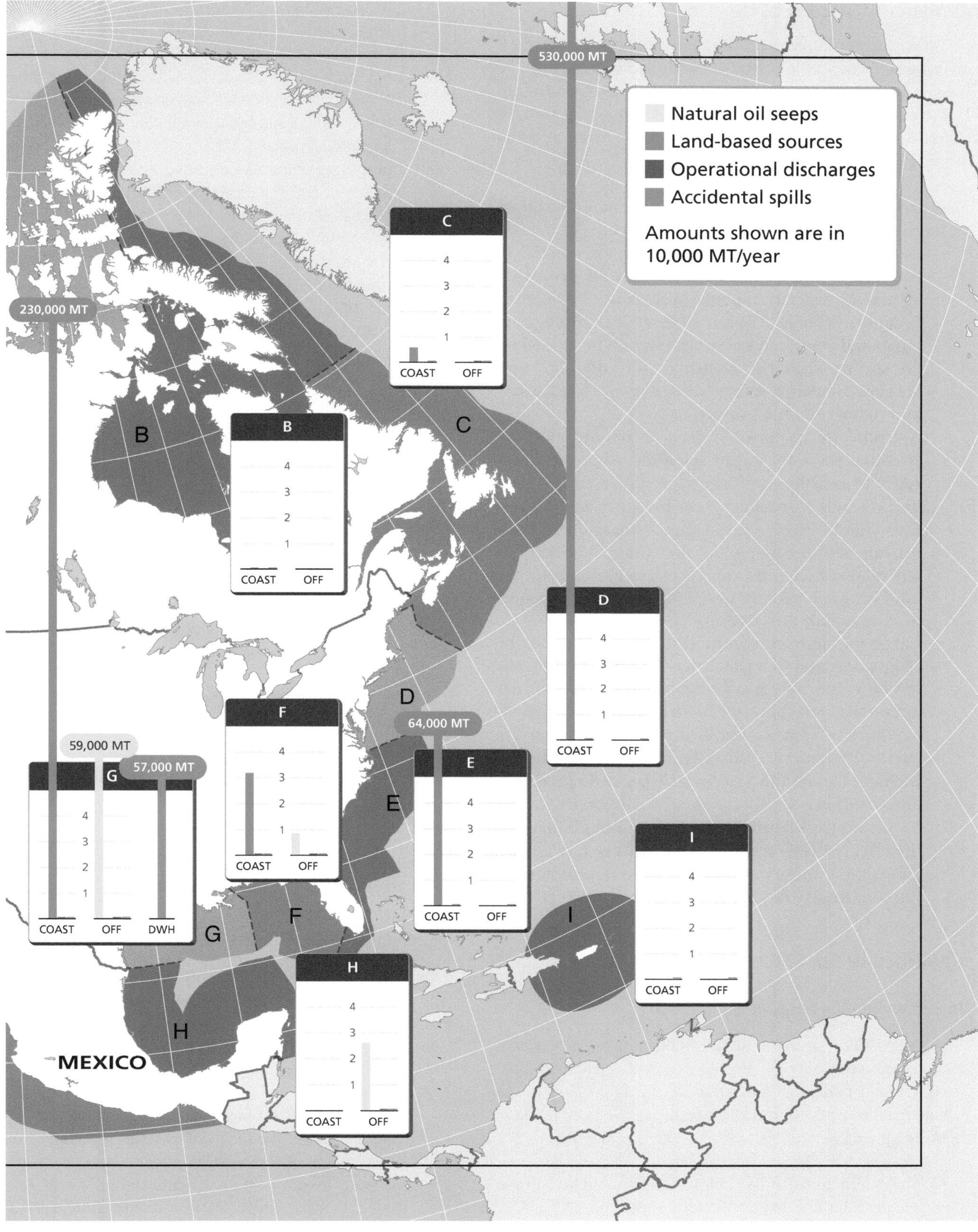

Natural oil seeps
Land-based sources
Operational discharges
Accidental spills
Amounts shown are in 10,000 MT/year
530,000 MT
230,000 MT
64,000 MT
59,000 MT
57,000 MT
B
C
D
E
F
G
H
I
4
3
2
1
COAST
OFF
DWH
MEXICO

MINIMIZING THE EFFECTS

While much is done to prevent and reduce amounts of oil entering the sea, accidental spills occur and will continue so long as there is offshore production and transportation of oil. In the event that a spill does occur, it is important to have tools available, and plans in place to most effectively respond to the spill and to protect the vulnerable.

Recommendation: The following steps should be taken to minimize the effects of oil entering the sea:

- **Regulatory mechanisms should be introduced to encourage evaluation, permitting, and deploying of new advanced response techniques when they become available. The use of these techniques during actual emergency response should be guided by the specific scenario to ensure they add value and maximize health, safety, environmental, cultural, and socioeconomic protection.**
- **Appropriate responsible agencies should plan for effective ways for rapid scientific response to oil spills to enable scientists to mobilize to the field quickly and in concert with response operations. This should involve rapid communication and approvals between all parties to define the operationally relevant scientific direction and gather relevant information for future decision making with respect to minimizing the effects of the spill.**
- **Large oil spills contaminating productive marine ecosystems and shorelines may inflict mass mortalities on vulnerable species such as seabirds, marine mammals, and shoreline biota and disrupt the ecology in that location. Following a spill, appropriate environmental specialists (e.g., a specialist in the Incident Command System [ICS]) should promptly identify these species, ecological linkages should be promptly identified, and their abundances in the affected region should be monitored, to enable detection of these indirect effects on populations at the community level and ensure their protection.**

ADVANCES IN OIL SPILL SCIENCE

Understanding the Fates of Oil in the Sea

In the past two decades, there has been major progress in understanding fundamental physical, chemical, and biological processes and reactions that influence the fate of oil in the marine environment. Significant advances have been made in the following areas:

- The interactions of live oil and gas spilled within the ocean water column are better understood.
- Gas bubble sizes and oil droplet breakup processes for oil released into the sea, critical to determining their fate and ultimately determining the affected environments, can be more accurately predicted.
- A new phenomenon, "tip streaming," responsible for creating micro-scale droplets, has been identified for droplets treated with chemical dispersant, allowing threads of oil to leak into and break within the droplet wake.
- Ongoing advancements in data storage platforms and computing capabilities have enabled profound advances in computational fluid dynamics modeling of oil spills in the oceans.
- Dissolution, long known as a relevant fate process for oil spilled at the sea surface, has been recognized as potentially significant for subsea oil spills, such as accidental oil well blowouts.
- Research on optimal design, implementation, and efficacy analysis of subsea dispersant injection (SSDI) was employed as a response option during the DWH oil spill. These analyses showed that even modest changes in oil droplet size distributions may lead to significant changes in the fate and effects of the spilled oil and may significantly benefit human health by reducing volatile organic carbon emissions to the atmosphere from subsea spills.
- There is renewed appreciation of the importance of photo-oxidation by solar irradiation of oil at the sea surface and near-surface as a major pathway for oil degradation. Numerous direct and indirect photo-oxidation reaction products likely influence the fate and toxicity of spilled oil.
- Marine oil snow sedimentation and flocculant accumulation (MOSSFA), a phenomenon observed in the Gulf of Mexico, has been proposed as a mechanism to explain observed oil transport to the seafloor during and after the DWH oil spill.
- There is an expanded appreciation of processes and oil fates unique to cold water and sea ice in deep sea and Arctic environments, and new recognition that oil biodegradation can occur at rates faster than previously assumed at near-freezing temperatures.

Understanding the Effects of Oil in the Sea

Understanding of exposure to and ultimately the acute and chronic effects of oil pollution on marine and estuarine habitats and resident organisms has significantly expanded and improved since the publication of *Oil in the Sea III* in 2003. Highlights include:

- The identification of new exposure scenarios (i.e., routes and/or specific chemical components involved), toxicity mechanisms of action, target sites and biological effects, mixture toxicity, environmental modifiers of toxicity (e.g., UV light, pressure), new biological receptors, impacted habitats and species, and long-term implications from direct or indirect exposure to oil spills.

- There have been significant advances in resolving the complexity of determining or predicting the effects of petroleum hydrocarbons in the marine environment within a changing ecosystem and multiple co-stressors using field, laboratory, or predictive modeling-based approaches.
- New knowledge of oil toxicity with "natural" lower-level exposures via respiration/absorption includes the chronic effects on the hypothalamic-pituitary-adrenal axis (at least in marine mammals), immune system in many higher-level vertebrates, and an organism's microbiome (external and/or internal), which may influence the overall health of the exposed organism.
- Biotechnology 'omics[1] tools have also been developed to detect target sites and mechanisms of action for oil toxicity and as biomonitoring tools for use in baseline studies, damage, and recovery assessments.
- There has been tremendous growth in the examination of the potential effects of oil spills on human health, including mental and behavioral effects; and considerations of socio-economic impacts and community resilience as well as the potential direct toxicity to humans of crude oil, its components, and derivatives.

COMMON THEMES FOR ADVANCING OIL SPILL SCIENCE

Complex and intrinsic interdependencies exist among inputs, fates, and effects and the measures to reduce them, either from the source or by response. Recognition of this interdependency accompanied by best practices with respect to oil spill science in general led to the emergence of several repeated themes throughout the report for advancing understanding of oil spill science, all with the end goal of minimizing negative impacts to the environment by better informing prevention, response, and restoration activities.

Long-Term Funding for Inputs, Fates, and Effects of Oil in the Sea

The Gulf of Mexico Research Initiative (GoMRI), a 10-year program initiated after the DWH oil spill, resulted in an extraordinary output of both discipline-specific and multidisciplinary research by funding a mix of field, laboratory, mesocosm, and test facility science and related modeling. Advancement of oil spill science has a history of being hindered by a boom-and-bust funding cycle and, consequently, the inability to sustain research and the scientific expertise to conduct the research.

Recommendation: As recommended in *Oil in the Sea III*, there remains a need for long-term, sustained funding focused on oil in the sea to support multi-disciplinary research projects that address current knowledge gaps, including those listed as Research Needs throughout this report. Research is needed to address new regulatory requirements and to improve response capabilities. The application of new data and technologies to advance interdisciplinary knowledge of fates and effects of oil in the sea will require a longer funding commitment than is currently typical.

[1] Meaning genomics and related biomolecular analyses.

Open Water Experimentation and Use of Spills of Opportunity

Government agencies, industry, private companies, non-governmental organizations, and academia are continually looking into improvements to response technologies, identifying the effects of oil on the environment, developing better ways to monitor oil in and on the water, and looking for ways to increase the efficiency of response operations and organizations. These projects generally take place in a laboratory or test tank. It is not possible to simulate all the complexities and variability of field conditions in a small-scale or large-scale laboratory setting alone. Field experiments with real oil and studies conducted during actual response events are critical for understanding the fate and behavior of oil in realistic conditions and for the development, testing, and improvement of response techniques.

Recommendation: As recommended in previous reports, controlled in situ field trials using real oils should be planned, permitted, and funded to incorporate multi-disciplinary research focused on important processes as well as response techniques that do not accurately scale from in vitro or ex situ experiments to in situ conditions. Additionally, funding and systemic mechanisms should be set in place by appropriate agencies to enable rapid deployment of qualified scientific personnel during actual oil spill events (i.e., spills of opportunity) to conduct appropriate, time-critical research in situ, outside the Natural Resource Damage Assessment process, while having minimal or no interference with spill response activities.

Baseline Knowledge and Data

After a spill has occurred, assessment and research efforts often do not have appropriate or requisite pre-spill data for comparison with post-spill observations and assessment of remediation. This limits the ability to assess the inputs, fates, and effects of oil in the sea.

Recommendation: There is a need to review how pertinent knowledge and data from numerous sources are most effectively assembled, made available, and archived, given the advances and gaps in understanding noted in this report.

- **The review should assess what is needed for baseline knowledge and data with recognition that both**

natural and anthropogenic influences (other than inputs of oil) result in baselines that are dynamic in space and time.

- **Funding should be established for appropriate baseline data acquisition and curation in locations of particular interest, such as coastal areas, areas with offshore energy exploration and production, and marine transportation routes.**
- **Data collections would include aspects such as physical oceanography, biogeochemical processes, contaminant source surveys, critical species (e.g., endangered, abundant, and vulnerable, or of commercial importance) and marine biodiversity, and pertinent metrics of human health and well-being.**
- **As a corollary, guidelines should be developed for collecting and analyzing baseline data immediately after and in the midst of a spill from neighboring, unaffected (control) areas, where possible.**
- **U.S. Interagency Coordination Committee on Oil Pollution Research (ICCOPR) member agencies, in cooperation with relevant agencies from Canada and Mexico and other interested parties, should convene a series of regional workshops and studies to inform the most efficient process of defining and assembling the evolving baseline knowledge and data. These workshops should gather relevant knowledge from stakeholders such as Indigenous peoples; diverse rural, suburban, and urban coastal communities; government agencies; business and industry; nonprofit groups; and the academic sector.**

Big Data and Interdisciplinary Research

Enormous streams of data have been generated from advances in analytical techniques, particularly in petroleum and environmental chemistry and in 'omics. Archival and maintenance of this "big data" to make it universally accessible is essential for meaningful interpretation of the data and to support interdisciplinary research linking oil chemistry to the fates and effects of the oil in the environment to inform response.

Recommendation: A free, central, universally accessible and curated repository should be formed for information pertinent to oil in the sea in order to better manage the enormous data sets being generated through advanced chemical analyses, 'omics techniques, geoscience surveys (among others), and especially field and laboratory studies pursuant to oil spills. Optimum use of such archives will require development of data analytics, data quality control, and reporting standards for associated metadata to enable integration and interpretation by, and training of, interdisciplinary teams.

Oil in the Arctic

Marine traffic in Arctic waters is increasing with seasonal decrease in ice cover, and increased offshore oil production is a possibility in the future. Both of these factors could lead to higher risk of oil pollution in the Arctic. Yet, oil spill science in Arctic waters and shorelines has lagged behind study of more temperate and accessible marine ecosystems. Field experiments in Norway, Canada, Alaska, Svalbard, and Greenland have uncovered many complex processes affecting oil in Arctic environments. However, utilizing this information in modeling or response requires additional work.

Recommendation: In agreement with previous reports, there should be a concerted effort to gather information about the fate of oil in Arctic marine ecosystems, with and without ice cover, in advance of further development of this region. This would include baseline surveys (geophysical and biological); efficacy of response and mitigation options; data acquisition on natural attenuation and active remediation strategies, including biodegradation kinetics at low temperature; and effects on higher organisms, populations, and ecosystems in Arctic waters and on shorelines.

New Fuels

New requirements for low sulfur fuel oils (LSFOs) for marine shipping came into effect in 2020, but studies on these oils are currently extremely limited. The few very-low sulfur fuel oil (VLSFO) and ultra-low sulfur fuel oil (ULSFO) samples studied to date differ chemically from traditional marine fuel oils and from each other. To date, insufficient research has been conducted to determine transport and weathering behavior, biodegradability, and toxicity of different LSFO formulations under diverse environmental conditions.

Recommendation: Government should fund research needed to study the composition, toxicity, and behavior of new types of marine fuels (e.g., LSFO, VLSFO, biofuels) and petroleum products (e.g., diluted bitumen) so that fate and effects of these products can be understood and response operations can be planned and executed most effectively to reduce impacts.

Human Health

The human aspect of oil in the sea permeates each of the decisions made in cleanup activities and endpoints. To that end, it is paramount to understand how oil, as well as oil spills and their responses, may affect human health and welfare. Each response to an oil spill is focused on minimizing harm to the environment. The ultimate clean-up goal is to ensure a healthy ecosystem for a sustainable future, and this includes both safety measures for those working, living, and recreating along its shores and the sustainability of marine resources such as food, energy, and transportation.

Recommendation: The governmental agencies involved in responding to an oil spill should upgrade the priority and attention given to individual and community mental and behavioral effects and community socioeconomic disruptions in the ICS decision-making and response processes. The inclusion of community-based human health assessment and mitigation measures into the ICS is needed to provide a more holistic approach regarding both human and ecosystem health.

OIL SPILL SCIENCE RESEARCH NEEDS

In addition to identifying overarching recommendations for advancing oil spill science, the report also highlights specific research necessary to better understand the fate and effects of oil in the marine environment, as well to advance response capabilities. Specific research needs are identified for the topical areas of oil spill response and the fates and effects of oil in the sea.

CONCLUDING REMARKS

Oil will continue to be a part of the global energy mix, though its share is likely to decrease as the use of alternative energy sources increases. New energy sources and fuels, such as biofuels, ammonia, and hydrogen, can introduce new safety and pollution concerns. The regulators, the research community, and the industry are encouraged to proactively review and address any potential adverse effects from these transitions. The recommendations and research gaps outlined in this report take into account a changing energy landscape. They are critical to address while this transition is taking place, and thereafter, to ultimately reduce impacts of oil on ocean ecosystems and human health and move toward a restorative and sustainable state following both human-caused chronic and episodic releases of oil into the sea.

1

Introduction

INTRODUCTION

Petroleum (sometimes referred to as crude oil or simply "oil") is a fossil fuel that was formed in the Earth's crust from the remains of plants and marine organisms. Oil, as a source of energy and for chemical synthesis, is critical for the technological and economic growth in the world. Although oil is a finite resource, new sources continue to be discovered. Oil has high energy density and is chemically complex, making it suitable for multiple uses. Although it is available in large quantities in many parts of the world, in recent decades global attention has turned to reducing greenhouse gas emissions, and alternative energy sources are also being discovered, developed, and scaled to play larger roles in the global energy economy. Even with an expanding portfolio of energy sources and the global movement toward decarbonization, oil will remain part of the energy and petrochemical mix until other alternatives are available at sufficient scale and are cost competitive. The pace of transition will depend on many factors including government policies and regulatory developments.

In some cases, oil naturally finds its way into the sea through fissures and faults. Often, while the collective volumes are large, individual seeps generally release petroleum slowly enough over hundreds to thousands of years to allow surrounding organisms to avoid, adapt to, and, in some instances, even thrive in their presence (NRC, 2003). In other cases, oil finds its way into the sea in less intrinsic ways, due to anthropogenic activities such as exploration, production, and global transport of oil. Oil is transported many miles through subsea pipelines and on transport vessels to support global fossil fuel demands as well as more general consumer needs.

Oil as it exists naturally in our environment does not always lead to damaging effects. However, oil present in locations, concentrations, timespans, or molecular forms outside its natural existence presents a different story. Recognizing the sources, quantities, and composition of natural and anthropogenic inputs of oil into the sea, and understanding the fate and effects of that oil in the marine environment, are important in order to minimize potential impacts. The application of this knowledge provides the mechanisms to evaluate and respond to potential environmental impacts of oil on ocean ecosystems and move toward a restorative state following both chronic and human-caused episodic events.

Oil is encountered in a variety of forms in daily life. Crude oil and other hydrocarbon liquids are refined into petroleum products used for many different purposes: to power vehicles, heat buildings, lubricate parts, and produce electricity. The petrochemical industry uses petroleum as a raw material, or feedstock, for a long list of products; these products represent important contributions to modern societal needs and wants. As energy reliance shifts to alternative sources, it is not clear how production of petroleum for these other uses may change, but consumption of hydrocarbons for purposes other than energy is small relative to energy use. The U.S. Energy Information Administration (EIA) estimates that the non-combustion uses of fossil fuels account for approximately 7% of total fossil fuel consumption in the United States,[1] and the International Energy Agency (IEA) estimates non-energy use of oil at approximately 17% and natural gas at approximately 12% globally (IEA, 2021a). Therefore, energy consumption is the main driver for the current and future demand of petroleum.

1.1 PRESENT AND FUTURE ENERGY NEEDS

1.1.1 World Energy Needs

World energy supply and demand was steadily growing until 2020, when the COVID-19 pandemic resulted in lockdowns and travel restrictions and, consequently, a drop in energy consumption. Following the reduction in demand,

[1] See www.eia.gov, "Today in Energy," April 6, 2018.

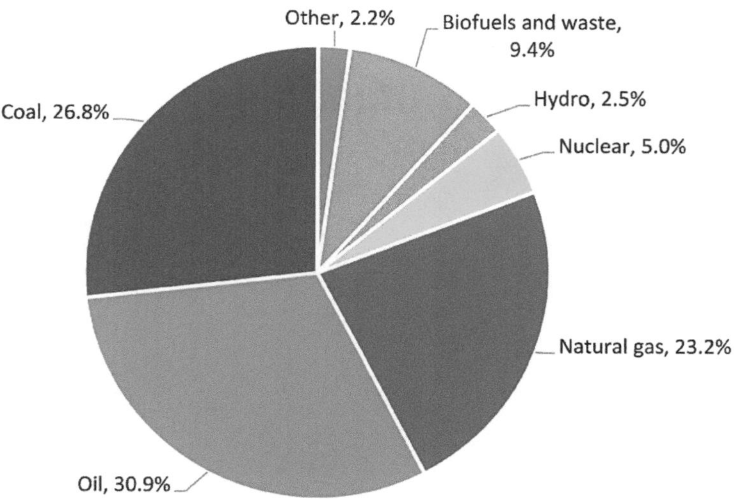

FIGURE 1.1 Global share of total energy supply by source, 2019.
NOTE: EJ = exajoules (10^{18} joules).
SOURCE: IEA, 2021.

production of worldwide crude oil and natural gas liquids dropped ~7% from 4,617 million tonnes in 2019 to 4,296 million tonnes in 2020 (IEA, 2021).

In 2019, oil and natural gas represented more than 50% of all energy supply and, with the addition of coal, the total hydrocarbon share was 80% of the worldwide energy supply (see Figure 1.1).

In 2020, the United States remained the top producer of both oil and natural gas. Other top oil producers were Russia, Saudi Arabia, Canada, and the United Arab Emirates, and the top natural gas producers were Russia, Iran, China, and Canada (IEA, 2021).

Transportation is the largest consumer of oil globally, using 56% of the total energy share in 2019 (see Figure 1.2).

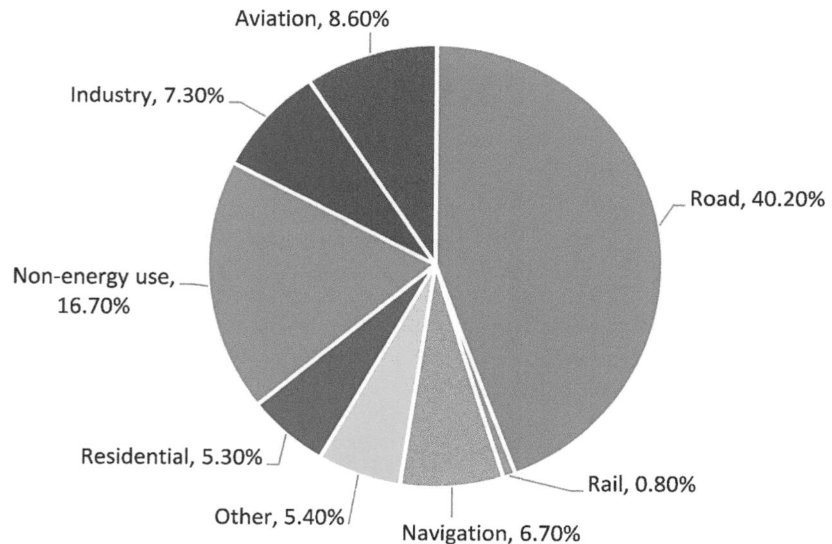

FIGURE 1.2 Global oil consumption, 2019.
SOURCE: IEA, 2021.

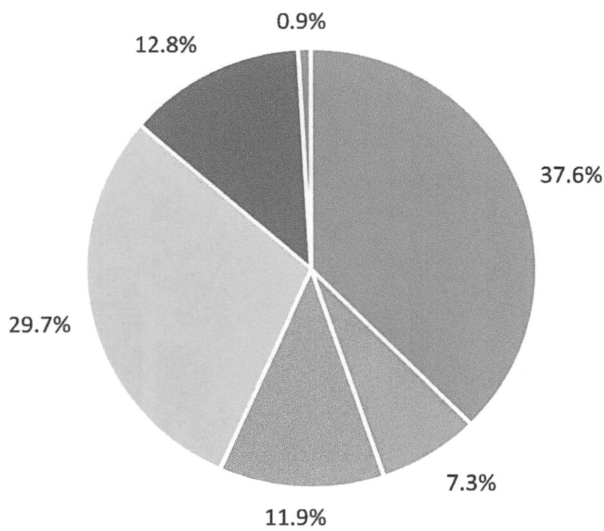

FIGURE 1.3 Global natural gas consumption, 2019.
SOURCE: IEA, 2021.

Industry and residential use are the largest consumers of natural gas, accounting for 67% of total in 2019 (see Figure 1.3).

The global energy landscape has been reconfigured considerably since the publication of *Oil in the Sea III* in 2003. A major driver for the change has been increased shale oil and gas production in the United States. Offshore oil and gas production share of the total oil and gas production has been declining (see Figure 3.18). Deep-water exploration is expected to further decline as the energy companies' investment strategies adapt to shifting energy requirements driven by decarbonization efforts.

The worldwide marine transportation flows have changed as a result of U.S. shale production, increased Asian demand, and the introduction of new refinery locations, mainly in the Middle East and Asia. Figures 1.4 and 1.5 illustrate the complex global flows of oil and gas.

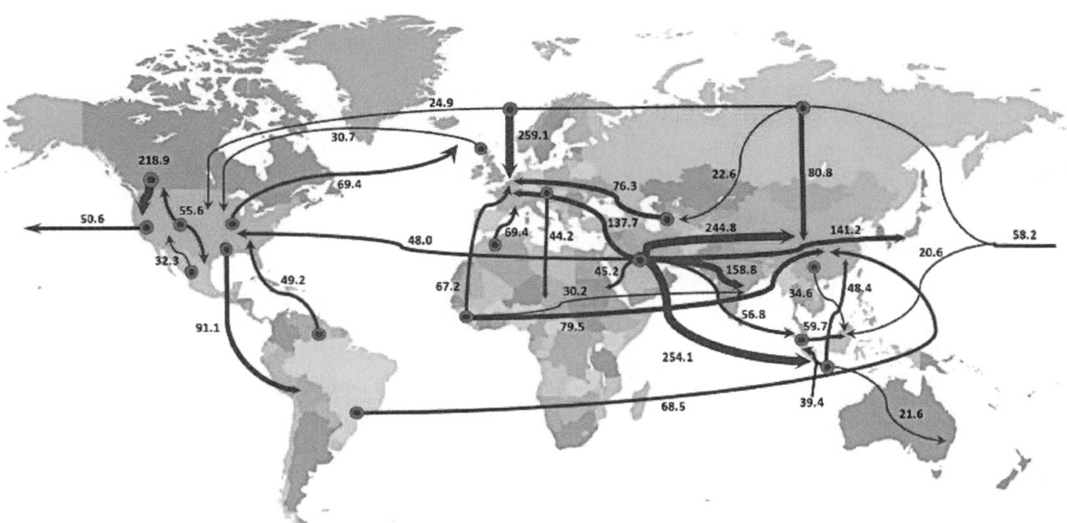

FIGURE 1.4 Oil trade flow diagram in million metric tonnes.
NOTE: Width of arrows represents relative volume transported; purple dots represent point of oil origination.
SOURCE: Adapted from BP, 2020.

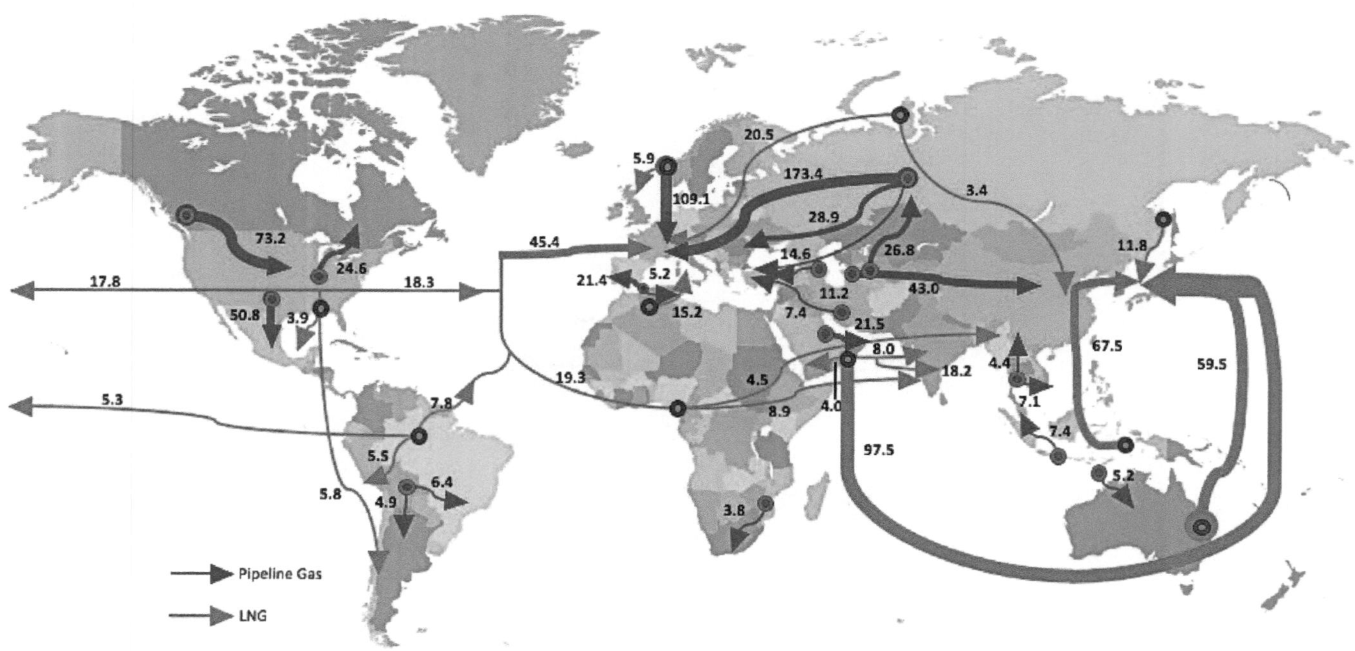

FIGURE 1.5 Gas trade flow diagrams in million MT.
NOTES: Width of arrows represent relative volume transported. LNG = liquified natural gas.
SOURCE: BP, 2020.

1.1.2 Energy Needs of North America

Petroleum is currently the largest North American energy source. As reported in the *BP Statistical Review of World Energy* (BP, 2021), in 2020 North American petroleum consumption averaged about 20.77 million barrels per day (bbl/d), or 2,967 tonnes per day (t/d). This was the lowest consumption rate since 1995 and the largest annual decrease in petroleum demand on record.[2] This anomaly can be attributed to the COVID-19 pandemic and is not necessarily reflective of the current trend. The annual differences were most pronounced in jet fuel, which decreased by 40% from 2019 to 2020. For this reason, 2019 numbers (from BP, 2020) are discussed in this section. In 2019, North American petroleum consumption averaged about 23.54 million bbl/d (3,362 t/d), with the transportation sector (including gasoline, distillate fuel, HGLs, and jet fuel) accounting for the largest share (68%) of U.S. petroleum consumption. The breakdown by sector for the United States is shown in Figure 1.6.

In North America, the most consumed petroleum product is gasoline (42% of total consumption), followed by distillate fuel oil (23% of total consumption), hydrocarbon gas liquids (HGLs) (18% of total consumption), and then jet fuel (6% of total consumption). Distillate fuel oil includes both diesel fuel (used to power farming and construction equipment, boats, trucks, trains, generators, etc.) and heating oil (used for residential and commercial heating and for producing energy in power plants). HGLs include propane, butane, ethane, and others used in the production of plastics, cooking fuel, lighter fuel, and in gasoline.[3]

1.1.3 Energy Outlook

The global energy demand continues to grow due to the population growth and the improving standard of living around the world (see Figure 1.7). The biggest increase in demand is predicted in Asia (EIA, 2019).

At the same time that global energy demand is increasing, societal concern over climate change is driving decarbonization efforts around the world to reduce greenhouse gas emissions. The need for a decarbonized future to mitigate the potential future financial and human costs of climate change is well accepted globally (IPCC, 2021). The transition from the current energy production, consumption, and transport to the decarbonized future is challenging to imagine, and equally challenging to forecast. To capture the breadth of

[2] See www.eia.gov, "Monthly Energy Review," April 2022.

[3] See https://www.eia.gov/energyexplained/oil-and-petroleum-products/use-of-oil.php.

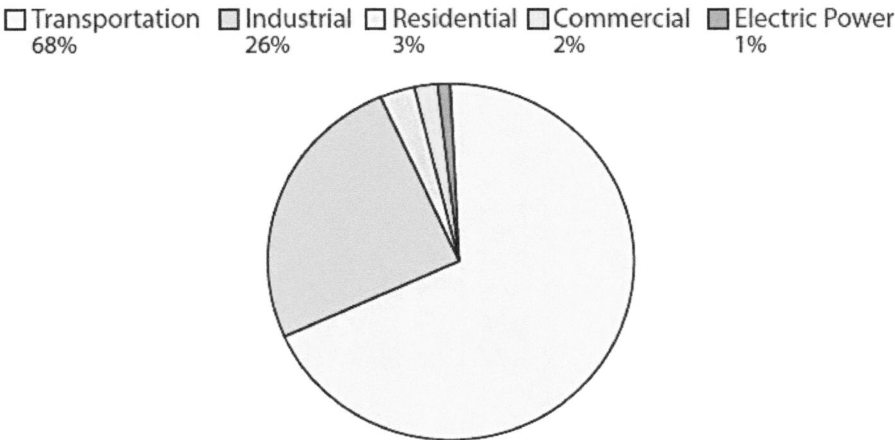

FIGURE 1.6 U.S. petroleum refined product consumption by end-use sectors' share of total in 2019.

future possibilities, Greene Economics (2021) considered three different projections:

1. The Reference case, or business-as-usual scenario: Continuation of current trends.
2. The Net Zero by 2050 (NZ2050) case: Reducing global carbon dioxide (CO_2) emissions to net zero by 2050 is consistent with efforts to limit the long-term increase in average global temperatures to 1.5°C. (IEA, 2021b).
3. Partial Transition: Halfway point between the Reference case and the NZ2050 case (ABS, 2020).

The share of oil and gas as the energy source globally differs considerably among the three scenarios (see Figure 1.8). In all scenarios, renewables have the largest rate of growth.

The following figures illustrate the crude oil (see Figure 1.9) and gas (see Figure 1.10) production in the United States, Canada, and Mexico for the three scenarios.

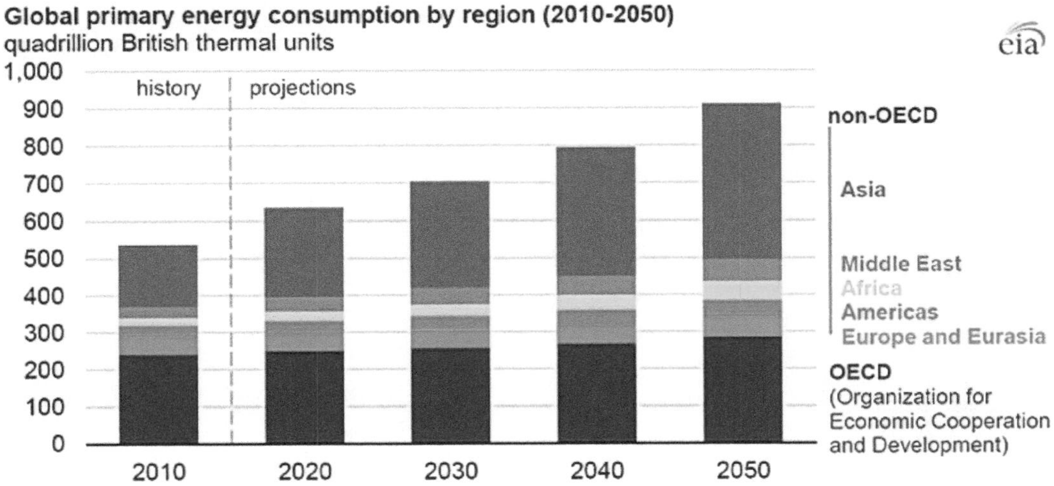

FIGURE 1.7 Global Energy Consumption. Projections represent the "Reference" case, which assumes continuation of current trends (see below).
SOURCE: EIA, 2019.

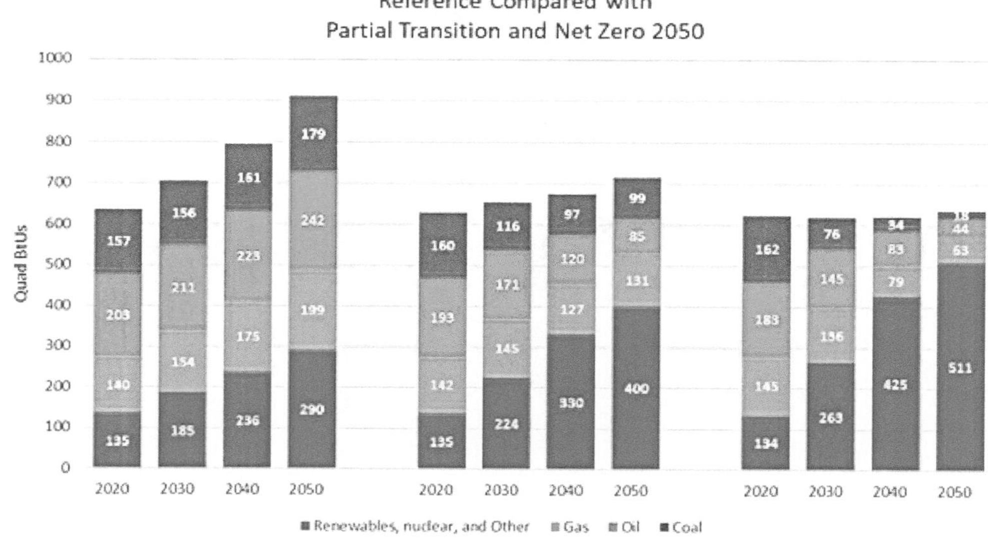

FIGURE 1.8 Global energy consumption projections for the reference, partial transition, and NZ2050 cases.
NOTE: Quad BTU = quadrillion British thermal unit.
SOURCE: Greene Economics, 2021.

In the Partial Transition and NZ2050 cases, the peak oil and gas production in the United States has already been reached. In the Reference case, the peak oil production would be reached around 2030, followed by relatively flat production thereafter. Gas production follows similar trends.

The energy outlook projections in Figures 1.9 and 1.10 show a large spread between the three scenarios, reflecting the uncertainty in the future political actions by governments, availability and cost of alternative energy and green fuels, and technological advances in carbon capture, large capacity

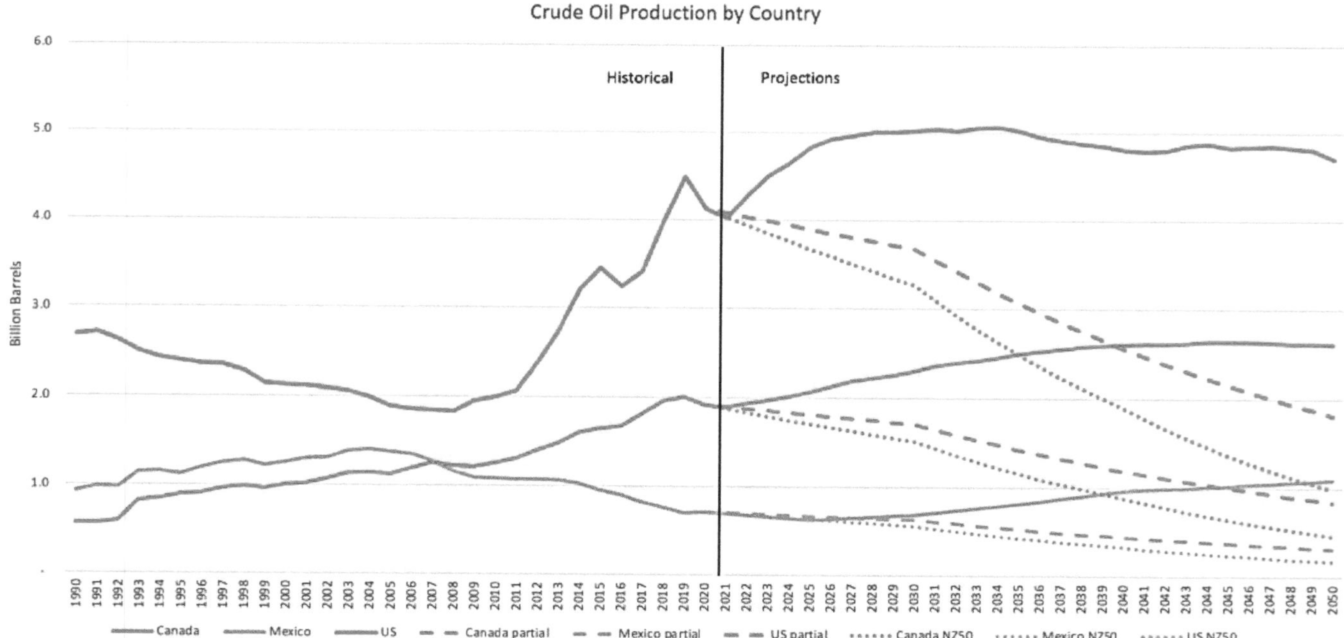

FIGURE 1.9 Crude oil production in the United States, Canada, and Mexico (billion bbl).
SOURCE: Greene Economics, 2021.

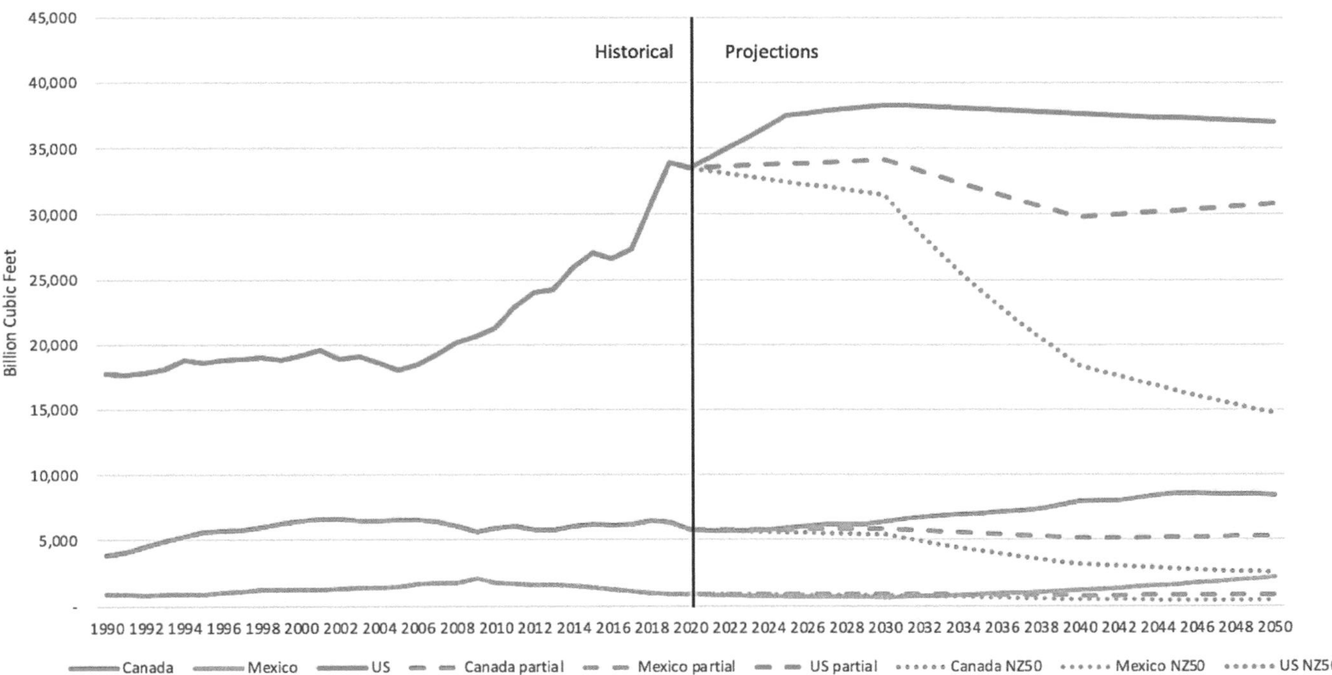

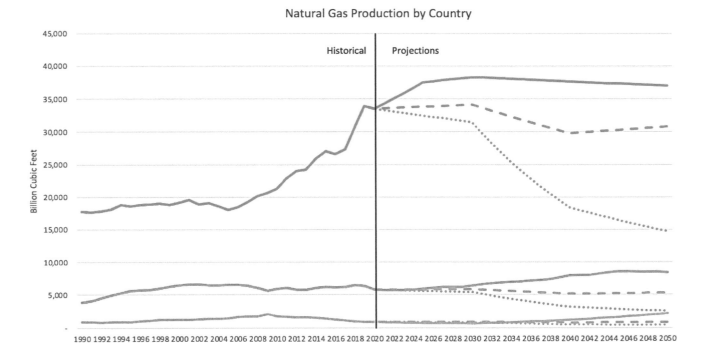

FIGURE 1.10 Natural gas production in the United States, Canada, and Mexico (billion bbl).
SOURCE: Greene Economics, 2021.

batteries and fuel cells. Increase in the use of electric vehicles will have an impact on gasoline demand and consequently on oil demand, production and transportation; however, to achieve zero carbon life cycle for electric vehicles, alternative energy sources such as wind, hydroelectric, and solar are needed.

Hydrocarbons will continue to be a part of the energy mix, but their share is likely to decrease if the use of alternative fuels continues to increase at a faster rate than oil production increases. This will introduce a major transition to both the energy industry and consumers. New energy sources and fuels, such as biofuels, ammonia, and hydrogen, can introduce new safety and pollution concerns, and the regulators, the research community, and the industry are encouraged to proactively review and address any potential adverse effects from these transitions. While energy transition is taking place, continued attention to environmental concerns associated with fates and effects from human-caused chronic and episodic inputs of petroleum and fossil fuels remains critical.

1.2 STUDY RATIONALE

This study is the fourth in a series by the National Academies on inputs, fates, and effects of petroleum-based hydrocarbon mixtures in the sea. The first study, *Petroleum in the Marine Environment*, was published in 1975. As understanding of the science surrounding oil in the sea advanced, new reports were released. Table 1.1 lists the report series and sponsors.

Nearly two decades have passed since *Oil in the Sea III* was released. Since the last report, there have been significant advances in technology and science in general. There are almost 20 additional years of research on long-term

TABLE 1-1 *Oil in the Sea* Report Series

Year	Title	Sponsor(s)
1975	*Petroleum in the Marine Environment*	U.S. Environmental Protection Agency (U.S. EPA), U.S. Coast Guard (USCG), Office of Naval Research, The Rockefeller Foundation, American Chemical Society
1985	*Oil in the Sea: Inputs, Fates, and Effects*	USCG
2003	*Oil in the Sea III: Inputs, Fates, and Effects*	Mineral Management Services, U.S. Geographical Survey (USGS), USCG, U.S. Department of Energy, U.S. EPA, National Oceanic and Atmospheric Administration (NOAA), U.S. Navy, American Petroleum Institute (API), National Ocean Industries Association
2022	*Oil in the Sea IV: Inputs, Fates, and Effects*	API, Bureau of Ocean Energy Management, Bureau of Safety and Environmental Enforcement, Fisheries and Oceans Canada, Gulf of Mexico Research Initiative (GoMRI), National Academies of Sciences, Engineering, and Medicine's Presidents' Circle Fund

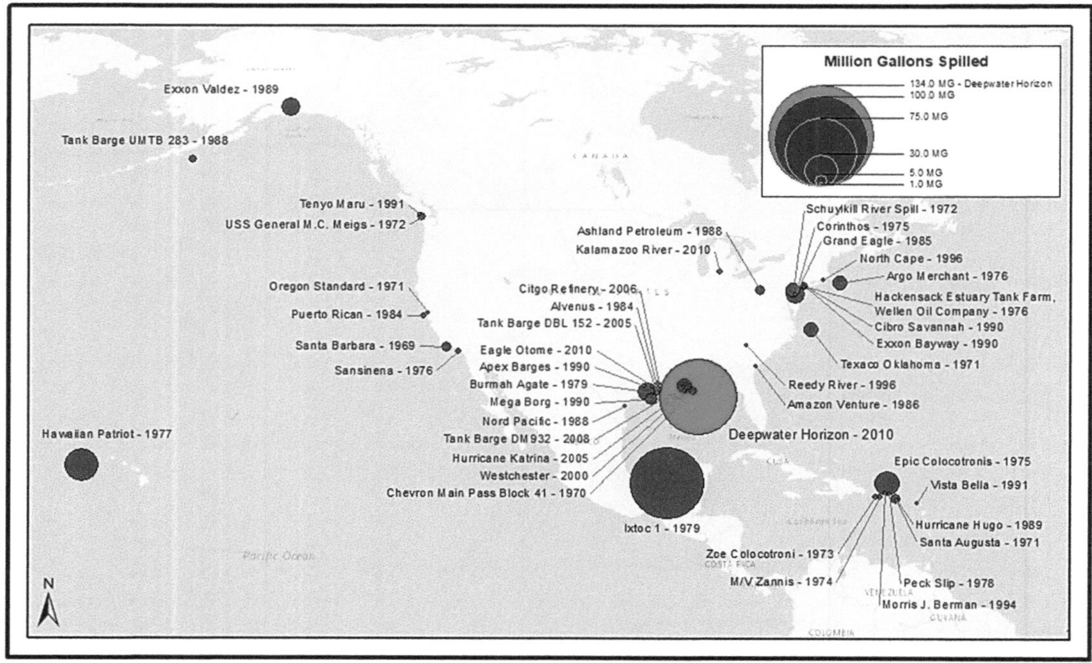

FIGURE 1.11 Largest oil spills affecting U.S. waters (1969–July 17, 2020), reported in million gallons (MG).
SOURCE: Office of Response and Restoration, National Oceanic and Atmospheric Administration.

effects of oil spills on the environment from accidents such as the *Exxon Valdez*. Since 2003, there have been six large (more than 10,000 bbl) spills in North American waters (see Figure 1.11), including the largest maritime spill in North American history, the *Deepwater Horizon* explosion and oil spill (see Box 1.1), referred to throughout this report as the DWH oil spill. Financial support for science issued directly from BP following the spill along with the civil and legal penalties, and enactment of the Natural Resource Damage Assessment (NRDA),[4] led to a wealth of new research and literature on oil spill science. This surge in knowledge was a strong impetus for the committee's undertaking of *Oil in the Sea IV*.

It should be noted that several other relevant studies have been conducted by the National Academies since the 2003 report to better understand inputs, fates, and effects of oil in the sea:

- *Oil Spill Dispersants: Efficacy and Effects* (2005)
- *An Ecosystem Services Approach to Assessing the Impacts of the* Deepwater Horizon *Oil Spill in the Gulf of Mexico* (2013)
- *Responding to Oil Spills in the U.S. Arctic Marine Environment* (2014)
- *Spills of Diluted Bitumen from Pipelines: A Comparative Study of Environmental Fate, Effects, and Response* (2016)
- *The Use of Dispersants in Marine Oil Spill Response* (2020)

1.2.1 Statement of Task

The committee's Statement of Task is included in Box 1.2. The committee's deliberations and report writing were informed by review of scientific literature and by a series of public meetings and presentations, drawing in expertise from academic, governmental, and non-governmental organizations.

In designing the study, the committee first reached consensus on what was and was not to be included in the report by defining four key terms included in the Statement of Task:

Oil: fossil fuel hydrocarbon; organic compounds containing primarily hydrogen (H) and carbon (C); originating from natural gas and petroleum present in geological formations; in their natural forms or as their human-produced hydrocarbon-containing derivatives (refined and modified products) in different physical states (gas, liquid, solid)

Marine environment: oceans, coastal ecosystems, and estuaries

Marine life and ecosystems: organisms directly in or in contact with the marine environment, at least part of the time

[4] See https://darrp.noaa.gov/what-we-do/natural-resource-damage-assessment.

BOX 1-1
Deepwater Horizon

The *Deepwater Horizon* (DWH) oil spill was not only a catastrophic event that ended in the loss of 11 lives, but also the largest oil spill in U.S. waters. An estimated 535,000 tonnes of oil were released over a period of 87 days from sources at depths exceeding 1,500 meters in the deep Gulf of Mexico (McNutt et al., 2012). The explosion and fire on the DWH drilling platform led to the sinking of the platform, which resulted in breakage of the submarine risers. The time line of oil leakage and interventions was reviewed by French-McCay et al. (2015) and is summarized here. After multiple failures of the blowout preventer (NAS, 2012), oil initially leaked predominantly from the end of the broken riser, which rested on the seafloor. The riser immediately above the blowout preventer was severely bent and started leaking as early as April 28, 2010. Over the following 30 days, the flow through the bent riser increased and oil continued to spill from the end of the riser.

During this period, dispersants were injected using the riser insertion tool at the end of the broken riser, but their usage was highly variable. Between May 26 and 29, 2010, BP attempted to stop the flow with drilling mud through the top-kill operation. When the top-kill failed, the broken riser was removed, which was completed on June 3, and the oil spill emanated from the top of the crippled wellhead. From this point onward, subsea dispersant injection (of varying volumes) was used as a mitigation technique to minimize the amount of oil transported to the surface. A "top-hat" device was installed over the exposed riser to help collect spilled oil and pump it to the surface to be recovered. The top-hat device ramped up collection from an initial rate of about 11,000 bbl/d on June 4 to 22,000 bbl/d by June 13, continuing through July 10; oil and gas not collected by the top hat were vented through ports on the sides of the top hat. During this time, engineers designed and built a capping stack device, which could shut in the well. The top hat was removed starting July 10, 2010, with the capping stack installed and the well shut in by July 15, 2010.

The fluids that leaked from the crippled Macondo well included gas and liquid phases. Because of the high pressure deep in the ocean and the high temperature of the release fluids, the liquid phase included a large fraction of dissolved gases, and the gas phase contained a large fraction of higher-molecular-weight compounds. As the discharged fluids rose in a plume above the wellhead, they entrained ambient seawater, eventually forming a flow of seawater, dissolved hydrocarbons, and gas bubbles and oil droplets that intruded into the ocean water column near 1,100 meters deep. Almost all of the methane released from the wellhead entered this subsurface intrusion layer along with significant fractions of other dissolved compounds and some tiny (order 100 microns) oil droplets (Lehr et al., 2010). The oil remaining in larger droplets rose to the sea surface, forming slicks.

Subsea dispersant injection was a unique response option used during the DWH oil spill. Dispersant was mixed with the oil and gas at the release point in an effort to disperse the oil that was escaping collection. Dispersant acts to reduce the interfacial tension between oil and water, allowing smaller oil droplets to form and be dispersed in the water column. While a significant fraction of oil still reached the sea surface and formed slicks (where targeted areas were then treated with surface dispersant), subsea dispersant injection significantly improved air quality by promoting greater dissolution of low-molecular-weight, volatile compounds from the oil during transit through the deep sea (Gros et al., 2017; Zhao et al., 2021).

While the oil was still flowing, BP announced they would provide \$500 million to support a 10-year research program on the impact of the spill and improving capabilities to respond to future spills. This funding, which formed the Gulf of Mexico Research Initiative (GoMRI), supported more than 2,500 scientists and was a catalyst for an unprecedented era of oil spill science advancement (Colwell et al., 2021). In 2013, with \$500 million in criminal settlement funds from the companies involved in the DWH oil spill, the National Academies' Gulf Research Program (GRP) was created. The studies, projects, and activities conducted by the GRP will advance and apply science, engineering, and public health knowledge to reduce risks from offshore oil spills and will enable the communities of the Gulf to better anticipate, mitigate, and recover from future disasters. Additional funds of more than \$19 billion were distributed from settlement of civil, criminal, natural resource, and administrative claims to support RESTORE Act programs, the Oil Spill Liability Trust Fund, Natural Resource Damages, and economic claims to Gulf of Mexico states.

North American waters: includes the countries of Canada, the continental United States and Hawai'i, and Mexico; extending seaward of these land masses to the exclusive economic zone (EEZ) established by the 1982 United Nations Convention in the Law of the Sea. In regions where the seaward boundary of the EEZ terminated before the base of the continental slope, North American waters extend to the base of the slope.

There are several important distinctions between this study and previous ones in the series. "Oil" is referred to in several ways in the Statement of Task, including as "hydrocarbons," "fossil fuel hydrocarbons," and "petroleum." The committee chose to define "oil" in this report to mean "fossil fuel hydrocarbons." Whereas the 2003 report's definition was limited to the liquid state, this committee's definition incorporates the gas, liquid, and solid states of fossil fuel hydrocarbons. Including the gaseous state allows the committee to consider hydrocarbon inputs into the water column such as those dissolved from natural seeps.

By opening up the definition of fossil fuel hydrocarbons to the solid state, flocculants and tar balls are also addressed. The committee also discussed possible inclusion of plastics composed of hydrocarbons synthesized from petrochemicals in the inputs section, as they are a source of fossil fuel hydrocarbons in the sea, but (1) the committee felt plastics, which

BOX 1.2
Statement of Task

The study will serve as an update of the previous report's (*Oil in the Sea III: Inputs, Fates, and Effects*, 2003) inventory of the sources, composition, and quantity of hydrocarbon inputs to the marine environment and assessment of the state of the science on the fate and effects of fossil fuel hydrocarbons in the marine environment. To the extent possible, Oil in the Sea IV will also identify, categorize, and quantify these sources of hydrocarbons (and their chemical composition) with an emphasis on North American waters. The committee will examine worldwide data in an effort to place numbers derived for North American waters into a global context.

Specifically, the committee will:

1. Examine natural and anthropogenic sources of hydrocarbons entering the marine environment, including but not limited to:
   - natural hydrocarbon seeps;
   - extraction of petroleum (spills from production and drilling platforms, deposition from platform air emission, produced water and gas condensate discharges—disaggregated by platform type [e.g., deepwater, ultra-deep, shallow waters, leaks associated with platform decommissioning]);
   - transportation of petroleum (pipeline spills, tanker spills, operational discharges, coastal facility spills, deposition from tanker exhaust and volatile organic compound [VOC] emissions); and
   - consumption of petroleum (land-based runoff, recreational vehicle discharge, spills from commercial vessels <100 GT, operational discharges from commercial vessels, atmospheric deposition from land-based sources, aircraft dumping during emergencies).
2. Identify and evaluate, to the extent possible, sources of quantitative information about approaches for estimating the volume of hydrocarbon input to the marine environment worldwide from all sources. Based on these sources, develop, and summarize quantitative estimates of hydrocarbon inputs to the marine environment with an emphasis on North American waters and provide estimates of the upper, lower, and most probable values for each subcategory.
3. Assess and discuss the physical and chemical characteristics and behavior of these hydrocarbons, the transport and fate of various hydrocarbon mixtures in the marine environment, and the effects of these mixtures on marine life and ecosystems.
4. Characterize the risk posed to the marine environment by fossil fuel hydrocarbon components or type of input, given the range of organisms or ecosystems likely to be affected.
5. Review progress in implementing the recommendations from the 2003 report regarding inputs, fates, and effects and identify priority recommendations that have yet to be implemented.
6. Provide recommendations to improve estimates of inputs and identify focus areas for reducing hydrocarbon inputs from human activities.
7. Provide recommendations to improve understanding of the fates and effects of hydrocarbon inputs from human activities and strategies for reducing the more harmful effects.

include more than hydrocarbons (e.g., cellulose acetate), were beyond the spirit of the task and (2) a National Academies report quantifying the United States' contribution to global marine plastic waste (NASEM, 2021) was under way concurrently with the *Oil in the Sea IV* study and should be referenced to understand the additional volume, transport, and fate of plastic pollution in the sea.

Additionally, the committee included a chapter (Chapter 2) to further define the commonly used but complex term "oil" and to review the basic physical chemical properties of gas, oil, and gas and oil mixtures. This chapter is foundational to later discussions on fates and effects of oil in the sea in Chapters 5 and 6. Geographically, this study encompasses North American waters, which the committee defined to include waters surrounding the continental United States, Hawai'i, Canada, and Mexico. The committee used the same geographic zones as presented in *Oil in the Sea III*, which are also included in Appendix A of this report.

In terms of defining marine life and ecosystems, the committee determined that humans are part of the marine ecosystem and therefore examined the effects of oil in the sea on human health that had not been thoroughly evaluated in previous reports in this series. This evaluation included the potential direct effects of crude oil and of crude oil components and derivatives through inhalation, ingestion, and skin contact on response workers and community members. The committee also evaluated the growing literature on the socioeconomic impact of oil spills on the mental and behavioral health of local community members.

Fates and effects of oil in the sea are not only coupled in many ways, they are also dependent on the particular oil spill response. Each response effort, from natural attenuation to complex operations, influences the fate and effects of oil in the sea. For this reason, a chapter on source control and response is also included (see Chapter 4).

1.2.2 Progress on *Oil in the Sea III* Recommendations

The committee's charge includes identifying progress made on recommendations included in the *Oil in the Sea III* report and, specifically, highlighting recommendations that have not been acted on, but remain a priority. Table 1.2 summarizes the committee's opinion on the status of each recommendation in *Oil in the Sea III*. Many of the recommendations are not

TABLE 1-2 Recommendations from *Oil in the Sea III: Inputs, Fates, and Effects*

Key:

Significant progress has been achieved

Recommendation partially addressed

Recommendation not addressed

#	Recommendation (with cross-reference to information supporting the rating, within the current report)
1	Federal agencies, especially the U.S. Geological Survey (USGS), the Minerals Management Service (MMS), and the National Oceanic and Atmospheric Administration (NOAA) should work to develop more accurate techniques for estimating inputs from natural seeps, especially those adjacent to sensitive habitats. (Chapter 3)
2	To refine estimates associated with non-point sources, federal agencies, especially the U.S. Environmental Protection Agency (U.S. EPA) and the USGS, should work with state and local authorities to routinely collect and share data on the concentration of petroleum hydrocarbons in major river outflows and harbors in storm and wastewater streams. (Chapter 3)
3	The comprehensive port control regime, administered by the U.S. Coast Guard (USCG), cooperative programs with ship owners and the boating community, and active participation of the International Maritime Organization (IMO) in developing effective international regulatory standards have contributed to the decline in oil spills and operational discharges. These efforts and relationships should be continued and further strengthened where appropriate. (Chapter 3)
4	Federal agencies, especially the USCG, should work with the transportation industry to undertake a systematic assessment of the extent of non-compliance. If the estimates of noncompliance assumed in this report are essentially correct, more rigorous monitoring and enforcement policies should be developed and implemented. (Chapter 3)
5	Federal agencies, especially the U.S. EPA, should continue efforts to regulate and encourage the phase-out of inefficient two-stroke engines, and a coordinated enforcement policy should be established. (Chapter 3)
6	The USCG should work with the IMO to assess the overall impact on air quality of VOC from tank vessels and establish design and/or operational standards on VOC emissions where appropriate. (Chapter 3)
7	Federal agencies, especially the Federal Aviation Administration, should work with industry to more rigorously determine the amount of fuel dumping by aircraft and to formulate appropriate actions to limit this potential threat to the marine environment. (Chapter 3)
8	Federal agencies, especially NOAA, MMS, the USCG, and the USGS, should work with industry to develop and support a systematic and sustained research effort to further basic understanding of the processes that govern the fate and transport of petroleum hydrocarbons released into the marine environment from a variety of sources (not just spills). (Chapter 5)
9	Federal agencies, especially the U.S. Coast Guard, NOAA, and U.S. EPA, should work with industry to develop a more comprehensive database of environmental information and ambient hydrocarbon levels, and should develop and implement a rapid response system to collect in situ information about spill behavior and impacts. (Chapters 5 and 6)
10	Federal agencies, especially the USGS, NOAA, U.S. EPA, and MMS, should develop and support targeted research into the fate and behavior of hydrocarbons released to the environment naturally through seeps or past spills. (Chapter 5)
11	Federal agencies, especially the USGS and U.S. EPA, should work with state and local authorities to establish or expand efforts to monitor vulnerable components of ecosystems likely to be exposed to petroleum releases. (Chapter 6)
12	To assess the impacts attributable to different sources including oil spills and nonpoint sources, federal agencies, especially the USGS and U.S. EPA should work with state and local authorities to undertake regular monitoring of total petroleum hydrocarbon (TPH) and polycyclic aromatic hydrocarbon (PAH) inputs from air and water (especially rivers and harbors) to determine background concentrations. (Chapter 3)
13	Federal agencies, especially the USGS, MMS, NOAA, and U.S. EPA, should work with industry to develop or expand research efforts to understand the cumulative effects on marine organisms. Furthermore, such research efforts should also address the fates and effects of those fractions that are known or suspected to be toxic in geographic regions where their rate of input is high. (Chapter 6)
14	Federal agencies, especially the USGS, U.S. EPA, and NOAA, should work with state and local authorities and industry to implement comprehensive laboratory- and field-based investigations of the impact of chronic releases of petroleum hydrocarbons. (Chapter 6)
15	In areas of sensitive environments or at-risk organisms, federal, state, and local entities responsible for contingency plans should develop mechanisms for higher levels of prevention, such as avoidance, improved vessel tracking systems, escort tugs, and technology for tanker safety. (Chapter 3)
16	The U.S. Departments of the Interior and Commerce should identify an agency, or combination of agencies, to develop priorities for continued research on the following: • the structure of populations of marine organisms and the spatial extent of the regions from which recruitment occurs; • the potential for cascades of effects when local populations of organisms that are key in structuring a community are removed by oiling; and • the basic population biology of marine organisms, which may lead to breakthroughs in understanding the relationship between sublethal effects, individual mortality and population consequences. (Chapter 6)
17	The federal agencies identified above, in collaboration with similar international institutions, should develop mechanisms to facilitate the transfer of information and experience. (Chapter 6)

Oil in the Sea: Inputs, Fates, and Effects

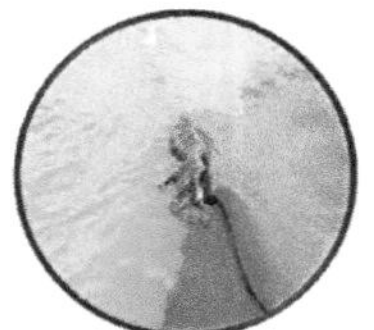

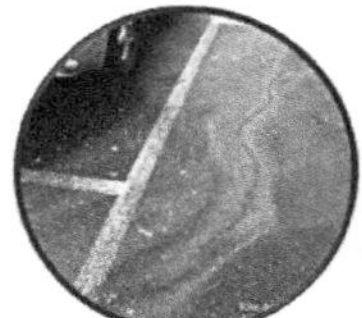

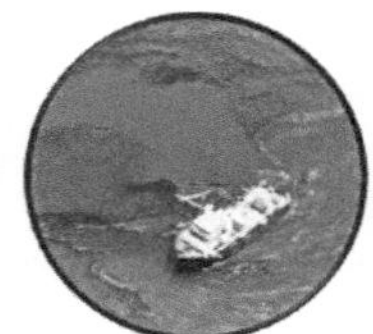

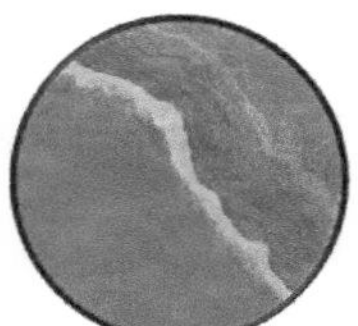

What is Oil? (Chapter 2)
Where does oil originate? (Chapter 3)
What can be done? (Chapters 3 & 4)
Where does the oil go? (Chapter 5)
What harm could the oil do? (Chapter 6)

FIGURE 1.12 *Oil in Sea IV: Inputs, Fates, and Effects* report organization.
PHOTO CREDIT (left to right): NOAA, Alan Mearns, NOAA, NOAA, NOAA.

written in a way in which success is measurable; in some cases the agencies called out have changed (or the agencies' responsibilities have changed); and in many cases there is no funding source to support the recommendations. For these reasons, the committee categorized the recommendations by whether or not (1) significant work has been accomplished, (2) the recommendation has been partially addressed, or (3) the recommendation has not been acted upon.

In the committee's assessment, significant progress has occurred toward addressing three of the recommendations. Recommendations partially addressed in the areas of fates and effects are primarily due to the surge in research following the DWH oil spill; the ranking is not necessarily indicative of specific agency or interagency response. Most notably, significant progress was made on recommendations aimed at reducing oil inputs into the sea (Recommendations 3, 5, and 6)—the committee commends these efforts. Recommendations 1, 2, 4, 7, and 12, which were aimed at better quantification of oil inputs into the sea were largely not addressed. Recommendation 8 regarding sustained funding and effort to advance understanding of fates, and Recommendations 14 and 16, regarding the effects of oil in the sea were not addressed outside the context of the DWH-funded research. Details on recommendations that have not been fully addressed but remain a priority are called out in the appropriate sections within the body of this report.

Throughout the study process, the committee acknowledged the remarkable and unprecedented advancements seen in oil spill science in the past 20 years, especially following the DWH oil spill.[5] Research funding sustained for over a decade was a key factor leading to the wealth of knowledge described in this report. This is not to say that unrelated efforts have not also been impactful, but it is clear what can be done with a continued dedicated funding stream, as seen over the past decade and as called for in previous *Oil in the Sea* recommendations.

This report aims to augment, but not repeat the information provided in *Oil in the Sea III*. During the writing of this report, more published information became available and will become available after this compilation is completed. The committee has aimed to bring the most recent published research to light.

Reiterating past recommendations, and to keep the momentum expanding understanding of the fates and effects of oil in the sea in the context of a constantly evolving fossil fuel landscape (i.e., changes in exploration, production, transportation, consumption of oil as well as the chemical composition of oil being transported and consumed), and to reduce and mitigate oil's impact on the marine environment, the committee developed the following overarching conclusion and recommendation, which is echoed throughout the report.

Conclusion: Long-Term Funding Enables Great Advances in Knowledge

The establishment of, and 10 years of funding for, the GoMRI after the DWH oil spill resulted in an extraordinary output of discipline-specific research and, more importantly, of multidisciplinary research by funding a mix of field, laboratory, mesocosm, and test facility research, and related modeling. Similar periods of advancement were seen following other oil spills such as the *Exxon Valdez* and the *Hebei Spirit*. Advancement of oil spill science has a history of being hindered by a boom and bust funding cycle, which led to the inability to sustain continuity of oil spill research along with the scientific community conducting that research.

1.2.3 Report Organization

The Statement of Task aligns with the narrative of how oil interacts with the sea, and this is how the report is organized, shown in Figure 1.12. Chapter 2 sets the stage by delving into

[5] There are thousands of scientific publications and reports pertinent to the committee's task; where practicable recent published reviews and a selection of relevant specific references were cited—it was impractical to cite all papers in the report's subject areas published since 2003.

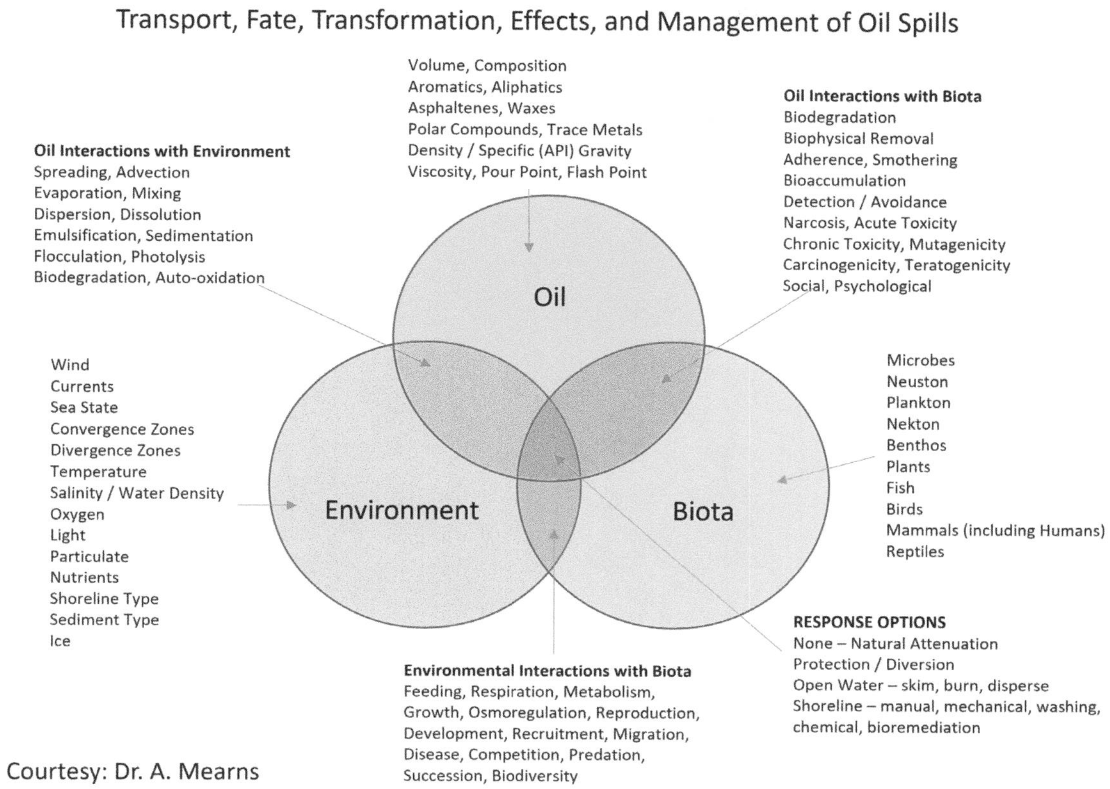

Transport, Fate, Transformation, Effects, and Management of Oil Spills

Volume, Composition
Aromatics, Aliphatics
Asphaltenes, Waxes
Polar Compounds, Trace Metals
Density / Specific (API) Gravity
Viscosity, Pour Point, Flash Point

Oil Interactions with Environment
Spreading, Advection
Evaporation, Mixing
Dispersion, Dissolution
Emulsification, Sedimentation
Flocculation, Photolysis
Biodegradation, Auto-oxidation

Oil Interactions with Biota
Biodegradation
Biophysical Removal
Adherence, Smothering
Bioaccumulation
Detection / Avoidance
Narcosis, Acute Toxicity
Chronic Toxicity, Mutagenicity
Carcinogenicity, Teratogenicity
Social, Psychological

Oil

Wind
Currents
Sea State
Convergence Zones
Divergence Zones
Temperature
Salinity / Water Density
Oxygen
Light
Particulate
Nutrients
Shoreline Type
Sediment Type
Ice

Environment

Biota

Microbes
Neuston
Plankton
Nekton
Benthos
Plants
Fish
Birds
Mammals (including Humans)
Reptiles

RESPONSE OPTIONS
None – Natural Attenuation
Protection / Diversion
Open Water – skim, burn, disperse
Shoreline – manual, mechanical, washing, chemical, bioremediation

Environmental Interactions with Biota
Feeding, Respiration, Metabolism,
Growth, Osmoregulation, Reproduction,
Development, Recruitment, Migration,
Disease, Competition, Predation,
Succession, Biodiversity

Courtesy: Dr. A. Mearns

FIGURE 1.13 Venn diagram depicting the connections between oil, the environment, biota, and response options.
SOURCE: Adapted from Mearns, 1997, by Dr. Alan Mearns, National Oceanic and Atmospheric Administration.

the details of exactly what oil is, incorporating important advancements in analytical chemistry and models to predict oil behavior. The journey begins in Chapter 3 with the origin of oil that ends up in the sea. Where is the oil coming from? What are the quantities of oil input into the sea? How have the quantities changed since *Oil in the Sea III*? How do we expect the volumes to change into the future and what other forms of potential inputs might be realized in the coming years? What can be done to prevent oil from entering the sea? Chapter 4 then discusses accidental spill mitigation measures through both source control and response in terms of what can be done to reduce the quantities of oil entering the environment and also to minimize the negative effects of oil on the marine ecosystem. Chapter 5 continues the story of what happens to the oil once it reaches the sea, assimilating significant advancements in understanding of transport and fate of oil in the sea. The story ends with Chapter 6, which discusses the harm the oil can have on the marine environment and marine organisms over its journey from source to contact, synthesizing a vast amount of literature focused on effects of oil on both the ecosystem and on humans.

Within each subsequent chapter, findings are identified in bold text. Synthesis of those findings is included in the last subsection of each chapter, described through conclusions and a list of specific research needs to advance understanding of the particular subject.

Although the committee has divided the report into distinct chapters, complex and intrinsic interdependencies exist among inputs, fates, effects, and the measures to reduce them—either from the source or by response. Because of the interconnected nature of oil spill science, overall recommendations are not specifically tied to one topic (or chapter) or another and are therefore presented in the final chapter of this report (see Chapter 7).

The Venn diagram in Figure 1.13 depicts the complicated relationships among oil parameters, environmental conditions, and biological components. These interactions illustrate the multitude of parameters adding to the complexity of changes over time and space in determining the potential effects from an oil spill and appropriate response strategies. While the listing of items in the diagram is not exhaustive, it clearly depicts the numerous combinations of possible variables. Thus, this simple diagram helps to serve as a visualization of the aspects covered in each of the chapters in this report.

2

Petroleum as a Complex Chemical Mixture

2.1 COMPOSITION OF PETROLEUM AS A COMPLEX MIXTURE OF CHEMICALS AND PROGRESS IN ANALYSES OF THIS COMPLEX MIXTURE

2.1.1 Introduction

This chapter provides a brief overview of the classification of petroleum oils used by industry as it produces, transports, and refines them by means of distillation and petroleum cracking; and the classifications of petroleum oils relevant to oil spill response. Petroleum and many of its distillation products (e.g., fuel oils) and residues such as asphalt are complex mixtures of individual chemical compounds. A few molecular structures of the thousands of chemicals involved are provided as illustrations of this complexity. Analytical chemical methods used in the forensics of connecting environmental occurrence to sources of inputs, and for studies of the fates and effects of inputs are described. Examples of significant advances in analytical chemical methods used in petroleum pollution research are presented and discussed. This informs recommendations for adoption for applications in oil spill response, for assessment of fates and effects, and for further developments of analytical methodologies.

The basic physical chemical properties of gas, oil, and gas and oil mixtures are reviewed to set the scene for Chapter 5, Fates, and Chapter 6, Effects.

"Oil" is a three-letter word used in everyday discussions. This simple term unfortunately masks, for many people, that the chemical composition of petroleum is a complex mixture of thousands of chemicals. Furthermore, each oil has a unique combination and proportion of these chemicals. These chemical composition differences translate to both profound and subtle, but important, differences in each chemical mixture's (each oil's) fate and effects in the environment. The chemical composition differences underpin distinguishing one oil from another, that is, the forensics of sources of inputs or spills.

Chemical structures impart properties such as volatility, solubility in water, ease or difficulty of dispersion, environmental persistence, and the ability to react in various ways. These are important to understanding the fates and effects of oil in the environment and in formulating detection, response, mitigation, and cleanup plans.

2.1.2 Definitions and Classifications of Petroleum

The fact that petroleum is a complex mixture of thousands of chemicals has been known for many decades. Many of its components are hydrocarbons that are composed only of hydrogen and carbon, whereas other petroleum chemicals include additional elements such as nitrogen, sulfur, and oxygen (N, S, and O) and thus are not strictly hydrocarbons, even though common use may lump all of these components together as hydrocarbons. A few of the chemicals also contain metals such as nickel and vanadium. This complex chemical composition varies within and between oil reservoirs. This is the result of (1) the sources of the organic matter from which geological processes (designated generally as maturation) generate petroleum in source rock sedimentary formations, (2) migration and accumulation processes in reservoir rock sedimentary formations, (3) in situ biodegradation, and (4) the varied influence of these processes in different geological formations. The details of gas and oil generation and accumulation in reservoirs, as they influence petroleum's complex composition, are beyond the scope of this report and have been described in several publications including Tissot and Welte (1984), Hunt (1996), Peters and Fowler (2002), Peters et al. (2005), and Philp (2018).

The complexity of petroleum's chemical composition and the variations between reservoirs necessitated, especially in the early years of the production and use of petroleum, operational definitions for material in reservoirs or obtained from reservoirs. Hunt (1996) has a useful glossary that includes these operational terms and reference to a 137-page detailed illustrated glossary by Miles (1989).

An example is the definition of petroleum: "Petroleum is a form of bitumen composed principally of hydrocarbons and existing in the gaseous or liquid state in its natural reservoir" (Hunt, 1996). In the strictest definition, the term "bitumen" refers to any organic material extracted from ancient sediments by organic solvents, where the exact organic solvent or solvent mixture and conditions of extraction are not specified. "Bitumen petroleum" is a term describing a collective of natural gas, crude oil, and asphalt components at the Earth's surface temperature and pressure (e.g., Hunt, 1996). However, the term "bitumen" is also used in various ways to describe the heaviest crude oils, some of which are solid or semi-solid rather than liquid in their natural reservoirs. The glossary of a report on spills of diluted bitumen (NASEM, 2016) states "Bitumen—A mixture of hydrocarbons that is too viscous to flow under ambient conditions. Commercial quantities are recovered by thermal processes." Radović et al. (2018) provide a useful summary explanation: "Technically, under reservoir conditions, we can differentiate 'heavy' oils (gravity 10^0–20^0 American Petroleum Institute [API], viscosity >100 centipoises, cp) which can flow; and non-flowable (*sic*) "bitumen" (gravity 100 API; viscosity >10,000 cp). However, since they are part of the same physicochemical continuum, the two terms will be used interchangeably in this chapter." We use the approach by Radović et al. (2018) in this report with respect to the term "bitumen" rather than that of Hunt (1996).

Crude oil is defined as "petroleum that is removed from the earth in a liquid state or is capable of being so removed" (Hunt, 1996). There are various classifications of crude oils, ranging from those used in the economic arena to gauge prices and investments in stock to those used in the production and processing or refining of crude oils (Hunt, 1996; Wnek et al., 2018; API, 2022, among others). In the latter situation, the crude oils are often classified by their density and chemical composition (see discussion of chemical composition in a following section) as light, medium, heavy crude oils, paraffinic or naphthenic crude oils, and sour or sweet crude oils (Hunt, 1996; Wnek et al., 2018). A sour crude oil or fuel oil is an oil containing quantities of sulfur compounds such as hydrogen sulfide or sulfur-containing compounds such as mercaptans that are noxious to smell. Sweet oils do not contain noxious quantities of those compounds (Hunt, 1996).

There are also designations of *live oil* and *dead oil*. Live oil still contains gases, mainly methane, but also ethane, propane, and butane (see Section 5.2.1); dead oil does not contain these gases. Generally, oil well blowout accidents such as the *Deepwater Horizon* (DWH) involve live oil, and this is important because of the influence of gas bubbles on the initial fate of the discharged petroleum. In comparison, oil spills from oil tankers do not contain methane, ethane, propane, or butane, unless the spill or release is from a liquefied natural gas (LNG) tanker or coastal facility. LNG is mainly methane (85–95%) with 5–15% ethane, propane, butane, and nitrogen (DOE, 2005).

An amalgamation of data from many dead crude oils, showing composition related to percentage of molecular types, general classification of fractions of a crude oil, the molecular type of the groups of chemicals using names common in the oil and gas industry, and boiling point in °C, is presented in Figure 2.1.

Oil classification schemes most relevant to our report are classifications by oil groups (see Box 2.1) and by crude oil density (see Table 2.1). Oil groups are classifications of refined products, which are distinguished by their composition and viscosity. Crude oils are often described by their density through a specialized form of the specific gravity.

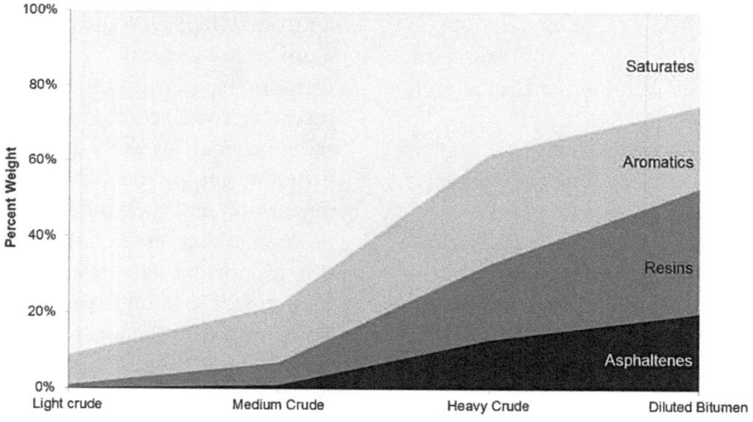

FIGURE 2.1 Components of typical crude oil types.
NOTES: Names of the crude oil types on the x-axis and percentages on the y-axis of the four major commonly used categories of molecules in crude oils: saturates, aromatics, resins, and asphaltenes (identified as SARA within the petroleum industry and the scientific communities involved in responses to oil spills and for oil inputs, fates and effects). See text of this chapter for further explanation.
SOURCE: NASEM, 2016.

BOX 2.1
Description of Refined Petroleum Products by Group Classification and the U.S. Code of Federal Regulations

Group 1: Non-Persistent Light Oils (e.g., gasoline, condensate)

- Highly volatile (should evaporate within 1–2 days).
- Do not leave a residue behind after evaporation.
- High concentrations of toxic (soluble) compounds.
- Localized, severe impacts to water column and intertidal resources.
- Cleanup can be dangerous due to high flammability and toxic air hazard.

Group 2: Persistent Light Oils (e.g., diesel, No. 2 fuel oil, light crudes)

- Moderately volatile; will leave residue (up to one-third of spill amount) after a few days.
- Moderate concentrations of toxic (soluble) compounds.
- Will "oil" intertidal resources with long-term contamination potential.
- Cleanup can be very effective.

Group 3: Medium Oils (e.g., most crude oils, IFO 180)

- About one-third will evaporate within 24 hours.
- Oil contamination of intertidal areas can be severe and long-term.
- Oil impacts to waterfowl and fur-bearing mammals can be severe.
- Cleanup most effective if conducted quickly.

Group 4: Heavy Oils (e.g., heavy crude oils, No. 6 fuel oil, bunker C)

- Little or no evaporation or dissolution.
- Heavy contamination of intertidal areas likely.
- Severe impacts to waterfowl and fur-bearing mammals (coating and ingestion).
- Long-term contamination of sediments possible.
- Weathers very slowly.
- Shoreline cleanup difficult under all conditions.

Group 5: Sinking Oils (e.g., slurry oils, residual oils)

- Will sink in water.
- If spilled on shoreline, oil will behave similarly to a Group 4 oil.
- If spilled on water, oil usually sinks quickly enough that no shoreline contamination occurs.
- No evaporation or dissolution when submerged.
- Severe impacts to animals living in bottom sediments, such as mussels.
- Long-term contamination of sediments possible.
- Can be removed from the bottom of a water body by dredging.

SOURCE: NOAA (2021).

Specific gravity of oils is often reported in units of API gravity, a parameter that evolved and has been standardized over the decades of the petroleum era. API gravity is related to the specific gravity of an oil, by the formula API gravity = (141.5/specific gravity at 15.6°C) – 131.5. An API gravity of 10 is neutrally buoyant in freshwater; lower values would sink, and higher values would float. Table 2.1 provides a few examples of API gravity for some oil types and with some examples connected to geographic locations.

2.1.3 Oils or Fuels of Emerging Importance

There are some categories of petroleum and/or oils that have emerged recently or are of emerging importance for a variety of factors including ease of fulfilling demand for petroleum (diluted bitumen), meeting new environmental regulatory standards (low sulfur fuel oils), improving performance (newer lubricating oils), and reducing net carbon inputs to the environment (biofuels). These are described briefly in the following sections.

TABLE 2.1 Crude Oil Classifications by API Gravity (dead oil composition)

Oil Type	API Gravity*	Examples (API)	References
Condensate	>45°	Agbami, Nigeria (48°)	Speight (2014)
		West Texas Intermediate (40°)	Speight (2014)
Light oil	35°–45°	Macondo (40°)	Reddy et al. (2012)
Medium oil	25°–35°	Alaska North Slope (32°)	Speight (2014)
Heavy oil	15°–25°	Venezuela Heavy (17°)	Speight (2014)
		Tar sands: Orinoco, Venezuela (8°–12°);	
Extra heavy oil	<15°	Oil sands: Athabasca, Canada (6°–10°)	Tissot and Welte (1984)

NOTES: For comparison, freshwater has an API gravity of 10°. * API gravity = (141.5/specific gravity at 15.6°C) – 131.5.
SOURCE: Adapted and modified with permission from Rullkötter and Farrington, 2021.

2.1.3.1 Diluted Bitumen

Diluted bitumen, a petroleum product derived from oil sands, is increasing in production. It is transported primarily inland in North America by pipelines and rail tank cars, although increased shipping by tanker to Asia is forecast. Semi-solid bitumen extracted from oil sands is diluted with lighter molecular weight petroleum distillates such as condensates, naphtha, diesel, or synthetic crude oil, creating blends such as "dilbit" and "synbit," respectively, that enable the blend to be transported as a liquid. Diluted bitumen, being a two-component blend, has bi-phasic characteristics if spilled: the very light diluent volatilizes rapidly (although not completely) leaving a viscous residue.[1] A report by the *National Academies, Spills of Diluted Bitumen from Pipeline: A Comparative Study of Environmental Fate, Effects, and Response*, contains pertinent, important information (NASEM, 2016). See also Chapter 5, Fates, and Section 5.3.2.4.

2.1.3.2 Heavy Fuel Oils and Low Sulfur Fuel Oils

Heavy fuel oil (HFO) is a generic term to describe fuels primarily used to power marine vessels. It is a residual fuel from the distillation of crude oils that is blended with other lighter molecular weight fuels oils to have contents of sulfur in the 3.5% range. They contribute to air pollution in general and especially in port areas. Intermediate fuel oils (IFOs), designated as marine diesel oils by some organizations, are distilled fuels blended with higher proportions of heavy fuel oils. According to a narrower definition, marine diesel oil is a blend of distillate fuels and a very low proportion of heavy fuel oil[2].

Recent national and international regulations have phased out HFOs with high sulfur content, that is, high sulfur fuel oils (HSFOs) with 3.5% sulfur, replacing them with low sulfur fuel oils (LSFOs—0.5% sulfur), and ultra-low sulfur fuel oils (ULSFOs—0.1 % sulfur).[3] The LSFOs and ULSFOs are blends of various types of both heavy oils and light oils (Sørheim et al., 2020). Burning HSFOs is still allowed if the emissions are scrubbed with an exhaust gas cleaning system, often referred to as a scrubber, to the equivalent sulfur oxide limit as achieved by burning LSFOs (see Section 5.2.3.1).

2.1.3.3 Lubricating Oils

Petroleum-related lubricating oils have evolved over the decades from initially being mainly composed of higher molecular weight cyclic alkanes from petroleum to the present mixtures of those hydrocarbons with synthetic chemical additives, or composed almost entirely of synthetic chemicals including biosynthetic or biological materials-based lubricants (Bolina et al., 2021; Murru et al., 2021, and references therein). One example, among many, are polyalphaolefin-based products, a synthetic lubricant produced from alpha olefin (see Section 2.1.3.5) and used as a major lubricant in wind-power turbines, including offshore wind turbines.[4]

2.1.3.4 Biofuels

Since the NRC 2003 report, the production, transportation, and use of biofuels has increased. There are two main biofuels at present: ethanol and biodiesel. Ethanol is produced mainly from corn and similar crops and biodiesel is a liquid fuel consisting of new and/or used vegetable oils and animal fats, which significantly differ chemically from petroleum oils, comprising primarily natural lipids such as esters of fatty acids (e.g., triglycerides—also known as triacylglycerols) rather than hydrocarbons. There is substantial research ongoing in production of biofuels from various sources of biomass, including algal cultures grown for that specific purpose and converted waste biomass. The details of production, release, fates and effects of biofuels are beyond the scope of this report, and furthermore, the chemical compositions and spill behavior of biofuels are substantially different from those of petrochemical fuels. It should be noted that the biological origin of "biofuels" does not mean, a priori, that there will be no adverse effects if such biofuels are spilled or released as chronic inputs to the marine environment and, it is therefore important to keep track of chemical compositions and potential fates and effects of such spills or chronic inputs.

2.1.3.5 Olefins

Olefins are unsaturated hydrocarbons (also named alkenes or cyclic alkenes) with C=C double bonds in various places in the molecule, and with overall structures similar to the n-alkanes and branched alkanes, and cyclic alkanes. They are both biosynthesized naturally occurring compounds and are synthesized for a variety of purposes, including incorporation into drilling muds. The discharge of the untreated synthetic-based drilling muds into the marine environment is not allowed in the United States, Canada, and Mexico. However, past uses of olefins mean that they may be found in sediment environments near previous petroleum drilling operations, (e.g., see Stout and Litman, 2022).

2.1.4 Chemical Compositions

There are various ways that are accepted within the scientific community to depict molecular structures. This brief introduction is provided to capture the range of molecular size,

[1] Further relevant definitions and explanations can be found at https://www.oilsandsmagazine.com/technical/product-streams#streams.

[2] See https://www.mabanaft.com/en/news-info/glossary/details/term/marine-diesel-oil-mdo-intermediate-fuel-oil-ifo.html.

[3] See https://www.oiltanking.com/en/news-info/glossary/details/term/heavy-fuel-oil-hfo.html.

[4] For example, see https://klinegroup.com/wind-turbine-lubricants-an-important-segment-of-the-industry.

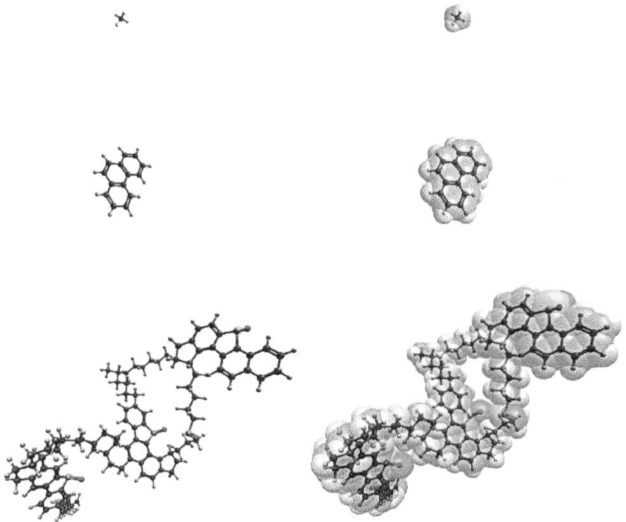

FIGURE 2.2 Depictions in three dimensions of example gas and oil molecules in relative size scale.
NOTES: Top row is methane. Middle row is the three ring aromatic compound phenanthrene. Bottom row is an example of a proposed structure of an asphaltene molecule. Left column: molecules shown without illustration of a rough approximation of electron clouds around the atoms. These are also known as ball and stick models. Right column: rough approximation of electron clouds around the individual atoms as they contribute to the overall size of the molecule in three dimensions. Black spheres represent carbon atoms, white spheres represent hydrogen atoms, yellow spheres represent sulfur atoms, and red spheres represent oxygen. The asphaltene molecule is a structure of one asphaltene molecule proposed by Yang et al. (2015).
SOURCE: Heather Dettman, Rafal Gieleciak, Natural Resources Canada.

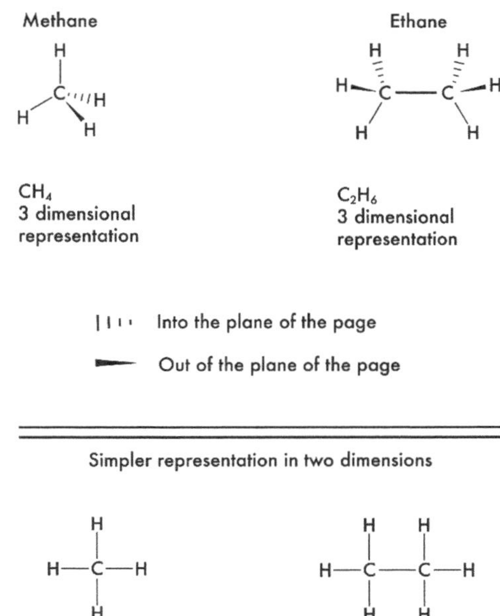

FIGURE 2.3 Three-dimensional, two-dimensional, and carbon–carbon bond representation of the gases methane and ethane.

examples of the classes of chemicals, and the complexity of various classes of chemicals found in oil and the important differences between molecules that contain the same number of elements of mainly carbon and hydrogen arranged in different spatial configurations. The differences in sizes and spatial configurations govern the fates and effects of the molecules in the environment.

Figure 2.2 provides a three-dimensional illustration of the comparison in relative size (to scale) between the simplest of hydrocarbons (methane), the three aromatic ring hydrocarbon phenanthrene, and one of the latest proposed structures for a relatively large asphaltene molecule. The information in this figure should be used as a guide when viewing the more traditional representation of molecular structures used in other figures in this report and in the scientific literature related to inputs, fates and effects of gas and oil chemicals in the marine environment. Also note that the representation of the asphaltene molecule is one of many such molecules in the asphaltenes. These molecules can assume various three dimensional configurations depending on their exact elemental compositions and chemical bonding, and interact with each other to form nanostructures as discussed briefly in a later section of this chapter.

A more familiar or traditional, longer used depiction of selected molecules is shown in the other molecular structure figures in this chapter. A few representative structures are depicted here, beginning with two and three dimensional skeletal configurations of methane and ethane. Figure 2.3 depicts the simplest structures, methane and ethane, and illustrates three-dimensional depictions, two-dimensional depictions, and a shorthand notation most often used to depict petroleum hydrocarbons. The shorthand notation for hydrocarbon composition is Cx, where x is the number of carbon atoms and the number of hydrogen atoms may or may not be included. Figures 2.4 and 2.5 present further example structures of increasing complexity.

2.1.4.1 Gases, Volatile Organic Compounds, and BTEX

The simplest manner to begin discussion of the composition of petroleum is with those constituent component chemicals that are volatile (i.e., quickly transform from the liquid to the gaseous phase) at average ranges of atmospheric pressure and temperature conditions at the Earth's surface. In general, these include chemical compounds of carbon and hydrogen atoms with relatively simple, or branched, short-chain or cyclic molecules of low molecular weight.

Gas in petroleum comprises mostly methane with varying additional quantities of ethane, propane, and butane. These are simple chemical structures (see Figures 2.3 and 2.4). There is overlap in the scientific literature of chemicals designated as gases and those designated as volatile organic compounds (VOCs) from petroleum. Generally, the overlap

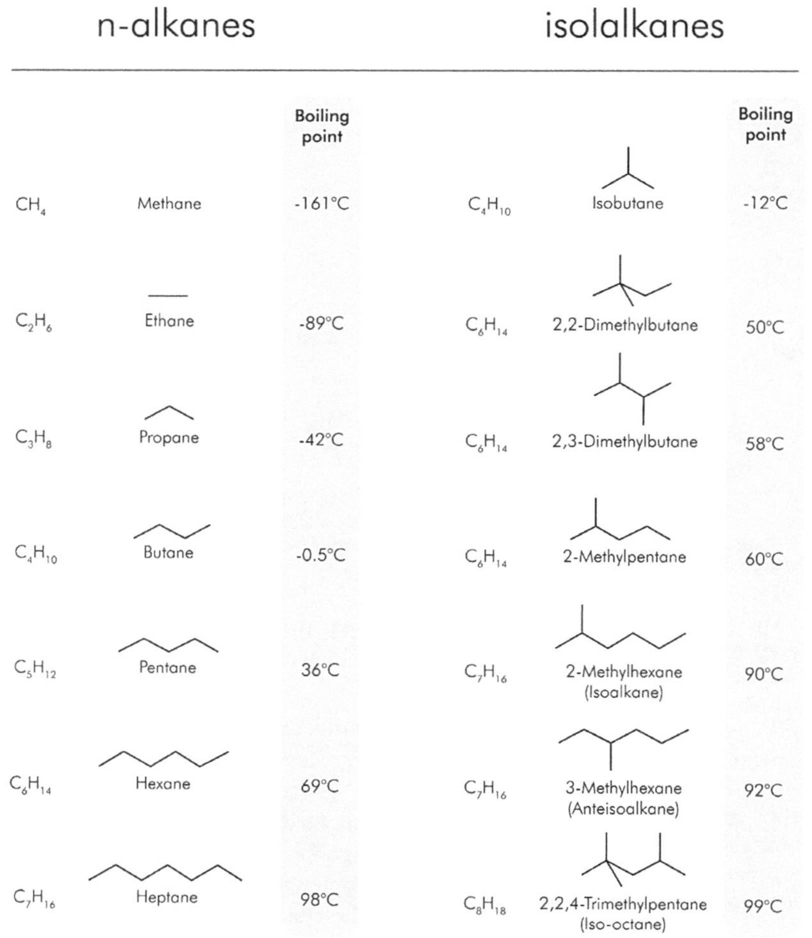

n-alkanes		Boiling point	isolalkanes		Boiling point
CH_4	Methane	-161°C	C_4H_{10}	Isobutane	-12°C
C_2H_6	Ethane	-89°C	C_6H_{14}	2,2-Dimethylbutane	50°C
C_3H_8	Propane	-42°C	C_6H_{14}	2,3-Dimethylbutane	58°C
C_4H_{10}	Butane	-0.5°C	C_6H_{14}	2-Methylpentane	60°C
C_5H_{12}	Pentane	36°C	C_7H_{16}	2-Methylhexane (Isoalkane)	90°C
C_6H_{14}	Hexane	69°C	C_7H_{16}	3-Methylhexane (Anteisoalkane)	92°C
C_7H_{16}	Heptane	98°C	C_8H_{18}	2,2,4-Trimethylpentane (Iso-octane)	99°C

FIGURE 2.4 Carbon–carbon bond (or skeletal) structures of *n*-alkanes and *iso*-alkanes (branched chain alkanes) found in petroleum (gas and oils). As indicated by the boiling points, these are gases or volatile compounds at atmospheric pressure.
SOURCE: Reprinted with permission from Hunt, 1996. Copyright 1997 American Chemical Society.

is in the designation of propane and butane. Examples of chemicals in the petroleum VOCs are the *n*-alkanes (straight chain hydrocarbons) *n*C3 to *n*C10 or propane, butane, pentane, hexane, heptane, octane, nonane, and decane. There are also branched chain alkanes having one or more alkyl substituents in different positions such as 2-methylbutane, 2-methylpentane, and 3-methylpentane (see Figure 2.4 for some examples). Cyclic alkanes include cyclohexane, and methylcyclopentane. A more complete list is presented by Wnek et al. (2018).

Among the petroleum VOCs are the mono-aromatic hydrocarbons benzene, toluene, ethylbenzene, and xylene isomers—commonly referred to as BTEX (see Figure 2.5). These are very important compounds from the perspective of human health concern during oil spills, especially for oil spill responders for at-sea oil spills (see Chapter 4) or the general public for nearshore and shoreline spills (see Chapter 6). They are also of concern regarding potential toxicity to marine mammals that may be in the vicinity of an oil spill (see Chapter 6).

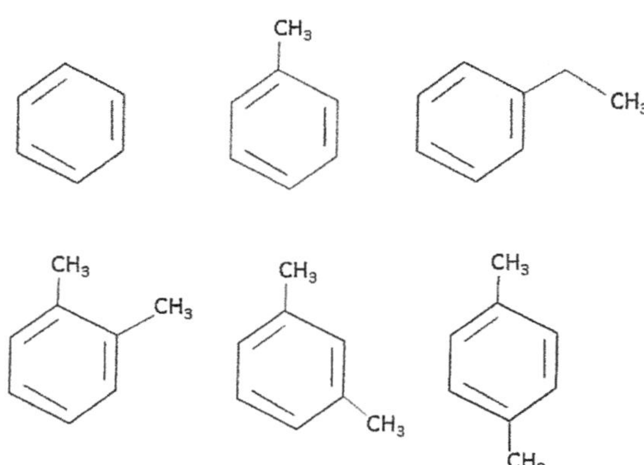

FIGURE 2.5 Chemical structures of the group of monoaromatic hydrocarbons collectively known as BTEX (benzene, toluene, ethyl benzene, and xylenes). Pi-bond electrons resonate within the ring structure. The structure is often depicted as a circle within the hexagon. If viewed edge-on, this would show clouds of electrons of these C=C bonds above and below the planar hexagon C skeleton.
SOURCE: Montero-Montoya et al., 2018. CC BY 4.0.

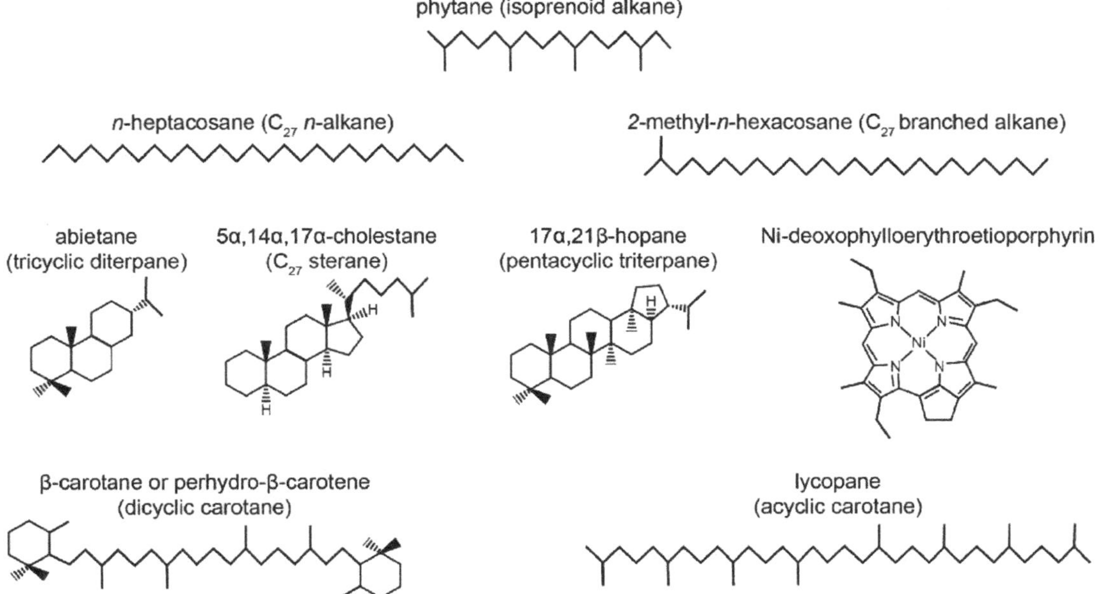

FIGURE 2.6 Examples of a saturated normal-alkane and common iso-alkane and cycloalkane petroleum biomarkers (see Box 2.2) and a nickel porphyrin found in crude oils.
NOTES: The dotted wedge-like bonds indicate C–H bonds into the plane of the page and the solid wedges indicate C–H bonds out of the plane of the page for four of the molecules depicted. (See Figure 2.1 for a simple example for methane and ethane.)
SOURCE: Rullkötter and Farrington, 2021. CC BY 4.0.

It is important to note that the U.S. EPA and similar other national and international agencies use the term "volatile organic compounds" (VOCs) to designate a much wider variety of volatile compounds and differentiate between indoor VOCs and outdoor VOCs.[5]

2.1.4.2 Saturates, Aromatics, Resins, and Asphaltenes (SARA) and Other Chemicals in Petroleum

One of the most common and useful ways to describe the composition of crude oil and its distillation products is an approach that is coupled to the analytical procedures discussed in the next section of the chapter. Petroleum can be divided into four fractions according to solubility and chemical composition: saturated hydrocarbons (alkanes comprising only carbon and hydrogen, with no unsaturated bonds), aromatic hydrocarbons (comprising unsaturated cyclic structures of carbon and hydrogen, with or without saturated side chains), resins (typically cyclic structures containing one or more heteroatoms in addition to carbon and hydrogen), and asphaltenes (similar in chemical composition to resins but having larger, more complex molecules such as that shown in Figure 2.2). The shorthand used is SARA. Functionally, the first three fractions (i.e., SAR) are soluble in alkane solvents such as *n*-pentane, *n*-hexane or *n*-heptane and are sometimes referred to as maltenes to distinguish them from the asphaltenes that are insoluble in light alkanes but soluble in toluene. Chemically, the saturated and aromatic hydrocarbons comprise only carbon and hydrogen atoms. Resins contain some hydrocabons and chemical compounds with additional elements such as N, S, and O, as is the case for asphaltenes. There is overlap in the solubility definition because some small resins behave like aromatic hydrocarbons.

Commonly, saturates represent the major fraction of oil. Chemical structures of alkanes range from simple methane (CH_4), a gas at atmospheric pressure, to volatile, liquid or waxy alkanes and intricate cyclic structures such as steranes and hopanes. Alkanes include three subgroups, with examples shown in Figure 2.6: (1) *n*-alkanes (or normal alkanes) are straight chains of carbon atoms with hydrogen bonded to the carbon atoms (e.g., *n*-heptacosane, *n*-C_{17}). (2) *iso*-alkanes are branched alkanes where one or more carbon atoms is bonded to a straight carbon chain (e.g. 2-methyl-*n*-hexacosane, also named 2-methylhexacosane). A special subgroup of branched alkanes is the isoprenoid branched alkanes, so named because of their structural relationship to a repeated unit of the isoprene moiety (e.g., phytane, β-carotane, and lycopane). (3) Cycloalkanes (also called cycloparaffins, naphthenes, alicyclic hydrocarbons) typically range from one ring of five or six carbon atoms to many fused rings, often with alkane substituents on the rings. Some specific examples are depicted in Figure 2.6—a tricyclic diterpane, a sterane, and a pentacyclic triterpane. Although these structures are flat on the page, in reality they are three dimensional "bent" structures as exemplified in the three dimensional example of Figure 2.7 for cholestane,

[5] See https://www.epa.gov.

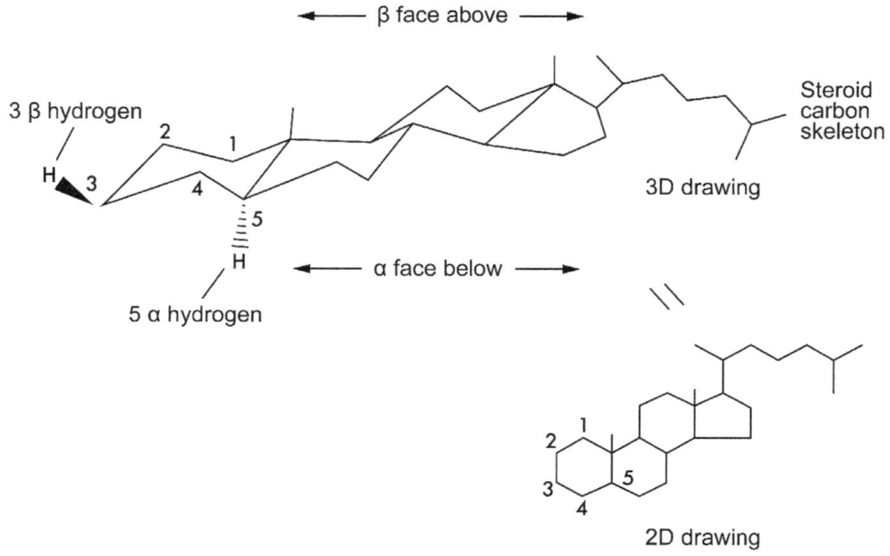

FIGURE 2.7 Three-dimensional rendering of the petroleum biomarker compound cholestane and comparison to two-dimensional drawing of the same molecule.
NOTE: The positioning of the carbon-hydrogen bond at the three position points above the plane of the molecule is designated 3β and that at the five position points below the plane 5α.
SOURCE: Adapted with permission from Hunt, 1996. Copyright 1997 American Chemical Society.

a sterane. The exact three-dimensional structures of these petroleum compounds are key to the use of steranes, triterpanes, and related compounds as petroleum biomarkers, as will be discussed in more detail in subsequent sections.

Aromatic hydrocarbons range from monocyclic aromatics to fused ring compounds comprising two or more rings (polycyclic aromatic hydrocarbons; PAH or PAHs). Benzene is a single-ring, six-carbon-atom compound as noted previously (see Figure 2.5), whereas naphthalene is a two-ring PAH, phenanthrene has three fused rings, and chrysene has four (see Figure 2.8). The rings are co-planar, and in addition to the carbon–carbon bonds, the ring carbons share a cloud of electrons resonating above and below the ring, a characteristic of aromatic rings, as noted earlier for BTEX compounds. These types of compounds were initially designated aromatic because the first few isolated and characterized had a distinctive aroma; it also is a reminder that smaller aromatic hydrocarbons are volatile.

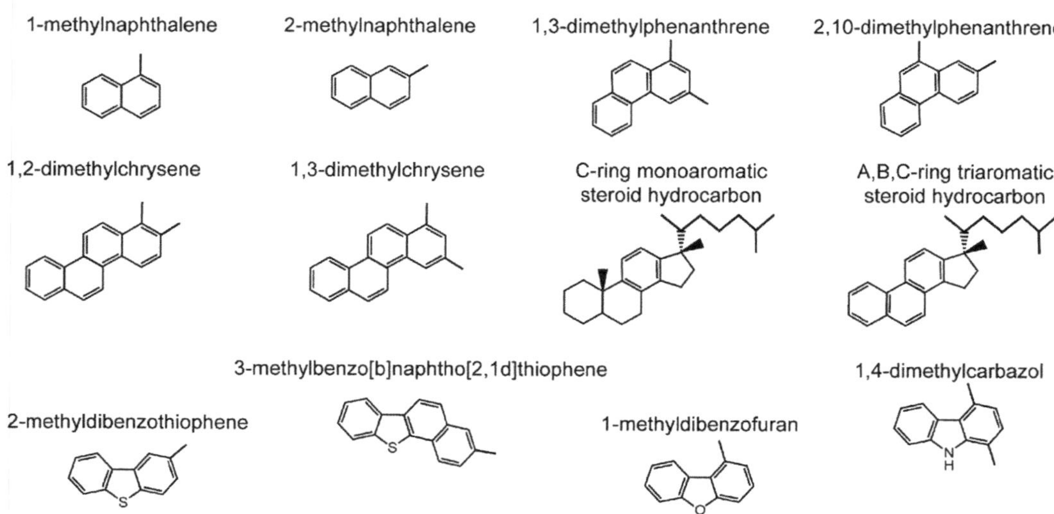

FIGURE 2.8 Examples of structures of alkylated aromatic hydrocarbons, aromatic hydrocarbon petroleum biomarkers, and alkylate heterocyclic aromatic hydrocarbons.
SOURCE: Adapted from Rullkotter and Farrington, 2021. CC BY 4.0.

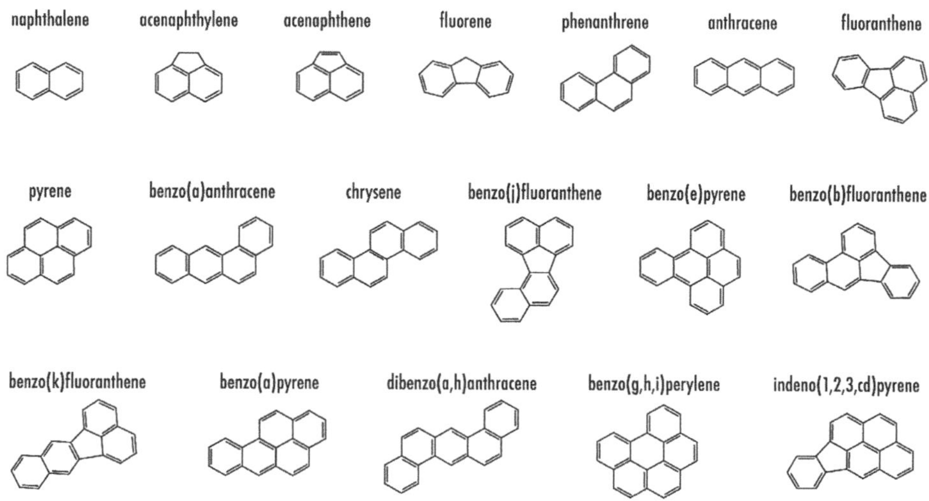

FIGURE 2.9 The 16 parent PAHs currently on the U.S. Environmental Protection Agency (EPA) Priority Pollutant List, plus benzo[f] fluoranthene and benzo[e]pyrene.
SOURCE: Rullkötter and Farrington, 2021. CC BY 4.0.

The aromatic rings can be fused in different specific geometries for PAHs of three or more rings, as depicted in Figure 2.9. For example, compare the three-ring compounds phenanthrene and anthracene, and the four-ring compounds pyrene, benzo[a]pyrene, and chrysene. Some aromatic hydrocarbons have a five-membered carbon ring as well as six-membered rings in their structures (e.g., acenaphthylene and fluoranthene, Figure 2.9). Aromatic hydrocarbons with two or more cyclic rings are designated by the shorthand notation polycyclic aromatic hydrocarbons (PAH or PAHs).

When nitrogen (N), oxygen (O), or sulfur (S) are substituted for a carbon atom in these compounds (see the thiophenes, dibenzofuran, and carbazole in Figure 2.8 as examples) the compounds are no longer strictly hydrocarbons and are known instead as polycyclic aromatic compounds (PACs). Hydrocarbon compounds for which one or more of the carbons has been replaced by the elements O, N, or S are also known as heteroatom compounds. These simple PACs commonly have chemical and physical properties similar to their PAH counterparts.

The various configurations of the fused rings influence physical and chemical properties of the compounds such as volatility, solubility, and also the chemical and biological reactivity of the specific PAH or PAC. A much discussed example is the comparison of benzo[a]pyrene, classified as a strong carcinogen, with its isomer benzo[e]pyrene, which is not classified as carcinogen. Both structures are depicted in Figure 2.9.

PAHs and PACs in petroleum are often "decorated" with substituent alkyl groups of different lengths and in various positions (i.e., alkanes or branched alkanes substituted on the ring structures), forming series of related chemicals having the same aromatic skeleton. Examples depicted in Figure 2.8 include 1- and 2-methylnaphthalene; 1, 3-, and 2, 10-dimethylphenanthrene; and 1, 2- and 1, 3-dimethylchrysene.

There are also aromatic hydrocarbons with a combination of saturated cyclic rings coupled with aromatic rings such as acenaphthene (see Figure 2.9) and the monoaromatic and triaromatic steroid hydrocarbons depicted in Figure 2.8.

The chemical structures depicted in Figures 2.2 through 2.9 are only a few examples of the several thousand individual chemical structures present in petroleum.

The resins fraction is not defined by chemical structure but rather by solubility, being insoluble in liquid propane; the structure of individual resins is generally not well known. The resins fraction comprises compounds with heteroatoms where N, S, and/or O replace C in one or more positions (see Figure 2.8). As noted previously, some simple PACs fall into the chemical definition of a resin, but their solubility and reactivity align them functionally with the PAHs.

The asphaltenes fractions, like the resins, are defined by solubility: they are petroleum chemicals that are not soluble in alkane solvents but do dissolve in toluene. They are the compounds that impart a dark color to crude oil. Asphaltenes are high molecular weight compounds typically of undefined structures, although some tentative structures or parts of structures have been proposed. A recent review of asphaltene structure and function by Schuler et al. (2020) suggests that previous reports of average molecular weights in the thousands of Daltons for model asphaltene molecules are overestimates, having been inferred from measuring nanoaggregates of smaller average asphaltene moieties (say, ~600–800 Daltons) (e.g., see Figure 2.2 for a proposed structure for an asphaltene molecule, Yang et al., 2015) that tend to self-assemble into macromolecular clusters.

Naphthenic acids (NAs) are composed of cyclic and non-cyclic carboxylic acids with varied alkyl substituents on these oxygen-containing compounds. They are similar to the resin fractions in being polar. There is concern that some NAs are toxic to aquatic animals, and some do not biodegrade easily (Lee et al., 2015). While examples of some structures have been noted or proposed, NAs are an example of a class of oil compounds in need of more accurate definition of structure by using advances in analytical chemistry. Other chemicals present in minor quantities in petroleum are metal-containing organic compounds such as the petroporphyrins containing Ni (nickel) and V (vanadium) (Chacón-Patiño et al., 2021, and references therein). See the nickel porphyrin structure in Figure 2.6 as an example. Elemental sulfur and H_2S (hydrogen sulfide) are present in some crude oils—the "sour crude oils."

2.1.4.3 Petroleum Biomarkers

There is a special group of isoprenoid hydrocarbons, cyclic alkanes, and aromatic hydrocarbons commonly referred to in the organic geochemistry and petroleum chemistry literature as "biomarker compounds" or "molecular biomarkers." It is important to note that these are not the same as "molecular biomarkers" that have come into the lexicon of molecular biology and environmental toxicology. (See Box 2.2 for an explanation of the differences.) To differentiate between toxicological and geochemical/petrochemical usages, the term "petroleum biomarkers" is used to indicate the latter (see Box 2.2). Among the most common of these compounds are the subclasses isoprenoid hydrocarbons, steranes, and hopanes (see examples in Figure 2.6), so named because their molecular structures are derived from biologically produced precursor compounds (e.g., the phytol side chain of chlorophylls, carotenoids, steroids, and hopanoids) that were deposited in sediments with other organic matter of biological origin. Over geologic time through diagenesis and catagenesis, this organic matter was the source of the petroleum that contains these petroleum biomarkers whose backbone chemical structure is the same as or similar to the isoprenoids, carotenoids, steroids, and hopanoids.

These petroleum biomarker compounds are important because their presence and exact proportions to one another in given crude oil or fuels oil can be and have been used to provide a unique, or nearly unique, identification for a given oil. Early in the oil spill literature of the late 1960s and into the 1970s these petroleum biomarkers or molecular signatures were explained in laypersons' terms as providing the "fingerprint" of an oil. The term "fingerprinting oils" became part of the lexicon of oil spills even though the unique nature of a given oil's fingerprint was not in the same statistically unique category as a human fingerprint. Since that time, the term "oil spill forensics" has also been used (e.g., Stout and Wang, 2018). Importantly, many of these petroleum biomarker compounds do not readily biodegrade nor are as chemically reactive as many of the other chemicals in petroleum. Thus, they are often used as reference compounds to discern recent alteration of spilled oil (e.g., through biodegradation, volatilization, dissolution, etc.; see Chapter 5). Petroleum biomarkers of oil are noted in various examples and discussion in the Chemical Methods section.

2.1.5 Physical Chemical Properties of Petroleum Hydrocarbons

The molecular weights and various molecular configurations of petroleum hydrocarbons (and any molecule for that matter) control the fundamental physical chemical properties of the individual compounds and of mixtures of compounds. These properties include density, volatility, solubility, and viscosity, among many important properties. As an example, Table 2.2 is a combination of data excerpted from May (1980). Note that molecular weight ranges over about a factor of 3 and the solubilities range over a factor of approximately 10^6. Table 2.2 also contains the Setschenow constants and the equation using those constants that adjust for solubility reductions as a function of salinity. The reduction of solubility as a function of increasing salt concentrations was noted first in 1889 by Setschenow that can influence hydrocarbon solubility as explained in May (1980). Relatively large changes in salinity are needed to significantly change solubility in contrast to greater sensitivity to small changes in ambient temperature (Whitehouse, 1984). We have noted salinity here as an important factor because large salinity gradients are present in estuarine and many coastal waters.

The presence of dissolved salts is among several parameters such as the presence of other organic chemicals that can influence the solubility of hydrocarbons in the aqueous phase as discussed in May (1980) and Schwarzenbach et al. (2016). These latter authors provide an extensive appendix of solubility data and other physical chemical data for a suite of hydrocarbons and other organic chemicals and reasonably up to date reference citations for these types of data.

There are numerous other properties and parameters of compounds that can be measured or calculated and are useful for understanding the fates and effects of petroleum compounds as discussed in Chapters 5 and 6. One example is the K_{iow} (also noted in some papers as K_{ow}) or octanol water partition coefficient, a measure of a compound's partitioning between two liquids of different polarity (i.e., octanol and water). This partitioning coefficient is used to predict partitioning between dissolved chemicals in water and the lipids in marine organisms, and between water and an organic phase (such as an organic coating on a water column or sediment particle or organism exuded polymer). Much has been learned about the fundamentals related to these partitioning processes as noted in the comprehensive text by Schwarzenbach et al. (2016).

BOX 2.2
Resolving the Confusion About the Term "Biomarker"

It is important to resolve to the extent practicable for this report, and the subject of oil pollution in general, confusion that could arise with the use of the term "biomarker" (short for biological marker) or "molecular biomarker." One definition of the term *biomarker* in *Merriam-Webster Dictionary* online (July 18, 2021) is as follows: "a distinctive biological or biologically derived indicator (such as a metabolite) of a process, event, or condition (such as aging, disease, or oil formation)." Note that the definition spans subjects from biological processes to oil formation. Expert scientists understand their use of the term within their respective disciplines. When discussions, papers, and reports (such as this one) encompass several scientific disciplines, it is important to clarify the terminology to avoid confusion and misunderstandings.

Petroleum Geochemistry Use Definition of Biomarkers

There is a long history of many decades of the use of the term "biomarker" in the subject area of petroleum geology and geochemistry. Hunt (1996) provides a review and concise explanation:

"The organic compounds in sediments, rocks, and crude oils whose carbon structures or skeletons can be traced back to a living organism are called biomarkers. They are micro fossils generally less than 30 nm in diameter and are highly variable in their stereochemistry, that is, the spatial arrangements of the atoms and groups within their molecules. Because of this variability, fossil biomarkers frequently can be linked directly to the specific group of plants, animals, or bacteria from which they originated."

A recent instructive paper by Philp (2018) in a collection of papers edited by Stout and Wang (2018), "Oil Spill Environmental Forensics Case Studies," provides an informative introduction and overview of biomarkers in petroleum geochemistry and organic geochemistry. Quoting Philp (2018): "The ability to relate molecules present in crude oils and source rocks today with those precursor molecules deposited millions of years ago, provides a wealth of information for the petroleum explorationist in reconstructing the origin and history of the oil. The fingerprints that can be obtained from multiple biomarkers present in crude oils and source rocks can also be used for oil/oil and oil/source rock correlations. This is where the crossover with environmental forensics enters since precisely the same approach can be applied to the correlation of a spilled oil with its suspected source." Note: "fingerprinting" is another term that could be misunderstood because the statistical probability of matching sample to source has yet to be shown to be as well-established as the forensic use of human fingerprints.

A more detailed and comprehensive discussion of the use of "biomarkers" in petroleum geochemistry and organic geochemistry in general is "The Biomarker Guide" (Peters et al., 2005).

Biological and Medical Use Definition of Biomarker

The National Institute of Environmental Health Sciences (NIEHS) provides this introduction to biomarkers.[a]

> A biomarker (short for biological marker) is an objective measure that captures what is happening in a cell or an organism at a given moment. Biomarkers can serve as early warning systems for your health. For example, high levels of lead in the bloodstream may indicate a need to test for nervous system and cognitive disorders, especially in children. High cholesterol levels are a common biomarker for disease risk.

In the environmental health field the term "biomarker" has been expanded to include specific types of biomarkers, for example, NIEHS supports the development of biomarkers that measure or provide information regarding:

- Exposure—what are the levels of environmental chemicals inside the body?
- Response (or effect)—are there biological indicators of adverse health effects?
- Susceptibility—do variations in genes place a person at higher risk to health effects from an environmental exposure?

This definition applies when expanded to organisms in general in terms of their responses to environmental natural chemicals and environmental contaminants. Biomarker responses occur in individual organisms at various levels of organization, from the molecular level (i.e., changes in gene regulation), biochemical, cellular, physiological, and developmental through behavioral levels). Responses at higher levels of organization (i.e., population, community, and ecosystem levels) are termed bioindicators.

RESOLUTION. In this report the term "petroleum biomarkers" is used to refer to organic molecules in oil or in oil contaminated samples that are useful in correlating oil or oil contaminated samples with a suspect source of oil input to the environmental area(s) that are the focus of response, investigation, and/or research. In addition, petroleum biomarkers are useful in discriminating aspects of physical, chemical (e.g., photo-oxidation), and biological (biodegradation) weathering processes.

The term "biomarker(s)," without qualifying terminology, will be used in the sense defined by NIEHS for medical use and extending to all biological organisms and systems. See Chapter 6, Effects.

[a] See https://www.niehs.nih.gov.

TABLE 2.2 Solubility of Some Example Aromatic Hydrocarbons in Pure Water at 25°C and the Corresponding Setschenow Constants Used to Adjust for the Effect of Salinity in Sea Water That Reduces Solubility

Chemical Name	Molecular Weight (Daltons)	Solubility at 25°C (mg/kg pure water)	Setschenow Constant K_s (liters/mole)
Benzene	78.1	1791 +/– 10	0.175 +/– 0.006
Naphthalene	128.2	31.69 +/– 0.23	0.213 +/– 0.001
Fluorene	166.2	1.685 +/– 0.005	0.267 +/– 0.005
Anthracene	178.2	0.0446 +/– 0.0002	0.238 +/– 0.004
Phenanthrene	178.2	1.002 +/– 0.011	0.275 +/– 0.010
2-Methylanthracene	192.3	0.0213 +/– 0.0003	0.336 +/– 0.006
1-Methylphenanthrene	192.3	0.269 +/– 0.003	0.211 +/– 0.018
Fluoranthene	202.3	0.206 +/– 0.002	0.339 +/– 0.010
Pyrene	202.3	0.132 +/– 0.001	0.286 +/– 0.003
Benzanthracene	228.3	0.00094 +/– 0.0001	0.354 +/– 0.002
Chrysene	228.3	0.0018 +/– 0.00002	0.336 +/– 0.010
Triphenylene	228.3	0.0066 +/– 0.0001	0.216 +/– 0.002

NOTES: Solubility data are means +/– one standard deviation. The equation for Setschenow constants is $\text{Log } S_0/\text{Log } S_s = K_s C_s$ where S_o is the concentration in freshwater and S_s is the concentration in seawater. K_s is the Setschenow constant and C_s is the molar salt concentration.
SOURCE: Excerpted and adapted from May, 1980.

2.1.6 General Sources of Hydrocarbons in the Marine Environment

The main sources of hydrocarbons in the marine environment need to be described briefly as this is relevant to understanding how marine ecosystems might be predisposed to the presence of certain hydrocarbons and/or mixtures of hydrocarbons such as those found in spilled or chronically released oil. It is also important to set the scene for analytical methods and data interpretation used to distinguish sources of hydrocarbons in forensics and in fates and effects studies.

1. Biosynthesis and biochemical transformations. Specific *n*-alkanes and alkenes are synthesized by land plants and marine organisms, as has been documented for decades (NRC, 1975, 1985). Biochemical transformations of biogenic compounds to hydrocarbons have also been well documented. One striking example is the biochemical transformation of the phytol side chain of chlorophyll present in its phytoplankton food to pristane by *Calanus* copepods (Avigan and Blumer, 1968). *Neocalanus* copepods in Prince William Sound biosynthesize pristane in a similar manner, and it is transferred through the food web (Short, 2005). Various species of diatoms biosynthesize highly branched C25 alkenes (Belt et al., 2019; Gao et al., 2020; and references therein).

   The significance of C_{15} and C_{17} *n*-alkanes by cyanobacteria *Prochlorococcus* and *Synechococcus* (Lea-Smith et al., 2015; Love et al., 2021) with an estimated yield of ~308–771 million tons of these hydrocarbon inputs to the global ocean from this source.
2. Microbial biochemical and geochemical transformations of natural organic matter in soils and surface sediments (often designated as early diagenesis) yield several specific steranes and aromatic hydrocarbons derived from biological precursor molecules, as reaction products of the early diagenesis of sterols in some deposition environments and in sinking particulate matter in the water column (e.g., Wakeham and Canuel, 2016). In particular, the PAHs retene and perylene are often present in analysis of marine sediments and are of known biological/early diagenesis origin (Lima et al., 2005).
3. Erosion of ancient sediments containing organic matter that has been transformed by diagenesis into material that is oil shale or near to oil shale type material or coal.
4. Natural seepage at the subsea floor and at the sediment-water interface releases both natural gas and oil to the marine water column (MacDonald et al., 2015; Ruppel and Kessler, 2017). Natural seepage is ubiquitous on the continental slopes and margins in many areas of the world, and natural seep sites are persistent at the annual and decadal time scales (see Section 3.2). Depending on the dynamics of the seep such as the rate of seepage and subsurface interactions with sea water and nutrients, the composition of the petroleum may be altered from that in the reservoir due to the development of microbial communities associated with a seep. (See Chapter 5 for a discussion of microbial degradation of petroleum.)
5. Human-mobilized coal that may end up in surface sediments as a result of losses due to routine handling or shipwrecks (e.g., Tripp et al., 1981; Hostettler et al., 1999). These authors noted that extraction of surface sediments in the present-day environment that contained coal particles and analysis by gas chromatography methods used to analyze petroleum hydrocarbons (see Analytical Chemistry Methods section) yields results that could be mistaken for the presence of petroleum hydrocarbons.

6. Combustion is a source of PAH. Grassland and forest wildfire along with human use of fossil fuels (coal, oil, and gas) and biogenic material (e.g., wood) for heating and power generation are major sources of PAHs in the environment in modern times (e.g., Lima et al., 2005). Grassland and forest fires have been a source for PAHs in the environment before human activities (e.g., Karp et al., 2020). The processes of reactions in combustion that likely yield PAHs and also yield soot and black carbon have been outlined recently by Johansson et al. (2018). Notably, these pyrogenic PAHs differ from oil-derived (petrogenic) PAHs in terms of the relative abundances of parent (i.e., non-alkylated) PAHs and related alkylated PAH homologs and other compositional differences (NRC, 1985; Lima et al., 2005).
7. There are various sources of groups of hydrocarbons or individual hydrocarbons that are synthesized for specific uses in various products utilized by general public consumers in everyday life. Many of these are disposed via wastewater streams and eventually are discharged in varying amounts to the estuarine and coastal marine environment. A complete review of these is beyond the scope of this report. Two illustrative examples suffice: decalin and linear alkyl benzenes.

   Decalin is a bicyclic saturated hydrocabon (formal name bicyclo[4.4.0]decane) and can be present in two conformations—*cis* and *trans*—in petroleum. Also decalin is synthesized as a jet fuel additive because of its energy density and high thermal stability (e.g., Wang et al., 2019), and as solvent used in various manufacturing processes (ScienceDirect, 2004).

   Linear alkylbenzenes (LABs) are involved in the production of linear alkylbenzenesulfonates (LAS) that are used as anionic surfactants in domestic laundry detergents and for washing of dishes. There are 26 congeners (i.e., similar molecular structures) consisting of 10, 11, 12, 13, and 14 carbon atoms arranged in various branched alkane configurations substituted on the benzene molecule. As such they overlap in the jet fuel–fuel oil range of molecular weights for petroleum chemicals. A portion of LABs are not sulfonated in the production and are included in the detergents. The large amount of detergent use results in considerable amounts of LABs in domestic wastewaters, particularly in large urban areas. Despite various wastewater treatment methods, significant amounts of LABs are discharged into the receiving streams or accumulated in sewage sludge (e.g., Eaganhouse et al., 1983; Takada and Ishiwatari, 1990; Sherblom and Eaganhouse, 1991; Takada et al., 1992, 1994; Eaganhouse and Sherblom, 2001; Gustafsson et al., 2001; Macias-Zamora and Ramirez-Alvarez, 2004; Martins et al., 2008, among others). The overlap of molecular structures, physical chemical properties, and biogeochemical fates of LABs with petroleum hydrocarbons leads to LABs being included in hydrocarbons extracted and isolated from marine samples for determination of petroleum hydrocarbons (see the previously cited references). It is incumbent on analysts of samples for assessment of oil spills and chronic inputs of oils, especially in urban harbors receiving wastewater discharges, to be cognizant of the potential presence of LABs in the samples.
8. Petroleum. Human-mobilized petroleum releases to the environment, either as spills or chronic inputs, will be discussed later in this report. At least two of the sources of chronic releases may contain mixtures of petroleum hydrocarbons and combustion source hydrocarbons. These are (1) motor oils that are dribbled into the environment on roads and driveways by leaking crankcases of cars, trucks, and similar vehicles. The leaking oil contains the original crankcase oil with its mostly higher molecular weight cycloalkanes plus PAHs from combustion of the fuel that have slipped across the rings in the combustion chamber to accumulate in the crankcase oil. (2) A similar process occurs for exhausts of outboard engines and inboard marine engines for boats and ships.

Another category of petroleum that is increasing in production and transport, especially by pipeline and rail tank car, is diluted bitumen. (See previous discussion).

In summary, these key factors are relevant to this report:

1. Hydrocarbons from biosynthesis or early (modern) diagenesis in the water column, surface sediments, and soils have a composition that is relatively simple compared to the complexity of chemicals in petroleum. Nevertheless, it is reasonable to posit that these hydrocarbons have stimulated the evolution of microbes that can break down and metabolize these types of hydrocarbons in the marine environment (Section 5.2.7).
2. Hydrocarbons, resins, and asphaltenes present in natural oil seeps have a compositional complexity that is the same as or similar to that of oil inputs from oil spills and several types of chronic releases. Depending on each natural hydrocarbon seep, differences in composition could be a function of physico-chemical dissolution and microbial decomposition processes active during the seepage processes pre-release in the seafloor and at the sea floor.
3. Combustion sources of PAHs (pyrogenic PAHs) generally have complex compositions similar to PAHs in petroleum. However, the parent PAHs are more abundant than alkylated PAHs in the same grouping of parent and alkylated PAHs (e.g., naphthalenes, phenanthrenes, and chrysenes). In general, the higher the efficiency of the combustion process, the higher the relative abundance of the parent PAHs. In comparison, PAHs in petroleum

(petrogenic PAHs) have greater relative abundances of the various alkylated PAHs compared to combustion sources. Assessing these relative abundances of parent and alkylated PAHs has been used in numerous instances to ascertain the relative contributions of pyrogenic and petrogenic PAHs to a given sample (e.g., see NRC, 1985; Lima et al., 2005, among others).

2.1.7 Sampling and In Situ Observations and Analyses

The most sophisticated laboratory procedures and analytical chemistry instruments will not provide useful data unless: (1) appropriate precautions are taken to obtain field samples that are not contaminated or compromised by the sampling process, (2) samples are labeled in accord with specifications for the intent of the sampling and analyses (e.g., U.S. Natural Resource Damage Assessment protocols or their equivalent in other countries), and (3) samples are preserved appropriately during transportation and storage prior to analysis. Furthermore, all of the preceding coupled with the appropriate analytical methods will provide the most useful data if there is attention to the details of appropriate sampling plans to answer the questions about responses to oil spills and inputs, fates, and effects of oil spills or chronic inputs (individually or in combination) discussed in the following chapters. This includes familiarity with the physical, chemical, biological, geological, and ecological processes of the ecosystem or ecosystems being sampled.

Several appropriate online documents have been prepared that describe planning and sampling within the context of oil spills (e.g., NOAA, 2014b; IPIECA, 2020, and references therein). These are updated periodically to take into account new knowledge gained from responses and research connected with oil spills. For example, the Royal Society of Canada report (Lee et al., 2015) and this NASEM report (and references herein) provide updated reviews of new knowledge of relevance. In addition, reports such as that by Payne and Driskell (2015) as part of the Natural Resource Damage Assessment (NRDA) studies of the DWH oil spill document sampling equipment and measurement instruments deployed to advantage in the field, including helpful photographs.

There have been advances in refinement and utilization of in situ observations of hydrocarbons within the ocean water column for chronic inputs and episodic spills. These instruments can be mounted on hydrocast-CTD (conduct-temperature-density) rosettes or on underwater vehicles (remotely operated, human-occupied, and autonomous). An important advance is the more extensive use of ultraviolet (UV)-induced fluorescence instruments for detection of aromatic hydrocarbons in seawater in assessment of oil spills. These fluorescence observations are often coupled with other sampling equipment, including water samplers and other analytical platforms. Use of a laser fluorosensor to detect oil on water and substrates such as shoreline, plants, and ice—among other substrates—has been reviewed by Fingas and Brown (2018). A coastal mounted sensor using UV induced fluorescence to detect oil harbors has been reported by Hou et al. (2018). As noted by Part et al. (2021), care must be taken to calibrate the sensors to avoid interference or false positives from the presence of natural colored (or chormophoric) dissolved organic matter (CDOM) and algae-derived chemicals such as chlorophyll A. A brief description of some other advances in UV-fluorescence analytical methods is provided in Section 2.1.8.

Coupling of an underwater mass spectrometer with an autonomous underwater vehicle has expanded capabilities beyond the fluorescence detectors to a broader range of dissolved hydrocarbons. These and other advances with in situ measurement devices such as advanced camera systems for imaging particle sizes and shapes have been summarized in an overview manner in Dannreuther et al. (2021) with relevant references therein. Some of these systems are noted with figures and photos in Payne and Driskell (2015).

A more thorough review of this topic is beyond the scope of this report because of the rapid advances of the past two decades in towed underwater vehicles, remotely operated underwater vehicles, autonomous underwater vehicles, and associated sensor developments.

The preceding applies not only to oil spill response and damage assessment but also to research activities focused on fates and effects.

2.2 PHASES AND STATES OF PETROLEUM FLUIDS IN THE SEA

2.2.1 Gas- and Liquid-Phase Petroleum

As discussed in this chapter, petroleum fluids are a complex mixture of hydrocarbon- and non-hydrocarbon-containing molecules. Depending on the temperature and pressure, they may occur in the gas, liquid, or solid phase of matter. In this report, the focus is on the gas and liquid phases because they are the most often encountered in the marine environment and more problematic. In everyday lives, petroleum fluids are normally used at atmospheric pressure; hence, some hydrocarbon molecules are commonly referred to as gases or liquids. For example, methane, ethane, and propane are gases at standard conditions, and benzene is a liquid. However, propane at 15°C and 1.7 MPa (megapascal, equivalent to about 150 m depth in the ocean) would be in the liquid phase. Because there is concern with petroleum fluids throughout the ocean water column, this report uses the terms *gas* and *liquid* when appropriate to refer to the in situ phase of the fluid of interest, rather than describing propane, for example, simply as a gas.

Commonly, the term *oil* is used to refer to the liquid-phase petroleum since crude oil and refined oil products are normally experienced as liquids at standard conditions. Here, this usage of the term *oil* may also be followed when appropriate and the phrase liquid petroleum may be used whenever the generic term *oil* would be ambiguous. However, it is important to remember that crude oil is a complex mixture that will have a different composition within the gas and liquid phases under different thermodynamic states, and that the term *oil* is not a precise, scientific description of either the state or the composition of a petroleum fluid.

There are different definitions of *standard thermodynamic conditions* in different fields of chemistry and engineering. In this chapter, the definition of standard conditions is used from the Society of Petroleum Engineers (SPE), which defines standard temperature to be 15°C (288.15 K or 59°F) and standard pressure to be 100 kPa (kilopascal, also equal to 1 bar). We use this definition since quantitative metrics describing oil spills, especially as related to offshore oil and gas exploration or hydrocarbon transport, are likely to be expressed using this standard.

Within a petroleum reservoir, all of the crude oil components are present as a mixture in equilibrium with the ambient pressure and temperature (McCain, 1990). Depending on the composition and ambient conditions, this mixture may or may not contain a gas phase. When no gas phase is present, all compounds are present in the liquid-phase petroleum, and some of the lightest components, such as methane, may be thought of as fully dissolved in the liquid phase. When a gas phase is present, all of the components of the whole petroleum will equilibrate between the gas and liquid phases. This means that a large fraction of the methane may be found dissolved in the liquid phase and that some of the higher molecular weight hydrocarbons, such as pyrene, with molecular weight of 202.25 g/mol, will partially partition into the gas phase. As an example, over half of the methane released from the DWH wellhead during the spill was released in the liquid-phase petroleum (Gros et al., 2016). The phase partitioning will be at equilibrium in the petroleum reservoir and may be in disequilibrium for some time as crude oil is conducted from the reservoir to surface storage.

The gas-to-oil ratio (GOR) is a measure of the fraction of compounds that would be gases at standard conditions within a crude oil. In petroleum engineering, GOR is commonly expressed as the number of cubic feet of gas that can be extracted at standard conditions to the number of stock barrels of liquid crude oil that would remain. This form of the GOR of a reservoir is very important in petroleum engineering as it explains what types of petroleum products can be extracted through a well and the volume flow rate of produced gas and liquid petroleum. The GOR is evaluated as an equilibrium condition for the petroleum mixture in a closed system (i.e., not in equilibrium with the atmosphere). Hence, the light, highly volatile and volatile components of the original petroleum fluid would still be present in the petroleum mixture for the purposes of evaluating the reservoir GOR. The GOR may also be expressed using in situ conditions or different units. For example, the GOR of the DWH crude oil at the sea floor conditions of the release was 29–44% gas and 56–71% liquid (Gros et al., 2016). When expressed this way, using the same volume units for oil and gas, the reported GOR may be unitless. In whatever way the GOR is specified, it is always critical to know the thermodynamic conditions and units used to make the evaluation.

For a surface spill of a crude oil or refined product, as from a tanker or ship accident, the original gaseous components of the oil would no longer be expected to be present, and the GOR would not be important. For submarine oil spills, especially for oil well blowouts, the GOR is a critically important parameter and must be considered in the context of the thermodynamic state for which it is evaluated. As oil and gas traverse the ocean water column, the thermodynamic state (notably, temperature and pressure) is continually changing, and the composition of the oil and gas may be changing, both directly through phase changes as well as via fate processes such as dissolution, biodegradation, and other interactions with seawater.

Because liquid petroleum is immiscible in water, it may also be referred to as a non-aqueous phase liquid (NAPL). NAPLs include hydrocarbons that exist as a separate, immiscible phase when in contact with water and/or air; they may either be crude oils or refined products. NAPLs are typically classified as either light nonaqueous phase liquids (LNAPLs), which have densities less than that of water, or dense nonaqueous phase liquids (DNAPLs), which have densities greater than that of water (Newell et al., 2015). Normally, liquids with an API gravity less than 10°API would be classified as DNAPLs. (See Table 2.1 for comparison to other API gravity of some other oil types.)

Coal tar is an example of a multicomponent DNAPL, which consists of polycyclic aromatic hydrocarbons; phenols; benzene, toluene, ethylbenzene, and xylene; and other compounds. Produced for medical and industrial uses, coal tar is a by-product of the production of coke and coal gas from coal. Coal tar, creosote, and No. 6 fuel oil DNAPL mixtures generally have very low bulk solubility and may result in small to negligible dissolved plumes when spilled. However, these DNAPL mixtures often contain soluble constituents, such as naphthalene, that dissolve in water creating plumes after a spill.

2.2.1.1 Live Oil and Gas

When a petroleum mixture is in equilibrium with standard conditions, most of the gases originally dissolved in the liquid-phase petroleum will have escaped, and the remaining compounds would remain in the liquid phase. This state of liquid petroleum is referred to as *dead oil*. Petroleum mixtures in most other states of equilibrium would be referred to as *live oil*. The term *live* is used to indicate that some gaseous compounds, such as methane, are present in the liquid-phase petroleum at concentrations such that some gas would evolve out of the liquid phase if brought to standard conditions.

Both the gas and liquid phases of a petroleum fluid can be termed as live gas or oil whenever they are out of equilibrium with standard conditions. Fluids released from a subsea oil well blowout, for example, would be live gas and oil as the composition of the gas and liquid phase petroleum fluids would be that at the high pressure of the release point, generally including much more of the light hydrocarbons, such as methane, within the liquid phase than will remain at surface conditions. The disequilibrium of the released fluids relative to standard conditions is important both for predicting bulk properties, such as density, and for evaluating the composition of each phase of petroleum. These parameters are critical for evaluating fate, effects, and safety issues, including explosivity.

2.2.1.2 Weathered Oil

Historically, the general term *weathering* has been used to describe any process that changes the composition of an oil from its release conditions. This term came into use when many early studies applicable to surface floating oil considered processes linked to the weather conditions. Such processes may include evaporation, natural dispersion, or emulsion formation. Today, oil spill scientists may use weathering to describe a much wider array of fate processes occurring throughout the ocean water column, including dissolution, photo-oxidation, biodegradation, sorption, sedimentation, and so on. Correspondingly, *weathered oil* is any oil whose composition has changed by any fate process since its release to the environment. Weathering or fate processes are described in detail in Chapter 5.

2.2.1.3 Gas Hydrates

Gas hydrates consist of gas molecules surrounded by cages of water molecules that form a crystalline lattice (cage) via hydrogen bonds under certain pressure and temperature conditions (Sloan and Koh, 2008; Hassanpouryouzband et al., 2020). Gas molecules small enough to fit inside the lattice structure are encaged, or enclathrated, within the water lattice. Natural gases that may form hydrates in seawater include methane, ethane, propane, isobutane, *n*-butane, nitrogen, carbon dioxide, and hydrogen sulfide, among others (Sloan and Koh, 2008). Natural gas hydrates can concentrate and store large quantities of natural gas within their water cage cavities. When heated or depressurized, they become unstable and dissociate into water and natural gas.

Gas hydrates are important in petroleum engineering because they can cause blockages of wells and pipelines. Methods to understand and prevent such hydrate formation are the topic of flow assurance (Sloan and Koh, 2008; Koh et al., 2011). In marine oil spills, hydrates may play a role in both response and petroleum fluid fate when spills occur within the hydrate stability zone, the region of the oceans for which the temperature is low enough and pressure high enough for hydrate stability. Hydrates cause issues for response because they may clog collection systems, freeze equipment, or otherwise interfere in sensors or operations. Gas hydrates may alter the fate of spilled oil and gas by creating a barrier between the free oil and gas and water. Because hydrate forms as a crystalline matrix of gas and water, it will grow from the gas–water interface. Natural gas molecules may be supplied from either the gas or liquid phase of the spilled fluids. Warzinski et al. (2014a) show the nature and dynamics of hydrate armoring on methane and natural gas bubbles in a high pressure water tunnel. Though hydrate itself is soluble in seawater, it is less soluble than the free gas; hence, hydrate armoring may interfere with dissolution of natural gas bubbles in seawater. The implications of hydrates on marine oil spills are discussed in more detail in Section 5.3.3, on deep-water processes for acute marine oil spills, and Section 5.4.1, on natural seeps.

Naturally formed gas hydrates occur beneath the seafloor, in permafrost areas, and beneath some ice sheets where temperature and pressure conditions for formation of methane hydrates are favorable (see Figure 2.10). Except on upper continental slopes (depths less than 300–700 m water depth), the seafloor of most of the oceans is within the hydrate stability zone. About 99% of the world's gas hydrates are in the uppermost hundreds of meters of marine sediments at water depths greater than ~500 m and close to continental margins (Ruppel and Kessler, 2017).

2.2.2 Advances in Analytical Chemistry Methods

There have been significant advances in chemical methods applied to analyzing gas and oil chemicals released to the environment since the NRC (2003) report. That report did not address analytical chemistry methods in detail because of the large amount of other information presented. Thus, advances since the NRC 1985 report are reviewed briefly, but the focus of this section is on the significant advances since the NRC (2003) report. More details of the analytical chemistry procedures are provided in Stout and Wang, 2018, and Wise et al., 2022, and specific references cited in this section.

2.2.2.1 Ultraviolet Fluorescence Analyses as a Sample Screening Method and Related Advances

The initial analyses of samples and extracts often involve ultraviolet fluorescence spectrometry to assess the presence of aromatic hydrocarbons and other aromatic compounds in petroleum—mostly the low-to-medium molecular weight compounds that have greater solubility in seawater than the higher molecular weight PAHs and PACs. This method has been used for decades as a quick scanning method to assess the presence or absence of these petroleum compounds (NRC, 1985).

Over the past several decades, the instrumentation and methodology has advanced to the point where there is now a robust excitation–emission matrix spectroscopy (EEMS) methodology that provides a reasonably rapid analyses of water samples and some extracts (Bugden et al., 2008). This

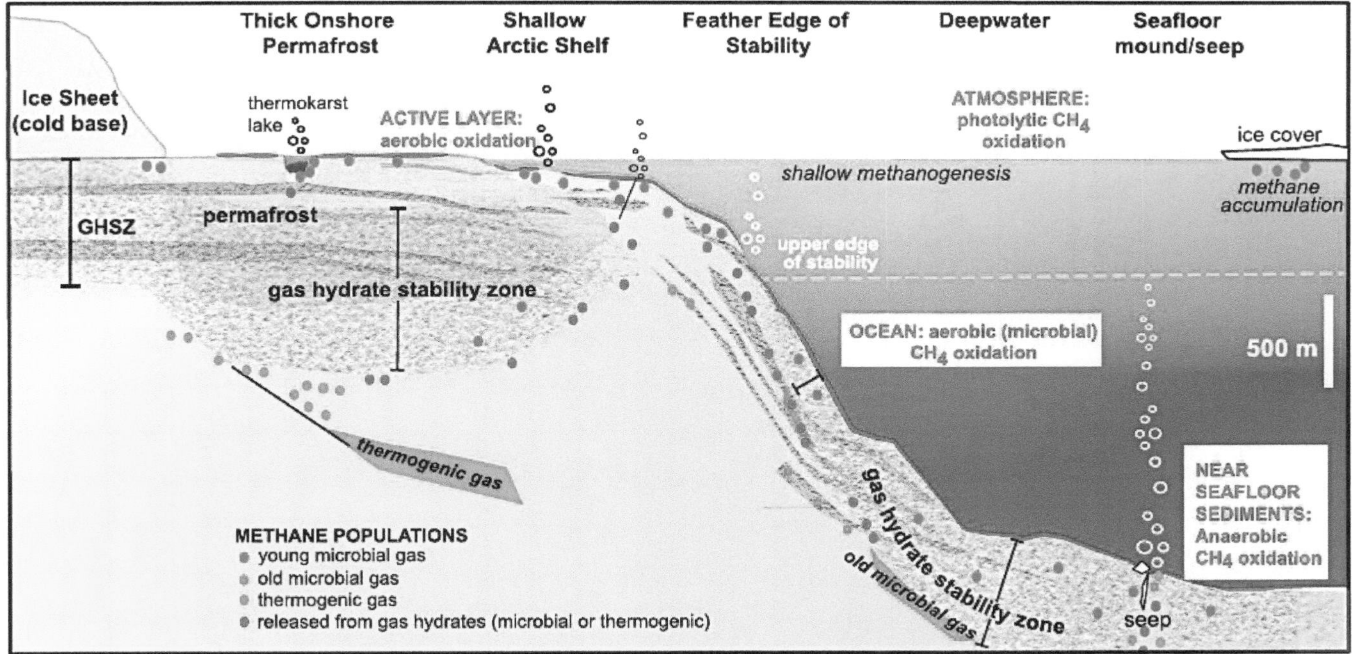

FIGURE 2.10 Schematic of the occurrence and dynamics of methane hydrate in terrestrial and marine systems.
SOURCE: Adapted from Ruppel and Kessler, 2017.

has been enhanced by coupling EEMS with parallel factor analysis (PARAFAC) to enhance the specificity of the methodology (e.g., Miranaghia et al., 2018; Araújo et al., 2021; Matheus et al., 2021; Oliveira et al., 2021). This methodology has also been used to provide insight into the fate of petroleum aromatic compounds (PACs) in sea water 2 years after the DWH oil spill (Bianchi et al., 2014).

Depending on the circumstances and the objectives of the analyses, these UV-fluorescence methods can have an important role in providing a first screening and identification of samples to be subjected to a progressive cascade of more detailed analyses as noted in the following section.

The basics of the flow schemes of detailed analyses are summarized in outline form in Figure 2.11. Oil samples or samples from the environment (e.g., air, water [dissolved and particulate fractions], sediments, organisms) are analyzed by (1) progressive physical and/or chemical separation of oil constituents from other organic materials in the samples; (2) solution-based separation of the resulting isolated oil into groups of chemicals, such as the SARA fractions previously described, by using chromatographic means such as column or thin-layer chromatography and high performance liquid chromatography; (3) further physico-chemical separation and quantification by gas chromatography (GC) based on properties of volatility and polarity of the individual constituent chemicals corresponding to their molecular weight and structure; (4) mass spectrometric (MS) analysis of compounds separated by GC or high-performance liquid chromatography (HPLC) (often via coupled GC-MS or HPLC-MS); or (5) direct analysis by MS of extracts, isolated fractions, or oil or weathered oil samples (see weathering in Chapter 5).

This analytical methodology of glass capillary gas GC and glass capillary gas chromatography–mass spectrometry (GC-MS) coupled to computer systems for data processing had become more routine just at the time of the 1985 NRC report. Since that time, there has been extensive application and demonstration of its practicality and utility. An informative, brief history of GC-MS computer systems is provided by one of the pioneers of these methods (Hites, 2016).

Recommended GC-MS methods for analyzing oil or oil-contaminated samples in environmental settings have been set forth by national agencies, for example, Lauenstein and Cantillo (1998), Olson et al. (2004), and NIST (2021).[6]

There are hundreds if not thousands of scientific papers and reports on uses of GC-MS relevant to inputs, fates, and effects of oil in the marine environment since the NRC 2003 report. An instructive collection of recent examples are provided by Stout and Wang (2016, 2018) and the recent review by Wise et al. (2022).

Advances in mass spectrometry leading to different types of commercially available and easier to use mass spectrometers coupled to GC or HPLC have enabled widespread adoption and significant technical advances, such as GC-orbitrap

[6] See https://www.canada.ca/en/environment-climate-change/services/canadian-environmental-protection-act-registry/agreements/related-federal-provincial-territorial/standards.html.

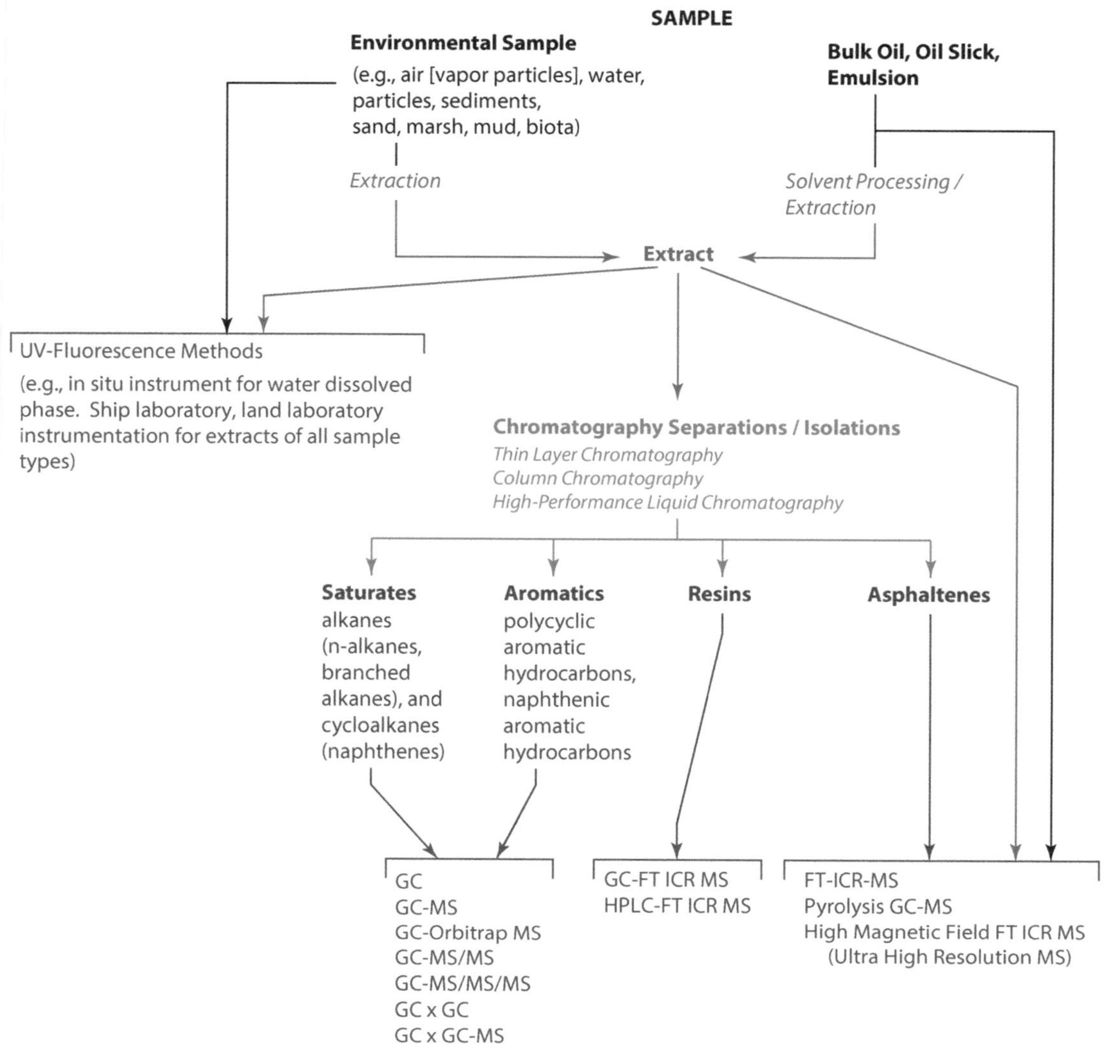

FIGURE 2.11 Simplified flow diagram of extraction and chemical analysis of bulk oil, oil slick, emulsions, and environmental samples emphasizing long established methods of instrumental analysis and newer instrumental analysis methods.
NOTES: Not all newer and potentially useful instrumental analysis methods are shown. Black text and arrows describe various environmental samples and their handling, blue text and arrows show the pathways of different extraction processes, green text and arrows show different types of separations or isolation, brown text and arrows describe the main types of chemical compounds in a petroleum sample (saturates, aromatics, resins, and asphaltenes), and purple text describes the chemical analysis systems.

MS (Heska et al., 2021), multiple mass spectrometers connected in sequence such as GC-triple quadrupole MS (e.g., Szulejko and Solouki, 2002; Adikari et al., 2017). Advances in this field of analytical chemistry are documented by numerous papers in the scientific literature (e.g., in the journal *Mass Spectrometry Reviews*, Wiley online). A recent review by Wise at al. (2022), although focused on advances since the DWH oil spill, provides references from numerous studies other than those focused on the DWH.

Several of the advances in methodology have focused on a progressively more detailed analysis of the GC detected portion of petroleum, that is, what is termed the GC-amenable fraction (see Box 2.3). It is important to emphasize that most of the compounds in resins and all of the asphaltenes are not GC-amenable. This is emphasized because the GC-amenable fraction of an oil might represent only a portion—and sometimes a minor portion—of some spilled and chronic oil inputs. In general, the proportion of the GC-amenable fraction decreases as the asphaltene fraction increases over the continuum of light—medium—heavy oils.

Given the complexity of the mixture of GC-amenable compounds in fuel oils and crude oils, analysis by capillary GC will not completely resolve all the compounds present. Some of the compounds present in greater relative abundance will be recorded as peaks and can be quantified and identified by GC-flame ionization detector (FID) signal and/or mass spectrometry in a GC-MS system. As time increases, many

BOX 2.3
GC-Amenable Fraction of Oil

To be detected using routine GC methods, the compound must have a molecular weight range and polarity that permit it to be introduced to the GC column as a gas (either directly or dissolved in a non-polar, volatile solvent such as *n*-hexane) via injection. Generally, the molecular weight range of GC-detected petroleum components ranges from C_5 to C_{35} (sometimes to C_{40}) saturated (alkanes and cycloalkanes) and aromatic (including PAHs) hydrocarbons and includes some simple heteroatom compounds with sulfur, nitrogen, or oxygen substituted for a C atom. Larger and/or more polar (heteroatom-containing) components such as asphaltenes and large resin molecules are not detected using routine GC analyses.

other compounds of increasing molecular weight and boiling point elute. Because of the similarity of their structures, the signals as recorded by the detector, overlap with one another. This gave rise to the term "unresolved complex mixture" (UCM) for this signal as explained by Farrington and Quinn (2015). Even routine analyses with GC columns could not resolve that signal into individual compounds or peaks in many of the gas chromatograms. The UCM is especially prominent as a feature in some weathered and biodegraded fuel oils and crude oils.

An example is depicted in Figure 2.12 that shows glass capillary GC of Macondo crude oil from the DWH accident, a surface slick from June 2010, a beach sand pattie (see Chapter 5) from April 2011, and a rock scraping from June 2011 on the shoreline (Aal from Wise et al., 2022). This time sequence from fresh to weathered and biodegraded oil is typical of a spilled oil when lower molecular weight compounds are lost to evaporation, and the resolved peaks representing compounds such as *n*-alkanes and branched alkanes are biodegraded, leaving behind the UCM compounds of cyclic and branched cyclic alkanes. It is important to note that if the detector signal for the Macondo Well oil was expanded on the y-axis so that the resolved peaks were off the top of the scale, then the UCM would be apparent in that chromatogram. That is, those UCM compounds are present in the original oil but only become more apparent in the signal as the other compounds are removed by weathering and biodegradation processes.

GC-MS and GC-MS-MS, GC/FT-ICR-MS methods allow for identifying and quantifying some of the compounds that make up or co-elute with the UCM. Comprehensive two-dimensional gas chromatography (GC×GC) has resolved

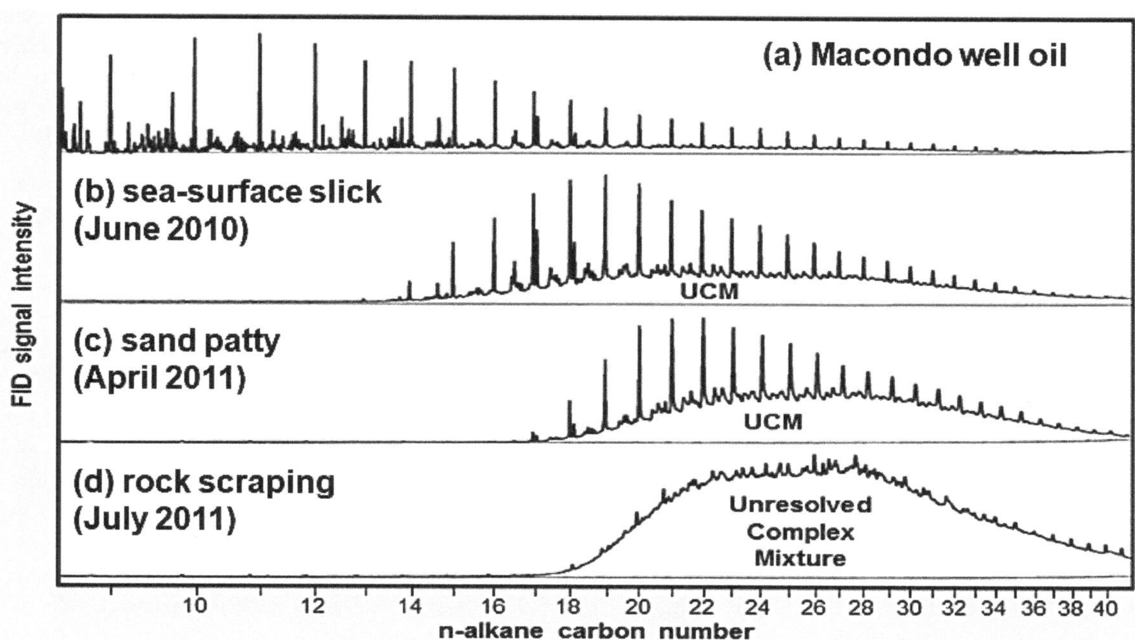

FIGURE 2.12 Capillary gas chromatograms of oil samples after the DWH oil spill: (a) original Macondo Well oil, (b) surface slick June 2010, (c) sand patty on beach April 2011, (d) rock scraping July 2011.

NOTES: The x-axis is increasing molecular weight as indicated by *n*-alkane carbon number. The y-axis is GC FID (flame ionization detector) signal intensity.

SOURCES: Courtesy of C. R. Reddy and R. K. Nelson, Woods Hole Oceanographic Institution (Wise et al., 2022).

the components as will be discussed below in more detail as an example of the advances in analytical methods relevant to the task and scope of this report.

The higher molecular weight and more polar fractions of crude oils and fuel oils are not amenable to conventional gas chromatography, that is, many of the resins and the asphaltenes had not been analyzed to an appreciable extent until developments in the past two decades involving Fourier transform-ion cyclotron resonance- mass spectrometry (FT-ICR-MS).

Examples of important advances in analytical chemical methodology relevant to the charge to this committee are summarized in Box 2.4 and explained briefly in the following paragraphs. A more detailed description of the methods is presented in Wise et al. (2022) and papers in Stout and Wang (2018).

2.2.2.2 Two-Dimensional Gas Chromatography (GC×GC) and GC×GC Mass Spectrometry (GC×GC-MS)

The analytical methodology of two-dimensional gas chromatography was introduced by Liu and Phillips (1991). Within a few years, Frysinger et al. (1999) had utilized the method to analyze benzene, toluene, ethylbenzene and xylenes, and total aromatic compounds in gasoline, Frysinger and Gaines (1999) had reported GC×GC use in analyzing petroleum, Gaines et al. (1999) had utilized the method for oil spill identification, and Frysinger and Gaines (2001) reported on the separation and identification of petroleum biomarkers.

The use of GC×GC methodology provides a significantly more comprehensive separation analysis of samples for petroleum chemicals compared to GC with glass capillary columns alone or interfaced with mass spectrometry. When combined with mass spectrometry, GC×GC-MS provides greater certainty in some cases of forensic analyses of sources of spilled oil (e.g., Lemkau et al., 2010; Nelson et al., 2016). In fates and effects assessments and research, the method provides significantly better chemical composition data to unravel or support important aspects of these assessments and research.

The basic principle of GC×GC is to utilize the resolving power of two sequential GC columns, the first of which has a nonpolar coating, to separate the compounds progressively mainly by boiling point/molecular weight. During that process, and with rapid periodicity, the effluent of that column is trapped by a freezing process that is then followed rapidly by a heating process. The heated compounds are "switched' into a second GC column having a more polar interior coating that then progressively separates the groups of trapped compounds from the first column by their polarity properties, derived from their specific molecular structure. The effluents of the second column pass to a GC detector or, most often, to an interfaced mass spectrometer with data collected by an interfaced computer. This is explained in more detail by Nelson et al. (2016).

Examples of GC×GC separations of the samples shown in Figure 2.12 are depicted in Figure 2.13, in which all or the vast majority of the individual compounds have been resolved. Admittedly, these depictions are currently less familiar to many oil spill responders and non-chemistry researchers. The power of these sets of data is that various expansions of the two-dimensional data can be faithfully generated from the data sets. For example, in Figure 2.14, the GC×GC-FID (flame ionization detector) analyses data for the DWH or Macondo Crude oil are shown in both three-dimensional plots Figure 2.14A and in a two-dimensional plot Figure 2.14B, with color codes indicating the intensity

BOX 2.4
Examples of Advances in Analytical Chemistry Methods

- Advances in ultraviolet fluorescence analysis for in situ screening of water samples and ex situ screening of water, organism, and sediment extracts for the presence of polycyclic aromatic compounds.
- Advances in several gas chromatography–mass spectrometry methods using several types of mass spectrometers and mass spectrometers in sequence (e.g., GC-MS-MS-MS)
- Comprehensive two-dimensional gas chromatography (GC×GC) and coupling of GC×GC with higher resolution mass spectrometry (GC×GC-MS) of various types and configurations.
- Fourier transform ion cyclotron resonance–mass spectrometry (FT-ICR-MS). Can be interfaced with glass capillary gas chromatography or high-performance liquid chromatography (i.e., GC-, or HPLC-).
- Thermal slicing pyrolysis—gas chromatography mass spectrometry.
- High magnetic field Fourier transform–ion cyclotron resonance–mass spectrometry (high magnetic FT-ICR-MS) or ultra high resolution mass spectrometry (Szulejko and Solouki, 2002).
- Gas chromatography with stable isotope (i.e., $\delta^{13}C$) mass spectrometry.
- Isolation of individual chemicals by gas chromatography and subsequent trapping followed by determination of $\delta^{14}C$.

All analytical methods are coupled with up-to-date computerized detector signal and data processing systems. This enables extensive data archiving and facilitates interpretation.

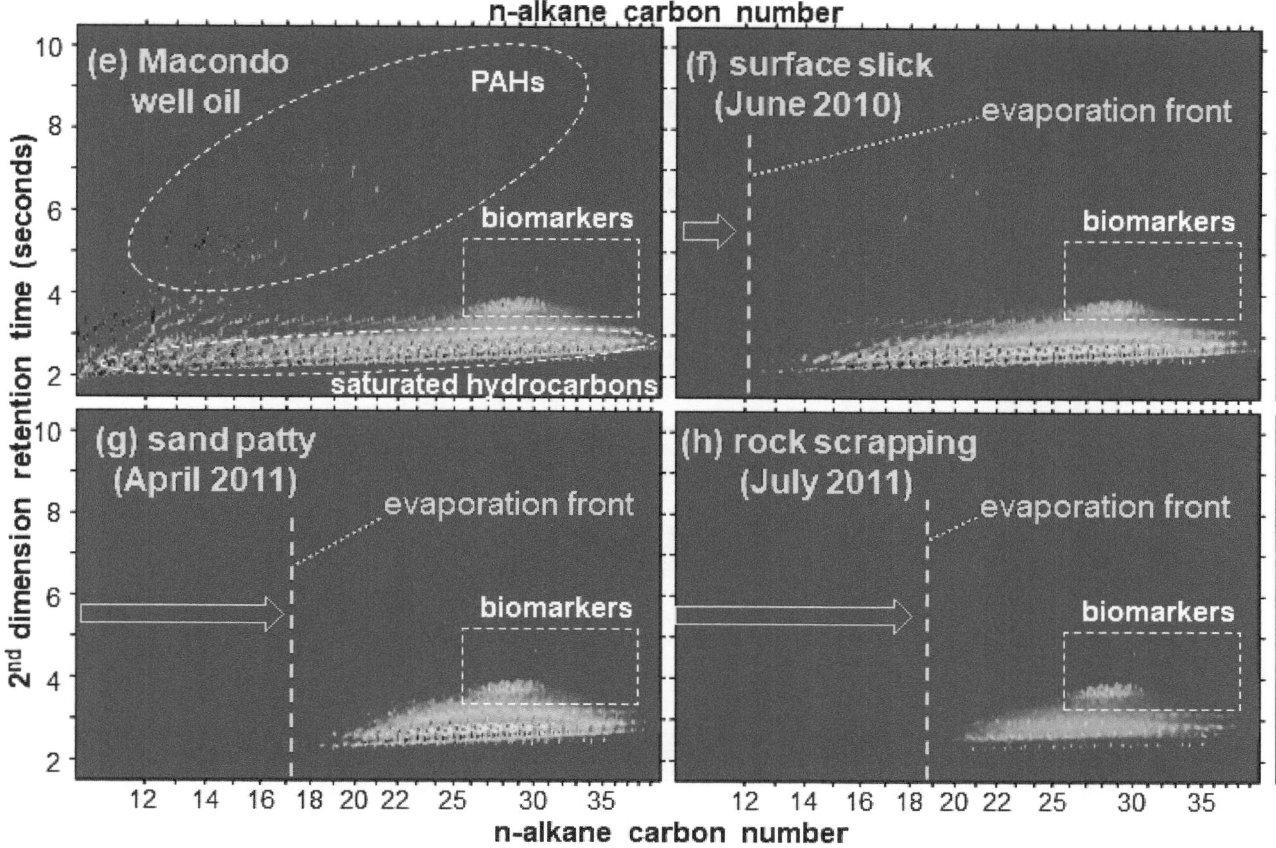

FIGURE 2.13 Comprehensive GC×GC chromatograms of samples shown in Figure 2.12.
NOTES: The *x*-axis shows separation in the first dimension: molecular weight, depicted by *n*-alkane number. The y-axis is separation in the second dimension: polarity (a function of separation by a slightly less non-polar phase coating the second GC column), given as retention time in seconds. Evaporation front refers to loss of petroleum compounds of greater volatility to the left of the line by initial and ongoing evaporation.
SOURCES: Courtesy of C. R. Reddy and R. K. Nelson, Woods Hole Oceanographic Institution (Wise et al., 2022).

of the signal. Sections of these types of GC×GC data can be expanded to a zoomed-in view to provide more detail as is the case in Figure 2.15 A for hopane petroleum biomarkers (A) and sterane petroleum biomarkers (B). Further explanations of shorthand notations in petroleum biomarker nomenclature are given by Philp (2018) and Peters et al. (2005).

Despite the high resolution and identification power of GC×GC and GC×GC-MS there have been only a few notable applications of the methodology to routine identification of spilled oil and to fate and effects assessments and research (e.g., the M/V *Cosco Busan* oil spill [Lemkau et al., 2010], the *Hebei Spirit* oil spill [Yim et al., 2012], and the DWH oil spill as noted in the Figures 2.13, 2.14, and 2.15 above). The reasons are explained in more detail by Górecki (2021) who addressed the lack of wider adoption of GC×GC in general in analyses for a multitude of chemicals in a variety of samples—not specifically the subject of this report. Górecki notes that a major impediment to wider use in general is the need to incorporate GC×GC and GC×GC-MS advances in standardized protocols. This applies in the case of responses to spilled oil and assessments of the accompanying fates and effects. This convinces and enables a wider range of laboratories in all sectors—government, industry, commercial analytical laboratories, spill response, environmental companies and academia—to invest in, and utilize, GC×GC and GC×GC-MS.

2.2.2.3 Fourier Transform Ion Cyclotron Resonance Mass Spectrometry (FT-ICR-MS)

Fourier transform ion cyclotron resonance mass spectrometry (FT-ICR-MS) is a powerful analytical instrument method for a variety of organic chemicals (e.g., Marshall et al., 1998). There has been substantial progress in the past two decades in the evolution of instrumentation and its applications. This has involved various configurations of varying powers of magnets used to generate the magnetic fields (e.g., see the review by Cho et al., 2014). These instruments have been interfaced in various configurations with GCs (GC-FT-ICR-MS) or

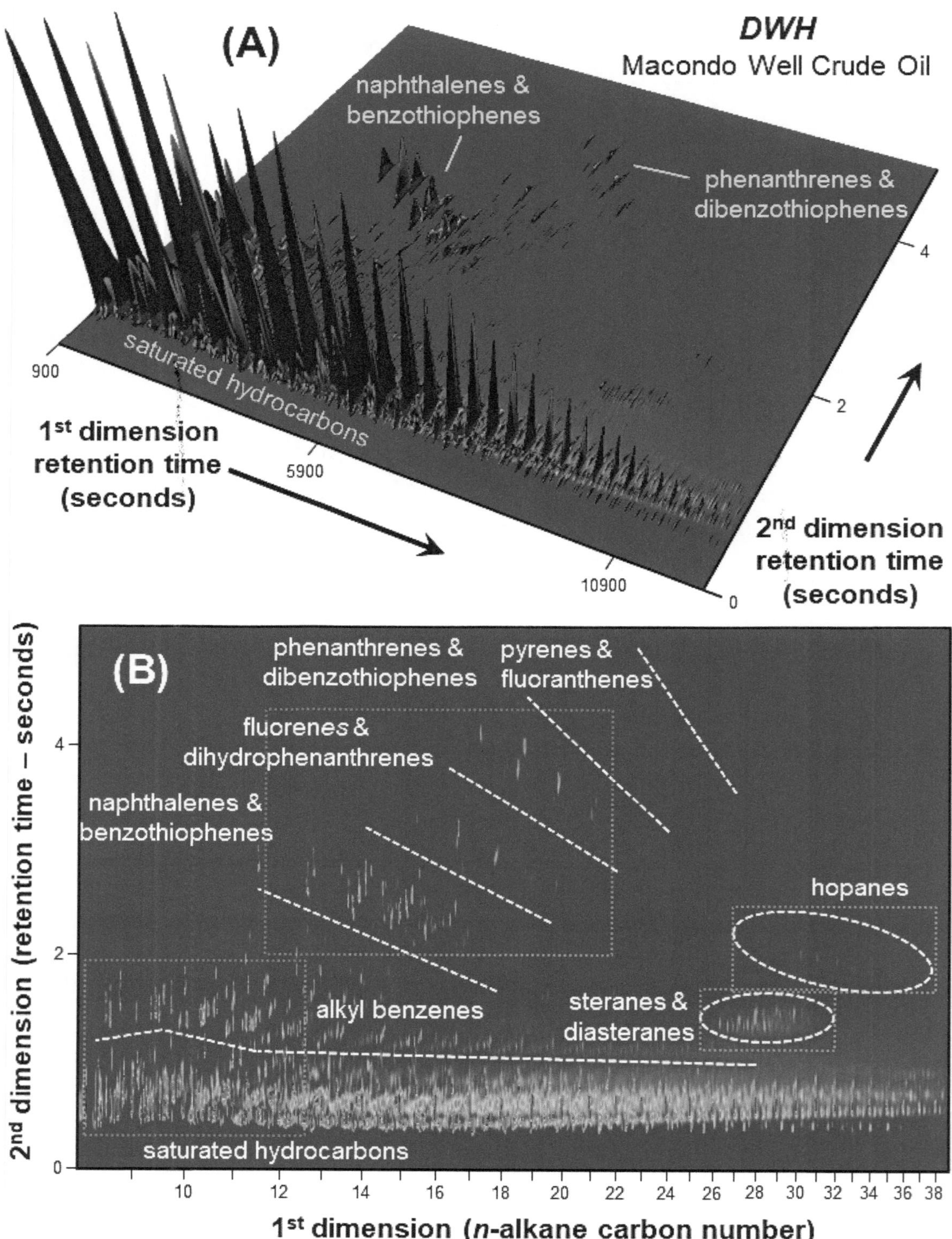

FIGURE 2.14 Comparison of three-dimensional surface rendering (mountain plot) (A) and a color contour plot plan view (B) for GC×GC-FID (flame ionization detector) chromatograms from the analysis of Macondo crude oil.
SOURCES: Courtesy of C. R. Reddy and R. K. Nelson, Woods Hole Oceanographic Institution (Wise et al., 2022).

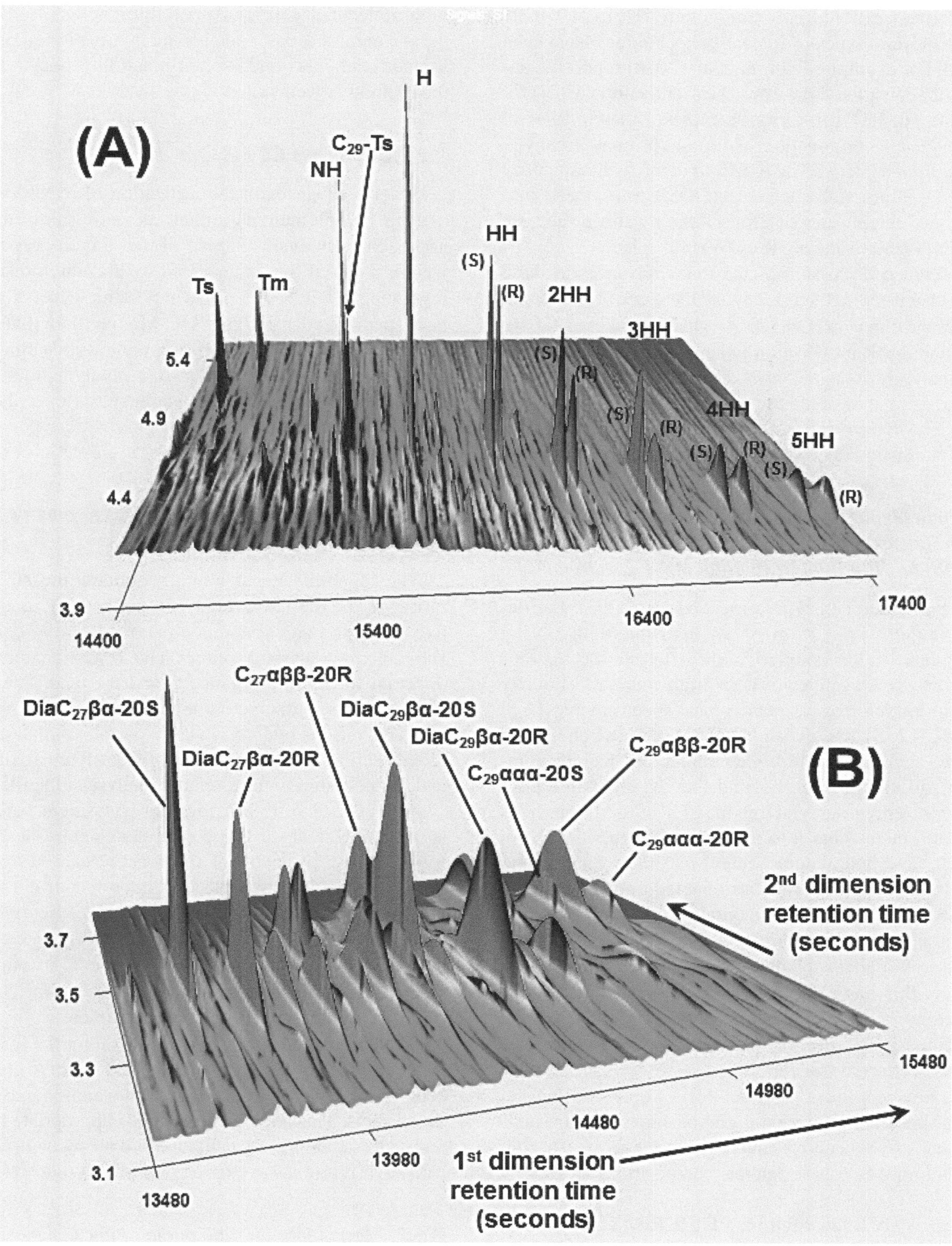

FIGURE 2.15 GC×GC FID chromatograms of petroleum biomarkers from Macondo crude oil. A zoomed in view of a GC×GC FID Mountain plot (see Figure 2.14) of the region where the hopanoid petroleum biomarkers are shown in panel (A). A zoomed-in view of the diasterane/steranes petroleum biomarker region is shown in panel (B).

SOURCES: Courtesy of C. R. Reddy and R. K. Nelson, Woods Hole Oceanographic Institution (Wise et al., 2022).

high performance liquid chromatography (HPLC-FT-ICR-MS). Also, several types of ionization techniques have been developed. For example, Benigni et al. (2016) report the application of atmospheric pressure laser ionization (APLI) for the screening of PAHs in various samples. Kujawinski et al. (2011) utilized a combination of liquid chromatography—mass spectrometry and FT-ICR-MS to identify and quantify the dioctyl sodium sulfosuccinate (DOSS) introduced into the Gulf of Mexico water column plume by subsurface use of COREXIT 9500A during the DWH oil spill.

More detailed discussion of the many advances in these analytical chemistry methods is beyond the scope of this report. References cited in Cho et al. (2014) and Benigni et al. (2016) provide a reasonable introduction and overview for the advances in these analytical methodologies. The next section discusses briefly the importance of the use of one of the higher magnetic field FT-ICR-MS instruments to advance our knowledge related to inputs, fates, and effects of oil in the marine environment.

2.2.2.4 High Magnetic Field Fourier Transform-Ion Cyclotron Resonance–Mass Spectrometry (High Mag FT-ICR-MS) or Ultra High Resolution Mass Spectrometry

Since publication of *Oil in the Sea III* (2003), the development and application of an instrument involving high magnetic fields coupled with different ion sources (e.g. positive or negative ions) in high magnetic Fourier transform-ion cyclotron resonance-mass spectrometry (High Mag FT-ICR-MS) has expanded the analytical chemistry window to encompass the higher molecular weight compounds of oil composition beyond the GC-amenable fraction (described previously). High mag FT-ICR-MS provides high resolution insights into elemental composition from C_{10} to C_{100+} and heteroatom content of sulfur, oxygen and nitrogen (0–20+ atoms). This has enabled a major advance in our understanding of the elemental composition of higher molecular weight components of oils especially resins, asphaltenes, and also asphalt. Of equal, or perhaps greater importance, this has also enabled a major advance in our understanding of the processes important in the fates and effects of inputs. Specific advances are described in the fates and effects section of this report.

These advances have also set forth some challenges. While high resolution elemental compositions of thousands of compounds have been presented, the exact molecular configurations linked to exact chemical compositions are largely unknown.

There are only a few high-mag FT-ICR-MS instruments available worldwide. The National High Magnet Laboratory Facility[7] at Florida State University–Tallahassee is funded as a user facility by the National Science Foundation, currently advancing research on fates and effects of spilled oil, among other research foci. There may be special cases of very high molecular weight oils and dilbit spills, used motor oil inputs, natural seeps, and chronic inputs of road runoff with asphalt chemicals and asphalt reaction products for which the method is well suited.

[7] See https://nationalmaglab.org.

2.2.2.5 Pyrolysis GC-MS

Several advances in the utilization of pyrolysis GC-MS offer a complementary quantitative or semi-quantitative approach to analysis of some of the same higher molecular weight material that is analyzed by ultra high resolution or high mag FT-ICR-MS. Thermoslicing pyrolysis GC-MS is a specialized pyrolysis GC-MS method that provides interesting insights into those compounds that may be trapped or absorbed into the resin/asphaltene fraction during weathering. It also provides insights into molecular components of asphaltene and asphalt polymeric-type substances (e.g., Kruge et al., 2018; Seeley et al., 2018).

2.2.2.6 Carbon Isotope (^{13}C, ^{14}C) Measurements of Environmental Samples

The capability exists now to routinely measure isotopic ratios of the stable carbon isotopes $^{13}C/^{12}C$ and radioactive ^{14}C in individual compounds or groups of compounds. Biogenic hydrocarbon sources have $\delta^{13}C$ signatures that are different for marine organisms and for land plant sources. There are some distinctive signatures for $\delta^{13}C$ among and between various hydrocarbons in petroleum that assist with identification of the presence of petroleum hydrocarbons and, on occasion, with forensics analyses of spilled oils.

Analyses of $\delta^{14}C$ provides an assessment of whether a specific chemical (or group of chemicals) has a source in a fossil fuel (i.e., is devoid of a radioactive carbon signal due to decay over geological time) or is modern in origin (e.g., from recent biosynthesis). This also allows for assessing whether carbon from a fossil fuel source has become incorporated into the food web by assaying either $\delta^{14}C$ of bulk organic material or $\delta^{14}C$ specific classes of biological compounds such as carbohydrates, proteins, and lipids.

Similar approaches are now available for the utilization of stable isotopes of nitrogen, oxygen, and sulfur to elucidate reactions involving the fates of petroleum compounds in marine ecosystems. This adds to the methods appropriate for tracing and understanding biodegradation and photochemical reaction pathways, discussed in later sections of this report (Chapter 5).

2.2.2.7 Expanding the Utilization of the Above Advancements to Assessment of Fate, Effects, and Forensics of Inputs

There is understandable concern about rapidly introducing newer analytical methods into a portfolio of standard methods which have a proven track record in a regulatory and legal framework. There have been examples with

GC×GC-MS where the use of this newer methodology has enabled a more complete forensic type analysis of spilled oil (e.g., Nelson et al., 2016) or sources of a collection of pelagic tar balls (Green et al., 2018).

It is recognized that there are trade-offs in the use of any given analytical methodology, such as instrument cost and availability, time of analysis, ease of data interpretation including requisite software for the interfaced computers, and other factors. The issue of analysis time and ease of data interpretation are often important factors for forensic analysis after sampling of oil spill samples and potential sources. For example, a recent exercise using a combination of GC-MS and ion mobility spectrometry-mass spectrometry (IMS-MS) in a table-top exercise demonstrated the feasibility of utilizing IMS-MS (minutes for analysis) compared to GC-MS (1 to 2 hours) to obtain comparable results (Roman-Hubers et al., 2021).

The advances in analytical chemistry methods described previously require judicious choices, depending on the questions being asked related to the samples to be analyzed, such as how rapidly answers are needed, whether the question is related to forensics, other regulatory activities, or research related to inputs, fates, and effects. For the arenas of forensics and regulatory activities, appropriate regulatory authorities should conduct periodic reviews every few years to ascertain which of the newer methods should be introduced into the portfolio of recommended methods for oil spill forensics and other regulatory activities. This should be accompanied by increased availability of appropriate standard reference materials and quality assurance and quality control inter-laboratory and intra-laboratory exercises.

2.2.2.8 The Challenges with Reporting of Petroleum and Other Hydrocarbon Concentrations in Environmental Samples

The reporting of hydrocarbon concentrations or sums of hydrocarbons in oils and environmental samples is common in oil pollution studies. The difficulties arise when terms such as *total petroleum hydrocarbons* (TPH), *total alkanes*, or *total polycyclic aromatic hydrocarbons* (PAHs) are used without a clear specification of the identities of the compounds included and their individual concentrations. This becomes a very important problem when attempting to compare studies of fates and effects of the same oil and the same oil at various stages of weathering and biodegradation (see Chapters 5 and 6) in which concentrations of TPH, total alkanes, and total PAHs are reported that contain summations of different mixtures of analytes. The problem becomes even more complicated when comparing fates and effects for different oils, or for differentiating the presence of given concentrations of PAHs from petrogenic or pyrogenic sources, or mixtures.

One approach that some studies have taken is to specify "target analytes," for example a given set of alkanes, cyclic alkanes, and/or PAHs that the analysts suspect could be present in the samples and can be separated and quantitatively assessed by the analytical methods employed. There is a need for general agreement among relevant agencies, organizations, and groups of scientists with respect to the chemicals to be measured and what group of compounds to include in summations (i.e., total of some specific group of chemicals). Such agreements should maximize the contributions to understanding and assessing fates and effects of spills and inputs and enable advantageous use of analytical chemistry methods for identification and tracking of spills and inputs within a forensic setting. The forensics needs may emphasize different sets of chemicals and therefore different types of measurements than assessments of fates and effects. All of the preceding most likely will vary depending on the type of oil spilled or variations in other types of input.

Irrespective of what methods are used and which compounds or summations of compounds are reported, there is a need for adequate quality assurances and quality control (QA/QC) for sampling and analytical methods within and between laboratories.

Quality Control and Quality Assurance

The need for QA/QC in applications of the analytical chemistry methods have been recognized for decades and implemented in recommended standard methods. Recent reviews have set forth the state of knowledge (e.g., Murray et al., 2015; Litman et al., 2018; Wise et al., 2022, and references therein).

The preparation and curation of standard reference materials, their utilization in QA/QC interlaboratory comparison exercises for the analyses of selected petroleum oils and petroleum and fossil fuel combustion PAHs in bivalve molluscs and surface sediments have been ongoing since the late 1970s (Kimbrough et al., 2008, and references therein; Wise et al., 2022, and references therein). Similar types of samples and/or interlaboratory comparison exercises have been ongoing within the international community during that same period of time. An example is an International Council for the Exploration of the Seas/International Oceanographic Commission (ICES/IOC) intercomparison exercise (Farrington et al., 1988, and references therein).

Interlaboratory comparisons and instrument calibration can be done using standard reference materials. These are mixtures of compounds with accepted composition and properties. Repositories of standard reference materials are maintained by various entities. See NIST (2021), among others, for details on the importance and usage of standard reference materials.

This has become standard protocol for official assessments such as those for forensics in oil spill identification and for activities such as NRDA by groups reporting to official trustees for a given spill in the United States. This is also the case for similar activities in Canada. It is becoming more common among academic and other independent researchers as is appropriate.

Authentic Samples of Sources of Spills and Inputs

An essential aspect of forensics and regulatory activities, and research studies of the fates and effects of any oil input to the environment, is to collect samples of original spilled oil or other sources of oil to the extent practicable. These samples are useful in forensic analyses to trace the geographic and temporal extent of a given input using petroleum biomarker compounds. Collections of potential sources of a mystery spill can also be used in forensic analyses to tie the mystery spill to a specific source or, at the very least, eliminate some potential sources. Stout and Wang (2016) provide an interesting, relevant collection of examples.

"Big Data" Issue

Sampling and analytical methods for forensic analyses and for assessing the fates and effects of oil spills and other inputs are now generating large sets of data because of the increased ability to analyze for more of the myriad compounds of interest. As with other sciences, oil pollution studies have entered the realm of "big data." It is not only in the analytical chemistry data, but data related to "omics" (see Chapter 5) and Effects (see Chapter 6). A related issue is bringing the most relevant "big data" together for the same samples. The inclusion of metadata such as sampling location, time, sample type, and other related environmental information into these data bases is essential.

While expense is often cited for the reasons for not bringing the most appropriate chemistry and biological analysis methods together for the same samples, the inefficiency connected with the "cost" of lost knowledge should enter the considerations in a more appropriate manner.

The need for appropriate data archives, readily accessible for all interested users is clear (e.g., McNutt et al., 2016). NOAA's DIVER database (NOAA, 2021e) and the recent Gulf of Mexico Research Initiative Information and Data Cooperative (GRIIDC) database (Gibeaut, 2016; GoMRI, 2021) are examples of a response to this need. Science and technology are poised for important expansion of content and interactive data archives for chemical analyses interfaced with archives of inputs, fates and effects data as noted in Chapters 5 and 6.

2.3 THERMODYNAMICS OF MIXTURES OF OILS

As described previously, oils are complex mixtures of molecules, each molecule having its own unique properties. At the same time, these molecules mix together into fairly homogeneous fluids that have their own properties, including mixture density, viscosity, interfacial tension, and others. These properties can be measured when samples are available. However, there are many instances where these properties need to be estimated, as for example in predicting the behavior of a spilled oil during its transport and fate in the marine water column.

The purpose of a model for the thermodynamic behavior of an oil is to predict its bulk properties and its interactions with seawater. Important bulk properties include density and viscosity; some models may also predict the gas–liquid phase partitioning. When released in seawater, it is also critical to predict interfacial tension, potential for hydrate formation, and the solubility and diffusivity of individual components of the oil. Each of these properties is thermodynamic, meaning that they depend on the composition of the oil–seawater system and its thermodynamic state, typically defined by the temperature and pressure. Some of these properties are important themselves for determining oil behavior; others are inputs to additional process models for predicting fate effects, such as breakup of oil and gas into droplets and bubbles, mass transfer from oil or gas to a dissolved state in seawater, and many other processes (see Chapter 5).

Irrespective of the thermodynamic model used, oil composition is normally described by quantifying the abundance of both individual molecules and groups of molecules having similar behavior. When oil composition is simplified by grouping molecules together, these groups are called pseudo-components. To make predictions, oil property models require additional data for several physical properties of each compound or pseudo-component. These properties may include molecular weight, boiling point, or critical point properties, among others, depending on the needs of the model. Using the analytical methods described previously, the molecular composition of oils can be known in great detail. Because of the added burden of also quantifying the physical properties of each of these compounds, models of oils may be based on individual molecules up to about n-C8 at most (Gros et al., 2016). For longer-chain hydrocarbons, pseudo-component groups become more efficient than itemizing all possible chemical structures individually.

Pseudo-components are most often composed of molecules having a similar fate. Solubility, octanol and K_{ow}, biodegradation rate, and boiling point are common properties used to distinguish them (McKay, 2003). For example, these groups may be distinguished by boiling point and SARA identification. More recently, Gros et al. (2016) identified pseudo-components by common regions in two-dimensional gas chromatography (GC×GC) spectrograms. They also present methods to compute several important properties for each pseudo-component using property correlations and group contribution methods. However, it is not always necessary to group molecules having similar fates in the aquatic environment. For example, the equation of state for DWH crude oil developed by Zick (2013) and used in the court case of the United States against BP (British Petroleum) grouped benzene and n-hexane together into a single pseudo-component despite these compounds having solubilities that differ by a factor of 151. Rather than predicting the oil fate, the purpose of the Zick (2013) model was to predict fluid phase density and gas–liquid phase equilibrium, for which this pseudo-component selection was appropriate. Irrespective of how pseudo-components are defined or their properties estimated, an oil mixture is defined by stating the compound and pseudo-component groups and quantifying the mass fraction of the whole petroleum fluid within each group.

Property models for oil normally fall into two broad categories: thermodynamic equations of state (EOS) or correlation methods. Thermodynamic EOSs have developed significantly since the 1970s within the petroleum chemistry and engineering fields, where they are mostly applied within systems having limited exposure to seawater (McCain, 1990). Common practical models are variations of the cubic equations of state (see the review by Valderrama, 2003). Recently, these models have also been adapted to predict the behavior of natural gas (McGinnis et al., 2006) and oil in the oceans (Gros et al., 2016, 2017; Dissanayake et al., 2018). Though powerful, cubic equations of state require significant input data for each component and are limited to predicting some of the fundamental properties of a fluid mixture, including density and component fugacity, hence, solubility. The alternative correlation methods include group-contribution methods and property correlations. Group-contribution methods predict oil properties based on discrete contributions of individual molecular components within a compound. As these methods provide component properties, these may be used to estimate the input data required by the cubic equations of state (e.g., Gros et al., 2016). The Poling et al. (2001) method for predicting the critical point pressure of a compound is a good example. Property correlation methods similarly develop empirical relationships between the properties of a given molecule or pseudo-component, for example, the molecular weight and boiling point, and another property, such as the vapor pressure. These methods are especially important for estimating oil properties not predicted by the cubic equations of state, such as viscosity, interfacial tension, and diffusivity in seawater.

Current oil spill models differ mostly in their approaches to predicting density, phase equilibrium, and component solubility of an oil mixture: some use thermodynamic EOS and others use property correlations for density and solubility with database lookup methods for phase equilibrium. An example using the thermodynamic EOS approach is presented in Gros et al. (2016). They applied the Peng–Robinson (PR) cubic EOS (Peng and Robinson, 1976) with volume translation (Péneloux et al., 1982) to various compositions of the DWH oil. The PR EOS predicts the density of the gas and liquid phases of the mixture and the fugacities of each component in the mixture given the component properties of the molecular weight, acentric factor, critical point temperature, pressure and volume, and the binary interaction coefficients. The gas–liquid phase equilibrium is determined by adjusting the estimated partitioning of components into the gas and liquid phases at a given thermodynamic state and iterating until the fugacities of each component in each phase converge (Michelsen and Mollerup, 2007).

Using the alternative property correlation method, the density of the liquid-phase petroleum is reconstructed from estimates of the component densities and their mass fractions in the whole oil. The gas-liquid phase equilibrium would normally be read from a thermodynamic table of properties, developed by way of another EOS, such as the Gros et al. (2016) or the Zick (2013) model of a crude oil (e.g., French-McCay et al., 2015) or based on laboratory measurements.

Using the EOS approach, the solubility of each component is derived by applying the modified Henry's law (King, 1969; Dhima et al., 1999) in which the partial pressure is replaced by the fugacity coefficients. The Henry's constant for each component at standard conditions is adjusted to the in situ thermodynamic state using standard corrections for temperature, pressure, and salinity that depend on the component properties of the partial molar volume at infinite dilution, the enthalpy of transfer from the gas phase to the liquid phase, and the Setschenow coefficient (Gros et al., 2016). Gros et al. (2016) reported each of these required model inputs for 131 individual molecules identified in the DWH oil and for 148 additional pseudo-components defined from the GC×GC-FID chromatogram and simulated distillation data. To estimate the properties for many of these components, they present several group-contribution and property correlations. Similarly, Gros et al. (2018) present a method to estimate the required parameters from distillation data. Through rigorous comparison to available laboratory data, they conclude that this model and its parameter estimation techniques is valid for pressures corresponding to ocean depths ≤2,500 m, temperatures between −2 and 30°C, and salinities up to 35%.

An alternative and more direct approach to estimate solubility using property correlations was introduced by Mackay and Leinonen (1977). There, the solubility of each component of an oil is given by the product of the solubility of the pure component in seawater, the mole fraction of that compound in the whole oil, and the solubility enhancement factor that accounts for non-ideal behavior of the petroleum liquid. Mackay and Leinonen (1977) report four values of the solubility enhancement factor, corresponding to components containing alkanes, cyclic alkanes, aromatics, and olefins, respectively. In situ effects of temperature, pressure, and salinity are captured through the pure-component solubility estimates; non-ideal, real-fluid effects are contained in the solubility enhancement factors. Many other similar methods exist. For example, Lehr et al. (2002) predict pure-component solubility for >C6 aromatics based on the molecular weight and component density. They then estimate the solubility of each component from the whole petroleum liquid using the oil–water partition coefficient, which they compute from a correlation with the pure-component solubility. An implicit assumption of these methods is knowledge of the oil-phase composition, which may have to be linked to an EOS model when the release involves oil in equilibrium with gas.

Because pure compounds, such as benzene, have a constant chemical activity regardless of thermodynamic state, their solubilities in seawater can be reported in thermodynamic tables, for instance as a function of temperature and pressure. When these compounds are present in a petroleum mixture, however, their activity and hence, solubility, are altered as described previously, and solubility data become composition-dependent. The situation is made even more complicated when considering reservoir fluid versus dead oil (see Section 2.2). The composition of fluids within a petroleum reservoir are in equilibrium with the local temperature and pressure, both of which are typically very high. In this state, a significant light fraction of the mixture, for example, methane, ethane, propane, and so on, may be dissolved in the liquid petroleum phase. Such a liquid is commonly termed *live*. *Dead* oil, by contrast, is the liquid petroleum composition after coming to equilibrium with standard conditions in a closed system; in this state, little of the gas fractions remains dissolved in the oil. Whether the oil enters the ocean as dead or live oil, hence, can significantly affect the solubility of all fractions of the oil. Moreover, the composition of the liquid phase depends on the chain of gas–liquid separations occurring after release. For example, once liquid oil droplets separate from gas bubbles, further evolution of the liquid phase composition will depend on this discrete composition and not on the composition of the whole oil, upstream of the initial separation into droplets and bubbles. Phase separation is discussed in more detail in Section 2.2 and the diversity of fate processes in Chapter 5. The effect of these processes on the solubility of benzene from various oil compositions is illustrated in Box 2.5.

BOX 2.5
Solubilities of Benzene in Diverse Petroleum Mixtures

The solubility of any given component from a complex petroleum mixture into seawater depends on the composition of the petroleum and the thermodynamic state. Different models exist to predict component solubility, each with different input data requirements and levels of complexity. This box presents predictions from two different modeling approaches (equations of state and property correlations) in their predictions of the behavior of the relatively soluble compound benzene.

The saturation concentration of benzene in freshwater at 25°C and atmospheric pressure is 0.0224 mol/l (1750 ppm) (Schwarzenbach et al., 2016). When benzene is present as one compound in a complex petroleum mixture, the solubility of benzene into water from the mixture will be reduced from this value. Louisiana light sweet crude oil, as described in Gros et al. (2018), is used to illustrate these effects. The dead-oil composition of the petroleum at 25°C and atmospheric pressure has a density of 841 kg/m^3 and a benzene mass fraction of 1.3%, taken as the pseudo-component "aromatics" in Gros et al. (2018), which has a boiling point range of 353.2 K to 356.5 K, hence, including benzene, which has a boiling point of 353.25 K (Schwarzenbach et al., 2003). The mole fraction of benzene in this mixture is 3.47%.

Equations of State Approaches

The cubic equations of state model of Gros et al. (2016) predicts a solubility of benzene from this dead-oil mixture into seawater at 25°C, 34.5 psu salinity, and atmospheric pressure of 0.00053 mol/l (41 ppm). Considering the effects of pressure only, this same system at 15 MPa has a benzene solubility in seawater of 0.00056 mol/l (44 ppm); hence, there is a negligible effect of pressure on this dead-oil petroleum liquid. Cooling this system to 5°C and maintaining 15 MPa pressure, the benzene solubility from this mixture into seawater is predicted to be 0.00059 mol/l (46 ppm). Thus, temperature also has a minimal effect on the solubility of benzene.

Correlation Equations Approaches

Correlation equation methods to compute benzene solubility from petroleum mixtures include using the octanol-water partition coefficient K_{ow} or the pure-compound solubilities with mole fractions and enhancement factors. The measured K_{ow} value of benzene is 148 (Schwarzenbach et al., 2016). Assuming that benzene partitions in the same way from crude oil as it does from octanol, K_{ow} may be used to predict solubility from crude oil mixtures. For the Louisiana light sweet crude composition defined above, the K_{ow} method predicts a solubility of benzene into seawater of 0.0010 mol/l (77 ppm), which is within a factor of 2 of the equations-of-state method of Gros et al. (2016).

The approach in Mackay and Leinonen (1977) gives similar estimates for benzene solubility. Using the enhancement factor for aromatic compounds of 2.2 in MacKay and Leinonen (1977) with the mole fraction of benzene given previously, their correlation equation predicts a benzene solubility in seawater of 0.00171 mol/l (130 ppm).

Effects in Live-Oil Mixtures

To demonstrate the solubilities from live-oil mixtures, hypothetically, Louisiana light sweet crude defined by Gros et al. (2018) is mixed with a natural gas composed of 93.3% methane, 4.2% ethane, 1.84% propane, 0.3% iso-butane, and 0.3% n-butane by mass fraction. The GOR is defined using standard petroleum engineering units. Fixing the temperature at 5°C and the pressure at 15 MPa and considering a GOR value of 2000 scf/sbbl, the Gros et al. (2016) model predicts a gas fraction by volume at equilibrium of 44% with a gas density of 148 kg/m^3. The corresponding liquid-phase density is 747 kg/m^3, reduced from the dead-oil density by the inclusion of dissolved gases. The mass fraction of benzene in the liquid-phase petroleum is 0.6%, as a significant amount of benzene has partitioned to the gas phase, which has a benzene mass fraction of 2.5%.

If the gas-phase and liquid-phase petroleum are in equilibrium, then the activity of benzene will be the same in both fluid phases; hence, the solubility of benzene from the gas or liquid phase is the same. For these conditions, the equations in Gros et al. (2016) predict the benzene solubility of 0.00029 mol/l (22 ppm). Thus, the presence of gas has reduced the solubility of benzene. This effect can also be seen from the correlation equation methods: the mass fraction of benzene in this gas-liquid mixture is 1.1% by mass fraction and 0.9% by mole fraction—lower than in the dead-oil mixture due to addition of natural gas so that solubilities predicted by these equations will also be lower.

These examples show the ranges of predicted benzene solubility from a light sweet crude oil using different modeling approaches. Despite the complex thermodynamics of the gas–liquid equilibrium, each of the methods presented here gives reasonable estimates for the benzene solubility over a wide range of relevant ocean thermodynamic states.

Because of the recent advances in predicting oil properties and their importance to fate and effects (e.g., evolving oil compositions, solubilities of complex mixtures, phase equilibria, and so on), databases of more complete oil composition data are needed. These could include elemental composition to n-C8 or lighter compounds and complete SARA analysis. Results of GC or GC×GC output in standardized formats could also be housed within the repository. The committee recognizes that these data may not be available for spills from exploration wells and that some of these data are proprietary for producing wells. However, there are existing databases of oil properties for well-known oils, and where these more comprehensive analyses can be made public, they should be stored in standardized data formats in dedicated public repositories.

2.4 CONCLUSIONS

Conclusion—Analytical Chemistry Methodology: There have been significant advances in analytical chemistry methodology utilized in forensics pertaining to oil spills and other inputs of petroleum to the marine environment and assessing and understanding the fates and effects of such inputs. Previous advances of the 1980s–1990s such as glass capillary–mass spectrometry (GC-MS) have been utilized routinely. Recent advances in analytical chemistry methodology such as glass capillary two-dimensional gas chromatography–mass spectrometry (GC×GC-MS), GC Orbitrap MS, GC-MS-MS-MS, FT-ICR-MS, and high magnetic field FT-ICR-MS have been tested in several studies and are ready for more routine use to provide much needed more detailed chemical compositional data pertinent to the questions being asked in forensic analyses for spilled oil and chronic inputs, and understanding fates and effects of various inputs of oil to the environment.

Conclusion—Elemental Composition: Remarkable insights have been gained into the elemental compositions and classes of compounds in resins, asphaltenes, and asphalt, as well as reaction products of processes governing the fate of oil spills and other petroleum inputs by utilization of FT-ICR-MS and in particular by ultra high resolution high magnetic FT-ICR-MS. Despite these advances, the exact molecular structures of these thousands to tens of thousands of chemicals have yet to be elucidated and are one of the most important challenges of analytical chemistry relevant to the subject of this report.

Conclusion—Big Data: Utilization of the advances in analytical chemistry methods yield large databases of concentrations and relative abundances of petroleum oil chemicals and reaction products resulting from physical, chemical, and biological processes acting on such chemicals in the environment. There are existing databases such as those maintained by NOAA and Environment and Climate Change Canada (as noted in this chapter) that might be expanded to meet the need for archiving these datasets in an accessible and useful manner. It is importance to incorporate appropriate metadata for the samples in these larger databases.

Conclusion—Reporting of Chemical Composition: The reporting of concentrations of individual chemicals, and summations of chemicals of different compositions, varies among and between scientific papers and reports. This can lead to inappropriate comparisons of different compositions of chemicals when elucidating fates and effects results between and among different studies and assessments.

Conclusion—Modeling: Models to predict the properties of petroleum fluids, including mixtures with gas, have improved recently, buoyed partly by the better elucidation of the composition of petroleum mixtures using advanced analytical methods.

Conclusion—New Fuels: Continued attention to the chemical compositions of biofuels and other evolving fuels as their production and use expands will inform response actions and the potential fates and effects of such spills or chronic inputs.

3

Input of Oil to the Sea

The life cycle of "oil in the sea" starts with the input of petroleum hydrocarbons, referred to throughout this report as "oil," to the marine environment. Oil enters the sea from natural seepages and from anthropogenic sources that can originate from routine human activity or from accidental releases. Estimating the volumes of oil inputs is subject to a degree of uncertainty but the relative orders of magnitude and general trends can be established. The rate of discharge of oil must also be considered. Natural seepage—as well as runoff and atmospheric deposition from land-based activity—typically takes place at a low rate, either continuously or intermittently. These chronic or continuous inputs have very different effects on the environments than do accidental spills. Accidental releases garner the most attention when they occur at higher volumes in a relatively short time frame; however, they can also occur at a range of volumes and time frames, and in some instances the releases can continue for months or even years. The volume and the rate of discharge, as well as other factors, are important for the fate and the effects of the oil, as will be discussed in detail in later chapters in this report.

As illustrated in Figure 3.1, this chapter discusses inputs and provides estimates (as possible) from the following sources:

- natural oil and gas seeps
- land-based runoff and atmospheric deposition
- operational discharges from
  - extraction of hydrocarbons
  - marine transportation
  - recreational vessels
  - aircraft fuel jettison
- accidental spills from
  - extraction of hydrocarbons
  - aging infrastructure and decommissioning
  - marine transportation by ships and pipelines
  - coastal storage facilities
  - sunken wrecks

In this report, *operational discharges* refers to water produced as a by-product from oil extraction and to discharges occurring in a controlled manner during routine machinery and cargo operations on ships. These discharges are governed by international and domestic regulations. *Spills* refers to all other discharges from marine operations and offshore oil extraction, accidental or intentional, not covered by the controlled routine discharges. For example, discharge of oily water in areas where discharges are prohibited or in quantities exceeding allowable limits, are considered illegal spills in this report. Natural seeps as well as runoff and atmospheric deposition from land-based activity are considered in a category of their own and are not included in the spill volumes.

The objective of this chapter is to provide an update of estimates from the previous Oil in the Sea reports. Therefore, the inputs are estimated using the same basis as in the NRC 2003 report, as much as is practical, to provide a comparison of the relative input volumes and to identify trends since *Oil in the Sea III*.

- The *Oil in the Sea III* accidental spill statistics covered a 10-year period from 1990 to 1999. This report covers the period of 10 years from 2010 to 2019.
- The geographic zones used are the same as in the *Oil in the Sea III* report and they are described in Appendix A.
- The input magnitudes are provided in metric tonnes (MT) for consistency with the previous report.

This report has added sources of oil not included in *Oil in the Sea III*. Natural gas seeps are estimated in addition to oil seeps. Potential spills from aging infrastructure, decommissioning leakage, and sunken wrecks are evaluated. Oil spills caused by extreme weather events have become a concern, which is covered in this chapter as well.

As described in Chapter 1, to the extent possible with available data, the analysis is of North American waters, which includes the economic exclusive zones (or out to the

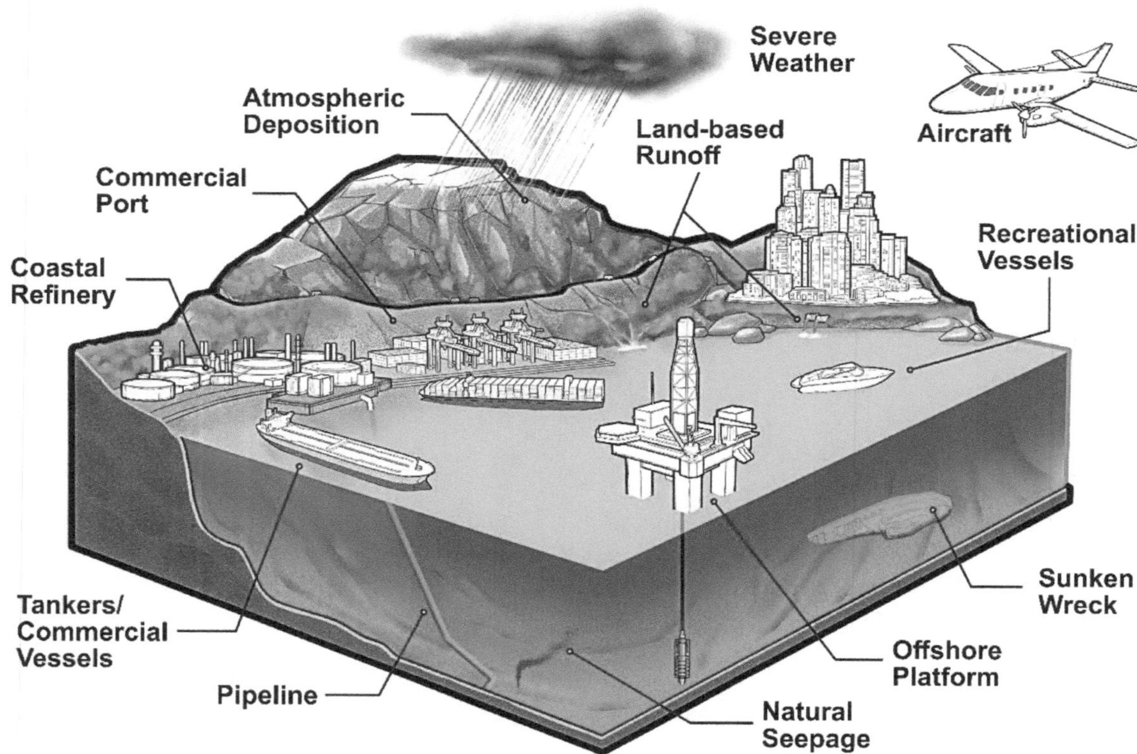

FIGURE 3.1 Sources of fossil fuel hydrocarbons entering the marine environment.
SOURCE: Image provided courtesy of the American Petroleum Institute, produced by Iron Octopus Productions, Inc.

continental shelf, whichever is further offshore) surrounding the United States, Canada, and Mexico.

The past 20 years have brought many changes to how oil is extracted, transported, and consumed. The North American energy landscape has changed with the production of shale oil and gas both in the United States and Canada. In the past, the United States was an importer of oil and gas, but now, because of the increased production and changes in legislation allowing export of crude, the United States has become an exporter of oil and gas. This, as well as Arctic ice melt, has had an impact on terminals and on shipping routes and the frequency of oil movement.

Increase in urban populations in coastal areas, consumer behavior, improved fuel efficiency of vehicles, and introduction of electric cars are all changes impacting land-based consumption and, although more difficult to quantify, inputs of oil into the sea.

The past two decades have been marked by a warming climate and associated extreme weather (NASEM, 2016). This introduces new potential for accidents and also, as sea ice cover continues to decrease, opens up more of the Arctic region for exploration, production, and transportation of oil as well as shipping and tourism (see Box 3.1). Other major changes since the *Oil in the Sea III* include new regulations on design and operation of ships carrying oil as cargo or as fuel, and on engines used on recreational vessels. Following the *Deepwater Horizon* (DWH) incident, major changes were introduced to the regulations governing offshore oil and gas operations (some of which vary in the United States from one administration to another). New performance measures and enforcement mechanisms have been introduced to improve pipeline safety. In general, there is an overall recognition of the importance of human factors and human error in the safety across sectors and there has consequently been increased focus on improving the safety culture in the oil and gas exploration, production, and transportation industries.

This chapter begins with an overview of the committee's findings regarding magnitude and trends of oil volumes entering the sea.[1] The following sections then provide detail on each type of input, describing the changes since the *Oil in the Sea III* report (many of which are regulatory in nature), the data and methodology used for the estimates, and a summary and a comparison of the input volumes. The chapter provides the best estimate and discusses the uncertainty associated with each type of input. However, because the available data does not support rigorous statistical analysis, the chapter does not use statistical approaches to provide the

[1] It should be noted that the Oil in the Sea IV committee's analysis incorporated information available through 2021.

BOX 3.1
Sources and Challenges of Oil Pollution in the Arctic

Although the Arctic Ocean comprises only 3.4% of the world ocean surface area, it includes 21% of the world's continental shelf (Menard and Smith, 1966). It is on these continental shelves that most of the estimated recoverable marine oil deposits are located. Recovery potential is greatest off the northern coast of Alaska in the Chukchi and Beaufort Seas, where an estimated 14–47 billion barrels of recoverable oil may be located (some of which is on land, see Figure 3.2). Currently offshore Alaskan oil is accessed by drilling rigs based on land or on artificial islands using directional drilling to tap oil reservoirs. These artificial islands are within 20 km of the shore, in less than 12 m of water depths. Other offshore oil production platforms currently operating in the Arctic Ocean include Prirazlomnoye, located in 6 m of water about 60 km offshore Russia in the southeastern Pechora Sea with ~500 million barrels of recoverable reserves[a] (and the Goliat oil field located in 175 m of water about 80 km offshore Norway in the Barents Sea with ~175 million barrels of recoverable oil.[b] Oil produced by these two platforms is loaded onto tankers for delivery to refineries. Also, the Snøhvit gas field located in 250–345 m of water about 80 km offshore Norway in the Barents Sea has ~170 million m^3 of recoverable natural gas and oil condensates.[c] Snøhvit production is transported through a pipeline buried in the seafloor to the mainland for processing. Overall, the great majority of oil reserves and current production in the Arctic Ocean is associated with shallow (<15 m) water depths on the continental shelf off the northern coast of Alaska.

While accidental offshore oil discharges from oil exploration, development, production, and transportation have the potential to cause the largest oil spills, discharges resulting from other causes cannot be dismissed. Commercial fishing, shipping, and tourism vessels carry substantial amounts of refined petroleum products as fuels, ranging from gasoline to heavy bunker oils and crude oils. Fuel capacities of large container vessels can exceed 10,000 m^3, and release of even half of this following a collision would constitute a major oil spill.[d] In the U.S. Arctic, commercial fishing has been forestalled indefinitely under the U.S. Arctic Fisheries Management Plan,[e] but in the Barents and Kara Seas off Norway and Russia commercial fishing has been conducted for more than a century. Currently commercial tourism, as with commercial fishing, is more developed along the northern coasts of Greenland, Norway, and Russia than in the North American Arctic.

[a]See https://www.gazprom.com/projects/prirazlomnoye.
[b]See https://www.norskpetroleum.no/en/facts/field/goliat.
[c]See https://www.norskpetroleum.no/en/facts/field/snohvit.
[d]See https://response.restoration.noaa.gov/about/media/how-much-oil-ship.html.
[e]See https://www.npfmc.org/arctic-fishery-management.

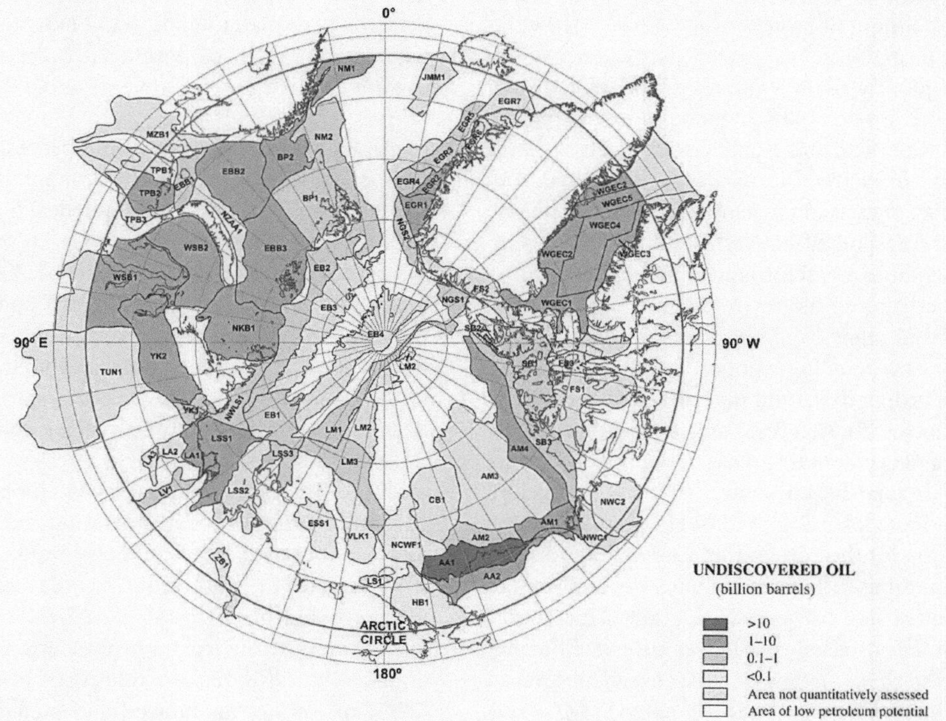

FIGURE 3.2 Map showing the assessment units (AUs) of the circum-Arctic resource appraisal is color-coded for mean estimated undiscovered oil.
NOTES: Only areas north of the Arctic Circle are included in the estimates. Black lines indicate AU boundaries.
SOURCE: From Gautier et al. (2009).

upper, lower, and most probable values for each category of inputs. The estimates provide insight into the order of magnitude and general trends of the inputs as well as their relative contribution to oil in the sea.

The chapter concludes with a discussion of future predictions on spills based on the current spillage rates and the assumption that the current energy usage trend will continue. It should be noted that the predictions do not account for potential changes in transportation and extraction practices, and they do not assume a future energy transition from oil to a mix of energy sources. They also do not take account for world events that have occurred since the start of the study.

3.1 OVERVIEW OF OIL INPUTS

Although data are lacking for accurately estimating many inputs of oil into North American waters, enough data do exist to at least understand the trends, and in many cases to provide more precise estimations of annual volumes. Overall, looking at oil input to the sea from 2010–2019, as compared to previous reported estimates from 1990–1999, if excluding the DWH event from the 10-year annualized statistics, it is evident that regulatory changes, advances in science and technology, and (for the most part) attention to safety would have helped to make North American waters less polluted with oil, if the DWH event did not happen. Although this trend is likely to continue, there is also the risk of a "black swan" type event, a failure of a system, as seen in 2010 with the DWH: a low-probability occurrence with detrimental effects and consequently, with volumes of oil that dwarf other sources. There are also new potential sources for oil pollution in North American and global waters, such as those related to the aging infrastructure in the sea and along the coasts; sea-level rise; increased intensification and frequency of storms; use and transport of new types of oil; changes in shipping routes including travel through arctic regions; shifts in workforce, expertise and asset ownership in the energy industry; warfare; and others.

Figure 3.3 shows the total inputs of oil in the sea, by geographic region, according to natural inputs (see Section 3.2), land-based sources (see Section 3.3), operational discharges (see Section 3.4), and accidental spills (see Section 3.5). The annual estimates of land-based sources, by far, outweigh other sources; this includes using the DWH oil spill and the worst-case projection for the Mississippi Canyon 20 oil spill in the estimates. Land-based sources include only land-based runoff; the committee did not include estimates for atmospheric deposition. Land-based runoff was calculated using the same approach as *Oil in the Sea III* and is based on several assumptions. The increase in estimates compared to *Oil in the Sea III* is within the range of uncertainty, and it is therefore not possible to determine if there has been a quantitative increase. These estimates are in line with global estimates (which are closer to 30 times as high as those reported in *Oil in the Sea III*), and it is plausible that there has been an increase in land-based runoff over the past two decades corresponding to an increase in the number of vehicles being used.

The second highest input is from natural seeps. These estimations were derived using more advanced techniques, such as satellite data, than in previous Oil in the Sea studies. The committee's estimates of natural seeps are roughly one-third lower than reported in *Oil in the Sea III*, and reflect updated estimates for the Gulf of Mexico; new data was not available for other North American regions.

Operational discharges presented in this report include produced water and discharges from machinery operations on commercial vessels. In the absence of non-compliance data, full compliance of oily discharges from tank vessels and non-tank vessels is assumed and illegal discharges are regarded as spills. Regulations prohibit discharges of oily water from cargo areas in tank vessels in North American waters, and therefore there is no discharge to report. Assuming full compliance, all discharges from commercial vessel operations are small, less than 10 MT per year. Because *Oil in the Sea III* discharge amounts assumed non-compliance, a direct comparison is not meaningful. Although it can be safely assumed that discharges from recreational boating greatly decreased with regulatory actions to ban sales of two-stroke engines, there is no data available for estimating actual discharges of oil in the marine environment from this source. Estimates are also not included for aircraft fuel jettison, as (1) the fuel volumes are not reported and (2) more research is needed to understand how much (if any) fuel jettisoned enters the sea. Produced water estimates are higher in total than they were 20 years ago, reflective of increased oil production.

Annualized estimates of spills represent the third highest input even if one includes a 10-year annualized estimate from the DWH spill. It should be noted, the estimates of annual input from the Mississippi Canyon-20 spill vary greatly and the estimates for the time period of 2000–2019 will more accurately reflect annual spillage if and when this volume is published. Spills occurred more frequently in offshore waters than nearshore waters and predominantly occurred in the Gulf of Mexico. Over 20 years, the volume of spills decreased significantly for pipelines, tank vessels, non-tank vessels, and coastal refineries.

Table 3.1 includes the total oil inputs broken down by natural, extraction, transportation, and consumption for direct comparison with estimates reported in *Oil in the Sea III*. Estimates of natural sources are roughly one-third lower than reported in *Oil in the Sea III*. Amounts of oil entering the marine environment by extraction have more than doubled but still remain relatively low, except when the DWH oil spill is annualized and included in the statistics. Spills and discharges related to transportation of petroleum are more than 10 times lower than they were 20 years ago. Then there is the category of consumption: oil entering the sea through consumption of oil has potentially increased significantly over the past 20 years; however, this estimation is heavily weighted by an estimate of land-based runoff that is largely based on assumptions (similar to *Oil in the Sea III* calculations), as appropriate measurements do not exist for providing data-based estimates for land-based runoff. The same is true for other input sources. An important factor in calculating land-based runoff for a region is number of vehicles; however, trends such as increased fuel efficiency, increased use of electric or hybrid vehicles, and motor oil recycling are not factored in; therefore, estimates are likely higher than actuals. If including estimates of land-based runoff, overall, the inputs of oil into the marine environment are estimated to have been five times as high from 2010–2019 as they were from 1990–1991. If excluding the category of consumption, which includes inputs in *Oil in the Sea III* that were not estimated in *Oil in the Sea IV* and therefore skew comparison, annual inputs are roughly 35% lower in the 2010s than they were in the 1990s.

3.2 NATURAL SEEPS

Natural seepage is a significant source of gaseous and liquid fossil fuel hydrocarbons introduced into the marine environment without human interference (Kvenvolden and Cooper, 2003). Oil and gas from deep subsea reservoirs move through faults and cracks in the seafloor or through sediments into the water column. In the water column, oil and gas compounds are subjected to dissolution and biodegradation according to their solubility, bioavailability and the environmental conditions (temperature, salinity, nutrient availability; refer to Chapter 5).

Natural oil seeps have been reported worldwide and their locations continue to be explored to determine locations that might be feasible for oil extraction. Regions with considerable inputs of oil from natural seepage, such as the Gulf of Mexico, have persistent surface slicks (Johansen et al., 2017), some of which are visible from space and are routinely detected through remote sensing methodologies such as satellite imagery (MacDonald et al., 1993; Garcia-Pineda et al., 2010). Other seeps may be more challenging to detect due to their low flux, the episodic nature of their releases, co-location with other oil sources, or the lack of a surface slick due to a wide range of reasons, including weather, sea state, flow rate, and subsea fate. Current estimates of the number of natural oil seeps are thought to be underestimated due to the remoteness and difficulty in accessing potential seep locations in the Southern Ocean, the Arctic Ocean, and the deep sea (Byrnes et al., 2017, for a detailed discussion of these limitations in the Gulf of Mexico).

Natural gas seeps are also ubiquitous along the North American continental margins (Ruppel and Kessler, 2017). Natural gas seeps release only gaseous hydrocarbons, whereas oil seeps may release a mixture of both gaseous and liquid hydrocarbons. Because of the large density difference between gas and water, natural gas bubbles are visible in images of acoustic backscatter from low-frequency (18 to 74 kHz), long-range (1,000 to 500 m) sonar (Weber et al., 2012). As a result, recent advances such as using water-column backscatter in acoustic multibeam data have significantly improved our ability to identify natural gas seeps and have led to many new discoveries (Römer et al., 2012; Skarke et al., 2014; Wang et al., 2016). These recent advances enabled gas seep flux to be included as a new hydrocarbon input in the *Oil in the Sea IV* analysis.

Total estimates of fossil fuel hydrocarbon inputs from oil and gas seeps in North American waters are discussed in the following subsections.

3.2.1 Oil

Estimates of the input of oil via natural seepage have varied considerably over time, primarily due to changes in the methods of estimation. Data from a limited number of seeps combined with a comprehensive survey of potential seep locations were first used to make the first estimates of oil inputs from natural seepage (NRC, 1975). This estimate was later revised to include consideration of the amount of crude oil known that could seep into the ocean over periods of time (NRC, 1985). Advances in remote sensing techniques, natural seep detection and assessment in regions including the Gulf of Mexico, offshore southern California, and offshore Alaska have enabled seepage rates to be calculated more accurately and recently be included as a component of the estimates of inputs of oil from natural seepage. The best estimate of the annual input of oil into North American waters from natural seeps in *Oil in the Sea III* was based on the available technology of the time and was determined to be 1.1 million barrels (160,000 tonnes)—the largest source of oil inputs to U.S. waters (NRC, 2003). The *Oil in the Sea III* global estimate of annual natural seeps was determined to be 4.2 million barrels (600,000 tonnes). Subsequently, an international assessment of oil inputs into the sea estimated the range of annual inputs of oil from natural seepage to be 0.14–14.0 million barrels (20,000–2,000,000 tonnes) annually (GESAMP, 2007).

Since these previous estimates were made, remote sensing and hindcasting methodologies, as well as our understanding of oceanographic constraints on natural seeps, have expanded. A recent estimate for the Gulf of Mexico was made between 1991 and 2019 (2021 report to the National Academies from MacDonald[2]), consisting of data collected from satellites using synthetic aperture radar that can detect oil slicks under all sunlight and cloud conditions (Brekke and Solberg, 2005; Leifer et al., 2012; Fingas and Brown, 2014; MacDonald et al., 2015). Synthetic aperture radar imaging was used to identify over 32,000 natural seep oil slick origins (as in Garcia-Pineda et al., 2010), which can be determined when the oil slick elongates and the flow

[2]Copy of report can be provided upon request from the National Academies Public Access Records Office.

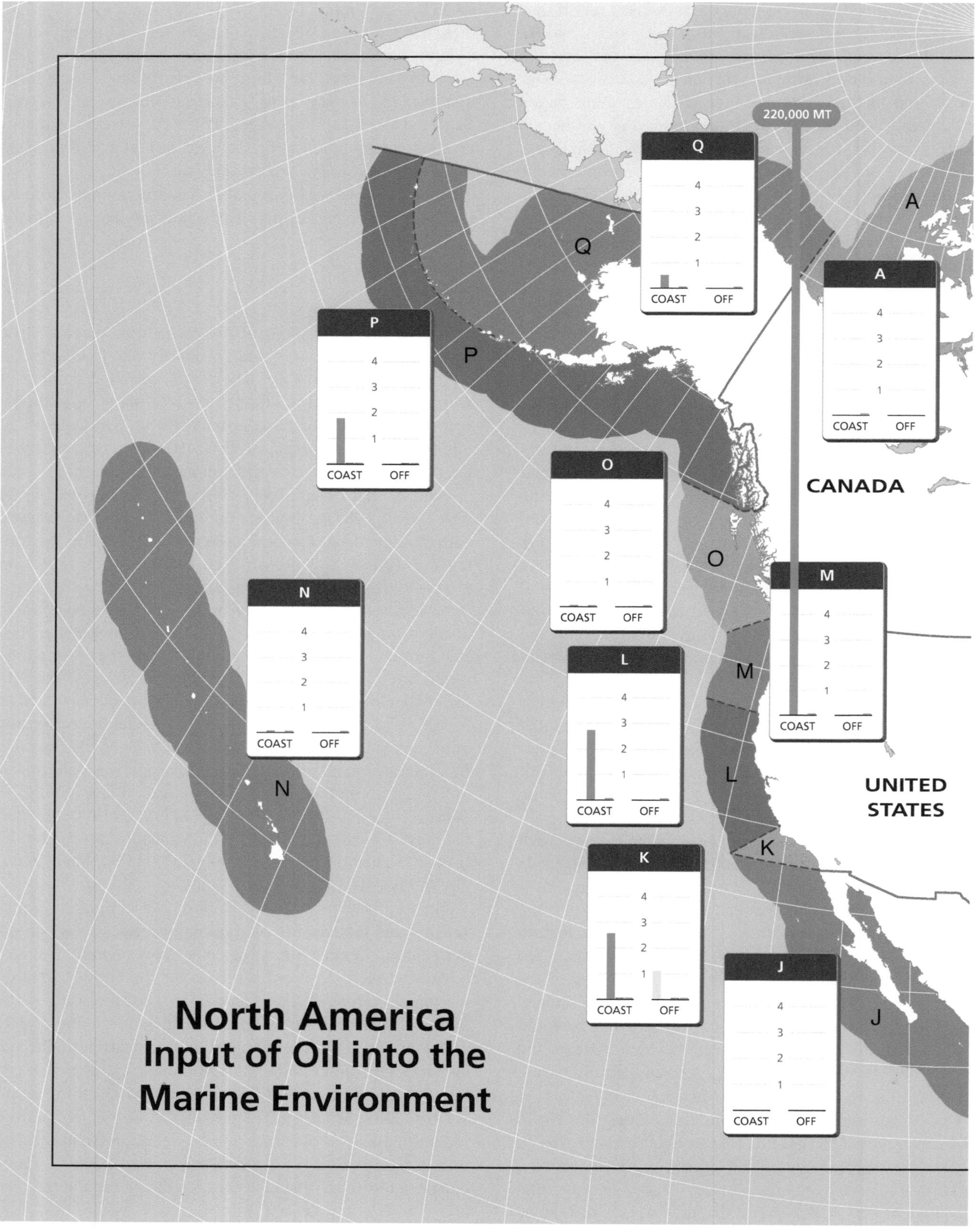

FIGURE 3.3 Map of North America showing estimates of oil in the sea through natural sources, land-based sources, operational discharges, and accidental spills.

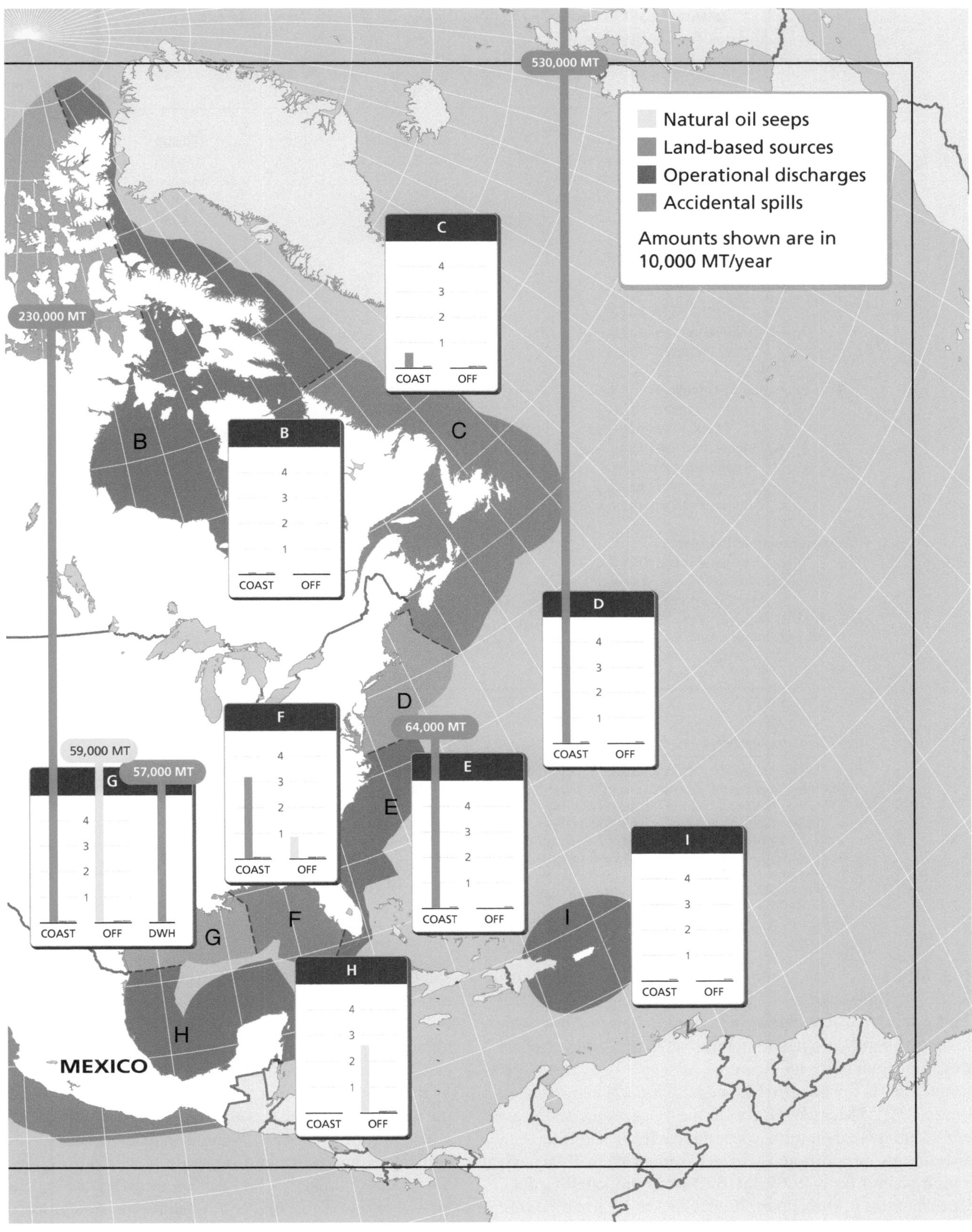

Natural oil seeps
Land-based sources
Operational discharges
Accidental spills
Amounts shown are in 10,000 MT/year
530,000 MT
230,000 MT
59,000 MT
57,000 MT
64,000 MT
B
C
D
E
F
G
H
I
COAST
OFF
DWH
MEXICO

TABLE 3.1 Estimates of Oil Inputs as Reported in *Oil in the Sea IV* with *Oil in the Sea III* Estimates for Comparison

	Oil in the Sea IV (2010–2019) (MT/yr)	*Oil in the Sea III*[a] (1990–1999) (MT/yr)
Natural Sources (total not including natural gas)	**100,000**	**160,000**
Oil Seeps	100,000	160,000
Gas Seeps	2–9 Tg[b]	Not reported
Extraction of Petroleum		**2,980**
excluding DWH	**9,500**	
including DWH	**66,500**	
Platforms	1,100	160
MC-20	1,600[c]	
DWH	57,000	
Atmospheric Deposition	Not reported	120
Produced Waters	6,800	2,700
Transportation of Petroleum	**818**	**9,209**
Pipeline Spills	380	1,900
Tank Vessel Spills	200	5,300
Commercial Vessel Spills	8	99
Coastal Terminal Spills	220	1,900
Coastal Refinery Spills	10	Included with terminal spills
Atmospheric Deposition	Not reported	10
Consumption of Petroleum	**1,200,399**	**83,520**
Land-based Runoff	1,200,000	54,000
Recreational Marine Vessels	Not reported	5,600
Spills (non-tank vessels)	390	1,200
Op. Discharges (Vessels >100 GT)	9[d]	100
Op. Discharges (Vessels <100 GT)	0[d]	120
Atmospheric Deposition	Not reported	21,000
Aircraft Jettison	Not reported	1,500
Total[e]	**1,400,000**	**260,000**
Total[e] *(Excluding Consumption and DWH)*	**110,000**	**170,000**

[a] *Oil in the Sea III*'s "Best Estimate" is shown in this column—the numbers vary from the totals shown by summing the regional inputs shown in the later tables within this chapter.
[b] Gas volumes reported in teragrams (Tg).
[c] Value shown reflects value of 1,600 MT/yr (30 bbls/day)—the amount oil collected from the site since April 2019. This value may be over- or underestimate of annual discharge for the full 2010–2019 timeframe. See Table 3.5 for range of published estimates.
[d] Assuming full compliance with discharge regulations.
[e] Reported values and Totals rounded to 2 significant figures.

is guided away from the origin by winds and surface currents (De Beukelaer et al., 2003; MacDonald et al., 2015). This comprehensive dataset was used in combination with hindcast methodologies to determine spatial and temporal trends in abundance and location of seep zones, and to estimate the fluxes of oil from seeps that had been detected in the Gulf of Mexico. Uncertainties in previous and present estimates of natural seep fluxes from satellite data are directly proportional to the assumed slick thickness, which cannot be directly inferred from the satellite data. Uncertainties in slick intermittency and residence times are improving with more satellite coverage and with ongoing efforts to identify slick signatures in the wealth of SAR satellite data (NOAA, 1996). The annual flux of oil from natural seepage in the Gulf of Mexico was estimated to be 0.2–0.7 million barrels (0.03–0.1 million tonnes), which agrees with previous annual flux estimates of MacDonald et al. (2015). These estimates are constrained by the accuracy of the estimated value of the average oil thickness and by the ability of satellite methods to identify all natural seeps, which may be influenced by anthropogenic sources from oil and gas exploration and production in the Gulf of Mexico. Table 3.2 summarizes oil seep input estimates developed for *Oil in the Sea IV* by geographic region for comparison. **The total estimated annual inputs from natural seeps is roughly 35% lower than estimated in *Oil in the Sea III*.**

TABLE 3.2 Annual Flux Estimates of Oil Seeps in North American Waters

Oil in the Sea Geographic Zone	Seep Annual Flux Estimate (MT)
F: Eastern GoM	8,000
G: Western GoM	60,000
H: Mexican GoM	30,000
K: California Pacific	10,000
P: South Alaska	1,000
Total	109,000

3.2.2 Methane

Methane is the dominant gaseous hydrocarbon that is emitted from seafloor hydrocarbon seeps. Methane may be released in dissolved form with emitted natural seep fluids and/or as gaseous bubbles. Depending on the depth, some of the released methane may contribute to seafloor hydrate formations, and released gas bubbles may form hydrate skins (Warzinski et al., 2014b). The majority of methane released as gas bubbles dissolves into the ocean water column (Ruppel and Kessler, 2017). Methane released from shallow seeps that reaches the sea surface either as gas bubbles or dissolved methane is emitted to the atmosphere.

Methane inputs to the ocean and atmosphere are quantified by two general approaches known as bottom-up or top-down methods (described in Kessler and Weber, 2021). Bottom-up approaches are useful for providing accurate flux estimates for a specific area at a specific time, whereas top-down approaches are preferable when estimating emissions of methane over larger areas and longer timescales. Top-down approaches consider the input as the amount of methane required to maintain the seawater inventory of methane while accounting for the loss of methane to the environment. Thus, for estimates of methane inputs made by this approach, information regarding methane concentration, aerobic methane oxidation, sea-to-air gas exchange, advection from the coastal environment, and other processes are required. This presents a variety of challenges to the top-down approach primarily due to the sparsity of data regarding the sinks of methane as well as the ability to determine seep versus non-seep sources of methane.

To estimate the emission of methane (all phases) from seafloor seeps into the coastal waters (not including emissions to the atmosphere) of North America, a top-down approach with novel methods was used to overcome the challenges of sparse datasets and source determination of methane (Kessler and Weber, 2021). Briefly, the fate of methane (sink) was estimated from datasets using artificial neural network models, and radiocarbon constraints were employed to estimate the seep contribution (source) of methane. Estimates were obtained for three case studies (Gulf of Mexico, North American Pacific margin, North American Arctic margin), which were then extrapolated to all North American regions. The case studies revealed that the Gulf of Mexico had significantly higher seepage rates than the North American Pacific and Arctic margins, which were relatively similar to one another. Given this, the North American Pacific and Arctic margin seepage rates were used to define a "low" methane source range and the Gulf of Mexico seepage rate defined a "high" source range. The regionally average seafloor methane seepage fluxes determined by this approach were 0.0042–0.087 moles of methane per meter squared per year, consistent with previous estimates (Kessler and Weber, 2021, and references therein). **A range for the total methane source in North American coastal waters was determined to be 4–20 Tg/yr, with 2–9 Tg/yr coming from seafloor seepage. Estimates of methane inputs to the sea provided by Kessler and Weber (2021) were not previously described in *Oil in the Sea III* (NRC, 2003); thus, this represents a new input term.**

3.3 LAND-BASED SOURCES

This section examines both runoff and atmospheric depositions as land-based sources of oil in the sea.

3.3.1 Runoff

Runoff describes excess water, typically from stormwater or melting snow, that flows across land into receiving water bodies such as lakes, rivers, and groundwater. Runoff can carry particles and dissolved chemicals that have been deposited on land surfaces together with airborne particulates and volatile organic compounds (VOCs) that are redeposited with the rainwater. Runoff contains suspended solids (Westerlund and Viklander, 2006), excess fertilizers, pesticides, oil, grease, salts, biochemical oxygen demand (BOD), and chemical oxygen demand (COD), as well as trace metals in both solid and liquid fractions (Huber et al., 2016; Pitt et al., 2018). Particles (or suspended solids) transported with stormwater also carry pollutants (e.g., fuel oil). Depending on the flow and particle characteristics, particles can remain in suspension or be deposited in the receiving water bodies.

Seasonal changes can influence the amounts of chemicals present in runoff including fertilizer chemicals (e.g., nitrates and phosphates) used in agriculture during growing seasons, as well as salt and deicing compounds (e.g., ethylene glycol) used on roads, parking lots, driveways, and vehicles during winter months and during snowmelt in early spring. Seasonal flooding can cause trash and debris to be washed from land. This trash and debris may contain petroleum-based contaminants either by sorption or as part of their composition (e.g., printing ink). During warmer months, roads and other impervious surfaces heated by sunlight may increase solubility and mobilization of some petroleum-derived compounds present in asphalt.

The primary sources of oil in runoff are derived from urban environments and automotive or transportation-related activities (Klimaszewska et al., 2007; Müller et al., 2020). Oils, grease, and other hydrocarbons are frequently found in highway runoff, and a prior estimate indicated that approximately 50% of solids and 70% of the polycyclic aromatic hydrocarbons (PAHs) found in receiving waters can be attributed to highway runoff sources (Ellis, 1986). Transportation-related sources such as automotive fluid leakages (Markiewicz et al., 2017), tire wear (Muschak, 1990), vehicle washing (Sörme et al., 2001; Björklund, 2010), and road abrasion (Hvitved-Jacobson and Yousef, 1991) are common inputs of petroleum hydrocarbons to stormwater (Brinkmann, 1985; Müller et al., 2020). Vehicle exhaust emissions from internal combustion engines also contribute particulate matter (e.g., soot), PAHs (Markiewicz et al., 2017), and benzene, toluene, ethylbenzene and xylene (BTEX) compounds (Liu et al., 2018), although some of the chemical components of vehicle exhaust pollution are partly controlled by catalytic converters (Rauch et al., 2005). Strong correlation exists between increased urbanization and traffic density and the levels of PAHs and *n*-alkanes detected in urban runoff (Hewitt and Rashed, 1990; Bomboi and Hernandez, 1991; Moilleron et al., 2002). A review of stormwater runoff quality monitoring data from industrial facilities has shown that PAHs are the most common in roof runoff (due to asphalt coating), parking area runoff, and vehicle service area runoff (Pitt et al., 1995).

Quantifying the inputs of oil to the sea from land-based runoff continues to present a significant challenge due to the absence of the data spanning both geographic and temporal scales that is needed for this purpose. A review of oil and grease and petroleum hydrocarbon data available in the Water Quality Portal[3] (WQP) sponsored by the U.S. Geological Survey (USGS), the U.S. Environmental Protection Agency (U.S. EPA), and the National Water Quality Monitoring Council (NWQMC) indicates that while a large amount of data have been collected, data from major inland river basins and coastal cities is limited, and there continues to be no data collected in the 2000s for the Alabama-Tombigbee, Altamaha, Brazos, Colorado (Texas), Copper (Alaska), Rio Grande, St. Lawrence, Santee, or Yukon river basins. The water data collected for the major inland rivers during the 2000s and reported in the WQP are still primarily oil and grease data, with some petroleum hydrocarbon data, and no PAH data (see Appendix C). **It is evident that the recommendations presented in *Oil in the Sea III* to comprehensively quantify the concentration of petroleum hydrocarbons (oil) and PAHs from major river outflows, harbors, urban runoff, and municipal wastewater effluents and streams in urban coastal cities have not been acted upon.**

Previous estimates of the inputs of petroleum hydrocarbons from land-based sources for the United States and Canada reported in *Oil in the Sea III* were based on 1990s oil and grease data for the Delaware and Mississippi Rivers. For estimates between 2010 and 2020, data from the Potomac River were used instead of the Mississippi River, which only has four total observations (see Appendix C). In the 2000s, the largest datasets are for the Columbia, Potomac, and Savannah River basins, but the data from these locations correspond to individual projects with associated research objectives that vary from one another and are not generalizable. The data from the Columbia river basin, for example, is from a storm sewer in the Portland Harbor Superfund Site[4]; the data from the Potomac River basin are from a stream being monitored to assess changes in water quality associated with the construction of a nearby planned unit development[5]; and the data from the Savannah River basin are being collected to compare structural devices for managing highway runoff (Conlon and Journey, 2008). Although these datasets are important in relation to specific projects, they cannot be used to either update or refine estimates using the same methodology as that previously presented, and due to the specificity of the projects to which they are related, were the types of studies that were excluded from *Oil in the Sea III* estimates.

Given the absence of oil, grease, and petroleum hydrocarbon concentration measurements in rivers, inputs of oil to the sea from land-based sources must be estimated using proxies such as changes in urban land area, population, and petroleum consumption via the numbers of vehicles per capita (as in *Oil in the Sea III*). This assumes that inputs from land are from predominantly urban areas and that vehicle operation and maintenance are the primary inputs of oil and grease to the land (as previously described). While this does not consider differences from changes in fuel efficiency, the use of all-electric vehicles, or distances traveled, it can provide some understanding of changes to land-based inputs and worldwide estimates of oil inputs to the sea.

Land-use patterns and river flows also affect land-based runoff inputs of petroleum hydrocarbons, both of which have seen changes over the past 20 years. An overall increase of 15% is calculated for major inland rivers in North America, but there are large differences between different rivers ranging from a 37% decrease in flow for 1980–1999 compared to 2000–2020 for the San Joaquin River to a 40% increase from 1980–1994 to 2000–2020 for the Mississippi River (see Appendix C). With respect to land use, both urban land area and populations for major inland rivers have increased with the exception of Trinity River, which has seen a 5% decrease in its urban land area (see Appendix C). Increases in urban land area between the 1990s and the 2010s range from 1% for the Susquehanna River to 560% for the Potomac River, echoed by changes in the range of population increases of 10% to 700% for the Susquehanna and Potomac Rivers respectively (see Appendix C).

In order to evaluate changes and to compare findings to *Oil in the Sea III*, the same approach was used to determine

[3] See https://www.waterqualitydata.us.

[4] See https://cumulis.epa.gov/supercpad/SiteProfiles/index.cfm?fuseaction=second.docdata&id=1002155.

[5] See https://frederickcountymd.gov/DocumentCenter/View/327149/FINAL-Peter-Pan-Run-Report-2019.

the average annual loads of oil and grease, petroleum hydrocarbons, and PAH to the sea. A reasonable estimate of the low and high ranges of the calculated oil and grease and PAH values was also calculated (methodological details and tables are provided in Appendix C).

The annual loading of oil and grease to the sea for North America is estimated to have increased to 5.8 million tonne/yr, and globally to 20.1 million tonne/yr (see Appendix C). The best estimate of land-based petroleum hydrocarbon to the sea was 1.2 million tonne/yr for North America and 4.0 million tonne/yr globally. This global estimate is about 28 times as large as *Oil in the Sea III* (NRC, 2003; see Appendix C) and is primarily due to an increase in the number of vehicles being used in North America, Europe and Asia.

The estimates provided here are based on the methods used in *Oil in the Sea III*, and therefore have the same range of uncertainty of four orders of magnitude.This uncertainty is primarily due to (1) the lack of available data; (2) the differences in methods used for measuring and reporting the data; and (3) the fact that the proportion of petroleum hydrocarbons and PAH calculated from oil and grease measurements is only an estimate. The increases described previously with respect to the inputs of petroleum hydrocarbons are within this range of uncertainty and thus it is not possible to determine whether there are any quantitative changes to these inputs. **Given that there have been increases in urban land area, population and vehicle ownership, it is plausible that land-based inputs of petroleum hydrocarbons ("oil") have increased over this time, but it is unclear by how much.** In addition, any estimate provided should be considered an upper limit as reductions due to personal choices and behaviors are not accounted for. **Smaller inputs would arise from reductions in vehicle usage (e.g., due to car pooling, increased use of public transportation), improvements in fuel efficiency, and increased usage of electric vehicles, as well as improvements to care, maintenance, and the replacement of older vehicles.**

3.3.2 Atmospheric Deposition

Land-based inputs of oil can be transported via the atmosphere and deposited in the ocean via wet deposition, dry particle deposition, and air–water gas exchange (dissolution, degassing) as summarized in Figure 3.4. The primary inputs of fossil fuel hydrocarbons are not in the form of liquid oil, but are combustion-derived products from combustion engines (vehicles), fossil fuel burning power plants, industrial

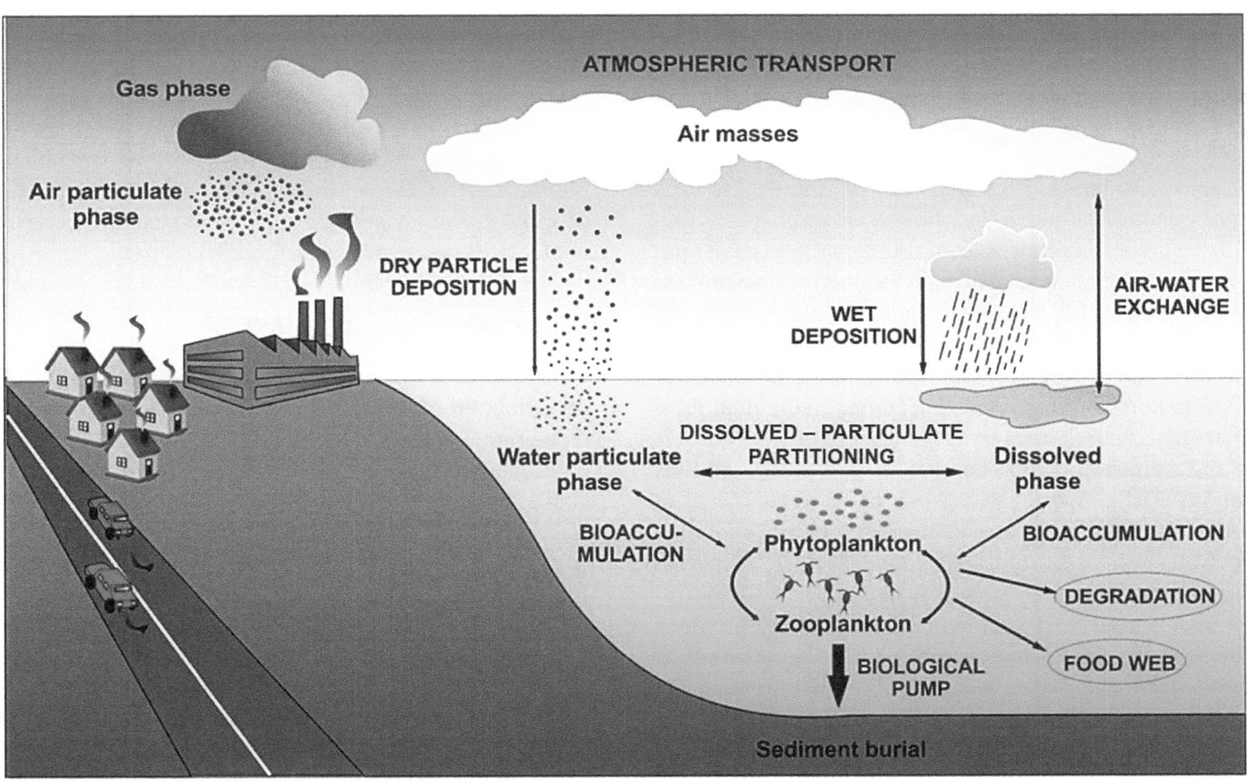

FIGURE 3.4 Atmospheric deposition of land-based sources into the sea, including fate of the deposited emissions.
SOURCE: Jordi Dachs; presented to the National Academies committee on March 11, 2021.

manufacturing facilities, the domestic burning of coal, and other urban emissions. Once deposited from the land to the ocean, several processes influence their transport and fate (refer to Chapter 5), including interactions with the ocean carbon cycle and food webs (see Figure 3.4).

Atmospheric deposition of oil from land-based sources was estimated in *Oil in the Sea III* to be 4.2% of the total input of oil to the sea and was therefore considered to be a significant input of petroleum, especially PAHs, to the marine environment (NRC, 2003). When considering land-based atmospheric inputs, it is important to remember that these oil inputs are applied over large surface areas of the global ocean.

To estimate the atmospheric inputs of petroleum hydrocarbons ("oil"), *n*-alkanes with carbon lengths C_{10}–C_{33}, and PAH compounds are typically examined. These compounds make up a large fraction of petroleum, and information about their concentrations in marine atmospheres and surface waters, as well as their physical properties, is available. Prior estimates of total inputs of these compounds made in *Oil in the Sea III* acknowledge that focusing on these compounds alone underestimates the total mass of inputs, and this bias is considered to be less than two-fold. The aforementioned hydrocarbon compounds fall under the definition of *oil* if they are derived from sources such as petroleum, and coal. These hydrocarbon compounds could also originate from non-fossil inputs such as marine plankton, volcanic eruptions, plant waxes, wood burning, and wildfires and these other sources must be carefully considered in any quantitative estimates of input to the sea. It is possible to estimate the relative percentages of *n*-alkanes and PAHs that are derived from petrogenic, pyrogenic, and biogenic inputs. This is especially important in inputs from certain regions of the world where biomass burning can contribute up to two-thirds of combustion-derived carbon to the atmosphere (Xu et al., 2012). Stable carbon isotopic composition and natural radiocarbon abundance have been measured to apportion the source of PAHs in the following studies: Mandalakis et al., 2005; Gocht et al., 2007; Zencak et al., 2007; and Gustafsson et al., 2009. Other methods for source apportionment of PAHs include examining wind trajectories, individual PAH compounds as source markers (e.g., Ma et al., 2013), or organic and elemental carbon and sulfate and potassium ions (Crimmins et al., 2004).

There have been numerous sampling campaigns to measure petroleum hydrocarbons, specifically PAHs, in marine atmospheres and surface waters in recent years. This increase in data covers a broad geographic area and provides insights into the inputs and sources of PAH compounds to marine surface waters (see Appendix C). Prior to the early 2000s, data on quantities of atmospheric hydrocarbons in the marine atmosphere that are deposited and present in surface waters were sparse, and geographic locations used for prior estimates instead included data from an urbanized North American coast (California), a less-impacted urbanized coast (Denver as proxy), a rural coastline (east and west coasts), and an offshore rural coastline. Data represented conditions in the early to mid-1990s. **The influx of new studies and the associated data present an opportunity to re-examine the atmospheric deposition of oil from land-based sources, as well as to improve understanding of the fates of these hydrocarbon compounds via net air-water flux calculations.**

Net air-water fluxes of PAHs are calculated from concentrations of PAHs in marine atmospheres and surface waters to determine which regions of the ocean exhibit overall volatilization of PAHs (e.g., Nizzetto et al., 2008; Lohmann et al., 2011), compared to others where deposition and absorption of PAHs into surface waters are occurring (Jernelöv, 2010; Castro-Jiménez et al., 2012; Lohmann et al., 2013; González-Gaya et al., 2016). Volatilization versus deposition of individual PAH compounds varies according to their chemical and physical properties as well as seasonally (Cabrerizo et al., 2014; Casal et al., 2018). By calculating air-water fluxes for the sum of all PAH compounds, estimates of the overall global atmospheric input of PAHs to the oceans can be made. The global atmospheric input of PAHs calculated from measuring the sum of 64 individual PAHs input to the oceans is estimated to be 0.9 Tg C y^{-1} (González-Gaya et al., 2016). Assuming that a maximum of 65% of these PAH is from biomass burning, this estimate would be ~0.5 Tg C y^{-1} or ~550,000 tons C y^{-1}, which is approximately 10 times over previous estimates (NRC, 2003). Global emission estimates of the 16 U.S. EPA priority PAHs for the year 2004 are 520,000 tonnes y^{-1} (Zhang and Tao, 2009), with the most significant contributions from biofuel (56.7%), wildfire (17.0%) and consumer product usage (6.9%). Power plants, open biomass burning, road transport (mostly diesel), industrial processes, and air and sea transport also contribute (Keyte et al., 2013, and references therein).

3.4 OPERATIONAL DISCHARGES

Operational discharges include oil discharges as a result of extraction of oil and gas, transportation of oil and gas, recreational vessels, and aircraft.

3.4.1 Extraction of Oil and Gas

3.4.1.1 Emissions

Emissions of fossil fuel hydrocarbons during the extraction of oil and gas were previously considered to be small in quantity and only significant in the context of the local air quality (NRC, 2003). **There are no new studies or data to suggest that this has changed and that these operations should be considered as an input of oil to the oceans.** We note that oil production is sometimes associated with a large amount of waste hydrocarbon gas, especially in regions that lack a subsea pipeline system that could bring the gas to market. Current practice is either to pump the gas back into the oil reservoir, use the gas to produce power, or burn the gas in a flare.

3.4.1.2 Produced Water

Produced water can be brought to the surface along with hydrocarbons and may include formation water from the reservoir, condensation water, and/or water that was injected into the reservoir to enhance oil recovery. Offshore, produced water is typically reinjected back into a formation to improve production, injected into a disposal well, or treated to remove excessive oil and discharged into the sea. Produced water management techniques include produced water minimization, treatment, reuse, recycling, and disposal (Veil et al., 2004; Abdulredha et al., 2018; Liang et al., 2018; Veil, 2020). To minimize the collection of produced water, various mechanical and well construction techniques can be used to block the water from entering the well including the use of dual completion wells to separate production of oil and water, the injection of gel-like materials to stop water migration, and the use of downhole or subsea oil/water and gas/water separators.

Produced water must be treated to remove harmful components before being discharged into the environment. Several studies provide an overview of produced water treatment options (Fakhru'l-Razi et al., 2009; Igunnu and Chen, 2012; Zheng et al., 2016; Jimenez et al., 2018; GWPC, 2019; Liu et al., 2021). Produced water treatment strategies for offshore platforms may require different approaches compared to onshore facilities due to space and weight constraints, which necessitate the use of more compact physical and chemical systems. Offshore locations also have limited options for the beneficial use of treated produced water. Reservoir fluids first go through a bulk separation where oil, gas, and water streams are separated. Next, the water undergoes primary treatment to remove free oil and larger oil droplets via gravity separators, skimmers, hydrocyclones, corrugated plate separators, or centrifuges. Primary treatment typically achieves concentrations of less than 100 mg/l of dispersed oil and 50–100 mg/l of total suspended solids (TSS). Secondary treatment to remove dispersed/dissolved oil and suspended solids may consist of flotation cells, centrifuges, filters, adsorption units, ion exchange, or organic extraction. Secondary water treatment typically achieves dispersed oil concentrations of less than 50 mg/l and less than 25 mg/l of TSS. Secondary water treatment is usually sufficient to reduce oil content to the regulatory required offshore discharge levels. If necessary, tertiary or polishing techniques could be used. These may include filtration with different media, adsorption/absorption or coalescing units, liquid-liquid solvent extraction, ultrafiltration membranes, advanced oxidation, and others. Depending on the technique, the tertiary treatment could achieve 1–10 mg/l of dispersed oil and 1–10 mg/l of TSS. After a desired level of treatment is achieved, produced water is discharged into the marine environment.

Regulatory Regimes

Several approaches are implemented globally to regulate produced water discharges into the marine environment. Most commonly used is the application of discharge standards for oil-in-water concentrations, total petroleum hydrocarbons, oil and grease, or dispersed oil. The United States and Canada use a risk-based approach based on Whole Effluent Toxicity (WET) testing. This establishes a field-specific dilution factor which should be achieved after discharge in order to minimize the risk of potential negative effects. In Europe, in the North-East Atlantic, a combination of oil-in-water standards and risk-based approaches is used. However, within the European risk-based approach there is a greater emphasis on chemical characterization of individual produced water constituents.

Operational discharges from exploration and production operations in Canada are regulated by the Offshore Waste Treatment Guidelines (National Energy Board, 2010). For produced water, a 30-day volume weighted average oil-in-water (or "oil and grease") concentration, the maximum allowable concentration is 30 mg/L. The maximum 24-hour average oil-in-water concentration, as calculated at least twice per day, is 44 mg/L. The deck drainage, ballast, bilge, and storage displacement water should not have a residual oil concentration that exceeds 15 mg/L.

In the United States, produced water discharges offshore are regulated by the Clean Water Act established by Congress in 1972 and managed through effluent limitations guidelines published in the Code of Federal Regulations (C.F.R.) at 40 C.F.R. Part 435. For the offshore installations, "oil and grease" effluent reduction attainable by the application of the best available technology economically achievable (BAT) is "the maximum for any one day shall not exceed 42 mg/l; the average of daily values for 30 consecutive days shall not exceed 29 mg/l." Effluent reduction attainable by the best practicable control technology currently available (BPT) and the best conventional pollutant control technology (BCT) is "72 mg/l maximum for any 1 day and 48 mg/l for 30 consecutive days." In coastal areas, produced water is not allowed for discharge except for Cook Inlet, Alaska, which has the same limits as offshore wells.

Factors Affecting Produced Water Inputs into Marine Environments

Produced water typically represents the largest byproduct by volume in oil and gas extraction operations, and its quantities tend to increase over time as the reservoir depletes (Reynolds and Kiker, 2003; Veil, 2020). While there is a notion that global rates for produced water discharges generally increase (Dal Ferro and Smith, 2007), local trends

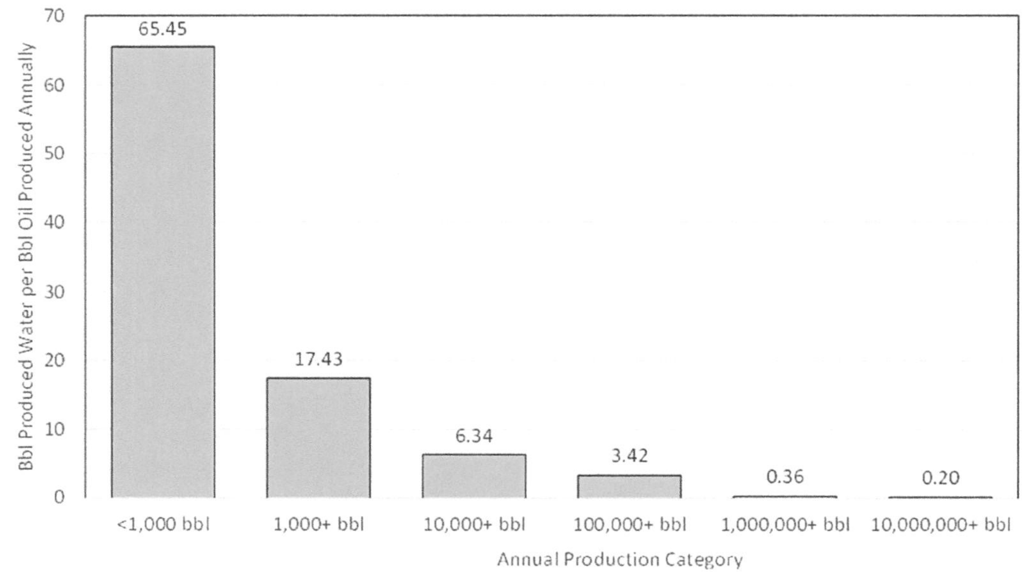

FIGURE 3.5 Average bbl of produced water per annual production by lease category in the U.S. Gulf of Mexico Outer Continental Shelf. NOTE: The water-to-oil ratio (WOR) is highlighted at the top of the bar for each category.

in recent years in the United States and Canada indicate reduction of produced water volumes even with increased oil production,[6] which is most likely due to better produced water management practices and addition of new prolific oil fields that produce significantly less volumes of water. If global oil production rates continue to decline, so will the volumes of discharged produced water.

The total volume of oil entering the marine environment depends on:

1. Total volume of oil and gas production in the area and either measured or estimated volume of total produced water volumes generated along with oil and gas.
2. Final concentration of dissolved and dispersed oil in produced water after all the treatment techniques.

Produced Water as a Function of Total Oil Production

The ratio of produced water to produced oil equivalents (water-to-oil ratio; WOR) and ratio of produced water to produced gas equivalents (water-to-gas ratio; WGR) depends on many factors and varies significantly. Some of the factors affecting produced water volumes are overall oil production volumes, type, age, properties and integrity of the reservoir, type of well (horizontal versus vertical), location of a well within the reservoir structure, type of well completion, type of produced water management techniques implemented, whether water injection is used to enhance oil recovery, etc. (Reynolds and Kiker, 2003; Veil et al., 2004; Clark and Veil, 2009; Veil, 2015, 2020).

Figure 3.5 illustrates the dependency of produced water volume on volume of produced oil.

Appendix D illustrates the variability among regional values for WORs in study regions calculated as the average, the mean of all the data as well as the mean with the outliers removed. This is due to great variability in the WORs among individual wells. Despite this variability, the "average world-wide" WOR was historically estimated to be around 2 to 3 (Khatib and Verbeek, 2003; Veil et al., 2004; Clark and Veil, 2009; Neff et al., 2011; Liang et al., 2018). Individual reported values range from zero for new reservoirs to more than 100 for depleted fields (Veil et al., 2004; Neff et al., 2011; Veil, 2015; BOEM/BSEE database). Veil conducted a systematic analysis of produced water generated in the United States, both onshore and offshore, and published comprehensive reports in 2009, 2015, and 2020. He also highlighted the challenges related to collection and interpretation of these data. Using 2017 data for the United States, Veil (2020) reported WORs ranging from 6.5 bbl/Mmcf to 5,490 bbl/Mmcf, with the weighted average being 76.4 bbl/Mmcf. He reported that the WOR ranges between 0.25 and 57.7 with weighted average 4.8, although he highlighted that due to dataset limitations, this weighted average may not be representative. He offered an alternative expression for WOR, which incorporates natural gas volumes converted to barrels of oil equivalents (BOEs) and combined with oil volumes. In this case this parameter was equal to 2.76 for the 2017 data. This is lower than a value of 3.4 calculated using 2007 data, indicating a reduction of generated produced water volumes (Veil, 2020).

[6]BSEE: www.data.bsee.gov; Canada-Newfound and Labrador Offshore Petroleum Board: https://www.cnlopb.ca; Canada-Nova Scotia Offshore Petroleum Board: https://www.cnsopb.ns.ca.

Table 3.3 summarizes the annual oil and gas production as well as produced water volumes generated in 2020 in the United States, Mexico, and Canada. 61,037,08 bbl of produced waters have been generated by the offshore facilities in Eastern Canada, 535,637,088 bbl in offshore and nearshore U.S. waters, and estimated 1,551,250,000 bbl in Mexican waters of the Gulf of Mexico. Because no data were available for the exact volume of produced water generated in Mexican waters, available data for oil production were used, with 2.5 WOR recommended by Veil (2011).

International Association of Oil & Gas Producers (IOGP) member companies in North America reported that less than 7% of produced water from offshore assets was reinjected back into the reservoir rather than released into water (IOGP, 2020b). This is comparable to the numbers reported by Clark and Veil (2009) and IOGP (2009). Veil (2020) suggested that it may be due to a risk of produced water injection resulting in hydrogen sulfide creation in the formation. Hence, when water injection is required to support production of offshore fields, the use of seawater treated to remove elements that may cause corrosion may be preferable.

General trends indicate a reduction of produced water generated per unit of oil for offshore United States and Canada in recent years, even at a time of increasing

TABLE 3.3 Estimates of Produced Water for Oil and Gas Operations[a]

	Annual Production[b]			Annual Oil Input in Metric Tonnes (MT/yr)		
Oil in the Sea Geographic Zone	Oil (bbl)	Gas (mcf)	Produced Water (bbl)	Assuming 20 mg/L (best estimate)	Assuming 15 mg/L (minimum estimate)	Assuming 29 mg/L (maximum estimate)
C: E Canada Offshore	104,005,970	1,615,773	61,037,083	194	146	281
F: E GOM Nearshore[c]	25,884	48,541,716	1,280,832	4	3	6
F: E GOM Offshore	34,991,045	78,403,905	11,811,098	38	28	54
G: W GOM Nearshore[d]	3,847,596	11,299,809	9,618,990[e]	31	23	44
G: W GOM Offshore	626,892,869	831,100,299	398,644,277	1,267	951	1,838
H: Mexican GOM[f] Offshore	620,500,000	73,000,000	No data[g]	4,932	3,699	7,151
K: CA Pacific Nearshore[h]	1,496,500	393,470,000	16,242,500	52	39	75
K: CA Pacific Offshore[i]	4,383,418	2,702,654	42,233,459	134	101	195
P: South Alaska[j] Nearshore	3,253,102	23,610,069	35,371,703	112	84	163
P: South Alaska Offshore	0	5,915,233	42,926	0	0	0
Q: North Alaska Nearshore	11,580,775	256,693,367	20,391,303	65	49	94
Total[k]	1,410,977,159	1,726,352,825	587,055,181	1,866	1,400	2,706
Total[l]	1,410,977,159	1,726,352,825	2,147,924,171	6,829	5,122	9,902

[a] Based on data available for 2020. Note that only nearshore (but not inland) and offshore oil and gas operations were included. Inland operations are outside of the scope of the *Oil in the Sea IV* study.

[b] U.S. estimates calculated based on publicly available data from BSEE (available at www.data.bsee.gov). Eastern Canada estimates for produced water for gas production based on value for exclusive gas production in Nova Scotia coupled with produced water data from the same area. (Data from Canada-Newfoundland and Labrador Offshore Petroleum Board and Canada-Nova Scotia Offshore Petroleum Board.) Estimates for Alaska based on State of Alaska data on well gas and oil production and associated produced water. Calculations conducted for gas production wells and oil production wells. Oil production wells generally also produce some gas.

[c] Alabama nearshore: https://www.gsa.state.al.us/ogb/production.

[d] Louisiana data from Louisiana Department of Natural Resources Office of Conservation (data for 2020). Texas data from Texas Railroad Commission (data for 2020) (http://webapps.rrc.texas.gov/PDQ/generalReportAction.do).

[e] Estimated based on 2.5 bbl water per bbl oil per Veil (2011).

[f] Mexican production data from U.S. Energy Information Administration (https://www.eia.gov/international/analysis/country/MEX).

[g] Estimated based on 2.5 bbl water per bbl oil per Veil (2011).

[h] California nearshore (state waters) data for Platforms Eva, Emmy, and Esther from California State Lands Commission. The water:oil ratios are relatively high because the production rates are now relatively low, totaling less than 1.5 million bbl per year for all three platforms.

[i] Data from BSEE Pacific. The water:oil ratios are relatively high because the production rates are now relatively low, totaling less than 4.4 million bbl per year for 11 leases.

[j] Data for Cook Inlet offshore production from State of Alaska (http://aogweb.state.ak.us/DataMiner3/Forms/Production.aspx).

[k] Totals for produced water without Mexico and Western GOM Nearshore because no data are available.

[l] Totals for produced water including estimates for produced water for Mexico and Western GOM Nearshore based on an estimate of 2.5 bbl water per bbl oil per Veil (2011).

oil production. Condensate-producing wells and wells with low-volume oil production (less than 1,000 bbl per year) tend to generate larger volumes of produced water than wells with higher oil production. Wells in deep waters tend to generate less produced water than wells in shallow waters. This is not related to the reservoir conditions per se, but instead is most likely driven by the production economics, which makes deep-water operations profitable if they are focused on producing oil from prolific non-depleted reservoirs not requiring water injection to facilitate oil recovery and do not generate significant produced water volumes.

Concentration of Oil in Produced Water Produced water physical and chemical properties vary significantly depending on the age, depth, and type of the reservoir, as well as chemicals added during production. Typically, produced water contains salts, dissolved and dispersed oil and other organic substances, residues from production chemicals (biocides, corrosion inhibitors, H_2S scavengers, clarifiers, emulsion breakers, hydrate inhibitors, etc.), and low levels of naturally occurring radioactive materials and heavy metals. (Neff, 2002; Lee et al., 2005; Veil et al., 2005; Burridge et al., 2011; Neff et al., 2011). Dissolved components in produced water typically include BTEX, phenols, aliphatic hydrocarbons, carboxylic acid, and low molecular weight aromatic compounds. Dispersed components typically include PAHs and alkyl phenols (Henderson et al., 1999; Røe Utvik, 1999; Veil et al., 2004; Neff, 2011; Liang et al., 2018). Reported concentrations of total saturated hydrocarbons in produced water range from 17 to 30 mg/L, total BTEX from 0.068 to 578 mg/L, and total PAH from 0.04 to 3 mg/L and consist primarily of the 2- and 3-ring PAHs, such as naphthalene, phenanthrene, and their alkylated homologs (Neff, 2002; Faksness et al., 2004; Johnsen et al., 2004; Neff et al., 2011). Factors that affect the concentration of oil in produced water include physical and chemical properties of extracted oil and type and efficiency of chemical treatment, as well as type and efficiency of the physical separation equipment (Veil, 2011).

Based on reports by IOGP member companies for North America, the quantity of oil discharged per unit of oil production offshore ranged from 13.7 mg/L in 2017 to 17.6 mg/L in 2019 (IOGP, 2020).

In Canada, monthly concentrations of oil in produced water in recent years have ranged between 18 mg/L and 30 mg/L (Stantec, 2018; Amec Foster Wheeler, 2019; Husky Energy, 2019).

In *Oil in the Sea III*, 29 mg/L was used as the default value (maximum), 20 mg/L was used as the best estimate and 15 mg/L was used as the minimum estimate. For consistency and comparability, the same values were used in this study for calculations of oil volumes discharged through produced water into marine environments. Table 3.3 summarizes the annual estimated input of oil into the marine environment in different study regions.

3.4.2 Transportation

3.4.2.1 Marine Transportation

Regulatory Framework Specific to Marine Transportation

The International Convention for the Prevention of Pollution from Ships, MARPOL 73/78, regulates ship-generated pollution through the six Annexes of the Convention. Parties to the Convention consist of 160 of the current 174 International Maritime Organization (IMO) Member States, including the United States, Canada, and Mexico, covering 98.8% of the world tonnage. Enforcement of the Convention is the responsibility of the governments of Member Parties.

Routine machinery operations on ships generate oily residues, which are discharged either to shore-based facilities or to the sea. MARPOL prohibits any discharge of oily mixtures from ships into the sea except when all of the following conditions are satisfied (MARPOL Annex I Reg. 15):

1. The ship is proceeding en route (the vessel is under way between ports);
2. The oily mixture is processed through approved oil filtering equipment;
3. The oil content of the effluent without dilution does not exceed 15 parts per million (ppm, mg/L);
4. The oily mixture does not originate from cargo pump-room bilges on oil tankers; and
5. The oily mixture, in the case of oil tankers, is not mixed with oil cargo residues.

The United States has implemented MARPOL 73/78 by the Act to Prevent Pollution from Ships and by U.S. Coast Guard (USCG) regulations found in 33 C.F.R. 151.09. The Vessel General Permit for Discharges Incidental to the Normal Operation of Vessels (VGP) is a permit issued by the U.S. EPA first in 2008 and reissued in 2013 under the Clean Water Act National Pollutant Discharge Elimination System (NPDES) that provides authority for discharges incidental to the normal operation of non-military and non-recreational vessels. All discharges of oil, including oily mixtures, from ships subject to Annex I of the MARPOL 73/78 must have concentrations of oil less than 15 mg/L before discharge. The discharge of motor gasoline and compensating effluent must not have oil in quantities that may be harmful as defined in 40 C.F.R. 110.3, which includes discharges resulting in a visible sheen, or an oil concentration that exceeds 15 mg/L.

Tankers carrying oil can have additional oily residues from pump-room bilges and from cargo tank cleaning operations. Discharges of oily waters from the cargo area of an oil tanker are allowed provided that the tanker has a slop tank and an oil discharge monitoring and control system (MARPOL Annex I Reg. 14). Additionally, the discharges are allowed only if the tanker is under way between ports outside of 50 nautical miles from the nearest land, and the instantaneous rate of discharge of oil content does not exceed

30 liters per nautical mile (MARPOL Annex I Reg. 34). However, these discharges are not allowed in Special Areas as defined in the Annex.

There are no legal discharges of oily waters from the cargo area of tankers within the North American territorial waters or in the zone contiguous to the territorial waters. Any accidental or intentional discharges would be classified as spills.

Oil residues and oily mixtures that cannot be discharged in compliance with the regulations shall be retained on board for subsequent discharge to reception facilities. Ports and terminals in which ships have oily residues to discharge are required to have reception facilities with adequate capacity to receive and process the oily waste generated by the vessels calling in the facility. In January 2021, the IMO GISIS Port Reception Facility database[7] listed 2,179 reception facilities in the United States, 2,273 in Canada, and 591 in Mexico. The enforcement of the requirements in the United States and Canadian waters is rigorous but a few cases of illegal discharges still occur annually; when identified, the USCG refers them to the Department of Justice.

As a result of the regulatory changes, as discussed later, **the amount of oily water discharges resulting from tank vessel cargo operations has been reduced significantly worldwide.** In 2014, Peters and Siuda (2014) reported on the decreased incidence of tar balls in the Sargasso Sea in the North Atlantic Ocean since the 1960s and 1970s, which they recognized as a long-term result of the implementation of international conventions that reduced inputs from oil tanker cargo tank cleaning and discharges of tank washings along with the continuation of floating tar removal process (Peters and Siuda, 2014).

The reduction of operational discharges is closely connected with the evolution of the regulatory requirements for the operation, design, and construction of tankers, which has gradually reduced or eliminated mixing of water with cargo.

The first significant operational change was crude oil washing of cargo tanks. Crude oil washing (COW) is a system whereby the oil tanks on a tanker are cleaned out between voyages not with water but with crude oil (the cargo). It is discharged with the rest of the cargo (any remaining sludge must be disposed of to a shore facility). Crude oil washing was made mandatory for new crude oil tankers in 1978 and for all crude oil tankers delivered after 1 June 1982.

In October 1983, MARPOL added a requirement for new oil tankers to be fitted with segregated ballast tanks to eliminate the need to carry ballast water in cargo tanks. **This was followed by the requirements for double-hull tankers, the second significant operational change resulting in decreased operational discharge.** The final year to operate single-hull tankers was 2010, with some exceptions that allowed operation until 2015.

[7] See https://www.imo.org/en/OurWork/Environment/Pages/Port-reception-facilities-database.aspx.

In 2020, the U.S. EPA, under the authority of the Vessel Incidental Discharge Act (VIDA), proposed to establish national standards for discharges incidental to the normal operation of primarily non-military and non-recreational vessels 79 feet in length and above into the waters of the United States or the waters of the contiguous zone. The proposed rule covering oil management reflects the acceptability of the currently permitted discharges (40 C.F.R. 139).

Tank Vessels This section covers operational discharges associated with tank vessel cargo operations. Operational discharges from machinery operations on tankers and other commercial vessels are basically the same, and they are discussed under the following section on commercial vessels.

In the *Oil in the Sea III* report, estimates of operational discharges included assumptions on the level of MARPOL compliance. **Because there is no quantitative data available on the current non-compliance levels,** the estimates in this report are based on full compliance acknowledging that even with rigorous enforcement illegal discharges still occur. The objective is to demonstrate the order of magnitude of discharges if full compliance is achieved. Illegal discharges are counted as spills, if they are reported.

In 2003 there were still pre-MARPOL tankers without segregated ballast tanks in operation. These vessels have now been phased out and all ballast water is segregated from the cargo, consequently the estimated oily discharges from the cargo area of tankers worldwide in 2020 is less than in 2003.

If the non-compliant discharges are excluded, **the worldwide 2003 cargo oil discharge from tankers was estimated at 7,400 tonnes per year; the estimate for 2020 is 1,730 tonnes per year, more than a 75% decrease.** The estimate is based on the discharge estimates for double hull and segregated ballast tank tankers in the 2003 study extrapolated to account for the fleet size of 16,978 tankers greater than 100 gross tons (GT) in 2020 (UNCTAD, 2020). These estimates are assumed to be higher than actuals because they are based on the full slop discharge into the sea at the maximum allowed concentrations whereas product tankers often discharge slops into shore-based facilities.

The regulations do not allow any discharges of oily waters from the cargo area of tankers within the North American territorial waters or in the zone contiguous to the territorial waters.

Sludge is discharged to shore-based facilities or incinerated. No sludge is allowed to be discharged into the sea.

VOCs, discussed in Chapter 2, are released from tankers during loading operations and during the voyage. At loading, the pressure in the cargo tank increases with the rising cargo level and VOCs are vented into the atmosphere unless the terminal and the vessel have a vapor control system in place. During the voyage evaporation from the cargo surface increases the pressure and VOCs may be released to the atmosphere. VOC releases from tankers are characterized as methane and non-methane (NMVOC). Methane and other lighter components contribute to the greenhouse effect, whereas the heavier components are pollutants (Oil Companies International Maritime Forum [OCIMF]).

Factors impacting VOC emissions from crude oil include (MEPC.1/Circ.680):

1. Vapor pressure of the crude oil
2. Temperature of the liquid and gas phases of the crude oil tank
3. Pressure setting or control of the vapor phase within the cargo tank
4. Size or volume of the vapor phase within the cargo tank

Since 2010, MARPOL (MARPOL Annex VI Reg. 15.6) has required crude oil tankers to implement a VOC management plan to prevent or minimize the release of VOC emissions by operational procedures and technical measures. Operational measures include using optimum loading procedures and target operating pressures for cargo tanks. Technical measures include vapor emission control systems (VECS), which are not mandatory. Ports and terminals can implement requirements for VECS. Various VOC emission control methods and control systems have recovery rates from 30% to 99% (OCIMF).

The NRC *Oil in the Sea III* estimated the atmospheric deposition of VOC from tankers in North American waters at 5 tonnes per year based on conservative assumptions on the VOC emissions and the report concluded that the input is significant only in terms of its impact on local air quality. No studies are available on the impact of the IMO regulations on the VOC emissions, but a reduction since 2010 can be assumed. Therefore, **although VOC emissions from ships remain a concern, the input into the sea is not considered significant.**

Commercial Vessels This section covers operational discharges from machinery operations on commercial vessels including tankers.

As with the discharges from cargo tank operations this report assumes full compliance while acknowledging that even with rigorous enforcement, illegal discharges still occur. Illegal discharges are counted as spills, if they are reported.

If the non-compliant discharges are excluded, the 2003 worldwide estimate for machinery bilge oil discharges is 240 tonnes per year corresponding to a total of 86,817 vessels greater than 100 gross tonnes (GT). The estimate for the North American waters is 8 tonnes per year for vessels greater than 100 GT.

By extrapolation, the machinery bilge oil discharges in 2020 are estimated at 270 tonnes per year corresponding to the total of 98,140 vessels greater than 100 GT (UNCTAD, 2020). At the same rate of increase, **the estimate for machinery bilge oil discharges in the North American waters is 9 tonnes per year, slightly higher than in 2003.** These are approximate estimates, accurate within an order of magnitude of the operational discharges, assuming full compliance.

Another operational oil leakage can occur from oil-lubricated stern tubes and other lubricated equipment with oil-to-sea interfaces, such as controllable pitch propellers, rudder bearings, and on-deck equipment water interfaces. A study by D. S. Etkin (2010) estimated that daily stern tube lubricant consumption rates can range up to 20 liters per day, and the operational lubricating oil discharges from vessels are estimated between 37 million liters and 61 million liters of lubricating oil into the marine port waters annually. Today these discharges are treated as oil pollution and all vessels operating in the United States, covered under the VGP, must use environmentally acceptable lubricants (EALs)[8] in all oil-to-sea interfaces unless technically infeasible.[9]

3.4.2.2 Recreational Vessels

The *Oil in the Sea III* report (NRC, 2003) emphasized how oil inputs from recreational engines had been unintentionally overlooked in previous editions of this publication and recommended that emissions from the then ubiquitous conventional two-stroke outboard engines operating in coastal waters were considerable and deserved to be accounted for in the calculations for total inputs of oil into the sea. The 2003 report concluded that the numbers of recreational two-stroke outboards (the predominant propulsion system on small gasoline-powered boats operating within the marine environment) had increased significantly within the previous decades and that the amount of fuel, lubricating oil, and additives was contributing an estimated 5,600 tonnes per year, approximately 6% of the total load of petroleum into North American waters. These figures suggest that inputs from recreational marine engines were the second highest contribution to oil in the sea within the "Consumption of Petroleum" category, behind land-based and river runoff, and the fourth highest contributor of total oil contribution when all categories were compared.

The original two-stroke engine designs allowed for a light and powerful engine but one that emits more pollutants into the environment than engine designs utilizing other available technologies. The U.S. EPA issued regulations that prescribed a 75% reduction of emissions from small marine engines sold by 2005. This was to be done through new design features such as direct injection and four-stroke technologies. Direct injection engines reduce emissions by approximately

[8] EALs are biodegradable and minimally toxic and are not bioaccumulative.

[9] For purposes of the EAL permit condition, *technically infeasible* means that no EAL products are approved for use in a given application that meets manufacturer specifications for that equipment, products that come pre-lubricated (e.g., wire ropes) have no available alternatives manufactured with EALs, EAL products meeting a manufacturer's specifications are not available within any port in which the vessel regularly calls, or change over and use of an EAL must wait until the vessel's next drydocking (epa.gov).

80% and some four-stroke engine designs currently exceed 90% emissions reduction over the U.S. EPA 1998 standard.[10]

The recreational boating market continues to increase in developing and first-world countries. In Australia, between 1999 and 2009, boat ownership grew by 36.4% (Burgin and Hardiman, 2011). Similar trends have been noted in the United States.[11] According to a Global Markets insights report, the recreational outboard market was estimated to be at $9.1 billion (U.S.) in 2020 and was expected to grow at a rate of 5% from 2021 to 2027. Although two-stroke outboard motors are no longer sold in the United States, they continue to hold a world share of the market. These sales are estimated to still be valued at around $3 billion (U.S.) in 2027. This share of the market is maintained predominantly by advantages of the engines' lighter weight, higher torque, and lower maintenance cost relative to four-stroke motors, and by their excellent power-to-weight ratio.[12]

As predicted, **the pervasiveness of the newer, cleaner four-stroke and direct injection engines has produced much cleaner recreational engines that emit far less oil directly into the marine ecosystem; however, in the United States, many of the existing older technology engines still exist and continue to contribute oil to those waters.** Also, as previously mentioned many other countries, particularly the developing nations, continue to utilize the older technology systems due to lower costs, weight and operational constraints. Some national parks in the United States such as Lake Mead and parts of the Colorado River have banned the use of engines that do not meet the U.S. EPA 2006 standards (36 C.F.R. § 7.48[f][3]).

3.4.2.3 Aircraft Fuel Jettison

Aircraft deliberately dump, or "jettison," fuel to reduce the aircraft's weight in emergency situations to allow the aircraft to land safely without sustaining structural damage. Such emergency situations include a return to the airport shortly after takeoff, compromised aircraft performance, and an emergency landing at an unintended destination.

In the United States and Canada, regulations on fuel jettison are aligned with the standards and recommended practices of the International Civil Aviation Organization (ICAO)—a United Nations agency aiming to foster international consensus on aviation activities. ICAO recommendations include advising air traffic control regarding the onset of dumping, and requiring control to coordinate the route, altitude, and duration of dumping with the flight crew to prioritize unpopulated areas (preferably over water and away from expected or reported storms[13]).

Oil in the Sea III recommended that "federal agencies, especially the Federal Aviation Administration (FAA), should work with industry to more rigorously determine the amount of fuel dumping by aircraft and to formulate appropriate actions to understand this potential threat to the marine environment." **This recommendation from 2003 has not been acted upon.**

The Transport Canada Civil Aviation (TCCA) agency does not maintain statistics on the frequency of fuel jettisoning. The FAA describes the frequency of emergency jettisoning as "extremely rare,"[14] and has neither required nor routinely recorded incidences of commercial aircraft fuel jettisons as they occur (Clewell, 1983). The *Oil in the Sea III* report (NRC, 2003) estimated that 1,500 tonnes of petroleum per year are released over the open ocean in North America, and 7,500 tonnes per year worldwide from this activity based on assumptions about the probability of fuel dumping occurring for civilian and military aircraft and experimental deposition studies estimating constraints on the amount of fuel that typically reaches the sea surface (NRC, 2003). Given that U.S.-system air traffic has tended to increase since the completion of the *Oil in the Sea III* report, which estimated an approximate total of 1,000 flights per day (700 North Atlantic, 100 North Pacific, 200 Southern hemisphere) for its calculations, an input of **1,500 tonnes of petroleum per year to North American waters may be an underestimate.** Following a low-altitude fuel dumping incident in January 2020, the FAA was quoted to be aware of 47 events globally in the past 3 years involving a U.S. airline.[15] Previous estimates of the frequency of fuel-dumping incidents are derived from U.S. Air Force assessments of military jettisoning, and cited in *Oil in the Sea III*.

3.5 ACCIDENTAL SPILLS

Accidental spills are oil spillages occurring during exploration and production of oil and gas, spills due to aging infrastructure, and spills from transportation, including pipelines, tank vessel spills, non-tank vessel spills, transportation by rail, and spills associated with coastal refineries and storage facilities.

3.5.1 Exploration and Production of Oil and Gas

Significant oil spillage related to oil extraction may occur due to a blowout (the uncontrolled release of oil and gas from a well to the environment or "loss of well control"). A *blowout*, also known as a *loss of well control*, is an uncontrolled flow of formation or other fluids, and can occur at the surface or underground. A blowout usually does not involve the release of much oil. However, it may, on occasion, lead to the release of large volumes of oil (such as in the case of the DWH spill). The volume of oil released in a blowout depends

[10] See https://www.bts.gov/content/federal-exhaust-emissions-standards-newly-manufactured-marine-spark-ignition-outboard-0.

[11] See https://www.nmma.org.

[12] See https://www.gminsights.com/industry-analysis/outboard-engines-market.

[13] See ops.group, icao.int.

[14] See www.faa.gov/airports/airport_development/omp/faq/general_concerns#q23.

[15] See popsci.com/story/technology/airplane-fuel-dumping-landing-weight.

on the flow rate, which is in turn dependent on the reservoir properties, well characteristics, and the duration of the flow (i.e., how long it takes for the oil to stop flowing naturally through natural bridging or through a human intervention, such as a blowout preventer, capping stack, or relief well).

Smaller spills may also occur during routine operations. As reported in *Oil in the Sea III*, spills related to oil extraction contributed a relatively minor input of oil to marine and estuarine waters—totaling an estimated 168 MT of oil per year with slightly more than half coming from offshore operations. This picture changed dramatically when, in April 2010, the Macondo MC252 well experienced a blowout event during drilling operations from the drilling rig *Deepwater Horizon*. Although the oil spilled from the Macondo MC252 well, the incident is generally referred to as the Deepwater Horizon or DWH oil spill. This event resulted in the release of 4 million bbl (about 571,000 MT) of oil into the Gulf of Mexico over the course of 86 days. **Except for the DWH outlier event, overall, despite an increase in volume of oil produced, the inputs of oil from oil extraction-related spills from 2010–2019 have been minor and of a similar scale to those reported in *Oil in the Sea III*. However, there has been a shift to more spillage in offshore waters and less spillage in nearshore/coastal waters.**

3.5.1.1 Estimation of Oil Inputs from Spills Resulting from Exploration and Production

In a 2009 report prepared for the American Petroleum Institute (API), the 95% reduction in spillage from U.S. offshore oil platforms was described as a significant achievement (Etkin, 2009). Although there had been other oil spills resulting from blowouts in North American waters during the 1960s and 1970s, including the 1969 Alpha Well 21 Platform A incident off Santa Barbara, California, in which 100,000 bbl (14,285 MT) were released, there were no incidents that approached the magnitude of the Ixtoc I incident in Mexico. In the Ixtoc I blowout, between 3.3 million and 10.2 million bbl of crude oil (471,000 to 1.46 million MT) spilled into the Gulf of Mexico at Bahia del Campeche, Mexico, over the course of 290 days in 1979 and into 1980 (Boehm and Feist, 1982; Dokken, 2011). The next largest blowout of 2 million bbl (286,000 MT) occurred off the United Arab Emirates in 1973. In Canada, the largest blowout, which occurred in 1984 off Nova Scotia, amounted to a release of 1,500 bbl (215 MT). In the year before the DWH incident, there was a blowout at the Montara platform offshore Australia in which 28,600 to 214,300 bbl had spilled (Commonwealth of Australia, 2011).

While it will never be possible to definitively establish the exact volume of oil released during DWH (refer to Box 3.2), in 2015, the U.S. District Court's findings of fact showed that approximately 4 million bbl of oil were released and that, given the oil collected at the source (810,000 bbl), a total of 3.19 million bbl of oil were released into the Gulf of Mexico. With regard to impacts, the amount released into the Gulf would be of greatest relevance, although for the purposes of evaluating potentials of releases from blowouts and responses to these events, the amount of oil collected at the source should not be discounted.

The 10-year time period selected for the analysis of data to produce an estimate of oil spill inputs corresponds to the 10-year period for spills just prior to the 2000 commencement of the *Oil in the Sea III* study (1990–1999); the fact that

BOX 3.2
Estimating Inputs from the *Deepwater Horizon* Spill

The total volume of the DWH spillage, the flow rate at different times during the 86-day period, and the amount of oil contained at the wellhead were in dispute for some time as part of multidistrict litigation (*U.S. v. BP et al.*, 2014). Estimates of the volume of oil released to the Gulf of Mexico 66 km from shore varied from 3,260,000 bbl (Fitch et al., 2013) to 5,140,000 bbl worst case (Lehr et al., 2010), further described in Box 1.1 and Section 5.2.9.3.

Estimates of the DWH oil flow rate were initially derived using physical and optical methods applied during the spill. These values were subsequently refined, and an official estimate of oil flow rate was published (McNutt et al., 2011). Video and acoustic data of the oil plume acquired by remotely operated vehicles (ROVs) were used as it exited the well in water 5,067 feet deep at the wellhead, and yielded relatively consistent flow rates between 25,000 and 60,000 bbl/day.

Ryerson et al. (2012) used combined atmospheric, surface, and subsurface chemical data and transport pathways to estimate the quantity of subsurface release from the leaking well during the DWH incident. Subsurface oil composition, dissolved oxygen, and dispersant data were used to estimate the quantity of oil released from the leaking well. The extent and nature of alteration attributable to dissolution and evaporation over time along different transport pathways were estimated by comparing the measured hydrocarbon compositions of atmospheric and subsurface DWH plume samples with the composition leaking from the Macondo well. The estimated quantity of $(7.8 \pm 1.9) \times 10^6$ kg of oil leaking on June 10, 2010, accounted for about three-fourths of the total leaked mass on that day. The average environmental release rate of $(10.1 \pm 2.0) \times 10^6$ kg/d was estimated using atmospheric and subsurface chemical data and agrees well within uncertainties with the official average leak rate of $(10.2 \pm 1.0) \times 10^6$ kg/d derived using physical and optical methods.

The volume of oil released from the Macondo well most commonly used is 4 million barrels, the volume determined by the U.S. District Court in the 2015 damage asessement proceedings.

the DWH incident coincidentally occurred at the beginning of the 10-year time frame being analyzed to determine estimates of "average annual oil inputs" from various types of spill, including oil extraction operations, presents a challenge for the *Oil in the Sea IV* study input estimation analysis. **The volume of 4 million bbl (571,430 MT) is assumed in this *Oil in the Sea IV* study. If this amount is "averaged" over 10 years (2010–2019), the "annual" input from oil extraction would come to more than 57,000 MT. This amount not only exceeds the entire annual spillage reported in *Oil in the Sea III* (7,941 MT) by seven times but is also the equivalent of more than half of the total inputs from all anthropogenic sources as estimated in *Oil in the Sea III*.** Given that the DWH incident can reasonably be considered an extreme outlier case (Ji et al., 2014), it is factored in separately in the *Oil in the Sea IV* analysis. The effect of including and excluding the DWH incident from the overall extraction spill estimates is shown in Table 3.4.

The estimated 160 MT of annual spillage from offshore and coastal/nearshore oil exploration and production activities represents the relatively small spills that occur during routine operations. These spills include the spills of hydraulic oils, diesel and other fuels, petroleum-based synthetic drilling muds, and occasionally crude oil that occur during exploration drilling and production activities. For Canadian offshore operations, spills of crude oil have averaged less than 6 bbl (0.9 MT) per year. There have been some spills of drilling muds that have all been less than 100 bbl each. In the United States, for the past decade, excluding the DWH incident, spills from oil and gas extraction have all involved 500 bbl or less; 77% have involved less than 10 bbl. (No data on individual incidents were available for Mexico.)

Prior to the 2010 DWH spill, U.S. offshore spillage per unit production had decreased by 87% since the 1970s and 71% since the 1990s (Etkin, 2009). This appeared to indicate increased safety measures or "Standard of Care" (i.e., implementation of accident prevention measures, inspections, maintenance, and operator training) in operations. However, the DWH incident brought up significant concerns regarding safety and spill prevention measures for offshore exploration and production operations, which were echoed in Canada, Mexico, and other parts of the world (see Box 3.3).

With respect to the statistical analysis of oil inputs from extraction operations for *Oil in the Sea IV*, DWH is included as a separate and distinct input of oil. The fact that the incident occurred during the 10-year *Oil in the Sea IV* time frame of 2010–2019 does not mean that this is a once in 10 years event. In fact, its occurrence once in 20 years (back to *Oil in the Sea III*) or longer (to the beginning of offshore oil exploration in the mid-1950s) does not have any bearing on

TABLE 3.4 Estimated Annual Spillage Inputs from Oil Extraction[a]

Geographic Zone	Subzone	Average Annual Spillage from Oil Extraction (MT/year)[b]: Excluding DWH	Including DWH	*Oil in the Sea III* Estimate for 1990–1999 (MT/year)[c]
C: E Canada	Coastal	0	0	ND
C: E Canada	Offshore	0.9*	0.9*	28
D: N Atlantic	Coastal	0	0	Trace
F: E GOM	Coastal	0	0	Trace
F: E GOM	Offshore	1.4*	1.4*	Trace
G: W GOM	Coastal	70.8	70.8	90
G: W GOM	Offshore	18.1	57,161.00	50
H: Mexican GOM	Coastal	0	0	ND
H: Mexican GOM	Offshore	1,016.70	1,016.70	61
J: Mexico Pacific	Coastal	0	0	NA
J: Mexico Pacific	Offshore	0	0	NA
K: CA Pacific	Coastal	0.1*	0.1*	Trace
K: CA Pacific	Offshore	1.2*	1.2*	Trace
L: Mid-Pacific	Coastal	NA	NA	Trace
P: South Alaska	Coastal	0.1*	0.1*	Trace
P: South Alaska	Offshore	0.3*	0.3*	NA
Total	Coastal	71	71	90
Total	Offshore	1,038.60	57,291.20	139
Grand Total		1,109.60	57,362.20	229

[a]Current estimates are based on data analyses conducted by Environmental Research Consulting and Greene Economics on the U.S. Coast Guard Marine Information for Safety and Law Enforcement (MISLE) database for U.S. zones, on data from the Canada-Nova Scotia Offshore Petroleum Board and Canada-Newfoundland and Labrador Offshore Petroleum Board for Canadian zones, and data provided by Petróleos Mexicanos (Pemex) for Mexican zones.

[b]Estimates that would have been classified as "trace" in the *Oil in the Sea III* study are shown with an asterisk.

[c]"Trace" indicates that estimated input is less than 10 MT or 70 bbl/year.

NOTE: NA = not applicable; ND = not determined.

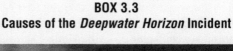

BOX 3.3
Causes of the *Deepwater Horizon* Incident

FIGURE 3.6 Response to the DWH explosion.
SOURCE: USCG.

On 20 April 2010, the *Deepwater Horizon* drilling rig, which was conducting temporary well-abandonment drilling operations at the Macondo oil well about 50 miles offshore of Louisiana, lost control of the well, causing a blowout. On the rig, released hydrocarbons ignited, resulting in explosions and fire (see Figure 3.6) that led to the deaths of 11 persons and serious injuries to 17 others on the rig. The remaining 115 workers were evacuated from the rig before it sank (USCSHIB, 2016a). The oil spill led to the release of 4 million bbl of oil into the Gulf of Mexico.

As detailed in several studies investigating the cause of the accident (National Commission on the BP Deepwater Horizon Oil Spill and Offshore Drilling, 2011a,b,c; NRC, 2012; USCSHIB, 2016a), the Macondo well blowout is a tragic example of a failure of a system—a complex series of interconnected technical, human, regulatory, and organizational factors.

> Offshore drilling, especially in deep water, is an inherently hazardous activity. Construction of deepwater wells like Macondo is a complex process. Sophisticated equipment is used, such as the Deepwater Horizon drilling rig, which must operate in a highly coordinated manner in areas of uncertain geology, often under challenging environmental conditions, and subject to failures from a variety of sources including those induced by human and organizational errors. (NRC, 2012)

The series of events and failures that occurred during the explosions and fire on the *Deepwater Horizon* drilling rig and the release from the Macondo well, as well as analyses of the precipitating causes of the incident, were covered in full in the 2012 NRC report *Macondo Well Deepwater Horizon Blowout: Lessons for Improving Offshore Drilling Safety*. In chronological order, the following findings, from NRC (2012) briefly describe the facts that led to the explosion:

1. The flow of hydrocarbons that led to the blowout of the Macondo well began when drilling mud was displaced by seawater during the temporary abandonment process.
2. The decision to proceed to displacement of the drilling mud by seawater was made despite a failure to demonstrate the integrity of the cement job even after multiple negative pressure tests. This was but one of a series of questionable decisions in the days preceding the blowout that had the effect of reducing the margins of safety and that evidenced a lack of safety-driven decision making.
3. The reservoir formation, encompassing multiple zones of varying pore pressures and fracture gradients, posed significant challenges to isolation using casing and cement. The approach chosen for well completion failed to provide adequate margins of safety and led to multiple potential failure mechanisms.
4. The loss of well control was not noted until more than 50 minutes after hydrocarbon flow from the formation started, and attempts to regain control by using the BOP were unsuccessful. The blind shear ram failed to sever the drill pipe and seal the well properly, and the emergency disconnect system failed to separate the lower marine riser and the *Deepwater Horizon* from the well.

5. The BOP system was neither designed nor tested for the dynamic conditions that most likely existed at the time that attempts were made to recapture well control. Furthermore, the design, test, operation, and maintenance of the BOP system were not consistent with a high-reliability, fail-safe device.
6. Once well control was lost, the large quantities of gaseous hydrocarbons released onto the *Deepwater Horizon*, exacerbated by low wind velocity and questionable venting selection, made ignition all but inevitable.
7. The actions, policies, and procedures of the corporations involved did not provide an effective system safety approach commensurate with the risks of the Macondo well. The lack of a strong safety culture resulting from a deficient overall systems approach to safety is evident in the multiple flawed decisions that led to the blowout. Industrial management involved with the Macondo well–DWH disaster failed to appreciate or plan for the safety challenges presented by the Macondo well.

the expected future occurrence of a similar well failure or oil spill of the magnitude of the DWH oil spill.

There are potential worst-case discharge scenarios for which operators and regulators are conducting contingency planning in the Gulf of Mexico, as well as in other offshore oil exploration and production areas in North America (notably Eastern Canada and possibly Mexico) that could hypothetically result in a blowout that would dwarf DWH (e.g., Buchholz et al., 2016). The likelihood of blowouts of this magnitude are affected by flow rates and the time by which the flow of oil and gas is stopped, either by natural bridging or by active source control measures (e.g., relief wells, capping stacks, or top hat installations) implemented by responders.

A number of studies have looked into the likelihood of blowouts with varying results.

Worldwide, there have been about 50,000 exploratory wells drilled with two large blowouts–the 1979 Ixtoc I well blowout, and the 2010 Macondo MC252 well blowout. That suggests an incidence rate of about 1 large blowout for every 25,000 exploration drilling operations (Imperial Oil Resources, 2013). However, this rate assumes that offshore operations have equivalent safety standards worldwide.

The most comprehensive data on offshore blowouts are maintained by SINTEF. Based on analyses of 607 offshore well blowouts and releases that have occurred worldwide since 1955, Holand (2013) has calculated blowout probabilities that range from about 4.0×10^{-6} to 1.4×10^{-4} (1 in 250,000 to 1 in 7,000) per well or per well-year, depending on the well depth and phase of drilling operations (development, production, or completion) (reviewed in Etkin, 2015). Blowouts are more likely to occur in wells in deeper water (Holand, 2013). Note that a blowout (or loss of well control) does not necessarily imply a large oil release. When an oil spill results from a blowout, there is a 56% likelihood of it lasting 2 days or less (i.e., bridging naturally with sediments filling the wellbore), and only a 15% chance of it lasting more than 2 weeks (Holand, 2006, 2013); however, the Macondo well in the DWH incident flowed for 85 days until it was capped.

In a study conducted for the U.S. Bureau of Safety and Environmental Enforcement (BSEE), loss of well control (LOWC) events in the U.S. Gulf of Mexico outer continental shelf (OCS) were analyzed for the purpose of developing a risk model to estimate risk for future LOWC events (Holand, 2016). One factor that was found to contribute to the probability of a large spill (of at least 500 to 5,000 bbl) is exploration drilling (as opposed to development or production drilling), especially from a floating vessel (drilling rig) rather than from a fixed platform.

The frequency (per well) of LOWC events or blowouts for development drilling resulting in at least 500 bbl of spillage was estimated to be significantly higher in the U.S. Gulf of Mexico OCS than in regulated areas in other parts of the world, including offshore East Canada (1 in 390, compared to 1 in 1,560). The greatest likelihood of a LOWC event is during the exploration drilling phase when the characteristics of the reservoirs are not well known. Here again, the U.S. Gulf of Mexico was calculated to have a higher probability of an LOWC event (1 in 160 compared to 1 in 600). The U.S. Gulf of Mexico has a higher reported rate of "kicks," although according to Holand (2016) there is no clear explanation for this. Possible reasons include the greater depth of the wells in the substrate (as distinct from water depth) which take a longer time to drill, or that there are more complicated formations to deal with. A *well kick* occurs when the pressure found within the drilled rock is higher than the mud hydrostatic pressure acting on the borehole or rock face, which can force formation fluids into the wellbore. The forced fluid flow is called a "kick."[16]

These calculations were based on historical data. The probabilities for future blowouts will be affected by the degree to which known mitigation and prevention measures are implemented (including the capping stack developed for the DWH response), as well as any new technologies that are developed. One research team (Caia et al., 2018), conducted analyses based on newer intervention technologies developed after the DWH incident. These newer interventions, such as Rapid CUBE (Rapid Containment of Underwater Blowout Events) (Andreussi and DeGhetto, 2013) and Blowstop injections (Caia et al., 2018), were shown to reduce the duration

[16] See https://petrowiki.spe.org/Kicks.

of flow by 30% to 60%, resulting in much lower volumes of release. Most of these technologies are designed to intervene in spills that originate from a broken riser above the seafloor. In the case in which a well failure results in subsurface damage and leakage through the near seabed sediments, the leak area will be diffuse, and rapid containment would be much more challenging and likely would be achieved through a relief well.

From a preparedness and response perspective, the ability to reliably control the source of the oil release (the well) as quickly as possible to reduce or stop the flow of oil will be the most effective way to reduce or mitigate impacts of any future blowout or large release (Buchholz et al., 2016). Section 4.1.2 includes additional detail on source control measures for offshore wells.

Offshore Exploration and Production Safety

Following the Macondo incident, a comprehensive review was conducted by the U.S. oil and gas industry, the U.S. Department of the Interior and the Presidential Oil Spill Commission, the National Research Council (NRC, 2012), and the National Chemical Safety Board—all of which resulted in a series of recommendations and actions to improve offshore safety, safety culture, and the regulatory framework. Since then, government and industry have been focused on developing regulations and standards as well as ensuring availability and readiness of well intervention and spill response capabilities.

The U.S. government, through the BSEE and the Bureau of Ocean Energy Management (BOEM), has made significant changes to the regulations governing offshore oil and gas operations. Among them are requirements for the safety and environmental management systems (SEMS), drilling safety, well control, blowout preventer systems, production safety, and spill response. The regulations for well design, integrity, and safety now have extensive requirements that include (among other things):

- Requirement for two independent barriers between hydrocarbon-bearing zone and the environment
- Requirements for the testing, inspection and certification of well completion activities as well as subsea well control equipment including blowout prevention equipment
- Real-time monitoring capabilities for high-risk drilling activities
- New requirements for firefighting systems and high pressure/high temperature (HP/HT) well equipment
- New reporting requirements

The USCG is another regulatory agency overseeing offshore oil and gas operations in relation to safety of life, property, and navigation as well as protection of the environment. During a maritime oil spill response, the U.S. Coast Guard acts as the designated federal Incident Commander. The USCG has also revised some of its regulations, guidance, inspections, and training practices including, among others:

- Requirements for the independent testing and certification of electrical equipment in hazardous locations on new mobile offshore drilling units (MODUs), floating offshore facilities, and vessels other than offshore supply vessels that engage in offshore activities
- Guidelines for fire and explosion risk analysis
- Guidelines for lifesaving and firefighting equipment, training, and drills onboard manned offshore facilities
- Updated inspection protocols for vessels in offshore operations
- Improved collaboration with the BSEE on regulatory oversight, inspection, and spill response

In the United States, the API develops industry standards for safe oil and gas operations. Since 2010 API has published more than 250 new and revised exploration and production standards including standards for wells, blowout preventers, subsea and well capping equipment, integrity management, and spill response, among others. Development of standards is underpinned by continuous innovation and development of new technologies to increase safety and environmental protection in all segments of offshore operations. Latest efforts have resulted in significant improvements in information management systems, large data analytics for well planning and design, equipment manufacturing, complex well completion tools, and a variety of other new methods and techniques.

In 2011, the industry formed the Center for Offshore Safety (COS) to help improve the safety performance of the offshore oil and natural gas industry. The COS is focused on a development of audit tools, standards, and best practices as well as protocols for industry data collection, analysis, reporting, and knowledge sharing among companies and regulators. The COS has developed SEMS tools for auditor qualification, conduct of audits, and accreditation of audit service providers (API, 2019a). The development of consistent and transparent audit tools and processes assists in effective implementation of SEMS and facilitates broad sharing of audit learnings.

Another significant achievement of the offshore oil and gas industry was the creation of well intervention and containment consortiums that provide containment and response capabilities for well intervention thousands of feet below the water's surface. These companies maintain specially designed equipment for well control in strategic locations around the globe and complement capabilities of commercial well control companies. More than 20 capping stacks (see Figure 3.7) are now available for well control. This ensures quick deployment of equipment tailored to specific incident conditions to any location in the world. Source

FIGURE 3.7 Capping stack.
SOURCE: Oil Spill Response, Limited.

control procedures and equipment are being continuously upgraded with new technologies and best practices to keep pace with industry operations and ensure global response readiness. Regulators in many countries now require oil and gas companies to demonstrate availability of equipment and trained staff to cap a well or capture uncontrolled flow of hydrocarbons prior to commencement of operations.

In Canada, following the Macondo incident, several regulatory reviews and audits were conducted. In 2010, the commissioner of the Environment and Sustainable Development branch of the Office of the Auditor General of Canada evaluated whether Transport Canada, the Canadian Coast Guard, Fisheries and Oceans Canada, and Environment Canada have implemented measures to prepare for and respond to pollution from ships in Canada's marine environment (The Office of the Auditor General of Canada, 2010[17]). In 2012, the commissioner evaluated whether existing regulations and agencies appropriately manage the environmental risks and impacts from offshore oil and gas activities (The Office of the Auditor General of Canada, 2012). These reports and subsequent work by the agencies resulted in several notable improvements:

- The Canadian Coast Guard established a new Environmental Response Branch within the Maritime Services Directorate. The Director, Environmental Response position was created and staffed, with additional personnel to provide a strategic focus for the program, and to ensure that the Coast Guard can fulfill its mandated obligations related to marine pollution preparedness and response. The Coast Guard also reviewed its Response Management System to ensure that it can support a multi-party response to a major oil spill in Canadian waters.
- Transport Canada, in their Report to Parliament 2006–2011 on Marine Oil Spill Preparedness and Response Regime (Transport Canada, 2014), reported improvements in interdepartmental governance and coordination. The Interdepartmental Marine Pollution Committee was formed, co-chaired by Coast Guard and Transport Canada, and reporting to the Assistant Deputy Ministers' Emergency Management Committee. Transport Canada, Environment Canada,

[17]See https://publications.gc.ca/collections/collection_2011/bvg-oag/FA1-14-2011-eng.pdf.

Natural Resources Canada, Aboriginal Affairs and Northern Development, National Defence, and Public Safety Canada are members of the committee. The committee supports the Government of Canada's objectives related to marine pollution, focusing on interdepartmental collaboration to strengthen Canada's ability in prevention, preparedness, and response and recovery capabilities regarding marine pollution events.

- Canada-Newfoundland and Labrador Offshore Petroleum Board (C-NLOPB) established a Special Oversight Team and defined Special Oversight Measures for critical wells. The measures are applied prior to and during the well approval review process, and extend throughout the execution phase and into the plugging and abandonment of the well.

Special Oversight Measures are focused on well control protocols, equipment, and competencies, blowout prevention, and oil spill response contingency plans. There is a heightened focus on kick prevention, maintaining dual well barrier envelopes, assessing the extent of means in place for primary and secondary well control, and assessing the integrity and functionality of primary and secondary BOP systems. There is also heightened focus on assessing plans and processes for oil spill response readiness, relief well drilling arrangements, and arrangements for capping stack/subsea containment systems. (C-NLOPB, 2018)

- In 2013, the governments of Canada, Nova Scotia, Newfoundland, and Labrador implemented changes to further strengthen the Canadian regimes to reinforce the "polluter-pays principle" in legislation, increasing "absolute" liability (no-fault) limits in the offshore from $30 million to $1 billion (CA).
- A regulation under the Canada Oil and Gas Operations Act added Corexit® EC9500A and Corexit® EC9580A to Schedule 1, enabling the use of the spill-treating agents when they are likely to achieve a net environmental benefit.
- The Multi-partner Research Initiative, part of the national Oceans Protection Plan, was launched in 2016. The Initiative provided $45.5 million over 5 years to support collaboration among leading national and international experts on oil spill research and response to:
    - identify knowledge gaps and research priorities;
    - improve understanding of how oil spills behave in water and their impacts on fish and other aquatic organisms;
    - develop new technologies and protocols to select the best methodologies for oil spill clean-up; and
    - support science-based decisions that will aim to minimize the environmental impacts of oil spills and enhance habitat recovery.

Pipeline Safety

In 2017, the U.S. Department of Transportation Pipeline and Hazardous Materials Safety Administration's (PHMSA's) Office of Pipeline Safety (OPS) Division worked with the Pipeline Safety Trust, the API, and the Association of Oil Pipe Lines and develop performance measures for pipeline systems transporting crude oil, refined petroleum, and biofuel. Safety metrics were defined, by individual operator and by safety program, that uses incident data for serious incidents (as rate per mile and causes), accidents impacting people or the environment (as rate per mile and volume spilled per barrel-mile transported and for all causes in terms of integrity inspection target, and operations and maintenance target), and miles inspected (as miles inspected by inspection method).[18] PHMSA monitors compliance through field inspections of facilities and construction projects; programmatic inspections of operator management systems, procedures, and processes; incident investigations; and direct dialogue with operator management. PHMSA has established enforcement mechanisms to require that operators take appropriate and timely corrective actions for violations, and that they take preventive measures to prevent future failures or non-compliant operation of the pipelines.[19]

In the United States for the past 5 years, the total number of onshore and offshore pipeline incidents impacting people or the environment decreased by 36% despite a 10% increase in pipeline mileage and barrels (10% increase in "barrel miles," or barrels of liquid moved one mile) delivered (API, 2019b). Leak prevention is accomplished by robust pipeline integrity management, which involves, among other things, risk identification and measures to mitigate these risks (API, 2019c). Industry's strategic plan to further improve pipeline safety includes four main goals:

- Promote and improve safety culture by implementing safety and integrity management systems (ANSI and API, 2015; API, 2019c), analyze and share learnings from incidents and implement mitigation measures;
- Improve safety through technology and innovation—for example, by developing and implementing a holistic leak detection program and advanced leak detection methods including corrosion identification and mitigation techniques (API, 2015);
- Increase stakeholder awareness and engagement to reduce potential for excavation damage for onshore assets (API, 2010); and
- Enhance emergency preparedness and response tools and practices (API, 2015).

[18] See https://www.phmsa.dot.gov/data-and-statistics/pipeline/national-pipeline-performance-measures.

[19] See https://primis.phmsa.dot.gov/comm/reports/enforce/Enforcement.html.

3.5.2 Spills Caused by Natural Hazards

Besides blowouts and more usual operational spillage, there is also a potential for oil spillage to occur from offshore oil platforms, wells, and other installations due to damage from earthquakes, hurricanes, and storm-related currents and mudslides. Storm damage may include wind damage to topside structures; topside damage by greenwater (waves that break over the deck); wave slamming on the underside of floating platforms; loss of mooring systems due to disconnection in high seas, breakage in high currents, or loss of anchors in sediment failures; and structural damage caused by submarine mudslides and shoreline erosion, among others. Loss of mooring systems can result in capsizing or uncontrolled drift, which may also result in damage to subsea equipment due to dragging the mooring system or risers with the rogue platform. Studies conducted in the Gulf of Mexico document damage to offshore platforms and structures during Hurricanes Andrew, Lili, Ivan, Gustav, and Ike (Wang et al., 2005; Energo Engineering, 2006, 2010). The high winds, storm surge, and associated seafloor mudslides during Hurricane Ivan toppled a number of offshore platforms in September 2004, including that of Taylor Energy at Mississippi Canyon (MC-20) (Bryant et al., 2020; see Box 3.4), causing the spillage of an undetermined amount of oil at the time of this study. Estimates from model calculations, reported by the National Oceanic and Atmospheric Administration's (NOAA's) National Centers for Coastal Ocean Science, range between 9–108 bbl of oil released daily into the Gulf of Mexico for, at the time of this study, 17 years (and counting). The amount of oil collected by the containment system installed in April 2018 is in the vicinity of 30 bbl per day. Oil from this incident is still being released periodically from the seafloor (Bryant et al., 2020).

The potential for future spillage related to damage from hurricanes, particularly in light of increase in intensity and frequency of severe storms and hurricanes related to global warming, represents a considerable risk of future oil inputs into the marine environment (Wang et al., 2005; Webster et al., 2005; Elsner et al., 2008; Bender et al., 2010; Knutson et al., 2010; Wijesekera et al., 2010).

3.5.3 Aging Infrastructure and Decommissioning Leakage

Offshore oil and gas exploration activities began in the late 1800s off Santa Barbara, California followed closely by exponential expansion of offshore exploration in the marshes, bays and nearshore waters of Louisiana. Shortly thereafter, exploration expanded to the nearshore and offshore waters throughout the Gulf of Mexico at an expeditious rate.[20] To date, this activity has produced some 55,000 oil and gas wells drilled, and roughly 23,000 miles of pipelines installed in waters of the U.S. outer continental shelf (OCS) associated with oil and gas exploration. Of the wells drilled, approximately 53% have been permanently or temporarily decommissioned.[21] The two largest oil producing states, Texas and Louisiana, each report a growing number of decommissioning and abandonments. Many wells and pipelines that have ceased production have gone into abandonment (orphaned) and potentially impart additional risks for leakage of oils into the environment. Even wells that have been properly plugged and abandoned (P&A) hold a significant long-term risk of release or total failure of the containment system.

Actively producing wells have been decreasing in number, particularly in shallow water. Over the past two decades, the BSEE reported more than a 50% decrease in active wells, from a peak in 2001 of approximately 4,000 to less than 1,908 remaining in 2017. Wells are decommissioned for a variety of reasons, but most often it is a matter of economics (costs of the operation surpassing the revenues from the production) or resource depletion. The decommissioning of an oil and gas well and associated infrastructure may include many processes but must include the plugging of each well bore by the use of means that meet regulatory standards. BSEE regulations also mandate that any well in an idle condition for more than three years, and platforms that have been inactive for 5 years, or platforms that have been toppled, or those for which there are no further plans for production must be listed on the BSEE "Idle Iron"[22] list and be physically removed and/or otherwise decommissioned within 5 years of inclusion on the list. The fate of the decommissioned material may include scrapping, or use as artificial reef material, per BSEE's Rigs-to-Reefs policy. Legislation regarding federal waters allows regulators to hold any previous owner responsible for necessary plug and abandonment and removals. Such legislation may not apply to waters not under 43 U.S.C. 1334 and 30 C.F.R. 250, Subpart Q, Decommissioning Activities.[23]

Oil and gas wells—whether active (in service), inactive, suspended, orphaned, or abandoned—have been sources of oil and gas released into the environment since the first wells were drilled. Releases occur during all phases of a well's existence (see Figure 3.9), including the exploration, extraction, and abandonment phases. Structural components of the well tubing, valves, and fittings, and even the materials

[20] American Oil & Gas Historical Society editors, "Offshore Petroleum History." January 10, 2010; updated November 9, 2020. American Oil & Gas Historical Society. https://aoghs.org/offshore-history/offshore-oil-history.

[21] Wesley C. Williams, PhD, PE, "Leakage During Decommissioning and from Decommissioned Wells" National Academies presentation. Advanced Reactor Systems Oak Ridge National Laboratory.

[22] See https://www.bsee.gov/notices-to-lessees-ntl/ntl-2018-g03-idle-iron-decommissioning-guidance-for-wells-and-platforms.

[23] 43 U.S.C. § 1334 and 30 C.F.R. Pt. 250, Subpart Q, Decommissioning Activities, https://www.govinfo.gov/content/pkg/CFR-2015-title30-vol2/pdf/CFR-2015-title30-vol2-part250-subpartQ.pdf.

BOX 3.4
MC-20 Oil Spill

In the aftermath of Hurricane Ivan (a Category 3 hurricane) in September 2004, an underwater mudslide caused the Mississippi Canyon 20 (MC-20) oil and gas platform to collapse onto the floor of the Gulf of Mexico approximately 11 miles off the Louisiana coast. The structure toppled and dragged to the south, snapping and then burying in sediment the cluster of 28 pipes drilled to carry oil to the surface[a] (see Figure 3.8), resulting in a slow release of oil from the broken pipe collection about 500 feet below the surface.

For 14 years, the broken pipelines leaked oil and the owner of the platform made attempts to deal with the releases. The MC-20 site has since been the source of persistent plumes of oil and gas and surface oil slicks. These surface slicks are visible from ships and by aerial and satellite remote sensing and have been used to measure the volume of oil leaking from the site.[b] A comprehensive analysis of these slicks and their potential sources is presented in Bryant et al. (2020).

Published estimates of the volume of oil that has entered the sea from the MC-20 site are summarized in Table 3.5. The inherent challenges and uncertainties of oil flow estimations as well as differences in the methodologies employed by researchers resulted in great variability of the reported released volume values. Caution is advised in using and interpreting these numbers even when published in the peer-reviewed literature.

In 2018, the USCG federalized the response and contracted a group to design, manufacture, install, and operate a recovery structure to reduce the amount of oil entering the waters of the Gulf of Mexico. This was a unique situation requiring a first-of-its-kind source control solution. The innovative subsurface oil containment and recovery system was installed in April 2019.

The containment system consists of a collection dome supported from the downed jacket lying on the seabed. The dome (shown in Figure 3.8(B) in purple) was placed on a structure cantilevered off the jacket leg directly above four oil plumes rising from the seabed. Once collected, the water is separated from the oil is in a subsea separator and the oil is stored in a subsea oil containment vessel. A submersible hydraulic

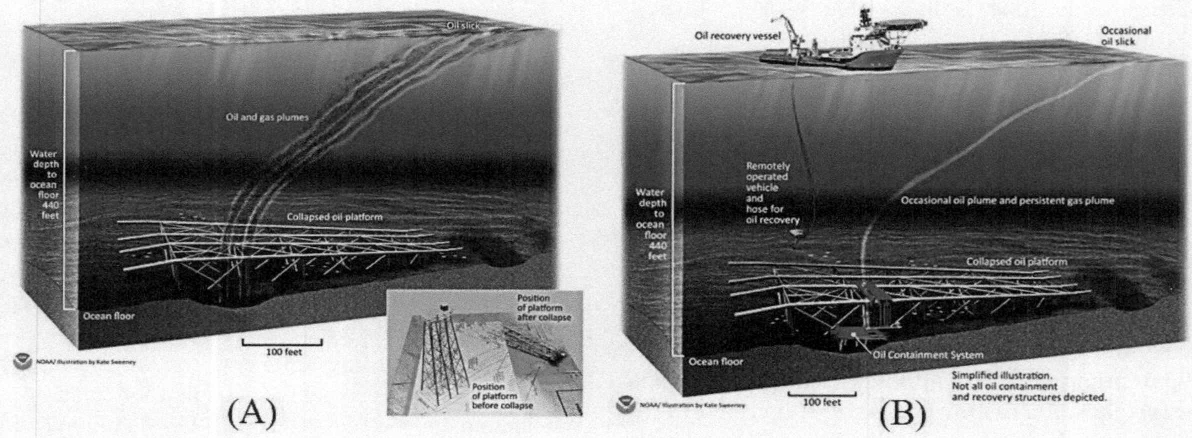

FIGURE 3.8 (A) Mississippi Canyon 20 Saratoga oil platform collapsed after Hurricane Ivan in 2004. Oil and gas discharge into the Gulf of Mexico and rise to the ocean surface creating a visible oil sheen. (B) The USCG directed the installation of an oil containment system (pictured in purple), which is designed to contain and recover as much oil as possible. An oil recovery vessel periodically connects to the oil containment system to remove oil collected in the system and transport the oil to shore.
SOURCE: NOAA, Kate Sweeney.

TABLE 3.5 Published Estimates of Annual Oil Seepage from Mississippi Canyon-20

Estimate Source(s)	Low MT per Year	High MT per Year
BSEE, 2015	9.8	85.3
Reddy, 2018	25	514.1
Windham, 2014	156.4	557.9
Fears, 2019; Couvillion, 2020	328	1,246.30
MacDonald et al., 2019; Mason et al., 2019	983.9	5,575.50
Pineda-Garcia, 2018	12,987.50	36,404.50
Sun et al., 2018	2,525.40	89,896.10

pump then offloads the oil from the containment vessel on a monthly basis to a surface vessel, which then transports the oil to shore, where it is recycled.[c]

The apparatus collects approximately 30 bbl (1,260 gallons) of oil per day.[d] As of 10 August 2021, the apparatus had collected approximately 830,000 gallons since becoming operational.[e] The collection system is not a complete solution; however, while a more permanent solution is planned, the system is capturing a majority of oil being released and has reduced the amount of oil entering the marine environment.

[a]See https://www.eenews.net/articles/feds-record-oil-leak-100-times-greater-than-company-says.

[b]See https://coastalscience.noaa.gov/project/mc20report.

[c]See https://couvillionmc20response.com/#technology.

[d]See https://coastalscience.noaa.gov/news/mc20report.

[e]See https://couvillionmc20response.com.

used to plug a well, are susceptible to corrosion, decay, tectonic shifting, underwater landslides, and accidents such as allisions by passing vessels, but more often fail due to well design and the materials used in their construction and containment efforts. **Although the age of the wells and associated infrastructure are not always the underlying reasons for an oil release into the environment, age can be, as has been established, a contributing factor and cause for concern.** Age ranges of pipelines and wells for the northern Gulf of Mexico are shown in Table 3.6. BSEE has

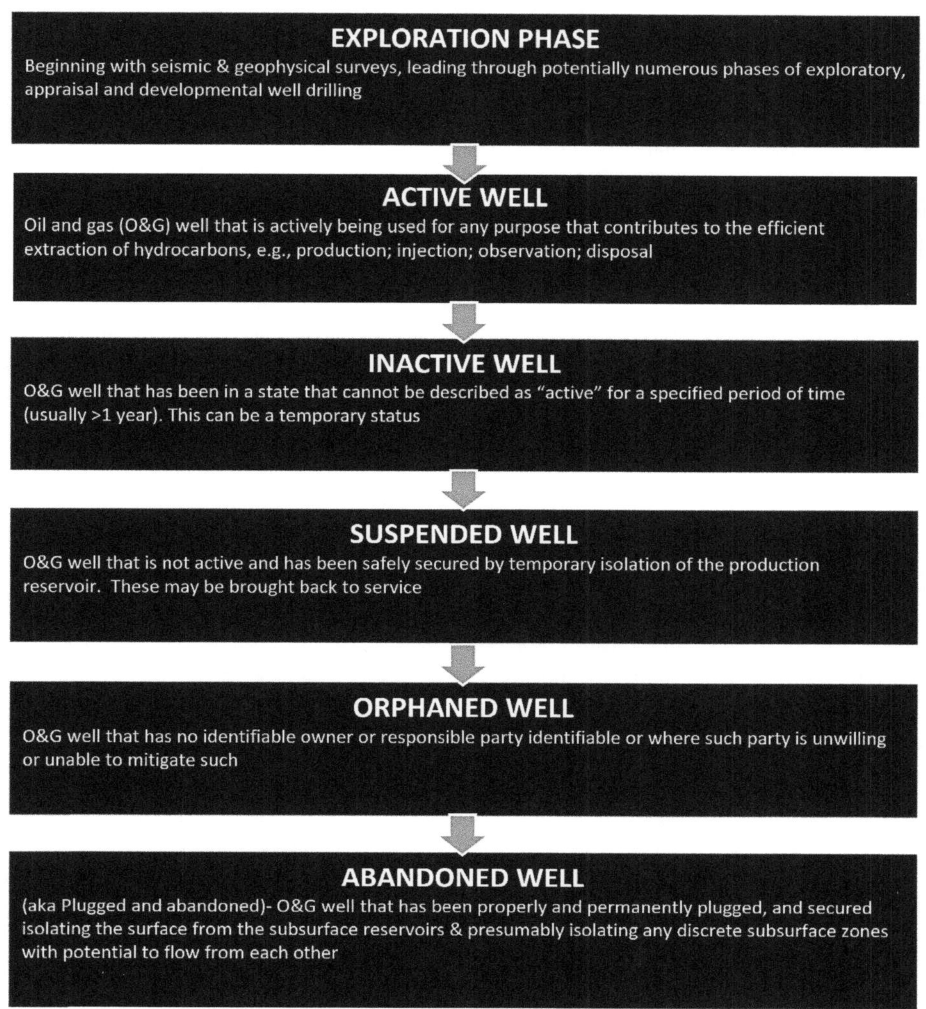

FIGURE 3.9 Phases of an oil and gas (O&G) production well.

TABLE 3.6 Age Categories of Pipelines and Wells in the Northern Gulf of Mexico

	Total Number of Pipelines and Wells (by Age) in Northern Gulf of Mexico								
Age	<10 years	10+ years	20+ years	30+ years	40+ years	50+ years	60+ years	70+ years	Total
Total Pipelines	2,293	4,396	3,280	1,409	2,799	898	109		15,184
Total Wells	1,349	4,205	7,209	7,514	8,149	6,677	1,948	69	37,120

SOURCE: Data based on BSEE data available as of November 30, 2020.
NOTE: Totals do not include sidetracks.

indicated there are more than 8,500 actively producing wells in the Gulf of Mexico alone with ages greater than 50 years.[24]

Wells that have been properly plugged and abandoned (P&Aed) continue to pose a threat to the environment. Several leak paths potentially exist (see Figure 3.10), including loss through the casing-annulus, cement micro-annulus, plug casing annulus, bulk flow through cement, casing penetration, and micro channeling through annulus cement and annulus cement-formation micro-annulus. Recently developed sealing alternatives such as bismuth alloys, thermite and resins have been replacing cements for the P&A process, but no data exists to date that any design currently in use will meet the conventional "eternal, 3,000 years or greater" expectation. Some research suggests that P&Aed wells are still under high pressure and that use of current P&A practices may have a statistical life of only a few hundred years. Some studies suggest that between 1.9 to 75% of drilled wells may eventually exhibit integrity failure of containment, resulting in a potential of more than 10 million bbl of oil that could be released into the environment, given the current number of wells drilled to date (Ingraffea et al., 2014; Boothroyd et al., 2016; Willis et al., 2019).

Pipelines associated with oil and gas activities have exhibited accidental spills emanating from failures due to allisions to the pipeline associated with vessel groundings, anchor drags and setting, and ruptures due to human oversight such as over-pressuring (Kaiser and Narra, 2017). Steel pipelines in saltwater marine environments are particularly susceptible to corrosion, which has contributed to a growing number of spills in older systems such as in the marshes of Louisiana and in Texas bays. Cathodic protection measures and coating are standard practice but have limitations to the extension of a pipeline's usable life. Pipelines are either trunklines or gathering lines. Trunklines are typically large in diameter and may carry products from wells, fields and lease blocks to market. Gathering lines typically are shorter segments of smaller diameter lines that transport liquids and gases between facilities or to a trunkline. Normal procedures for the retirement of an existing pipeline include full or partial removal off the pipeline, or the pigging (a process that may include internal inspection and purging of the line) and cutting of the ends of the line, as well as ensuring proper burial depth and abandonment, after which pipelines pose little threat for spillage. In the Gulf of Mexico, a total of 45,310 miles of pipeline have been installed and more than half have been decommissioned (Kaiser and Narra, 2019). Data suggest that the number of spills from pipelines has increased dramatically. From 1968 to 1977, an average of 47 pipeline spills per year were reported. Decadal averages increased to 188 per year in 1978–1987 and to 228 per year in 1988–1997 (GESAMP, 2007). Some papers suggest that the increase in releases from pipelines may be attributable not only to the increase in number and length of lines, but also to the aging of the existing lines (Jernelöv, 2010).

Well infrastructure and pipeline spill data are not collected in a manner that denotes a cause due to aging; therefore, the ability to quantify the amount of oil entering the seas due to such releases cannot be discerned.

3.5.4 Transportation of Oil and Gas

Between crude oil extraction (exploration and production processes) and the use of oil as a fuel or material for the production of chemicals, oil is transported and handled many times.

Crude oil and refined petroleum products are transported from one place to another in a number of ways: by tank vessel (tanker or tank barge); by pipeline; and, more recently, by

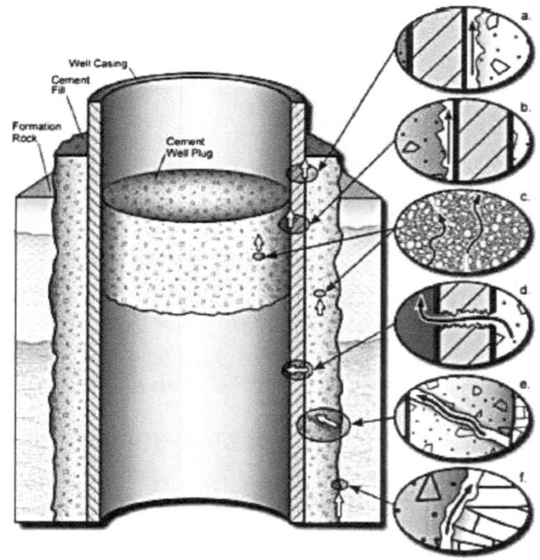

FIGURE 3.10 Potential leakage points in a casing plug.
SOURCE: Willis et al. (2019).

[24] See https://www.data.bsee.gov.

TABLE 3.7 Estimated Annual Spillage from Pipeline Oil Transportation[a]

Geographic Zone	Subzone	Average Annual Spillage from Pipeline Oil Transportation 2010–2019 (MT/year)[b]	Average Annual Spillage from Pipeline Oil Transportation *Oil in the Sea III* Estimate for 1990–1999 (MT/year)[c]
D: N Atlantic	Coastal	24	150
E: Mid-Atlantic	Coastal	3.7*	36
F: E GOM	Coastal	11.4	Trace
F: E GOM	Offshore	0	Trace
G: W GOM	Coastal	296.1	890
G: W GOM	Offshore	2.1*	60
H: Mexican GOM	Coastal	0	Trace
H: Mexican GOM	Offshore	0	ND
I: Puerto Rico	Coastal	0	Trace
J: Mexico Pacific	Coastal	0	ND
J: Mexico Pacific	Offshore	0	ND
K: CA Pacific	Coastal	42	39
K: CA Pacific	Offshore	0	Trace
L: Mid-Pacific	Coastal	0	Trace
M: NW Pacific	Coastal	0	Trace
N: Hawaii	Coastal	1.0*	Trace
P: South Alaska	Coastal	0	Trace
Total	Coastal	378.3	1,115.0
Total	Offshore	2.1*	60
Grand Total		380.4	1,175.0

[a] Current estimates are based on data analyses conducted by Environmental Research Consulting on the Pipeline and Hazardous Material Safety Administration (PHMSA) database for U.S. zones. There are currently no offshore or coastal pipelines in Canada. No data were publicly available for Mexican zones.
[b] Estimates that would have been classified as "trace" in the *Oil in the Sea III* study are shown with an asterisk.
[c] "Trace" indicates that estimated input is less than 10 MT or 70 bbl/year.
NOTE: NA = not applicable; ND = not determined.

rail. Each time oil is handled, temporarily stored, transferred, or transported, there is the potential for spillage to occur. Spills that occur from tank vessels, from coastal or offshore pipelines, and conceivably from trains that transit along coastal routes would affect the marine waters of concern in the *Oil in the Sea IV* study. In addition, spills may occur from coastal storage terminals and refineries.

In the *Oil in the Sea IV* analysis, the spillage related to the "transportation" phase is considered to take place from the conclusion of the offshore (or onshore) oil extraction phase through the transport by tank vessel, pipeline, or rail to storage terminals, and then through the transport to consumption facilities. Only spills that occur into marine and estuarine waters of North America are included.

3.5.4.1 Pipelines

Pipelines are a vital part of the oil transportation infrastructure of North America. Crude oil is transported through offshore pipelines to terminals and to connect with coastal/inland pipeline systems. Crude oil and diluted bitumen (dilbit; see Section 2.1) are transported from inland production sites in the United States and Canada through pipelines to inland and coastal terminals for distribution by tank vessel, rail, or tanker truck. Some pipelines run directly to refineries. Refined petroleum products are often transmitted through pipelines to terminals and end-use consumption facilities, such as power plants and airports.

Although tens of thousands of pipeline spills have been reported in the last 50 years in the United States and Canada (Etkin, 2014, 2017a), only a small percentage of those spills have affected the marine and estuarine waters of North America as covered by the *Oil in the Sea IV* study.

The estimated annual inputs of oil to North American marine and estuarine waters are shown in Table 3.7. **There has been a 68% reduction in the amount of oil spilled from pipelines between the 1990–1999 *Oil in the Sea III* time frame and 2010–2019. More than 99% of the spillage has occurred in nearshore or coastal waters.**

The significant reduction in coastal and offshore pipeline spillage has mirrored the continuing trend seen in inland pipelines of the United States over the past 50 years (Etkin, 2017a). Many of the safety improvements made to inland pipeline operations have also been applied to offshore and coastal pipelines. These measures include increased inspections and monitoring, as well as replacement of older lines. However, the aging infrastructure of the offshore pipeline system in the Gulf of Mexico is of concern, as described in Section 3.5.2. **Of the more than 15,000 miles of oil and gas pipelines in the Gulf of Mexico, over one-third are at least 30 years old** (see Table 3.8).

TABLE 3.8 Age of Active Pipelines in Northern Gulf of Mexico[a]

Product Type Transported	Total Pipeline Mileage (by Age)							
	<10 years	10+ years	20+ years	30+ years	40+ years	50+ years	60+ years	Total
Oil	1,393	2,253	1,534	525	788	242	48	6,783
Gas	879	2,056	1,138	653	1,797	537	61	7,120
Condensate	0	0	12	0	0	0	0	12
Gas/Oil	0	1	14	45	40	24	1	126
Gas/ Condensate	21	86	582	186	174	94	0	1,144
Total	2,293	4,396	3,280	1,409	2,799	898	109	15,184

[a] Based on analyses of BSEE data conducted by Environmental Research Consulting.

TABLE 3.9 Estimated Annual Spillage from Tank Vessel Oil Transportation[a]

Geographic Zone	Subzone	Average Annual Spillage from Tank Vessel Oil Transportation 2010–2019 (MT/year)[b]	Average Annual Spillage from Tank Vessel Oil Transportation *Oil in the Sea III* Estimate for 1990–1999 (MT/year)[c]
C: E Canada	Coastal	0.2*	0
D: N Atlantic	Coastal	9.5*	740
D: N Atlantic	Offshore	0	17
E: Mid-Atlantic	Coastal	0.2*	14
E: Mid-Atlantic	Offshore	0	Trace
F: E GOM	Coastal	2.3*	140
F: E GOM	Offshore	0	10
G: W GOM	Coastal	84.8	770
G: W GOM	Offshore	91	1,500.00
H: Mexican GOM	Coastal	ND	80
H: Mexican GOM	Offshore	ND	ND
I: Puerto Rico	Coastal	0	Trace
I: Puerto Rico	Offshore	0	490
J: Mexico Pacific	Coastal	ND	ND
J: Mexico Pacific	Offshore	ND	Trace
K: CA Pacific	Coastal	0	150
K: CA Pacific	Offshore	0	0
L: Mid-Pacific	Coastal	0	Trace
L: Mid-Pacific	Offshore	0	12
M: NW Pacific	Coastal	0.2*	10
M: NW Pacific	Offshore	0	0
N: Hawaii	Coastal	0	Trace
N: Hawaii	Offshore	0	Trace
O: Pacific Canada	Coastal	9.9*	0
O: Pacific Canada	Offshore	0	0
P: South Alaska	Coastal	0.2*	20
P: South Alaska	Offshore	0	Trace
Q: North Alaska	Coastal	0	Trace
Q: North Alaska	Offshore	0	Trace
Total	Coastal	107.4	1,924.00
Total	Offshore	91.1	2,029.00
Grand Total		198.5	3,953.00

[a] Current estimates are based on data analyses conducted by Environmental Research Consulting and Greene Economics on the U.S. Coast Guard Marine Information for Safety and Law Enforcement (MISLE) database for U.S. zones, and on the Marine Safety Information System (MARSIS) data for Canadian zones. No data were publicly available for Mexican zones.

[b] Estimates that would have been classified as "trace" in the *Oil in the Sea III* study (less than 10 MT) are shown with an asterisk.

[c] "Trace" indicates that estimated input is less than 10 MT or 70 bbl/year.

NOTE: NA = not applicable; ND = not determined.

For coastal and inshore pipelines, 28% of spills are caused by corrosion, which is greater in older pipelines. For offshore pipelines, corrosion is identified as the main cause of pipeline spills 14% of the time. However, these pipeline incidents tend to involve smaller volumes as corrosion and associated leakage are generally detected during inspections. Larger pipeline spills are more likely to occur from natural forces or other outside damage, which for offshore and coastal pipelines can include hurricanes (Energo Engineering, 2010) and seismic activity, anchor dragging, dredging, or excavation.

Pipeline spill prevention measures that would address these types of incidents include regular inspections and maintenance, and clear marking of locations of pipelines on navigational and coastal maps.

3.5.4.2 Tank Vessel Spills

The category of "tank vessels" includes all the tankships (tankers) and tank barges that carry crude oil or petroleum products as cargo. These vessels also carry oil as fuel and for lubrication, as with all non-tank vessels.

Average annual spillage by geographic zone and subzone is shown in Table 3.9 in MT for the current *Oil in the Sea IV* analysis, which encompasses the years 2010–2019 and for *Oil in the Sea III*, which encompassed 1990–1999. Only spills of one bbl (0.14 MT) or more are included in the analysis. The *Oil in the Sea III* analysis excluded vessels less than 100 gross tonnage (GT). There were a total of 94 tanker spills of at least one bbl—less than 10 spills per year. (Final estimates, as shown in Table 3.9, are expressed in MT to be consistent with the *Oil in the Sea III* results.) The locations of the spills are shown in Figure 3.11.

While tankers were of the greatest concern as spill sources in the *Oil in the Sea III* study and other risk assessments at that time, the picture is considerably different now. **There was a nearly 95% reduction in the overall volume of tank vessel spillage in North American waters during 2010–2019 compared to the *Oil in the Sea III* time**

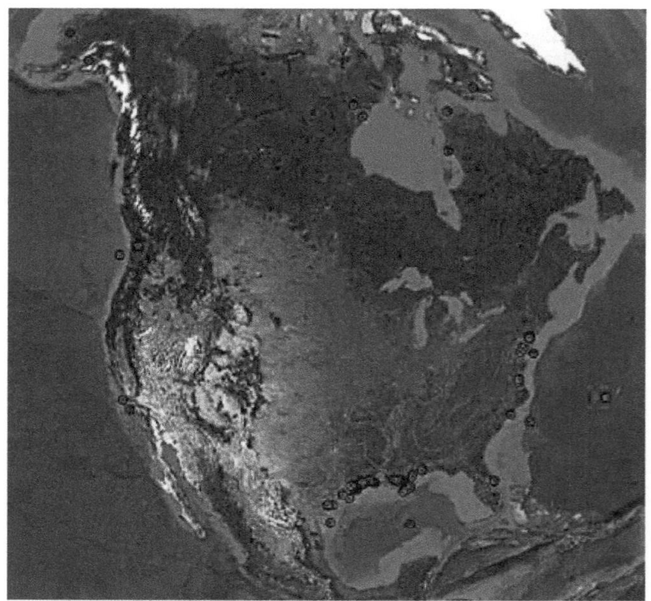

FIGURE 3.11 Locations of tank vessels spills in North America 2010–2019.
SOURCE: Courtesy of Environmental Research Consulting.

TABLE 3.10 Global Tanker Spill Trend

Time Frame	Average Annual Tanker Spills Worldwide (>7 MT)	Reduction in Numbers from Previous Decade	Average Annual Amount of Oil Spilled by Tankers (MT)	Reduction in Spill Amount from Previous Decade
1970s	76.8	—	319,500	—
1980s	45.4	41%	117,500	63%
1990s	35.8	32%	113,400	4%
2000s	18.1	49%	19,600	83%
2010s	6.3	65%	16,400	16%

SOURCE: Based on ITOPF (2021).

frame. Nearly 62% of the spills involved less than 10 bbl. Only 2% involved 1,000 bbl or more.

This reduction in tank vessel spillage mirrors the international trend. The International Tanker Owners Pollution Federation (ITOPF) reported that the numbers and total volumes of worldwide tanker spills has decreased significantly over five decades, as shown in Table 3.10. These data only include tanker spills of 7 tonnes (MT) (about 50 bbl) or more. Overall, there has been a nearly 92% reduction in tanker spill numbers since the 1970s, and an 82% reduction since the *Oil in the Sea III* time frame of the 1990s. There has been a 95% reduction in the amount of spillage since the 1970s, and an 86% reduction since the 1990s. Tanker spillage worldwide decreased even with an increase in the amount of oil transported by tanker.

Likewise, in the United States, oil spillage by tank vessels in relation to the amount of oil transported in U.S. waters had been decreasing since the late 1970s, as shown in Table 3.11. (Note that a comparison between the current data and previous time periods on the basis of ton-miles is not possible, as ton-mile data are no longer available from the U.S. Army Corps of Engineers.) A separate analysis comparing oil spillage per tonnage of oil transported for the *Oil in the Sea III* and *Oil in the Sea IV* time frames is shown in Table 3.12. There has been a 97% reduction in the amount of oil spilled from tank vessels in U.S. waters since the *Oil in the Sea III* time frame.

TABLE 3.11 Average Annual Tank Vessel Spillage in Relation to Ton-Miles Transported in U.S. Waters

Time Frame	Average Annual Spillage (MT)	Average Spillage per Billion Ton-Miles Oil Transport	Reduction from Previous Decade
1978–1987	16,177	27.4	—
1988–1997	9,297	18.22	34%
1998–2007	1,290	5.28	71%

SOURCE: Based on Etkin (2010).

TABLE 3.12 Average Annual Tank Vessel Spillage in Relation to Oil Transported in U.S. Waters

Time Frame	Average Annual Spillage (MT)	Average Oil Transport/Year (Million MT)	Average Spillage per Million MT Oil Transport
Oil in the Sea III (1990–1999)	3,861.00	313.1	12.331
Oil in the Sea IV (2010–2019)	188.2	497.1	0.379

The significant reduction in oil tanker spillage that has occurred over the past few decades can be attributed to a number of factors:

- Overall increase in accident prevention measures, inspection and maintenance processes, and safety programs, sometimes called "Standard of Care," by the oil industry in light of significant costs and damages related to major tanker spills, especially the 1989 *Exxon Valdez* tanker spill, and, in the United States, the Oil Pollution Act of 1990's more stringent requirements for spill liability (Homan and Steiner, 2008). Ships (300 GT or more) transporting oil and operating in the U.S. waters are required to have the Certificate of Financial Responsibility to ensure that the owner/operator of the vessel has the financial means to meet the requirements of the law.
- Reduction in the number of major impact accidents (groundings, collisions, and allisions) due to increased implementation of measures such as improvements and establishment of vessel traffic systems, which employ Automatic Identification System (AIS) and Geographic Information System (GIS) mapping capabilities, traffic separation schemes, compulsory pilotage, and escort tugs in busy ports (see Table 3.13);
- Improvements in safety measures taken during oil transfer and lightering procedures (Etkin, 2006);
- Certificates of Financial Responsibility, improved inspections and audits, and other measures to assure the quality of tankers entering U.S. waters (Etkin and Neel, 2001; Homan and Steiner, 2008);
- Internationally, the IMO adopted the International Convention on Oil Spill Preparedness, Response and Cooperation (OPRC) in 1990, providing a global framework to facilitate international cooperation and mutual assistance in preparing for and responding to major oil pollution incidents. The convention entered into force in 1995. Ships are required to carry a shipboard emergency plan (SOPEP), and they are required to report pollution incidents to coastal authorities (Article 3 of OPRC and MARPOL Annex I Reg. 19); and
- Requirements for the phase-out of single-hulled tankers by 2015.

TABLE 3.13 Effectiveness of Accident Mitigation Measures for Tankers

Mitigation Measure	Estimated Effectiveness	Data Source
Traffic Separation Scheme in Ports	60% reduction in collisions	DNV, 1999
Vessel Traffic Service (VTS)	84% reduction in collisions	DNV, 1999
Compulsory Pilotage	75% reduction in collisions	NRC, 1994; Young 1994, 1995
Compulsory Pilotage with VTS	88% reduction in collisions	NRC, 1994; Young 1994, 1995
Tug Escorts	varies	Gray et al., 2005

Beyond the regulatory changes, the tanker industry has initiated a number of programs that have contributed to safer oil transportation.

- ITOPF, a not-for-profit membership organization, provides services in the areas of spill response, claims analysis and damage assessment, contingency planning, training and information. Members of ITOPF are tanker owners through their P&I Clubs and other oil pollution insurers who pay the ITOPF annual dues on behalf of the members.
- The Oil Companies International Maritime Forum (OCIMF) focuses on promoting best practices in the tanker industry (ocimf.org). In 1993 OCIMF launched the Ship Inspection Report Program (SIRE) to address concerns about substandard ships. The SIRE database stores inspection reports for ships inspected in accordance with the SIRE inspection protocol. The information is used by charterers, ship and terminal operators, and government bodies to assess the quality of tankers. Oil company vetting of tankers based on the SIRE protocol has become a prerequisite for tanker eligibility in worldwide chartering.
- INTERTANKO, an industry body representing the tanker industry, promotes safe, environmentally sound, and efficient seaborne transportation of oil, gas, and chemical products. The organization's main focus areas are (1) safety and technical, (2) human element, (3) environment, (4) quality operations, and (5) commercial sustainability.[25] The organization has a non-governmental organization status in the IMO and it provides input regarding regulatory development.

Papanikolau and Eliopoulou (2008) analyzed tanker accidents for tankers larger than 60,000 deadweight tonnage (see Figure 3.12). The analysis shows a significant decrease in large tanker accidents in the post-1990 period and credits the reduction to the introduction of a series of regulatory

[25] See intertanko.com.

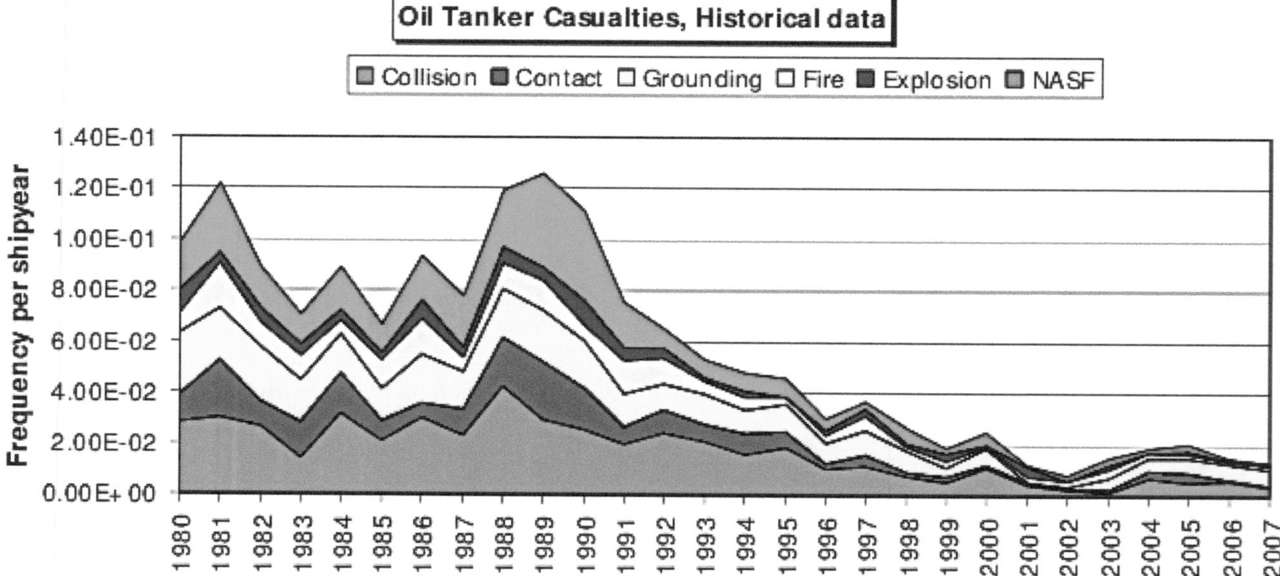

FIGURE 3.12 Historical trends in oil tanker casualties.
NOTE: Casualties include accidents in the following categories: (1) collisions of two vessels, (2) contacts of a vessel with a floating or a fixed object, (3) groundings of a vessel touching the sea bottom or shore, (4) fires as the first initiative event, (5) explosions as the first initiative event, and (6) non-accidental structural failures (NASFs) where hull cracks or fractures affect the ship's structural integrity and seaworthiness.
SOURCE: Papanikolaou and Elipoulou, 2008.

measures, changes in ship design, and overall improvement of the safety culture of the maritime industry.

Effectiveness of Double Hulls for Mitigating Vessel Spills

The significant decrease in tanker oil spills has been observed worldwide since the 1970s (ITOPF, 2021), largely attributed to the reduction in tanker spills to double hulls. While double hulls have played an important role in reducing tanker oil spills, they are not the only mitigation factor at play. Reductions in tanker spills were seen even before widespread implementation of double hulls. Overall changes in the standard of care taken by the tanker industry—as well as the effects of the large costs of oil spills, especially in the United States with the enactment of the Oil Pollution Act of 1990 (OPA 90), which set high liability limits for response and natural resource damages—also had an effect.

Double hulls on tankers were required by a 1992 amendment to MARPOL that was further amended in 2001 to accelerate the phase-in schedule so that all tankers of 5,000 deadweight tonnes and above would be outfitted no later than 2015. This requirement has had a significant effect on reducing the very large tanker spills due to groundings, collisions, and allisions. Having a double hull on the cargo tanks of a tank vessel reduces the likelihood that there will be an outflow of oil in the event of impact. Impact accidents have caused about 50% of all tanker spills over the past 50 years, and they have been implicated in 62% of the largest incidents (ITOPF, 2021). Double hulls help to reduce the likelihood of spillage (Yip et al., 2011b). The probability of having no outflow of oil in the event of an accident with a double-hull tanker was calculated to be 16% to 25% of the probability of no outflow with a single-hull tanker (Michel et al., 1996). Several NRC studies have evaluated the effectiveness of the double-hull design (NRC, 1991, 1998, 2001) and concluded that although no design is superior in all conditions the advantage of the double-hull design has been demonstrated (NRC, 2001).

Double hulls on tankers have no effect on the likelihood of spills caused by engine fires, bunkering or cargo transfer errors, mechanical defects, or operational mishaps. However, these types of spills tend to have lower volumes of release.

While the regulations addressed prevention of and preparedness for oil spills from tankers, a concern about fuel tank spills emerged as cargo vessels became larger and bunker fuel quantities grew. To mitigate the risk, a new IMO regulation entered into force in 2007 to require double hulls for fuel tanks in new ships with large oil-fuel capacity (MARPOL Annex I Reg. 12A).

Double hulls on bunker tanks reduce, but do not eliminate, the likelihood of an oil spill in the event of an impact-related accident. However, they have no effect on reducing the amount of oil released in the event of a breach of both hulls.

The degree to which double hulls on either cargo or bunker tanks are effective in preventing the breach of both hulls to release oil in a specific accident case depends on a number of factors including:

- The speed of the vessel(s) at the time of contact;
- The weight of the vessel(s), which determines the forces involved;
- The angle and relative height of the point of contact; and
- The condition of the vessel(s).

Overall, double hulls on tankers are about 90% effective in reducing oil spill incidence in groundings and 75% in collisions (Keith, 1993). Statistical and modeling studies conducted on the effectiveness of double hulls in reducing oil spillage have shown that for a double-hull tanker, the probability of an oil spill in an impact-related accident is still about 0.15 (i.e., a 15% chance of a spill with each accident). The likelihood of a tanker spill in which at least half of the oil in the cargo tanks is released is about 1 in 67,000 per accident (based on NRC, 1998, 2001; Rawson et al., 1998; Etkin and Neel, 2001; Etkin and Michel, 2003; and Yip et al., 2011b).

For bunker tank spills, the likelihood of a spill is reduced by 60% with double hulls. The likelihood of the outflow of the entire bunker fuel capacity is about 1 in 1,250 per accident (Michel and Winslow, 2000; Etkin and Michel, 2003; Herbert Engineering and Designers & Planners Inc., 2003; Barone et al., 2007). Unlike double hulls for cargo tanks, double-hull bunker tanks are not yet universal. In 2020, about 50% of the world's vessel bunker tanks are double-hulled.

Given the degree to which vessel tonnage and maneuverability may affect the likelihood of accidents and hull breaches, it is important to consider that the overall size of tankers and cargo ships, particularly container ships, is increasing.

The chances of a large tanker spill may be escalating despite the full implementation of double hulls and other prevention measures because of the increased capacity of tankers (Statistica, 2021), as well as the increasing size of other cargo vessels, especially container ships, that may collide with them in busy ports and shipping lanes. Larger vessels are less maneuverable and have a greater impact due to the force of their tonnage in the event of a collision.

While the probability of major tank vessel spills has been reduced significantly in the past 20 years, there is still a possibility that such a spill might occur. Accident prevention measures and oil outflow mitigation from double hulls do not entirely eliminate the risk of spills from tankers. Although not presently a concern in the North American waters, piracy and terrorist attacks on tankers in some areas of the world create dangerous situations for the crews and generate a risk of oil spills. As long as oil is transported by tankers and tank barges, there is a possibility of a major spill, and a continuing need to remain prepared and vigilant.

Prevention of Lightering and Oil Transfer Spills

Oil spills may occur during operations in which oil is transferred from one vessel to another or between a vessel and a facility through loading arms and hoses. Oil transfers routinely occur when:

- Oil is transferred or lightered from a larger tank vessel to a smaller one (e.g., when a portion of the oil cargo in a sea-going tanker is lightered to a smaller tanker or tank barge that will deliver the oil to a coastal or inland river facility;
- Oil is transferred from a tank vessel to a facility during offloading operations (e.g., a delivery of crude oil to a storage terminal or to a refinery);
- Oil is transferred from a facility to a tank vessel during loading operations (e.g., from a refinery to a product tanker that will deliver the refined product to another facility); and
- Oil is transferred during bunkering or fueling operations from a storage tank at a facility or between a bunkering barge and a vessel.

In the United States, spills occur less than about once in 2,500 transfer operations (a probability of 0.0004). With the implementation of strict standards to reduce spillage during transfer operations, the spillage rate may be reduced to about a 0.00026 probability of a spill during every transfer operation, or one spill every 3,850 transfer operations. Between 1985 and 2004, there was a 96% reduction in the number of transfer-related spills in the United States, but this rate could be reduced further (Etkin, 2006).

One of the approaches to mitigating spills from transfer operations is to pre-boom the vessel: that is to place an oil spill containment boom around the vessel so that any spilled oil is contained within the boomed area. With oil spill removal equipment (pumps and skimmers) on standby, a relatively high percentage of the oil may be removed from the water surface before it can spread. This does not prevent the spill as safety protocols, monitoring, and alarm systems employed during the operation do, but it may reduce the impacts of a spill. Pre-booming is only effective when the velocity of the current does not exceed about 0.7 knots. Higher currents will allow some of the oil to begin entraining (going under) the boom (Etkin et al., 2007).

The spills that occur during lightering and oil transfer operations are generally quite small. About 72% of the spills amount to less than one barrel (bbl); 90% are less than 10 bbl.

3.5.4.3 Non-Tank Vessel Spills

The category of "non-tank vessels" includes all the ships and boats that use oil only as fuel and for lubrication but excludes those that carry crude oil or refined petroleum products as cargo. Spills from tank vessels that carry oil as cargo are included under Transportation (see 3.4.2). In this study, the non-tank vessels are divided into two size subcategories: less than 100 gross tons and 100 GT and over. The larger category includes cargo ships, industrial vessels, passenger vessels, and larger fishing vessels. The smaller category includes smaller fishing boats, recreational boats, and miscellaneous smaller vessels.

Average annual spillage by geographic zone and subzone is shown in Table 3.14 in MT for the current *Oil in the Sea IV* analysis, which encompasses the years 2010–2019 and for *Oil in the Sea III*, which encompassed 1990–1999. Only spills of one barrel (bbl) (0.14 MT) or more are included in the analysis. The *Oil in the Sea III* analysis excluded vessels less than 100 GT. From 2010 to 2019, there was a total of 1,010 spills of at least one bbl—or about 100 spills per year, 32 from larger non-tank vessels and 68 from vessels of less than 100 GT. (Final estimates of volume of oil spilled, as shown in Table 3.14, are expressed in MT to be consistent with the *Oil in the Sea III* results.)

Considering only spills from non-tank vessels of 100 GT and over, there was a 70% reduction in spill volume for the time period 2010–2019 compared with the 1990–1999 time period covered in the *Oil in the Sea III* study. That study did not include analyses of smaller vessels, which were found in the current study to contribute about as much oil pollution as larger vessels. It is reasonable to assume that there were likely to have been similar contributions to the overall non-tank vessel pollution from smaller vessels 20 years ago.

Another notable trend for non-tank vessels is a shift from offshore oil spillage to a greater proportion of spillage occurring in coastal and nearshore waters. Nearly 66% of the spillage during 2010–2019 occurred in coastal and nearshore waters compared to 45% during 1990–1999.

Spill volumes from non-tank vessels tended to be low during the past decade, as was the case in previous time periods. For larger (≥ 100 GT) non-tank vessels, 67% of the spills were of less than 10 bbl. (Note that spills of less than one bbl were not counted in this analysis.) Only 0.6% of the spills were of 1,000 bbl or more.

The reduction in spillage from non-tank vessels may reflect the overall increase in diligence of vessel operators to reduce spills and measures to reduce accidents, particularly for larger cargo ships, through vessel traffic services and AIS in busy ports.

Although the volumes of oil contained in cargo ships are generally smaller than the volumes of cargo oil carried by large tankers, spills from these vessels can cause significant damage. In 2007, the container ship *Cosco Busan* allided with a support tower on a bridge in San Francisco Bay, California, spilling 1,262 bbl (180 MT) of heavy fuel oil into the bay. In an allision, the moving vessel strikes a stationary

TABLE 3.14 Estimated Annual Spillage from Non-Tank Vessels[a]

Geographic Zone	Subzone	Average Annual Spillage 2010–2019 (MT/year)[b] Non-Tank Vessels <100 GT	Non-Tank Vessels ≥100 GT	*Oil in the Sea III* Estimate for 1990–1999 for Non-Tank Vessels > 100 GT[c] (MT/year)[d]
A: NW Canada	Coastal	0	0	NA
A: NW Canada	Offshore	0	0	NA
B: N Canada	Coastal	0	0	NA
B: N Canada	Offshore	0	0	Trace
C: E Canada	Coastal	0.4*	0	Trace
C: E Canada	Offshore	0.7*	0	Trace
D: N Atlantic	Coastal	10.2	6.7*	88
D: N Atlantic	Offshore	1.3*	2.4*	Trace
E: Mid-Atlantic	Coastal	5.5*	8.5*	22
E: Mid-Atlantic	Offshore	0.8*	0.3*	40
F: E GOM	Coastal	2.1*	12.7	30
F: E GOM	Offshore	0.3*	1.2*	70
G: W GOM	Coastal	21.3	28.4	100
G: W GOM	Offshore	24.9	18.3	120
H: Mexican GOM	Coastal	0	0	Trace
H: Mexican GOM	Offshore	0	0	Trace
I: Puerto Rico	Coastal	0.5*	0.8*	Trace
I: Puerto Rico	Offshore	0.1*	0.2*	10
J: Mexico Pacific	Coastal	0	0	ND
J: Mexico Pacific	Offshore	0	0	Trace
K: CA Pacific	Coastal	3.7*	1.3*	28
K: CA Pacific	Offshore	0.5*	0	Trace
L: Mid-Pacific	Coastal	2.1*	0.1*	Trace
L: Mid-Pacific	Offshore	0	0.1*	16
M: NW Pacific	Coastal	6.8*	5.9*	35
M: NW Pacific	Offshore	0.8*	0	13
N: Hawaii	Coastal	38.3	19	26
N: Hawaii	Offshore	2.6*	0.1*	87
O: Pacific Canada	Coastal	24.9	21.4	Trace
O: Pacific Canada	Offshore	10	0	52
P: South Alaska	Coastal	6.4*	9.9*	70
P: South Alaska	Offshore	0.2*	4.4*	30
Q: North Alaska	Coastal	0	4.6*	Trace
Q: North Alaska	Offshore	0	35.1	50
Total	Coastal	122.1	119.3	399
Total	Offshore	42.1	62.2	488
Grand Total		164.2	181.6	887

[a] Current estimates are based on data analyses conducted by Environmental Research Consulting and Greene Economics on the U.S. Coast Guard Marine Information for Safety and Law Enforcement (MISLE) database for U.S. zones, and on the Marine Safety Information System (MARSIS) data for Canadian zones. No data were publicly available for Mexican zones.
[b] Estimates that would have been classified as "trace" in the *Oil in the Sea III* study (less than 10 MT) are shown with an asterisk.
[c] Gross tons (GT).
[d] "Trace" indicates that estimated input is less than 10 MT or 70 bbl/year.
NOTE: NA = not applicable; ND = not determined.

object, such as the bridge tower. In a collision, the moving vessel strikes another moving vessel. (This incident is not included in the data analysis for *Oil in the Sea IV* because it occurred prior to the 2010–2019 time frame selected as representative of the current situation.)

The volumes of bunker fuel carried by large cargo ships can be considerable. For example, the container ship *Benjamin Franklin* that visits the Port of Seattle, with a capacity of 18,000 20-foot equivalent units (TEUs), contains more than 107,000 bbl of bunker fuel.[26] This volume is nearly half the amount of oil spilled from the tanker *Exxon Valdez* in Prince William Sound in 1989. The potential for a major spill of fuel oil from a large cargo ship continues to be a concern.

One significant improvement since the *Oil in the Sea III* time period is the introduction of double hulls on

[26] See https://response.restoration.noaa.gov/about/media/how-much-oil-ship.html.

TABLE 3.15 Cargo Ship Oil Spillage per Dry Cargo Shipments in U.S. Waters

Years	Dry Cargo Shipment (million short tons)	Annual MT Oil Spilled	MT Oil Spilled per Million Short Tons Shipped[a]	Source
1978–1987	1,057	969	0.9	Etkin, 2010
1988–1997	1,256	402	0.32	Etkin, 2010
1998–2007	1,382	229	0.16	Etkin, 2010
2010–2019	2,340	160	0.07	Current *Oil in the Sea IV* data

[a] Shipment data from U.S. Army Corps of Engineers.

bunker tanks. Similar to double hulls on cargo tanks as present on tank vessels, double hulls on bunker tanks reduce the likelihood of an oil release in the event of an impact accident—a collision, grounding, or allision (such as occurred with the *Cosco Busan*). With the inclusion of a double hull on bunker tanks, the probability of a release of oil due to an impact accident decreases by 60% from 0.05 per accident to 0.02 per accident (Michel and Winslow, 1999, 2000; Etkin and Michel, 2003; Herbert Engineering and Designers & Planners Inc., 2003; Barone et al., 2007). Double hulls on bunker tanks do not provide any additional protection against other types of incidents, such as operational or bunker fuel transfer errors. Currently, approximately 50% of cargo ships have double-hull bunker tanks.

The trend in the reduction of oil spillage from non-tank vessels in U.S. waters was noted even prior to the time period of *Oil in the Sea III*. Even with a 30% increase in dry cargo shipments, there was a 76% reduction in bunker spills from cargo ships: that is, an 82% reduction in the oil spilled per cargo shipped between the decades of 1978–1987 and 1998–2007, and a 50% reduction between 1987–1997 and 1998–2007 (Etkin, 2010). **The rate of spillage in U.S. waters during 2010–2019 showed an even greater reduction, 92% since 1978–1987 and 56% since 1998–2007** (see Table 3.15).

3.5.4.4 Transportation by Rail

Although rail tank cars have been used to transport fuel and petrochemicals for many years in most parts of the world with extensive freight railroad systems, the use of trains to transport large quantities of crude oil is a relatively new phenomenon in the past decade and one that is, at least currently, limited to the United States and Canada. In both the United States and Canada, rapidly increasing inland shale oil production in the early 2010s exceeded the capacity and availability of pipelines for transport to refineries and terminals. Key trains and unit trains of 20 to 120 tank cars carrying crude oil were utilized as "moving pipelines" of "crude-by-rail." Movement by rail made it easier to change routes utilizing existing rail lines and oil could be efficiently loaded at production facilities and unloaded at refineries and terminals.

In 2005, about 17 tank cars of oil per day (each one holding about 714 bbl) were moving on U.S. railroads. By 2015, more than 1,400 tank cars per day were on the railroads. After a peak in about 2017, crude-by-rail movements began to decrease. In 2019 and 2020, there was some increase in crude-by rail traffic (see Figure 3.13). In some cases during the past decade, there have been temporary switches to the use of rail when there have been pipeline or refinery closures. For example, during Hurricane Harvey in 2017, when Texas refineries were temporarily

FIGURE 3.13 Crude-by-rail traffic in the United States and Canada.
SOURCE: EIA (2021b).

TABLE 3.16 Estimated Annual Spillage from Coastal Oil Storage Facilities[a]

Geographic Zone	Subzone	Average Annual Spillage from Coastal Oil Storage Facilities 2010–2019 (MT/year)[b]	Average Annual Spillage from Coastal Oil Storage Facilities *Oil in the Sea III* Estimate for 1990–1999 (MT/year)[c]
C: E Canada	Coastal	0	Trace
D: N Atlantic	Coastal	14.6	500
E: Mid-Atlantic	Coastal	0.2*	79
F: E GOM	Coastal	174.3	10
G: W GOM	Coastal	14.3	740
I: Puerto Rico	Coastal	0	130
J: Mexico Pacific	Coastal	ND	Trace
K: CA Pacific	Coastal	7.4*	62
L: Mid-Pacific	Coastal	0	26
L: Mid-Pacific	Offshore	0	12
M: NW Pacific	Coastal	0.3*	10
N: Hawaii	Coastal	0.1*	31
P: South Alaska	Coastal	16.5	30
Q: North Alaska	Coastal	0	10
Total	Coastal	216.4	1,697.00
Total	Offshore	0	12
Grand Total		216.4	1,709.00

[a] Current estimates are based on data analyses conducted by Environmental Research Consulting and Greene Economics on the U.S. Coast Guard Marine Information for Safety and Law Enforcement (MISLE) database for U.S. zones, and on the Marine Safety Information System (MARSIS) data for Canadian zones. No data were publicly available for Mexican zones.
[b] Estimates that would have been classified as "trace" in the *Oil in the Sea III* study are shown with an asterisk.
[c] "Trace" indicates that estimated input is less than 10 MT or 70 bbl/year.
NOTE: NA = not applicable; ND = not determined.

shut down, increased operations at Northeast U.S. refineries required eight trainloads of oil to be brought to Philadelphia via a route that went along the Hudson River in New York.

The greatest concern about the crude-by-rail trains was the possibility of an accident that could cause an explosion, such as occurred in 2013 in Lac-Megantic, Quebec, where there were 47 fatalities. There were a number of other accidents in the United States and Canada, which created a significant public outcry and spurred changes in regulations to increase safety (Etkin et al., 2015; Etkin, 2017b).

With respect to the current *Oil in the Sea IV* study, there are no reports of spillage to marine and estuarine waters that should be considered in estimating rail transportation-related oil inputs. However, there is the possibility of future spills that may affect these waters.

3.5.4.5 Coastal Storage Facilities

Coastal storage facilities included under the Transportation category of the *Oil in the Sea III* and *Oil in the Sea IV* studies include large oil terminals at which oil transported by tank vessel or pipeline is loaded and offloaded and stored. These facilities do not include smaller consumer-oriented fuel storage facilities (local home heating oil storage or marinas) or facilities involved in the consumption of oil (e.g., power plants and manufacturing facilities).

Estimates of oil inputs from this source are summarized in Table 3.16. **For the 2010–2019 time frame, 56 MT (395 bbl) of spillage occurred annually. This represents an 87% reduction in spillage since the 1990s and *Oil in the Sea III.***

With the storage of large quantities of oil at coastal terminals (individual storage tanks may contain as much as 300,000 bbl of oil), there is a possibility of a significant facility spill in the future. Spills that occur from storage tanks at coastal terminals will usually be contained within required secondary containment. However, there are circumstances when this containment, which is designed to hold more than the volume of the tanks, may be breached, causing some or all of the spilled oil to enter a waterway. This may occur during heavy flooding conditions and during hurricanes or other storms. There is also the possibility that flood waters may swamp the containment area, in which case secondary containment is generally ineffective and oil may enter marine waters.

In the United States, Hurricanes Katrina and Rita in 2005 caused significant damage to coastal terminals in Louisiana and Texas and the release of nearly 132,000 bbl (19,000 MT) of oil. One 12,000-bbl tank was lifted and moved 3 miles in a Texas marsh by the tidal surge during Hurricane Rita (USCG, 2006).

In 2017, Hurricane Harvey caused damage to more than two dozen aboveground storage tanks in Texas, spilling more than 24,000 bbl of oil and chemicals. The largest spillage came from two storage tanks that released 12,000 bbl (1,714 MT) of gasoline. This event, in addition to the incidents that occurred during Hurricanes Katrina and Rita, clearly exposed the vulnerability of aboveground storage tanks to damage from severe storms and flooding (Bernier et al., 2017).

With the potential for increasing intensity and frequency of hurricanes due to climate change and sea level rise (Webster et al., 2005; Elsner et al., 2008; Bender et al., 2010; Knutson et al., 2010; Kang and Elsner, 2015), **the risk for major spills from coastal terminals may be increasing.**

3.5.4.6 Coastal Refineries

Average annual spillage by geographic zone and subzone for refineries is shown in Table 3.17 in MT for the current *Oil in the Sea IV* analysis, which encompasses the years 2010–2019. *Oil in the Sea III* did not include coastal refineries as a separate category. Only spills of one barrel or more are included in the analysis. **The average annual amount spilled from coastal refineries was less than 11 MT (75 bbl).** However, because refineries often contain aboveground storage tanks, they may be subject to similar damage during earthquakes, hurricanes, and flooding.

TABLE 3.17 Estimated Annual Spillage from Coastal Refineries[a]

Geographic Zone	Subzone	Average Annual Spillage from Coastal Refineries 2010–2019 (MT/year)[b]	*Oil in the Sea III* Estimate for 1990–1999 (MT/year)[c]
D: N Atlantic	Coastal	0.3*	ND
F: E GOM	Coastal	0.1*	ND
G: W GOM	Coastal	10.4	ND
Total	Coastal	10.7	ND
Total	Offshore	0	ND
Grand Total		10.7	ND

[a] Current estimates are based on data analyses conducted by Environmental Research Consulting and Greene Economics on the U.S. Coast Guard Marine Information for Safety and Law Enforcement (MISLE) database for U.S. zones, and on the Marine Safety Information System (MARSIS) data for Canadian zones. No data were publicly available for Mexican zones.

[b] Estimates that would have been classified as "trace" in the *Oil in the Sea III* study (less than 10 MT) are shown with an asterisk.

[c] "Trace" indicates that estimated input is less than 10 MT or 70 bbl/year.

NOTE: NA = not applicable; ND = not determined.

Combined Spillage from Transportation of Oil

Oil spillage from transportation-related sources has been reduced significantly in the past 20 years (see Figure 3.14 and Table 3.18). The greatest reduction has been in tank vessel spills. Notably, tank vessel spillage was about one-half of the spillage from coastal and offshore pipelines. A summary of transportation inputs by geographic zone and subzone is shown in Table 3.19.

3.5.5 Potentially Polluting Sunken Wrecks

The 2003 *Oil in the Sea III* study (NRC, 2003) mentioned a few historic shipwrecks still containing oil as ongoing and potential future sources of pollution, but did not provide details regarding their location or the quantity of oil that could be released. The study briefly mentioned the particular risk of chronic or potentially significant oil releases from the thousands of World War II-era wrecks of tankers (see Box 3.5), merchant ships, and military vessels due to their large oil contents and 60 years of corrosion at the time of the study. The study cited the example of the *USS Mississinewa*, an oil tanker sunk in 1944 in Ulithi Lagoon, Caroline Islands, as leaking heavy fuel oil after a storm in 2001 (Gilbert, 2001; Gilbert et al., 2003).

In the two decades since *Oil in the Sea III*, there has been significant progress in documenting the larger wrecks and in developing programs for conducting risk assessments for future oil pollution potential. At the 2005 International Oil Spill Conference, an issue paper and panel discussion of experts estimated that **worldwide there are about 8,600 wrecks of tankers and other large vessels (at least 400 gross tons) that in total may still contain as much as 2.5 to 20.4 million tonnes (17.5 million to 143 million bbl) of oil.** Three-quarters of the identified large wrecks are World War II–related (Michel et al., 2005), making them about 80 years old at present.

There have also been non-war wrecks that have caused environmental damage through leakage, notably the S.S. *Jacob Luckenbach* that sank off central California in 1953 (see Box 3.6). After this wreck was reported to be causing continuing oiling of beaches and birds, an oil removal operation was conducted (Hampton et al., 2003; Moffat

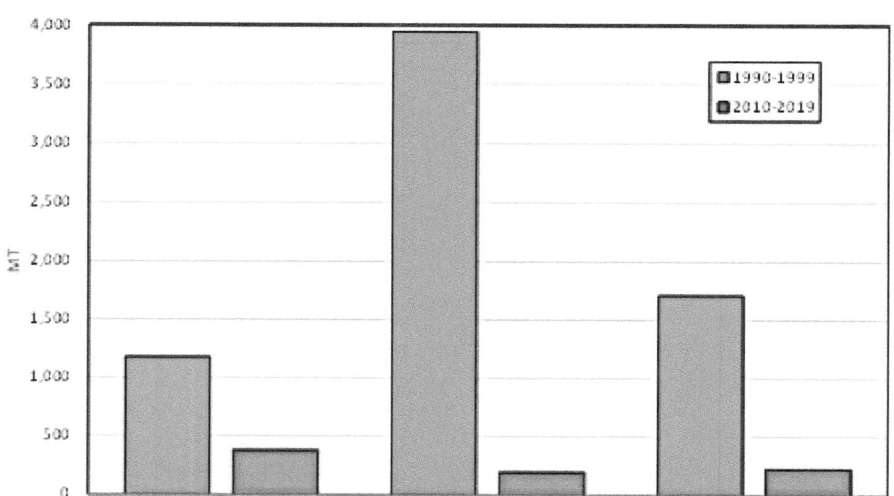

FIGURE 3.14 Average annual oil spillage from transportation (*Oil in the Sea III* versus *Oil in the Sea IV*).

TABLE 3.18 Comparison of *Oil in the Sea III* and *Oil in the Sea IV* for Oil Transportation Spillage

	Oil in the Sea IV (2010–2019)		*Oil in the Sea III* (1990–1999)		
Source	Annual Spillage (MT)	% Total	Annual Spillage (MT)	% Total	Reduction Between *Oil in the Sea III* and *Oil in the Sea IV*
Pipelines	380	47.20%	1,175	17.20%	67.60%
Tank Vessels	199	24.60%	3,941	57.70%	95.00%
Coastal Terminals	227	28.20%	1,709	25.00%	86.70%
Total	806	100.00%	6,825	100.00%	88.20%

TABLE 3.19 Estimated Annual Spillage from Oil Transportation[a]

		Average Annual Spillage for 2010–2019 (MT/year)[b]					*Oil in the Sea III* Estimate for 1990–1999 (MT/year)[c]				
Geographic Zone	Subzone	Pipelines	Tank Vessels	Coastal Terminals	Refineries	Total	Pipelines	Tank Vessels	Coastal Terminals	Refineries	Total[d]
A: NW Canada	Coastal	0	0	0	0	0	NA	NA	NA	ND	0
A: NW Canada	Offshore	0	0	0	0	0	NA	NA	NA	ND	0
B: N Canada	Coastal	0	0	0	0	0	NA	NA	NA	ND	0
B: N Canada	Offshore	0	0	0	0	0	NA	NA	NA	ND	0
C: E Canada	Coastal	0	0.2*	0	0	0.2*	0	0	Trace	ND	0
C: E Canada	Offshore	0	0	0	0	0	NA	NA	0	ND	0
D: N Atlantic	Coastal	24	9.5*	14.6	0.3*	48.4	150	740	500	ND	1,390.00
D: N Atlantic	Offshore	0	0	0	0	0	NA	17	0	ND	17
E: Mid-Atlantic	Coastal	3.7*	0.2*	0.2*	0	4.1*	36	14	79	ND	129
E: Mid-Atlantic	Offshore	0	0	0	0	0	NA	Trace	0	ND	0
F: E GOM	Coastal	11.4	2.3*	174.3	0.1	188.1	Trace	140	10	ND	150
F: E GOM	Offshore	0	0	0	0	0	Trace	10	0	ND	10
G: W GOM	Coastal	296.1	84.8	14.3	10.4	405.6	890	770	740	ND	2,400.00
G: W GOM	Offshore	2.1*	91	0	0	93.1	60	1,500.00	0	ND	1,560.00
H: Mexican GOM	Coastal	0	ND	ND	0	0	Trace	80	ND	ND	80
H: Mexican GOM	Offshore	0	ND	ND	0	0	ND	ND	0	ND	0
I: Puerto Rico	Coastal	0	0	0	0	0	Trace	Trace	130	ND	130
I: Puerto Rico	Offshore	0	0	0	0	0	NA	490	0	ND	490
J: Mexico Pacific	Coastal	0	ND	ND	0	0	ND	ND	Trace	ND	0
J: Mexico Pacific	Offshore	0	ND	ND	0	0	ND	Trace	0	ND	0
K: CA Pacific	Coastal	42	0	7.4*	0	49.4	39	150	62	ND	251
K: CA Pacific	Offshore	0	0	0	0	0	Trace	0	0	ND	0
L: Mid-Pacific	Coastal	0	0	0	0	0	Trace	Trace	26	ND	26
L: Mid-Pacific	Offshore	0	0	0	0	0	0	Trace	12	ND	12
M: NW Pacific U.S.	Coastal	0	0.2*	0.3*	0	0.5*	Trace	10	10	ND	20
M: NW Pacific U.S.	Offshore	0	0	0	0	0	0	0	0	ND	0
N: Hawaii	Coastal	1.0*	0	0.1*	0	1.1*	Trace	Trace	31	ND	31
N: Hawaii	Offshore	0	0	0	0	0	0	Trace	0	ND	0
O: Pacific Canada	Coastal	0	9.9*	0	0	9.9*	0	0	69	ND	69
O: Pacific Canada	Offshore	0	0	0	0	0	0	0	0	ND	0
P: South Alaska	Coastal	0	0.2*	16.5	0	16.7	Trace	20	30	ND	50
P: South Alaska	Offshore	0	0	0	0	0	NA	Trace	0	ND	0
Q: North Alaska	Coastal	0	0	0	0	0	NA	Trace	10	ND	10
Q: North Alaska	Offshore	0	0	0	0	0	NA	Trace	0	ND	0
Total	Coastal	378.3	107.4	216.4	10.7	712.8	1,115.00	1,924.00	1,697.00	ND	4,736.00
Total	Offshore	2.1*	91.1	0	0	93.2	60	2,017.00	12	ND	2,089.00
Grand Total		380.4	198.5	216.4	10.7	806	1,175.00	3,941.00	1,709.00	ND	6,825.00

[a] Current estimates are based on data analyses conducted by Environmental Research Consulting and Greene Economics on the U.S. Coast Guard Marine Information for Safety and Law Enforcement (MISLE) database for U.S. zones, and on the Marine Safety Information System (MARSIS) data for Canadian zones. No data were publicly available for Mexican zones.

[b] Estimates that would have been classified as "trace" in the *Oil in the Sea III* study are shown with an asterisk.

[c] "Trace" indicates that estimated input is less than 10 MT or 70 bbl/year.

[d] Totals assume that "trace" values equal zero.

NOTE: NA = not applicable; ND = not determined.

BOX 3.5
Oil Wars

Oil, as a valuable natural resource and because of its role in society, has been the impetus for aggression and war to enable countries to prosper and grow. Between 1912 and 2010, countries fought 180 times over territories that contained, or were believed to contain, oil or natural gas resources.[a] In 1918, during World War I, the British dealt Germany a decisive blow by stopping Germany from capturing the Baku oil fields in Azerbaijan. The British had plans to destroy the Baku facility if necessary. In World War II Japan's and Germany's aggression was fueled by the pursuit of areas with oil production, so as to decrease their reliance on external entities and to expand their idealisms. The fuel oil produced on the east coast of the United States was able to supply crucial petroleum needs to Europe and Africa (Yergin, 2008). This was one of the major logistical advantages the Allies were able to maintain over the Axis powers.

The same situation spurred Iraq's military aggression in the Middle East. Starting in January of 1991, Iraqui forces strategically released large quantities of oil into the Persian Gulf in order to impede U.S. troops from attempting beach landings. This use of a natural resource—and specifically pollution—as a defense tactic resulted in more than 240 million gallons of crude oil being intentionally released into the Persian Gulf from tankers, pipelines, and storage facilities. The effects of the oil spill on marine life and on the shores was extensive and the region has not recovered. Loss of biodiversity and uninhabitable coastlines persist.[b]

Terrorist attacks on pipelines (as have occurred intermittently in Colombia and Nigeria for some time) are mainly inland spill inputs, not marine or estuarine. However, in February 2021, the source of an oil spill off the coast of Israel was suspected to be from a tanker that spilled oil smuggled from Iran into Israeli waters, a possible incidence of "environmental terrorism."

Although no specific conflict-related oil spills were reported in North American waters within the 20-year period covered by this report, the possibility cannot be ruled out in the future. Instead, in North American waters the major conflict-related oil sources include the thousands of shipwrecks resulting from sea battles during World Wars I and II that released oil in the past or could in the future, as discussed and quantified in Section 3.5.5.

Wars described as "oil wars" include Vietnam War, Iran–Iraq War, Gulf War, Iraqi no-fly zones conflicts, and the Iraq War.

[a] See https://www.lawfareblog.com/exaggerated-threat-oil-wars.

[b] See http://www.environmentandsociety.org/tools/keywords/gulf-war-oil-spill-man-made-disaster.

et al., 2003; Luckenbach Trustee Council, 2006; Henkel et al., 2014). Another example is the wreck of the U.S. Army Transport ship *Brigadier General M.G. Zalinski* that grounded and sank in the Grenville Channel, British Columbia, in 1946. This wreck presented a continuing environmental threat until the remaining fuel oil was removed in 2014 (Elliott and DeVilbiss, 2014).

An important outcome of the 2005 study was the recognition that, in addition to the sheer magnitude of the wreck problem, there were many significant legal and financial obstacles to conducting proactive oil removal projects even for the most threatening wrecks. The financial and legal liability and responsibility of the vessel owners was a particular issue of concern, especially for the war-related wrecks. The need to evaluate the benefits of removing the oil from a wreck in relation to the intervention costs was identified (Girin, 2004).

In addition to the costs involved, and the potential lack of funding to conduct proactive oil removal operations, the logistical and safety challenges of wreck oil removal operations can be formidable (Findlay, 2003). However, in the past decade, salvage masters and crews have conducted increasingly complex operations on various wrecks. There are also a number of significant technological developments that have improved safety and efficiency of these operations, such as the use of remotely operated vehicles (ROVs), remote sensing, saturated diving systems, and hot-tapping (Elliott and DeVilbiss, 2014).

The application of systematic risk assessments to prioritize the wrecks with respect to the likelihood of leakage and degree of potential environmental impact is another significant development in recent years (Etkin, 2019). Advances in monitoring of sunken wrecks through remote sensing technologies, as well as models to simulate hypothetical releases (French-McCay et al., 2012) and models to determine more accurately the probability of discharges from particular vessels based on vessel architecture and condition, disturbance potential, and environmental factors (Landquist et al., 2013, 2014, 2017) have greatly improved the capabilities for desktop risk assessments.

The most comprehensive wreck risk assessment conducted to date is the NOAA Remediation of Underwater Legacy Environmental Threats (RULET) project (Symons et al., 2013). Ultimately, the study team identified 87 wrecks for more detailed analysis with respect to wreck condition and the potential for impacts to ecological and socioeconomic resources in the water column, on the water surface, and on the shoreline based on thresholds of concern for oiling (French-McCay, 2016). The overall analysis and prioritization of the wrecks with respect to risk provides the USCG with specific criteria that may be applied on a district basis to access the nation's Oil Spill Liability Trust Fund (OSLTF) Emergency Fund for monitoring or removal operations. The OSLTF funds are derived from a tax on imported and domestic oil. These funds are available to finance spill response operations. Typically, an identified responsible party (spiller) would have to reimburse the OSLTF, but funds may also be used

BOX 3.6
S.S. *Jacob Luckenbach*

FIGURE 3.15 Left: S.S. *Jacob Luckenbach* (Image credit: NOAA). Right: Oiled common murre (Image credit: California Department of Fish and Wildlife, Steve Hampton).

On July 14, 1953, the S.S. *Jacob Luckenbach* (see Figure 3.15), bound for South Korea and loaded with railroad parts and 10,880 bbl of bunker fuel, collided with its sister ship, the *Hawaiian Pilot*, and sank in 180 feet of water, 17 miles west-southwest of San Francisco and just southeast of the Farallon Islands, a National Wildlife Reserve and seabird nesting colony—specifically, of the common murre, a relative of the puffin, which is the most common bird on the Farallon Islands.[a]

Starting as early as 1973, the California coastline experienced a series of spills of undetermined source. These episodic events were most common in winter months and associated with strong winter storms. The spills were most notable due to hundreds of oiled birds—mainly the common murre (see Figure 3.15)—that regularly washed up on the California coastline every winter. For years, these events were blamed on chronic oil of unknown origin such as illegal discharges (Hampton et al., 2003).

Following the San Mateo mystery spill in 2001, which led to months of oiled birds washing ashore and 14 months of continual active rescue and rehabilitation efforts for oiled birds by California's Oiled Wildlife Care Network, a federal response was launched to determine the source of the oil. Oil fingerprinting was used to determine that the oil on the bird feathers collected during these episodic events was originating from the same source. Synthetic Aperture Radar was used to look for surface slicks and narrow down the search to several sunken vessels off of the San Francisco coast. Consultation with wildlife experts was used to understand the most likely origin of where murre oiling was occurring. Finally, divers were able to collect samples from the *Luckenbach*, confirming the source of these winter storm mystery spills (Hampton et al., 2003).

Further investigation tied the episodic winter return of these events to significant wave height; starting in 1996, any time the significant wave height exceeded 7 m, an oiling event occurred. Though little is known of the bottom currents in this area, anecdotal evidence supports a correlation between the large swells and intensification of bottom currents near the vessel (Hampton et al., 2003).

Clean up operations started in the summer of 2002; dive teams were able to salvage approximately 2,380 bbl of bunker fuel, and a remaining 690 bbl of oil were sealed in the wreckage. No responsible party was identified for the *Luckenbach* spill. The National Pollution Funds Center funded the cleanup, response, and restoration efforts through use of the Oil Spill Liability Trust Fund (Hampton et al., 2003).

Over an estimated 30 years, approximately 7,810 bbl of fuel were released into California's coastal waters, killing more than 50,000 seabirds. Restoration projects totaling $23M are still under way from the Kokechik Flats of Alaska to the Baja peninsula.[a]

[a]See https://wildlife.ca.gov/OSPR/NRDA/Jacob-Luckenbach.

if there is no identified responsible party, or the spiller is unable to pay. In the case of a potentially polluting wreck where there is no actual spill at the time, the USCG needs to establish that there is a "substantial and imminent threat" of an oil spill in order to be able to procure funds from the OSLTF to perform any monitoring, surveying, or oil removal operations.

For the current *Oil in the Sea IV* study, documented wrecks of tankers and larger non-tank vessels of 400 gross tons and larger in the Environmental Research Consulting databases (as used in the 2005 study) and the RULET study results for North America and the Caribbean were evaluated to estimate the potential volumes and types of oil present by geographic zone. Note that these wrecks generally only present a potential for leakage in the future, although some episodic releases are occurring presently as some wrecks are currently releasing oil on occasion. The analytical results are shown in Table 3.20 in MT, and in the map in Figure 3.16.

TABLE 3.20 Estimated Potential Oil Content in Sunken Wrecks by Geographic Zone

		Crude Oil (MT)		Heavy Oil (MT)		Light Oil (MT)		All Oils (MT)	
Zone[a]	Subzone	Max	Min	Max	Min	Max	Min	Max	Min
B: N Canada	Offshore	0	0	535	54	0	0	535	54
C: E Canada	Coastal	0	0	1,453	145	0	0	1,453	145
C: E Canada	Offshore	19,491	1,949	138,105	13,811	64,936	6,494	222,533	22,253
D: N Atlantic	Coastal	0	0	0	0	1,714	171	1,714	171
D: N Atlantic	Offshore	11,134	1,113	46,140	4,614	59,165	5,916	116,438	11,644
E: Mid-Atlantic	Coastal	0	0	1,981	198	0	0	1,981	198
E: Mid-Atlantic	Offshore	82,783	8,278	81,590	8,159	39,498	3,950	203,871	20,387
F: E GOM	Offshore	12,857	1,286	11,706	1,171	714	71	25,278	2,528
G: W GOM	Coastal	0	0	27,450	2,745	0	0	27,450	2,745
G: W GOM	Offshore	33,999	3,400	57,566	5,757	9,947	995	101,512	10,151
H: Mexican GOM	Offshore	0	0	51,991	5,199	0	0	51,991	5,199
I: Puerto Rico	Coastal	0	0	7,032	703	0	0	7,032	703
I: Puerto Rico	Offshore	0	0	0	0	30,994	3,099	30,994	3,099
J: Mexico Pacific	Coastal	0	0	3,401	340	0	0	3,401	340
K: CA Pacific	Coastal	0	0	0	0	1,120	112	1,120	112
L: Mid-Pacific	Coastal	0	0	0	0	1,786	179	1,786	179
L: Mid-Pacific	Offshore	0	0	25,926	2,593	0	0	25,926	2,593
M: NW Pacific	Coastal	0	0	245	24	929	93	1,173	117
M: NW Pacific	Offshore	0	0	14,930	1,493	561	56	15,491	1,549
N: Hawaii	Coastal	0	0	0	0	3,286	329	3,286	329
N: Hawaii	Offshore	0	0	0	0	1,714	171	1,714	171
O: Pacific Canada	Offshore	0	0	3,852	385	1,036	104	4,888	489
P: South Alaska	Coastal	0	0	1,000	100	0	0	1,000	100
P: South Alaska	Offshore	0	0	1,722	172	0	0	1,722	172
Total	Coastal	0	0	42,562	4,256	8,835	883	51,397	5,140
Total	Offshore	160,263	16,026	434,063	43,406	208,566	20,857	802,892	80,289
Grand Total		160,263	16,026	476,625	47,662	217,401	21,740	854,289	85,429
Caribbean	Coastal	16,438	1,644	43,489	4,349	2,708	271	62,634	6,263
Caribbean	Offshore	30,253	3,025	259,055	25,905	48,157	4,816	337,464	33,746

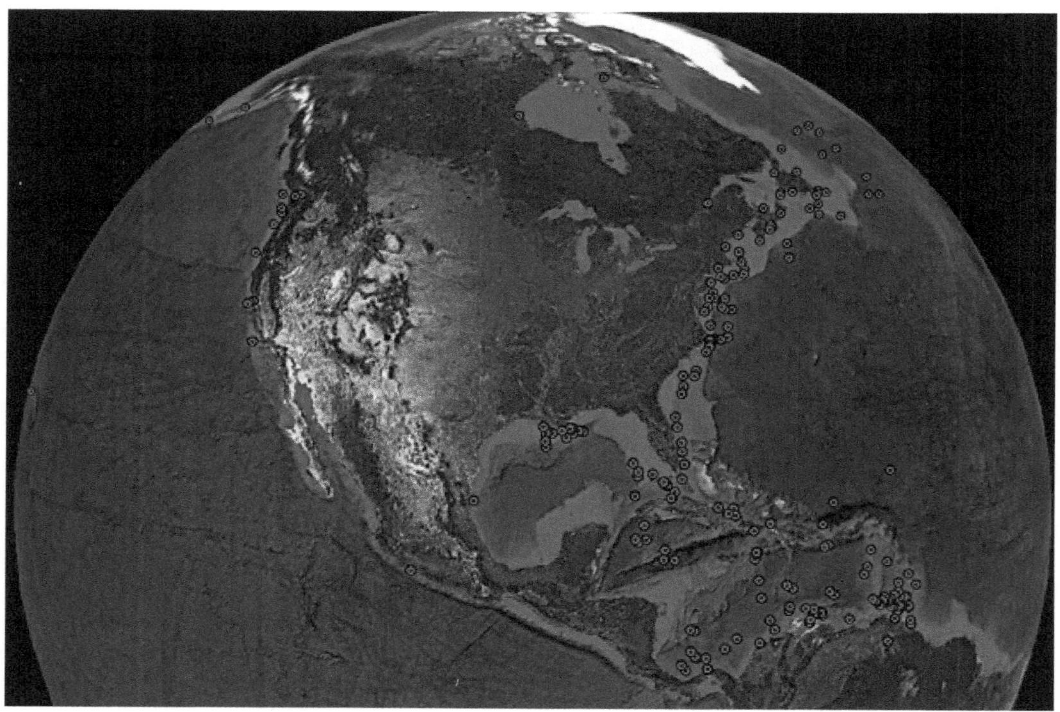

FIGURE 3.16 Map of largest potentially polluting shipwrecks in North American waters.
SOURCE: Courtesy of Environmental Research Consulting.

In North American waters, there are at least 236 large wrecks containing an estimated 600,000 to 6 million bbl (85,000 to 850,000 MT) of oil that may present a threat of pollution in the future. There are an additional 89 wrecks with 280,000 to 2.8 million bbl (40,000 to 400,000 MT) of oil that may threaten Caribbean waters. Oil pollution from these wrecks may come in the form of chronic smaller discharges similar to natural seeps. There is also the possibility of episodic discharges and even large release events when the vessel is disturbed by storms, landslides, tremors, anchors, or fish trawling nets, or when corrosion becomes extensive. Many of the World War II wrecks may contain particularly toxic fuel oils that were often used during that era (Faksness et al., 2015).

3.5.6 Projections of Future Oil Spillage

The estimates of oil spillage into marine and estuarine waters of North America from oil extraction, transportation, and consumption activities as estimated for *Oil in the Sea IV* (based on spill data for the years 2010–2019 to represent current spillage), were compared to historical data from *Oil in the Sea III* (1990s) in addition to future projections through 2050 (energy outlook based on data provided to the committee from Greene Economics—data sources are included as Appendix B).

The projections for oil spillage were based on the fundamental assumption that current spillage rates (i.e., amount spilled per amount extracted, transported, and consumed) would apply in the future. In other words, the rate of spillage would not be further decreased by prevention measures and source control beyond those measures that are currently in place. In reality, there may well be more spill prevention measures put into place over the next decades that are not currently anticipated or imagined. This approach also assumes that the spillage rates would not be increased due to negligence or decreases in the current practices that prevent accidents and associated spills. The projections also do not account for increases in spillage that could occur due to natural hazards, aging and changing infrastructure, changes in transportation routes, or other potential sources of oil in the sea. The oil spillage projections merely track the projections for oil extraction, transportation, and consumption in the future.

Three separate oil energy outlook scenarios were included in the analysis (Greene Economics, 2021):

- The Reference scenario is the business as usual (BAU) case based on the EIA Energy Outlook (2021), which assumes a continuation of energy usage trends (and petroleum usage as part of that usage).
- The Partial Decarbonization scenario is based on the Greene Economics (2021) report in which the forecasts for 2020–2050 by fuel type are the midpoint between the Reference scenario and the Net Zero 2050 scenario with the exception of the shipping data, which are based on previous estimates by an American Bureau of Shipping (2020) study.
- The Net Zero 2050 scenario is based on projections made for a decarbonized future by the International Energy Agency (2021).

Spillage rates were calculated separately for:

- Oil exploration and production (oil exploration and extraction activities in offshore and state marine waters);
- Oil transportation (tank vessels, pipelines, coastal storage facilities, and refineries); and
- Oil consumption (oil used as fuel in non-tank vessels and coastal facilities).

In addition, total spillage rates were calculated.

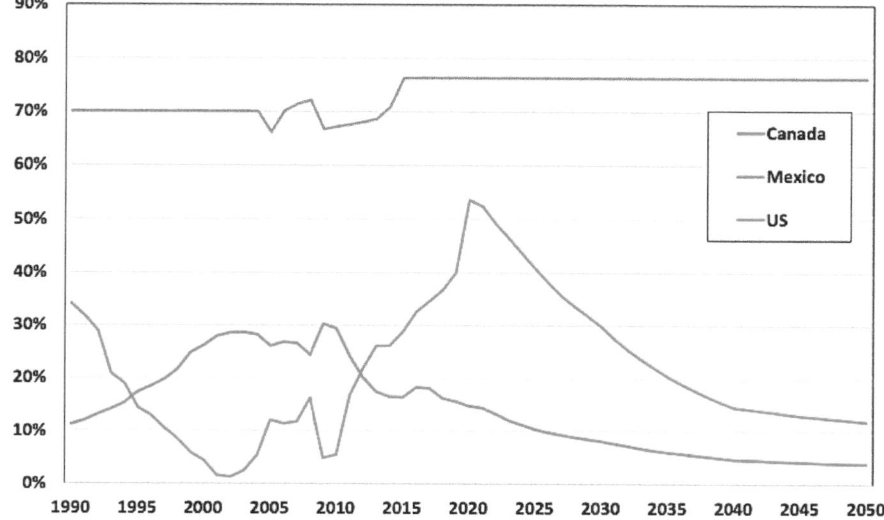

FIGURE 3.17 Proportion of total oil production conducted offshore (reference case).

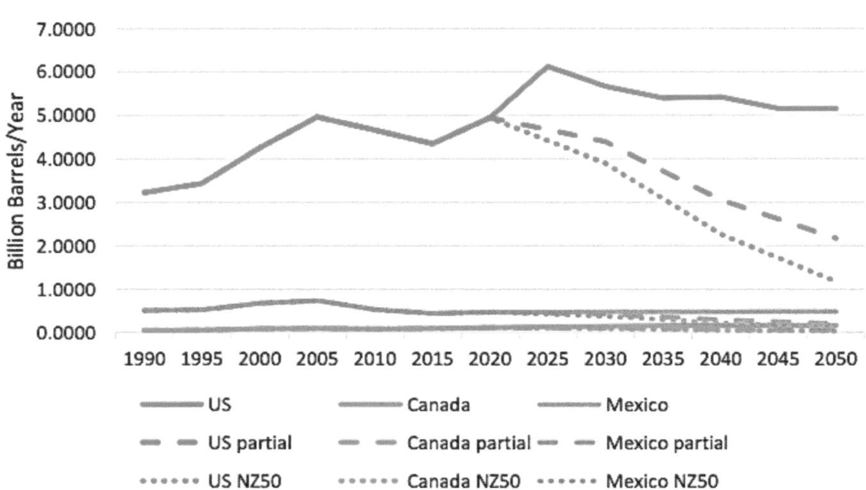

FIGURE 3.18 Seaborne oil transportation by nation in reference (solid line), partial decarbonization (dashed line), and NZ2050 (dotted line) scenarios.

All annual data (MT/year) were based on averages of the decade shown (e.g., 2020s = 2020–2029). For the 2050s, the value was based on the estimate for the year 2050.

Because the *Oil in the Sea IV* study is limited to input to marine and estuarine waters and do not include spills to inland areas, only oil extracted from offshore and nearshore (state) waters was included as the denominator for spillage rates. The future oil production data presented in the Greene Economics report (2021) included all oil extraction activities, including for inland areas of the United States, Canada, and Mexico (see Figure 1.9). The relative percentage of offshore production out of total production for each nation was assumed to follow the same downward trajectory (i.e., a lower proportion of offshore production in the future) in the two decarbonization scenarios (Partial Decarbonization and Net Zero 2050) as for the Reference case (see Figure 3.17). Oil spillage from extraction does not include outlier well blowouts such as the DWH incident.

For oil transportation, seaborne transportation data were applied (see Figure 3.18). Although these data only directly address the transport of imported and exported oil by tank vessels, it was assumed that the same oil would also be stored in coastal facilities (terminals) and transported to refineries. Oil offloaded at offshore oil ports would be transported landward by pipeline.

The projected oil consumption data in the three future scenarios as presented in Greene Economics (2021) were global estimates. The relative proportion of oil consumed

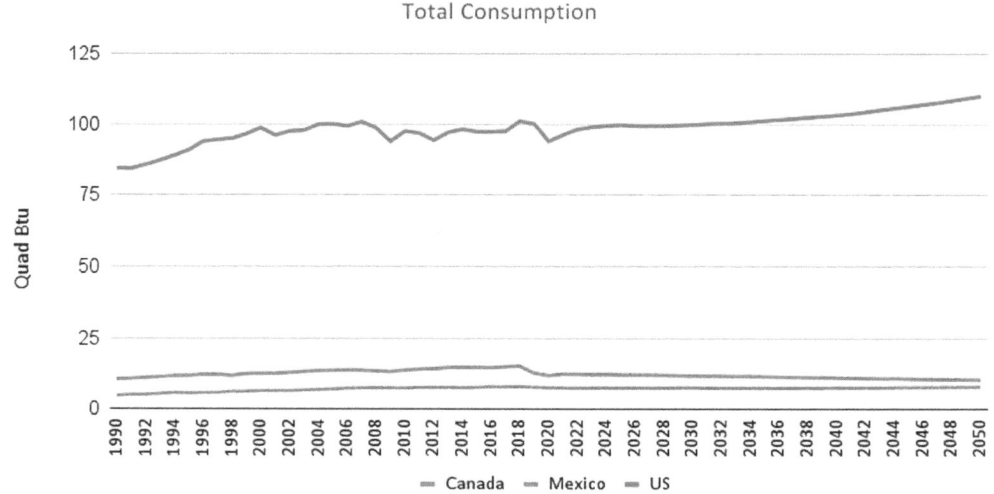

FIGURE 3.19 Oil consumption by nation in reference scenario.

TABLE 3.21 Oil Extraction-Related Spilling (excluding outlier events)

		Average Annual Input by Nation (MT/year)			
Energy Scenario	Time Period	USA	Canada	Mexico	Total
Historical Data	1990s (OITS III)	140	28	61	229
	2010s (OITS IV)	92.0	0.9	1,016.7	1,110
Reference Scenario	2020s	82.8	1.6	708.9	793
	2030s	51	1	863.9	9169
	2040s	33.1	0.7	1,094.8	1,129
	2050s	29.6	0.6	1,171.2	1,201
Partial Decarbonization Scenario	2020s	71.3	1.5	720.5	793
	2030s	32.2	0.6	577.2	610
	2040s	15.2	0.3	405.1	421
	2050s	11.5	0.2	333.9	346
Net Zero 2050 Scenario	2020s	66.9	1.4	542.5	611
	2030s	26.5	0.5	378.3	405
	2040s	9.9	0.2	210.1	220
	2050s	6.3	0.1	144.4	151

by the three nations under the two decarbonization scenarios was assumed to be the same as for the reference scenario (see Figure 3.19).

For oil extraction-related spillage, the estimated historical and projected spillage amounts are shown in Table 3.21 and Figure 3.20. Note that these do not include outlier blowouts such as the DWH incident.

For oil transportation-related spillage, the estimated historical and projected spillage amounts are shown in Table 3.22 and Figure 3.21. Though the general trend in tanker spillage has been downward (ITOPF, 2021), a large tanker spill in the future could change the spillage volumes significantly. The chances of a large tanker spill may be escalating, despite the full implementation of double hulls and other prevention measures, because of the increased capacity of tankers (Statistica, 2021), as well as the increasing size of other cargo vessels, especially container ships, that may collide with them in busy ports and

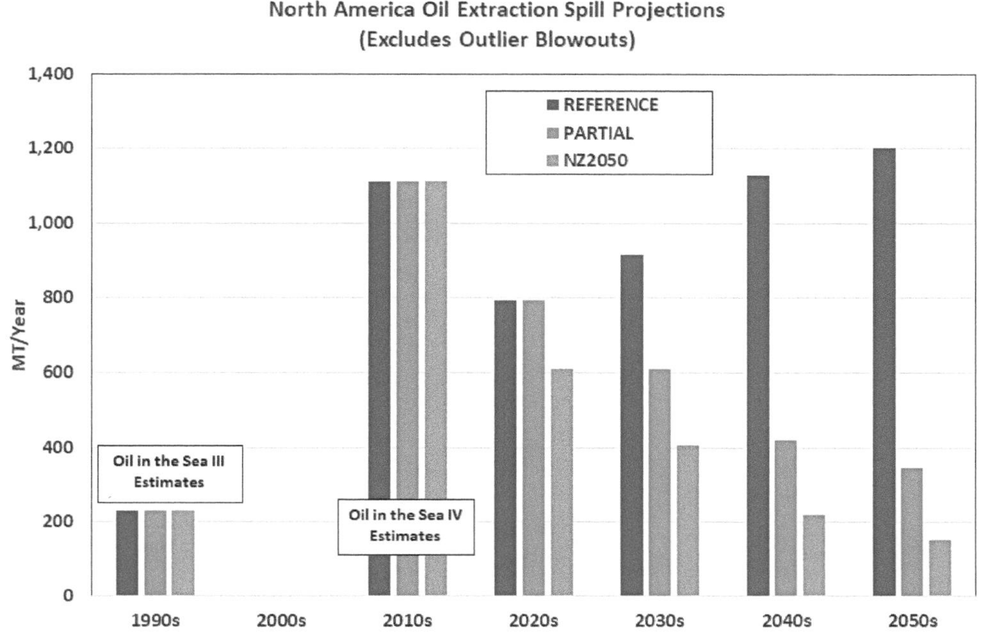

FIGURE 3.20 North America oil extraction spill projections (excluding/outlier blowouts).

TABLE 3.22 Oil Transportation-Related Spillage

Energy Scenario	Time Period	Average Annual Input by Nation (MT/year)			
		USA	Canada	Mexico	Total
Historical Data	1990s (OITS III)	6,676	69	80	6,825
	2010s (OITS IV)	796	10	ND	806
Reference Scenario	2020s	976	14	ND	990
	2030s	975	18	ND	994
	2040s	932	20	ND	952
	2050s	909	20	ND	929
Partial Decarbonization Scenario	2020s	848	13	ND	860
	2030s	717	11	ND	728
	2040s	501	7	ND	509
	2050s	384	6	ND	390
Net Zero 2050 Scenario	2020s	826	12	ND	838
	2030s	617	9	ND	626
	2040s	353	5	ND	358
	2050s	209	3	ND	212

NOTE: ND = no data.

shipping lanes. Larger vessels are less maneuverable and have a greater impact due to the force of their tonnage in the event of a collision.

For oil consumption-related spillage, the estimated historical and projected spillage amounts are shown in Table 3.23 and Figure 3.22. The probability of a significant spill from a consumption-related source (fuel) is unlikely, as the oil volumes contained in these sources are considerably lower than in large tankers or outlier well blowouts.

The total spillage from all sources is summarized in Table 3.24 and Figure 3.23. Again, outlier blowouts are excluded from the analysis.

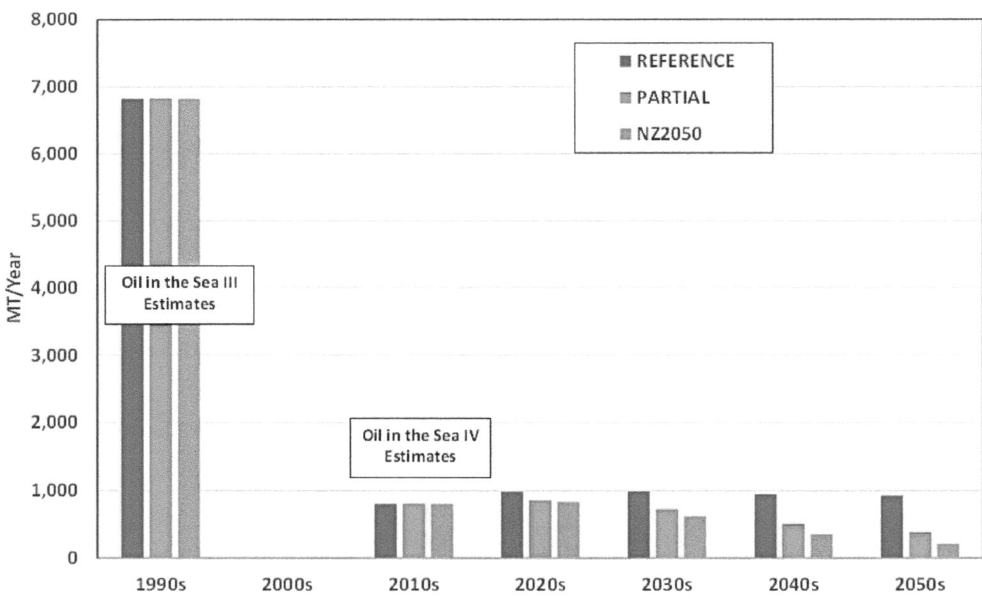

FIGURE 3.21 North America oil transportation spill projections.

TABLE 3.23 Oil Consumption-Related Spillage

		Average Annual Input by Nation (MT/year)			
Energy Scenario	Time Period	USA	Canada	Mexico	Total
Historical Data	1990s (OITS III)	835	52	ND	887
	2010s (OITS IV)	337	57	ND	394
Reference Scenario	2020s	335	50	ND	385
	2030s	341	44	ND	385
	2040s	350	39	ND	389
	2050s	360	35	ND	395
Partial Decarbonization Scenario	2020s	318	47	ND	365
	2030s	281	42	ND	323
	2040s	198	29	ND	227
	2050s	141	21	ND	162
Net Zero 2050 Scenario	2020s	302	52	ND	354
	2030s	239	41	ND	280
	2040s	138	24	ND	162
	2050s	73	13	ND	86

NOTE: ND = no data.

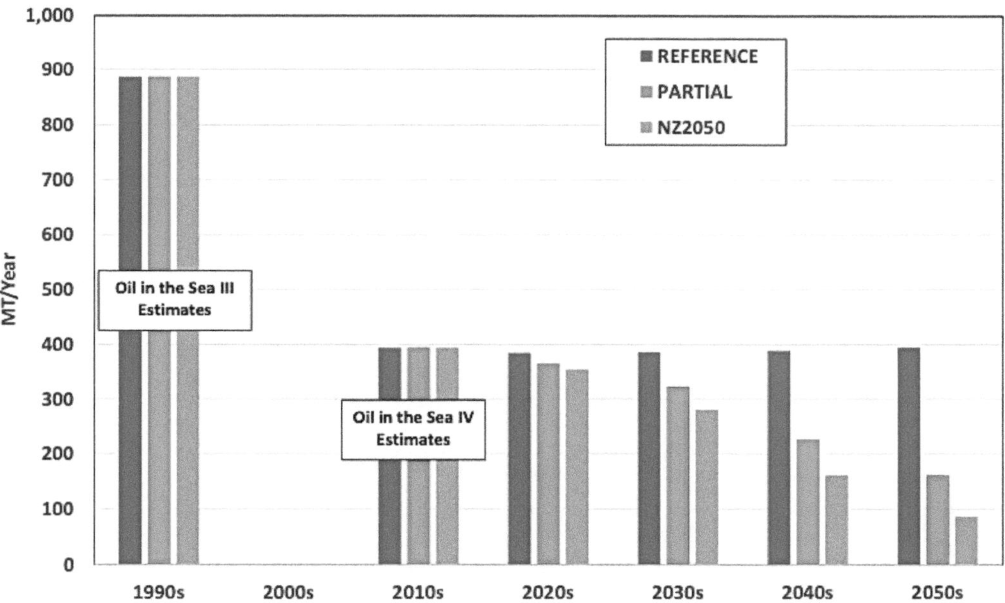

FIGURE 3.22 North America oil consumption spill projections.

TABLE 3.24 Total Oil Spillage (excluding outlier blowouts)

		Average Annual Input by Nation (MT/year)			
Energy Scenario	Time Period	USA	Canada	Mexico	Total
Historical Data	1990s (OITS III)	7,651	149	141	7,941
	2010s (OITS IV)	1,224	69	1,017	2,310
Reference Scenario	2020s	1,394	65	709	2,168
	2030s	1,367	63	864	2,295
	2040s	1,316	59	1,095	2,469
	2050s	1,298	55	1,171	2,525
Partial Decarbonization Scenario	2020s	1,237	61	721	2,018
	2030s	1,030	53	577	1,660
	2040s	714	37	405	1,156
	2050s	536	27	334	896
Net Zero 2050 Scenario	2020s	1,195	65	543	1,802
	2030s	882	51	378	1,311
	2040s	501	29	210	740
	2050s	289	16	144	449

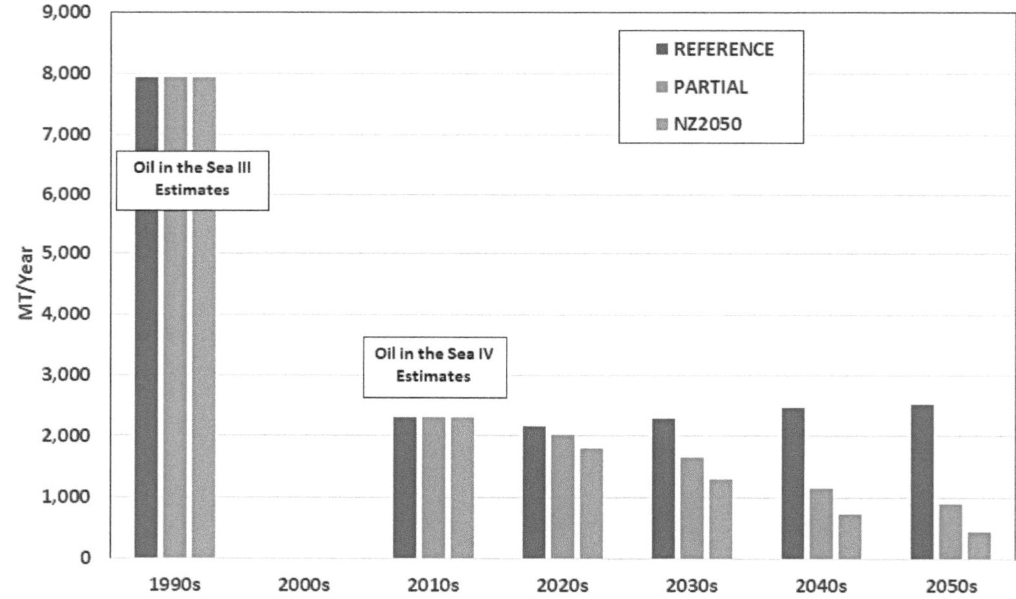

FIGURE 3.23 North America oil spill projections for all sources.

3.6 CONCLUSIONS

Conclusion—In agreement with the 1975 National Research Council report titled *Petroleum in the Marine Environment* (the first in the series of Oil in the Sea reports), this report also recognizes "that there was a significant lack of systematic data collection concerning petroleum hydrocarbon discharges entering the oceans." As was stated in *Oil in the Sea III* (2003) "the report, lacking significant quantitative data, was based on estimates and in some instances, educated guesses." Although the *Oil in the Sea IV* committee found more specific data on some inputs (spill volumes, permitted discharges), other values in input categories remain vague (land-based runoff, sewage discharge, two-cycle engines, atmospheric deposition, and emergency dumping) to nonexistent.

Conclusion—Energy transition from hydrocarbon fossil fuels to alternative (renewable, nuclear) energy will ultimately reduce the input of oil in the sea from production, transportation, and consumption. New energy sources (e.g., wind farms) and fuels (e.g., biofuels, ammonia, and hydrogen) can introduce new safety and pollution concerns and the regulators, the research community, and the industry are encouraged to proactively review and address any potential adverse effects from these transitions. While energy transition is taking place, continued attention to environmental concerns associated with fates and effects from human-caused chronic and episodic inputs of oil remains critical.

Natural Seeps

Conclusion—Oil in the Sea III recommended that "[f]ederal agencies especially the USGS, MMS, and NOAA should work to develop more accurate techniques for estimating inputs from natural seeps, especially those adjacent to sensitive habitats." Some progress has been made on identification of seep locations, but not much work has advanced quantification of oil and gaseous hydrocarbons released from natural seeps, particularly outside the Gulf of Mexico in areas such as the North American Pacific margin and North American Arctic margin. The geographic and temporal expanse of natural seeps could be better understood through use of existing techniques, along with exploration of new technologies for identification and quantification of natural seeps.

Land-Based Sources

Conclusion—Oil in the Sea III recommended that "[i]n order to assess impacts attributable to different sources including oil spills, and nonpoint sources, federal agencies, especially USGS and EPA should work with state and local authorities to undertake regular monitoring of TPH and PAH inputs from air and water (especially rivers and harbors) to determine background concentrations." It is evident that this recommendation has not been acted upon. In order to quantify inputs from nonpoint sources and thus understand the effects these inputs may have on the ecosystem, regular monitoring of rivers and harbors has to occur. Oil and grease and petroleum hydrocarbon data was collected in the 2000s, but that data corresponds to individual and unique projects aimed at monitoring runoff in specific locations. These projects do not allow for temporal or spatial coverage or extrapolation, or improved insights into baseline data regarding the composition of oil compounds present in runoff from land.

Conclusion—Personal and industrial practices for controlling nonpoint pollution, such as motor oil recycling, and reduction in fossil fuel usage via improved fuel efficiency and electrification of transportation, will have the added benefit of reduction in land-based runoff.

Conclusion—Further research is required to refine estimates of land-based inputs of oil that is transported via the atmosphere to the sea. This would include (1) expanding the geographic and temporal coverage of data collection; and (2) refining understanding of the source of hydrocarbons measured in marine atmospheres and surface seawater by improved structural elucidation of compounds of interest coupled to robust source determination via the methods previously mentioned in this section.

Operational Discharges

Conclusion—Operational discharges from exploration and production operations are permitted as long as they do not exceed maximum weighted average concentration of oil-in-water (or "oil and grease"). Under this approach the exact concentration of oil in produced water is not reported, preventing accurate estimation of oil inputs into the marine environment.

Conclusion—Regulatory changes had a positive impact on decreasing operational discharges from exploration, production, and transportation, in the North American waters. With compliance, the operational discharges are small relative to potential spill volumes. Illegal discharges still occur and the best prevention is rigorous and effective enforcement of the regulations. Technologies such as drones and satellites can be used to improve surveillance and enforcement. Additionally, the corporate culture increasing the standard of care for environmental stewardship within the industry has grown and continues to aid in the reduction of illegal activities.

Conclusion—Illegal discharges from ships can have a significant impact on the total input volumes in the locations where they occur. The *Oil in the Sea III* report recommended: "federal agencies, especially the U.S. Coast Guard, should work with the transportation industry to undertake a systematic assessment of the extent of noncompliance." The extent of noncompliance with illegal-discharge regulations in North American waters remains unknown.

Conclusion—Oil in the Sea III recommended that "federal agencies, especially the U.S. EPA, should continue efforts to regulate and encourage the phase out of inefficient two-stroke engines, and a coordinated enforcement policy should be established." The committee commends the U.S. EPA's efforts to address this important recommendation. Continued efforts to reduce use of two-stroke engines in North American waters are encouraged.

Conclusion—Oil in the Sea III recommended that "[f]ederal agencies, especially the Federal Aviation Administration (FAA), should work with industry to more rigorously determine the amount of fuel dumping by aircraft and to formulate appropriate actions to understand this potential threat to the marine environment." It was found that this recommendation has not been addressed and remains an important gap in the understanding of oil inputs into the marine environment. Additionally, obtaining accurate estimates of fuel released into the seas would require reporting of jettison dumping by the defense sectors, in addition to the private sector.

Accidental Spills

Conclusion—Excluding the DWH incident in annualized estimates, in U.S. and Canadian waters, the spillage from oil extraction activities is nearly the same as that estimated in *Oil in the Sea III* (1990–1999), although there is a shift away from coastal/nearshore spillage to more offshore spillage. This reflects the increase in offshore oil exploration and production activities in deeper offshore waters both in the United States and Canada. However, the actual rate of spillage per amount of production has decreased in the past two decades.

Conclusion—The DWH incident was unprecedented with regard to the total volume of the release. It represents an "outlier" case in that it is expected to be a very low-probability–high-volume event. Its occurrence during the time period considered as the "sampling period" for the estimation of average annual inputs of oil into the sea (2010–2019) from oil extraction activities is treated separately from the annual estimates while recognizing its impacts during the period reviewed in this report.

Conclusion—Extensive investigations in the aftermath of the DWH incident resulted in numerous positive changes by government agencies, industry, and regulators to prevent such a disaster from happening again. This includes technological improvements to detect anomalies that may indicate potential issues with well integrity during offshore drilling operations and to intervene with source control measures in as timely a fashion as possible to prevent large releases and blowouts. Continued support of technological and organizational improvements is critical to being better prepared for the next event, should it occur.

Conclusion—Governmental agencies and industry have improved pipeline systems' reliability as well as spill prevention and early detection techniques and processes, resulting in a decrease of spills from onshore and offshore pipelines despite the increase in pipeline mileage and barrels delivered.

Conclusion—The potential for future spillage related to storm damage of offshore structures, pipelines, and coastal facilities, particularly in light of increased intensity and frequency of severe storms and hurricanes related to climate change, represents a considerable risk of future oil inputs into the marine environment.

Conclusion—Double hulls on tankers and on fuel tanks of other types of vessels help reduce the likelihood of spillage from low-impact groundings and collisions, but they will not guarantee zero pollution in all incidents. Port congestion and increased vessel size make maneuvering more challenging and create a risk of collisions and allisions. Robust systems for monitoring and managing vessel traffic decrease the risk of pollution incidents.

Conclusion—While the probability of major tank vessel spills has been reduced significantly in the past 20 years, there is still a possibility that such a spill might occur. Future advancement in oil spill prevention would benefit from improved connectivity via satellites, available automatic information system data, machine learning, and other new technologies that would help both the onboard crew and onshore personnel to identify potentially dangerous situations before they occur.

Conclusion—Aging infrastructure (i.e., abandoned wells and pipelines, coastal facilities, etc.) poses potential sources of oil into the marine environment. Appropriate federal, state, or tribal entities should take inventory of the existing remnants of oil storage, transport, or production activities that still remain to be identified, triaged for potential losses, and as appropriate salvaged or capped.

Conclusion—There is an increased potential for future spillage from sunken wrecks as those wrecks age. The majority of the larger wrecks are World War II–related casualties and thus more than 70 years old. As such, they are subject to significant corrosion in addition to possible damage incurred in battle. U.S. Coast Guard and environmental authorities from the United States, Canada, and Mexico are encouraged to take advantage of the availability of data from existing risk assessment studies to prioritize wrecks for monitoring, surveying, and potential oil removal operations. The availability of advanced tools for conducting more effective monitoring and remote sensing survey operations, as well as the technological capability to conduct effective oil removal operations may help to reduce this threat. Although the number of wrecks may seem a daunting task, a strategy of systematic risk assessment and prioritization using existing tools will focus efforts on the wrecks that present the greatest threats to marine waters and sensitive ecological and socioeconomic resources.

4

Accidental Spill Mitigation

Highlights

This chapter covers the efforts to prevent and minimize the volume of oil entering the marine environment, as well as tools and best practices in mitigation measures aiming to reduce the impacts of accidental spills. Important highlights over the past 20 years include:

- Salvage operations for distressed vessels have proven an important mitigation step in preventing or reducing oil outflow.
- Significant advances in blowout prevention and source control technologies for offshore wells have been made, in addition to greater emphasis on human factors and strengthening of regulatory oversight.
- Improvements have been made in onshore and offshore pipeline integrity monitoring and leak detection technologies improving leak prevention as well as reduce their volume and consequences.
- The Incident Command System (ICS) has become the standard for emergency management across the United States and internationally.
- The response toolbox concept was adopted widely as an approach for optimizing recovery efficiency and maximizing protection for people and the environment.
- Several risk assessment methods were developed to assist response decision-makers with selection of response options with the greatest potential to mitigate spill impacts taking into account efficiency, benefits, and additional impacts response options may have under unique environmental conditions.
- Response and damage assessment have benefited from significant progress in oil spill modeling, monitoring, and big data collection, processing, and computation over the past two decades.
- Due to the advances in information technologies, sensors, and automation, the ability to monitor field activities and environmental impacts has improved over the past decades.
- Great advances in analytical equipment and bacterial genomics improved understanding of natural attenuation and bioremediation processes and allow their optimization and monitoring in the field.
- Oiled wildlife response has evolved into a comprehensive program fully integrated with oil spill response efforts.

Oil inputs from accidental spills can be substantially reduced through prevention and source control. It is important to have tools available to effectively respond and minimize volume of oil spilled and the impact on people and on the environment should there be a need. This chapter discusses operational consideration of the various options available to respond to accidental oil spills and the rationale for the selection of response techniques, the pros and cons of each measure, a realistic assessment of their effectiveness under various conditions, and suggestions for future improvements.

4.1 SOURCE CONTROL

4.1.1 Salvage as Source Control for Vessel Spills

When a vessel is involved in a major accident—a collision, allision, or grounding—there is a risk of oil being released into the environment. Although double hulls on cargo and bunker tanks add a level of protection that reduces the likelihood and potential volume of a release in an impact accident (also see Section 3.2.4.1) (Rawson et al., 1998; Michel and Winslow, 2000; NRC, 2001; Herbert Engineering Corp. and Designers & Planners Inc., 2003; Barone et al., 2007; Yip et al., 2011b), there is still a chance of an oil release. Furthermore, as discussed in Chapter 3, as the capacity of both tankers and cargo ships increases, the vessels potentially become less maneuverable and the impact force, should a collision occur, potentially increases.

In the event of a vessel accident in which there is a risk of an oil release, the vessel operators exercise their vessel response plan (VRP) and alert not only the appropriate coast guard authorities, but also the salvage and marine firefighting (SMFF) organization named on their plan or recommended by authorities. The SMFF providers are equipped and trained to extinguish fires and to secure stability and structural strength of the vessel to prevent further damage to the vessel and to prevent or reduce oil outflow from the cargo and bunker tanks. Much of this work is accomplished with large cranes and machinery, as well as with trained divers. In some cases, the source control work of salvage teams involves lightering the vessel—or removing the oil into another vessel or storage tank. Preventing releases or at least reducing the amount of oil released, in the event of a vessel accident is the first level of mitigation in preventing the effects of large oil spills.

Ensuring that VRPs and SMFF response capabilities are appropriate to meet the needs of existing and future vessel traffic will be an important mitigating factor for large tanker and non-tank vessel spills (GAO, 2020).

4.1.2 Offshore Wells

In the pursuit of readiness to respond to a potential oil spill from an offshore exploration and production (E&P) facility, the primary approach of both the regulatory community and the industry community is prevention of and preparedness for incidents. **Great advances have been made in recent years in blowout prevention and source control.** Prevention can be achieved through adherence to drilling standards and safety regulations combined with a culture of safety and proactive risk management (see Section 3.5.1 for more details). During drilling operations, well parameters are continuously monitored and warning systems exist to alert operators of a potential threat to well integrity. This real-time well monitoring information is often displayed simultaneously in several locations, and examined by different teams of experts to ensure that any deviation from normal operations is immediately detected and mitigated. Among other parameters, these systems can monitor the condition of well barriers. A *barrier* is defined as a system or a device that can be used to contain fluid or pressure within the well. These barriers may include high-pressure wellhead housings, multiple casing strings cemented in place, blowout preventers, and weighted drilling fluids. Drilling standards require at least two independent barriers to be maintained at any given time (API, 2018b,c; IOGP and IPIECA, 2019) to prevent a single failure from leading to a loss of well control and a resulting spill. Barriers are tested prior to and after installation, as well as at regular intervals during operations. Should the prevention systems fail, resulting in an uncontrolled release of oil to the environment, a variety of well intervention activities are designed to take place.

The primary well control barrier is the weighted drilling fluid that maintains hydrostatic pressure in the wellbore at a higher level than the pressure in the reservoir, thus preventing hydrocarbons from entering the wellbore. Its composition and density are designed for specific well conditions and are continuously monitored throughout drilling operations. A blowout preventer (BOP) is a device installed subsea on a wellhead or at the rig and is specifically designed to prevent uncontrolled release of gas and fluids from the well (API, 2018b). It serves as a secondary well control barrier. The BOP consists of a series of devices to close off the wellbore in various conditions, including cutting jaws (called "shear rams") to slice through the pipe and seal off the well in case of an emergency. A BOP must be able to cut the drill pipe and seal the well under the maximum anticipated pressure. In the United States, the Bureau of Safety and Environmental Enforcement (BSEE) requires a BOP to be certified and tested to ensure readiness of all components. If an early warning system indicates a potential loss of well containment, the BOP can be activated to seal the well before oil escapes, either remotely or using a remotely operated underwater vehicle with the help of a subsea intervention skid.

In the event that the primary and secondary well control measures fail, additional source control activities can be initiated (API, 2006). A new well can be drilled to intersect and kill the well experiencing loss of containment. Relief well operations are effective but may take time, during which escaping oil may negatively affect the environment. This issue was addressed by the development of the subsea capping stack and containment equipment technologies. Subsea capping involves installing a capping stack (see Figure 3.7) onto the well and then closing its valves to shut off the flow of hydrocarbons and seal the well. Capping stacks should be capable of withstanding the maximum anticipated pressure at the wellhead. Some of them weigh almost 400,000 pounds and operate in temperatures as high as 400°F, pressures of 20,000 psi, and water depths up to 15,000 ft water depths (IOGP and IPIECA, 2019). The main advantage of the capping stack is its ability to isolate and stop the flow in a relatively short period of time. A Global Subsea Response Network (GSRN) was formed to leverage collective expertise of leading source control companies and ensure that best equipment and practices in source control are available worldwide.

If it is not possible to safely shut in a well with a capping stack, containment methods can be used to prevent the environmental impacts while a relief well is being drilled. The goal of the containment operation is to capture wellbore fluids and hydrocarbons exiting the well, and divert them to processing vessels. This process involves subsea infrastructure and surface processing equipment, as well as the connection of flow lines and riser systems to create a temporary subsea production system. The capping stack may potentially serve as the interface to connect containment system lines that will divert the flow to the capture vessel. Otherwise, a containment structure ("top hat" or a "containment dome") could be placed over the well to collect escaping hydrocarbons. The current industry containment systems are designed to handle oil flow rates up to 100,000 BOPD (barrels of oil per day) and 200 MCFD (million standard cubic feet per day) of gas. Hydrate inhibition chemicals could be used to prevent the formation of hydrates that can plug flowlines transporting the oil to the surface (see Section 2.2). Collected and processed liquids are then offloaded from the surface production vessel to crude tankers and taken to shore for utilization and

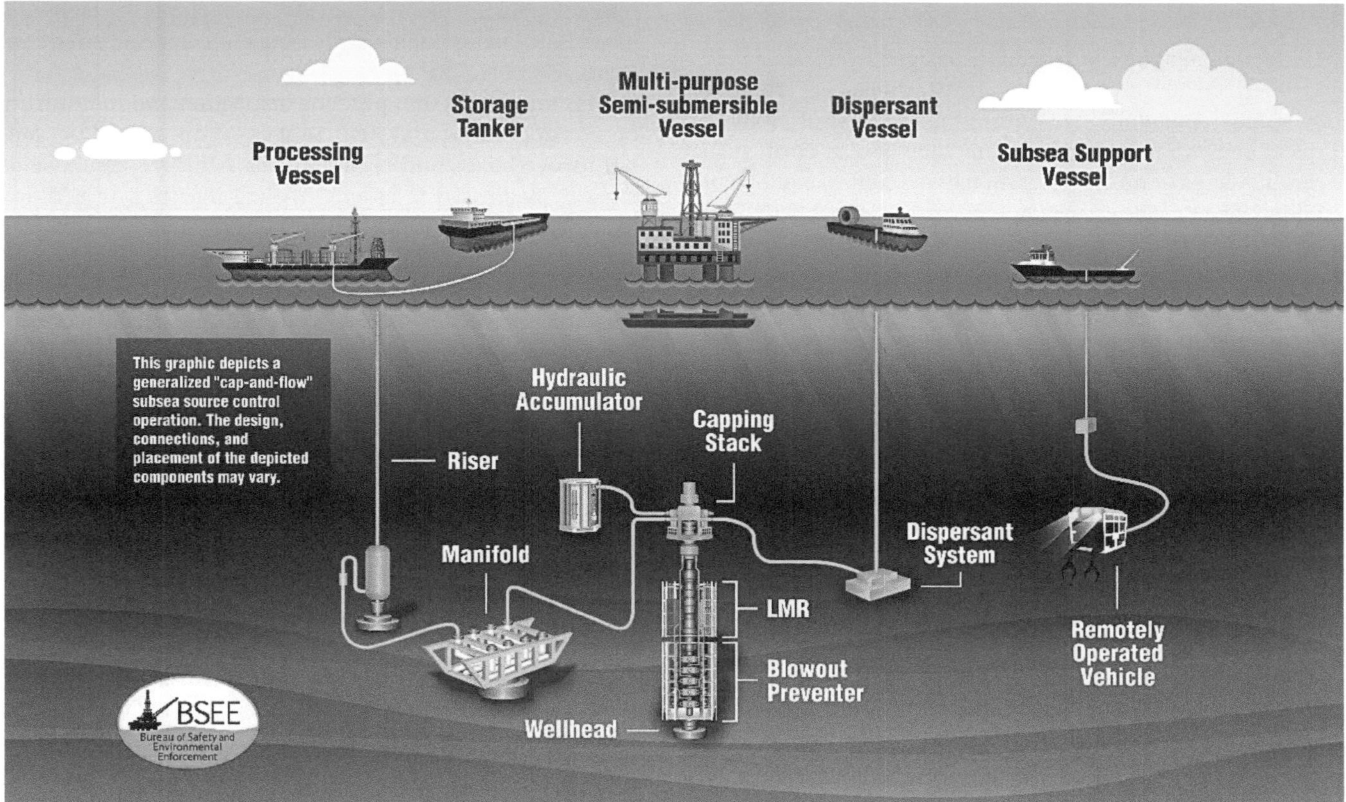

FIGURE 4.1 An illustration of source control activities.
SOURCE: BSEE.

disposal. An example of use of a containment structure to capture oil during the ongoing Mississippi Canyon, Block 20 (MC-20) oil spill is detailed in Box 3.4. Various components of the source control operations are illustrated in Figure 4.1.

As a permanent source control solution, a relief well could be drilled in parallel with capping and containment activities to intersect the incident well at some predetermined distance below the seabed and seal the well by injecting heavy mud and cement. A relief well is drilled in the same way as a regular well and is positioned at an appropriate distance from the incident wellsite to allow safe drilling operations and avoid interference with the source control activities. The distance between surface locations for the blowout well and the relief well can range between 500 feet and 3,500 feet. When the relief well intersects the target well, drilling mud is pumped into the well to increase hydrostatic pressure and regain control of the well. When the well is no longer flowing, cement can be pumped into the well to seal it. Izon et al. (2007) analyzed the Mineral Management Service (MMS) data for blowout wells in the outer continental shelf of the United States in a period from 1971 to 2006. They found that out of 39 blowouts that occurred during that period, relief wells were initiated for two of them, because they were controlled by other means prior to completion of the relief wells. During the Macondo incident, well control was also regained through installation of a capping stack before completion of a relief well. Since then, there has been progress in making well control devices, such as capping stacks, more readily available for incident response.

Human factors are an important aspect of safe drilling operations and effective source control. In the United States and Canada, regulators require professional certifications as well as frequent drills and exercises to ensure personnel qualification, drilling safety, and response readiness. These include exercises with drilling crews to test their ability to detect and mitigate simulated influx of hydrocarbons, a requirement for rig supervisors to take a certified well control course every 2 years, conduct safety seminars, and Drill the Well on Paper (DWOP) exercises prior to the start of a drilling program. An important human factor component to safety on the rig is a broadly accepted industry practice, called "stop work authority," where anyone has the right to stop the work if he/she feels that it is unsafe, without repercussion. This enables rig personnel to immediately stop unsafe operations without delays that may be caused by the need for managerial approvals. Additional discussion on human factors related to offshore operational process safety can be found in the proceedings of the 2018 National Research Council (NRC) workshop, *The Human Factors of Process Safety and Worker Empowerment in the Offshore Oil Industry: Proceedings of a Workshop* (NRC, 2018).

4.1.3 Pipelines

Although pipelines are specially designed, built, installed and operated to safely move hydrocarbons, an integrity failure may potentially occur resulting in a loss of product. Typical causes for such a failure include human error, such as anchor drags or misinterpretation of the control equipment readings and/or alarms, pipeline corrosion, process defects during installation, and flaws occurring during the manufacturing process, as well as geophysical external factors. Pipelines associated with oil and gas exploration are regulated by the U.S. Department of Transportation's (DOT's) Pipeline and Hazardous Materials Safety Administration (PHMSA) as well as by the BSEE. Industry employs pipeline safety management systems (SMS) (ANSI and API, 2015; API, 2019b) to significantly reduce the risk of incident occurrence as well as an incident's impacts on people and environment. This approach goes beyond operational leak detection and also includes risk management, safety assurance, incident investigation, lessons-learned sharing and continuous improvement, staff competence maintenance, and emergency preparedness and response, among others.

Pipeline integrity monitoring allows an early detection of structural issues and repair or replacement of the impacted segment before loss of integrity and a potential leak. Leak detection techniques allow quick shut off to stop hydrocarbon discharge and reduce the lost volume and environmental consequences due to the pipeline failures. A variety of pipeline monitoring and leak detection techniques and approaches have been developed over the years, resulting in a decrease of pipeline-related incidents and their consequences. Some of them aim to detect the leak at the exterior of the pipeline, while others use devices, sensors, and computational algorithms (API, 2007) to monitor pipelines from the inside.

For offshore use, leak detection techniques include acoustic detection, fiber-optic sensors, pressure point analysis, rate of change / conditional rate of change, dynamic modeling, vapor sampling, infrared thermography, digital signal processing and mass-volume balance technique, among others (DNV, 2016; Adegboye et al., 2019). Additional techniques can be employed onshore; for example, the use of ground-penetrating radars, spectral scanners, or specially trained dogs that can detect a leak as deep as 15 feet (API, 2016a). Pipeline integrity can be assessed by smart pigging, an in-line monitoring technique employing sensor-equipped probes that are propelled through pipelines for cleaning and inspection activities. Additionally, remote leak monitoring can be conducted by sensors installed on autonomous underwater vehicles (AUVs), remotely operated vehicles (ROVs), or drones, or by using sensor networks. Recent advances in using AUVs and ROVs for subsea pipeline inspection and monitoring have significantly expanded offshore monitoring capabilities and reduced the risk of human exposure and errors. These systems can be operated remotely making them suitable for inspections in potentially hazardous environments such as deep-water and Arctic regions (Ho et al., 2019).

It is important that pipeline inspection and monitoring be considered as a system that is composed of personnel, procedures, and technologies. All three components must be addressed to assure pipeline integrity. In addition to the development and deployment of robust monitoring and detection technologies, development and implementation of human controls is also important. These may include performance metrics and key performance indicators, control center procedures, competencies and responsibilities of personnel, training and exercises, among others (API, 2015c, 2019c; 49 C.F.R. pt 195).

4.2 RESPONSE

4.2.1 Introduction

Oil spill scenarios, their potential impacts, suitable response options, and the effectiveness of those options vary greatly from spill to spill. Over the past several decades, the trend has shown the volume and the number of spills have been declining over the years (see Chapter 3). Most spills are fairly small, are in the industrial areas close to response equipment depots, and can be effectively responded to (e.g., by mechanical recovery) without the need for complicated equipment and comprehensive analysis. Fewer spills—for example, large spills in offshore or coastal areas or spills in particularly sensitive environments—may require a larger number of Incident Command System (ICS) managers and numerous responders, focused engagements with stakeholders, a detailed environmental and operational analyses, and the use of complex response strategies and monitoring techniques. This section aims to review various oil spill response tools that have a potential to reduce the volume and impacts of the released product and which could be considered for response in marine environments. It is important to note that no single response technique is absolutely effective, safe, or even applicable or necessary in every situation. The benefits, scenario-specific effectiveness, operational challenges, and any potential additional environmental and socioeconomic impacts of any tool use have to be carefully considered using best available expertise and information before a decision to proceed is made.

4.2.1.1 Response Structure

In the 1970s, California was devastated by a series of catastrophic forest fires encroaching on urban turf. As part of an effort to determine the cause of this disaster, case histories were examined and it was discovered that incident failures were far more likely to result from inadequate management than from lack of resources, faulty tactics, or any other factor.

It was from this insight that ICS was born. ICS represents organizational "best practices," and has become the standard for emergency management across the United States (HSPD 5, 2003). ICS provides a common management structure through administrative support and oversight in a response to create time- and resource-efficiency between management personnel from various agencies. ICS is interdisciplinary and agile by design so as to meet the needs of various agencies and various incidents. ICS has been tested in more than 30 years of emergency and nonemergency applications, by all levels of government and in the private sector (NPS, 2017).

Walker et al. (2015) define formal authorities involved in the response structure as including agency representatives having oil spill authority at multiple levels of government and potential parties responsible for making and implementing preparedness and response decisions to mitigate impacts to the ecosystem and the ecosystem's resources, users, and property owners. Stakeholders are identified as "part of a community and broadly defined as those groups that have a stake, interest, or right in an issue or activity (e.g., an oil spill) and those that will be affected either negatively or positively by decisions about the issue or activity" (Krick et al., 2005).

In the early 1990s the U.S. Coast Guard (USCG) modified the original ICS into its *Incident Management Handbook* (IMH) for use on oil spills, chemical releases, and other environmental and emergency responses. This became the de facto management system for response in the United States and has been adopted by many other nations and companies. Through training and certification, individuals are able to move from incident to incident and fill designated positions in the management structure (USCG, 2020).

The essential philosophical modification from strict ICS to a unified management system aligns with the Oil Pollution Act of 1990 (OPA 90) legislation where the responsible party (RP) is mandated to be the entity charged with source control, oil removal, and environmental restoration. This provides a regulatory role for the RP, along with the federal and state agencies involved in environmental protection. In an ICS chart, this would form the Unified Command triangle on the top of the ICS pyramid. As opposed to a Stafford Act response, where the government is wholly in charge of response activities to a natural disaster, an oil spill is more cooperatively managed, with the government providing oversight to the RP to direct cleanup activities.

Over the decades, after the *Exxon Valdez* (1989), with the incorporation of this modern response management system, response activities have generally become more cooperative and less combative at the corporate/governmental levels. To achieve the optimum outcome from a response, ICS helps ensure best management practices, streamlined organization structure, minimally environmentally intrusive tactics, and economically viable and socially accepted practices to most effectively plan for, respond to, and restore the environment after an oil spill occurs.

4.2.1.2 Common Operating Picture and Information Management Systems

The U.S. Department of Homeland Security's (DHS's) definition of a common operating picture (COP) is a continuously updated overview of an incident compiled throughout an incident's life cycle from data shared among integrated communication, information management, and intelligence and information sharing systems (DHS, 2008). The goal of a COP is real-time situational awareness across all levels of incident management and across jurisdictions. This need was called out in the NRC's *Oil in the Sea III* (2003) report recommendation for "federal agencies ... to work with industry to develop and implement a rapid response system to collect in situ information about spill behavior." Several COP systems have been developed over the years.

One such system, developed by the National Oceanic and Atmospheric Administration (NOAA) and the University of New Hampshire, with the U.S. Environmental Protection Agency (U.S. EPA), the USCG, and the U.S. Department of the Interior, is the Environmental Response Management Application (ERMA®) (NOAA, 2014a). This is an online mapping tool designed to act as a common operating picture that integrates both static and real-time data, such as Environmental Sensitivity Index (ESI) maps, ship locations, weather, and ocean currents, in a centralized format for environmental responders and natural resource decision makers. ERMA® enables a user to quickly and securely upload, analyze, export, and display spatial data in a Geographic Information System (GIS) map. ERMA® provides decision makers with the information needed to make informed decisions for response, damage assessment, and recovery and restoration efforts.

4.2.1.3 Classification of Coastal Environments and Environmental Sensitivity Index

The process of mapping of coastal environments and ranking their relative sensitivity was first used in 1976 for Lower Cook Inlet, Alaska (Michel et al. 1978). Since then, the coastal environment ranking methodology to aid/expedite response decision making by providing information about resources at risk has been refined and expanded to cover shoreline types throughout the world. ESI maps have been used for oil spill contingency planning and response since 1979. ESI maps provide a succinct visual summary of resources, such as birds, shellfish, drinking water intakes, corals, and coastal recreational areas, that are at risk if oil is spilled in that geographic region. ESI maps help responders and planners determine protection priorities in view of sensitive biological resources and human-use resources using specific symbol sets, icons, and hatch patterns (NOAA, 2019) (see Figure 4.2; see Appendix G for ESI map definitions).[1]

---

[1] See https://response.restoration.noaa.gov/resources/environmental-sensitivity-index-esi-maps.

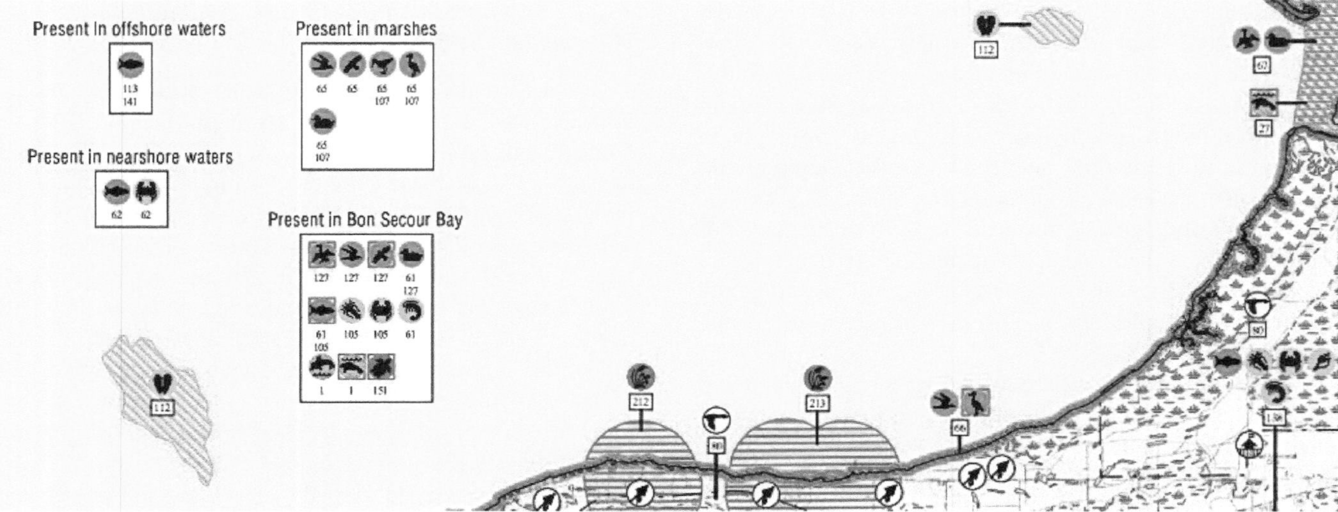

FIGURE 4.2 Example of ESI map.
NOTES: Shorelines on ESI maps are color-coded by sensitivity to oil. Symbols mark localized areas for biological and human-use resources.
SOURCE: NOAA, 2019.

In addition to the NOAA standards, modified ESI mapping methods have been developed in Europe (IOGP and IPIECA, 2012) by a joint effort of IPIECA (the global oil and gas industry association for environmental and social issues), the International Maritime Organization (IMO), and the International Association of Oil and Gas Producers (IOGP).

To archive and disseminate data after an incident, NOAA developed the Data Integration Visualization Exploration and Reporting (DIVER) system (NOAA, n.d.). This system allows users to search and download a wide range of environmental description and project planning data by geographic areas or incident activities. The DIVER tool provides natural resource trustees and the public with the ability to access, query, visualize, and download vast data on environmental pollution, sampling, and restoration efforts. ERMA® provides direct access to DIVER for data query and download.

The DIVER tool is an application for the integration and distribution of Natural Resource Damage Assessment (NRDA)-related impact assessment and restoration data. It also contains historical data collected from oil spills and hazardous waste sites around the country. The DIVER Explorer query tool allows public users to search, filter, access, and download available data. *Deepwater Horizon* (DWH) restoration projects and monitoring data are also incorporated into DIVER.

Significant increases in information gathered and generated during emergency response and natural resource damage assessment phases led to the need to store and process large volumes of data and corresponding metadata. Since the publication of *Oil in the Sea III*, there has been considerable improvement in information and computational technologies; further development is expected to progress at a fast pace in the coming years. The preceding examples are government-provided assets. Additionally, industry-developed electronic COPs and data management solutions used to augment or replace those mentioned earlier are also available. Please refer to Chapter 5 for the more detailed discussion on this topic.

4.2.1.4 Response Toolbox

There is a long list of response techniques and technologies that may be employed during an oil spill (see Figure 4.3). These methods include mechanical cleanup to remove oil, dispersing agents, in situ burning, various shoreline cleanup techniques, etc. (NOAA, 2013a; see Sections 4.2.2, 4.2.3, and 4.2.4). Using an analogy of a toolbox, one must choose the right tool(s) for the right job. If a piece of wood has to be trimmed, a hammer or screwdriver would not be the first choice of tools to undertake the chore. However, the toolbox would also include a saw, clamps, file, knife, etc., to accomplish the task efficiently. Similarly, depending on the type of oil, the location of the spill, water depth, water and air temperature, wind speed, time of day, precipitation, etc., the specific tools that are correct for the response will vary.

The other concept that must be factored into the choice of the best response strategies are environmental and socioeconomic concerns. Ideally, the response techniques chosen will do no further harm and will contribute to mitigating further impacts. The processes of net environmental benefit analysis (NEBA) (IPIECA, 2015) or spill impact mitigation assessment (SIMA) (IPIECA and IOGP, 2018) are used in the ICS Planning Section by the Environmental Unit to identify which tools in the toolbox are most suited to the specific situation (refer to Section 4.2.5 for more details). Monitored natural attenuation is usually a starting point

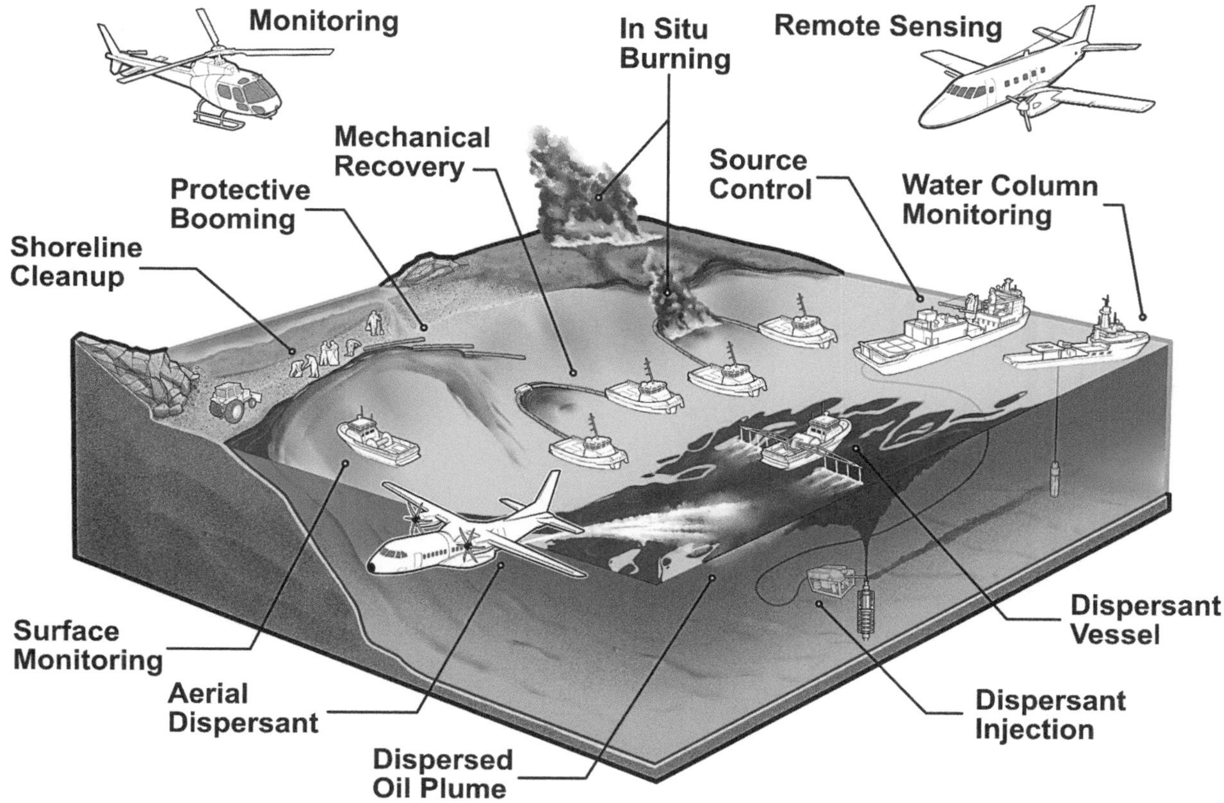

FIGURE 4.3 Oil spill response techniques and technologies.
SOURCE: Image provided courtesy of the American Petroleum Institute, produced by Iron Octopus Productions, Inc.

for weighing appropriateness of the actions, and can be the most appropriate approach. For example, a small to medium size spill of light fuel oil in an offshore area will probably evaporate quickly and may not require further response actions. Allowing the oil to naturally evaporate and biodegrade without application of additional response techniques could, under some circumstances, be the best-case solution.

To demonstrate the concept of tradeoff decision-making in identifying appropriate response actions, Figure 4.4 illustrates the generic "normal" fluctuations over time in the environment under conditions in the absence of a spill, and the potential gradual recovery time after an incident with no cleanup response actions, shortened recovery time with the application of cleanup options 1 and 2, and potential longer recovery time after cleanup option 3. Using a net environmental benefit analysis, Option 3 would most likely not be implemented. Eventually, with time and possible restoration activities, the response recovery lines will approach the conditions in the absence of a spill (blue line).

Oil spill response equipment may be owned by individual companies or by federal, state or local governments, but the majority of it is held by the oil spill removal organizations (OSROs). Following the OPA 90 and its requirement for certain U.S. facilities to have plans and equipment to respond to spills, various OSROs have been established. These organizations are voluntarily classified by the USCG. They are classified by the Captain of the Port (COTP) Zones by spill size, removal capacity, response times, and by different operating environments (e.g., rivers and canals, inland, great lakes, and oceans). These organizations are responsible for holding, maintaining, and deploying appropriate response equipment and trained personnel at the time of response as requested by their stakeholders. The USCG uses the Response Resource Inventory[2] (RRI) to maintain a comprehensive list of spill removal equipment held by OSROs. It contains information about response resources (including ownership and location) in 12 equipment categories (boom, pumps, vessels, skimmers, dispersants, dispersant delivery systems, vacuum systems, beach cleaners, portable storage, oil/water separators, fire-fighting equipment, and logistical support equipment) and also tracks response-trained personnel. The RRI aims to improve the effectiveness of deploying response equipment to an oil spill, and may be used to develop contingency plans.

[2] See https://cgrri.uscg.mil/logon.aspx?ReturnUrl=%2f.

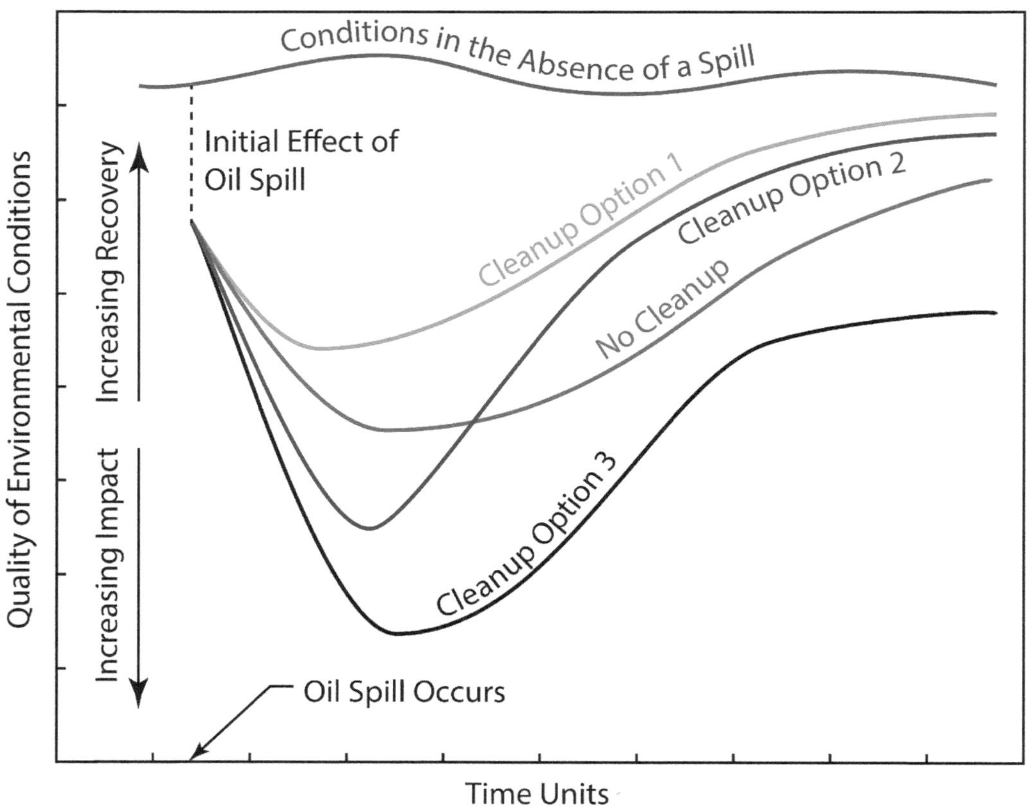

FIGURE 4.4 A conceptual model of potential impacts and recovery over time as derived from modeling and experience considering several cleanup options as compared to pre-spill conditions or no response actions.
SOURCE: Adapted from NOAA.

4.2.1.5 Response Research and Development

The field of oil spill response has similarities to the medical profession. Both have professionals with advanced degrees and many years of practice. Both rely on science and experience for decision-making and advancement of knowledge. In each case, for operational application best restorative practices using the most appropriate available technologies are implemented. Researchers (academic, government, and industry) are at the forefront of discovery and innovation. Their laboratory results have to be tested and assessed prior to being implemented on a broad level in the field or general population. Each relies on a prescribed theory of triage to help prioritize the locations and types of treatment. There is a large cadre of individuals involved to assist at all levels of care to ensure the best outcome for the patient or environment from the activities pursued. There are numerous treatment options (tools in the toolbox) to choose from and while each of them has their benefits and tradeoffs, only options that are anticipated to produce positive results with minimal negative effects are implemented. Stakeholders (trustees or patients) have a voice in the process as to what may work best in a particular situation. Each individual and situation is unique and the treatment options vary from place to place and time to time as knowledge and technology improve.

The preferable time for doing oil spill response research and development is not during an actual emergency but rather in preparation for one. Government agencies (NOAA, USCG, the Bureau of Ocean Energy Management [BOEM], BSEE, Environment and Climate Change Canada [ECCC], Department of Fisheries and Oceans Canada [DFO], Texas General Land Office [TGLO], etc.), industry, private companies, nongovernmental organizations (NGOs), and academia have well-developed and administered programs: they are continually looking into improvements for response technologies, identifying the effects of oil on the environment, developing better ways to monitor oil in and on the water, and looking to increase the efficiency of response operations and organizations. These projects generally take place in a laboratory or test tank. **To verify laboratory experiments in real-world conditions, in addition to modeling and reviewing data from previous incidents, field testing of different response countermeasures is needed.** While laboratory experiments can be conducted with crude oils and petroleum products, receiving a regulatory approval for a field experiment with hydrocarbons has been a challenge.

Researchers and responders attempted to simulate an oil spill in the field using various surrogates—popcorn, oranges, soy and canola oil mixtures, tree barks, food dyes, etc., but they found that the complexities of oil's chemistry, fate and behavior discussed in Chapters 2 and 5 cannot be simulated using these substitutes. Field experiments with real oils and petroleum products are needed to further advance oil spill science and optimize oil spill response techniques. Use of novel experimental techniques are not generally approved during emergency response situations. However, in advance of an accidental oil spill, planning bodies (e.g., Area Committees and Regional Response Teams in the United States) could have mechanisms in place to allow "small science" projects to occur on a portion of a release to be able to make informed scientific measurements and conclusions on different products in collaboration with industry and academia.

To establish a uniform and impartial method to determine the level of development of new response technology, and to determine when it is ready for field use, BSEE has developed technology readiness levels (TRL) (see Table 4.1). These range from TRL 1, with basic science exploration for future technology applications to TRL 9, where the technology is deployed for a real spill (Panetta et al., 2016).

In order to help sift through the many new ideas and suggestions that may be presented to an incident commander during a response, the Alternative Response Tool Evaluation System (ARTES) (API, 2013a; NOAA, 2019b) may be implemented. ARTES allows a special response team to swiftly review proposed response tools and provide a recommendation to the on-scene coordinator. This system allows people who submitted response ideas to track their proposals and progress through the assessment process. During the DWH incident, 123,000 individual ideas were submitted and tracked; 470 made the initial cut, 100 of those were officially reviewed, and about 30 were executed during response field operations resulting in incremental improvements. Of the original 123,000 submissions, there were about 80,000 ideas submitted for subsea response and about 43,000 ideas for surface oil slick response. It should be noted that this process took tremendous effort and a significant number of personnel. Such evaluations require careful consideration in the context of specific response needs and resources available to make sure that evaluation efforts do not reduce overall efficiency of the response. Development of oil spill response science requires cooperative actions from industry, governmental agencies, and academia internationally. Many projects are conducted collaboratively through joint projects to ensure incorporation of the best available and diverse expertise and to facilitate the widest dissemination and adoption of new knowledge and technologies. There are a number of standing conferences, regular workshops and committees that bring together all interested parties and facilitate information sharing. The International Oil Spill Conference (IOSC) has taken place since 1969 and served as a primary venue for sharing and documenting oil spill research and best operational practices. All of the proceedings are available online free of charge at the International Oil Spill Conference Proceedings website.[3] More recently, additional conferences moved to fill the tri-annual cycle with venues rotating among the United States (IOSC), Europe (Interspill), and Australia (Spillcon). Another long-standing venue is the Arctic and Marine Oilspill Program (AMOP). AMOP began in 1977 and still serves as a meeting place for oil spill and environmental scientists to share their experiences and ideas. The Industry Technical Advisory Committee (ITAC) has been organized annually since 1996 to bring together scientists, regulators, and spill response practitioners to discuss new developments in the field and facilitate their practical adoption in oil spill response operations.

The Interagency Coordinating Committee on Oil Pollution Research (ICCOPR) is a 15-member body established by the Oil Pollution Act of 1990 to "coordinate a comprehensive program of oil pollution research, technology development, and demonstration among the federal agencies, in cooperation and coordination with industry, universities, research institutions, state governments, and other nations, as appropriate, and shall foster cost-effective research mechanisms, including the joint funding of the research" (ICCOPR, 2015). The American Society for Testing and Materials (ASTM) F20 Committee on Hazardous Substances and Oil Spill Response was formed in 1975 and is still developing and updating documents and standards relevant to hazardous substances and oil spill response. Other meetings and information-sharing sessions are organized through various research programs and agencies such as BSEE, USCG, TGLO, NOAA, the Coastal Response Research Center,

TABLE 4.1 Oil Spill Technology Readiness Levels (TRLs) Used by BSEE for Research and Development Prioritization

TRL	Title
Technology Research and Development	
1	Basic principles observed or reported
2	Technology concept and speculative application formulated
3	Technology proof of concept demonstrated
Technology Advancement, Development, and Demonstration	
4	Technology prototype demonstrated in laboratory environment or model scenario
5	Technology prototype tested in relevant environments
6	Full-scale prototype demonstrated in relevant environments
Technology Implementation in Operational Environments	
7	Integrated technology tested on a large scale or in open water
8	Final integrated system tested in real or relevant environment
Technology Deployment in Real Spill Environment	
9	Final integrated system deployed in real spill environment

SOURCE: Panetta and Potter (2016).

[3] See https://meridian.allenpress.com/iosc.

the Oil Spill Preparedness Regional Initiative, ECCC, DFO, the Canadian Association of Petroleum Producers, API, IPIECA, ITOPF, the American Chemical Society, U.S. Navy, etc. Additionally, there are focused multi-year, multidisciplinary research programs with a fixed timeline for funding, such as the recently completed Gulf of Mexico Research Initiative program in the United States and the ongoing Multi-partner Research Initiatives program in Canada. An *X Prize* competition, such as the Wendy Schmidt Oil Cleanup XCHALLENGE, is another example of a funding mechanism fostering innovation on a specific response topic. **All of these programs have generated invaluable advances in oil spill response knowledge and technologies and facilitated global sharing and deployment of this information.**

4.2.2 Monitoring and Assessment

Over the past two decades, significant advances in sensing instrumentation have occurred. These smarter, more sensitive, broader-spectrum, smaller, and more varied devices can now track spills from above and below the ocean surface as the oil spreads and moves through the environment. We consider monitoring and assessment platforms to be the systems that maneuver sensors through the environment. Sensors themselves make and record observations. In general, multiple sensors can be installed in any observing platform, and many sensors are suitable to multiple platforms. This section is organized by considering the different platforms deployed for monitoring and assessment. Platforms available in the toolbox for deploying sensors for oil visual and chemical observations include moored instruments, equipment casts from the vessels, subsurface and surface vessels (including manned and autonomous), aircraft, and satellites (API, 2013a; IPIECA, 2016; Fingas, 2018; IPIECA, 2021).

A new essential component of oil spill response is remote sensing. It is the science of obtaining information about areas or specific objects from a distance, typically from aircraft, vessels or satellites. The public expectation is that remote sensing now allows for the precise mapping of oil spill extent and location. However, in a practical sense, the ability to accurately map oil either on or in water is not fully achievable, especially in the water column. Whether remote or in situ sensors are used, the observations must be geolocated. Response personnel can use location information to implement countermeasures to minimize the effect of pollution. Remote sensing can also provide information about illegal discharges from ships (Fingas and Brown, 2018). For a given mission, a combination of remote sensing systems may be needed. Different data end-uses, including documentation of spill location, enforcement or cleanup support, or documentation of affected resources, may require a specific and differing characteristics of the data such as resolution. Remote sensing data collected during the response phase can also be very valuable for the NRDA phase.

Figure 4.5 illustrates a variety of remote sensing platforms that may be involved in oil spill surveillance.

4.2.2.1 Surface Oil Detection and Monitoring

Aerial and surface remote sensing is used in spill response for several types of missions:

- Oil on Water Observations:
  - Mapping and documentation of the area, slick thickness, and percentage cover in time and space
  - Verification/"ground-truthing" of satellite imagery and modeled trajectories
- Support to Tactical and Strategic Countermeasures:
  - Identify most suitable areas of the slick for mechanical recovery, in situ burning, and surface dispersants
  - Monitor effectiveness of response strategies including monitored natural attenuation
- Resources at Risk Observations:
  - Monitor and document presence and absence of wildlife in the area
  - Monitor environmental, socioeconomic, and cultural resources at risk
- Shoreline Cleanup Assessment Technique (SCAT) Support:
  - Aerial surveillance of the shoreline to document baseline conditions
  - Aerial surveillance of the shoreline to document and quantify oil on the shorelines

In addition to response activities, remote sensing could also be used to gather information related to the NRDA; for example, documentation of pre-impact baseline conditions at reference sites, documentation of impacts, as well as progress of restoration projects. These above missions could be delivered through a variety of remote sensing platforms.

Aircraft

A trained observer in an aircraft, either a helicopter or fixed-wing platform, can be rapidly deployed to provide feedback concerning the location, surface coverage, movement, physical description of the product, surrounding environmental and wildlife concerns, and qualitative assessment of volume. Limitations on this method include the platform operating parameters (weather, altitude, light, etc.), ability to see a limited area at a time, darkness, cloud cover, and sun glint. With the current technology, aircraft can plot exact locations (using GPS) of observed oil and other features of concern, map the outline of slicks, transmit the observations (in near real time), and give a human analysis of the situation.

Cameras with GPS can capture images for display in the command post to show examples of the types and degree of oiling and other concerns. These images can also be used as evidence in post-incident legal matters. Both still and video images are useful. Light-enhanced (night-vision) cameras can be used in the dark to help identify oil.

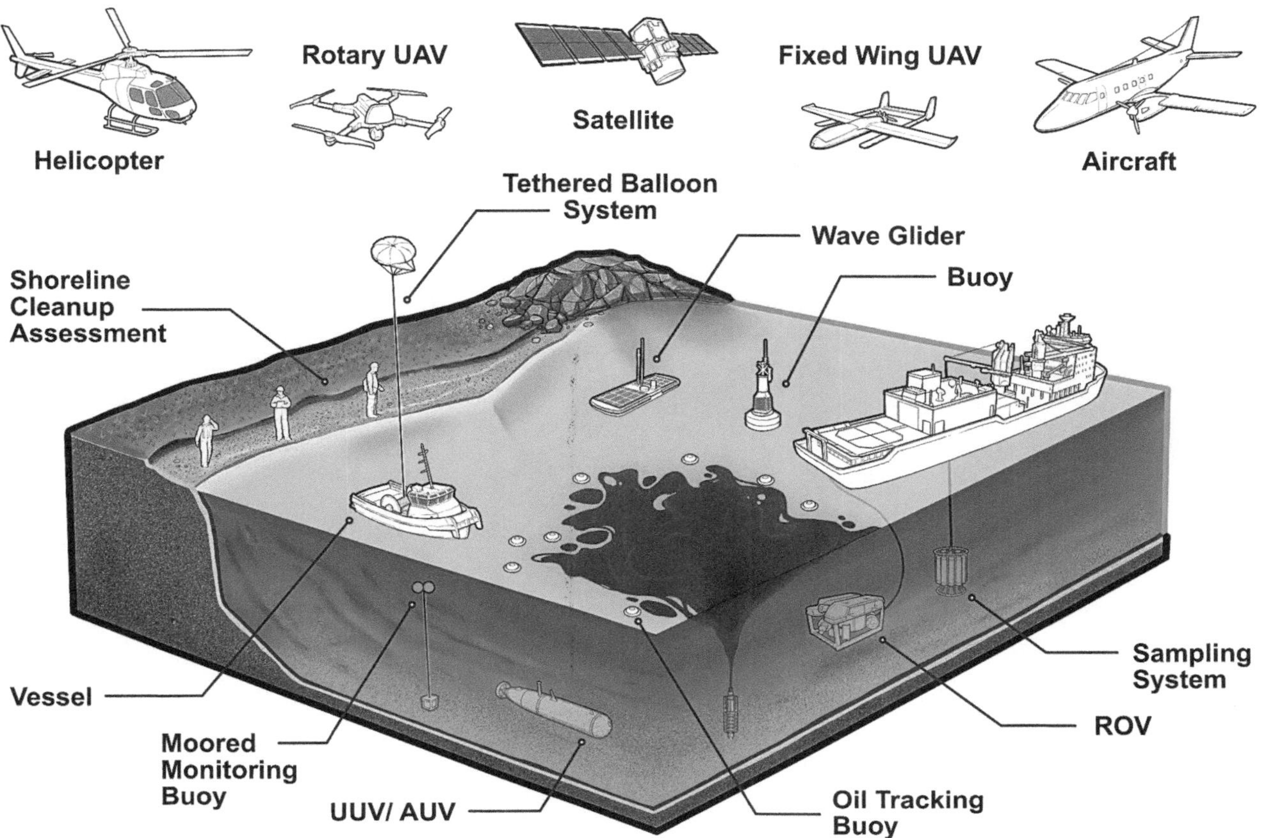

FIGURE 4.5 Remote sensing platforms where sensors can be placed for oil spill surveillance.
NOTE: AUV = autonomous underwater vehicle; ROV = remote operated vehicle; UAV = unmanned aerial vehicle; UUV = unmanned underwater vehicle.
SOURCE: Image provided courtesy of the American Petroleum Institute, produced by Iron Octopus Productions, Inc.

Sensing instrumentation used from an aircraft typically include synthetic aperture radar (SAR), Side Looking Airborne Radar (SLAR), U-cameras operating in ultraviolet (UV), visible and infrared (IR) ranges, and laser-induced fluorosensors. The API Planning Guidance for Remote Sensing in Support of Oil Spill Response (2013) and the paper by Fingas and Brown (2017) provide detailed overviews of various sensors, their strengths, limitations, and operational conditions. Sensors such as IR sensors are referred to as "passive" sensors because they passively detect slick radiation, as opposed to "active" radars or laser fluorosensors that emit energy for slick detection. The difference in oil slick emissivity, heat capacity, and thermal conductivity compared to the surrounding water allows observers to distinguish it using UV/IR sensors during day or night. Figure 4.6 illustrates the difference in slick mapping done by IR and UV sensors. UV sensors tend to map the entire extent of the slick including thin sheen, which is helpful for the environmental impact assessment, where IR sensors map thicker portions of the slick, which helps to direct response resources to the areas where they can be most efficient.

Radar techniques can operate day and night and under cloud cover. They detect the capillary waves on the sea surface and any dampening effect produced by the floating oil. As there are multiple substances that can cause this dampening effect, these images must be verified to confirm presence of a slick. Although not frequently used due to logistics constraints, laser fluorosensors can detect oil by emitting laser light and detecting the fluorescence emitted by the hydrocarbons. Because different substances emit fluorescent light at different wavelengths, this technique allows one to discriminate between different types of oil and other materials that may be present on the water surface or shoreline.

There are modern advanced multispectral and hyperspectral systems that integrate the feed from multiple sensors for maximum detection accuracy. Because no one sensor is 100% effective in every spill scenario, it is important to have access to multiple remote sensing tools for optimal data collection and interpretation suitable for various missions. Several government-owned and commercial planes offer integrated solutions with a variety of sensors that allow tailoring of data acquisition and processing to a specific scenario and data needs.

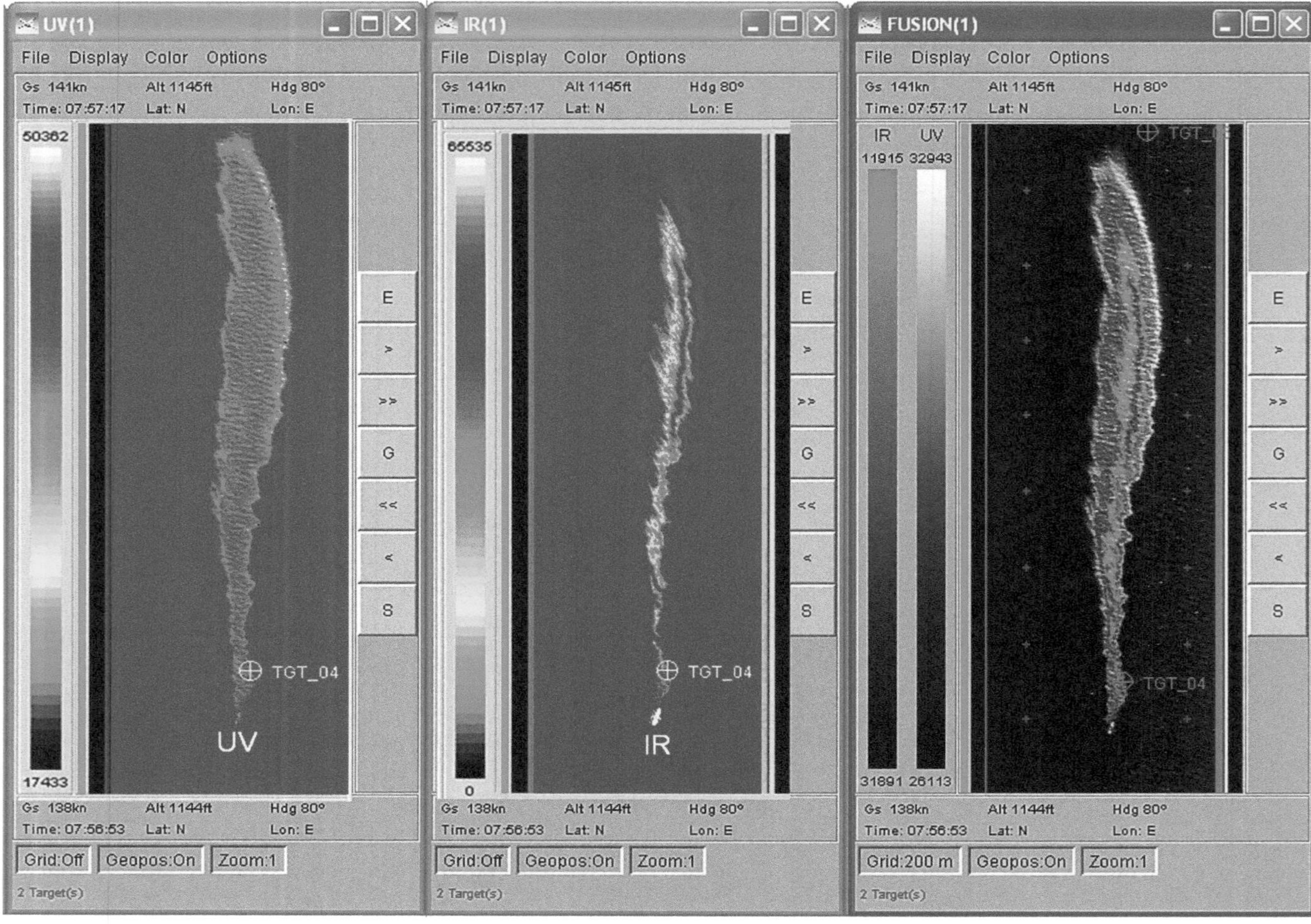

FIGURE 4.6 Examples of oil slick mapping using UV sensor (left), IR sensor (middle), and a fusion of both images (right).
SOURCE: Fototerra Aerial Survey, LLC.

Aerostat or Balloon

Usually tethered to an oil spill response vessel (OSRV), an aerostat or a balloon may carry passive remote sensors similar to those on an aircraft, including optical and thermal infrared camera systems. These systems may improve the OSRV's response ability to locate and position the vessel to best contain and remove the thickest oil, thereby maximizing response and recovery operations.

Unmanned Aerial System or Drone

These platforms have been used for both shoreline surveys and open water reconnaissance. Currently, in the United States, drone flights have to remain line-of-sight operations (except with waivers from the Federal Aviation Administration [FAA] or for the U.S. Department of Defense). This may be an impediment to their larger operational use. Depending on the platform size, the weight and dimensions of the attachable instrumentation may be limited. They may be fitted with one or multiple of the previously described instruments.

Vessel

Some of the X-band marine navigation radars used on the vessels can be reconfigured and accompanied with data processing software to map a slick in proximity to the vessel. Just as other radars do, it detects the areas with dampened capillary waves due to slick presence.

Satellite

Satellite remote sensing tools and best practices have been reviewed in several publications (Partington, 2014; IPIECA, 2016). Satellite imagery can provide the outline or footprint of the extent of surface oil. However, having the outline does not provide detailed information about oil coverage, as oil slicks are patchy with great variability in the thickness of the oil. The outline of the oil slick does inform on scene observers to collect additional information. Analyses using accrued knowledge of slick behavior and visual interpretation of images are starting to allow remote sensing to define not only the extent but also identify the thickest parts of the slicks (Garcia-Pineda et al., 2020).

The application of visual satellite images for oil spills has been undertaken over the past few decades. QuickBird, WorldView I and II, GOES East (GOES-16), and GOES West (GOES-17) now provide recurrent satellite coverage of the Earth's surface. Various wavelength data from multi-spectral satellites (MODIS and MERIS [Medium Resolution Imaging Spectrometer]) provide additional, new, means of Earth observation. Detecting oil spill from remote sensors in the visible spectrum depend on weather conditions, oil types, and view angles. Cloud cover and sun glint inhibit detection in visible imagery. Sun glint is often severe and can obscure an entire scene; however advances to remove sun glint have been made. Several visual imaging systems were employed during the DWH spill. A multi-spectral image derived from MODIS was corrected by an automated classification system to improve the image and classify oil on water. (Fingas and Brown, 2018).

The Marine Pollution Surveillance Report (MPSR) (see Figure 4.7) is a product package, generated in the NOAA National Environmental Satellite, Data, and Information Service (NESDIS) Satellite Analysis Branch (SAB), when a marine anomaly is identified in or approaching U.S. waters and believed to be the result of an accidental or intentional oil discharge. Most often potential oil slicks are detected through the analysis of multispectral satellite imagery and synthetic aperture radar, but are sometimes identified through other surveillance means such as aerial photography. Anomalies are identified based on visual inspection, and through the use of various auxiliary datasets including an automated oil spill mapping tool. Visible imagery is advantageous because different color combinations can be analyzed to distinguish oil from vegetation. Radar imagery is advantageous because it can "see" the surface of the ocean during night time hours and through clouds.

Satellite SAR (image resolution of order 10 m) has been used to define areas containing oil and as "oil free," additionally it allows for identification of oil emulsions within an oil slick and rapid classification of oil types.

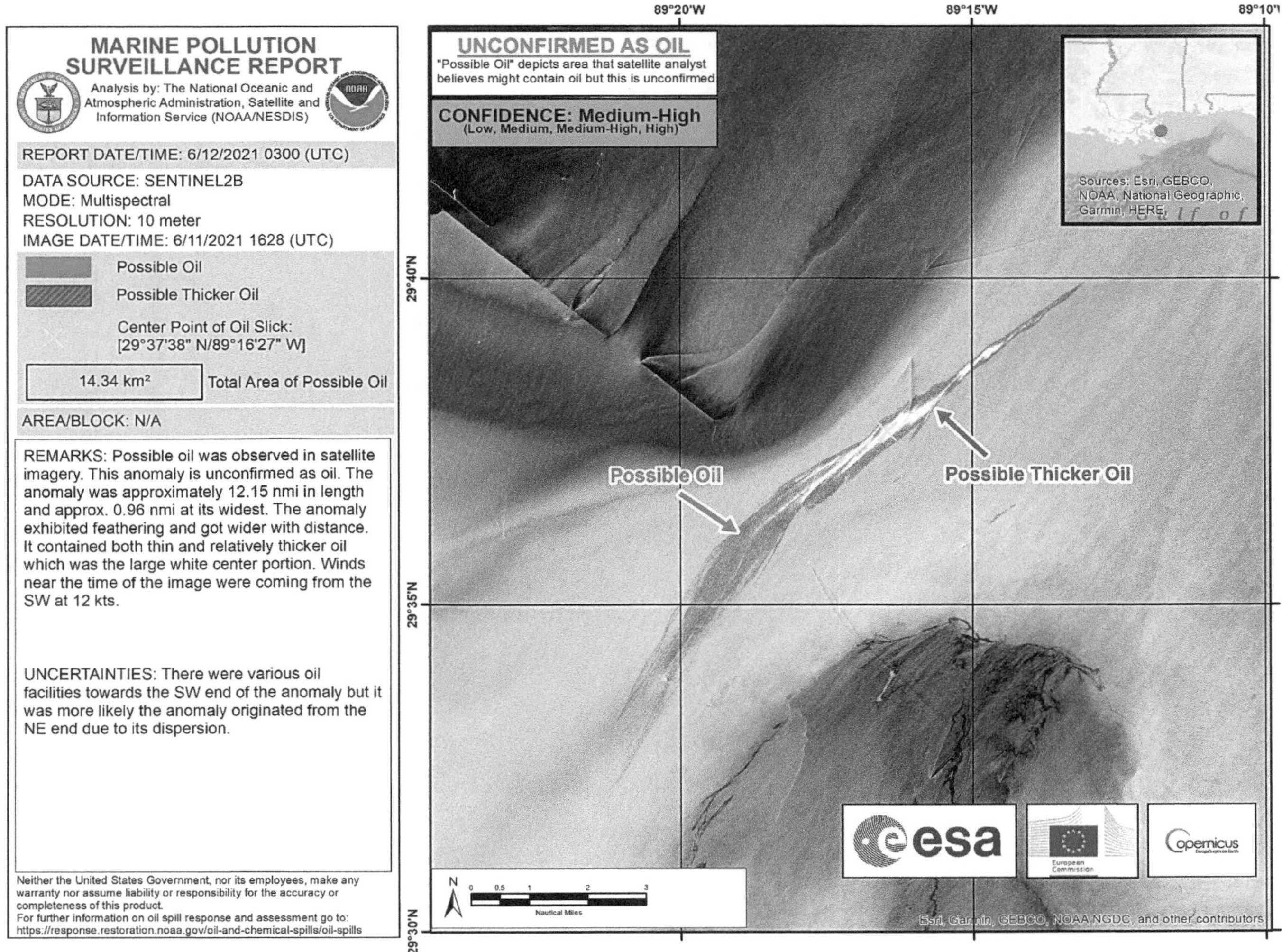

FIGURE 4.7 Satellite image with annotations.
SOURCE: NOAA NESDIS.

SAR imagery provides information about thick oil and oil emulsions (i.e., "actionable" oil) that can quickly aid responders in the field. At the Ohmsett (the National Oil Spill Response & Renewable Energy Test Facility in New Jersey) experiments determined that during certain viewing conditions, differences can be detected between thick stable emulsions and non-emulsified oil using a single polarization SAR image when taken from a sensor at elevation (Garcia-Pineda et al., 2020). Novel remote sensing methods have the potential to be used for oil spill responses by providing fast information to emergency responders. It is important to note that the data outputs are instrument-dependent because the sensitivity of the instruments is highly dependent on the analysis of backscatter in different frequency bands and different methods for noise elimination. At this time, these experimental technologies allow the comparison of relative oil thickness, but are not sensitive enough to quantify specific volumes.

Aerial images and satellite images can be used for tracking the oil as it moves at sea. The time lapse imagery from the MODIS (Moderate Resolution Imaging Spectroradiometer; resolution ranging from 250 m to 1,000 m) instrument, on board NASA's Terra and Aqua satellites, was used after the DWH spill for tracking the oil slick (which appears grayish beige in Figure 4.8 and is altered by changing weather, currents, and use of oil dispersing chemicals). Terra MODIS and Aqua MODIS view Earth's entire surface every 1 to 2 days, acquiring data in 36 spectral bands, or groups of

FIGURE 4.8 Satellite images of oil after the DWH spill, acquired on May 24, 2010.
NOTES: These images were captured by the multi-angle imaging spectroradiometer (MISR) instrument. The left-hand image contains data from MISR's vertical-viewing camera. It is shown in near-true color, except that data from the instrument's near-infrared band, where vegetation appears bright, has been blended with the instrument's green band to enhance the appearance of vegetation. The right-hand panel was constructed by combining data from several MISR channels. In this false-color view, oil appears in shades of inky blue to black; silt-laden water due to runoff from the Mississippi River shows up as orange, red, and violet; and land and clouds appear in shades of cyan.
SOURCES: NASA/GSFC/LaRC/JPL, MISR Team. https://www.nasa.gov/topics/earth/features/oil20100602.html.

wavelengths. Recently, a new breed of tiny, modular satellites (Cubesat) offers a wide variety of imagery, depending on the location, with resolutions up to 5 m.

4.2.2.2. Oil Detection and Monitoring in the Water Column

A comprehensive monitoring program is critical to effective oil spill response (Payne and Driskell, 2015). Several recent publications, including the IPIECA/IOGP Good Practice Guide for in-water surveillance of oil spills at sea (IPIECA, 2016), describe tools and processes used to monitor oil spill fate and behavior in the water column (see Section 2.1.7). Generally speaking, water column monitoring involves the following tools deployed from different platforms, as shown in Figure 4.5.

Vessels

Vessels may house both sensor systems and other sensor platforms. Sensing systems may include visual reconnaissance and camera and video systems to observe surface floating oil, air quality monitors, continuous flow-through water quality analysis, and acoustics-based observing systems, such as sonar (split-beam and multibeam) and acoustic doppler current profilers (ADCPs). Sensor platforms fixed to the boat include conductivity, temperature, and depth (CTD) profilers and ROVs, discussed separately later in this chapter. Vessels thus provide a mobile hub for bringing experts, sensing systems, and sensors to the spill site and the affected area.

Split-beam and multibeam sonar are traditionally used to measure seafloor bathymetry, subsurface properties, and large pelagic organisms (e.g., fish). In oil spill science, sonar may also be used to observe oil droplets or gas bubbles as they traverse the water column. Oceanographic sonar is characterized by its insonification frequency, with lower frequencies being able to traverse longer distances. Lower frequency sonar also has longer wave-length sound forms. Objects smaller than the acoustic wave length typically are not observable by sonar and are considered acoustically transparent. Gas bubbles are an exception to this rule as the bubble-water interface is excited by a wide range of typical sonar frequencies, emitting a loud signal despite the bubbles being smaller than the acoustic wavelength (Weber et al., 2014). In contrast, oil droplets, which have densities similar to sea water, remain acoustically transparent to sound waves with wavelengths greater than their diameters. Hence, multibeam sonar has become a major tool for observing natural gas seepage in the oceans. Oil droplets may be observable in short-range, high-frequency sonar, but are less evident in long-range, low-frequency data (see Section 5.3.1 for more on natural gas seeps).

ADCPs are used to measure water currents in integrated bins over a profile under a ship. ADCPs can be mounted in a vessel hull or incorporated into other observing platforms (e.g., ROVs or moorings, described in the following text). ADCPs differentiate themselves predominantly by frequency, with the lowest-frequency ADCPs penetrating to about 1,100 m depth. Data are reported within fixed depth bins. Some floating oil and gas exploration and production platforms also include ADCPs, and these data are often uploaded in real time to public repositories. Especially in the nearfield of a subsea release, measured currents using ADCP are extremely beneficial to response modeling.

Aromatic hydrocarbons, which are highly fluorescent, are often a major constituent of typical crude oils; therefore, fluorescence techniques can be used to detect oil. These techniques typically excite oils using ultraviolet wavelengths (300–400 nm) causing fluoresce in the visible wavelength range from 400–600 nm. The specific fluorescent traits are determined by the individual compound's carbon structure. The addition of additives will also alter the fluorescence characteristics of oils.

Ambient ocean properties are commonly measured using a sensor package that records conductivity, temperature, and depth (pressure) (CTD). These sensor platforms often also house a wide array of other sensors and may be deployed with additional sampling equipment. The CTD package is deployed from a wire connected to a spool on the ship deck and controlled by a deck computer. Conductivity, temperature, and depth are observed to produce density profiles of the ocean water column. Other sensors often integrated with the CTD include oxygen sensors, fluorometers calibrated to observe aromatic hydrocarbons, dissolved organic matter, phytoplankton, or other fluorescent constituents, and transmissivity sensors, among others. These sensors may be deployed together with a rosette system for collecting water samples in specialized bottles (e.g., Niskin bottles) throughout the water column.

Buoys

Buoys can be either stationary or free-floating. Stationary buoys provide fixed, moored platforms for deployment of various sensor packages. Many buoys are operational, providing continuous, real-time or near real-time monitoring of ocean properties. Others may be deployed strategically as part of a spill response. Generally, a stationary buoy consists of an anchor weight and a buoyancy package providing upward tension on a line between the anchor and buoy. A buoy cable may extend to the sea surface or may be suspended anywhere in the water column. Most sensors attached to buoys are located at a fixed point, attached to the buoy or buoy wire. Sensors include those described earlier for CTDs, ROVs, or AUVs. They may also include point current meters, wave sensors, and weather stations. ADCPs may be mounted near the surface profiling downward or in the water column profiling either upward or downward. Some complex buoys also include a wire walker that continuously moves a CTD up and down through the ocean water column.

Drifters

Floating buoys or tracking buoys (drifters) are designed to float together with an oil slick. They may be powered by a battery or solar power and may include satellite telemetry to help track an oil slick trajectory over time. Two broad classes of drifters are utilized in ocean sensing: surface floating drifters and sub-surface drifters. Subsurface drifters achieve neutral buoyancy at a selected isopycnal depth and rise to the surface occasionally to transmit their observed data. More than 2,000 such floats are continuously monitored around the globe. Surface drifters have a surface float and data communication system, sometimes attached by a tether to a subsurface sail, designed to track surface slicks (no tether and sail) or the currents at a given depth near the sea surface (with tether and sail), usually within 1 to 2 m water depth. Economic, biodegradable surface float technology has grown significantly since the DWH oil spill through research focused on circulation in the Gulf of Mexico (Özgökmen et al., 2016; Dannreuther et al., 2021).

Remotely Operated Vehicles

ROVs are deployed from a ship and continuously powered and controlled from a ship-board control room connected to the ROV over a wire tether. ROVs come in a wide array of sizes and are built for a multitude of purposes. In oil spill response, ROVs may be used downstream of the spill to monitor the ocean water column or within the response zone to conduct subsea operations, including subsea dispersant application. ROVs used in monitoring normally carry a high-definition video camera and other sensor systems similar to those installed on CTDs or ship vessels, including sonars and ADCPs. Because ROVs may be large and provide significant power subsea, they may also be integrated with more sophisticated sensors, including mass spectrometers and gas chromatography/mass spectroscopy (GC/MS) systems, though these sensors can generally be integrated with any platform that can provide enough space and power (Chua et al., 2016). Two advantages of an ROV over a CTD or AUV (see next heading) are that the operator has real-time video footage of the sample location and that the sensor platform can be piloted to enable sampling of specific locations.

Because ROVs can be maneuvered using real-time video feeds, they are an ideal platform for installing sensors designed to observe oil droplet and gas bubble size distributions. During the DWH oil spill, oil droplet sizes were measured by water samples collected from CTD Niskin bottles on a vessel deck and using an under-water holographic camera in the subsurface intrusion several kilometers from the wellhead (French-McCay, 2015). Niskin samples were analyzed using bench-top laser in situ scattering and transmissivity (LISST) instruments. This general technology uses the scattering of light by small particles to infer a particle size distribution. Current instruments can be deployed to great ocean depths and measure particles 500 microns in diameter and smaller. The LISST does not include a visual image; hence, additional observations are needed to ensure that the measurement corresponds to oil droplets. Other size distribution measurement methods utilize cameras. To avoid parallax error, a recent advancement utilizes collimated light to produce silhouette images with a camera (Brandvik et al., 2021). This has the added advantages that it may observe droplets or gas bubbles larger than 500 microns, does not assume that fluid particles are spherical, and can distinguish among gas, oil, and non-hydrocarbon particles (e.g., marine snow). Although an ROV may be the best platform to maximize effectiveness of these instruments, they can be deployed on CTD, AUV, and other platforms during response.

Autonomous Underwater Vehicles

AUVs include a wide range of platforms designed to carry out their missions autonomously, not connected to a ship or controlled in real time during their sampling. AUVs are mainly distinguished by their propulsion mechanism. Gliders move by adjusting their buoyancy such that they alternate between sinking and floating and by hydrofoils that produce horizontal motion from their vertical descent and ascent. Wave gliders likewise use the upward and downward motion of ocean waves to produce horizontal motion. Propeller-driven systems rely on battery power for most of their propulsion. These systems may carry sensors similar to those on CTDs and ROVs. AUVs are further outfitted with inertial navigation systems, communications systems, and the ability to carry out autonomous surveys. Over the past decade, advancement in autonomous vessels has increased rapidly, including improvements to navigation control and safety (Gu et al., 2021). Recent advances employ machine learning and on-board sensors to produce adaptive surveys. Long-range AUVs, such as gliders, receive new instructions over remote communications at intervals. Short-range AUVs conduct a survey and then surface, transmitting their location for pick-up.

The development of AUV technology for sea patrol and environmental monitoring included structural design, hardware and software design equipped with oceanographic sensors, surveillance cameras, and underwater cameras for several applications. Wireless communication with 2.4 GHz frequency has been utilized over short distances for monitoring, control, and real-time communication between a base station and the AUV.

Multiple AUVs were deployed during the DWH oil spill to observe and track the subsurface intrusion layer that formed near 1,100 m water depth predominantly to the west-southwest of the wellhead (Camilli et al., 2010; Zhang et al., 2011). Unique sensors are increasingly being integrated with AUVs, including sensors like the in situ mass spectrometer deployed with the Sentry AUV during the DWH oil spill (Camilli et al., 2010).

Hundreds of oil and gas industry structures in the marine environment are approaching decommissioning (see Section 3.5.2). As decommissioning of these aged structures occurs environmental assessment and monitoring must be conducted and potentially lasting over the life of any assemblies left in place. One solution to the major oversight challenge is to use marine autonomous systems (MASs) to monitor the decommissioned structures. The acoustic, visual, and oceanographic sensors installed on MAS provide necessary data for decommissioning oil and gas structures (Crabb et al., 2019). MAS provide both the considerable potential for cost savings and a dramatic improvement in the temporal and spatial resolution of environmental monitoring. MAS offer viable alternatives where a direct match for the conventional decommissioning monitoring approach is not possible (Jones et al., 2019).

Bottom Samplers

Sampling platforms may also be installed at the seafloor. Some are transported by an ROV, such as a push-core sampler that retrieves seafloor sediments. Others are set in place and capture marine snow and sediment as it reaches the seafloor (e.g., a sediment trap).

Sunken Oil Detection

A specific case of oil monitoring in the water column is detection and mapping of heavy oil that is submerged in the water column or has sunk to the bottom. API developed a technical report and an operational guide (2016a,b) describing techniques for sunken oil detection and recovery. Some of the detection techniques include diver observations, observations with the camera from underwater vehicles, sonar systems, acoustic camera, towed and stationary sorbents, bottom samplers, laser fluorosensors, and water column sampling. Recent experiments have also evaluated the use of marine-induced polarization for oil detection in the water column (Wynn et al., 2017).

4.2.2.3 Oil Spill Detection Above and Under Ice

Arctic environment presents some unique challenges for oil detection and monitoring, such as short daylight hours during winter months, prolonged periods of fog, strong wind gusts, low temperatures, and presence of the snow and ice cover for a significant portion of the year. All these factors complicate and sometimes reduce the efficiency of remote sensing operations. Oil spills in the Arctic or ice-prone environments behave differently from oil spills in open water (see Chapter 5 for more details), which affects remote sensing techniques and strategies. Some of the tools that are used to monitor oil in open water can also be used in the Arctic to detect oil in open water, between, or on top of the ice. Presence of different ice types requires special considerations and different tools for oil detection. Puestow et al. (2013) provided a detailed review of the above-ice remote sensing techniques that can be used to map oil in low visibility and ice. Wilkinson et al. (2014) described the techniques that could be used for oil detection under ice using AUVs. Watkins et al. (2016) developed an operational guide for oil detection in ice-covered waters integrating information from the first two reviews.

Some of the unique tools used for oil detection under ice include ground penetration radar (GPR) deployed from the ice surface as well as a newer method to use it from a helicopter. Detection of oil under ice and snow using trained dogs has also been successfully demonstrated. Oil can also be detected from below the ice using sensors deployed from AUVs or ROVs. Figure 4.9 illustrates these techniques. Watkins et al. (2016) provide additional information on the applicability and effectiveness of these tools for various ice and oil distribution scenarios.

Some of the more experimental detection techniques include nuclear magnetic resonance (NMR), marine induced polarization (IP) (Wynn and Flemming, 2012), as well as C-band scatterometer (Firoozy et al., 2017). Experiments conducted at the U.S. Army Corps of Engineers Cold Regions Research Laboratory showed that marine IP could detect oil under and within frozen ice, floating on the water surface, and floating with broken ice. Marine IP sensors must be deployed in water to observe oil; hence, they may be deployed on moorings, integrated with CTD, ROV, or AUV platforms, or towed behind ships. Fingas and Brown (2015) provide additional information on detection techniques in ice and snow.

4.2.2.4 Special Monitoring of Applied Response Technologies

Since the early 1980s, the response community recognized a need for procedures to monitor response technologies used during oil spills. The increased reception to using dispersants and in situ burning (referred to as applied or advanced response technologies) in most regions of the United States came about because of technological advances. Throughout the United States pre-approval zones for dispersant and in situ burn operations are in place with pre-approval conditions, and requirements for monitoring rules. The Special Monitoring of Applied Response Technologies (SMART) establishes a monitoring system for rapid collection and reporting of real-time, scientifically based information, to aid the Unified Command with directing in situ burning or dispersant operations. SMART recommends monitoring methods, equipment, personnel training, and command and control procedures that manages the counterpoise between rapid response efforts and informed decision making (SMART, 2006).

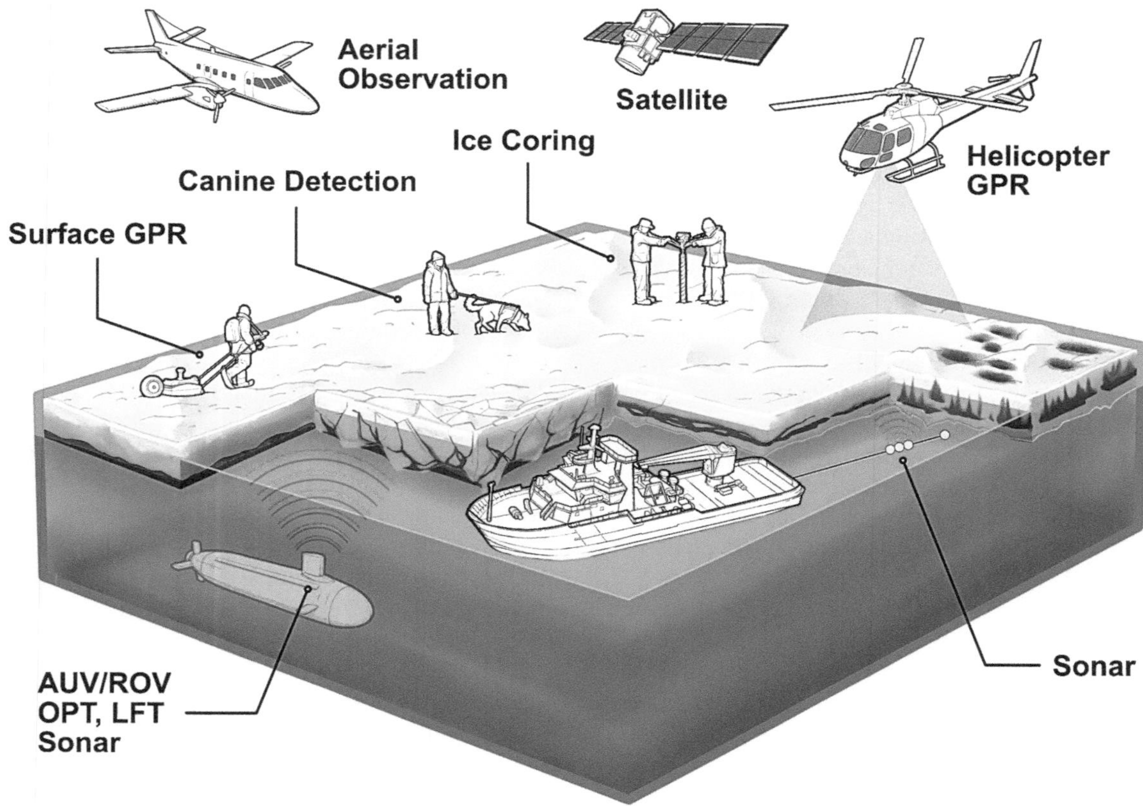

FIGURE 4.9 Platforms and sensors for oil detection in ice.
NOTE: AUV = autonomous underwater vehicle; GPR = ground penetrating radar; OPT/LFT = optical and laser fluorosensor; ROV = remote operated vehicle.
SOURCE: Image provided courtesy of the American Petroleum Institute, produced by Iron Octopus Productions, Inc.

The SMART program was used to monitor the effectiveness of sea surface dispersant use during the DWH oil spill. To quantitatively measure dispersant effectiveness, samples of chemically and naturally dispersed oil where the ratios of total petroleum aromatic hydrocarbon (TPAH) and total petroleum hydrocarbon (TPH) (see Section 2.1.2) were compared, the results showed good agreement with SMART field assessments of dispersant effectiveness. The SMART analyses generated data about acute biological effects of value to the larger scientific community, in addition to the primary goal of providing near real-time effectiveness data to the response (Bejarano et al., 2013).

Monitoring a heterogeneous plume of dispersed oil droplets in very dynamic ocean conditions presents a number of challenges. First, the access of monitoring vessels to the location of aerial dispersant application may be limited by the exclusion zones set around spraying operations, so vessels may be challenged to quickly access the precise location requiring monitoring; second, a dispersed plume may be transported in a different direction than any remaining surface slick thus complicating its tracking; third, by the time surface dispersants are usually applied, aromatic compounds targeted by fluorometers could have evaporated or dissolved from the slick; fourth, in some situations dispersion may occur rapidly before a monitoring vessel arrives on location, or could be delayed due to high oil viscosity or low ocean turbulence. It is important that the readings obtained as a result of the SMART monitoring efforts are put into appropriate context and interpreted by experts. These and other challenges prompted the ongoing development of updates to the SMART equipment and monitoring protocols to optimize its effectiveness and incorporate latest technologies.

4.2.2.5 Shoreline Cleanup Assessment Technique

Shoreline Cleanup Assessment Technique (SCAT) is a systematic method for surveying an affected shoreline after an oil spill to document shoreline oiling and monitor effectiveness of response techniques. The SCAT method originated during the response to the 1989 *Exxon Valdez* oil spill, when responders needed a systematic way to document the spill's impacts on many miles of affected shoreline (NOAA, 2013b).

The SCAT approach documents shoreline oiling conditions by using standardized terminology. The use of standard oiling condition terminology supports decision-making for shoreline cleanup. SCAT is flexible in the scope of its surveys and detail of datasets collected. SCAT is a regular part of oil spill response.

An initial assessment of the shoreline conditions require a SCAT surveys early in the response (see Appendix G), and ideally continuing in advance of operational cleanup. Throughout the response surveys continue to verify shoreline oiling and cleanup effectiveness, and eventually, to conduct final evaluations of shorelines to ensure that they meet cleanup end goals. SCAT survey teams are also trained to look for subsurface oil by digging trenches in locations where oil burial is likely.

The SCAT process includes eight basic steps (NOAA, 2013b):

1. Conduct reconnaissance survey(s).
2. Segment the shoreline.
3. Assign teams and conduct SCAT surveys.
4. Develop cleanup guidelines and endpoints.
5. Submit survey reports and shoreline oiling sketches.
6. Monitor effectiveness of cleanup.
7. Conduct post-cleanup inspections.
8. Conduct final evaluation of cleanup activities.

According to Michel and Ploen (2017) many natural materials could be mistaken for oil (e.g., algal blooms, suspended sediments, bacterial sheens), leading to the need for ground truthing aerial observations. A non-petroleum sheen and petroleum sheens act differently when disturbed, which cannot be conducted from aerial observation. Disturbing a bacterial sheen causes it to separate into small platelets or break like broken glass. A petroleum sheen, on the other hand, swirls and quickly reorganizes after being disturbed. Other techniques to differentiate non-petroleum sheens include:

1. Hexane test, where the sheen is collected with a sheen net. The net is inserted into a glass vial containing hexane, shaken, and allowed to stabilize. A petroleum sheen dissolves in hexane, causing the hexane to discolor. Biogenic sheens do not dissolve in hexane, so there is no change in color.
2. Ultraviolet test, in which hexane vials are viewed under ultraviolet light. Petroleum oils fluoresce, whereas biogenic sheens do not (EPA, 2016).

Final SCAT data and documentation should be input and archived into robust computer databases such as the ERMA® and the DIVER system.

Canine SCAT

Since the publication of *Oil in the Sea III* (2003), the use of canine detection of oil buried in the sediment or under ice has demonstrated improved effectiveness and efficiency of oil spill assessment surveys and leak detection. The methodology continues to be refined and improved as it is used in real oil spill situations, and as our understanding of how and what the dogs are actually detecting increases (API, 2016a).

A canine detection team can be deployed in a variety of field roles, which increase the speed and efficiency of SCAT surveys, improve the level of confidence in oil detection and delineation and provide more timely information for the planning and direction of response efforts. For more in-depth information and an overview that includes 2020 field trials, see Owens and Bunker (2021) and Owens et al. (2021).

Underwater SCAT

There have been several incidents (i.e., DWH, *M/T Athos I* [see Box 4.1], *TB Morris J. Berman*, and *TB Vista Bella*) where oil has become submerged in the nearshore area. Specially trained and equipped SCAT teams have been deployed to map the extent and concentration of oil. This methodology is referred to as underwater or snorkel SCAT. From the information gained thereby, cleanup strategies have been developed and deployed to areas where oiling warranted removal.

Electronic SCAT

The process of performing SCAT surveys has evolved over the decades, and in keeping up with modern practices several variations of electronic SCAT (e-SCAT) have emerged. These have been spearheaded by both governmental agencies and private enterprises. It is now possible to use digital pads or smartphones on shoreline surveys to mark locations, take photographs, fill in forms, read help screens, and upload data to command posts for decision-making determinations and archiving. The speed of data entry and increased accuracy of these digital platforms is a large step forward in the SCAT data management arena.

Drone SCAT

Over the past two decades the ability to remotely operate unmanned aerial vehicles (UAVs) or drones and the degree of precision and high-resolution imaging have greatly improved. Just as remote aerial surveillance can be used for tracking oil at sea (see Section 4.2.2.1), the use of drones to fly shorelines and identify oil has become more prevalent (Allen et al., 2008; Prascal et al., 2014; Tarpley et al., 2014; Muskat, 2021). Drones can access areas quickly, go where people may not be able to access, record images, and—with the aid of various lenses, filters, and artificial intelligence (AI)—be able to identify and interpret items that the human eye may miss. The use of drones is regulated by the FAA and is not appropriate in all situations. As the approach becomes less pioneering, the state of the art continues to evolve, and use becomes more commonplace, the integration of drones will probably become part of the standard SCAT team makeup. See Section 6.5.3.4 for additional examples by Kenworthy et al. (2017) for use of drones for ESI mapping.

BOX 4.1
***M/T Athos I* Tanker Spill, Delaware Bay**

FIGURE 4.10 *M/T Athos I* with an 8° list to port in the federal anchorage off of the CITGO dock in the Delaware River 1 day after the initial oil release.
SOURCE: Ed Levine, NOAA.

On November 26, 2004, the U.S. Coast Guard was notified that the *M/T Athos I* was spilling oil. The *M/T Athos I* was a 750-foot-long, Cypriot-flagged tank ship with a single bottom, double-sided hull built in 1983 (NOAA, 2004). The tank ship was inbound with approximately 13 million gallons of Bachaquero Venezuelan crude oil. The characteristics of Bachaquero crude oil are marginally buoyant, highly viscous and sticky. It is a cargo that is heated, has a high asphalt content, and weathers slowly and can easily form into tar balls.

Due to a "blowout" tide (strong winds blowing from the north caused the river level to be significantly below tide table predictions) the ship's bottom was close enough to the river bottom to come in contact with submerged debris (using side scan sonar, this debris was determined to be a U-shaped pipe, part of a large dredge pump housing, and was later removed). The vessel's hull had been breached by the contact, which caused the vessel to take on an eight-degree list to port (see Figure 4.10), this caused an automatic shutdown of its engines.

Within a few hours, thick oil covered the river extending 6 miles north of the incident, and began to spread. After several days the vessel was stabilized, then tank gauging was possible, and the worst-case estimate of oil released into the river was approximately 473,500 gallons.

It was determined by a dive survey that the vessel had sustained a hole in a center cargo tank, as well as a port ballast tank. The bulkhead between the cargo and ballast tank had been compromised, allowing an unknown amount of cargo to flow into the ballast tank and then spill out into the river. A vessel salvage plan was developed, initially lightering the vessel and then removing the vessel for repairs.

The Delaware River and Bay had 214 miles of shoreline (including creeks) affected by this oil release, in Pennsylvania, New Jersey, Delaware, and Maryland. All of this shoreline was surveyed by SCAT teams with participation from the affected states plus federal and responsible party representatives. Each oiled segment received cleanup recommendations for responders and final sign-off occurred about one year after the spill. Environmental concerns were equally balanced with economic issues through close cooperation between the Environmental and Economic Units in the Planning Section. An extensive wildlife capture and rehabilitation program was established. Additionally, endangered species of eagles and short-nosed sturgeon had to be considered in response operations.

Due to the proximity of the ship's bottom to the river bed (~18 inches), oil released from the hole was injected into the sediment, causing a change in specific gravity. Sediment laden oil became neutrally and negatively buoyant, whereas samples of oil from inside the cargo hold tested positively buoyant. It took several days to discover that there was oil floating below the surface and on the river bottom.

The submerged oil proved to be a challenge for the response operation. Out of precaution, the Salem Nuclear Power Plant shutdown two reactors upon the discovery of oil in the cooling water intake debris screen. This instituted a first-of-its-kind submerged oil tracking and recovery operation.

Cleanup activities and final signoff by three state and federal partner assessment teams of all affected shorelines took more than one year to achieve. Following the OPA 90 protocols, a NRDA was performed and several years later a restoration plan was agreed to and implemented.

4.2.3 Offshore Response

4.2.3.1 Monitored Natural Attenuation and Biodegradation

Fate and behavior of hydrocarbons in the environment depend on their physical and chemical properties as well as conditions of the environment in which they were released. *Natural attenuation* (sometimes called intrinsic bioremediation) is a general term that refers to a combination of natural processes that decrease concentration of hydrocarbons in the environment, ultimately transforming and removing harmful components from the ecosystem (Pequin et al., 2022). These processes include spreading, evaporation, sedimentation, photooxidation, natural dispersion, and biodegradation (see Section 5.2.8). When combined with a regimen of sampling and measurement, it is called monitored natural attenuation. Biodegradation, which relies on microorganisms to convert hydrocarbons into less harmful components, is the ultimate natural attenuation process for removing hydrocarbon components from the environment. In general, light products such as gasoline, diesel, and condensate are not persistent in the environment, because of their high volatility, solubility, and natural dispersion and degradation rates. Typically, small spills of light crude oils are not very persistent, but their fate and behavior are strongly affected by environmental conditions. In warm climates and high seas conditions they may evaporate or naturally disperse within days. In cold climates, however, they could remain in the environment for weeks to months, especially if they freeze into ice or snow (see Section 5.3.5). Medium and heavy products are more persistent in any type of environment, and often require response efforts to mitigate their potential impacts.

The selection of an optimal combination of oil spill response options is made based on the environmental conditions, expected behavior of a spilled product, and resources at risk in the area. In some situations, it may not be safe or feasible to deploy spill response countermeasures. For example, a spill of a light product in high seas may evaporate or disperse naturally, not requiring an active cleanup. Even in calm conditions, light products may spread into very thin sheens that would not be feasible to clean up using mechanical recovery equipment, dispersants or in situ burning. These sheens would be expected to dissipate and degrade in a relatively short time frame. As long as they stay offshore and away from sensitive resources, natural dissipation and biodegradation may be the best "response" option. Although no active recovery operations would then be taking place in the field, identification of resources at risk, forecasting of surface and dispersed oil trajectory, remote sensing monitoring, sampling, and wildlife protection efforts could still be undertaken.

Natural attenuation processes are typically more effective in transforming oil at the water surface and in the water column rather than on the shoreline or in sediments. This is due to high dilution potential and abundance of oxygen and nutrients required for effective biodegradation (see Chapters 2 and 5). If hydrocarbons reach the shoreline or benthic sediments, they can potentially accumulate at higher concentrations in an environment less favorable for the effective removal of hydrocarbons. Even in those situations, it sometimes can be more ecologically sound to leave an oil-contaminated site to recover naturally rather than to conduct an active response. Some examples of such cases include spills at remote or inaccessible locations when oil properties and environmental conditions are suitable for effective natural attenuation or when conditions are too hazardous to risk human health and safety.

The safety of the public and responders is always the highest priority in any response operation. Depending on the spill scenario and environmental conditions, it may be impractical or unsafe to conduct containment and recovery operations. Not all products can be or should be recovered mechanically. Spills of very volatile products or blowouts with gas releases can create hazardous environments for the responders requiring them to leave the area or utilize personal protective equipment (PPE) and strategies that may reduce effectiveness of the response. Very light hydrocarbons (e.g., condensate and gasoline) evaporate rapidly and are typically allowed to do so naturally. They usually spread too thin to contain and could create a safety and exposure hazard if contained by a boom. If safe to do so, light, medium, and heavy hydrocarbon products could be recovered mechanically. Mechanical recovery has a wide window of opportunity and can continue to be effective even as an oil slick weathers over time. A variety of strategies and equipment have been developed specifically for the recovery of heavy and viscous products. Special recovery strategies have also been developed for the containment and recovery of heavy oils that may sink to the bottom (API, 2016d), although these operations would be even more challenging than oil recovery at the water surface.

Studies and field observations have shown that the relationship between oil and fine mineral particles in the nearshore area can play an important role in natural cleaning of contaminated marine shorelines (Lee et al., 2003b). The formation of micro-aggregates between oil and small suspended particles reduces the adhesion of oil to intertidal shoreline substrates and facilitates their removal by tidal action and currents through dispersion, dilution, burial in sediments, and biodegradation. In contrast, oil interaction with large sediment particles, such as sand, may form macro-aggregates and potentially cause oil to settle in nearshore waters as was evidenced during the DWH incident (Gustitus and Clement, 2017) (see Box 5.10 for more details). The latter scenario may require more active clean-up measures.

A spill of a light product that reaches a remote rocky shoreline with an energetic wave environment is not expected to persist and will degrade naturally as wave exposure increases both physical removal and weathering processes. It would not be feasible or safe to deploy recovery equipment and personnel under these circumstances. Another example is a spill of a light product in a marsh area. Deployment of a large amount of equipment and personnel into this sensitive environment will likely cause greater impact by driving hydrocarbons into the marsh soil and destroying the root system of vegetation. It may be better to let attenuation processes remove hydrocarbons naturally rather than to conduct an invasive cleanup (Hoff, 1996). Natural attenuation as a shoreline cleanup method is typically still monitored to assess and document the effectiveness of this process.

Similar considerations are applied for the evaluation of natural attenuation potential of spills in cold climates. Under some circumstances, the presence of extended daylight time and local oil-degrading bacterial communities adapted to cold temperature could facilitate natural attenuation of oil at rates comparable to those in more temperate regions (see Section 5.3.5.2). This is especially true for light hydrocarbons and ice-free environments. Cold temperatures and ice filled waters can significantly affect the fate and behavior of medium and heavy hydrocarbons as well as the rate of natural attenuation processes, but can potentially create favorable conditions with an extended window of opportunity for other response options (Sørstrøm et al., 2010).

While bioremediation is sometimes used to remediate oil spills on shorelines, it is generally recognized that biostimulation (adding fertilizers) or bioaugmentation (adding lab-generated bacteria) are not appropriate methods for remediation of oil spills in the open marine environment (see Appendix E). It is well established that indigenous oil degraders, which exist in every marine ecosystem examined to date are adapted to the local conditions at any given spill site, whereas introduced microbes are at a disadvantage and may not compete with indigenous microbes (Hazen et al., 2016; McGenity, 2018). Also, adding laboratory-grown bacteria to the natural environment may be prohibited by local environmental regulations. Marine environments typically have sufficient quantities of oxygen and nutrients to support effective biodegradation of hydrocarbon products, as long as those products are present in small concentrations and offer substrate suitable for bacterial colonizations (e.g., very small droplets rather than thick emulsified slicks) (Zhu et al., 2001).

4.2.3.2 Mechanical Recovery

The containment and recovery of oil are often effective when responding to relatively small spills especially in calm waters or areas close to large stockpiles of equipment. Rapid deployment of equipment and personnel is critical for the success of mechanical recovery. Oil recovery at sea is always a race against time and natural physical processes which complicate the response. As soon as oil is released into water, it begins to spread and can form very thin sheens, thinner than a sheet of paper. It also breaks up into patches or windrows making it more difficult to collect and recover (see Section 5.2.2.1). Oil spill response vessels have to maneuver between these slicks and collect them into thicker layers to allow for more effective recovery by skimmers. Oil encounter rate—defined as the amount of oil accessed by a skimming system per unit of time—often determines the feasibility and effectiveness of mechanical recovery. Modern skimmers can process large volumes of oil, but it is the ability to collect enough oil in the boom and make it available for skimming that often determines how much oil can be collected in the field.

Mechanical recovery removes spilled oil from the water surface by relying on a complex system of devices and strategies, shown in Figure 4.11. First, remote sensing techniques are used to identify areas of recoverable slicks with a relatively high thickness. Oil on water does not spread uniformly and its thickness in different areas of a slick can vary greatly. These thicker areas generally occupy a relatively small portion of the total contaminated area yet contain the most oil volume. Mechanical recovery equipment deployed in the thickest patches of surface oil will be most effective and can potentially recover 100 times as much oil as can be recovered working on thin sheens. The next step of recovery operations is corralling and containing the oil. Because even the thickest areas of a free-spreading slick are too thin for effective recovery by oil skimmers, specially designed floating fences (booms) are towed by the vessels through the slick to corral oil into layers several inches thick. Skimming devices located in the thickest oil at the apex of the boom then separate oil from water and move recovered fluids into a temporary vessel storage before it is transferred to the shore for recycling or disposal. **Mechanical recovery should be viewed as a complex multi-component system; removal efficiency depends on all components.** This view is reflected in several calculators that aid in estimation of mechanical recovery system capacity and identification of any "bottlenecks" that could negatively affect its efficiency.[4]

Mechanical recovery was used for many years around the world under a variety of conditions. Most accidental releases of oil from a facility or vessel are small spills of less than 100 gallons (refer to Chapter 3 for more details). These spills often take place in industrial areas, ports, marinas, shipping lanes, and areas of general proximity to the response equipment depots. Depending on the type and volume of oil, the location and time of year, as well as resources at risk, the most appropriate response method is selected in consultation between regulators and the responsible party. For spills that are determined to require countermeasures beyond natural attenuations and monitoring, the responsible party will be required to activate its spill response plans and mobilize OSROs. OSROs will bring the required equipment

[4]Response Options Calculator (ROC): https://response.restoration.noaa.gov/oil-and-chemical-spills/oil-spills/response-options-calculator-roc; BSEE Response System Planning Calculators: https://www.bsee.gov/what-we-do/oil-spill-preparedness/response-system-planning-calculators.

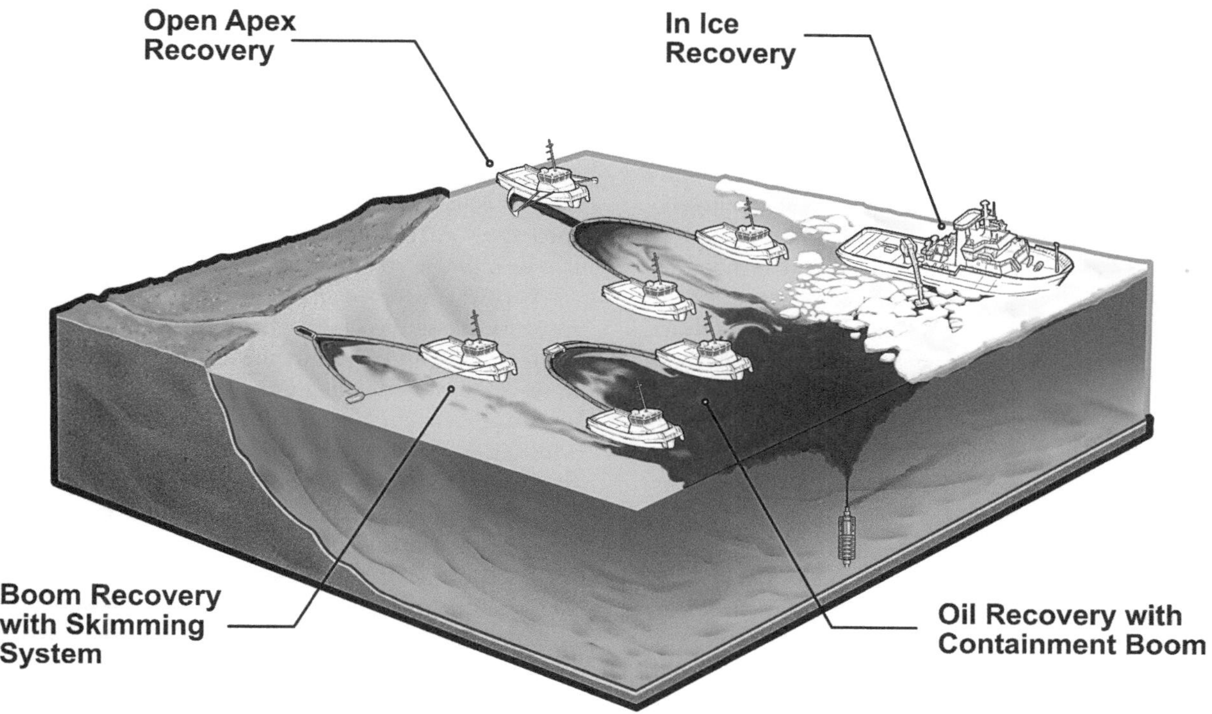

FIGURE 4.11 Mechanical recovery configurations.
SOURCE: Image provided courtesy of the American Petroleum Institute, produced by Iron Octopus Productions, Inc.

and personnel to clean up the spill while being monitored by the appropriate federal and state agencies. This could include sorbent materials, booms, skimmers, manual labor, vacuum trucks, etc. The incidents are usually dealt with and completed by a signoff from local, state, and federal authorities within days to weeks. Mechanical recovery can be very effective on small spills in calm waters and in areas where response can be assembled quickly, and will remain a preferred response technique for such situations.

Offshore conditions, large-volume spills, and remote areas present several challenges for mechanical response. Mechanical recovery is often favored by the general public based on the notion that a successful outcome of mechanical recovery is oil "removed" from the environment. In reality, and especially in when spills occur far offshore, or when winds or sea states are high, mechanical recovery may not be effective in recovering large volumes of oil. Historically, the "rule of thumb" has been that up to 10–30% of an offshore spill can be collected by mechanical recovery. A review of historical offshore oil spills found that mechanical recovery was only effective at removing between 2% and 6% of the total volume of oil spilled (Schmidt Etkin and Nedwed, 2021). During the Macondo response in the Gulf of Mexico, mechanical recovery removed only 3% of the spilled volume, despite an unprecedented number of personnel, recovery equipment, and vessels involved in the response (The Federal Interagency Solutions Group, 2010). **The reliance on mechanical recovery alone for cleaning up large widespread offshore slicks may result in significant exposure of marine environment and its resources to potentially large volumes of unrecovered oil that continues to drift into new areas and may eventually reach sensitive nearshore and shoreline areas.** Also, often overlooked are the impacts associated with the transport, treatment, and disposal of recovered fluids and debris (e.g., additional air contamination resulting from the waste transportation and disposal, potential contamination from the temporary storage sites, additional exposures of humans and wildlife to hydrocarbons, etc.).

Sea state is another major factor determining the feasibility of mechanical recovery. It can operate in short-period wind waves of up to 3–4 feet or in more developed "swell" waves not exceeding 5–6 feet. Higher wave heights cause equipment deployment and retrieval challenging, reduce the effectiveness of containment and recovery, and may reduce the safety of working conditions. In this situation, natural or chemically enhanced dispersion may become a valuable response tool. Because dispersion is most effective in the presence of waves, a significant portion of a slick is likely to be dispersed naturally under these conditions.

Because mechanical recovery efficiency is determined by the encounter rate of oil by a skimming system and the ability to recover it with minimal volumes of free water, recent technological improvements have focused on new boom designs allowing for faster oil collection and improved skimmer designs, allowing the collection of larger volumes of oil with as little free water as possible. Conventional containment booms are limited to tow speeds of about 1 knot. Several new boom types use innovative designs to allow them to corral oil at higher speeds (Jensen et al., 2012). These systems modify the flow of oil and water in the containment area, allowing them to collect oil at 3 knots in flat water and 2 knots with light to moderate waves. Another innovation in booming optimizes mechanical recovery by using containment booms with fewer boats (Hansen, 2000). When deployed in water, this system behaves as a horizontal kite that pulls the end of the boom into the current and holds it at a fixed location relative to the towing vessel. This allows for one-vessel boom operation rather than needing two vessels to tow one boom.

Just as with car designs, some mechanical response systems resemble their earlier versions while incorporating major engineering and design improvements resulting from years of response experience and research. Recent improvements in skimming systems were focused on optimizing the texture and geometry of skimming surfaces to improve oil adhesion and water repellency. For example, smooth skimming surfaces have been enhanced with a specially designed material resembling a seal skin or a series of grooves to increase skimming surface area and improve recovery of oil from the water surface (Meyer et al., 2012; Miller et al., 2014). The Wendy Schmidt Oil Cleanup X CHALLENGE, initiated to address mechanical recovery challenges observed during the DWH response brought about the development of several novel skimming techniques (Meyer et al., 2012). Other advances in skimming systems include improved oil and ice processing, heating systems to facilitate recovery in cold weather conditions, and improved processing of viscous oils and emulsions. Researchers developed advanced pumping equipment to allow transfer of cold and viscous oil/water/ice mixtures (Hvidbak, 2001; Fleming and Hyde Marine, 2003).

In developing countries, where access to modern recovery equipment may be limited, booms and recovery devices have historically been made from locally available materials: nets, bamboo, floating polypropylene hawsers, foam blocks, plastic bottles and drums, etc. (Guena, 2012). In-land, sorbents and solidifiers are often used to absorb released products from the ground, prevent its further spreading, and facilitate its collection. Various types of chemical solidifiers as well as chemical and natural sorbents have been developed (Merlin and Le Guerroue, 2009). Sorbents could be used in a bulk form as fibers, powder, or granules, or in a confined form of pads, rolls, pillows, booms, or mops. They typically generate a much greater volume of waste requiring disposal than the actual volume of the product that they recover. Attention should also be paid to prevent their sinking in the water or redistribution in the environment. Collection, recovery, and disposal of contaminated sorbents is a critical and often most challenging component of response operations. Because of these reasons, **use of sorbents in a loose unconsolidated form is not preferred for responses in marine environments due to additional risks that unrecovered contaminated sorbents may present to marine life.** Conventional skimming systems are not designed for collection, processing, and transfer of oil-contaminated loose sorbents or solidifiers. In marine environments, sorbents can have a niche application in recovering light products when used as a sorbent boom placed inside a hard boom to facilitate its ultimate recovery and disposal.

A critical factor for an effective offshore containment and recovery operation is the availability of sufficient storage for the recovered oil/water mixture and the ability to transfer it to the shore for recycling or disposal. Even effective skimmers can recover considerable volumes of free water along with oil, especially in offshore conditions. In some cases, recovered products can contain as much as 90% water. The limited storage available on the offshore vessels may get overwhelmed resulting in a delay of recovery operations while recovered product is transferred to another vessel and taken ashore. The decanting of free recovered water in specified areas can potentially be used to optimize storage capacity and reduce downtime in recovery operations. In Canada, this method is not allowed by existing regulations, but in the United States, decanting is accepted as an integral part of response operations and is covered in area contingency plans. The MARPOL Convention also has provisions for decanting operations. If regulatory approval is granted for this procedure, free water separated from oil is drained from the bottom of a storage tank into a boomed area next to the vessel. Occasionally, this process may require the onboard chemical treatment of recovered fluids to facilitate oil and water separation. This process is always monitored and documented to prevent any additional oil from being released into the environment.

The development of special response tools and strategies for mechanical recovery in cold climate and ice conditions came about because of decades of experience. **Under certain conditions, when snow, ice, or cold temperatures reduce oil spreading and allow it to remain in thick slicks and pools, mechanical recovery can be even more effective than in open water where oil can quickly spread over a very large area** (Sørstrøm et al., 2010). The type and concentration of ice cover often determines the selection of a suitable containment and recovery system in cold weather conditions. A variety of techniques to contain and recover oil on—and under—solid stable ice have been developed and practiced for more than 40 years (ACS, 2015). Several types of ice-capable response vessels, that contain built-in and over-the-side recovery equipment, are used in Arctic regions (Wilkman et al., 2014). Azimuth Stern Drive (ASD) vessels have high maneuverability in ice and are very valuable for Arctic oil spill response and supporting logistics.

Conventional open water containment and recovery techniques can be used with concentrations of drift ice up to 10%. At higher ice concentrations, the opening of a containment boom can be adjusted to maneuver around individual ice floes. Short sections of boom connected to an ice-strengthened skimming vessel by "outrigger arms" are used instead of long vessel-towed booms when drift ice concentrations are 10–70%. These narrower systems are easier to maneuver around ice floes and can be lifted, as needed, to avoid ice pileup inside the boom that may damage the equipment. These shorter sections of a boom direct oil toward a skimming system capable of processing oil and ice mixtures, which is built into the hull of a vessel. The ice itself acts as a containment to inhibit oil from spreading with high enough concentrations when booms cannot be used. At drift ice concentrations greater than 70%, specialized "over-the-side" skimmers can be deployed from the ice-strengthened response vessels to recover oil collected between ice floes (Potter et al., 2012). High capacity Arctic skimmers have been developed and tested with oil and ice at low temperatures (Ross, 2010; Sørstrøm et al., 2010; Meyer et al., 2014; Wilkman et al., 2014). These skimming systems were designed to process ice pieces and oil/ice mixtures. Some of them have their own propulsion systems, allowing them to maneuver between ice floes. Others may be lifted and positioned in a desired location by a crane aboard a vessel.

Decades of response experience show that mechanical recovery operations can be successfully used especially on relatively small spills in calm conditions. Mechanical recovery typically requires a greater amount of personnel, equipment and complex logistical support over a longer period of time than any other response technique. This consideration may become a critical factor when responding to spills in offshore locations or remote areas that may not be suited to host the necessary amounts of equipment and personnel for the extended period of time without negative impact on local communities and environment. **Responsible use of mechanical recovery methods could be informed by a "cradle to grave" analysis evaluating potential additional environmental impacts and human exposure associated with recovery itself,** the presence of a potentially large number of vessels and associated air and noise impacts; impacts associated with the transfer of oil to a disposal facility that may be located a significant distance from the response site; impacts from the recovered product disposal, possibly including incineration, recycling, or a landfill; and other activities required to support this response option.

4.2.3.3 In Situ Burning

Controlled ignition of hydrocarbon emissions to prevent their accidental ignition and reduce associated risks to human health and safety is a much-used technique. For example, a well blowout with significant volumes of gas could be ignited to mitigate a risk of accidental ignition as well as to reduce human health exposure, especially for releases with H_2S, and ultimately allow safer access to the site to implement source control measures. Combustion will consume a significant portion of liquid and gaseous hydrocarbons, reducing their input into marine environments. *In situ burning* is a process of controlled burning of an oil slick in the field and it has been used as a response countermeasure technique for more than 60 years. Fingas (2018) provides the most recent review of this technique. The first recorded in situ burning was conducted in 1958 during a pipeline spill in the Mackenzie River, Northwest Territories (McLeod and McLeod, 1972). More recently, around 400 individual controlled burns were safely conducted during the DWH response, removing an estimated 220,000 to 310,000 barrels of oil from the Gulf of Mexico sea surface (Allen et al., 2011). **Decades of research and field responses have proven that under favorable conditions, in situ burning can be a valuable response tool for removing large volumes of oil quickly, safely, and effectively with minimal environmental impact.** Successful in situ burning operations eliminate the need to collect, store, transport and dispose of recovered oil as would be required in the case of mechanical recovery.

During slick combustion, liquid oil itself is not burning. The heat from an ignition source converts liquid hydrocarbons into vapors, which are then consumed by the fire. Once a small area of the slick is burning, heat from the flames will radiate to the adjacent oil, vaporize it into gas, and sustain the combustion process. The oil removal rate is a function of the oil type and its weathering degree, burning area size, slick thickness, and environmental conditions. The rule of thumb is that for fresh crude oil slick fires of more than 3 m in diameter the burning can remove oil by reducing slick thickness at a rate of about 3.5 mm/min (USCG, 2003; Buist et al., 2013a). Diesel and jet fuel fires burn at a slightly faster rate of 4 mm/min. At these rates, more than 500 bbl of oil can be eliminated in less than an hour with 500 feet of fire boom. Removal efficiency of in situ burning of crude oils has been reported as high as 98%; this means that up to 98% of collected oil can be eliminated through combustion, not that 98% percent of a total spilled volume will be burned.

The critical factor for successful burning operations is oil slick thickness. At sufficient slick thickness, oil acts as insulation from underlying cold water and maintains high temperature at a slick surface. As burning removes oil volume, a slick will eventually thin out to the point that surrounding water will cool it below the temperature required for hydrocarbon vaporization and burning will cease. Extensive laboratory and field experiments determined that minimal thickness for burning fresh crude oil on water is about 1 mm, 2–3 mm for unemulsified crude oil and diesel fuels, and about 10 mm for heavy fuel oils (USCG, 2003). Oils with water contents up to about 12.5% burn at rates similar to those of fresh oils. Stable emulsions with water contents above 25% are difficult to ignite and burn because water has to be evaporated first for the combustion process to begin (Buist et al., 2013a). If permitted by regulations, chemicals called "emulsion breakers" can sometimes be used to separate oil and water emulsions and facilitate more effective burning.

The minimum thickness required for in situ burning is typically much greater than the thickness of a free-spreading oil slick in the field. To make oil combustible, it must be captured, thickened, and isolated from the rest of the slick with special types of booms, similar to how oil is collected for mechanical recovery. These operations require relatively calm conditions with wind speeds not exceeding 10 m/s (20 nautical miles) and wave heights lower than 3–4 feet. In a more turbulent environment, the oil may be dispersed in a top portion of the water column making it difficult to contain the oil by a boom. Safe deployment and retrieval of booms will also become a greater challenge in a more dynamic environment. Several types of fire-resistant boom designs have been developed over the years (Allen, 1999; Fingas, 2018) and tested during the DWH oil spill (Mabile, 2010). There are two main types of fire-resistant booms. One uses metallic or ceramic floats that are permanent and solid and are covered with or attached to a fire-resistant material. The other has inflatable buoyancy chambers covered with felted material and with water distribution lines that are used to continuously pump sea water to saturate and cool the material, hence reducing its direct contact with the fire.

Special safety procedures have been developed to ensure that in situ burning operations are conducted safely (USCG, 2003; API, 2015b, 2016c, 2018a; Fingas, 2018). Booms with burning oil are always towed into the wind, so that the burn area is as far from the vessels as possible with the smoke plume dispersing away from the vessel. The burn area inside the boom can be expanded or contracted by slowing down or increasing the speed of the vessel. Should the burn require extinguishment, the vessel's speed can be increased, allowing oil to escape under the boom into open water, or one end of the containment boom can be released to allow the oil to spread naturally. In both cases, the burning slick will spread out to thicknesses below the threshold required for sustaining combustion and fire will extinguish.

Recently, the use of oil-herding chemicals was proposed as an alternative to booms for thickening slicks for in situ burning in open and ice-covered waters. *Herders* are low-toxicity surface-active chemicals that can be applied in very small quantities around the perimeter of a slick to reduce surface tension of the water and allow an oil slick to contract into a smaller area and higher film thickness. Herders have existed since the 1970s, but several recent laboratory and field-scale research projects specifically evaluated their effectiveness in containing the oil and facilitating in situ burning (Buist et al., 2010, 2017; Ross and Danish Centre for Energy and the Environment, 2015; Byrne et al., 2018; Bullock et al., 2019; Tomco et al., 2022). These experiments proved that herders can effectively contract and contain oil slicks in temperate open water as well as brash and slush ice concentrations of up to 70% ice coverage and low temperatures. The burn efficiencies measured for the herded slicks were comparable to those expected from burns in conventional containment booms (Buist et al., 2010b; Kalimov et al., 2021). Herder toxicity was also evaluated: research indicated that the effects of concentrations for the tested species are several orders of magnitude greater than those expected in the field (Buist et al., 2017).

Several types of igniters are available to responders (Buist et al., 2016; Fingas, 2018). The Heli-torch was originally developed for forest-fire control, but was adopted for the use of in situ burning in the mid-1980s and was refined over the years. It works by releasing a stream of gelled fuel from a container underslung beneath a helicopter; the fuel is ignited as it leaves the device. The globules of burning gelled fuel are then dropped on a slick contained inside the boom and provide the initial ignition source to vaporize hydrocarbons and initiate the combustion process. A variety of handheld igniters that could be deployed from a vessel or a helicopter are also available (Buist et al., 2016). Most of these devices have fuses that provide sufficient delay time to assure safe deployment and ignition. In the absence of specialized ignition equipment, ad hoc devices such as weed burners, torches, and sorbent pads soaked in fuel have also been used to ignite small contained spills.

The most recent technological development combines mechanisms for application of a herder and a gelled igniter in a remotely operated surface vehicle (ROSV) built on the base design of a jet ski (Nedwed et al., 2021). The system is intended to be deployable from a ship or helicopter and operate autonomously in transit to spill locations; control is transferred to a remote operator once the slicks are located. An ROSV can travel at speeds up to 100 kph (depending on sea states), have at least an 800-km range, and at least 12 hours of operation before refueling. Once on site, the ROSV would first apply herder around the slick and then ignite the slick with a stream of gelled igniter, eliminating the need for conventional vessel-towed fire booms and exposure of personnel to hydrocarbons or burn by-products. Although this system has not yet been tested under field conditions, it holds promise for significantly improving speed and efficiency of in situ burning operations. See Figure 4.12 for a graphic summary of in situ burning tactics.

The temperatures generated during crude oil burns on water were measured to be as high as 900°C to 1200°C (Koseki, 1993; Guenette et al., 1994). Most of this heat is radiated upward and radially, while the water temperature just a few inches underneath the slick remains at ambient temperature. The flame temperatures are so high that the relative effect of ambient air temperature on the burning process is negligible whether it is +20°C or −20°C, meaning that in situ burning can be successfully conducted in cold-weather conditions. In fact, more than 40 years of research have shown that in situ burning is even more effective under cold Arctic conditions than it is in temperate climates (Buist et al., 2013a,b). This is due to the fact that burning is most effective on fresh, unemulsified oil containing light fractions, at high film thickness. At cold temperatures, evaporation slows down and oil viscosity increases, resulting in reduced spreading, smaller

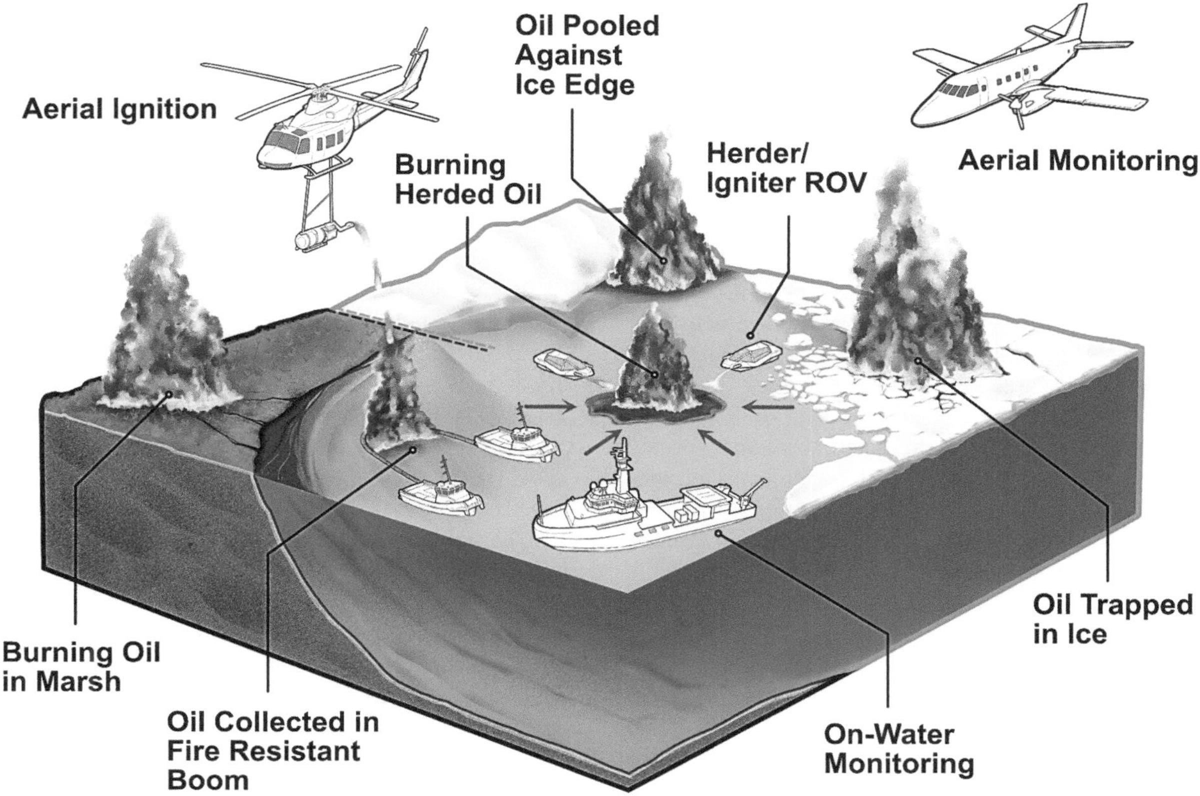

FIGURE 4.12 In situ burning tactics as an oil spill response technique.
SOURCE: Image provided courtesy of the American Petroleum Institute, produced by Iron Octopus Productions, Inc.

contaminated area, and higher oil slick thickness than typically would form in warm water. The presence of ice and snow can further reduce spreading and maintain slick at higher thickness. Natural dispersion and emulsification may also be reduced due to the wave-dampening effect of ice floes and as a result, reduced mixing energy. **Thicker slicks, smaller affected areas, and fresh, unemulsified oil conditions result in an extended "window of opportunity," the time during which in situ burning could be used and can increase its effectiveness** (Sørstrøm et al., 2010; Buist et al., 2013a,b).

In relatively open water with up to 10% drift ice concentrations, standard containment tactics using fire-resistant booms or herders can be used. In drift ice concentrations between 10% and 70%, when the deployment of booms becomes challenging due to potential risk of ice damage, herders could be used to corral and concentrate oil slicks. In higher ice concentrations, ice itself can act as a barrier, preventing oil spreading and concentrating it into higher film thickness. Under these conditions, oil can be burned between ice floes without the need for additional containment. Oil could also be burned on solid ice or frozen ground in pools or mixed with snow. In situ burning was shown to be effective in burning oil in concentrated drift ice off the Canadian East Coast in 1986 and the Norwegian Barents Sea in 2009 (Buist and Dickins, 1987; Sørstrøm et al., 2010), oil that surfaced in melt pools during spring time after being trapped inside the ice sheet in the winter (NORCOR Engineering and Research Ltd., 1975; Dickins and Buist, 1981; Brandvik et al., 2006), and oil mixed with small ice pieces and brash ice collected in a fire-resistant boom (Potter and Buist, 2010; Potter et al., 2012). Buist et al. (2013b) and Fingas (2018) provide the most comprehensive recent summaries of the history of in situ burning in the Arctic and cold-weather conditions.

Numerous studies evaluated composition and concentrations of emissions and residues produced as a result of in situ burning, as summarized in Sholz et al. (2004), Fingas et al. (2010), Buist et al. (2013a,b), Fingas (2018), and CTEH (2019). On average, about 85–95% of the burned oil is converted mostly into carbon dioxide and water vapor with small concentrations of nitrogen dioxide, sulfur dioxide, and other gases; 1–10% of the oil is converted to particulates (mostly soot) and the rest, 1–10% will remain on the water surface as an unburned residue. Overall, emissions from burning crude oil can be similar to those from agricultural and wildfire burns (CTEH, 2019). The black smoke made up of solid particles (soot), and liquid material (mists, fogs, sprays) may persist suspended in the air long enough to possibly be inhaled by response personnel or the public. The size of the particles plays a critical role in how long they will remain in the air, how far they can be transported, and whether they can affect human lungs. Particles of 10 and 2.5 microns are usually small enough to stay suspended in the air and present inhalation risks. If inhaled at high concentrations or for a long period of time, they can cause respiratory problems. Concentration and transport of the soot particles along with emitted gases are always monitored when burning of slicks is conducted. (Refer to Section 4.2.2.4 on SMART monitoring.)

The effects of burning polycyclic aromatic hydrocarbons (PAHs) results in their complete or partial destruction or conversion into higher molecular weight (HMW) PAHs. HMW PAHs are regarded as less acutely toxic, and they are typically found in low concentrations in both soot and residue. A few of these HMW PAHs are known or suspected carcinogens and thus are monitored by chemical measurement of particulate matter around in situ burning operations. Available measurements indicate that the concentration of PAHs in emissions is typically low or barely detectable (Barnea, 1995; Fingas et al., 2001; Middlebrook et al., 2012). Because more PAHs are destroyed by in situ burning than are created, the quantity of PAHs in the smoke plume is less than in the original oil (ASTM, 2014). Human exposure to any type of burn by-products presents an inherent risk and should be carefully considered and monitored, especially for inland burns (e.g., burning oil in marshes).

Extensive studies and field tests have demonstrated that in a field application, the concentrations of smoke particulates and gases typically quickly dilute to below levels of concern (Fingas et al., 1995, 2001; Walton et al., 1995; Sholz, 2004; Buist et al. 2013a; Fingas, 2018; CTEH, 2019). Extensive research conducted by Canadian and U.S. specialists in the 1990s, involving burns with real oil, progressed the understanding of smoke components, concentrations, and downwind transport. These studies evaluated several medium-scale burns at fire test facilities in Alabama, a set of burns in Alaska (McGrattan et al., 1993, 1995), and the highly recorded Newfoundland Oil Burn Experiment known as NOBE, a large-scale burn at sea off the Canadian East Coast (Fingas et al., 1995; Fingas, 2018). The NOBE experiment offered an invaluable opportunity to conduct a full-scale burn with real oil in the field and carefully measure multiple parameters related to smoke concentration and composition (including carcinogens and PAHs), residue toxicity, and impacts on the upper water column (Fingas et al., 1995). This experiment confirmed that **when in situ burning is conducted according to best practices, surface-level particulates and hazardous gas concentrations during in situ burns fall well below human health levels of concern.** To ensure protection of public and responders during these operations, in situ burns are conducted at a predetermined safe distance (e.g., 3 miles) from the general public and using appropriate personal protective equipment for responders (API, 2015b, 2016c, 2018a; USCG, 2003).

Dioxins, which can have a negative impact on human health and the environment, are formed from the incomplete combustion of organic matter in the presence of chlorine. Several studies evaluated whether dioxins may present a risk to human health as a result of in situ burning. Aurell and Gullett (2010) described measurements conducted by the U.S. EPA during the DWH response and found that the concentrations of dioxins and dibenzofurans were either at background level or in the range similar to residential woodstove emissions, which is much lower than the emissions that result from burning residential waste or forest burns. Schaum et al. (2010) evaluated the risk of exposure to workers and residents and also found it to be low. Burning of oil on water is not dissimilar to burning hydrocarbons in a furnace or a car, but due to limited oxygen access it is not as efficient and produces black soot particulates that appear as black smoke. Several projects attempted to address this issue by introducing new devices aimed at improving efficiency of combustion and reducing production of soot particles (Buist, 2013a; Tuttle et al., 2017).

The unburned residue remaining after oil combustion is typically a semi-solid tarry substance that will initially float and could be collected using sorbents, but may potentially become heavier than water and sink after cooling off. Residues produced from burning in a boom towed at sea, even if allowed to sink, are expected to scatter at low concentrations at the bottom. In rare cases when a significant volume of unburned residue sinks in a relatively small area, localized suffocating of benthic habitats and fouling of fish nets are the highest concern. This was observed during the Haven spill in Italy in 1991 and during the Honam Jade spill in South Korea in 1983. There were also reports of burn residue adversely impacting the royal red shrimp (*Pleoticus robustus*) fishery during the DWH response.

Burn residues are less acutely toxic and less bioavailable to aquatic organisms than the original oil. This is because the combustion process eliminates the lightest, most bioavailable and most acutely toxic oil components with boiling points lower than 200°C and reduces the number of components with boiling points between 200°C and 500°C, including light and medium-weight PAHs (Garrett et al., 2000; Fritt-Rasmussen, 2012; Fingas, 2017, 2018). As a result, burn residues do not contain light hydrocarbons and have lower concentrations of total PAHs. For example, the burn residues from the NOBE field experiment had 70–75% less PAHs compared with the parent oil (Blenkinsopp et al., 1996). Because burning can remove up to 98% of the original oil volume collected in the boom, along with the light- and medium-weight PAHs, the residual volume will have a relatively higher proportion of the high molecular weight PAHs, pyrogenic components, and metals compared to the fresh oil. Although these components alone may have a chronic toxicity, in the field, they are a part of the dense residue matrix and have low aquatic bioavailability, unless directly ingested. Due to these physical and chemical changes, burn residue has low acute toxicity to indicator species in saltwater and freshwater. Studies by Daykin et al. (1994); Blenkinsopp et al. (1997), and Gulec and Holdway (1999) found little or no acute toxicity of burn residues in sand dollars, oyster larvae, and inland silversides; no acute aquatic toxicity in fish (rainbow trout and three-spined stickleback) and sea urchin fertilization; no acute toxicity in amphipods; and very low sublethal toxicity (burying behavior) in marine snails. The greater risk from the residue would be the smothering of marine organisms if the burn residue was deposited to the benthic environment in high concentrations at one location.

Human health risks as well as environmental risks associated with in situ burning in the context of the risk presented by the oil spill itself are discussed in Chapter 6. **The impact of a temporary reduction in air quality from burning and any potential impacts of oil residue should be weighed against the impact of an untreated oil in the environment.** As much as 30–50% of the spilled volume may evaporate naturally, so a reduction in air quality could be expected regardless of the response method involved. Responders involved in in situ burning work under robust safety plans that address the risks and exposures associated with the response (API, 2015b, 2016c, 2018a; USCG, 2003). Exposure to the general public is minimized by conducting burns at a considerable distance from populated areas (e.g., 3 miles in most U.S. regions) and only under favorable conditions that allow dilution of the smoke plume below levels of concern. In situ burning operations in the field are always accompanied by a monitoring program (see Section 4.2.2.5) involving visual observations, measurements of gases and particulate composition and concentration, and air sampling. Modeling can also be used to predict concentration of particulates in the plume and potential trajectory of the plume and particulates (Walton et al., 2003). Additional observations are conducted to estimate the volume of oil that was eliminated through in situ burning (Mabile, 2013).

4.2.3.4 Dispersants

Several recent publications provide a more detailed overview of dispersant science as well as impacts of dispersants and dispersed oil on the environment: A publication by the National Academies of Sciences, Engineering, and Medicine, *The Use of Dispersants in Marine Oil Spill Response* (2020); The Canadian Science Advisory Secretariat report *State of Knowledge of Chemical Dispersants for Canadian Marine Oil Spills* (DFO, 2021); and the GAO-22-104153 report *Offshore Oil Spills: Additional Information Is Needed to Better Understand the Environmental Tradeoffs of Using Chemical Dispersants* (GAO, 2021). This section focuses on operational aspects of dispersant use in marine environments as one of the components of the response toolbox.

Surface Dispersants

The rapidly changing offshore environment require access to all appropriate response options to ensure maximum response effectiveness and environmental protection for an effective response. Response techniques such as mechanical recovery and in situ burning using fire-resistant booms require deployment of a large number of field personnel and on-water equipment and are challenged by the need to collect and concentrate spread-out patchy oil slicks. Though they can be effective in responding to small spills in calm waters relatively close to equipment stockpiles, they may not be as effective in recovering large offshore spills in remote areas or in high wind and sea conditions (Schmidt Etkin and Nedwed, 2021). In these situations, an aerial application of dispersants may be a critical response tool. Dispersant planes are able to arrive at a spill site quickly, treat larger volumes of oil, and operate in higher sea and wind conditions with limited risk and exposure to a smaller number of personnel involved than other response techniques. In a surface application, dispersants can also be sprayed from helicopters and from vessels either using fire cannons equipped with spray nozzles or specially designed application equipment (see Figure 4.13).

Dispersion of oil is a natural process facilitated by wave action, which breaks the oil into small droplets and dissipates the droplets into the water column (Delvigne and Sweeney, 1988; NASEM, 2020; see Section 5.3.1.3). The properties of the oil combined with extent of wave energy at the surface determine the effectiveness of this process. In most cases, lower viscosity oils are more inclined to natural dispersion than those with higher viscosity. Also, higher wave energy generates greater natural dispersion. Oil droplets of less than 100 microns in diameter (roughly the diameter of a human hair) generally tend to stay suspended in the water column, dilute, and eventually biodegrade; larger droplets are more likely to float back to the surface and re-coalesce into a slick. A portion of any spill in the marine environment is expected to disperse naturally. Several spills in rough sea conditions were completely dispersed by the natural wave action; an example is the *North Cape* spill (Michel et al., 1997).

Chemical dispersants are mixtures specifically designed to enhance natural dispersion of oil in marine environments. Dispersants typically consist of surface-active agents (surfactants) and a solvent (NASEM, 2020). The surfactant molecules consist of two parts: an oleophilic part that tends to stay in oil and a hydrophilic part that tends to stay in water. Upon contact with oil, these molecules orient themselves at an oil-water interface and reduce the interfacial tension, making it easier for waves or other turbulence to create small oil droplets and dissipate them in the water column (Clayton et al., 1993). Monomolecular surfactant "coating" around the droplets also minimizes re-coalescence and general stickiness of the droplets. The solvent reduces the viscosity of the surfactant, allowing it to be sprayed from the plane in the form of small droplets of about 600 to 800 microns in diameter and facilitates penetration of the surfactant into oil. Dispersing an oil slick into water as droplets creates greater surface area at the oil-water interface, thus increasing availability for natural biodegradation (NASEM, 2020; Pequin et al., 2022) (see Section 5.2.8.3).

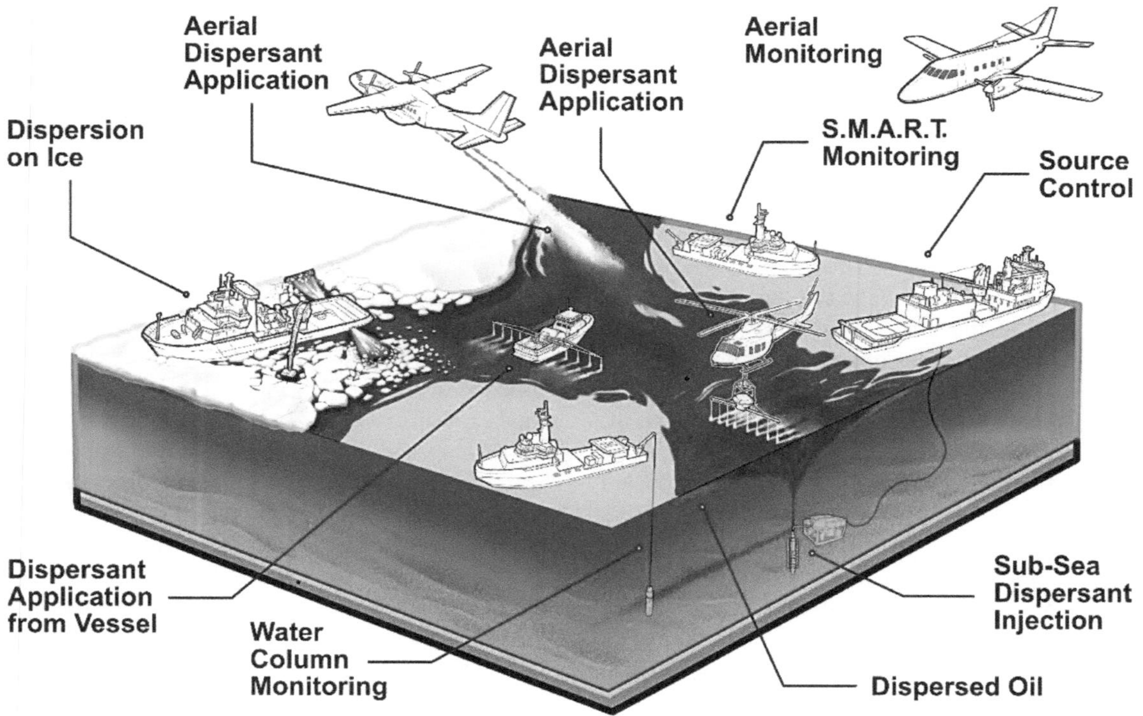

FIGURE 4.13 Dispersant application methods and monitoring techniques.
NOTE: S.M.A.R.T. = Self-Monitoring, Analysis and Reporting Technology System.
SOURCE: Image provided courtesy of the American Petroleum Institute, produced by Iron Octopus Productions, Inc.

The first application of dispersant-like chemicals in an oil spill response took place in 1967 during the *Torrey Canyon* oil spill in the United Kingdom (see Section 6.5.3.3). The chemicals used to disperse the oil were industrial degreasers, which were not intended for use in marine environments. Since then, several generations of modern dispersants have been developed using low-toxicity ingredients also used in household products, food, and cosmetics (Hemmer et al., 2011; Word et al., 2015; NASEM, 2020). Although the exact composition of most commercial chemical dispersants is proprietary, the composition of Corexit EC9500A was disclosed during the DWH spill. It contains the surfactants DOSS, Tween 80, Tween 85, and Span 90 as well as glycols and dipropylene glycol n-butyl ether (DGBE) in a hydrotreated light petroleum distillate (Parker et al., 2014; Fingas, 2017). All these ingredients are approved by the U.S. Food and Drug Administration and have other household uses. For example, the anionic surfactant DOSS (dioctyl sulfosuccinate sodium salt; sodium dioctyl sulfosuccinate) is a common ingredient used in a variety of applications such as a wetting and flavoring agent in food, a cosmetics ingredient, and an over-the-counter laxative. Dickey and Dickhoff (2011) provide a detailed overview of Corexit ingredients, their uses, and their environmental toxicity. Recently, researchers explored the use of biosurfactants and natural surfactants as alternatives to chemical dispersants (Quigg et al., 2021a; Pequin et al., 2022). While showing some promise at a laboratory scale, further research and development are needed to optimize these concepts and formulations for various spill scenarios and demonstrate their feasibility for field use, as well as to evaluate their industrial-scale production and distribution.

In 1996, 445 tons of modern-formulation dispersants were used nearshore during the *Sea Empress* oil spill in an effort to protect sensitive coastal resources from the impacts of drifting oil; this prevented at least 36,000 tons of oil from washing ashore (Harris, 1997). During the 1979 Ixtoc 1 spill in the Bay of Campeche, Mexico, response efforts over a five-month period applied approximately 9,000 tons of dispersants (Jernelöv and Lindén, 1981). When the Montara wellhead blew out in 2009 in Western Australia, 48,000 gallons of a combination of seven different dispersants were used (NASEM, 2020). During the DWH incident approximately 53,000 tons of dispersants were applied through a variety of deployment resources, including aerial, vessel, and subsea methods. Prior to the DWH incident, between 1968 and 2007, dispersants were (reportedly) operationally used globally 213 times (Steen and Findlay, 2008). Over the past 40 years, dispersants were used in the United States 27 times (Helton, 2021).

Chemical dispersants are most effective when applied during or quickly after a spill or sub-sea release incident. Natural processes such as dilution, weathering, and emulsification of the oil can quickly reduce the effectiveness of surface dispersants, creating a limited "window of opportunity" (the time window during which application of surface dispersants can be successful). The effectiveness of surface dispersants is influenced by the efficiency of the employment process (encounter rate), properties of the slick at the time of application (e.g., viscosity, pour point, slick thickness), and the sea conditions (e.g., wave energy).

Encounter Rate

The encounter rate for surface dispersant deployment is determined by the speed of the delivery vehicle, the amount of dispersant that it can apply, and the width of the spray pattern. A Boeing 727 dispersant aircraft can travel at speeds of more than 500 miles/hour compared to a transit speed for a response vessel of 8 miles/hour. A Boeing 727 can arrive at a spill site quickly, before an oil slick spreads out over a large area or breaks into small patches. Another dispersant plane, a C-130A with an internal spray system can treat around 7,000 bbl of oil with a dispersant-to-oil ratio of 1:20 in a 12-hour day, compared to the 1,000 bbl that could be recovered by a large oil spill response vessel even under optimal conditions (API, 2015a). The encounter rate of aerial dispersants is the highest on continuous slicks. Once oil spreads out, thins out, and breaks into smaller slicks (patches) aerial application may have reduced efficiency.

Slick Properties

Not all oils are suitable for dispersion. Properties that determine oil dispersibility include chemical composition, density, viscosity, pour point, and the degree of weathering. As discussed in Chapter 2, very volatile hydrocarbons (Group I) evaporate rapidly and are typically allowed to do so. Very heavy products (Group V) that have density higher than water are also not suitable for dispersion. An oil cooled below its pour point (the temperature at which oil remains liquid) may become very viscous and lose its dispersibility. Dispersants can be used on slicks much thinner than those that could be collected for mechanical and in situ burning operations (typically higher than 0.1 mm), but are typically not applied on very thin sheens of less than 0.05 mm, as dispersant droplets could penetrate through the sheen and into the water column without interaction with the oil. Surface dispersants are most suitable for liquid hydrocarbons that form dark/black slicks on the water surface and are not expected to attenuate naturally. Dispersants are most effective on fresh oils. As a result of weathering processes (see Chapter 5), oil viscosity increases over time due to evaporation of lighter fractions. In addition, many crudes and some refined products can form stable, difficult-to-disperse emulsions when, over time, wave action mixes them with water. The use of dispersants can disrupt or prevent formation of stable water-in-oil emulsions and, in some cases, even undo already formed emulsions. The time window during which surface dispersants are effective can be as short as one to three days, after which the oil could become too viscous or emulsified. The type of oil, as well as freshness, affects its dispersability, newly spilled light to medium crude oils are considered readily dispersible whereas highly viscous oils are not. In the late 1970s it was assumed that dispersants were not effective on oils with a viscosity greater than 2,000 cP, which is similar to honey at room temperature (Martinelli and Cormack, 1979). Studies that are more recent have shown modern dispersants are more effective at dispersing weathered and emulsified oils, and are effective on oils with a viscosity as high as 20,000 cP, which is similar to chocolate syrup at room temperature (Lewis et al., 1995; Strøm-Kristiansen et al., 1997; Lessard and DeMarco, 2000; Belore et al., 2008; Nedwed et al., 2008). As a rule of thumb, oils with a viscosity of up to 10,000 centipoises are considered potentially dispersible (Daling and Strøm, 1999), but whether or not they could actually be dispersed at sea, depends on the sea conditions.

Sea Conditions

Dispersants can be used over a wider range of sea conditions than other response options, and this is the only response technique that can be used in high waves and winds. An important factor for surface dispersant application is the sea state, it affects both the mixing energy available for breaking slicks into small droplets and the effectiveness of oil plume dilution in the water column. At low mixing energies, dispersion may be ineffective and any droplets that were dispersed may resurface back to the slick. Oil slick can be treated with dispersants in calm conditions so that it can disperse when wave energy increases in the following days. If wave energy is too high, oil may be dispersed naturally in the water column and not be available for treatment. Generally the upper limits for spraying dispersants from aircraft are gale force winds (speeds > 35 knots [18 m/s]) and wave heights of 5 meters, there are instances where dispersants have been applied from aircraft in winds greater than 50 knots (ESGOSS, 1994; IPIECA, 2015). High wind and wave conditions affect not only dispersant application efficiency, but also the safety of surface spraying operations.

Another important limitation for surface dispersant application is visibility. Sufficient visibility to accurately track the oil slick (e.g., cloud ceiling of 1,000 feet and daylight) are necessary to perform aerial dispersant applications. Spraying operations from the vessel could potentially proceed at night with support from specialized remote sensing equipment. During an actual event, the operational use of dispersant may have to be preceded by a small-scale field test to ensure that the specific dispersant will work on the specific oil under the specific weather and field conditions.

While the dispersants efficiency measurements from actual spill responses are limited, measurements obtained from the field scale tests and experiments in large test tanks report high effectiveness of dispersant applications (Belore et al., 2008, 2009; Ross, 2011; NASEM, 2020). In contrast, standard laboratory tests used to evaluate effectiveness of dispersants for the purpose of regulatory approvals typically report dispersion efficiency in the range of 40–60% (EPA, n.d.a). This is because simulating dispersion under actual field conditions have not been the objective of standard laboratory tests. Rather, they were designed to screen candidate dispersants and are intentionally conducted at low mixing energy to allow better differentiation between candidate dispersants. Assessing the effectiveness of oil dispersants under more realistic sea state conditions experiments should be conducted in large test tanks—and even then there are inherent limitations which may affect dispersant performance (Nedwed and Coolbaugh, 2008). When dispersion tests have been conducted in large test tanks results exceeded those observed in small-scale tests and demonstrated high effectiveness of dispersants. For example, Corexit 9500 and Corexit 9527 were 85–99% effective in dispersing both fresh and weathered Alaskan oils in a test tank even when tested at cold temperatures (Belore et al., 2009). Test tank experiments also showed good dispersibility of viscous oils compared to bench-scale experiments (Belore et al., 2008). Figure 4.14 illustrates the effect of oil type and viscosity (shown in parentheses), as well as test method, on the results of the effectiveness tests conducted by Ross in 2011. Tested oils included Alaska North Slope (ANS), Harmony, and PXP-02 (Irene commingled crude), as well as intermediate fuel oils (IFOs). Ohmsett is a very large outdoor test tank that can generate turbulence comparable to that in marine environments. Other test methods are bench scale with various levels of mixing/turbulence. The Exxon Dispersant Effectiveness Test (EXDET) was designed to generate relatively higher turbulence to better represent ocean turbulence than used in the swirling flask test or Warren Spring Laboratory used for screening of dispersants for regulatory certification purposes. The baffled flask test is a modification of the swirling flask test that allows testing at higher turbulence, but is still not directly translatable to field conditions. Higher viscosity oils are generally less dispersible, but even they could be dispersed in the field, test tanks, or bench scale tests with sufficient level of turbulence compared to the bench scale tests with limited mixing. **The direct use of the dispersant efficiency values generated for the purpose of regulatory certification in low-turbulence conditions is not appropriate for the prediction of dispersant efficiency in the field and operational decision-making.** This may prevent the use of dispersants that could have been effective under field conditions, which most of the time have higher levels of natural mixing and dilution potential than those simulated in the small-scale laboratory tests. A pilot spray test is often conducted in the field to confirm the effectiveness of dispersant use under specific field conditions.

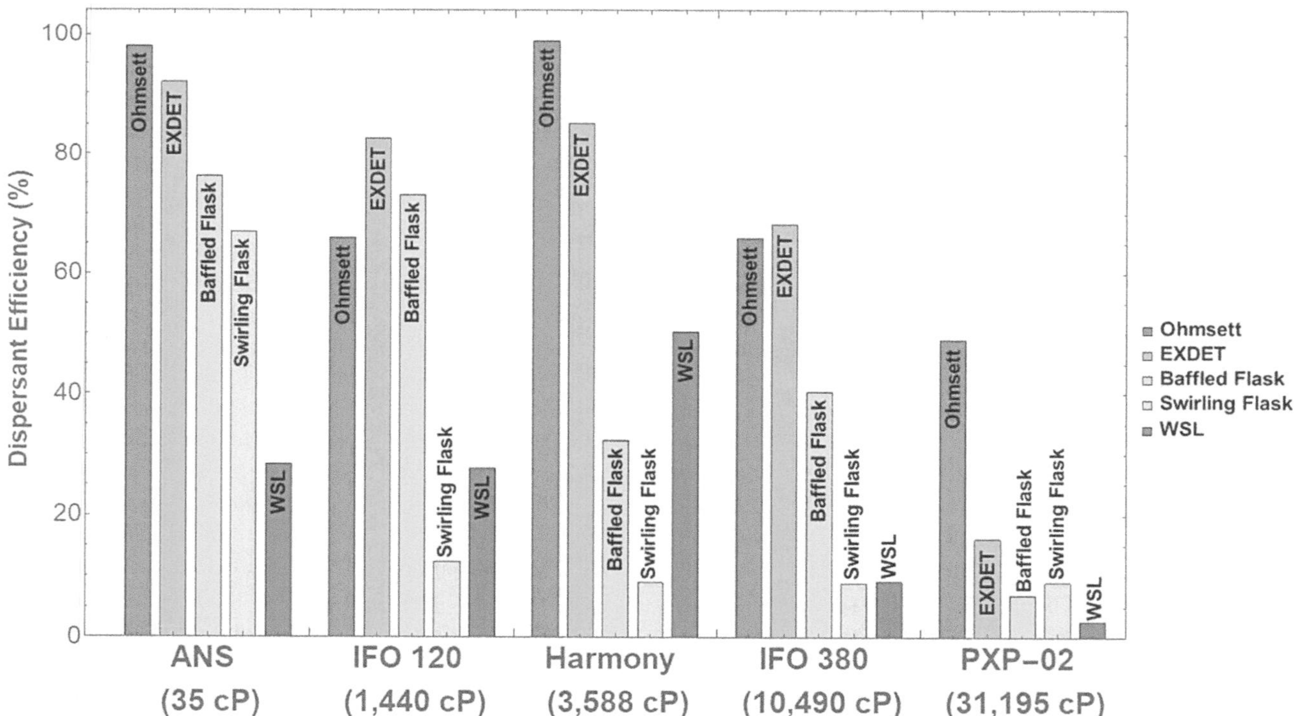

FIGURE 4.14 Comparison of dispersants efficiency estimates as a function of oil viscosity (highlighted in parentheses) and the type of the test.
NOTE: cP = centipoise.
SOURCE: Adapted from SL Ross (2011).

Several studies have shown that the dispersability of many oils have not been significantly reduced by cold temperatures (Brown and Goodman, 1996; Owens and Belore, 2004; Lewis and Daling, 2007; Sørstrøm et al., 2010; Lewis, 2013; Lewis and Prince, 2018), and most oils remain dispersible until they are cooled below their pour point (Daling et al., 1990, Brandvik et al., 1995; Nedwed et al., 2006). Although cold temperature can increase oil viscosity, this change is usually not as drastic as the viscosity increase resulting from oil evaporation and emulsification. These weathering processes can proceed quickly in warm climates, potentially limiting the window of opportunity for surface oil dispersion to one to three days. In contrast, Arctic and cold-weather conditions can increase the period of effectiveness for dispersant use by reducing oil weathering (Brandvik et al., 2010; NRC, 2014). An oil slick in cold weather conditions will have lower spreading and evaporation rates, especially if contained between ice floes. Under these conditions, the oil may remain fresh and dispersible from days to weeks. Ice-covered marine environments typically have low mixing energy conditions due to absence of breaking waves, although some studies found that the motion and contact between broken ice floes under certain conditions could generate surface turbulence sufficient for dispersion (Owens and Belore, 2004; Faksness et al., 2017). The icebreaker-facilitated dispersion concept was introduced to overcome the wave-dampening effect of ice and provide the necessary turbulence. This technique uses the mixing energy from the propeller of an ice breaker or support boats to disperse dispersant-treated oil (Spring et al., 2006; Nedwed et al., 2007; Daling et al., 2010). A special hydraulic arm operating from the side of the vessel was developed for targeted application of dispersants between ice floes. A 2009 field experiment in the Arctic demonstrated the effectiveness of this technique on oil that was weathered between ice for 6 days (Sørstrøm, 2009). Experiments have also shown that oil can remain dispersible during spring melt after being encapsulated into ice for as long as 3 months (Nedwed et al., 2017). Although most commercial dispersants were formulated for use in marine environments and have reduced effectiveness in waters with lower salinities, several dispersant types are available for use in fresh waters.

Under offshore conditions, dispersed oil is expected to dilute in the top portion of the water column and be carried away and dissipated by the currents. Because currents typically run parallel to the shore, oil dispersed offshore is not expected to arrive at the shore, as surface slicks driven by the winds often do. Dispersants are typically applied in marine environments at water depths greater than 10 meters and 3 nautical miles away from a shoreline. This is to allow sufficient dilution of the dispersed plume and reduce potential exposure to aquatic organisms. **Application of dispersants in shallow waters can be suitable under some circumstances but should be evaluated on a case-by-case basis to examine potential benefits, such as protecting sensitive or economically important nearshore or shoreline areas versus potential impacts in the water column.** Effective dispersant use increases concentration of oil in the water column, but because slicks are dispersed into the water immediately underneath, oil is typically dispersed into the area already contaminated with hydrocarbons. This volume will experience the highest initial loading of dispersed oil before ultimate plume dilution. Dispersants are typically not used in shallow waters with limited water circulation and dilution potential or in waters with high sediment load, typically found close to the shore and in estuaries. Chemically or naturally dispersed oil in waters with high turbidity and organic content could potentially result in formation of marine snow. Brakstad et al. (2018c) conducted a critical review of marine snow literature in the context of oil spills and dispersant treatment, but concluded that the exact contribution of dispersant to the formation of marine oil snow events during the DWH oil spill could not be determine from available published laboratory studies, as these studies were performed at very high oil concentrations and not representative of the rapid dilution in the open ocean. A more detailed discussion can be found in NASEM (2020) and Section 5.3.2.2.

Experiments and field observations have shown that initial dilution of dispersed oil in the field is rapid with maximum measured dispersed oil concentrations being 10–50 mg/L (NASEM, 2020). Trudel et al. (2009) showed that, even in closed wave tanks, concentrations of dispersed oil are rarely higher than 100 mg/L. This initial concentration is reduced through dilution and advection to 1 to 2 mg/L in less than 2 hours (Cormack and Nichols, 1977; McAuliffe et al., 1980, 1981; Lunel, 1994; Daling and Indrebo, 1996; Strøm-Kristiansen et al., 1997; NASEM, 2020). **With time, dispersed oil plumes continue to dilute and offshore concentrations of dispersed oil are estimated to fall below a threshold for acute impacts for aquatic organisms in less than a day** (Cormack and Nichols, 1977; McAuliffe et al., 1980; French-McCay and Payne, 2001; French-McCay et al., 2006; Judson et al., 2010; BenKinney et al., 2011; Coelho et al., 2011; Hemmer et al., 2011; Bejarano et al., 2013). This explains why during the DWH response, only 33 samples had TPH concentration higher than 10 ppb among 2,779 individual samples collected in that area (Lee et al., 2011b, 2013). An extensive federal and state monitoring program initiated after the DWH oil spill to address public concerns about the consumption of seafood tainted from the oil, dispersants, and dispersed oil found that dispersed oil levels were well below levels of concern (LOC) for human health risk (FDA, 2011; Gohlke et al., 2011; Ylitalo et al., 2012). **Additional research into the topic of seafood safety has identified improvement opportunities for future monitoring programs:** for example, the need to include alkylated polycyclic aromatic hydrocarbons into human health assessment for contaminated seafood (see Farrington [2020] and Chapter 6 for more details).

Dispersants alone are also present in low concentrations in the water. The typical dispersant-to-oil ratio for surface application of modern dispersants is generally around 1:20, and a standard aerial application rate is 5 g/acre (47 l/ha). If this volume of dispersant was applied to water, the resulting initial concentration of dispersant in the top meter of seawater would be around 5–13 ppm and would dilute to less than 1 ppm within minutes to hours (Lewis and Prince, 2018; NASEM, 2020), which is well below the toxicity threshold for most aquatic species (NASEM, 2020). After application, dispersants are subjected to dilution and degradation processes, including biodegradation and photodegradation at the surface. The early literature on surfactants, based on laboratory experiments, provided evidence both for and against inhibition of biodegradation of dispersed oils, but some of these studies used unrealistic incubation conditions including excessive concentrations and dispersant:oil ratios (Prince and Butler, 2014). More recent literature generally has demonstrated that dispersants do not inhibit oil biodegradation (Prince et al., 2013) but rather promote biodegradation (e.g., Bælum et al., 2012; Sun et al., 2019), although contradictory results are still reported from laboratory studies (e.g., Kleindienst et al., 2015; Langenhoff et al., 2020); **diverging results are likely due to differences in experimental design** (NASEM, 2020). Laboratory experiments measuring Corexit EC9500A loss have shown that the surfactant compounds DOSS, Tween 80, and Tween 85 and the petroleum distillates biodegrade on the order of days (Garcia et al., 2009; Campo et al., 2013; Brakstad et al., 2018b; NASEM, 2020). Although the general consensus seems to be that dispersant constituents are biodegradable, DOSS biodegradation may be delayed in the presence of oil and its metabolite ethylhexyl sulfosuccinate (EHSS) may transiently accumulate (Gofstein et al., 2020). It appears that dispersant components generally do not persist at high concentrations in marine waters, especially at the surface where photo-oxidation can occur (Ward et al., 2018a), but may persist when sequestered in sediments (White et al., 2014).

Prince (2015) reviewed the literature and proposed that the overall environmental benefits of using oil spill dispersants likely outweigh any potential negative environmental impacts, while listing numerous questions that remain regarding dispersant use. **When effectively applied, dispersant use can reduce the potential of the public to come in contact with oil by preventing oil from coming ashore.** On-scene health hazard evaluations performed by the National Institute for Occupational Safety and Health (NIOSH) found that standard personal protective equipment with exposure monitoring (if deemed necessary) was adequate to protect oil spill responders including during dispersants operations (NIOSH, 2010; King and Gibbins, 2011). Dispersants also reduce the potential of workers to be exposed to oil and oil fumes while conducting clean up and recovery efforts at sea or on the shoreline. Some researchers have evaluated whether application of dispersants may increase aerosolization of oil into atmosphere (Afshar-Mohajer et al., 2020), although further studies are needed to translate these laboratory findings into field conditions and integrate them into larger scale dispersion and exposure models. The comparison of emissions for the total time that untreated oil slick is present at the water surface versus emissions from the limited duration that a treated slick is present on the surface before it disperses into water column will provide valuable information for this comparison. A more detailed discussion on the effects of dispersants and dispersed and undispersed oil on human health can be found in NASEM (2020).

Dispersant application in the field is typically accompanied by a variety of dedicated monitoring programs (e.g., SMART; see Section 4.2.2.4) as well as more comprehensive water column and air quality sampling and monitoring activities. It is critical to collect field data on the actual dispersant effectiveness, water column concentrations, and transport to validate assumptions of the Net Environmental Benefit Analysis (see Section 4.2.5) and better assess environmental and socio-economic impacts of oil and dispersed oil.

Point-Source and Subsea Dispersant Injection

When an oil spill occurs at a localized source and involves a continuous release, chemical dispersants may also be applied directly at the spill source (see Figure 4.15). For cases

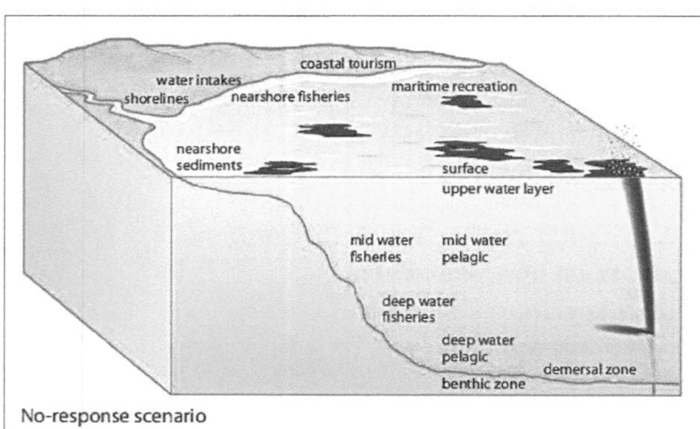

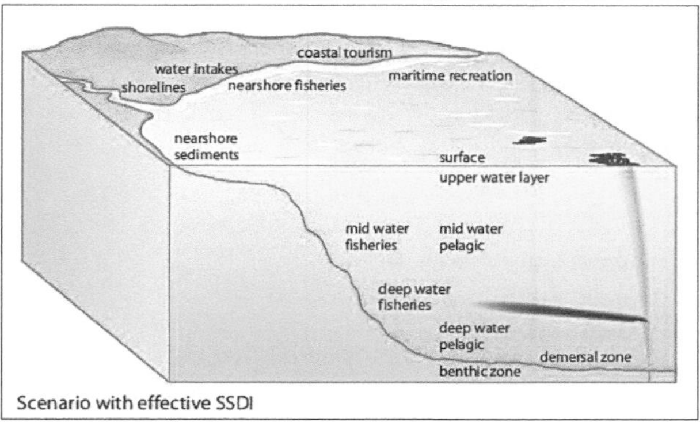

FIGURE 4.15 Illustration of the effect of use of subsea dispersants on the fate of released oil.
SOURCE: IPIECA.

involving an accidental oil well blowout or a leak from a disabled pipeline near the sea floor, this type of dispersant application is termed *subsea dispersant injection* (SSDI). This technique is not used exclusively in deep waters, and could potentially be applied to shallow or even surface releases. In this case, it would be considered a "point-source" injection aiming to premix dispersants into uncontrolled release so that oil can later be dispersed by waves at a sea surface.

SSDI was first used as a major response option during the DWH oil spill. As with all dispersant applications, it is only used when mechanical collection and recovery are not feasible to collect all the released oil and the oil is already entering the marine environment. The purpose of SSDI is to enhance natural dispersion of the released oil into the water column and, especially for subsea spills, to limit the amount of liquid hydrocarbons and volatile compounds reaching the sea surface where they may pose a hazard to humans, marine mammals, birds, and other marine life.

SSDI is not always an appropriate response technique. For example, a blowout of a condensate may naturally disperse in the water column as well as disperse and evaporate once it reaches the water surface, not requiring additional intervention. Under some circumstances, it may be more advantageous to let hydrocarbons evaporate into atmosphere than enhance their dispersion into water column. SSDI could potentially be considered for blowouts that result in a formation of a large surface slick threatening environmental resources if the spill cannot be addressed by other response measures, or to protect source control responders from harmful hydrocarbon vapors. The potential effectiveness of SSDI, its additional impacts, logistical feasibility, complexity of the regulatory approvals, and stakeholders' concerns should all be carefully evaluated before the decision to proceed is made.

Subsea dispersant injection typically involves several steps:

- A vessel with specialized dispersant injection equipment and stockpiles of dispersants arrives at the site.
- An ROV positions the nozzle of a flexible line connected to the vessel as close to the release point as possible to allow direct injection of dispersant into the oil stream. Alternatively, dispersant could be injected through special ports on a well-capping structure.
- Dispersant is pumped at a controlled rate from the deck of the vessel.
- Dispersion efficiency is monitored and injection rate is adjusted accordingly.

One of the primary reasons for SSDI injection is to reduce the volume of oil and gas surfacing near the release site to create less hazardous working conditions for source control vessels working on closing the well and stopping the release of hydrocarbons into the environment. Another reason to consider SSDI is that lower dispersant volumes may be required, due to lower dispersant-to-oil ratios (DORs) for equivalent effectiveness, and more oil may be treated than by surface dispersant spraying. Several studies have evaluated the effectiveness of SSDI (Brandvik et al., 2017; Gros et al., 2017; Zhao et al., 2021; see Chapter 5). Zhao et al. (2021) analyzed 91,566 volatile organic compound (VOC) measurements from the DWH spill collected near the well site and concluded with statistical confidence that SSDI reduced airborne VOC concentrations at the well site and enhanced the safety and health conditions of the responders. Measured VOC concentrations were lower during subsea dispersant use, and incidents of peak concentrations exceeding 50 ppm VOC that could have been of immediate concern to worker health were reduced by a factor of ~6 to 19 when dispersants were delivered at the intended rate. A study by Gros et al. (2017) also discussed the effectiveness of SSDI in dispersing oil at depth and the resulting changes in hydrocarbon concentrations at the water surface. Those authors reported that dispersant injection decreased the size of the droplets, which increased dissolution of hydrocarbons in the water column. Higher dissolution of light hydrocarbons in the water resulted in a 2,000-fold decrease in emissions of benzene at the surface, which lowered health risks for source control workers. Section 5.3.3.5 includes a more detailed discussion on the effect of subsea dispersants on formation and fate of oil droplets as well as its overall effectiveness. As with other response options, SSDI effectiveness in the field depends on many factors and it is not anticipated to be 100% effective, yet, as discussed in Section 5.3.3.5, even modest changes in oil droplet sizes could potentially result in considerable changes in their behavior and corresponding transport and impacts of the spill at large.

The advantages of SSDI, as compared to application of dispersants at a sea surface, include (IPIECA and IOGP, 2015):

- SSDI is suitable for almost any weather condition and around-the-clock operations. Other response techniques often have to cease during the night and are much more sensitive to limitations imposed by environmental conditions.
- The degree of control over the dispersant application process is higher when dispersants are injected at one manageable location directly into oil rather than sprayed from planes on patchy weathered surface slicks.
- Large oil volumes can be treated immediately at the source. The fresh nature of the oil combined with potentially high mixing and turbulence at the release point create advantageous conditions for effective dispersion.
- SSDI can reduce the amount of oil reaching the surface, forming slicks, affecting marine mammals and birds, or drifting toward sensitive areas and shorelines where high densities of organisms at sensitive life stages may be present.
- SSDI can reduce the need for other response measures including surface recovery, in situ burning, and surface

dispersant operations, which also reduces the possibility of exposure or accidents during these operations.
- SSDI can reduce/eliminate challenges associated with handling, storage, and disposal of large volumes of waste that could otherwise be generated by on-water mechanical recovery and shoreline clean-up operations.
- SSDI can reduce the amount of oil exposed to sunlight that may potentially increase their overall toxicity and impacts.
- SSDI requires a lower dispersant-to-oil ratio compared to surface spraying operations (1:100 subsea versus 1:20 at the surface).
- Larger water volume is available for dispersed oil dilution compared to surface application.

As with all response tools, the benefits of SSDI must be evaluated against potential operational challenges and additional environmental impacts. Subsea injection of dispersant in deep waters is a complex procedure requiring specialized injection and monitoring equipment. In a high flow rate oil and gas blowout, a portion of released oil may naturally disperse and dissolve in the water column. SSDI further increases the concentration of hydrocarbons in the water and hence potential exposure to marine organisms in certain areas (see Chapters 5 and 6 for more details). During the DWH spill, higher concentrations of hydrocarbons were observed in subsea intrusion layers, such as the deep plume between 1200 and 900 m water depth. Gros et al. (2017) estimate that on June 8, 2010, 1.5 times as much dissolved petroleum fluids by mass entered the subsea intrusion than would have occurred without SSDI. It should be noted, though, that concentrations of hydrocarbons in this plume were mostly at the ppm level. Hydrocarbon concentrations were consistently well below 5 ppm measured at about 1 km (0.6 mi) from the wellhead at 1,200 m depth (3,937 ft) (BenKinney et al., 2011; Coelho et al., 2011). About 84% of more than 20,000 water column samples had oil concentrations below 1 ppb, even though sampling was focused on locations where hydrocarbon concentrations were expected (Wade et al., 2011, 2016).

Dissolved and dispersed hydrocarbons are in a suitable form to be degraded by in situ bacteria, and several studies have documented the effectiveness of biodegradation in the DWH subsea plume (Valentine et al., 2010; Kessler et al., 2011; Hazen et al., 2016). Depending on the location of a spill globally, this may have important implications for oxygen concentration in the deep ocean as aerobic bacteria consume dissolved oxygen along with hydrocarbons. Indeed, the subsea plume associated with the DWH oil spill could be identified by dissolved oxygen anomalies in the CTD profiles collected during the spill. For this region of the Gulf of Mexico, oxygen concentration remained well above critical levels. However, at other locations around the globe, such as offshore Western Africa, where background oxygen concentrations are already low, dissolved oxygen may fall to hypoxic levels as the dissolved material from a subsea blowout is degraded.

The effectiveness and efficiency of SSDI application in the field will vary and depends the number of factors such as oil properties, the application method, the release conditions, and others. In situ monitoring is needed to evaluate the effectiveness of SSDI and confirm its applicability. Several documents were developed by both governmental agencies and industry to guide monitoring of SSDI (NRT, 2013; API, 2020; EPA, 2021; NRT, 2021). The purpose of operational SSDI sampling and monitoring described in detail by API (2020) is to:

- Determine dispersant efficacy.
- Characterize the dissolved oxygen and oil droplet sizes for subsea, or near surface, dispersed oil plumes including background samples.
- Assess potential ecological effects as they relate to operational response decision-making.

Operational SSDI monitoring is organized in three phases of increasing complexity:

- Phase 1: Confirmation of dispersant effectiveness near the discharge point and reduction in surfacing VOCs (e.g., improvement of air quality).
- Phase 2: Characterization of oil droplet size near plume and dispersed oil concentrations at depths in water column.
- Phase 3: Detailed chemical characterization of water samples.

The National Response Team (NRT) 2013 Monitoring Guide describes a comparable program encompassing:

- Site characterization;
- Source oil sampling;
- Water sampling and monitoring; and
- Sediment sampling and monitoring.

Although the API guide is focused on operational sampling and monitoring that can assist with real-time decision-making, the NRT recommends more detailed analysis of potential environmental impacts. Whether as a part of an SSDI monitoring program, or as a separate program, additional sampling and monitoring activities are always conducted to assess environmental impact as it relates to NRDA.

In 2021, the U.S. EPA published an updated rule, *Dispersant Monitoring Provisions Under Subpart J of the National Contingency Plan*, addressing dispersant use in response to major oil releases and other specific uncommon dispersant use situations in the navigable waters of the United States and their coastlines (EPA, n.d.).

The amendments establish dispersant monitoring requirements for the following scenarios:

- Use of dispersants in subsurface settings;
- Surface use of dispersant over prolonged periods, specifically in excess of 96 hours after initial application; and
- Surface dispersant use in response to oil discharges greater than 100,000 U.S. gallons in a 24-hour period.

The monitoring elements in the final rule include:

- Source Characterization and Information on Dispersant Application—Flow rate or volume of oil discharged, type of dispersant, dispersant-to-oil ratio, application rates, and total amount of dispersant needed.
- Water Column Sampling—In situ water column samples of background, baseline, and dispersed oil plume with recorded oil droplet size distribution, fluorometry and fluorescence, total petroleum hydrocarbons, dissolved oxygen (subsurface only), methane (subsurface only), heavy metals, turbidity, water temperature, pH, and conductivity.
- Oil Distribution Analysis—Description of dispersant efficiency and oil distribution.
- Ecological Characterization—Description of potential ecological receptors and habitats, and their accompanying exposure pathways.
- Immediate and Daily Reporting—Immediate reports for alterations to specified application plan, and daily reports of water sampling and data analyses to the On-Scene Coordinator and the Regional Response Team.

Subsea Mechanical Dispersion

An alternative dispersion method was recently proposed (Brandvik et al., 2021) for use under some circumstances when dispersion of a subsea blowout is desirable, but not achievable, either because chemical dispersants are not allowed or not available, or if a blowout does not have sufficient turbulence to break the plume into small droplets. Researchers tested various methods to facilitate the formation of small oil droplets by introducing mechanical mixing within the rising oil plume. After evaluating rotating mixing blades, ultrasonic cavitation, and high-pressure water jetting they concluded that creation of additional turbulence by introducing a powerful jet of water across the plume is the most promising technique, especially because it can be delivered using the equipment typically available at offshore facilities. Although this technique still has to be verified under field conditions and different blowout scenarios, it may become a viable response option in some situations. Since subsea mechanical dispersion relies on the power of a water jet to generate smaller droplets by cutting across the blowout, it may have limitations during high-volume blowouts and with certain oil types. In some scenarios, this technique could potentially facilitate subsea blowout dispersion when conventional SSDI is not possible or to increase the effectiveness of conventional SSDI, if needed.

4.2.3.5 Summary of Offshore Response Techniques

Earlier sections described a number of techniques that are available to respond to offshore oil spills. No technique is 100% effective under every response scenario, and there are advantages, challenges, and optimal conditions for the use of each technique. The conditions specific to each spill inform response decision-making so that the optimal combination of response techniques can be selected that will result in maximum protection of sensitive resources. Table 4.2 summarizes benefits, challenges, and potential additional impacts for different on-water response techniques.

4.2.3.6 Submerged and Sunken Oil Response

Spilled oil and hydrocarbon products may become submerged in the water column or sink to the bottom either if their initial or weathered density exceeds the density of seawater or if they form heavy agglomerate after mixing with suspended sediments (refer to Chapter 5 for more details). The *API Technical Report* and *Operational Guide on Sunken Oil Detection and Recovery* (2016a,b) provide a recent overview of recovery methods that could be used in such situations, along with their advantages and limitations. The effectiveness of these techniques will be strongly influenced by the response conditions (e.g., depth, visibility, sediment load, accessibility, presence and absence of ice, current speed, etc.) as well as product properties and behavior (e.g., sunken to the bottom or floating in the water column, in a shape of large mats or small aggregates, solid or liquid, concentration, areal distribution, etc.). Large concentrated oil mats are typically easier to recover than scattered small aggregates. Depending on the conditions, this could be accomplished by a variety of techniques: divers with pumps or vacuums, heavy machinery such as a dredge or an excavator, nets, trawls, and sorbents can be used. Table 4.3 summarizes applicability of available recovery techniques under various conditions.

4.2.4 Shoreline Protection and Cleanup

One of the results of discharges of oil into the environment is that it may eventually strand and, depending on the location, may affect open-water facing beaches and shores, rocky or vegetated shorelines, tidal flats, or manmade structures, such as rip-rap, jetties, or bulkheads. Appendix G, as well as Section 5.3.4.1, describe various shoreline types and potential oil fate and behavior in these unique environments. Various response methodologies have been developed to protect, clean up, and mitigate the detrimental effects of such oiling events. These strategies are often specific to shoreline types, environmental conditions, oil types, and even stakeholder

TABLE 4.2 A Summary of On-Water Response Options

Benefits	Challenges	Potential Impacts
	Monitored Natural Attenuation	
Does not cause additional environmental impact from response activities	• Only suitable for relatively light products when there is no threat to human life or sensitive environment • Public perception that nothing is being done • Requires dedicated monitoring efforts • If conditions change, calling for other response options, the slick could be weathered and inhibit their effective application	Slick could be transported to uncontaminated or sensitive areas
	Mechanical Recovery	
• Removes oil from the water surface • Widely accepted first response option • Works on most fresh and weathered oil types (large window of opportunity) • Good availability of equipment and expertise • Recovered product may be reprocessed	• Requires collection of oil by booms, which is not effective in high seas, for spread-out thin slicks, or large-volume spills • Recovery of oil in ice is challenging • Challenges with resource mobilization to remote offshore location before oil spreads too thin • Only limited oil volume can be concentrated by booms and made available for recovery • Slow collection and recovery process • Very labor and equipment intensive • Large volumes of free water are often recovered together with oil • Requires storage for the recovered product • Challenges with transport and disposal of recovered product in remote regions • Operations are often limited to daylight	• Additional environmental impacts (e.g., air quality, noise) due to the need for large number of vessels for extended periods of time • Oil that was not collected and recovered will continue to drift into clean areas and can potentially reach shorelines or sensitive environments • Impacts associated with transport, storage, and disposal of recovered oil and water, especially in remote and limited infrastructure regions • Exposes responders to hydrocarbons and risks involving personnel safety • Impacts to seagrass, corals, and sensitive benthic environments during nearshore or shallow water response
	In Situ Burning	
• Can quickly remove up to 98% of oil that was collected within booms or by herders • Reduction of hazardous/flammable vapors • Lower logistics and equipment requirements than mechanical recovery • Storage and waste disposal are typically minimal or not required • Minimal additional environmental impact in most cases • Effective at low temperatures and for oil in ice • Ignition of the well could eliminate large volumes of oil and gas directly at the source	• Requires specialized equipment and expertise for application and monitoring • Requires regulatory approvals and may generate public concerns • Typically conducted offshore away from populated areas • Requires collection of oil by booms or herders, which is not effective in high seas or on spread-out thin slicks • Only limited oil volume can be concentrated by booms. Slow collection process • Works better on fresh oil. Not effective on emulsified oil, resulting in limited window of opportunity • Daylight operations only	• Localized and temporary decrease in air quality • High temperature impacts in the water column are minimal and only in already affected area/volume • Residue may adversely affect sensitive benthic environments and thus have to be recovered in some cases • Oil that was not collected and burned will continue to drift into clean areas and may reach sensitive habitats • Safety protocols and proper PPE should be used to reduce exposure of responders to fire and emmissions
	Dispersants: Surface Application	
• Dispersant planes can get to a spill site and begin response quicker than other response techniques • Can remove large volumes of oil from the surface quickly, preventing it from coming to the shore and affecting sensitive environments • Enhances natural dilution and biodegradation • No storage or disposal issues • Minimal equipment and manpower requirements • Aerial use eliminates exposure of personnel to hydrocarbons or the need for labor- and equipment-intensive on-water recovery operations • Effective in turbulent sea conditions that would hinder effectiveness of mechanical recovery and in situ burning (e.g., wave height greater than 4–6 ft) • Icebreakers can create mixing energy to disperse oil in ice	• Requires specialized equipment and expertise for application and monitoring • Daylight operations for aerial spraying; vessels could potentially spray at night • Requires regulatory approvals and may generate public concerns • Requires wave action to create oil droplets and mix them into a water column. • Cannot be used in very high winds which prevent safe aerial operations and accurate targeting of the slick with dispersants (e.g., wind speed over 35 knots) • Limited window of opportunity (most effective on fresh light or medium products at temperatures above pour point) • Requires sufficient volume of water for dilution; use in shallow waters requires NEBA/SIMA analysis • Public perception issues	• Localized and temporary decrease in water quality • Exposes organisms in the top portion of water column to higher concentrations of hydrocarbons • In waters with high sediment and organics content, high concentrations of hydrocarbons could potentially result in marine oil snow formation and transport of degraded oil components to benthic environment • Application from the vessel may result in exposure of responders to dispersants; appropriate PPE should be used • May result in formation of aerosols, creation of which should be weighed against impacts of natural evaporation of undispersed slick • Undispersed oil will continue to drift into clean areas and may reach sensitive habitats

TABLE 4.2 Continued

Benefits	Challenges	Potential Impacts
	Dispersants: Subsea Application	
• Large oil volumes can be treated immediately at the source with high effectiveness • Prevents oil from reaching the surface, forming slicks and clouds of VOCs; enables source control operations • If efficient, in some scenarios can eliminate/reduce the need for surface recovery, in situ burning, and surface dispersant operations, thereby reducing the potential for exposure and accidents during these operations • Reduces/eliminates challenges with handling, storage, and disposal of large volumes of waste that could otherwise be generated by on-water mechanical recovery or shoreline cleanup • Keeps hydrocarbons away from sensitive habitats and areas with high densities of organisms at sensitive life stages (exception is coral reefs if present) • Keeps hydrocarbons away from sunlight that may increase their overall toxicity and impacts • The only response option suitable for any weather and 24/7 operations • Requires reduced dispersant-to-oil ratio compared to surface spraying operations • High degree of control over dispersant application process • Sufficient water volume available for dilution • Reallocates oil to areas with smaller biological density than at the water surface or at near-shore/shoreline areas	• Requires regulatory approvals and may generate public concerns • Requires specialized equipment and expertise for application and monitoring • Public perception issues • Relies on a blowout turbulence for generation and mixing/dilution of droplets	• Significantly increases volume of dispersed oil in the water column, including in the trap layer; increases exposure of some aquatic organisms to hydrocarbons and potentially reduced oxygen levels • Can expose benthic organisms to dispersed oil • In waters with high sediment and organics, high concentrations of hydrocarbons could potentially result in marine oil snow formation and transport of degraded oil components to benthic environment • Undispersed oil will continue to drift into clean areas and may reach sensitive habitats.

interests. Spill responders and oil spill scientists recognize that there is no single cleanup method that is effective in every situation and that all cleanup activities are a tradeoff, in response to an already existing detrimental pollution event.

Oil spill history has shown that the greatest impact to people and to the environment happens when oil reaches sensitive nearshore and shoreline areas (refer to Chapters 5 and 6 for more details). This is because these areas have higher densities of organisms than offshore or deepwater areas, and those organisms are often at sensitive life stages. Oil arriving to shallow waters can be present in relatively high concentrations, can accumulate on shorelines over time, and can be quite persistent. This overlap of high biological density and high concentrations of persistent oil should be avoided to the extent possible. Cleanup activities, however effective, can create additional impacts, as they often involve large numbers of people and equipment for long periods of time and generate large volumes of waste. Exclusion, deflection, diversion, and collection shoreline booming strategies are deployed to reduce the consequences of an oil spill reaching the shore and provide protection to critical habitats.

Unfortunately, **due to the nature of booming operations, booms can only protect relatively small sections of the shoreline.** Important tradeoffs must be evaluated when deciding where and when to place booms. Booms take time to deploy; once deployed, they take additional time to tend and relocate. Booms have to be managed from vessels and placed at an angle, rather than perpendicular to the oil slick trajectory; booming will not be effective if not placed at an angle as oil will be pushed over or under the boom. If left unattended, booms can be affected by tides and currents, lose function, and be dislocated and pushed into sensitive shorelines resulting in an additional environmental impact. This was illustrated during the DWH response when long stretches of boom deployed parallel to the shore were not able to stop oil from coming to shore and created additional environmental impacts when they were pushed into marshes requiring complicated retrieval operations. Exclusion and deflection booming strategies are not designed to remove oil from the environment, only to divert the slick to another location. Hence, they are used to protect relatively short stretches of the most sensitive locations such as inlets into back water or small estuaries (API, 2014, 2016b). Attention should be focused on the location to which the oil is diverted to prevent additional harm to the environment. Diversion booming aims to direct oil to a dedicated collection location, but it is often more suitable for responses in rivers, when oil moves parallel to the shore, rather than the offshore environment when an

TABLE 4.3 Applicability and Likely Effectiveness of Sunken Oil Recovery Techniques

Red = not likely effective; yellow = may be effective; green = most likely effective

	Suction Dredge	Diver Vacuum	Diver Pump	Excavator	Grab Dredge	Environmental Clamshell	Sorbents/V-SORS	Trawls and Nets	Manual Removal Shallow Water	Manual Removal by Divers	Agitation/Refloat
Water Depth (ft)											
— <5ft											
— 5 to 40 ft											
— 40 to 80 ft											
— >80 ft											
Water Visibility											
— >5 ft											
— <5 ft											
Water Current											
— <1 kt											
— 1 to 2 kt											
— >2 kt											
Water Height (ft)											
— <2 ft											
— >2 ft											
Availability											
Oil Pumpability											
— Fluid											
— Not Fluid											
Oil Distribution											
— <10 %											
— 10 to 50 %											
— >50 %											
Oil Patch Size											
— <0.1 ft^2											
— 0.1 to 1 ft^2											
— >1 to 10 ft^2											
— >10 ft^2											
Substrate Type											
— Sandy											
— Muddy											
Bottom Obstructions											
Buried Oil											
Sensitive Habitat											
Removal Rate*											
Waste Generation**											
Environmental Impact**											
Cost **											

*Classified as rapid, medium, or slow.
**Classified as low, medium, or high.

SOURCE: API Technical Report 1154-2, First Edition, February 2016. Reproduced with permission from the American Petroleum Institute.

oil slick approaches the shoreline as a large front. *Shoreline Protection Guide for Sand Beaches* (API, 2013) describes additional tactics such as dams, dikes, and barriers that could be implemented to protect and reduce the spread of oil on sandy beaches as well as response and environmental considerations that should be considered in selection of these strategies.

When an oil slick arrives to shore, it may affect environmentally sensitive areas as well as areas sociologically, economically and culturally important to humans, such as areas of recreation, fishing, industry, and tourism. Depending on the characteristics, type, and sensitivity of the shoreline and the physical and chemical properties of the oil itself (amount, type, and degree of weathering), various methodologies may be employed to mitigate an oiling event (NOAA, 2010). Cleanup activities can range from the allowance of natural attenuation of the oil, to intrusive manual and mechanical removal of the oil. Careful consideration should be given to any additional environmental and socioeconomic impacts that could result from the implementation of these techniques (Hoff, 1996; Martínez et al., 2012; Michel and Ruherford, 2014; Michel et al., 2017). Numerous cleanup methodologies have been developed, all of which are part of the available response toolkit. The predominant cleanup methods and responses, and their efficacy, their advantages and disadvantages are discussed in Appendix E, Shoreline Cleanup Methodologies. Oil deposited on a surface may be removed by manual methods such as a hand rake and shovel, sorbents, or by mechanical methods such as maintainers/road graders and small front-end loaders, skimmers, and vacuum systems. Purpose-built mechanical beach cleaners and sand sifters can sometimes be utilized; some of those are designed to minimize the removal of valuable sediment material along with the oil (API, 2013; Michel et al., 2017). Oil that has adhered to a substrate may necessitate the use of chemicals for removal or may be removed by the use of water washing utilizing varying degrees of pressure and temperatures. Other response strategies also exist, including bioremediation, debris removal, and tilling.

The choice of oil removal response technique(s) is always a tradeoff among efficacy, expedience, safety, and environmental impact of the cleanup, which must be considered in each event. For example, complete removal of the oil contamination is often not a desirable endpoint, as such a degree of removal may often cause more damage to flora and fauna through direct secondary impacts or through stress to vulnerable species. Also, even minimal human cleanup activities may affect feeding behavior of certain species, such as threatened or endangered foraging birds, and thus are subject to provisions under the Endangered Species Act. These provisions require the overseeing federal entities to engage with the U.S. Fish and Wildlife and National Marine Fisheries Services, dependent on the identified species that may be affected, in a consultation process (U.S. Congress, Senate Committee on Environment, 1983).

The physical intrusion into an area, for the purposes of an oil spill assessment or cleanup, can have unintentional and potentially detrimental consequences that can worsen an already bad situation. Spill response equipment and even response personnel can create secondary—sometimes greater—impacts than would have been seen had the intrusion not occurred. Heavy equipment, response measures, and even foot traffic can damage the established biota of a locale through trampling or flattening, particularly by heavy tracked equipment or in areas of soft substrate. Equipment movement and response strategies, such as berms and barriers, can change the topology of an area, causing impediments to organisms that depend on the area, such as nesting sea turtles, or alter the area's natural hydrology, potentially causing long-term changes to an entire area. Intrusive response strategies such as building sand berms, inlets restrictions, and freshwater diversion into coastal marshes should be carefully evaluated, as they can result in greater environmental impacts than the oiling itself (Martínez et al., 2012). Refer to Section 4.2.5 for the best practices in comparing response options and their potential consequences. Oil that was once on the surface, subject to natural degradation processes, can be driven into the substrate where the lack of oxygen can cause the material to remain for many years (Beland et al., 2017). Additionally, shoreline response typically generates a large volume of waste that has to be collected, stored, transported, and disposed of. This can create additional environmental impacts and logistical challenges. These and many other considerations must be thoughtfully analyzed by knowledgeable and experienced response personnel to ensure that the most appropriate treatment strategy is selected.

Marine shoreline ecosystems that may be affected by spilled oil are highly variable in geophysical structure as well as resident and transient morphological composition (refer to Chapter 5 and Appendix G for more details on oil fate and behavior on shorelines). Oil stranding on a fine-grained sand beach will require different protection and cleanup methodologies than oil affecting a rocky shoreline or marsh environment. The fate and behavior of oil, once it comes ashore, must also be considered in response and protection strategies. The oil spilled and affecting a shoreline in tropical or temperate climates may require a different cleanup strategy than the same oil in an Arctic environment. As in the open ocean marine environment, natural biodegradation and chemical composition continually alter an oil's makeup, which may in turn necessitate changes in the response strategies chosen. Oil that was once amenable to one strategy can change, sometimes rather quickly, to a point that the initial strategy no longer works. Most common cleanup methods and response techniques are illustrated in Figure 4.16 and described in Appendix E. Table 4.4 summarizes benefits, challenges, and potential impacts of various methods.

4.2.4.1 Advanced Shoreline Cleanup Techniques

Although every shoreline cleanup technique presents certain challenges and tradeoffs, several of them require special

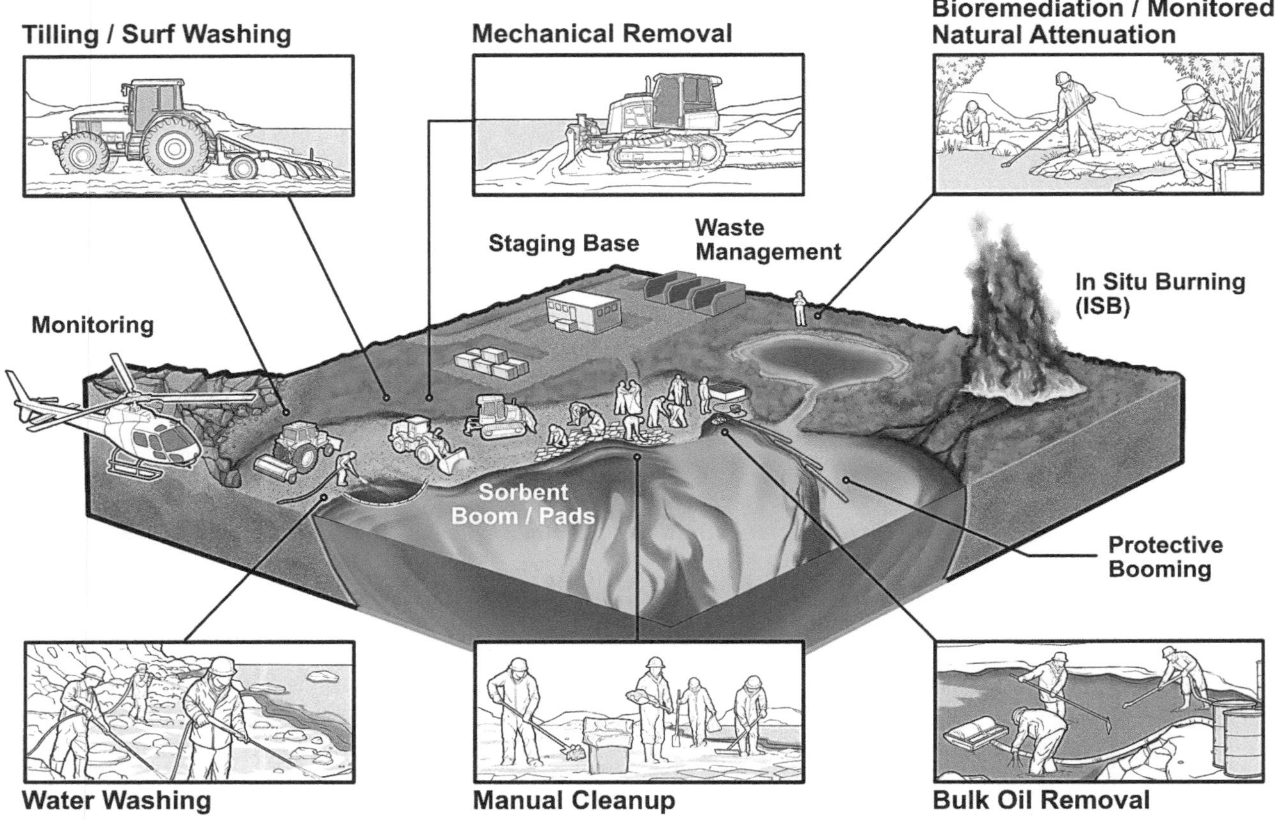

FIGURE 4.16 Examples of shoreline oil spill response operations.
SOURCE: Image provided courtesy of the American Petroleum Institute, produced by Iron Octopus Productions, Inc.

regulatory approvals, scientific information, and additional considerations (see Section 4.2.5).

Surface Washing Agents

Surface washing agents (SWAs) are specially designed chemicals, composed of a surfactant and sometimes solvent, that are used to loosen heavily coated or stubborn oils on surfaces; the oils can then be more easily removed by wiping with sorbents or water washing techniques that move the oil into another location for removal by other methods. SWAs used in marine oil spill response fall into two categories: "lift and float" and "lift and disperse." As with all chemical countermeasures, use of SWAs must be approved by the designated agencies. In the United States, Subpart J of the National Oil and Hazardous Substances Pollution Contingency Plan (National Contingency Plan), which lists approved surface washing agents does not distinguish between the categories of "lift and float" or "lift and disperse," but only surface washing agents shown to float the oil after application are typically allowed for use in the marine environment. The method allows the floating oil to be corralled and collected for removal and disposal. Subpart J also requires that the overseeing federal jurisdictional entity, usually the U.S. EPA or USCG, request permission from their local Regional Response Team prior to use. Some industrial locations have gone through the preauthorization process for use of SWAs, expediting cleanup operations in locations where environmental impacts have been considered and deemed an acceptable risk.

Burning of Oil in Marshes

In situ burning of oil in marshes involves the controlled incineration of an oil product that has affected a salt or brackish marsh (Fingas, 2018). Often oil that has affected marsh vegetation becomes trapped within it and becomes inaccessible for mitigation by other cleanup measures. The dense cover of some marsh inhibits evaporative processes on the oil, which may have a long retention time especially if it penetrates into the marsh sediments. Many species utilize marsh habitat for forage and refuge, thus creating the possibility of secondary oiling from this remaining oil. Though burning of oil in marshes can have very high efficacy rates (up to 70% reported by NOAA in some cases), there are numerous tradeoffs and precautions that must be taken when

TABLE 4.4 Benefits, Challenges, and Potential Impacts of Shoreline Cleanup Methods

Benefits	Challenges	Potential Impacts
	Debris Removal	
• Removal of debris before impact lessens the amount of hazmat material that potentially may have to be disposed • Allows for an unobstructed work zone for cleanup and response activates to occur	• Very labor-intensive and time-consuming • May place responders in situations of potentially hazardous conditions causing concerns about human health and safety	• Removes material that may be used as forage or shelter for local biota • Removes material that may stabilize an area from erosion • Equipment and personnel can inflict secondary impacts on flora and fauna
	Monitored Natural Attenuation	
• Does not cause additional environmental impacts from response activities or generate a waste stream	• Time-consuming and not suitable for all oil products • Requires emendable environmental conditions to be an effective response option	• Potential environmental impacts from persistent oils • Possible remobilization to uncontaminated or sensitive areas
	Tilling/Aeration	
• Often used to augment natural attenuation • Allows oil to be more rapidly and naturally broken down by natural processes by increasing accessibility to nutrients and oxygen and by increasing the surface area of the oil for colonization of microbes	• Requires mobilization of light, medium, or heavy equipment to remote locations • Time-consuming strategy, that relies on natural processes to remove the oil	• Equipment and personnel can inflict secondary impacts on flora and fauna • Potential of disturbance of local fauna from noise and activity • Process moves oil to new locations and can potentially bring previously non-exposed organisms into contact with the oil • Potential disruption of subterranean fauna
	Bioremediation (Microbial, Nutrient Enrichment, Enzymatic)	
• Enhances natural biodegradation and lets microbes convert the oil into inert byproducts • Low physical disturbance of affected area • Ease of use	• Slow process of removal and not suitable for all oil products • Requires emendable environmental conditions to be an effective response option • May require addition of nutrient enrichment, which may have secondary effects • May be used in conjunction with tilling (aeration) • Oxygen and nutrient addition are typically more effective and accepted than bacterial seeding	• Slow process that may allow for secondary impacts to organisms by contact or ingestion • Possible remobilization to more sensitive locales by being transported to uncontaminated or sensitive areas
	Surfwashing	
• Removes oil from contaminated sand and gravel by relocating the material into nearby waters, utilizing the natural wave and water flow to help dislodge and naturally remove the contamination • Allows for large area of contaminated shoreline to be remediated • No additional waste is generated	• Requires mobilization of light, medium, or heavy equipment to remote locations	• Additional environmental impacts to substrate and flora should be expected due to damage from equipment mobilization and use • Potential of disturbance of local fauna from noise and activity • Process moves oil to new locations and can potentially bring previously non-exposed organisms into contact with the oil • Potential disruption of subterranean fauna • Potential for removal of valuable non-contaminated substrate and material by inexperienced responders
	Manual Removal	
• Can be an effective, surgical removal option with minimal secondary impacts when only the contamination is removed • Generates a low volume of waste material	• Very labor-intensive and time-consuming • May require responders to be in remote areas, in close contact with potentially hazardous materials, causing concerns about human health and safety	• Can inflict secondary impacts to flora and fauna • Risks removal of valuable non-contaminated substrate and material by inexperienced responders • Can generate additional waste requiring storage, transportation, and disposal

continued

TABLE 4.4 Continued

Benefits	Challenges	Potential Impacts
	Vegetation Cutting	
• Removes oiled vegetation and allows access to areas of oil accumulation beneath the canopy for removal • Eliminates secondary oiling of biota that may utilize the area for forage or shelter	• Very labor-intensive and time-consuming • May require responders to be in remote areas, in close contact with potentially hazardous materials, causing concerns about human health and safety • Consultation with professionals should dictate use of this method as many tradeoffs must be considered • Generates a high volume of waste material that must be properly disposed	• May cause long-term habitat loss due to the area of plant removal • Equipment and personnel can inflict secondary impacts on flora and fauna

this type of response is considered.[5] The NOAA/API *Guide on Oil Spills in Marshes* (2013) identified and described 30 oil spills, three field experiments, and three laboratory studies concerning in situ burning conducted in marshes. Of the 27 oil spills reviewed, 23 were light to medium crude oils and 4 were light refined products. The NOAA guide also discussed various parameters that will affect feasibility and effectiveness of burning and concluded that if burning is conducted following appropriate guidelines, the wetland vegetation is expected to recover within 5 years, but for many spills within one to two growing seasons.

Federal, local, and state agencies involved in emergency response and protection of the air quality must be consulted and authorize the burning technique. These entities will consider the amount of air pollution impacts including particulate matter that will be emitted and potentially affected areas of populations downwind (see Section 4.2.3.3). Other entities such as federal, state, and tribal authorities, along with historical and cultural authorities, should be consulted. Threatened and endangered species as well as other local flora and fauna must be identified and protected using numerous methodologies. If the ignition of the oil in the marsh requires the use of a chemical accelerant or ignition agent, this may require additional approval.

Numerous considerations must be addressed when evaluating oil burning in marshes (NOAA and API, 2013). This may include time of the year, the type and weathering condition of oil, soil type and degree of oil penetration, wind speed, water depth (if applicable to protect vulnerable vegetative root systems), containment of the burn, etc. Containment of a burn in a marsh is difficult, as the installation of protective fire boom is typically not possible due to the vegetation coverage. Fire spreading is typically prevented by creating corridors of open water separating out the burn area. Numerous response organizations have developed checklists and guidance documents for undertaking safe burning operations (API, 2015b).

Bioremediation

Bioremediation, biostimulation, and bioaugmentation are response strategies that utilize naturally occurring or introduced microbial biodegraders to consume the oil, resulting in breakdown of some of its components, predominantly into carbon dioxide and water (API, 2014). Bioremediation has been and still is a tested form for spill response on land spills where there may be a limitation of microbes or nutrients and there is sufficient time and proper conditions for the processes to work. On marine shorelines bioremediation is most often incorporated as a "polishing" step or response to augment the natural degradation processes after gross oil contamination has been removed and continued cleanup would cause more harm than benefit.

Three types of bioremediation methodologies are discussed here; each aligns within one or both categories of biostimulation or bioaugmentation:

1. Nutrient enrichment/addition—stimulates naturally occurring bacterial communities by the addition of limited essential nutrients to accelerate the degradation process.
2. Microbial addition—involves the addition of petroleum hydrocarbon-degrading bacteria/biodegraders directly to an oil that has been released into the environment for the purpose of exponentially increasing their numbers to expedite the degradation process.
3. Enzyme addition—this process is meant to accelerate the natural degradation processes by the addition of a catalyst that speeds the chemical reaction rate of a particular reaction, thereby expediting the natural breakdown processes.

Biostimulation involves the judicious addition of chemical supplements (typically sources of nitrogen and/or phosphorus, often in commercial formulations) that assist indigenous

[7] See https://www7.nau.edu/itep/main/HazSubMap/docs/OilSpill/USGSOSRinFastCurrents.pdf; https://oilspillprevention.org/-/media/Oil-Spill-Prevention/spillprevention/r-and-d/inland/swift-water-spill-response-guide-april-2.pdf; https://www.itopf.org/fileadmin/uploads/itopf/data/Documents/TIPS_TAPS_new/TIP_3_Use_of_Booms_in_Oil_Pollution_Response.pdf; and https://www.osti.gov/etdeweb/biblio/20103189.

hydrocarbon-degrading microbes to metabolize oil components by balancing the in situ carbon:nitrogen:phosphorus ratio. Addition of nutrients does not usually increase the mass or range of hydrocarbons eventually degraded, but rather accelerates the rate of biodegradation (reviewed by Prince, 1993). Biostimulation, where permitted, may be most useful on nutrient-poor beaches where the applications can be controlled and not immediately be diluted by wave and tidal action. Several types of nutrient application, including solid and slow-release oleophilic fertilizers, were used successfully on shorelines after the *Exxon Valdez* spill in Alaska; some were found to increase biodegradation rates three- to five-fold (Atlas, 1995). However, in field trials of other near-shore environments (e.g., Delaware shoreline and salt marshes; Mearns et al., 1997; Zhu et al., 2001), the application of nutrients was found to be unnecessary because terrestrial runoff supplied sufficient nitrogen and phosphorus. Biostimulation using oleophilic fertilizers may be particularly useful on beaches for accelerated degradation of buried oil in sandy intertidal (aerated) beach sediments (Pontes et al., 2013), but whether biostimulation of fine-grained marine sediments would enhance anaerobic biodegradation is untested in the field. More recently, in vitro trials with seawater microcosms combined nutrient supplementation with application of biosurfactants, resulting in enrichment of native hydrocarbon-degrading species and greater biodegradation than that achieved using either nutrients or biosurfactant alone (McKew et al., 2007). However, **biostimulation for remediation of oil spills in the open ocean is unlikely to be successful because of dilution factors and therefore has not been rigorously field tested.**

Bioaugmentation ("seeding") is the addition of exogenous microbes to an environment. In the case of oil spills, bioaugementation involves introducing a known hydrocarbon-degrading species or cocktail of species previously enriched during growth on oil in a laboratory or commercial facility. Although bioaugmentation can be successful at small scale in vitro (e.g., McKew et al., 2007) and large scale ex situ (Hassanshahian et al., 2014), **bioaugmentation remains controversial and currently is not implemented in situ for marine oil spills, for several reasons.** First, introducing exogenous microbial cultures into a natural environment is not permitted in some jurisdictions. Second, the survival and transport of introduced microbes cannot be controlled once they are released into an open environment. Third, it is well established that the indigenous oil degraders that exist in every marine ecosystem examined to date are adapted to the local conditions and predators at any given spill site, whereas introduced microbes are at a disadvantage and may not compete with indigenous microbes. Finally, introduced organisms would be subject to the same environmental limitations as the native organisms and may still require biostimulation. New approaches—for example, immobilization of bacteria in pellets of biodegradable floating support material (Luo et al., 2021)—may revive interest in testing this strategy for confined oil slicks.

Enzyme addition involves distributing a formulation of one or more cell-free enzymes to the environment. However, several inherent limitations make this option unlikely to be effective in situ. Among these are the facts that (1) hydrocarbon oxidation requires multiple enzymatic steps that cannot be achieved by a single enzyme, and suspensions of enzyme mixtures cannot be coordinated into an efficient pathway in a dilute environment; (2) many enzymes require sensitive co-factors that would have to be supplied in a protected matrix; and (3) enzymes are proteins and therefore likely to be consumed as nutrients by native microbiota before they are able to effect a change in the oil.

Arctic Shoreline Response

The Arctic environment presents some unique challenges, especially for the shoreline cleanup. Limited daylight hours during winter, frequent days with fog, strong wind gusts, and low temperatures can make any outdoor activities very difficult. Limited infrastructure presents another obstacle to field operations. Above the Arctic circle in North America, there are no rail connections, two deep-water ports, only one of which (Tuktoyaktuk, in the western Northwest Territories of Canada) is connected to the road system, and four airports open to regularly scheduled jet aircraft service, three of which are in Alaska. Travel from these airports to more remote field sites usually requires small chartered aircraft. There are very few settlements in the Arctic and they are not likely to accommodate large number of workers that may be required for the shoreline cleanup. Challenging environmental conditions, limited transportation options, difficulties with accommodating large number of personnel and equipment in the remote and often sensitive areas makes planning and implementing Arctic shoreline response very difficult. Arctic shoreline types vary from rocky shores and gravel and sandy beaches that are similar to more temperate regions to some unique shoreline types such as tundra cliffs and peat shorelines. **The sensitive nature of Arctic shorelines require special considerations for oil spill assessment and cleanup.** Under some circumstances, monitored natural attenuation may be the most appropriate response method, especially considering logistics challenges in remote regions. In addition to the challenges of bringing equipment and personnel to the remote beaches and accommodating them for long periods of time without existing infrastructure, the negative impacts of large volumes of waste that would get generated and additional disturbance from human presence has to be carefully considered as they may outweigh benefits of cleanup. One positive aspect of the Arctic environment is that landfast ice (ice cover growing from the shore toward the sea) forms first and melts last, providing shorelines with natural barriers from offshore oil spills for a significant portion of the year. An oil spill heading toward the shore will accumulate against the ice edge where it can be effectively recovered or burned using the techniques described in earlier sections. The frozen nature of the ground during significant portions of the year also prevents penetration of pollution deep into the soil and may allow for a successful recovery from the surface. Emergency Prevention, Preparedness and Response (EPPR) Guide to Oil Spill Response in Snow and Ice Conditions (2015) provides a comprehensive overview of Arctic shoreline types, expected oil behavior, and recommended response strategies.

4.2.5 Comparison of Response Options for Decision-Making

As discussed in earlier sections, no response option is completely effective under every condition or completely risk free. Every response action, as well as inaction, can result in additional impacts, which must be carefully considered before the decision to proceed is made. In the preparedness phase or at the time of an oil spill, decision-makers need to evaluate possible response strategies—their advantages, challenges, and additional impacts they my cause—and ultimately select those that would be optimal for a specific spill scenario. Without this consideration, the response actions could result in greater impacts than the oiling itself and delay the environmental and socioeconomic recovery of the affected areas (Hoff, 1996; Martínez et al., 2012; Michel and Ruherford, 2014; Michel et al., 2017).

Depending on the scale and complexity of the spill scenario and the phase of planning or response, this analysis may range from a mental health assessment and a brief discussion to a comprehensive analysis including formal ecological and socioeconomic assessment and involving a variety of stakeholders. The essence of the principle "minimize the harm" is integrated into all steps of contingency planning and response even if a formal analysis is not conducted. Some of the examples include:

- Contingency planning, incorporating the entire toolbox of response options and allowing flexibility of utilization of individual tactics at the time of response;
- Focus of the initial response on stopping the spill source and minimize oil spreading;
- Setting response objectives aiming "to minimize harm to the environment" and "protect resources at risk";
- Development of response technique checklists and decision trees by regulatory bodies to expedite decision-making process at the time of the response; and
- Pre-authorized and designated "no use" dispersant zones in offshore United States areas that have been established, based on consensus, from previously conducted analysis.

In fact, many small spills or responses, where only one response method can be used (e.g., mechanical recovery) do not require comparative analysis; responders can use best operational practices and existing contingency plans to respond to a spill of this nature. In the case of a larger spill, when several response options are viable, a comparison between response options can be conducted in several ways, all comparative analyses are built on the same principles and aim to engage with key stakeholders, minimize impacts on people and the environment, and assist with the most efficient and effective recovery of the ecosystem (including the local community). The primary concern of any response is the safety of the public and responders. Once action is taken to protect human health and safety, comparison of response options can be analyzed. This process can be applied with different degrees of involvedness while maintaining the same "minimize the harm" principle based on the nature of the event and the timing of the decision (during contingency planning or the response phase). NEBA is a comparison tool historically used by stakeholders and the response community to analyze response options and create a response strategy that minimizes the effect of an oil spill on both the environment and local community.

It is important to note that a formal NEBA process may be time- and resource-consuming and when possible, should be conducted as a part of the contingency planning process guiding the response strategies used for planning scenarios. It may be difficult to conduct a comprehensive formal NEBA analysis from scratch at the time of a spill without causing unnecessary delays in spill response. During a response, the NEBA process (formal or informal) is generally performed by the environmental unit and used to ensure that response strategies and tactics are tailored to the evolving spill-specific conditions and offer maximum protection to the resources at risk. The environmental unit specialists can use NEBA scenarios generated during planning phase and adjust them for the specifics of a spill.

Whether conducted formally or informally, the NEBA process typically consists of the following steps (IPIECA, 2015):

1. Compile and evaluate available information.
2. Develop response scenarios, including a "no response" scenario as a base case. Evaluate their effectiveness and resulting changes in the oil fate and behavior.
3. Identify environmental and socioeconomic resources that may be affected under different response scenarios.
4. Evaluate potential impacts to the resources and their recovery potential under different response scenarios, including potential additional impacts of response activities.
5. Compare potential outcomes and select response option(s) resulting in a higher degree of environmental and socioeconomic protection and fastest ecosystem and economic recovery.

There are several methods for implementing NEBA principles in a formal assessment of response options (NASEM, 2020). These are:

- Consensus Ecological Risk Assessment (CERA)
- Spill Impact Mitigation Assessment (SIMA)
- Comparative Risk Assessment (CRA)

All these methods are based on similar considerations:

- Safety of public and responders is addressed first.
- A critical assumption is made that oil is already in the environment and response options will change the oil's presence in varying environmental conditions. Impacts of both untreated slicks and slicks treated through various response tools need to be weighed.
- All feasible response options are considered (Response Toolbox approach).
- No response methods will be completely effective and all response efforts pose some level of risk.
- Realistic effectiveness of response techniques are considered under conditions unique to each spill.
- Consider all affected resources and use an ecosystem-based approach to assess impacts to populations and habitats together, rather than assessing impacts to individual organisms.
- Incorporate evaluation of socioeconomic and cultural resources, if needed.
- Take a holistic view to appraise long-term impacts on the public and the affected ecosystem and consider recovery rates of affected resources to their pre-spill condition.
- Rely on best available science and location-specific information when available.
- Encourage stakeholder engagement and transparency of decision-making. Local and Indigenous knowledge on topics such as baseline conditions, environmental patterns, resources at risk and protection priorities, temporal and spatial population variability, recovery rates, etc., should be sought and incorporated into the analysis to ensure appropriateness and transparency of response decision-making.

4.2.5.1 Consensus Ecological Risk Assessment

The USCG sponsored 17 CERA workshops from 1998 to 2012 (Aurand et al., 2000). These workshops brought together a variety of stakeholders who conducted a formal NEBA using comparative risk methodology to evaluate regional oil spill response options in a planning environment. More recently, two additional CERA workshops were held: one focused on potential transportation-related Bakken and dilbit crude oil spills in Delaware Bay (sponsored by USCG; published in 2016) and one in Hawaii (sponsored by Oceania RRT; published in 2018). The CERA workshops present a valuable mechanism to bring together and build working relationships among key stakeholders, educate them on and involve them into the decision-making process, compile available regional information and best practices, and facilitate timely selection of response options at the time of response. The outcomes of these workshops may need to be periodically reviewed to ensure that they are still relevant for the ever-changing spill scenarios, environmental conditions, best available science, and resources protection priorities.

In the CERA framework, a response scenario is developed with specific detailed information on oil composition, toxicity, weathering, and trajectory. The scenario, response options and evaluation of the impact of the response options are developed by a group of diverse stakeholders to evaluate the impacts of different response options, including no response. The affected environment is divided up into a number of compartments (e.g., water, shoreline, socioeconomic resources, etc.). Each compartment is defined by its representative population or resources and its relationship to the other defined compartments. This results in a risk-ranking matrix in which the affected resources are ranked by the percent of a population/habitat affected and the rate of its recovery. The most affected and slower-to-recover populations and habitats receive higher risk scores; the populations and habitats affected to a lesser degree and capable of rapid recovery receive lower risk scores. This comprehensive analysis, conducted in coastal regions in the United States led to the designation of the pre-authorized offshore zones for dispersants and in situ burning as well as the development of checklists and decision trees that could be used in an emergency. The CERA method proves challenging because of the necessary amount of scientific information and time required for analysis. It may not be appropriate for short response efforts or when detailed information is not available. The SIMA method was recently developed to address this need.

4.2.5.2 Spill Impact Mitigation Assessment

SIMA underscores the importance of stakeholder engagement and depends on practical experience, best available data, and local knowledge to expeditiously evaluate response options, even with limited information. It also specifically addresses the socioeconomic, cultural, and special value resources of local stakeholders (IPIECA and IOGP, 2018). In the contingency planning phase, this approach could be used with the same intricacy and thoroughness as the CERA methodology, but during a time-critical response phase, it could be done much faster. Similar to CERA, SIMA identifies compartments within the potentially affected environment and assesses the impacts with and without response measures. When the impacts are assessed, the process uses consensus opinion of experts and stakeholders to assign relative impact ranking for different compartments. Then the group evaluates the response options and for each compartment determines whether a response option improves the situation or creates additional impacts, as well as the degree of improvement or additional negative effects. These assessments are based on expert knowledge and best available information and are translated into a numerical matrix by first assigning an initial impact coefficient and then multiplying it by a mitigation factor. Scores for each response technique are then summed up across all compartments. A higher positive SIMA score indicates the likely ability of a response technique to mitigate impact from the spill. Negative SIMA scores indicate that a response technique may result in greater damage than an unmitigated slick. This analysis helps responders and decision-makers to select response techniques that would be most suitable under a specific spill scenario.

It is important to note that, under different response options, the numerical value of SIMA scores may be misunderstood as having greater precision than was intended by this methodology. In a simplified form, the SIMA converts professional opinions into numeric scores; these numbers should not be taken as absolute values, but rather used as a relative comparison metric. Small single digit differences between impact mitigation scores do not indicate that one technique is better than the other. Generally speaking, positive values indicate response techniques that have a potential to result in a reduction of spill impacts and negative scores highlight the techniques that may result in additional environmental impacts.

4.2.5.3 Comparative Risk Assessment

The CRA is a new, computationally advanced approach to the comparison of response techniques and their potential effects (Bock et al., 2018; French-McCay et al., 2018; Walker et al., 2018). This assessment is similar in process to CERA in terms of identifying compartments and representative populations within each compartment. However, a CRA takes a more quantitative approach that involves complex three-dimensional numerical modeling to calculate the volume of water and the area of the water's surface that is affected by the oil concentrations above the impact threshold. The results are then overlapped with the relative density distributions of valuable ecosystem components (VECs) across environmental compartments. A risk coefficient is then assigned based on the population's sensitivity, vulnerability, and recovery rates. As in most risk assessments, the organisms that are slowest to reproduce and those that are long-lived appear to be at greatest risk and come up at the top of protection priority (Bock et al., 2018; French-McCay et al., 2018).

The international community may use other numerical modeling methods, which are not typically used in North America.

Summary

During oil spill preparedness, planning, and response, the response community and stakeholders use NEBA, CERA, SIMA, CRA, and other techniques to develop a response strategy. Using these approaches, either formally or informally, ensures the response strategy compares the impact mitigation potential of each response option while minimizing the net impact of an oil spill on the environmental, socio-economic, and cultural resources at risk. Conducting a formal NEBA, CERA, SIMA, CRA process is not always feasible or practical, yet there is always a need to base response decisions on the best available information in a timely way. Lessons learned from the earlier assessments in a particular region, assessments administered for similar scenarios in other areas, and experiences from oil spill responses and restoration projects, globally, could be used to facilitate timely, informed, and transparent decision-making. Responders can use an earlier formal analysis when a spill takes place in an area that is already covered by a NEBA-type of analysis. If the actual spill scenario and NEBA-type analysis are reasonably similar, a significant portion of the earlier analysis and conclusions can be used to expedite the decision-making process (with some quick adjustments to reflect specifics of the new scenario).

It should be noted that the NEBA-type processes are quite distinct from the NRDA process. The NEBA framework attempts to integrate all available knowledge about the impacts of oil in different compartments and evaluates the ability of response techniques to mitigate them. It starts with an assumption that oil is already in the environment and seeks to find an optimal response solution, recognizing that none of the response techniques are 100% effective and that all of them come with their own risks. This analysis can be performed reasonably quickly using high-level information at a population and ecosystem level and does not require extensive scientific analysis or information about individual organisms. NRDA, in contrast, compares the impacts of a specific spill in a specific location to specific resources to their unimpacted state. It typically requires an extensive and detailed analysis of information about oil properties, toxicity, distribution, and impacts specific to that individual scenario. It does so without evaluating the alternatives, and usually takes many years to complete. The type and level of detail of information used, the timescale, the purpose of analysis, stakeholder engagement mechanisms, and other factors are very different for NEBA and NRDA and they should not be confused.

4.2.6 Wildlife Rescue and Rehabilitation

Over the past 50 years, the practice of oiled wildlife response has evolved from an activity conducted by small individual nonprofit organizations with minimal external support to a fully integrated part of the overall spill response effort performed in a professional and coordinated manner. In the United States, this progress has corresponded to that seen after the *Exxon Valdez* oil spill in 1989, with a more ordered and measured response effort as dictated by OPA 90 performed under a structured ICS (see Figure 4.17). While there are some differences in specific protocols and operational activities in active response efforts to different taxa involved (e.g., birds versus marine mammals versus sea turtles), the overall concepts of wildlife rescue and rehabilitation can be applied holistically to all species.

Wildlife response typically falls under the Wildlife Branch within the Operations Section, but works closely with other sections, particularly the environmental unit within the

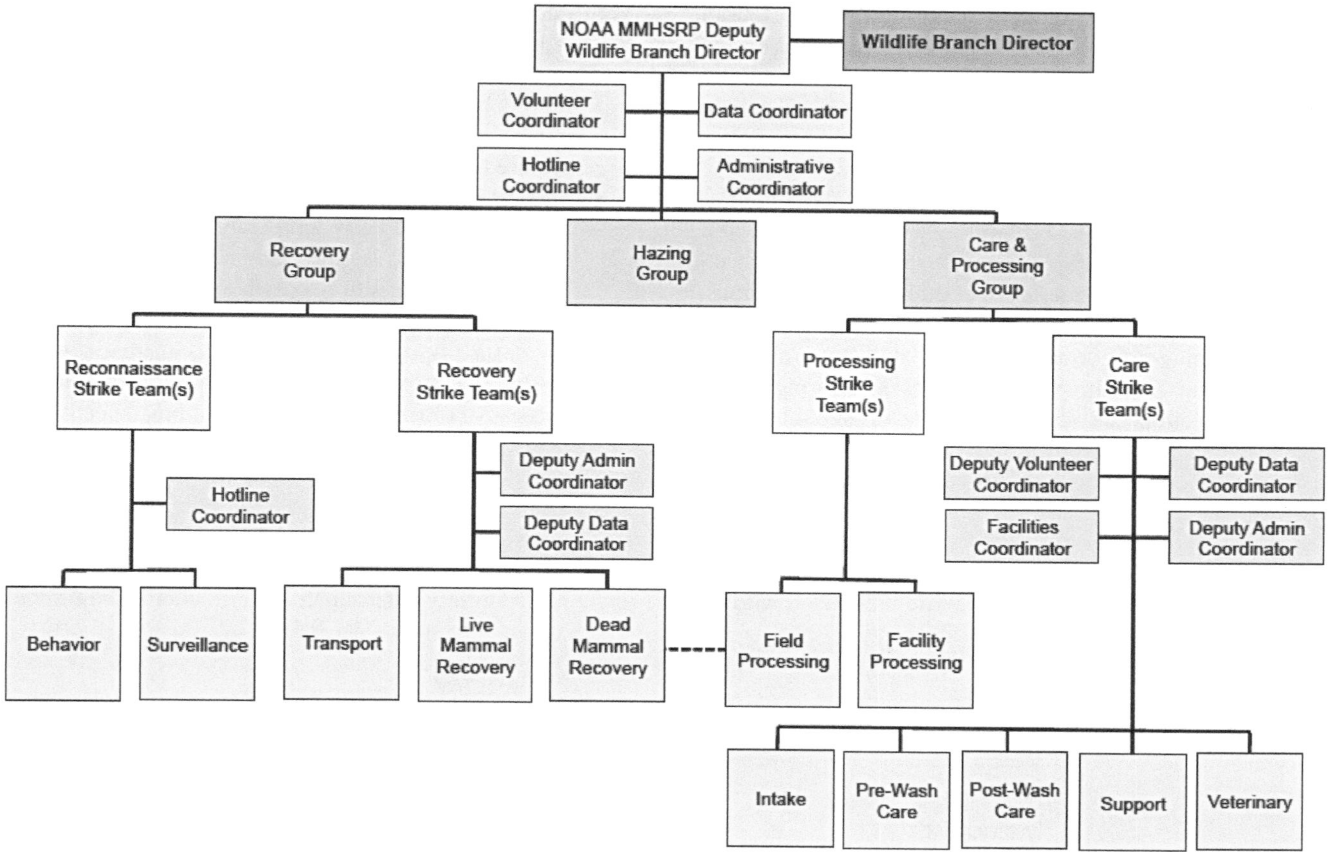

FIGURE 4.17 Example organizational chart for U.S.-based marine mammal responses under NOAA's National Marine Fisheries Service.
NOTE: The Wildlife Branch is a branch of the Operations Section within the larger ICS structure.
SOURCE: Ziccardi et al. (2015).

planning section. Under the guidance of the Wildlife Branch Director (WBD), the main objectives of the Wildlife Branch are typically to:

- Conduct all operations in such a way that are safe to both people as well as animals;
- Minimize injuries to wildlife and habitats from the contamination, cleanup effort, and animal recovery operations;
- Follow legal guidelines when collecting all data, samples, and animals;
- Document the immediate impacts to wildlife, and report in a timely and complete manner all relevant data and information necessary;
- Support efforts in spreading information to the media, public, and other interested parties; and
- Provide the best possible care to affected and/or threatened wildlife.

In general, oiled wildlife response can be divided into three different response strategies: primary, secondary, and tertiary. Primary response tactics include those actions that "keep oil away from wildlife." The most effective means by which oil is kept from animals is through mechanical, physical, or non-mechanical cleanup (e.g., in situ burning and chemical/biological response). However, reconnaissance of at-risk populations is also critical: both the collection of *a priori* historical information as well as rapid, real-time reconnaissance of animal presence in (and adjacent to and anticipated [incoming migrations]) the spill area. Additionally, collection of oiled carcasses can be considered a primary response action, as these can act as a risk of oiling (and associated toxicological effects) to scavengers as well as a key component of environmental cleanup actions.

Secondary oil spill response tactics include efforts to "keep wildlife away from oil." These actions fall into two main categories: deterrence (or hazing) and pre-emptive capture of wildlife. Birds and other animals can be scared away from an oiled area through use of visual or auditory devices, deterred from areas by use of exclusion devices (e.g., fences, netting), or attracted to less risky regions through bait or environmental manipulation. Capture of unoiled, at-risk species (and subsequent handling, transportation, short-term holding, and release of unoiled wildlife) has been done in select incidents, including the relocation of tens of thousands of African penguins during the *Treasure* oil spill in South Africa and the capture and long-term holding of New Zealand dotterels during the M/V *Rena* oil spill.

Tertiary response efforts are those typically thought of as the "standard wildlife rehabilitation efforts" during oil spill events (see Figure 4.18). Such recovery and rehabilitation activities are not always undertaken (e.g., in certain situations, recovery and rehabilitation of animals may not likely result in their returning to "normal" after care), but comprehensive assessment for the elimination of animal suffering (often through collection and humane euthanasia) should always be considered even when full-scale response cannot occur. Tertiary response actions fall into several main categories:

- ***Capture:*** During spill events, government agency personnel or trained and experienced wildlife responders from rehabilitation groups capture most oiled wildlife during search and collection efforts. Areas where oiled wildlife are likely to be found must be systematically covered, additionally, responders should accurately record field data (such as date, time, location, and name) as well as start additional legal requirements if required (such as chain of custody procedures).
- ***Stabilization:*** Stabilization and initial first aid (e.g., return to normal body temperature, fluids, and initial brief assessment) should be administered to live oiled animals either in situ if possible, or by being quickly taken to a field stabilization site. At this point, initial triage (or prioritization and sorting of treatment) can be done, including the determination of likelihood of success of rehabilitation and early humane euthanasia using veterinary-approved means when rehabilitation is highly unlikely to be successful.
- ***Intake:*** On arrival at the facility, a standardized intake protocol should be followed, to ensure legal evidence is obtained and documented for possible NRDA injury determination. Each oiled animal should

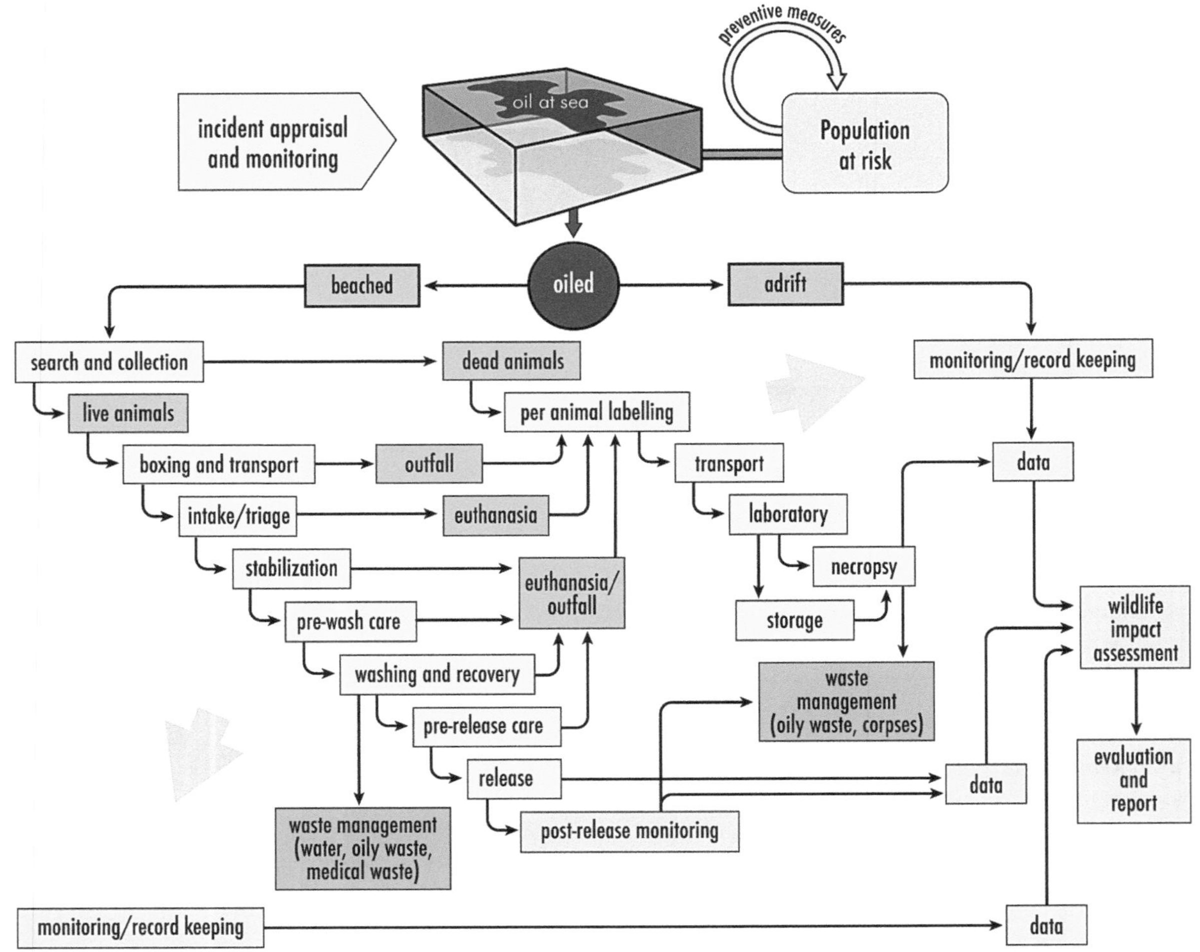

FIGURE 4.18 Tertiary oiled wildlife response options and activities.
SOURCE: From *A Guide to Oiled Wildlife Response Planning*, IPIECA Report Series Volume 13.

be individually identified, a feather sample or fur swab collected for proof of oiling, and a photograph taken. Animals undergo thorough physical examinations based on triaged prioritization (e.g., threatened/endangered status or requiring immediate care going first), noting especially the extent of oiling, and signs of hypothermia, dehydration, and malnutrition. Blood samples are drawn to evaluate, as a minimum, anemia and low serum protein, and all data are recorded in individual records for each oiled animal.

- ***Pre-wash care:*** The washing of oiled birds and heavily furred mammals is a rigorous and stressful process and should not be done on newly admitted animals until they are determined to be physiologically stable. They should only undergo this procedure when they meet certain standard criteria (typically stabilization with warmth, fluids, and nutritive support). Other species (such as pinnipeds aside from fur seals and sea turtles), however, may be able to be cleaned immediately upon intake if blood work and initial exam show no abnormalities beyond physical coating. Animals should be provided nutritional and hydration support (e.g., birds gavage-fed high-calorie nutritional slurries alternating with rehydrating solutions up to eight times daily, with volumes dependent on species, size, and health status) and regularly re-examined to determine when they are medically stable enough to move to cleaning.
- ***Cleaning:*** Standard wash procedures include the use of dishwashing detergent (diluted for birds and furred mammals, full strength in other species) in water heated to, and maintained at, physiologically normal temperatures. This solution is manually agitated around, or massaged into, the oiled pelage, until oil is completely removed. Monitoring of core body temperature and/or close observation of animals for distress should be done at all times. Animals are then rinsed with water heated to physiologically normal temperatures using an adjustable high pressure (40–60 psi) nozzle until all detergent is removed from the feathers or underfur. For heavily furred mammals and birds, it is also recommended that rinse water be 2–5 grains hardness, as minerals in the water will bind with microscopic amounts of detergent and can cause calcium carbonate crystals to form. After animals are completely clean, birds and furred mammals are then allowed to dry using towels, pet groomer dryers, heat lamps, or ambient conditions in their pen, depending on species.
- ***Pre-release conditioning:*** After cleaning, animals should be moved into appropriate enclosures (typically outdoor pools, aviaries, or pens) that are of sufficient size to allow accurate determination of behavior, feeding habits, and waterproofing status. Pen and water quality must be excellent to prevent recontamination of feathers/fur by oily feces and fish, and animals should be continually observed for signs of inadequate waterproofing. Excellent nutritional support should be provided, and regular medical and waterproofing checks should be done to plot progress. Once animals meet established release criteria (usually, at a minimum, normal behavior, good body weight, waterproof, and normal physical exam, hematology, and serum chemistry values), they can be marked with a permanent tag, metal leg band, or other means of post-release monitoring, and released into an appropriate clean environment.

Since the release of *Oil in the Sea III*, numerous advances in oiled wildlife rehabilitation have occurred, both at a systemic level and a scientific level. Internationally, the importance of oiled wildlife efforts as an integral part of overall spill response has been recognized, particularly after the DWH incident in 2010. The International Petroleum Industry Environmental Conservation Association (IPIECA)-International Association of Oil and Gas Producers (IPIECA-IOGP) Oil Spill Response Joint Industry Project (OSR-JIP) recognized oiled wildlife response involving specialist personnel as one of the 15 key capabilities necessary for effective preparedness and response. In 2015, as part of Phase 2 of the OSR-JIP effort, an ambitious wildlife response preparedness project was initiated. This Global Oiled Wildlife Response System (GOWRS) project, which involved 11 leading wildlife response organizations from seven countries (including four leading organizations from the United States), aimed to develop an international framework for oiled wildlife response as well as encourage the further development of wildlife response preparedness by industry and other stakeholders. This effort has led to the development of numerous globally acceptable planning and response documents; animal care standards; a system of readiness and response encompassing training; equipment, and personnel readiness; and a source for the best available information on effective planning needs (IPIECA, 2014, 2017; Ziccardi, 2021).

In the United States, as a consequence of the DWH incident, NOAA's National Marine Fisheries Service embarked on a significant effort to establish national protocols and procedures for cetaceans, pinnipeds, and sea turtles to support both integrated response efforts, but also better organization of NRDA efforts (NOAA, 2015). This national effort has also led to specific efforts in key regions of the United States, with Arctic, Gulf of Mexico, and Pacific Island regional plans being developed to focus on key risk issues in those sensitive habitats. On the scientific front, many investigational and laboratory studies following DWH focused on attempting to better understand the overarching effects of oil on wildlife (see Chapter 6), and these have led to significant advances in how to better collect and care for animals and are now being more fully incorporated into the medical protocols of all professional oiled wildlife response organizations.

Probably the most contentious issue surrounding the recovery and rehabilitation of oiled wildlife is whether the effort is a waste of resources focused on individual animals (versus populations), whether survival of rehabilitated animals is poor, and whether the resources expended on such would benefit wildlife more if directed to other conservation efforts (Estes, 1998; Jessup, 1998). There is considerable variability among post-release survival studies in the literature, dependent on species differences, characteristics of the spill including product spilled and speed of response, or specifics of rehabilitation methods (summarized in Henkel and Ziccardi, 2018). However, despite this variation, more recent cases have clearly shown survival far in excess of estimates generated prior to 1993 (as detailed in Sharp, 1996), and in several studies the survival of rehabilitated and control animals showed no discernable differences (Golightly et al., 2002). Recent studies have also shown that rehabilitated oiled wildlife can successfully re-enter the breeding population and reproduce (Whittington, 1999; Sievwright, 2014).

In summary, wildlife rescue and rehabilitation can be a highly successful operational effort during oil spills—one of the most positive, publically visible outcomes of such emergencies. **To achieve high release rates and best possible post-release recovery rates, robust planning for oil spill response is critical, as is conducting ongoing research into, and incorporating the results from, the latest and best available science in protocols for veterinary care and rehabilitation.** Only with this focus on readiness and applying "lessons learned" can rapid capture and best achievable care of oil-affected wildlife be realized.

4.3 CONCLUSIONS AND RESEARCH NEEDS

Conclusion—Source Control: Second only to prevention, an effective source control is the key strategy that can reduce the volume of hydrocarbons entering the marine environment as well as its potential environmental and socioeconomic impacts. Advanced source control measures have been developed for wells, pipelines, and vessel salvage. The use of proven source control techniques continues to reduce potential hydrocarbon volumes that may enter North American waters, and their continuous advancements should be enabled through research and development efforts.

Conclusion—Incident Command System: The value of the Incident Command System (ICS) in effective incident management, integrating interests of diverse groups of stakeholders, and ensuring effective coordination and deployment of response resources to maximize protection of human health and safety as well as resources at risk has been documented since its adoption.

Conclusion—Response Toolbox: No single response technique is effective under every environmental condition and every spill scenario. They each have their advantages, challenges, and optimal operational conditions for use. Greatest effectiveness of response activities is achieved when various response techniques are available (a response toolbox concept) and their use is tailored to a specific response scenario based on safety, effectiveness, ability to protect environmental and socioeconomic resources and facilitate fastest recovery to the pre-spill conditions. If required, the NEBA tools and processes (CERA, SIMA, and CRA) are available to compare oil spill response options and select an optimal combination of activities that could result in maximum protection of environmental, socioeconomic, and indigenous resources at risk.

There are some barriers that may challenge the integration of best available science and new technologies as well as utilization of available response techniques and monitoring tools in spill response. Some of these barriers include inability to conduct full-scale field tests with real oil in the United States and significant challenges with conducting them in Canada, which prevents verification and optimization of new response technologies in the field and gathering of field-scale scientific information; challenges with commercialization and regulatory approvals of new technologies; insufficient knowledge of decision-makers and stakeholders of best available techniques and practices; extended time required to make decisions at the time of response; communication challenges of response details and decision-making process with interested stakeholders; and much more. Joint education and outreach efforts are needed from the agencies and industry to increase understanding and acceptance of these tools among specialists and the general public, as well as to facilitate involvement of local and indigenous representatives.

Conclusion—Funding for Oil Spill Research: Significant progress has been made in advancing response techniques since *Oil in the Sea III*. There are a number of standing research programs as well as focused multi-year, multidisciplinary research programs with a fixed timeline for funding. All of these programs have generated invaluable advances in oil spill response knowledge and technologies. Continuous investment into oil spill research and development and deployment of verified technologies in the field should be encouraged.

Conclusion—Extrapolation of Research Results: Researchers who clearly indicate response scenarios/conditions being simulated during laboratory, test tank or modeling experiments, and explain the process for extrapolation of the results into field conditions, enhance the relevance of their work and make it easier to integrate their findings into real-life responses. This allows for a better comparison between different studies and eases their integration into the response decision-making. Great caution is advised in extrapolating conclusions from smaller-scale experiments to field conditions when the optimal process for doing so is not suggested by the researchers.

Conclusion—Field Experiments: It is not possible to simulate all the complexities and variability of field conditions in even the largest test tanks. Field experiments with real oil and spill-treating agents and studies conducted during actual response events are critical for the development, testing, and improvement of response techniques under realistic conditions.

Conclusion—Remote Sensing, Monitoring, Modeling and Other Information/Computation-Related Technologies: Since the *Oil in the Sea III* report, significant progress has been made in spill modeling, monitoring, remote-sensing, big data collection and processing, and other information/computation-related technologies. These techniques have become an integral part of the response and natural resource damage assessment activities. Their progress and advancements are expected to continue at a fast pace in the coming years. Further progress and adoption of relevant and verified technologies will improve response effectiveness.

Conclusion—New Fuel Types: New requirements for low sulfur fuel oils (LSFOs) for marine shipping came into effect in 2020 but studies on these oils, which have properties that are combinations of light distillates and heavier distillates, are currently extremely limited. The few very low and ultra-low (VLSFO and ULSFO) samples studied to date differ chemically from traditional marine fuel oils and from each other. Their fate and behavior are also likely to be different, including a potential to become semi-solid under some circumstances. The semi-solid state of a slick can significantly limit the effectiveness of available response options and require special response strategies.

Conclusion—Health and Safety Risks to Response Professionals: Protection of the health and safety of the general public as well as responders is the primary priority during emergency response. Numerous best practices for safe operations and use of personal protective equipment by responders have been developed, implemented in the field, and integrated in the incident management processes. Still, response operations, especially those involving mechanical recovery and shoreline cleanup where responders have to remain in close proximity to the oil, as well as operations involving heavy equipment under offshore conditions, have inherent risks.

RESEARCH NEEDS

Continued research, technology development, and implementation efforts aimed at prevention of spills, early detection, and limitation of spill volume if a spill occurrs is encouraged. More specifically, the research included in Table 4.5 would benefit future oil spill response efforts.

TABLE 4.5 Research Recommended to Advance Oil Spill Response and Minimize Effects

4.1	**New Fuel Types and Oilfield Production Products:** The effectiveness of various response techniques and their windows of opportunity in responding to hybrid fuel oil, particularly low sulfur fuel oils, and to diluted bitumen ("dilbit") oilfield products, must be carefully evaluated.
4.2	**Lifecycle Analysis of Oil Spill Based on Response Scenarios:** A continuous collaboration among academic, government, and industry scientists and response practitioners is needed to develop a comprehensive, multifaceted, and realistic analysis of "cradle to grave" oil slick fate and effects for various response scenarios, including monitored natural attenuation. This analysis should include all possible variations of response scenarios of a single event so as to give decision-makers a complete and comprehensive picture of the decision outcomes in each of the scenarios. At present, it is challenging to create a complete and comprehensive picture from the results of multiple uncoordinated studies with a narrow focus, as these studies are conducted with different goals and conditions in mind and do not lend themselves to a seamless integration.
4.3	**Health and Safety Risks to Response Professionals:** Research into health risks and psychological impacts to response personnel involved in various types of response operations should be conducted. This information should be integrated into response decision-making.
4.4	**Response Tools—Mechanical Recovery:** Mechanical recovery technologies would benefit from the research efforts aimed at improving encounter rates, specifically the volume of oil entering the containment devices and available for recovery as well as optimization of a particular technology's efficiency under various environmental conditions and spill scenarios (e.g., recovery of submerged oil). Mechanical recovery should also be viewed as a multicomponent system involving equipment mobilization, oil collection, recovery, storage, transfer, and disposal. Analysis of potential bottlenecks in this system under different response scenarios will help to identify potential areas for improvements.
4.5	**Response Tools—In Situ Burning:** Research focused on improving efficiency of and expanding a window of opportunity for in situ burning should continue and include various scenarios. Such scenarios could include burning in conventional booms, use of herders, inland burning, and burning under Arctic conditions.
4.6	**Response Tools—Chemical Dispersants:** Continuous research efforts focused on increasing natural dispersion processes through use of chemical dispersants (including new formulations and natural materials), and on mechanical dispersion techniques for selected offshore blowout scenarios are recommended.
4.7	**Response Tools—Arctic Conditions:** Although significant progress has been made in our understanding of the applicability and efficiency of various response techniques under Arctic conditions, these efforts should continue, given the great diversity of potential response scenarios as well as new formulations of fuels that may be encountered in that region.
4.8	**Oiled Wildlife Management:** Additional research into long-term impacts, survival rates, and return to normal function of treated and released animals would be beneficial to refine oiled wildlife management methods.

5

Fates of Oil in the Sea

Highlights

In the past two decades, due in large part to the depth and breadth of research that occurred after the *Deepwater Horizon* (DWH) oil spill, there has been major progress in understanding fundamental physical, chemical and biological processes and reactions that influence the fate of oil in the marine environment. Highlights in this chapter include:

- Building on advances in analytical chemistry, petroleum fluid models, and new data from spills and laboratories, significant progress has been made in understanding the interactions of live oil and gas spilled within the ocean water column.
- Significant advances have been made in understanding and predicting gas bubble sizes and oil droplet breakup processes for oil released into the sea, critical to determining their fate and ultimately the affected environments.
- A new phenomenon, "tip streaming," responsible for creating micro-scale droplets has been identified for droplets treated with chemical dispersant, allowing threads of oil to leak into and break within the droplet wake.
- Little information exists about the eventual fate of gas phase or aerosol pollutants. Future work should consider the fluxes into and out of the air–seawater interface.
- Ongoing advancements in data storage platforms and computing capabilities have enabled profound advances in computational fluid dynamics (CFD) modeling of oil spills in the oceans, critical to understanding the behavior of oil in the sea.
- Dissolution, long known as a relevant fate process for oil spilled at the sea surface, recently has been recognized as potentially significant for subsea oil spills, such as accidental oil well blowouts. New research has focused on mass transfer of bubbles and droplets, including analysis of gas hydrate effects and the fates of gas released from natural hydrocarbon seeps.
- Subsea dispersant injection (SSDI) became a major response option during the DWH oil spill and spurred research on optimal design, implementation, and efficacy analysis of SSDI. These analyses showed that even modest changes in oil droplet size distributions may lead to significant changes in the fate and effects of the spilled oil and may significantly benefit human health by reducing volatile organic compound (VOC) emissions to the atmosphere from subsea spills.
- There is renewed appreciation of the importance of photo-oxidation by solar irradiation of oil at the sea surface and near-surface. Numerous direct and indirect photo-oxidation reaction products, likely influence the fate of spilled oil.
- The terminology used to describe microscopic and macroscopic aggregates of oil with mineral and organic particles in the sea and stranded on-shore has been refined, addressing confusing terms used in the literature and among responders.
- MOSSFA (marine oil snow sedimentation and flocculant accumulation), a phenomenon observed in the Gulf of Mexico, is a mechanism proposed to explain observed oil transport to the seafloor during and after the DWH oil spill.
- Revolutionary advances in high-throughput DNA sequencing, bioinformatic software, and sequence databases ('omics) in the past two decades have enhanced the ability to detect and analyze microbial communities, including those that affect the fate of oil in the ocean through biodegradation and MOSSFA.
- There is an expanded appreciation of processes and oil fates unique to cold water and sea ice in deep sea and Arctic environments, and new recognition that oil biodegradation can occur at rates faster than previously assumed at near-freezing temperatures.
- Important advances have been made toward integrating and interpreting data acquired by different scientific specialties. Because the ultimate fate of oil is an aggregate of complex interactions involving many physical, chemical, and biological processes, meaningful understanding and prediction of the ultimate fate of oil requires synthesis of multi-disciplinary knowledge.

5.1 INTRODUCTION

Once oil has entered the marine environment, its chemical composition, physical properties, and behavior immediately begins to change due to combinations of dynamic processes that ultimately determine the fate of its components over time. Various combinations of processes dominate in different circumstances and locations in the marine ecosystem, discussed in detail in this chapter (see Figure 5.1). The aggregate of these processes determines the fate of oil and its diverse components, including the transport of bulk oil from one marine compartment to another (e.g., moving from a surface slick to dispersion in the water column or stranding on the shoreline), transformation of oil components to partially oxidized products (e.g., by photo-oxidation or biodegradation), and/or selective removal of oil components from the ocean (e.g., by volatilization or biodegradation). It should be clear that the "ultimate fate" of oil is dependent on time, as the processes described below act over time spans of hours (e.g., evaporation) to geological time (e.g., burial in sediments) and affect different proportions of different oils, ranging from minor changes to residual oil (e.g., ultra-heavy oils) to nearly complete removal from the ocean (e.g., jet fuel).

5.1.1 Major Advances in the Past 20 Years

Since publication of *Oil in the Sea III* report (NRC, 2003) there have been major technological advances in our ability to monitor and predict the fates of spilled oil in the ocean. In the previous report the chapter describing the behavior and fate of oil focused on physico-chemical weathering of spilled oil, including transport mechanisms, primarily at or near the surface. With the exception of photo-oxidation, our understanding of these fundamental processes has not changed substantially since then, yet our appreciation of how these processes work and how to predict them has improved substantially. The past two decades have provided tremendous advances in analytical methodology for detection and identification of oil constituents and their oxidized products (reviewed in Chapter 2), enabling sophisticated tracking, monitoring and forensic identification of spilled oil. Research into the *Deepwater Horizon* (DWH) spill provided additional information about the fate of subsurface spills, use of subsurface dispersant injection, modeling of gas-oil mixtures at depth, and behavior of oil plumes in deep waters. Whereas the 2003 report did not thoroughly examine biodegradation as a fate of spilled oil, extraordinary technological and conceptual advances made in analysis of microbial communities in the ocean ('omics techniques) were buoyed by broad progress in DNA sequencing and bioinformatic software development and applied to research into the DWH spill. Furthermore, anaerobic hydrocarbon biodegradation pathways, which were not considered in the previous report, have been increasingly recognized as the major fate of oil sequestered in anaerobic marine sediments. These biological/biochemical advances, combined with the analytical chemistry developments, have taken marine oil spill research into the realm of "big data." In addition to technological advances, there have been surprising insights into important processes associated with the fate of marine oil spills that arose from the DWH spill, including recognition of the importance of photo-oxidation and the large-scale but transient role of marine oil snow sedimentation and flocculant accumulation (MOSSFA) in oil removal and sedimentation, described in this chapter. In parallel, significant advances have been made in understanding the formation and transport of oil:mineral aggregates in the near-shore and on beaches. The DWH spill was also a driver for refining models of bubble and droplet formation

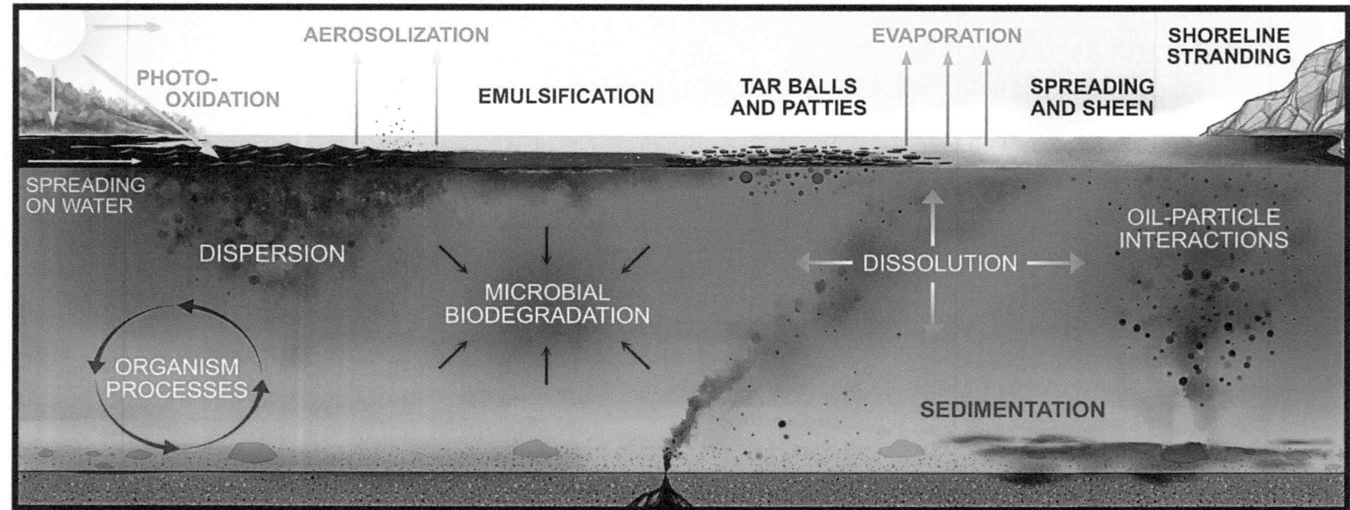

FIGURE 5.1 Overview of the fate of oil in the marine environment.
SOURCE: Image provided courtesy of the American Petroleum Institute, produced by Iron Octopus Productions, Inc.

and understanding the consequences of subsurface dispersant injection. Finally, the previous report concentrated on continental U.S. waters, whereas there is increased concern about oil spills in cold regions; the current chapter describes recent advances into the fate of oil in Arctic marine environments, particularly interactions of oil and ice.

5.1.2 Chapter Structure and Caveats

This chapter will convey to the reader the myriad interacting processes that affect the transport, transformation and ultimate fates of oil components, and emphasize the complexity of spill modeling. However, it is important that the reader appreciates some limitations to understanding, monitoring, and predicting the fate of oil in the sea. Three caveats in particular are noteworthy. (1) As Figure 5.1 suggests, at any given marine location there potentially are numerous interacting processes that can transport and transform oil over broad timespans. Each spill is unique because most oils are themselves unique and furthermore change dynamically over time in the diverse marine environments that they impact. Responders quip that they can "never respond to the same spill twice" because each individual spill is idiosyncratic and the processes affecting it are constantly changing, and because different spills experience unique combinations of processes in different parts of their impacted ecosystem. (2) Observing and measuring the fate of oil in the sea requires repeated sampling over time and in geographical space. This can be more difficult than it sounds because the marine environment is heterogeneous over many size and time scales, from micrometer-sized oil droplets in millimeter-sized pore spaces in beach sand to many square miles of evaporating oil slick on the water surface. Sampling a water column at exactly the same geographical location and depth on successive days may reveal very different oil concentrations, depending on where the currents, tides, eddies, wind, and so forth have transported the oil. Some sample variation might be overcome by exhaustive sampling, but that may challenge analytical capacity and be very expensive. Therefore, observing, modeling, and predicting the fate of oil in a specific situation requires inexact (but experienced) extrapolation from previous oil spill observations and from literature describing laboratory research. (3) Laboratory experiments can provide guidance for predicting oil fates in the field but, unfortunately, some of the published literature is based on experimental conditions that, while academically interesting, do not scale up to actual oil spills. For example, sometimes the type of oil tested or its experimental concentration are inappropriate; or the experimental conditions do not capture field parameters due to difficulty simulating natural processes in the laboratory (e.g., tidal action, mixing by wind and wave energy, oil leaking from sunken wrecks [see Box 3.6] ambient pressure and biological activity rather than high hydrostatic pressure, lack of solar radiation, and so on); sometimes the experimental time scale is too short to capture meaningful data on long-term fates; and sometimes the experimental biota within the test are inappropriate, having been cultivated in the laboratory rather than collected from the spill site, or are incubated with nutrients at concentrations not found in the ocean. In fact, it is difficult even for experts to define "environmentally relevant experimental conditions" (see Chapter 6, Box 6.4), given the first two caveats on dynamics and heterogeneity. Nevertheless, these limitations fundamentally influence calculation of oil budgets (see Section 5.2.9), modeling of oil spill fates (see Section 5.5), and monitoring of oil spills (see Chapter 4).

With these limitations in mind, the descriptions of processes in this chapter, assembled from a combination of laboratory and field measurements, summarize our best current understanding of the fates of oil in the sea. Section 5.2 is an overview of the fundamental physical, chemical, and biological processes and reactions that influence the fate of oil in the marine environment, regardless of specific geographical location and oil type. It also provides examples of "spill budgets" that estimate the proportional fates of spilled oil. Section 5.3 discusses the fates of episodic oil spills (typically from a single, finite event) in specific marine systems, from surface and near-surface waters, through the water column to deep water and deep-sea sediments, along shorelines including beaches and estuaries, and describing the fate of oil in Arctic conditions as a special case. Section 5.4 describes the fates of oil from continuing ("chronic") oil inputs, where their typically lower concentrations and more diffuse sources impact the relevant fate processes in important ways. Section 5.5 summarizes the uses of oil spill models for predicting the fates of marine oil spills. Section 5.6 summarizes conclusions and research needs arising from the literature review. Studies conducted during and after the DWH oil spill generated enormous amounts of information about the fate of Macondo 252 oil in the Gulf of Mexico; highlights of this research are integrated throughout the chapter without allowing this wealth of literature to dominate the knowledge and research needs of the broader marine environment. Notably, research and assessment of smaller spills in U.S. waters and elsewhere in the world also have contributed to advancing our knowledge of oil fates relevant to the focus of this chapter and the Statement of Task for this report.

5.2 FUNDAMENTAL TRANSPORT AND WEATHERING PROCESSES

This section provides an overview of the physico-chemical and biological processes that affect oil in the sea, regardless of geography and oil type. It describes the different states of oil and gas, formation of bubbles and droplets in water, transport in the ocean environment, volatilization of hydrocarbons to the atmosphere, photo-oxidation at the ocean surface, dissolution of hydrocarbons into water, emulsification of oil in water and water in oil, biodegradation fundamentals and the interactions of these processes. For context, Figure 5.2 illustrates the relative persistence of oil fractions in the

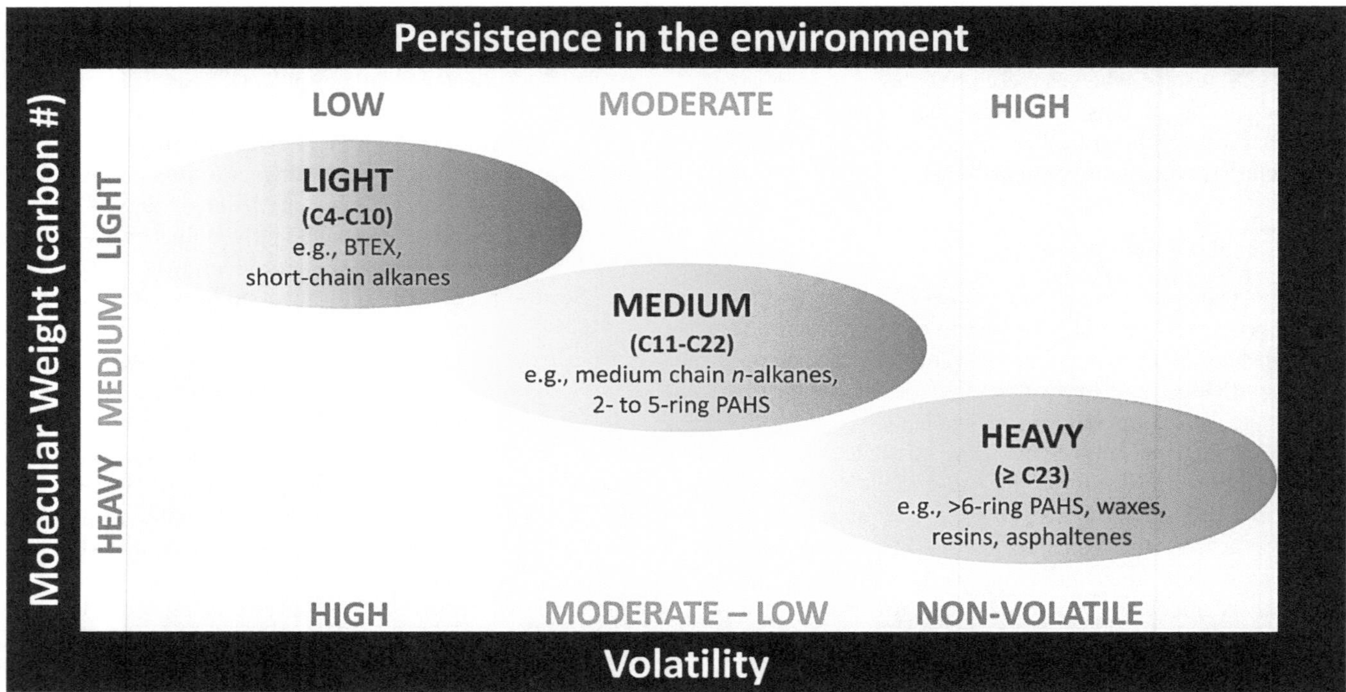

FIGURE 5.2 High-level overview of the relative persistence of oil components in the environment in relation to molecular weight and volatility.

environment, which is affected by the fate processes summarized in Box 5.1 and Table 5.1 and described throughout this chapter.

5.2.1 Phases and States of Petroleum Fluids in the Sea

As discussed in detail in Section 2.2, petroleum may occur in the natural environment as a solid, liquid, or gas. In this section, we focus on liquid and gas as these are the most common forms of petroleum in the marine environment. Oil and gas properties depend on the composition of the petroleum fluid and on the thermodynamic state, normally defined by a given temperature and pressure. Following the approach in Chapter 2, we define standard conditions in this chapter following the Society of Petroleum Engineers (SPE), with standard temperature given as 15°C and standard pressure at 100 kilopascal (kPa).

Several important terms describing petroleum states and the various stages of transformation will be used in this chapter. As explained in Section 2.2, mixtures of liquid- and gas-phase petroleum may be in thermodynamic equilibrium within the petroleum reservoir, with a large fraction of the low molecular weight compounds dissolved in the liquid oil. As these fluids are released, either naturally through seeps, purposefully during production, or accidentally in an oil spill, the temperature and pressure of these fluids may change, possibly resulting in their phase and composition change. Liquid oil or gas at equilibrium in the reservoir or at equilibrium or disequilibrium with any state other than standard conditions is considered to be *live oil* and *live gas*, the term "live" indicating that the composition would evolve if brought to standard conditions, usually by release of gaseous compounds still dissolved in the liquid-phase oil. Likewise, *dead oil* refers to liquid petroleum that has released enough of the dissolved gases that it is in equilibrium with its gaseous headspace at standard conditions. Dead oil still contains some of its volatile components. **Because the oil and gas from the DWH oil spill was emitted as a live mixture, oil spill science now recognizes the important implications of the live oil state, and significant new research has been conducted at a wide range of temperatures and pressures to understand the interactions of live oil and gas with the sea.** *Weathered oil* is used to describe a petroleum liquid that has an altered composition from that with which it was released. Numerous processes alter the composition of petroleum fluids, and these are the topics of this chapter. We use the term *weathering* to refer generally to any process that alters the composition of oil or gas in the environment; we will also refer to many specific weathering processes by their various names, for example, dissolution, biodegradation, and photo-oxidation.

Because we commonly refer to liquid-phase petroleum as oil, we use this term throughout the report wherever it is not ambiguous. When the particular phase of matter must be specified for clarity, we will use the terms *liquid oil* or *liquid petroleum*. Some petroleum compounds are also commonly

BOX 5.1
What Happens to Oil in Water

The fate of oil in the sea involves complex interactions of gas and liquid petroleum phases, seawater, floating particles, sediment, and the atmosphere through physical, chemical and biological processes as well as horizontal and vertical mixing. These processes are dynamic and their relative significance changes over time depending on the type of oil spilled, spill location, and environmental conditions. Moreover, as these processes change the chemical composition of the petroleum fluids, the properties of these fluids (e.g., density, viscosity, interfacial tension) also change, in turn altering the fate processes. Table 5.1 lists the major processes, cites the section(s) in which they are described, provides a brief definition, and lists conditions favorable for increasing the role of a given process in petroleum fate.

TABLE 5.1 Summary of Processes Affecting the Fate of Oil in the Sea

Process; Chapter Section	Definition	Promoting Conditions
Surface spreading; Sections 5.2.2.1, 5.3.1	Movement of oil on the sea surface that creates thin, floating pools of slicks and sheens	Low viscosity oils spread more quickly than those with high viscosity
Bubble and droplet formation; Sections 5.2.2.2–5.2.2.4, 5.3.3.2	Breakup of immiscible oil and gas into droplets and bubbles suspended in the ocean water column	Turbulence, shear, low interfacial tension, mixing of oil or gas with water
Mixing, dispersion, and dilution; Sections 5.2.2.3, 5.2.4, 5.3.1.3, 5.3.2	Changes in concentrations of spilled hydrocarbons by incorporation of those compounds in different volumes of seawater	Concentrations decrease when hydrocarbons are mixed with greater volumes of seawater, caused by turbulence and ocean currents. Concentrations increase when oil droplets accumulate at the sea surface, caused by oil droplet rise and boundary interaction with the sea surface (see also surface spreading)
Evaporation; Section 5.2.4	Transfer of the lighter or more volatile substances from oil slick to air by vaporization	• High surface area exposed to air • High fraction of volatile (or lighter) compounds in oil • Warm temperatures • Wind and wave action (turbulence)
Aerosolization; Section 5.2.4	Transfer of liquid petroleum from the sea surface to the atmosphere by particle formation	• Waves and white capping • Wind
Atmospheric redeposition; Section 5.2.4	Transfer of petroleum chemicals and particles from the atmosphere to the sea surface	Rain and snow
Photo-oxidation; Section 5.2.5	Chemical reactions occurring in liquid-phase petroleum and weathered petroleum as a result of sunlight in the presence of oxygen	• Oil at sea surface, in the photic zone, and on shorelines/beaches • Sunlight • Formation of free radical which in turn reacts with oxygen to produce reactive oxygen species
Dissolution; Section 5.2.6	Transfer of water-soluble compounds in oil into the surrounding water	• High fraction of water-soluble compounds in oil spilled • Turbulence • Large interfacial area between petroleum fluid and water
Emulsification; Section 5.2.7	Formation of a mixture consisting of small droplets of oil and water	• Oil at sea surface • Wind and wave action (turbulence) • Use of dispersants
Biodegradation; Section 5.2.8	Metabolism and breakdown of organic compounds by microorganisms such as bacteria	• Presence of nutrients such as nitrogen and phosphorus • Presence of wide range of microorganisms
Sorption to mineral particles; Section 5.3.2.1	Attachment of hydrophobic organic chemicals onto mineral surfaces by physico-chemical interactions	• High hydrophobicity • High mineral content • Small particle size (for increased surface availability) • Low K_{ow}
Sorption and entrainment into marine snow; Section 5.3.2.2	Entrainment of hydrophobic organic chemicals in marine snow by physico-chemical interactions	• High hydrophobicity • Small oil droplet size • Small particle size (for increased surface availability) • Low K_{ow}
Sorption to plastics; Section 5.3.2.3	Attachment of hydrophobic organic chemicals onto plastic surfaces by physico-chemical interactions	• High hydrophobicity • Presence of plastics • High surface area for attachment • Low K_{ow}

continued

TABLE 5.1 Continued

Process; Chapter Section	Definition	Promoting Conditions
Deep sea bubble or droplet breakup; Section 5.3.3.2	Formation of immiscible bubbles or droplets by breaking larger patches of oil or gas into smaller fluid particles	• Turbulence • Mixing into water • Lower interfacial tension and viscosity of petroleum fluid
Submergence; Sections 5.3.2.4, 5.3.3.6	Settling of oil fragments as they become heavier than water	• Loss of lighter compounds by volatilization and/or solubilization • Attachment of particles with mineral content
Sinking; Sections 5.3.2.4, 5.3.4	Settling of oil due to the adhesion to sediment particles or organic matter	• Shallow waters loaded with suspended solids • Oil trapped on sandy shorelines, mixed with sand and other sediments

referred to as gases because they are in the gas phase at standard conditions, for example, methane. We will likewise refer to these compounds as gases, specifying their state only if they are present in the liquid phase, as may occur at low temperature and high pressure. For more details about the states and compositions of petroleum fluids, see Chapter 2.

5.2.2 Immiscible Dynamics of Oil and Gas in Seawater: Sheens, Slicks, Bubbles, and Droplets

Spilled oil and gas interact with the marine environment through their immiscible interfaces. For oil on the sea surface, these interfaces result in thin oil sheens and slicks. For submerged oil and gas, these interfaces take the form of oil droplets and gas bubbles. The sizes of these sheens, slicks, droplets, and bubbles critically affect the fate of petroleum fluids in the oceans because they set the available interfacial area for exchange. For droplets or bubbles, which may form either due to a subsurface release or due to entrainment of surface floating oil into the water column, their sizes also control their residence time in the water column and their trajectory. Smaller droplets or bubbles normally rise slower than larger ones, and because ocean velocities vary with depth, their longer rise times will translate into very different lateral trajectories compared to larger droplets or bubbles. Hence, **the extent and thickness of sheens and slicks and the size distributions of oil droplets and gas bubbles are key parameters controlling the fate of spilled oil in the marine environment, ultimately determining the affected communities and their exposure concentrations.**

While the initial spreading of oil on a quiescent interface may be well understood, the formation of sheens and slicks and the breakup of oil and gas into droplets and bubbles is a complex process. For slicks and sheens, their organization into patches of varying thicknesses depends on the oil properties, which depend on the origin of the oil and which change over time, and on the surface-ocean dynamics, including waves, Langmuir cells, Lagrangian coherent structures, wind, and surface-ocean currents and turbulence. In the real ocean, these processes are stochastic and interacting so that only a statistical prediction of slick dynamics is possible. Likewise, oil droplet and gas bubble formation is also stochastic and complex. In marine oil spills, oil droplets normally originate either by entrainment into the water column from a surface slick or directly by turbulent breakup from a subsurface source; gas bubbles are normally only associated with subsurface sources, as in a pipeline leak or oil well blowout. Generally, breakup of droplets or bubbles continues until a maximum stable size is reached for which the internal forces of the oil droplet or gas bubble resist the external forces of the turbulent flow at the droplet or bubble scale. This is a major reason why generalized theories of breakup for oil and gas are not available: turbulence is a property of the flow field and not a property of the fluid (Tennekes and Lumley, 1972). **Thus, formation of slicks and sheens and breakup of droplets or bubbles will be different for different releases and in different environments,** such as in the turbulent field of the upper mixed layer of the ocean, in a buoyant jet, from a subsea blowout, or in the low-energy turbulence of the deep ocean, as that surrounding individual bubbles or droplets rising from a subsea leak or natural seep.

In this section, we introduce the types and descriptions of sheens and slicks and some methods to estimate their thicknesses. For droplet and bubble breakup, we discuss the fundamental dynamics occurring at the bubble or droplet scale, including the parameters affecting breakup and the effects of chemical dispersants. Here, the discussion is thus limited to universal behavior of oil and gas interacting with the sea. In Sections 5.3 and 5.4, we apply these fundamental mechanisms to understand dynamics pertinent to specific spill locations or natural release scenarios, where the turbulent properties of each unique situation will be applied.

5.2.2.1 Surface Oil Spreading

Liquid-phase petroleum released at the ocean surface or sub-sea, once it reaches the ocean surface, will spread on water to form a sheen or slick. The reason a slick appears "slick" is due to the dampening effect on capillary waves by the oil on the surface. The visual appearance of such oil can

BOX 5.2
Terminology Describing Oil Slick Appearance

The Open Water Oil Identification Job Aid for Aerial Observation released by the National Oceanic and Atmospheric Administration (NOAA, 1996) and the U.S. Coast Guard describes oil slicks for spill response operations using the following terminology and descriptions for the appearance of oil layers (NOAA, 2016). Representative photographs are shown in the main text.

Light Sheen: A light, almost transparent layer of oil. Sometimes confused with windrows and natural sheen resulting from biological processes. Sometimes referred to as transparent sheen.

Silver Sheen: A slightly thicker layer of oil that appears silvery or shimmers. Occasionally called gray sheen.

Rainbow Sheen: Sheen that reflects colors.

Brown Oil: Typically, a 0.1- to 1.0-mm thick layer of water-in-oil emulsion. Thickness can vary widely depending on wind and current conditions. Maybe referred to as heavy or dull colored sheens.

Mousse: Water-in-oil emulsion often formed as oil weathers: colors can range from orange or tan to dark brown.

Black Oil: Area of black colored oil sometimes appearing with a latex texture. Often confused with kelp beds and other natural phenomena.

Streamers: Oil or sheen oriented in lines, windrows, or streaks. Brown oil and mousse can be easily confused with algae scum collecting in convergence lines, algae patches, or mats of kelp or fucus. Sometimes called streaks, stringers, or fingers.

Emulsified: Water-in-oil mixture that appears as various shades of orange, brown, and/or red.

Tar Balls: Weathered oil that has formed a pliable ball. Size may vary from pinhead to about 30 cm. Sheen may or may not be present.

Tar Mats: Non-floating mats of oily debris (usually sediment and/or plant matter) that are found on beaches or in shallow water just offshore.

Pancakes: An isolated patch of oil shaped in a mostly circular fashion; pancakes can range in size from a few meters across to hundreds of meters in diameter. Sheen may or may not be present.

be used as an indicator of oil properties as well as state and thickness of the oil slick (Fingas, 2021; see Box 5.2). Most crude and fuel oils are dark brown or black. Diesel fuel is sold in three varieties as clear and dyed red or blue, the color indicating different usages subject to different taxes.

When spread on the water surface to form a slick or changed by mixing with water (e.g., emulsions) oils take on other appearances depending on their interaction with light, viewing angle, atmospheric conditions, wind effects, solar illumination, and water conditions. This affects the detection of spill extent and monitoring during response and remediation efforts (see Chapter 4).

Thin sheens have a very small amount of oil. For thicker oil layers, it is difficult to estimate the thickness of the oil layers exactly (see Figure 5.3). Slick thicknesses also vary over several orders of magnitude, from thin sheens of a few

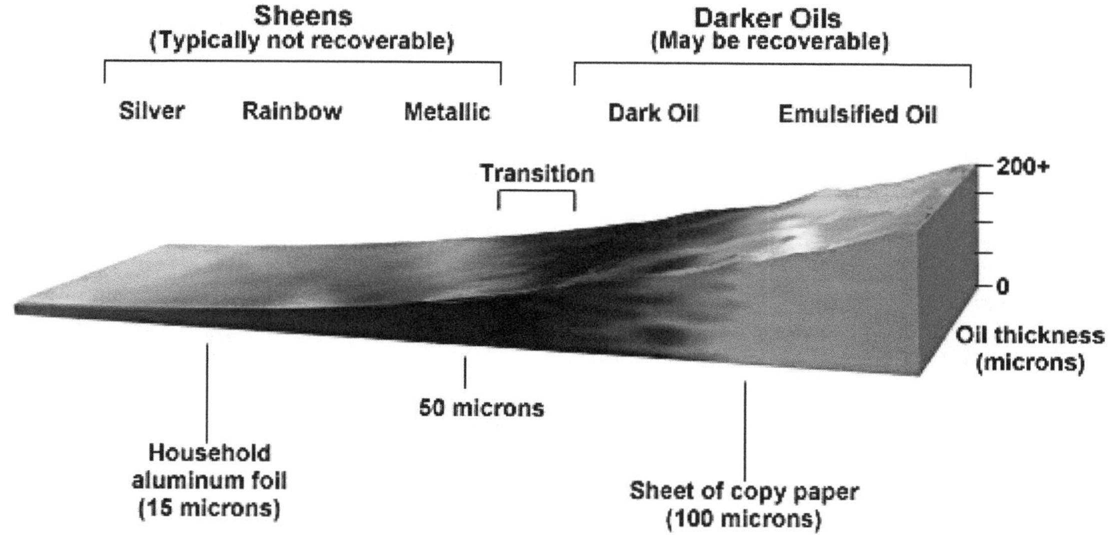

FIGURE 5.3 Relative thickness of different colors of oil slicks.
SOURCE: Comet Program, 2014.

micrometers to dark and emulsified oil that may be hundreds to thousands of micrometers in thickness—though still thin, as 1,000 micrometers is only one millimeter thick.

Different color codes have been developed for responders to estimate the thickness of oil slicks. The color codes are generally consistent for thin slicks (<3 μm) but not for thicker portions (>3 μm) that comprise a greater oil volume. For thin oil slicks (thinner than a rainbow sheen; <3 μm), the appearance of oil depends on the thickness of the slick as the optical phenomena involved in oil coloration can be applied. Hence, the appearance of thin slicks is relatively consistent. However, for thicker oil slicks (>3 μm), the appearance of the oil slick is not correlated with thickness because different physical factors predominate and affect the appearance of the slick (e.g., absorption and attenuation of light).

The Bonn Agreement Oil Appearance Code provides a standard method to assess the volume of oil on water based on appearance (Bonn Agreement, 2012). The code classifies the oil thickness into several classes as: sheens (silver/gray), rainbow, metallic, discontinuous true oil color, and continuous true oil color (see Figure 5.4).

Other information that can be obtained from the appearance of an oil slick include coverage and distribution on the surface, formation of water-in-oil emulsions (Lu et al., 2020), indication of oil-in-water emulsions (IPIECA, 2015), rate of emulsion formation (Sicot et al., 2015), measurement of subsea discharge rates (Fingas et al., 1999), and measurement of the oil geometry on the sea (De Padova et al., 2017).

Color also indicates changes due to weathering and response applications. When water-in-oil emulsions form, they often appear reddish in color, depending on the properties of the oil (see Figure 5.6). If dispersant is used as a response method (see Chapter 4), formation of a coffee-colored plume in the water column is a sign of dispersant effectiveness (Cedre, 2005; IPIECA, 2015) and depends on disappearance of oil on the water surface to block the transmission of light into and out of the water column. The brown color develops after the application of dispersants, then slowly dissipates—it is the result of the light reflection from the 5–50 μm droplets dispersed in the water column (Fingas, 2011a). The dispersed plume in the water column may transport in a different direction than the surface slick due to different transport mechanisms (e.g., no wind effect).

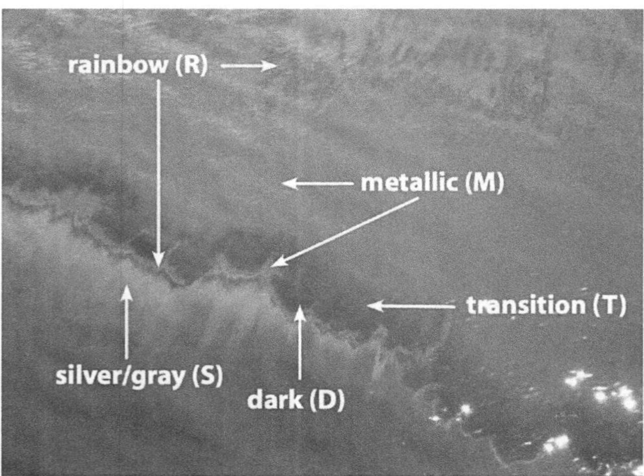

FIGURE 5.4 Aerial photo of oil spill and descriptions of oil slick appearance.
NOTES: Silver sheen (S) is typically <1 μm thick; Rainbow (R) is 1–3 μm thick (see also Figure 5.5); Metallic (M) 3–10 μm thick; slicks of greater thickness and emulsions have inconsistent colors (Fingas, 2021).
SOURCE: From the Open Water Oil Identification Job Aid. Image credit: NOAA.

5.2.2.2 Gas Bubble Breakup

Gas released subsea will form bubbles, and bubbles breaching the sea surface will rapidly enter the atmosphere. Hence, gas-phase petroleum does not form slicks, and the sizes of individual bubbles determine their fate in the water column.

Breakup of gas bubbles is fundamentally different from that of oil droplets due to the low dynamic viscosity and low density of gas compared to oil. As early as Hinze (1955), it was known that it is difficult to disperse gases in liquids due to the high contrast in dynamic viscosities between liquids and gases. This remains true for petroleum gases in seawater, which have a dynamic viscosity ratio of seawater to gas of order 100, indicating that significant energy is required to form smaller bubbles. For most gas releases, the energy available to create a dispersion of bubbles comes from the release rather than ambient ocean currents. For low gas flow rates, bubbles pinch off as their buoyancy lifts them from the release. Even for a very high gas flow rate, the energy input is small due to the low density of gas, which yields a low momentum flux. Wang et al. (2018) observed these facts

FIGURE 5.5 Rainbow sheen from an incident in the Gulf of Mexico spotted near a leaking natural gas well off the Louisiana coast.
SOURCE: NOAA.

FIGURE 5.6 Aerial view of emulsified oil at the Bayou Perot well blowout changing to dark oil as it weathers.
SOURCE: Gary Shigenaka/NOAA.

from experiments on gas jets into a large laboratory tank. Gas jets with large volume fluxes formed large bubbles near the release because the momentum of the gas was insufficient to penetrate very far into the receiving water and form a dispersion. Instead, breakup occurred near a packet of gas surrounding the release. Hence, breakup of gas bubbles is limited in the aquatic environment, and the maximum stable sizes of gas bubbles can be quite large (Clift et al., 1978; Grace et al., 1978).

The dynamics of gas breakup change when a liquid phase is discharged with the gas. This may be a petroleum liquid phase or co-released water. In this case, the higher density of the liquid phase in the release provides significant momentum, allowing the mixture of gas and liquid to penetrate the receiving seawater, generating stronger turbulence and mixing, and leading to greater breakup of gas bubbles. Measurements of bubble sizes for air jets co-released with water are available in Lima-Neto et al. (2008). Data for pure gas jets and a theoretical approach to predict gas bubbles sizes in both types of release are presented in Wang et al. (2018). Wang et al. (2016, 2020) also measured gas bubble sizes distributions for several natural seeps in the Gulf of Mexico. All these studies, which span small to large gas fluxes with and without co-flowing liquids, observe gas bubble sizes in the 2 mm to 5 mm diameter size range, with maximum sizes on the order of 10 mm.

Recent work has adapted theoretical models to predict gas bubble sizes in pure gas releases and for gas released with co-flowing liquid. Zhao et al. (2016) applied a model for gas bubble and oil droplet breakup in a jet that includes the momentum flux of the gas and co-released fluids. Importantly, Zhao et al. (2016) also include the energy input of the bubbles due to their buoyant motion as a contribution to the overall turbulent kinetic energy of the jet. This effect of turbulence production by gas bubbles has been observed in several previous studies. Recently, Lai and Socolofsky (2019) quantify a comprehensive turbulent energy budget for a bubble plume, also showing a significant contribution from the energy input of individual bubbles. Hence, turbulence modulation by bubbles at the bubble scale is an important factor contributing to bubble breakup in large gas-flux releases, such as from accidental marine oil well blowouts. **Overall, because of the low viscosity and density of gas, bubble sizes for a wide range of release scenarios fall in a similar, millimeter-scale range and are influenced by the dynamics caused by any co-released fluids.**

5.2.2.3 Oil Droplet Breakup and Dispersion

Liquid-phase petroleum, while it is suspended subsea, will form droplets of various sizes. Droplets may form at a subsea release point in a process normally called breakup or as surface floating oil is entrained into the water column through a process normally referred to as dispersion. In either case, turbulent motion in the seawater is responsible for the droplet formation.

Droplet formation of oil dispersed in seawater primarily occurs due to pressure fluctuations at the droplet scale resulting from turbulent eddies in the flow field (Hinze, 1955). To understand how this small-scale turbulent motion relates to the larger turbulent flow, we introduce a few key concepts of the canonical model of turbulence (Pope and Pope, 2000). Turbulent kinetic energy is produced at large scales of the flow, and this energy is transferred to smaller and smaller eddies until reaching the dissipation range, where the length scales are small enough that viscosity (i.e., fluid friction) damps the turbulent energy and converts it to heat. The largest scales of turbulence are highly situationally dependent, controlled by the geometry of the boundaries and the energy input (Tennekes and Lumley, 1972). In the statistically stationary case of steady forcing, the total rate of production of turbulent kinetic energy at the large scales is balanced by the rate of dissipation of turbulent kinetic energy at the smallest scales. Within scales smaller than the production scale but larger than the dissipation scale, Kolmogorov hypothesized that the characteristic length and time scales of the turbulence would depend only on the kinematic viscosity of the fluid and the dissipation rate of the turbulent kinetic energy; this region of the turbulence spectrum is called the inertial subrange, and the turbulent eddies within this region may be considered three-dimensional and isotropic (Pope and Pope, 2000). Because the largest stable droplet sizes of oil are usually small compared to the scales of turbulence production, they may be expected to fall in the inertial subrange of typical environmental flows, where the kinematic viscosity and turbulent dissipation rate fully describe the turbulence. In this way, droplet breakup may be described in terms of these parameters, independent of the larger flow dynamics. However, because the dissipation is intimately linked to the large scales through the production term, the actual rate of turbulent kinetic energy dissipation, and hence the droplet breakup process, remains specific to each flow situation.

Oil dispersed in a turbulent flow continues to break up until forces within a given oil droplet are large enough to resist further breakup by the turbulent pressure fluctuations. The two properties of oil that resist breakup are the interfacial tension and viscosity. Interfacial tension is the force per unit length along the oil–water interface. As droplets get smaller, the turbulent pressure fluctuations affecting the droplet also become smaller and may eventually become comparable to the interfacial tension forces. When the interfacial tension is very low, however, the viscosity, a form of fluid friction, may also act to limit breakup. Hence, droplet breakup for oil in the sea normally depends on the viscosity and turbulent dissipation rate of the seawater and on the interfacial tension and viscosity of the oil.

There are two basic approaches to predicting droplet breakup within turbulent flows. In the empirical approach, the characteristic scales of the turbulence dynamics and droplet properties are combined through dimensional analysis to yield predictive equations. Box 5.3 applies this approach to define the Weber and viscosity numbers, two key parameters for droplet breakup. Equations to predict characteristic droplet sizes using these parameters must be fitted to experimental data, and are expected to have different fit coefficients in

BOX 5.3
Parameters Controlling Droplet Breakup

Empirical Parameters: Hinze (1955) utilized scaling laws from the canonical model of turbulence with properties of immiscible liquids to develop an empirical relationship for the maximum stable droplet size D_{max} in a turbulent flow field (see Appendix F for additional details). He considered the case that the interfacial tension σ is the dominant droplet force resisting breakup and applied scaling relations within the inertial subrange of the turbulence to define a droplet Weber number We_p, given by

$$We_p = \frac{\rho \varepsilon^{2/3} D_{max}^{5/3}}{\sigma} \qquad \text{(B.1)}$$

where ρ is the density of the continuous phase (seawater) and ε is the dissipation rate of the turbulent kinetic energy. For breakup in a turbulent flow, the maximum stable droplet size is predicted by We_p equal to a constant. Hinze (1955) also proposed a correction when oil dynamic viscosity μ_p becomes important to resist droplet breakup. Wang and Calabrese (1986) modified the original viscosity group proposed by Hinze (1955) based on laboratory data in Calabrese et al. (1986), leading to the viscosity number Vi given by

$$Vi = \frac{\mu_p U}{\sigma}\left(\frac{\rho}{\rho_p}\right)^{1/2} \qquad \text{(B.2)}$$

where U is a characteristic velocity scale of the turbulence and ρ_p is the density of the dispersed phase (e.g., oil). Following the same turbulence scaling arguments as used by Hinze (1955), U should scale as $(\varepsilon D_{max})^{1/3}$ at the scale of droplet breakup. Applying this scaling, Wang and Calabrese (1986) proposed the empirical equation

$$\frac{D_n}{L_C} = f(We_L, Vi) \qquad \text{(B.3)}$$

where D_n is a characteristic droplet size of the droplet size distribution, L_C is a characteristic length scale of the flow field, and We_L is a Weber number defined by substituting L_C for D_{max} in Equation B.1. Thus, the Weber and viscosity numbers are the key parameters describing oil droplet breakup in a turbulent flow, and empirical equations predicting characteristic oil droplet sizes should encapsulate some version of Equation B.3, adapted to the specific scales of turbulence in a given flow field.

Population Balance Modeling Parameters: In the population balance modeling approach, the rate of change of the number of droplets of a given size is balanced by (1) the rate of creation of droplets of this size by breakup of larger droplets, (2) the rate of loss of droplets of this size by breakup into smaller droplets, (3) the rate of creation of droplets of this size by coalescence of two droplets of other sizes, and (4) the rate of loss of droplets of this size by coalescence with another droplet. In predicting these rates, population balance models consider physical processes occurring at the droplet scale. Breakup is modeled following the dynamics described by Hinze (1955) but using a probabilistic approach that considers the frequency at which the dispersed phases encounter a turbulent eddy of similar size and the probability that this encounter would generate breakup. In the Zhao et al. (2014a) model, breakup is resisted by both surface tension and viscosity, which are combined in an overall breakage efficiency. Surface tension is included by evaluating the formation energy of the new droplets from the parent droplet. For viscosity, Zhao et al. (2014a) observed that the process depended on time: a droplet may be able to resist breakup by viscosity over a short time, but this resistance diminishes over time. Zhao et al. (2014a) fitted a time-dependent viscous energy term, which allowed the model to predict both surface tension- and viscosity-resisted breakup regimes. Likewise, the coalescence terms consider the probabilities of droplet collisions and of collisions resulting in coalescence. For these models to predict field and laboratory observations, breakage and coalescence fitting parameters need to be calibrated and validated. These parameters adjust the overall breakup and coalescence probabilities by a constant measure. To date, the breakage and coalescence parameters remain empirical and depend on dimensional correlations specific to each spill type. **There is a research need to measure and understand the probability distributions of particle–eddy encounter rate and of fluid particle breakup per encounter to make these models more general.**

each type of turbulent flow field. Moreover, these models must assume a probability distribution and width parameter to predict the whole size distribution (see Appendix F). An alternative approach to droplet size modeling involves simulation of the droplet-scale physics of particle breakup and turbulent eddies to predict the time evolution of the whole population of droplets sizes in the droplet size distribution (Zhao et al., 2014a; Nissanka and Yapa, 2016). Key concepts of these population dynamics models are also highlighted in Box 5.3.

The population balance models differ from the empirical equations approach in three main ways. First, because they track the interactions of a full spectrum of droplet sizes, population balance models predict the size distribution directly, without having to assume a probability density function. Second, they consider the time-evolution of the fluid-particle breakup and interaction with turbulence. Hence, they may be applied in cases where the turbulent field is evolving in time, and they can predict the time-dependence of the size distribution in steady and unsteady turbulence. Third, the model equations are based on physics relations at the particle scale. When these scales can be related to the larger-scale turbulent flow, the models can be adapted to a wide range of breakup scenarios, as for example mixing tanks with constant turbulence (Zhao et al., 2014), intermittent turbulence of breaking waves (Cui et al., 2020c), and steady, non-uniform turbulence of blowout jets (Zhao et al., 2014b, 2015, 2016, 2017; Nissanka and Yapa, 2016; Aiyer and Meneveau, 2020). Empirical equations can also be adapted to these flow cases (e.g., Wang and Calabrese, 1986; Johansen et al., 2013, 2015), and in both of these modeling approaches, experimental data are required to calibrate and validate the model predictions.

While bubble and droplet breakup has been a topic of chemical engineering for years, the past 10 years have seen a tremendous increase in models adapted to oil spill scenarios and in laboratory data relevant to oil spills which can be used to calibrate and validate such models. Specific oil spill models and experimental observations in the context of different spill types are reviewed in Sections 5.3 and 5.4. Additional details about the theoretical foundations of empirical equations and models for droplet breakup in turbulent flows are given in Appendix F.

5.2.2.4 Effects of Chemical Dispersants on Droplet Breakup

A recent National Academies committee has comprehensively reviewed the usage of chemical dispersants as response agents for marine oil spills (NASEM, 2020). Here, we briefly explain what dispersants are and how they affect droplet breakup (see also the discussion of dispersants in Sections 4.2.3, 5.3.1.3, and 5.3.3.5).

Chemical dispersants are mixtures of surfactants that are dissolved in one or more solvents. Surfactants have active groups with affinity for oil (i.e., oleophilic or hydrophobic) and affinity for water (i.e., hydrophilic). The orientation of the surfactant at the oil–water interface reduces the interfacial free energy, reducing the interfacial tension. This reduction in the interfacial tension has two main effects on droplet breakup. First, by reducing the interfacial tension, the droplet resistance to breakup is reduced, and smaller droplets will form under the same turbulent conditions as compared to untreated droplets, that is, those naturally dispersed. This effect holds in the primary break-up phase of dispersion in which droplets are formed by interactions with turbulent eddies down to the inertial scale of the turbulence. Second, fluid motion at the oil droplet-water interface can further concentrate dispersant at convergence points at the lee of a rising droplet, leading to singularities in the interfacial tension, that is, zero interfacial tension, at the wake separation points (Gopalan and Katz, 2009, 2010). Oil may then leak through these separation points, forming very thin oil threads, with diameters on the order of a few microns. **This effect is a new phenomenon identified since the previous *Oil in the Sea* report (NRC, 2003) and is known as tip streaming** (see Sections 5.3.1.3 and 5.3.3.5). Tip streaming has been observed for oil droplets suspended in homogeneous, isotropic turbulence (Gopalan and Katz, 2010), for droplets rising to the sea surface after passage of a breaking wave (Li et al., 2017), and for droplets stabilized in a counter-flowing water tunnel (Davies et al., 2019). The oil threads leaking from these chemically treated droplets eventually break up by sinuous-wave instabilities along the oil thread, forming droplets with diameters of order 1 micron down to potentially 100 nanometers (Li et al., 2017).

As explained in Chapter 4, dispersants are used to promote smaller droplet sizes because smaller droplets help increase biodegradation rates in the water column and reduce the amount of oil on the water surface, thus reducing the amount of oil that can reach the shoreline. Dispersants may be applied at the sea surface to promote breakup of floating oil slicks into dispersions of suspended droplets (see Section 5.3.1.3). Or, they may be applied locally at the spill source to promote formation of smaller droplets in the primary breakup zone of a release (see Section 5.3.3.5). We discuss the dynamics associated with each of these use scenarios in the cited sections.

5.2.3 Transport and Dilution of Oil and Gas in the Sea

Hydrocarbons released into the sea occur as two different types of tracers: dissolved hydrocarbons are passive, being transported and mixed much like dissolved oxygen; gas bubbles and oil droplets are active, being affected both by the local ocean currents and their own buoyancy and immiscibility. The processes by which passive and active tracers are transported and mixed by ocean processes is the topic of environmental fluid dynamics. A classic treatment of the subject is presented in Fischer et al. (1979); a modern, comprehensive exposition is presented in Fernando (2013a,b).

Transport is a technical term referring to the movement of a dissolved or suspended material with the local currents. Mixing results when a parcel of water is homogenized with another parcel of water. Mixing normally reduces the concentrations of tracers in the combined parcel and reduces the differences between the maximum and minimum concentrations across a tracer cloud. Hence, mixing normally results in dilution of concentrations and homogenization of the concentration field (Fischer et al., 1979). Mixing mechanisms in the ocean include molecular diffusion, turbulent motion, and fluid motion resulting from unstable density fields, among other apparently random processes.

For hydrocarbon transport and mixing in the oceans, all mechanisms and scales of dynamics are present. At the droplet–water and bubble–water interface, molecular diffusion limits transfer of liquid and gaseous hydrocarbons to the aqueous phase (see Section 5.2.6). Outside the bubble or droplet, a chemical boundary layer forms, which is affected by the fine–scale turbulence surrounding the bubble or droplet and its wake. Once outside the concentration boundary layer, dissolved hydrocarbon is affected by ocean currents (advection) and turbulence. Advection is a deterministic transport process that moves dissolved constituents with the local fluid flow. Turbulence causes a random kind of advection that is normally approximated by an enhanced turbulent diffusion process. Diffusion is a random process that moves material from high-concentration regions into low-concentration areas, reducing the concentration of local constituents as they are diffused. Because there are many different scales of turbulent eddies, the turbulent diffusion coefficient summarizing the mixing caused by turbulence scales with the size of the tracer cloud—larger clouds are mixed by larger eddies, giving larger apparent turbulent diffusion coefficients compared to smaller clouds mixed by smaller eddies. Experiments from centimeter- to multiple kilometer-scale are observed to obey the Richardson 4/3-power law, which was summarized for ocean mixing in the Okubo diagram (Okubo, 1972; Fischer et al., 1979). This diagram predicts the apparent turbulent diffusion coefficient as a function of the size of tracer cloud being diffused. Hence, dissolved hydrocarbon diffuses more and more rapidly as the cloud of dissolved material grows in size.

In the oceans, turbulent diffusion normally differs in the horizontal and vertical directions. Because of gradients in the ocean salinity and temperature profiles, the water column is density stratified, with lighter water near the surface and denser water at depth. This density stratification stabilizes the water column and limits mixing in the vertical direction. The total movement of a dissolved compound as a result of diffusion is due to the combined effect of the turbulent diffusion coefficient and the concentration gradient. If the concentration is spatially uniform, diffusion will not be active. In the oceans, concentration gradients are usually smaller in the horizontal direction than the vertical direction, largely due to the density stratification, which promotes lateral motion and inhibits vertical motion. Hence, vertical diffusion, owing to the persistent vertical concentration gradients, is normally the dominant mode of tracer mixing in the oceans.

The net effect of diffusion is to dilute or reduce constituent concentrations. In a turbulent flow, eddies smaller than a tracer cloud erode away at the edges, producing a diffusive effect, but eddies larger than the tracer cloud cause advection, or transport, of the cloud. Eddies by nature are three-dimensional and tend to strain (deform) fluid parcels. Thus, eddies larger than a tracer cloud can tear it into pieces, creating filaments of high concentrations surrounded by dilute or pristine water. This process is called turbulent stirring, and concentrations only reduce after these high concentration filaments diffuse into the surrounding low-concentration waters. Ledwell et al. (2016) report on observations of an inert tracer injected at about 1100 m into the deep Gulf of Mexico near the DWH spill site. They found enhanced mixing near the continental slope and that mixing occurring near the slope resulted in intrusion of mixed fluid into the interior of the Gulf of Mexico. The tracer was clearly identifiable over 12 months after injection, but peak concentrations had reduced by 10^8 times after 12 months compared to the initial release. These very large dilutions were attributed to stirring of the tracer by boundary currents, mesoscale eddies, and three-dimensional turbulent eddies followed by turbulent diffusion across the high-concentration filaments and ultimately molecular diffusion at the smallest scales of tracer gradients. These observations were for a quasi-instantaneous release, however similar mixing occurs across wider scales for more episodic or continuous injections, such as the DWH oil spill. Hence, ocean currents are effective at reducing concentrations by diffusion, and the resulting concentration cloud is not a homogeneous concentration field but rather a complex, stirred tracer field that may be distributed over large areas.

Many discussions of ocean mixing use the term *dispersion* to account for mixing by turbulent motions. Dispersion, initially identified by Taylor, refers to the combined effects of diffusion and velocity shear, much like the preceding example on turbulent stirring. Velocity shear stretches concentration patches into elongated forms and sets up concentration gradients between the filaments and surrounding water. Diffusion, whether molecular or turbulent, mixes the sheared concentration patch into the surrounding water. The net effect of dispersion is to spread tracer over a much larger area than diffusion alone, since shear advection is working to stretch concentration patches. It is important to keep in mind that dispersion, though, is the combination of two processes, an advection or transport step caused by a sheared velocity field followed by a diffusion step across the sheared concentration gradient. Here, we reserve the word *dispersion* to refer to suspensions of gas bubbles and oil droplets in seawater (see Box 5.4) and avoid its use in reference to ocean mixing, using instead *turbulent stirring* and *diffusion*.

Gas bubbles and oil droplets can also be mixed by ocean currents, much like tracer clouds of dissolved material. The major difference is their active nature: gas bubbles and oil droplets normally rise through the ocean water column and remain immiscibly dispersed as bubbles or droplets. As a result, gas bubbles and oil droplets may not remain spread out or diluted after a mixing event. For example, gas bubbles rising out of a subsea layer may form into a plume, converging together into a narrow column of rising bubbles. Also, oil droplets that reach the sea surface will accumulate there, not passing entirely into the atmosphere nor resuspending immediately into the ocean water column. Hence, oil and gas may merge back into high concentration layers after they are initially mixed. This is not possible for passive tracers acted on by diffusion since diffusion is an irreversible process. Thus, the active nature of buoyant oil and gas, as well as ocean particles, is important to keep in mind when considering their mixing. Oil that accumulates on the ocean surface (see Section 5.2.2) often remains floating or suspended near the sea surface, and it is transported and mixed by near-surface currents. Surface transport processes as they apply to floating oil have recently received significant attention, especially through several new field experiments involving large numbers of floating drifters (D'Asaro et al., 2020). Floating material tends to accumulate along lines where two water masses meet and downwelling occurs (D'Asaro et al., 2018; Özgökmen, 2018). These convergence zones can spread oil over long distances but also keep it concentrated along the convergence lines. These submesoscale fronts are important at large scales, but Langmuir currents also predominate at small scales. The distribution of oil throughout the water column in the presence of Langmuir cells is strongly dependent on the droplet sizes of dispersed oil (D'Asaro et al., 2020). In fact, fluctuations in ocean flows below the scales of 10 km are the dominant mechanisms for the initial spread of floating tracer clouds, and neither operational circulation models nor satellite altimeters capture the scales of these flows (Poje et al., 2014). Thus, the complex convergent and divergent field of the ocean surface is critical to understanding the spread of floating oil.

In the remainder of this chapter, we will consider all of the processes discussed here to be summarized by the terms *transport* and *mixing*, where transport refers to advection with the currents and mixing to any process that changes concentrations by interactions with local ambient water. Because transport determines the affected environment and mixing alters the concentrations, these mechanisms are critical to an understanding of the fates and effects of oil and gas in the seas. A more exhaustive treatment of mixing and transport are presented in the cited reference monographs (Fischer et al., 1979; Fernando, 2013a,b).

5.2.4 Routes to and from the Atmosphere: Evaporation, Aerosolization, and Atmospheric Re-deposition

The sources of oil in the sea are varied, and their route to the atmosphere is dependent on chemical and physical properties of the compounds found in oil. Gas phase compounds readily enter the atmosphere. The ability of the oil to partition to the gas-phase (evaporate), react with existing atmospheric compounds (oxidize), and form condensed phase airborne particles (aerosolize, also referred to as atmospheric aerosol) is complex. The following sections review what is currently known about the generation and potential deposition of atmospheric gas-phase and aerosol pollutants to and from marine oil sources.

5.2.4.1 Primary Atmospheric Pollutants

Hydrocarbons are a primary gas-phase pollutant. Atmospheric hydrocarbons are mainly derived from evaporated oil and contribute the greatest atmospheric mass of oil pollutants (Middlebrook et al., 2012; French-McCay et al., 2021). If the amount of oil at the sea surface is known, one can estimate the evaporation of hydrocarbons by knowing the volatility of compounds in oil and the environmental conditions (e.g., temperature, pressure, volume). Specifically, there are models available to calculate the mass of gas-phase material from condensed phase compounds. Here we briefly describe available models, key assumptions, and additional considerations to estimate the rate of evaporation of oil in the sea.

The evaporation of specific condensed phase components (e.g., liquid or solid) is fundamentally a function of vapor pressure and mass transfer coefficients, both of which depend on temperature, pressure, and volume. Thus, the amount and rate of evaporated oil can be derived by coupling thermodynamic equations of state, vapor–liquid equilibrium models, and mass transfer equations. Evaporation has been long known to change oil composition on water surfaces; evaporation removes lower boiling point and lower molecular weight components from the liquid phase into the gas phase. The most volatile compounds evaporate in as quickly as an hour (McAuliffe, 1989). Experimental work since the 1970s has measured rates of evaporation from refined and diesel oils (Blumer et al., 1973; Mackay and Matsugu, 1973; Regnier and Scott, 1975) and provided some of the earliest data regarding the evaporation of oil components. The most widely used model to describe the evaporation of oil hydrocarbons and petroleum mixtures is the work of (Stiver and Mackay, 1984). The Jones (1997) model is more advanced and assumes a pseudo-component evaporation model. The Jones model employs components representative of benzene, toluene, ethylbenzene and xylene (BTEX), polycyclic aromatic hydrocarbon (PAH) fractions, volatile aliphatics and two semi-volatile aliphatic fractions. Each component evaporates according to its binned vapor pressure, diffusivity and molecular weight. To date, state-of-the art evaporation models applied to oil spills have been validated and extended the number of components used in Jones to better represent the complexity of oil compositions (Lehr et al., 2002; McCay and Rowe, 2004; Spaulding, 2017; French-McCay et al., 2018). However Jones and subsequent evaporation models rely on parameterizations of a mass transfer coefficient and simplifications of fuel formulations. Thus, it should be noted that simplified empirical evaporative models, such as the Fingas model have also been applied to oil in marine environments. The Fingas model (Fingas, 1999) is specific for oil types and depends on temperature and time. Regardless of the model, as fuel formulations advance, the evaporative models must keep up to date with changes in new formulations. Additional details of evaporative models for oil in the sea can be found in the recent review by Keramea et al. (2021).

Gas-phase oxidized sulfur (SOx) and organosulfur compounds are emitted into the air from the combustion of sulfur-containing fossil fuels. The higher the sulfur fuel content the larger the potential airborne emissions of combusted SOx. Much research has considered the fates of atmospheric sulfur emissions from varied shipping fuel formulations (Streets et al., 2000; Corbett, 2003; Endresen, 2003; Perraud et al., 2015; Abdul Jameel et al., 2017; Peng et al., 2020; Pei et al., 2021). One of the largest uncertainties from atmospheric SOx emissions has been attributed to the international shipping operations in marine environments (Smith et al., 2011). The International Maritime Organization regulations have gradually reduced the allowable sulfur content of ships' fuel oil. From 1 January 2020 the global upper limit was reduced from 3.50% to 0.50% mass. Higher sulfur content is allowed if a vessel operates an exhaust gas cleaning system that results in SOx emissions equivalent to burning 0.5% sulfur content fuel. A stricter sulfur limit of 0.10% mass has been applied in Emission Control Areas (ECAs) since 2015. The North American coastal waters were designated as an ECA in 2010, and the waters around Puerto Rico and the U.S. Virgin Islands were designated as an ECA in July 2011; ECA and a timeline of sulfur content requirements are shown in Figure 5.7.

ISO-8217 is a specification of marine fuels by the International Organization for Standardization (ISO). The standard applies to High Sulfur Fuel Oil (HSFO) as well as 0.5% sulfur fuels, generally referred to as Very Low Sulfur Fuel Oil (VLSFO). It is currently under review to update it to reflect the quality changes resulting from the introduction of VLSFOs, recognizing that there are wide variations depending on how the fuel is produced or blended.

It is also noted that fuel sulfur content is positively and linearly related to primary particulate matter emissions (Streets et al., 2000; Corbett, 2003; Endresen, 2003; Perraud et al., 2015; Abdul Jameel et al., 2017; Kim and Seo, 2019; Peng et al., 2020; Pei et al., 2021); high sulfur content fuels contribute to higher particulate matter (PM) concentrations. Thus, several countries have reduced the sulfur content of shipping diesel fuels to ultra low levels of 10–15 ppm (ultra low sulfur diesel) to reduce ship atmospheric emissions that

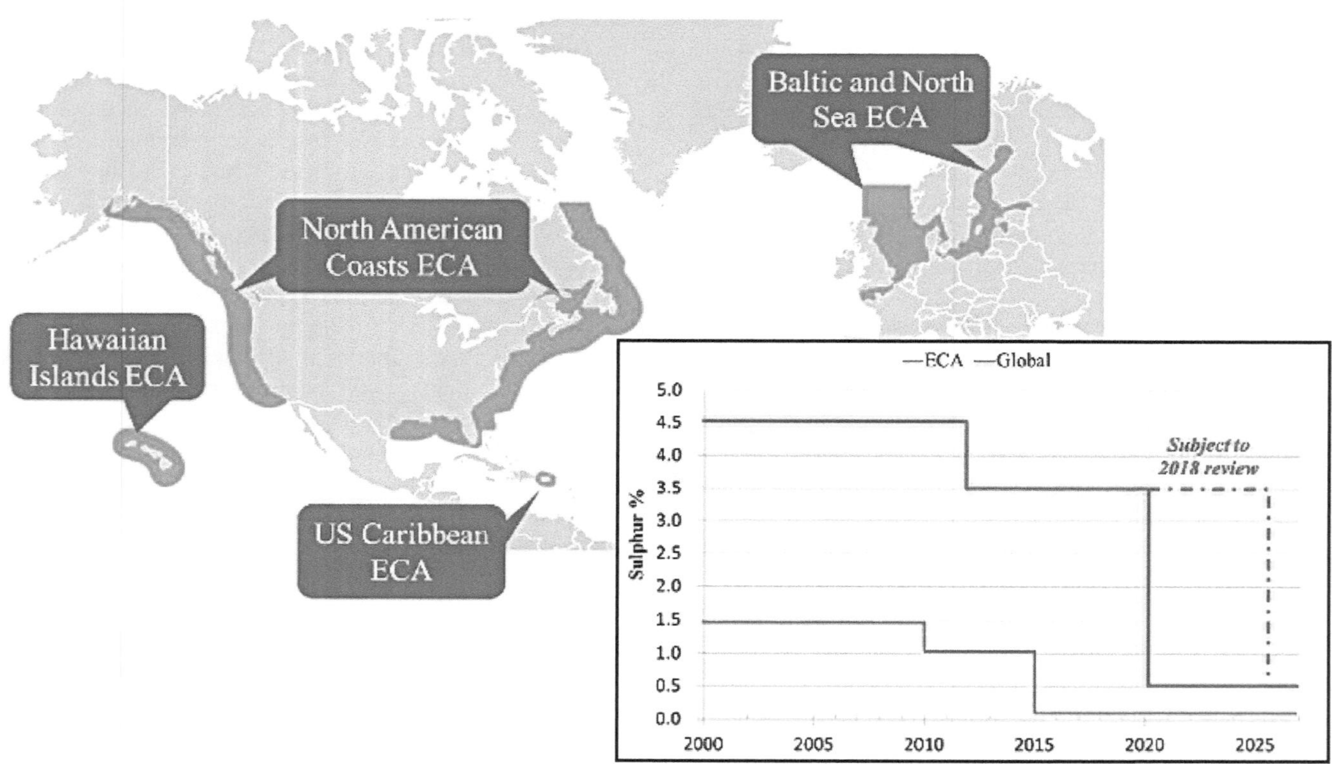

FIGURE 5.7 Emission Control Areas and timeline of IMO sulfur content requirements.
NOTE: The red line indicates global requirements, and the blue line indicates ECA requirements.
SOURCES: Gu et al. (2016); http://dx. doi.org/10.2139/ssrn.2870407.

significantly affect local port communities (Eyring et al., 2005). Furthermore, the use of sulfur-containing biodiesel fuels in marine transportation is of growing interest (Lin, 2013; Price et al., 2017; Mohd Noor et al., 2018; Svanberg et al., 2018; Zhou et al., 2020; Deng et al., 2021).

Indeed, the emissions of oil in the sea may be directly expelled as solid or liquid condensed-phase material (particulate matter or PM). High winds, wave crests, bubble bursting on marine surfaces, and raindrops can generate particles in scales from nanometers to hundreds of micrometers (Blanchard and Woodcock, 1957; Monahan et al., 1983; O'Dowd and de Leeuw, 2007; Ryerson et al., 2011; Murphy et al., 2015). For example, the bubble bursting of surface oil has been shown to directly emit oil droplets into the atmosphere (Ehrenhauser et al., 2014; Sampath et al., 2019) and the use of dispersants can modify the amount of primary gas and aerosol phase material emitted into the atmosphere (Afshar-Mohajer et al., 2018).

Oil that undergoes combustion can also directly emit black carbon or soot-like aerosol into the atmosphere. For example, during the DWH oil spill, black carbonaceous aerosol was measured directly downwind of the source (Perring et al., 2011). Furthermore, extensive literature has measured soot aerosol in marine environments directly attributed to anthropogenic shipping and oil operations (Agrawal et al., 2008, 2010; Moldanová et al., 2009, 2013; Popovicheva et al., 2009; Ault et al., 2010; Zheng et al., 2010; Khan et al., 2012; Lack and Corbett, 2012; Browse et al., 2013; Gaston et al., 2013; Tao et al., 2013; Buffaloe et al., 2014; Cappa et al., 2014; Celo et al., 2015; Kleinman et al., 2016; Betha et al., 2017; Streibel et al., 2017; Corbin et al., 2018; Fingas and Lambert, 2018; Jiang et al., 2018; Zhang et al., 2019). Many of the aforementioned studies also quantify the co-emission of regulated pollutants, such as carbon dioxide and nitrogen dioxide that contribute to global warming and the oxidation of secondary pollutants. Indeed, carbon dioxide and its greenhouse gas considerations can be primary and secondary atmospheric pollutants as considered in the following section.

5.2.4.2 Formation of Secondary Pollutants

The chemical reactivity of hydrocarbons can lead to the formation of secondary gas-phase and aerosol pollutants. Natural gas operations are sources of hazardous pollutants and photochemical ozone precursors that can produce secondary ozone (Kemball-Cook et al., n.d.). These aged organic vapors may undergo fragmentation (breaking of carbon–carbon bonds) and functionalization (e.g., the addition of polar functional groups), and the contribution of oxidation products to secondary pollutant concentrations continues to be of scientific interest. Several semi-empirical parameterizations have been applied (Robinson et al., 2007; Hodzic et al., 2010; Shrivastava et al., 2015), but more is needed to address the complex chemistry. Additionally, the chemical reactivity of intermediate volatile organic compounds (VOCs) (C14–C18) can lead to the formation of higher molecular weight products to create secondary organic aerosol, SOA (de Gouw et al., 2011). Aerosol scientists use knowledge of chemical reactions (e.g., photochemistry) and thermodynamic models to estimate the formation of SOA from hydrocarbons. The volatility basis set (Donahue et al., 2006, 2011, 2012), is an empirical model that assumes that volatility of the gas and condensed phases materials is in equilibrium for a concentration range.

It also should be noted that the combustion and oxidation of reactive hydrocarbons ultimately leads to the formation of atmospheric CO_2. Thus, for the formation of greenhouse gases (natural and anthropogenic) from oil sources, it should also be considered (McAlexander, 2014) whether they are primary (e.g., methane) or secondary (e.g., CO_2, N_2O) greenhouse gas atmospheric pollutants.

More research is needed to understand the formation of secondary air pollutants immediately downwind and transported long range from oil in the sea sources.

5.2.4.3 Deposition of Atmospheric Pollutants in the Marine Environment

Currently, the deposition of gas-phase and aerosol pollutants derived from oil sources into the marine environment is not well quantified. In this section, we review key papers that provide information regarding oil sources from the atmosphere and the biogeochemical cycle into the marine environment.

The mass transfer at the air–sea interface is complex. Much of the work addressing this exchange is derived from our understanding of PAHs (see Section 3.3.2). PAHs are ubiquitous in the atmosphere, can be transported over long distances, and are compounds of interest to quantify the biogeochemical cycles from oil in the marine environment. PAHs are natural and anthropogenic in origin; they are mainly derived from the incomplete combustion of fuels. Atmospheric PAHs in the marine environment have been measured in coastal areas and across several oceans; the measured PAH concentration varies in different seas (Ding et al., 2007; Nizzetto et al., 2008; Castro-Jiménez et al., 2012; Wang et al., 2013; Ke et al., 2017; Pegoraro et al., 2020). Recent work has measured and characterized the dry and wet deposition of PAHs into seawater (Castro-Jiménez et al., 2012; Lammel et al., 2016; Everaert et al., 2017; Chen et al., 2021) but more work is required to quantify the contributions of oil derived PAH sources back into the sea. Oceans are considered a major sink for long-range transport air pollutants (e.g., CO_2, and PAHs) (Wania and Mackay, 1996; Dachs et al., 2002; Lohmann and Belkin, 2014). **Long-range transport air pollutants can be long-lived and therefore specific air pollutant tracers from oil industries must be identified to directly assess the impact of oil in the sea. The identification of such tracers are critical for understanding the air-water exchange, and the overall biogeochemical cycle of pollutants.**

5.2.5 Photochemical Reactions

There has been a significant advance over several decades in our understanding of the role of photochemical reactions in the fate of spilled oil, and perhaps by extension to some of the other types of petroleum or oil inputs to the marine environment. **Renewed appreciation of photochemical reactions has resulted in a paradigm shift, causing us to consider photochemical reactions as one of the major factors influencing the fate of spilled oil at the sea surface.** In addition, there are implications for better understanding the effects of oil photochemical reaction products in concert with other processes influencing the fate of oil inputs as will be discussed below. Furthermore, this new knowledge has important implications for understanding effects of oil on marine organisms to be discussed in Chapter 6 of this report.

Photo-oxidation, sometimes referred to as photochemistry in some papers and reviews, was recognized in oil spill research of the late 1960s and early 1970s as one of the processes contributing to the fate of spilled oil slicks and sheens (NRC, 1975). An influential paper by Burwood and Speers (1974) aptly noted the importance of photo-oxidation as a factor in the dispersal of crude oil slicks. The importance of photo-oxidation was further emphasized in the literature reviewed by Payne and Philips (1985) and the report *Oil in the Sea* (1985). The process of photo-oxidation is also of concern because of research demonstrating that such reactions can produce products that are toxic (NRC, 1975, 1985, 2003; Lee, 2003, and references therein) as will be discussed in Chapter 6. As noted in the preceding references, initial attention was focused on photo-oxidation of aromatic hydrocarbons and aromatic ring compounds containing nitrogen due to experiments that indicated a few of these types of reactions yielded compounds with toxicity to marine organisms.

Photo-oxidation is often used as the overarching term in oil pollution literature and may appear in the citations and discussion that pertain in this report. However, it is important to recognize that at least three processes are recognized to occur, or have the potential to occur, during photochemical reactions of oil as suggested by Overton et al. (1980) and more recently demonstrated through research during the past 10 years (Rodgers et al., 2021; Freeman and Ward, 2022):

- Direct/indirect photo-oxidation
- Photo-induced polymerization
- Photo-cracking of large molecules

One type of photochemical reaction that might occur is photosensitized oxidation as outlined in simplified form in Figure 5.8. This depicts a compound other than the compound of interest, the simple hydrocarbon n-hexadecane, being activated by light energy to a triplet state and then subsequently a series of resulting follow-on reactions. Figure 5.9 depicts other mechanisms of photo-oxidation of petroleum hydrocarbons including those involving singlet oxygen.

$$X + h\nu \rightarrow X^*$$
$$X^* + RH \rightarrow XH\bullet + R\bullet$$
$$XH\bullet + O_2 \rightarrow X + HO_2\bullet$$
$$R\bullet + O_2 \rightarrow RO_2\bullet$$
$$RO_2\bullet + RH \rightarrow RO_2H + R^\bullet$$
$$RO_2\bullet + XH\bullet \rightarrow RO_2H + X$$
$$RO_2H \rightarrow RO\bullet + \bullet OH$$
$$RO\bullet + RH \rightarrow ROH + R\bullet$$
$$RO_2H + R\bullet \rightarrow RO\bullet + ROH$$

FIGURE 5.8 Type 1 Photosensitized oxidation of 1 mm thick film of *n*-hexadecane containing 40mmol/Liter xanthone layered over water in a laboratory experimental cell.
NOTES: X is xanthone; *hv* is light; X* is the xanthone triplet; RH is *n*-hexadecane; XH• is the hydrated xanthone radical; R• is the *n*-hexadecane free radical; O_2 is molecular oxygen; HO_2• is the hydroperoxy radical; RO_2• is the *n*-hexadecane peroxy radical; RO_2H is *n*-hexadecanoic acid, RO• is the oxygenated *n*-hexadecane radical; •OH is the hydroxyl radical; ROH is *n*-hexadecanol. Xanthone is an example of one of several photosensitizers that could be encountered in the environment of an oil spill.
SOURCE: Adapted with permission from Gesser et al. (1977). Copyright 1997, American Chemical Society.

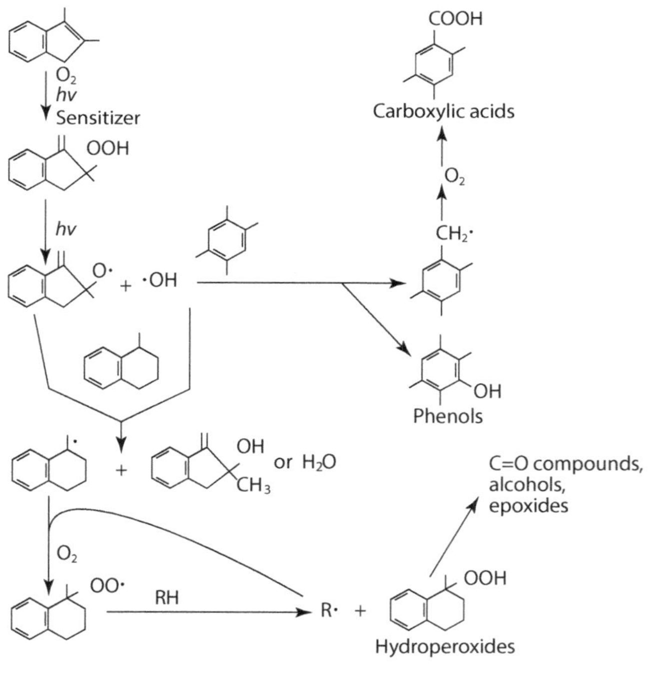

FIGURE 5.9 Example mechanisms proposed for photo-oxidation of some fuel oil compounds.
SOURCE: Adapted with permission from Payne and Phillips et al. (1985). Copyright 1985, American Chemical Society.

These are representative reaction diagrams discussed by Payne and Phillips (1985). Given the thousands of hydrocarbons in crude oils, as discussed in Chapter 2, it was not difficult to imagine that there might be many thousands of photo-oxidation reaction products. Applications of advances in analytical chemistry to field samples and samples from laboratory experiments post-DWH spill have shown that there are indeed many thousands of reaction products. This presents new challenges as will be discussed.

Research since the NRC (2003) report explores various aspects of the photochemical reactions involving oil and oil compounds (e.g., Aeppli et al., 2012; Corea et al., 2012; Ray and Tarr, 2014a,b; Cao and Tarr, 2017; Ward et al., 2018b; Wang et al., 2020). A reasonably comprehensive review of research by different groups of investigators, including relevant papers from the 1970s onward, of both laboratory experiments and field observations (mainly from the DWH related research 2010 to 2020) of photo-oxidation as it relates to the fate of spilled oil in the marine environment is presented by Ward and Overton (2020). They build upon the early research summarized in reviews such as that of Payne and Phillips (1985), and NRC (1985). They note that the need to incorporate photo-oxidation as a component of oil spill fate and trajectory modeling was stated by Spaulding (1988).

Inclusion of photochemical reactions in oil spill modeling has begun in part with the paper by Ward et al. (2018b). They combined results from laboratory experiments, field sampling and analysis, field observations, and modeling to demonstrate that for the DWH oil spill photo-oxidation of oil compounds in the surface slick converted many compounds to reaction products and was a significant quantitative fate for these compounds. The reaction products may then undergo further degradation by microbial or other biological processes, although this has yet to be determined. This finding is at odds with the prevailing wisdom or assumptions pre-DWH. It is a significant update to what was stated in *Oil in the Sea III* (NRC, 2003).

The *Oil in the Sea III* report contained a statement in the section "Photo-oxidation in Sea Water" in the Behavior and Fate of Oil chapter, pages 94–95: "(Parker et al., 1971, cited in Malins, 1977) Photo-oxidation is unimportant from a mass balance consideration: however, products of photo-oxidation of petroleum slicks may be more toxic than those in the parent material (Lacaze and Villedon de Nevde, 1976)." In hindsight that statement is perplexing. The latter citation does not deal with mass balance consideration. It is focused on photo-oxidation products that cause toxicity. The statement about "from a mass balance consideration" seems to be based on a conceptual model that emphasized other processes such as evaporation, dissolution, and biodegradation.

A similar concept or assumption was incorporated into various field response guidance manuals (NOAA, 2013a; ExxonMobil, 2014; Ward and Overton, 2020). As is often the case in scientific research, new findings require revisions to previous assumptions. Informed by new research, the current understanding, based on recent published research as noted above, is depicted in the paradigm of Figure 5.10 (Ward et al., 2018b). These are generalized representations

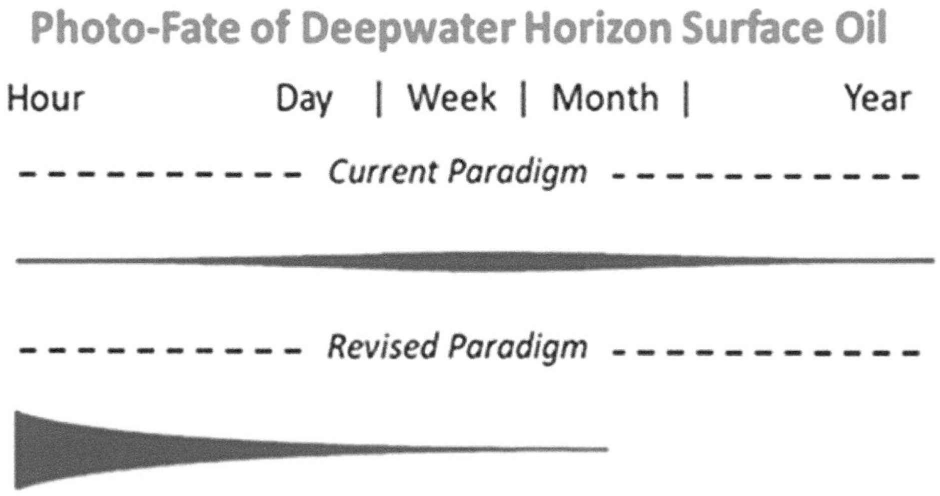

Photo-Fate of Deepwater Horizon Surface Oil

Hour | Day | Week | Month | Year

---------- *Current Paradigm* -----------

---------- *Revised Paradigm* -----------

FIGURE 5.10 Current paradigm refers to interpretations post NRC (2003) report up to circa 2018.
NOTES: Revised paradigm refers to post-DWH research and revisiting photo-oxidation of oil slicks circa 1971 onward (e.g., Ward et al., 2018b; Ward and Overton, 2020). Intensity of photochemical action on the surface oil slick begins earlier and is more intense than previously depicted and comparable to evaporation as a weathering process in published paradigms similar to this figure (see Ward and Overton, 2020). As noted in this report in Chapter 2 and this chapter, each set of chemical constituents in the various types of oil have specific relative fates with their own time-dependence and depending on the geographic/ecosystem location.
SOURCE: Ward et al. (2018b). Copyright 2018, American Chemical Society.

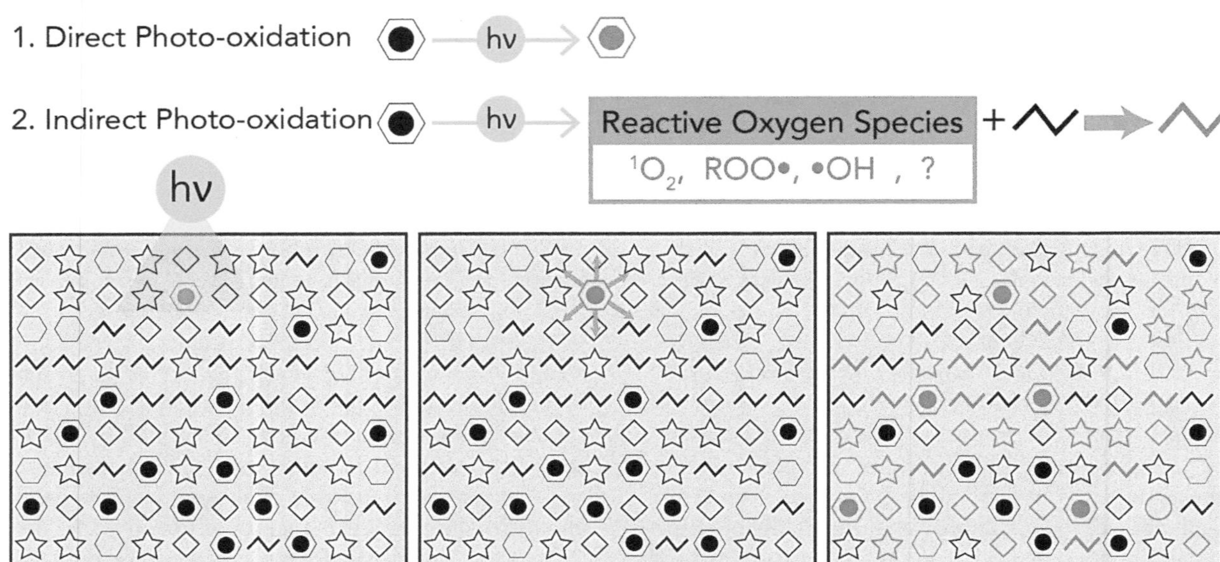

FIGURE 5.11 A conceptual model depicting direct and indirect photo-oxidation.
NOTES: In direct photo-oxidation (Reaction 1) a light absorbing molecule (depicted as a black aromatic ring) is partially oxidized into a new molecule (depicted as an orange aromatic ring). Indirect photolysis occurs when the absorption of light (Reaction 2 and left panel) leads to the production of reactive oxygen species (middle panel), like singlet oxygen, peroxyl radicals, and hydroxyl radicals. These reactive intermediates can oxidize a wide range of compounds, not just the compounds that directly absorb light (right panel).
SOURCE: From Ward and Overton (2020) with permission.

(NOAA, 2013a; ExxonMobil, 2014) that must be evaluated in terms of type of oil spilled and the climate regime (e.g., insolation, temperature) and ecosystems involved. Nevertheless, it is clear that photo-oxidation has re-emerged as an important factor in understanding fates of oils spilled in the marine environment.

The evidence for both direct and indirect photo-oxidation of oil components is summarized by Ward and Overton (2020). Their Figure 5 is presented here as Figure 5.11. As noted in the section of this report on advances in analytical chemistry (see Section 2.3), the ability to characterize reaction products by high magnetic field Fourier-transform ion-cyclotron-resonance mass spectrometry (FT-ICR-MS), and other high-resolution mass spectrometric methods, in several ionization modes has enabled new understanding of oil photo-oxidation reactions and reaction products and their fates (e.g., Niles et al., 2019, 2020). Considering the thousands of petroleum hydrocarbons and other less prominent compound classes containing heteroatoms of oxygen, nitrogen and sulfur (O, N, and S) that can be present in crude oils, fuel oils of various types and lubricating oils, among other fossil fuel oils, both direct and indirect photo-oxidation reactions can yield a myriad of reaction products. Hydrocarbons can yield photo-oxidation products that contain one to multiple oxygen atoms. In addition, N and S-containing compounds in crude oils yield photo-oxidation products with multiple oxygen atoms.

Zito et al. (2020) report the formation of water-soluble products from photo-oxidation of oil that form interfacial material (IM) at the oil–water interface in laboratory experiments. The photo-oxidation products then progressively become part of the dissolved organic matter (DOM). Figures 5.12 and 5.13 provide a glimpse of the complexity of reaction products produced by photo-oxidation of a Macondo surrogate oil spread on a thin film of pre-irradiated sea water and exposed to simulated sunlight in a laboratory experiment (Zito et al., 2020). The heteroatom class of reaction products and how photo-oxidation reactions produce a sequence of products proceeding over time from oil is depicted in Figure 5.12. Figure 5.13 comes from the same samples and further delves into the heteroatom composition in a van Krevelen plot of O/C versus H/C (O = oxygen, H = hydrogen, C = carbon). Each dot in Figure 5.13 represents a molecular formula. The elemental formulas for the reaction products are known; however, some of the general pathways of photochemical reactions are known (e.g., see Figure 5.11 and also Ray and Tarr, 2014a,b). Except for relatively few specific examples, the details of the exact molecular structures of both the reaction intermediates and reaction products are not known at this time.

By extension from natural organic compounds, a reasonable assumption is that some of the photo-oxidation reaction products are susceptible to microbial degradation. Likewise, uptake and metabolism by at least some multicellular marine organisms is likely. However, this has yet to be explored in any substantive manner. Moreover, the interactivity of the photochemical reaction products

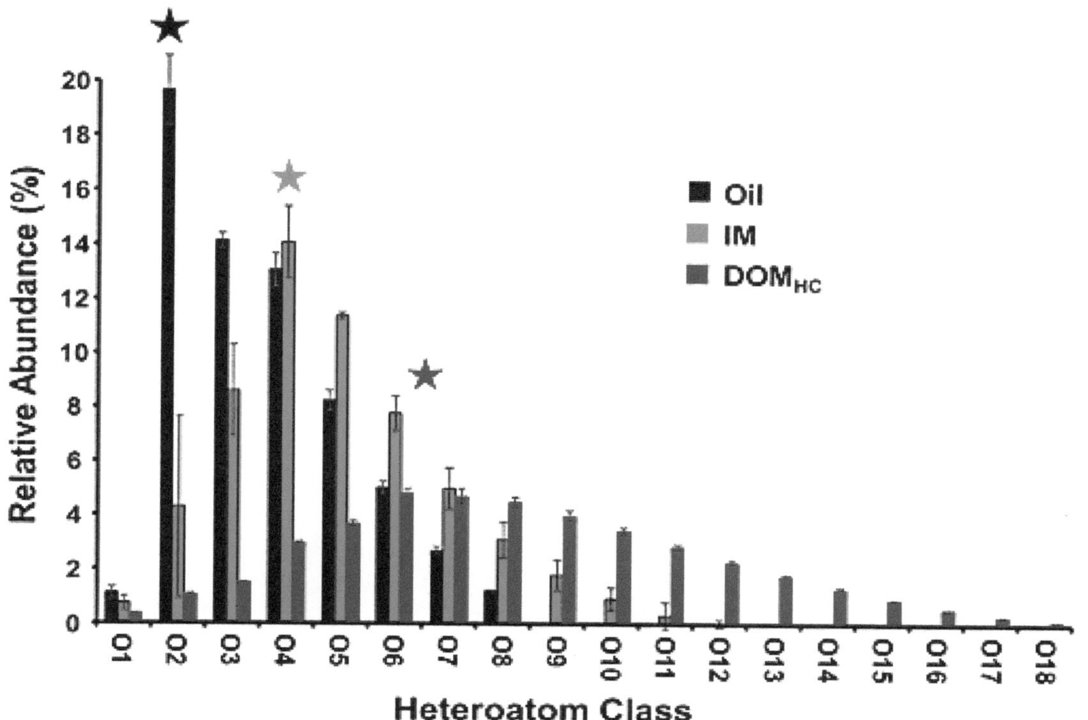

FIGURE 5.12 Heteroatom oxygen class graphs for the FT-ICR-MS data obtained in the triplicate analysis of molecules dissolved in the oil phase (black), interfacial-intermediate layer material (IM) at the oil–water interface (red), and dissolved organic molecules (DOM_{HC}) in the water phase (blue) fractions after simulated sunlight exposure.
NOTES: Black star indicates the most abundant species in the O_2 class in the oil soluble species contain 1–8 oxygens. The interfacially active species contained 1–12 oxygens and had an increased tendency to bind water and remained associated with oil. The most abundant IM components contain four oxygens per molecule (red star). DOM_{HC} are the most water soluble species and had 1–18 oxygens per molecule with the most abundant having 6–7 oxygens per molecule (blue star).
SOURCE: Zito et al. (2020). Copyright 2020, American Chemical Society.

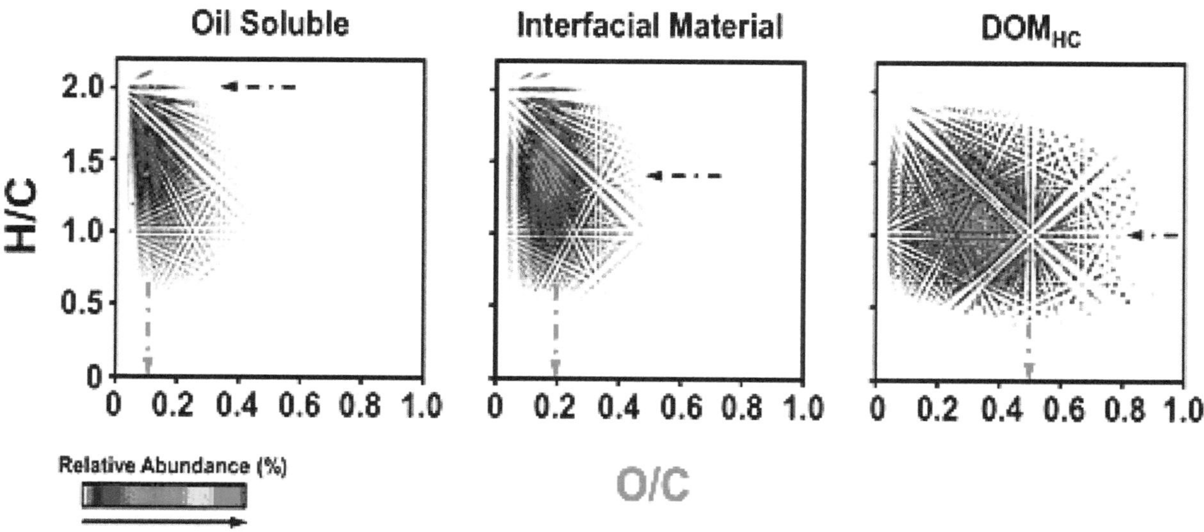

FIGURE 5.13 van Krevelen diagrams of the molecular formula unique to each fraction (oil soluble, interfacial material, dissolved organic matter or DOM_{HC}) showing the progression of higher oxygenated species from each fraction after simulated sunlight exposure.
NOTES: Black and red arrows highlight the most abundant H/C and O/C values, respectively. C(carbon), O(oxygen), H(hydrogen).
SOURCE: Zito et al. (2020). Copyright 2020, American Chemical Society.

in sorption/desorption processes with particulate matter from mineral particles to larger marine snow complex assemblages of particles (see Section 5.2.2) is a reasonable hypothesis yet to be tested in a substantive manner.

The photochemical reaction products formed at the sea surface and subsequently part of the oil slick or oil sand mixture coming ashore on beaches results in a conglomerate of resin-like and asphaltene-like material along with photo-oxidation products that are residual oil soluble and associated with the sediment-oil-agglomerates (SOAs) and oiled sand patties (e.g., John et al., 2016; White et al., 2016; Harriman et al., 2017, 2018; Aepelli et al., 2018; Bostic et al., 2018; Bociu et al., 2019; see Section 5.3.2).

Given that there is evidence that these SOAs can last as long as 32 years (Bociu et al., 2019) it is important to ascertain what specific reaction products are present, their fate, and potential for bioavailability to young children (toddlers) as noted in Chapter 6.

Marshes are another area of shoreline research that has documented the importance of the resin-like and asphaltene-like photochemical reaction products in the persistence of tar mat-like materials for at least several years (e.g., Lin et al., 2016).

In summary, research of the past 10–15 years has documented unequivocally the importance of photochemical reactions in the mass balance loss and fate of spilled oil at the air–sea interface in slicks and in films in temperate and subtropical regions. Variations of intensity of sunlight (various wavelengths) and corresponding variations in photo-oxidation reaction rates can be expected depending on sunlight intensity (insolation and angle of sunlight to the surface), cloud cover, day length and temperature, particularly in the polar regions (e.g., see Freeman and Ward, 2022).

Ward et al. (2018a) present compelling evidence and arguments that new findings of the photochemical chemical reactions of the oil at the sea surface have important implications for response, damage assessment, and restoration activities related to oil spills. For example, many of the photo-oxidation products have various oxygenated functional groups and free radicals that enhance interactivity of water and oil (as noted above), perhaps causing emulsions and/or interfering with and decreasing the efficacy of present-day dispersants, herders, or emulsion treating agents. **Moreover, the emergence of the greater importance of the photochemical reaction processes requires more extensive and intensive research focused on interactions of the photochemical reaction products with other aspects of the fates and effects of oil inputs such as slick thickness, emulsification, and biodegradation. This is important for all geographic regions.**

5.2.6 Dissolution

Dissolution is the process by which components in a gas- or liquid-phase petroleum fluid are transferred to an aqueous dissolved state in seawater. The solubility is the maximum amount of a given compound that may be dissolved in water at equilibrium with the dissolving mixture; the corresponding component concentration is the saturation concentration. Solubility for pure compounds and petroleum mixtures has been described in detail in Section 2.4. Dissolution occurs as long as the water phase has not reached the solubility limit. Hence, dissolution is inherently an unsteady (time-evolving) process with a rate controlled by the dissolution kinetics. In the oceans, ambient concentrations are low and the saturation concentration is rarely encountered so that dissolution is a continuous process for submerged petroleum fluids.

Directly at the interface between a petroleum fluid and water, dissolution is rapid, and the saturation concentration is expected to occur. As long as the concentration in the bulk water phase far from the interface is below saturation, dissolved components will diffuse away from the interface. This diffusive flux is matched by ongoing dissolution at the petroleum–water interface. Using the analytical solution for diffusion from a constant-concentration interface, one may write the rate of dissolution, or mass transfer dm_i/dt, of a component i from the petroleum to the aqueous dissolved state as

$$\frac{dm_i}{dt} = -A\beta_i\,(C_{s,i} - C_{b,i}) \tag{5.1}$$

where A is the area of the interface, $C_{s,i}$ is the solubility of compound i, $C_{b,i}$ is the concentration of compound i in the bulk water, and β_i is a mass transfer coefficient for component i, also called the mass transfer velocity (Clift et al., 1978).

The mass transfer coefficient encapsulates the thermo- and fluid-dynamic processes occurring in the thin concentration boundary layer near the petroleum–water interface. This includes molecular diffusion of component i away from the interface, convective or turbulent transport of component i through the concentration boundary layer, and the thickness of the concentration boundary layer itself. Because these properties are different at different water temperatures, for different interface types, for example bubbles or sheens, and for different conditions of fluid motion, the value of the mass transfer coefficient depends on the in situ conditions where dissolution is occurring. General behavior of dissolution mass transfer from floating oil and suspended droplets and bubbles is discussed below. More details related to specific spill scenarios are discussed in Sections 5.3 and 5.4.

5.2.6.1 Dissolution Mass Transfer from Floating Oil

Oil floating on the sea surface is exposed to both seawater at its bottom interface and the atmosphere at its top interface. Mass transfer from the floating oil to the atmosphere is by evaporation, or volatilization; this is discussed in Section 5.2.4. At the same time, dissolution may be occurring at the petroleum–water interface of the surface slick and for oil droplets suspended in the water column by natural dispersion. Dissolution from suspended droplets is considered below. MacKay and Leinonen (1977) suggest using Equation 5.1 for dissolution mass transfer from a surface slick, and they proposed a single value of the mass transfer coefficient for all compounds in the oil of $\beta_i = 2.36 \times 10^{-4}$ cm/s, which was experimentally derived for experiments on oil slicks in ponds. This approach was also used by McCay (2003) and French-McCay (2004) to assess potential damages of surface spills. For different wind and sea states, the mass transfer coefficients may be quite different (MacKay and Leinonen, 1977); however, because all soluble components in an oil are also volatile, there is a competition between evaporation and dissolution for mass transfer out of the liquid-phase petroleum of the surface slick. Because mass transfer by evaporation in some of the light molecular weight hydrocarbons is up to 10 times that by dissolution, dissolution has been a less important process for mass balance calculations of surface floating oil (MacKay and Leinonen, 1977). Toxicologically, dissolution is always important, and for Arctic oil spills where ice coverage may restrict or eliminate evaporation, dissolution from surface floating oil under ice may be particularly important.

5.2.6.2 Dissolution Mass Transfer from Suspended Gas and Oil

Unlike for floating oil, suspended oil droplets and gas bubbles are not exposed to the atmosphere, and all mass transfer from the petroleum to the aqueous phase must be by dissolution—evaporation, or volatilization, does not occur. **As a result, dissolution may be a significant fate process for the mass balance of petroleum fluid spilled subsea.** When oil and gas are released subsea, they break up into droplets and bubbles (see Section 5.2.2). Hence, dissolution will occur following Equation 5.1 with mass transfer coefficients applicable to droplets or bubbles. The rate of dissolution changes as the surface area, mass transfer coefficient, and solubility change. These parameters in turn change due to changes in seawater temperature and pressure as bubbles or droplets rise as well as by the evolving composition of the droplet or bubble as they dissolve.

Oil may also be found subsea in the form of various oil-particle aggregates (OPAs; see Box 5.9 later in this chapter). Mass transfer coefficients from solid particles are similar to those for droplets, but in the case of aggregates, not all of the oil will be exposed to water. Because mass transfer rates are not uniform over the surface of a suspended particle, the dissolution rate cannot simply be adjusted by the fraction of exposed oil. Instead, case-specific mass transfer rates would have to be developed for each oil–particle mixture.

Clift et al. (1978) reports on classical understanding of mass transfer from bubbles, drops, and particles. Mass transfer depends on the shape of a fluid particle and whether there is circulation inside the particle. Shapes are classified as spherical, ellipsoidal, or spherical cap, and depend on the fluid particle volume, density, and interfacial tension and on the density and viscosity of the continuous phase seawater (Clift et al., 1978; Zheng and Yapa, 2000). Circulation may occur inside a fluid particle as a result of fluid motion at the particle–water interface. In very clean systems, the interface will move due to the applied drag force from the continuous phase water, setting up a circulation cell within the fluid particle. This circulation enhances convection through the concentration boundary layer, resulting in higher values of the mass transfer coefficient for clean fluid particles. In contaminated systems, Marangoni forces result from concentration gradients of surfactants occurring at the interface of the fluid particle and resist the motion forced by the continuous phase drag. This effect shuts down the circulation within the droplet or bubble, resulting in less convection within the concentration boundary layer. As a result, mass transfer coefficients for contaminated, or dirty, fluid particles are lower than those for clean bubbles or droplets. This effect is expected both for naturally occurring surfactants and for surfactants mixed with the bubbles or droplets by chemical dispersant addition. Clift et al. (1978) provides correlation equations for mass transfer coefficients for fluid particles in a wide range of conditions; Johnson et al. (1969) provides a unified equation for mass transfer coefficients of clean bubbles across the full scope of bubble shapes.

Because of the ubiquitous nature of surfactants in the environment, dirty bubble dynamics are expected in most natural systems; however, several studies in the oceans indicate that clean bubble mass transfer rates may be appropriate early after release of a droplet or bubble. Rehder et al. (2002, 2009) report on measurements of methane bubbles released from a remotely operated vehicle in the deep Monterey Bay, California. They observed bubble size over time, allowing estimation of the shrinkage rate. Shrinkage rates over the first few minutes after release were consistent with mass transfer rates for clean bubbles. During this time, some bubbles shrank by up to half their diameter, which is equivalent to 83% of the volume. Thus, clean bubble mass transfer rates may be important for the initial dissolution of bubbles or droplets released into the oceans.

Likewise, Olsen et al. (2019) released methane gas from diffusers at 100 m and 300 m depth in Trondheimsfjord, Norway. Bubble sizes were initially in the range of 5 mm to 7 mm, in agreement with the observations reported in Section 5.2.2.2. They tracked the bubbles, measuring their diameters until they shrank to 2 mm. Similarly to Rehder et al. (2002, 2009), Olsen et al. (2019) observed that bubble shrinkage rates correlate with clean bubble mass transfer coefficients initially and transition to dirty bubble shrinkage rates at sizes between 3.5 and 4.5 mm. McGinnis et al. (2006) similarly suggested 3.5 mm as the transition size for switching from clean to dirty bubble mass transfer rates using the data in Rehder et al. (2002). In the Olsen et al. (2019) experiments, hydrate formation would not be expected owing to the shallow depths; whereas, for Rehder et al. (2002, 2009), hydrate skins were observed to form on the bubbles released within the hydrate stability zone. Like the effect of surface contamination, hydrate skins would create a rigid bubble–water interface, similar to the behavior for dirty bubbles. Based on these studies, it appears that clean bubble mass transfer coefficients are appropriate for predicting mass transfer from bubbles in the ocean initially after their release. After some time, mass transfer reduces to dirty bubble mass transfer rates, either due to surfactant contamination (Olsen et al., 2019) or also hydrate armoring (Rehder et al., 2009; Wang et al., 2020).

Although the 3.5 mm transition scale used by McGinnis et al. (2006) agreed with the experiments presented by Rehder et al. (2002), some of the results of Rehder et al. (2009) show different transition points. Wang et al. (2020) considered the Rehder et al. (2009) data and postulated that the hydrate growth rate should depend on the hydrate subcooling and that the time for hydrate growth to complete should be proportional to the initial bubble surface area. Using these parameters, they fitted an empirical model that could predict the transition time from clean- to dirty-bubble mass transfer rates as observed in the Rehder et al. (2002, 2009) datasets. They validated this model to observations of rise height for natural seeps in the deep Gulf of Mexico. We discuss this aspect of their work in Sections 5.3.3.4 and 5.4.1. This approach is also consistent with Vasconcelos et al. (2002), who correlated contamination time with bubble surface area. Despite some uncertainty in contamination mechanisms and transition times, mass transfer coefficients for clean and dirty bubbles have been well studied in the literature, and reliable correlations (e.g., Clift et al., 1978) for each bubble type are well known and tested. **Significant new data exist for bubble dissolution rates in the oceans. These data show initial clean-bubble mass transfer behavior followed within minutes or seconds by dirty bubble mass transfer rates. There remains a research need to understand the mechanisms modulating bubble mass transfer rates in the oceans and to collect new observations of mass transfer of bubbles co-released with oil, which may act as a surfactant and inhibit any clean-bubble mass transfer observed in pure-gas plumes.**

5.2.7 Emulsification

Oil–water emulsions at sea form by wave action and affect the weathering processes and cleanup options after oil spills. Emulsions consist of a mixture of small droplets of one liquid that remain dispersed in another liquid; see Box 5.4 for definitions. When oil and water are mixed, for example by wind and/or wave energy, two types of emulsions may form depending on the relative quantity of one liquid in the other: water-in-oil emulsions and oil-in-water emulsions (see Figure 5.15). Water-in-oil emulsions, which are also referred as "chocolate mousse" due to their appearance, have water droplets dispersed throughout the oil (Zafirakou, 2019). Water-in-oil emulsions can contain as much as 80% water and can remain floating or submerge (Overstreet and Galt, 1995). Different oils have different susceptibility to emulsion formation and different stability of emulsions that do form, some persisting for a few hours or days and other remaining emulsified for decades. Oil in this form is difficult to degrade because of its high viscosity and reduced surface area. In addition, emulsified oil exhibits reduced evaporation, dissolution, entrainment into the water column and transport (reviewed by French-McCay et al., 2021).

Oil-in-water emulsions, on the other hand, have oil droplets dispersed throughout the water column by the action of agitation, such as wind and wave activity. Dispersing floating oil as droplets (i.e., forming oil-in-water emulsion) increases the oil–water interface and therefore, enhances dissolution of water-soluble fractions and availability of hydrocarbons for biodegradation. These types of emulsions are considered in more detail in Section 5.3.1.3 (natural dispersion).

Droplet sizes of oil-in-water emulsions are related to viscosity and turbulence levels at sea. Lower viscosity and higher turbulence increase formation of droplets and exposure of oil to water column, while entraining smaller droplets. The droplet diameter in emulsions can be estimated from the Weber and Reynolds numbers (Boxall et al., 2012). Stable emulsions have droplet diameters of 8 microns or less and they start to become unstable at droplet sizes of about 50 microns (Opedal et al., 2009).

The interface between the incorporated water and the oil needs to be stabilized in order to form an emulsion (Fingas and Fieldhouse, 2014, 2015). For oil-in-water emulsification, resins facilitate emulsion formation, and they also act as solvents for asphaltenes that support emulsion stability. Although earlier studies reported that as an emulsion forms, water content increases exponentially (MacKay and Zagorski, 1982), later studies showed that characteristics and composition of oil are important for formation of stable emulsions. An emulsification stability index (SI) was developed by Fingas and Fieldhouse (Fingas and Fieldhouse, 2009, 2011; Fingas, 2011a,b) based on stabilizing oil components of asphaltenes and resins, combined with general oil properties of density and viscosity. The emulsification SI can be used to determine whether an oil is *unstable*, *entrained water-in-oil state*, *meso stable*, or *stable*. Certain microbial species, including those that degrade hydrocarbons, can enhance emulsification and stabilize emulsions through production of extracellular biosurfactants such as rhamnolipids and glycolipids (Gutierrez et al., 2013; Das et al., 2014).

Understanding the processes that lead to emulsification allows better predictions as to when emulsification will occur (Bacosa et al., 2015). In this report, the term "emulsion" is used to refer to oil-in-water emulsions. Processes that promote emulsification include changes in oil characteristics caused by evaporation, photo-oxidation, spreading, and dissolution; and increase in oil viscosity. In flume studies with Macondo oil, over the period of a week, the emulsion viscosity increased, becoming stable in the first 3 days, with

BOX 5.4
Emulsion and Dispersion Terminology

There are some variations in the use of terminology to describe oil-water mixtures in the literature and usage by different disciplines (e.g., petroleum, food industry, pharmaceuticals). The definitions provided below are specific to the use of terminology for oil spills at sea.

Emulsification is generally defined as the incorporation of water droplets into an oil matrix (water-in-oil) or conversely oil droplets in a water matrix (oil-in-water) achieved by the action of agitation, such as wind and wave activity (Lee et al., 2015). In oil spill science and oil spill response, emulsification of crude oils typically refers to the process of incorporation of water droplets into oil matrix to form a water-in-oil emulsion. This mixing is promoted by the turbulence generated by waves at the sea surface and can increase as oil weathers and becomes more viscous allowing it to retain water droplets within its structure.

Emulsion is the mixture formed by mechanical mixing of oil and water that do not naturally mix. Different oils exhibit different tendencies to emulsify, and emulsification of surface oil spills (slick) is likely to occur under high energy conditions (strong winds and waves) (NOAA, 2021a). Emulsions are classified by the nature of the dispersed phase and dispersing medium as **oil in water (o/w)**, where oil droplets are dispersed in water and **water in oil (w/o)**, where water droplets are incorporated in oil. In oil spill literature, "emulsion" typically refers to the water-in-oil emulsion, sometime termed *chocolate mousse*, because of its appearance. These types of emulsions are more viscous and persistent than the original oil and their formation may cause the volume of the slick to increase as much as four-fold. This process changes oil properties, affects other weathering processes, and can complicate the response. These emulsions can be unstable, separating into oil and water phases soon after formation, or stable for months or years (Lee et al., 2015). In oil spill literature, the formation of oil-in-water emulsions is typically referred to as entrainment or dispersion, but in other fields of science they can be referred to as emulsions of various types.

In **macroemulsions**, oil is dispersed in water as drops with droplet diameters generally exceeding 1000 nm. Macroemulsions are thermodynamically unstable; therefore, surfactants or small particles can be used to stabilize them. **Miniemulsions** are formed when oil is dispersed in another phase as droplets with diameters between 100 and 1000 nm. They can be thermodynamically unstable. **Microemulsions** have smaller droplet sizes than 100 nm, making them appear transparent. Microemulsions are thermodynamically stable and usually require a surfactant and a cosurfactant (e.g., short chain alcohol).

Pickering emulsions refer to oil-water mixtures stabilized only by solid particles positioned at the oil–water interface (i.e., substituting solid particles for surfactants).

Dispersion (or entrainment) is the movement of spilled oil from the water surface down into the upper layers of the water column in the form of droplets (oil-in-water emulsion), caused either by natural wave action or by the application of chemical dispersants made for this purpose (NOAA, 2021a).

Dispersants are chemical agents (similar to soaps and detergents) that help break up the oil slick into small droplets by reducing the surface tension between the water and oil (NOAA, 2021a). While this does not remove the spilled oil, smaller oil droplets are easier to be colonized and biodegraded by bacteria. The terms *dispersant* and *surfactant* are often used interchangeably; however, from a scientific perspective, the difference is that a dispersant improves the separation of particles in a suspension whereas a **surfactant** lowers the surface tension between two phases (oil and water).

What Happens to Dispersed Oil at the Sea Surface?

Initially, dispersed oil moves down into the water column to depths ranging from 1 to 10 meters (about 3 to 30 feet). To avoid contaminating the sea floor, most dispersant use to date has been restricted to waters deeper than 10 meters (about 30 feet). Figure 5.14 shows concentrations of dispersed oil at different depths (estimated from field studies), during the first few hours after dispersants have been applied. These concentrations drop within hours as currents and waves disperse and dilute the oil further (NOAA, 2021b).

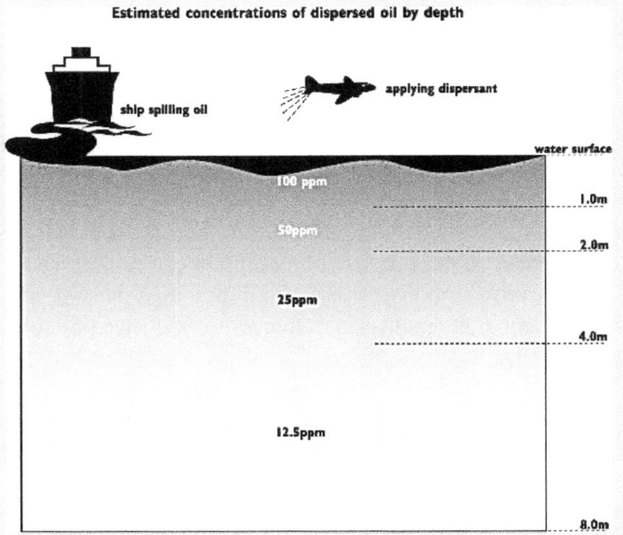

FIGURE 5.14 Estimated concentrations of dispersed oil by depth in the upper water column (≤ 8 m depth).
SOURCE: NOAA (2021).

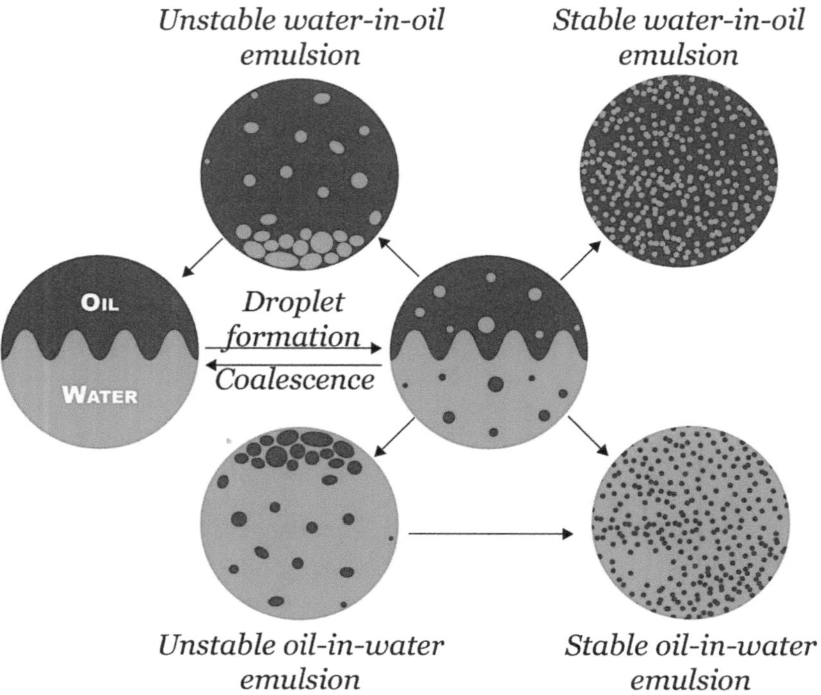

FIGURE 5.15 Emulsification of oil and water. Generation of stable and unstable water-in-oil (w/o) and oil-in-water (o/w) emulsions in seawater. SOURCE: Lee et al., 2015. This content has been reproduced with the permission of the Royal Society of Canada.

the viscosity increasing from 1,000 mPa to over 11,600 mPa over 12 days (Daling et al., 2014). As the emulsification process continues, some emulsified oil sinks; and oil spreading and vertical mixing is reduced (Xie et al., 2007). Recent research in emulsified oil has focused on the effects of photo-oxidation on the initiation of emulsion formation (e.g., Zito et al., 2020), as photo-oxidation of some hydrocarbons makes them more water soluble, thus facilitating formation of an oil–water emulsion. Furthermore, UV irradiation leads to a reduction of micelle size of surfactant molecules beyond that achieved using visible light, and increases the viscosity and stability of the emulsion to freeze–thaw cycles (Genuino et al., 2012).

Daling et al. (2003) provided the findings from four oil release experiments in the North Sea and Norwegian Sea with surface and subsurface oil releases and laboratory studies that involved emulsification. The results of laboratory experiments and full-scale field experiments revealed that both the physico-chemical properties of the oils and the release conditions control the initial film thickness and rate of emulsion formation (Daling et al., 2003). Simulated underwater pipeline release experiments indicated that, within a limited area, the oil droplets surface, forming oil slicks with adequate thickness to support emulsion formation. At moderate depths (300 m), simulated underwater blowout experiments resulted in the gas-bubble plume surfacing, bringing with it entrained water. In deep waters (>500 m), the simulated blowout showed that the plume may be confined to the water column. In this case, the oil droplets will rise to the surface within a more limited area, leading to initial oil film thicknesses that are appropriate for emulsion formation. These studies were used to estimate the timing for dispersant use after the DWH spill (Zito et al., 2020).

Because emulsification has such profound effects on the behavior of oil, French-McCay et al. (2021) persuasively articulate that additional research is needed to improve prediction of oil emulsification, including formation, stability, and behavior, beyond the existing model algorithms.

5.2.8 Microbial Biodegradation of Oil

5.2.8.1 Microbes in the Sea: Who Is Out There and How We Know

Marine systems harbor diverse microbes in astonishing abundance: the world's oceans have been estimated to contain 10^{29} microbial cells (Whitman et al., 1998), perhaps exceeding the number of stars in the universe 100,000-fold. A single milliliter of seawater may harbor more than 100,000 bacterial cells (Hazen et al., 2010), and marine viruses probably number 10-fold greater still (Suttle, 2007). The most abundant and geographically widespread living organism known to date, *Pelagibacter ubique* in the cryptic SAR-11 clade of bacteria (Giovannoni, 2017) inhabits ocean waters, and microbes abound even in the permanently dark and cold ocean depths (Orcutt et al., 2011) including long-lived cells in ancient deep-sea sediments (Jørgensen, 2012) and abysses such as the Mariana Trench (Liu et al., 2019). Niches for marine prokaryotes (bacteria and archaea) and eukaryotes (including phytoplankton, zooplankton, and fungi) include the sea surface microlayer (neuston); the full depth of the water column; shallow and deep sediments of the seafloor; hydrothermal vents and natural oil and gas seeps; habitable oil reservoirs beneath the seafloor; and beaches, estuaries, and salt marshes. Microbes form associations with all forms of plant and animal marine life in these niches.

Diversity, abundance, and metabolic activity in the ocean are dominated by the kingdom Bacteria (Azam and Malfatti, 2007) whereas the metabolic contributions of marine fungi are not yet fully known (Richards et al., 2012; Grossart et al., 2019). Bacterioplankton, phytoplankton, and microscopic zooplankton predominate in the neuston and photic zones (the upper water column where light still supports photosynthesis) where they drive primary productivity (photosynthetic fixation of carbon dioxide to organic carbon) and underpin the marine food web. Primary productivity in the ocean is chiefly attributed to a few key cyanobacterial genera (Flombaum et al., 2013) and exceeds that of land-based primary productivity. Heterotrophic prokaryotes—those that consume dissolved and particulate organic carbon (DOM and POM) rather than fixing CO_2—dominate the deeper water column and sea floor sediments, sustained by detritus sinking from the photic zone and by light-independent geochemical reactions and contribute to biological processes in salt marshes, estuaries, and intertidal beach ecosystems. Microbes degrade an enormous range of organic substrates including ancient plant material such as components of petroleum and its refined products. Viruses, despite being numerically dominant and very important in carbon cycling, are non-living entities that have no metabolism outside a suitable host and are not included here in discussions of microbial numbers and metabolic activity.

Marine microbes may exist as free-living single cells suspended in water but are far more likely to form multi-species communities by attaching via biopolymers to each other, to higher organisms, and/or to detritus and minerals, forming community aggregates and biofilms (e.g., Sanli et al., 2015) (see Section 5.3.2.2). Permanent and transient members of these dynamic microbial communities flourish or perish; compete or cooperate through cross-feeding in metabolic tasks (McGenity et al., 2012; Zengler and Zaramela, 2018); may parasitize or establish symbiotic relationships with higher organisms such as mussels and corals; and consume and contribute to organic carbon and other nutrients in the food web. Unseen microbial life controls the marine biosphere and therefore influences global cycles.

The classical method for identifying and enumerating microbes is to collect samples and cultivate individual species on artificial media as clones (i.e., colonies on agar plates). This is a slow, labor-intensive, and highly biased procedure that we now know isolates perhaps <1% of the total microbes in a sample: only those capable of growing on laboratory media in pure culture. Astounding advances in DNA sequencing and software technologies for describing and enumerating microbial communities have developed since *Oil in the Sea III* (NRC, 2003) and continue to evolve. The greatest value in the new techniques is that microbial identity and activity now can be studied without having to cultivate organisms or separate them from their community partners. Speed is also a factor: with the advancement of high-throughput sequencing methods and, recently, miniaturized hand-held DNA sequencers (e.g., MinIon), analyses that once required sample transport, significant laboratory infrastructure, and days or weeks of data acquisition can now be accomplished in hours (e.g., Joye and Kostka, 2020) and made fieldable for nearly real-time analysis of community structure and activity, thereby permitting repeated temporal and spatial surveys of dynamic environments with statistical replication.

High-throughput (next-generation) sequencing and other analytical technologies that are used to describe evolutionary relationships and metabolic activities of individual organisms or communities are collectively termed 'omics. Those that apply to microbial communities are described in Appendix H, including:

1. Sequencing selected "marker" genes amplified from cultivated isolates or community DNA to infer taxonomic identity or metabolic potential, often without additional genomic context for interpretation;
2. Sequencing the total DNA (genome) of a single species to determine its taxonomy, biochemical potential, evolutionary history, and more (genomics). The DNA may be isolated from a cultivated colony or from an uncultivated single microbial cell physically teased out of a community in a sample (single cell–assembled genomes [SAGs]);
3. Sequencing the total DNA of a microbial community, then using software to assemble discrete metagenome-assembled genomes (MAGs) of individual species or to analyze selected marker genes in the community (metagenomics);
4. Isolating and sequencing total messenger RNA from a species (transcriptomics) or community (metatranscriptomics) to discern metabolic activity in comparison to a basal metabolic state; and
5. Isolating and chemically analyzing cellular macromolecules and small molecules, namely: proteins (proteomics), lipids (lipidomics), or biochemical pathway products (metabolomics) to derive metabolic pathways.

Prior to the DNA sequencing revolution, a commercial DNA chip technology was developed based on hybridization of sample DNA to small pieces of known reference DNA for rapid taxonomic (PhyloChip) and metabolic (GeoChip) analysis. These microarray technologies were used early on during the DWH oil spill for rapid, fieldable estimation of microbial community composition and metabolic potential (e.g., Beazley et al., 2012). However, because the output of such screening tests was limited by the range of known DNA samples on the chip, it likely missed novel organisms or gene variants. 'Omics now allow novel microbes previously uncultivated by classical methods to be detected in the environment. These "virtual microbes" are taxonomically identified by comparison of a marker gene sequence (usually fragments of 16S or 18S ribosomal RNA genes) to other sequences archived in public databases, which themselves may be known only as a short DNA sequence. Many new branches of microbial life have been discovered in this way, and Pallen et al. (2020) have suggested that millions of microbial lineages remain to be described and named, many of them currently known only as DNA fragment sequences. Notably, detecting a DNA sequence or genome in an environmental sample does not necessarily indicate whether the organism is active, recently dead, or dormant, whereas transcriptomics is better for revealing which organisms are actually alive and active by discerning the complement of genes that they are expressing in response to their environment.

An essential component for application of 'omics to oil spills is the development, curation, and accessibility of public databases. Several comprehensive databases of taxonomic gene sequences exist, having different degrees of rigor and curation (e.g., GenBank, SILVA, RDP, Greengenes). Newly acquired DNA sequences are compared to ever-expanding databases to assign provisional identities at different taxonomic levels and with varying confidence. General analytics include the U.S. Joint Genome Institute Integrated Microbial Genomes and Microbiomes site[1] (IMG/M) for functional gene identification and UniProt, LipidMaps, and National Metabolomics Data Repository sites for macromolecules. The Tara Oceans database[2] (Sunagawa et al., 2015) comprises four gene catalogs of microbial community sequences acquired from global ocean water surveys, currently providing access to more than 110 million functional and taxonomic gene sequences that enable study of microbial biodiversity and ecosystem activity. Hydrocarbon-specific databases such as GROS (Genome Repository of Oil Systems; Karthikeyan et al., 2020) and gene sequence analysis and annotation pipelines (e.g., CANT-HYD; Khot et al., 2022) have recently emerged to facilitate analysis of such gene repositories.

As spectacular as the 'omics revolution is, numerous caveats must be applied to acquisition and interpretation of the data. Reagents and methods (e.g., for DNA extraction, amplification, sequencing platforms, and annotation pipelines) are constantly changing as methodological biases are removed and analytical platforms improve. The availability and succession of new methods challenge quality assurance and quality control measures within each laboratory even over short time frames and make interlaboratory comparisons exceedingly difficult. Furthermore, the number of genes available for analysis expands daily as new sequences are input by scientists across the globe (scientific journals typically require that sequences cited in publications be accessible in open repositories), so that gene identification can change over time depending on which sequences of what quality are available for comparison. Likewise, sequence analysis software is constantly evolving in response to advances in bioinformatic developments, so that the same sequence may be identified differently by successive versions of the same software or by new software applications. Such changes can quickly make archived data outdated or impossible to analyze using new bioinformatic platforms. Finally, microbial taxonomy is also in flux as new classes, genera, and species are recognized, so a sequence identified as "Microbe X" last year may be named "Microbe Y" this year, sometimes confounding the literature.

Data acquisition, input, storage, accessibility, and management represent a large proportion of research effort and expense when 'omics methods are applied; a moderate-size project can quickly generate terabytes of data for analysis and archiving. To extract meaning from the 'omics dataset, comprehensive metadata describing the sample are essential, including site chemistry, meteorology, and geographical and oceanographical information. Furthermore, interpretation of 'omics for biodegradation also requires information about changes in oil chemistry and/or metabolite formation. However, when corresponding data from new analytical chemistry techniques (see Section 2.1.7) are also available, their coupling further adds to the burden of integrating "big data" sets, and requires a multidisciplinary approach to achieve a comprehensive understanding of the sample.

Although these caveats might appear to discredit 'omics output, environmental microbiologists have learned not to consider sequence identification as a permanent label but rather as the best interpretation at the time of analysis, particularly for uncultivated environmental microbes. **Despite the fluidity of information and analysis, 'omics-based studies can yield astonishing and unparalleled levels of insight into environmental processes provided that researchers follow rigorous procedures, heed caveats, and cautiously interpret information.**

5.2.8.2 Biodegradation: Why Microbes Are Important to Oil in the Sea

Biodegradation is the biological process of breaking down organic matter. Among the myriad microbial lineages in ocean waters and sediments, certain ubiquitous microbes—particularly bacteria—can biodegrade specific petroleum compounds, using them as high-energy substrates for growth (reviewed by Head et al., 2006; Hazen et al., 2016). Microbial communities have been called the "first responders" to oil

[1] See https://img.jgi.doe.gov.

[2] See http://ocean-microbiome.embl.de/companion.html.

and, given appropriate conditions, can begin degrading the oil within hours or days. They commonly are also the 'final responders" or "polishers," continuing to biodegrade susceptible oil components that remain after natural processes (see Sections 5.2.1–5.2.7) wane and human interventions (see Chapter 4) are complete. To put the importance of biodegradation in perspective, most natural processes only transport, transform, dilute, or sequester the oil whereas, in theory, biodegradation can remove oil components from the ocean by completely oxidizing them to gas and water. In practice, however, oil biodegradation is rarely complete but instead comprises mineralization and transformation (see Box 5.5), yielding carbon dioxide (CO_2), water, and cells (biomass) plus partially oxidized compounds that either associate with the oil or sediments, or dissolve in the water column. Biodegradation leaves behind high molecular weight components such as large PAHs, resins, and asphaltenes (see Section 2.1.3) plus unaltered whole oil that is not bioavailable (see Box 5.5). In addition, some of the reaction product intermediates from biodegradation processes may combine with other organic molecules in the environment to form biogeopolymers that are then less susceptible to further biodegradation. These compounds may be incorporated into sediments and become buried over time as geological deposits. Microbes may mineralize from 10% to 90% of the oil mass, depending on the oil's chemical composition and environmental conditions. The residual oil remaining after biodegradation is depleted in "labile" chemicals and enriched in "refractory" compounds.

The Relative Abundance of Microbial Hydrocarbon-Degrading Species Is Dynamic and Responsive

Hydrocarbon-degrading species are typically present in low abundance in the environment in the absence of oil and form part of the "rare biosphere": taxa present at <1% abundance in the local microbial community. Incursion of oil provides a selective advantage to those species, temporarily permitting them to become abundant by out-competing species that cannot utilize the oil for growth, but only until those high-energy hydrocarbon substrate(s) have been depleted, after which their advantage is lost. For example, shortly after the DWH oil spill, hydrocarbon-degrading bacterial species that were minor components of the background microbiota increased to >90% of the bacteria detected in the dispersed oil plume during the period of active biodegradation, then diminished again (e.g., Redmond and Valentine, 2012; Kleindienst et al., 2016). This pattern of "bloom and bust" is common after oil spills.

A succession of dominant taxa is typical during the bloom phase because different species utilize different oil components: some specialize in saturates and others in aromatics (see Appendix I). They usually degrade their respective substrates at different rates, leading to temporal succession of species in the microbial community that echoes

BOX 5.5
Biodegradation: Mineralization Versus Transformation

Use of the term *biodegradation* often implies incorrectly that a given substrate is completely removed from an environment. However, the extent of biodegradation, particularly of a chemically complex substrate such as oil, depends upon the organism(s) present, the chemical composition of the substrate (its susceptibility to enzymes), its bioavailability,[a] the prevailing environmental conditions (whether conducive or limiting), and the timeframe.

Biodegradation of individual oil components may be complete or partial, both in quality and quantity. Qualitatively, some substrates such as *n*-alkanes and small PAHs may be **mineralized** (completely oxidized to carbon dioxide gas, water, and biomass) through a series of enzymatic reactions. More complex components may be only **transformed** (chemically altered) by introducing oxygen to the compound by adding chemical groups (e.g., hydroxyl, carboxyl) to the hydrocarbon skeleton to form alcohols and organic acids and/or break ring structures. In this case, some or most of the original chemical structure remains, even though the product may no longer be detected directly using conventional analytical techniques (see Box 5.8 later in the chapter for the analytical consequences of transformation). Conditions that interrupt the mineralization process (e.g., depletion of oxygen or nutrients) can shift biodegradation toward transformation. The partially oxidized products may become more water soluble than the original compound and therefore partition into the water, or may remain associated with the oil or with organic coatings on particulate matter. In some cases, the transformed products may become substrates for other microbes to metabolize or further transform (see Box 5.7 later in the chapter).

Quantitatively, rarely are all the diverse components of an oil biodegraded, either completely or partially; condensates or light fuels comprising simple hydrocarbons may be exceptions. Typically, the most susceptible hydrocarbons (see Table 5.2) are depleted and recalcitrant oil components remain, either partially oxidized or completely unaltered. Thus, unfortunately, the simple term "biodegradation" used in the literature covers a range of fates from mineralization to transformation and from minor depletion of components to almost complete removal of a mixture. The reader should be wary of assuming the best-case scenario.

[a] Note that the term *bioavailability*, when used in the context of microbial biodegradation, is synonymous with "accessible to microbial cells," whether dispersed or emulsified in the water column, sorbed to sediments or organic material, or partitioned into the tissues of higher organisms.

the order of oil component depletion (e.g., Valentine et al., 2012; Dubinsky et al., 2013; Rodriguez et al., 2015; Hu et al., 2017). Furthermore, some species donate and/or accept genes that enable hydrocarbon degradation, thus spreading the capability amongst the microbial population in response to the selective pressure of oil incursion. Such blooms may temporarily sequester nutrients like nitrogen, phosphate, and iron and deplete dissolved oxygen locally, but their metabolic waste products can be consumed by other members of the microbial community, and the bloom attracts predators such as viruses and zooplankton that re-cycle the nutrients (see Box 5.6). When the petroleum substrates are exhausted, the degraders are less competitive, their dead biomass becomes a substrate for non-hydrocarbon-degrading species and they return to being a small proportion of the total community. In this way the petroleum molecules that increase the biomass of hydrocarbon-degrading microbes contribute to higher organisms' biomass and enter the food web (e.g., Chanton et al., 2012; Fernández-Carrera et al., 2016). The ubiquitous presence of hydrocarbon-degrading marine microbes, adapted to local conditions, obviates the introduction of commercially grown microbial inocula for bioremediation purposes ("bioaugmentation"; see Appendix E).

In terrestrial environments almost all hydrocarbon-degrading bacteria are generalists that grow on a variety of organic substrates and so persist in soil even in the prolonged absence of oil. Individual cells may relinquish the ability to degrade hydrocarbons by shedding "accessory genes," while a small proportion of the community retains the genetic ability and can disseminate those genes when oil is present.

BOX 5.6
Microbial Interactions with Oil and the Marine Food Web

Figure 5.16 schematically shows the roles of marine microbes when surface or subsurface oil intersects with the marine biological pump that transports organic matter from the photic zone to the seafloor (Azam and Malfatti, 2007; Louvado et al., 2015). Although the diagram focuses on carbon cycling, other essential nutrients follow the same paths when sequestered in biomass. The "short" and "long" hydrocarbon cycles (Valentine and Reddy, 2015) refer to the depth and duration of the cycles, not the size of hydrocarbons involved.

FIGURE 5.16 Intersection of the marine biological pump with natural sources of oils and oil spills (not to scale).
NOTES: **Blue lines** represent processes comprising the biological pump, independent of oil. Photosynthesis by phytoplankton (green and yellow circles) in the photic zone consumes dissolved carbon dioxide (CO_2), produces oxygen (O_2), and generates biomass (cells). Phytoplankton are consumed by zooplankton, grazers, and predators, sustaining the marine food web. Water-soluble products of metabolism contribute to DOM in the water column. Dead cells, fecal pellets, and organic detritus comprising POM may remain suspended temporarily before eventually sinking. Marine viruses (not shown) are essential agents in converting POM to DOM by lysing bacterial cells and killing higher organisms. Heterotrophic (non-photosynthetic) bacteria metabolize DOM and POM aerobically and form aggregates (marine snow; see Section 5.3.2.2) by adhering to extracellular polymeric substances (EPS; white lines) synthesized by phytoplankton and heterotrophs and/or released by dead cells. As marine snow settles through the water column over days to weeks, aerobic metabolism and predation cycle DOM and POM and the refractory remnants contribute to seafloor sediments. Over the short term of weeks to years, aerobic and anaerobic metabolism of residual marine snow in sediments produce $CO_2 \pm$ methane (CH_4) gases, DOM, and POM. With burial, anaerobic biodegradation, and geological time, the organic carbon may mature to form crude oil and/or methane deposits. **Orange lines** represent processes in the short hydrocarbon cycle. Some plankton, particularly certain phytoplankton species, synthesize and accumulate specific alkanes (e.g., *n*C15, *n*C17, pristane). The thin orange lines indicate that the biosynthesized hydrocarbons are a fraction of total photosynthetic carbon. As these cells die or associate with heterotrophic bacteria to form marine snow, the hydrocarbons are readily metabolized, consuming O_2 and producing CO_2, DOM, and POM. The biosynthesized hydrocarbons do not accumulate in the water column because they are easily biodegraded, but they do contribute to organic carbon sedimentation via marine snow formation. The cycle operates independently of petroleum incursion from spills or natural seeps. **Gray lines** represent processes in the long hydrocarbon cycle. Introduction of oil at the surface or from the subsurface produces dispersed oil droplets in the water column. Hydrocarbon-degrading bacteria colonize the droplets to form marine oil snow (MOS) aggregates. Additionally, existing marine snow particles may sorb oil while sinking through dispersed droplets, forming MOS. The presence of oil as a substrate shifts the MOS microbial community composition as hydrocarbon-degrading species are enriched (community succession) and preferred oil components are depleted and dissolved. As MOS settles over days to months, microbes degrade the oil, consume O_2, and generate DOM, POM, and CO_2. Residual oil becomes depleted in labile hydrocarbons and enriched in refractory compounds, with transit time influencing the degree of degradation; in shallow waters, undegraded but physically altered oil may still remain when MOS reaches the sea floor. After sedimentation, burial, anaerobic biodegradation, and geological time, the carbon may be transformed to oil and gas to complete the long cycle. The contribution of the long hydrocarbon cycle to the biological pump depends on the mass of introduced oil.
SOURCES: Committee generated, based on Azam and Malfatti (2007) and Valentine and Reddy (2015).

Some marine taxa similarly are opportunistic generalists ("facultative" oil-degraders), but others are considered "obligate hydrocarbonoclastic bacteria" (Yakimov et al., 2007), so specialized that they grow almost exclusively on specific oil constituent classes such as alkanes (see Appendix I). An obvious question is how they persist in the ocean in the absence of spilled oil. The selective pressure likely is continuously low concentrations of hydrocarbons from chronic sources. These include gaseous and/or liquid hydrocarbons and chemical analogs released from natural oil and gas seeps (reviewed by Ruppel and Kessler, 2017; Joye, 2020), which are ubiquitous on the continental margins of North America (see Chapter 3); chronic pollution such as in shipping lanes (Weiman et al., 2021); and/or PAHs from forest fires deposited from the atmosphere (see Section 5.2.4) These diffuse long-term sources of hydrocarbons, in addition to "pyrogenic" PAH deposited via the atmosphere via forest fires and early diagenetic PAHs (see Section 2.1.5) can sustain the genetic potential of a small proportion of hydrocarbon-degrading species in a community. Another natural continuous source of hydrocarbons is contemporary biological production in the "short hydrocarbon cycle" (see Box 5.6 and Section 2.1.5). Strains of the ubiquitous and abundant marine cyanobacteria *Prochlorococcus* and *Synechococcus* (Flombaum et al., 2013) annually synthesize millions of tonnes of the alkane *n*-pentadecane (nC15) and somewhat less nC17 (Lea-Smith et al., 2015; Love et al., 2021) that accumulate in thylakoid and cytoplasmic membranes (Lea-Smith et al., 2016) and which facultative and obligate hydrocarbon-degraders can access and consume (Chernikova et al., 2020). Similarly, certain

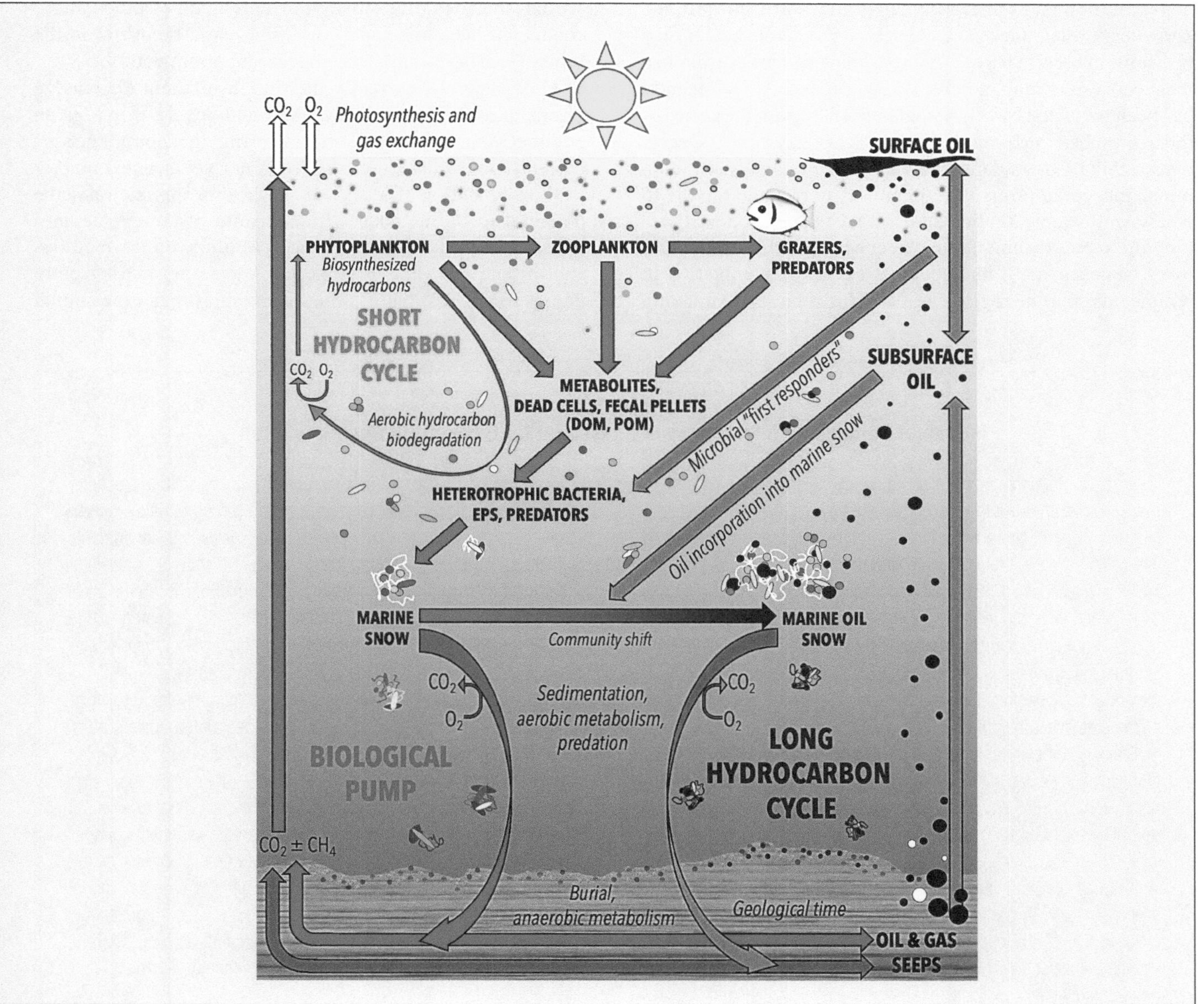

zooplankton, phytoplankton, and benthic algae synthesize the *iso*-alkane pristane, some *n*-alkanes, and olefins (Blumer et al., 1963, 1971; Clark and Blumer, 1967), some of which appear to pass up the food chain to the liver oils of whales and sharks. Some eukaryotic plankton produce isoprenes and monoterpenes (Shaw et al., 2010; McGenity et al., 2018) and recently have been shown to associate tightly with obligate bacterial hydrocarbon-degraders (Thompson et al., 2020), presumably sustaining the bacteria; plant-derived waxes and natural aromatic compounds are also present in some marine environments. Thus, hydrocarbon-degrading microbes are ubiquitous and persist in marine environments at low abundance even in the absence of spilled oil and can multiply rapidly after an oil spill. Some species even detect and swim toward oil droplets (using chemotaxis) to maximize their selective advantage over non-degrading cells (e.g., Dewangan and Conrad, 2020; Gregson et al., 2020).

Even though hydrocarbon degraders are ubiquitous, knowing the abundance and identity of indicator species may be useful in several ways. Their presence and proportion in a local ecosystem may indicate whether that site is or recently has been impacted by oil. Abundance of "sentinel" species—those enriched early during a spill—or "keystone" species that sustain biodegradation may indicate previous petroleum inputs that could prime the community to respond rapidly to a new spill event. Krolica et al. (2019) have proposed employing 'omics techniques to detect key microbial bioindicators ("genosensing") that could reveal oil contamination in remote cold marine regions as a tool for chemically tracking covert petroleum discharges. In contrast, the presence of a "naïve" community where degraders are still part of the rare biosphere may foretell a long lag time until hydrocarbon-degrading species are recruited and enriched. "Accessory" species that degrade co-metabolites or by-products of oil metabolism, or contribute to distributed metabolism (see Box 5.7) by sharing metabolic pathways within a community (Weiman et al., 2021) might serve as indicators of the in situ biodegradation trajectory (Lozada et al., 2014), especially when such information is paired with chemical analysis of residual oil. Finally, the return of the microbial community to approximate the pre-spill composition might serve as an indicator of ecosystem recovery and clean-up success, although this proxy has not yet been used formally for closure. Spill sites may have several ecological niches (e.g., water column, beaches, marshes), each having their own baseline and responding communities (Liu and Liu, 2013) so that both spatial and temporal surveys of community composition, conducted pre-, mid- and post-spill, could be useful in the future for determining the biodegradation trajectory.

Unfortunately, currently there is insufficient knowledge to create a checklist of key microbial species for a given environment, and simply monitoring the abundance of currently known sentinel species is not yet a robust marker of biodegradation efficacy. However, this approach has the potential to be included within a suite of complementary indicators, especially when incorporating 'omics methods and oil analysis. For example, an interesting but currently unproven application of 'omics is monitoring marine animal

BOX 5.7
Microbial Community Interactions Are Essential for Efficient Oil Biodegradation

Aerobic biodegradation of an individual hydrocarbon or suite of related hydrocarbons can be accomplished by a single microbial species using one or more biochemical pathways to capture carbon and energy from the substrate. Species tend to specialize in certain classes and/or size ranges of hydrocarbons, and each has a finite substrate range. Therefore, biodegradation of a chemically complex substrate like oil requires a diverse assembly of microbes. The community of degraders shifts in composition ("community succession") as labile oil components are degraded and the oil composition changes, and species interactions affect the ultimate biodegradation achieved.

Species compete for access to the oil, for example at the oil–water interface of dispersed oil droplets, and for nutrients and electron acceptors (e.g., O_2) in the environment, but they also co-operate in various ways. For example, Species A may grow by mineralizing a specific suite of hydrocarbons but may inadvertently partially oxidize analogous compounds to products that it cannot utilize ("co-metabolism"). If these co-metabolic products accumulate, they may become inhibitory to Species A, but if Species B can utilize those products for growth, it benefits both species. This is one example of "distributed metabolism," in which individual species do not have the entire complement of genes to achieve complete metabolism of a substrate, but by sharing biochemical abilities the community benefits, whether through complementary biochemical pathways, or sharing of water-soluble vitamins or nutrients, or inter-species transfer of hydrogen (H_2), or production of extracellular polysaccharides such as in marine snow formation. In some cases, the biochemically intimate partnership ("syntrophy," meaning "eating together," a common phenomenon in anaerobic hydrocarbon biodegradation) is essential to survival of one or both species.

Recruitment and assembly of a competent community in the most efficient proportions and physical orientations at the correct phase of oil biodegradation may take some time. This process may explain in part the acclimation period ("lag time") often observed before biodegradation becomes apparent, which may last days to weeks in oxygenated water or months to years in anaerobic sediments; a complementary explanation is that microbes need time to access the oil, replicate, and begin to express the requisite genes before multiplying to achieve sufficient numbers that their metabolism has an observable effect on the oil chemistry. The model of Valentine et al. (2012) that couples physical mixing with metabolic components accommodates these variables.

microbiomes (the total microbial complement in and on an organism) as an indicator of oil exposure: Walter et al. (2019) detected putative hydrocarbon-degrading bacterial species in the gastrointestinal microbiome of wild Atlantic cod caught in Norwegian waters that correlated with exposure of the fish to low oil concentrations. **In the future, monitoring microbial community composition via 'omics techniques may become a component of oil spill site assessment and recovery. However, research is required to achieve and validate this approach to creating a practical bioremediation tool linking community composition with hydrocarbon biodegradation, concomitant with acquisition of baseline data for comparison.**

5.2.8.3 The Physical State of Oil Influences Its Bioavailability

The physical state of oil—whether gaseous, liquid, (semi-)solid, free-floating, or sorbed to minerals or organic matter—influences biodegradation by affecting bioavailability of the oil components through dissolution, dispersion, attachment, and emulsification.

Gaseous alkanes (≤C4, including methane, ethane, propane, and butane) enter marine water and/or submerged sediments from methane hydrate deposits, gas seeps, or subsurface gas releases as free gas bubbles and/or dissolved in liquid oil, depending on hydrostatic pressure (see Section 5.2.1). Bioavailability is influenced by the kinetics of gas dissolution from bubbles into water, by partitioning into water, and by duration of bubble rise. Notably, methanotrophs (microbes that specialize in utilizing methane) have been observed to be transported from seafloor sediments to the water column by attachment to gas bubble surfaces during ebullition (Schmale et al., 2015), suggesting that close association with the bubbles assists methane metabolism.

Bioavailability of liquid oil, whether present as droplets suspended in water or in a surface slick, is influenced by the oil–water interfacial area. Natural dispersion expedites dissolution (the smaller the droplets, the greater the total surface area) and governs physical access to the oil. Microbes live in the water phase and access dissolved and liquid oil using various biochemical and physical strategies including energy-dependent uptake of dissolved hydrocarbons (Miyata et al., 2004), emulsification and pseudosolubilization of liquid oil (reviewed by Wang et al., 2020a). Some respond to diffusion gradients by sensing and swimming toward oil via chemotaxis (Zhou et al., 2017) where they may adhere to the droplet surface (Godfrin et al., 2018), forming biofilms (see Section 5.2.8.7).

Chemical dispersion of oil slicks during spill response (see Chapter 4) generates fine droplets under appropriate conditions and would be expected to enhance bioavailability compared to a slick, and therefore oil biodegradation (Prince, 2015). However, for decades the literature has documented contradictory reports of chemical dispersants inhibiting, enhancing, or having neutral effects on oil biodegradation. Much of this controversy is due to widely differing in vitro experimental conditions, including excessive initial masses of oil and/or dispersant in closed-system cultures (where dilution cannot occur) and/or unrealistically high ratios of dispersant:oil (reviewed by Lee et al., 2003a); using unsuitable inoculum sources, such as laboratory cultures instead of natural seawater; providing nutrients at unreasonably high concentrations that skew community activities (Prince et al., 2016b); and using ineffective means of generating and maintaining fine dispersions of oil droplets, among others (see also Figure 4.17). Incidentally, Prince (2017) has developed a simple apparatus to maintain dispersions for biodegradation studies using "environmentally relevant" concentrations of oil, dispersant, and nutrients to address some of these variables.

As an example of conflicting observations, Brakstad et al. (2015a) prepared Corexit 9500A dispersions of Macondo oil using near-surface Norwegian seawater in the laboratory and found, as predicted, that the smaller the droplet, the greater the biodegradation: dispersions of 10 μm oil droplets degraded significantly faster than 30 μm droplets. However, despite using the same oil, dispersant, and method to generate 10- and 30-μm oil dispersions, when Wang et al. (2016) used deep (~1,200 m) Gulf of Mexico seawater, they found no statistical difference in biodegradation rates between the two droplet sizes. The main difference was the water source, suggesting that the chemical and/or biological composition of the deep waters influenced the observed effects more than the microscopic droplet size. In separate studies that also used deep Gulf of Mexico water samples incubated with Macondo oil, the presence of Corexit 9500A enriched different microbial communities and did not enhance biodegradation rates (Kleindienst et al., 2015). This implies that chemical dispersion of oil droplets is not the main factor limiting oil biodegradation in deep Gulf waters, and that other nuances such as nutrient availability, community composition and/or metabolic status of the inoculum may be more important. To address the latter possibility, Rughöft et al. (2020) simplified an in vitro experiment by using a single key hydrocarbon-degrading species growing with pure *n*-hexadecane. They found that the effect of Corexit 9500A on biodegradation was influenced by the metabolic state of the inoculum, that is, whether the cells were starving or replete: starving cells were inhibited by the presence of dispersant.

The National Academies issued a report on dispersant use in oil spills, discussing the complexities of evaluating dispersant effects on biodegradation and particularly focusing on whether subsurface dispersant injection used during the DWH spill enhanced deep sea oil biodegradation (see Box 2.2 in NASEM, 2020). That report determined that the in situ conditions precluded clear-cut conclusions about the effect of Corexit 9500A on biodegradation in the DWH dispersed deep plume. Furthermore, the report pointed out that increased surface area will not accelerate dispersed oil biodegradation if other parameters such as nutrient concentration are limited, as inferred above. It is important to note that dispersants are usually applied to oil at the surface and deep subsea injection such as that implemented during the DWH spill was anomalous. However, controversy about the effects of chemical dispersion on biodegradation of near-surface oil persists. **The current divergent conclusions about positive, negative, or neutral effects of chemical dispersion on biodegradation of liquid oils necessitate more methodical experimentation using carefully selected, environmentally relevant conditions to resolve the persistent controversy about biodegradation of chemically dispersed oil. Because chemical dispersion is an important response option for oil slicks, its impacts on biodegradation of spilled oil, whether at the surface or subsurface, should be resolved.**

Solid and semi-solid oils such as naturally heavy oils and weathered dilbit (see Chapter 2), or conventional oils that form tar balls, mats, or patties after severe weathering (see Section 5.3.4) present numerous physical barriers to efficient biodegradation. First, the formation of the (semi-)solid oil resulted from prior abiotic weathering and/or biodegradation, either over geological time in the reservoir or subsequent to entering the sea. Such oil is depleted in labile substrates and enriched in chemicals recalcitrant to further biodegradation. (Fresh or lightly weathered dilbit is a special case because the lighter components comprising the hydrocarbon diluent in the dilbit blend are generally biodegradable [Schreiber et al., 2019] whereas the bitumen fraction predominantly comprises resins and asphaltenes that are not significantly biodegradable; see Section 5.3.2.4.) The outer surface of tar balls, etc., typically is enriched in refractory asphaltenic and resin components (see Section 2.1) that diffuse within the oil structure to assemble at the oil:water or oil:air interface as a water-insoluble, non-biodegradable "rind," along with oxygenated hydrocarbons from photo-oxidation (e.g., White et al., 2016) that may also be poorly susceptible to biodegradation. Additionally, on the shoreline the oil surface may be physically occluded by mineral particles (e.g., silt, sand) that further reduce microbial access and photo-oxidation. These surface features decrease interfacial area and reduce water penetration, limiting diffusion of soluble nutrients and oxygen and decreasing bioavailability and biodegradation rates (reviewed by Gustitus and Clement, 2017). Limited, selective biodegradation of stranded tar balls near the beach surface has been observed, albeit with estimated half-lives of years to decades (Harriman et al., 2017; Bostic et al., 2018; Bociu et al., 2019), whereas other spill sites still had weathered asphalt-like "pavements" on beaches 30 years post-spill (Lee et al., 2003a).

5.2.8.4 The Chemical Composition of Oil Influences Its Biodegradation

Of the four SARA analytical classes (see Section 2.1), measurement of oil biodegradation historically has focused on the saturates and aromatics—the two classes that are most susceptible to aerobic microbial attack and are amenable to gas chromatographic analysis. Not all constituents within these two classes can be biodegraded, due to individual chemical and biological properties including toxicity, water solubility and biochemical reactivity. Instead, microbial enzymes exhibit substrate specificity according to molecular size, complexity, isomeric arrangement of side groups or heteroatoms, and even stereoisomeric configurations, leading to recurring patterns of biodegradation observed using GC-MS (see Table 5.2). The general patterns of susceptibility are that small, simple molecules are more labile than complex, high-molecular-weight compounds; unsubstituted (parent) hydrocarbons are more susceptible than alkyl-substituted members within that chemical family (and different alkyl positions or increasing alkylation affect biodegradability [e.g., Lamberts et al., 2008]); the presence of heteroatoms may decrease susceptibility. There are exceptions to each of these general statements. For example, the three-ring PAH phenanthrene is far more labile than its isomer anthracene;

TABLE 5.2 General Susceptibility of Oil Constituents to Aerobic and Anaerobic Biodegradation, Based on Data from Relatively Short-Term Observations in Water and Soil, and After Geological Time in Oil Reservoirs

Overall Susceptibility	Class (Section 2.1.3)	Relative Susceptibility to Biodegradation Within Class
Most	*n*-Alkanes	$C_3 \sim C_8–C_{12} > \sim C_{12}–C_{15} > C_{15+}$
	i-Alkanes	Lesser > greater alkyl substitution
	Isoprenoids	Lower molecular weight (e.g., C_{10}) > pristane, phytane > higher molecular weight (e.g., > C_{20}); acyclic > polycyclic
	Aromatics	Monocyclic > polycyclic; 1-ring > 2-ring > 3-ring > 4- and 5-ring PAHs
	Alkyl-substituted aromatics	Unsubstituted > alkyl-substituted; methyl and dimethyl > trimethyl or more
	Cyclic alkanes (alicyclic hydrocarbons; naphthenes) and steroids	Simple cycloalkanes (e.g., methylcyclohexane, decalin) > complex cyclic alkanes (hopanes > steranes > diasteranes) > aromatic steroids
	Resins	Simple (e.g., carbazole, dibenzothiophene) > complex (e.g., porphyrins)
	Asphaltenes	Smaller hydrocarbons that co-purify with asphaltenes may degrade but asphaltene biodegradation is unlikely
Least		

NOTES: No individual microbial species can attack all classes; species tend to specialize within groups or sub-groups of hydrocarbons. Some compounds may only be partially oxidized or may be co-metabolized only when a suitable hydrocarbon that sustains growth is also present (see Boxes 5.5 and 5.7).
SOURCES: Huesmann (1995); Van Hamme et al. (2003); Prince and Walters (2007).

the alkyl-substituted monoaromatic toluene (methylbenzene) is more degradable than its parent, benzene; very small hydrocarbons may elude biodegradation at high concentrations because they are toxic membrane solvents and can inhibit even the microbes that can degrade them; PAHs may be degraded in preference to alkanes (e.g., phenanthrene before *n*-hexadecane; Foght et al., 1990) depending on the microbial community composition.

The wealth of published data about biodegradation of SARA components (see Chapter 2) is summarized in text below, but three points are notable: (1) unless specifically mentioned, the summary below refers to aerobic biodegradation because much less is known about the anaerobic counterpart (see Section 5.2.8.8); (2) much of the information was obtained using pure cultures of terrestrial organisms rather than communities of marine microbes and/or using pure hydrocarbons rather than whole oils. These experimental conditions likely skew presumptions of actual substrate susceptibility in situ, particularly regarding bioavailability; (3) this brief summary cannot present the nuances and exceptions to general rules that have been reported in the literature for specific circumstances.

Despite being chemically very stable, *n*-alkanes are the most biodegradable class in the saturate fraction. Within that family, biodegradation of gaseous alkanes (≤C4) has been discussed above and is relevant only for subsea conditions; C5–C8 short-chain *n*-alkanes are volatile, may evaporate before they are biodegraded, and can have some toxicity as membrane solvents; medium-chain *n*-alkanes (~C10–C30) are usually the most readily biodegraded; and longer *n*-alkanes (>C30) are waxy and less susceptible to biodegradation, possibly due to difficulty crossing the cell membrane or to limited surface area compared with liquid alkanes (Lyu et al., 2018). Beta-oxidation is the most common biochemical pathway for alkane metabolism (Van Hamme et al., 2003), analogous to lipid metabolism. Alkyl substitution of the alkane backbone (e.g., di- and tri-methyl-alkanes, or ethyl-alkanes) to form *iso*-alkanes reduces biodegradability, possibly due to steric hindrance at the enzyme reactive site. Multiply-alkylated *iso*-alkanes such as pristane and phytane are much less biodegradable than their straight-chain homologs and some may be used as "petroleum biomarkers" (see Sections 2.1.3 and 5.2.8.6) because they are biodegraded slowly, if at all. Little is known about biodegradation of unsubstituted cycloalkanes, particularly in the ocean, although decalin (two fused cyclohexane rings) can be co-metabolized (see Box 5.7) when provided with an *n*-alkane as a growth substrate (Kirkwood et al., 2008). Alkyl-cycloalkanes may be partially degraded if the alkyl chain is long enough to be attacked, producing a naphthenic acid. Olefins (unsaturated hydrocarbons such as alkenes) are uncommon in crude oil (Speight, 2014) although they are components of some synthetic-based drilling muds (Reddy et al., 2007) that may enter the marine environment. Olefin biodiodegradation is not well documented, but selective biodegradation or persistence of different olefins has been noted in Gulf of Mexico sediments (Stout and Payne, 2017). Complex fused-ring aliphatics such as hopanoids and sterols also tend to resist biodegradation and may serve as petroleum biomarkers. Similarly, petroleum-based lubricating oils that comprise complex long-chain *iso*-alkanes and condensed cycloalkanes resist biodegradation, driving the development of biodegradable plant-based lubricant oils (Mobarak et al., 2014).

Mono-aromatics (e.g., BTEX: benzene, toluene, ethylbenzene, and m-xylene isomers) are biodegradable and are somewhat water-soluble but can be toxic to microbes at high concentrations and are severely depleted or absent from heavy oils and weathered oils. PAHs and alkyl-PAHs are often less biodegradable than mono-aromatics, in part due to their lower water solubility. PAHs comprising two or three rings are generally degraded in preference to 4- and 5-ring PAHs (Kanaly and Harayama, 2010), which may persist until the smaller PAHs are depleted. Alkyl-PAHs with one or more alkyl groups are differentially degraded with no discernable pattern of susceptibility (e.g., Wammer and Peters, 2005; Lamberts et al., 2008), and mixtures of PAHs biodegrade differently than individual substrates (e.g., Knightes and Peters, 2006). Some of the differences between biodegradation potential of PAHs may be due to bioavailability in situ rather than inherent recalcitrance to enzymatic attack (Wammer and Peters, 2005).

Some simple resin compounds that are not, strictly speaking, PAHs (e.g., N-, S-, and O-heteroaromatic compounds from the carbazole, dibenzothiophene and fluorenone families; see Section 2.1) are considered here with the PAHs because they have similar chemical properties (i.e., they fractionate with the aromatics) and their pattern of biodegradation preference (parent compound over alkyl-substituted homolog) is similar to that of their PAH analogs. More complex resins either are not detected or are not resolved using routine gas chromatography. Furthermore, although simple naphthenic acids may be biodegraded, the more complex members of the class (e.g., tetrameric forms) appear to be highly recalcitrant to biodegradation since they are detected in otherwise-biodegraded crude oils (reviewed by Barros et al., 2022). Because of this analytical limitation and the fact that their structures may not be known, it is difficult to measure the biodegradation potential of most resins. FT-ICR-MS studies (see Section 2.1.7) are beginning to reveal changes to N-, S-, and O-containing oil compounds as they decrease or accumulate during biodegradation, but assigning structure remains difficult. It is important to note that some metabolites (e.g., naphthenic acids, aromatic epoxides, and alcohols; see Section 5.2.8.8) produced by partial biological oxidation of other hydrocarbon classes are, by definition, resins and may appear transiently as they are produced then further degraded, or may accumulate if they are dead-end products and sufficiently non-polar to partition with the oil, thereby contributing to the total resins fraction in the altered oil. Naphthenic acids are natural components of heavy crude oils such as bitumen, and spills of dilbit (see Chapter 2) may introduce this class of potentially toxic organic acids into receiving waters (Monaghan et al., 2021).

Asphaltenes comprise a solubility class of complex molecules with diverse chemical composition and dynamic intra-molecular arrangements, making analysis very difficult (Scott et al., 2021). Although a few reports of asphaltene biodegradation have been published, almost all suffer from flaws in experimental design and/or rigorous analysis (Gray, 2021). Asphaltenes are the most recalcitrant of the oil classes, being too large to be transported across cell membranes for intracellular metabolism, and too chemically diverse to be susceptible to the types of substrate-specific enzymes that degrade saturates and aromatics. Because of their complexity and low aqueous solubility (i.e., bioavailability), they tend to be persistent but inert and of little toxicological concern. Small asphaltenic components such as some petroporphyrins (see Figure 2.5) are susceptible to limited oxidation by non-specific extracellular enzymes of soil fungi (reviewed by Hernández-Lopez et al., 2015), but such an attack usually only transforms the molecules or releases coordinated metals from the structures but does not fully mineralize them. The resistance of resins and asphaltenes to biodegradation is one of the reasons they persist over geological time and accumulate in heavy oils that are otherwise considered severely biodegraded (Prince and Walters, 2007). It is also why asphaltenes commonly are used in paving roads and making long-lasting roofing material.

Finally, toxic oil components can influence biodegradation potential, such as the presence of hydrogen sulfide (H_2S) in sour crude oils and naphthenic acids in heavy and ultra-heavy crude oils. Evaporation and dissolution in situ may remove these toxicants and enable subsequent biodegradation of the oil.

In summary, the overall biodegradability of a given oil and the sequence of its degradation, aside from any environmental considerations, is a function of the oil's chemical composition (e.g., the ratio of susceptible to recalcitrant or biodegradation-resistant constituents and the presence of toxic components); its bioavailability to microbes; and the composition of the microbial community attacking it. The biodegradation potential of oils and refined products then can be considered as a continuum corresponding to their chemical compositions, beginning with light oils including condensates, refined products like jet fuel, and West Texas intermediate (>~40° API), through medium oils like Alaska North Slope and North Sea oils, to shale oils and waxy crudes (~25–40° API) and to heavy and ultra-heavy oils like fuel oils and bitumen (<~25° API) (Lee et al., 2015). **The biodegradability of new classes of very low sulfur fuel oils and ultra low sulfur fuel oils has not yet been tested so their place on the biodegradability continuum is currently unknown.**

5.2.8.5 Biodegradation Kinetics: Measuring Rates of Biodegradation

An important metric describing observations of biodegradation is the rate at which oil components or pseudo-components are transformed by biological processes. As explained earlier, the term *biodegradation* is often misused to describe any biologically mediated transformation process that converts a given oil component into another compound (see Box 5.5). In this section and in much of the cited literature, biodegradation in the context of reaction kinetics usually refers to the transformation of a certain oil component. Socolofsky et al. (2019) review recent literature of biodegradation rate studies for oil and the approaches used in oil fates modeling. In these studies, quantification of biodegradation rates involves determining the reaction kinetics of biological transformation processes and determining their rate constants. In most studies, multiple chemical components are tracked within a whole oil, and the observed loss rate of each chemical is taken as the net transformation rate of that component. The transformation chain is not considered, and source terms for one component are not evaluated with respect to the loss terms of another.

Although some transformation reactions have been observed to be higher order, the most common kinetics law used in oil biodegradation analysis is first order. In this model, the loss rate of a given component is proportional to the concentration of that component in the environment, with the rate constant k being the proportionality constant. The solution to this model is an exponential decay, with a half-life equal to $-\ln(1/2)/k$. This model conforms well to our understanding of the mechanisms of biological transformation of dissolved chemical species, and the concentration used in the model would be the dissolved concentration of the chemical of interest.

When liquid petroleum is present as droplets or floating on the surface, the mechanisms of biodegradation and the appropriate concentration to use in the first-order rate law are less obvious. This is due to the heterogeneous nature of the oil–water mixture, as discussed in Section 5.2.2. Oil-degrading bacteria live in the water phase and primarily degrade oil by colonizing the oil–water interface, though they have also been found suspended within the liquid oil. In most biodegradation studies, the first-order rate equation for liquid petroleum is expressed as a function of the total mass of oil compounds in the system rather than the oil concentration (Socolofsky et al., 2019). This combines the liquid and dissolved petroleum into a single measure. This approach ignores the surface-area dependent nature of biodegradation of oil droplets (Brakstad et al., 2015; Thrift-Viveros et al., 2015; Wang et al., 2016). However, the method is convenient for models that track the whole oil, and biodegradation rate constants evaluated using this approach appear to have some consistency with measured rates across multiple studies (Socolofsky et al., 2019).

One phenomenon often observed in laboratory incubations is an initial lag period between the start of an experiment when excess petroleum is added to the system and the onset of rapid biological transformation. When evaluating transformation rate constants relevant to oil spills, this lag period is normally ignored and only the exponential phase of biodegradation is considered (Socolofsky et al., 2019). Lag periods can be variable for similar experiments. Hypotheses explaining the lag phase include (1) the lag period may be a recovery time associated with the stress experienced by the biological community due to changes in temperature, pressure, and light following field collection and subsequent use in the laboratory; and/or (2) the lag period may reflect the low initial concentrations of oil-degrading bacteria in a sample and the time needed for the bacterial community to assemble or adjust to an artificial oil input. This may be particularly true of anaerobic biodegradation (see Section 5.2.8.8). Certainly, community changes were observed in the field during the DWH oil spill (Valentine et al., 2010), but whether the time-scale of the lag period in the field is similar to those observed in the laboratory remains an open question. **Hence, there remains a need for field studies to determine the importance of and the mechanisms controlling lag periods of petroleum biodegradation in response to marine oil inputs (whether ongoing or episodic). These studies would need to span different ocean water depths and geographic regions to uncover the complicated roles of temperature, pressure, native community composition, and background state. Unless these lag periods are very short (hours to days) compared to typical biodegradation rates (days to weeks), they may be very important to the fate of oil released in the sea.**

Aside from the issue of using environmentally relevant concentrations, biodegradation rate studies lack a consensus on appropriate best practices (e.g., Prince et al., 2017). This can make it difficult to compare results among different studies or to identify appropriate biodegradation rates for a given spill scenario. Biodegradation rate studies can be done either in laboratory or in situ mesocosms (Socolofsky et al., 2019) or by evaluating the concentration field of natural or accidental releases (Thessen and North, 2017). Experiments may also be conducted for isolated cultures or for natural marine-water samples. Differences in the approaches used in biodegradation studies include how the oil–water system is stirred or agitated, how often stirring is conducted, what droplet sizes are present when liquid oil is involved, what temperatures and light conditions are used, and whether and how nutrients or oxygen are added to the system (Socolofsky et al., 2019). The type and frequency of stirring or agitation is particularly important in studies that include liquid oil because this may affect the oil droplet size, hence, the surface area of the oil–water interface. Rate constants are known to vary with temperature, incubation conditions, and oil constituents. An example of the variability of literature values for rate constants is given by Socolofsky et al. (2019), who report values for the biodegradation rate constants of extended SARA components for two different oils using a database of literature values for individual oil compounds. These are also compared to the rate constants used in two common oil spill trajectory models.

While the order of magnitudes of reported rate constants generally agree for a given pseudo-component, significant variability is present in these data, even for experiments conducted using similar conditions. Moreover, how to represent the surface area of suspended oil in a first-order biodegradation rate law remains unclear. The only known studies using the same oil and methods, including carefully controlled and measured droplet sizes (Brakstad et al., 2015a; Wang et al., 2016), show different trends with droplet size: Brakstad et al. (2015b) found faster degradation rate constants for fresh Macondo crude oil with smaller droplet sizes; Wang et al. (2016) reported faster rates for the larger droplet sizes (although, notably, the seawater inoculum differed in the two experiments; see Section 5.2.8.3). Further complicating the issue of estimating biodegradation rate constants are observations that oil droplet surfaces can become occluded by microbes and/or their polymers, mineral particles such as silt, and possibly photo-oxidation products, the combination of which influence degradation rates and are site- and event-specific. **Hence, it remains important to establish best practices for conducting degradation experiments with liquid petroleum, including the importance of controlled and measured droplet size, and to intercompare results among similar studies, conducted both by single groups and by multiple groups. These studies will help elucidate the mechanisms of biologically mediated transformation for liquid oil and allow better simulation of biodegradation processes for ocean oil spills.**

5.2.8.6 Biodegradation Changes the Chemical Composition of Oils

Selective biodegradation alters the chemical composition of the residual oil (described below) and therefore its physical properties (described in Section 5.2.8.7). Chemical changes include complete or incomplete removal of specific oil components and generation of partially oxidized metabolites that may continue to associate with the oil (see Box 5.5). As simpler components are removed the residual oil becomes enriched in resins and asphaltenes that are less biodegradable.

Because microbes most commonly attack saturates and aromatics, which are GC-amenable (see Box 2.2), gas chromatography is most commonly used to qualitatively or quantitatively assess the extent of biodegradation over time. There are caveats to interpretation of gas chromatograms, however (see Box 5.8).

Using Petroleum Biomarkers to Discern Changes to Oil Composition

Biodegradation targets some of the same small hydrocarbons as the abiotic weathering processes discussed earlier in this chapter, although rates and duration may differ. Careful selection and use of petroleum biomarkers intrinsic to the oil (see Box 2.1) that resist abiotic and biotic fates can allow discrimination between abiotic processes and biodegradation as oil weathers (Wang et al., 1998). The highly branched *iso*-alkanes pristane and phytane have been used widely because they are often prominent peaks in GC analysis of the saturates fraction (see Box 5.8). The peak area ratios of *n*-heptadecane (nC_{17}) to pristane or of *n*-octadecane (nC_{18}) to phytane (each

BOX 5.8
Using Gas Chromatography (GC) to Monitor Biodegradation, and Some Caveats for Interpretation

Microbial depletion of saturates and aromatics in oil routinely is observed by applying gas chromatographic methods such as gas chromatography plus mass spectrometry (GC-MS; see Section 2.1.7) to in situ and in vitro samples. The characteristic profiles of chromatographic peaks identified using mass spectrometry (see Figure 5.17) can be compared to the source oil, if available, for forensic purposes. For environmental monitoring purposes, profile changes over time or from different sites can be discerned by comparison to internal petroleum biomarkers such as the *iso*-alkanes pristane and phytane.

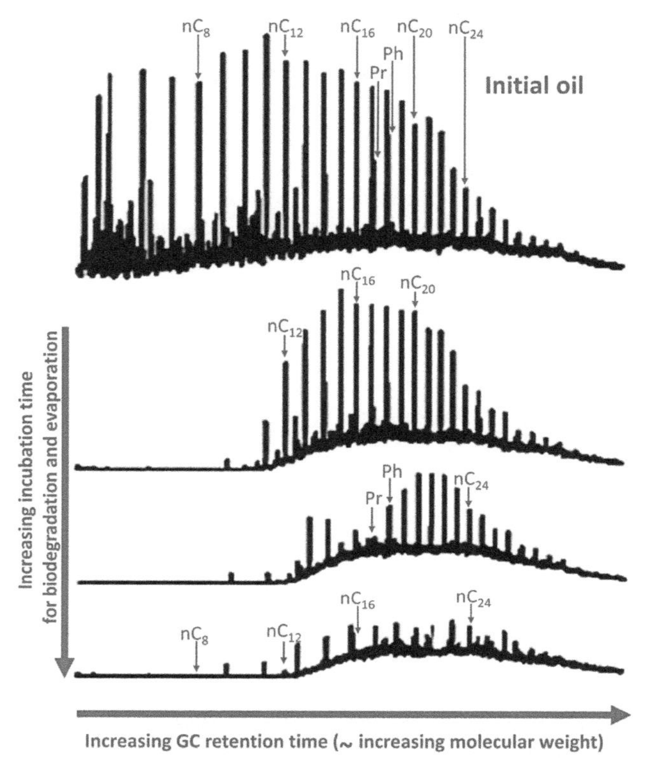

FIGURE 5.17 A time series of GC-MS chromatograms of Alaska North Slope crude oil incubated in open mesocosms in the laboratory and analyzed at intervals.

NOTES: Chromatograms are aligned vertically so that the relative height of each peak can be followed with time. Each peak represents a different oil component; most of the largest peaks are a regular series of *n*-alkanes increasing in carbon number with GC retention time; smaller peaks are *iso*- and cycloalkanes. Some *n*-alkane peaks are labeled for reference; Pr and Ph are pristane and phytane. The top panel shows the initial oil, slightly pre-weathered by evaporation. With increasing incubation time (bottom three panels) more peaks become smaller or disappear. Evaporation likely accounts for losses of the smallest peaks at the shortest sampling time, and biodegradation for subsequent losses; inclusion of a sterile control incubated under the same conditions would have discriminated between abiotic and biological removal. Peaks that remain after the longest incubation period represent compounds that may eventually biodegrade or may persist. Decreased GC peak area is commonly misinterpreted as demonstrating that the entire molecule has been removed from the sample. This is true for dissolution and evaporation processes, but not necessarily true for photo-oxidation or biodegradation. For the latter two oxidative events, all that is required for the peak to decrease or disappear from the chromatogram is the introduction of a functional group (transformation; see Box 5.5). Most or all of the carbon skeleton may remain either associated with the oil but no longer detectable using routine GC-MS, or dissolved in the water where it requires modified solvent extraction techniques and an additional chemical reaction ("derivatization") to be detected by GC. Thus, caution is required when using routine GC methods to infer the degree of loss of oil components. An in vitro alternative or adjunct to GC is the sensitive analysis of metabolites from individual hydrocarbons labelled with stable (^{13}C) or radioactive (^{14}C) isotopes, using methods described in Section 2.1.7.5. For example, measurement of $^{14}CO_2$ production ("radiorespirometry") indicates the extent of mineralization of the labeled hydrocarbon.

SOURCE: McFarlin et al. (2014). CC-BY 4.0. Significantly adapted by the committee.

pair being adjacent in the chromatogram and having similar volatility, to self-correct for evaporation losses) have been used for short-term incubations, for example, until the *n*-alkanes have been mostly depleted. However, both pristane and phytane may be biodegraded slowly or after a lag, confounding analysis in the later stages of incubation or with prolonged environmental oil exposure. The complex alicyclic saturates in the hopanoid and steroid classes (e.g., triterpanes and steranes, respectively) are more resistant to biodegradation and can be useful internal petroleum biomarkers even in severely degraded oil, even though some hopanes also are biodegradable (e.g., Bagby et al., 2016) and others are susceptible to photo-oxidation (Aeppli et al., 2014). In the aromatics fraction, multiply-alkylated monoaromatics (e.g., trimethylbenzene) and/or (alkyl) PAHs with more than three rings (e.g., benzanthracenes, chrysenes) or aromatic steroids (Yang et al., 2013) tend to resist biodegradation and may be suitable markers.

In laboratory and ex situ samples analytical chemical markers can be added to the experimental oil sample before analysis. Such chemicals typically do not exist naturally in the oil or are present in the oil in very small amounts but behave similarly to natural oil components during extraction and chromatography. These markers can be added before incubation in an experimental system ("exogenous internal standards," which must resist biodegradation), or after incubation but before solvent extraction ("surrogate standards," which must have solvent partitioning properties similar to the hydrocarbons of interest), or before GC analysis ("external standards," which must have chromatographic properties similar to the analyte hydrocarbons). A few examples include squalane, chrysene, *o*-terphenyl, hexamethylbenzene, thianthrene, and hydrocarbons enriched in stable isotopes (e.g., multiply deuterated phenanthrene). Inclusion of multiple markers, when possible, enables more rigorous interpretation of GC results.

5.2.8.7 Biodegradation Changes the Physical Properties and Behavior of Oils

Changes to the chemical composition of oil due to biodegradation, discussed in Section 5.2.8.6, affect the physical properties and behavior of the oil. Analogous to evaporation, depletion of low-molecular-weight hydrocarbons during the initial stages of bioremediation increases oil viscosity, which reduces surface spreading and dispersion; increases residual oil density, which decreases buoyancy; and may increase the "stickiness" of oil as the resins and asphaltene proportions increase, which may promote formation of tar balls, etc. (see Sections 5.3.2 and 5.3.4).

With further biodegradation, interfacial effects also occur, affecting dispersion and emulsification. In the water column, association of microbes and their extracellular polymers can produce visible string-like threads of oil and/or affect the droplet size of dispersed oil in various ways. First, hydrocarbon-degrading species can adjust their cell surface to become more hydrophobic (Dorobantu et al., 2008; Godfrin et al., 2018), allowing them to physically associate with oil droplets or slicks at the oil-water interface (Abbasnezhad et al., 2011). This produces a biofilm on the droplets that alters the mechanical properties of the interface (Kang et al., 2008a; Omarova et al., 2019) and stabilizes dispersions by impeding coalescence of droplets Figure 5.18 (top). A second microbial strategy is to produce cell-associated or excreted biosurfactants that increase dispersion of oil as very fine droplets, thereby increasing hydrocarbon bioavailability through increased surface area (Ron and Rosenberg, 2002; Van Hamme et al., 2003; Quigg et al., 2021a). Even in the absence of biosurfactants or added chemical dispersants, the cells themselves can enhance and stabilize oil-in-water and water-in-oil emulsions (Dorobantu et al., 2004; Kang et al., 2008b) (see Figure 5.18 [bottom]), although biodegradation of oil in stable emulsions ("mousse") may be limited by poor accessibility of microbes to oil in the interior of the mousse. For example, limited biodegradation was noted by Liu et al. (2012) when characterizing *n*-alkane and alkyl-PAH profiles in oil mousse samples from surface waters of the DWH spill.

5.2.8.8 General Environmental Factors Affecting Biodegradation

Numerous environmental factors affect whether and how quickly biodegradation brings about the physical and chemical changes to oil described above. These include aerobic versus anaerobic conditions, nutrient supply, salinity, temperature, hydrostatic pressure, toxicity, and predation, discussed below.

Aerobic Versus Anaerobic Biodegradation

Higher life forms exclusively use O_2 to respire aerobically. Many microbes likewise are restricted to using O_2, but others respire using dissolved inorganic molecules such as nitrate, sulfate, iron, or CO_2, or ferment using organic compounds such as simple organic acids, alcohols, and/or sugars. Some facultative species readily switch among these types of metabolism depending on the available electron acceptor(s). Less chemical energy is available from anaerobic processes, so O_2 is usually preferred for metabolism, if tolerated, but strictly anaerobic microbes can be killed even by minute concentrations of O_2.

Aerobic and anaerobic hydrocarbon biodegradation are biochemically dissimilar processes, involving different microbes, pathways, metabolites, end products, kinetics, and somewhat different substrate ranges and preferences. Both processes involve a specific substrate undergoing sequential enzymatic steps in a coordinated pathway that, ideally, mineralizes the substrate. This is the biochemical equivalent of combustion, with the added parallel of occurring more efficiently in the presence of molecular oxygen. (Of course, a major difference between biodegradation and combustion is that, during biodegradation, a percentage of the substrate

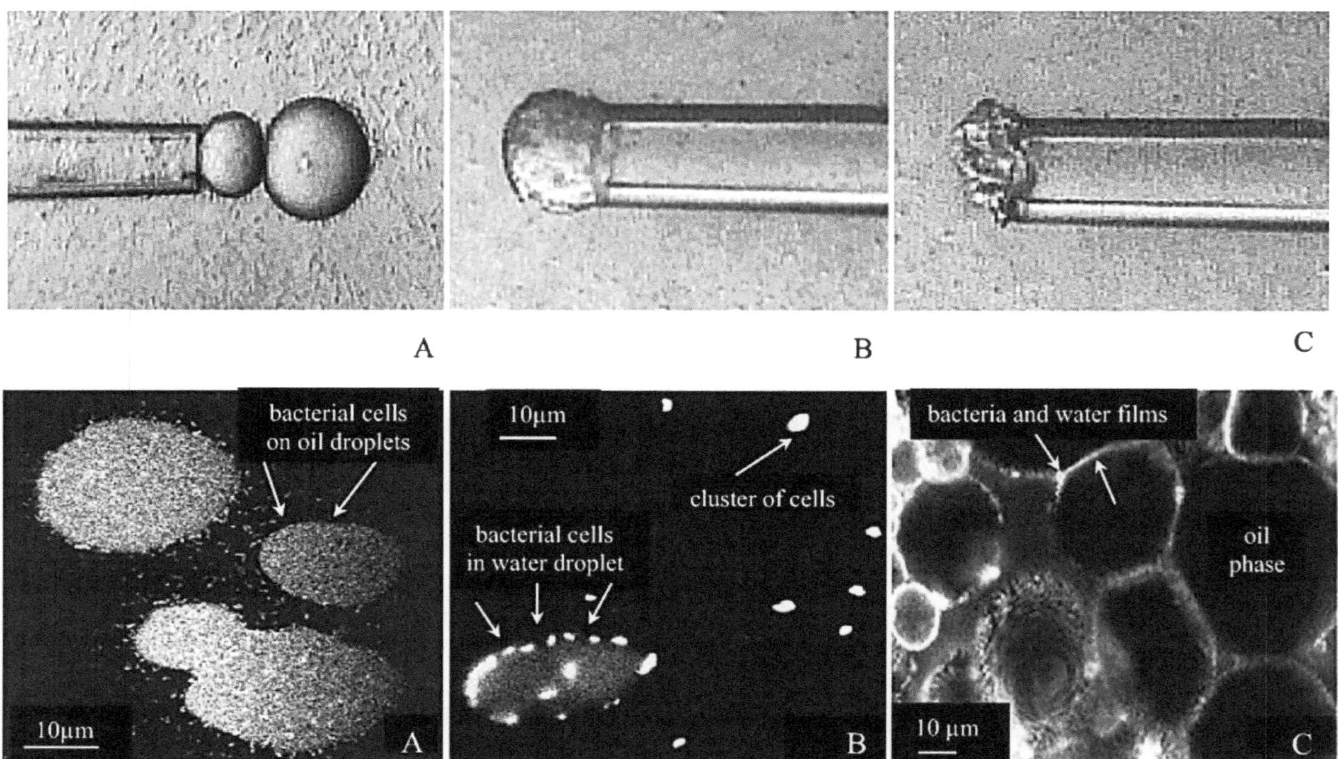

FIGURE 5.18 Localization to and assembly of microbes at hydrocarbon:water interfaces. Top panel: (A) Micrograph of two *n*-hexadecane droplets being extruded from a micropipette (tip diameter 14 μm) into an aqueous suspension of bacterial cells (small dots, each ~1 μm). The droplets fail to coalesce because cells adhering at the oil–water interface serve as steric barriers. (B) and (C): As an extruded droplet is partially withdrawn into the micropipette, its surface deforms and wrinkles due to attached bacterial cells changing the oil surface rheological properties. Bottom panel: Confocal micrographs of emulsions generated using *n*-hexadecane and aqueous suspensions of bacterial cells. (A) Oil-in-water emulsion of bacterial cells, the small white particles suspended in the dark aqueous solution and covering the outer surface of *n*-hexadecane droplets; (B) water-in-oil emulsion with bacteria (fluorescent particles) associating with the inner surface of an aqueous droplet (gray) at the interface of a continuous *n*-hexadecane phase (black); (C): a stable "emulsion gel" showing bacteria concentrated at the aqueous interfaces surrounding dark *n*-hexadecane droplets, stabilizing the emulsion by decreasing coalescence.
SOURCE: All images from Dorobantu et al. (2004).

is used to synthesize new macromolecules and biomass by using chemical energy for metabolism, with less energy lost as heat than in combustion.)

Aerobic hydrocarbon biodegradation occurs in oxygenated water, the uppermost layer of subsurface sediments, and intertidal beach sediments. Diffusion-limited seafloor sediments, wetland, estuarine, and fine-grained beach sediments and muddy tidal flats that are continuously anaerobic support various anaerobic processes. Because seawater has a relatively high concentration of dissolved sulfate, sulfate reduction is common in anaerobic marine environments, generating end products including hydrogen sulfide (H_2S), hydrosulfide (HS^-), elemental sulfur (S°), pyrites, and/or metal sulfide precipitates. Where sulfate is depleted, the energetically less-favorable processes of fermentation and methanogenesis can occur.

Aerobic microbes commonly degrade hydrocarbons without a partner species, and some exclusively metabolize hydrocarbons (obligate hydrocarbon-degrading bacteria; see Appendix I). In contrast, anaerobic hydrocarbon biodegradation, which currently is far less well understood, is often achieved through "distributed metabolism" (see Box 5.7) involving two or more species that are biochemically and/or physically linked, so as to overcome marginally favorable thermodynamic reactions. Aerobic hydrocarbon degradation is considered to be more rapid than anaerobic, and the known substrate range of biodegradable hydrocarbons is considerably larger under aerobic conditions. These factors are important for oil in the sea because it is relatively easy and fast for free-living aerobic hydrocarbon-degraders to begin metabolizing oil near the ocean surface within hours or days, but it may take weeks, months, or years for an anaerobic community to recruit and assemble suitable partners (physically and/or biochemically), then finally begin to biodegrade oil in anaerobic sediments, particularly when limited to fermentation and methane production. Lag times of several years have been observed for methane production from alkanes and aromatics in vitro (Siddique et al., 2015) but are unexplored in situ.

Degradation of methane provides a good example of contrasting microbial strategies: aerobic methanotrophs (not to be confused with methanogens that are anaerobic methane producers) that are commonly detected near natural gas seeps consume methane without requiring other microbial partners. Conversely, anaerobic methane consumption commonly requires a slow-growing syntrophic partnership between two anaerobic members, for example, an anaerobic methane-oxidizing archaeon in close physical contact with a sulfate-reducing bacterium, forming a microscopic consortium (reviewed by Knittel and Boetius, 2009). These consortia tend to live in anaerobic sediments near natural seeps in the transition zone where the chemical gradients of methane gas and dissolved sulfate overlap (Knittel et al., 2005). Despite the extremely small energy yield of anaerobic methane oxidation, this activity in marine sediments exerts a major control over global methane flux from the ocean to the atmosphere (Reeburgh, 2007). Larger gaseous alkanes (ethane, propane, butane) released from subsurface gas seeps can be degraded aerobically by microbes other than methanotrophs. For example, Valentine et al. (2010) deduced that respiration of ethane and propane was an early driver of microbial response to the DWH event, and subsequent studies have suggested that *Cycloclasticus*, a bacterial genus previously thought to be an obligate PAH-degrader, may have utilized ethane, butane, and propane during active gas release early in the spill, then shifted to PAH degradation after the gas release ended (Rubin-Blum et al., 2017). Anaerobic degradation of non-methane gaseous hydrocarbons by sulfate-reducing marine bacteria has been observed both with (Singh et al., 2017; Chen et al., 2019) and without an archaeal partner (Kniemeyer et al., 2007).

In contrast to the special cases of gaseous hydrocarbons, biodegradation of liquid oil components is a widespread and well-studied phenomenon in marine systems, and metabolism of saturates and aromatics by diverse species has been known for decades (Van Hamme et al., 2003). The well-described aerobic biochemical pathways for saturates and aromatics differ. *n*-Alkanes typically are oxidized via beta-oxidation using pathways analogous to lipid metabolism, with fatty acids and alcohols as transient intermediates (Rojo, 2009). Neither *n*-alkanes nor their aerobic metabolites tend to accumulate in the environment or contribute to toxicity, and even highly branched pristane and phytane can be degraded by the ubiquitous species *Alcanivorax borkumensis* (Gregson et al., 2019) or depleted via co-metabolism (see Box 5.7) with long-chain *n*-alkanes (Deppe et al., 2005). Other *iso*-alkanes and cycloalkanes are less readily degradable, likely due to enzyme specificity, and suites of partially oxidized organic acids that fit the chemical definition of potentially toxic naphthenic acids (Clemente and Fedorak, 2005) may be produced and persist either in the oil or dissolved in surrounding water. In anaerobic systems fewer alkanes have been shown to biodegrade, the pathways are poorly known, and metabolites have proven difficult to identify and isolate (Mbadinga et al., 2011). Notably, anaerobic hydrocarbon research primarily has been done in terrestrial systems rather than marine sediments, but the processes are likely analogous. The anaerobic pathway for saturates requires activating the molecule via enzymatic addition of fumarate followed by beta-oxidation and central metabolism (Callaghan, 2013; Abu Laban et al., 2015; Jaekel et al., 2015). Under anaerobic but non-methanogenic conditions (e.g., sulfate reduction), a limited group of bacterial species is known to degrade alkanes without a partner. Until recently, hydrocarbon biodegradation under methanogenic conditions was thought to require the syntrophic association of a methanogenic archaeon with one of several bacterial species (reviewed by Gieg et al., 2014) as a complicated example of distributed metabolism (see Box 5.7). However, a methanogenic archaeon now has been inferred to oxidize long-chain alkanes and alkyl side-chains without a partner (Zhou et al., 2022).

Aromatic hydrocarbons are more chemically reactive than saturates, and numerous aromatics are mineralized or transformed aerobically by bacteria and eukaryotes (described below). Bacterial aerobic pathways for oxidizing (alkyl) monoaromatics and (alkyl)PAH are lengthy, sometimes requiring ≥14 enzymes (Elyamine et al., 2021) and usually involve addition of one or two oxygen atoms to the aromatic ring or the alkyl side chain, followed by ring opening and further degradation to alcohols, aldehydes, and carboxylic acids that can be funneled into central metabolic pathways (Jindrova et al., 2002; Haritash and Kaushik, 2009; Ghosal et al., 2016). However, the accumulation of polycyclic aromatic acids from partial biodegradation of alkyl-PAH may occur, with lower molecular weight metabolites (e.g., naphthoic acids) being transient but larger metabolites (e.g., fluorene carboxylic acid) persisting longer, especially if dissolved oxygen is limiting (Kristensen et al., 2021). The toxicity of such mixtures of aromatic acids to marine fauna is not yet known, as dilution in the water column or distributed metabolism may reduce any potential effects. In contrast to aerobic attacks, the anaerobic susceptibility of numerous alkyl-PAH is not yet fully known (Foght, 2008), nor are the substrate ranges for different types of anaerobic communities, although new 'omics techniques are yielding some insights (Laczi et al., 2020). Some bacterial anaerobic pathways for aromatics require initial enzymatic activation via addition of fumarate, methyl, hydroxyl, or carboxyl groups prior to ring opening and/or saturation. Anaerobic depletion of alkanes, low-molecular-weight PAHs, and high-molecular-weight alkyl-PAHs in situ recently has been inferred from Gulf of Mexico deep-sea sediments (Bagby et al., 2016; Shin et al., 2019a), although the completeness of the oxidation is not known. Novel sulfate-reducing PAH-degrading bacteria and distinct metabolic pathways were discovered after the DWH spill (Shin et al., 2019a), indicating that we have yet to fully catalog the diversity of anaerobic oil degradation in the environment. **Anaerobic biodegradation of hydrocarbons is not yet fully understood, and although the range of substrates is believed to be much narrower and more selective than for aerobic attack, general rules for anaerobic susceptibility have not yet been described. This hampers understanding and prediction of ecosystem recovery from oil that impacts seafloor sediments.**

Eukaryotes, including some fungi and the organs of some animals, can biodegrade or transform hydrocarbons, exclusively aerobically. The biochemical pathways differ somewhat between prokaryotes and eukaryotes, with the former primarily being used for mineralization and cell growth and the latter being employed for detoxification purposes. That is, prokaryotic pathways have evolved to achieve maximum oxidation of the substrate for carbon and energy, whereas most eukaryotes only transform the hydrocarbon enough that it is less toxic and/or can be excreted because it has been made more water-soluble; alkane- and PAH-utilizing fungi are a notable exception. Importantly, some eukaryotic processes inadvertently increase hydrocarbon toxicity and carcinogenicity. The first steps of aerobic PAH oxidation are a good example. Bacterial enzymes commonly initiate aromatic degradation by introducing both atoms of O_2 into the PAH structure via dioxygenase enzymes, some of which are quite substrate-specific and define the range of PAHs that each species can grow on. The products are funneled into central metabolism to be mineralized. Eukaryotes, in contrast, often use non-specific peroxidases or mono-oxygenases like P_{450} cytochromes (CYP1A; see Section 6.3.4.2) to oxidize hydrocarbons as well as other chemical classes including steroids, drugs, and carcinogens. Another eukaryotic option is glycosylation that chemically marks the compound for active excretion. Unfortunately, some transformed products have greater cell toxicity and/or carcinogenicity than the parent hydrocarbon and may accumulate as "dead-end" products that subsequently cause mutations or cell toxicity (Guengerich, 2008).

In summary, the ultimate outcome of combined aerobic and anaerobic oil biodegradation is the removal of most *n*-alkanes and monoaromatics and the depletion of selected *iso*- and cycloalkanes, PAHs up to five rings including certain alkyl-PAH isomers, and N-, S- and O-substituted aromatic resins (PACs). Bacterial attack with or without oxygen typically results in mineralization or substantial transformation of susceptible oil components to simple gases, biomass, and relatively innocuous water-soluble pathway products and may render the residual oil less acutely toxic. Aerobic eukaryotic metabolism may deplete some alkanes, mono-aromatics, and PAHs, with production and excretion of potentially toxic and/or carcinogenic products that can pass up the food chain.

Nutrient Availability

Efficient biodegradation requires sufficient nutrients to balance the high carbon content of oil (reviewed by Vergeynst et al., 2018). Water-soluble forms of nitrogen and phosphate are the most frequently measured bioavailable nutrients assessed for oil degradation and the Redfield ratio of 106:16:1 C:N:P (Anon, 2014) historically was used as a rule of thumb to estimate the N and P demand for efficient attenuation of oil spills (see Section 4.2.3.1 for discussion of bioremediation). Concentrations of dissolved nitrate, nitrite, ammonium and phosphate from terrestrial runoff may be adequate for oil biodegradation in near-shore waters and shorelines, but when these nutrients are sequestered into biomass (e.g., blooms of marine snow; see Section 5.3.2.2) they become less bioavailable, and dissolved concentrations may become limiting. Nitrogen compounds can be continuously supplied by some species of marine bacteria, particularly bacterioplankton that fix N_2 gas, and by some N_2-fixing heterotrophs that also degrade hydrocarbons (although seldom simultaneously; Foght, 2018). For example, N_2 fixation by suspended particles in the Gulf of Mexico was substantial during the first few months after the DWH spill in parallel with methane consumption (Fernandez-Carrera et al., 2016), but it is not clear whether the same organisms were responsible for both activities simultaneously, or whether N_2 fixation occurred aerobically (e.g., by cyanobacteria) or anaerobically (e.g., by consortia of archaeal methanotrophs and bacterial sulfate-reducers). In anaerobic seafloor sediments microbial nitrogen-cycling has been linked to PAH exposure (Scott et al., 2014). Thus, nutrient demand for oil degradation may be met by natural sources, and their concentrations generally are adequate, although not necessarily optimal, for hydrocarbon biodegradation in most marine environments.

Addition of exogenous nutrients (biostimulation; see Section 4.2.4) has long been known to enhance microbial activity and diversity in the presence of oil, as observed by in situ and in vitro experimentation, for example, in samples of Arctic and Gulf of Mexico surface waters (Sun and Kostka, 2019) and on beaches with sediment–oil aggregates (Shin et al., 2019b). Such amendment is justified only where permitted in locations where nutrients are limiting and amendments will not be diluted or washed away. Unfortunately, laboratory studies often provide nutrients at concentrations many-fold greater than the micromolar concentrations commonly available in situ, and therefore may generate unrealistic observations of activity and community response.

Trace elements including iron, copper, nickel, and rare earth elements (Shiller et al., 2017; Mehaja et al., 2019) may also be necessary for optimum biodegradation, functioning as co-factors for hydrocarbon-oxidizing enzymes.

Salinity

The salinity of seawater is sufficient to restrict microbial growth to species that are halotolerant, yet significant oil biodegradation is accomplished in the sea by microbes adapted to marine salinity. Higher salinity due to evaporation may occur in warm shallow coastal waters, wetlands, and beaches where it not only affects the composition of the microbiota but decreases water-solubility of some oil components, particularly PAHs, and limits biodegradation (Geng et al., 2021). Elango et al. (2014) determined that hypersaline conditions in the intertidal zone limited biodegradation of stranded tar balls, and Chen et al. (2010) likewise found that increasing salinity in mangrove sediments (from 5 to 25 ppt) decreased PAH biodegradation. Salinity may also be a factor in Arctic sea ice, where brine channels containing oil and microbes have higher salinity than the ice (see Section 5.3.5.3). Salinity gradients found in estuaries and some continental shelf environments can influence petroleum compound solubilities (see Chapter 2, Section 2.1.5). This could interact with the aforementioned microbial species responses to salinity differences and have an overall influence on microbial degradation of petroleum compounds.

Temperature

Temperature is a factor in all microbial activities but distinct microbial communities are selected by and adapt to ambient temperatures where they perform optimally. They inhabit subsea hydrothermal vents, hot tropical beaches, temperate waters and wetlands, constantly cold deep-sea waters and sediments, and sub-zero polar sea ice, the latter being discussed in detail in Section 5.3.5. Biodegradation typically occurs over a range of ~30°C for individual species, with different communities prevailing over a sliding range from sub-zero to ≥60°C, given accompanying suitable conditions (reviewed by Margesin and Schinner, 2001).

Pressure

Hydrostatic pressure is an environmental co-stressor with low temperature in deep-sea water and sediments impacted by oil (reviewed by Louvado et al., 2015). Oil biodegradation was observed in the DWH dispersed plume at ~5°C and depths >1,000 m (Hazen et al., 2010) and oil biodegradation was inferred in seafloor sediments at ~2,000 m (Bagby et al., 2016). However, quantifying the effects of pressure on oil biodegradation is difficult because of the logistical challenges of sampling in situ and technical challenges of handling samples in the laboratory when simulating deep-sea conditions. This has hampered in situ and in vitro study of oil biodegradation, and our current understanding is very limited. Laboratory studies at high pressure and low temperature have varied considerably in their methodology, including the hydrostatic pressure applied (6–50 Mpa [megaPascal]) where 0.1 MPa is atmospheric pressure and 15 MPa corresponds to ~1,500 m depth). Some studies have used pure isolates or near-surface communities not previously adapted to pressure (e.g., Schwarz et al., 1975; Schedler et al., 2014; Scoma et al., 2016a), calling into question the relevance of the observations. Others have used piezotolerant microbial communities from geographically diverse deep waters (e.g., Prince et al., 2016a; Calderon et al., 2018; Marietou et al., 2018) or sediments (Calderon et al., 2018; Nguyen et al., 2018; Noirungsee et al., 2020) that necessarily were depressurized during sampling and manipulation, with unknown effects on cell viability before being repressurized. Some studies have amended the samples with pure hydrocarbons (Schedler et al., 2014; Scoma et al., 2016a) or a mixture of pure hydrocarbons (Calderon et al., 2018) whereas others used crude oil, either with (Prince et al., 2016a; Noirungsee et al., 2020) or without dispersant (Nguyen et al., 2018). Some studies have supplemented the cultures with nutrients (Schedler et al., 2014) and others have not (Barbato and Scoma, 2020); this is relevant because starving microbes may be more sensitive to oil and/or pressure.

Given the diverse parameters of these disparate studies, it is difficult to draw broad conclusions, but in general moderate pressure (6–15 Mpa) results in modest inhibition of oil biodegradation under laboratory conditions (Prince et al., 2016a; Nguyen et al., 2018), with different hydrocarbon classes possibly being more susceptible than others. Microbial community composition changes with pressure but the significance is not fully known. Notably, none of the studies have examined anaerobic biodegradation, which is most relevant in deep-sea sediments. Furthermore, although the solubility of hydrocarbons in liquid oil exhibits little variability with pressure, modeling predicts that high pressure may slightly increase water solubility, potentially increasing bioavailability and biodegradation. This contrasts with modeling of pure solid PAHs (not dissolved in oil) whose solubility decreases with increasing pressure (Oliveira et al., 2009). Unfortunately, measured effects of pressure on hydrocarbon solubility in seawater have not been reported in the literature, especially in combination with low temperatures experienced in the deep sea (Louvado et al., 2015). Understanding the synergistic effects of pressure, temperature, nutrients, oil type, and presence of dispersant will require improved experimental design using combinations of in situ and in vitro observations (Scoma et al., 2016b). **Numerous gaps exist in understanding the effects of deep-sea conditions on aerobic and anaerobic oil biodegradation. The enormous volumes of deep waters and large areas of deep-sea sediments that may be affected by oil from natural sources and spills warrants the technical efforts that high-pressure studies demand.**

Predation

Predation is a factor in all marine environments, affecting microbial communities and therefore oil biodegradation. Marine oil snow, for example, is a hotspot of microbial activity and therefore also concentrates microscopic predators (e.g., zooplankton) and bacterial viruses (bacteriophage; Suttle, 2007). The effects of predation, whether negative due to mortality or positive due to nutrient cycling, have not been rigorously studied, in part because they cannot be controlled during in situ experiments or faithfully replicated in vitro.

To summarize Section 5.2.8.8, no single environmental factor controls oil biodegradation in the sea. Rather, unique combinations of factors that reflect the specific environment and abiotic weathering progression, in concert with the microbial community composition and oil characteristics, influence the ultimate rate and extent of oil biodegradation in the sea.

5.2.9 Examples of Oil Spill Budgets

As demonstrated in Sections 5.2.1–5.2.8, oil in the sea experiences multiple fates depending on its composition, the site, prevailing conditions, and timeline. Section 5.2.9 describes how the various fates can be assigned proportions during a spill. This bridges Section 5.2, which describes fundamental processes affecting oil fate irrespective of the site, and Section 5.3 that describes oil fates in specific marine environments. At a fundamental level, an oil spill budget reports the amount of spilled oil entering different environmental compartments (e.g., water column, sediments, sea surface, atmosphere, etc.) and being removed or altered by different fate processes (e.g., dissolution, biodegradation, volatilization, etc.). Spill budgets are developed for different purposes at different times, each dependent on different amounts and quality of data. The definition of the various compartments and processes should be suited to the purpose of the budget. Early during a spill, the main purpose of a spill budget is to direct the response and inform the public and decision makes. Following a spill, budgets are determined through the NRDA process as part of the penalty assessment. During and following a spill, detailed scientific budgets may also be developed to support research. Here, we give a few examples of spill budgets, their uses, uncertainties, and variability.

5.2.9.1 Caveats on Oil Budgets

During an oil spill response the Unified Command, stakeholders, and general public require reporting on the mass balance/oil budget from mitigation activities and natural processes. These spill accounting volumes help estimate the percentage of cleanup completed and how much is left to respond to. It helps measure the pace of response activities and gives a sense of how long the incident will take to resolve. Under the Incident Command System, Form 209 collects these data. Amounts entered include volume spilled/released, recovered oil, evaporation/airborne, natural dispersion, chemical dispersion, burned, floating contained, floating uncontained, onshore, and total oil accounted for. As will be seen in the discussion below, each of these volume categories has a wide range of potential values, so the oil budget is never an absolute single number, but generally represented by a range of values.

Oil budget calculations for a particular incident may include:

- initial amount and type of oil released (if known);
- amounts remaining as floating, evaporated, naturally dispersed, and/or beached oil;
- amounts removed—by collecting at the source, skimming oil from the water, chemically dispersing the oil (if used), and/or in situ burning (if used); and
- amount of oil recovered from the shoreline.

As described in Chapter 2, oil is a mixture of many different chemicals, with varying toxicity and behavior in the environment which can affect the behavior of the oil itself. When oil spills, several factors determine how the oil will behave, how the oil will interact with the environments, and the eventual fate of the oil:

- The volatility of the oil components determine how quickly the oil will evaporate when spilled.
- Oil components that are more soluble in seawater will partition more to the water phase.
- The biodegradability of the oil will determine how easily the oil is broken down by microbes.
- Sunlight can transform oil components.
- Chemical changes in the oil affect the physical features like density and viscosity.

These physical and biological processes results in what is referred to as the "weathering" of the oil. Weathering describes changes in the oil's chemical composition and physical characteristics over time.

While laboratory studies of oil weathering processes can be rigorously performed, studied, and measured to very precise percentages of the different weathering compartments, extrapolating controlled laboratory measurements to field conditions is not possible to the same degree of precision due to the heterogeneity of the ocean and degree of uncertainty of each of the input categories. In a laboratory the initial conditions (input quantity, type of oil, temperature, agitation, etc.) can be strictly controlled and measured. These data can be incorporated into mathematical algorithms that are used in computer models to estimate weathering in the field. The programs may provide a "best guess" answer and are designed to run on as little information as possible, especially using the type of information that can be estimated quickly or obtained in the field. The programs may incorporate environmental properties such as salinity and temperature of the water, wind speed, and/or wave height. Spill properties include the type of oil spilled along with the rate and duration of the release. Spill properties may be very uncertain, especially early during a spill. Hence, it is valuable to remember the general rule of thumb with any computer modeling program: "Garbage in = Garbage out." Hence, the greater the uncertainty of your data inputs, the greater the uncertainty in your model outputs.

For many incidents it is very difficult to measure the exact amount of oil initially released; this estimate affects subsequent calculations. The composition of the spilled fluids may also be uncertain. Thereafter, one of the greatest uncertainty inputs for calculating an oil budget lies in the weather conditions and forecasts, both in time and space. Whereas, in a laboratory you can control the wind speed, duration, and temperature, in the field these are changing moment to moment and from location to location.

Due to the heterogeneity of environmental conditions and oil properties, some portions of the slick will form a sheen, naturally disperse and evaporate, and some portions will stay thick and emulsify. Ocean circulation may concentrate some of it and dissipate other areas. Mass balance calculations would ideally identify 100% of the oil spilled. In fact, some processes (physical recovery and biodegradation) eliminate oil, and some others—such as emulsification—increase the volume of the slick floating on the surface, which is the oil that is being observed, measured, and recovered. Therefore, the volume of the petroleum product is constantly changing in both competing directions, as it incorporates water or loses components, resulting in the slick areas varying significantly. The amount of oil stranded on a shoreline is also extremely varied in thickness (quantity), and calculations attempt to average out these volumes over the distance covered.

In general, evaporation is very well understood and can be fairly well extrapolated to field conditions. Dissolution is next in line with a fairly high degree of confidence from lab studies as it is related to the amount of "light ends" (low-molecular-weight components) in the parent oil, as is evaporation. In the subsurface the problem increases three-dimensionally as the currents are not well characterized and we are not able to directly see the oil so that it is difficult to track, and there is also the difficulty of predicting the droplet size distribution. Estimates of natural dispersion and droplet formation are very uncertain as they depend significantly on the energy from wind and waves on the surface being transferred to the floating oil. Although natural dispersion and droplet formation are mechanisms that remove oil from the water's surface, this is not truly a weathering process in terms of any chemical transformation—it is more of a transport mechanism with the oil remaining physically the same, just in smaller droplets. Additionally, naturally dispersed oil droplets may recoalesce at the water's surface when wind and turbulence decrease. Emulsion formation prediction is probably the least reliable calculation as it too is related to the type of parent oil and the physical energy it encounters in the field, with the physical mechanism of action greatly different from laboratory waves to real ocean waves. Running many statistical analyses to compute the uncertainties and reduce them with comparisons to real-world observations is one method to help produce a "best guess" estimate to within an order of magnitude.

Oil budgets are produced by experts in their fields using the best available data for input into calculations and models. This is followed by a process of comparing the calculated results with observable situations and finding approximate fits to the solutions. There is almost always a spread in the results and through a process of informed negotiations and best professional judgement consensus on the closest approximations of the range of results is issued. **Oil budgets should be viewed as scientifically derived qualitative information rather than precise numbers** for the oil volume partitioning between different compartments. There are many challenges described in this report related to the estimation of the released volume, oil volume present on the water surface, as well as estimations of weathered and recovered oil volumes sited in the mass balance table. The focus on these numbers without proper understanding of associated uncertainties has created issues, misunderstandings, and delays during past responses and exercises. The response community is encouraged to describe a changing spill situation by focusing on the objective and verifiable numbers important for response and impact assessment. For example, surface area affected by the spill that can be estimated and documented using remote sensing techniques or the length of the affected shoreline that could be estimated and documented by SCAT programs assisted with remote sensing if needed.

One example of a computer system to estimate oil movement and weathering is NOAA's WebGNOME trajectory modeling software (see Figure 5.19). ADIOS2® (Automated Data Inquiry for Oil Spills) is the oil weathering model component. It is an oil spill response tool that models how different types of oils weather (undergo physical and compositional changes) and calculates the amount of evaporation, floating, dispersion, sedimentation, water content, density, and viscosity of the oil in the marine environment. Working from a database incorporating laboratory data of more than a thousand different crude oils and refined products and given an estimate of the composition of a spilled oil, ADIOS2 quickly estimates the expected characteristics and behavior of spilled oil (NOAA, 2021c).

Properties compiled in the database include the density, viscosity, and water content of an oil or refined product, and the rates at which it evaporates from the sea surface, disperses into the water column, and forms oil droplets that become emulsified, or suspended, in the water (NOAA, 2021c). The database was compiled from a variety of sources, including Environment and Climate Change Canada, the U.S. Department of Energy, and industry. The predictions are designed to help decision-makers develop cleanup strategies. All of the outputs need to be understood with the limitations to precision and the range of variability there may be from the low to high estimates. Quoting from the WebGNOME manual, "Uncertainty is the only certainty there is."

5.2.9.2 Exxon Valdez *Oil Spill Budget*

From the report by Wolfe et al. (1994) (see Figure 5.20), it is estimated that over time by day 1,000+:

- 11% of spilled oil was estimated to remain in the sediment
- 5% had been stranded on beaches
- 12% had been mechanically recovered
- 2% was dispersed in the water column
- 50% had been biodegraded (to an unknown degree)
- 20% had evaporated or photo-oxidized

It can be seen in the graph that over time the estimated percentages of each compartment varied. *Evaporated and photolysis products* were a continuing process throughout the incident. *Floating* oil was gone about two months after the initial release. Oil *recovered or disposed of* began soon after the release and grew throughout the response. *Biodegradation products* started within the first month and continued throughout. Naturally *dispersed* oil peaked in the first month and then waned. Oil incorporation into the *sediments* occurred after a few months and then continued. The graph does not show any ranges in the values on the estimates for each compartment that do appear in the text of the report.

5.2.9.3 Deepwater Horizon *Oil Spill Budget*

These estimates were supplied as part of the oil budget: Oil Budget Calculator Technical Documentation (2010), a peer-reviewed report over 200 pages that presented the formulas used and updated the percentages in the original budget (see Figure 5.21).

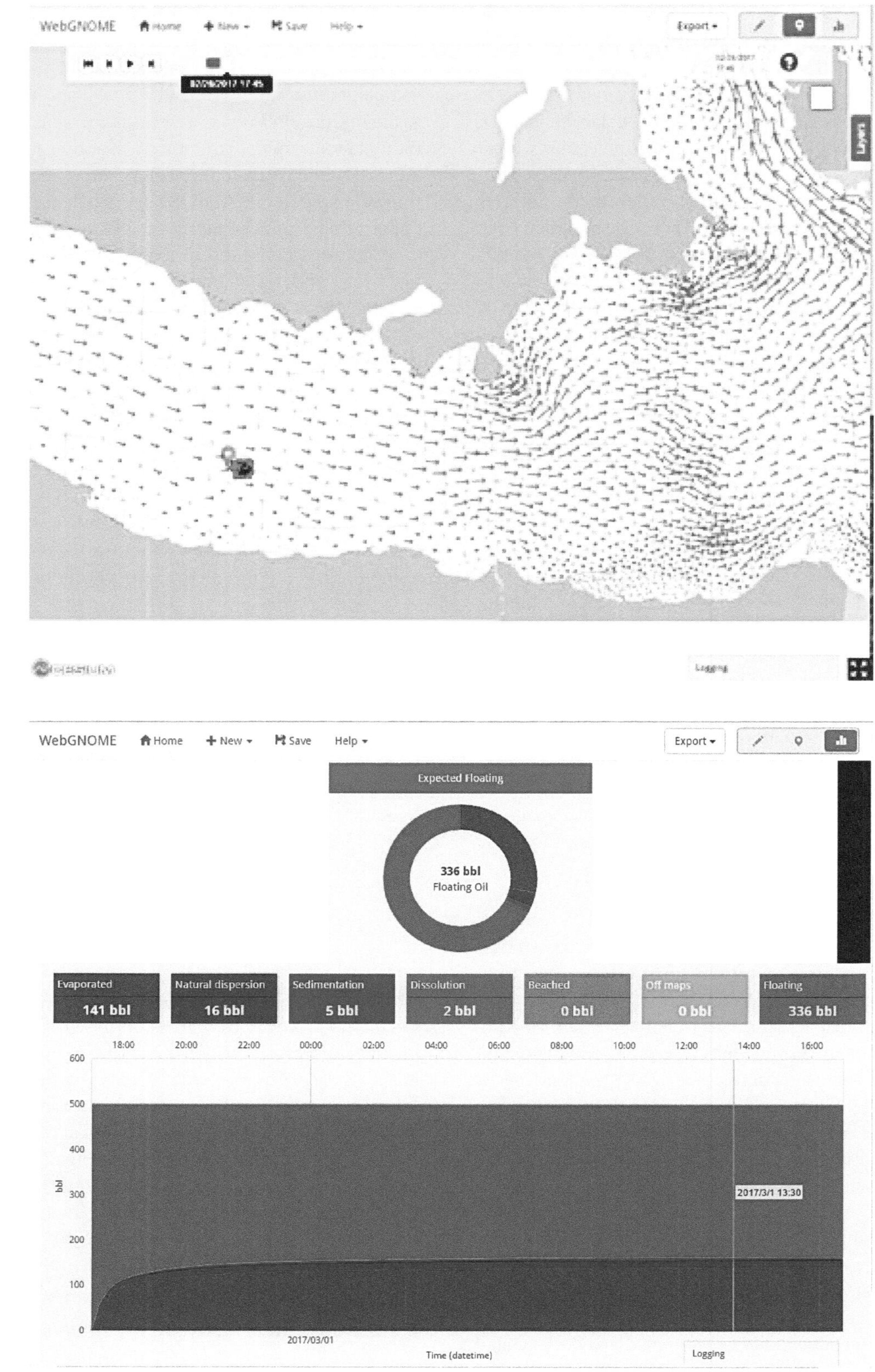

FIGURE 5.19 NOAA's WebGNOME trajectory modeling software. Top: A map view from WebGNOME, showing currents moving the oil in a simulated spill. Bottom: Oil fate (weathering) view from WebGNOME, showing an oil budget for a simulated spill.
SOURCES: NOAA; ADIOS (2021).

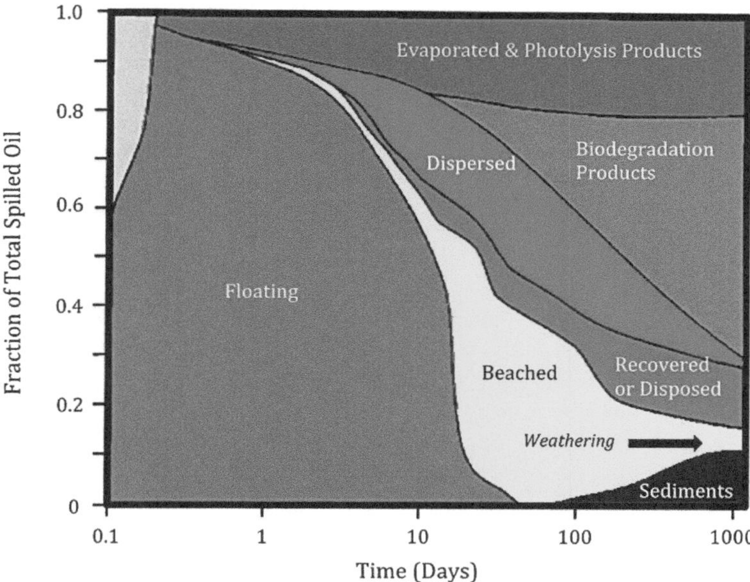

FIGURE 5.20 Oil spill budget compiled from observations and measurements of the *Exxon Valdez* oil spill.
SOURCE: Reprinted with permission from Wolfe et al. (1994). Copyright 1994, American Chemical Society.

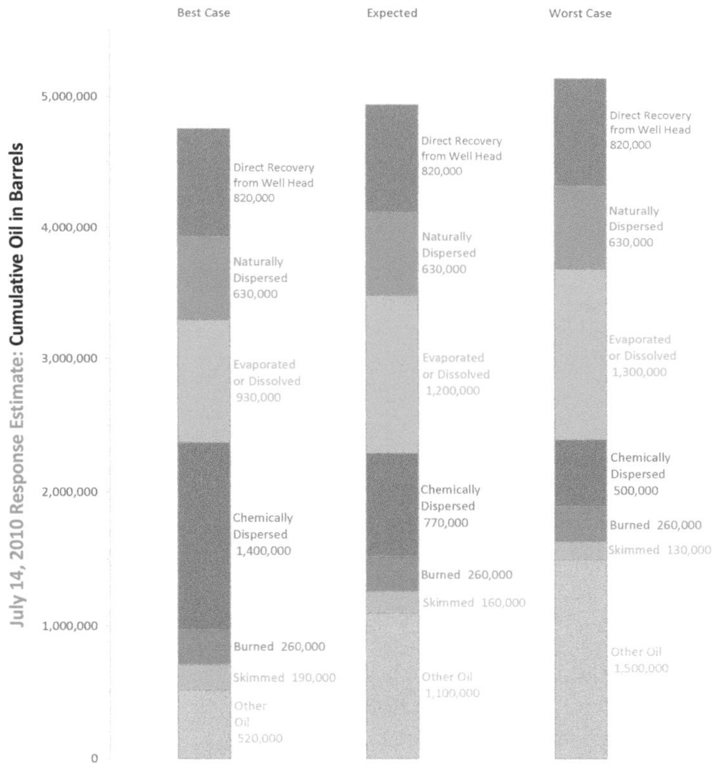

FIGURE 5.21 Response estimates produced by the DWH Oil Budget Calculator showing best case, expected, and worst-case volumes of the seven different categories that the calculator tracks individually, shown as cumulative volume of oil discharged through July 14, 2010.
NOTES: These estimates served solely as a guide for the national response to the DWH MC252 gulf incident. The best and worst cases defined in the report are the combinations of values of the seven variables depicted in each stack that correspond to the lower and upper endpoints of a 95% confidence interval for the volume of "Other Oil." The Oil Budget Calculator does not quantify the volume of oil that forms tar balls or surface slicks, sinks due to sedimentation, remains in the surf zone, or impacts the shore and is subsequently cleaned up. These are instead grouped together as "other oil." The amount of oil listed as "other" is quite large in the DWH oil budget.
SOURCE: Federal Interagency Solutions (2010).

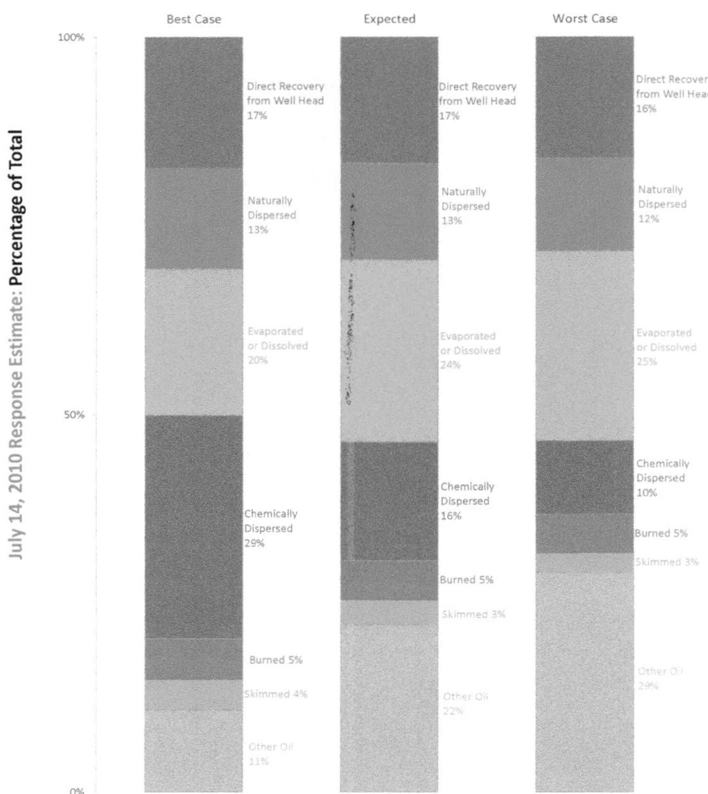

FIGURE 5.22 Response estimates produced by the DWH Oil Budget Calculator, expressed as percentages of the volumes shown in Figure 5.21.
SOURCE: Federal Interagency Solutions (2010).

The oil budget (see Figures 5.22 and 5.23) indicated that response and containment operations collected, eliminated, or dispersed in the expected case about 41% of the oil, with containment ("direct recovery from wellhead") being the most effective method, and chemical dispersant applications at surface and subsurface dispersing a substantial fraction (National Commission, 2011a). Subsurface application of dispersant resulted in dispersion of the oil before it reached the surface, limiting the amount of surface oil that could be skimmed, burned, or dispersed at the surface. Roughly 8 percent of the oil in the expected case was removed through skimming or burning, with burning operations considered to be more successful than skimming despite the resources directed to skimming operations (National Commission, 2011a). Comparing the three cases—best, expected, and worst—it can be seen that the results may vary several-fold, depending on the assumptions.

Lessons Identified by the Federal Interagency Solutions Group (2010, p. 38):

The experience in developing the calculator pointed to areas needing future research and planning:

1. Protocols for surface and subsurface sampling: While oil samples were collected for impact assessment, few samples were properly collected and categorized for response.
2. Dispersed oil droplet size: A major improvement in estimating dispersant efficiency would be possible if practical operational tools and methods existed to characterize droplet size distribution of subsurface oil.
3. Basic models for longer-term processes: While longer-term processes such as biodegradation often happen outside the time frames of the response, understanding and being able to predict such longer-term changes may be useful in making response decisions.
4. Estimation of collected shoreline oil: For a complete mass balance, procedures should be implemented that estimate the fraction that is oil on the oiled debris gathered from shoreline cleanup.
5. Expanded modeling capabilities: Many of the team members that assisted with the Oil Budget Calculator were also part of a working group of spill experts developing the specifications for the next generation of oil spill model. These specifications need to be translated into real code.
6. Revised interface: A better interface is necessary to more properly display the intrinsic uncertainty in the calculator.

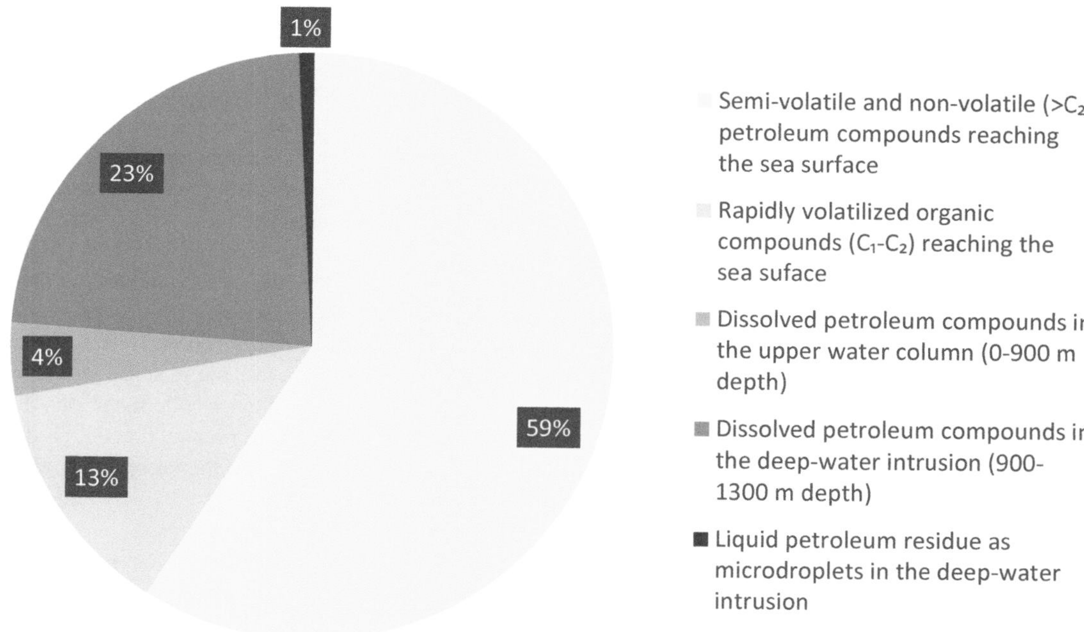

FIGURE 5.23 Mass balance for the fate of oil discharged from the DWH oil spill for the fluids released on June 8, 2010, based on model simulations by Gros et al. (2017).
NOTES: The shaded regions report the percentages of the total discharge resident in each category and environmental compartment (see inset legend for details). Simulation results are for the base-case scenario of the most likely fate with the applied subsea dispersant injection of June 8, 2010.
SOURCE: Gros et al. (2017).

Gros et al. (2017) and French-McCay et al. (2021a) reported mass balance predictions for the DWH oil spill based on oil fate and trajectory modeling (see Figures 5.23 and 5.24). The authors report the mass balance for the fate of fluids discharged to the ocean water column, not including the fluids collected directly at the wellhead. Gros et al. (2017) reported fates for different compound groups and environmental compartments and performed simulations only for June 8, 2010; French-McCay et al. (2021b) analyzed the total mass entering each environmental compartment. From Figure 5.23, Gros et al. (2017) reported that 72% of the petroleum released on June 8 reached the sea surface, 13% of which rapidly volatilized to the atmosphere, 27% dissolved into the ocean water column, and about 1% became trapped in the water column as tiny oil droplets. From Figure 5.24, French-McCay et al. (2021b) reported, up to June 8, 2010,

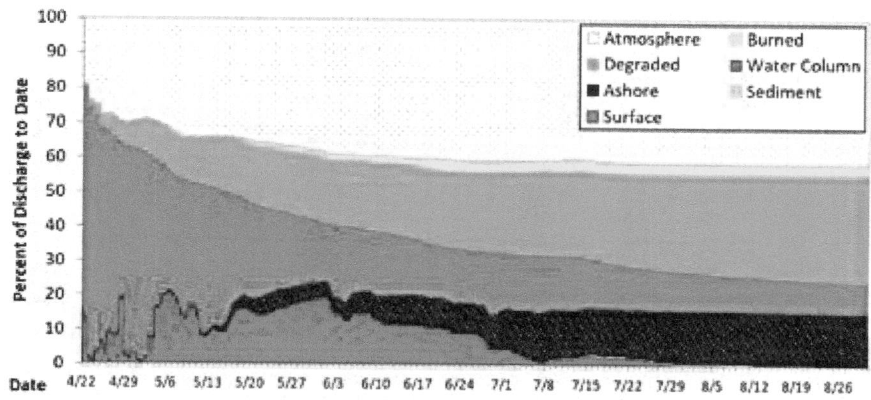

FIGURE 5.24 Cumulative mass balance (percentage of discharge to date) for the DWH oil spill based on model simulations reported in French-McCay et al. (2021b).
NOTES: The shaded regions report the percentage of discharge represented in each environmental compartment over time (see inset legend for details). Simulation results are for the base-case scenario of the most likely fate. This is the only model that considered biodegradation; the model did not consider photo-oxidation.
SOURCE: French-McCay et al. (2021b). CC-BY ND.

that 14% of the petroleum discharged since April 22, 2010, was floating on the sea surface, 8% had stranded ashore, 17% had become trapped in the water column, 20% had been biodegraded, 2% was burned, and 39% had partitioned to the atmosphere. Figure 5.24 also shows that this partitioning varied in time as different intervention methods were applied and because the various fate processes have different time scales.

As can be seen by these different approaches to calculating oil budgets: it is difficult to compare different types of calculations and models; there are very large/rough/round numbers used; the "Other" category may encompass a large volume and vary in definition; there are many different combinations of ways to view the numbers (not every assumption will hold true for low, mean, high supposition—some may be high, some may be low, etc.); the results may differ by millions of gallons or be "unaccounted for;" and the time scales vary for each scenario.

5.2.9.4 M/T Athos I *Oil Budget*

During the *M/T Athos I* (see Box 4.1) response, the Unified Command requested the Environmental Unit to calculate a mass balance/oil budget for the recovery activities. As can be seen in Figure 5.25, there is a wide spread in the values calculated depending on the assumptions made, as there were very few absolute values to use. The most reliable number was the total amount released, as that was able to be calculated from the volume loaded into the hold and the amount pumped out after the spill. This number was also rounded to the nearest thousand. Due to the complex system that the oil was released into, the resultant calculations are highly variable in space and time. Each of the categories in the table show a wide range of values except for evaporation, as there was consensus that this particular heavy oil had negligible light ends and therefore nearly zero evaporation.

***M/T Athos I* Incident Response – Mass Balance/Oil Budget**

Total Spill Volume (Gallons) 264,000

Total Spill Volume = Evaporation + Dissolution + Emulsification + Dispersion + Removed while on Water* + Deposited on Shoreline +Oil Unaccounted For

	Gallons of Oil		
	Low Estimate	Midpoint Estimate	High Estimate
Evaporation estimates are based on ADIOS model runs conducted by NOAA and verified by ENTRIX. Evaporation is estimated as a percent of *Total Spill Volume* minus the estimated volume of pooled oil in trench #1 and #2. Estimates range from 3 to 13 percent with a midpoint estimate of 8 percent. Evaporation is the process by which material in a liquid state enters a gaseous state and oil on the surface is transferred into the atmosphere.	7,818	20,840	33,851
Dissolution estimates are based on ADIOS model runs conducted by NOAA and verified by ENTRIX. Dissolution is estimated as a percent of *Total Spill Volume*. Estimated dissolution is 0 to 1 percent with a midpoint estimate of 0.5 percent. Dissolution is the process by which soluble components of the oil dissolve in the water column and become indistinguishable from the water.	0	1,320	2,640
Emulsification estimates are based on ADIOS model runs conducted by NOAA and verified by ENTRIX. Emulsification is estimated as a percent of *Total Spill Volume*. The model run suggests that no oil was emulsified. Emulsification is the process by which water droplets are forced into oil forming a "water-in-oil-emulsion" called a mousse. This process is driven by the physical turbulence of the surrounding water. It is the opposite of dispersion.	0	0	0
Dispersion estimates are based on ADIOS model runs conducted by NOAA and verified by ENTRIX. Dispersion is estimated as a percent of *Total Spill Volume*. Estimated dispersion is 0 to 1 percent with a midpoint estimate of 0.5 percent. Dispersion is the process by which oil on the surface is broken down into small droplets and dissipated into the water column.	0	1,320	2,640
Removed While on Water* estimates are based on the ICS 209 - Incident Status Summary from January 13, 2005. These reports indicate that 79,131 gallons of Oily Liquid were recovered. The % composition of oil in this oil-water mixture was assumed to range from 2–6.5% with a midpoint estimate of 4.25%.	1,583	3,363	5,144
Deposited on Shoreline estimates are based on a combination of SCAT data collected during the spill and back-calculation of shoreline oiling based on waste stream estimates and % completion of bulk oil removal. Calculations used to estimate *Deposited on Shoreline* are detailed in subsequent worksheets.	202,785	218,383	219,725
Oil Unaccounted for is estimated as the difference between Total Spill Volume and the sum of *Evaporation, Dissolution, Emulsification, Dispersion, Removed While on Water, and Deposited on Shoreline.*	51,814	18,774	0
Total	264,000	264,000	264,000

*Includes submerged pooled oil removed from trenches at the collision site

INTENDED FOR INFORMATIONAL PURPOSES ONLY

The mass balance/oil budget estimates provided herein is a semi-quantitative analysis based on the best available information compiled from multiple sources. Numerous assumptions were required to complete the computations. While the range of assumed values is considered both appropriate and realistic, significant uncertainty still exists. When compounded through multiple calculations, the final results vary widely, even though the range of values for specific parameters is narrow. For these reasons, the mass balance estimates should be used for general information purposes only to provide a general indication of the relative fates of the spilled oil. They are not intended to serve as benchmarks for gauging removal efficiency, and their use for this or similar purposes is inappropriate given the inherent uncertainty.

FIGURE 5.25 Oil budget from the *M/T Athos I* incident using low, medium, and high estimates to calculate each of the weathering and removal categories.
NOTE: This is intended for informational purposes only.
SOURCE : Levine, E. Unpublished data.

Depending on the assumptions for each variable, the remainder in the "oil unaccounted for" varies greatly (from 20% to 8% to 0%). These calculations have merit for framing the pace of the response, but not as an absolute accounting.

5.3 OIL FATES IN SPECIFIC MARINE ENVIRONMENTS: EPISODIC INPUTS

Section 5.2 introduced the fundamental parameters and processes influencing the fate of oil including slick formation, gas bubble and oil droplet breakup, and dispersion, regardless of geographic location and marine system. In Section 5.3 we use these universal mechanisms to describe how the properties of specific marine systems affect the fate of oil from episodic inputs (i.e., typically single, finite events). Section 5.4 considers chronic inputs.

This section begins by describing oil in surface and near-surface waters (to ~10 m depth); then in the underlying water column from the photic zone (up to ~50 m depth) and below (<1,000 m depth); then deep-sea waters and sediments (>1,000 m depth); and finally oil-impacted shorelines including beaches and estuaries. The fate of oil in the Arctic is presented as a special case.

5.3.1 Sea Surface Processes Affecting Oil Fate

Oil may reach the ocean surface either directly as a surface spill or from a subsea petroleum source. This section addresses processes immediately at the ocean surface or originating at the ocean surface and potentially affecting transport over the upper mixed layer of the ocean, extending to depths of order ~10 m. Most oils are lighter than seawater, so newly spilled oil tends to rise toward the surface and accumulate, at least temporarily, where it is affected by a wealth of weathering and transportation processes. Oil floating at the sea:air interface is exposed to the atmosphere, where it may evaporate or photo-oxidize, and also interacts with the near-surface water column, where it may be dispersed or dissolved. Wind and waves together with ocean currents transport the oil and may promote dispersion of the oil from a floating slick to submerged oil droplets or emulsions. These processes are described in the sections below.

5.3.1.1 Initial Spreading of Oil on Surface Waters

Oil at the surface initially spreads under gravitational forcing to form a slick or sheen. In the open ocean, where there are no physical boundaries, oil spreads out by gravitational forcing and eventually pools in slicks and sheens (described in detail in Section 5.2.2.1 and Figures 5.3–5.5). These layers of oil can be extremely thin, approaching a thickness of a few micrometers. Gravitational spreading typically is important on length scales of meters to hundreds of meters and ceases when the interfacial tension of the oil layer begins to resist further spreading. After initial spreading, wind and wave action may cause the oil to undergo dispersion, evaporation, dissolution, or emulsification. Importantly, oil floating at the surface is more susceptible to photo-oxidation than submerged oil (see Section 5.2.5). The ultimate fates of various photo-oxidation products (whether via dissolution, sedimentation, biodegradation, etc.) are not yet fully known, but amphiphilic photo-oxidation products that increase hydrophobic interactions and aggregation of oil droplets (by altering surface activity or stickiness) can enhance both emulsification (Ward et al., 2018a; Ward and Overton, 2020) and the formation of marine oil snow, discussed below (see Section 5.3.2.2). In contrast to photo-oxidation, oil slicks are less susceptible to biodegradation than submerged oil droplets because the surface area available to microbes is smaller in a slick for a given oil volume (Prince and Butler, 2014).

There is a significant knowledge of the accumulation of natural organic material such as biosynthesized lipids, natural marine biopolymers (e.g., exopolymeric substances; see Section 5.3.2.2) and natural surfactants in oily-like surface films at the air-sea interface. The physical chemical properties of these substances indicate that such materials will also accumulate with petroleum oil slicks and films. Human synthesized and human mobilized contaminant/pollutant chemicals with physical chemical properties such as chlorinated pesticides, polychlorinated biphenyls, and other persistent organic pollutant and organic pollutants of emerging concern with physical chemical properties similar to petroleum compounds are likely to be partitioned into the petroleum oil slicks. Discussion of these interactions and subsequent fates and effects is beyond the scope of this report.

5.3.1.2 Wind Drift and Stokes Drift

Oil transport modelers have observed for decades that floating oil tends to move by a combination of the near-surface ocean currents and the wind vector. This added velocity component from the wind is called wind drift; typical values are in the range of 1% to 3% of the wind velocity. It is also well known that there is a nonlinear component of the surface wave orbital velocity that contributes a non-zero net transport. This wave effect is known as Stokes drift, and its direction and magnitude are a function of the surface wave energy spectrum (D'Asaro et al., 2020). Oil floating on the sea surface or suspended within the orbital velocity field of surface waves will be acted on by wind drift and Stokes drift. Stokes drift can be included through analytical expressions; wind drift depends on the mechanisms causing its effect.

The motion within a surface oil slick will be in balance with the water and wind forcing so that the friction, or shear stress, experienced on either side of the oil–water or oil–air interfaces is the same. For relatively thick surface oil, this means that the shear stress between the air and oil could differ from that between the oil and water so that the wind could accelerate the oil separately from the acceleration transferred to the water column. Surface texture of thick oil layers could also result in different stresses applied to the oil and water. Either effect could cause a wind drag on the oil that pushes the oil in the downwind direction, adding a wind drag to the oil transport and causing the oil not to purely follow the surface ocean currents.

Another mechanism explaining the apparent wind drift derives from the variation of current velocity with depth in the near-surface of the ocean. At the air–water or air–oil interface, the friction, or shear stress, on the water or oil side of the interface will be equal to that applied by the winds. Moving away from the air–water or air–oil interface, currents will adjust to match general ocean currents at depth that may not be aligned with the winds. In the oceans, this adjustment will further occur in the surface Ekman boundary layer, the region coupling the rotation-dominated ocean interior, which may be in geostrophic balance with currents perpendicular to the pressure gradient, to the surface boundary conditions, in which currents are parallel to the pressure gradient. Because of these competing dynamics, recent measurements in the northern Gulf of Mexico (Laxague et al., 2018) confirm that significant velocity shear may be confined to the upper few centimeters of the ocean surface. Numerical models of ocean circulation must solve for very large domains so that the upper-most cell of a numerical model may be of order 1 m or more in depth. While numerical simulations are likewise in equilibrium with the applied surface stresses, they integrate these fine-scale variations in currents over a comparatively large depth. Because oil slicks are very thin, oil is likely to always be transported at the same speed as the real ocean currents at the sea surface. However, because numerical models predict average currents over a relatively thick surface grid cell, fine-scale effects of the winds need to be added to the oil transport floating on the top of these grid cells. Because ocean circulation models typically resolve the surface layer with similar order-of-magnitude length scales, the wind drift factor of 1% to 3% of the wind is a fairly stable parameter in oil trajectory modeling. As ocean circulation models become more resolved or include subgrid scale parameterizations for the surface boundary layer, the wind-drift parameter may become less important or eliminated altogether. Hence, apparent wind drift is an artifact of low model resolution at the air–water interface that generally should adapt to the resolution of the model as circulation models improve.

5.3.1.3 Natural and Chemical Dispersion of Oil in Near-Surface Waters

Entrainment of oil from the sea surface into near-surface water occurs when oil droplets separate from the oil slick and enter the water column—a process commonly referred to as natural dispersion when dispersants are not used and chemical dispersion when they are used (NASEM, 2020) (see Box 5.4 and Section 5.2.2.4). Oil entrainment is caused naturally by surface mixing via wind, waves, and currents and can be enhanced by the application of chemical dispersants. When floating oil is dispersed, its surface area to volume ratio increases, providing greater interaction of oil with seawater and allowing other natural processes to occur at faster rates (e.g., dissolution, biodegradation). To the extent that these oil droplets shade others or become dispersed below the photic zone, photo-oxidation for entrained oil may be reduced or arrested (Xiao and Yang, 2020).

The processes of dispersion of oil are complex. Floating oil may form submerged oil droplets when fluid motion at the oil–water interface entrains oil below the sea surface and turbulence in the upper mixed layer of the ocean breaks the entrained oil into droplets of various sizes. Whereas larger oil droplets can rise back to the water surface, smaller droplets remain in suspension in dispersed form in the water column. Entrainment of oil floating on the sea surface into the ocean interior requires both downward currents and encapsulation of oil within these currents. Breaking waves are the most notable source of entrained oil (Delvigne and Sweeney, 1988). Oil entrainment has also been observed for regular, non-breaking waves (Li et al., 2008), and may result from downwelling between the windrows of Langmuir circulations. Langmuir cells consist of alternating rows of helical vortices resulting from nonlinear interactions of the winds and waves (Yang et al., 2015). These complex processes set the source rate of oil entering the upper ocean turbulence. Turbulence within this domain also depends on a range of processes occurring near the air-water interface. These include wind stress at the sea surface, wind-induced currents, currents and turbulence resulting from ocean circulation, a broad spectrum of breaking and non-breaking waves, and Langmuir circulations (D'Asaro et al., 2020). Turbulence within this complex system is highly variable in space, and turbulence generated by periodic forcing, such as waves and wind gusts, can be heterogeneous and intermittent, resulting in strong temporal variability. Hence, breakup of oil droplets or gas bubbles in the upper ocean is a complex process, highly variable in both space and time, dependent on the local wind and sea state and on the oil properties of viscosity and surface tension, which set the interaction dynamics between entrained oil and the ocean currents and turbulence.

The entrainment rate of oil as droplets is faster for light oils with low viscosity and low interfacial tension. Sometimes chemical dispersants are sprayed on floating oil to reduce interfacial tension and promote dispersion. Low air and water temperatures, as well as most weathering processes, increase the viscosity of oil, thereby reducing the rate of natural dispersion. The rate of natural dispersion may also decrease with increasing ice coverage. The presence of ice on water reduces water surface and wave activity; thus, mixing and energy dissipation could be very low. Ice coverage may also reduce the interaction of oil with the atmosphere. Together, these processes may slow weathering, allowing fresh oil to persist longer. **Interactions of oil with ice are complex and how oil is dispersed under partial or complete ice coverage remains a major challenge for predicting oil trajectories during response.**

Classical understanding of natural oil dispersion from the sea surface originated with Delvigne and Sweeney (1988), who showed, based on small- and large-scale laboratory experiments, that floating oil could be entrained as droplets due to the effect of breaking waves on the sea surface. Their measurements in two laboratory wave flumes showed two important conclusions. First, surface oil is entrained to a characteristic depth that depends on the wave height at breaking. Second, the maximum droplet size observed within the entrained oil depends on the characteristic depth of oil entrainment, the rise velocity, or slip velocity, of the formed droplets, and the time since wave breaking. Some of the entrained oil failed to break up into small droplets and rose to the sea surface nearly immediately following breakup. As the time since breaking increased, the droplet sizes remaining in the water column, therefore, decreased. Delvigne and Sweeney (1988) further related the volume flux of entrained oil per unit area of the ocean to the dissipated breaking wave energy per unit surface area and the fractional coverage of the sea surface by oil and by breaking waves. Though these correlations are highly empirical and must be fitted to each type of oil, the general understanding remains valid: Breaking waves entrain oil to a characteristic depth, and droplets remaining in the water column must be small enough not to resurface in the time between breaking events.

Recent insight on natural dispersion of oil by waves grew out of new experimental techniques that allow measurement of the in situ turbulent kinetic energy dissipation rate and the evolving oil droplet size distribution. An important set of experiments using regular and breaking waves and involving oil and oil plus dispersant were conducted by Li et al. (2007, 2008, 2009a,b,c, 2010). This work and other historical contributions are carefully reviewed in the more recent experimental work of Li et al. (2017). The Li et al. (2007, 2008, 2009a) studies showed that, while breaking waves are much more effective at dispersing surface oil, oil entrainment does occur under regular, non-breaking waves. They also showed that chemical dispersants can significantly reduce the droplet sizes of oil droplets forming under waves and increase the volume of entrained, dispersed oil. Through their laboratory measurements, Li et al. (2008) introduced a model that they refined in Li et al. (2009b,c) for the kinetics of oil droplet breakup under waves. They also quantified the effects of the turbulent eddy dissipation rate under waves. The focus in these studies was on the droplet size distribution. Li et al. (2010) used oil concentration measurements to further define the dynamic dispersant effectiveness, which is a measure of the mass fraction of oil suspended in the water column a certain time after a wave passage or breaking event. Using results for a heavy fuel oil, they concluded that less than 15% of the oil is dispersed at low temperature (<10°C) or for non-breaking waves; whereas better than 90% efficiency could be achieved using chemical dispersants at 16°C. This illustrates the importance of oil properties, especially viscosity and surface tension, and the surface flow dynamics, including wave characteristics, on dispersion of oil from the surface ocean. Li et al. (2017) used cutting-edge fluid dynamics observations to further elucidate the dynamics of oil droplet formation under breaking waves with and without chemical dispersant addition. They quantified the turbulent eddy dissipation rate in space and time for various laboratory simulations of deep-water breaking waves and measured the dynamic oil droplet size distribution. During the breakup process, they did not observe any coalescence events but did observe two forms of droplet breakup. Larger droplets form by primary breakup caused by the droplet-scale turbulence. When dispersant is added, droplets further undergo tip streaming, resulting in long micro-threads of oil which eventually break up into micron- and submicron-sized droplets (Gopalan and Katz, 2010; Davies et al., 2019).

Because Li et al. (2017) directly measured the time-evolving turbulence field, these measurements could also be compared to droplet size predictions using the Hinze (1955) model. Using the theory expressed in that paper (specifically, Equation F.1 in Appendix F), Li et al. (2017) showed good agreement between predicted and observed maximum droplet sizes in most experiments when using initial values of turbulent fluctuating velocities observed immediately following wave breaking. This is in agreement with their observation that larger droplets form rapidly by primary breakup in the turbulent field of the breaking wave. Li et al. (2017) also applied the Hinze theory using empirical relations for turbulent eddy dissipation rate under the roller of a breaking wave. These likewise showed good agreement in most experiments with the observed maximum droplet sizes. These Weber-number type predictions following Hinze (1955) are less accurate for the experiments with high dispersant to oil ratio (e.g., 1:25). In those experiments, it was more difficult to identify the characteristic droplet sizes in the measurements, and breakup by oil threads via tip streaming becomes important—a process that does not depend on the turbulent dynamics of the breaking wave and is not predicted by the dynamics described by Hinze (1955). **Hence, when breakup occurs by turbulent eddies, predictions using the Hinze (1955) analysis do agree with observations, but when surface tension is low, as following chemical dispersant addition, other processes of droplet breakup dominate.**

Two recent empirical equations (Johansen et al., 2015; Li et al., 2017) attempt to predict the droplet sizes of naturally dispersed oil. Both are based on an approach similar to that of Wang and Calabrese (1986) in which both Weber number and viscosity groups are involved and where the droplet-scale turbulence is parameterized by bulk, flow-scale variables. Johansen et al. (2015) define a modified Weber number, following the approach of Johansen et al. (2013), in which a bulk Weber number and viscosity number are combined. In their approach, the characteristic length scale is the oil slick film thickness, and the turbulent eddy dissipation rate is considered proportional to the kinetic energy of the breaker. In the Li et al. (2017) approach, they combine a bulk Weber number with the Ohnesorge number, another non-dimensional parameter that compares viscosity effects to surface tension forces and here, equivalent to the Hinze (1955) viscosity group. In their equations, the characteristic length scale is taken as the maximum stable droplet size for a drop rising in quiescent water (Clift et al., 1978). Both of these equations show good agreement with measured droplet size distributions after calibration of their fit parameters. The Li et al. (2017) approach was explicitly calibrated for experiments with and without dispersant. The Johansen et al. (2015) approach, though appropriately including the viscosity number, has not yet been validated to experiments with chemical dispersant treatment. Overall, these approaches appear promising for predicting droplet size distributions for natural dispersion of oil at the air–sea interface.

The transport of dispersed oil in the upper mixed layer of the ocean is also a complex process. Cui et al. (2018) simulated transport of oil droplets under a deep-water breaking wave, and Cui et al. (2020a,b) coupled the under-wave transport with a population balance model for predicting the oil droplet size distribution. The advection and dispersion field within the region affected by wave breaking is highly heterogeneous and complex, containing downwelling currents, entrained air and oil, and highly fluctuating vorticity and eddy dissipation fields, among other complex processes. These affect oil dispersion in two key ways. First, the heterogeneous turbulent field-controlled droplet breakup. Second, the non-uniform turbulence and current velocity fields create non-uniform diffusivity fields. To model the transport of oil droplets using a Lagrangian particle tracking approach in a non-uniform diffusivity field requires a random displacement model (Boufadel et al., 2018; Cui et al., 2020b). Sophisticated laboratory data (e.g., Li et al., 2017) and numerical simulations (e.g., Cui et al., 2018, 2020b) are helping to understand these processes and develop appropriate field tools. **At the same time, little field-scale data are available to study the dispersion process or to validate models that predict droplet size distributions under different dynamic ocean conditions.**

5.3.1.4 Other Processes Affecting Surface and Near-Surface Oil

Photo-oxidation, dissolution, evaporation, emulsification, and biodegradation affect oil at and near the water surface, and have been described in detail in Sections 5.2.3–5.2.8. Photo-oxidation is most effective at the surface and rapidly diminishes with water depth, refraction off suspended particulates and self-shading by dispersed droplets. Thus, slick thickness, suspended sediment content, and depth of dispersed oil in the water column affect the impact of photo-oxidation. Dissolution occurs between the petroleum–water interface of surface floating oil and suspended water droplets. Because evaporation is faster than dissolution, dissolution is less significant for predicting the mass balance of surface, floating oil slicks. However, natural dispersion and dissolution from the suspended oil droplets together remain a similar order of magnitude to evaporation, and should be considered (MacKay and Leinonen, 1977). Natural dispersion is assessed as described above, and dissolution is analyzed using the approaches described in Section 5.2.5 for suspended bubbles and droplets. Emulsions arise through physical forces but may be enhanced by biological polymers and stabilized by microbial cells and/or their extracellular products (see Section 5.2.8.3). Whereas oil-in-water emulsification may increase the interfacial area available for biodegradation, particularly in the case of naturally produced microbial biosurfactants (Ron and Rosenberg, 2002), water-in-oil emulsification may decrease biodegradation due to limited diffusion of nutrients in seawater to the interface of internal water droplets. As previously discussed, biodegradation of near-surface oil is likely to be exclusively aerobic (see Section 5.2.8.8), and any predation of oil-degrading microbial blooms by predators and grazers (viruses, zooplankton) may decrease the rate of biodegradation of oil components. The role of marine oil snow in oil sedimentation and biodegradation is discussed in Section 5.3.2.2.

Oil chemistry affects the fate of surface oil. Notably, little is known yet about the in situ weathering properties of low sulfur fuel oils (LSFOs) and very low sulfur fuel oils (VLSFOs), new classes of marine fuel oils with diverse compositions. A December 2019 spill of a LSFO in Korea (Song et al., 2021) with a high saturates content (32%) and long-chain *n*-alkanes (~C20–C31) resulted in the oil solidifying at the water surface and on the shoreline at ambient cold temperature. Evaporation was inferred to be the main weathering process explaining losses of polycyclic alkanes, but little photo-oxidation of PAHs occurred during the 24–48 hr period of post-spill sample collection. Analysis of a single sample of VLSFO oil spilled from the MV *Wakashio* in Mauritius in 2020 indicated that this particular fuel oil experienced evaporative losses of low molecular weight *n*-alkanes at the warm sea surface within days of the spill (Scarlett et al., 2021). Its relatively high proportions of monoaromatic hydrocarbons and low proportions of PAHs, including the naphthalene and phenanthrene families, volatilized and/or dissolved in predictable patterns. Some emulsification was observed on-site; however, no studies have yet been published on additional fates of this VLSFO at the surface or in the water column, including photo-oxidation, dispersion, biodegradability, or sedimentation. **Studies of LSFO and VLSFO weathering and behavior should be conducted under conditions relevant to the many different marine environments that may be impacted by global use of these heterogeneous fuel oils.**

5.3.2 Processes Affecting Oil in the Water Column

After oil has been dispersed below the surface and near-surface water (see Section 5.3.1), it is subject to additional processes affecting its fate and transport through the water column. Most oils, even after weathering, are less dense than seawater and therefore float or have neutral buoyancy, remaining suspended at the surface or near-surface unless they interact with mineral particles and/or organic aggregates. Such interactions may increase the density of aggregates sufficiently to cause them to submerge, sink, and/or sediment to the seafloor, during which time the oil may change composition.

Here we describe processes that typically begin in the epipelagic zone of the sea (<200 m depth), specifically interactions between oil and suspended particulates: mineral particles, microbial cells and organic detritus, and plastics. These interactions lead to formation of oil-particle conglomerates and may result in sinking of oil through the mesopelagic zone (200–1,000 m depth). We conclude this section by considering the special case of submergence and potential sinking of heavy oils and semi-solid oils. Deep waters (>1,000 m depth) and deep seafloor sediments are discussed in Section 5.3.3.

5.3.2.1 Sorption of Oil to Mineral Particles

Interactions of oil with suspended particles have long been known. Unfortunately, the terminology describing the microscopic and macroscopic aggregates of different sizes and compositions that form has evolved over the past few decades and the literature can be confusing. Box 5.9 defines the most common terms in use and presents a scale for size reference. Both are pertinent to the section below on water column fates and to Section 5.3.4 describing oil on shorelines and sediments.

Oil–mineral aggregates (OMAs) are microscopic particles comprising distinct oil and mineral phases that are stable over periods of weeks in seawater. The surface properties of most minerals in marine sediments allow ionic and polar organic solutes to attach to the mineral surface, thus making their surface at least slightly lipophilic. Therefore, OMAs form when the suspended mineral particles attach to dispersed oil droplets by hydrophobic bonding (Stoffyn-Egli and Lee, 2002). The finer the oil dispersion, the greater the surface area available for OMA formation. Therefore, use of chemical dispersants that promote formation of small droplets can result in greater aggregation of oil droplets.

There are three different structural types of OMA: droplet, solid, and flake (Lee and Stoffyn-Egli, 2001; Stoffyn-Egli and Lee, 2002) (see Figure 5.28). Droplet OMAs consist of one or more oil droplets with mineral particles attached to their surface only. The size of the oil droplets ranges from less than 1 μm to tens of μm. The larger sizes are usually found as floating droplet OMAs. The number of mineral particles attached to the oil droplet are highly variable. Buoyant OMAs contain fewer mineral particles than the sinking OMAs. Solid OMAs are typically non-spherical in shape with irregular contours around mineral particles. The shape of the oil in solid OMA depends on the shape of the minerals included in it, whereas in droplet OMA the mineral particles are arranged around the exterior of the oil droplets. The solid OMAs often are elongated and curved or branched, and their sizes vary and can reach 250 μm in length. Flake OMAs resemble membranes, usually floating or neutrally buoyant. They can be several millimeters wide. Their microstructure is highly organized, exhibiting a dendritic or feather-like arrangement.

The oil-to-mineral ratio is the primary factor that controls the buoyancy of OMAs (Stoffyn-Egli and Lee, 2002). For droplet OMA, because the number of particles on an oil drop is limited by its surface area, buoyancy depends on the volume of the oil droplet. For a given size of mineral particle, aggregates with larger oil drops are relatively more buoyant. Once OMAs have formed, the oil droplets do not break down into smaller ones because the mineral coating protects the integrity of the droplets; below a certain size, oil drops are not broken down further by water turbulence. OMAs can become colonized by marine microbes to form OPAs (Lee et al., 1997; Weise et al., 1999). The additional reactive surface area afforded by the minerals enhances biodegradation compared with dispersed oil droplets.

The presence of mineral particles and temperature affect the rate of formation of aggregates and fate of oil (Passow, 2016). The aggregation process occurs faster at higher temperatures (≥20°C) incorporating higher fractions of oil compounds (Henry et al., 2020). This may be due to reduced viscosity of water and higher energy of the particles which allow weak interactions between particles. Temperature also affects the aggregate sedimentation characteristics. Experimental results show that below 20°C, aggregates without mineral particles did not exhibit sinking/sedimentation characteristics but required mineral particles to promote aggregate sinking (Henry et al., 2020). It is possible that chemical dispersant may contribute to MOS formation in cold sub-Arctic waters (Suja et al., 2019), but additional study is required to verify this possibility.

5.3.2.2 Sorption of Oil to Organic Particles: Marine Oil Snow (MOS) and MOSSFA

Formation and Significance of Marine Snow in the Absence of Oil

Because MOS derives from marine snow, a brief introduction to marine snow is warranted. Marine snow refers to organic matter aggregates that form naturally in the upper water column even in the absence of oil, settle by gravity into the deeper waters or seafloor, and are integral to the ocean ecosystem (see Box 5.6). Marine snow aggregates range in size from <0.5 mm to tens of centimeters (Daly et al., 2016) and comprise organic and inorganic particles (e.g., living microbial cells, dead or dying plankton, fecal pellets, sand, soot, and other inorganic dust, degraded larvacean houses, concentrated nutrient sources, and semi-stable substrates such as excreted polysaccharides and proteins [Tansel, 2018]). These components are integrated into a highly hydrated organic matrix of extracellular polymeric substances (EPS) excreted by aggregate- (or floc-) forming microbes.

BOX 5.9
Oil-Particle Terminology and Scale

For microscopic aggregates, the term *oil mineral aggregate* (OMA) was adopted in the 1990s. OMAs consist of solid aggregates that are a mixture of oil and minerals blended into microscopic bodies of various shapes. On average, OMAs are less dense than mineral-only aggregates and can remain in suspension longer and be dispersed further than unoiled suspended sediment. *Oil sediment aggregate* (OSA), a term adopted around 2010, refers to smaller aggregates associated with sediments. The term *oil–particle aggregate* (OPA) was introduced in 2015 to represent aggregates that form from mineral and organic particles. OPA terminology is used, generally, in reference to the particles associated with marine oil snow (MOS) aggregates (see Section 5.3.2.2 and Figures 5.26 and 5.27). Large patches of floating oil can interact with sediments to produce large sediment-oil mats (SOMs) that can reach up to several meters in length (Gustitus and Clement, 2017). Until the DWH oil spill, SOAs were considered under the broad category of tar balls and were referred to as tar mats. After the DWH spill, the terms *surface residual ball* (SRB) and *submerged oil mat* (SOM) were introduced to describe these agglomerates. SOA refers to sediment-oil aggregates that are between 1cm and 10 cm. Oil–sediment aggregates that are 10–100 cm (i.e., larger than an SRB and smaller than an SOM) have been referred to as surface residue patties (SRPs). Since the term "surface" does not refer to sediments, the *sediment–oil patties* (SOPs) terminology is used to describe aggregates that are 10–100 cm in sediments with relatively fresh oil at their core. The terms SRB, SRP, and SOM are almost exclusively used to describe DWH residues.

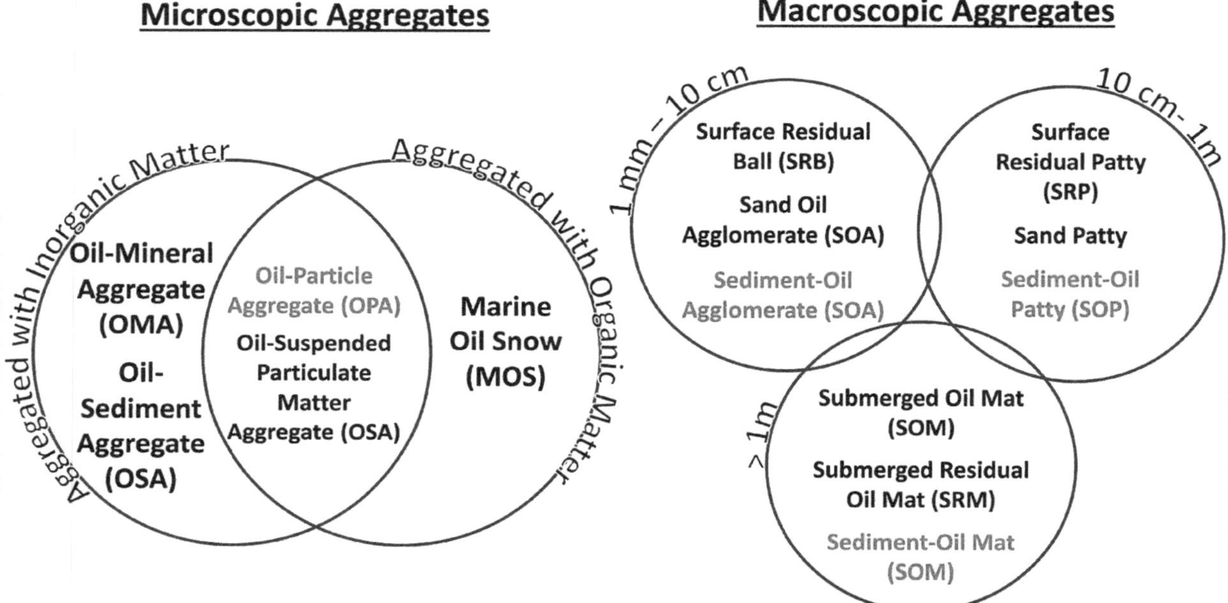

FIGURE 5.26 Terminology used for microscopic and macroscopic OPAs and agglomerates in the marine environment.
NOTES: Terms highlighted in orange are those recommended by Gustitus and Clement (2017) in their review to refer to the microscopic aggregates and macroscopic agglomerates. Compositions are outlined below and described in detail in the main text in Sections 5.3.2 and 5.3.4.
SOURCE: Gustitus and Clement (2017).

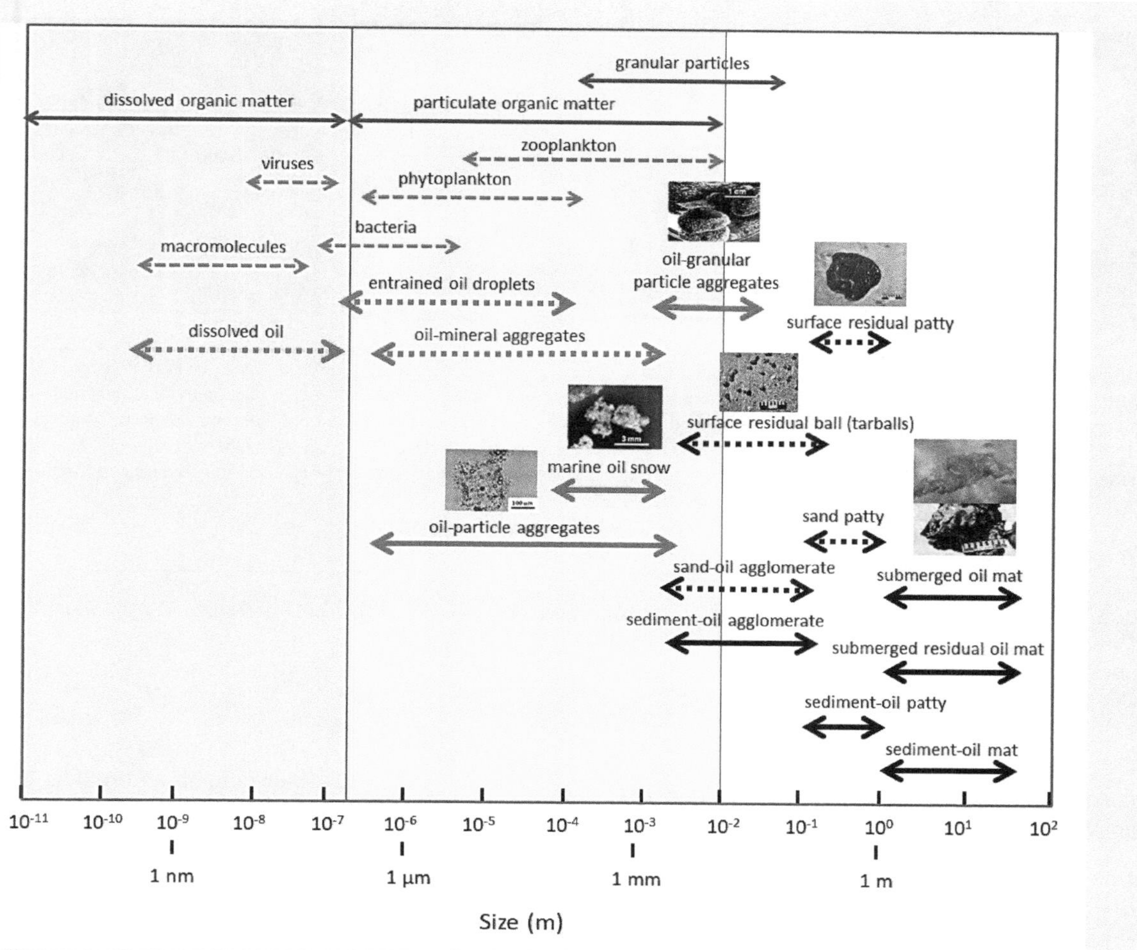

FIGURE 5.27 Size scale of OPAs.
NOTES: Solid purple arrows: natural state of particles present in water that are not associated with live organisms; dashed blue arrows: biological particles; dashed red arrows: oil state; red solid arrows: OPA; dashed black arrows: macroscopic surface aggregates; black solid arrows: macroscopic submerged and sediment aggregates. Figure does not include larger geologic structures such as tar mounds and asphalt volcanoes.
SOURCE: Adapted with modifications from Figure 7 in Boglaienko and Tansel (2018).

The EPS enables the bacteria to adhere to surfaces, including other cells, while providing a protective layer. Cells that do not possess a gelatinous matrix may also exhibit flocculation ability (especially under stress), thus contributing to floc formation (Tansel, 2018). Marine snow aggregates are densely colonized (up to 10^8–10^9 cells per ml; Alldredge et al., 1986) by microbes and their predators. Their combined metabolic activity converts the organic matter into non-sinking DOM, forming plumes of DOM and soluble nutrients (nitrogen, phosphorus, and iron) in the upper mixed layer of the ocean to support primary productivity, as well as creating sinking POM that transports carbon, nitrogen, phosphorus, iron, and silicon from near-surface waters into the deeper layers of the water column and sediments to support the pelagic and benthic food webs (Azam and Malfatti, 2007). These microscale interactions influence global biogeochemical processes (e.g., carbon storage and the regulation of carbon flux). Formation and abundance of marine snow varies seasonally and with ocean currents due to the changes in light (i.e., primary productivity) and nutrients available to the organisms in the upper water column and, furthermore, responds to the presence of dispersed oil and chemical dispersants. The color of marine snow varies from nearly white to dark brown depending on the nutrient content, types of bacteria available, density of particle aggregates, oxygen availability, and age of aggregates (Tansel, 2018).

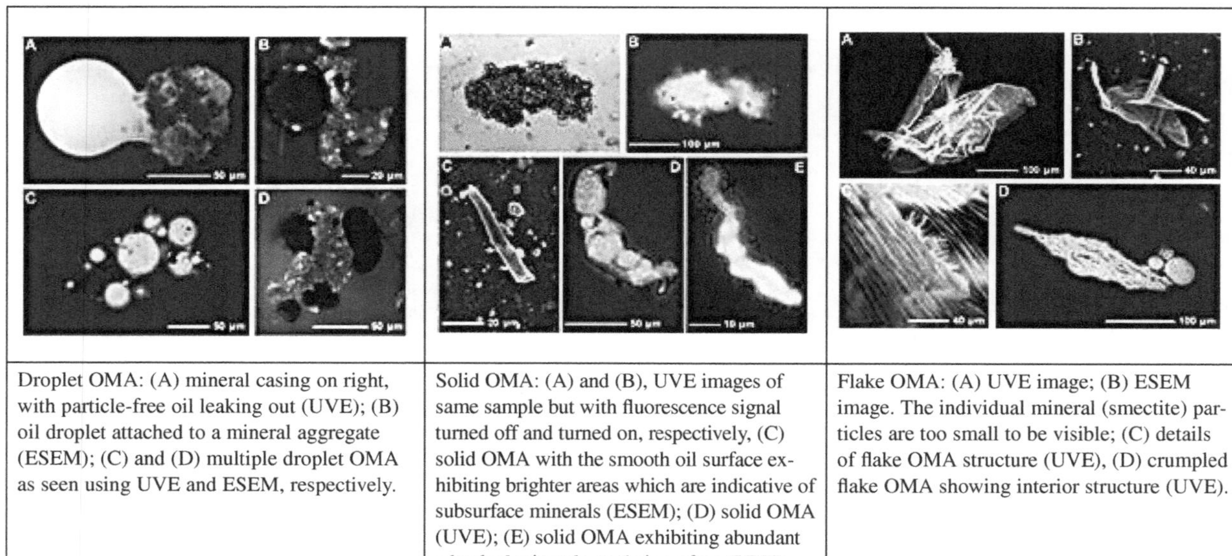

Droplet OMA: (A) mineral casing on right, with particle-free oil leaking out (UVE); (B) oil droplet attached to a mineral aggregate (ESEM); (C) and (D) multiple droplet OMA as seen using UVE and ESEM, respectively.	Solid OMA: (A) and (B), UVE images of same sample but with fluorescence signal turned off and turned on, respectively, (C) solid OMA with the smooth oil surface exhibiting brighter areas which are indicative of subsurface minerals (ESEM); (D) solid OMA (UVE); (E) solid OMA exhibiting abundant adsorbed minerals on their surface (UVE).	Flake OMA: (A) UVE image; (B) ESEM image. The individual mineral (smectite) particles are too small to be visible; (C) details of flake OMA structure (UVE), (D) crumpled flake OMA showing interior structure (UVE).

FIGURE 5.28 Micrographs showing droplet, solid, and flake structural types of OMA, obtained using UV epifluorescence microscopy (UVE) or environmental scanning electron microscopy (ESEM).
NOTE: Scale bars are shown in each image.
SOURCE: Adapted from Stoffyn-Egli and Lee (2002).

Mechanisms of Marine Oil Snow (MOS) Formation and Its Characteristics

Marine oil snow (MOS) forms when naturally or chemically dispersed oil droplets attach to marine snow and mineral particles (Brakstad et al., 2018c) (see Figure 5.29). MOS particles, defined as microhabitats <0.5 mm in size and rich in organic matter (Burd et al., 2020; Trudnowska, 2021), may form via two general mechanisms that are not mutually exclusive: (1) pre-formed aggregates interacting with oil, or (2) oil acting as a nucleus for microbial biofilm growth and floc formation (i.e., bacteria–oil aggregate or BOA) (Passow and Overton, 2021). Within these mechanisms are various processes and components that contribute to MOS formation including physical coagulation of marine particles, microbially mediated formation of marine snow, synthesis of biopolymers by algae and bacteria due to oil and dispersant exposure, and incorporation of oil into mucous feeding webs followed by ingestion of oil by zooplankton and/or attachment of oil to fecal pellets (MOSSFA, 2013). From the perspective of aggregation theory, oil droplets and marine particles collide and stick to form larger aggregates with the particle size and concentration determining the rate of particle collisions (Lambert and Variano, 2016).

The presence of non-biological particles such as suspended sediment from riverine sources or high-energy beaches, or black soot from in situ burning of surface oil also can contribute to MOS formation. Furthermore, application of chemical dispersants to oil slicks can stimulate EPS production and MOS formation (Passow, 2016). Dispersants allow oil to enter the water column as droplets by reducing the interfacial surface tension between the oil and seawater (see Section 5.3.1.2) so the presence of small but numerous oil droplets in water results in increased collision rates between droplets and marine particles and, therefore, increases the rate of MOS formation. However, experiments conducted with oil, diatoms, and the dispersant Corexit 9500A showed that Corexit inhibited the formation of MOS aggregates. Therefore, the net effect of Corexit 9500A on sedimentation rates of oil could not be quantified (Passow et al., 2017). Although the relative importance of dispersant, oil, and marine snow to the MOSSFA phenomenon observed during the DWH spill is difficult to ascertain in situ, controlled laboratory studies (Fu et al., 2014) noted that oil, Corexit EC9500A dispersant and suspended particulate matter, including living indigenous microbes, were all required to enhance MOS formation. Their hypothesis is that, when chemical dispersion increases the volume of small oil droplets in the water column compared with naturally or mechanically dispersed oil, the finely dispersed oil provides increased surface area and substrates for enhanced production of EPS and growth of microbial flocs, as well as facilitating physical interactions with natural suspended solids to form MOS having high oil content. Both sinking and rising of MOS were observed in the experiment (Fu et al., 2014); extrapolation to in situ conditions suggests that MOS sinking would generate a MOSSFA event. The importance of suspended matter in MOS production and growth suggests that deepwater oil plumes are less likely to generate MOS than surface waters. Note, however, that due to the inherent limitations of

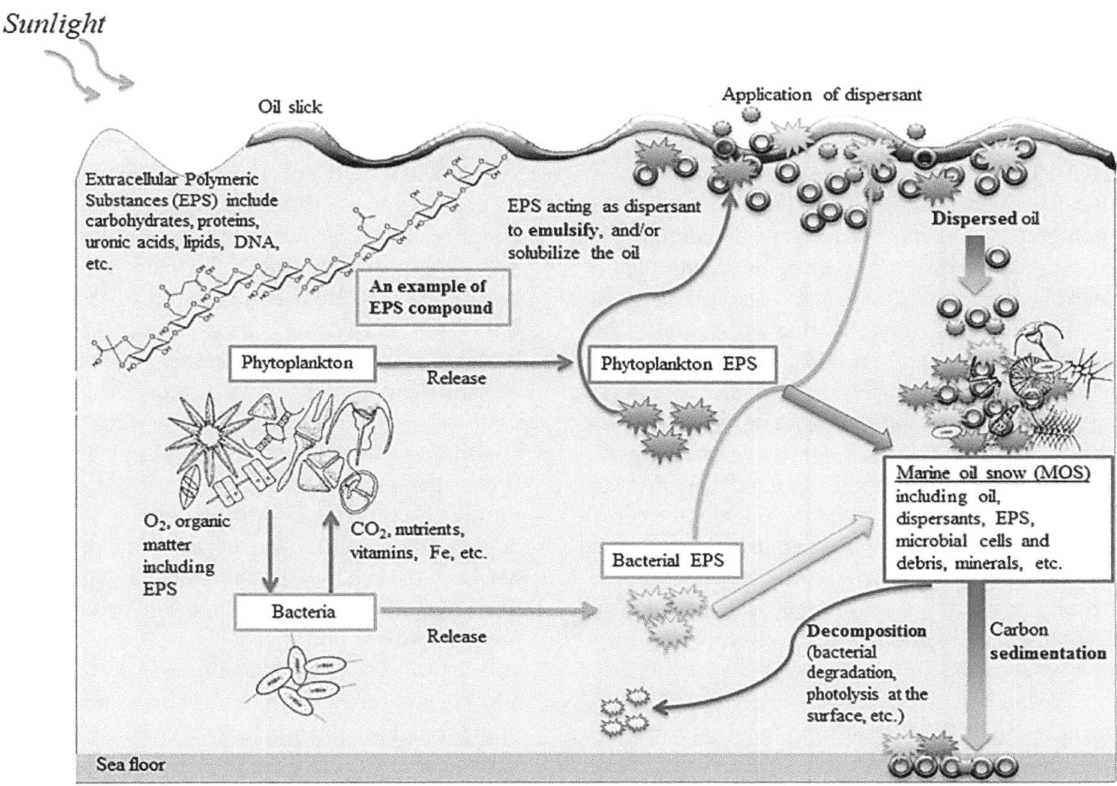

FIGURE 5.29 Interactions between microbes, extracellular polymeric substances (EPS), oil, and chemical dispersant when present to form MOS.
NOTES: EPS adsorbs and disperses oil, thus enhancing its solubility and bioavailability; provides a physical structure for microbes to assemble and degrade oil efficiently; and traps colloidal and particulate matter via coagulation, aggregation, and cross-linking processes. Thus, MOS aggregates may include EPS, oil, dispersant, cells or cellular debris, entrained solutes, particulates and other materials from the water column. Similar processes affect oil in the absence of dispersant when the oil is entrained by wind and waves into the surface water column.
SOURCE: Quigg et al. (2016). CC BY.

laboratory experiments, direct extrapolation to MOSSFA formation in the field may be challenging. The results should be viewed in the context of realistic concentrations, droplet size distributions, natural dilution rates, and other parameters relevant to field conditions.

Living organisms are a key component of MOS, affecting its formation and biochemical activity. Changes in the surface water microbial community composition that occur after oil spills (see Section 5.2.8.2) can affect the production of exopolymers and, subsequently, MOS formation and composition. Primary producers (phytoplankton) as well as heterotrophic bacteria that thrive in oiled water, such as strains of *Alteromonas*, produce large quantities of EPS that further enhance MOS formation, whereas strains of other hydrocarbon-degrading bacteria such as *Pseudoalteromonas*, and *Cycloclasticus* produce relatively small quantities of EPS (Gutierrez et al., 2018). Other biochemical activities also influence MOS characteristics: the patterns of polysaccharide- hydrolyzing enzyme activities in MOS differ significantly from the surrounding water matrix, and the enhanced lipase activity indicates increased biodegradation rates within the aggregates (Doyle et al., 2018; Kamalanathan et al., 2018). Oxygenation patterns in MOS extracts show degraded oil compounds similar to those of environmentally aged oil (Wozniak et al., 2019). That is, MOS aggregates include multifunctional microbial communities involved with oil biodegradation either directly as primary and secondary oil-degraders (e.g., by mineralizing, transforming, and/or assimilating hydrocarbons and degradation products), or indirectly (e.g., by emulsifying oil with EPS). In contrast to obligate hydrocarbon-degrading bacteria that specialize in biodegradation of petroleum components, most MOS species can utilize a broad range of substrates and do not depend exclusively on carbon sources originating from petroleum (Arnosti et al., 2016). Because the communities are dynamic, their composition can shift as the particles sink and oil components are biodegraded. Ultimately, the residual oil deposited with MOS on the seafloor is chemically different from the original spilled oil (Bagby et al., 2016), discussed in Section 5.3.3.6, and the diverse organic matter associated with MOS can be further degraded in seafloor sediments, contributing to the benthic carbon cycle. Marine snow does not only occur in subtropical gyres but also has been reported in the continental shelf off North Carolina and in a fjord of the San Juan Islands (Washington), and in the Baltic Sea (Turner, 2015). Large accumulations of marine snow at a fjord in the San Juan Islands had discrete thin layers of diatoms. Photographs of dense marine snow accumulations in the Baltic Sea show copepods feeding upon particles of marine snow. There is ample evidence of the advances in the past two decades of in situ collection and visualization instrumentation that now facilitates documentation of the presence of marine snow/extracellular polymeric material as well as other particles accumulating on density gradients (Trudnowska et al., 2021). Dynamic interactions between the marine snow/extracellular polymeric organic material and other particles, feeding by some zooplankton, and microbial degradation of natural organic matter contribute to spatial and temporal variability of marine snow deposits.

Fundamental physical chemical properties of medium to higher molecular weight material predict adsorption/absorption to marine snow/extracellular polymeric organic material (see Chapter 2) and has been demonstrated in laboratory studies especially after DWH. It is important that future responses to oil spills, and consideration of the fate and effects of petroleum compounds from oil spills and other inputs, consider marine snow as a potentially important factor for fate of oil.

Environmental and Biological Controls on MOS Formation and MOSSFA Events

Formation of marine snow with incorporation of oil to generate MOS with subsequent settling to the seafloor (i.e., MOSSFA) is a mechanism proposed to explain the fate of some Macondo oil during and after the DWH spill (Quigg et al., 2020). However, it is difficult to quantify how much of the oil released during DWH was transported by MOS to sediments because some oil-containing flocculant material that settled was resuspended from the sediment–water interface and transported with underwater currents. Due to resuspension, sea floor sediment samples did not consistently detect oil but, conversely, oil-associated flocculant material indicative of MOS deposition was found in deep sea coral reefs during and after the DWH oil spill (Passow and Ziervogel, 2016).

Occurrences of MOSSFA events were reported in the literature at different geographical locations before the DWH spill (Quigg et al., 2020), although not by that acronym. However, pre-DWH literature mostly described shallow waters, near-shore environments (where there are high concentrations of mineral particles that promote formation of both MOS and OMA) and small-scale observations.

The phenomenon observed during DWH provided a new perspective for conditions that can result in a MOSSFA event due to (1) a very large quantity of oil entering the marine environment, (2) deep water (offshore) conditions, and (3) possibly the use of subsurface dispersants (Fu et al., 2014). Research efforts after DWH allowed better spatial and temporal resolution for observations and analyses of data to study the MOSSFA event (e.g., Romero et al., 2021).

Formation of MOS aggregates is affected by several combined factors including: timing and location of spill and of response measures, the state (i.e., degree of weathering) and chemical composition of oil, influences from the ocean and coastal environment (including interaction with riverine and shelf processes), and the type of marine biota present and their related processes (Daly et al., 2016; Gregson et al., 2021) (see Figure 5.30). Individually these conditions do not necessarily lead to significant oil deposition on the seafloor in MOSSFA events, such as that observed during the DWH oil spill (see Box 5.10). The deposition, accumulation and biogeochemical fate of oil and associated MOS on the seafloor are influenced by the dynamic interactions of benthic fauna (bioturbation, resuspension, feeding); presence of oil and dispersants (petrogenic hydrocarbons, smaller oil droplets, pyrogenic material generated from in situ burning processes); and riverine and terrestrial inputs of clays and organic matter (MOSSFA, 2013). For example, surface application of Corexit 9500A during the DWH spill was associated with unusually large in situ accumulations of MOS followed by MOSSFA events, described in Box 5.10 and Figure 5.29.

At present, the DWH oil spill is the most-studied example of MOSSFA contributing significantly to the fate of oil through sedimentation to the seafloor. Retrospective analyses suggest that MOSSFA events may have occurred during the 1979–1980 Ixtoc 1 spill in the Gulf of Mexico and other spills (Vonk et al., 2015; Schwing et al., 2020a), but currently it is unknown whether such large-scale events are common or significant outside the Gulf of Mexico and other temperate marine waters. **The possible global significance of MOSSFA events should be examined so that, if relevant, they may be incorporated into response plans, oil spill budgets, and oil spill models, as needed.**

5.3.2.3 Sorption to Plastics

Plastics are produced from petroleum hydrocarbons and enter marine environments; however, they are not considered in the same manner as oils (liquid or semiliquid) as their transport and fate are different from those associated with oil spills at sea. Plastics entering the marine environment and their fate are considered in a new study by NASEM (2021). Examples, as described by the Smithsonian Institution's Ocean Portal Team,[3] are provided in Box 5.11.

Plastics enter the marine environment through activities associated with fishing and aquaculture, intentional disposal

[3] See https://ocean.si.edu/conservation/pollution/marine-plastics.

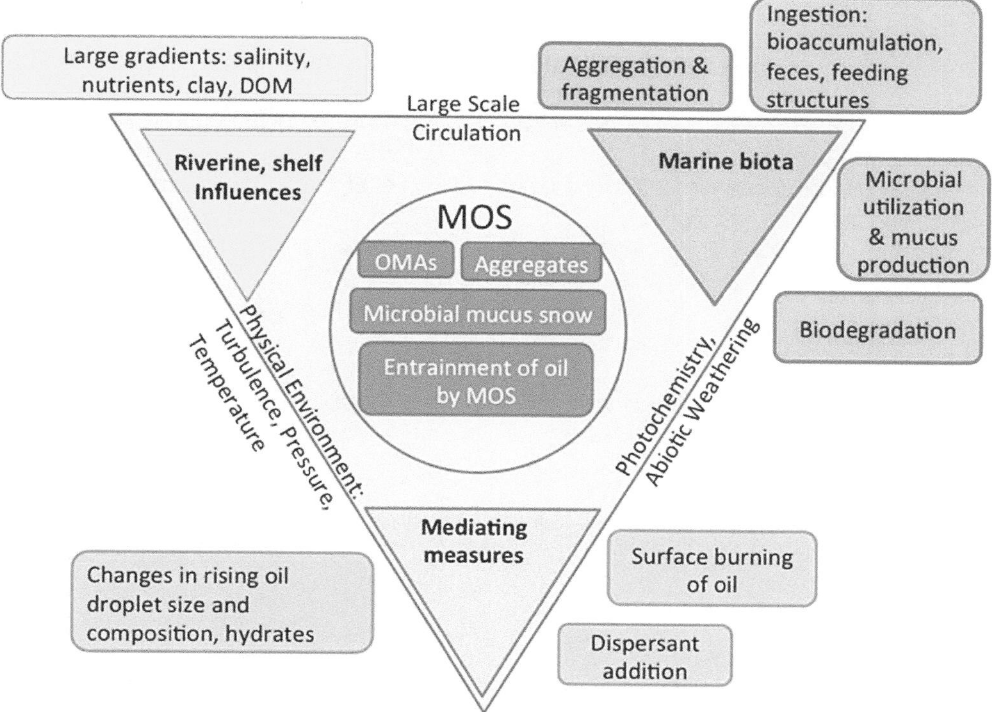

FIGURE 5.30 Environmental and biological factors that influence formation and modification of MOS and potentially to MOSSFA events. SOURCE: Daly et al. (2016). CC BY NC ND.

BOX 5.10
MOS Formation and MOSSA Events During the DWH Oil Spill

Unusually rapid and abundant MOS formation occurred in the oil-contaminated surface waters of the Gulf of Mexico during and shortly after the DWH spill and had a major role in the fate of the spilled oil and its transport to the seafloor. This phenomenon, called MOSSFA (marine oil snow sedimentation and flocculant accumulation), was the first such significant event documented. The presence of thick patches of surface oil promoted the formation of a large masses of MOS with highly variable hydrated densities, suggesting that the aggregates comprised a wide variety of constituents, including weathered and photo-oxidized oil components (MOSSFA, 2013).

Three mechanisms may have led to high concentrations of MOS near the Gulf of Mexico surface after the spill: (1) production of mucous webs by oil-degrading bacteria associated with the floating oil layer to enhance marine snow formation; (2) the coagulation of chemically dispersed and/or photo-oxidized oil with suspended matter and marine snow; and (3) the coagulation of phytoplankton with oil droplets incorporated into aggregates (Passow et al., 2012). In addition, a deliberate elevated and extended discharge from the Mississippi River during the spill generated high concentrations of suspended sediments in the water column. This enhanced phytoplankton growth, suspended particle concentrations, zooplankton grazing and EPS formation (Daly et al., 2016), and may have contributed to subsurface formation of OPAs, especially near the shelf.

As the MOS particles lost buoyancy (due to oil biodegradation within the MOS aggregates and/or to MOS interaction with mineral particles), their porosity and density became similar to those of diatom or miscellaneous aggregates (Passow et al., 2012) and their sinking velocity was high. Thus, notable MOSSFA events occurred due to combinations of oil–particle associations including (1) enhanced marine snow production due to oil and chemical dispersant at the surface; (2) formation of MOS both near the surface and during passage of marine snow through the dispersed subsea plume of oil; (3) enhanced formation of OPA due to high suspended sediment load.

A significant fraction of the oil from the DWH spill submerged, some of which was entrained in MOS, and was deposited on the seafloor and on corals during MOSSFA events (Passow and Overton, 2021). The sedimented oil therefore represented a proportion of subsurface oil that reached the (near-) surface but was not reclaimed at the surface, plus an unknown proportion of the subsea dispersed oil plume that sedimented as MOS and OPAs. The oil that reached the sea floor was chemically altered compared with the original oil, and continued to change after burial (Bagby et al., 2016).

BOX 5.11
Types of Plastics Entering the Marine Environment

1. Polyethylene terephthalate (PET or PETE): water and soda bottles, microwavable meal trays, and clothing.
2. High-density polyethylene (HDPE): grocery bags, garbage bags, shampoo bottles, and some bottles and caps.
3. Polyvinyl chloride (polyvinyl, vinyl or PVC): including raincoats, shower curtains, plumbing materials, garden hoses, and window frames.
4. Low-density polyethylene (LDPE): plastic wrap, bags, squeeze bottles, toys, and gas and water pipes.
5. Polypropylene (PP): medicine containers, diapers, rope, and outdoor furniture.
6. Polystyrene (PS):
   - Solid polystyrene: CD cases and cassettes, cups
   - Styrofoam (PS in foam state): egg cartons, packing materials, take-out containers and disposable plates are made out of foamed polystyrene
7. Other (O):
   - Polycarbonates: Compact discs for DVDs, eyeglasses, and "glass" panels of greenhouses
   - Polylactic acid: industrial compostable containers and cups
   - Nylon: clothing, car tire components, and rope
   - Acrylonitrile butadiene styrene (ABS): Legos and toys

Plastic Debris by Size

Particle Category	Diameter Range (mm)	Examples of Debris
Nanoplastics	<0.0001	Fragments from larger items
Small microplastics	0.0001–1.00	Fragments from larger items
Large microplastics	1.00–4.75	Buttons, thumbtacks, eyedroppers, string, "nurdles"
Mesoplastics	4.76–200	Plastic bottles, plastic cups, beach balls, combs, plastic spoons
Macroplastics	>200	Flip flops, balloons, umbrellas, gas canisters

SOURCES: Eriksen et al. (2014); https://www.vox.com/2016/5/23/11735856/plastic-ocean.

from vessels, and storm-related debris (Erickson et al., 2014; Hale et al., 2020). Plastics and construction debris are deployed in water bodies for the creation of artificial reefs to create habitats for finfish and shellfish. Episodic events such as storms and hurricanes in coastal areas result in floodwaters that carry significant quantities of plastics to coastal waters. Depending on the treatment processes used, wastewater discharges also may contain microplastics (MP). "Nurdles," the preproduction building blocks for nearly all plastic goods, may enter the oceans through shipping mishaps. The pellets, weighing about 20 mg each, usually float or have neutral buoyancy. They absorb toxic chemicals from the water column and are often mistaken for food by animals.

Floating plastics in the marine environment are transported by wind, wave action, and ocean currents. Plastics can adsorb hydrophobic organic pollutants and metals, and the adsorption capacity increases with weathering and aging of the material (Yu et al., 2019; Li et al., 2020). Boom material used for oil absorption of oil spills is made of plastic compounds. As a result, once oiled, such booms may become another part of the plastics waste stream. Salinity, temperature, and pH can affect the adsorption characteristics of hydrophobic compounds on plastics. The petroleum-based hydrophobic compounds adsorbed on plastic debris also can be transformed and/or transported by photo-oxidation, biofouling of surfaces (microscopic, algae), marine growth (macroscopic), sinking/sedimentation, deposition on sediments, and consumption by marine organisms (e.g., fish, birds). It is estimated that atmospheric deposition of synthetic fibers (such as from laundering synthetic fleece) is between 3 and 10 metric tonnes each year, and on a 2,500 km^2 area can deposit 355 particles m^{-2} day^{-1} with the majority consisting of small fibers 7–15 μm in diameter (Dris et al., 2016).

The circulation patterns in the oceans create accumulation zones on the ocean's surface including plastic debris in subtropical gyres (Howell et al., 2012; Maximenko et al., 2018; Li et al., 2020). Physical oceanographic phenomena (e.g., Ekman transport, geostrophy) explain why plastic debris accumulates within the more calm areas of oceanic gyres, while more complex processes transport marine plastics in less obvious locales, including deep sea sediments and ice sheets (Hale et al., 2020).

Sedimentation rates and characteristics of microplastics are different from deposition characteristics of natural sediment. Microplastics have been found in deep-sea sediments cores from the Arctic Central Basin, North and South Atlantic, Southern Ocean, Mediterranean Sea, and Indian Ocean (Erni-Cassola et al., 2019; Kanhai et al., 2019). Microplastics are found in fish and other organisms, coastal sediments including beaches, shorelines (Browne et al., 2010), and coastal lagoons (Vianello et al., 2013). The most common polymers identified in the intertidal sediments from urban and semi-natural southwest Atlantic estuaries included polyethylene, polyethylene terephthalate/polyester, polyvinyl chloride, and polypropylene. Different microplastic characteristics among the estuarine environments suggests

TABLE 5.3 Fate and Transport Mechanisms of Plastics That Affect Fate and Transport of Oil in Marine Environments

Sources of Plastics	Interactions of Plastics with Oil at Sea
• Atmospheric deposition • Stormwater runoff • River discharges • Ship debris • Fishing lines/nets • Artificial reefs	• Adsorption/absorption of oil and other hydrophobic compounds • Weathering and solubilization of hydrocarbon fragments • Breaking into smaller pieces/particles • Aggregating (e.g., during MOS formation, oil-particle aggregation) • Photo-oxidation • Decomposition/denaturing (release of hydrocarbon fragments) • Biofouling (microscopic, algae) • Marine growth (e.g., macroscopic, barnacles which can metabolize oil) • Sinking/sedimentation (e.g., during MOSSFA) • Deposition on sediments (affecting sediment characteristics) • Biota transport as consumption by marine organisms (e.g., transport of adsorbed oil on plastics with fish and birds)

different anthropogenic sources (Díaz-Jaramillo et al., 2021). High concentrations of denser polymers (e.g., polyester, polyamide, and acrylic) are found in intertidal and subtidal sediments, as well as locations with periodic polystyrene inputs (Erni-Cassola et al., 2019).

At sea, these plastics can interact with oil and other hydrophobic compounds. Depending on the size of plastic particles and type of the polymer, different types of plastic particle–oil interactions can occur (see Table 5.3). Plastics can become covered by oil, microorganisms, algae, invertebrates, and other organic and inorganic materials that then secondarily are exposed to the oil. Biofilms on plastic debris may also support microorganisms that expel extracellular enzymes which may degrade oil and/or polymers (Dang and Lovell, 2016).

Oil adsorbed on macroplastics can be transported long distances with ocean currents. Microplastics can be incorporated into aggregates that form during coagulation of organic matter and oil droplets. Marine snow can transport microplastics of different shapes, sizes, and polymers and enhance their bioavailability to benthic organisms (Zhao et al., 2018). Mass sedimentation episodes of marine aggregates (e.g., MOS) can capture slowly sinking particles, including microplastics (Passow and Stout, 2020). Sinking rates of microplastics increased when incorporated into marine snow and also increased microplastic bioavailability for mussels (Porter et al., 2018).

Plastics are often ingested by marine organisms including micro- and nanoplankton species and oil adsorbed on the plastics can be metabolized (Andrady, 2011) (see Figure 5.31).

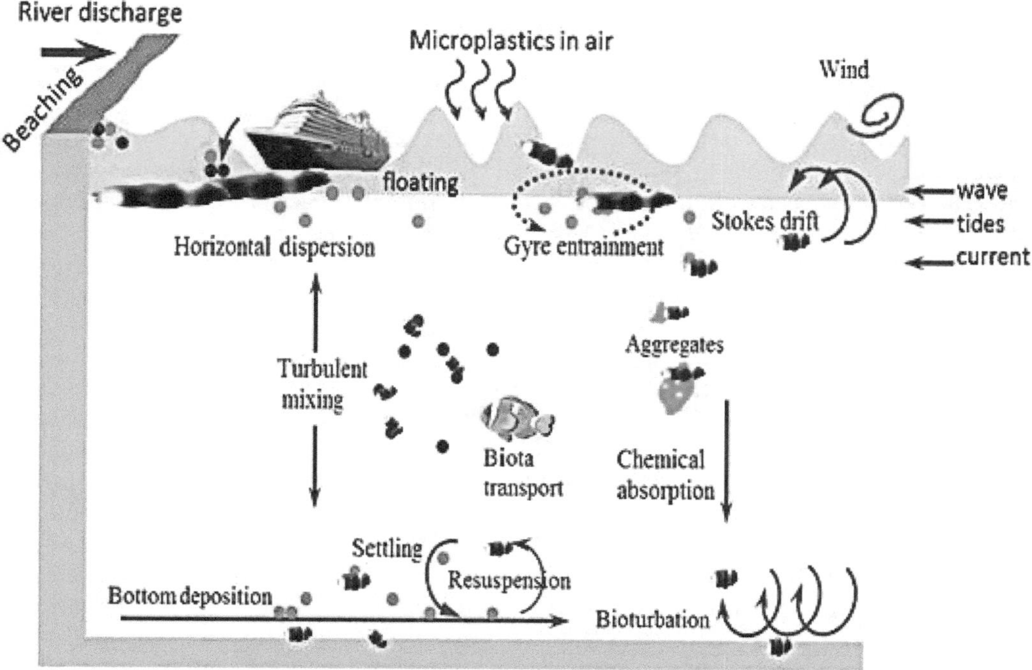

FIGURE 5.31 Schematic of the transport pathways of microplastics in the ocean.
SOURCE: Modified from Li et al. (2020).

5.3.2.4 Submergence and Sinking of Heavy Oils and Semi-Solid Oils

Conventional oils are less dense than seawater and float or remain suspended in seawater even after weathering, whereas some unconventional oils have densities greater than that of freshwater (1.00 g/cm^3) and therefore may sink in freshwater. Stated in terms of American Petroleum Institute (API) gravity, oils having >10° API will float and those of <10° API will sink in freshwater (NASEM, 2013). This behavior will differ in seawater, which has a typical density ≥1.02 g/cm^3. Therefore, Bunker C (No. 6 Fuel Oil), ranging in density from 0.95 to ≥1.03 g/cm^3 and with API gravity ~12° (see Chapter 2), may float or sink in seawater. Furthermore, interactions with mineral particles and/or organic aggregates as described in Section 5.3.2.1 may cause otherwise buoyant oil types to sink. For example, the inland Refugio pipeline spill in 2015 near Santa Barbara, California, released a diluted heavy crude oil that flowed over earth, sand, and gravel before entering the ocean. Much of the oil floated and oiled nearby beaches, but a portion submerged and sank, likely because it adhered organic and mineral particles from the beach and surf zone, and affected near-shore reefs and kelp beds (NASEM, 2016a).

Natural (undiluted) bitumen has a density of 0.977–1.016 g/cm^3 and API gravity <10° (Fingas, 2015), in principle allowing it to sink in fresh or brackish water but not seawater. However, it is not transported (or spilled) as neat bitumen but rather as dilbit, a blend of bitumen and light hydrocarbon diluent (see Chapter 2). The diluent decreases the density of the blend sufficiently to enable spilled dilbit to float, at least initially, but floating results in rapid evaporation of a significant proportion of diluent, enabling the residual dilbit to sink in freshwater, as observed in laboratory studies (e.g., Stoyanovic et al., 2019) and in the Kalamazoo River spill in 2010 (NASEM, 2016a). The potential for sinking in seawater and brackish water has been debated and, in the fortunate absence of significant marine spills of dilbit, a variety of laboratory and ex situ tests using outdoor flumes and tanks has been conducted to gain insight into sinking potential under simulated spill conditions. The relevance of some observations has been criticized because some experiments failed to include one or more environmental factors such as solar irradiation (precluding photo-oxidation effects), in situ water temperatures (affecting oil viscosity and evaporation), relevant mixing energy (affecting dispersion of oil), oil confinement (slick thickness, affecting evaporation), and presence/absence of suspended sediments (OPA formation; see Section 5.3.2.1). Nonetheless, trends can be discerned from the combined results. In laboratory simulations, Hua et al. (2018) found that the weathering state of dilbit and the particle size and type were major factors influencing interaction and buoyancy in synthetic seawater. Sediment loads similar to seasonal maxima at coastal river outflows formed OPAs when mixed artificially with fresh or lightly weathered dilbit (0% or 16% mass loss) and subsequently sank, whereas heavily weathered dilbit (25% mass loss) had less interaction with sediment and formed submerged oil balls, some of which later resurfaced but did not sediment. Sand particles had almost no interaction with dilbit in comparison with clay and benthic sediment particles (Hua et al., 2018).

The general consensus appears to be that weathered dilbit can submerge (but not sink) as discrete oil balls in brackish water and even in sediment-free coastal seawater (King et al., 2014), especially in fjords or near river outfalls if heavy rainfall decreases the salinity of the upper 1–3 m of water (Johannessen et al., 2020). Light or mild weathering of dilbit could increase OPA formation, given sufficient sediment load and mixing energy, but would also maintain buoyancy so that lightly weathered dilbit should not sink in coastal seawater with seasonal densities of ≥1.02 g/cm^3 (Ortmann et al., 2020) and therefore would not sink in higher salinity offshore seawater. Heavily weathered dilbit is more dense, but interacts less with suspended sediment and therefore is more likely to form suspended oil balls than to sink. The possible contribution of dilbit to MOSSFA events (see Section 5.3.2) has not been examined, but because weathered dilbit is poorly available, with only very selective biodegradation having been documented in situ (Schreiber et al., 2021), theoretically it should not support marine snow formation to enhance sedimentation. **Although research into the fate of heavy oils and diluted bitumen in the ocean has accelerated in the past decade, some questions remain regarding weathering properties, behavior (particularly the sinking potential), and biodegradability of such oils in the sea; additional ex situ research under relevant conditions would help address this knowledge gap.**

5.3.3 Deep Sea and Deep Sediment Processes

Accidental subsea releases of oil and gas include oil well blowouts, pipeline leaks, and releases from sunken ships. Near-surface processes (see Section 5.2.1) are mostly limited to the upper mixed-layer of the ocean. Here, we consider processes below 1,000 m depth, extending to the deep seabed. Leaks from accidental oil well blowouts or pipeline ruptures are expected to be localized, energetic inputs of oil, gas, or mixtures of oil and gas. The Ixtoc I and *Deepwater Horizon* oil well blowouts are historical examples. These may form an atomized spray of oil droplets and gas bubbles from the release orifice that combine with entrained seawater to form buoyant jets, which rise as buoyant plumes through the ocean water column, and depending on the depth, form lateral, subsurface intrusion layers (Socolofsky et al., 2011). Above an intrusion layer, individual bubbles or droplets may continue to rise by their own terminal rise velocity (Dissanayake et al., 2018). Alternatively, if the well failure causing an oil well blowout occurs subsurface, it is possible that oil and gas would flow into the sub-bottom geologic formation and percolate diffusely through the seabed sediments. This may result in much less rigorous releases distributed over a larger area such that droplet and bubble formation follow different dynamics, and a buoyant plume stage of transport may not form. Similarly, releases from sunken ships would be expected to have much less energy at the source, forming oil droplets by sinuous-wave breakup (Masutani and Adams, 2000). Entrainment of ambient seawater and formation of a buoyant plume would not be expected. Like the bubbles or droplets escaping the intrusion layers of the more rigorous blowout plumes, droplets released from diffuse seabed leakage and sunken ships rise at their individual terminal velocities. Hence, the main differences among these types of subsea releases are in the droplet formation dynamics and whether or not a buoyant plume stage may be expected.

As noted earlier, most of the fate processes for oil in the sea depend on the interfacial area of the droplets or bubbles, which increases per unit volume of oil as the droplet size decreases. Smaller droplets also rise slower and are transported along different trajectories than larger droplets. One process that accelerates the vertical rise of droplets independent of their sizes is a buoyant jet or plume. When a plume forms, it can transport small oil droplets or gas bubbles at plume speeds of about 0.5 to 1.0 m/s (Gros et al., 2020). Eventually, oil and gas plumes transiting the ocean water column are arrested by the ambient density stratification and will form a lateral intrusion at a level of neutral buoyancy. Bubbles and droplets may rise out of this intrusion at their own terminal rise velocity, rising toward the surface without the assistance of subsequent plumes. Point-source releases from broken pipelines or a crippled oil well are expected to have a buoyant plume stage of transport. Releases from sunken ships or diffuse seepage over a wide region of the seabed leads to weak plume coherence; hence, the plume stage of transport would not be expected (Wang et al., 2019, 2020), and oil and gas would rise at their terminal velocity throughout their ascent from the source.

5.3.3.1 Plume Dynamics and Intrusion Formation

For an oil well blowout or pipeline leak through a localized orifice with sizes of order 1 m in diameter or smaller, the released oil and gas would be expected to mix with ambient seawater and form a buoyant jet at the release that would quickly transition to a vertically rising buoyant plume. The dynamics of this buoyant plume will depend on the total buoyancy flux of oil and gas released at the source, and droplets and bubbles engulfed within the plume would rise at the plume velocity plus their terminal rise velocity (Socolofsky and Adams, 2005; Dissanayake et al., 2018). To understand the breakup dynamics causing formation of oil droplets and gas bubbles, we need to consider the turbulent dynamics in the buoyant jet and plumes stages of transport. To predict the rise of these droplets and bubbles in the ocean water column, we must also understand the buoyant plume dynamics and its interactions with ambient ocean currents and density stratification.

There are three stages of turbulent dynamics moving from the release point to the buoyant plume. At the release, the turbulent properties may be characterized by the turbulence upstream of the release, normally some form of pipe-flow turbulence. Once released to the ocean, this stream of oil and gas rapidly mixes with seawater through a zone-of-flow-establishment (ZFE) to form a buoyant jet. When gas, which carries high buoyancy, is released the jet, which is dominated by the released momentum, quickly transitions to a plume, which is dominated by the released buoyancy. See Lee and Chu (2003) for a comprehensive review of these dynamics applied to single-phase (e.g., wastewater) discharges. For the DWH oil spill, the release at the end of the broken rise transitioned from the pipe flow to a buoyant plume (vertically rising) within one frame of the ROV camera, or a few pipe diameters downstream. Bubble and droplet breakup will follow different dynamics in the ZFE, jet, and plume stages of a blowout because the dissipation rate of turbulent kinetic energy follows different scaling laws within each of these regimes. Within the ZFE, the turbulent dissipation rate is constant and depends on the release velocity and orifice diameter. In the jet stage, the turbulent dissipation rate rapidly decreases, scaling with inverse distance from the source to the fourth power. The decrease in turbulent dissipation rate is less in the plume stage, scaling with inverse distance to the second power, owing to energy input from the buoyancy driving the plume. These scaling laws are likewise reviewed for single-phase releases in Lee and Chu (2003) and applied to breakup of oil droplets by Zhao et al. (2014).

In multi-phase flows, involving immiscible bubbles or droplets like releases of oil and gas into seawater, the behavior alters from that of single-phase plumes due to the slip velocity of the bubbles and droplets relative to the entrained seawater. Two main consequences are important here. First, bubbles and droplets are not required to follow the entrained seawater but instead may separate from the plume (see Figure 5.32). This may occur in lateral currents that bend over the plume in the downstream direction (see Figure 5.32a)—the vertical rise velocity of the bubbles or droplets then causes their trajectories to escape on the upstream edge of the entrained seawater plume (Socolofsky and Adams, 2002; Murphy et al., 2016). This may also occur when ocean density stratification prevents the entrained seawater from rising above a level of neutral buoyancy (see Figure 5.32b), where the entrained seawater plume would be expected to be arrested by the stratification and intrude laterally (Socolofsky and Adams, 2003, 2005). In the deep oceans, both currents and density stratification may be important. Figure 5.32c depicts the case of a stratification-dominated plume in a weak crossflow; separation dynamics and empirical equations for predicting the occurrence and scales are discussed in more detail in Socolofsky et al. (2011).

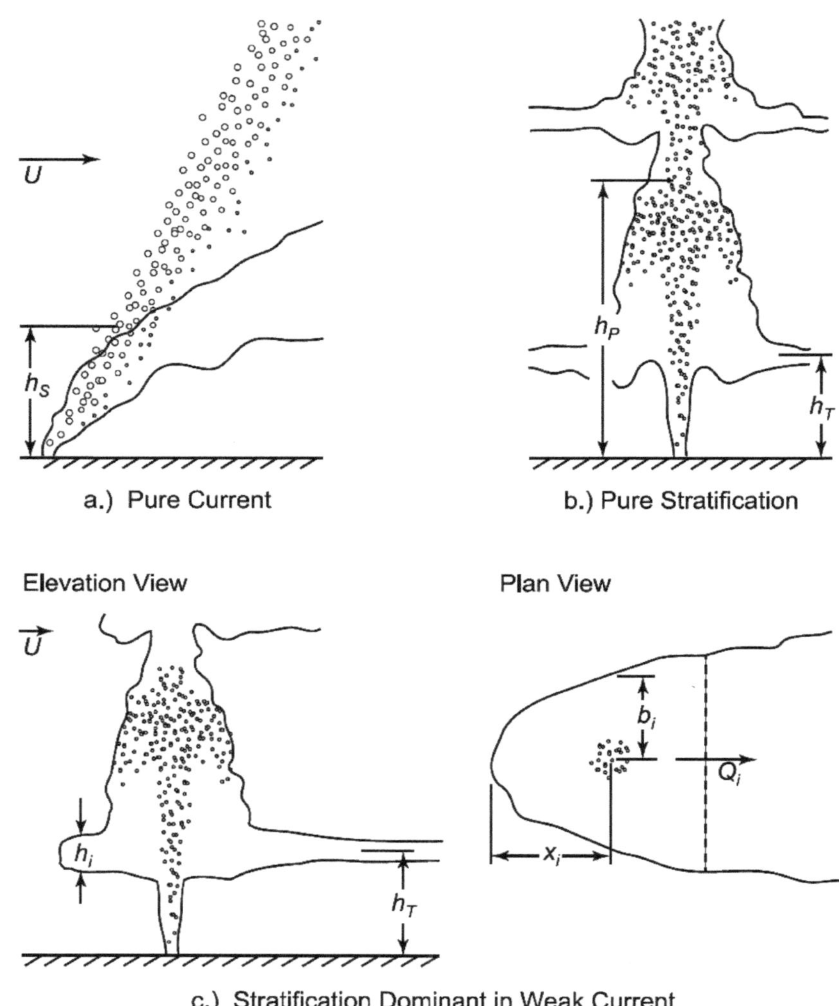

FIGURE 5.32 Schematic of a multiphase plume in (a) pure current, (b) pure density stratification, and (c) a weak current in which density stratification dominates the plume dynamics.
SOURCE: Adapted from Socolofsky et al. (2011).

Second, the presence of bubbles or droplets in the flow breaks down the assumptions used to derive solutions for single-phase buoyant; namely, self-similarity is no longer strictly valid (Socolofsky et al., 2008). This means that turbulence dynamics in the flow are altered somewhat from their single-phase analogues. Plume spreading may be affected by the breakdown in self similarity (Seol et al., 2009), and the classical entrainment coefficient is no longer constant along the plume trajectory (Milgram, 1983; Seol et al., 2007). This is important for the development of one-dimensional integral models based on the entrainment hypothesis; however, as for single-phase plumes in density stratification, models based on self-similarity remain robust even for significant deviations from the self-similarity requirements (Turner, 1986). An important consequence of this fact for understanding oil well blowouts is that the turbulent eddy dissipation rate may not follow correlations for classical jets and plumes, with important contributions to the turbulence production term arising from the wakes of the oil droplets and gas bubbles (Fraga et al., 2016; Zhao et al., 2016, 2017; Lai et al., 2019).

For more details on the plume dynamics as they relate to accidental oil well blowouts, see the reviews by Socolofsky et al. (2016) and by Boufadel et al. (2020a,b).

5.3.3.2 Bubble and Droplet Breakup from Subsurface Leaks

Since the 2003 *Oil in the Sea* report and following the DWH oil spill, there has been tremendous advancement in data and models to predict the droplet and bubble sizes of oil and gas released subsea. Here, we consider both slow or diffuse releases as well as oil well blowouts. Oil droplet size is important in subsea releases because it determines the rise velocity of the droplets and sets the interfacial area for dissolution and biodegradation fate processes. For deep releases, these fate processes may represent a significant fraction of the overall release mass balance.

Oil droplet formation for subsea releases may occur by different mechanisms, depending on the flow rate and source geometry. Masutani and Adams (2000) reported on these breakup processes based on observations of different types of oil into water. They showed that at slow releases, droplet breakup is by pendent drop formation at the lowest release rates followed by sinuous wave breakup as the release rate increases. At higher flow rates, an atomization breakup regime is found. The boundary defining the onset of atomization occurs at a jet Weber number of 324 (Johansen et al., 2013). Droplet sizes for the pendent drop formation and sinuous wave breakup scale with the smaller of the orifice diameter or the maximum stable droplet size for a drop rising in a quiescent ambient fluid (Clift et al., 1978). Droplet sizes in the atomization stage depend on the turbulence dynamics at the release and in the ensuing jet or plume. Slow or diffused subsurface releases fall outside the atomization region; accidental oil well blowouts likely fall in the atomization region.

Slow or Diffused Subsurface Releases

For oil releases from sunken wrecks, natural seeps, or diffuse seepage through the seabed from a subsurface oil well blowout, the droplet formation process follows the pendent drop or sinuous wave breakup modes. For a wide range of oils, the maximum stable droplet size is about 10 mm in diameter (Li et al., 2017). Maximum stable gas bubbles sizes are more complicated to estimate, requiring an iterative solution to a stability analysis problem (Grace et al., 1978). For typical natural gases in seawater, maximum stable bubble sizes can be over 100 mm in equivalent spherical diameter. Hence, bubble sizes scale more often with the orifice diameter. For example, for a wide range of natural gas seeps in the deep Gulf of Mexico and for laboratory experiments using air in a 16 m deep tank, gas bubble size distributions ranged from 1 to 10 mm, with median diameters of about 5 mm and maximum sizes up to 15 mm (Wang et al., 2016, 2018, 2019, 2020). These sizes are similar to the expected oil droplet size from weak sources, assuming no chemical dispersants are used, which is reasonable for these slow or distributed sources. As a result, for diffuse or weak sources, millimeter-scale droplets and bubbles are expected to form via non-atomizing breakup processes.

Strong Buoyant Jet Releases

For releases from broken pipelines or accidental oil well blowouts, the droplet formation process is expected to follow dynamics in the atomization regime of breakup. There, the turbulence in the released fluids and entrained seawater act to achieve breakup, with breakup following the Weber number and viscosity number parameterizations given by Hinze (1955) and Wang and Calabrese et al. (1986) (see Section 5.2.2.3 and Box 5.3). To apply these scaling relationships in empirical equations for droplet size, the dissipation rate of the turbulent kinetic energy must be evaluated for the buoyant jet dynamics in the primary breakup regime of the release. As noted above, strong buoyant jet releases include a ZFE, jet, and plume stage of development, and different scaling relationships for turbulent dissipation rate apply within each of these zones. Approaches using empirical equations normally take the turbulence characteristics from relations valid within the ZFE (Johansen et al., 2013; Li et al., 2017; Brandvik et al., 2021). Population balance models, which can consider time-varying turbulence properties, may consider all three stages from the ZFE to the plume transport stage (Zhao et al., 2014; Nissanka and Yapa, 2016). Appendix F includes a brief summary of how these approaches arrive at their model equations. Whichever approach is used, the equations and models must be calibrated and validated to measured droplet size distributions for experiments involving localized releases into water.

When conducting laboratory experiments in fluid dynamics, it is normally required to identify the governing nondimensional variables, here the Weber number and one of either the viscosity, Reynolds, or Ohnesorge numbers (see Box 5.3), and to match their values between the laboratory experiments and the field-scale prototype. This approach is called dynamic similitude. For oil jets into water using crude or refined petroleum products as the dispersed phase, it is not possible to match the Weber and viscosity numbers between the laboratory and the field without conducting the laboratory experiment at full scale or changing the scales of the jet to plume transition. As a result, existing observations must be extrapolated to the field scale.

Figure 5.33 shows an example of the laboratory and field data available during the Model Intercomparison Study reported by Socolofsky et al. (2015), extended with data at higher modified Weber number from Brandvik et al. (2021). The data are plotted using the modified Weber number, defined by Brandvik et al. (2013), which combines the Weber and viscosity numbers into a single parameter. Symbols show the laboratory and DeepSpill field observations. The parameter space of the DWH oil spill is also shown as the shaded region in the figure along with extrapolated predictions for the droplet size of untreated DWH oil using the empirical equations of Johansen et al. (2013) and Li et al. (2017) and the population balance prediction of Gros et al. (2017). These model predictions for the DWH span more than an order of magnitude in the non-dimensional droplet size: clearly, extrapolating the observations to the scale of the DWH is difficult. One solution might be to use different fluids with properties yielding the field-scale parameters, but to have an accurate physical effect of the potential subsea dispersant injection, real oils and dispersants must be used in the experiments. This difficulty in making relevant laboratory measurements at adequate scales has been noted

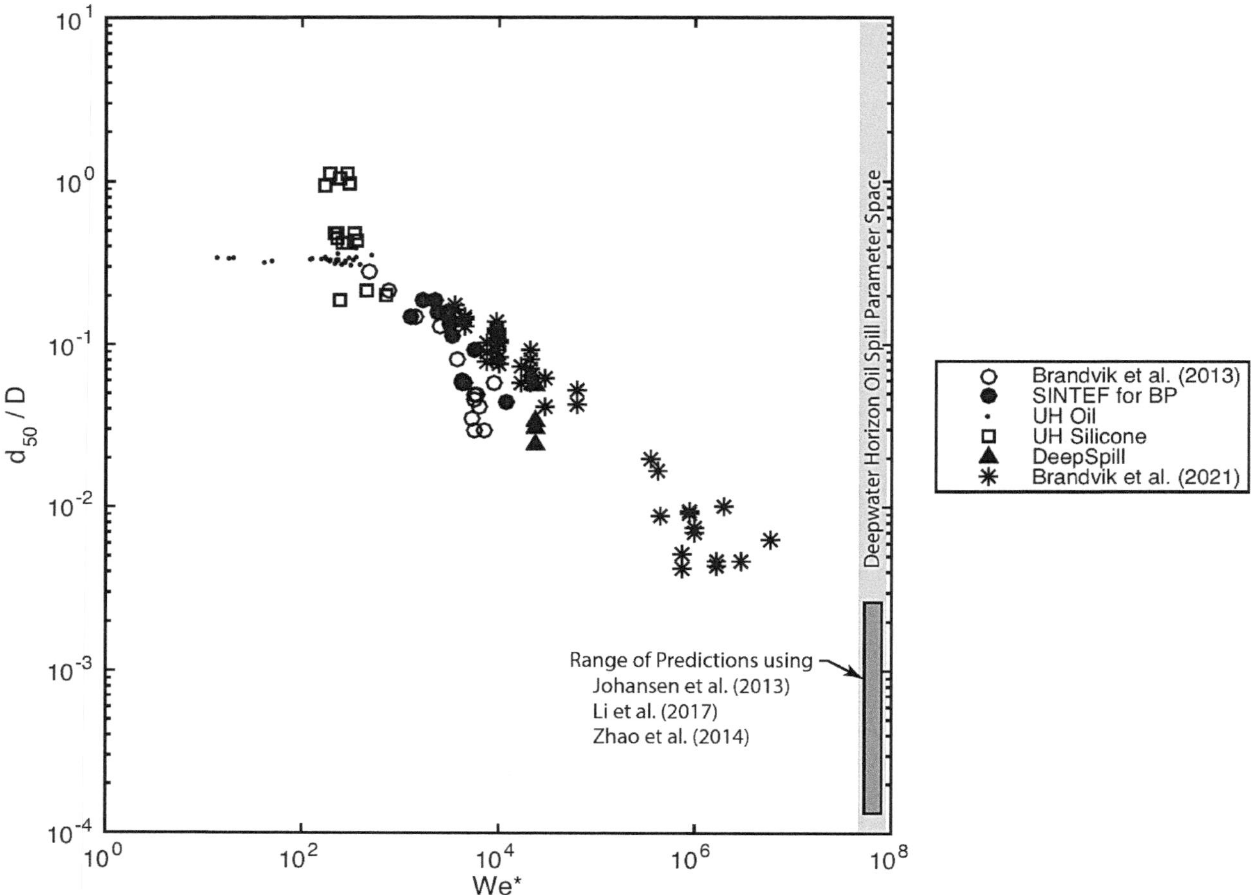

FIGURE 5.33 Plot of the available laboratory data for volume median droplet size (d_{50}) normalized by the orifice diameter (D) from a straight orifice versus to modified Weber number (We*) as defined in Brandvik et al. (2013).
NOTES: Laboratory data include data published in Brandvik et al. (2013) using the SINTEF Tower Basin, additional data collected by SINTEF in the Tower Basin and provided to British Petroleum, data reported in Masutani and Adams (2000) for experiments at the University of Hawaii for crude oil and silicone oil, and data reported by Brandvik et al. (2021) for experiments in the Ohmsett facility. The results of the DeepSpill field experiment (Johansen et al., 2003) are also shown. The grey region on the right-hand-side of the figure depicts the parameter space relevant to the DWH using flow rate and property data reported by Gros et al. (2017) for June 8, 2010. The highlighted gray bar spans the range of droplet size predictions obtained using the empirical equations from Johansen et al. (2013) and Li et al. (2017) and the VDROP-J model (Zhao et al., 2014), each computed using the flow and property values reported by Gros et al. (2010) for June 8, 2010. Note that the field-scale parameter space is at nearly two orders of magnitude higher in the modified Weber number than the data; hence, extrapolation is required to make predictions. Moreover, the model predictions span a full order of magnitude. The smallest droplets are predicted by the Li et al. (2017) model; the largest droplets are predicted by VDROP-J (Zhao et al., 2014); and the Johansen et al. (2013) predictions using the update parameter calibration falls near the middle of the plotted span of data. Whether laboratory trends may be extrapolated linearly in log-log space to make the field-scale predictions requires additional data for validation.
SOURCE: Brandvik et al. (2013).

numerous times previously (NASEM, 2020) and remains a problem for validation of droplet size equations and models today. **Because complete dynamic similitude is not achieved between existing jet break-up experiments and field prototypes, large-scale field or laboratory experiments are needed to demonstrate the reliability of existing droplet-size prediction equations to the field scale. This can be achieved either by full-scale experiments or by demonstrating the prediction results at smaller scales become independent of the non-matched scaling parameters. These data are needed for oil droplet breakup with and without co-flowing gas and with and without subsea dispersant injection.**

One approach to bridge this gap is to utilize physics-resolving numerical models. Aiyer et al. (2019) developed a numerical approach that combines the population balance equations within a large eddy simulation (LES) fluid dynamics model to solve for the evolution of the droplet size distribution in space and time owing to the resolved and subgrid scale turbulent energy predicted by the LES. The population balance model follows the approach of Zhao et al. (2014a), but includes breakup within the viscous subrange; the LES treats the dispersed phase using an Eulerian description (Yang et al., 2015). Aiyer and Meneveau (2020) adapted the coupled population balance and LES model to an oil jet into water, using a one-dimensional approximation of the model at the orifice to resolve the smallest scales of jet and breakup dynamics. This approach allows the ability to study the non-uniform and unsteady breakup dynamics. Importantly, they were able to quantify the evolution of characteristic diameters of the distribution, including the total surface area of the whole size distribution, given by the Sauter mean diameter of the simulated distribution. Although the actual size distributions showed strong dependence on the Weber number, the shapes of the lateral profiles of normalized properties of the size distribution did not depend on the Weber number. This indicates that the breakup process is largely self-similar and that the approach in VDROP-J, which uses uniform distributions of velocity and size distribution, is appropriate.

Other scaled experimental observations also study details of the breakup physics, including the effects of pressure, live oil, and the fine-scale details of droplet formation and turbulence in the immediate vicinity of an oil jet release. Malone et al. (2018) released dead and live oil mixtures of Louisiana sweet crude (LSC) and *n*-decane, using methane as the gaseous chemical species for creating the live oil liquids in the live oil experiments. They simulated the breakup of droplets from a circular orifice and measured the size distributions. For both LSC and *n*-decane, the measured droplet size distributions had larger median droplet sizes for the live oil cases than for the dead oils by about a factor of 2 despite the live oils having lower viscosities and only slightly higher interfacial tensions to the corresponding dead oils (Malone et al., 2018). Aman et al. (2015) also used live oil in high-pressure experiments for a mixing tank (i.e., not a jet into water). Although they did not observe dead oil, droplet sizes observed for the live oil experimental conditions agreed with predictions based on empirical models fit to literature data on water-in-oil mixing systems at atmospheric pressure. This suggests that the breakup of live oils can be predicted using equations fit to data for dead oils, provided the live oil properties are used with the prediction equations. Indeed, experiments reported by Brandvik et al. (2019) for live oil and gas jets into water at high pressure showed good agreement between predictions using the Johansen et al. (2013) equation with in situ live-oil properties. They found no effect of pressure. The effect of co-released gas was to reduce the oil droplet sizes compared to experiments with the same pure oil flow rate owing to the higher exit velocity of oil when released with gas. A mild effect of pressure in oil and gas releases was also observed due to the increased density of gas with higher pressure, resulting in greater momentum flux at the release. **In summary, it appears that droplet size distribution models developed using experiments for dead oil at atmospheric conditions accurately capture the relevant physics and may be used for live oil at high pressure provided the relevant release dynamics, including in situ buoyancy, momentum fluxes, and oil and gas properties, are evaluated.**

Laboratory experiments utilizing measurements at the droplet scale of droplet formation from a jet were reported by Xue and Katz (2019). They used a silicone oil jet into sugar water in order to match the index of refraction in the dispersed oil and continuous water phases; they visualized the breakup process immediately above the nozzle and to distances of 30 nozzle diameters using planar laser induced fluorescence (PLIF). Sample results for three different jet Reynolds numbers are shown in Figure 5.34. The experimental images clearly show the engulfment of ambient water by entrainment into the oil phase and the later breakup of the oil into droplets. A distinguishing feature of the formed droplets was the inclusion of smaller water droplets within individual oil droplets for the experiments with jet Reynolds numbers of 1358 and 2122. These compound droplets had been observed in previous droplet breakup studies, but not previously for jet breakup. The included water increases the bulk density of the oil–water mixture of a compound droplet compared to the density of pure oil. The included water also increases the total surface area of oil exposed to seawater. For some of the larger oil droplets, multiple water droplets were included, sometimes as a cascade of Russian dolls, with water inside oil inside water inside oil. The droplet breakup process was observed to continue to at least 30 nozzle diameters downstream, indicating that primary breakup occurs beyond the ZFE of the buoyant jet and out into the self-similar region of these multiphase plumes.

Similar fine-scale experiments were also reported by Xue et al. (2021), including also velocity measurements of the oil and water phases separately using high-resolution particle image velocimetry (PIV) simultaneously with PLIF. The final spacing between PIV velocity vectors was as small as 128.5 μm. From these data, Xue et al. (2021) could compute the time-average velocity and fluctuating statistics in both the oil and water phases, as well as the turbulent kinetic energy production. They found higher turbulence levels than in a single-phase jet, though this difference decreased with distance from the nozzle. They also found differences between the turbulence magnitude and characteristics for the oil-phase and water-phase turbulent fluctuations. The water had higher turbulent kinetic energy than the oil near the nozzle. At distances greater than six nozzle diameters from the release, the turbulent kinetic energy is higher in the oil than the water. These effects demonstrate the transfer of energy from the oil to the water via entrainment near the source and the subsequent production of turbulent kinetic energy in the oil by the water downstream of the jet orifice. Turbulent kinetic energy production is higher in the oil along the edges of the jet, and higher in the water along the jet centerline, both effects owing to the different mechanisms of turbulence production in the oil and water. These data will be important to further validate numerical models of oil jets into water and to evaluate the assumptions and turbulent quantities used

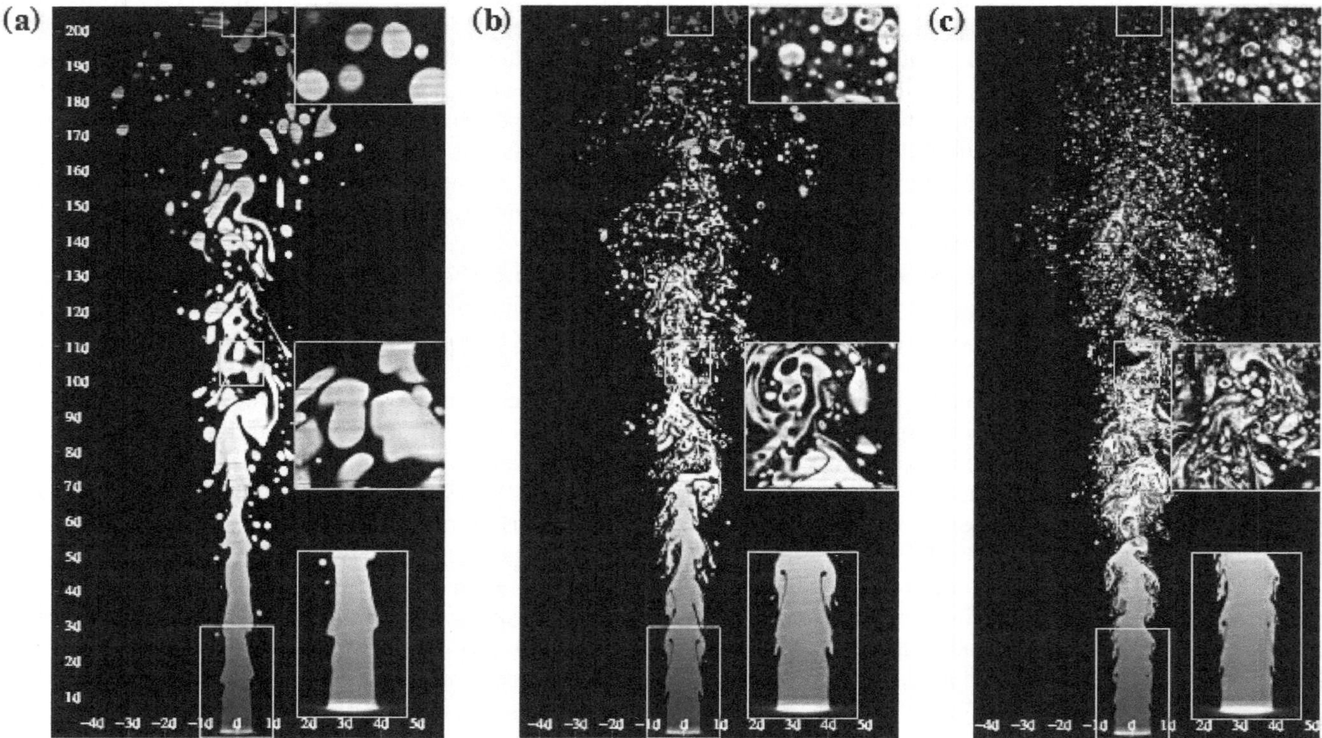

FIGURE 5.34 Breakup of a silicone oil jet into sugar water and visualized by planar laser induced fluorescence; axes labels are scale in millimeters.
NOTES: Jet Reynolds numbers are (a) 594, (b) 1358, and (c) 2122. Insets on the right-hand side of each image correspond to the smaller, highlighted region in the main jet. These images show the entrainment of ambient water (black) into the oil jet (gray-scale), creating filaments and blobs of oil and subsequently break up into individual droplets. The insets show the details of the formed oil droplets, which include many compound droplets composed of water droplets suspended in oil droplets and sometimes cascading as in a set of Russian dolls. These data help to elucidate the mechanisms of droplet breakup and the structure of the formed droplets for different release conditions.
SOURCE: Xue and Katz (2019).

by equations and models to predict droplet breakup. **High resolution imaging and data collection technologies are providing better insight for advancing our understanding of subsea oil release phenomena to develop and refine appropriate mathematical models.**

5.3.3.3 Formation of Natural Gas Hydrates

Natural gas hydrates are crystalline structures of water and gas molecules that are thermodynamically stable at the high pressure and low temperature of the deep oceans (Sloan and Koh, 2008)—at depths of order 100 m or more, depending on the gas composition and temperature profile. Hydrates can cause problems in petroleum production by blocking pipelines and seizing equipment. During response to deepwater oil spills, hydrates may interfere with response equipment or form shells on released gas bubbles. Hydrate shells on bubbles may change their mass transfer coefficients.

In the laboratory, gas hydrates are observed to form at the interface of gas and water after the ambient water phase becomes saturated with dissolved gas relative to the hydrate solubility limit (Warzinski et al., 2014b). Anderson et al. (2012) used this criterion to assess the conditions within an accidental oil well blowout plume under which hydrates may be expected to form. Due to higher temperatures in the deep oil reservoir and the expected rapid rise time from the reservoir to the leak orifice, the released oil and gas are expected to be very warm. For example, temperatures measured during DWH were as high as 105°C (Reddy et al., 2012). Anderson et al. (2012) used an integral model to predict the evolving temperature and dissolved gas concentration in the plume resulting from oil and gas dissolution and entrainment of cold, ambient water. They concluded that by the time the plume cooled to a temperature favorable for hydrate formation, the dissolved gas concentration was diluted by entrained water adequately that hydrates were unlikely to form for the typical blowout cases simulated.

This is not to say that hydrates play no role in accidental subsea oil well blowouts. On the contrary, they were a major factor complicating the response to the DWH accident. As Anderson et al. (2012) also point out, hydrates do form rapidly wherever water and released gas collect, as under and within subsea and response equipment. A good example is the failure of the coffer dam intervention during DWH, which rapidly became clogged with hydrates (McNutt et al., 2011). Hence, the top hat device used to eventually shut in the DWH leak was carefully engineered to avoid hydrate clogging. Hydrate formation has also been studied for natural gas releases from seafloor seeps; this is discussed in conjunction with dissolution in Section 5.3.3.4.

5.3.3.4 Gas Ebullition and Dissolution

Oil and gas released subsea, whether from an oil well blowout, pipeline leak, or diffuse seepage through the seafloor, is expected to enter the ocean water column as live petroleum droplets and bubbles. The soluble components within these fluid particles will immediately begin to dissolve and, depending on their initial conditions and rise speed, may experience phase changes. Gas bubbles may evolve out of the liquid petroleum phase *via* ebullition; gas bubbles may condense to liquids after dissolution of their lighter components. Whether these phase changes occur depends on the kinetics of dissolution and ebullition, the degree of initial supersaturation of gases in the liquid phase petroleum, the evolving mixture composition, local thermodynamic state, and the rise rate of droplets and bubbles.

For an accidental discharge directly from a crippled wellhead or pipeline leak, supersaturation of gas in the liquid phase may occur. As reservoir fluid is transported through the subsurface pipeline to the leak point, the pressure will be decreasing, and gas will be evolving out of the liquid phase. If the ebullition kinetics are not fast enough for the gas–liquid petroleum system to be at equilibrium with the temperature and pressure at the release, the liquid petroleum would be super-saturated with gas, in equilibrium with a higher, down-hole pressure and temperature. Observations of the in situ fluid phase equilibrium at a full-scale release of live petroleum fluids has not been made; hence, conclusions that may be drawn are tentative, based on limited laboratory observation and modeling.

For shallow releases, significant amounts of free gas may survive to the sea surface. When the gas arrives at sufficient atmospheric concentration, the gas may ignite and burn. This was the case for the gas released in the Ixtoc I blowout and for a recent gas pipeline leak in the Gulf of Mexico. Figure 5.35 shows a photo of the burning gas for that pipeline leak as it exited the water column near an offshore platform. Hence, predicting the fire hazard of gas released subsea requires accurate estimation of both the subsea gas plume dynamics and gas dissolution. If the majority of gas is dissolved subsea and is sequestered in a subsea intrusion layer, surface exposure or fire risk is low, as was the case for the DWH oil spill. When free or dissolved gas does reach the atmosphere in sufficient volumes, explosion and fire hazards are expected.

In oil and gas releases, the gas components must be tracked in both the gas and liquid phases of the petroleum. In a high-pressure water tunnel, Pesch et al. (2018) studied single, live-oil droplets under simulated deep-ocean conditions. They observed the droplet size evolution and considered

FIGURE 5.35 A gas leak from an underwater pipeline in the Gulf of Mexico.
SOURCES: NOAA ORR, https://upload.wikimedia.org/wikipedia/commons/a/a6/IXTOC_I_oil_well_blowout.jpg.

cases with depressurization corresponding to high rise speeds, as may be experienced in an energetic plume of oil, gas, and entrained seawater, and slower rise speeds, closer to the rise rate of the observed droplets. Droplets remained of similar size over the majority of their depressurization, growing significantly in the final few tens of meters of simulated rise. Dissolution of methane from the droplet into the recirculating water was evident by gas bubbles that formed in the water below on the side-walls of the apparatus at system pressures below 5 bar.

Gros et al. (2020) applied the TAMOC model to simulate the dissolution and ebullition dynamics of the Pesch et al. (2018) experimental dataset. Gros et al. (2020) used the same model set up as for their hindcast of June 8, 2010, of the DWH and reported in Gros et al. (2017). In those simulations, phase equilibrium and dissolution are modeled simultaneously for each gas bubble and oil droplet. For the Pesch et al. (2018) data, only oil droplets were considered. If free gas is predicted to exist in equilibrium with the petroleum mixture at any stage of the experiment, the TAMOC model immediately creates an attached gas bubble—ebullition kinetics were, thus, ignored, and degassing was assumed instantaneous. Over the stage of the experiments where the droplet sizes remained steady, TAMOC predicted dissolution kinetics to be faster than degassing. That is, supersaturation was not observed due to the rapid dissolution of methane into the circulating water. Only in the final 5 bar of depressurization of the simulations did gas appear, and this occurred largely due to saturation of dissolved methane in the recirculation water, preventing further dissolution of methane from the oil droplet. Corresponding theoretical simulations in an open ocean with background methane concentration showed that degassing is a negligible process for oil droplets over a wide range of initial conditions for all but potentially the final few meters of rise through the ocean water column (Gros et al., 2020).

To understand the role of hydrate formation on dissolution from individual gas bubbles, Warzinski et al. (2014a) conducted experiments for suspended gas bubbles in a high-pressure water tunnel (HPWT) at the U.S. National Energy Technology Laboratory. The full suite of experiments is reported in Warzinski et al. (2014) and Levine et al. (2015). They used high-resolution and high-speed video imagery to identify hydrates. They observed that gas bubbles in the millimeter diameter size range initially formed thin hydrate shells, and that these shells became thicker and more rigid as hydrates are more thermodynamically favorable (i.e., at higher pressures and lower temperatures). The effect of the hydrate skins was observable in the high-speed imagery in which the wave oscillations on the bubble–water interface were damped and then frozen by the hydrate formation. As in previous laboratory experiments (e.g., Masutani and Adams, 2000), hydrates did not form in the HPWT until the background gas concentration reached saturation of the liquid–hydrate system. Once hydrate formed, it remained stable even for different dissolved gas concentrations as long as the hydrate was thermodynamically stable. Because the background gas concentration was high in their experiments and the experiment durations were relatively short, bubble shrinkage rates owing to dissolution processes could not be accurately quantified. Nonetheless, these experiments demonstrate the key phenomenology of hydrate effects of natural gas bubbles.

Wang et al. (2016, 2020) utilized a similar approach to Warzinski et al. (2014a) to observe natural gas hydrates on bubbles released from natural seeps in the deep Gulf of Mexico. They developed a high-speed, stereoscopic video system (Wang et al., 2015), which they integrated with a remotely operated vehicle (ROV) to observe bubble size distributions and hydrate effects. The high-speed imagery confirmed that hydrate shells were forming on the bubbles based on the immobilization of the bubble–water interface after formation, which generally occurred within the first few meters of rise from the seabed. Parallel measurements included analysis of the gas composition from the source and quantification of the dissolved gases in samples collected within the gas bubble column. Although dissolved gas concentrations in the bubble stream were low compared to hydrate solubility, hydrates rapidly formed. This may have been due to nucleation of the hydrate formation from solid hydrate present in the marine sediments, but no definitive mechanism for the rapid hydrate formation at low ambient dissolved gas concentration was determined.

Wang et al. (2020) observed the rise heights of bubbles from natural seeps using acoustic multibeam sonar, mounted both on the hull of the research vessel and on the ROV. They simulated the gas bubble evolution using the measured bubble size distributions at the sea floor using TAMOC and applied acoustic models to convert the simulated bubble properties to acoustic backscatter. By comparing the modeled acoustic behavior to the measured multibeam data, they could validate the model predictions for gas bubble dissolution. As explained in Section 5.2.2, they found that initially, gas bubbles dissolve at rates predicted by clean-bubble mass transfer coefficients. This was also corroborated by analysis of the isotopic signature of the dissolved gas samples (Leonte et al., 2018). After hydrate formation, the mass transfer switches to rates given by dirty bubble mass transfer coefficients. The hydrate formation time was predicted by an empirical equation using the hydrate subcooling and the initial surface area of the bubble and calibrated to data in Rehder et al. (2009). Dirty bubble mass transfer rates were appropriate due to the immobilization of the bubble–water interface by the hydrate. Wang et al. (2020) also found that it was the free gas inside the bubble and not the hydrate shell itself that was most responsible for the bubble shrinkage. This is consistent with the observation that hydrate skins crack and mend as bubbles expand due to the decrease in pressure with height (Warzinski et al., 2014b).

Gros et al. (2017) likewise simulated the effects of hydrate shells for a hindcast of June 8, 2010, of the DWH oil spill using dirty bubble mass transfer rates and dissolution kinetics given by the free gas solubility rather than that of hydrate. Gros et al. (2017) applied dirty bubble mass transfer coefficients directly from the release due to the fact that chemical surfactants were being applied subsea on June 8. For simulations assuming no surfactant injection, it was assumed that oil released with the gas would act as a contaminant, leading to dirty bubble mass transfer coefficients. The good performance of the Gros et al. (2017) model compared to the measured data (see Section 5.2.3.4) supports the conclusion that hydrate dynamics within accidental subsea blowout plumes may be modeled using dirty bubble mass transfer coefficients and dissolution related to the free gas and oil solubilities at the bubble- or droplet-water interface; hence, providing guidance for simulating potential future, deepwater oil spills.

5.3.3.5 Effects of Subsea Dispersant Injection

Like surface application of chemical dispersants, dispersants may be used subsea to affect the formation of oil droplets (NASEM, 2020). As explained in Chapter 4, this method is normally applied locally, near the release source, and is referred to as subsea dispersant injection (SSDI). The effect of SSDI is to reduce the interfacial tension between the treated liquid petroleum and seawater (Socolofsky et al., 2015). The amount of interfacial tension reduction depends on the properties of the released oil, the amount of injected dispersant, and the mixing between dispersant and oil. Depending on the oil and injection methods, interfacial tension reduction factors could span from 300 times (Brandvik et al., 2013, for Oseberg blend in the laboratory) to 5.4 times (Gros et al., 2017, for DWH oil under field conditions). Brandvik et al. (2013, 2018) tested different injection methods. The consensus to date is that dispersant injection near the release orifice is the most effective, achieving the greatest reduction in interfacial tension for a given dispersant volume. Hence, most SSDI contingency plans call for application by a dispersant wand at the end of a supply line and maneuvered into the released fluids by an ROV.

By reducing the interfacial tension of the released oil, smaller droplets may be generated. As explained in Section 5.2.2, turbulence in the seawater acts to break up droplets, and oil interfacial tension and viscosity resist droplet breakup. By reducing the interfacial tension, smaller droplets may form. This is especially true for SSDI when the dispersant is injected upstream of the high turbulence region of the release. For example, most of the oil droplet breakup occurs within 50 to 100 times the release diameter for a blowout or pipeline leak (Zhao et al., 2015). By treating the oil directly at the release, smaller droplets may be expected to form in this primary breakup region (Zhao et al., 2014, 2015). At some point, further reduction in interfacial tension no longer results in smaller droplets as the oil viscosity will dominate the breakup resistance, and viscosity is not affected by chemical dispersants. This was especially evident through experiments conducted on simulated SSDI for oil jets under diverse conditions in the laboratory (Brandvik et al., 2015). Hence, for each crude oil and spill release scenario, there may be a different optimal dispersant injection that will minimize the droplet size, though this dispersant to oil ratio is expected to be on the order of 1:100.

Droplets treated by dispersant through SSDI may further undergo breakup as they traverse the ocean water column. This secondary breakup results from a process called tip streaming. See Section 5.2.1.1. Depending on the amount of dispersant contained in a droplet and its rise time, a significant amount of oil may leave the droplet through this process, resulting in a dispersion of many, micron-sized droplets (Gopalan and Katz, 2010; Davies et al., 2019). As a result, SSDI affects droplet sizes by both reducing the average interfacial tension throughout the primary breakup zone near the source and by allowing some droplets to continue to break up into a chain of micron-size droplets through the process of tip streaming.

As has been carefully discussed in this chapter, the fate and effects of an oil droplet in the ocean are critically dependent on its size. This is the case because droplet size determines the available surface area for exchange (both for dissolution and biodegradation); sets the rise velocity, hence the residence time, of a droplet in the ocean water column; and dictates the trajectory of droplets rising through unsteady, non-uniform ocean currents. These processes critically determine the effects of the spilled oil in the ocean (see Chapter 6). **Hence, if SSDI alters the droplet size, it will change the fate and effect of the spilled oil.**

For example, by reducing the droplet size, SSDI acts to increase the total amount of dissolved hydrocarbons entering the deep sea, thereby reducing the amount of volatile compounds reaching the sea surface and the atmosphere (Gros et al., 2017; Zhao et al., 2021). When used effectively, SSDI may also result in droplets that spread out over wider areas, resulting in thinner sheens and slicks than for untreated releases. Gros et al. (2017) considered the effectiveness of SSDI during the DWH oil spill using a numerical simulation for June 8, 2010, a period for which intensive observations were available and SSDI was conducted at the end of the severed well head. Because of limitations in dispersant availability and application tools, the dispersant to oil ratio was not optimal on June 8, achieving an average dispersant to oil ratio of 1:250 and interfacial tension reduction estimated to be 5.4-fold. This resulted in oil droplets that were simulated to be 3.2-fold smaller than untreated droplets, though still having a volume median diameter of 1.3 mm. Because the droplets remained in the millimeter size range with SSDI application, the location of the surfacing oil and the total mass flux of liquid oil reaching the sea surface was minimally affected. However, this modest reduction in droplet size significantly affected the dissolution of the light, soluble volatiles. For example, benzene mass fluxes to the atmosphere were reduced by over 2,000 times with this SSDI operation compared to simulations without SSDI. This resulted in significantly improved air quality within the response zone. This has recently been corroborated by an analysis of VOC alarm data for the entire DWH incident, which showed significantly improved air quality during times that SSDI was being used (Zhao et al., 2021). **Hence, even when droplet trajectories are weakly affected, modest reductions in droplet size may have significant benefit to human health by reducing VOC emissions from subsea spills.**

The increased dissolution promoted by SSDI will also result in higher hydrocarbon fluxes into subsea intrusion layers, such as the deep plume between 1,200 and 900 m water depth observed during the DWH spill. Gros et al. (2017) estimate that on June 8, 2010, 1.5 times as much dissolved petroleum fluids by mass entered the subsea intrusion as would have occurred without SSDI. This dissolved hydrocarbon is in an ideal form to be degraded by in situ bacteria, and several studies have documented the effectiveness of biodegradation for the dissolved methane, ethane, propane, and butane in the DWH subsea plume (Valentine et al., 2010; Kessler et al., 2011; Rubin-Blum et al., 2017). Depending on the location of a spill globally, this may have important implications for oxygen concentration in the deep ocean. Most of the biodegradation within the water column is by aerobic bacteria. Indeed, the subsea plume associated with the DWH oil spill could be identified by dissolved oxygen anomalies in the CTD profiles collected during the spill. For this region of the Gulf of Mexico, oxygen concentration remained above anoxic levels (Du and Kessler, 2012). However, at other locations around the globe, such as offshore western Africa, where background oxygen concentrations are already low, dissolved oxygen may fall to anoxic levels as the dissolved material from a subsea blowout is degraded.

Because the effects of SSDI depend on the oil, the application method, the release conditions, and the receiving environment, future applications of SSDI to subsea oil spills should be accompanied by in situ monitoring to demonstrate effectiveness. Section 4.2.3 summarizes three phases of SSDI sampling and monitoring, which includes observation of the source conditions and surface slicks, characterization of the oil droplet size distributions near the source, and chemical characterization of the source oil and affected water column and sediments. Measurement of the oil droplet size distribution is especially valuable for predicting and evaluating the fate of the released oil, and in situ observations will significantly reduce uncertainty in this important initial condition.

As explained in Section 5.3.3.2, data to validate droplet size predictions at field scale are currently lacking. This remains true for releases with or without SSDI and with or without co-flowing gas releases. Moreover, the U.S. Environmental Protection Agency has issued new monitoring requirements for oil droplet size distribution when SSDI is used. These requirements include observation of the oil droplet size distribution for sizes between 2.5 micron to 2 mm and quantification of the volume or mass median diameter (40 CFR Part 300). While technology exists to meet this requirement, some of it is experimental, and work is needed to integrate these observations with subsea response to maximize data quality and minimize impact of the response. Hence, it is important to ensure that tools to monitor oil droplet sizes in conjunction with SSDI are integrated with existing response infrastructure that will be deployed at future spills. **Importantly, and in agreement with the recent National Academies dispersant study (NASEM, 2020), we conclude that there is a critical research need to collect large-scale experimental data to validate field-scale prediction of droplet sizes with and without SSDI, and to develop methods to assess the interfacial tension reduction for a spilled oil as a function of the dispersant to oil ratio and SSDI method.**

5.3.3.6 Sedimentation and Burial of Oil

Very deep ocean basins contain cold waters that are older than surface waters and have ancient sediments comprising particles that have sedimented through thousands of meters of water column and are buried over time. The deep-sea sediments below these waters represent the Earth's largest oxygen-depleted ecosystem: they cover ~65% of its surface (Danovaro et al., 2014, 2016) and harbor large numbers of living microbes, predominantly prokaryotes, to subsurface depths of several kilometers below the seafloor (Jørgensen, 2012; Orsi, 2018), highlighting the importance of understanding the fate of oil impacting deep sea waters and sediments.

Oil may impact the deep subsea through natural seeps (see Chapter 3) and by sedimentation in OMAs or MOS (see Section 5.3.2), eventually becoming buried when unoiled material subsequently covers the oiled material. There has been debate in the literature about the importance of MOSSFA in sedimentation of oil spilled during the DWH event. In part the divergent views may be due to a time lag of several years between initial entrainment of oil in suspended particles and their deposition in deep sea sediments, as well as sediment resuspension and transportation events (Diercks et al., 2021). Analysis of early sediment samples suggested that MOSSFA occurred during the first 4–5 months after the spill (Brooks et al., 2015), but that oil–particle sedimentation did not stabilize in the region until at least 2013–2016 (Larson et al., 2018). Thus, samples collected during and shortly after the spill may have underestimated the impact of mechanisms facilitating oil sedimentation such as MOSSFA. Romero et al. (2021) subsequently used a broad suite of chemical measurements from sediment samples collected in 2010–2011 to determine the source of hydrocarbons (i.e., subsea plume versus surface slick) and coupled them with data from samples collected in 2018 to understand biological and physical changes in the oil composition. (The oil had diagnostic ratios of alkanes and biomarkers that matched the Macondo oil spilled from the DWH event.) The oil composition in sediments collected in 2018 enabled discrimination among different sources of oil: whether it traveled from the wellhead to the sea surface where it became entrained in MOS or OPA, then settled to the seafloor, or resulted from particles such as marine snow encountering oil as they settled through the deep-sea intrusion layer or through non-plume deep water containing dispersed oil. Each path would transport oil to the deep sea, with different degrees of weathering and biodegradation occurring before sedimentation. The Romero et al. (2021) analyses revealed different oil concentrations and proportions of surface areas impacted by these three sources of weathered oil.

Despite the physical and chemical limitations to biodegradation in deep sea waters (low temperature, high pressure, and patchy microbial density associated primarily with marine snow; see Section 5.1.7.4), oil biodegradation does occur as and after oil sinks. Thessen and North (2017) calculated or inferred first-order degradation rate constants for 54 selected hydrocarbons using data collected in situ during and after the DWH spill at >700 m water depths. They reported that biodegradation of the selected hydrocarbons in Macondo oil occurred in the order toluene > methylcyclohexane > benzene > C1-naphthalene, and the slowest hydrocarbons to degrade were long-chain *n*-alkanes (C26–C33), which are also the least soluble of those target compounds. Bracco et al. (2020) noted that the microbiome of the deep Gulf waters changed composition during this period of intense oil biodegradation, rapidly shifting from methane, ethane, and propane gas utilization during the active spill to biodegradation of liquid oil components post-spill. In addition to MOS that formed near the ocean surface, microbial biomass also increased in the deepwater plume (Hazen et al., 2010), generating MOS that incorporated water-insoluble hydrocarbons such as high molecular weight alkanes and PAHs from suspended droplets while sorbing soluble low-molecular weight aromatics (Wirth et al., 2018). OMA also contributed to oil sedimentation during the DWH spill, as flushing of the Mississippi River during the spill increased the suspended sediment load in the northern Gulf (Langenhoff et al., 2020). Thus, the chemical composition of oil deposited at the deep seafloor usually differs from the original oil and may depend on whether the oil sedimented as part of a mineral aggregate or in an organic MOS particle. In the first case, the oil will likely experience little biodegradation or abiotic weathering while settling through the water column, but in the latter case the oil may undergo substantial biodegradation during sedimentation, as noted above.

Oil that reaches deep sea sediments as a component of MOS is more likely to be recalcitrant to further biodegradation than oil in newly formed flocs for three reasons: (1) the most labile (biodegradable) components have already been depleted by bacteria in the MOS aggregates, leaving the more refractory compounds such as high molecular weight PAHs, resins, asphaltenes, and petroleum biomarkers (Stout and Payne, 2016; Romero et al., 2021); (2) onset of oil utilization by sediment microbes may be delayed by co-sedimentation of biological polymers (e.g., EPS and cell detritus from MOS particles) that typically are metabolized in preference to hydrocarbons and furthermore may alter the sediment biogeochemistry through redox reactions (Hastings et al., 2020) and metabolite production (e.g., carbon dioxide and organic acids) (Joye and Kostka, 2020); and (3) preferential utilization of the biopolymers may deplete dissolved oxygen (forcing slower anaerobic degradation) and sequester the nutrients and trace elements needed for hydrocarbon biodegradation in biomass, further retarding oil degradation (Shiller and Joung, 2012; Shiller et al., 2017).

Supporting these inferences, Bagby et al. (2016) determined that biodegradation of Macondo oil after the DWH spill occurred in two main phases: initially a rapid loss of suspended and settling oil components (presumably via aerobic biodegradation in MOS), followed by slower loss after deposition to the seafloor, but within 4 years of the spill (presumably predominantly via anaerobic biodegradation). Interestingly, the biodegradation patterns observed by Bagby et al. (2016) were analogous to those routinely observed in other environments: simpler compounds before larger complex chemicals, and parent PAHs before their alkylated series members. They even found evidence that numerous petroleum biomarkers had been extensively biodegraded, suggesting that abyssal microbes have evolved or been selected for the ability to utilize recalcitrant molecules.

Aerobic oil biodegradation may occur after deposition of OMA and MOS and during burial but will become limited by diffusion of dissolved oxygen to replenish that consumed during oxidation of hydrocarbons. Oxygen penetration into marine sediments can be on the order of millimeters to centimeters in shallow coastal areas with high microbial respiration (Revsbech et al., 1980) to tens of meters in oligotrophic areas such as deep sea sediments of the South Pacific Gyre (D'Hondt et al., 2015). Although bioturbation can replenish oxygen locally in anoxic sediments, this produces patchy and transiently aerobic conditions in near-surface sediments (Sørensen et al., 1979). Thus, the duration and depth of aerobic oil degradation is site-specific, generally being lesser in carbon-replete shallow sediments and greater in offshore deep sea sediments. Likewise, diffusive replenishment of soluble nutrients in buried sediments is limited, potentially leading to suboptimal conditions for hydrocarbon biodegradation unless biomass turnover occurs. As expected, the microbial communities and characteristic metabolic products in newly deposited aggregates and buried sediments reflect succession from aerobic to anaerobic metabolism (Kimes et al., 2014; Yang et al., 2016). Microbial sulfate reduction is the major anaerobic process in shallow (recently buried) seafloor sediments because of the relatively high concentration of sulfate in seawater but, once it is depleted, poor diffusion limits its replenishment as a terminal electron acceptor. This leads to dominance of fermentation and methanogenesis deeper in the sediments. Sulfides (hydrogen sulfide gas, soluble hydrosulfide, and/or metal sulfide precipitates) that are end products of microbial sulfate reduction may accumulate and have toxic effects on benthic fauna. Recent studies applying 'omics and biogeochemistry have raised the possibility of previously unknown biodiversity and hydrocarbon degradation potential in deep-sea sediments (Dong et al., 2019).

Anaerobic hydrocarbon biodegradation is often considered to be slower and more selective than aerobic processes (see Section 5.2.8), but long-term retention of oil in buried sediments can eventually lead to biodegradation of even some recalcitrant compounds including petroleum biomarkers like hopanes, as observed by Bagby et al. (2016). Furthermore, even slow biodegradation in sediments reduces the "reservoir" of residual oil that may be re-mobilized by bioturbation or currents. Whether or not anaerobic biodegradation significantly reduces toxicity to benthic organisms is not fully known, but likely will depend on the extent of biodegradation and production of metabolites. Notably, some high molecular weight, multiply-substituted components and petroleum biomarkers of Macondo oil were still detectable in surface seafloor sediments at least four years after the DWH oil spill (Stout et al., 2015; Bagby et al., 2016) whereas the microbial sediment communities had returned to nearly pre-spill composition within approximately 2 years (Overholt et al., 2019) showing that, in the short term, biodegradation leaves residues enriched in recalcitrant molecules.

5.3.4 Shorelines and Near-Shore Sediments

Oil-impacted shorelines can be categorized as supratidal, intertidal, subtidal, ice areas and on-water areas (NOAA, 2010). Each area has distinct air–water–land interaction characteristics and habitats that affect oil deposition and decomposition characteristics (see Appendix G). Coastal processes and landforms are affected by waves, tides, and wind. Ocean waves generated by wind blowing over the ocean surface provide about half of the energy to do work at the coastlines. In coastal zones, the beach morphology (or shape) is transformed through shoaling, breaking, and swash of ocean waves as they interact with the seabed (Short, 2012). In general, shorelines with coarse-grained sediments have a higher wave energy environment than shorelines with fine-grained sediments.

The tides and tidal currents also provide about half of the energy delivered to the coastlines as shown in Figure 5.36 (Short, 2012). High and low tides as well as tidal currents can shift the shoreline and transfer sediment. Tidal currents can run either parallel to the shoreline or perpendicular to inlets, creating current through coastal inlets.

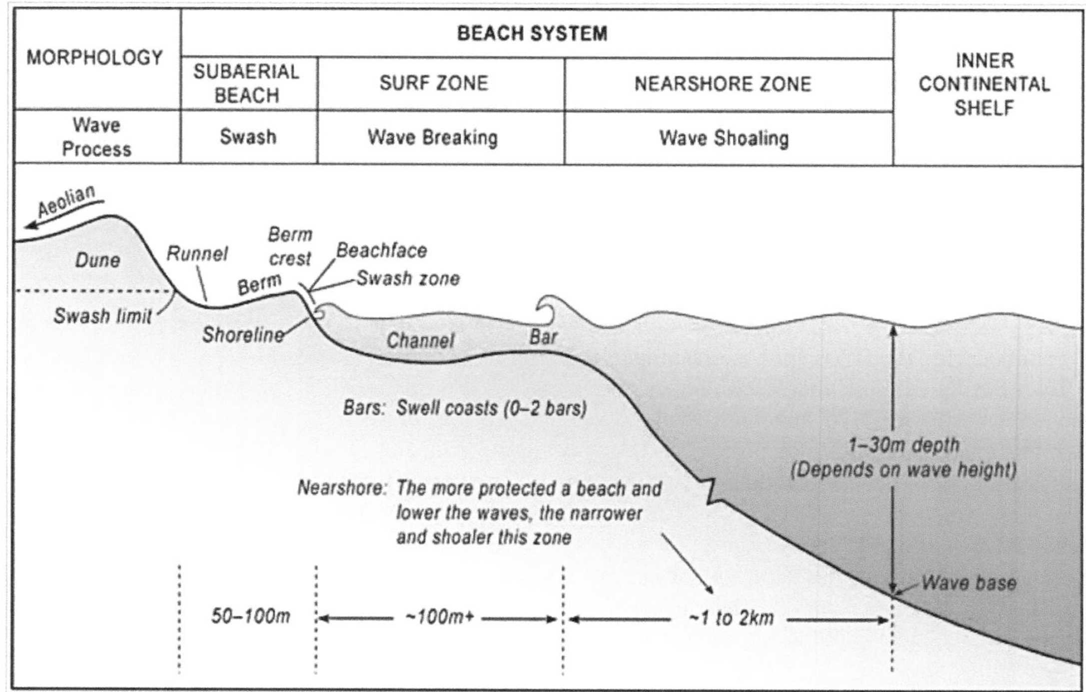

FIGURE 5.36 Different zones of a beach impacted by waves.
NOTE: Cross-sectional schematic of a beach system including a swash zone backed by a "dry" beach; an energetic surf zone with breaking waves and currents; and nearshore subtidal zone extending to the wave base.
SOURCE: Short (2012).

5.3.4.1 Behavior of Oil in Shorelines

Weathering rates of oil in coastal areas depend on oil type, chemistry, and physical properties (e.g., viscosity and pour point), volume of oil spilled, weather and shoreline conditions, location of oil (on water or stranded, in surface or deep water), and physical energy levels of the shoreline (marine and coastal processes) (Etkin et al., 2007; Michel and Rutherford, 2014; Environment and Climate Change Canada, 2018).

The coastline represents a boundary to the ocean basin, and as such, ocean currents tend to deflect parallel to coastlines as they approach these boundaries. As a result, oil may only reach the coast under specialized conditions. Normally, this requires both an on-shore wind and a falling tide. Once in contact with the coastline, several new processes affecting oil transport and fate come into play.

Significant post-spill monitoring activities have documented the persistence of oil and recovery of ecosystems at shorelines that have been impacted by two major spills (*Exxon Valdez* and DWH). Characteristics and examples of different types of intertidal shorelines as categorized by NOAA (NOAA, 2017; Petersen et al., 2019) and predicted oil behavior at these settings are provided in Appendix G. At intertidal areas with exposed shorelines including rocky shores, rocky banks, solid man-made structures, and rocky cliffs with boulder talus bases, oil is generally held offshore by waves or rapidly removed from the exposed surface by wave action. However, oil can penetrate to the wet rock surface at shorelines with exposed wave-cut platforms, and shelving bedrock shores (NOAA, 2017; Petersen et al., 2019). Oil can accumulate and be buried in and on sand and gravel beaches. At tundra cliffs, oil can be stranded on-shore during summer months when there is no ice. At shorelines with tidal flats, oil moves with the tide; however, biological damage can be severe. Oil can remain stranded on shorelines with sheltered scarps and sheltered man-made structures. In salt and brackish water marshes, oil adheres to the intertidal vegetation, and especially light oils can penetrate and persist in the sediments. At shorelines with mangroves, oil can be trapped in the root zone, adhere to the roots, and penetrate into the sediments. The recovery of ecosystems in intertidal shorelines are discussed in detail in Chapter 6.

The persistence and character of stranded oil on coarse sediment beaches depend on oil character, oil amount, shoreline type, location with respect to tidal water levels, location with respect to mobile sediments, and interference by man and by nature (Owens et al., 2008) (see Figure 5.37). Washover events occurring during storm surge from cold fronts, high tides, tropical storms, and hurricanes can mobilize and deposit sand. Storm-driven transport of MC 252 oil released into the Gulf of Mexico from the DWH spill reaching low-relief beaches of sand and shell aggregates in Louisiana, United States, showed that these aggregates can be mobilized from the subtidal and intertidal zone of the beach during more energetic events such as storms and high tides (Curtis et al., 2018). Curtis et al. (2018) further showed

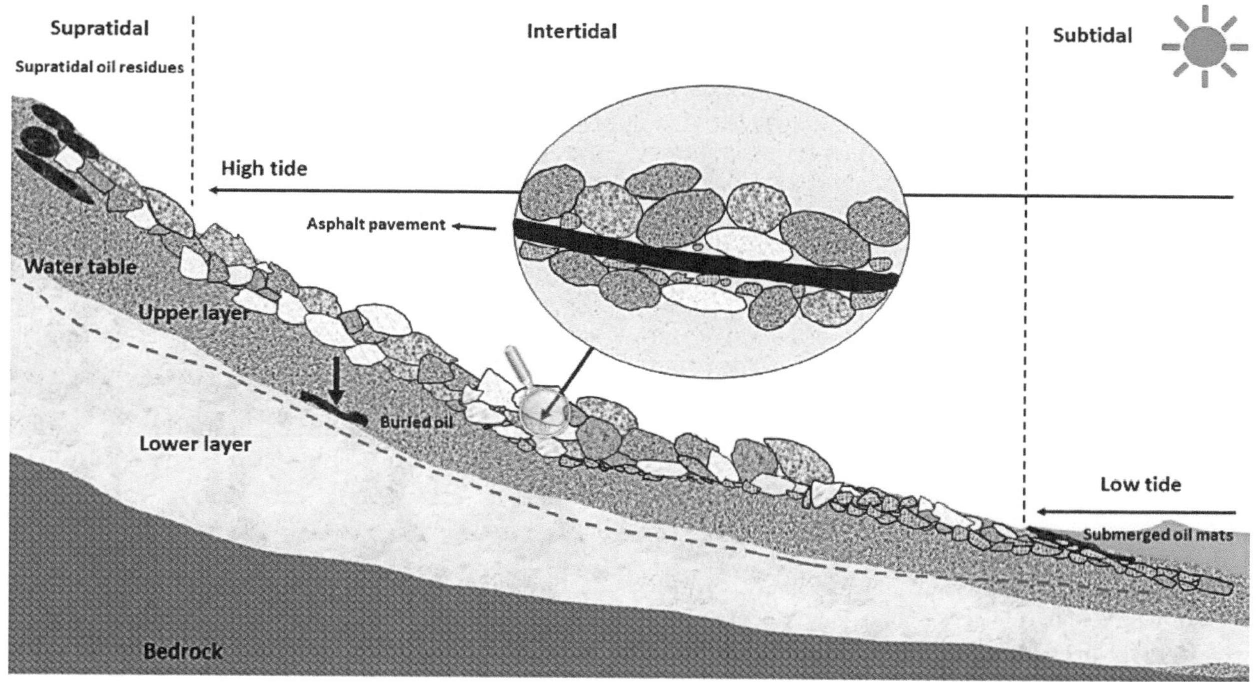

FIGURE 5.37 The distribution and behavior of stranded oil on shorelines affected by the beach topography and environmental factors. SOURCE: Wang et al. (2020).

that washover suspension of oil resulted "in the deposition within washover fans and channels in the supratidal portion of the beach." These results show that crude oil aggregates could move across the beach by physical transport processes as well oil transport from intertidal and subtidal sediments.

On coarse sediment beaches, small amounts of oil can remain for decades in the intertidal zone. Oil that survives natural attenuation on the scale of weeks to months will likely remain on the beach in the form of tar mats, asphalt-like pavements, or as veneers on coarse particles or hard substrate until the environmental conditions shift (Curtis et al., 2018). Subsurface oil residues can penetrate into the pore spaces of the particles until it reaches limiting layers such as fine-grained sediment, the water table, or bedrock. Small fractions of the residual oil stranded within the protected residues can continue to degrade (Curtis et al., 2018).

The hydrodynamic conditions in coastal areas exposed to recurrent contamination result in long-term persistence of deep oil spills from wrecks by burying and resurfacing the oil in the intertidal zone (Bernabeu et al., 2013). Beach geomorphology contributed to the persistence of subsurface oil in a tidally influenced gravel beach of Prince William Sound (Alaska) polluted by the 1989 *Exxon Valdez* oil spill; the oil plume moved alongshore, which was attributed to the gradual slope and smooth substrate of the beach in that particular area (Xia and Boufadel, 2011). Oil biodegradation is sensitive to both nutrient concentration and biomass concentration (see Section 5.2.8). The surface area of sediments is a key parameter for microbial growth in sediments (Torlapthi and Boufadel, 2014): fine-grained sediments with little porespace for oxygen and nutrient diffusion can slow down the biodegradation processes in near shore environments (Pardue et al., 2014).

Residual oil from the DWH spill in the shallow surf-zone in the northern Gulf of Mexico was found primarily in two forms—as SOMs (Michel and Bambach, 2020) and surface residual balls (SRBs; see Box 5.10). Mats formed when weathered oil at the surface reached a shallow location that was energetic enough for waves to mix the sediment with the oil creating a sand and oil mixture (Plant et al., 2013). The mousse interacted with the sediment-oil mixture near the shoreline and sank to the bottom, forming SOMs. Over time, SOMs were buried, exposed, and broken apart through natural coastal process dynamics, and appeared as SRBs in the beach system, with sizes ranging from a few millimeters to several centimeters (Hayworth et al., 2015; Gustitus and Clement, 2017). SRBs (often referred to as "tar balls") that formed after the DWH event, and found on northern Gulf of Mexico beaches, were different from traditional tar balls. DWH SRBs were a brownish, sticky substance with a strong odor of petroleum as opposed to the more extremely weathered tar ball (dark, either rubbery or hard, and without odor) more commonly seen (Hayworth et al., 2011; OSAT-2, 2011; Michel et al., 2013; Mulabagal et al., 2013; Yin et al., 2015).

Re-surveying the shoreline 13 years after the *Exxon Valdez* oil spill at 39 sites in Prince William Sound showed that, despite evidence of oil weathering, the natural weathering rates were slow both at the surface and in the subsurface (Taylor and Reimer, 2008; Li and Boufadel, 2010). The slow weathering was due to oil residue being mixed in with finer sediments and remaining sheltered on the shoreline from wave dynamics and other active weathering processes (e.g., the oil was isolated by boulders and outcrops or shallow bedrock formations). Persistence of subsurface *Exxon Valdez* oil extending into the biologically productive middle and lower intertidal zones after 12 years confirmed the potential for long-term biological effects on beaches most heavily impacted by the spill (Short et al., 2004, 2006). Surveys conducted after 16 and 20 years showed similar results, namely that stranded oil in subsurface sediments of exposed shores and subsurface oil may persist for decades with little change (Short et al., 2007; Boehm et al., 2014). In contrast, the half-lives of aliphatic and aromatic hydrocarbons in Macondo oil from the DWH spill that impacted Pensacola Beach (Florida) sands were 25 and 22 days, respectively, and aerobic biodegradation removed oil to background concentrations within one year (Huettel et al., 2018). Gros et al. (2014) determined that the dominant alkanes remaining 12–19 months after the spill were >C22, with the smaller alkanes presumably having been weathered and/or degraded in situ. Saturates in the C22–C29 range were partially depleted in the order *n*-alkanes > methylalkanes and alkylcyclopentanes + alkylcyclohexanes > cyclic and acyclic isoprenoids, consistent with other oiled aerobic environments. Even components of buried oiled sand patties, known to resist biodegradation, were being metabolized and incorporated into microbial biomass 5 years post-spill (Bostic et al., 2018). Bociu et al. (2019) estimated that golf ball–sized SOAs (see Box 5.10) embedded in sandy beaches were being weathered and aerobically degraded on north Florida beaches. They estimated that, although decomposition of the SOAs would take at least 32 years, in the absence of sediment contact the SOAs would persist for more than 100 years; therefore, incorporation of sand accelerated oil removal because of the porosity afforded by incorporated sand. Collins et al. (2020) similarly determined that the half-life of PAHs in Macondo oil buried in the intertidal zone depended on the flux of oxygen into the sediments and the permeability of the sediments. Nutrient permeation in some beach sediments may be augmented by natural microbial nitrogen fixation (Shin et al., 2019b), and the availability of nutrients (particularly nitrogen and phosphate) in oiled beach sand may influence which hydrocarbon-degrading species dominate—alkane-degrading *Alcanivorax* or PAH-degrading *Cycloclasticus* (Singh et al., 2014). Other factors in bioremediation on beaches are moisture and salinity: Elango et al. (2014) observed that PAH and alkane biodegradation rates were negatively affected in the intertidal zone where salinity was high and oxygenation was low, whereas in the supratidal zone nutrients and moisture limited biodegradation of stranded oil:sand aggregates.

The DWH spill posed challenging shoreline oiling characteristics (Michel et al., 2013). The normal erosional and depositional processes of the beach cycle and seasonal wind patterns caused the oil to become buried, exposed, and remobilized. Oil became stranded on beaches in three zones (i.e., supratidal, intertidal, and intertidal/nearshore subtidal zones). In the supratidal zone (see Figure 5.37), oil was stranded in patches by storm waves. In the intertidal zone, SRBs and SRPs became buried (0.1 m in places). Tropical storm Lee and Hurricane Isaac resulted in extensive beach erosion and release of oil residues. As described by Michel et al. (2013), two different patterns of oil accumulation occurred in the lowest intertidal/nearshore subtidal zone: (1) Along the more heavily oiled sand beaches along the northeast Gulf of Mexico, some of the oil/sand mixture accumulated in the nearshore subtidal areas, forming extensive SOMs (between the toe of the beach and the first offshore bar) which were recurrently buried and exposed by sand transport. As the SOMs broke up, they became persistent sources of SRBs/SRPs on the neighboring shoreline (see Figure 5.38); (2) Along most of the marshes, the stranded oil spread into the marsh with tidal currents due to high density of vegetation and the high viscosity residual oil (Michel et al., 2013). Salt marsh sediments are typically anaerobic below the surface and the presence of oil plus sulfate from tidal water supports anaerobic biodegradation through sulfate reduction (Natter et al., 2012), leading to production of high sulfide concentrations that can negatively affect marsh plants (Mills and McNeal, 2014). However, biodegradation of weathered Macondo oil was observed in sediment cores 18–36 months after the DWH oil spill, indicating that recovery of salt marsh sediments was occurring (Atlas et al., 2015). This observation was supported by detection of progressively oxygenated compounds assumed to represent metabolites from biodegradation of Macondo oil (Chen et al., 2016). 'Omics surveys of marsh grass sediments suggested that the rhizosphere communities (microbes associated with plant roots) may have contributed to oil biodegradation in Gulf of Mexico salt marshes (Beazley et al., 2012). However, in coastal environments these oxygenated

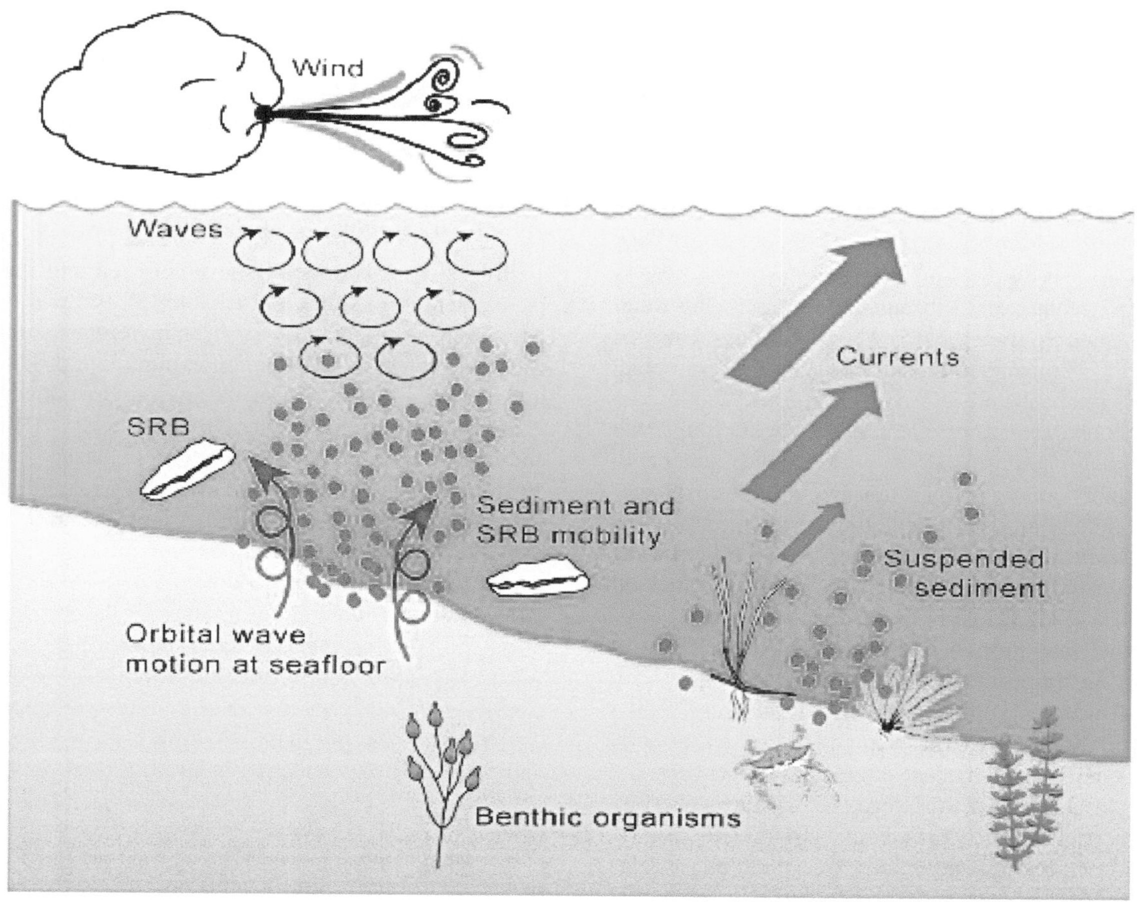

FIGURE 5.38 Resuspension and mobilization of sand, SRBs, and other material from the sea floor by wave- and current-induced shear stress along the coastline.
SOURCES: Plant et al. (2013); USGS.

hydrocarbons accumulate and may be recalcitrant to further degradation (White et al., 2016), emphasizing that simply monitoring the "disappearance" of parent hydrocarbons does not mean the partially oxidized metabolites or photochemical by-products have been removed from the environment. **The fate of oxygenated hydrocarbon species and their effects in coastal ecosystems require additional study.**

Beach oiling from the DWH spill showed oil contamination on the surface and buried below surface sands in the form of tar balls, tar patties, tar cakes, oil sheet, and stained sand (Wang and Roberts, 2013). The cross-shore distribution of surface oil was bound landward by the maximum high-tide wave run-up, which was, in turn, controlled by the incident wave condition. The dynamic and continual swash motion on the foreshore prevented preservation of surface oil deposition. The burial of oil contaminants was driven by the same processes as the initial surface deposition resulting in layers of oil contaminants that varied in both thickness (up to 15 cm) and depth (up to 50 cm). As the beach erodes, these buried oil contaminants can naturally resurface; alternatively, buried oil can also be removed through excavation.

The rate of PAH weathering decreases significantly once the oil becomes buried in sand or trapped within the SRBs. Chemical data indicate that submerged oil containing heavy PAHs (e.g., parent and alkylated chrysenes) can remain in the beach system for several years (Yin et al., 2015). Numerical analyses confirm that spatial distribution of infiltration flux due to waves was dependent on the large-scale hydraulic gradient at the beach and high landward water table reduced wave-engendered seawater infiltration. The decrease in seawater infiltration can have adverse effects on chemical transformation processes (e.g., nutrient recycle and redox condition) in beaches, and subsequently on receiving water bodies (Geng and Boufadel, 2015). Using measurements that integrated oil chemistry, hydrocarbon degrading microbial populations, nutrient and DO concentrations, and fundamental beach characteristics, Geng et al. (2021) found that "intrinsic beach capillarity along with groundwater depth provides primary controls on aeration and infiltration of near-surface sediments, thereby modulating moisture and redox conditions within the oil-contaminated zone." Hypersaline sediment environments in beach pore water inhibited oil decomposition along the Gulf shorelines.

The knowledge gained since the *Exxon Valdez* and DWH spills on the interaction of mineral particles with stranded oil was a significant step in understanding the behavior and fate of oil in different coastal environments, particularly with low physical energy levels (Owens and Lee, 2003; Michel and Rutherford, 2014; Tarpley et al., 2014; Evans et al., 2017; Curtis et al., 2018; Owens et al., 2018). The interaction of especially fine mineral particles with stranded oil in coastal areas decreases the potential for oil to adhere to solid surfaces such as sediments or bedrock. The formation of stable, micron-sized oil droplets that disperse into the water column makes the oil more favorable for biodegradation, allowing the oil to be removed naturally in very sheltered coastal environments such as those with little to no wave action.

5.3.4.2 Tar Balls

Tar balls, tar mats, and tar patties are types of marine tar residues and can range in size from millimeters in diameter (tar balls) to several meters in length and width (tar mats) (Warnock et al., 2015; also see Figure 5.27). The term "marine tar residue" refers to different types of weathered oil conglomerates that can be found on beaches, the open ocean surface, and the seabed. Tar residues from the ocean surface or seafloor can be transported to the shore by waves and currents and deposited on sediments and beaches. Tar ball deposition on a beach depends on several factors such as a recent oil spill, ocean currents, winds, natural seeps, and tanker traffic. The appearance of new tar balls may indicate the occurrence of a recent oil spill.

Tar balls and patties can be transformed by several factors and it is difficult to determine how long they will retain form.[4] Tar balls can break open from turbulence in the water such as beach activity, or directly though interaction with animals, exposing less weathered centers (more fluid and sticky). Tar balls also become more fluid and sticky as air and water temperatures rise. The amount of sediment present can have the opposite effect by adhering to the surface and hardening the tar ball, making it more difficult to break open.

5.3.5 Arctic Marine Systems and Sea Ice

Polar oceans historically have not been impacted greatly by accidental oil spills (see Chapter 3), but with decreased Arctic sea ice extent there has been increased shipping, tourism and interest in off-shore oil exploration, production and transportation in Arctic waters around Greenland, Canada, Alaska, Norway, and Russia (Brakstad et al., 2018a; Lewis and Prince, 2018). The possibility of increased frequency of oil spills coupled with vulnerability to global climate change (de Sousa et al., 2019), rapid ocean warming (Grossart et al., 2020), and a dearth of baseline information about the unique Arctic marine environment (reviewed by Lee et al., 2015) makes it imperative to consider the impact and fate of oil spills in this generally pristine environment. The U.S. Arctic Research Commission[5] and the international Arctic Council[6] are stakeholders in this endeavor. Studies conducted under the international Joint Industry Programme[7]; and SINTEF[8] have included in situ mesocosm experiments

[4] See https://response.restoration.noaa.gov/oil-and-chemical-spills/oil-spills/resources/tarballs.html.

[5] See https://www.arctic.gov.

[6] See https://arctic-council.org/en.

[7] See http://www.arcticresponsetechnology.org/research-projects.

[8] See https://www.sintef.no/projectweb/jip-oil-in-ice/publications.

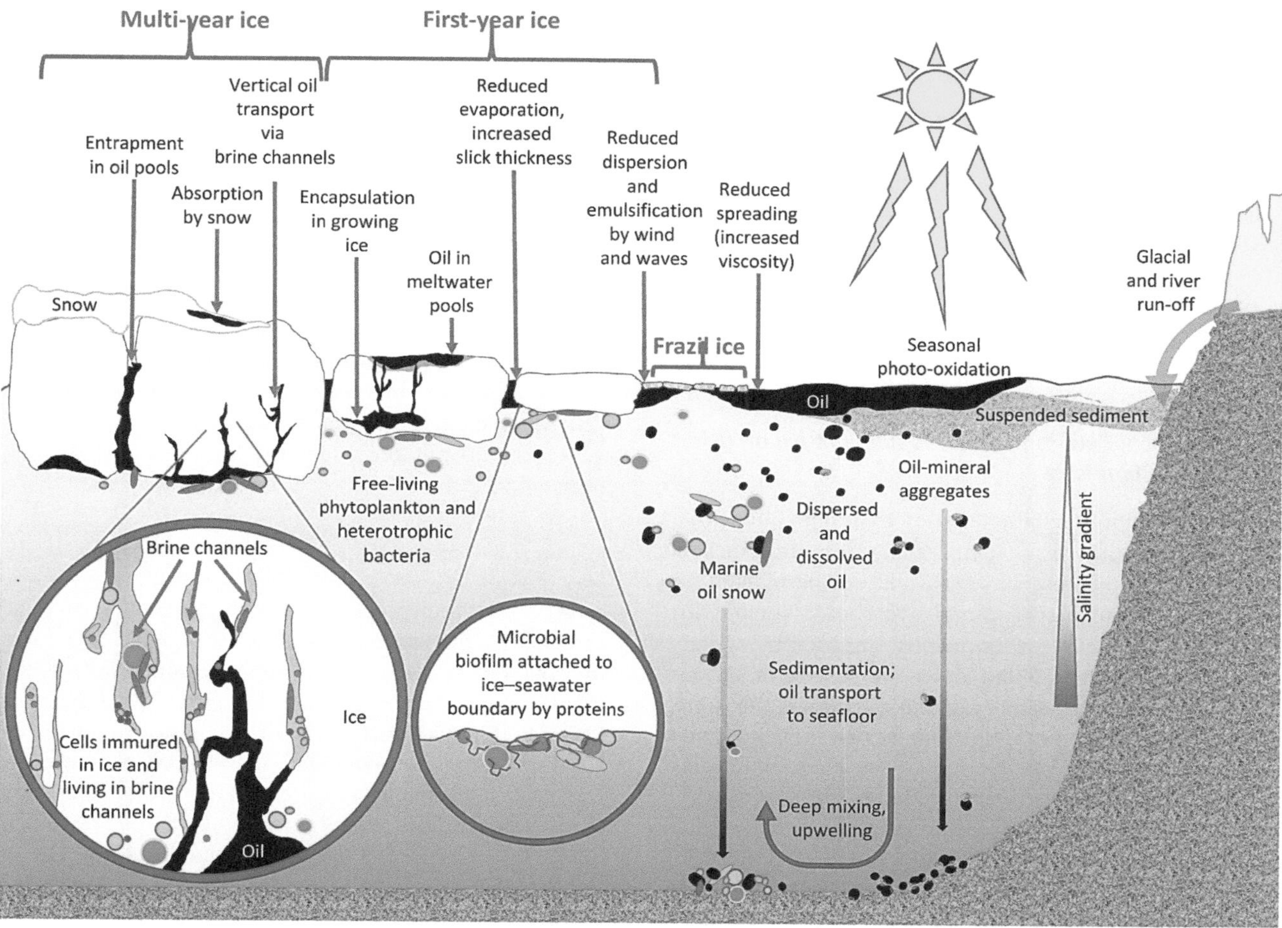

FIGURE 5.39 Physical, chemical and microbial interactions of oil with Arctic marine systems.
NOTES: Sea ice affects oil slick thickness by reducing surface spreading, by "herding" oil into pools and by encapsulating oil within ice masses; it also reduces mixing by waves and wind, decreasing natural dispersion of oil into suspended droplets. At the surface, oil can spread on meltwater pools or be absorbed by snow cover where it is subject to photo-oxidation. Glacial and river runoff provide fresh water, establishing a salinity gradient in the water column, and suspended sediments that may form oil-mineral aggregates (see Section 5.3.2) and sink to the seafloor. Deep mixing and upwelling can replenish nutrients from deep water. Seasonal solar irradiation photo-oxidizes floating oil and supports free-living phytoplankton and cyanobacterial communities that form marine snow and can interact with dispersed oil to form marine oil snow (MOS) that eventually sinks to the seafloor. Bacteria and plankton form biofilms on the underside of sea ice by attaching via extracellular polymers, and also may become entrained in brine channels. They may encounter oil in the water column, at the seawater:ice interface, or within ice masses. Bulk oil may migrate vertically through sea ice, especially during melting as it displaces brine in channels, and also may travel laterally when immured in drifting ice, and dissolved oil components may diffuse via brine channels.
SOURCES: Committee generated; adapted from NRC (2014) and Vergeynst et al. (2018).

and field trials addressing various aspects of Arctic oil spill research and response.

Most of the same factors that affect the fate of spilled oil at lower latitudes also affect polar marine environments, but some are magnified and some are unique to polar oceans (see Figure 5.39). Briefly, these include (1) seasonally low air temperatures and constantly low water temperatures that affect oil behavior and chemistry by increasing its viscosity; (2) seasonal ice cover that reduces mixing of water by wind and wave action, thus indirectly decreasing dispersion and emulsification of floating oil (see Sections 5.2.2 and 5.2.7), and also resulting in stratification of water layers by salinity (density) more so than by temperature; (3) extreme seasonal differences in solar input that directly affect oil photo-oxidation of oil (see Section 5.2.5) and also influence seasonal phytoplankton blooms that can accelerate oil sedimentation (see Section 5.3.2); (4) seasonally low dissolved nutrient concentrations affected by sequestration in phytoplankton blooms and salinity stratification; (5) seasonal increases in suspended sediment plumes from glacial and river runoff that affect sedimentation of oil associated with particulates (see Section 5.3.2); and (6) generally pristine conditions in which hydrocarbon-degrading microbes are relegated to the "rare biosphere" (see Section 5.2.8.2) with some exceptions, such as natural bitumen seeps on the Mackenzie River that runs to the Arctic Ocean (Carey et al., 1990) and point-source natural sub-sea seeps (see Section 5.4.1). These factors and their consequences for oil properties and fates are discussed in greater detail below, first by describing the effects of temperature and sea ice on oil behavior and chemistry (see Section 5.3.5.1), their effects on biodegradation (see Sections 5.3.5.2 and 5.3.5.3), and other pertinent factors in the water column (see Section 5.3.5.4) and in Arctic seafloor sediments and shorelines (see Section 5.3.5.5).

5.3.5.1 Effects of Low Temperatures and Sea Ice on Oil Behavior and Chemistry

The most obvious characteristics distinguishing polar marine environments from temperate oceans are constant low temperature and presence of seasonal and permanent sea ice (frazil ice versus first- and multi-year ice). Notably, the Arctic water column is predominantly stratified by salinity rather than temperature. Wind and wave mixing of surface waters (which have low salinity due to river inputs, seasonal sea ice, and glacial melt runoffs) with denser, higher-salinity deep waters is reduced under ice cover. Thus, unlike the steep and seasonal temperature gradients characteristic of temperate oceans, the Arctic ocean temperature has a shallow temperature gradient that does not change markedly with either season or depth. Surface water temperatures typically range from −1.8°C in winter (the freezing point of seawater) to ~8°C in summer and from −1.2–4°C in deep Arctic waters (> 500 m depth) (Vergeynst et al. 2018b).

Temperature Effects

Low temperature affects oil behavior and chemistry (reviewed by Dickins, 2011; Lee et al., 2011a; Vergeynst et al., 2018) by increasing oil viscosity, with complicated consequences. Greater viscosity slows oil spreading on water and produces thicker films, resulting in a smaller impacted surface area than in temperate open waters. (Ice and snow further reduce oil spreading, as discussed below.) The thicker slicks have decreased photo-oxidation potential that may be offset by high seasonal insolation at high latitudes. Thicker slicks and cold air temperatures decrease evaporation rates, allowing toxic low molecular weight components to persist longer in the oil where they may be available to dissolve into the water column. However, dissolution is balanced by greater viscosity that reduces oil droplet size (a balance between dispersion and coalescence) and decreases both dissolution and biodegradation due to smaller oil:water interfacial area. Additional temperature effects on dissolution include the water solubility of hydrocarbons, which generally decreases until a critical temperature range is reached (usually between 0°C and 30°C), then slightly increasing. Temperature also affects molecular dispersion: a drop in water temperature from 10°C to 0°C decreases diffusion of low molecular weight alkanes (≤C5) by ~20% (Michalis et al., 2016). Crude oils with high wax content can gel at low temperatures, changing spreading and evaporative behavior (reviewed by Potter et al., 2012).

Thus, cold temperature overall decreases loss of bulk oil by physical weathering processes so that water temperatures near freezing may prolong exposure of local marine biota to toxic low molecular weight water-soluble oil components such as aromatic hydrocarbons compared with temperate waters. There has been speculation that dispersants would be ineffective in cold Arctic waters, but Lewis and Prince (2018) concluded that water temperatures as low as 0–2°C generally do not preclude dispersant effectiveness (see Section 4.2.3.4).

Ice Effects

Although transportation of both floating oil and ice are affected by the same driving forces of wind and currents, ice coverage affects oil behavior and chemistry in additional ways (reviewed by Fingas and Hollebone, 2003; Buist et al., 2008; Potter et al., 2012): (1) sea ice physically restrains oil from spreading on water (sometimes by orders of magnitude; Vefsnmo and Johannessen, 1994), and can concentrate oil in leads between floes, thereby reducing the area of slicks and decreasing evaporative losses of low molecular weight components from slicks; (2) sea ice decreases wave energy and surface water mixing (Brandvik and Faksness, 2009), which in turn decreases emulsification and dispersion, increases droplet size in the water and thereby decreases bioavailability; (3) sea ice, whether present as frazil, first-year, or multi-year ice, influences light penetration and photo-oxidation, with consequences for oil chemistry, toxicity, biodegradation, and algal growth; and (4) sea ice can entrain floating oil and transfer it both vertically through the ice, as described below, and horizontally with pack ice, often moving in a different direction from the underlying currents.

Sea ice comprises pure ice, gas bubbles, brine-filled channels, solutes and entrained particulates including living and dead microorganisms and their polymers (EPS), organic detritus and mineral particles (Meiners and Michel, 2017). Unlike freshwater ice, sea ice is porous, having an internal network of fluid-filled channels as a consequence of brine exclusion while the ice is forming. The bottommost pores are directly in contact with the underlying seawater and permit exchange of fluids (including oil) between the water column and ice. As described by Wang et al. (2017), sea ice acts as "a dynamic and porous 'lid' between the atmosphere and the ocean [acting] as both a temporary storage and an effective transporter for contaminants, moving them in space (vertically between the atmosphere and the ocean, and laterally

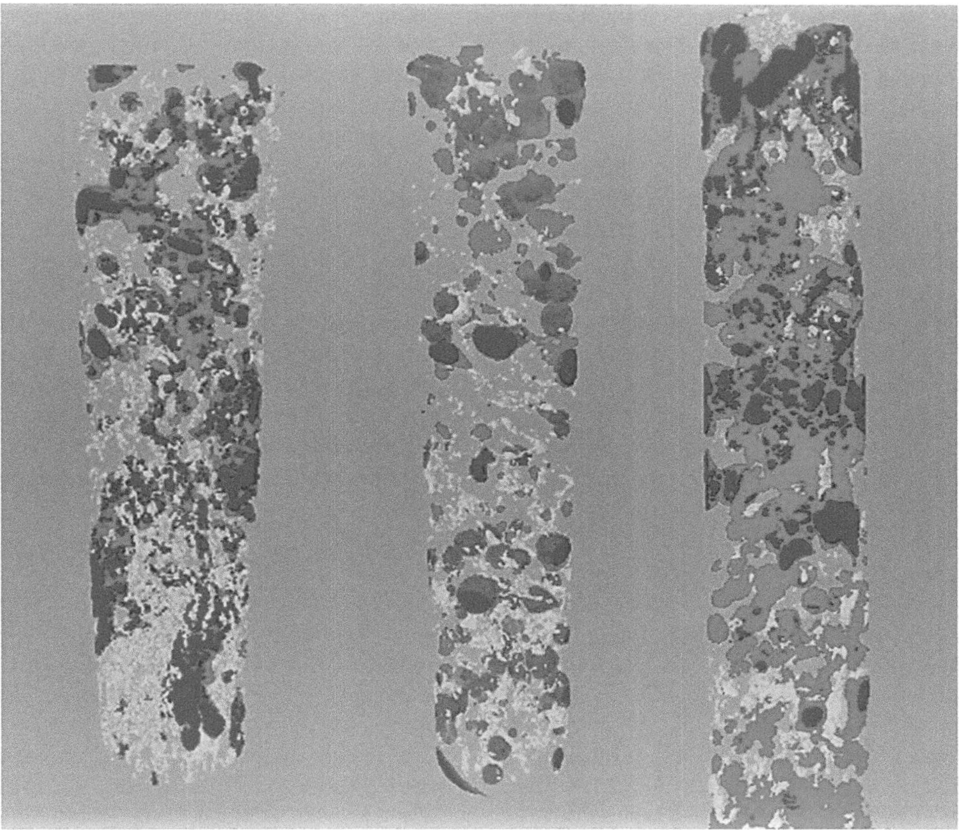

FIGURE 5.40 X-ray microtomography scan of mesocosm oil-in-sea ice cores at different temperatures in which oil (red), air (blue), and brine (white) signals are overlaid. Left: Ice temperature −15°C; middle, −10°C; right, −4.8°C.
SOURCE: Desmond et al. (2021b), supplementary information.

as ice drifts) and in time (seasonal storage and release via freeze and melt)." The extent of oil movement within the ice differs according to the age and temperature of the ice. Recent x-ray micro-computerized tomography (CT) studies of oil, air and brine distribution in artificial sea ice cores revealed that bulk oil injected beneath the ice migrated toward the surface and became occluded within the ice, sometimes as pockets and often surrounding the perimeter of air inclusions (Desmond et al., 2021b; see Figure 5.40). In contrast, the oil had minimal interfacial contact with brine, suggesting that oil progressively displaced and replaced brine as it percolated through the channels and became encapsulated in the ice. If this mesocosm study translates to natural sea ice, physical separation of oil and brine likely has consequences for biodegradation, since bacteria are associated with the brine. Microbial contact with oil then would primarily occur at the base of the ice or in melt ponds rather than within the ice mass, or in warm ice (> −5°C) when channels are larger. The separation of oil and brine also reduces dissolution of hydrocarbons into water due to their reduced interfacial areas, affecting biodegradation potential within cold ice (Desmond et al., 2021a). In addition to movement of bulk ice, dissolution and diffusion of oil components also occurs: in an in situ mesocosm study in Svalbard, Boccadoro et al. (2018) noted that PAHs, as opposed to alkanes, in a light crude oil applied to first-year ice surface migrated to the underlying water column via dissolution. The commissioning of the new Ocean–Sea Ice Mesocosm (OSIM) facility at the Churchill Marine Observatory (CMO) on Hudson Bay[9] in 2022 will provide infrastructure for examination of oil–sea ice interactions.

Frazil ice comprises newly forming, randomly oriented ice crystals, occurring usually in supercooled water under turbulent conditions and often where pack ice meets open ocean and where microbial primary productivity is greatest (Lofthus et al., 2020). In the process of consolidation of frazil into first-year ice, solutes, cells, and particles can become encapsulated rapidly—within hours to days (Potter et al., 2012)—as can floating oil or suspended oil droplets near the surface. Floating oil can also penetrate the bottom of pre-existing first-year ice. Petrich et al. (2013) estimated from field samples that < 2 L crude oil might be entrained per square meter of cold first-year ice in winter, and 5–10 L/m^2

[9] See https://umanitoba.ca/environment-earth-resources/earth-observation-science/marine-observatory.

entrained in warmer ice in spring. The oil overridden by ice can then be transported vertically via brine channels to the ice surface and atmosphere during spring melt. Oil spilled under multi-year ice can be retained by the rough ice at the water:ice boundary, creating relatively thick pools (reviewed by Potter et al., 2012) with little penetration into the ice. Conversely, oil that is encapsulated during ice growth subsequently can be exposed and released to seawater during ice melt (ablation) as brine channels drain (Oggier et al., 2020). The extent of entrainment and transport depends on the type and thickness of ice and particularly its porosity (percent of volume occupied by brine channels), both of which are influenced by ambient air temperature (Faksness and Brandvik, 2008a). Warmer ice is typically more porous and, as the ice deteriorates in spring, the channels within may exchange brine for floating oil. Colder ice may simply encapsulate oil until seasonal warming enables migration, or may transport oil vertically as much as 30 cm (Oggier et al., 2020). Oil may persist in multiyear ice for up to 5 years or seasonal melting may release the oil from first-year ice to impact the food web at a distance from the original location, since sea ice movement can be a major pathway for long-distance cryptic transport of entrained oil (Wang et al., 2017). Recent modeling of the transport and weathering fates of the PAH naphthalene predicted a low potential for incorporation into a hypothetical Arctic food web (Oliveira et al., 2019), but the authors caution that there is limited knowledge of the ecological effects of oil spills in ice-covered waters.

Oil incorporated into sea ice contacts hyper-saline brine, which decreases water solubility of aliphatic and aromatic hydrocarbons by 3- to 60-fold (reviewed by Vergeynst et al., 2018; Table 2 in Saltymakova et al., 2020) and influences the hydrocarbon content of brine released to the ocean during ice melt. This can lead to chromatographic separation of water-soluble oil components as oil migrates through brine channels and air bubbles in the ice (Faksness and Brandvik, 2008b; Desmond et al., 2019; Saltymakova et al., 2020) and may also affect biodegradation (see Section 5.2.8). Boccadoro et al. (2018) additionally determined experimentally that PAH migrated through the sea ice more effectively when dispersant was present, likely improving the otherwise-limited bioavailability. Sea ice cover thus serves as a reservoir for retention and migration of oil as well as particulates and solutes that support microbial activity, discussed below.

While the studies cited above provide detailed understanding of the fine-scale interactions of oil and ice and the possible breadth of these interactions and their impact on oil fate processes, predicting response-scale transport of oil within ice-infested waters is exceedingly difficult at present. Accurate predictions would require knowledge of the detailed under-ice topology, the ice porosity, density of surrounding seawater, and many other small-scale aspects of the ice. This is especially difficult since ice and the surrounding seawater may be moving at different velocities and because the ice is dynamic, undergoing a myriad of changes over time. Current models for oil transport assume that oil transports with the drift ice field when ice coverage is 70% to 80% or greater, that oil moves as it would be in open water at drift ice coverages of 20% to 30% or less, and the oil moves at a linear interpolation of these two extremes for intermediate ice coverages. Some models also use these percentages to scale fate processes for interaction with the atmosphere. The real situation may be much more complex, dependent on the actual in situ ice properties and coverage and their interactions with local currents. Hence, **though there is a good understanding of the fine-scale interaction of oil with ice, models have imprecise ice predictions and must average over large scales so that simplified approaches are required to predict oil–ice behavior. Development of these algorithms require validation data and will likely rely on real-time data integration of observing systems during future spills.**

5.3.5.2 Effect of Low Temperature on Oil Biodegradation

In the event of an oil spill, the remoteness of some Arctic regions would delay and alter spill response options some of which, furthermore, would be less effective due to local conditions (see Chapter 4). Biodegradation (via natural attenuation) might then be the primary or the only feasible fate of spilled oil, and onset of that activity might be delayed because most Arctic sites are historically pristine. The Arctic Monitoring and Assessment Programme (2010) has estimated that 95% of hydrocarbons in Arctic oceans (~4×10^4 tons crude oil per year) are associated with point-source natural seeps rather than with spills. Therefore, aside from localized exposure to hydrocarbons at seeps (Cramm et al., 2021; see Section 5.3.5.5) and associations with hydrocarbon-synthesizing phototrophs (see Section 5.2.8), there is little selective pressure to enrich or maintain hydrocarbon-degrading bacteria for recruitment to spilled oil, and they are typically part of the "rare biosphere" (see Section 5.2.8.2). This paucity might increase the lag time before biodegradation becomes significant, although the rates of biodegradation may then be relatively fast, as discussed below. In addition to their importance for oil removal, microbes are at the base of the Arctic food web (reviewed by Kellogg et al., 2019) and any effects on their metabolism will impact higher organisms. Unfortunately, due in part to the difficulty of accessing remote northern sites for scientific sampling, until recently **the application of 'omics techniques to polar marine regions has lagged behind studies of temperate marine environments, and further research effort is needed to augment understanding of Arctic ecosystem responses to oil** (Lauritano et al., 2020; Zhong et al., 2020). Current Canadian endeavors in Arctic 'omics include GENICE (Microbial Genomics for Oil Spill Preparedness in Canada's Arctic Marine Environment) and the MPRI (Multi-Partner Research Initiative; Lee, 2021).

Temperature affects all biological processes although, surprisingly, it may have less impact on Arctic oil biodegradation than expected because (1) O_2 solubility increases with decreasing water temperature and (2) the temperature gradients observed in temperate water columns are not as steep in polar oceans and the native microbiota are adapted to constant cold temperature throughout the water column. Broadly accepted terms describing microbes adapted to cold are: "psychrotolerant" or "psychrotroph" for those that grow optimally at temperatures >15°C but tolerate and grow at lower temperatures; and "psychrophile" for organisms that die at >20°C, grow optimally at ≤15°C, and can be metabolically active below 0°C (Moyer and Morita, 2007). Psychrotrophs are quite common in diverse environments and many of the recognized oil-degrading microbes in Arctic oceans are psychrotolerant, whereas true psychrophiles are rarer and restricted to constantly cold environments such as deep oceans and polar regions. Oil-degrading genera that are commonly detected by polar marine studies during low-temperature biodegradation include members of the family Oceanospirillaceae such as the psychrophilic obligate hydrocarbon-degraders *Oleispira antarctica*, *Oleiphilus*, and *Thalassolituus*; psychrotolerant heterotrophs *Polaribacter*, *Colwellia*, and *Cycloclasticus* and the recently detected *Zhongshania* and *Paraperlucidibaca* (see Appendix I). 'Omic surveys of microbiomes in Canadian Arctic seawater, sediment, shorelines, and deep seeps have revealed ubiquitous distribution of hydrocarbon-degrading species in the rare biosphere (C. Hubert, personal communication) that are enriched when incubated with oil in vitro. Comparison of sequences to the global Tara Oceans database (Sunagawa et al., 2015) indicates that some key hydrocarbon-degrading taxa, such as the latter two listed above, are endemic to polar marine environments but rare in lower latitudes (Murphy et al., 2021); conversely, other keystone species are more common in temperate marine environments. Recognition of key endemic species could inform response decisions and interventions might be tailored to enrich polar-specific oil-degrading species.

When temperature is invoked as a limiting factor for microbial metabolism, the concept of Q_{10} based on the Arrhenius equation is often applied to whole organism (or community) metabolism. Briefly, the Q_{10} rule of thumb states that, below the optimum temperature for an activity (an individual enzyme or microbial cell growth), a reduction of 10°C results in a 2–3-fold reduction in the rate of that activity. This generalization is incorporated into some numerical models for the fates of oils. However, cold-adapted microbial communities may not conform to this rule, which was developed primarily to describe in vitro enzyme reaction rates at moderate temperatures. Therefore, the relationships between temperature and rate of oil biodegradation or pattern of susceptibility to biodegradation are not straightforward. Because temperatures are low but constant in Arctic marine waters, the magnitude of the Q_{10} effect is dampened because the microbiota are adapted to low temperature; therefore, extrapolating Q_{10} values from temperate water data may be invalid (Vergeynst et al., 2018). Furthermore, applying a universal Q_{10} factor to different oil spill conditions is questionable because the metabolic rates of key genera (e.g., *Alcanivorax* and *Cycloclasticus*) have not been determined experimentally under temperature and nutrient concentrations relevant to cold marine environments (Bagi et al., 2014). Meta-analysis of 10 published datasets indicated that models using correction factors based on the Arrhenius equation underestimated in situ oil biodegradation rates and that lag time before onset of biodegradation (which is accommodated differently by different models) was inversely influenced by temperature (Brown et al., 2020). More specifically, Nordam et al. (2020) examined Q_{10} scaling for biodegradation of individual oil components at four temperatures spanning −2°C to +13°C. They found that conventional Q_{10} scaling was adequate for predicting biodegradation rates for the more water-soluble, lighter oil components, but were less accurate for poorly water-soluble, higher molecular weight oil components, possibly because Q_{10} does not capture temperature effects on oil behavior (e.g., increased viscosity and decreased water solubility of PAHs at low temperature; reviewed by Margesin and Schinner, 2001). Based on these and other reports, best practices would then require that experiments be conducted at temperatures relevant to in situ conditions, rather than being extrapolated from data collected at moderate temperatures, and that alternative approaches be developed to predict low-temperature biodegradation rates.

Unsurprisingly, there are numerous reports that oil biotransformation rates are slower at cold than at moderate temperatures (e.g., Prince et al., 2013) but there are also reports of unexpectedly rapid biodegradation at low temperature. Laboratory and in situ experiments using cold deep waters and warmer surface waters collected near the DWH site found that PAHs degraded slightly faster at 4°C than at 24°C and *n*-alkane degradation was only marginally faster at the higher temperature (Liu et al., 2017). Interestingly, the deep-sea microbial community outperformed the surface community at both temperatures. Thus, temperature may not be as important in cold ecosystems as co-existing factors like seasonal nutrient concentrations (see Section 5.3.5.4) and oil bioavailability (see Box 5.5). However, even this generalization is inadequate, as temperature may affect degradation of aliphatics differently than aromatics in some cases. Kristensen et al. (2015) examined crude oil biodegradation in cold Arctic and warmer North Sea water samples and found that the order of biodegradation of oil classes differed at the two sites, with aliphatics being preferentially degraded before PACs in Greenland seawater incubated at 5°C versus the opposite pattern of faster PAC biodegradation in Danish seawater incubated at 15°C. That is, preferences were site-specific, providing a cautionary lesson about generalizing hydrocarbon biodegradation susceptibility in cold waters. Scheibye et al. (2017) also observed biodegradation of *n*-alkanes (C_{13}–C_{30}) in deep Greenland seawater at 2°C, but losses of (alkyl) PACs were attributed to abiotic dissolution rather than biodegradation. Lofthus et al. (2018) incubated thin films of oil adsorbed to a support fabric in seawater at temperatures from 0°C to −20°C and found that the *n*-alkane biotransformation rate decreased at lower temperature, likely due to physical changes in the oil. Biodegradation has been observed in seawater even at sub-zero temperatures: chemically dispersed diesel fuel was incubated with seawater from Norway (after enhancing salinity to depress the freezing point) at −2°C and −6°C (Dang et al., 2020). Although chemical dispersion of diesel would not be a response option in situ, the experimental observation of significant degradation of *n*-alkanes (*n*C10–*n*C36), decalin, and 2–5-ring PACs at these low temperatures expands the permissive temperature range for hydrocarbon biodegradation.

5.3.5.3 Biodegradation Within and Below Sea Ice

Sea ice, and particularly liquid brine channels within the ice, presents an extreme habitat for microbial life, imposing combinations of stressors including limited diffusion of nutrients, seasonally fluctuating salinity from near freshwater to hypersaline (200 ppt) and fluctuating low temperatures from −1 to < −20°C. Nonetheless, sea ice harbors communities of microbes and viruses recruited from, yet distinct from, those in sea water (Fernández-Gómez et al., 2019; Zhong et al., 2020) and provides a niche for assembly of microbial biofilms both within and under the ice (see Figure 5.41). Furthermore, sea ice is not merely a refuge for dormant microbes: bacterial metabolism has been detected within Arctic sea ice at temperatures as low as −20°C (Junge et al., 2004), particularly in microbes physically associated with other cells such as algae or with inorganic particles or ice crystal faces. Because of these physicochemical stressors, microbial communities in bulk sea ice are less dense and less diverse than in sea water (Yergeau et al., 2017) whereas microbes, solutes, and other particles are concentrated in brine channels by exclusion as ice forms from seawater (Zhong et al., 2020), during which cold-adapted microbes are enriched (Deming and Collins, 2017). Reported densities of bacteria in sea ice range from 10^3 to 10^7 per mL of bulk ice, or up to 10^8 cells per mL of brine fluid, although not all of the cells may be active at sub-zero temperatures. Cooper et al. (2019) visually counted ~10^5 cells per mL of sea ice brine, ~5% of which were dividing; virus-like particles were also noted at virus:cell ratios of 0.7–3. Yergeau et al. (2017) used metagenomics (see Section 5.2.8.1 and Appendix H) to determine that distribution of microbes and genes was uneven in sea ice, with greater microbial numbers being detected at the bottom of the ice columns near the ice:water boundary. Virus (particularly bacteriophage) densities range from 10^4 to 10^8 units per mL in bulk ice (Deming and Collins, 2017) but their activity in situ is not well known. Notably, most studies to date have been conducted in frazil and first-year ice rather than multi-year ice (reviewed by Deming and Collins, 2017), highlighting another knowledge gap.

The value of in situ experiments is demonstrated by a pair of Arctic field studies examining the fates of aliphatic and aromatic components of three oils: marine gas oil (a diesel distillate), a blended North Sea crude oil, and a high-sulfur intermediate fuel oil. In the first study (Vergeynst et al., 2019a) the oils were sorbed to mesh supports and incubated in Greenland fjord water during the summer months. The oil was depleted by a combination of evaporation,

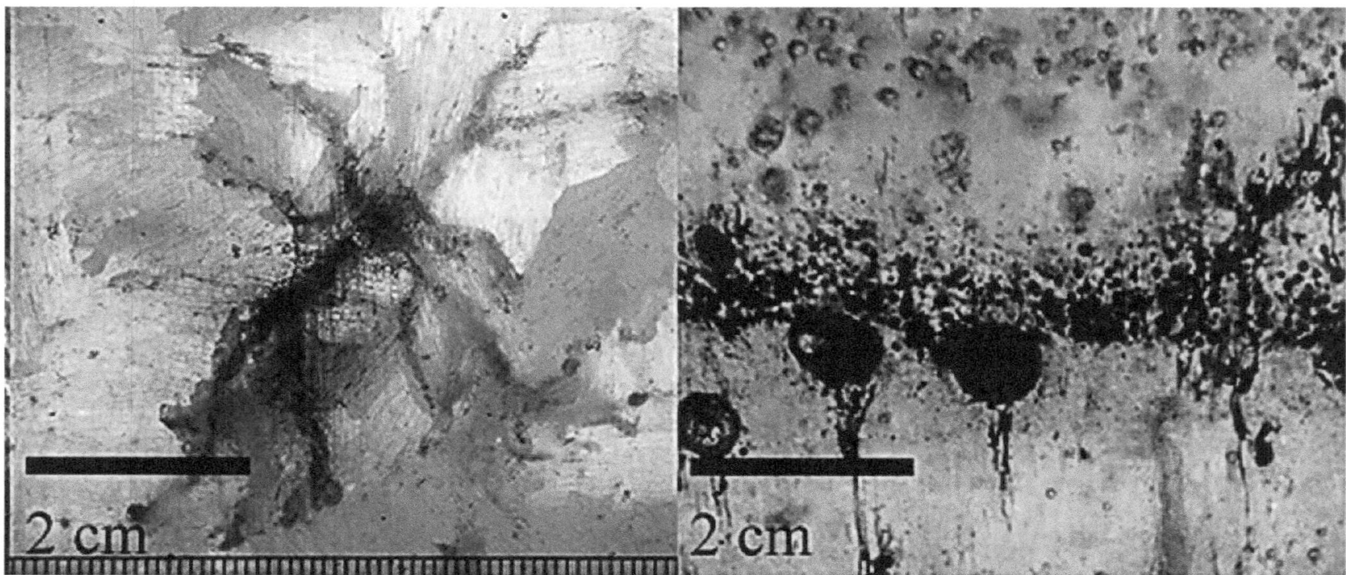

FIGURE 5.41 Photographs of oil migrating through columnar sea ice grown in experimental tanks. Left panel: oil distribution in a brine channel ~10 cm above oil:water interface, with fanning into "feeder: brine channels; Right panel: vertical section of ice showing mechanically dispersed oil droplets entrained in the bulk ice. Scale bars are given in each photograph.
SOURCE: Adapted from Oggier et al. (2020).

photo-oxidation, dissolution and biodegradation, with the latter being the dominant fate for alkanes and PACs concurrent with development of an enriched bacterial biofilm on the mesh. Among the abiotic fates, evaporation accounted for early losses of short-chain alkanes, dissolution accounted for early depletion of two- and three-ring PACs and photo-oxidation depleted high-molecular weight PACs. Overall, dissolution and photo-oxidation of PACs were more important fates for marine gas oil and the crude oil, whereas dissolution and biodegradation processes dominated PAC depletion from the fuel oil. In the companion study (Vergeynst et al., 2019b), mesh supports with sorbed marine gas oil were frozen into first-year ice, extending into the underlying sub-zero Greenland fjord seawater. Rapid biodegradation of n-alkanes in the sub-zero seawater occurred at rates comparable to those in temperate waters but degradation within the sea ice was negligible. Genomic analysis of the biofilms revealed that *Oleispira antarctica* (see Appendix I) was present in 25–100-fold greater abundance in seawater than in sea ice and likely contributed to the different *n*-alkane biodegradation outcomes. PAC depletion differed in seawater and sea ice, with dissolution occurring 3–6.5 times faster in seawater. Photo-oxidation was important in ice and the upper water column, increasing with increasing PAC alkylation and ring number. The experiment showed that sub-zero temperatures permit oil biodegradation in seawater, but severely limit biodegradation in sea ice at a similar temperature. The two studies highlight the need to measure oil weathering fates and rates in different compartments of ice-covered marine systems and to study different oil types separately.

Total microbiological activity associated with sea ice occurs primarily along the bottom of the ice (Boccadoro et al., 2018), and is dominated in spring and summer by photosynthetic microalgae (reviewed by Lee et al., 2011a; Bowman, 2015). Microbes attached to and living in ice near the ice-water contact zone produce metabolites that "sculpt" the ice, enhancing biofilm attachment and growth of the algae, and facilitating access to nutrients in brine channels. Thus, sea ice is a dynamic reservoir of active microbes, both photosynthetic and heterotrophic, and their associated viruses, originally sourced from seawater then concentrated and selected during brine channel formation.

Due to their location in the ice and water column, algae are likely to be adversely affected by floating oil, although the susceptibility of different species varies and some diatoms are more resistant to the effects of oil (Lee et al., 2011a); zooplankton that graze on the algae may be affected secondarily. Arctic phyto- and zooplankton experienced sub-lethal effects when exposed to water-soluble oil components at concentrations of 0.07–0.55 mg L^{-1} both with and without additional photo-oxidation (Lemcke et al., 2019) (see Section 5.2.5). Seawater below melting ice during the winter-summer transition in the Greenland Sea was found to harbor a naturally high abundance of hydrocarbon-degrading bacteria (*Marinobacter* and *Alcanivorax*; see Appendix I), perhaps sustained by biofilms on the underside of sea ice that comprise bacteria and marine algae, some of which synthesize alkanes. Thus, the underside of ice and the water in contact with it may enrich hydrocarbon-utilizing microbiota.

Microbial biofilms associated with the underside of sea ice (inset, Figure 5.39) are likely mediated by ice-binding proteins (IBP) that have been identified in various polar eukaryotic algae, fungi and bacteria. These extracellular microbial IBPs, unlike the intracellular antifreeze proteins of higher organisms, alter the structure and behavior of ice crystals, protect the cells against freeze-thaw damage, and facilitate attachment of secreting cells to ice surfaces and to IBP-free cells such as diatoms and microalgae, thus promoting the formation of symbiotic photosynthetic biofilms at the sea ice:seawater interface (Guo et al., 2017). By sequestering the cells near the top of the water column, they are optimally placed to access oxygen and nutrients from photosynthesis (Bar Dolev et al., 2016) and/or to contact floating oil or newly ice-embedded oil. Biofilm formation on the ice undersurface may also facilitate close interactions between alkane-synthesizing cyanobacteria and hydrocarbon-degrading bacteria as a source of competent microbes for recruitment in the event of an oil spill.

Within bulk sea ice, 'omics has revealed the presence of putative hydrocarbon-degrading bacteria and associated genes (Bowman and Deming, 2014) even in the absence of oil, and exposure of ice to oil shifts the microbial community composition as key species increase in abundance (e.g., Garneau et al., 2016). In a winter field experiment at Svalbard designed to demonstrate emergence of selected species in response to oil, Brakstad et al. (2008) overlaid an oil slick with an ice slurry, which froze in place. The presence of oil increased the total microbial abundance 5-fold compared with unoiled ice, and the community composition of the clean and oiled ice diverged over several months as the key hydrocarbon-degrading bacterium *Colwellia* was enriched. Analysis of oil recovered from the ice did not conclusively demonstrate biodegradation in situ but did suggest slow biotransformation where the oil concentration was low (Brakstad et al., 2008). This field study demonstrated that bacterial communities exposed to oil could become enriched even in ice at sub-zero temperatures, even if biodegradation within the ice was slow or negligible.

The latter observation is interesting. Despite the presence of appropriate species and apparent community shifts in response to oil exposure, Lofthus et al. (2020) noted that no experiments to date have demonstrated substantial biodegradation of bulk oil within sea ice. Instead, losses of specific oil components have been detected in the bottom-most portions of ice that are in contact with seawater; such losses might be attributed to a combination of dissolution and biodegradation (Boccadoro et al., 2018; Vergeynst et al., 2019a). The extent and pattern of losses also depended upon the type and concentration of oil in the ice and whether the oil had been weathered before becoming embedded, thereby removing the most water-soluble and bioavailable components such as short-chain alkanes and monoaromatics; that is, perhaps only hydrocarbons dissolved in liquid brine channels are bioavailable in ice. Another factor proposed by Garneau et al. (2016) is that oil biodegradation in sea ice might be limited by a combination of low nutrient concentrations and relatively higher concentrations of labile dissolved organic matter, such as exopolymers produced by algae in the sea ice, which might be degraded in preference to the oil components. Yet another factor potentially limiting oil biodegradation in bulk ice might be access to the oil itself rather than survival, since Garneau et al. (2016) observed good biodegradation of oil by melted sea ice (indicating that appropriate organisms and sufficient nutrients were present in the ice), and in agreement with speculation by Desmond et al. (2021a,b) about partitioned distribution of oil, air, and brine in ice (see Section 5.3.5.1). Frazil ice with its high water:ice contact area differs physically from first- and multi-year ice, and might provide greater opportunity for oil biodegradation if it entrained floating oil while forming. Lofthus et al. (2020) examined the fate of weathered, chemically dispersed North Sea crude oil in frazil ice formed from seawater and held at –2°C. Over 125 d the presence of frazil ice enhanced *n*-alkane degradation and decreased two- to three-ring PAH degradation compared with parallel ice-free seawater controls. Despite depletion of some oil components, there was negligible loss of total oil regardless of ice because the biodegradable fraction of hydrocarbons represented only a small proportion of the bulk North Sea oil applied.

Thus, selective biodegradation of oil components may occur within cold sea ice at very slow rates that are difficult to distinguish from dissolution and evaporation. Given that biodegradation can occur at moderate rates in seawater at the same temperature, it is clear that other factors reduce biodegradation rates within the ice, such as decreased surface area (oil bioavailability); limited nutrient and/or oxygen diffusion; the presence of alternate labile carbon sources like exopolymers from biofilms; limited diffusion of substrates and/or metabolic products in brine channels for efficient distributed metabolism; and possibly hypersaline brine. However, as the ice warms, melts, and liberates the microbes and oil, biodegradation can occur in near-surface seawater or meltwater ponds on the ice or in expanded brine channels. Further laboratory and in situ studies are needed to examine this possibility.

5.3.5.4 Other Factors Affecting Biodegradation in the Arctic Water Column: Bioavailability and Nutrient Concentrations

As discussed in Section 5.3.5.1, low temperature increases oil viscosity and makes dispersion more difficult, leading to larger droplet sizes with smaller total oil:water interfacial surface area that in turn affects biodegradation rates by limiting bioavailability. As noted in Section 5.2.8.3, Brakstad et al. (2015a) showed shorter lag times and faster degradation of *n*-C10–21 alkanes and 2–3 ring PACs in 10 μm oil droplets than in 30 μm droplets at 5°C, whereas biodegradation of higher molecular weight alkanes (*n*-C22–36) and 4- to 5-ring PACs was slower but comparable in both droplet sizes. That is, as expected, greater surface area increased bioavailability of the more water-soluble oil components. Besides affecting dissolution, the interfacial area of droplets affects microbial community dynamics. In one experiment, larger droplets exhibited delayed succession of oil-degrading species in biofilms associated with the oil (Brakstad et al., 2015b). In another study, biodegradation proceeded only after biofilms had formed on the surface of the oil droplets; in fact, agitating the culture to decrease adherence delayed microbial growth. Once biofilms had formed, droplet size decreased and droplets disappeared with longer incubation time and disappeared (Deppe et al., 2005). Therefore low temperature, in addition to presence of sea ice damping surface mixing in Arctic waters, can indirectly affect oil biodegradation by influencing oil behavior.

Another parameter particularly affecting biodegradation in many Arctic waters is insufficient concentrations of nutrients (e.g., nitrogen) required for effective oil biodegradation. There are several reasons for this phenomenon. First, stratification occurs when low-salinity meltwater at the surface is not effectively mixed with underlying layers due to density and the seasonal damping effect of sea ice that decreases mixing by wind and waves: nutrients may remain at depth rather than at the surface where oil is spilled and degraded. Second, seasonal phytoplankton blooms may sequester nutrients in cell biomass during photosynthesis, even if they fix nitrogen (N_2). As an example, total nitrogen (mainly nitrate) concentrations in deep Greenland waters (>100 m depth, below the photic zone) are commonly 10–18 μM year-round (reviewed by Vergeynst et al., 2018), but in the surface waters nutrient concentrations undergo a yearly cycle. In the photic zone (surface to 25–50 m depth), phytoplankton deplete dissolved nitrate to concentrations of ~0–5 μM in the spring and summer. In winter nitrate is replenished from below. This cycle may be amplified in polynyas and marginal ice zones (Vergeynst et al., 2018). Despite such sequestration, nutrient concentrations in seawater may be adequate to enable biodegradation of low concentrations of oil in the water column: McFarlin et al. (2014) incubated Alaska North Slope crude oil with Alaskan seawater at –1°C, without nutrient amendment but also with or without chemical dispersion, and measured significant biodegradation within 60 d. Subsequently, McFarlin et al. (2018) reported that surface seawater collected from the Chukchi Sea and amended with low concentrations of nutrients (approximating natural background conditions) not only aerobically biodegraded 36–41% of North Slope oil within 28 d at 2°C, but also significantly biodegraded the surfactant component of Corexit 9500. *Colwellia*, *Polaribacter*, and *Oleispira* were implicated in both oil and Corexit biodegradation and may in fact have degraded dispersant components in preference to oil under the study conditions. Similarly, Gofstein et al. (2020) determined that crude oil components (in the order *n*-alkanes > *iso*-alkanes > PAHs) and the surfactant component of Corexit 9500 were degraded by Arctic seawater microbes, with some taxa specializing in one or the other substrate and others apparently able to degrade both.

5.3.5.5 Arctic Seafloor Sediments, Deep Seeps, and Shorelines

In a survey of seafloor sediments from a transect of the Arctic Ocean, Dong et al. (2015) detected 16 U.S. EPA priority PAHs in all assayed cores. The concentrations decreased with sediment sample depth and with location from southernmost to northernmost sites. Taxonomic analysis of 16S rRNA genes revealed the widespread presence of known hydrocarbon degraders commonly detected at lower latitudes, including *Cycloclasticus*, *Alcananivorax*, *Colwellia*, and *Dietzia* (see Appendix I), with the latter being most abundant. Notably, these genera degrade oil aerobically, whereas subsurface sediments typically are anaerobic; it is therefore unclear whether these cells were active or dormant in situ, even though they were still viable and could be isolated and/or enriched in the laboratory. Ferguson et al. (2017) incubated subarctic deep-sea sediments at 0°C and 5°C with a model oil comprising 20 hydrocarbons (representing saturates, monoaromatics, 3–4 ring PAHs and resin components >C8) in the presence and absence of the marine oil dispersant SuperDispersant 25. The sediment microbial community composition shifted during hydrocarbon degradation at both temperatures, and the dispersant had variable effects on biodegradation, from insignificant effects at 5°C to enhanced biodegradation at 0°C. The research concluded that the effect of dispersant was ambiguous, needing further investigation under cold conditions. In contrast to the aerobic PAH study, Gittel et al. (2015) specifically surveyed cold marine sediments for the presence of anaerobic alkane degradation genes, which are represented by some marker sequences in databases. They found the genes to be ubiquitous, along with nanomolar concentrations of short-chain alkanes even in sediments considered pristine. Taken together, the studies suggest that the potential for both aerobic and anaerobic alkane biodegradation is widespread in cold seafloor sediments, and there is cosmopolitan distribution of species and genes associated with oil degradation.

Cold hydrocarbon seeps support cold-adapted hydrocarbon-degrading bacteria that may be recruited to Arctic marine oil spills. A hydrocarbon seep recently sampled in Scott Inlet, Baffin Bay in the high Arctic, supported seafloor microbial mats and colonies of higher organisms, and the water column above and down-current of the seep were enriched in putative methane- and hydrocarbon-oxidizing bacteria, the latter including the genera *Polaribacter* and *Colwellia* (Cramm et al., 2021) commonly enriched during oil spills in cold seawater. Both aerobic and anaerobic methane-consuming microbes can exist at gas seeps. An Arctic cold gas seep in the Barents Sea harbored macroscopic biofilms, located at the sulfate–methane transition zone, comprising a single anaerobic methane oxidizer and several members of sulfate-reducing bacteria (Gründger et al., 2019). Thus, key hydrocarbon-degrading species may be sustained by natural hydrocarbon inputs at pristine sites to be recruited in the event of an oil spill.

Oil degradation on Arctic shorelines has been studied by applying oil directly to the intertidal zone or by using solid surfaces as a proxy for a rocky substratum. For example, Gustavson et al. (2020) applied crude oil or heavy fuel oil to slate tiles placed throughout the tidal zone, with or without sunlight, on a West Greenland shoreline in the summer. Because the petroleum biomarker ratios were nearly constant during the 95-day experiment, biodegradation was deemed to be "fairly unimportant" in the field trial. This may have resulted when biofilms forming on the oiled surfaces were sloughed off due to wave action and/or inhibited by solar irradiation at the high latitude. In contrast, results from the Baffin Island Oil Spill (BIOS) long-term shoreline study in the Canadian Arctic, summarized in Box 5.12, illustrate that effective shoreline bioremediation can occur slowly at high latitudes.

In conclusion, it appears that temperature is likely not the most important factor controlling oil biodegradation in Arctic oceans compared with other limitations such as nutrient supply and replenishment and oil bioavailability. Key microbial oil-degrading species are present in Arctic marine environments as members of the rare biosphere, but combinations of limiting environmental factors may slow or restrict their response to and subsequent biodegradation of spilled oil, requiring prolonged time for oil removal. Whereas oil biodegradation may not be significant within cold bulk ice, microbes are present and can begin to degrade the oil once the ice is warmer or in the ocean surface when the ice melts.

5.4 FATES IN SPECIFIC MARINE ENVIRONMENTS: CHRONIC INPUTS

In addition to the episodic oil spills described in Section 5.3 that have a defined onset and finite duration of oil input, there are numerous long-term sources of oils that enter the marine environment (see Chapter 3). The inputs may be intermittent or uninterrupted, and may be diffuse or point sources; often they comprise small volumes released over a long time. Continuing inputs include natural sources such as oil and gas seeps as well as anthropogenic inputs arising from resource development and use, such as wastes from offshore drilling, commercial shipping, wrecks, recreational boating, and urban runoff or discharges into rivers. In many cases the concentration of oil entering the ocean from such releases is more dilute and more diffuse than in an acute oil spill, making it more difficult to monitor and ascertain the

BOX 5.12
A Long-Term Arctic Experimental Eite, 1980–2019: The Baffin Island Oil Spill (BIOS)

The BIOS experimental oil spill comprised a suite of multidisciplinary field studies conducted at Cape Hatt, Baffin Island, Canada (72°3'1 N, 79°50'W) from 1980 to 1983 (Boehm et al., 1985; reviewed by Lee et al., 2015; Sergy and Blackall, 1987, summarizing papers cited therein). The main objectives were to determine the fate and environmental effects of oil spilled with or without a chemical dispersant in Arctic near-shore waters during the ice-free season, and to determine the efficacy of shoreline remediation approaches for stranded neat oil and an oil-water emulsion. A lightly weathered Venezuelan Lago Medio oil (a sweet medium gravity crude) was applied to near-shore water either as neat oil at the surface or premixed 10:1 with the dispersant Corexit 9527 and applied subsurface as an emulsion in sea water (Dickins et al., 1987). The oil was allowed to strand naturally on the shoreline via tide, wind, and wave action. Independent shoreline applications of oil were conducted on low-energy beaches away from the near-shore spill by spraying slightly weathered oil or a 1:1 oil:water emulsion directly onto plots in the intertidal and supratidal zones (Owens and Robson, 1987). Pre-spill baseline data and control plots provided comparisons with experimental plots. Water and sediment samples were collected yearly from 1980 to 1983, intermittently until 1993, again in 2001 (Prince et al., 2002) and in 2019 (GENICE, personal communication) to document long-term shoreline recovery.

The main conclusions about the short-term fate of oil were that (1) application of a chemical dispersant to oil spilled in pristine Arctic nearshore waters was not contraindicated and, in some situations, could be beneficial (e.g., the impacts of dispersed oil on subtidal sediments were found to be less than the impacts of undispersed oil; Boehm et al., 1987); and (2) natural processes (e.g., abiotic weathering and natural washing via moderate wave action) rapidly dispersed fresh intertidal oil at the shoreline surface, rendering intervention (i.e., mechanical mixing or chemical dispersion) unnecessary (Sergy and Blackall, 1987). Some of the neat oil applied at the surface stranded on the beach and a portion remained over the short term, whereas the dispersed oil rapidly disappeared from the water column and the beaches by tide and wave action. In the shoreline application experiment, oil retention was influenced by the type of substratum (e.g., sand versus gravel/pebble), porosity, water table height, and whether the oil was neat or emulsified, but generally the bulk of spilled oil was rapidly removed from the shore by tidal action within 1–6 weeks. Dispersant washing of some beach plots immediately reduced surficial oil content. A soft "asphalt pavement" developed in some parts of the upper intertidal zone where oil remained in patches on the shoreline surface, particularly in low-energy beaches and backwater shorelines (Owens et al., 1987). One year later, oil was detected only high up in the intertidal zone sediments, and by summer's end was undetectable. In the water column, no significant effects of neat or dispersed oil were detected on microbial numbers and general metabolic activity using techniques available at the time (Bunch, 1987).

Characterization of residual oil collected from the beach four years later revealed biodegradation of *n*-alkanes, isoprenoids, and low molecular weight PAH (Wang et al., 1995) but small-scale differences in microbiota and/or local conditions influenced the fate of oil components, as well as the location of the residual oil—whether exposed at the beach surface or present as liquid oil embedded in sediment.

Twenty years later most of the beach was oil-free, with occasional subsurface patches of relatively fresh oil (chemically similar to the original oil) and a few samples characteristic of extensively weathered, photo-oxidized, and biodegraded oil that had experienced loss of more than 87% of the original hydrocarbons, alkyl-chrysenes, and certain terpane and sterane petroleum biomarkers (Prince et al., 2002).

In 2019, almost 40 years post-spill, samples of water and subtidal, intertidal, and supratidal sediments were collected for oil chemistry and 'omics analysis. The shore sampling campaign was not exhaustive because in most cases the original plot outlines were no longer evident, having been naturally attenuated. Preliminary results from samples collected at two shoreline sites revealed that total petroleum hydrocarbons (TPHs) had decreased from 1983 measurements, with the subsurface samples retaining more TPH than surface samples. Analyses showed evidence of continued but slight weathering of alkanes in surface sediments, with a few supratidal samples still having oil that was depleted in low molecular weight alkanes and enriched in partially oxidized hydrocarbons (determined using FT-ICR-MS; see Chapter 2). PAHs were still detectable at the two sites, with surface samples having experienced proportionally greater losses than subsurface samples (G.A. Stern, University of Manitoba, personal communication). Metagenomic analyses of sediments showed that genes associated with aerobic biodegradation of alkanes and PAHs remained elevated at a few oiled shoreline sites compared to pristine control sites (L. Schreiber, National Research Council, Canada, personal communication).

This long-term resampling with chemical and genomic characterization demonstrates that substantial natural attenuation of stranded medium-weight crude oil is possible even under typical Arctic conditions, and that chemical and genomic markers of oil biodegradation can persist for decades.

ultimate fates of the contaminants in the sea. For the most part, the fates of oil from continuing inputs are assumed to be similar to those of episodic spills (refer to Section 5.3 for episodic inputs and Section 5.2 for fundamental hydrocarbon fate processes in the oceans).

5.4.1 Fates of Oil and Gas from Natural Seeps

Natural seeps are ubiquitous on the continental margins and are sources of natural gas and liquid-phase petroleum to the ocean water column (Ruppel and Kessler, 2017; see also Chapter 3). Natural seep sources discharge through cracks and fissures on the seafloor, and petroleum fluids enter the water column as oil droplets or gas bubbles. Because of the low turbulence normally associated with weak natural seep discharge, droplets and bubbles are in the millimeter diameter size range (Wang et al., 2016, 2020; Romer et al., 2019).

The fate of oil and gas discharged from natural seeps is controlled by droplet and bubble processes, notably advection and dissolution (see Sections 5.2.2 and 5.2.5), and by biodegradation of dissolved compounds (see Section 5.2.8). Oil from seeps is mostly present in the ocean water column, following processes similar to those for episodic spills (see Sections 5.3.2 and 5.3.3), but with lower overall inflow rates. Natural gases are predominantly methane; biogenic sources may be exclusively methane, and authigenic sources may include fractions of other, light hydrocarbons, including ethane, propane, and butane, along with atmospheric gases, including nitrogen and carbon dioxide (Wang et al., 2016, 2020; Ruppel and Kessler, 2017; Leonte et al., 2018). Because of their slow seepage through the marine sediments before release allowing solute equilibrium with the subsurface water, liquid oil is not expected to contain significant amounts of gaseous compounds (e.g., methane, ethane, or propane).

Because the natural gases are quite soluble at depth, dissolution is a major process altering the composition of natural gas bubbles released from natural seeps (Rehder et al., 2002, 2009). Because natural seep bubbles are normally depleted in some of the atmospheric gases, they may also strip dissolved gases from the ocean water column (McGinnis et al., 2006). Wang et al. (2020) showed that, although bubbles remained on the order of 1 mm in diameter after 400 m of rise from the seafloor, their model simulations predicted that over 99.9% of the initial methane had dissolved out of the bubbles, and that the bubbles were dominated by atmospheric gases, stripped from the surrounding seawater. **Hence, the presence of bubbles in the water column above a natural seep is not proof that hydrocarbon gases remain an important component of these bubbles.**

Because of the large density difference between gas bubbles and seawater, even at deep ocean depths, they can be easily observed by long-range acoustic multibeam sonar (Römer et al., 2012, 2019; Skarke et al., 2014). This observation is possible because gas bubbles resonate and have high backscatter levels at sizes of order 1 mm in diameter or greater for acoustic excitation down to 18 kHz; higher frequency multibeam penetrates less deep into the ocean water column, but can visualize smaller bubbles (Weber et al., 2014). **This has led to many new discoveries of natural gas seeps through analysis of the water column backscatter of multibeam bathymetric surveys.**

By contrast, dead oil released at natural seeps undergoes little dissolution on transiting the ocean water column. The oil often reaches the sea surface and, depending on the sea state, may form a distinct slick that extends many kilometers from the seep source. Unlike gas bubbles, the similar density of oil to seawater makes oil droplets acoustically transparent to most common multibeam sonar frequencies. Hence, oil droplets are seldom identified in multibeam surveys. Instead, their surface floating oil signatures are visible in satellite imagery, which is a major means of discovering new oil seep sites (MacDonald et al., 2015).

Aerobic and anaerobic biodegradation of methane from natural seeps has been documented (see Section 5.2.8.8) and recent studies applying ‘omics and biogeochemistry to hydrocarbon seep sediments revealed a diverse community of microbes that are inferred to use a range of electron accepting processes including sulfate reduction and methanogenesis (Dong et al., 2019, 2020). Such microbes support diverse communities of higher organisms including bivalves, tube worms, corals, and crustaceans, some as symbionts (Rubin-Bloom et al., 2017) and others as the base of the food chain at natural seeps (Cramm et al., 2021).

5.4.2 Offshore Produced Water

Produced water from offshore production facilities, described in detail in Section 3.4.1.2, comprises various volume ratios of oil, gas, and water containing dissolved chemicals; its composition varies between oil fields and with time. Oil, gas, and water are separated at the surface facilities and may undergo chemical and/or physical treatment before being discharged to the ocean (Ahmadun et al., 2009). The discharge includes unrecovered finely dispersed oil and dissolved hydrocarbons plus, commonly, other water-soluble organic compounds such as volatile fatty acids and naphthenic acids, biocides, corrosion inhibitors, metals and inorganic solutes (Barman Skaare et al., 2007; Neff et al., 2011; Harman et al., 2014). Produced water also often harbors viable microbes that survived the injection, production and separation processes or were displaced from biofilms that developed in the reservoir and/or within production infrastructure (Gieg et al., 2011).

The unrecovered oil in discharged produced water is assumed to be subject to the same sub-surface physico-chemical and biological fates discussed in Section 5.2.2, including dispersion, dissolution, biodegradation, and sedimentation (the fates of other chemical additives is beyond the scope of this report) and may form sheens at the surface (King et al., 2016) where it would undergo evaporation and photo-oxidation. Dilution is a major factor in determining local concentrations of produced water hydrocarbons in the water column (see Section 3.4.1.2) and would also lessen any deleterious effects of biocides discharged with the produced water. Localized enrichment of native hydrocarbon-degrading bacteria and/or introduction of hydrocarbon-degraders with discharged water (Yeung et al., 2011) might amplify aerobic biodegradation by free-living and MOS microbes in the water column, in addition to serving as living markers of produced water plume movement via ‘omics (see Appendix H). Similarly, oil transported to the seafloor by MOS sedimentation and subsequently buried eventually would be subject to anaerobic conditions. In either case, in theory hydrocarbon biodegradation kinetics would be affected by nutrient availability, temperature, depth, and oil composition (see Section 5.2.8), but there are few published studies supporting these assumptions.

A rare laboratory study of produced water biodegradation potential incubated in a North Sea oilfield produced water with Norwegian fjord seawater in the dark (Lofthus et al., 2018b). Measured PAH half-lives ranged from 8 to >100 d (median 16 d) after lag periods of 6–12 d; other organic solutes (alkylated phenols) likewise were biodegraded. Significant growth and attachment of bacteria to OMAs and depletion of dissolved oxygen during incubation in closed containers were observed, indicating that components of produced water from this offshore field are biodegradable relatively rapidly under simulated in situ conditions. Additional studies of oil fields from different geographical regions are needed to capture the scope of produced water biodegradability and its consequences, especially since produced water discharge can continue for many years during offshore well operation. Such surveys might indicate whether biological "polishing" of produced water at surface facilities prior to discharge would be beneficial (Nilssen and Bakke, 2011; Camarillo and Stringfellow, 2018; Deng et al., 2021; Nepstad et al., 2021).

Few models that describe the dispersion, physico-chemical fates, and transport of produced water include biodegradation parameters (e.g., Reed and Rye, 2011; Nepstad et al., 2021). Questions that arise include direct effects on MOS formation (either positively by supplying viable microbes and cell detritus, or negatively through inhibitory solutes) and indirect ecosystem effects of contributing particulate and dissolved organic carbon to the biological pump, thereby increasing transport of biomass and oil to seafloor sediments in the long hydrocarbon cycle (see Box 5.6) and potentially depleting dissolved oxygen locally. A report by Klaise et al. (2014) found that offshore platform structures enhance local biodiversity and marine heterotrophic productivity by providing habitable surfaces, but the contribution of produced water to productivity was not addressed. A recent study detected hydrocarbon-degrading microbiota associated with the infrastructure of decommissioned offshore production wells (Vigneron et al., 2021), indicating the widespread effects of offshore oil production on local biodiversity. **Research is needed to quantify the fates and effects of prolonged produced water discharge particularly in areas with concentrated off-shore oil production such as the Gulf of Mexico and North Sea. Models describing the physico-chemical fates of produced water should include factors for biodegradation and MOS sedimentation, which would require acquisition of laboratory and field data.**

5.4.3 Fates of Oil from Ship Discharges

Discharges of oily wastes from oil tankers have been described in Section 3.4.2, and discharges from recreational vehicles and facilities in Section 3.2.3. Because these discharges take place at or near the surface, oil components will be subject to the same physico-chemical and aerobic biodegradation fates described for the water column in Sections 5.3.1 and 5.3.2. As with riverine sources (see Section 5.4.4 below), these discharges often contain other chemicals such as metals and organic chemical co-contaminants that may influence the specific fates of the discharges. As discussed in Chapter 3, operational discharges from tankers have been reduced with the evolution of the regulatory requirements for the operation, design and construction of tankers, which has gradually reduced or eliminated mixing water with cargo. In contrast, oil from sunken wrecks (described in Section 3.5.4) may impact the water column as the oil rises and/or the sediment, depending on the position of the wreck. In this case the fate of the oil could also include aerobic and anaerobic seafloor processes described in Sections 5.2.8.8 and 5.3.3.6.

In harbors and marinas, small oil spills from commercial and recreational boating incrementally become entrained in anaerobic fine-grained sediments. In this case, anaerobic biodegradation prevails (typically via sulfate reduction and/or methanogenesis; see Section 5.2.8.8), but may be positively or negatively influenced by co-contamination with other organic wastes.

5.4.4 Fates of Oil from Riverine Sources

The contributions of oil to the ocean via runoff are described in Chapter 3. Many of the urban inputs via stormwater runoff and water treatment plant effluents, discussed below, are complex mixtures (e.g., include non-hydrocarbon fat, oils, and greases [FOG] in addition to petroleum components), and furthermore their concentrations have a range of uncertainty of four orders of magnitude (see Section 3.3.1), making it difficult to quantify their fates during transport and upon reaching the sea. In general, the assumption is that the fates of any oil-related compounds in river outflows will be similar to those from oil spills and natural seeps described earlier in this chapter, being dependent on the physical and chemical properties of the compounds themselves and of the coastal areas receiving the inputs, including salinity gradients, suspended sediment loads, nutrient concentrations, co-contaminants, water temperature, mixing energy, and so on.

Particulate and dissolved contaminants that intermittently wash from the atmosphere, soil, and urban surfaces into rivers via stormwater constitute a constant but relatively dilute source of hydrocarbons (see Appendix B). The low concentrations of hydrocarbons in these dilute inputs may have decreased bioavailability (e.g., through sorption to particulates; Alexander, 1985) and furthermore may be insufficient to induce microbial biodegradation.

FOG, which includes animal fats and oils as well as petroleum, in wastewater streams originate from food processing and cooking activities, oil mills and refineries, as well as runoff or direct disposal to sewer lines (Collin et al., 2020).

Although >90% of the FOG entering wastewater treatment plants is removed through physical, biochemical, and chemical processes (e.g., gravity separation and skimming, dissolved air flotation, de-emulsification, coagulation, flocculation, biodegradation), the effluents still contain some oil and grease composed of animal and vegetable origin that will biodegrade quite rapidly in the marine environment and do not require the presence and activity of specialized microbial species, unlike some petroleum components. Vegetable oils, animal fats, and some greases are mainly non-hydrocarbons, containing only minor amounts of biogenic hydrocarbons such as plant waxes, n-alkanes, and carotenes. They differ chemically from petroleum oils and can be distinguished from petrogenic and pyrogenic hydrocarbons using routine chemical analyses described in Chapter 2. However, the typical analyses conducted on wastewater treatment plant effluents do not readily distinguish oils and greases originating from food sources (plant- or animal-based oil and grease) from those with petrogenic origin; such methods should be updated to enable discrimination of the categories.

5.4.5 Fates of Oil in Coastal Ecosystems: Monitoring PAH Profiles in Sediments and Bivalves

One of the fates of petroleum and other hydrocarbon inputs, such as pyrogenic PAH (see Section 2.1.5), in the sea is accumulation in muds (mainly silt-clay and organic matter) and surface sediments, which are impacted by oil that sediments from the water column or is re-mobilized from beaches. Direct measurement of hydrocarbons in surface sediments is useful for assessing the current status and longer term trends of contamination of coastal and estuarine ecosystems, but is subject to three main limitations: (1) bottom-dwelling organisms stir (bioturbate) the sediments, smearing the time record over many years for typical coastal and estuarine muddy sediments; (2) physical processes resuspend, transport, and redeposit oil-impacted sediments and are another factor smearing the record in space and time; and (3) chemical extraction of sediments and analyses of those extracts yield hydrocarbons that might be sequestered in, or tightly adsorbed to, sediment particulates (NRC, 2003) and thus do not necessarily provide an accurate assessment of hydrocarbon bioavailability to higher organisms. For this latter purpose, bioavailability can be assessed by analyzing the hydrocarbon content of various water column and benthic organisms. To assess geographical trends, it is best to utilize cosmopolitan species inhabiting a significant geographical range of coastal and estuarine areas. Bivalves, specifically mussels and oysters, have been the species of choice not only for petroleum hydrocarbons and pyrogenic PAH, but also for other chemicals of environmental concern such as chlorinated pesticides, PCBs, mercury, lead, and tributyltin. Bivalves bioaccumulate such chemicals from their surrounding habitat by factors of several orders of magnitude up to 106 compared to concentrations in water, depending on the chemical involved, and are good indicators of PAH persistence.

The need for regional, national, and international monitoring programs to assess the status and trends of chemicals of environmental concern, including the fates of petroleum hydrocarbons, was recognized in the late 1960s and early 1970s. The use of mussels and oysters in such programs on a wide geographic scale was proposed in 1975 (Goldberg, 1975). Examples of progress in various regions, countries, and internationally, and the advantages and limitations of bivalve mussel monitoring programs noted since then have been reviewed by Farrington et al. (2016). That review, references therein, and citations below are the source of the following information. Bivalves bioaccumulate hydrocarbons such as PAH from their surrounding habit, rapidly taking up hydrocarbons within hours to a day after exposure. If the exposure source is removed rapidly by shifting currents or other means, resulting in decreased water concentrations, the bivalves release hydrocarbons back to the surrounding water. If the exposure is prolonged, then there is evidence that hydrocarbons are distributed to various tissues within the bivalve. Once the source of hydrocarbons is reduced, release of hydrocarbons back to the water will proceed, often in a multi-step process with more rapid release of some followed by slower release of others. This is best explained by considering that different tissues within the bivalves have different uptake and release rates and are interconnected by the bivalve's circulatory fluids. The longer the exposure, the longer the release time once the exposure is reduced or eliminated. The uptake and release are mainly via filter feeding of particulate matter and from or to water via the gill surfaces. There may be some hydrocarbon metabolism by the bivalves themselves or biodegradation by microbes associated with the gills and other tissues, but this is small compared to enzymatic activity induced in various species of fish and crustacea by the uptake, for example, of PAHs (NRC, 1985, 2003; Chapter 6 herein).

The U.S. NOAA National Status and Trends Program (NS&T) (Kimbrough et al., 2008) annually has measured a suite of PAHs in sediments and mussels and oysters sampled in U.S. coastal areas, beginning with 145 sites in 1986 and increasing to nearly 300 sites as the program progressed. The last comprehensive reporting on a national basis for the Mussel Watch portion of the NS&T Program was by Kimbrough et al. (2008). As noted in Chapter 2, PAHs enter the coastal environment from various sources such as combustion of fossil fuels; forest and grass fires; creosote on wood pilings; and oil inputs from spills; and chronic inputs from rivers, municipal and industrial effluents, land runoff, and natural oil seeps. The NS&T Mussel Watch Data provided both contemporary geographic (i.e., Status) and long-term assessments (Trends) for mussel species at 108 sampling stations in the Northeast, Southwest, Northwest, and Alaska coastal areas and for oyster species at 105 sampling stations in the Southeast and Gulf of Mexico coastal areas (plus freshwater mussel species at 23 sampling stations in Great Lakes coastal areas).

There was a significant decreasing trend for the sum of measured PAH concentrations at 33 of 236 sampling stations and increasing concentrations at only 2 of 236 sampling stations (Kimbrough et al., 2008). Generally, the highest concentrations were in samples near urban areas and included both pyrogenic and petrogenic PAHs. A rich data set of individual PAH measurements remain to be explored further that may disclose separate trends for petrogenic and pyrogenic PAH. In comparison, data for total PCBs, a class of chemicals of environmental concern that has properties similar to PAHs, showed decreasing trends at 46 of the 236 sampling stations and only one sampling station with an increasing trend.

To illustrate the importance of the NS&T Mussel Watch Program, baseline data from the program were important in assessing coastal contamination of several geographical locations as a result of the DWH oil spill (Apeti et al., 2013). Regional and local programs of baseline monitoring and trends assessments have built around the NS&T Program, such as the Gulfwatch Contaminants Monitoring Program in the Gulf of Maine that involved sampling and measurement of a suite of PAHs by Canadian and U.S collaborators from 2005 to 2010 (Chamberlain et al., 2018). A more local example built on the NS&T model is a program based in Puget Sound, Washington, using mussels from an aquaculture source transplanted to various stations in Puget Sound that measured a suite of PAHs plus other chemicals of environmental concern. (Lanksbury et al., 2017). Such long-term analyses provide insight into the persistence of PAHs in marine sediments.

The NOAA NS&T Program transitioned about 10 years ago from being a national collection and analysis program that included measurements of a suite of PAHs to a program that seems to emphasize joint funding collaborations with regional and local government entities. This program, proven useful for assessing the status and trends of PAHs in the coastal environment, appears to be at an end with respect to U.S. nationwide sampling and analyses on an annual basis.

5.5 MODELING THE TRANSPORT AND FATE OF SPILLED OIL

Many different models have been developed to predict and study the fate and effects of oil spilled in the marine environment, including governmental, commercial, and research models. These models normally include similar types of components, with the degree of model complexity depending on the model type and intended usage. The most common model application is for contingency planning; models used for real spills include forecasting models to help direct the response and injury assessment models to help understand the impacts of different response decisions and the ultimate injury to the environment resulting from the spill. Here, we summarize these aspects of models and review some of the major advancements that have occurred since the previous NRC report on *Oil in the Sea III*.

5.5.1 Model Components

Oil spill models simulate different processes of oil transport, fate, and effects through various sub-models that handle each process. Many of the algorithms used in these sub-modules are described in Chapters 5 and 6 of this report. Here, we briefly describe the chain of processes simulated by most oil spill models and some of the strategies used to include these algorithms within realistic oil spill simulations.

Models simulating the transport, fate, and effects of oil in the oceans must solve a coupled set of differential equations, which predict the time-evolution of the spill dynamics. These equations rely on *boundary conditions* to provide the external forcing and source *initial conditions* to provide the release information. As explained in Section 5.2.9.1, many of the inputs required by these models have large uncertainties, especially early on during a spill; hence, model predictions should be viewed as being only as certain as the certainty of their inputs.

Boundary conditions in oil spill models include the hydrodynamic motions, density profiles, and background chemical concentrations of the oceans and atmosphere. Because the atmospheric and ocean fluid dynamics are normally considered to be independent of, or unaffected by, the oil spill, models to predict the coupled ocean–atmosphere dynamics are run separately from the oil spill model. There are a wide number of operational ocean, atmosphere, and coupled ocean-atmosphere models (Ainsworth et al., 2021). We define an operational model to be one that regularly runs and posts model results on web servers that may be accessed either publicly or by spill modelers during an actual oil spill. Most of these models simulate historic data up to the current date and then make model forecasts for several days into the future. To predict the true ocean–atmosphere dynamics most accurately, these models assimilate measured data up to the real-time simulation point. In the oceans, these assimilated data include satellite observations of ocean temperature and water level (i.e., altimetry), which allow the models to predict the meso-scale eddy dynamics (rotating vortices of order 10 km in diameter and larger) and, in the Gulf of Mexico, to also predict the Loop Current. Where available, full and partial water-column profiles of temperature, salinity, and currents have also been assimilated. Since assimilation data are not available for future times, forecasts are limited to a few days in the coupled ocean–atmosphere models, or for the ocean-side alone, a week to a month. Some models handle uncertainty of future model predictions by running several ensemble forecasts, each forecast assuming a slightly different set of model initial conditions or parameter values. To predict the trajectories and fates of spilled oil, the currents, temperature, and salinity dynamics of these ocean–atmosphere hydrodynamic models are used.

Initial conditions for an oil spill include the flow rate, composition, and location of the spilled fluids, and the geometry of the release. These values may be known at varying levels of certainty during a real or simulated spill event. This information is normally synthesized by an initial conditions sub-model that predicts the bubble and droplet size distributions for subsurface spills or the initial spreading of a surface slick for a surface spill. The oil spill model then tracks the evolution of the spilled petroleum fluids as they are transported by the ocean currents.

The oil transport models are normally composed of different modules for the near- and far-field dynamics of the spill. The spill near-field is a region close to the spill in which the source geometry and the spill momentum and buoyancy affect the transport. For a subsea oil well blowout, this includes the orifice, rising plume of oil droplets, gas bubbles, entrained seawater, and the fate and transport of this plume, including dissolution, until it is no longer controlled by the collective buoyancy of the spilled oil and gas. For a surface spill, this includes the release geometry, potential droplet formation, and initial spreading to form a slick on the sea surface. The near-field is normally contained within a radius of order 1 km or smaller around the spill source. Because these are localized, small domains of the ocean, the hydrodynamic ocean boundary condition data can be provided either by operational hydrodynamic models or measured ocean profiles—there is no need to know the full dynamics of the ocean basin. Once oil droplets, gas bubbles, dissolved components, or floating slicks leave the near-field, their subsequent transport is modeled by far-field trajectory models.

Far-field transport models solve the governing advection–diffusion equation, or transport equation, with the advection of oil and gas given by the superposition of the ocean currents and the rise velocity of individual oil droplets or gas bubbles. Generally, two modeling approaches are available to solve the transport equation. Ocean circulation models use the Eulerian approach to simulate the transport of temperature and salinity. This results in model predictions made at each grid-cell of the model domain. This works well for these variables where high concentration gradients are not expected—the temperature and salinity fields are fairly smooth at the resolution of the numerical model. Pictures of spilled oil show a different characteristic: oil slicks can be highly localized compared to the kilometer-scale resolution of ocean circulation models, breaking up into slicks a few 100 meters in diameter to patches of dispersed oil droplets on scales set by the surface breaking wave dynamics and Langmuir cells (around 1 to 10 m). If the Eulerian approach were used in oil trajectory modeling, oil slicks could not be resolved at scales less than several grid cells—or of several kilometers for most operational spill models. To avoid this problem, oil spill models use a Lagrangian approach to solve the transport equation. There, numerical particles called Lagrangian parcels are initialized in the model at the end of the near-field.

The Lagrangian transport approach uses numerical parcels that contain information about the amount of oil present and its distribution in the water column, either as droplets or a surface slick. The model then interpolates the gridded hydrodynamic data from the ocean circulation model to the exact location of the Lagrangian parcel and then solves for the transport using a deterministic advection step that depends on the ocean currents and oil droplet rise velocity and a random walk component, calibrated to the expected local turbulent diffusion in the ocean. Where the turbulent diffusivity is varying in space, as near the ocean surface, random displacement models that accurately include spatially varying diffusion are required. Using the random-walk or random-displacement approach to spreading, Lagrangian transport models avoid numerical diffusion and can predict sharp gradients in oil concentration and fine-scale spill structures (order a tens to hundreds of meters). The end-result of the far-field model is a simulation tool capable of tracking the individual, time-varying trajectories of hundreds to thousands of Lagrangian parcels, each initialized at the end of the near-field region and each tracking the unique state of the oil for that parcel within the ocean water column.

Oil fate and effects processes are simulated along the trajectories of each Lagrangian parcel. Historically, fate processes were termed oil weathering since they were mainly considered for surface spills, where oil fate is dominated by processes linked to the weather. However, any process that alters the mass or composition of oil within a Lagrangian parcel is a fate, or weathering, process. The dominant processes include dissolution, evaporation, biodegradation, and photo-oxidation. One may also consider sedimentation or formation of OMAs as types of fate processes as they affect the properties of the oil within a Lagrangian parcel.

To predict the effects of oil on the environment, models must convert the Lagrangian parcel data into composition and concentration information (i.e., the exposure level) and this must be coupled to models predicting the distribution of biological resources and an estimate of the effects, or toxicity, that would be experienced by these biological resources if encountering these predicted exposures. Toxicity and effects modeling is summarized in Chapter 6. To estimate concentration, Lagrangian parcels usually contain information about the total amount of oil considered, its centroid location, and its spatial extent. Because oil may be transported in different directions from the surrounding ocean water, concentrations of dissolved chemical species within a Lagrangian parcel are not directly predicted. Moreover, since the Lagrangian parcel is not considered to occupy a fixed volume of seawater, the concentration of liquid oil represented by a Lagrangian parcel is not a normal model state variable. Hence, a concentration sub-model is needed to convert the Lagrangian parcel information into field exposure levels.

For oil spill models to accurately track the evolving mass of oil in the ocean water column, they must also include submodels for different response options, such as skimming, in situ burning, aerial application, and subsea use of chemical dispersants, or other response actions that change the amount or dispersion of oil in the environment. Techniques to keep sensitive resources away from injury, such as bird hazing, are also used. These processes are normally represented as sink and transformation terms in the oil transport models and in modeling the oil spill response itself, with simulated booms, skimmers, dispersants, etc. Moreover, to improve accuracy, models require a feedback look from real-time observations to reset model predictions and improve the next output time series.

Oil may also leave the ocean-domain of the model by deposition on the shoreline or seafloor or by evaporation or transport into the atmosphere. Once deposited on the seafloor, further transport of the oil is normally ignored, and only fate and effects processes would be considered. As volatile and gaseous compounds leave a surface slick and enter the atmosphere, they may be simulated by atmospheric dispersion models. These models work similarly to the Lagrangian parcel models of the ocean-side transport except that the volatilized compounds are miscible in the air; hence, there is no longer a need to track liquid droplets separately from the atmospheric dynamics. As a result, atmospheric dispersion models do track component concentrations and can be directly used to estimate air quality.

5.5.2 Types of Models

Although nearly every oil spill model is unique, most are built using well-known numerical algorithms in computational fluid dynamics. Recent advances in oil spill models include better sub-models for the fate and effects algorithms and, because of the increase in computing power available in recent years, the application of more complex modeling tools to oil spill scenarios.

Here, we arbitrarily separate models by their speed and domain size. We consider integrated oil spill models to be those that can simulate the full extent of the spill and run at speeds adequate to inform spill response or injury assessment. Typically, responders must make forecasts in a matter of a few hours. Likewise, injury assessment must consider the full model domain and must be efficient enough to consider many spill scenarios. This class of models generally covers all response models, injury assessment tools, and contingency planning models. The other class of models we consider are research models. These have a greater diversity of model algorithms, but their main characteristic is that at present time they are largely limited to simulating a smaller domain than the full-scale spill. Research models are normally used to study high-resolution dynamics of specific aspects of the spill dynamics, such as the release conditions, the near-field plume, oil slick dynamics in the ocean mixed layer, or other regional domains of a spill.

5.5.2.1 Integrated Oil Spill Models

Because integrated oil spill models must be efficient, their algorithms are more limited. Initial conditions are normally predicted using either empirical equations for bubble or droplet size distributions or by running population balance models to predict the equilibrium size distribution at the end of the primary break-up zone. If a near-field buoyant plume forms, as for a subsea blowout, integral models are used to solve a one-dimensional set of conservation equations along the bent trajectory of the plume. An important aspect of all integral models is that they are steady-state models, which means they do not consider time-varying dynamics. To obtain an unsteady result, these models must be run successively under different conditions for each time-step—but the solution obtained for each model run is a steady-state solution. In the far-field, operational models all utilize Lagrangian parcel models, with model differences stemming from the way ocean circulation data are interpolated and the numerical algorithm used to solve the advection step. Models also differ by the random walk algorithm used, with the most flexible models using random displacement algorithms appropriate for non-uniform diffusivity fields. Because these models keep track of the oil mass and its characteristic state within the ocean water column (i.e., number and size of oil droplets), operational models easily include all fate and effects processes through sub-models that utilize algorithms normally expressed at the droplet or slick level (see the wide array of mechanisms discussed in Chapters 5 and 6 that support this model development). Hence, integrated oil spill models are mostly differentiated by the number of fate and effects processes they consider and the particular algorithms available within the model to simulate each process.

The main advances occurring recently for operational models are through the development of new and improving sub-models for each of the processes considered in these models. The overall skeleton of a droplet-size model feeding an integral model linked to a far-field Lagrangian parcel model has not changed much in the past 20 years. It is the key advances in the understanding of the fate and effects processes, detailed throughout Chapters 5 and 6, that have contributed the most to the ongoing development of integrated spill models. **At the same time, there remains an urgent need to convert our new insights into operational algorithms for oil fate and trajectory modeling, to include these algorithms in models, validate their predictions, ideally to field-scale observations, and to integrate oil spill models with the enormous stream of observation data that may be part of a spill response. Some of these new insights include new findings (warranted by spill observations) such as photo-oxidation, MOSSFA, temperature effects on biodegradation kinetics, and anaerobic biodegradation, among others. While some companies may make model improvements as part of their competitiveness profile, research is needed to define and test algorithms and funding is needed to allow this work to be completed ahead of the next spill scenario.**

5.5.2.2 Research Models

Research models have a much greater diversity since each model is normally built to investigate a different process.

However, only a few canonical model types are available for the three-dimensional simulation of multiphase flow. In our definition, research models are computational fluid dynamics (CFD) models that solve either a two- or three-dimensional version of the governing fluid dynamics equations, the Navier–Stokes equations. Because oil spills and ocean currents are turbulent, these models must address the turbulent nature of the flow. And, because oil and gas are immiscible, these models must consider more than one phase (gases and liquids) in the simulation.

There are two major branches of numerical modeling approaches for turbulent flow. The earliest, and most common approach is the Reynolds-Averaged Navier–Stokes (RANS) solution. In this approach, the governing equations are time-averaged, and the model only solves for the time-average, mean velocity profile. The model state variables are the time-average velocity components and the pressure. None of the turbulent motions are resolved by the model. In the process of taking the time-average of the governing equations, several new unknowns are also generated. These are various products of the turbulent velocity and pressure fluctuations and are commonly expressed in terms of Reynolds stresses. The algorithm used to relate these new unknowns to the time-average velocity and pressure field is called the turbulence closure model. Several closures are available, including algebraic and dynamic closures. Many commercial CFD codes provide the common turbulence closures. Because turbulence is a property of the flow field and not a property of the fluid, however, whenever a RANS model is applied to a new flow type, the turbulence closure model must be calibrated and validated to observations.

The other major approach to modeling turbulent flow is large eddy simulation (LES). In this approach, instead of applying a time-average to the model governing equations, a spatial average, or filter, is applied. This filtering has the effect of removing turbulent motions that occur below the filter scale, but allowing the model to resolve turbulent motion above this scale. Like the RANS approach, new unknowns are introduced in the model equations for products of the fluctuating velocity and pressure below the filter scale, and another type of turbulence closure is required. Ideally, the filter scale is set so that the production scales of turbulence and much of the turbulent cascade is resolved so that the closure model handles turbulent scales within the inertial sub-range, where the behavior approaches a universal law. Like RANS models, several closure models have been developed, and some are now available within commercial and open-source LES codes.

A third approach to turbulence modeling has been used in fluid dynamics studies, but has little application thus far in oil spills. This method of dynamic numerical simulation (DNS) does not apply any time- or spatial-averaging to the governing equations and instead solves for all of the turbulent motions in the flow. To date, model domain sizes are limited to be too small to have practical applications in oil spill modeling. Hence, in an oil spill research model, the continuous, or water, phase of the flow field is simulated using either RANS or LES methods.

There are also two major approaches to including the multi-phase nature of oil and gas flows in the oceans within a RANS or LES model. One approach treats the dispersed oil or gas as continuous distributions, and the equations are modeled using Eulerian transport equations. Because the RANS or LES model of the water is already an Eulerian model, this approach is called the Eulerian–Eulerian approach. The other major approach tracks each oil droplet or gas bubble separately using a Lagrangian particle approach, in models using an Eulerian–Lagrangian approach. Both methods couple the particle motions to the water dynamics by force coupling between the continuous-phase water and dispersed bubbles or droplets. And, both methods must deal appropriately with low concentrations of bubbles or droplets within each computational grid cell for the water flow.

As for turbulence models, there is also a third common approach to multiphase dynamics in CFD modeling that has had little application in oil spills. These are models that fully resolve the fluid dynamic motions in both fluid phases (e.g., water and oil) and then track the fluid interface. Because oil droplets and gas bubbles are small compared to the domains they are transported within, fully resolved multi-fluid models with interface tracking have not been effectively applied to oil spills.

Some of the largest advances in research modeling that has occurred in recent years is in development of new LES models for oil spill simulations. These advances have been made for both Eulerian-Eulerian (e.g., Yang et al., 2015) and Eulerian-Lagrangian (e.g., Fraga et al., 2016) approaches. The Eulerian-Eulerian models of oil spills have been applied to oil transport in the upper, mixed-layer of the ocean (Chor et al., 2018; Chamecki et al., 2019; D'Asaro et al., 2020), including the interactions of oil droplets with Langmuir circulations (Yang et al., 2014, 2015), and more recently to subsea oil well blowouts plumes (Yang et al., 2016; Daskiran et al., 2020), including the effects of subsea and aerial dispersant applications (Chen et al., 2018). These models have also integrated population balance models to dynamically track the oil droplet size distribution (Aiyer et al., 2019; Aiyer and Meneveau, 2020), and have developed numerical methods to include fate algorithms for aerosolization and atmospheric transport (Li et al., 2019), dissolution (Peng et al., 2020), and photo-oxidation (Xiao and Yang, 2020). Eulerian–Lagrangian models have been applied to simulate laboratory-scale bubble plumes (Fraga et al., 2016) and are helping to understand the detailed turbulent dynamics of multiphase plumes (Fraga and Stoesser, 2016), which in turn will aid calibration and validation of turbulent closure for integral and RANS models.

Though these research models may not simulate full-scale oil spills spanning the ocean basin scale, they are exceptional tools to study the detailed dynamics of oil spill processes within numerical domains that approach field scale. To the extent these simulations are validated, they can help bridge the gap between laboratory experiments, which are necessarily at reduced scale and contain idealized conditions, and field experiments, which are generally not permitted with oil at the present time. Even when field experiments are permitted, obtaining reliable and detailed observations is difficult. **Hence, insights from LES research models on multiphase oil spill dynamics are critical to help further our understanding of the behavior of oil in the sea. Research submodels are often the precursors for advancing operational and natural resource damage assessment models. These are especially valuable to making prototype-scale simulations where field trials are either prohibitively expensive or not permitted. Hence, the ongoing development of physics-based research models for understanding the complex behavior of oil fate in the real ocean is key to advancing response and injury assessment models, thus also benefiting decision making during and following real spills.**

5.5.2.3 Uses of Models

As inferred in the previous discussion, there are many different uses for oil spill models. Some require none or only a few sub-models related to oil fate; others require fully integrated operational models; and others involve specific, purpose-build research models. Here, we briefly summarize the most common uses of oil spill models for simulating the behavior of oil in the sea.

As an example of using just a selection of sub-models, weathering analysis studies the weathering properties of an oil assuming it is on the surface of the ocean and subject to idealized, constant weather forcing. Such models can predict the mass partitioning of oil for a surface spill into different fate categories, including natural dispersion, dissolution, evaporation, and emulsion formation. The U.S. NOAA Automated Data Inquiry for Oil Spills (ADIOS) model is a typical example. These models help establish the general weathering characteristics of an oil and are important for contingency planning and development of oil property databases. They are also useful in oil budget calculations, as described in Section 5.2.8.1.

During an actual oil spill, models are run to forecast the dynamics of the oil spill and help to direct the response. These types of model runs may include deterministic runs (single model scenario using the best available information) for two-dimensional surface fate and transport, deterministic runs for fully three-dimensional fate and transport, or stochastic runs of two- or three-dimensional fate and transport. For stochastic simulations, multiple simulations, or ensembles, are run, each potentially initialized with different conditions, each using potentially different circulation model output data, and each potentially based on different choices for the calibration parameters of the sub-models. Model output from several ensemble simulations is combined into a single forecast that can include uncertainty bands, such as maps with different probabilities of oil occurrence. Overlaying these trajectories over environmental sensitivity index (ESI) maps helps responders prioritize the daily needs of a response, for example, bird rookeries or coral reef areas may be prioritized for protection before more easily cleaned sand beaches. The Incident Command System brings together a variety of experts to work together to develop consensus for trajectory modeling and prioritize areas for protection and cleanup. The main feature of response model runs is that they involve forecasting future oil distributions based on the best available information on present oil amounts and locations and forecasted ocean currents. Rapid return of field observations to the response modelers is key for this to work well. Ideally, models should predict both the location and state of the oil (e.g., what response strategies may still be effective after a given amount of weathering) as well as predict the atmospheric air quality using atmospheric dispersion models.

Both during and after an actual oil spill, injury assessment models are also run to understand environmental and human health trade-offs of response options (e.g., during a spill response), and to quantify the environmental injury (e.g., after a spill and during the Natural Resource Damage Assessment [NRDA]) (refer to Chapter 6). Injury assessment modeling may rely on very similar simulations of fate and transport to those used in spill response modeling, but these simulation results are further analyzed to compute exposure and injury to organisms throughout the ocean environment.

Although models are often developed to help guide oil spill response and understand the impacts of an oil spill, the main way models are used is in contingency planning. Here, no actual oil spill has occurred, and models are used to understand the potential outcomes of a set of hypothetical oil spills, and these results are used to guide response planning and responder tabletop exercises.

Modeling activities could be integrated into contingency planning in several ways. A probabilistic modeling could evaluate possible trajectories of a worst-case discharge (typically a scenario with the largest potential release volume) under variable environmental conditions obtained from a historic dataset for that region. This modeling can indicate any preferential directions for oil spreading, illustrate seasonal differences in slick transport, determine minimum time for oil to reach sensitive areas, and identify worst case deterministic scenarios that could be considered for planning purposes.

Another form of modeling is directed at understanding the best response options for a given location or spill scenario. This type of modeling can support net environmental benefit analysis (NEBA), spill impact mitigation assessment (SIMA), or comparative risk assessment (CRA) (see Chapter 4 for details on NEBA, SIMA, and CRA). In this case, a selected deterministic spill trajectory is modeled with and without the use of various spill response techniques to evaluate their effect on changes in oil transport, and potential cultural, economic, human health, and environmental impacts. This provides a scientific basis for the analysis of response options tradeoffs for a given spill scenario.

5.5.3 Model Validation and Uncertainty

5.5.3.1 Model Validation

By definition, models are approximations for the real physics, chemistry, and thermodynamics controlling oil fate and transport. Hence, the accuracy of their approximations should be evaluated. Model validation is the process by which model predictions are compared to accepted, and true values are assessed in terms of the model appropriateness. These accepted values may be results of laboratory, mesocosm, or field experiments or observations of real events, such as the DWH oil spill. Each of these types of data are important, and each present different modeling challenges. For example, laboratory experiments are often better controlled, with more precise knowledge of inputs and outputs, but they are themselves approximations of field-scale events. On the other hand, field observations are made within the prototype system but are limited in their extent and often have forcing fields (e.g., wind, waves, currents) that are partly unknown. As a result, the observation databases against which models are compared have their own uncertainties in terms of the system dynamics (what processes were active to what degree during the experiment or event) and their outputs (how comprehensive is the database of observations). In some jurisdictions, open sea experiments involving oil are currently not permitted, and the only means to obtain prototype data is either during an accidental spill or at a natural oil or gas release. Spills of opportunity are rare and are controlled by the response; natural sources or oil or gas usually have slow release rates, not matching typical spill dynamics. Hence, there are many challenges first in obtaining relevant data to conduct model validation.

Assessment of model appropriateness should be done by comparing model predictions to these observations. Models should be run with inputs and conditions that match as closely as possible the conditions of the experiments. The model output data should be processed to create predictions that also most closely match the observations. Because the ocean is a turbulent, dynamic system and observations are rarely comprehensive in terms of spatial and temporal coverage, care must always be taken to compare values that accurately reflect the processes being evaluated. As an example, concentration measurements made in the field integrate both the fate processes that have altered concentration from the source to the sample (e.g., dissolution, biodegradation, emulsion formation) and the mixing that has occurred as a result of ocean currents (e.g., advection, turbulent diffusion, dispersion). One means to isolate fate processes from mixing processes is the use of fractionation indices (Gros et al., 2017). There, model predictions for the relative composition changes of modeled compounds are compared to similar fractional (i.e., percentage) changes in the sample. Whether a large amount or a little amount of dilution has occurred makes no difference in the fractionation indices.

A very important aspect of model validation is to quantify the degree of model correspondence with the data. Often, model validation is characterized by qualitative metrics, such as the terms *good* or *acceptable*. To the extent possible, these evaluations should be based on some quantitative metric. This is often difficult to achieve because natural processes undergo periodic variability over large ranges of absolute values. For example, daily solar insolation varies from a peak around noon to zero at night. Model processes that depend on sunlight will similarly range from high to low values. A single absolute error level may appear small during daytime predictions and very large during the night. Several model goodness-of-fit metrics are designed to handle large amplitude, periodic oscillation in model output, and these should be used where appropriate. Yet model validation using quantified metrics is critical to assessment of model performance, including model intercomparison and the evaluation of the appropriateness of individual models.

5.5.3.2 Model Uncertainty

While model validation may be used to determine whether a given model or algorithm adequately approximates a real system, model uncertainty remains. Model uncertainty arises due to approximations of the model algorithms and due to incomplete or uncertain knowledge of the model inputs or the algorithm parameters. Uncertainty due to errors in the model algorithms is usually uncovered and assessed through model validation (see previous sections). Even perfect models, though, will produce uncertain results because of uncertain inputs, both in terms of environmental data and model algorithm parameters.

In oil spill modeling, there may be large uncertainties in the environmental forcing (wind, waves, currents, density stratification, etc.), the spill location and volume, and the composition of the spilled fluids. Each of these uncertainties will contribute to an overall uncertainty in the model predictions. Model uncertainty tends to be ignored for two main reasons. First, model algorithms are rarely designed in such a way to propagate uncertainties in model inputs through to the equivalent uncertainty in model outputs. A 20% uncertainty in a model input may result in an output uncertainty that could vary from negligible to enormous, depending on the mathematical form of the model algorithm. Hence, models need to be built with explicit error propagation to make accurate predictions of model uncertainty. This is itself a field of study (probability and statistics), and few modelers have the expertise and time available to devote to this aspect of model performance. Second, it is generally assumed that accurate models (i.e., those that pass validation tests) will give the correct central tendency of the result. Models that deviate from the central tendency are considered biased. Hence, despite input uncertainties, an unbiased model would be expected to accurately predict the mean output field. Error bounds would then be symmetric about the mean, and the model estimate would be viewed as the most likely predicted value.

Unfortunately, models are rarely unbiased, and uncertainties in some input data (e.g., the spill flow rate, location, or composition) themselves introduce bias to the model predictions. One relatively easy method to estimate model uncertainty in this case is by means of computing several ensemble simulations. An ensemble is a set of simulations, each utilizing slightly different forcing or input data within the range of expected input values. Ensembles may also be constituted by running multiple models with the same set of inputs. The IPCC predictions for global temperature rise under different atmospheric CO_2 scenarios are a familiar example for ensemble simulations (*IPCC Sixth Assessment Report, Summary for Policy Makers, 2021*). Model uncertainty is then quantified by the spread in model results given this uncertainty in inputs and algorithms. Ensemble analysis can be time consuming when models are slow to produce results, but they are a robust means of estimating the uncertainty in model predictions. However model uncertainty is assessed, it is critical that decision makers have an understanding of the sources and levels of model uncertainty and the degree of certainty present in model predictions.

5.6 CONCLUSIONS AND RESEARCH NEEDS

Conclusion—Insights afforded by the DWH oil spill: The majority of oil spill observations and research prior to 2010 focused on the fate of oil spilled at the surface, but the DWH oil spill in the deep subsurface focused attention on additional processes affecting oil behavior and fate. Although future observations may establish that these insights pertain to the specific circumstances of a subsea blowout of light crude oil treated with subsurface dispersant injection in subtropical waters with historically high microbial activity, proximal to coastal systems that input nutrients and sediments, currently they highlight oil behaviors and fates that may warrant consideration in other circumstances.

1. **Appreciation of the physics of "dead oil" versus "live oil,"** when oil with dissolved gas is released from the deep subsurface. This phenomenon has led to significant research into interactions of live oil droplets and gas bubbles with the sea over a wide range of temperatures and pressures, and revealed the dynamic properties and behaviors of subsurface bubbles and droplets. Methods to connect equations-of-state with analyzed oil properties and models describing bubble and droplet breakup have evolved tremendously and laboratory data have been used to calibrate and validate these new models.
2. **The effects of subsurface dispersant injection (SSDI)** during the DWH oil spill. Implementation of this response method led to better understanding of deep-sea oil dispersion dynamics and demonstrated that SSDI can reduce VOC emissions at the sea surface. Further demonstration of its efficacy and suitability remains to be determined through laboratory research and modeling exercises under different conditions of well pressure, depth, type of gas–oil mixture, type and dosage regime of dispersant, and physical oceanography conditions.
3. **The importance of oil-mineral aggregates (OMAs)** in submergence and sinking of Macondo 252 oil from the DWH oil spill. Although the role of oil–particle interactions in causing oil to submerge has been known for decades, light oils have been considered to be non-sinking. The presence of high concentrations of suspended sand particles in the nearshore Gulf of Mexico and their interaction with weathered Macondo oil led to a significant proportion of oil being sedimented to the seafloor in the shallow nearshore waters.
4. **The role of marine snow in transporting spilled oil to the seafloor.** Although marine snow is a common natural phenomenon in the sea, the estimated magnitude of MOS formation associated with the DWH spill and the subsequent sedimentation and flocculant accumulation (MOSSFA) was unexpectedly large. This process potentially has implications for both surface and subsurface oil spills elsewhere, as it has not previously been considered to be a significant fate for oil. Its global importance and the role of subsurface dispersant injection on MOSSFA remain to be quantified.
5. **The magnitude of oil biodegradation in cold, deep ocean water.** Samples from the deep dispersed oil plume from the DWH spill revealed that the natural Gulf microbiota were capable of significant biodegradation of several low- to medium-molecular weight oil components under in situ conditions, including those of low temperature and high pressure.

Conclusion—The significance of photo-oxidation of oil at or near the ocean surface has received renewed appreciation. DWH oil spill observations and related experiments, rediscovery and reinterpretation of pre-2003 research, recent international studies and availability of sophisticated chemical methods for analyzing and quantifying photo-oxidation reaction products have led to a paradigm shift in appreciating the quantitative importance of photo-oxidation as a major factor early in the fate of slicks on water or oil coating sand and rocks on shorelines and vegetation in marshes, and has generated new questions about the quantity, identity, fate and toxicity of photo-oxidation reaction products.

Conclusion—"Big data" management and inter-disciplinary research: New analytical techniques, particularly in petroleum and environmental chemistry and in 'omics (both microbial and higher organisms) are generating enormous amounts of information or "big data." Although some federally funded data repositories already exist, long-term funding is essential for recruitment of personnel (e.g., discipline-specific curators) and for maintenance of infrastructure (e.g., computational ability, data storage) to archive data in repositories that are universally accessible. Concomitant with this support is the need to develop standards for data classification, quality control and reporting formats for each technique and 'omics-associated metadata that provide context for the analytical information, perhaps incorporating data and information using the geographic information system (GIS) for compiling location-specific data that can be accessed, mapped, and analyzed. Furthermore, meaningful interpretation of the data must include integration of large datasets, which is not currently achieved easily. This goal requires training of informaticians familiar with the different scientific fields generating the big data and associated metadata, as well as long-term funding to ensure that the software is constantly updated and the data are perpetually archived in accessible form. An adjunct to big data acquisition and integration is the need for a central data repository of oil properties, beginning with defining standard measurements and the appropriate cataloging platforms that are required for such an archive.

Conclusion—Baseline knowledge and data: After a spill has occurred, assessment and research efforts often do not have appropriate or requisite pre-spill environmental data for comparison with post-spill observations and assessment measurement of remediation. This limits the conclusions that can be drawn and affects prediction of spill recovery trajectories as well as recognition of ecosystem recovery in comparison to pre-spill conditions. Several programs have been reviewed by the National Academies, and others, over previous decades with accompanying recommendations for improvements. We applaud ongoing efforts to collect environmental data from marine sites, such as those supported by the U.S. Bureau of Ocean Energy Management, the U.S. Department of the Interior, the NOAA Coastal Ocean Observing System, and research programs funded by the Environment and Climate Change Canada, the U.S. National Science Foundation, the NOAA Sea Grant, and U.S. EPA.

Conclusion—Arctic studies: Marine traffic in Arctic waters is increasing with seasonal decrease in ice cover, and off-shore oil production is a possibility in the future, yet examination of the fate of oil in Arctic waters and shorelines has lagged behind study of more temperate and accessible marine ecosystems. Field experiments in Norway, Canada, and Alaska, and increasing international interest in the Arctic have uncovered many complex processes affecting oil in Arctic environments. However, utilizing this information in modeling or response still requires additional work.

Conclusion—Laboratory and mesocosm experiments, field studies, and modeling: Oil spill science relies on small-scale laboratory (in vitro) research, larger outdoor mesocosm (ex situ) experiments, in situ field studies, and modeling. Each contributes to our understanding of the behavior and fate of oil. All four types of research have strengths and limitations that should be acknowledged and considered when comparing measurements and conclusions; ideally, the results from all scales and approaches should be used to synthesize our understanding of the fate of oil in the sea. Despite adhering to "best practices" (see Box 6.4), laboratory and mesocosm experiments designed with the advantage of controlled conditions and replicated sampling may have shortcomings such as: inherent inaccuracy of scaling up results from lab flask to ecosystem scale; lack of agreement about environmentally relevant concentrations (see Box 6.4) and ratios of oil and/or chemicals (e.g., dispersants), along with inability to replicate dilution and concentration mechanisms that exist in the field; omission of natural illumination regimes (simulated daylight irradiation and exposure) to incorporate photo-oxidation effects or phytoplankton contributions; lack of representative source breakup dynamics resulting in bubble and droplet size distributions that may not scale to in situ behavior; spatial inability to accommodate multiple trophic levels of natural marine biota; short incubation times that cannot capture seasonal changes; and technical inability to simulate in situ wave and tidal action, hydrostatic pressure, temperature gradients, and suspended sediment loads, among others. Conversely, in situ studies during accidental oil spill studies and field trials are limited by ability to simultaneously and repeatedly sample the environment sufficiently to capture the dynamics of the system, and inherently lack the control afforded by in vitro and ex situ experiments. These potential limitations may explain in part why the literature sometimes presents and perpetuates conflicting conclusions compared to in situ observations. This sometimes leads to incorrect extrapolations of data for inappropriate spatial and time scales, or ignoring essential processes that were not measured. As well, field studies experience regulatory hurdles in some countries and are extremely expensive.

Models of various types are useful and important in connecting results from laboratory, mesocosm, and field studies, and projecting fates and effects of inputs. They can also assist in identifying processes or phenomena needing further study in situ or ex situ. At the same time, they are limited by our understanding of the processes simulated, the environmental conditions present during an event or experiment, and uncertainty in the oil and gas composition, among others (see Section 5.5). Moreover, models for some processes disagree in the literature, and models can only be as good as the data to which they are validated (see previous paragraph). Hence, though models are imperfect, they are an important tool for guiding response and damage assessments for oil spills.

Conclusion—Microbial ecology: The development and application of 'omics has revolutionized microbial ecology and understanding of the microbes that respond to and biodegrade oil in the sea. 'Omics techniques provide insight into the composition of microbial communities, their succession patterns, and both individual and composite biochemical activities which, together, influence the fate of oil in the sea.

Conclusion—New fuel types: New requirements for low sulfur fuel oils (LSFOs), came into effect in 2020 but studies on these oils are currently extremely limited. The few very low and ultra-low sulfur (VLSFO and ULSFO) samples studied to date differ chemically from traditional marine fuel oils and from each other. To date, insufficient research has been conducted to determine transport and weathering behavior, biodegradability, and toxicity of different LSFO formulations under diverse environmental conditions.

Conclusion—Oil spill budget shortcomings: Calculating and reporting on the mass balance/oil budget from mitigation activities and natural processes is often required during an oil spill response. These spill accounting volumes help estimate the percent of cleanup completed and how much oil remains to respond to. It helps measure the pace of response activities and gives a sense of how long the incident will take to resolve. Reported amounts include Volume Spilled/Released, Recovered Oil, Evaporation/Airborne, Natural Dispersion, Chemical Dispersion, Burned, Floating Contained, Floating Uncontained, Onshore, and Total Oil accounted for. As a matter of practicality, it must be recognized that some of these numbers are modeled, some are measured, and some are estimated using different approaches. In addition, each of these volume categories has a wide range of potential values and generally is not an accurate number. The parameters comprising an oil budget are generally represented by a range of values with varying degrees of accuracy and do not lend themselves to a precise breakdown of 100% of the originally spilled volume, especially as this volume changes over time as a result of weathering processes. The focus on these numbers without proper understanding of associated uncertainties has created issues, misunderstandings, and delays during past responses and exercises. Describing a changing spill situation through objective and verifiable numbers important for response and impact assessment would eliminate some of these issues. For example, surface area affected by the spill, which can be estimated and documented using remote sensing techniques or the length of the affected shoreline that could be estimated and documented by SCAT programs assisted with remote sensing if needed. The use of field-derived or modeled values should always be referenced with appropriate caveats and caution and used to glean insights into general trends rather than to obtain precise numbers.

Conclusion—Monitoring PAH profiles in sediments and bivalves: The NOAA National Status and Trends Program (Kimbrough et al., 2008) has provided a unique, nationwide assessment of the geographic status and trends over time of a suite of PAH concentrations in coastal sediments. Advances in chemical methods of analyses for an expanded suite of PAHs within this program would provide valuable data about petroleum contamination in the nation's coastal areas and could be expanded to a cooperative program with Canada as was the case with the Gulfwatch Contaminants Monitoring Program in the Gulf of Maine.

Research Needs to Better Understand Fates of Oil in the Marine Environment

Continued research to better understand and model the fate of oil in the marine environment is encouraged; more specifically, the research included in Table 5.4 would continue to advance this important component of oil spill science.

TABLE 5.4 Research Recommended to Advance Understanding of the Fate of Oil in the Sea

5.1	**Physical mechanisms affecting the fate of oil:** With new laboratory facilities and methods, significant progress has been made in measuring droplet size distributions for oil jet breakup and dispersion of floating oil. These data can be used to develop and test models for oil droplet size distribution; however, because the reduced-scale laboratory experiments do not match the field-scale parameter space, field scale data for oil and gas breakup and dispersion remains an important need. Experiments utilizing SSDI are particularly important to test dispersant mixing at field scale and because treated oil at the field scale falls further outside the parameter space of existing measurements than does untreated releases (see Sections 5.2.3.2, 5.2.3.3, 5.2.5.2, 5.3.1.3, 5.3.3.2, and 5.3.3.5).
5.2	**Chemical reactions affecting the fates of oil:** With the renewed appreciation of photo-oxidation as a significant process affecting oil chemistry, more research is needed to focus on interactions of photo-chemical products with the physical and chemical properties of oil, its behavior in the water column and on shorelines (e.g., emulsification and adherence to mineral surfaces), and its effect on biodegradation. The fate and effects of oxygenated hydrocarbons especially in coastal regions should be examined, as well as the effect of surface or subsurface dispersant addition on photo-oxidation and subsequent processes such as marine oil snow formation (see Sections 5.1.4 and 5.2.5.1).
5.3	**Biological effects on the fates of oil:** Aerobic biodegradation of oil components has been well studied for decades, but the range and kinetics of anaerobic hydrocarbon biodegradation, relevant to seafloor and estuarine sediments and fine-grained shoreline sediments, are less well known. Furthermore, it is not known how the phenomenon of the "lag phase" often seen in laboratory studies of anaerobic biodegradation is manifested in situ; this would affect the time scale of natural attenuation in anaerobic sites. Thus, further research is needed to better understand the kinetics and range of anaerobic biodegradation of oil in the sea (see Sections 5.2.8.5 and 5.2.8.8) as a component of natural attenuation assessment. A physical factor that is not well studied is high hydrostatic pressure, especially when combined with low temperature and limited nutrients such as in deep sea sediments, where it likely affects persistence of sedimented and buried oil. Because such conditions are difficult to achieve in the laboratory, technological developments are needed to conduct in situ experiments and/or to collect samples from the deep sea and subsequently manipulate them in the laboratory without depressurization (see Section 5.2.8.8). The effect of chemical dispersant addition on biodegradation of oil has been controversial in the literature, due at least in part to diverse laboratory conditions that do not mimic in situ spill response circumstances. This controversy should be addressed by adopting "best practices" for designing experiments relevant to spill conditions under which dispersant might be used, such as whether the oil type being evaluated is suitable for dispersion, the weathered state of the oil, oil concentration, dispersant-to-oil ratio and the mixing energy applied (see Section 5.2.8.3). MOSSFA was recognized as a significant transport mechanism for oil spilled during the DWH event (and possibly, in hindsight, during the Ixtoc I spill). However, parallel cases of extreme marine snow sedimentation and flocculation outside the Gulf of Mexico have not yet been documented. Possible reasons for the currently novel DWH observation include (1) DWH was a high-volume offshore spill in water having low concentrations of suspended mineral particles than previous, more common near-shore smaller-volume spills with more suspended particles; (2) DWH response involved an unprecedented magnitude of SSDI; and (3) in recent decades significant advances in field sampling and monitoring techniques and instruments (e.g., sediment traps/particle interceptor traps, core sampling of undisturbed surface sediments, underwater imaging) have provided means for observing MOS occurrence. It will be important to tap the potential of these techniques and instruments for future oil spills in areas where marine snow is a natural phenomenon that could lead to MOSSFA events. The presence of natural (oil-free) marine snow in marine ecosystems argues that MOS and MOSFFA may be found to be important at other locations, although the combined roles of deep sea oil release and implementation of SSDI in fostering other MOSSFA events needs to be ascertained. Because MOS formation involves physical, chemical, and biological processes (e.g., evaporation, adsorption and enzymatic reactions), such study should be interdisciplinary. Observation of "spills of opportunity" should include measurements of MOS abundance and consider the contribution of MOSSFA to oil sedimentation. If globally significant, oil budget models should incorporate MOSSFA terms (Ross et al., 2021; also see Section 5.3.2.2). The process of natural attenuation implies that no intervention is required to enable the native microbiota to biodegrade oil in situ. However, continued research is necessary to evaluate the efficiency and extent of natural attenuation of various hydrocarbon mixtures and response products in diverse environments (e.g., in Arctic versus temperate waters, at different water column depths, in various types of shorelines, benthic sediments, etc.). Such inquiries will provide better insight into the applicability of natural attenuation in different scenarios and will generate additional data for assessing oil fate and biodegradation potential. The power of 'omics techniques has not yet been fully implemented in oil spill research, but could contribute to baseline studies, prediction of natural attenuation potential, and monitoring of bioremediation trajectories. Considerable research is needed to translate 'omics data into meaningful information as a bioremediation tool for modeling oil fate and monitoring natural attenuation progress (see Section 5.1.7.2).
5.4	**Fates of oil in remote sites:** Some ecosystems have been under-studied due to their inaccessibility, such as the Arctic and deep sea; regarding the technical difficulties of Arctic research, see Chapter 6. Within the Arctic, there is a critical research need to collect new data to validate oil-in-ice transport algorithms; to correlate predictions of ice evolution models with mechanisms controlling oil fate and transport; to develop new, more process-oriented models of oil interaction with ice; and to propose observing systems that can be used during oil spill response to collect the data needed to make accurate oil fate and trajectory predictions. Interactions of oil with ice are complex, and determining how oil is dispersed under partial or complete ice coverage remains a major challenge for predicting oil trajectories during response. The new OSIM facility on Hudson Bay will provide opportunities to study these processes. Furthermore, the application of 'omics techniques to polar marine regions has lagged behind studies of temperate marine environments, and further research effort is needed to augment understanding of Arctic ecosystem responses to oil (see Section 5.3.5). A better appreciation of the relationship between oil biodegradation kinetics and temperature would benefit both Arctic and deep-sea studies.

continued

TABLE 5.4 Continued

5.5	**Behavior and fates of new or unconventional oils:** Two classes of unconventional oils are due to be transported by ship in increasing volumes within the next decade: diluted bitumen products and LSFO and VLSFO fuel oils. Whereas some research has been conducted on the submergence and sinking potential of dilbit in various environments, there has not yet been a major marine spill of this two-component blend and the fates of the diluent versus the weathered dilbit warrant further large-scale open-air experimentation to provide insight into potential behavior and fates (see Section 5.3.2). The newly mandated marine fuel oil classes are known to be highly variable in composition but very little is currently known about evaporation, gelling, dispersion, shoreline adherence, and so on. Because the fuels will be used globally, it is essential that laboratory and in situ experiments be conducted under different environmental conditions to increase knowledge and awareness of their potential fates (see Sections 5.2.8.4 and 5.3.1.4). A third possible class is biofuels, which have not been discussed extensively in this report but could become a significant transportation fuel. Within this broad category of fuels, the fate of individual components could be inferred from other knowledge but currently little is known about their composite fate in the sea.
5.6	**Refining models of oil behavior and fate:** As our understanding of oil fate and transport in the sea improves, the need to convert our new insights into operational algorithms for oil fate and trajectory modeling also emerges. This includes the need to develop new modeling algorithms, add these algorithms to models, and validate their predictions, ideally using in situ observations. Some of the new insights that are currently being developed or still need to be parameterized for oil spill models include photo-oxidation, MOSSFA, temperature effects on biodegradation kinetics, and anaerobic biodegradation, among others. There is also a present need to integrate oil spill models with the enormous stream of observation data that may be part of a spill response (see Section 5.5.2).

6

Effects of Oil in the Sea

HIGHLIGHTS

This chapter focuses on advances and changes in our understanding of the effects of oil in the marine environment based on new knowledge since the *Oil in the Sea III* report, including new exposure scenarios (routes and/or specific chemical components), toxicity mechanisms of action, environmental modifiers of toxicity (UV light, pressure), new affected habitats and species, and long-term implications of oil spills. Key advances include:

- Complexity of determining or predicting the effects of petroleum hydrocarbons in the marine environment within a changing ecosystem and multiple co-stressors, using field, laboratory, or predictive modeling based approaches.
- Understanding of the longer-term effects of petroleum hydrocarbons and effects at the population and community levels.
- Identification of common misconceptions regarding the impact of oil in the marine environment and different viewpoints regarding traditional concepts for modes of action of petroleum hydrocarbons (e.g., narcosis) and major impacts. Provision of examples of disconnections between the scientific data presented and the conclusions made.
- Examination of the potential effects of oil spills on human health, including mental and behavioral effects, and considerations of socioeconomic impacts and community resilience.
- Identification of important information gaps that exist in our understanding of the effects of petroleum hydrocarbons (and other petroleum constituents), their microbial degradation products, and metabolic and chemical reaction products on marine organisms.

6.1 INTRODUCTION

The variety of oil inputs to the marine environment challenges efforts to anticipate, respond to, understand, or even describe their potential biological and environmental effects. Although there are a number of ways oil may enter the sea (see Chapter 3), oil spills into the marine environment have received particular attention and public interest, in part due to images of oil-coated animals such as seabirds, marine turtles, and marine mammals. Oil may harm individual organisms, populations, or communities directly through adverse effects that impair survival or reproduction, and indirectly either through cascading consequences of direct effects or via impaired dependencies among different species, populations or trophic structure (see Figure 6.1). Effects from acute, short-term exposures may be limited, resulting in sublethal responses that may lead to mortality, or may involve longer-term or delayed responses on individuals, populations, communities, and ecosystems. Releases of oil into the sea, where smaller amounts are released over a protracted period (e.g., by natural seeps or leaking infrastructure) may also cause adverse effects for exposed organisms. Chronic exposure to oil may also occur after an oil spill in particular habitats and locations, such as armored beach sediments or entrainment in mangrove roots. The importance of long-term effects from acute and/or chronic oil inputs into the sea was recognized and a focus of *Oil in the Sea III*, and this chapter expands this further (see Highlights box and Figure 6.1), recognizing multigenerational, epigenetic, population, and community-level effects.

Since publication of *Oil in the Sea III* (NRC, 2003), molecular technologies and tools ('omics) have advanced significantly and are increasingly being employed to study the presence, fate, and effects of environmental contaminants, including oil and its constituents. Omic approaches (i.e., genomics [DNA], transcriptomics [RNA], proteomics [proteins], lipidomics, and metabolomics), coupled with other disciplines have been used to provide a systems-wide approach

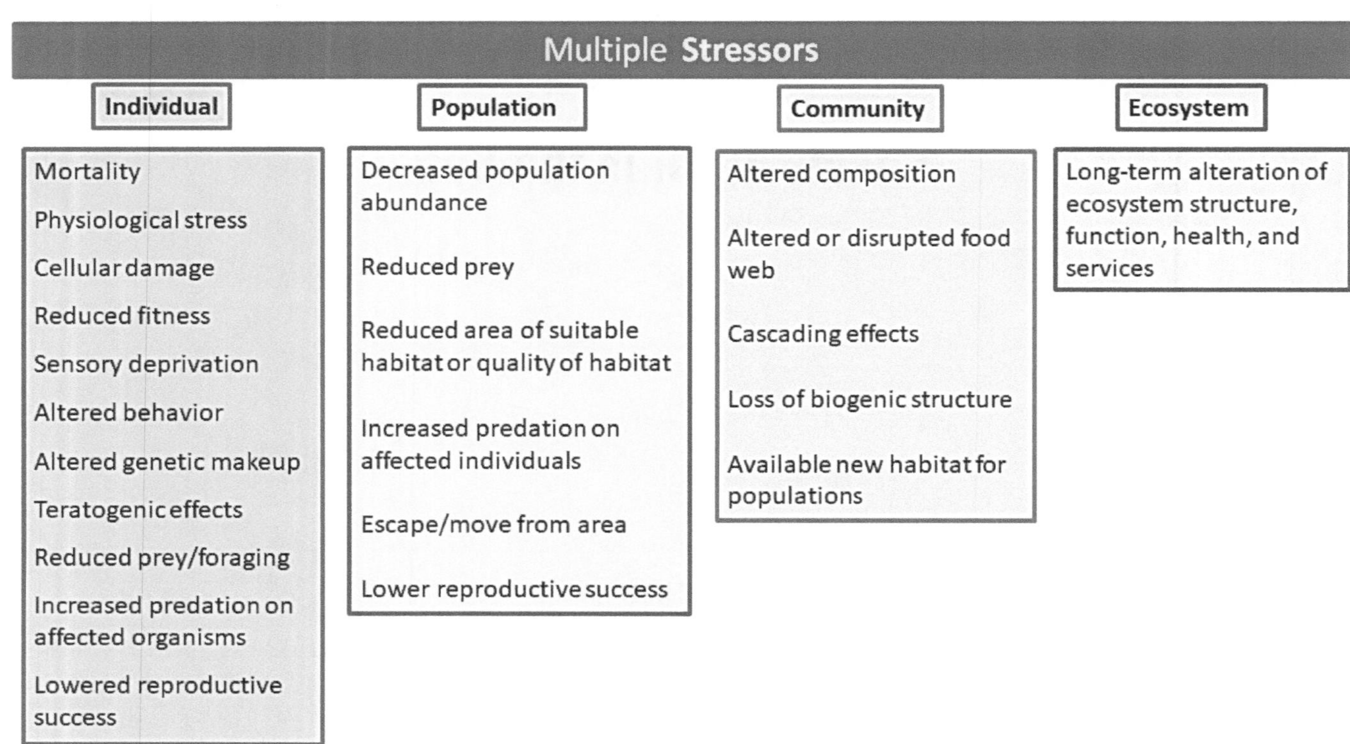

FIGURE 6.1 Potential adverse effects of oil spill exposure within levels of ecosystem organization within the context of multiple stressors.

to monitor and assess ecosystem health and function (Beale et al., 2022; see Box 6.1). For example, microbes have been used as bioindicators of oil and in studying fate and impact for utility in pre-spill response planning, in oil spill response, restoration and in a predictive capacity (Harik et al., 2022; see Chapter 5). In higher organisms they have been used to confirm traditional and identify new mechanisms of actions (MOAs) of oil constituents (e.g., AhR receptor and cardiac toxicity mechanisms; see Box 6.1). Despite advances in 'omic tools, there are limitations and challenges, so further developments are required before these techniques can be recognized for their full potential and utility in oil spill science.

Each oil spill or input is unique, although we understand a series of general groupings of types of effects, for different types of organisms and populations, for oil in general and for certain oil types. **Continuing discovery of the importance of exposure routes and indirect effects, new mechanisms of toxicity, the influence of environmental co-stressors in modifying toxicity, and effects at the community level or long-term effects after major oil spills confirms our incomplete knowledge of how oil can harm ecosystems, while spills of national significance (SONS) (e.g., *Deepwater Horizon* [DWH]) offer rare opportunities for substantial investigation.**

Studies addressing the impact of oil have also used genomic tools at broader scales, such as using environmental DNA (eDNA) and metagenomics (i.e., the sequencing of genes to look for all organisms or targeted taxonomic groups in an environmental sample) to study biodiversity, and community composition shifts and/or to assess organisms' metabolic potential to conduct important ecosystem functions and processes (e.g., carbon degradation). These analyses can be conducted in space and time in non-oiled areas (i.e., baseline studies) for comparison with oil-contaminated locations, including monitoring changes that expose recovery and/or long-term effects. A number of studies have suggested that eDNA surveys may be useful tools in future biomonitoring and impact assessments for oil, although further research is required to assess their utility (Cordier et al., 2019). Similarly, metatranscriptomics involving the sequencing of active or expressed genes could enable analysis of ecosystem functions and activities, although much more research is needed to develop these tools for use in oil spill assessments.

Genomic and transcriptomic approaches do not necessarily reflect functional components; hence their utility would be improved through combination with metabolomics, proteomics, and other monitoring approaches, used in an integrated eco-surveillance framework (Beale et al., 2022). Further research is required to assess the relevance of transcriptomic changes to adverse outcomes in individuals and their relation to population-level consequences, especially given concentration- and temporally-related responses. Metagenomics approaches (i.e., eDNA), while revealing many of the organisms present in the ecosystem under study, may not represent all species (given differences in DNA recovery or rare samples) and also do not provide quantitative data for each detected species. eDNA metabarcoding near oil and gas extraction locations has shown correlations with oil constituents in some but not all studies

BOX 6.1
Use of 'Omics in Oil Spills

'Omic technologies include genomics (DNA), transcriptomics (RNA), proteomics (proteins), and metabolomics (metabolites, including lipidomics [lipids]; see Section 5.2.7.1 and Appendix I). Specifically, in oil spill science they have been used to detect and study the fate of oil in the environment (as described in Chapter 5), identify new toxicity mechanisms of action (MOAs) and determine the impacts of oil to individual organisms, populations, and various ecosystem functions (see Figure 6.2). These tools have significantly advanced the state of knowledge on how oil impacts organisms, although many require further development particularly in their relevance to adverse outcome pathways (AOPs) and population-level consequences. 'Omics technologies were used extensively during and following the DWH incident (especially microbial genomics), leading to many important discoveries, such as the identification of new oil-degrading bacteria, microbial metabolic pathways, and a further understanding of microbial communities and changes in space and time after oil input (see Chapter 5). As highlighted in Figure 6.3, 'omics have also been used to elucidate new MOAs in higher organisms, confirming established MOAs in oil and individuals exposed to chemically dispersed oil and furthering understanding of physiological responses and effects of oil. The role of epigenetics (i.e., modifications of the DNA molecule that affect its availability, such as DNA methylation) has been an important discovery highlighting the long-term consequences of oil exposure for future generations that were not directly exposed to oil. Furthermore, the importance of oil-induced changes to the transcriptome has also been highlighted, although this is still an area of development, particularly in understanding population-level implications (Portnoy et al., 2020). For example, 10 years after the *Exxon Valdez* oil spill, cytochrome P4501A expression was elevated in the absence population level declines in the fish species studied (Jewett et al., 2003). In addition to microbial composition changes in environmental matrices (see Chapter 5), other studies have found evidence for oil-induced changes in organism microbiomes (e.g., fish gut microbiomes) that may ultimately affect the organisms overall health (see Adamovsky et al., 2018; Grosell et al., 2020).

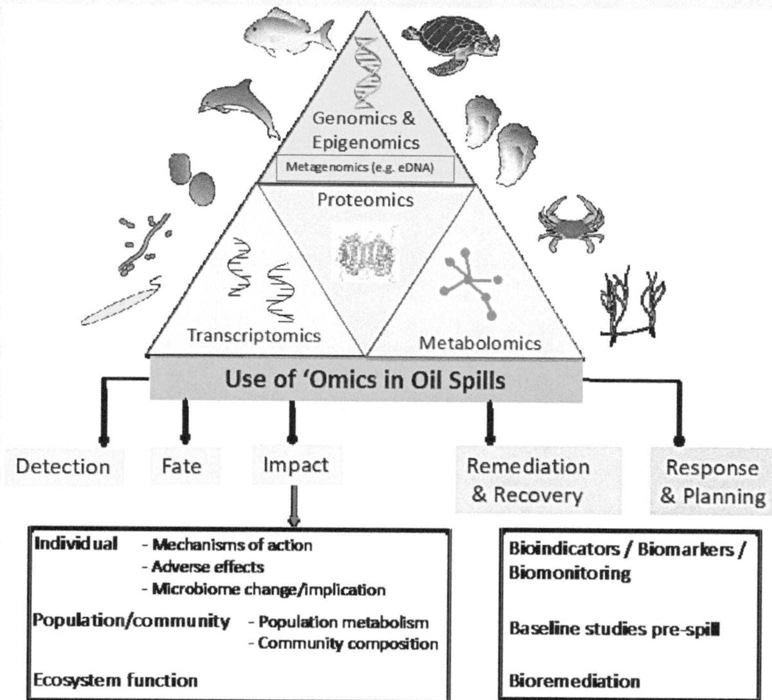

Examples of advances in understanding the impact of oil on organisms using 'omics:

1. Microbial food webs may be impacted by oil for years after a spill event (Rodgers et al., 2019; Chanton et al., 2020).
2. Using microbial community composition as *in situ* 'sensors' to identify oil release and utility in damage and recovery assessments (Krolicka et al., 2019).
3. Impact of oil on organism microbiomes (Brown-Peterson et al., 2015; Bayha et al., 2017) and their importance for organism health (e.g., gut microbiome, Adamovsky et al., 2018).
4. Confirmation of traditional MOAs of oil toxicity (e.g., the aryl hydrocarbon receptor [AHR]) and identification of new MOAs (e.g., developmental toxicity in fish showing various cardiotoxic mechanisms) (Incardona, 2017; Sørhus et al., 2017).
5. Epigenetic changes (5 methylated loci) between field-collected oiled and non-oiled salt marsh grass (Robertson et al., 2017).

FIGURE 6.2 Numerous 'omics tools have been used to study the occurrence, fate, and effect of oil in the environment.
NOTES: Genomics tools study genes (DNA) and by use of metagenomic approaches (e.g., eDNA) can provide information on biodiversity or community structure in the environment. Individual-based studies can provide information on changes to their microbiome, while epigenetic approaches show DNA modifications (e.g., DNA methylation) in response to present and past oil impacts that may result in long-term changes and inherited traits. Transcriptomics identify the up- or down-regulation of genes, proteomics enable an understanding of which proteins are made in response to oil exposure (e.g., bacterial hydrocarbon-degradation enzymes), and metabolomics are used to identify metabolites (e.g., from microbial oil biodegradation).
SOURCE: Graphic images are from the Integration and Application Network, University of Maryland Center for Environmental Science (ian.umces.edu/imagelibrary), CC BY-SA.

(Cordier et al., 2019; Lanzen et al., 2021). However, as detailed above, many advances have been made using ‘omics approaches and will continue to advance oil spill science and contaminant environmental assessments in the future (Martyniuk, 2018). Additional research efforts are needed to translate and adapt laboratory findings for field application, particularly for predictions and oil spill preparedness and for damage assessment and recovery tools in higher organisms.

As described in Chapter 2, oils may be refined products of pure or limited oil types, or even more complicated mixtures, such as crude oils. Effects from exposure to these compounds arise from the interaction among the various chemical compounds in released oils, together with the complexity of the ecosystem where the oil input or chronic release has occurred. Toxicity, which is understood to mean a harmful quality, may result from physical effects such as coating or smothering, or chemical effects involving any of a variety of distinct toxicity mechanisms. The effects of oil on organisms vary widely, depending on initial oil composition and its subsequent weathering state and fate, the mode of exposure, environmental conditions at the time of exposure, the species and life stage of exposed organisms and their habitat, the mechanism(s) of toxicity, and the exposure concentration and duration due in part to the heterogeneity of oil distribution. The relevant mechanism of toxicity depends strongly on the life stage of the exposed species, the environmental conditions of their habitat, and the mode(s) of exposure. For example, oil slicks present a serious threat to adult seabirds (life stage and species) when they come into contact (mode of exposure) with a floating slick oil (habitat and oil composition), through impairment of their buoyancy, thermoregulation, and mobility (toxic mechanism) that increases risk of death by starvation or consumption by predators. If the oil is highly weathered floating tar balls, however, this threat is much attenuated. In contrast, the threat presented by physical contact with oil to the mobility of fish inhabiting surface waters immediately beneath a floating oil slick is negligible, although threats are presented to these same fish by other modes of exposure, and threats presented by different components of oil acting through other mechanisms of toxicity may be more substantial particularly for fish early-life stages in surface waters.

As a result, the effects of oil on organisms are both complicated and complex. They are complicated because of the intricate interactions among the numerous important factors that determine which effects occur and how serious these effects may be, and complex because often these interactions are non-linear, or depend on the scale in space and time of the exposure incident, or both. These intricacies seriously limit the applicability of most generalizations regarding the effects of oil to marine organisms and ecosystems. A simplified summary of the complex effects of oil on the marine environment is shown in Figure 6.3; each component is explained in detail within this chapter.

Our understanding of the biological effects of petroleum released into the marine environment is informed by experimental laboratory and field studies, theoretical considerations, and observations of marine oil pollution incidents and natural oil seeps. Each of these have advantages and limitations. Studies of natural oil seeps provide insights into how communities of organisms adapt to chronic oil exposure. Laboratory and mesocosm experiments permit control of experimental conditions. However, accounting for differences between testing conditions in the laboratory or mesocosm and conditions in field settings is usually problematic, so these studies provide useful insights into what might happen but do not necessarily indicate that those events will occur during a particular oil pollution event. Laboratory studies are especially useful in identifying mechanisms of action and providing data for predictive models, but often fall short of providing environmentally relevant information useful for oil spill response or assessing the actual impacts under specific field conditions. Field studies provide wider integration of environmental factors, but extrapolating results to other combinations of environmental factors, oils or oil components, and species or life stages may be difficult. Research focused on theoretical considerations provides efficient testing and invaluable guidance for detecting adverse effects of oil pollution in the field, but predictions based on theory always require evidence for validation. Observations of effects during oil spills obviously provide reliable indications of harm to organisms and the environment, but establishing causal relationships with exposure to oil is often difficult. Moreover, each of these approaches to detecting effects of oil exposure is better suited for some organisms than others. Overall, the best combination of data for determining effects of an oil spill consists of oil composition and toxicity, and the response of organisms or habitats.

Environmental conditions of marine ecosystems around North America vary widely, as does our understanding of them. Extensive studies of the coastal waters of Mexico, the contiguous United States, and southern Canada provide a wealth of information about the physical and biological characteristics and functioning of these ecosystems, often including long-term data sets that are invaluable for detecting significant perturbations resulting from oil contamination. The Outer Continental Shelf Environmental Assessment Program during the 1970s provided the first extensive and detailed studies of the seas around Alaska, now augmented by studies prompted by the 1989 *Exxon Valdez* spill and continuing Arctic oil production in Alaska. Elsewhere, the marine ecosystems of the North American Arctic and sub-Arctic have received much less scientific study, even while trans-Arctic marine transportation and the attendant increased risk of oil contamination are expected to increase dramatically as global warming continues to shrink the Arctic ice cap. The Trans Mountain pipeline expansion from the Alberta oil sands to Vancouver, British Columbia, and the expected increased use of very low sulfur fuel oils in shipping opens up the potential for new types of oil spills (i.e., dilbit; NASEM, 2016a). Nonetheless, **across all of these regions, it is the sea surface and shoreline habitats that are most vulnerable to the adverse effects of oil spills and chronic oil discharges. Seabirds, marine mammals, sea turtles, and**

Relative Environmental Effects of a Generic Oil Spill

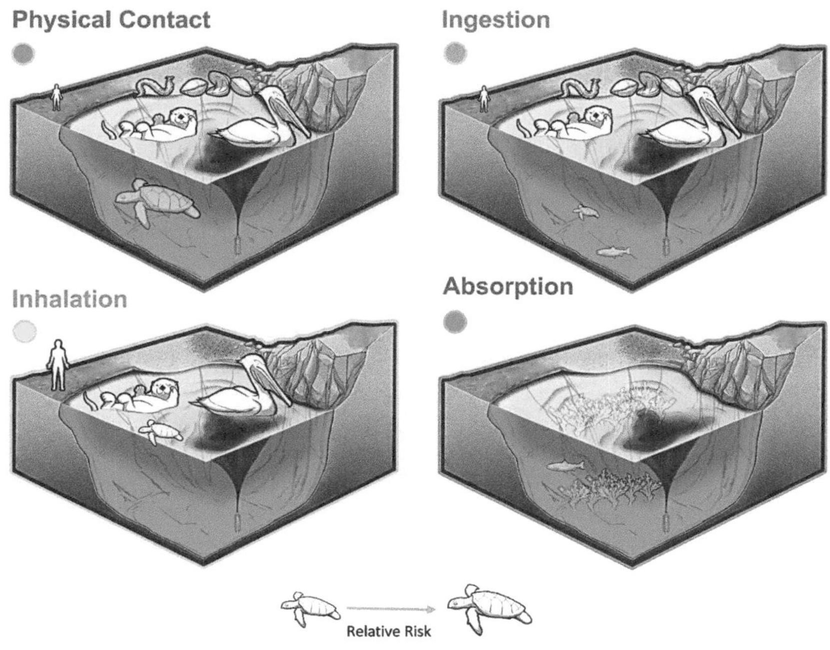

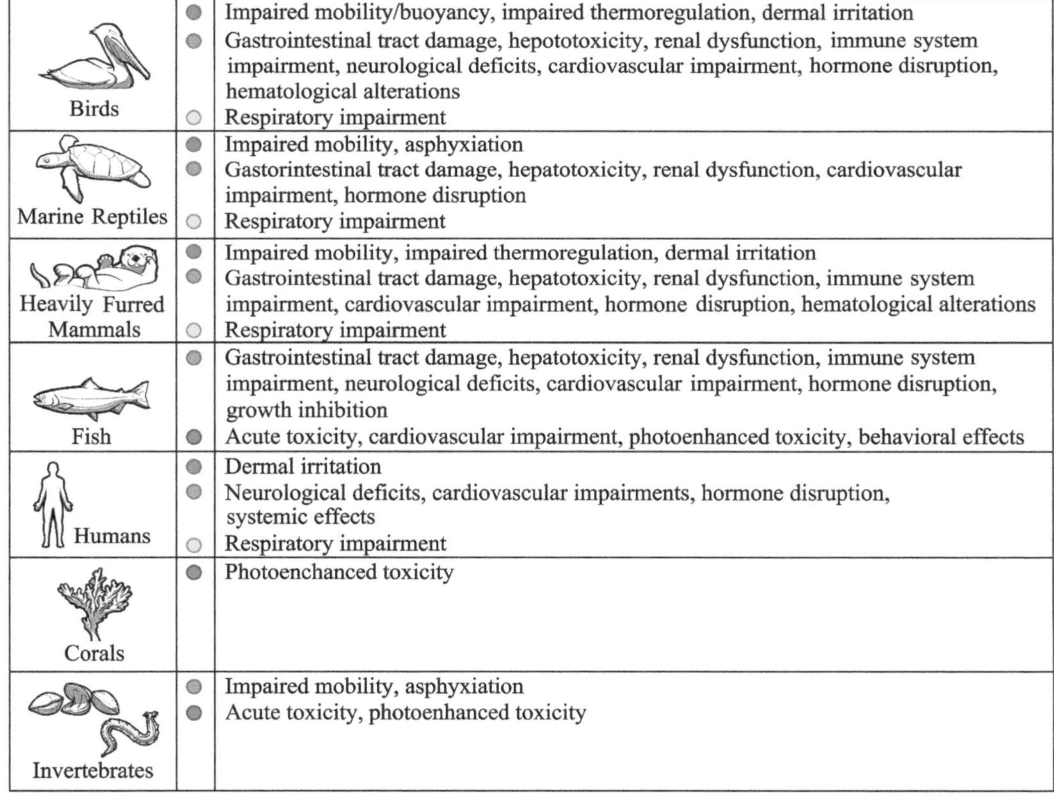

Birds	Impaired mobility/buoyancy, impaired thermoregulation, dermal irritation
	Gastrointestinal tract damage, hepototoxicity, renal dysfunction, immune system impairment, neurological deficits, cardiovascular impairment, hormone disruption, hematological alterations
	Respiratory impairment
Marine Reptiles	Impaired mobility, asphyxiation
	Gastorintestinal tract damage, hepatotoxicity, renal dysfunction, cardiovascular impairment, hormone disruption
	Respiratory impairment
Heavily Furred Mammals	Impaired mobility, impaired thermoregulation, dermal irritation
	Gastrointestinal tract damage, hepatotoxicity, renal dysfunction, immune system impairment, cardiovascular impairment, hormone disruption, hematological alterations
	Respiratory impairment
Fish	Gastrointestinal tract damage, hepatotoxicity, renal dysfunction, immune system impairment, neurological deficits, cardiovascular impairment, hormone disruption, growth inhibition
	Acute toxicity, cardiovascular impairment, photoenhanced toxicity, behavioral effects
Humans	Dermal irritation
	Neurological deficits, cardiovascular impairments, hormone disruption, systemic effects
	Respiratory impairment
Corals	Photoenchanced toxicity
Invertebrates	Impaired mobility, asphyxiation
	Acute toxicity, photoenhanced toxicity

FIGURE 6.3 Overview of relative effects of a generic oil discharge on the marine fauna, according to mechanism of exposure: physical contact, ingestion, inhalation, and absorption. The color scale (red to green) indicates the relative exposure risk: red, or physical contact, representing the highest exposure risk to the ecosystem and green, or absorption, representing lower exposure risk to the marine ecosystem. The size of the organism group icon in each diagram reflects the relative sensitivity of that organism group when exposed to oil through the exposure mechanism indicated.

other surface-dwelling organisms are vulnerable to direct contact with oil from spills or other discharges, and once oil contaminates shorelines, it may linger for decades, presenting a long-term threat through contact and other mechanisms of toxicity.

Oil spills or pollution abatement response efforts provide yet another dimension of effects to organisms, populations, and communities (see Chapter 4). These efforts aim first to protect human life, health, and property, and second to minimize ecological harm. No oil spill response option is without ecological consequences, and these consequences must be considered as effects of oil spills or contamination as well. For example, whereas under some circumstances the use of chemical dispersants may be a valuable response tool, it may adversely affect aquatic organisms near dispersed oil plumes. Because this topic is thoroughly reviewed in a recent National Academies of Sciences, Engineering, and Medicine report (NASEM, 2020), our discussions on this topic only summarize pertinent literature since the National Academies report.

Once oil contamination begins to decline and degrade, affected organisms, populations, and habitats begin to recover. Some organisms, populations, or communities may recover relatively rapidly, but other effects may extend over a decade or much longer depending on the specific habitat and species. Although aspects of ecosystem recovery may be measured and monitored, debate continues regarding when, if, or how much ecosystems have recovered. Identifying which changes are consequences of particular perturbations, such as oil spills, and distinguishing them from natural changes that may have occurred anyway is extraordinarily challenging, but nonetheless necessary for determining when rehabilitation efforts are no longer worthwhile. Thus, **determining when adverse effects of oil contamination have abated, or when the affected ecosystem has adapted, perhaps irreversibly, to such perturbations, remains an active aspect of research and management policy discussion regarding oil pollution effects.**

Oil in the sea, as well as subsequent cleanup activities, affects humans as well. We consider these relationships briefly under the One Health framework, emphasizing that these mutual dependencies are inseparable. Similar to the greater awareness of interconnectedness in infectious disease ecology, where land use, socioeconomic status, and climate resiliency are directly related to risks of emerging disease, it is clear that the effects of oil in the sea are much more complex and multifactorial than previously appreciated. For instance, oil contamination affects the health and well-being of spill responders, local inhabitants, and coastal communities—which, in contrast to earlier *Oil in the Sea* reports, we explicitly address here. This chapter specifically addresses human medical health harm, including seafood safety, but recognizes other human stressors as well, including economic and social issues. **Only through strong, collaborative science, bringing together experts from many different fields to work collectively, can we fully appreciate health-related impacts, and mitigate adverse consequences not only during a spill but also potentially, for effects in future events.**

Finally, we present the current state of oil pollution effects modeling in Section 6.7. **Although these models have advanced considerably over the past two decades, there remains considerable scope for improvements, in part in relation to the data required to build and validate these models.** Traditionally, laboratory toxicity tests have been used to try to mimic or replicate field conditions during a spill, which is not feasible; however, they have been useful in establishing toxicity thresholds for a number of diverse taxa that have been exposed to numerous types of oils (at differing weathering states), hydrocarbon mixtures, or single hydrocarbon components. These data have been used to develop and validate various biological effects and toxicity models used to predict toxicity (especially to new and understudied species) and have been of use both in the National Resource Damage Assessment (NRDA) process and in oil spill decision making to determine the best response option. How these tests are conducted and reported defines their utility; over- or under-estimations of toxicity can occur depending on how test media are made, chemically verified, and the experiment conducted and reported. These issues led to the development of a standardized protocol that was published a couple years before the *Oil in the Sea III* was released (i.e., Chemical Response to Oil Spills: Ecological Research Forum [CROSERF]; Singer et al., 2000). As highlighted in Section 6.4, new knowledge and technical advances in analytical chemistry warrant assimilation into better understanding of effects.

This chapter further explores the significant potential effects of oil spills in marine and estuarine habitats, highlighting what has been learned since *Oil in the Sea III* (NRC, 2003), and recognizing critical reports on effects that were not included in *Oil in the Sea III*. The numerous studies conducted during and after the DWH incident significantly expanded and improved our understanding of oil spill effects, not just for that specific spill but for oil pollution in general. However, the studies also highlighted and uncovered many new data gaps and information/research needs. Furthermore, there are new oil types (see Chapter 2), including very-low sulfur fuel oils (VLSFOs) and diluted bitumen (dilbit; NASEM, 2016a), for which very little information exists regarding their effects. The chapter begins with Sections 6.2 and 6.3 summarizing the modes of oil exposure to organisms (including humans) and the mechanisms of oil toxicity by which oil may harm organisms, recognizing similarities and differences across taxonomic groups. Section 6.4 follows with a discussion on the limitations and challenges in interpreting the toxicity data. Next, we review the effects of oil contamination on marine habitats, communities, and ecosystems in Section 6.5. We then review research and critical research needs in the Arctic in Section 6.6, oil effects modeling in Section 6.7, and the One Health framework (along with human health effects) in Section 6.8. We end this chapter with conclusions and identification of data gaps and research needs in Section 6.9.

6.2 MODES OF EXPOSURE

Oil can harm organisms through any combination of four major modes of exposure: physical contact, ingestion, inhalation of volatile components and oil droplets, or absorption of dissolved components (see Figure 6.3). The organisms are most susceptible to a particular combination of these modes of exposure depends on the species, life stage, habitat, initial oil discharge source, composition and subsequent weathering state, and environmental conditions. Here, we summarize the species, life stages and habitats most often affected by each mode of exposure.

6.2.1 Physical Contact

While physical contact with fresh or weathered oil most commonly occurs at the air-sea interface or on shorelines following a discharge of oil to marine waters, contact with organisms inhabiting the benthos, the water column, or the air may also occur. After an oil spill, the most visible hazard of oil is from oil slicks and sheens that may coat larger organisms that occupy or routinely traverse the oil-water interface—mainly seabirds, marine mammals, and marine turtles—and organisms that inhabit the intertidal surfaces of oiled shorelines, including humans (particularly oil spill response personnel and industry personnel working on oiled shorelines). These are the initial effects most commonly seen by the public and reported at the beginning of an oil spill. More energetic sea states promote entrainment of oil droplets into the water column and the air (see Section 5.2), promoting contact and possible coating of marine organisms inhabiting the mixed layer of the marine water column and perhaps seabirds such as petrels that may fly near breaking waves within oil slicks. As oil weathers, and possibly associates with inorganic sediments or organic matter in the upper water column, oil may sink to the benthos, contaminating coral reefs, epibenthic organisms, and eventually perhaps benthic infauna. However, **restriction of released oil to mainly two dimensions at the air-sea interface and on intertidal shorelines leads to especially severe and widespread adverse effects from physical coating of organisms occupying these habitats. Consequently, the most significant acute effects of oil spills or other surface discharges on larger vertebrates are those related to physical coating based on contact with the oil in a surface slick.**

The unique morphology of seabird feathers makes them particularly susceptible to contact with oil in surface slicks. The microscopic interlocking of barbules and barbicels creates a waterproof barrier that traps air next to the skin, providing critical insulation in species with high body temperatures (103–106°F), as well as buoyancy when in the water (Albers, 1995; Jessup and Leighton, 1996). Oil exposure causes the collapse of this microstructure (Hartung, 1967; Clark et al., 1968; Jenssen and Ekker, 1988). As a result, a comparatively lower surface tension can allow water to penetrate deeply into this insulative air layer (Stephenson and Andrews, 1997; Newman et al., 2000; O'Hara and Morandin, 2010).

In heavily furred aquatic mammals (e.g., sea otters, fur seals), the density and alignment of interlocking hair bundles can create an insulative air layer beneath the pelage in a manner very similar to a bird's feathers (Tarasoff, 1972; Williams et al., 1992), in contrast to with the blubber or fat layer used by many other marine mammal species to remain warm. Upon exposure to oil products, this coat loses its ability to repel water, thereby decreasing insulation and buoyancy.

Sessile organisms inhabiting the intertidal reaches of shorelines are vulnerable to smothering when oil accumulates in relatively thick surface layers. Porous shoreline sediments may allow these surface accumulations to penetrate beneath the surface, coating infauna and rooted vegetation by direct contact with oil, especially when the tidal excursion range allows for partial dehydration of subsurface sediments that permit surface oil to flow downward through aerated sediment interstices. Resuspension of surface and subsurface oil through bioturbation—for example, when sea otters or starfish encounter oiled sediments while excavating shoreline sediments in search of infaunal prey organisms—may also lead to oil coating the external surfaces of these predators.

Although oil that settles to the seafloor could also coat organisms residing there, oil loadings on the seafloor surface are usually modest. After the *Exxon Valdez* oil spill, transport of oil mixed with inorganic sediment particles from heavily oiled shorelines to the shallow (<6 m depth) subtidal sediments by wave action resulted in measured oil loadings consistently less than 0.2 mg oil/g sediment in the uppermost 2 cm of benthic sediments, and loadings were consistently even lower in deeper subtidal sediments, based on a total of 39 polycyclic aromatic hydrocarbons (PAHs) congener classes and the ratio of this total PAH to the mass of 30.5% weathered Alaska North Slope oil (O'Clair et al., 1996; Wang et al., 2003). Oil loadings of benthic sediments following the DWH oil spill provide an extreme example, with maximum loadings of ~15 mg oil/g sediment in the uppermost 1–2 cm of benthic sediments across ~30 km^2 in the vicinity of the well blowout on the seafloor, based on comparison of hopane measurements of seafloor surface sediments (Valentine et al., 2014) and the hopane concentration in 27.7% weathered South Louisiana crude oil (Wang et al., 2003). By comparison, a 0.5-mm thick oil slick on the sea surface at a density of 0.90 g/cm^3 above the uppermost 2 cm of the sea has an equivalent loading of ~22 mg oil/cm^3, and in the case of the DWH spill, slicks of this thickness or greater likely contaminated several hundred km^2 of sea surface. Measurements of shoreline oil loadings 12 years after the *Exxon Valdez* spill imply initial loadings of the order of 100 mg/cm^3 (Short et al., 2004). These comparisons imply that **oil loadings at the sea surface or on heavily oiled shorelines usually present much more serious and widespread contact hazards to marine organisms than oil that sinks to the seafloor.**

Oil droplets dispersed into the mixed layer of the marine water column may contact and coat a wide variety of pelagic marine organisms. Respiratory structures such as gills of fish and invertebrates are especially susceptible to such contamination, as are particulate-collection structures of suspension-feeding organisms. Concentrations of oil droplets entrained into the water column largely depend on the surface mixing energy supplied by breaking waves (see Section 5.2), the presence of dispersants, and the viscosity of the oil. High winds during or immediately after spills promote and entrain dispersion of oil droplets into seawater, increasing effects on fish and wildlife. The 1993 MV *Braer* and the 1996 *North Cape* spills provide examples of serious effects attributed to oil naturally dispersed into the air and water. The MV *Braer* spill grounded on 5 January 1993 in 100+ km/hr winds, eventually releasing nearly 85,000 t of light crude oil just off the south coast of the Shetland Islands (Conroy et al., 1994). Amid a winter storm near the coast of Rhode Island the barge *North Cape* discharged ~3,000 t of No. 2 fuel oil. Storm winds above 100 km/h and breaking waves higher than 5 m spread the oil along the coast and into inshore salt ponds, and dispersed the oil throughout the water column (Reddy and Quinn, 2001). Concentrations of 26 PAC (polycyclic aromatic compounds)[1] and of total petroleum hydrocarbons, existing as droplets of dispersed oil and dissolved compounds, in the water column reached 115 and 3,940 µg/L, respectively. These measurements are some of the highest concentrations of PAH in the water column ever recorded after an oil spill (Reddy and Quinn, 2001), causing substantial mortality to aquatic organisms (see Section 6.5.6.5).

6.2.2 Ingestion

Seabirds and sea otters may ingest oil while preening or grooming to remove oil from feathers or pelage. Sea turtles, other marine mammals, and particle-feeding fish and invertebrates may ingest oil directly while feeding. All these organisms, including sea turtles, may ingest oil through consumption of oil-contaminated prey organisms.

Preening or grooming by seabirds or heavily furred mammals is likely the second most important mode of oil exposure for these organisms. Hartung and Hunt (1966) estimated that ducks exposed to 7 g of oil would preen off 1.5 g in the first day, and Cunningham et al. (2017) calculated that a 20% covering in double-crested cormorants (the high limit to a "lightly oiled" category used in the DWH NRDA efforts) equated to 13 g of oil and, following previous work (Hartung, 1963), assumed that 50% of the oil would be preened off by day 8 of the trial.

Field studies have shown that sea turtles may consume oil-contaminated food (Hall et al., 1983; Camacho et al., 2013). Other suspension- or filter-feeding organisms also readily ingest dispersed oil droplets, including jellyfish; numerous shrimp, krill, and other crustaceans; sea butterflies (pteropods); barnacles; mussels; and oysters, among numerous other species. Deposit-feeding benthic infauna may also ingest oil (Gordon et al., 1978). Once contaminated, consumption of these organisms by higher-order consumer species provides a secondary route of exposure for these predators. For example, sea otters have an extremely high metabolic rate to maintain basal body temperatures, estimated at 2.4 times that of a comparable terrestrial mammal (Costa and Kooyman, 1982), and eat up to 25% of their body weight per day (Kenyon, 1969). This dietary intake can result in additional internal exposure to PAHs and petroleum compounds contained in prey species in affected environments (Neff et al., 1987; Jaouen-Madoulet et al., 2000; Bodkin et al., 2012).

6.2.3 Inhalation

Inhalation mainly involves fractionation of oil components into the air, so the composition of inhaled oil components is determined by components that have substantial partial pressures. These more volatile oil components include the BTEX (benzene, toluene, ethylbenzene, and xylene) and other alkyl-substituted monocyclic aromatic compounds, and alkane hydrocarbons containing 10 or fewer carbon atoms (see Section 2.1.3). Less frequently, microdroplets of whole oil may also be inhaled once oil is atomized by breaking waves (see Section 5.2.3) or by remediation methods such as high-pressure washing of oiled shorelines (see Section 4.2.4, Table 4.4). The addition of chemical dispersants can potentially increase the formation of aerosolized oil, with smaller droplet size distribution leading to greater droplet numbers (Afshar-Mohajer et al., 2018). The toxicity of the oil aerosols decreases with the use of chemical dispersant, with the dispersant creating a higher surface-to-volume ratio of the droplets, suggesting increased dissolution. Drozd et al. (2015) developed a composition-based model including evaporation, characterizing oil components out to a maximum of 30 carbons, though this model requires further examination before it is utilized this operationally.

Seabirds, marine mammals, and sea turtles are all susceptible to inhalation of volatile oil components in the air above oil slicks on the sea surface. Of these, cetaceans and deep-diving pinnipeds (e.g., elephant seals) are likely the most vulnerable, because they often breathe explosively immediately after returning to the sea surface following a dive, and show little inclination to avoid oil slicks (e.g., Smultea and Würsig, 1995). Inhalation is an important potential exposure pathway for humans too, especially oil spill response personnel and industry personnel working near accidental oil discharges.

6.2.4 Absorption

Oil components that dissolve into seawater may be absorbed by aquatic organisms, and this mode of exposure has been the most extensively studied. Like inhalation, absorption involves fractionation of oil components from the oil phase into the receiving medium, in this case seawater. At equilibrium, the composition of dissolved oil components is

[1] Although PACS and PAHs are sometimes used interchangeably in the literature, PAHs containing only carbon and hydrogen are a subset of PACs. See Section 2.1.3 for more details.

determined mainly by a particular compound's mole fraction in the discharged oil, and its partition coefficient K_D, which is the ratio of the chemical's equilibrium concentration in the oil and in the seawater phase. Following actual oil discharges, however, equilibrium conditions are almost never approached, so the composition of compounds in oil that dissolve into seawater is determined by their relative dissolution rates (see Section 5.2.5), and are always lower than the equilibrium concentrations. These dissolution rates are directly related by the relative surface area of the discharged oil in receiving waters, so that dissolution from naturally or chemically dispersed oil droplets is usually faster than dissolution from surface oil slicks, because the surface area of dispersed oil droplets is usually considerably greater than the surface area of surface oil slicks for an equivalent mass of oil in the two cases (see Section 5.2.5).

The compounds in oil that dissolve most readily include the BTEX compounds, other mono-, di-, and polycyclic aromatic compounds and their alkyl-substituted congeners (see Chapter 2). Concentrations of BTEX compounds that dissolve into surface waters tend to be ephemeral, owing to their high vapor pressures that favor evaporative losses to the atmosphere. Also, BTEX concentrations in the water column beneath oil discharges to the sea surface are often rapidly diluted within the mixed layer of the water column. For example, benzene is the most water-soluble hydrocarbon at 1,340 mg/L seawater (Mackay and Shiu, 1975), and has a partition coefficient of K_D (as approximated by the octanol-water partition coefficient K_{ow}) of about 135. Its concentration in South Louisiana crude oil is about 1.87 g/L. The equilibrium concentration of benzene after partitioning into a fixed volume of water from a fixed volume of oil may be approximately estimated as $C_w = C_o/(K_{ow} + V_w/V_o)$, where V refers to volume and the subscripts w and o refer to the water and oil phases (Shiu et al., 1988). A 1-mm surface slick of this oil floating on a 10-cm thick seawater column will result in an equilibrium benzene concentration of 7.9 mg/L in the seawater, whereas this same slick will result in an equilibrium benzene concentration of only 0.18 mg/L in a 10-cm thick mixed seawater layer. Even lower concentrations result from less soluble hydrocarbons in non-equilibrium conditions, especially the PACs. These results are consistent with measurements of dissolved PACs in surface waters contaminated by oil slicks following major oil spills. After the 1989 *Exxon Valdez* oil spill, the highest combined concentration of dissolved PACs measured in surface waters was less than 0.015 mg/L (Neff and Stubblefield, 1995; Short et al., 1996), and after the 2010 DWH spill this concentration was rarely exceeded even in surface waters above the well break.[2]

Dissolved oil compounds may be absorbed by almost all aquatic organisms that inhabit or come into contact with oil-contaminated waters. Absorption by most organisms is through respiratory or other gas-exchange organs, and secondarily through epidermal tissues. The ambient concentration of a chemical primarily determines the rate of absorption, although this may be modified somewhat by an organism's movement through the water or of its appendages or other structures.

[2] See https://www.diver.orr.noaa.gov/web/guest/diver-explorer.

6.3 MECHANISMS OF TOXICITY

Oil may harm biota through a variety of toxic mechanisms, involving both adverse effects from physical contact and poisoning from toxic compounds derived from oil. The vulnerability of organisms varies widely, depending on species, life-stages, their habitats, the mode(s) of exposure and the toxic mechanisms involved. Furthermore, environmental parameters (e.g., temperature, pressure, and UV light), the presence of other co-stressors (e.g., chemical contaminants), and complex toxicity relationships, such as, oil driven alterations to an organisms microbiome and ultimate impacts to health (i.e., immune system) increase the complexity of determining the effects of oil exposure on marine organisms. Groups of toxic mechanisms are strongly or even exclusively associated with particular modes of exposure. Therefore, the known toxic mechanisms of oil are reviewed together within each exposure mode category of Section 6.2, along with the most vulnerable species associated with each toxic mechanism.

6.3.1 Toxicity from Physical Contact

Organisms that come into contact with oil may suffer impaired mobility, impaired thermoregulation, dermal irritation and increased susceptibility to infection, and asphyxiation.

6.3.1.1 Impaired Mobility

Physical contact with oil may drastically reduce the mobility of affected organisms, impairing the ability of affected organisms to locate, capture and consume food, to avoid or escape from predators; and in some cases to avoid sinking and subsequent death by drowning. At sea, the most vulnerable organisms are those that inhabit or traverse the air-sea interface, such as seabirds, marine mammals, and sea turtles.

In birds, flight capability and capacity rely on the orderly structure of the remiges (e.g., flight feathers) to provide both lift and thrust. The physical presence of oil on these feathers interferes with the feathers' ability to interlock, thereby decreasing their ability to promote optimal flight dynamics as well as increasing body weight (Holmes et al., 1978; Leighton, 1993). These effects have been experimentally shown in Western sandpipers (*Calidris mauri*) to decrease takeoff speed by 29%, reduce takeoff angle by 10 degrees, require increased energy needs for flight (20% increase for lightly oiled, 41% increase for moderately oiled), and result in greater wingbeat frequencies and amplitudes (Maggini et al., 2017). Similarly, in experimentally oiled homing pigeons (*Columba livia*), it was found that oiling birds at 20% levels (on the wing and tail surface) resulted in a 1.6 times greater return time when compared to baseline flights (Perez et al., 2017). Furthermore, oiled birds flew significantly longer distances, at slower speeds, had more drastic elevation changes, and had greater maximal elevations reached, reflecting behavioral changes necessary to return to their original location. These impacts on flight capacity can lead to wild birds needing to expend increased energy stores and/or alter flight behaviors (e.g., increased wingbeat frequency/amplitude, flying at greater elevations and/or closer to ridgelines) to enable flight to occur. This loss of energy stores, in combination with that due to increased metabolic demands from hypothermia, can lead to significant loss of pectoral muscle mass that further affects normal flight characteristics. Alterations in flight capabilities can directly cause a number of different injurious outcomes, including an inability to evade predators (Burns and Ydenberg, 2002) and delayed arrival at breeding grounds. Also, the need to expend additional nutritive resources (in combination with increased heat loss from external oiling and decreased uptake of energy from food items due to internal exposure) can rapidly cause decreased fat and lean mass, leading to mortality. In addition, the removal of the air layer next to the skin can cause birds to lose the capability to swim or float in the water (McEwan and Koelink, 1973; Vermeer and Vermeer, 1975), leading to drowning at sea or, if they are able to make it to shore, increased vulnerability to dehydration, starvation, and/or predation.

6.3.1.2 Impaired Thermoregulation

Physical contact with oil may reduce the ability of seabirds and marine mammals to limit heat loss, leading to increased energy expenditure to maintain body temperature. These increased energy demands require increased food consumption needs, which may be more difficult to meet if contacted oil also impairs mobility as well as limiting their ability to remain in an aquatic habitat. Seabirds and heavily furred marine mammals that inhabit or traverse the air–sea interface are especially vulnerable to adverse effects from increased heat loss.

Oil penetration into the insulating air layer next to seabird skin results in increased heat loss from the skin and a much greater challenge to remain euthermic (e.g., tendency to become hypo- or hyper-thermic). Experimentally, in studies on double-crested cormorants (*Phalacrocorax auritus*) following the DWH oil spill, externally dosed birds were found to lose heat but able to maintain core body temperature (as opposed to orally dosed birds, which had difficulty maintaining internal temperature; Cunningham et al., 2017). Most often, birds exhibit a significantly greater basal metabolic rate to maintain core body temperature (Jenssen and Ekker, 1991; Jenssen, 1994), which was estimated recently to be as much as a 13–18% increase even in sublethal exposures to oil (Mathewson et al., 2018). However, should sufficient food stores be present, this increase in energy demand can be offset for some time (Oka and Okuyama, 2000) and can even lead to preservation of core body temperatures through increased foraging.

The decreased insulation, however, typically also increases risk of starvation because oiling increases the rate at which stored body fat is used (Hartung, 1967; Fry and Lowenstine, 1985) and, subsequently, muscle wasting (Leighton, 1993; Bursian et al., 2017; Perez et al., 2017). The degree and speed of morbidity/mortality associated with loss of insulation is dependent on, among other factors, species, degree of oiling, and environmental conditions. The timing of mortalities of birds due to external oiling is more likely extended in lesser oiled categories (e.g., trace to light) and for those species whose habits do not require being in water for foraging or other normal functions. However, the cumulative effects of these issues (in combination with many of the internal effects discussed in the following sections) eventually can exhaust body energy stores to a point that, even in trace to lightly oiled birds cannot maintain physiological function.

In heavily furred aquatic mammals (e.g., sea otters, fur seals), the density and alignment of interlocking hair bundles can create an insulative air layer beneath the pelage in a manner very similar to birds' feathers (Tarasoff, 1972; Williams et al., 1992), and in contrast to the blubber or fat layer used by many other marine mammal species to remain warm. Exposure to oil products causes this coat to lose its ability to repel water, which decreases the animal's insulation and buoyancy, and can lead to hypothermia and associated physiological problems similar to those seen in birds (Davis et al., 1988).

6.3.1.3 Dermal Irritation

Physical contact with oil may irritate the skin of most wildlife. In addition to reducing their ability to locate and capture prey, this may also increase their susceptibility to infection if their behavioral response to irritation leads to dermal abrasion or lesions. Oil contacting the skin or gills of fish may similarly induce lesions that also increase their susceptibility to infections. The physical presence of oil on vertebrates' skin, mucous membranes, and other sensitive tissues has been shown to cause irritation, burning, and permanent damage or loss of function to the skin and eyes in some species. These lesions have included the presence of inflamed, ulcerated, thickened, or sloughing skin and/or an inability to hear or see normally (Mazet et al., 2002; Tseng, 2006; Camacho et al., 2013). The skin of birds is particularly sensitive, being extremely thin and fragile over most of the body surface (Bauck et al., 1997). Petroleum products, depending on their constituent fractions, weathering, and other physical properties, can have a number of physical effects on tissues, causing both acute and chronic physical damage to the epidermis and the underlying layers. In particular, more highly refined products (e.g., gasoline, kerosene) can cause significant damage if not cleaned off—particularly in areas where bare skin may be present (such as the junction of the lower to upper leg in birds), regions where the product may accumulate (such as the patagium in birds and inner thigh or axillary areas), and sensitive tissues such as the corneum (Engelhardt, 1983; Massey, 2006; Helm et al., 2014; Cunningham et al., 2017). Effects to vertebrates relate not only to the specific elements of the product, but also to the adherence of the oil to the animal. Birds and furred mammals have been both experimentally and anecdotally proven to have oil adhere readily to their outer pelage (summarized in Engelhardt, 1982), where animals with no pelage or feathers (e.g., cetaceans, sea turtles) are more resistant. Similarly, experimental studies have shown cetacean skin may also bar petroleum compounds from causing adverse effects. In one study, dolphin skin was directly exposed to gasoline for 75 minutes with no observed acute effects. In addition, the healing ability of superficial cuts, when massaged with either crude oil or gasoline for 30 minutes, was not significantly different (Geraci and St. Aubin, 1982; Geraci, 1990). Exposure of the skin to petroleum in susceptible areas can also allow dermal absorption of BTEX compounds and some smaller PACs, leading to potential chronic health effects (Peakall et al., 1982, 1983; Pérez et al., 2008). In fish, prolonged exposure to oiled sediment followed by exposure to high titers of the pathogen *Vibrio anguillarm* caused dermal lesions through apparent immunosuppression (Bayha et al. 2017; see Section 6.3.2.4).

Exposure to oil from the DWH oil spill has been proposed as the cause of dermal lesions in fish (Murawski et al., 2014, 2021; Romero et al., 2018, 2020; Pultser et al., 2020), but corroborating chemical evidence of exposure to Macondo oil was not conclusive. The presence of a time series from 2010 to 2016 of apparently increasing and decreasing PAH concentrations with a petrogenic or mixed petrogenic and pyrogenic signature reported for mesopelagic fish and cephalopods in the northern Gulf of Mexico illustrates the need for a better understanding of the dynamics of sources, fates, and effects of PAHs/PACs, and other petroleum chemicals in deep water column biota and ecosystems.

6.3.1.4 Asphyxiation

Intertidal organisms that become covered by oil may be unable to respire, resulting in death by asphyxiation. Asphyxiation may also kill seabirds, shorebirds, marine mammals and sea turtles if their behavioral response to contact with oil leads to occlusion of their nostrils or airways by oil (Camacho et al., 2013). Necropsy findings during the DWH incident for heavily oiled animals collected during directed field capture efforts found asphyxiation by oil as the proximate cause of death (n=2/7), with oil found obstructing the glottis or in the trachea and bronchi in five of 10 dead turtles in the stranding, and significant amounts of oil in the mouth and esophagus (Stacy et al., 2012).

Oil smothering may also cause asphyxiation in plants growing in salt-water marshes or other intertidal habitats. Widespread mortality of mangrove forests following two large oil spills in Panama was attributed to asphyxiation and possibly to other toxic effects (Duke et al., 1997), although chemically toxic effects on intertidal plants have not been well studied. Early assessments for the DWH spill (in July 2010) in southeast Louisiana salt marshes clearly documented the dieback of all marsh vegetation in heavily oiled areas. The oiled marshes no longer contained living vegetation instead there were only dead stems layering the exposed, oiled sediments (Lin and Mendelssohn, 2012; Silliman et al., 2012; Zengel et al., 2015). As with the mangroves, it is not clear whether the ultimate mortality was due to asphyxiation or toxic compounds.

6.3.2 Toxicity from Ingestion

Attempts of contaminated animals to rid themselves of oil (feathers of seabirds or the pelage of marine mammals) often leads to ingestion of substantial amounts of oil. Sea turtles can also ingest harmful amounts of oil while breathing at the sea surface or through ingestion of oil-contaminated prey, as can fish through consumption of dispersed oil droplets, oil-contaminated prey, or for some species when gulping air at the sea surface. Ingestion of oil can cause numerous toxic effects, including damage to the gastrointestinal tract, liver, and kidney, and to the immune, neurological, cardiovascular, and hormonal (i.e., adrenal, hypothalamic, thyroid) systems, as well as causing anemia and inhibiting growth.

6.3.2.1 Gastrointestinal Tract Damage

Ingestion of oil via preening or grooming can initially cause significant alterations in gastrointestinal function, elimination of gastric microbiota, and direct damage to tissues of the gastrointestinal tract. These effects, due either to physical presence of oil or to direct damage to the gastrointestinal system, can first manifest in animals via watery stools, diarrhea, and wasting in the presence of increased food/water uptake (Rebar et al., 1995; Briggs et al., 1996; Massey, 2006; Cunningham et al., 2017). If the subsequent damage is severe, it has been shown to lead to gastric erosion/hemorrhagic enteritis and degeneration of intestinal villi (Hartung and Hunt, 1966; Fry and Lowenstine, 1985; Lipscomb et al., 1993; Camacho et al., 2013). Malabsorption and maldigestion of fluids and nutrients can lead acutely to cachexia, wasting, and severe dehydration (Briggs et al., 1996; Newman et al., 2000; Holmes and Cronshaw, 2013) and, should food again become available, can lead to "refeeding syndrome" in which altered electrolyte balances and increased extracellular fluid volumes can lead to tetany/seizures, hemolytic anemia, dysrhythmias/cardiac failure, and even death (Orosz, 2013; Fravel et al., 2016). Marine iguanas appeared to be especially sensitive to gastrointestinal tract damage following a small oil spill near the Galapagos Islands (Wikelski et al., 2002). Increasing evidence has highlighted the importance of organisms' relationships with bacteria, not just for symbiotic organisms like corals but also in an organism's microbiome (e.g., epithelial and GI tracts), alterations in which can have ramifications for the immune, metabolic, and other systems and ultimate toxicity of oil constituents through microbial metabolism (Adamovsky et al., 2018; Duperron et al., 2020). Recent studies have shown changes in gut microbiota from oil exposure in zebrafish (González-Penagos et al., 2020) and in southern flounder following exposure to DWH-oil contaminated sediments (Brown-Peterson et al., 2017); such changes have also been proposed for use as potential biomarkers for oil contamination (Walter et al., 2019).

6.3.2.2 Hepatotoxicity

Once petroleum-related compounds are absorbed via the gastrointestinal system, "first-pass metabolism," where compounds are transported via the portal vein to the liver, becomes important for their removal. The liver is the key organ responsible for xenobiotic metabolism, and oral exposure to oil has been shown in numerous studies to cause significant damage and alterations to this system; such studies included sea turtles in the Canary Islands (Camacho et al., 2013) and more recently a host of bird exposure studies stemming from the DWH spill (Bursian et al., 2017; Harr et al., 2017; Horak et al., 2017). Aryl hydrocarbon (Ah) receptor/cytochrome P450 enzymes, necessary for eliminating deleterious compounds from animals, have been proven to be activated in the presence of PAH congeners (Lee et al., 1985; Peakall et al., 1989; Trust et al., 2000; Schwartz et al., 2004a; Esler et al., 2010), but this metabolism can also lead to producing toxic and carcinogenic reactive intermediate compounds (Harvey, 1991) including oxygen radicals (Gutteridge and Halliwell, 2010), and it is unclear if it is the metabolic activity or the compounds themselves that cause pathological findings. Decreases in liver function, no matter if related to direct or indirect damage to the liver, can cause alterations in plasma protein levels and function (e.g., decreases in albumin and immunoglobulins), leading to immune dysfunction (Briggs et al., 1997; Newman et al., 2000); decreased protein synthesis and carbohydrate/lipid metabolism, leading to altered nutritive homeostasis (Hazelwood, 1986); decreased production of clotting factors, leading to increased health risk from injuries (Hochleithner et al., 2006); and impaired detoxification capacity, leading to an inability to eliminate PAH congeners (Leighton, 1993; Troisi et al., 2006). Additionally, the hemolytic anemia produced by oil exposure (see Section 6.3.2.9) can cause an accumulation of iron in the Kupffer cells of the liver, leading to hemosiderosis or hemochromatosis that, if severe enough, can decrease the liver's functional capacity (Fry and Lowenstine, 1985; Khan and Nag, 1993; Balseiro et al., 2005).

6.3.2.3 Renal Dysfunction

Once petroleum compounds enter the circulatory system, significant damage and alterations have been seen in the renal system, manifested by alterations in kidney metabolic function from both direct effects of PAHs leading to glomerulonephritis (Fry and Lowenstine, 1985; Couillard and Leighton, 1990) or other renal structural changes (Dean et al., 2017; Harr et al., 2017), or the presence/effects of oil and PAHs to the gastrointestinal system leading to intestinal inflammation and damage to villi, severe dehydration, and, thus, renal damage (Leighton et al., 1986). Decreases in kidney function can cause significant and deleterious alterations to blood electrolyte balances (e.g., hyperkalemia, hypochloremia), thereby affecting intra- and extracellular fluid volumes, blood pressure, and acidosis (and subsequent changes to cardiac function) (Tseng and Ziccardi, 2019), and can also lead to a decreased capacity to eliminate metabolic waste, reduced hemostasis, and generalized debilitation (Echols, 2006).

6.3.2.4 Immune System Impairment

Similar to the impacts of oil absorption via the gastrointestinal system on red blood cells, leukocyte (white blood cell) presence and composition can be seriously affected, leading to significant effects on immune function. Alterations in white blood counts and distributions of types of cells have been noted in numerous experimental exposures (Rocke et al., 1984; Briggs et al., 1997; Schwartz et al., 2004b; Troisi, 2013; Dean et al., 2017). These changes appear to be due to a number of different primary and secondary immunosuppressive factors, including a shift of emphasis in cellular production from leukocytes to erythrocytes (due to anemia), malabsorption of nutrients from the gut, abnormal concentrations of corticosteroids due to stress, and potentially immunosuppressive action due to PAH induction of the Ah receptor/Cytochrome P450 metabolic system producing reactive intermediate compounds. A depression in the number and distribution of leukocytes (specifically, decreases in lymphocytes and often a concomitant increase in monocytes/heterophils) in association with depletion in lymphoid tissues appear to be the most common findings (Leighton, 1986; Briggs et al., 1997; Schwartz et al., 2004b; Dean et al., 2017). However, it has also been postulated that these declines are more linked to non-specific reactions (driven by the multifactorial nature of the immune system and its interactions with nutrition, stress, and other biological factors) than direct reactions to oil exposure, leading to an inconsistency in results in the published literature. As discussed in the preceding gastrointestinal section, recent new knowledge has evolved regarding the importance of an organism's microbiome, with studies highlighting secondary consequences to the immune (and other) systems as a result of oil-induced changes to the organism's external and/or internal microbiome, including dysbiosis (Bayha et al., 2017; Tarnecki et al., 2022). In any event, immune dysfunction, if significant, can lead to significant morbidity and/or mortality due to the animal's inability to combat bacterial, fungal, viral, or parasitic infections, leading to increased energy demands.

6.3.2.5 Neurological Deficits

Neurological deficits have also been observed in live oiled animals (Massey, 2006; Helm et al., 2014), though it is unclear if the deficits noted were directly related to petroleum exposure or from other biomedical causes (e.g., hypoglycemia or from other nutritional causes, trauma-related, associated with liver dysfunction). If neuropathies were due to oil exposure, changes were likely due to either direct narcotic-type effects of single-ring aromatic hydrocarbons on the central nervous system, alterations in neurotransmitter levels in the brain (ATSDR, 1995), or direct morphological changes in neuronal tissues (Peterson et al., 2003). Alterations in behavioral function can lead to lack or avoidance of predators, inability to forage, decreased reproductive efforts, decreased migratory habits, and other secondary but significant effects.

6.3.2.6 Cardiovascular Impairments

The ingestion of oil may also be linked to alterations in cardiovascular function, manifested by visually observable cardiac abnormalities (e.g., flaccid heart musculature), changes to cardiac morphology (e.g., increased ejection velocities and volumes), and decreased perfusion/blood pressure (Harr et al., 2017). While this issue is just now becoming evident in birds and marine mammals, there is broad evidence of cardiac-associated pathology in developing fish species (Incardona et al., 2013; Whitehead, 2013; Incardona et al., 2014; see also Section 6.3.4.2). If these effects apply to bird and/or mammal species, it is currently unclear whether these changes might be due to direct effects of PAHs on heart muscle (Ou and Ramos, 1992), alterations in cardiac conduction (Brette et al., 2014), activation of the Ah receptor/cytochrome P450 causing ventricular remodeling (Incardona et al., 2004), secondary changes due to other oil-related pathology (e.g., hematological, renal, and gastrointestinal effects causing increased blood pressure needs) (Leighton et al., 1985), or a combination of the above. Cardiovascular impairment from PACs absorbed by fish have been particularly well studied (see Section 6.3.4.2).

6.3.2.7 Hormonal System Disruption

Changes in hormonal systems (similar to dysfunction noted previously above for dolphins) can also occur, appearing to be driven primarily through direct or indirect effects on the adrenal gland (Peakall et al., 1983), followed by increases in plasma corticosterone levels (Holmes et al., 1979; Rattner and Eastin, 1981; Lattin et al., 2014). Less direct evidence has been found on direct and/or indirect effects on thyroid function (Rattner et al., 1984; Jenssen et al., 1990), possibly due to direct effects of PAHs (and the metabolism of such compounds by cytochrome P450 systems) on the hypothalamus-pituitary-adrenal (HPA) axis (Fairbrother et al., 2004; Mohr et al., 2010; Schwacke et al., 2013) and/or effects of reactive metabolites (from hepatic activity) on these tissues (Rolland, 2000). Oil ingestion can also lead to direct alterations in reproductive function, manifested by changes in reproductive behavior, embryo/fetal mortality, teratogenesis, failed hatching/births, and increased chick/pup abandonment (well summarized regarding birds by Leighton, 1993).

6.3.2.8 Anemia

Should the metabolites of petroleum compounds pass into the blood, significant direct alterations to erythrocyte (e.g., red blood cell) presence and function, manifested by regenerative and non-regenerative anemias, can occur. Consistently during oiled wildlife response, birds and sea otters that have been collected and cared for have exhibited significant anemias (reflected by low packed cell volumes/hematocrits) (Rebar et al., 1995; Tseng, 1999), though the source of such anemias in oil spill settings is unclear (e.g., potentially lack of production due to nutritive issues, destruction due to damage to cells, or a combination of factors). Numerous (though not all) experimental studies have shown significant destructive anemias associated with oral oil exposure in birds (Leighton, 1985; Balseiro et al., 2005; Troisi et al., 2007; Harr et al., 2017) and in mink as a model for sea otters (Mazet et al., 2000; Beckett et al., 2002; Schwartz et al., 2004a). When destructive anemias occur, they appear to be driven primarily through oxidative damage to the cell membranes via exposure to oxygen radicals formed in the metabolism of PAHs (Leighton, 1986; Troisi et al., 2006; Harr et al., 2017), leading to the denaturation of hemoglobin and, in the case of birds, the formation of so-called Heinz bodies. Hemolytic anemias, however, are not universal in non-laboratory exposures, and are likely one component of a more complex host of factors (including lack of erythrocyte production due to stress and poor nutrition from reduced foraging and lack of absorption) that lead to significant challenges in oiled animals for oxygenation of tissues.

6.3.2.9 Growth Inhibition

Ingestion of oil or oil-contaminated food inhibits growth in fish (e.g., Carls et al., 1996) and birds (e.g., Butler and Lukasiewicz, 1979), and likely has comparable effects on marine mammal growth. Ingestion of alkane hydrocarbons retards growth in fish (Luquet et al., 1983, 1984), which may make them more vulnerable to consumption by predators (e.g., Craig et al., 2006). The ability of juvenile fish to avoid their predators is a strong function of their body size, so faster-growing juveniles spend less time reaching maturity, growing out of the most vulnerable smaller body sizes of younger life stages.

6.3.3 Toxicity from Inhalation

Nearly all of the toxicological effects of inhaled hydrocarbons on seabirds, marine mammals, and sea turtles are inferred from field studies, although a few laboratory studies have been conducted on seabirds (e.g., Bruner et al. 1984; Olsgard et al., 2008; Cruz-Martinez, 2015). Toxicity from inhaled hydrocarbons in field studies is usually inferred from a combination of pathological and chemical evidence, such as evidence of direct damage to pulmonary epithelial cells following known exposure to hydrocarbon vapors, perhaps coupled with analysis for hydrocarbons that shows the presence of volatile hydrocarbons in pulmonary tissues at relative concentrations consistent with a vapor phase fraction of a petroleum source (see Chapter 2). However, distinguishing the toxic effects caused by inhalation from those caused by ingestion is usually problematic, as both modes of exposure are often significant, and inhaled hydrocarbons rapidly enter the bloodstream where they cause the same suite of effects as hydrocarbons that enter the bloodstream, following ingestion of oil. We summarize toxic effects associated with inhalation of petroleum or fractions of petroleum, with reference to the toxicity mechanisms noted earlier for ingestion (see Section 6.3.2).

Marine mammals, and probably sea turtles, are especially vulnerable to toxic effects following inhalation of hydrocarbon vapors. Many marine mammal and sea turtle species inhale deeply when surfacing immediately after a protracted dive. If this happens when an animal surfaces through a relatively fresh oil slick that has hydrocarbon vapor pressures sufficient to substantially displace atmospheric oxygen, inhalation of the gas mixture above the slick may cause the animal to lose consciousness and drown. Although direct evidence of this is not available for marine mammals, accidental inhalation of high concentrations of hydrocarbon vapors among oil and gas extraction workers has led to sudden deaths attributed to oxygen deprivation and toxic effects (Harrison et al., 2016). Marine mammals and sea turtles that surface within a large and relatively fresh oil slick may be susceptible to similar risks; however, there is a paucity of data, especially for sea turtles, so future research should be directed at addressing this knowledge gap.

In sea otters, direct respiratory damage due to oil exposure was observed as one of the most significant findings in animals during the *Exxon Valdez* oil spill (Lipscomb et al., 1993). Overall, interstitial pulmonary emphysema was seen in 73% of heavily contaminated, 45% of moderately contaminated, and 15% of lightly contaminated animals necropsied during the event, with dyspnea and subcutaneous emphysema also diagnosed in live otters in the rehabilitation facility. The underlying etiology of this emphysema remains unclear, as it has not been reproduced in subsequent laboratory exposure trials on surrogate species, but otters may be anatomically predisposed due to well-developed interlobular septa in their lungs.

Similarly, following the DWH oil spill, a number of studies focused on the short- and long-term health impacts on coastal bottlenose dolphins (*Tursiops truncatus*) in heavily oiled Barataria Bay, Louisiana, due largely to exposure to oil and volatile compounds produced from the dispersing product. In 2011, 43% of the evaluated dolphins were considered "unhealthy" and 17% were given a poor or grave diagnosis, meaning they were not likely to survive (Schwacke et al., 2014). Furthermore, Barataria Bay dolphins were five times as likely to have moderate to severe lung disease, mostly described by substantial alveolar interstitial syndrome, lung masses, and pulmonary consolidation, compared to those found in a control group (Sarasota Bay, Florida). Additionally, strong evidence of adrenal compromise and an impaired stress response, leading to an overall increased susceptibility to infectious disease, was found in this dolphin population (reflected by low blood cortisol levels and associated low aldosterone values), as well as evidence of inflammation, hypoglycemia, and altered iron metabolism (Schwacke et al., 2014; Venn-Watson et al., 2015; Smith et al., 2017). **While it is unclear what role ingestion and subsequent metabolism of petroleum had in dolphins with impaired health, inhalation/aspiration and subsequent transfer of toxins from the respiratory system to the blood is strongly suspected as a driving factor** (see Figure 6.4). Necropsy results on dolphins found within the spill region revealed potentially lethal changes to their adrenal gland (33%; including unusually thin adrenal cortices) and primary bacterial pneumonia (22%) in agreement with earlier findings in live-sampled dolphins (Venn-Watson et al., 2015). Follow-up health assessments of this population in 2013 and 2014 indicated that, although overall improvements were seen in population health, pulmonary abnormalities (e.g., moderate to severe lung disease evidenced by pleural effusion, alveolar-interstitial syndrome, and pulmonary masses, nodules, and consolidation) and impaired stress responses continued for at least 4 years after the spill (Smith et al., 2017). This heightened risk of pulmonary effects (and adrenal compromise via subsequent transfer to the blood) is likely due to the uniqueness of cetacean physiology, with

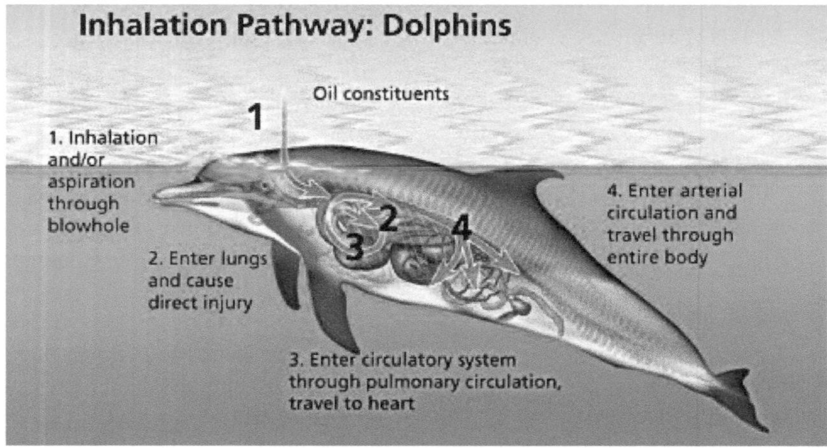

FIGURE 6.4 Illustration of how chemical components become available to cetaceans through inhalation and aspiration exposure pathways. SOURCE: NOAA.

the short trachea in *Tursiops* causing them to be explosive breathers that can exchange 70–90% of total lung capacity in 0.3 seconds, leading to rapid gas exchange and high air flow (Aksenov et al., 2014). Previous experimental studies hypothesized that cetaceans, on the whole, would avoid surface slicks (Geraci et al., 1983); however, in practice, this appears not to be the case in calm waters (Smultea and Würsig, 1995), leading to significant risk in these species.

Health impacts on marine mammals and sea turtles exposed at the air-sea interface, primarily those with rapid gas exchange and high air flow (cetaceans) or prolonged exposure (sea otters) manifest in short- and long-term effects (at least 4 years post spill for bottlenosed dolphins in the DWH oil spill). Lack of avoidance of oil (Smultea and Würsig, 1995; Stacy et al., 2017), **especially in calm waters, leads to significant risk in these species. Continued monitoring and determination of health effects require long-term studies for the assessment of population impacts.**

Changes in respiratory function of birds through inhalation are manifested by increased respiratory effort, physiological damage, and decreased ability to fly/ambulate/dive, leading to decreased oxygenation of the blood and necessary changes in normal foraging behaviors. Avian respiratory physiology is unique, with the lack of a diaphragm and air sacs requiring large volumes of air to move across respiratory surfaces to allow the needed oxygenation of blood. Additionally, due to very efficient gas exchange across thin respiratory tissues to the bloodstream (in tandem to the increased tidal volume), the transfer of lower molecular size PACs to the bloodstream as a preface to systemic effects is comparatively higher in birds (Duncker, 1974; Brown et al., 1997). Direct respiratory damage to birds is not frequently reported in the literature; however, pathological consequences are likely hidden by the more overt causes of morbidity and mortality due to external coating and ingestion from preening. Rehabilitated birds during oil spills have regularly been reported to have respiratory distress in captivity (Mazet et al., 2002), which can increase the propensity for significant fungal infections (e.g., aspergillosis). Experimental studies in laughing gulls (*Leucophaeus atricilla*) exposed orally to DWH oil found respiratory inflammation (pneumonia and/or air sacculitis) in nearly one-third of the subjects (Horak et al., 2017).

6.3.4 Toxicity from Absorption of Soluble Oil Components

Oil components that dissolve into the water column may harm organisms by causing acute (i.e., short-term) toxicity that can lead to death, cardiovascular impairment that decreases fitness, and for organisms with translucent tissues that inhabit the near (~1-m depth) surface layer of oil-contaminated waters illuminated by strong sunlight, photo-enhanced toxicity. As vertebrates, fish may also suffer the systemic effects of inhaled hydrocarbons experienced by seabirds, marine mammals, and sea turtles (Takeshita et al., 2021) (see Section 6.3.3).

6.3.4.1 Acute Toxicity

Bioassays typically measure acute toxicity as the concentration of a toxicant that will kill 50% of exposed aquatic organisms within some specified exposure period, often 96 hours, and typically summarized as 96-h LC_{50} (shorthand for "96-hour lethal concentration for 50% of the test organisms"). In these tests, test organisms are exposed to each of several different concentrations of the toxicant solution, and a concentration that would kill 50% of the test organisms is estimated from the number found dead at a specified time for each exposure concentration. The same approach is used to evaluate sublethal toxicity endpoints such as abnormalities associated with fish embryotoxicity (e.g., Turcotte et al., 2011; Lin et al., 2015), in which case the effective concentration causing 50% of the test organisms to display the response being evaluated is estimated, and denoted as EC_{50}.

Extensive laboratory tests conducted with the water-soluble fractions (WSFs) of many kinds of crude and refined oils have found that 96-h LC_{50} values usually exceed ~0.1 mg of total dissolved hydrocarbons measured per liter of seawater (Anderson et al., 1974; Rice et al., 1977; Fuller and Bonner, 2001; Mitchelmore et al., 2020a,b). Most of these dissolved hydrocarbons are aromatic compounds because of their higher water solubility relative to the saturated alkane, resin and asphaltene fractions of oil (see Chapter 2). Refined oils such as diesel fuels are somewhat more toxic than crude or heavier refined oils (Anderson et al., 1974; Rice et al., 1977; NASEM, 2016a; Adams et al., 2017; Hodson et al., 2019). In most organisms tested, the toxicity of aromatic compounds increases with the number of aromatic rings and the extent of alkyl substitution (Rice et al., 1977; Turcotte et al., 2011; Lin et al., 2015). Apparent sensitivity varies by more than two orders of magnitude among species, and within species, sensitivity may vary unpredictably among life stages (Rice et al., 1977; Mitchelmore et al., 2020b), although embryonic life stages of fish are especially sensitive (see Section 6.3.4.2).

Much of the wide variability in apparent sensitivity among species may be the result of differences in acute bioassay test conditions. Mixing conditions can greatly affect the proportions of compounds that dissolve from the test oil into seawater. Analysis methods used to characterize the composition and concentrations of these compounds vary widely, and quality assurance measures range from absent to extensive. Characterizing the effective doses of test solutions is particularly challenging, because the concentrations of the dissolved compounds usually decline with time because of volatility losses, microbial degradation (see Section 5.1.7), and possibly catabolism of accumulated compounds by the test organisms. Under static test conditions, where the exposure solutions containing the test organisms are left undisturbed for the duration of the exposure, concentrations of dissolved aromatic hydrocarbons may decline to less than half of the initial concentrations. These declines may be mitigated by "static renewal" tests, in which the test solution is replaced at intervals (usually daily), or flow-through or partition-controlled dosing systems that maintain nearly constant concentrations are used (e.g., Turcotte, 2011). Also, most acute toxicity bioassays reported in the literature have been conducted under incandescent or fluorescent illumination, so effects from photoenhanced toxicity (see Section 6.3.4.3) and from photo-oxidized compounds that result from exposure to the UV component of sunlight, now known to be important (Ward et al., 2018b; see Section 5.2.5), are largely precluded unless the tests are conducted outdoors. Taken together, these differences in experimental details may account for substantial proportions of the variability in toxicity reported for a given life stage of the species tested (see Section 6.4). **Moreover, evaluation of the toxicity of photo-oxidized products of compounds that dissolve from oil into seawater would assist in determining how much they contribute to acutely toxic effects on test organisms.**

6.3.4.2 Cardiovascular Impairment

Differences in the survival of pink salmon embryos in oiled compared with unoiled spawning habitat following the 1989 *Exxon Valdez* oil spill prompted laboratory studies confirming toxicity from exposure to water that had contacted oil (reviewed by Rice et al., 2001). *Oil in the Sea III* (NRC, 2003) noted that these findings remained controversial, in part because the concentrations of dissolved PAHs thought to be responsible for these effects were so low (~10 µg/L total of 39 PACs) compared with 96-h LC_{50}s, and because no known mechanism of toxicity could account for the high toxicity and the associated sublethal effects (Brannon et al., 2001). These sublethal effects include deformed jaws, missing or deformed fins, spinal curvature, and pericardial and yolk sac edema that appeared in larvae after exposure to oil had ceased. Results from experiments published 1 year after *Oil in the Sea III* (NRC, 2003) found that these sublethal effects resulted from impaired development of the embryonic heart following exposure to three ring PACs (Incardona et al., 2004). This insight was a major advance in oil toxicology, and led to a considerable body of ongoing research establishing the details and environmental ramifications of embryonic cardiac impairment following exposure to PACs, as well as independent confirmation of these effects in several species of fish by researchers in Canada (Hodson, 2017), China (Zhang et al., 2012), Korea (Jung et al., 2013), and Norway (Sørhus et al., 2015).

Two distinct general toxicological mechanisms accounting for embryonic cardiac impairment in developing fish embryos are now clearly established. The aryl hydrocarbon receptor-dependent (AhR-dependent) mechanism involves initial intracellular binding of alkyl-substituted PACs with three or more rings, or unsubstituted PACs having four or more rings (Barron et al., 2004). This initiates induction of cytochrome P450-1A (CYP1A), a PAC-detoxifying enzyme that oxidizes PACs to more water-soluble and excretable products, along with genetic transcriptional effects wherein up- or down-regulate genes associated with cardiac development or function, leading to impaired cardiac development and function in embryos that persist in later surviving life stages (see Figure 6.5A). A second AhR-independent mechanism involves direct interference in calcium and potassium ion cycling of excitation-contraction coupling in developing cardiomyocytes by un- and alkyl-substituted three-ring PACs, along with somewhat different genetic transcriptional effects of up- or down-regulating genes associated with cardiac development or function, again leading to persistent impaired cardiac development and function in embryos (see Figure 6.5B). Phenotypic expression of the AhR-dependent pathway (e.g., pyrene in Figure 6.5 and Box 6.2) is distinctly different from that of the AhR-independent pathway (e.g., fluorene, phenanthrene and dibenzothiophene in Figure 6.5 and Box 6.2).

BOX 6.2
Narcosis: A Paradigm Shift?

Many models of oil effects are underpinned by the concept of a "narcosis" mechanism of action for low molecular weight polycyclic aromatic compounds (PACs). However, recent insights highlight the inadequacy of this concept (e.g., in cells such as fish cardiomyocytes; see Incardona et al., 2021), which was rejected by the medical field (i.e., anesthesiology) in the 1990s but continues in aquatic toxicity studies to date. A review of the history behind these concepts illustrates the shortcomings of these models, noting that "not a single paper could be identified that unequivocally demonstrates an anesthetic/narcotic action of PAHs in any vertebrate" (Incardona et al., 2017).

In Veith et al. (1983) narcosis is described as "a reversible state of arrested activity of protoplasmic structures" and based on the narcotic mode of action of anesthetics in the human medical field. One of the most commonly used oil toxicity models is the Target Lipid Model, which is based on the assumption that PACs nonspecifically interact with membrane lipids (Di Toro et al., 2000). However, as discussed in Incardona et al. (2017), this assumption was based on science from the 1900s that correlated anesthetic potency with lipid solubility and has since been criticized (for reviews, see Veith, 1990; Franks and Lieb, 1994; Weir, 2006).

The effect that PAH naphthalene has on fish demonstrates this inconsistency with a narcotic mode of action, as exposure causes loss of equilibrium to fish at lethal concentrations and hyperactivity before death (Geiger et al., 1985; Spehar et al., 1999). As detailed in McKim et al. (1987), a fish exposed to a narcotic stops moving and rolls on its side. As an example, in a recent study, Incardona et al. (2021) observed robust dose-response relationships for PAC effects on fish hearts at low ΣPAC concentrations (i.e., ~1 µg/L). Tissue concentrations were lower by factors of 1,000 to 10,000 than the tissue residue concentrations that have been labeled "narcotic" or "baseline" toxicity because of acute or chronic exposures (McCarty and Mackay, 1993). Instead, these tissue concentrations are consistent with specific receptor interactions, as detected in cellular studies (Brette et al., 2014, 2017; see Figure 6.5). Therefore, observed correlations between lipophilicity and toxicity may be due to PAC interactions with specific (and hydrophobic) membrane protein targets in the heart (Incardona, 2017).

A common mechanism of PAC action involves aryl hydrocarbon receptor (AhR) binding, although in accordance with traditional beliefs about PACs as narcotics they were thought not to have other specific toxicity targets due to their lack of functional groups. However, the recently identified developmental-cardiotoxic mechanism in fish embryos mentioned earlier does not fit the AhR pathway or narcosis mechanisms (Incardona et al., 2017). Laboratory experiments have shown that a common target for chemicals with three aromatic rings is a key potassium channel, ERG (ether-a-go-go-related gene), which forms the basis of why early life stages of fish are affected by three-ring PACs (see Figure 6.5). The relationship between aromatic ring number and ERG effects looks very similar to the narcotic mode of action assumptions, in that there are relationships between molecular weight, lipophilicity, water solubility and bioavailability, and hence PAC toxicity. These examples highlight the limitations in applying the broad concept of narcosis in oil toxicity and associated "baseline toxicity" terminology.

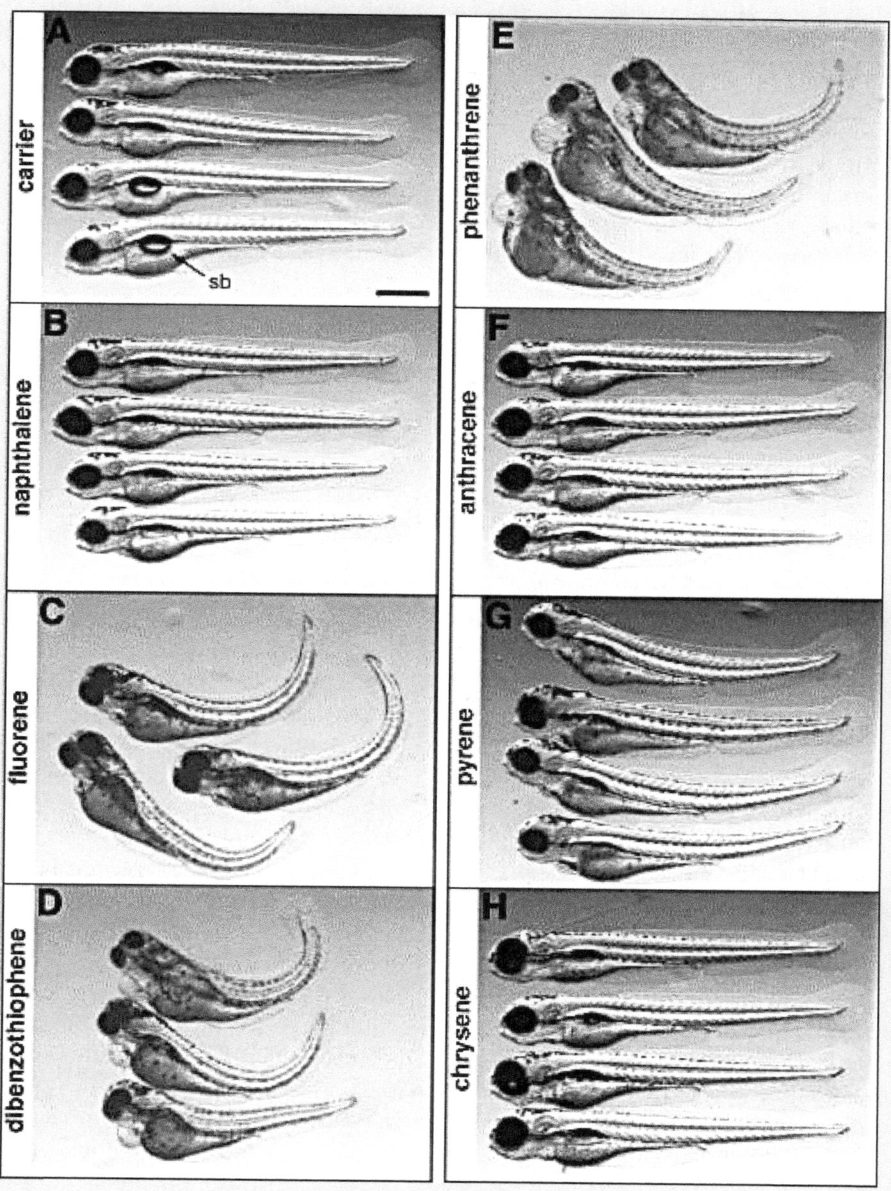

FIGURE 6.5 Images showing zebrafish development affected by individual non-alkylated PAHs. Displayed are four zebrafish larva after contact to (A) 0.2% DMSO (solvent control), (B) 10 mg/L naphthalene, (C) 10 mg/L fluorene, (D) 10 mg/L dibenzothiophene, (E) 10 mg/L phenanthrene, (F) 10 mg/L anthracene, (G) 1 mg/L pyrene, and (H) 2 mg/L chrysene. Swim bladder (*sb*) is identified in A. Scale bar is 0.5 mm. Larvae shown are examples from at least three replicative experiments and approximately 25–100 treated embryos. SOURCE: Incardona et al. (2004).

Although only a subset of PACs that dissolve from oil into water act through the AhR-dependent mechanism leading to CYP1A induction, once induced, CYP1A may act on other PACs as well, as most PACs are substrates for CYP1A (Incardona, 2017). This accounts for why PACs typically do not persist in vertebrate tissues (Meador et al., 1995). Though protective, induction of CYP1A may not entirely offset the combined effects of PAC mixtures that act through both the AhR-dependent and -independent mechanisms simultaneously.

Aqueous extracts of PACs from oils may be toxic at total PAC concentrations of less than 1µg/L, considerably lower than the potency of any of the individual PACs tested (e.g., Turcotte et al., 2011). While this may suggest synergistic interaction between the AhR-dependent and -independent mechanisms, PAC mixtures may be highly cardiotoxic without inducing CYP1A in endo- or myocardial cells (Jung et al., 2013; Sørhus et al., 2016). Harmful effects of these PAC extracts have been found at concentrations as low as 0.23 µg/L (Incardona et al., 2015), far lower than toxic thresholds associated with acute toxicity (see Section 6.2.4.1). See Figure 6.6.

Subtle interactions among PACs that act through either or both of the AhR-dependent and -independent mechanisms may also account for differences in effects stemming from genetic up- and down-regulation associated with fish that occupy different habitats. Jung et al. (2017) found considerable differences in the up- and down-regulation of genes of olive flounder (*Paralichthys olivaceus*) compared with those of spotted sea bass (*Lateolabrax maculatus*) exposed to PACs from Iranian heavy crude oil, and these differences also depended on the weathering state of the oil. From these and other related studies (summarized in Hodson, 2017, and Incardona, 2017), it is clear that the embryological effects of dissolved PACs on fish are intricate, complex, and dependent on numerous interacting factors, substantially complicating efforts to anticipate the results for particular exposures and species.

The <1 µg/L concentrations of three-ring PACs that can inflict substantial damage to developing fish embryos implies that these effects may be widespread after oil spills, or in association with chronic discharges such as stormwater runoff to marine receiving waters (e.g., McIntyre et al., 2016). Fish spawning and rearing habitats are especially vulnerable, including surface waters immediately beneath or near oil slicks for species that broadcast buoyant eggs (e.g., Jung et al., 2015), intertidal reaches of shorelines oiled by spills or oil-derived PACs from chronic discharges (e.g., pink salmon spawning streams after the *Exxon Valdez*; Bue et al., 1996), and perhaps the surface of the seafloor susceptible to contamination from produced water discharges for species that deposit their eggs there (see Section 6.5.6.6).

Fish, seabirds, and probably other vertebrates are susceptible to cardiovascular impairment caused by three and four ring PACs accumulated by any exposure route, including absorption through the skin and gastrointestinal tract, through respiratory surfaces of aquatic organisms, or by inhalation of hydrocarbon vapors (reviewed by Takeshita et al., 2021). Because at least some of the intracellular binding sites associated with the AhR-dependent and -independent toxicity mechanisms are highly conserved across vertebrates (Incardona, 2017), cardiotoxicity of PACs may be commensurately widespread. For example, Brette et al. (2017) suggest that environmental exposure to phenanthrene may account for some of the acute cardiac impacts of air pollution.

While the AhR-dependent and -independent cardiotoxicity effects of PACs may provide the basis for a unified toxicological model that accounts for cardiotoxic effects across wide ranges of oils, exposure conditions, habitats, species and life stages, it is likely that additional toxic effects of oil-derived PACs remain to be discovered. For example, naphthalene, a two-ring PAH, is toxic to early life stages of fish (Hodson, 2017), but not through either of the AhR-dependent or -independent mechanisms. **In light of the likelihood that results may have additional implications for human health, this is an important area of future research.**

6.3.4.3 Photoenhanced Toxicity

Photoenhanced toxicity refers to catalysis of the production of singlet oxygen by certain PACs absorbed within cells and exposed to UV light. Whereas most of the molecules within the Earth's biosphere have paired electrons that have zero net spin (called a "singlet" state, denoted as ^1X, where X is a molecule), molecular oxygen is highly unusual in having two unpaired electrons that have one of three spin states (called a "triplet" state, denoted as $^3O_2$), making molecular oxygen a diradical molecule. Conservation of angular momentum strongly inhibits reactions between singlet and triplet molecules. Absorption of UV light by PACs may lead to promotion of a PAC electron into a higher-energy triplet state, which can readily transfer this energy to triplet oxygen when the energy differences are similar, converting the triplet oxygen into higher-energy singlet oxygen. This exchange returns the PAC to its lower-energy singlet state, and the higher-energy singlet oxygen is extremely reactive with singlet organic molecules, allowing this process to repeat indefinitely as long as (singlet) oxygen is available and the PAC is not oxidized by it (see Figure 6.7). This pathway is usually denoted as Type II photosensitization (e.g., Wang et al., 2009).

In a separate pathway, denoted as Type I, the excited triplet PAC molecule may directly oxidize biological substrates such as lipids, amino acids, or DNA/RNA bases, producing a highly reactive PAC anion radical ($PAC^{\bullet -}$), which may then react with other biological substrates, or donate its radical electron to $^3O_2$, producing a superoxide anion ($O_2^{\bullet -}$), which can then induce generation or reactive oxygen species (ROS) such as peroxides or hydroxyl radicals (Wang et al., 2009) (see Figure 6.7). This pathway also returns the excited-state PAC back to its ground (singlet) state. Consequently, one PAC molecule may catalyze production of many

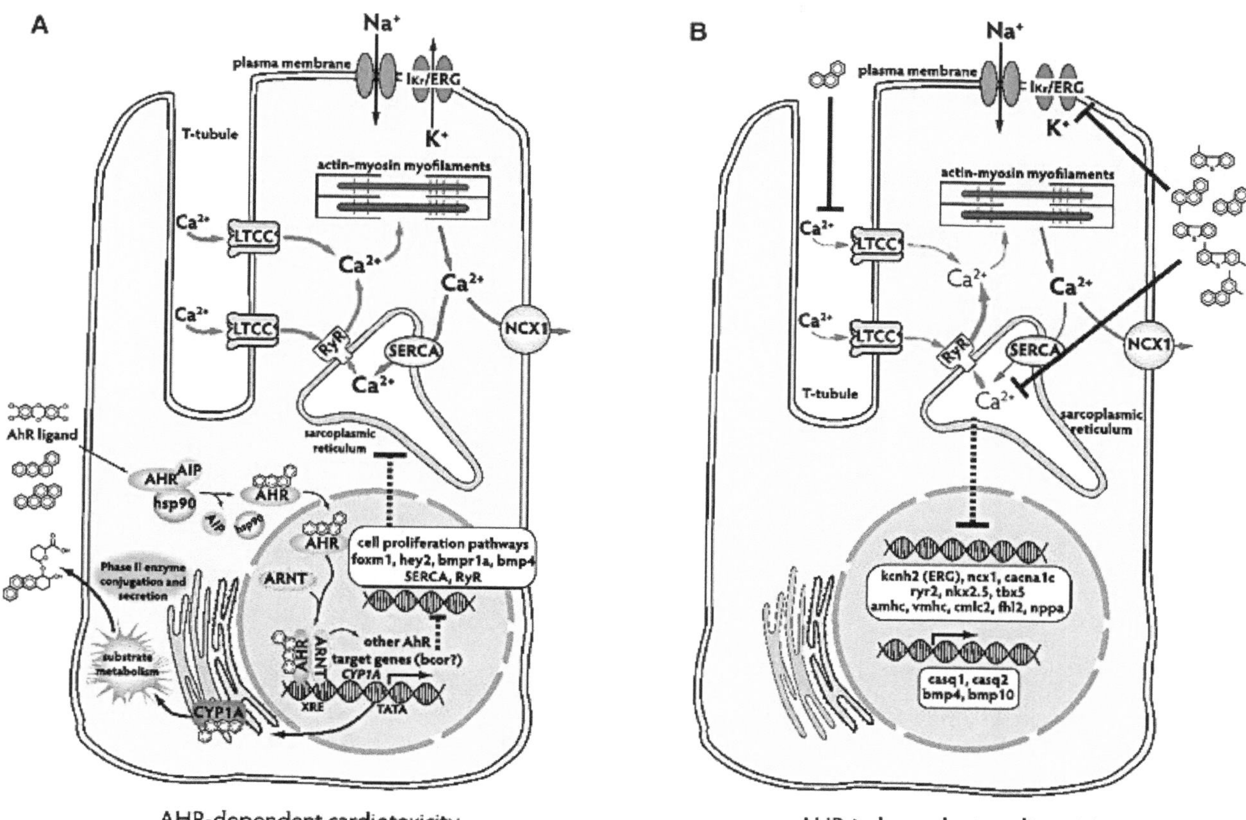

FIGURE 6.6 Aryl hydrocarbon receptor-dependent (a) and -independent (b) cardiotoxicity pathways. Schematics in both A and V represent gene expression and a simplified description of excitation–contraction (EC) coupling in developing cardiomyocytes. Cardiac action potentials are initiated by influx of Na^+ through voltage-gated channels in the plasma membrane, depolarizing the cell. Cells are returned to resting potential and readied for subsequent cycles by an outward K^+ flux (I_{Kr}), for which the erythroblast transformation-specific (ETS)-related gene (ERG) is the major channel subunit. Membrane depolarization also triggers the opening of L-type calcium channels (LTCCs), which are enriched in specialized plasma membrane invaginations called T-tubules that penetrate the cytoplasm, bringing LTCCs adjacent to the sarcoplasmic reticulum (SR), itself a region of the myocyte endoplasmic reticulum specialized for calcium storage. Extracellular Ca^{2+} entering through LTCCs activates the ryanodine receptor (RyR), releasing additional Ca^{2+} from the SR. Mobilized calcium ions then bind to myofilaments, driving contraction (*green arrows*). After release from myofilaments, Ca^{2+} levels are restored in the SR and extracellular space by SERCA and NCX1 (red arrows). AhR ligands bind a complex of AhR in the cytoplasm (containing heat-shock protein 90, hsp90, and AhR-interacting protein, AIP), which dissociates and translocates to the nucleus. Ligand-bound AhR then dimerizes with the AhR nuclear translocator protein (Arnt), binds DNA at xenobiotic response elements (XRE), activating transcription. For PACs this is the major detoxification pathway, with *cyp1a* a primary AhR target gene. CYP1A enzyme resides in the ER, and catalyzes the first step leading to creation of a more water-soluble derivative through Phase II enzymes. Metabolized PAHs are thus ultimately secreted from the cell. Persistent AhR activation by poorly metabolized compounds (organochlorines) or high concentrations of PAHs such as the indicated benz(a)anthracene and benzo(a)pyrene ultimately leads to down regulation of several indicated pathways and genes, presumably through the activity of other AhR target genes (*small dashed inhibition symbol*). At later stages of AhR-dependent cardiotoxicity, intracellular calcium handling by the SR is also inhibited (*large dashed inhibition symbol*). (B) Tricyclic PACs from crude oil directly block the efflux of K^+ ions through I_{Kr}, and disrupt SR calcium handling so that the SR Ca^{2+} stores are depleted and phenanthrene also blocks the LTCC (*large solid inhibitory symbols*). At later time points, through currently unknown molecular steps (*dashed inhibitory symbol*), select genes are down-regulated, encoding proteins that participate in K^+ and Ca^{2+} ion movement, cardiac morphogenesis and contractility. However, effects are complex with up-regulation of genes encoding SR calcium-buffering proteins (*casq1 and casq2*). In mahi mahi, *bmp4* expression was dynamically affected, being inappropriately higher and lower at different time points. In Atlantic haddock embryos, up-regulation of *bmp10* was an early response prior to visible malformation. With complex PAC mixtures, it is likely that both AhR-independent and -dependent mechanisms may be active in the same cells under some conditions.
SOURCE: Incardona (2017).

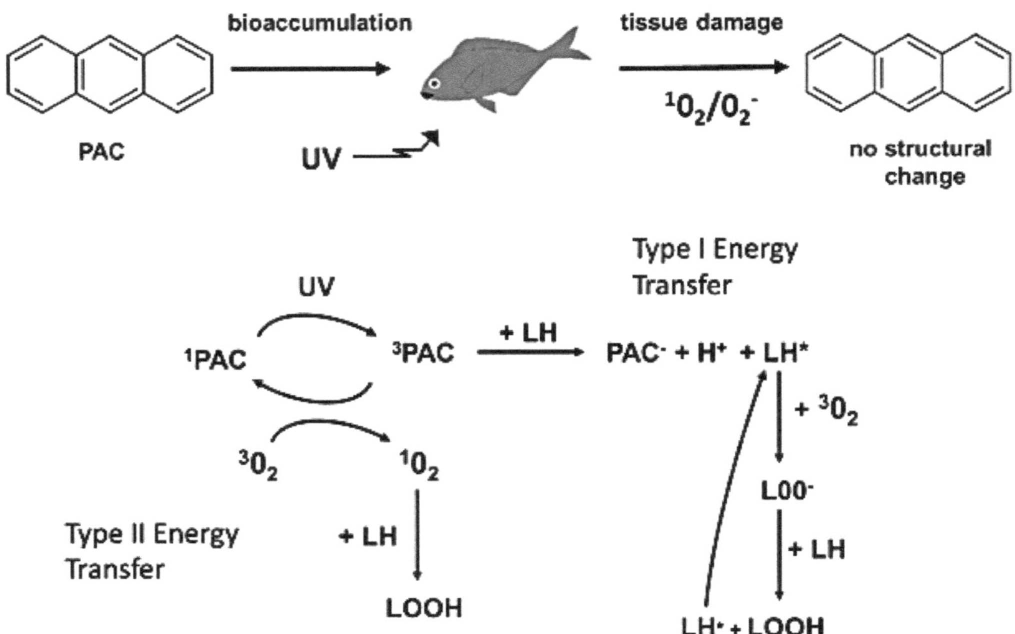

FIGURE 6.7 Photoactivation of a polycyclic aromatic compound (PAC) by ultraviolet light (UV) causing the creation of reactive oxygen species and a series of lipid (LH) peroxidation (LOOH).
SOURCES: Derived from Landrum et al. (1987), Choi and Oris (2003), Diamond (2003), and Wang et al. (2003).

$^3O_2$ molecules and ROS, which then in effect proceed to burn tissues from the inside out, along with occasional direct binding of the PAC involved with biological substrates including proteins, RNA, and DNA, the latter of which may lead to damage to the genome.

The requirement for similar energy gaps between the ground and excited states of PACs and $^3O_2$ limits which PACs can catalyze $^1O_2$ production. The most active PACs include parent and alkylated variants of anthracenes, fluoranthenes, chrysenes, pyrenes, benzochrysenes, and benzopyrenes, along with the oxygen, nitrogen, and sulfur analogs of these compounds such as dibenzothiophenes and acridines (e.g., Pelletier et al., 1997; Wiegman et al., 2001; Lee, 2003). This same requirement also limits the UV wavelengths to the "UVA" component of sunlight (320–400 nm). Tissue damage from photoenhanced toxicity in organisms requires absorption of the right PAC into tissues, exposure to the UVA component of sunlight during or immediately after PAC absorption, translucent tissues at least at an organism's epidermis, and oxygen availability. When these conditions are met, photoenhanced toxicity can be much more harmful compared with toxic effects when UVA radiation is low or absent. Depending on the PAC, such increases of toxicity range from factors of two to nearly a thousand (Willis and Oris, 2014). Consequently, small organisms with translucent bodies that inhabit the upper water column or intertidal reaches of shorelines are most vulnerable. **More than 30 aquatic species, often in their earlier life stages, have been shown to be sensitive to photoenhanced toxicity effects, including crustaceans, mollusks, oligochaetes, and fish (Boese et al., 1997; Spehar et al., 1999; Barron and Ka'aihue, 2001; Barron, 2007). Conversely, large-bodied organisms, especially those with extensive pigmentation or exoskeletal armoring, or organisms that inhabit deeper waters where UVA radiation does not penetrate, are unlikely to be susceptible to effects from photoenhanced toxicity. Given the toxicity range of two to nearly a thousand, understanding of photoenhanced toxicity effects would benefit from study characterization of additional organisms and numerous life stages.**

6.3.4.4 Immune System Impairment

Fish exposed to chemically enhanced water accommodated fractions (CEWAF) followed by exposure to high titers of the pathogen *Vibrio anguillarm* developed dermal lesions through apparent immunosuppression (Tarnecki et al., 2022; see Section 6.3.2.4).

6.3.4.5 Behavioral Effects

Laboratory tests clearly show that fish are able to detect and will avoid oil (Rice, 1973, 1977; Meinard and Weber, 1981; Martin, 2017; Claireaux et al., 2018), although repeated exposure may de-sensitize this response (Schlenker et al., 2019).

6.4 LIMITATIONS AND CHALLENGES IN INTERPRETING LABORATORY TOXICITY DATA

Laboratory toxicity tests are used for a number of reasons, first in a regulatory (environmental risk) framework to estimate the toxicity of single or mixtures of hydrocarbon constituents, oil and/or chemicals used in oil spill response (e.g., dispersants, herders). Second, they have been used with standard test species to establish relationships with and among oil types and to highlight species and/or life stages that may be more susceptible to oil exposure. The data, assuming they meet data quality requirements, can be used to populate and generate, calibrate, and/or validate toxicity tests and models to predict the impacts of oil and its constituents to various species and support decision-making during oil spill response (French-McCay et al., 2018; NASEM, 2020; see Section 6.7). Laboratory toxicity tests are also useful in establishing the specific mechanism(s) of toxic action (e.g., receptor sites and resulting effects) of oil and its constituent chemicals, and finally how oil effects can be influenced (i.e., become more or less toxic) when combined with other covariables that may occur in a field setting, such as UV light, temperature, salinity and pressure in addition to other chemical contaminants. Toxicity tests are also used following a spill event in NRDA activities and by researchers trying to establish the potential impact of a specific spill on resident species. Tests are conducted with both standard toxicity test organisms and with non-standard toxicity test species to provide a more accurate assessment of oil toxicity to exposed resident species. However, laboratory toxicity tests cannot mimic or replicate field conditions, and recently a paradigm shift was suggested in the recent National Academy dispersant report to conduct toxicity tests in such a way that the data generated would be useful in informing and validating toxicity models (NASEM, 2020).

Since the *Oil in the Sea III* report, new analytical molecular techniques and testing approaches have provided a wealth of new knowledge regarding drivers of toxicity (see Highlights box; NRC, 2003). These include specific chemical constituents involved in existing and new mechanisms of toxicity (e.g., cardiotoxicity in fish embryos; Incardona, 2017; Incardona et al., 2021), the influence of oil droplets (i.e., via additional exposure routes, enhanced bioavailability of dissolved constituents, or physical mechanisms) and environmental modifiers decreasing (i.e., pressure; Paquin et al., 2018) or increasing toxicity (i.e., UV light; Barron, 2017). However, how laboratory toxicity tests with complex mixtures of oil and oil/dispersants are specifically conducted influences the results obtained and hence their interpretations and in some cases decreases the utility of the data collected. Examples include inappropriate extrapolation of laboratory data to describe field conditions, over- or under-estimations of toxicity, and comparisons between oil and dispersed oil exposures based on the lack of or choice of analytical characterization metric (Bejarano et al., 2014; Redman and Parkerton, 2015; Bejarano, 2018; Hodson et al., 2019; NASEM, 2020; Mitchelmore et al., 2020a,b). Many of these issues were addressed in detail in the recent National Academy dispersant report (NASEM, 2020). This section aims to highlight some of the key points regarding the use, approaches and interpretation of laboratory toxicity tests since *Oil in the Sea III*. Dispersed oil and dispersant studies are not discussed in as much detail as oil-alone studies as these were recently reviewed in the National Academies dispersant report (NASEM, 2020).

6.4.1 Implications of the Variability in the Design, Execution, and Reporting of Toxicity Tests

Laboratory toxicity tests are conducted under controlled conditions and numerous standard protocols on how to conduct them exist, albeit for a relatively limited (but representative) number of standard test species (e.g., following U.S. Environmental Protection Agency [U.S. EPA] and Organisation for Economic Co-operation and Development [OECD] guidelines). However, the complexity of oil itself, and the additional use of spill response agents (e.g., chemical dispersants) combined with the influence of additional environmental variables, result in additional complications and considerations for these tests particularly concerning the variability observed and reproducibility of toxicant test exposure solutions. Standardization of protocols detailing how to prepare exposure solutions and how to conduct toxicity tests, including the use of appropriate analytical verification and choice of biological endpoints is critical. Without these standard methods, comparison across studies, data utility for inclusions and/or validation of toxicity models, or simply an accurate representation of toxicity is challenging (Redman and Parkerton, 2015; Bejarano, 2018; Hodson et al., 2019; NASEM, 2020; Mitchelmore et al., 2021a,b; Nordtug and Hansen, 2021). This need was recognized back in the 1990s and led to the development of the CROSERF Ecological Effects Forum, which standardized methods for preparing exposure media and conducting toxicity tests (Singer et al., 2000; Aurand and Coelho, 2005). Despite these guidelines, studies have continued to either not employ them or have modified them without providing all of the details required to allow toxicological interpretation. Many modified versions of the CROSERF protocol have been used since its initiation, in part due to new knowledge on processes that affect toxicity (e.g., UV light; phototoxicity) and the development of new exposure preparation methods (e.g., high energy water accommodated fraction [HEWAF]). These CROSERF protocols are now over 20 years old and new oil types and test species, an increased understanding of toxicological drivers, and new technologies for test media preparation require that these protocols be revisited and updated (see Table 6.1 for an overview and Boxes 6.3 and 6.4 for descriptions of terminologies used, an overview of exposure preparation methods, and characterization of solutions, respectively).

TABLE 6.1 Main Considerations of Parameters and Approaches That Can Lead to Different Results and Interpretations in Oil and Oil and Dispersant Toxicity Tests

Parameter/Approach	Considerations Related to Test Media and Effects
Choice of oil type, weathering state	Differing chemical composition and entrainment of droplets (number and size), dispersant efficacy.
Loading rate (oil-to-water ratio) and choice of variable loading or variable dilution	See Figures 6.10 and 6.11 that demonstrate the differences in chemical composition between the two approaches. Will also affect coalescence of oil droplets.
Mixing regime (extent) and dispersion technique (physical or chemical) and dispersant:oil ratio (DOR)	Low-, mid-, and high-energy methods, with or without chemical dispersants will alter the chemical composition, oil droplet size, and oil droplet quantity. DOR will alter the efficacy and/or extent of dispersant compounds free in the water or associated with oil droplets. Affects stability of oil and coalescence of oil droplets.
Mixing time and settlement period	Time to reach equilibrium for dissolved hydrocarbons? Settlement time will affect stability in physically/chemically dispersed oil (i.e., longer time results in smaller droplets as larger ones have risen to the surface).
Type and duration of the toxicity test	Chemical and physical composition of the test media is dynamic and changes over time (e.g., loss of volatiles, photochemical reactions, coalescence of droplets). Exposure depends on test type; static, static-renewal, spiked declining, constant, etc. Effect on organisms and environmental relevance depends on exposure concentration and time.
Analytical chemistry	Nominal reporting (loading rate, %, etc.) unacceptable for exposure concentration. Limitations in use of bulk parameters (TPH, THC, etc.); minimum use of TPAH (50–70 individual parent/alkyl PAHs), BTEX, quantitation of dispersant components. Bulk oil via fluorescence methods. Dissolved PAH via passive sampling. Determination of dissolved and particulate fractions.
Physical analysis (droplet volume, number, and size)	Analysis (laser in situ scattering and transmissivity [LISST], coulter counter) of particulate phase (oil droplets) for volume, number, and size.
Toxicological considerations	WAF: soluble and readily bioavailable hydrocarbons.
Other variables to consider	What are the questions/hypotheses for conducting this test? Mixing vessel/exposure chamber types: headspace/closed or open containers? Salinity, temperature, lighting and feeding regime. Types of endpoints and frequency of assessing biological endpoints (e.g. time to death considerations for acute tests); environmental relevance, choice of species/life stage.

NOTE: BTEX = benzene, toluene, ethylbenzene, and xylene isomers; PAH = polycyclic aromatic hydrocarbon; TPAH = total PAH; THC = total hydrocarbon content; TPH = total petroleum hydrocarbon; WAF = water-accommodated fraction of oil.

Numerous requests for further standardization of exposure test media preparation and toxicity test protocols have been made in various publications, including National Academies reports and other papers (Aurand and Coelho, 2005; NRC, 2005; Bejarano et al., 2014; Redman et al., 2015; Hodson et al., 2019; Mitchelmore et al., 2020a,b; NASEM, 2020). Key aspects of new protocols should include detailed quality assurance and quality control metrics and the reporting of key elements of the test procedures (including any deviations from the standard protocols) so that there is a minimum set of reporting requirements provided in each test, examples of which can be found in tables in the review by Mitchelmore et al. (2020a) and in Table 6.2.

As the solubility and/or partitioning behavior of oil constituents (and hence bioavailability to organisms) is heavily influenced by how exposure solutions are made up, the CROSERF standardized methods recommended in detail how to prepare exposure media. A focus for toxicity testing has been on the preparation and characterization of dissolved oil (and/or dispersant) constituent exposures, first because this fraction is considered bioavailable and second, its use in oil spill toxicity prediction models (Carls et al., 2008; MacKay et al., 2011; French-McCay et al., 2018, 2021b). The CROSERF protocols detail the type of container to use (glass aspirator bottle), the amount of headspace (in the sealed bottle), the mixing energy (i.e., low and no vortex for minimum droplet entrainment and medium, 25% vortex, when using chemical dispersants to result in oil dispersal and droplet formation) and the recommended oil to dispersant ratio when using chemical dispersants (Singer et al., 2000). Since the original method was published, additional considerations regarding how test media are prepared (such as whether to use variable dilution or loading of the oil or other preparation methods, such as passive dosing) have been discussed (NASEM, 2020; Parkerton et al., 2021).

One of the test preparation parameters that has received particular attention and discussion is that of using variable loadings of oil to prepare the concentration range versus the preparation of a high concentration stock solution from which dilutions are made to prepare each concentration. This issue has been discussed at length in previous National Academies reports (NRC, 2005; NASEM, 2020) and in reviews (Barron and Ka'aihue, 2003; Redman et al., 2015; Mitchelmore et al., 2020a) regarding each approach's influence on the concentration and composition of the resulting test exposure solutions. The arguments for and against each method and the ramifications of each are summarized in Figures 6.10 and 6.11 and pertain to considerations of the solubility of each chemical constituent and the role of oil droplets. In the study by Forth et al. (2017a,b), WAFs prepared by the variable dilution approach presented concentrations (linear with dilution) and compositions as expected in the measured prepared exposure solutions compared with what was estimated based on chemical solubility. Furthermore, in this study the role of

BOX 6.3
Terminology and Approaches Used to Prepare Oil Exposure Solutions

Oil is a complex mixture of many chemicals that vary in their physical and chemical properties influencing their solubility and partitioning behavior (i.e., dissolved versus particulate fractions). Therefore, toxicity is directly dependent on the specific test procedures used to prepare exposure solutions (see Figures 6.7, 6.8, and 6.9). Using standardized protocols to prepare test exposure solutions is critical for interpretation of the toxicity results generated and comparison across studies (Aurand and Coelho, 2005; NRC, 2005; Mitchelmore et al., 2020a; NASEM, 2020). Common preparation methods used to prepare water accommodated fractions (WAFs) of water:oil and water:oil and chemical dispersants involve mixing in glass containers for a certain period of time, decanting test solutions from the bottom (to avoid neat oil and resurfaced oil) and include the following (* highlights those detailed in original Chemical Response to Oil Spills: Ecological Research Forum protocols).

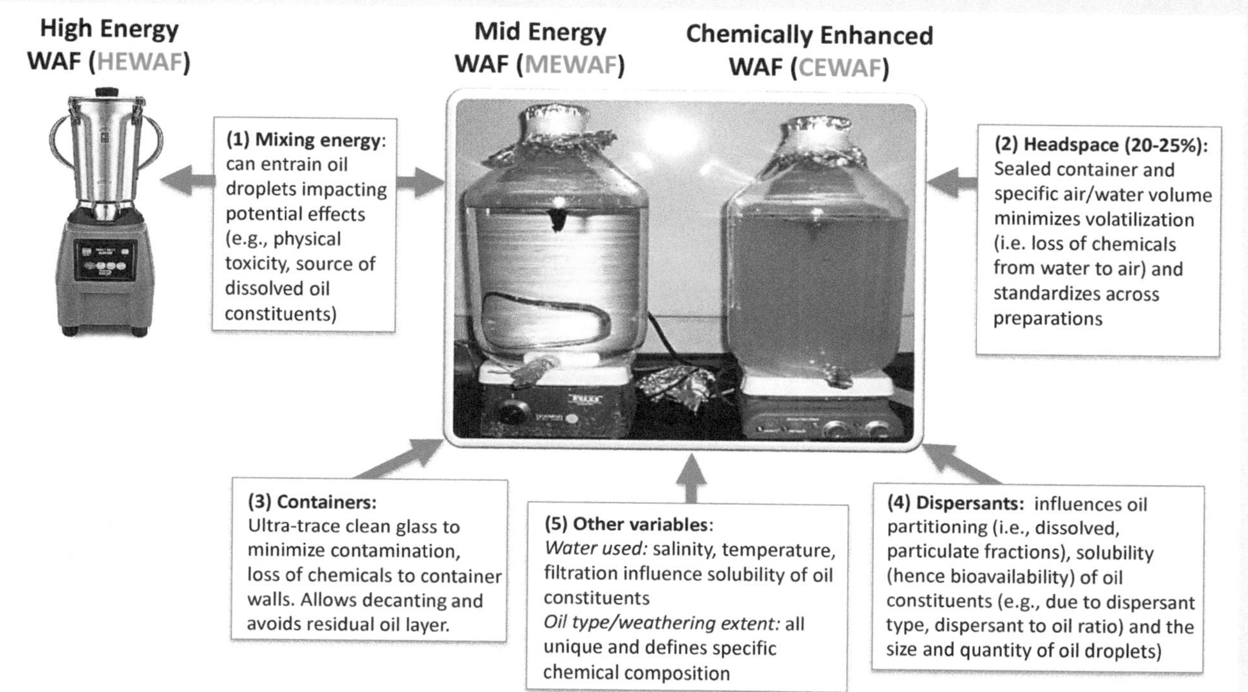

High Energy WAF (HEWAF)
Mid Energy WAF (MEWAF)
Chemically Enhanced WAF (CEWAF)
(1) Mixing energy: can entrain oil droplets impacting potential effects (e.g., physical toxicity, source of dissolved oil constituents)
(2) Headspace (20-25%): Sealed container and specific air/water volume minimizes volatilization (i.e. loss of chemicals from water to air) and standardizes across preparations
(3) Containers: Ultra-trace clean glass to minimize contamination, loss of chemicals to container walls. Allows decanting and avoids residual oil layer.
(5) Other variables: Water used: salinity, temperature, filtration influence solubility of oil constituents
Oil type/weathering extent: all unique and defines specific chemical composition
(4) Dispersants: influences oil partitioning (i.e., dissolved, particulate fractions), solubility (hence bioavailability) of oil constituents (e.g., due to dispersant type, dispersant to oil ratio) and the size and quantity of oil droplets)

FIGURE 6.8 Pictures depicting the preparation of HEWAF, MEWAF, and CEWAF and descriptions of some of the main variables during the procedures that would influence the concentration and chemical composition of oil /dispersant constituents in toxicity test exposure media.

LEWAF*: low energy WAF. Solution derived from low energy mixing (18–24 hours) of water and oil resulting in no vortex. No settlement time after mixing.

MEWAF: mid energy WAF. Solution derived from mid energy mixing (18–24 hours) of water and oil resulting in a vortex of ~20–25%. Settlement time usually 2–8 hours before decanting test solution.

HEWAF: high energy WAF. Solution prepared using a 30-sec blend at low speed in a blender. Includes a 1-hour settlement time in separatory funnel before decanting.

SHE-WAF: super high energy WAF. Similar to HEWAF, but a 120-sec mix time is used.

CEWAF*: chemically enhanced WAF. Solution derived from mid energy mixing (18–24 hours) of water, oil, and chemical dispersant resulting in a vortex of ~20–25%. Settlement time usually 2–8 hours before decanting test solution. Dispersant-to-oil ratios typically 1:10 or 20 for surface application and 1:50 or 100 for subsurface application.

HE-CEWAF: high energy CEWAF. Is the same as HEWAF except for the addition of dispersant.

NOTE: There are also a number of other exposure preparation methods, including oiled-sediment and passive dosing techniques.

BOX 6.4
Terminology and Approaches Used to Characterize Oil Exposure Evolutions

The physical (i.e., oil droplet quantity and size) and chemical characteristics of oil exposure solutions (i.e., specific chemical composition and concentration of oil and [if used] chemical dispersant constituents) are directly dependent on the specific test procedures used to prepare exposure solutions. Therefore, to be able to translate exposure and interpret the biological results from the toxicity tests, it is critical to appropriately characterize the exposure media, which was also a focus in the original Chemical Response to Oil Spills: Ecological Research Forum protocols. Given advances in analytical capabilities and understanding of the toxicity of oil and dispersants, numerous new approaches have also been suggested (see Redman and Parkerton, 2015; Mitchelmore et al. 2020a; NASEM, 2020). Common exposure metrics reported include total hydrocarbons (THCs), total petroleum (TPH), and a suite of individual PAHs combined for total PAH reporting (e.g., TPAH50, representing 50 parent and/or more recently also inclusive of alkylated PAHs). Also, % WAF dilution (with or without analytical characterization of stock preparations) and even nominal concentrations/loadings are reported, despite more than 20 years of recommendations against the nominal approach. The use of nominal concentrations/loadings is not an acceptable metric; these numbers have very limited utility in understanding the toxicity of the oil and prevent comparison to previous studies and/or any interpretation regarding field exposures (Mitchelmore et al., 2020a).

Using a composite metric approach (e.g., THC, TPH) does not account for any changes in composition and concentration of the individual constituents. Furthermore, even with detailed analyses (i.e., measurement of 50+ individual parent and alkylated PAHs [e.g., TPAH50]),

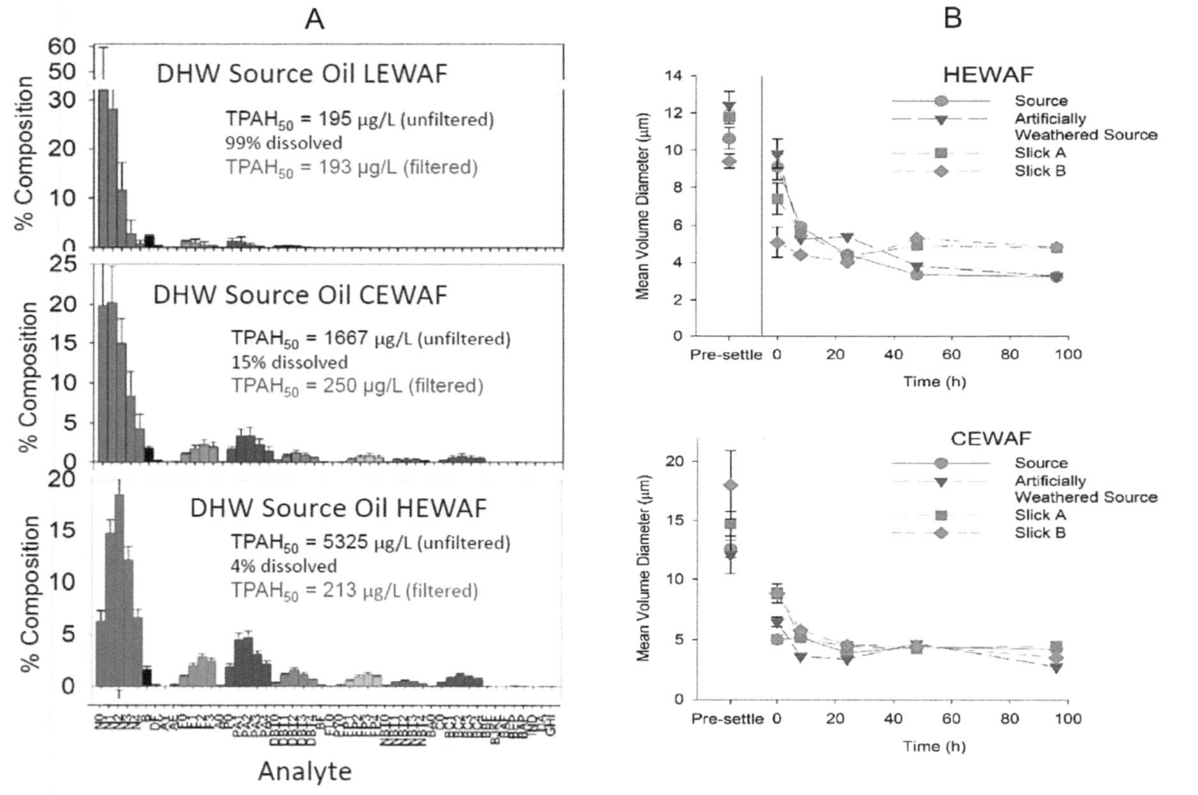

FIGURE 6.9 (A) Percentage composition of 50 parent and alkylated PAHs measured in DWH source oil exposure media (LEWAF, CEWAF, and HEWAF preparations using 1 g/L oil loading). TPAH50 concentrations, specific composition, and the percentage dissolved fractions (i.e., the amount of PAHs in the dissolved versus the total [unfiltered] exposure solution and hence the difference from 100% represents the extent of the PAHs in the particulate [oil droplet] fraction). In all preparations the dissolved fractions (i.e., filtered) are similar concentrations indicating equilibrium. The differences in concentrations of TPAH in the unfiltered solutions reflect the highly modulating role of oil droplets in analytical chemistry reporting and ultimately the interpretations of the toxicity of the test solutions. (B) Graphs of HEWAF and CEWAF preparations of various DWH oils depicting variability in droplet sizes over time in a 96-hour static toxicity test which is dependent on oil type and preparation method.
SOURCE: Forth et al. (2017a,b).

this approach does not account for the variation in the toxicity of the different hydrocarbon constituents, and it also does not identify or quantify other constituents (e.g., unresolved hydrocarbons, polar constituents, etc.) that may also contribute to toxicity (NASEM, 2020).

Toxicity is dependent on the specific composition and concentration of the oil constituents but also on form (i.e., dissolved versus particulate fraction). The importance of differentiating the dissolved from the particulate (i.e., oil droplet) fraction was highlighted in the NRC (2005) report and various methods have been used to investigate this, mainly including filtering exposure solutions through glass fiber filters (see Forth et al., 2017a,b). Differentiation is important for toxicological interpretation for a number of reasons. First, oil droplets may be toxic through physical processes, (e.g., block respiratory surfaces); second, they may enhance exposure (bioavailability) or uptake of oil into organisms. The quantity and size of oil droplets are important to measure especially with respect to direct organism uptake, as the organisms may select or filter only certain size classes. Using composite concentration metrics (e.g., TPH, TPAH) will not accurately reflect the exposure of bioavailable components to organisms. Preparations of oil:water solutions containing higher concentrations of particulate oil may underestimate the toxicity of oil (as depicted in Figure 6.8) for numerous reasons. Furthermore, many comparisons regarding the toxicity of CEWAF versus WAF preparations have been misinterpreted simply based on the choice of the exposure characterization metric used. This has led to a number of studies concluding that chemically dispersed oil is more toxic than physically dispersed oil. However, for most species/studies, chemical dispersants are not inherently more toxic; rather, differences are due to dispersants increasing the concentration and/or bioavailability of the oil constituents (NRC, 2005; NASEM, 2020). Similarly, comparisons of the toxicity of different chemical dispersants can also be misinterpreted based on the choice of analytical characterization by reflecting the efficacy of the dispersant rather than its inherent toxicity. This can lead to wrong conclusions that some dispersants are less toxic than others simply because they were not effective and simply entrained the oil constituents into the exposure media.

As we continue to evolve analytical capabilities and further understand the role of both chemical and physical attributes of oil exposure solutions, it is important that the procedures used to characterize test exposure media reflect toxicity drivers (i.e., individual chemical compositions and the quantity and size of oil droplets) so that more accurate interpretations regarding toxicity can be made.

microdroplets modulating WAF chemistry was highlighted emphasizing the need to differentiate between dissolved and particulate oil. Preparations of these fractions were achieved using filtration through a stacked GFF (0.3 μm) procedure resulting in a filtrate that was considered the "dissolved" fraction with the unfiltered fraction representing the "total." Test solutions were prepared using the same initial oil loading (i.e., 1 μl/L) and although LEWAF, CEWAF, and HEWAF preparations were similar in concentrations of dissolved TPAH50, the total TPAH50 concentration and composition in each preparation type varied widely (i.e., 195, 1667, and 5325 TPAH50 μg/L for unfiltered LEWAF, CEWAF, and HEWAF, respectively). Again, these data, as highlighted also by Redman and Parkerton (2015) show the importance of choosing and using appropriate analytical methods to quantify the concentration and specific composition of oil test exposure solutions.

The importance of oil partitioning (i.e., dissolved and particulate fractions) and its characterization including oil droplet sizes, are also new considerations from the original CROSERF methods highlighted by field research from the M/V *New Carissa* oil spill by Payne and Driskell (2003) and subsequently discussed in detail in the National Academies dispersant report (NRC, 2005). This need has been reiterated in subsequent reviews, reports and publications (Forth et al., 2017a,b; Redman et al., 2017; Sandoval et al., 2017; Mitchelmore et al., 2020a; NASEM, 2020). Partitioning of the oil constituents can influence oil fate processes, exposure routes, bioavailability, and ultimately toxicity. Oil droplets can cause physical toxicity, increase hydrocarbon bioavailability and uptake across membranes (i.e., due to dissolution; Ramachandran et al., 2004; Sørhus et al., 2015; Sørensen et al., 2017; Laurel et al., 2019) and can be ingested/filtered by certain species (i.e., zooplankton and oysters; Payne and Driskell, 2003; Hansen et al., 2018) resulting in species-specific impacts. **Despite being flagged for nearly 20 years, there still remain many data gaps hindering a full understanding of the implications and importance of oil droplets in driving toxicity; further research in this area is recommended including methods to characterize droplet concentrations and sizes.**

Many environmental variables that can influence the toxicity of oil (e.g., temperature and salinity) are controlled for and measured in toxicity tests. One variable that was not usually considered during toxicity testing is the influence of UV light (quantity and spectral quality). As discussed in Section 6.3.4.3, some oil constituents are more toxic to certain organisms under UV light compared to traditional laboratory lighting. The phototoxicity of certain oil constituents has been discussed at length in previous reports, but is still a relatively understudied area (Barron et al., 2004; NRC, 2005; NASEM, 2020).

One of the most important considerations and recommendations from the CROSERF working group was the requirement for detailed analytical verification and not the use of nominal exposures (i.e., a volume or mass of oil only), such as total petroleum hydrocarbons (TPHs) or total PAHs (TPAHs; Bejarano et al., 2014; Mitchelmore et al., 2020a). The focus and specific details required of the analytical methods have since been updated in various reports to reflect new analytical capabilities and also increased understanding regarding the importance of oil exposure routes and specific chemical compositions (NRC, 2005; Forth et al., 2017a; Mitchelmore et al., 2020a,b; NASEM, 2020). Comparisons across studies using different analytical measures is difficult

TABLE 6.2 Some Considerations on How Parameters Involved in Exposure Media Preparation, Conduct of Toxicity Tests, and Analytical Verification of Exposure Solutions Can Affect Test Results and Recommendations for the Minimum Reporting Requirement for Each

(1) Exposure Media Preparation		
Parameter	Considerations/Variables for Test	Suggestions for Minimum Reporting Requirements for Future Test Protocols
Vessel and laboratory conditions	Type of vessel, headspace to water volume (usually 25%), how sealed (usually closed to minimize loss of volatile components) and lighting (in the dark [i.e., dark room or covered with foil]) to reduce photolysis or photochemical reactions.	Specify vessel type used, volume of water, headspace volume, sealed/unsealed, and lighting conditions.
Type of oil, chemical dispersant	Environmental relevance (i.e., often more appropriate to use oil weathered to some degree and concentration) Dispersant: oil ratio depends on specific surface or subsurface application.	Detail specific oil used including weathering state and specific dispersant used (e.g., CAS #). Detail oil/water and dispersant/oil ratios used.
Exposure water type	Oil concentration/composition in exposure solution (dissolution/partitioning) depends on water quality (i.e., salinity, temperature, pressure, presence of potential modifying/binding substrates [i.e., organic carbon, etc.]).	Report water used (freshwater [FW], saltwater [SW]) and salinity, if filtration was used (e.g., glass fiber filter [GFF]) and for artificial seawater report salt source. If natural waters are used, ideally report water quality measures (e.g., total suspended solids [TSSs], dissolved organic carbon [DOC], particulate organic carbon [POC], etc.).
Test concentration preparation method (also see Figures 6.9 and 6.10)	Variable loading (i.e., each test concentration prepared individually with increasing volumes of oil) or variable dilution approach (i.e., high-concentration stock solution prepared and test concentrations are dilutions from the stock). Approach depends on specific test objectives, WAF or CEWAF, volume of test media, and cost of analytical verification.	Report the preparation method used in detail. Conduct appropriate analytical verification of test solutions (e.g., concentration and composition). Report if one (and holding times/conditions if used multiple times) or multiple preparations of test media were used.
Specific WAF preparation method	Report if WAF, LEWAF, MEWAF, CEWAF, HEWAF prepared and specific mixing time, mixing energy (e.g., 25% vortex) and settlement time used. The oil concentration and composition are dependent on these parameters.	Report mixing times/energy and settlement time. Report in detail how solutions are removed from the test preparation chambers, whether they are used immediately and, if not, how they are stored and methods used to resample to ensure a continued homogeneous mixture is prepared.
Preliminary abiotic test run	OECD test (2019) requires that for difficult to-work-with substances, investigations into test solution stability is required.	Conduct a prior test to determine the stability and reliability of the test solution preparations, report chemistry (although these can be simpler methods [such as fluorescence detection] than used for analytical verification of definitive tests).

(2) Toxicity Test Type Considerations		
Parameter	Considerations/Variables for Test	Suggestions for Minimum Reporting Requirements for Future Test Protocols
Specific test used	Acute or chronic test, standard method used (or based on [e.g., EPA, OECD, ASTM]). Test duration will affect toxicity. Inclusion of appropriate negative and positive controls so that results obtained reflect the test compound rather than a problem with the health of the species under study. Include solvent and test water-only controls if a solvent is used.	Detail test type and length of time conducted. For standard tests report method number/reference. State any and all modifications/deviations from the standard test protocol. Use a positive control to ensure reliability of the test and include two controls if a solvent is used.
Species and life stage used	Choice of test species and life stage critical. Some species/life-stages are more sensitive to oil constituents than others.	Detail the name, source (e.g., culture, supplier or wild-caught) and give details of field collections (location, time, etc.). Detail life stage, age, and, if possible, sex of organism.
Test conditions	Static (non-renewal and renewal), flow through constant exposures and spiked/declining or pulsed exposures. An organism's exposure will depend on the test type used given the dynamic nature of oil exposure media over time. Test concentration/composition will affect toxicity.	Report how the test is conducted and if and how exposure solutions are renewed over the time course of the experiment. Detail how the analytical verification of exposure solutions ties in with this and report nominal target and measured concentrations using an appropriate sampling plan to calculate average concentrations over time.

TABLE 6.2 Continued

(2) Toxicity Test Type Considerations		
Parameter	Considerations/Variables for Test	Suggestions for Minimum Reporting Requirements for Future Test Protocols
Additional toxicity test details	Specify the water used (e.g., FW, SW, filtered or non-filtered); temperature and pressure will alter oil concentration/composition. Size of exposure chamber: exposure solution volume and number of animals per chamber (inappropriate ratios of these may result in quickly declining concentrations of oil components and/or decline in water quality parameters [e.g., dissolved oxygen, build-up of ammonia] that may end up being covariables in the organisms' response observed). Appropriate lighting for the test species and questions being asked. Normal laboratory lighting underestimates toxicity compared to lighting mimicking natural spectral quantity/quality, as some chemical constituents are phototoxic (e.g., more toxic under UV lights); alternatively other components undergo photolysis and are degraded. Photochemical oxidation products have been measured and toxicity is largely unknown. Feeding of organisms may affect results due to binding of oil components to the food (causing co-uptake by the organisms) or sedimentation to the chamber bottom and removal from organism exposure.	Detail specific water used and salt sources. Also type of test chamber (material), size of test chamber, and exposure volume. Report number of replicate test chambers per treatment concentration and if appropriate the number of test subjects in a test chamber. Conduct appropriate water quality tests for the species under study (e.g., salinity, pH, conductivity, dissolved oxygen, temperature and for vertebrate species ammonia). Measure lighting (i.e., spectral quality and quantity). For water renewals, feeding, cleaning, quality/lighting parameters, etc., give details about the time of day measurements were taken; suggest keeping measurements to the same time each day. Report the day/night regime used. Report aeration and feeding regime used, how tanks are cleaned, and how cleaning (if any) occurs in relation to how solutions are renewed. Report when and if and how dead organisms are removed. For chronic tests, report how (if) test chambers are cleaned to prevent biofilm/fouling.
Preliminary range-finding versus definitive toxicity tests	Conduct a preliminary range-finding test using log doses over a wide range of concentrations. This will establish the approximate concentration range for effects. For a definitive test recommended to use a 2×–3× concentration range (for more accurate statistical analyses).	A preliminary test is recommended so that the concentrations used in a definitive test bracket the toxicity test thresholds appropriately.
Biological endpoints and statistics	Appropriate endpoints for the type of toxicity test (acute; LC50 or chronic; EC10, NOEC, etc.). TU approach for WAF and include time endpoints (e.g., LT50). Various statistical approaches may be taken and depend on the specific dataset.	Report appropriate toxicity thresholds for the type of test and what the test is being used for (regulatory, model data, etc.). Report raw data (use supplementary information in publications) and detail the specific statistical methods used. Report any problems with the dataset and why alternate methods (if any) are used.

(3) Analytical Approaches		
Parameter	Considerations/Variables for Test	Suggestions for Minimum Reporting Requirements for Future Test Protocols
Analytes quantified	The concentration and specific composition of oil has been reported in numerous ways (e.g., TPH or TPAH, using a variety of numbers of parent and/or alkylated PAHs). What is reported may have ramifications for the ability to compare across studies and for utility in toxicity models and may ultimately over- or underestimate toxicity (see Figure 6.11). The use of nominal concentrations of oil provides limited information with respect to toxicity of the resulting exposure solutions.	Analysis of VOCs (e.g., BTEX). Suggest to report as total PAH (50+) including alkyl homologues (rather than TPH). For CEWAF/HEWAFs, suggest titration (i.e., GFF) or centrifugation to determine chemical composition in the dissolved and particulate phases. If possible, also measure droplet size distribution (plus volume/quantity). The use of nominal exposures is unacceptable.
Oil droplet quantity and size	Oil droplets may be an exposure route for some organisms (e.g., filter feeders, incidental ingestion when drinking/feeding), some of which may feed on oil droplets depending on their size.	LISST/coulter counter analysis to determine droplet quantity (volume) and size distribution. Also important for comparison to field measurements.
Tissue residue and/or metabolites	Toxicity is defined by internal exposure and thus measuring TPAH in organism tissues relates internal exposure (body burden) to toxicity endpoints. In vertebrate species (e.g., fish), direct relationships between the extent of internal exposure to oil and bile metabolites have been shown in laboratory and field tests.	Measurement will depend on the objective of the test (e.g., bioaccumulation study) and the costs involved. Determining bile metabolites also depends on the questions (e.g., validation with field specimens to estimate exposure).

SOURCE: Modified and expanded from a table in Mitchelmore et al. (2020b).

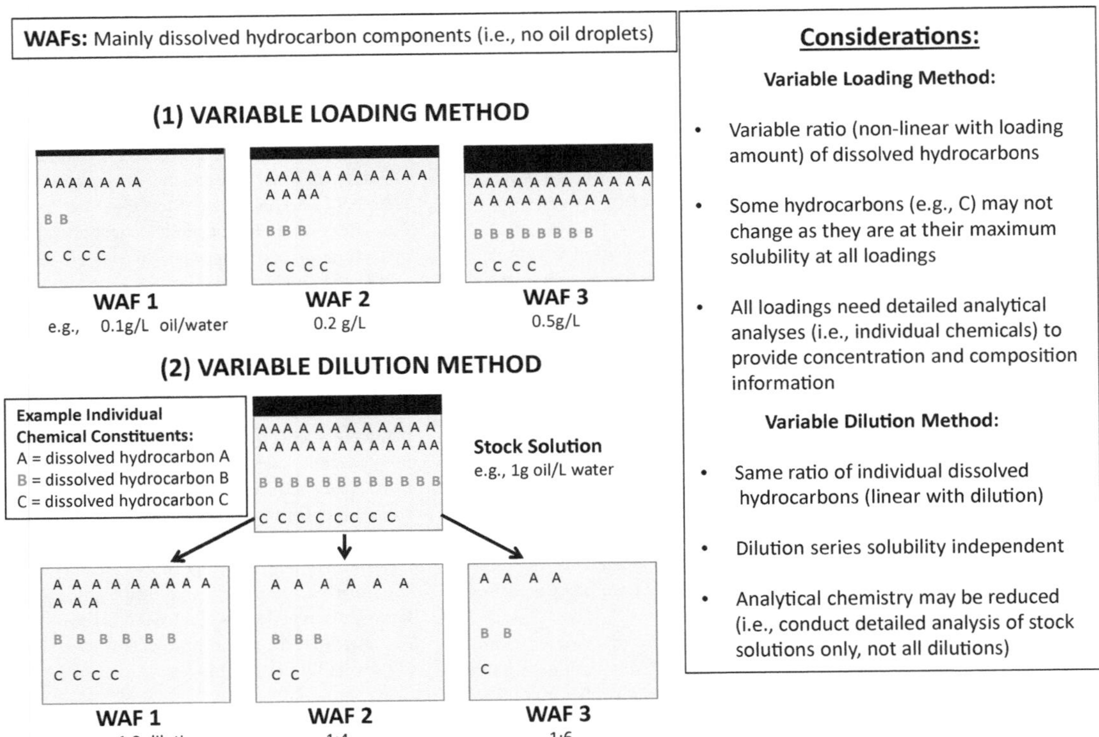

FIGURE 6.10 Comparison of variable loading and variable dilution methods for preparing WAFs (oil and water mixtures) for toxicity testing.

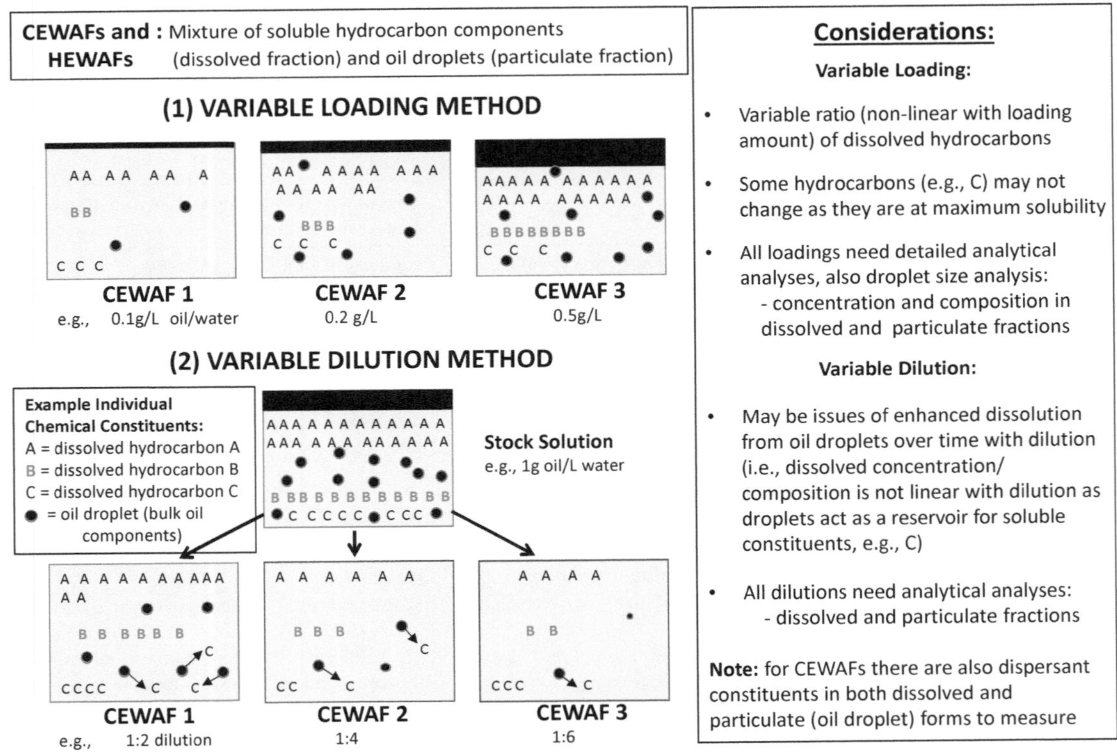

FIGURE 6.11 Comparison of variable loading and variable dilution methods for preparing CEWAFs (oil, water, and dispersant mixtures) and HEWAFs (high energy preparation of oil and water mixtures to entrain oil droplets) for toxicity testing.

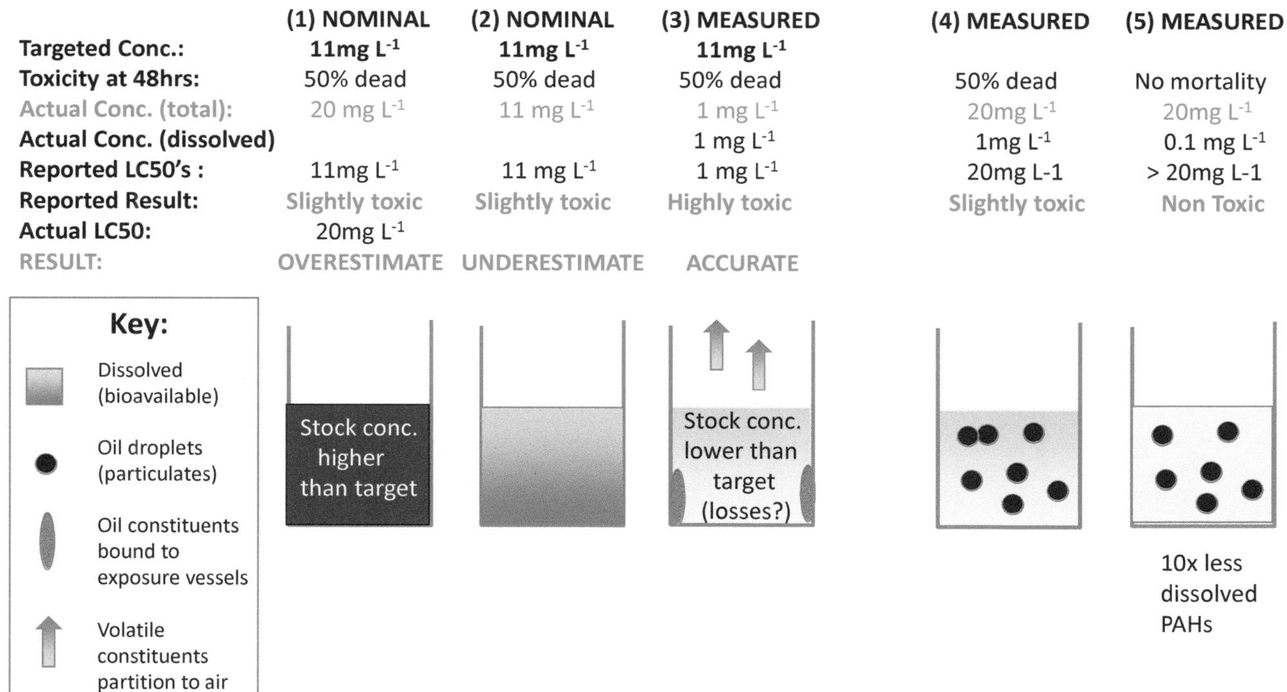

FIGURE 6.12 Example of misinterpretation of toxicity test data based on analytical approach (nominal versus measured exposure characterizations) in addition to the type of target analytes and their form (dissolved versus particulate) in a species where acute toxicity is due to dissolved PAHs (LC50 threshold of 1 mg/L) and analytical chemistry is reported in terms of total PAH (or TPH).
SOURCE: Figure modified and expanded from Figure 28.2 in Mitchelmore et al. (2020b).

(e.g., using TPH versus TPAH [and how many individual analytes are measured]) and many studies have only measured the parent and not the alkylated PAHs. Expansions of this analytical effort can also include the identification and quantification of the additional unresolved hydrocarbons that can contribute to toxicity.

As recently highlighted in the National Academies dispersant report, traditional exposure metrics (i.e., TPH and TPAH) do not account for the variation in the toxicity of individual hydrocarbons. The use of toxic units (TUs) has been proposed (NASEM, 2020; see discussion of modeling in Section 6.7). Even without these new insights into the importance of specific chemical composition, CROSERF methods significantly improved the ability to compare across studies—although nominal exposures or inappropriate analytical verification continue to be used and can result in both under- and over- estimations of toxicity (see Figure 6.12).

In addition to how exposure test media are prepared and characterized, and the results interpreted, consideration must also be given to the biological endpoints that are measured and when (see Figure 6.13). For example, acute toxicity tests are often conducted for 48–96 hours depending on the species, which may not be reflective of the typical exposure time observed in an oil spill event. Two approaches to address this include shorter-duration tests or assessments at multiple earlier time points than the terminal test duration. This latter approach is useful to determine time-to-death metrics. Technological advances in molecular biology have also resulted in a new understanding of mechanisms of action and impacts to organisms. The use of additional biological endpoints, besides just the typical metrics used in risk assessments (i.e., mortality, growth, and reproduction) should be investigated further particularly with respect to their underlying assumptions and relevance to individual-, population-, and ecosystem-level impacts.

6.4.2 Challenges Regarding the Environmental Relevance and Field Applicability of Laboratory Toxicity Tests

As summarized in Figure 6.14, a variety of studies are used to study the impacts of oil on marine organisms, and all vary in their complexity, ability to control variables, and field relevance (see Box 6.5). Although laboratory tests cannot replicate or mimic field conditions during an oil spill they are an important tool in estimating potential effects. The generation of reproducible and reliable toxicity test results is critical so that implications of exposure and its relationship to toxicity can be established and compared across studies to determine potential impacts from an oil spill, identify sensitive species, and calculate the relative toxicity of different oils and/or dispersant mixtures.

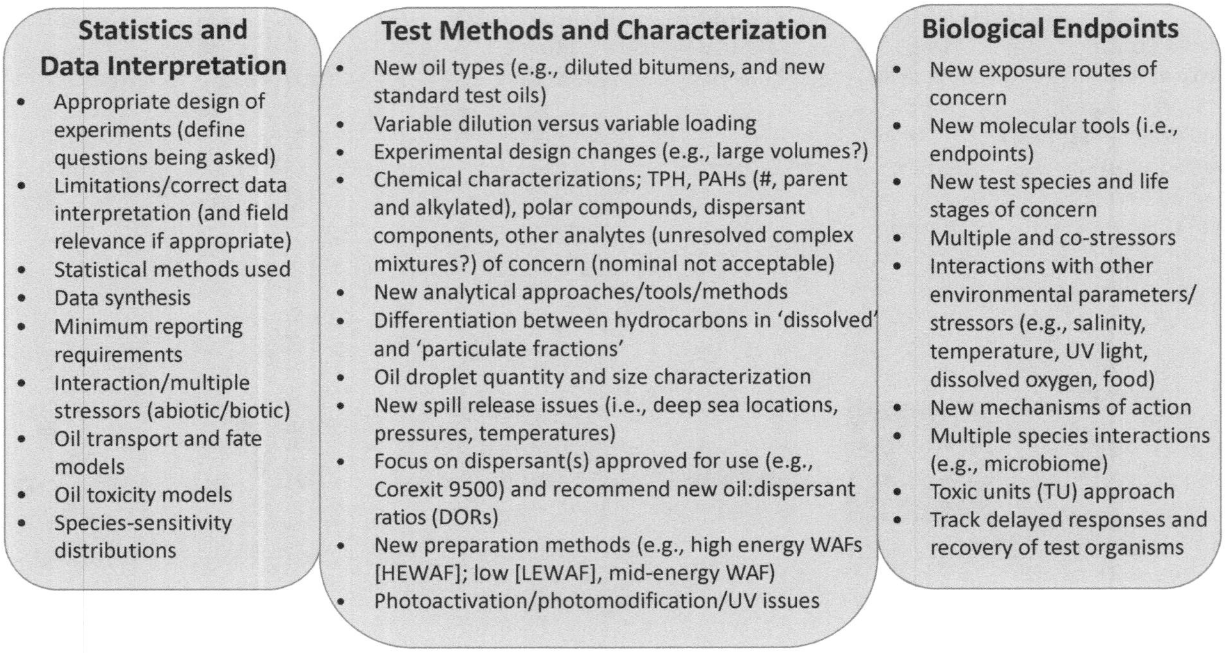

Statistics and Data Interpretation

- Appropriate design of experiments (define questions being asked)
- Limitations/correct data interpretation (and field relevance if appropriate)
- Statistical methods used
- Data synthesis
- Minimum reporting requirements
- Interaction/multiple stressors (abiotic/biotic)
- Oil transport and fate models
- Oil toxicity models
- Species-sensitivity distributions

Test Methods and Characterization

- New oil types (e.g., diluted bitumens, and new standard test oils)
- Variable dilution versus variable loading
- Experimental design changes (e.g., large volumes?)
- Chemical characterizations; TPH, PAHs (#, parent and alkylated), polar compounds, dispersant components, other analytes (unresolved complex mixtures?) of concern (nominal not acceptable)
- New analytical approaches/tools/methods
- Differentiation between hydrocarbons in 'dissolved' and 'particulate fractions'
- Oil droplet quantity and size characterization
- New spill release issues (i.e., deep sea locations, pressures, temperatures)
- Focus on dispersant(s) approved for use (e.g., Corexit 9500) and recommend new oil:dispersant ratios (DORs)
- New preparation methods (e.g., high energy WAFs [HEWAF]; low [LEWAF], mid-energy WAF)
- Photoactivation/photomodification/UV issues

Biological Endpoints

- New exposure routes of concern
- New molecular tools (i.e., endpoints)
- New test species and life stages of concern
- Multiple and co-stressors
- Interactions with other environmental parameters/ stressors (e.g., salinity, temperature, UV light, dissolved oxygen, food)
- New mechanisms of action
- Multiple species interactions (e.g., microbiome)
- Toxic units (TU) approach
- Track delayed responses and recovery of test organisms

FIGURE 6.13 Summary of changing procedures, new developments, and considerations for inclusion since publication of the CROSERF protocols for standardizing oil toxicity testing.
SOURCE: Modified and expanded from Mitchelmore et al. (2020b).

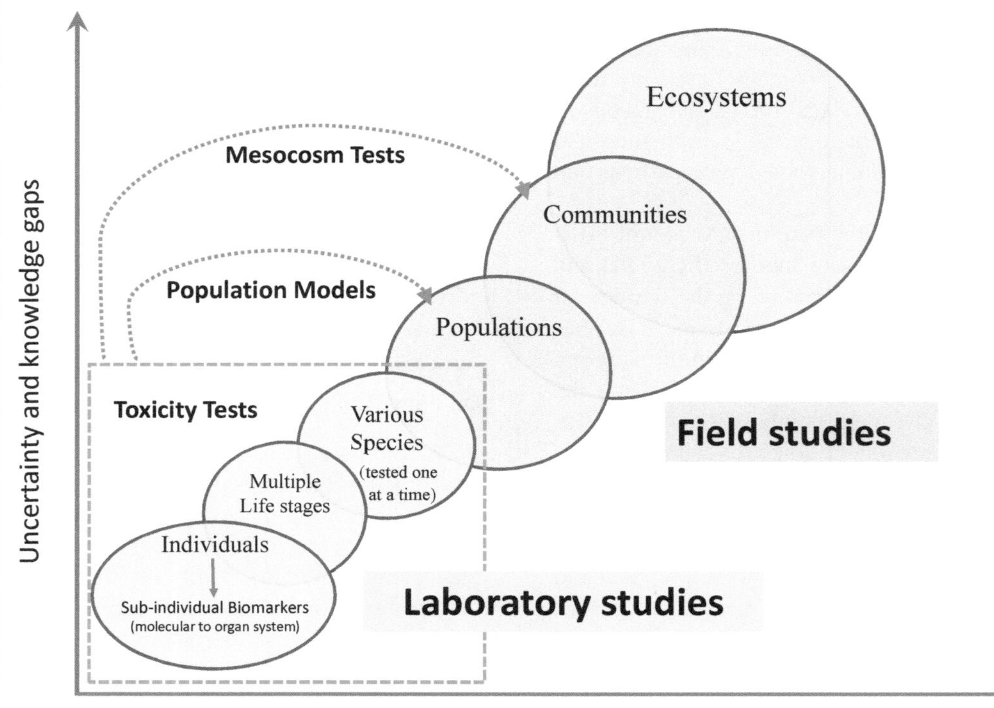

FIGURE 6.14 Highlighting the complexity in studying oil effects at multiple scales using laboratory, mesocosm, and field-testing approaches.

BOX 6.5
"Best Practices" and "Environmentally Relevant Concentrations"

Best Practices is a term used frequently in presentations and discussions of science and engineering subjects. Different groups of people, professional organizations, industrial associations, and/or government bodies may decide among themselves, or for their constituents, what constitutes best practices. For example, industrial associations, such as the American Petroleum Institute or the Interagency Coordinating Committee on Oil Pollution Research, will publish best practices for the industry as a whole, and individual companies might then use these as the basis of their own internal documents, with appropriate modifications for their unique circumstances. Government agencies and scientific associations have taken similar approaches, such as the National Oceanic and Atmospheric Administration[a] and the American Chemical Society.[b] Best practices for oil spill response (see Chapter 4) and for oil spill research on measurements, inputs, fates, and effects (also see Chapters 2, 3, and 5) may differ, but have the same goals: achieving the most appropriate and effective response, and acquiring the most robust data and deepest knowledge possible.

Best practices for oil spill response and damage assessments have been well defined by prior experience, pre-planning, practice, and information available about the spilled oil and factors such as worker safety, human health of nearby people, and ecosystems involved. For example, the government agency or agencies responsible for response to oil spills will promulgate response plans, guidelines, and practices based on their interpretations of up-to-date knowledge and technology. They are also informed through the input by and comments from interested groups. A non-exclusive set of examples of interested groups are: indigenous people of potentially affected areas, environmental nongovernmental organizations, industry associations, organizations of local and regional businesses, groups of the general public, and local governments. This process is designed to result in procedures that represent the most efficient or prudent course of action and have been shown by research and experience to produce optimal results with least damage to the environment and human health. Such procedures for oil spill response keep evolving and should be upgraded as new and better solutions are found or evolve from better awareness, new technologies, or simply different ways of looking at processes.

Best practices for oil spill research are underpinned by adherence to the scientific method; a nuanced explanation of the scientific method is contained in a recent report on *Reproducibility and Replicability in Science* (NASEM, 2019). Oil spill research is often guided by previously described recommended analytical and/or experimental procedures such as outlined by the Chemical Response to Oil Spills: Ecological Research Forum (CROSERF), described in this chapter. Research focused on unraveling poorly understood individual phenomena or complicated interactions in fates and effects issues often involves new experimental conditions or ways for collecting and processing data. Difficulties arise when experimental or observational conditions are not described in sufficient detail for others to replicate the results, or when the results are extrapolated, usually in the discussion section of scientific papers or reports, to claim relevance to other experiments, field observations, and past or potential future fates-and-effects scenarios. Moreover, there are different systemic limitations to in situ studies during accidental oil spill studies and field trials when attempting to replicate rapidly evolving situations in space and time, including the inability to sample and measure many locations simultaneously.

The following points form the basis of recommended "Best Practices" to be implemented in experimental design for oil spill science:

1. In vitro and ex situ experiments intended to inform in situ knowledge should be designed to reflect the marine ecosystem being studied as well as possible (i.e., applying appropriate scaling and matching physical, chemical, and biological parameters to the extent practicable between the experiment and field representation);
2. Extrapolation from laboratory and mesocosm experiments to the field should utilize accepted approaches to similitude, clearly stating when results are extrapolated outside the observed parameter space;
3. Field or laboratory studies wherein polycyclic aromatic compounds are known or suspected to be important stressor factors should include analysis results for alkyl-substituted congeners in addition to the un-substituted parent compounds, as limitation of these analyses to the U.S. Environmental Protection Agency priority pollutant polycyclic aromatic hydrocarbons is rarely appropriate for oil pollution studies;
4. Field studies aimed at estimating the extent of environmental damages from accidental oil discharges should use statistically-based sampling methods to estimate upper and lower bounds of the extent and intensity of shoreline oiling (including estimates of subsurface oil), estimates of numbers of animals killed, and other factors when appropriate, as these would permit meaningful comparisons among sampling periods to estimate temporal trends of oil persistence and recovery rates of adversely affected organisms.

One of the frequently quoted criticisms of laboratory experiment extrapolations is that they did not involve or do not pertain to environmentally relevant concentrations. A precise and concise definition of *Environmentally Relevant Concentrations* for oil in the sea is problematic because of the multiple, dynamic combinations of environmental conditions affecting oil concentrations in situ (e.g., oil type and volume, climatic conditions, physical and chemical processes, dilution, etc.) overlaid by temporal (e.g., oil changes due to weathering, chemical and biological reactions) and spatial factors (e.g., decreasing concentration with distance from the oil source, dilution with tidal action, changing oil droplet size with wave action, etc.). Few, if any, laboratory or mesocosm studies could hope to replicate the many possible combinations of conditions in situ. Adding biota exposure routes, concentrations, and spatial-temporal parameters to these scenarios further complicates experimentation. As a simplistic example, a school of fish passing through a dispersed plume of oil will have a short exposure duration, whereas plankton drifting with a surface slick will be exposed for a longer time, aside from their different exposure mechanisms and metabolic types. Furthermore, in some experiments researchers deliberately are not trying to replicate environmental conditions, but instead are emphasizing one or more factors to elicit a response or result that demonstrates a particular fate or effect. It is therefore incumbent on the scientific community to overtly describe the experimental conditions, state the scientific intent of applying those conditions, and indicate to what degree the observations and conclusions are applicable to the environment, so that laboratory and mesocosm results are not extrapolated incorrectly to subsequent oil spill response and modeling efforts.

[a] See https://response.restoration.noaa.gov/oil-and-chemical-spills.

[b] See https://www.acs.org/content/acs/en/greenchemistry/research-innovation/tools-for-green-chemistry.html.

Toxicity data are used for the development, calibration, and validation of predictive toxicity models (i.e., target lipid models) to provide predictions of oil toxicity (e.g., OilTox, PETROTOX; see Section 6.7) and species sensitivity distributions. As discussed earlier, toxicity tests provide information for basic mechanistic science, if realistic exposures (concentrations/time) under pertinent and applicable environmental conditions and (if operationally relevant) chemical dispersant concentrations are used, then toxicity results can be used for oil spill response and planning and ultimately to provide tools (models) for real-world applicability in operational decisions. In summary the choice of methodological approaches used in toxicity tests has limited the utility of many studies for inclusion in toxicity models and hindered assessments of comparability across studies and accurate predictions of toxicity of petroleum hydrocarbon constituents, oil mixtures, dispersants, and dispersed oil (Bejarano, 2018; Mitchelmore et al., 2020a,b).

More generally, NRDA studies provide most of the factual basis for quantifying the environmental damage caused by oil spills. The results are used to guide oil spill response and restoration efforts, and to assess any legal liability of the responsible party (RP). The scope of these studies is usually determined collaboratively by the RP and government trustee agencies, which then use scientifically rigorous methods to address the concerns identified within the agreed scope. Advantages of this approach include study designs informed by both industry and government expertise, engagement of both government and industry personnel and resources in the conduct of these studies, and avoidance of duplicative efforts associated with adversarial approaches. Disadvantages include the sometimes lengthy process required to reach agreement on study scope, objectives, and design, which may impair identification and exploitation of ephemeral opportunities to collect critically important data for the eventually identified scoping objectives; offer scant formal capacity for recognizing evidence of environmental injury from damage pathways that are not well-understood; and use study designs that may favor consensus-point estimates of environmental damage at the expense of quantifying the associated uncertainty, and that may be prone to unquantified underestimation bias. Consequently, conclusions of some NRDA studies may not be as accurate as conclusions informed by additional data from non-NRDA sources (including data collected by non-NRDA entities before, concurrently, or after NRDA study initiation), and by differing assumptions and methods used for the data analysis.

6.5 EFFECTS ON POPULATIONS, COMMUNITIES, AND ECOSYSTEMS

While oil pollution can affect organisms in a variety of ways, as summarized in Sections 6.2 and 6.3, the likelihood and severity of these effects may vary considerably during and after a particular oil spill or other oil discharge events. The most serious effects may significantly alter population-level fitness or growth rates, which in turn may perturb the ecological communities within which these populations interact. Conversely, other effects, such as mass fish mortalities from short-term acute toxicity, occur infrequently, even after very large spills such as the *Exxon Valdez* or the DWH.

Often the most serious effects of oil spills and discharges are obvious, such as widespread mortalities of heavily oiled seabirds, marine mammals, or sea turtles. Less obvious effects typically require carefully designed field studies to detect them, and are necessary to quantify any effects, obvious or not. Unfortunately, while oil spills, especially large ones, present unique opportunities to evaluate the actual effects that may result from possible toxicity mechanisms, exploiting these opportunities is exceptionally challenging. Oil spills usually occur without prior notice, and opportunities to collect crucial information for damage assessment and effects research studies are ephemeral. Initial field study designs are often crafted in a crisis atmosphere, and accorded lower priority than human safety and response efforts to limit property and environmental damage. The most qualified research personnel initially available may have little familiarity with the oil pollution effects literature, so initial data collection efforts may be less than optimal, although usually any sampling immediately following an oil spill or other discharge event is preferable to not sampling at all. **Yet studies of how ecosystems respond to the strong ecological perturbations caused by large marine oil spills present unparalleled opportunities to gain insights into marine ecosystem structure and function, in addition to how organisms and populations respond to the various ways that oil can cause toxicity. Research is needed that couples oil toxicity responses in organisms to populations.**

Recognizing the unusual and often stressful conditions that confront oil pollution response, damage assessment, and research personnel, we begin this section with a summary of the limitations and challenges involved when designing and executing field studies to evaluate oil pollution effects (see Section 6.5.1), in the hope that this will be useful for those faced for the first time with responding to future oil spills or evaluating their consequences. Following this, we summarize effects that have been clearly established during oil spills and discharges, beginning with the most vulnerable habitats, the sea-air interface and shorelines (see Sections 6.5.2 and 6.5.3, respectively), where oil is mainly compressed into the two horizontal dimensions. Following this are coral communities (see Section 6.5.4) because of their sensitivity to oil and benthic communities (see Section 6.5.5), where oil may again be largely constrained to two horizontal dimensions, but less likely to be as heavily oiled as the sea surface. Finally, we consider effects in the water column (see Section 6.5.6), where oil contamination is usually much lower than oiling at the sea surface and along shorelines. We conclude with a summary of ecosystem-level effects, which may occur when populations of species that strongly interact with oil-affected species are themselves strongly affected through these interactions.

6.5.1 Limitations and Challenges

One message from the *Oil in the Sea III* (NRC, 2003) report, which was not necessarily new then or now, is that **it becomes increasingly difficult to identify effects or recovery or both in a progression of individuals to population to community levels within an ecosystem** (see Figure 6.1). For example, the clearly identified toxic effects of oil from the DWH oil spill on marsh nekton in some field studies and most laboratory experiments (Whitehead et al., 2012) were not necessarily reflected in adverse negative effects on populations of these same fish within the marsh habitat (Fodrie et al., 2014; Able et al., 2015), when populations of marsh and other coastal fishes increased 1–2 years after the spill (Schaefer et al., 2016). Similarly, sea grass nekton were as abundant or increased in abundance within 1 year of the spill in oiled and unoiled sites, with no change in fish assemblages (Fodrie and Heck, 2011). It is even more difficult to extrapolate to ecosystems that will be identified as "stressed" or "recovered" after 5–10 years of potential recovery. Similar effects on changes in complex aquatic communities and ecosystems are confounded by other changes over time that may or may not be related to the initial hydrocarbon exposure (e.g., multiple stressors several years from the initial exposure).

6.5.1.1 Multiple Stressors

The inputs of petroleum into marine environments occurs with other environmental stressors, so the impact of oil will depend on the interactions of oil and influences of various potential co-stressors as a complex multiple stressor event. As an example, the 2010 DWH oil spill in 1500 m water depth off the Mississippi River delta occurred against a background of various other stressors, which cause chronic hydrocarbon contamination for the northern Gulf of Mexico. These stressors included natural deepwater oil and gas seeps, a heavily industrialized oil and gas production and transport system, coastal petroleum activity onshore, regular marine transportation of petroleum and non-petroleum products (Garcia-Pineda et al., 2014), and a recurring oil leakage from a damaged well (MacFayden et al., 2014). The area of the DWH oil spill is influenced by freshwater discharge, nutrient loads, and sediment flux from a major river, the Mississippi, high rates of sea-level rise, shoreline erosion, loss of coastal habitats, and coastal systems perennially exposed to petroleum from oil and gas production and transportation (Boesch, 2014). Laboratory experiments usually conclude in mortality, genetic impairments, or indirect effects, such as the inability to locate food. The ability to transfer laboratory experimental results to populations, communities, or ecosystems is not straight forward. Extrapolation of these experimental and field effects does not necessarily lead to accurate estimates or calculation of impacts on higher trophic levels or complexity of organismal organization. The ability to disentangle multiple stressors from petroleum impacts may be difficult (e.g., Turner et al., 2016).

6.5.1.2 Baseline and Long-Term Data

It is challenging to identify or acquire population-, community-, or ecosystem-level characterizations and ancillary chemical data prior to an oil spill, as well as to identify appropriate existing baseline data from existing databases for comparison to an ongoing spill or exposure to other oil sources.

Baseline data include:

1. numerical data, such as existing concentrations of petroleum hydrocarbons and their reaction products, and the density and diversity of potentially affected populations, prior to oiling, and
2. knowledge of biological, chemical, physical and geological ecosystem processes relevant to the responses to oil spills (see Chapter 4), and the fates and effects of oil inputs, including those relevant to human health (see Chapters 5 and 6).

Several U.S. agencies developed programs to collect baseline data prior to Outer Continental Shelf oil and gas drilling on continental shelves (e.g., the U.S. Department of the Interior, first Bureau of Land Management, Minerals Management Service, now Bureau of Ocean Energy Management) (e.g., Flint and Rabalais, 1981; Murray, 1998). There were appropriate long-term data sets from the Louisiana shelf low oxygen area that provided a comparison of low oxygen

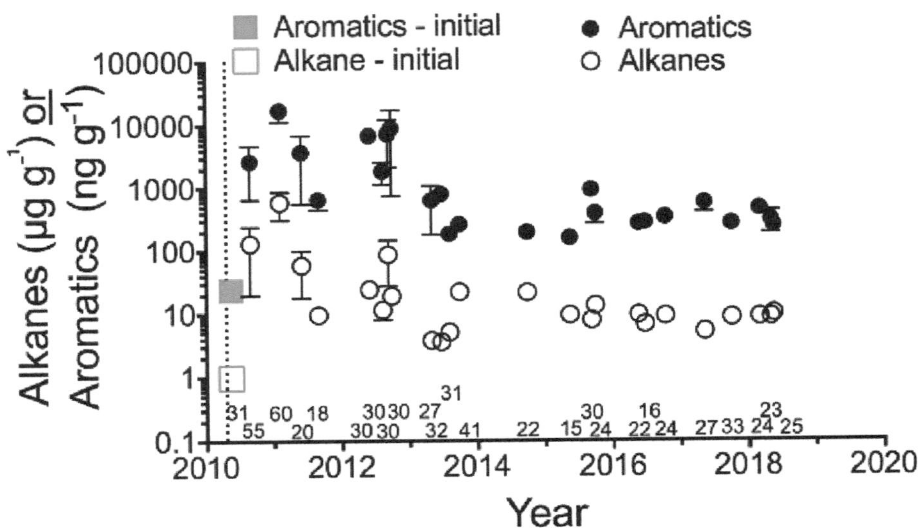

FIGURE 6.15 The concentration of total aromatics (filled symbols, ng/g ±1 SE) and total alkanes (open symbols, μg/g ±1 SE) in Louisiana salt marsh sediments for each sample from May 2010 to May 2018.
NOTES: The red squares are the samples taken before oil invaded the marsh. The vertical dotted line represents the start of the DWH oil spill offshore. The sample numbers averaged for each sampling trip are below each symbol (along the x-axis) and are the same for alkanes and aromatics. (The targeted 28 alkanes and 43 polycyclic aromatic hydrocarbons and their respective alkyl homologs [18 parent PAHs and 25 alkyl homolog groups] are provided in Tables 1 and 3 of the Supplemental Information.)
SOURCE: Turner et al. (2019), CC BY NC ND.

conditions in the spring and summer of the DWH oil spill (Rabalais et al., 2018). Appropriate historical hydrocarbon data were available from Gulf of Mexico coastal ecosystems (e.g., Wade et al., 1988). The U.S. EPA's Environmental Monitoring and Assessment Program[3] followed later by the U.S. EPA's Aquatic Resource Surveys,[4] collected some appropriate baseline data for assessing general environmental quality. Prior to the *Exxon Valdez* spill, baseline data documented regional background concentrations of hydrocarbons (Karinen et al., 1993). Additionally, the National Institute of Standards and Technology (NIST) maintains a bank of frozen marine mammal tissues collected from "natural" stranded animals throughout the United States such that comparative studies can be undertaken after an oil spill.

Baseline, and especially the term *benchmark* that was used at one time in the Outer Continental Shelf (OCS) programs, invoke interpretations involving a static situation. It is important that the concept of an evolving baseline of data and knowledge be adopted relevant to inputs, fates, and effects of oil inputs to the sea and to inform oil spill response and damage assessment activities. It is known that numerous ecosystems in coastal and continental margin waters are in a state of change as a result of changing climate; human activities, such as commercial and recreational fishing and waste disposal to coastal waters; and invasive species. These programs have evolved with time, in part in response to reviews and suggestions of the programs' Scientific Advisory Committees and also in response to reviews and recommendations set forth in reports of committees of the National Research Council (see, e.g., NRC, 1993).

Appropriate baseline data can be acquired prior to the interface of spilled oil within an environment, given enough response time and reasonable model predictions of where the oil may reach a shoreline. In anticipation of oil from the DWH spill reaching Louisiana marshes, a group of researchers collected a suite of samples from several salt marsh sites in southeastern Louisiana for typical marsh populations and hydrocarbon composition (Turner et al., 2019). Data for marsh sediment hydrocarbons provided a historical context, a measure of peak oil exposure, the peak sediment concentrations when DWH oil residue reached the marsh sediments, and a post-oiling period of reduced contamination (see Figure 6.15). The return to "pre-spill" hydrocarbon conditions, however, remained elevated for at least 8 years (i.e., a new baseline).

6.5.1.3 Inability to Achieve Appropriate Experimental Designs

An oil spill does not affect coastal and offshore waters and intertidal coastlines uniformly; there are differences in the amount of oiling (heavily oiled versus lightly oiled) and concentrations of various constituents and levels of

[3] See https://archive.epa.gov/emap/archive-emap/web/html/index.html.
[4] See https://www.epa.gov/national-aquatic-resource-surveys.

toxicity of the encroaching oil residue. Also, the duration, frequency, and mechanisms of exposure (see Section 6.3) will generate variable levels of contamination. There is also the complication of presumed exposure on communities such as phytoplankton and zooplankton near, or even within, an area of an oil spill, without adequate exposure data for the hydrocarbon residues. Just because an oil spill traversed a body of water at one time does not imply similar exposure or contamination of the plankton and nekton communities in the same or nearby locations or predicted new location (based on circulation, known or modeled). This understanding is critical for determining effects, potential effects, or no effects. The ability to reference long-term data may help alleviate some of these problems (Rabalais et al., 2018).

The lack of comparable habitats, other than oil exposure, which is not uniform, is a daunting complication for experimental design. There is seldom an adequate "control" for habitat comparisons following exposure to oiling. Placing a control site in a rocky intertidal habitat that was more quiescent with regard to waves or was less exposed compared to a high-energy rocky intertidal habitat that was heavily oiled is inappropriate (see the Environmental Sensitivity Index in Appendix G). A similar situation would exist with oiling of a salt marsh margin in a highly eroding environment subject to more frequent and higher wave characteristics compared to an opposite shoreline that is more sheltered with regard to meteorological conditions and has a lower erosion rate. These challenges can be mitigated by matching reference and oiled sites for important environmental factors, and by randomizing pairings between reference and oiled sites that have closely similar factors. In addition, sampling designs of field studies should ensure that there is adequate statistical power to detect effects that actually occur, because conclusions of no effects based on sampling designs that have low power to detect those effects may be misleading (e.g., Peterson et al., 2001).

6.5.1.4 An "Open" Ecosystem

Communities and ecosystems change substantially over time, and may not remain the same before and after exposure to an oil spill. There is also a notably high variability in physical and biological characteristics against which to define a "difference," even if the best data on exposure are available.

Motile organisms, such as marsh nekton, are difficult to monitor relative to effects of exposure to oil in high-exposure field conditions that lead to an observed decline in densities or change in the composition of the nekton community (Fodrie et al., 2014). They can move among marsh microhabitats such as marsh tidal creeks, marsh surface, or marsh ponds, depending on inundation levels, and their marsh microhabitats may not be related to the presence of oil or its residues. Other dynamics that might hide negative population impacts are high spatiotemporal variability, behavioral avoidance, compensatory pathways, and temporal lags. However, "positive" density responses can also be a warning of disruption if changes in population age structure releases juveniles from competition with or predation from older and larger individuals and species, such as appeared to have occurred with Gulf menhaden following the DWH oil spill (Short et al., 2017). In addition, results from subtle, long-term effects to ecosystem damage caused by loss of habitat and elevated oil contamination in sediments for decades may not yet be recognizable (Zengel et al., 2021).

Less mobile organisms, such as attached epifauna or benthic infauna, truly are exposed to hydrocarbon constituents of known or unknown toxicity. This is why they are often considered the "canary in the coal mine," because they cannot escape the exposure. Benthic infaunal communities are notoriously variable in their composition and relative abundance and require numerous time-consuming replications to detect statistical differences. In other instances, visual analysis of photographs or videos can capture these differences (see Box 6.6).

6.5.1.5 Before-and-After Controlled Experiments Versus Inferential Observations

A clear-cut experimental design that defines the effects of an oil spill on organisms or habitat based on conditions before and after the exposure, or in a control-versus-exposure comparison, is preferred over less definitive results or inferential observations. The latter are more likely than the former in the case of an oil spill under field conditions. Many efforts to detect environmental damage do not provide results with statistical significance to reject a null hypothesis of no difference—even when it is false—nor can these studies make the connection between observed decreases in densities following anthropogenic disturbance and a specific event.

The difficulty in identifying effects on higher organization of biological structure—populations to ecosystems—continues. Longer-term funded studies of populations, communities, and ecosystems that combine ecosystem-level processes and interactions over years to decades are necessary to better understand those relationships.

6.5.2 Air-Sea Interface

Some of the most immediate and serious effects of oil discharges occur when oil spreads across the sea surface as a sheen or oil slick. Seabirds are frequently the most vulnerable, as attested by the photographs of oiled birds that often accompany media reports of even small oil spills. Other vulnerable groups include marine mammals, sea turtles, fish that gulp air at the sea surface to inflate or maintain pressure in their swim bladders, and floating marine vegetation. Buoyant eggs released by some species of fish and other marine organisms that spawn at sea may rise to the surface, bringing them into contact with surface oil slicks when present.

BOX 6.6
Mearns Rock Update

Variability Matters When Determining Ecological Recovery from Oil Spills

How long does it take for an ecosystem to recover from an oil spill? What is full recovery and how is it determined? Ecological recovery following major perturbations such as large oil spills occurs within the context of natural variability, often highly dynamic, in affected populations. Evaluating this variability requires long-term monitoring studies that apply consistent methods on time scales of several years to several decades, which are rare. However, a long-term monitoring study at Mearns Rock, conducted after the 1989 *Exxon Valdez* spill (Mearns et al., 2017), provides a striking example of just how variable intertidal communities can be and how challenging answering the question of recovery is.

Photographic documentation of Mearns Rock, a boulder (see Figure 6.16) in an oiled but subsequently undisturbed location within

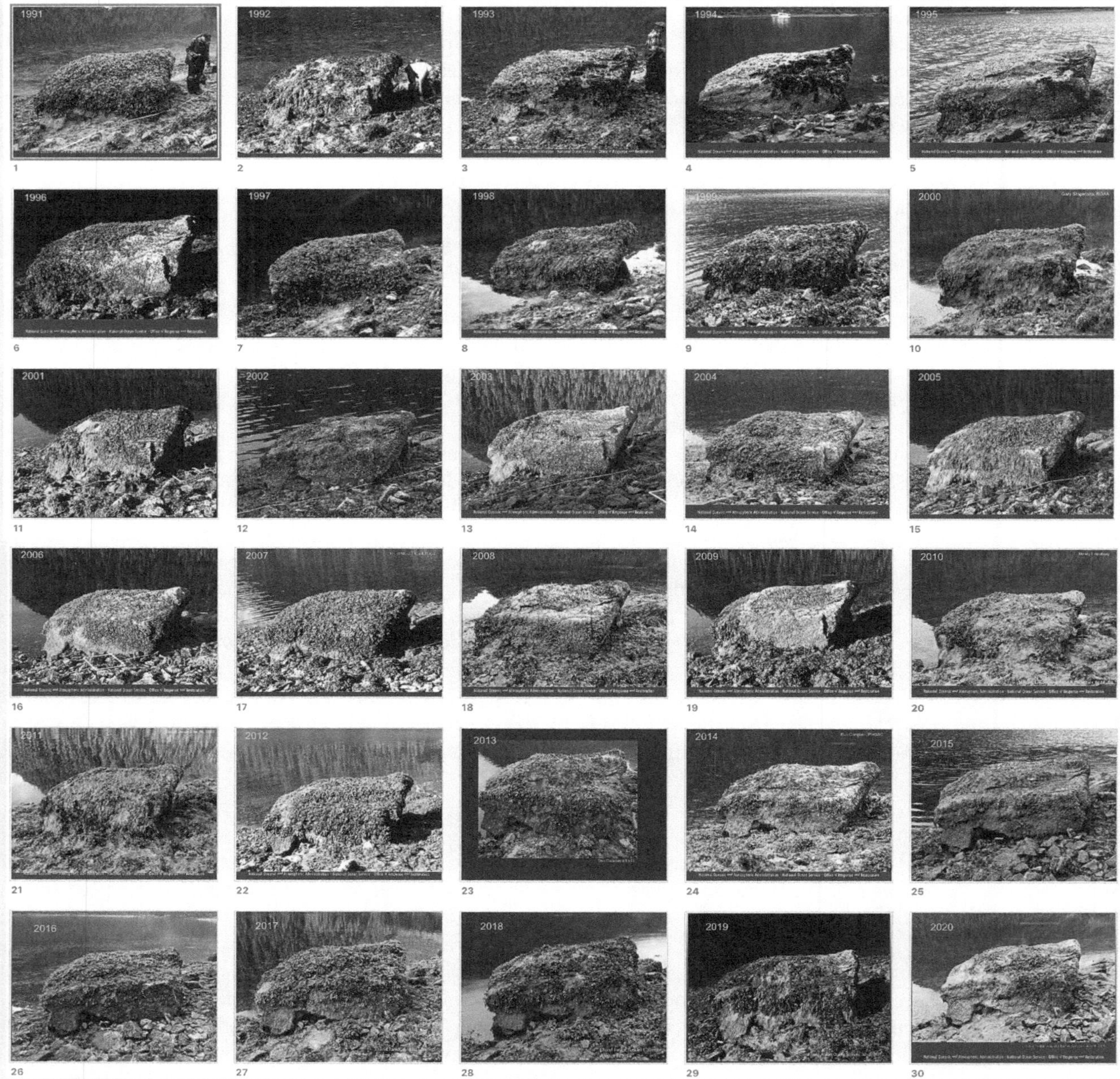

FIGURE 6.16 Photographs of Mearns Rock from 1991 to 2020.
SOURCE: Courtesy of Alan Mearns.

Snug Harbor in Prince William Sound, Alaska, during summer allowed semi-quantitative estimation of the proportion of the rock covered by rockweed (*Fucus* sp.), mussels (*Mytilus trossulus*), and barnacles (*Balanus* sp.) from 1989 through 2020 (NOAA Office of Response and Restoration.[a] Results show major 4- to 7-year fluctuations in rockweed and mussel coverages (see Figure 6.17). There were multiple-year periods when these organisms were rare, followed by one or more years of heavy cover followed again by declines in coverage. There have been five such "boom-and-bust" episodes for rockweed during the past 30 years, and three for mussels. These time-series photographs involved volunteer scientists and citizens and provide visual records of change, easily understood by the public and decision makers alike.

This dynamism presents serious challenges for determining when an area has "recovered." Due to interannual variability, impacts from an oil spill depend on when the spill occurs relative to their "stage" in their boom-and-bust cycles. Population variability of many species is considerably less than that typical of rockweed and mussels. Also, ecosystems respond to long-term secular change in response to climate cycles and global warming. Various standards of recovery have been proposed, including return to baseline conditions, population changes that are similar in nonaffect and oil-affected areas, and populations that return to numbers within the range of historical variability, among others. However, none of these has found widespread acceptance. In reality, some oil typically lingers on decadal time scales after major oil spills, and the differences between the state of an ecosystem following a spill are usually difficult—if not impossible—to distinguish from what the state would have been had the spill not occurred.

[a] See https://response.restoration.noaa.gov/oil-and-chemical-spills/significant-incidents/exxon-valdez-oil-spill/mearns-rock-long-term-study-ecological-recovery.html (revised November 9, 2020.)

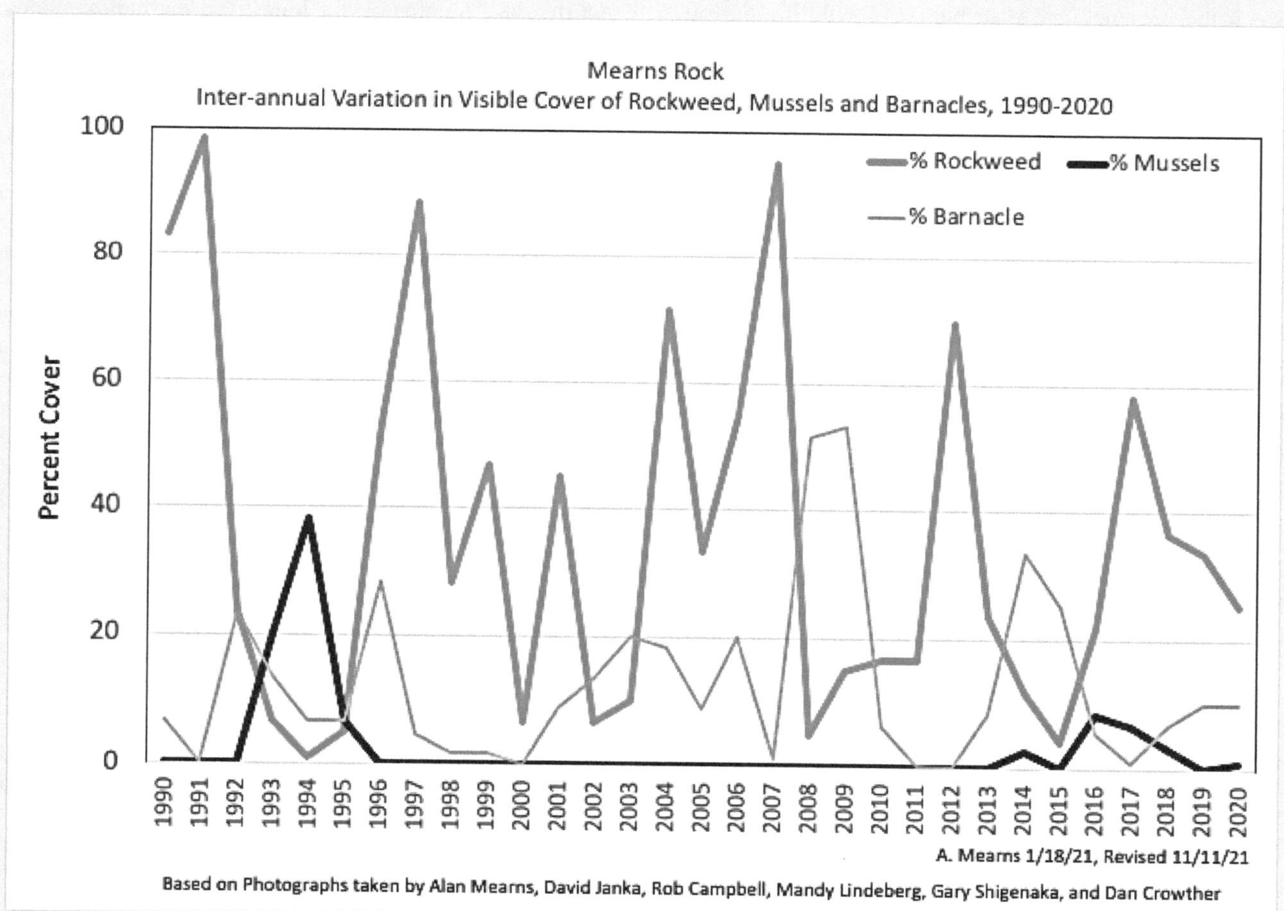

FIGURE 6.17 Inter-annual variation in cover of rockweed, mussels and barnacles on Mearns Rock from 1990 to 2020.
SOURCE: Personal communication, updated by A. Mearns, 2021.

Contact with oil released in large oil spills can cause mass mortalities of seabirds, marine mammals and sea turtles. The *Exxon Valdez* incident is a stark example of this: up to 375,000 seabirds and 5,500 sea otters may have died from acute exposure to oil (Bodkin and Udevitz, 1994; Ford et al., 1996). More recently, sea turtles were regularly observed trapped in the oily "mousse" entrained in the windrows of the Gulf of Mexico during the DWH oil spill (Stacy, 2012), and more than 400,000 surface-pelagic juveniles were estimated to have been oiled in that spill (McDonald et al., 2017). NRDA studies following the DWH oil spill estimated that up to 7,600 large juvenile/adult sea turtles and 166,000 small juvenile sea turtles were killed (see Box 6.7), 1,141 dolphins died from March 2010 through July 2014, and hundreds of thousands of birds across 93 species were killed by the spill (Haney et al., 2014a,b; DWH NRDA Injury Assessment, 2015). Although during an oil spill response efforts are made to capture, rehabilitate and document the overlay between the presence and distribution of wildlife and oil, direct observations of all potentially exposed and impacted organisms are limited both spatially and temporally. To address this limitation following the DWH incident, DWH trustees employed a number of different techniques to best estimate overall impact to the environment—including novel model-based approaches (DWH Natural Resource Damage Assessment Trustees, 2016). These injury determination efforts extended for many affected taxa, with the mission to

> quantify exposure and injuries to resources where direct measurement was infeasible given the scope of the incident. For example, numerical modeling was employed to quantify injuries to nearshore resources based on exposure to toxic concentrations of oil, and to quantify injuries to marsh fauna such as flounder and shrimp.

For higher vertebrates (e.g., birds, marine mammals, and sea turtles), very different approaches were employed for each. In birds, a combination of observational data, laboratory models, field studies, and three models (Shoreline Deposition, Offshore Exposure, and Live Oiled Bird models) were blended to get the best estimate possible for the "impact" in the various species in the Gulf. For marine mammals, historical distribution data were combined with stranding records and an extensive case-control study for dolphins in Barataria Bay to establish estimates of injury (see Section 6.3.3 for detailed findings). Due to the extent of these assessments, specific details of each cannot be adequately covered in a review such as this, and the reader is directed to the full DWH Final Programmatic Restoration Plan for details. However, as an example of the detailed modeling efforts that can be undertaken as a consequence of SONS-level incidents (and how those then inform scientific knowledge), Box 6.7 details the modeling accomplished for sea turtles, a taxa of particular concern in this region due to difficulty ascertaining observable injury coupled with their conservation value, especially in species like turtles (Shigenaka and Milton, 2003).

Although such large-scale petroleum releases can be catastrophic, it should also be noted that chronic, comparatively low-level releases into the marine environment can take a significant toll on wildlife populations. Wiese and Robertson (2004) estimated that, between 1998 and 2000, the illegal discharges of oil from ships caused an average of 315,000 ±65,000 alcid deaths annually in southeastern Newfoundland. Similarly, the S.S. *Jacob Luckenbach* (see Box 3.5), which sank off of San Francisco, California, in 1953 with 457,000 gallons of bunker fuel, was found to be leaking sporadically over the years and to have killed 51,569 birds and 8 sea otters between 1990 and 2003 (Luckenbach Trustee Council, 2006).

Impaired ability of seabirds, marine mammals, and sea turtles to find and capture food or to avoid predators resulting from sub-lethal effects of exposure to oil (see Section 6.3) may result in delayed mortalities, and these may lead to underestimates of population losses. Afflicted individuals may not die until months or, for long-lived marine mammals or sea turtles, years following oil exposure. These individuals may move considerable distances between exposure and death, so the place where they die may be far removed from the oil-contaminated region, which increases the likelihood that the death will not be attributed to oiling. More generally, attributing the cause of death in oil-exposed seabirds, marine mammals, and sea turtles is complicated by the possibility of several mechanisms of toxicity contributing to eventual mortality. These complications may also apply to fish such as herring and other physostomous fish that may ingest slick oil while gulping air at the sea surface (e.g., Price and Mager, 2020). **Widespread mortalities of seabirds, marine mammals and sea turtles may have population- and community-level effects (see Section 6.7); understanding of these effects could be improved by conducting studies focused on better estimation of these mortalities during future spills of opportunity.**

The pelagic brown alga *Sargassum* spp. forms biogenically structured habitat for high biodiversity (including endangered and threatened species) and productivity in surface waters of the western Atlantic Ocean, the Caribbean Sea, and the northern Gulf of Mexico. A large section of the Gulf of Mexico's floating (*S. natans* and *S. fluitans*, considered a single complex) was exposed to the immense pool of oil from the DWH oil spill. Theses oiled *Sargassum* mats were treated with aerially applied Corexit 9500 A dispersants. Aerial surveys each over the same 3,100 km^2 of ocean surface from the panhandle of Florida to the Chandeleur Islands, Louisiana, were completed in 2010; these documented extensive co-occurrence of oil/dispersant and *Sargassum* (Powers et al., 2013). The surveys documented that *Sargassum* abundance was less during the oil spill in 2010 compared to a 4-fold increase in abundance of *Sargassum* in 2011 and 2012. A delta-lognormal approach for aerial data from fishery-independent data was applied to determine abundance (Powers et al., 2013) from the aerial survey data, which take into account non-zero observations with positive sightings (Lo et al., 1992; Maunder and Punt, 2004).

BOX 6.7
Approaches Used to Estimate Injury to Sea Turtles Following the DWH Incident

Although active shore-based response and recovery of oiled turtles following the DWH incident began immediately after the accident, only visibly unoiled animals were collected in the weeks following the initial release. Once it became clear, a few weeks after the oil spill, that juveniles living in the *Sargassum* spp. were at greatest risk (and were not coming ashore but remaining entrained in the oiled substrate), dedicated boat-based activities (coupled with aerial reconnaissance to direct boats onto areas of greatest risk to turtles) versus passive beach-based recoveries were quickly organized and deployed. Activities continued until September 21, 2010 (before the wellhead was capped; Stacy 2012), focusing on direct observations and assessment of oiling extent of surface-pelagic juvenile sea turtles (see Figure 6.18).

Estimating injury to sea turtles following oil spills faces many challenges, including the impossible task of surveying all affected areas in space and time; unrecoverable (i.e., sinking) dead organisms; and, for some species, limited knowledge on the toxicological effects of oil (see Mitchelmore et al., 2017). Following the DWH incident as part of the NRDA injury estimation for turtles, a modeling approach was developed to estimate the probability that all sea turtles in the area would encounter oil (DHW Trustees, 2016b; Wallace et al., 2017; see Figure 6.18), especially heavy surface oil, for which it was determined a turtle would not survive contact (Stacy, 2012). This model utilized the data obtained from the rescue operations described previously combined with aerial surveys demonstrating locations (although not oiling status) of neritic juvenile and adult turtles. This approach spatially and temporally expanded the exposure and injury quantifications available (see Figure 6.19).

As turtles were recovered during the DWH incident, they provided examples of impacts relating to degree of observed external and estimated internal oiling (Stacy, 2012; McDonald et al., 2017). This enabled modification of the heavy-oiled model to provide estimations of the number of turtles likely to have encountered some level of heavy oil exposure, and to refine estimates of mortality of surface-pelagic turtles in the region (DWH Trustees, 2016b; McDonald et al., 2017). Besides physical oiling impacts, effects on turtles may be a result of numerous toxicological mechanisms of action resulting from one or numerous exposure routes of oil and its constituents (i.e., inhalation, direct or indirect ingestion, membrane coating), although data are limited (Shigenaka and Milton, 2003; Camacho et al., 2013; Mitchelmore et al., 2015).

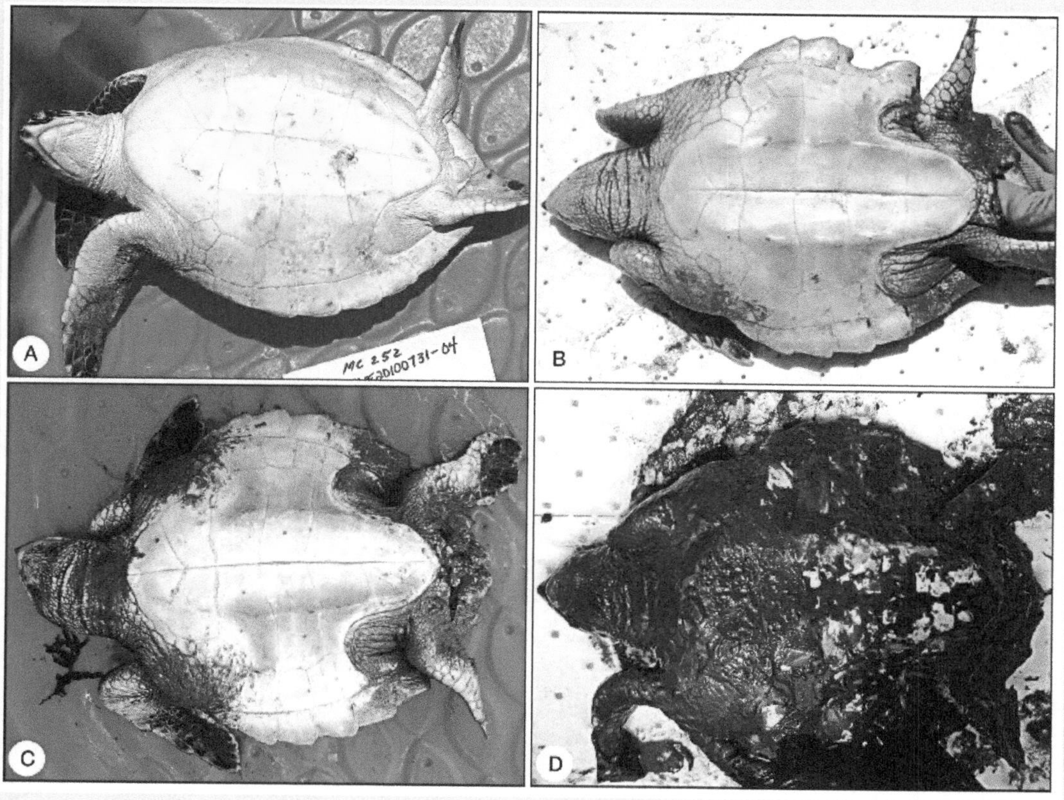

FIGURE 6.18 Photos depicting the oiling categories assigned to captured turtles during the DWH incident; (A) minimally oiled, (B) lightly oiled, (C) moderately oiled, (D) heavily oiled.
SOURCE: Stacy (2012).

To provide further estimates of injury to turtles from the DWH oil spill, a toxicological evaluation was conducted. This approach first estimated the potential exposure of turtles to oil using measured exposure metrics (i.e., degree of external/internal oiling [Stacy, 2012] and bile PAH metabolites [Ylitalo et al., 2017]) and models of exposure (Wallace et al., 2017). Both measured and modeled sources were used to calculate a turtle's oil exposure, which combined with mortality estimates based on oil/PAH toxicity data in reptiles and other species, provided an overall estimation of injury for turtles following the DWH incident in all oiling categories (except heavily oiled, as these were deemed to have 100% mortality) (Mitchelmore et al., 2017). These combined approaches estimated that an overall mortality of 30% would occur for all oceanic turtles within the footprint of the DWH oil spill in addition to those that already would have died from heavy oiling (DHW Trustees, 2015; McDonald et al., 2017; Mitchelmore et al., 2017).

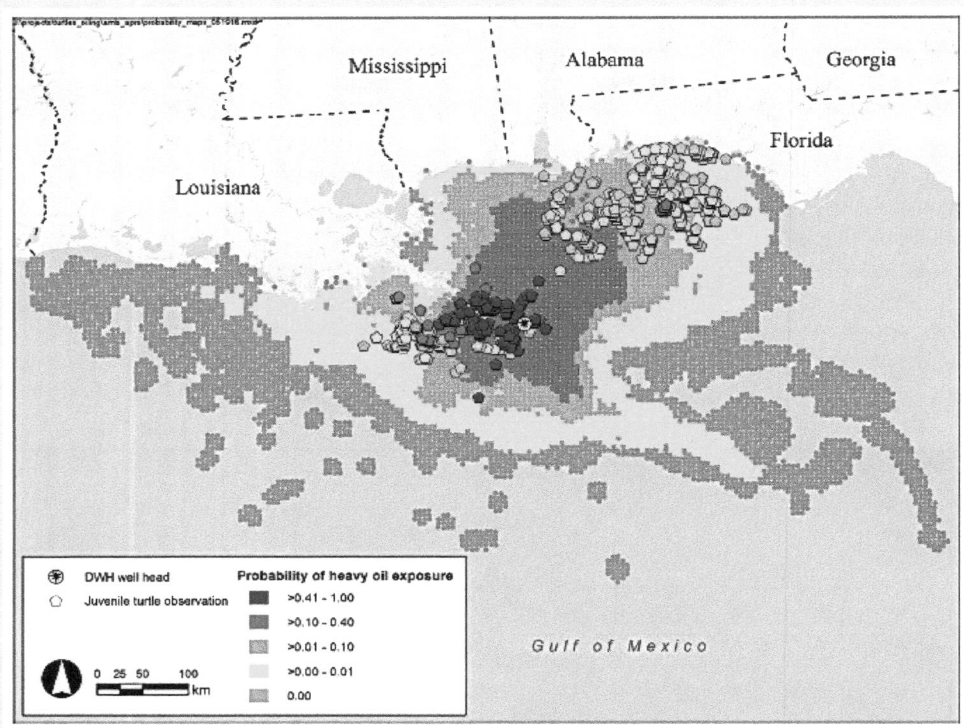

FIGURE 6.19 Modeled likelihoods of heavy oil exposure across the DWH oil spill area and period.
NOTE: Colors in pentagons specify the probability of heavy oiling calculated for each turtle collected or sighted during boat-based rescue operations.
SOURCE: Wallace et al. (2017), CC-BY Open Access article.

Buoyant eggs of fish and other marine organisms may directly contact surface oil slicks, promoting uptake of PACs dissolved in seawater immediately beneath the slick, or directly from the surface slick oil. Because fish embryos are extremely vulnerable to adverse cardiotoxic effects following exposure to dissolved PAC concentrations of less than 1 µg/L (see Section 6.3.4.2), even brief exposure may have lethal consequences for developing fish embryos.

Confirming evidence from field studies regarding the effects of surface oil ingested by fish or of cardiotoxic effects of PACs from surface oil on developing fish embryos is lacking; future spills of opportunity could be used to conduct such studies.

6.5.3 Shorelines

Regardless of whether an oil spill impinges on an intertidal shoreline from a surface spill or a submerged spill, these areas will show different hydrocarbon composition of the oil, depending on the weathering of the oil from its source to the shoreline. Weathering of the oil will vary with the length of trajectory times from a subsurface oil input to a surface expression, through the many fates outlined in Chapter 5 or through mitigation measures. Similarly, weathering of the oil at the surface, time to deposition at a shoreline, subsequent degradation in the exposed shoreline, and additional mitigation efforts will change the composition of the oil. Thus, there are many intertidal shorelines that may be affected by different contaminant compositions and levels of exposure.

There is an abundance of historical documentation of the effects of oil spills on coastal areas across a range of latitudes. Some garnered a high level of both fates and effects studies (e.g., West Falmouth, Massachusetts), because there were scientists with expertise, technical support, and scientific curiosity associated with the nearby Woods Hole Oceanographic Institution (see Box 6.8). Others were much larger in volume, but also located near a research institution, such as the *Galeta* oil spill near the Smithsonian Tropical Research Institution in Panama. Two were within the top five accidental marine oil spills in volume and also well funded for research: the *Exxon Valdez* tanker grounding in Prince William Sound, Alaska, in 1989; and the DWH catastrophic explosion and subsequent expulsion of oil from the uncapped wellhead in 2010.

The National Oceanic and Atmospheric Administration Office of Response and Restoration (NOAA, revised 2019) developed an Environmental Sensitivity Index (ESI) (see Section 4.2.1.3 and Appendix G). A rank of "1" indicates shorelines with the *least susceptibility to damage* by oiling. Such as steep, exposed rocky cliffs and banks. Waves and tidal action will quickly wash the oil off the rocks; the oil cannot penetrate into rocks. A rank of "10" indicates shorelines *most likely to be damaged* by oiling. These include protected, vegetated wetlands, such as mangrove swamps and salt marshes.

6.5.3.1 Salt Marshes (ESI 10)

The DWH oil spill represents the largest accidental marine oil spill in volume (McNutt et al., 2011) and as measured by shoreline length (Nixon et al., 2016). The broad scope of research, length of continued study, and funding by multiple agencies and industry (BP and other operators) generated immediate data on the effects of oil on coastal marshes, as well as longer-term observations. The National Science Foundation Rapid Response Research program and initial funding to the five Gulf states from BP placed researchers in the field to collect pre-spill salt marsh data before the oil came ashore. The Gulf of Mexico Research Initiative (GoMRI) developed and guided a 10-year, $500 million research program on the DWH oil spill (Zimmermann et al., 2021).

The oil residue-water mixture began coming ashore in mid May 2021 (Turner et al., 2019) after a 1-month period in offshore surface waters. "Fresh" oil continued to be deposited in salt marshes through September 2010, as seen in the sharp rise in total target aromatics concentrations in marsh sediments within Barataria Bay, compared to the previous three months (see Figure 6.15).

Several features of the marsh oiling from the DWH were that (1) pre-spill petroleum concentrations in sediment samples were determined and followed at the same locations for 9 years (see Figure 6.15), (2) consistent gas chromatography-mass spectrometry (GC-MS) analytical chemistry with the same total targeted aromatics and total targeted alkanes was applied to all marsh sediment samples (at least for the research conducted by the Coastal Waters Consortium, GoMRI [Turner et al., 2014]), (3) tropical storm activity resuspended and redistributed the oil, especially during Hurricane Isaac (Turner et al., 2014) (see Figure 6.20), and (4) recovery occurred at different intervals for different marsh and marsh community features, and some were not considered recovered even after 9 years (Zengel et al., 2021).

Researchers have been following the fate of No. 2 fuel oil from the grounded tanker barge *Florida* since 1969 (see Box 6.8). Study of the long-term fate of the oil, and elucidation of continuing effects, was undertaken 30 years after the spill. As with the original studies of the West Falmouth spill, these latter studies combined the latest in analytical chemistry methodology with the latest in biological/ecological studies.

6.5.3.1.1 Effects on Marsh Vegetation and Stability

Early assessments for the DWH oil spill in two southeast Louisiana salt marshes (Barataria Bay and Terrebonne Bay) in July 2010 found no signs of living vegetation, only dead stems above the exposed oiled sediments, which clearly documented the dieback of all marsh vegetation in heavily oiled marsh (Lin and Mendelssohn, 2012; Silliman et al. 2012; Zengel et al., 2015) (see Figure 6.22). Vegetation cover after 7 to 16 months continued at much lower levels than in the control sites (Silliman et al., 2012; Lin and Mendelssohn, 2012, respectively). The average live aboveground biomass combined of *Spartina alterniflora* (smooth cordgrass) and *Juncus romaeriaus* (black needle rush)—dominant species in the marsh—was significantly lower, almost none, in the heavily oiled marsh compared to reference marshes. Moderately oiled marshes fared better, with no significant difference between the combined biomass by weight of the dominant species with that in the reference marsh. However, the live aboveground biomass and stem density of *Spartina* was about 10 times as much as it was for *Juncus* in the moderately oiled marsh.

All metrics indicated initial impacts from oiling and most showed recovery time frames of several years or more for *Spartina*. *Spartina* stem density was the exception, with more rapid recovery due to possible stimulation by unoccupied space and perhaps residual oiling (inference); however, increased stem density was not leading to comparable increases in cover or biomass (Lin and Mendelssohn, 2012). In contrast to *Spartina*, *Juncus* was affected to a greater degree, with much slower or lack of recovery. In comparison to the marsh edge, the oiled marsh interior tended to have a lesser degree of impacts, at least initially. Impacts from oil were eventually detected in the interior, which saw similar recovery to the marsh edge. Complete vegetation recovery was not observed after 9 years, especially for marshes with a *Juncus* component and for belowground biomass (Zengel et al., 2021).

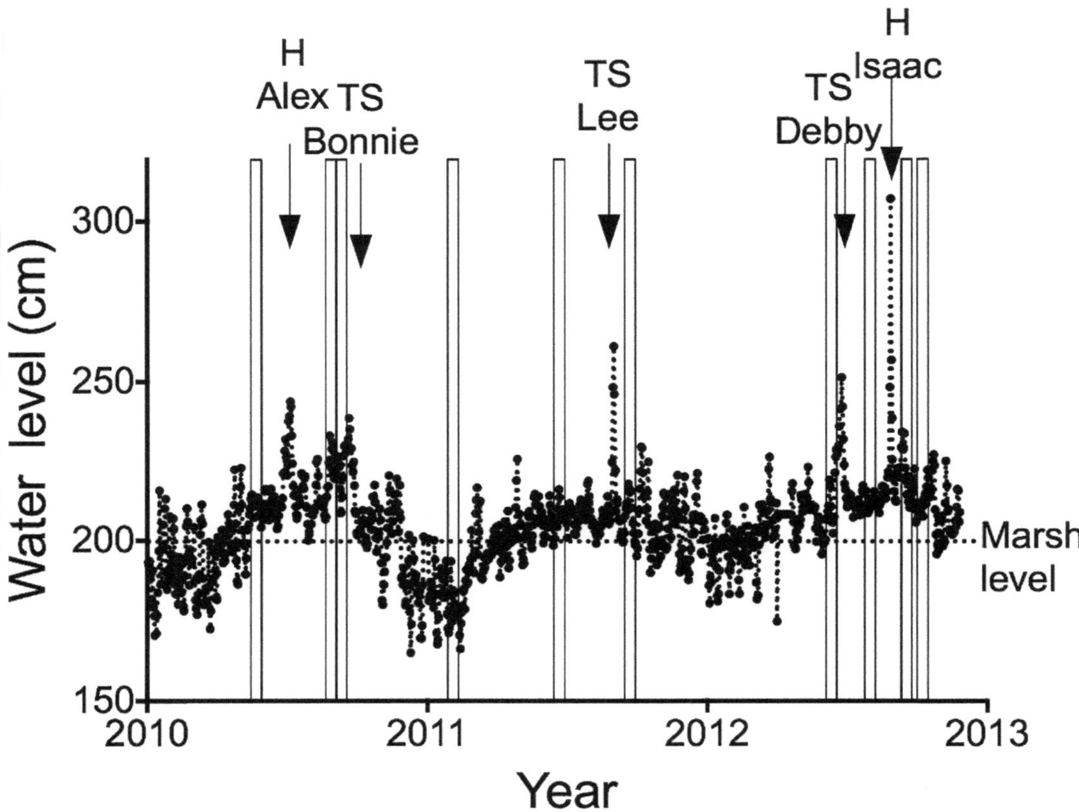

FIGURE 6.20 Marsh water level at the time of sampling trips in southeastern Louisiana salt marshes (vertical bars) and the times of tropical storms and hurricanes (Turner et al., 2014). Relatively fresh oil was redistributed into the marsh, especially during Tropical Storm Lee (at least 10 m into the marsh) and Hurricane Isaac, when oil was dispersed up to 100 m into the marsh from a 1-m shoreline mark (Turner et al., 2014).

BOX 6.8
1969 West Falmouth, Massachusetts, Oil Spill 35 Years Later

The oil barge *Florida* ran aground in a storm in Buzzards Bay, Massachusetts, and spilled an estimated 595,000 kg of No. 2 fuel oil in the nearshore coastal waters and coastal marshes and beaches of West Falmouth, Massachusetts. Max Blumer (organic geochemist), Howard Sanders (benthic ecologist), and John Teal (marsh ecologist) from the nearby Woods Hole Oceanographic Institution, and their laboratories began studies of the spill within a day or less, interacting with the Town of Falmouth Shellfish Warden, George Souza. This research is among the first, if not the first, to combine advanced analytical chemistry (e.g., GC-MS), knowledge of the biogeochemistry of organic chemicals in the marine environment, benthic biology/ecology, and marsh biology/ecology to understand the fate and effects of spilled oil in the marine environment. The initial studies continued over a 3- to 5-year period and have been noted and cited in previous *Oil in the Sea* studies (NRC, 1975, 1985, 2003); specific literature citations are in the follow-on studies noted later in this box.

A 20-year survey assessment of the status of concentrations and composition of the No. 2 fuel oil hydrocarbons in subtidal and intertidal marsh sediment in Wild Harbor River estuary, West Falmouth, Massachusetts, was reported by Teal et al. (1992). A more in-depth assessment of the recovery, long term fate of the oil, and elucidation of continuing effects was undertaken 30 years after the spill by Reddy et al. (2002), Frysinger et al. (2003), Peacock et al. (2005), Slater et al. (2005), White et al. (2005a,b) and Culbertson et al. (2007, 2008a,b). As with the original studies of this spill, these latter studies combined the latest in analytical chemistry methodology with the latest in biological/ecological studies.

A synopsis of their findings as of 2005–2006 follows. The general depth distribution in the marsh sediment and the composition of oil compounds remaining in the sediment, including the chemical composition of the PAH, had not changed significantly since 1989. This indicates that weathering is proceeding at a significantly slower rate, or is stalled (White et al., 2005). These persistent oil chemicals, and/or their degradation products, present in the sediment have continuing effects on some of the components of the salt marsh ecosystem. The ribbed mussel, *Geukensia demissa*, had reduced condition index (a measure of the health of bivalves), shorter shell lengths, and reduced filtration rates when compared to the same species from the nearby, unoiled control site—the Great Sippewissett Marsh. Mussels transplanted from that control site to the oiled Wild Harbor Marsh site showed similar reductions in filtration rates and conditions index after several months (July to October).

Fiddler crabs (*Uca pugnax*) from the oiled marsh areas only burrowed to the depth of the oiled layers in the marsh and then most burrows stopped or turned sideways (see Figure 6.21).

The fiddler crabs from the oiled marsh area attained lower population densities, lower feeding rates, and delayed escape responses compared to the control site.

There was lower stem density above and below ground for marsh grass biomass in the oiled area compared to the control. The resulting increased topographic variations can lead to loss of saltmarsh habitat.

There was a significant decrease of oil concentrations and total area where oil chemicals could be detected over the years, as previously noted. For the most heavily oiled marsh area, the intermediate (years) and long-term (decades) adverse impact on the salt marsh organisms and ecology was suspected as a strong possibility after the initial first few years' studies. These expectations and extrapolations into the future, though not proven at the time, contributed to coastal wetlands being ranked among the most sensitive to spilled oil in the Environmental Sensitivity Index rankings (see Appendix G).

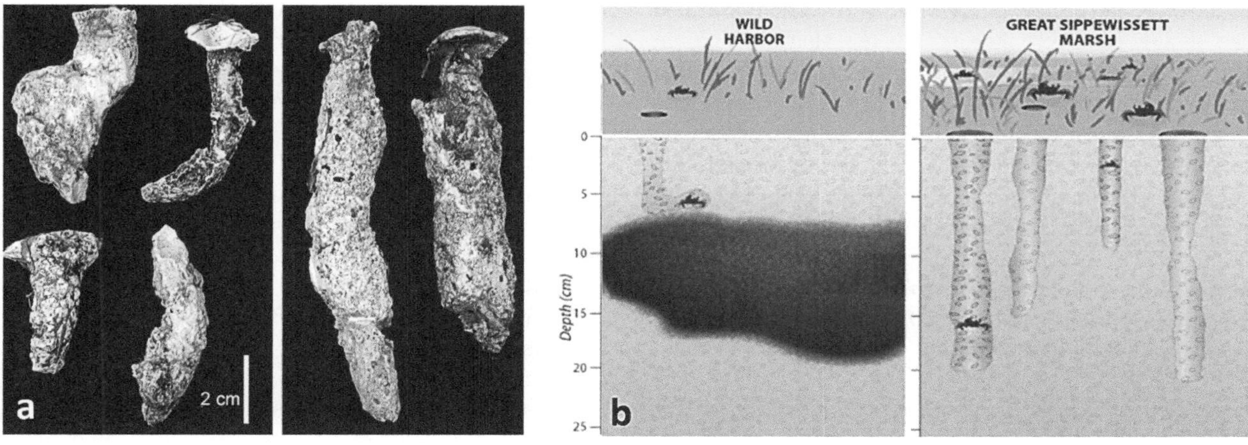

FIGURE 6.21 (a) Plaster of Paris casts of fiddler crab burrows from oiled Wild Harbor marshes approximately 35 years after oiling (left) and (right) similar casts from unoiled burrows in the Great Sippewissett Marsh; (b) Diagram of fiddler crab density and burrow depth from an oiled marsh at Wild Harbor (left) and an unoiled marsh in Great Sippewissett Marsh (right).
SOURCE: Woods Hole Oceanographic Institution, Oberlander and Kleindist.

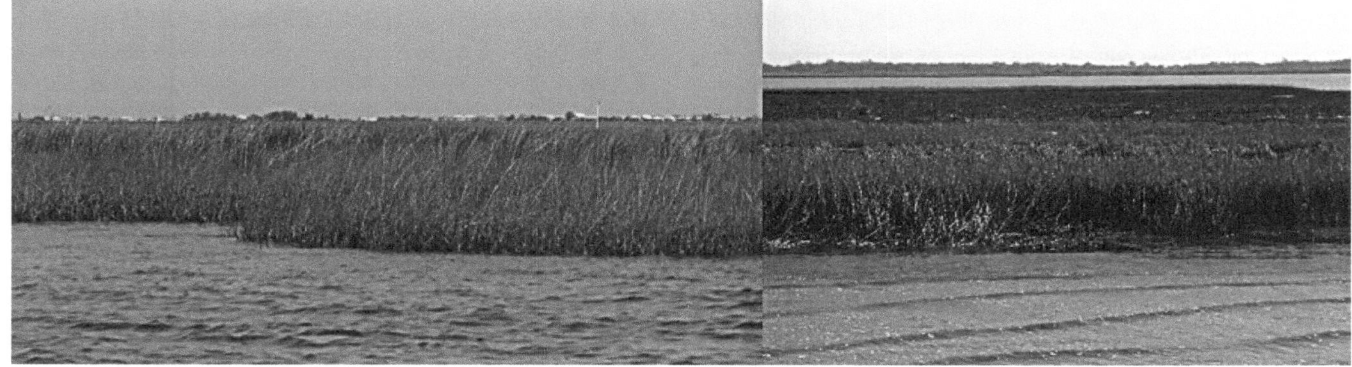

FIGURE 6.22 An unoiled marsh in southeast Louisiana at the time of the DWH oil spill (left) and an oiled marsh (right) with demonstrable die-off following oiling from the DWH oil spill.
SOURCE: Photos from Coastal Waters Consortium, Gulf of Mexico Research Initiative.

The common reed, *Phragmites australis*, is the dominant vegetation in the Mississippi River bird-foot delta and was exposed to DWH oil in 2010 (Hester and Willis, 2011; Shapiro et al., 2016). Field assessments were minimal and there was little exposure data other than qualitative categorization by shoreline cleanup assessment technique (SCAT). Shapiro et al. (2016) examined AVIRIS data for *Phragmites* marsh for 2010, prior to the oil spill and in 2011 post-spill. They documented minimal change in percent coverage of photosynthetic vegetation (an increase of 2.6% from 2010). In contrast, Zhu et al. (2013) indicated that there were detrimental effects to reed communities related to increasing hydrocarbon concentrations in an area of high oil production in the Yellow River delta. In greenhouse mesocosm experiments with application of weathered and emulsified Macondo oil (DWH) to *Phragmites* aboveground shoots, Judy et al. (2014) demonstrated that across a range of oil exposures and repeated shoot oiling, there was no major impact on overall plant growth. A factor in the lack of response is that the plants developed side branches that compensated for any vegetative stress. When the oil treatments were applied to the soils in the mesocosms, the total belowground biomass was highest in the controls and decreased with increasing oil dosage. However, there were no oil effects on dead root biomass, live rhizome, or live above- and belowground total biomass. At least for marsh vegetation in southeastern Louisiana exposed to DWH oil, there is a higher to lower sensitivity—from *Phragmites* to *Juncus* to *Spartina*.

6.5.3.1.2 Shoreline Erosion

Southeastern Louisiana is a location for high conversion rates of coastal land to open water, with the average land loss rate of 42.9 $km^2\ y^{-1}$ from 1985 to 2010 (Couvillion et al., 2011), although higher in prior decades. Marsh shoreline erosion is a significant factor in "recovery" following an oil spill because it results in emergent vegetation seldom being reestablished after oiling (Silliman et al., 2012; McClenachan et al., 2013; Zengel et al., 2015; Rangoonwala et al., 2016; Turner et al., 2016).

McClenachan et al. (2013) identified 30 sites along a north shoreline in Bay Batiste of the northern Barataria Bay, Louisiana, estuary that ranged from "low" oil (<200 $\mu g\ kg^{-1}$ PAHs, most without the Macondo oil chemical signature) to "high" oil (>20,000 $\mu g\ kg^{-1}$, all with the Macondo oil chemical signature). The location of these sites along a similar shoreline negated the differential effects of wind fetch and erosion potential that might affect erosion rates. Measurements of shoreline erosion, soil strength, percent vegetation cover, sediment PAH concentrations, and marsh overhang (distance from marsh edge under which the sediment has sloughed off into the adjacent water) taken in November 2010 to assess shoreline health. High oil sites showed significantly higher marsh overhang than observed at the low oil sites, with the exception of one location. The upper 50 cm of both high and low oil sights showed similar oil shear strength, but the shear strength in the high oil sites below 60 cm was much lower than the low oil sites. Soil shear strength demonstrates a decomposition of root structure in anaerobic, oiled sediments that would leave the marsh structure less strong and more likely to be eroded by wave energy (Turner et al., 2016). Although the marsh platform appeared healthy, it was falling apart below ground. During four of the five sampling time frames the percent of *Sporobolus alterniflora* vegetation cover showed no significant differences. After this period, the promontories at the low oil sites, produced by erosion of neighboring oiled sites, began to erode because of exposure to wave action. Moreover, the erosion rates attributed to DWH oil averaged greater than the long-term average for the area.

Turner et al. (2016) focused on the erosion rates at distinct oiled and unoiled islands in three estuaries of southeast Louisiana. In the first 6 months after oiling, erosion at oiled islands increased by 275% and for the first 2.5 years after oiling, were 200% of that of the unoiled islands. These erosion rates were 12 times as high as the average land loss in the deltaic plain of 0.4% per year from 1988 to 2011 (Turner et al., 2016).

Assessments of oil spill damage to marshes typically include percent cover by plants, species of vegetation, aboveground live biomass of vegetation by species, canopy height, and sometimes below ground living/dead vegetation biomass. This suite of measurements, along with hydrocarbon concentrations, may document damage and recovery of the dominant vegetation of marshes that were impaired by an oil spill (e.g., Lin and Mendelssohn, 2012; Silliman et al., 2012; Zengel et al., 2015). **These measurements, however, may not reveal the overall health of the marsh.** Live belowground biomass is an indicator of the structural support for the aboveground marsh biomass (Turner et al., 2011, 2016) that prevents the breakup of the marsh where erosion rates can be high in anaerobic sediments, due to waterlogging or an oil spill that results in marsh loss. Determining belowground live biomass is a tedious, time-consuming exercise. An alternative method of measuring the strength of the living belowground root structure is shear vane resistance in a vertical soil profile. In a study of oiled marsh shoreline erosion, Turner et al. (2016) demonstrated that while the soil strength within the upper layer of the marsh soil down to 50 cm was similar in both highly oiled and more lightly oiled salt marshes, the soil below 60 cm was less strong, indicating that the shoreline was susceptible to erosion—and it eventually did fall apart. **The protocols of natural resource damage assessment for marshes following exposure to an oil spill should incorporate additional measures of the "health" of marshes, including their structural integrity. Research into other measures or development of technological advances (e.g., portable photosynthesis systems for gas exchange and chlorophyll fluorescence measurements in plants) may also generate more universal representative indicators.**

Marsh-Dwelling Invertebrates

The obvious marsh inhabitants, such as fiddler crabs, marsh periwinkles, mussels, oysters, blue crabs, shrimp, and small fish (nekton), are often the foci of oil spill impacts. The same was true for the DWH, but the expanded research funding provided many opportunities to examine aspects outside of natural resource assessments. Fiddler crabs are easy to enumerate if one accepts the assumptions that (1) the number of burrows within a known area approximates their abundance, and (2) the width of their burrow approximates their size. Burrow densities were reduced by 39% in oiled sites, with effects of oiling and only partial recovery observed over 2010–2014 (Zengel et al., 2016). A return to the "reference" abundances was not complete by 2014. However, burrow diameters (~crab size) recovered by 2012, after being reduced from 2010–2011. Following oil deposition on the marsh surface, the proportion of *Uca spinicarpa* surpassed that of *Uca longisignalis* because of increase in largely unvegetated areas with a residual oil crust over the sediment surface (Deis et al., 2017). A reduction in burrow size initially and then an increase may be related to the altered species composition, with *U. longisignalis* being larger in carapace width and size of male claw compared to *U. spinicarpa* (Crane, 1975). Zengel et al. (2016) proposed that a return to species composition would likely follow the revegetation of *Spartina*.

The common salt marsh gastropod is the periwinkle (*Littoraria irrorata*), which feeds on benthic microalgae and algae on the stems of *Spartina*, both microhabitats subject to oiling. Post-spill surveys indicated significant losses of periwinkles in oiled habitat and a continuing slow recovery in both their abundance and size distribution attributed to habitat recovery (Zengel et al., 2016b). With longer-term data, neither density nor population size structure of periwinkles recovered at heavily oiled sites after 9 years, where snails were smaller and more variable in size structure. Likely linked to the lower total aboveground live plant biomass and stem density remaining over time in heavily oiled marshes. Periwinkle population rebound in moderately and heavily oiled sites may take one to two decades after the oil spill, respectively (Deis et al., 2020).

Benthic infauna (small animals living in the marsh sediments) are often used as indicators of pollution (in this case, an oil spill), because they cannot move away. In heavily oiled areas, total petroleum hydrocarbon (TPH) concentrations ranged from 50 mg TPH per gram sediment to 500 mg TPH per gram sediment compared to reference marsh levels of ~0.3 mg TPH g^{-1} sediment. These levels caused severe damage among meiofauna (animals > 63 μm but < 0.3 to 0.5 mm) similar to that of *Spartina* in heavily oiled areas. Over time, TPH degraded and Meiofauna began to recover following similar time courses of *Spartina* recovery, with considerable recovery of many organisms within 36 months of the spill. But, certain organisms such as, polychaetes, ostracods, and kinorhynchs, had not recovered to background levels in reference marshes 48 months post spill (Fleeger et al., 2015).

After 6.5 years, one community of 12 abundant taxa of meiofauna and juvenile macroinfauna had not fully recovered despite beginning to rebound from oiling in less than 2 years. The rate and speed of recovery of nematodes, copepods, most polychaetes, tanaids, juvenile bivalves, and amphipods were significantly positively linked to the recovery of *Spartina* and benthic microalgae (phytoplankton living on the sediment surface) (Fleeger et al., 2020). However, over time TPH concentrations remained high similar to the aromatics and alkanes, and live belowground plant biomass, bulk density, dead aboveground plant biomass, and live aboveground biomass of *Juncus* were not resilient. These conditions suppressed recovery of a kinorhynch, a polychaete, ostracods, and juvenile gastropods (Fleeger et al., 2020).

Overall abundances of the terrestrial arthropod insect community in oiled and unoiled *Spartina* marshes exposed to DWH oil were diminished by 50% at oiled sites in 2010, but by 2011 had largely recovered. Additionally, subguilds of predators, sucking herbivores, stem-boring herbivores, parasitoids, and detritivores all appeared to be suppressed at oiled sites by 25% to 50% in 2010 and recovered by 2011 (McCall and Pennings, 2012).

Studies of the greenhead horse fly (a top predator insect in marsh food webs of south Louisiana) were done, with biweekly monitoring in oiled and un-oiled areas from June 2010 through October 2011 (Husseneder et al., 2016). The population of horse flies crashed in oiled areas in 2010. The genetic makeup of six of seven oiled populations compared to six "pristine" sites indicated 10 polymorphic loci that identified genetic bottlenecks caused by fewer breeding parents, reduced effective population size, less family clusters and fewer migrants amid communities. The beauty of the experimental design was that it ranged from genetics to population levels on a keystone insect species with consistent oiled and unoiled results. Follow-up studies 4–5 years later in 2015 and 2016 by the same researchers (Husseneder et al., 2018) demonstrated signs of recovery of populations of the greenhead horse fly in formerly oiled areas, and previously detected genetic bottlenecks in oiled populations no longer exist. Husseneder et al. (2018) postulated that the greenhead horse fly larvae and adults followed, in succession, the regrowth of *Spartina* and recovery of 90% of the meiofauna, as documented by Fleeger et al. (2015).

Marsh-Dwelling Terrestrial Vertebrates

A dominant terrestrial vertebrate in Louisiana coastal salt marshes is the resident seaside sparrow (*Ammospiza maritima*). Those living in oiled marshes were potentially exposed to DWH oil directly, through inhalation and ingestion during preening and feeding while potentially experiencing reduced prey, especially insects (Pennings and McCall, 2014). Sparrows in areas with higher PAHs in sediments had elevated CYP1A gene expression, indicating metabolism of PAHs in oil. Carbon isotopic evidence further indicated DWH oil incorporated into seaside sparrow tissues as lower levels of ^{14}C were found in feathers and crop contents in sparrows in oiled versus unoiled sites (Bonisoli-Alquati et al., 2020). Comparing the transcriptomic response in the livers of seaside sparrows exposed to DWH oil with birds from a control site found 295 differentially expressed genes. These genes are critical in basic physiological attributes, such as energy homeostasis, including carbohydrate metabolism and gluconeogenesis, and the biosynthesis, transport and metabolism of lipids.

FIGURE 6.23 Seaside sparrow hatchlings in a nest within a salt marsh in southeastern Louisiana.
SOURCE: Phillip C Stouffer, Louisiana State University.

Furthermore, these genetic analyses offer molecular explanations for the long-standing observation of hepatic hypertrophy and altered lipid biosynthesis and transport in birds exposed to crude oil (Bonisoli-Alquati et al., 2020). Further research on visibly oiled seaside sparrows and those where UV light was used to determine oiling indicated that small amounts of external exposure to oil were associated with hemolytic anemia (Fallon et al., 2018, 2020). The multiple pathways for DNA damage and cell death support the observations of reproductive failures in seaside sparrows. Data from 2012 and 2013 (2 and 3 years after marsh oiling) of seaside sparrow nesting indicated that nests on unoiled sites were considerably more likely to fledge than those on oiled sites (Bergeon Burns et al., 2014).

The nest failure results were re-examined with further studies of the seaside sparrow in the same salt marshes through 2017 (Hart et al., 2021). The authors suggested that oil exposure may have initially reduced nesting success and that with further degradation of the oil nesting success would increase. Multiple factors affecting the salt marsh ecosystem were indicated in the follow-on research, including (1) redistribution of DWH oil during Hurricane Isaac in 2012 and continued oil degradation, (2) loss of salt marsh habitat across southeastern Louisiana, (3) loss of insect prey, (4) population dynamics of predatory marsh rice rats, (5) nest placement in microhabitats of high marsh grass stem density at nest height, and (6) location of some nest sites in areas of greater exposure to higher wind and wave energy and subsequent higher water levels. Hart et al. (2021) found that overall (2012–2017) nest survival was low (24%) and the majority of nests (76%) failed due to depredation (see Figure 6.23). There was no definitive effect of initial oiling or oiled sediment, or estimated predator abundance for years with those data. Nest success was greater in areas of less dense marsh grass, a factor that may have reduced depredation. More dense marsh grass could possibly provide small mammals with refugia or easier climbing access to nests. Predators identified (via deployed video cameras) were primarily the marsh rice rat and the American mink, and a sequence of the squareback marsh crab followed by a rice rat.

Although the effects of oiling on individuals was clear, the proposal that these would translate to population-level effects on reproductive success (Bergeon Burns et al., 2014) was not supported by the additional research (Hart et al., 2021). However, there was a decidedly better ecosystem-level understanding of interacting factors.

The longer-term study with multiple integrated features of the salt marsh ecosystem points to the need for these types of studies to integrate the multiple aspects of contamination, salt marsh ecology, trophic structure, predator–prey dynamics, understanding of microhabitats within the marsh, and an integrated approach ranging from genetics and enzymatic responses to ecosystem-level effects.

Marsh Nekton

Marsh killifish, or minnows, are ubiquitous across the temperate salt marsh platform, in tidal creeks, marsh ponds, and adjacent waterways across 1,000 km and are considered resident (movements of tagged fish within 100 m) (Fodrie et al., 2014). As expected, early in the arrival of DWH oil to salt marshes in Louisiana, the in situ observations by Whitehead et al. (2012) of killifish in lower Barataria Bay, Louisiana, near marshes that had been heavily oiled had elevated CYP1A gene expression, indicating a response to PAHs in their livers; this was not so for killifish in Mississippi and Alabama. Experimental results ranged across populations present in Gulf of Mexico marshes before oil landfall (May 1–9, 2010), during the peak of oil landfall (June 28–30, 2010), and after much of the surface oil was no longer apparent two months later (August 30–September 1, 2010). Whitehead et al. (2012) found that controlled laboratory exposures of developing killifish to water collected from heavily oiled salt marshes in lower Barataria Bay on June 28 and August 30, 2010, prompted CYP1A protein expression in larval fish. This response matched the location and timing of oil contamination, and showed that the remaining oil components dissolved at very low concentrations in lower Barataria Bay after landfall and were bioavailable and bioactive to developing fish.

Later field collections of marsh nekton from oiled and unoiled sites did not reveal any effects of oil exposure on population dynamics. These collections did not occur until 2012–2013, and hydrocarbon degradation may have rendered remaining DWH oil less toxic than that in the initial oiling. The mostly consistent results showed little indication of any effect from potential DWH oil exposure (Fodrie and Heck, 2011; Fodrie et al., 2014; Able et al., 2015). Another study compared the total fishes caught on seagrass beds from the Chandeleur Islands, Louisiana, to St. Joseph's Bay, Florida (all along oiled shorelines) in June–September of 2006–2009 with peak oiling in 2010. The total fishes caught across geographic areas between the 4 pre-spill years and the year of the spill showed no statistical difference (Fodrie and Heck, 2011; Moody et al., 2013).

These results leave a gap between toxicity of the initial oiling and physiological performance of killifish and population dynamics following the immediate time of the oil spill in the salt marshes (Fodrie et al., 2014). There is a possibility that gulf killifish distributions reflect extremely local conditions (<100 m). Meaning, this species can act as a site-specific indicator of disruption, however only individuals collected within the same year as the event were likely directly exposed (Jensen et al., 2019).

Marsh nekton distribution did not differ in oiled vs unoiled marshes after 2 years of oil spill exposure, because of many species-specific characteristics, population dynamics, and trophic dynamics (see Figure 6.19). Population responses did not reflect the immediate demise of heavily oiled killifish because of many of the characteristics described in Figure 6.24

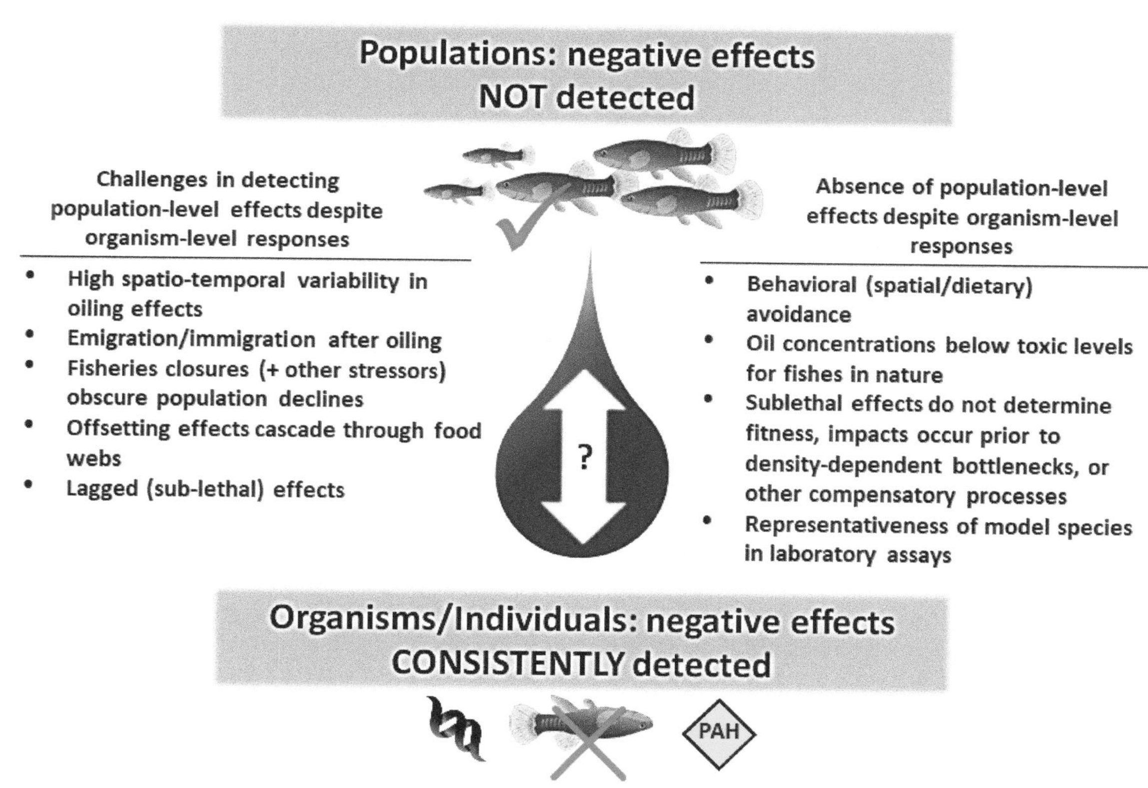

FIGURE 6.24 Potential pathways for the opposing results of organismal (genomic, physiological, developmental) and population-level (densities, assemblage structure) investigations describing the responses of fishes to the 2010 DWH oil spill.
NOTES: The list on the left highlights logistical and structural challenges in detecting population-level effects following large-scale disturbances. The list on the right concentrates on why oil pollution may not have affected estuarine fishes at the population (or community) level, despite known responses at the organismal level. PAH = polycyclic aromatic hydrocarbons.
SOURCES: The symbols are courtesy of the Integration and Application Network, University of Maryland Center for Environmental Science (http://ian.umces.edu/symbols). Figure modified by J. Fodrie, from Fodrie et al. (2014); permission for use granted by BioScience.

(Fodrie et al., 2014). Another possibility is that reduced mortality of these fishes as a result of the deaths of avian predators by direct oiling overshadowed fish mortalities from the directly toxic effects of oil exposure (Short et al., 2017).

Salt Marsh Recovery

For many spills, marsh recovery occurs within one to two growing seasons, even in the absence of any treatment (Michel and Rutherford, 2014), but that time period is often the length of a research program.

A meta-analysis by Zengel et al. (2021) of DWH-oiled marshes in southeastern Louisiana over 7 years post-spill concluded that the DWH oil spill effects were multi-year and that full recovery would likely exceed 10 years, especially in heavily oiled marshes where erosion may prevent full recovery or there is no recovery. In other words, there are no effects to be documented when the salt marsh no longer exists. All metrics (plant cover, stem density, vegetation height, and above- and belowground biomass) were tracked in 10 studies and 255 sampling sites. All plant metrics pointed to impacts from oiling, with 20–100% maximum reductions depending on oiling level and distance from the marsh edge. Heavily oiled sites at the marsh edge showed peak reductions of ~70–90% in total plant cover, total aboveground biomass, and belowground biomass, with *Juncus roemerianus* affected more than *Spartina alterniflora*. Most plant recovery metrics ranged from 3 years to at least 7 years post-spill (the length of the research programs). Belowground biomass reductions were particularly concerning, because the declines over time were longer than recovery of aboveground vegetation. The loss of belowground biomass and root structure as indicated by soil shear strength (Turner et al., 2016) over time was a strong indicator of continuing impact, limited recovery, and impaired resilience.

In experimental mitigation efforts within heavily oiled salt marshes covered by the DWH oil, manual treatment of oiled marshes lead to greater vegetation cover than mechanical treatment and no treatment, through 1 year (Zengel et al., 2015). The percentage of vegetation cover in both treated and untreated marshes did not reach the level of reference marshes for more than 2 years after the initial oiling. Planting allowed for quicker vegetation recovery and reduced shoreline retreat compared with areas of no planting.

Michel and Rutherford (2014) combined their assessment of marsh oil spills and known or predicted recovery rates (see Figure 6.25) with similar syntheses by Hoff (1996) and Sell et al. (1995) and provided the following summary.

Spills in the following conditions display the longest recovery time (example of spills):

- Cold environments (e.g., *Metula*, *Arrow*, *Amoco Cadiz*)
- Sheltered/protected locations (e.g., *Metula*, *Arrow*, Gulf War, Nairn pipeline, Mill River)
- Thick oil on the marsh surface (e.g., *Metula*, *Amoco Cadiz*, Gulf War)
- Light refined products with heavy loading (e.g., *Florida*, *Bouchard-65*, *Exxon Bayway*)
- Heavy fuel oils that formed persistent thick residues (e.g., *Arrow*)
- Intensive treatment (e.g., Aransas Pass, *Amoco Cadiz*, *Golden Robin*)

Spills in the following conditions display the shortest recovery:

- Warm environments (e.g., many spills in Louisiana and Texas)
- Light to heavy oiling of the vegetation only
- Medium crude oils
- Less-intensive treatment

6.5.3.2 Mangrove Communities (ESI 10)

Mangroves as intertidal flora range from short bushes (usually the black mangrove, *Avecinnia germinans*, ~1 m high) to more substantial genera, such as red mangroves (*Rhizophora mangle*) or white mangroves (*Laguncularia racemose*) that reach 6 m and 15 m, respectively. Mangroves are not a majority part of the intertidal shoreline habitats in the primary study area of this report, inclusive of Canada and the Mexico/Caribbean region, with the exception of the subtropical and tropical areas. The intricate structure of the mangals (mangrove swamps) provides habitat to a high diversity of organisms and stability to sediments. Mangroves as a shoreline habitat are much more developed in the southern Gulf of Mexico and throughout the Caribbean than in the northern Gulf of Mexico. Mangroves are affected variably by oil spills, depending on the type of oil and toxicity, the amount of oil, and the duration of weathering. Mechanisms of effects are via direct toxicity to the plants, via the prop roots and the pneumatophores through smothering and reducing the uptake of oxygen as avoidance to low oxygen conditions in the sediments, and via penetration of the oil into burrows of associated animals.

One of the more obvious indicators of the effects of oil on mangrove communities is the albinism of the plant leaves when exposed to xenobiotics, which is related to the translocation of aromatic hydrocarbons into their tissues (Getter et al., 1985). Trees heterozygous for chlorophyll-deficient alleles are fairly easy to distinguish in the mangrove species Rhizophora mangle (Klekowski et al., 1994) and are associated with albinism. This genetic mutation is often seen in red mangroves exposed to xenobiotics, and albinism in mangroves is recognized as an indicator of pollutant stress.

Getter et al. (1981) summarized the effects of oil spills at five sites in the Gulf of Mexico and the Caribbean Sea. Light, refined oils, such as No. 2 fuel oil, often penetrate the substrate through animal burrows and can be took up by the tree roots, with mortality in 24 to 48 hours in red mangroves

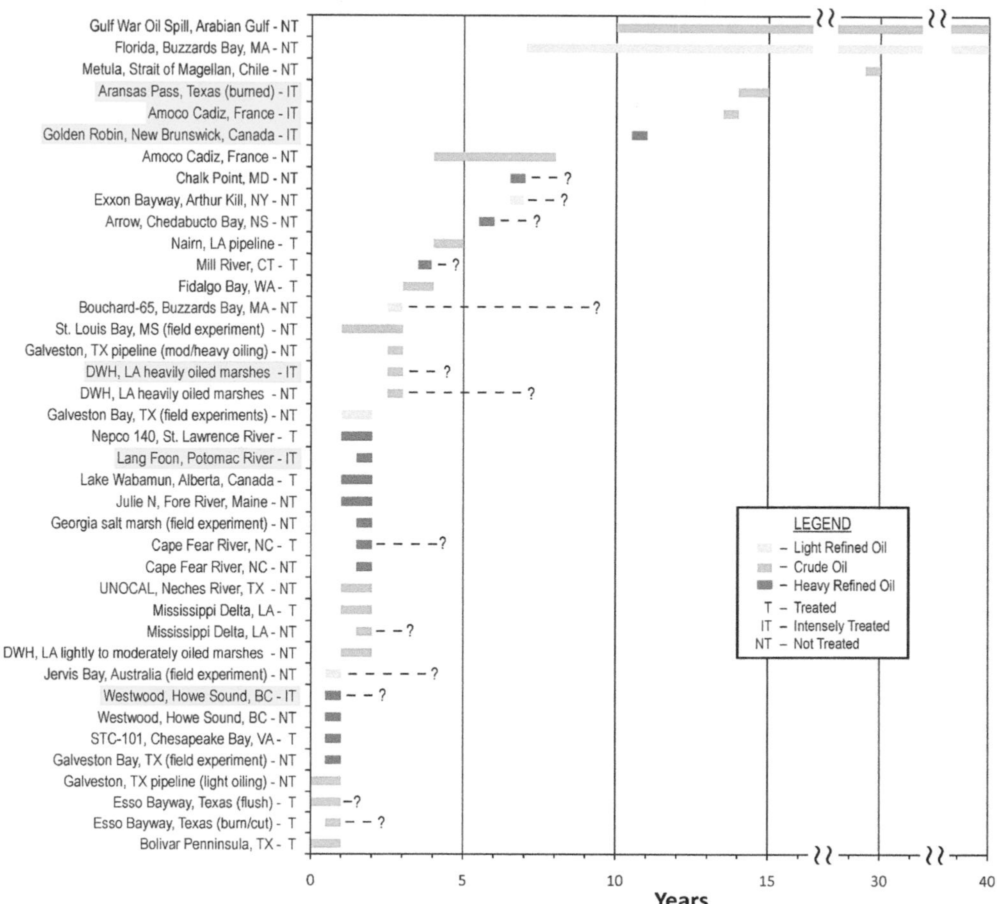

FIGURE 6.25 Years to recovery for spills and a few field experiments (n=32) in marshes (Michel and Rutherford, 2014), color-coded by oil group, from longest to shortest recovery.
NOTES: Recovery is defined as re-establishment of natural biota within the range of dominance, diversity, abundance, and zonation expected for the affected habitat (Sell et al., 1995), or return to comparable values for important ecological parameters for affected and nearby similar control habitats (Hoff, 1996). Yellow highlighting identifies spills with intensive treatment. Question marks represent potential time to recovery based on the results of the most recent data.
SOURCE: Used with permission of the authors.

and black mangroves. Alternately, crude oils and heavy refined products such as Bunker C coat the prop roots and pneumatophores, limiting the ability of the tree to exchange gases (see Figure 6.26). Crude oil contamination—both in the presence and in absence of dispersant in the water—causes a significant reduction in water flux through the red mangrove root tissues (Tansel et al., 2015). The root samples that were in the solutions contaminated with Louisiana crude oil showed dehydration and drying and separation between the epidermis and endodermis both in the absence and presence of dispersant after 7 days (see Figure 6.27). Heavy oils will have long-term persistence, especially with heavy accumulations resulting in leaf loss and possibly death. Oiled areas may also see reduced recruitment of seedlings (Getter et al., 1981).

In oil spills with massive and thick oil coverage on the roots and pneumatophores (Galeta spill in Panama), the epibiont and sessile invertebrate communities die off to < 5% of complete coverage (Burns et al., 1993). All groups of epibionts were present after 2 years, but corals, anemones, and tunicates were still scarcer at oiled versus unoiled sites. Oiled and unoiled sites continued to show significant differences through the 5-year post-spill study.

In August 2006, the tanker *SOLAR 1* sank in the Philippines' Guimaras Strait, spilling a substantial part of her 2,100-tonne cargo of intermediate fuel oil (Yender et al., 2008). The area is rich in mangroves, coral reefs, and seagrass beds, with a rugged shoreline of sand, pebble, and cobble beaches, into which seeped the oil. Cleanup focused on the shoreline with little cleanup in oiled mangroves. Mangrove prop roots and pneumatophores have lenticels, large pores for gas exchange, which if coated with oil lose the ability for gas exchange. After weeks of the oil spill the root surface remained oiled, however, these roots had increased the shedding of cells from lenticels, freeing them up for gas

FIGURE 6.26 Oiled red mangroves (*Rhizophora mangle*) from a spill in the eastern Gulf of Paria, Trinidad and Tobago (17 Dec 2013; 7,554 barrels of Bunker C fuel oil).
NOTES: The photo depicts oil adhered to the mangrove roots 3–4 weeks after the spill (January 2014). Some mangrove leaves turned from green to yellow in color, an indication of sublethal effects. The strings of pom-poms were deployed to prevent re-oiling.
SOURCE: With permission from Paul Schuler, Oil Spill Response Limited.

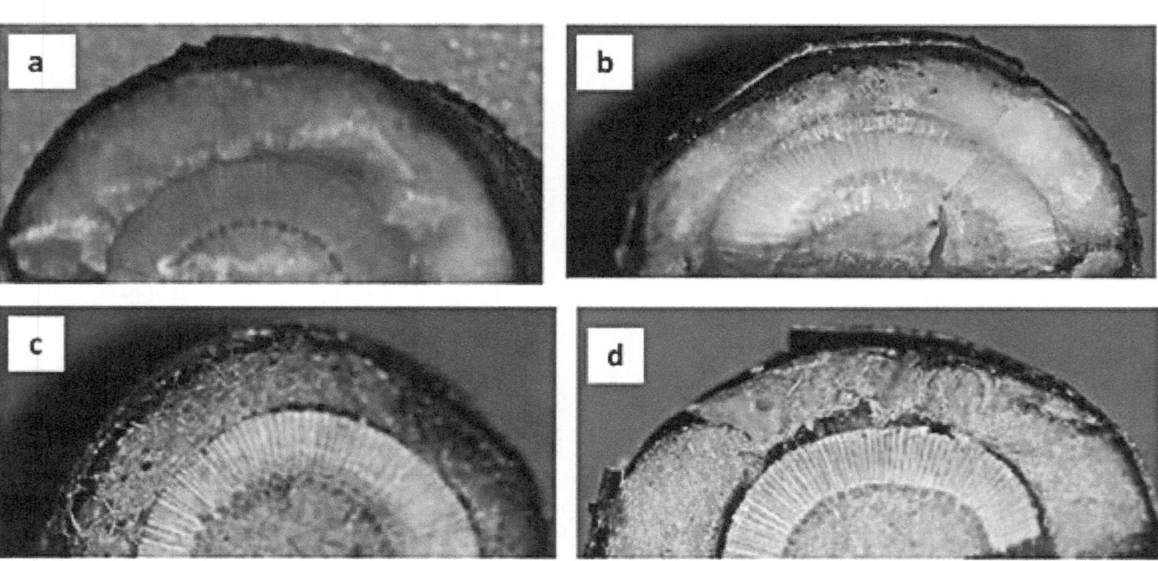

FIGURE 6.27 Cross-sections of the mangrove root segments: (a) fresh cut root, (b) root exposed to salt water for 1 week, (c) root exposed to salt water and crude oil (1 ml/150 ml), (d) root exposed to salt water, crude oil (1 ml/150 ml), dispersant at DOR=1:5.
SOURCE: Tansel et al. (2015).

exchange. Severely affected mangrove trees displayed signs of stress, including yellowing of the leaves, partial defoliation, and occasionally the spread of roots and trunk sprouts, among other symptoms. Areas where the oil remained trapped in sediments for several weeks or months after the spill saw highest adult tree mortality. Observations 1 year after the spill showed that oiled mangroves were recovering naturally and suffered only minor mortality.

Depending on the severity, results from oil exposure can range from initial defoliation and eventual recovery to mass mortality and complete loss of habitat. Many follow-up studies of mangrove recovery do not last long enough to determine the time to recovery. A spill from the Era tanker in southern Australia in September 1992 resulted in the oiling of approximately 100 ha of a monospecific stand of *Avicenna marina* mangroves (Connoly et al., 2020). Lightly oiled trees saw a full recovery; however heavily oiled trees experienced mass defoliation and death over several months after the oiling event. Aerial imagery analysis indicated that heavily oiled areas did not recover for 10 years following the oiling event. Seedlings had become established and canopy cover increased to 35% of pre-oiling cover within heavily oiled areas after between 10 and 25 years. Predictive modeling estimates based on trajectories of vegetation cover change over periods with adequate aerial imagery were used to model complete recovery of mangroves to pre-oiling cover, which was estimated to take 55 years from oiling (Connolly et al., 2020).

6.5.3.3 Sheltered Rocky Shores (ESI 8)

In 1967, the tanker *Torrey Canyon* wrecked while carrying 117,000 tons of crude oil off the southwest coast of the United Kingdom. Experts from the Plymouth Laboratory of the Marine Biological Association (MBA) of the United Kingdom mobilized in response to this environmental catastrophe. Staff from the MBA, A.J. and E.C. Southward (1978) studied the rocky shores affected by the spill and unaffected control sites; charting fluctuations of rocky shore fauna (particularly barnacles) and flora from the early 1950s, in relation to climate. This study provided a baseline to judge recovery of rocky shoreline from the beached oil and application of toxic engine room degreasers not designed for used in the marine environment. Subsequent follow-up studies in the 1980s and 1990s found that recovery took up to 15 years along the shore at Porthleven, which was subject to the most severe degreaser application. In contrast, at Godrevy where degreasers were not applied over the concerns about their impact on seals, recovery occurred in 2–3 years. The degreasers killed the dominant grazer, limpets of the genus *Patella*, which allowed a massive colonization by seaweeds. The resulting canopy of macroalgae ("rockweed" or "wrack") fostered a dense population of limpets, which then completely grazed the seaweeds until the starving limpets largely died off after migrating across the shore in search of food. The die off of limpets and now reduced grazing pressure prompted another bloom of algae. By the mid-1980s through 2016, normal levels of spatial and temporal variation of seaweeds and limpets fluctuations were charted at Porthleven. Lessons learned from observations over 60 years, both before and after the spill, for rocky shore monitoring highlighted the need for broad-scale and long-term monitoring to differentiate local impacts (such as oil spills) from global climate-driven change (Hawkins et al., 2017).

In the reports following the natural recovery of Mearns Rock after the *Exxon Valdez* oil spill in 1989 in Snug Harbor on Knight Island, Alaska, more than 30 years of observations have memorialized huge annual variations on the attached flora and fauna. Those crafting definitions of "fully recovered" needs to bear in mind these annual/seasonal fluctuations (see Box 6.6).

Spilled oil that contacts rocky armored shorelines presents a smothering hazard for diverse intertidal communities that inhabit the surface as well as subsurface sediments, which may require years to recover. Efforts to remove this oil may compound adverse effects by causing additional mortalities. For example, high-pressure washing of poorly sorted gravel beaches after oiling by the *Exxon Valdez* oil spill transported finer-grained sediments from the upper to the lower intertidal, and appears to have resulted in high mortalities of clams that survived the initial oiling. In fact, this may have reduced clam recruitment by removing finer-grained sediments and organic matter from these shorelines (Houghton et al., 1996). Seabirds, sea otters, and predatory invertebrates such as starfish that forage in the intertidal are especially vulnerable to adverse (and often lethal) effects of contact with oil. After the *Exxon Valdez*, population recovery and habitat reoccupancy of intertidally foraging seabirds and sea otters took well over a decade (Esler et al., 2018). Infauna that consume sediments are susceptible to ingesting oil, as are filter- and suspension-feeding invertebrates such as clams, mussels, and other organisms, which may be important subsistence food items for Indigenous communities.

Oil that contaminates hypoxic zones of porous, armored shorelines may persist for decades with little weathering beyond any that occurred prior to deposition, forming a long-term reservoir of oil that can contaminate aquatic biota (Nixon and Michel, 2018). After the 1989 *Exxon Valdez* spill, sea otters, sea ducks, and other species that excavate intertidal sediments in search of prey occasionally encountered subsurface shoreline oil for more than a decade following the spill (Short et al., 2006), oiling their pelage, feathers, or epidermal tissues and thereby often providing an oil ingestion pathway.

6.5.3.4. Seagrass Communities (no ESI category)

Healthy seagrass beds are usually located between supratidal habitats, such as mangroves or salt marshes, and coastal subtidal areas. The health of seagrasses depends on shallow, calm waters with low turbidity and lack of toxic pollutants and excess nutrients. During an oil spill, exposure may occur via coating of seagrass blades and seepage into sediments (Zieman et al., 1984). Seagrass habitats are often less studied than the more obvious salt marshes, mangroves, and coral reefs.

Kenworthy et al. (2017) documented shoreline oiling following the DWH oil spill via aerial and satellite imagery on the shallow back barrier portions of the Chandeleur Islands where seagrass communities existed in a patchy mosaic among unvegetated and vegetated areas of subtidal sediments. The method used by Kenworthy et al. (2017) was aerial photography assessment of the seagrasses in May 2010 before the spill and post-spill in October 2010, 2011, and 2012, where DWH oil exposure was confirmed. The investigation conservatively assessed a seagrass loss of 104.22 acres (42.18 ha). An estimated loss of 271 acres of seagrasses were lost due to immediate or delayed oil exposure and from outboard motor propeller scarring associated with spill response activities (Kenworthy et al., 2017). Oil exposure did not result in a catastrophe for seagrasses in this area. Similar results were found for the seagrass beds along the mainland coastline of eastern Mississippi Sound, with no short-term declines in their growth or abundance following the DWH oil spill (Cho et al., 2017). Results from other oil spills affecting seagrass beds demonstrate more negative effects, depending on the intertidal environment and relative differences in oil exposure. The most serious known effects of oil on seagrasses have been observed on communities that are intertidal or marginally subtidal and subject to occasional exposure, unlike seagrass beds that remain submerged during oil exposure (reviewed by Zieman et al., 1984).

One of the better-documented oil spills was that from the *Amoco Cadiz*, which occurred near Roscoff, on the Brittany coast of France, in March 1978. A massive oil slick washed ashore and portions of it covered a *Zostera marina* bed that had been studied continuously since 1976 by den Hartog and Jacobs (1980). Two seagrass beds at different tidal levels were studied: the upper one was exposed at nearly every low tide, whereas the lower station was about 0.5 m lower and located in an enormous tidepool that retained several centimeters of water during every low tide. The *Zostera* was "almost unaffected by the oil spill" (den Hartog and Jacobs, 1980). During the few weeks following the oiling, numerous leaves turned black and were lost, but this seemed to be only an acceleration of the normal leaf loss, and no other short-term damage to the plants was observed. den Hartog and Jacobs (1980) concluded that there were minimal effects on the seagrasses. Other baseline data on the fauna associated with the *Zostera* beds beginning the year before the oil spill provided comparative data for post-spill effects (Boucher, 1985; Dauvin, 1988). The influence of oiling left some animal groups unaffected, some with rapid recovery, but there was little re-establishment of a diverse amphipod community that was replaced by only two species of amphipods.

Fonseca et al. (2017) added other measurements of effect to oiled eelgrass in their study of oiling impacts to eelgrass (*Zostera marina*) from the 2007 *Cosco Busan* event in San Francisco Bay. At sites with pre-spill records, data collections included the shoot densities, reproductive status, and rhizome elongation of *Z. marina*, and post-spill measurements captured eelgrass photosynthetic efficiency. **SCAT oiling categories identified during shoreline cleanup assessments show no consistent relationship with variations in shoot densities and percent elongation of rhizome internodes formed after the oil spill. In addition, changes in seagrass photosynthetic efficiency were also not consistent with SCAT oiling categories. A comparison of SCAT oiling categories with chemical characterization of hydrocarbons would benefit subsequent field studies and comparison of effects results based primarily on the four-tier SCAT categories.**

6.5.3.5 Tidal Flats (ESI 7 and 9)

Heavy fuel oil was accidentally spilled in a drainage ditch near the Kaomei Wetland, south of the Tachia River estuary, Taiwan, in June 2010. Semi-diurnal tides of three to four meters quickly spread about eight metric tonnes of the oil over 300 ha of the reserve, including extensive tidal flats. The area is subtropical with average temperatures of 29°C in July and August. There were baseline data for the area, which had been monitored since 2008 for benthic communities and microbial processes (Lee and Lin, 2013). One site was selected for the polluted area; a similar tidal flat less than 400 m from the polluted site was established as a control site, which had not been oiled but likely was exposed to dispersed oil in the water column. CO_2 fluxes at the air-sediment interface during emersion at low tide were determined by benthic chambers. All parameters were measured five times within a 38-day period following the oil spill. Sediment bacteria increased in abundance and microalgal biomass decreased, indicating that the ecological functioning of the flats was compromised with the oil suppressing or stopping gross community production and increasing community respiration. Net community production became negative within 5 days, indicating that the community had switched to heterotrophy with suppressed microalgal photosynthesis. Net community production also became negative after 8 days at the control site, indicating that the oil had been dispersed by the tides from the polluted area. The contamination at the control site was not as great as at the polluted site because the gross community production remained positive. Macroinfaunal abundance was reduced shortly after the spill, and different fauna recovered at varying rates. Crabs and bivalves were observed after five days in the polluted site, but no macroinfauna were observed eight days post-spill, thereafter recolonizing at different times. Polychaetes and amphipods were observed only at the "control" site, but polychaetes reappeared at the polluted site only 38 days after the oil spills.

Paranaguá Bay in southern Brazil is one of the largest subtropical estuaries and best preserved in the southern hemisphere, but susceptible to oil operations that process mostly diesel fuel oil. Intertidal flats dominate in this estuary and in confined low-energy areas along the southeastern and southern Brazilian coasts. An experimental in situ diesel fuel oil spill was conducted in unvegetated tidal flats along the Cotinga Channel, a sub-estuary of Paranaguá Bay (Gonzalez Egres et al., 2012). The experiment was a multivariate before and after/

control and impact (M-BACI) model, with three levels of treatment contrasted with controls in 14 successive periods before and after the oil spill covering a total of 147 days for the full experiment. The monitored response was the composition of the benthic infauna. An acute effect was recorded immediately after the impact, but the recovery to pre-disturbance population levels occurred mostly complete within 30 days (i.e., within long-term variability of the community). There was an increase in the total density of the benthic community of small snails, oligochaetes, and ostracods after the disturbance, which was ascribed to background variability.

Nwipie et al. (2019) followed a series of oil spills in Bobo Creek (Niger delta) and related low-intertidal, soft-bottom infaunal macrobenthic invertebrates for 7 years after spills and compared changes in the benthic community after initial observations from Zabbey and Uyi (2014). Initially, the two major spills reduced macroinfauna by 81%, with two stations not supporting any benthos. Subsequent sampling on a bimonthly basis (spring and neap low tides) for 2 years indicated an initial increase in an opportunistic polychaete, then a polychaete-dominated but more diverse benthic community in 3 and 5 years after the spill. A recovery rate of only 9.7% of the benthic fauna abundance after 7 years (Nwipie et al., 2019) reflected remaining high levels of hydrocarbons, continued oiling from polluted tidal creeks, low dissolved oxygen and high biological oxygen demand from decomposing flora and fauna. There was no increase in an endemic characteristic chemosymbiotic lucinid bivalve population for at least 7 years post-spill.

6.5.3.6 Sandy Beaches (ESI 3 and 4)

Beach oiling can occur when oil is deposited by waves and coastal currents onto the shoreface of barrier islands. Subsequently the oil can be incorporated into the beach sands by tides and waves. With weathering of the oil residue and mixing by waves, the oil may be incorporated into subsurface oil mats along exposed sandy beaches, within burrows of organisms (Amos et al., 1983), or as tar balls (see Chapter 5, Section 5.3.2.1).

Oil spills on sandy beaches are seldom often thought about past the effects on tourism and the health of oil spill cleanup workers (see Section 5.2.5.2). However, the beach intertidal sands are home to many sediment-dwelling organisms that form the food base for shorebirds in residence and at important periods of migration cycles. Although sandy beach shoreline habitats are often the first affected by an oil spill coming ashore, the effects on their living inhabitants and communities are seldom studied at the same level of detail as are salt marshes, mangroves, and rocky shorelines. Analyses of the abundance and community composition of interstitial sediment dwellers are seldom accompanied by the makeup and concentrations of the oil residue hydrocarbons. Bejarano and Michel (2016, and references therein) synthesized the peer-reviewed literature and reported on the impacts of oil spills on sand beaches (63-μm to 1-mm grain size) with a focus on intertidal invertebrate communities and their recovery. Several complications exist in comparing results among beach intertidal infauna studies:

- Different ranges of sizes of organism categories (e.g., meiofauna are usually considered >63 μm and less than 0.5 mm, but are often described as >45 μm and less than 0.5 mm; macrofauna are usually those retained on a 0.5-mm sieve size, but some researchers put the lower limit at 0.3-mm sieve size for retention of macrofauna, especially juveniles).
- Inconsistent level of taxonomic enumeration (e.g., grouping of oligochaetes or nematodes versus specific identification of harpacticoid copepods, or specific identification of polychaetes versus counting as larvae versus adults).
- The groupings of invertebrates are often considered residents within positions along the intertidal interface of the beach between low- and high-water levels, for which there is overlap between the upper, middle, and lower intertidal communities.
- Infaunal beach communities shift in position on the beach forefront as water levels increase with tides or wind-induced water submergence of the beach and move to lower elevations of the beach with receding tides. Supratidal zones maintain more "permanent" residents such as ghost crabs (*Ocypode quadrata*).

Despite the complications already listed and limited publications that document effects of oil spills on sand beaches, Bejarano and Michel (2016), identified

> (1) an impact phase, where the associated invertebrate community experiences a measurable reduction in abundance and species diversity caused mostly by mortality and oil fouling; and (2) a recovery phase, where there is an increase in dominance of opportunistic species, followed by the return of species characteristic of the assemblage signaling the start of the recovery.

These findings are similar to those for other oiled habitats, with the exception that the organisms are mostly small infaunal meiofauna and macrofauna. Loss of abundance of beach infaunal organisms can be well over 50% and up to 100%. Groups with direct development, such as harpacticoid copepods and amphipods, are more affected than polychaetes with larval recruitment from ambient waters. Recovery may be rapid, such as within 1 year, or much longer, taking place over several years.

6.5.4 Coral Reefs (ESI 4)

Corals are found in almost all of the world's oceans: there are the shallow coral reefs that are restricted to tropical, warm, clear waters, but also deep or cold-water corals in temperate to Arctic locations. In tropical shallow-water locations, reef-building corals are symbiotic with photosynthetic zooxanthellae (e.g., *Acropora* spp.), but the common cold-water deepsea stony coral, *Lophelia pertusa*, does not contain these algal symbionts. Intertidal and subtidal shallow coral reefs in tropical regions of the world are often close to areas of oil extraction and tanker routes, and there have been a number of examples of acute oil spill events and chronic exposures from operational discharges or natural seeps (summarized in NRC, 2003). In a Net Environmental Benefit Analysis (NEBA), corals are often one of the organisms highest at risk and valued ecosystems listed for protection. This assessment was largely based on the results from one of the longest-lasting (currently over three decades) oil release field experiments (the 1984 TROPICS study, see Box 6.9; Renegar et al., 2017, 2021, 2022), which was conducted to determine if the use of a chemical dispersant would reduce the overall impact of oil to a coral reef and result in a quicker recovery (Ballou et al. 1987; Dodge et al., 1995). As coral reproduction and early life stages are especially sensitive to oil the timing of when an oil spill occurs is critical (see Nordborg et al., 2020, 2021). Furthermore, these early life stages of coral (i.e., gametes, embryos, and planula larvae) often reside in the ocean's surface where oil concentrations are typically higher. The effects may also be compounded by environmental factors and co-stressors such as ultraviolet radiation (UVR), elevated temperature, and specific phototoxicity reactions that enhance or exacerbate oil toxicity (see Section 6.3.4.3; also Kegler et al. 2015; Nordborg et al., 2018, 2020, 2021). The impact of oil and/or chemical dispersants on corals has also been investigated using laboratory toxicity tests, although much fewer than for standard toxicity test organisms given the complexities and difficulties with working with corals. Coral toxicity records (all chemical contaminants) in the CAFE database (see Section 6.7.2) represent only 0.1% of the total records across all taxonomic groups (Bejarano et al., 2016). Toxicity studies have been conducted exposing numerous coral species and life stages (i.e., adult fragments, gametes, and larvae) to single hydrocarbons, hydrocarbon mixtures, various fuels and oils, and chemical dispersants.

Although the vast majority of studies have focused on intertidal and subtidal species, the DWH incident highlighted the significant knowledge gaps regarding the toxicity of oil and chemical dispersants to deep-sea coral species, including whether these coral ecosystems would be more sensitive than their shallow-water counterparts. This led to a new research focus on understanding the exposure to and effects of oil on mesotrophic and deep-sea coral species, not just in the Gulf of Mexico but also in other areas of the world, as discussed later in this chapter (e.g., White et al., 2012; Silva et al., 2016; Girard et al., 2019; Bytingsvik et al., 2020).

Unlike for many other marine species, there are currently no standard toxicity test organisms or protocols for corals, which are notoriously challenging to obtain, sustain, and maintain health in laboratory settings (Mitchelmore et al., 2021). Typical metrics used in toxicity tests (i.e., mortality, growth and reproduction) are difficult to assess or achieve in typical testing time frames. Determining mortality in a coral is difficult given that polyps retract into the skeleton, and accurate assessments for many species may only be completed after assessment in control exposure media during recovery as discussed by Bytingsvik et al. (2020). Many corals are also slow-growing and hard to maintain in laboratory culture for extended periods making chronic toxicity studies challenging. There are, however, some species for which growth differences are possible in a suitable time frame (see Renegar et al., 2019; Renegar and Turner, 2021). Offsetting these limitations has been the inclusion of additional non-traditional acute and chronic endpoints, directed at both the host coral and its algal symbionts (if present). For example, Bytingsvik et al. (2020) used polyp retraction instead of mortality to calculate acute EC50s. However, the relevance of polyp behavior and other alternate non-traditional toxicity test biological endpoints to population-level consequences is not well established, and further investigation is needed (Mitchelmore et al., 2021).

This section discusses the implications of oil spills to both shallow-water and deep-sea coral reefs, highlighting studies that were not discussed or occurred after the *Oil in the Sea III* report (NRC, 2003). It includes results from field studies following oil spill events, in situ experiments and also laboratory toxicity studies.

6.5.4.1 Intertidal and Subtidal Coral Reefs

Intertidal reef flats are potentially at risk of higher impact from oil than subtidal reefs, as oil can coat surfaces in addition to the water-soluble fraction exposures typical for both reef types. The 1985 report on oil in the sea (NRC, 1985) focused on describing the effects to corals from a number of oil spills that had occurred in tropical habitats. The *Oil in the Sea III* (NRC, 2003) study highlighted the extensive additional field, in situ and laboratory studies that were done in the subsequent 20 years. The 1986 Galeta oil spill (Bahia las Minas, Panama) was highlighted, as long-term effects to corals were observed due to chronic oil exposure from the surrounding mangroves. In addition to the previous National Academies reports, many recent comprehensive reviews have summarized the effects of petroleum hydrocarbons on coral reefs, including field exposures during oil-spill incidents, planned field studies (in situ exposures), and the wealth of data stemming from laboratory exposures using single PAH compounds, mixtures of PAHs, and various fuels and oils (e.g., Shigenaka, 2001; NRC, 2005; van Dam et al., 2011; Renegar et al., 2017; Turner and Renegar, 2017; Kroon et al., 2020; NASEM, 2020; Negri et al., 2021). Studies have also included investigations of response options (e.g., chemical dispersants; Negri et al., 2018), covariables, and multiple stressors, including elevated UV light and temperature (Nordborg et al., 2018, 2020, 2021).

Responses to and the effect of oil and/or hydrocarbon exposures to corals in laboratory tests and from field and/or in situ studies vary from no impact to reproductive failures, increased mortality, reductions in coral cover and health, growth or skeletal differences, and injury, including bacterial infections. An array of sublethal effects have also been reported, many of which have shown correlations to and implications for individual or coral ecosystem health (Turner and Renegar, 2017 [summary tables detailing oil spills, in situ and ex situ exposures]). This wide array of impacts is due in part to the diversity of oil (and dispersant) types, specific environmental conditions including exposure concentration and duration, the specific coral species (and life stage), and other potentially confounding environmental variables. For example, chronic crude oil pollution in the Red Sea increased mortality and reduced coral reproduction, but a short-term exposure of dispersed oil showed little residual effect on growth (Rinkevich, 1977). Field surveys of chronically oiled sites in the Caribbean Sea showed a decline in coral cover (survival) and suggested that coral recruitment was impaired (Bak, 1987). In contrast, monitoring in 1992–1994 of the Gulf War oil spill of 1991 showed no detectable longer-term impacts to corals despite the large volumes of oil released (Vogt, 1995).

A number of oil spills have occurred around coral reefs globally, with the most well studied being the 1986 Bahia las Minas spill on the Caribbean coast of Panama (in 1986) and the DWH spill in the Gulf of Mexico (in 2010). Coral mortality was observed following the Bahia las Minas crude oil spill and reductions in coral cover and growth rates and other sublethal changes were observed 2 years post-spill, with reproduction and fecundity still affected 5 years after the spill, although longer-term (>10 years) chronic impacts could not be assessed due to confounding influences of additional stressors (summaries in Shigenaka et al., 2001; NRC, 2003, 2005; Turner and Renegar, 2017; Guzman et al., 2020). However, studies of other spill events have shown no detectable impacts to corals (see Turner and Renegar, 2017).

The most extensive and longest studies of oil and dispersant impacts on corals come from the Panama TROPICS study (in 1984) and the Arabian Gulf (LeGore et al., 1989) study, where controlled in situ oil spills (with and without the use of chemical dispersants) were conducted (see Renegar et al., 2017, 2021, 2022; see also Box 6.9). In the TROPICS experiment *Acropora cervicornis* and three other coral species showed a limited impact at the crude oil alone site but coral cover declined and growth was reduced at the chemically dispersed crude oil site during the first 2 years (Ballou et al., 1987). No residual effects of crude or dispersed crude oil were observed 10 years after the study, although mangroves continued to be impacted, even after 10 years at the oil-only site (Dodge et al., 1995; Ward et al., 2003). In contrast, in the Arabian Gulf experiment, no impact on *Acropora* spp. growth was observed following exposure to Arabian light crude oil and Corexit 9527 (LeGore et al., 1989). The greater short-term effects resulting from the 1986 Bahia las Minas spill compared to the 1984 TROPICS experiment that were observed in corals may be partially explained by species sensitivities to oil exposure, as the more sensitive *Acropora palmata* was largely absent at the TROPICS site, which was dominated instead by the more resilient *Porites furcata* species. A species-specific differential impact of oil was also noted in Bak (1987), where the branching *Acropora* spp. was found to be more sensitive than the massive species, *Montastrea* spp. Recent laboratory studies with single PAHs have also noted differences in coral species sensitivities, as discussed later in this section (Renegar and Turner, 2021).

Laboratory exposure studies have highlighted specific mechanisms of action of oil constituents and/or chemical dispersants and assessed exposure routes, toxicity thresholds, and species sensitivities in a number of species and multiple life stages from adult, juvenile, and larval stages. As with the field studies, a variety of biological repercussions have been demonstrated (see Shigenaka et al., 2001; Turner and Renegar, 2017). It is important to recognize that in many of these studies, the exposure concentration and durations used may be much higher than those expected and/or measured following oil spill events. Thus, they may not be environmentally relevant, and/or have limitations due to media preparation and/or validation, toxicity test designs or reporting (Bejarano, 2018; Mitchelmore et al., 2020a,b) as discussed in Section 6.4.

A number of studies have investigated the impact of hydrocarbons on coral reproduction, either gamete fertilization success or larval settlement, metamorphosis, and development. For example, exposure of *Pocillopora damicornis* to WAFs of natural gas condensate resulted in larval expulsion during early embryogenesis and early release of larvae in late embryogenesis (Villanueva et al., 2011). This is similar to the results of earlier studies, where corals showed reproductive failure as they aborted/released planula larvae following oil contact (Loya and Rinkevich, 1979). More recently, oil-contaminated seawater was shown to reduce settlement of *Orbicella faveolata* and *Agaricia humilis* (Hartmann et al., 2015). A number of studies have shown that crude oil inhibited fertilization (Negri and Heyward, 2000) and also metamorphosis (Te, 1991). A concentration-dependent reduction in settlement and survival occurred with larvae exposed to weathered oil, chemical dispersant, and chemically dispersed oil in *P. damicornis* and *O. faveolate* (Goodbody-Gringley et al., 2013). The use of dispersants has exacerbated reproductive damage (Epstein et al., 2000; Negri and Hayward, 2000).

Numerous other endpoints have also been assessed in both the coral host and algal symbiont, and presented either as individual measures (e.g., visual coloration, mucus production, polyp retraction/extension, tissue thinning/swelling, algal or chlorophyll a content, or photosynthetic efficiency) or summations of a number of endpoints to reflect a corals condition index (e.g., Renegar et al., 2021). Polyp behavior has been a common endpoint reported in numerous coral studies with other chemical contaminants, but the implication of this response, especially at the population level, is unknown and further studies should investigate the long-term repercussions of this to fully understand its utility in determining impact (Mitchelmore et al., 2021). Polyp retraction is also sensitive to changes in water flow/handling, time of day and lighting and the presence of food, so measurements must be conducted considering these variables (e.g., at set times of the day) (Bytinsvik et al., 2020; May et al., 2020). Coral exposed to HEWAF in both 96-hr static and pulsed exposures showed tissue regeneration and polyp behavior to be sensitive endpoints (May et al., 2020). Polyp retraction was also the metric used in a deep-sea coral study and highlighted to provide results suitable for an acute assessment (see following text and Bytingsvik et al., 2020).

Corals also represent a variety of species and forms (i.e., branching, massive) and life stages, all of which can significantly influence the toxicity thresholds reported (Shigenaka et al., 2001). Although differences in coral species sensitivity was highlighted by Bak (1987), very few studies have actually investigated this. One method to investigate a species' sensitivity to a chemical is to use the toxicity thresholds reported for that chemical in multiple species, producing a cumulative distribution which is termed a *species sensitivity distribution* or SSD (see Section 6.7 and Bejarano, 2016).

Species-sensitivity distributions for five Atlantic scleractinian coral species' exposure to 1-methylnaphthalene demonstrated that, similar to the Bak (1987) observation, *Acropora* spp. (in this case *Acropora cervicornis*) was the most sensitive species (Renegar and Turner, 2021). Interestingly, recent research using passive-dosing exposures to MC 252 surrogate oil and single hydrocarbons has indicated the relative resilience of some corals, likely due to their ability to produce excessive protective mucus, at least over short-term exposure periods (Renegar et al., 2017, 2019; Renegar and Turner, 2021). Corals have typically been thought of as one of the most sensitive marine species; however, in ranking coral toxicity endpoints with those of other marine species the corals were not the most sensitive, but considerably more resilient to this single hydrocarbon dissolved phase exposure (Renegar and Turner, 2021; also see Section 6.7 on SSDs). Of eight tropical species tested in acute exposures to weathered Ichthys condensate, *Acropora millepora* larvae were not the most sensitive species (a *Porifera* sponge larvae was), although they were second (Negri et al., 2021). The least sensitive species was adult fragments of *Acropora muricata* which were not significantly affected by the condensate WAF at the highest concentration used (2,031 μg/L TAH or 100% WAF; Negri et al., 2021). In comparing the toxicity of chemical dispersants in five coral species compared to other aquatic species, Bejarano (2018) highlighted that "[a]lthough it is commonly assumed that corals are among the most sensitive taxa, the sensitivity of five coral species fell within the lower to middle percentiles and were not clustered toward the lower percentiles."

Corals are exposed to many other natural and anthropogenic stressors, so hydrocarbon pollution often does not act in isolation. Many studies have shown that exposure to additional stressors results in elevated impacts from hydrocarbons, although this remains a relatively understudied area. Furthermore, many environmental conditions typical of coral reefs (i.e., high UV light, temperature stress) are often overlooked and rarely considered in coral oil toxicity tests and risk assessments. For example, impacts of diesel exposure in *Pocillopora verrucosa* were only apparent with co-exposure to elevated temperature (Kegler et al., 2015). One of the most studied co-stressors has been UV light due to observations of elevated PAH toxicity due to phototoxicity reactions (as discussed in Section 6.3) via photosensitization or photomodification/photo-oxidation reactions (Barron, 2017; Nordborg et al., 2018, 2020). For example, enhanced phototoxicity in low-dose, short-term exposures of fluoranthene to *Porites* spp. in natural sunlight was calculated to result in a 14× increase in toxicity by comparing upper and lower sides of the coral fragments. However, reduced effects were seen in replicate exposures kept under laboratory lighting, thereby highlighting the importance of using appropriate spectral quantity and quality in laboratory tests so that they do not underestimate toxicity in the field (Martinez et al., 2007). The inhibition of metamorphosis in *Acropora tenuis* larvae exposed to low concentrations of crude oil WAF (103 μg/L TPAH; suggested to be similar to levels that would be found following a spill) resulted in a 40% increased sensitivity of the larvae when co-exposed to UV light (Negri et al., 2016). Similarly, exacerbation of toxicity of a heavy fuel oil was observed in several early life stages of *Acropora millepora* following co-exposure with UVR, resulting on average in a 1.3-fold reduction of toxicity thresholds across life stages and endpoints (Nordborg et al., 2021). Co-exposure of two common marine fuels and UVR resulted in decreased larval settlement success in *Acropora tenuis* compared exposure to the fuels alone (Nordborg et al., 2018).

A review of all coral literature suggested UVR exposure could account for increases in toxicity up to 7.2-fold, leading the authors to conclude that UVR co-exposure should be accounted for in all future coral oil toxicity studies so that reliable toxicity thresholds can be determined for use in the development of credible oil spill risk models (Nordborg et al., 2020; see Figure 6.28). Although there are fewer data available to assess the influences of increased temperature or low pH on oil toxicity, increases, although more modest, were observed (i.e., 3- and 1.3-fold, respectively). **Further research is needed to assess the impacts that tropical environmental and climate change co-factors have on the impact of oil pollution in shallow reef ecosystems** (see Figure 6.28).

Other confounding stressors that may influence the impact of oil on corals are the health and disease status of the corals. Colonies of the reef-building coral *Orbicella faveolata* affected with Caribbean yellow band disease (CYBD) were

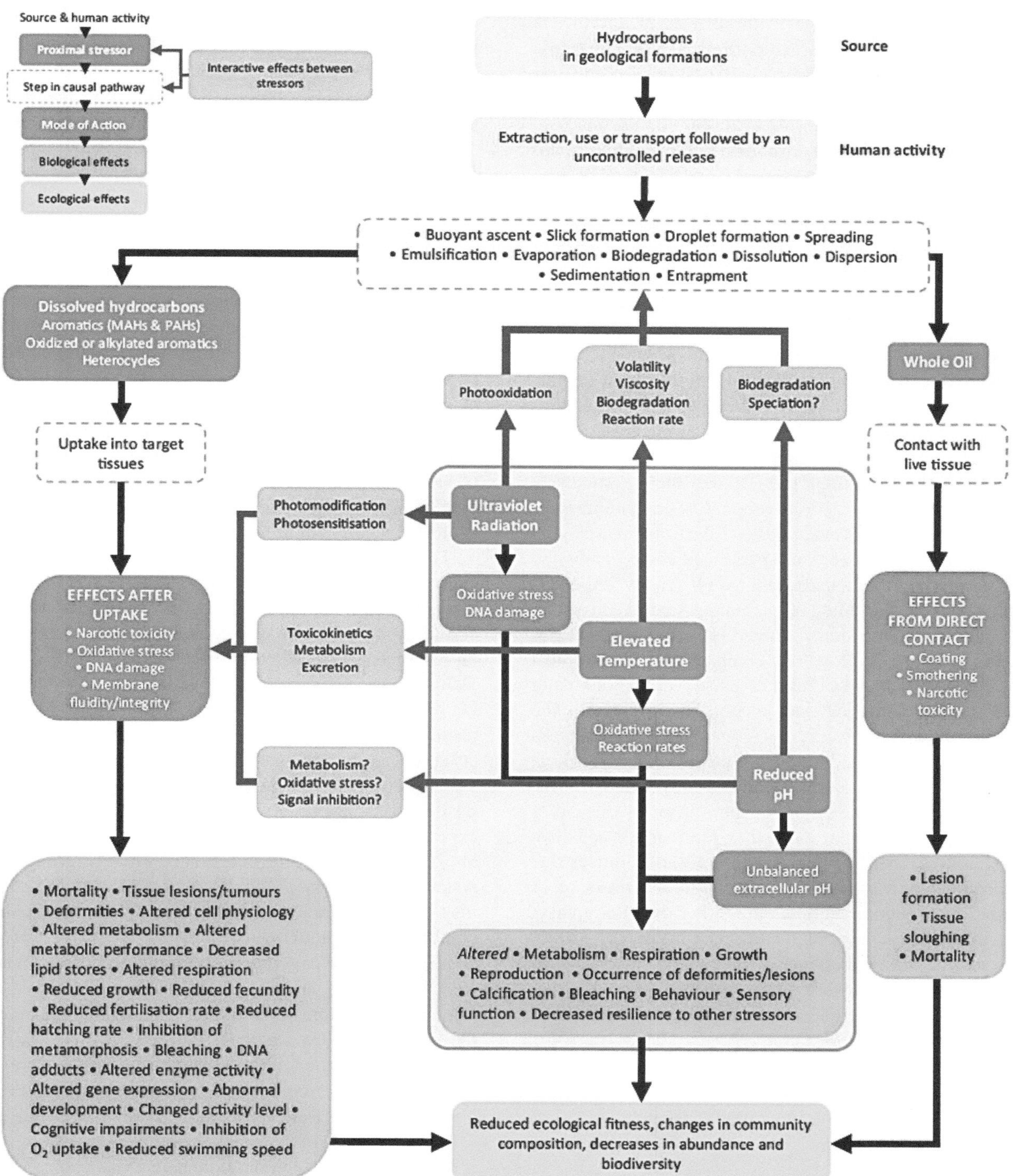

FIGURE 6.28 Conceptual figure highlighting exposure pathways and synopses of effects from exposure to oil pollutants with tropical environmental pressures (UVR, elevated temperature and reduced pH) on coral reef organisms.
NOTE: Oil toxicity can be affected by environmental co-factors (purple boxes) through direct co-exposure, and resulting interactions, or by affecting oil weathering, dissolution and reaction rates prior to or during uptake into target tissues.
SOURCE: Nordborg et al. (2020).

found to be more vulnerable to the effects of anthracene than healthy colonies, due to a compromised anti-xenobiotic response (Montilla et al., 2016). Alternatively, corals exposed to petroleum hydrocarbons may be less resilient to subsequent infections and levels of disease could increase. **Therefore, numerous environmental and coral health co-stressors, together with additional investigations of species and life-stage sensitivities, must be explored further to understand the impact of oil on these important coral reef ecosystems.**

6.5.4.2 Mesophotic and Deep-Sea, Cold-Water Corals

As highlighted earlier, deep-sea, cold-water corals are found globally over a wide range of latitudes and depths, and are some of the longest-lived deep-water organisms (i.e., 100 to >1,000 years old; Watling et al., 2011). This high longevity, along with their low metabolic rates, slow growth, and low recruitment rates, makes them particularly vulnerable and slow to recover from anthropogenic impacts (Risk et al., 2002; Girard et al., 2019). Like tropical inter- and sub-tidal corals, these deep-sea species are ecologically and economically important, harboring a high density and diversity of organisms, providing habitat and essential ecosystem services for many, including commercially important species (Cordes et al., 2016). However, compared to their tropical counterparts, the impact of oil and dispersant on cold water species is a significant data gap, recently highlighted by the DWH oil spill which affected abundant and diverse deep sea coral ecosystems in the Northern Gulf of Mexico—which contains 285 deep-sea coral species (Etnoyer and Cairns, 2017). Very little is known about the impact of oil and/or dispersants on these species. Are they more or less sensitive than shallow-water coral species? Are any differences due to environmental variables, such as increased pressure or decreased temperatures, influencing either the fate, uptake, or toxicity of petroleum constituents? Bytinsvik et al. (2000) concluded that, based on their results and the current literature, deep-sea species were similar in their acute toxicity sensitivities to hydrocarbons, oil, and dispersant compared with shallow-water species.

During the DWH oil spill, deep sea coral ecosystems were exposed to oil plumes, dispersed oil, and dispersant, and were also affected by sinking of oil-contaminated marine snow (Camilli et al., 2010; Passow et al., 2017). An impacted coral community covered in brown flocculant material containing DWH oil and dispersant constituents was discovered at 1370 m by the U.S. Bureau of Ocean Energy Management (BOEM; White et al., 2012, 2014). Corals, primarily the octocoral *Paramuricea biscaya*, showed visual signs of stress, including abnormal skeletal development, increased mucus production, tissue sloughing, and death (White et al., 2012). Subsequent studies have highlighted sublethal impacts and health declines and long-term impacts (7+ years), including branch loss, reduced growth, and hydroid overgrowth (indirect impact from oil/dispersant exposure), suggesting that decades would be needed for recovery to original status (see Hsing et al., 2013; Girard and Fisher, 2018; Girard et al., 2019). Girard and Fisher (2018) highlighted that the ongoing effect (branch loss) observed in 2016–2017 in DWH oil-injured corals could ultimately result in delayed mortality. Similar to other deep-sea ecosystems, the NRDA process highlighted that there was a lack of baseline information on the eco-toxicological vulnerability of these deep-sea species (Peterson et al., 2012).

Similar to deep-sea species, injury to mesophotic coral reefs (at depths of 65–75 m) was quantified in more than 400 colonies using pathological assessments. Commonly reported was a biofilm with a clumped or flake-like appearance, with more extensive injuries showing broken and loss of branches and bare skeletons (Etnoyer et al., 2016; Silva et al., 2016). Many of the injured gorgonian octocorals highlighted in 2012 had declined further in condition by 2014 (Etnoyer et al., 2016; see Figure 6.29). The presence of elevated tissue TPAH levels led the authors to conclude that the injuries observed in 2011 may have resulted from an acute event (Silva et al., 2016).

The observation of DWH-impacted deep-sea corals spurred laboratory toxicity experiments with oil and dispersant using octocorals to provide estimates of toxicity for comparison with other species and mechanistic information on responses at the physiological and molecular levels (see DeLeo et al., 2016; Frometa et al., 2017; Ruiz-Ramos et al., 2017; DeLeo et al., 2018, 2021). Many of the gene expression responses were common to those observed in shallow water coral species, including up-regulation of metabolic, immune, wound-repair and oxidative stress responses. However, opposite to typical responses, a reduction in cytochrome-P450 expression was observed following oil exposure (DeLeo et al., 2018). To answer the question of whether an Arctic cold-water coral species (*Lophelia pertusa*) was more sensitive than other species, acute toxicity tests with individual hydrocarbons were conducted and fit to the target lipid model to generate predictive models and determine species sensitivity (Bytingsvik et al., 2020). Although it appeared that the deep-sea coral was more sensitive to 1-methylnaphthalene than the tropical shallow-water *Porites* spp., this was attributed to differences in the biological endpoints each study used, as observations of narcotic effects were in agreement between the two studies (Renegar et al., 2017b; Bytingsvik et al., 2020). Responses of the deep-sea and tropical corals in these studies were also similar with both 2-methylnaphthalene and phenanthrene. A potential limitation of this work was that toxicity tests were conducted at ambient rather than elevated pressures, although (as discussed in Section 6.7) modeling efforts have shown that elevated pressures result in reductions in hydrocarbon toxicity (Paquin et al., 2018). Therefore, these results are probably conservative, although more empirical data are needed to confirm the acute toxicity of oil and dispersants to deep-sea species and to investigate delayed and chronic impacts.

FIGURE 6.29 Examples of healthy (left) and injured colonies (right) of the gorgonian octocorals *Swiftia exertia* (a, b), *Hypnogorgia pendula* (c, d), and *Placogorgia* sp. (e, f) similar in appearance to *Paramuricea*.
SOURCE: Etinoyer et al. (2016).

BOX 6.9
The TROPICS Field Study, Panama: Long-Term Study of Corals Exposure to Oil and Dispersants

In 1984, along the Caribbean coast in Bahia Almirante, Panama, the **TR**opical **O**il **P**ollution **I**nvestigations in **C**oastal **S**ystems (TROPICS) in situ experiment was conducted to provide information for oil spill preparedness and response, especially with respect to tradeoff decisions for the use of chemical dispersants in nearshore tropical intertidal and subtidal marine ecosystems. It is now one of the most comprehensive experiments examining long-term impacts of oil and dispersed oil in tropical nearshore ecosystems and is foundational to the concept of Net Environmental Benefit Analysis (NEBA) (see Figure 6.30). Oil or oil mixed with chemical dispersant (Corexit 9527) was added at two separate locations and compared with a nearby reference (non-treated) site. Assessments were made before oil release (i.e., baseline) and at multiple time points during the 2.5-year study (Ballou et al., 1987). To examine long-term impacts and recovery, re-assessments of the sites were made at seven subsequent time points (e.g., Ward et al., 2003; Baca et al., 2005; DeMicco et al., 2011) including extensive studies 10 (Dodge et al., 1995) and 32 years after the initial experiment (Renegar et al., 2017, 2021, 2022). The 32-year study was conducted because prior research had identified residual oil in mangrove substrates, leading to the hypothesis that chronic re-exposure to oil could cause long-term impacts. Therefore, in addition to health assessments of the coral, seagrass, and mangrove ecosystems, one of the main goals of the 2016 re-visit was to determine the presence/absence of oil in core samples using a new petroleum biomarker.

Overall, this experimental study demonstrated—and still continues to show—the significantly different damage and recovery regimes in coral reefs, seagrass, and mangrove ecosystems depending on whether they were exposed to crude oil or dispersant-treated crude oil, thus highlighting some of the tradeoffs that are considered in NEBA analyses in these locations. Significant impacts to percent coral and growth occurred in the dispersed-oil site immediately following exposure, which were not observed at the oil-only site (Ballou et al., 1987). Continued and long-term reductions in coral cover remained at the dispersed oil site for at least 2 years, but by 10 years the coral parameters measured were not statistically different from those at any of the three experimental sites. Similarly, in 2016, no significant differences in health metrics for hard or soft corals were found between sites. Furthermore, the hydrocarbon and petroleum biomarker assessments showed no indication of the original oil or its degradation products. In contrast to coral responses, significant reductions in seagrass density, mangrove canopy cover, and root density were found in the oil-only site in comparison to the dispersed-oil site, consistent with previous assessments (Dodge et al., 1995; Renegar et al., 2017, 2021, 2022).

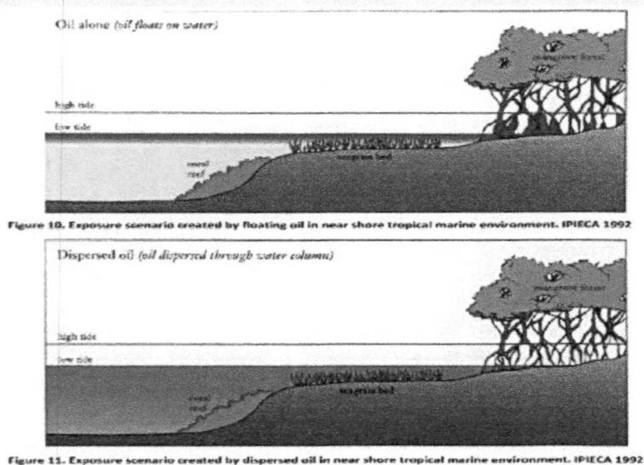

FIGURE 6.30 (A) Exposure scenarios in a near shore tropical marine environment created by (a) floating oil and (b) dispersed oil (IPIECA, 1992). (B) Location of the TROPICS experimental oil, dispersed oil, and reference sites within Almirante Bay, Panama (figure from Renegar et al., 2022). This study was limited due to the non-replication of the experimental sites, although results from this 32-year field study broadly demonstrated significant impacts of crude oil to mangroves with long-term implications for substrate stability. Dispersed oil resulted in fewer impacts to the mangroves but significant (though short-term) effects on coral cover and seagrass ecosystems, with total recovery, at least for the endpoints assessed, observed to have occurred within 10 years.

A number of studies have investigated the toxicity of Corexit 9500 on deep-sea corals (i.e., Arctic and Gulf of Mexico species), and found similar acute toxicities across the five species despite differences in the biological endpoints used (DeLeo et al., 2016; Frometa et al., 2017; Bytingsvik et al., 2020): 34.8 mg/L for *L. pertusa*, 7.9–35 mg/L range for *Paramuricea* type B3, *Callogoria delta*, and *Leiopathes glaberrima* and 70.3 mg/L for *Swiftia exserta*. These ranges are typical compared to other aquatic species tested, falling around the middle of the SSDs reported, and so we may conclude that these deep-sea species are not any more sensitive than other species.

These studies in mesophotic and deep-sea corals highlight the need for prior baseline studies of the health of benthic ecosystems, together with long-term follow-up studies on recovery, delayed mortality, and continued

declines in health in these species, particularly given their slow growth and lower recruitment compared to other marine species and hence potential for a protracted recovery period.

6.5.5 Benthic Communities

Benthos (adj. benthic) are the flora and fauna associated with the bottom sediments of an aquatic system. Benthos are often characterized by position related to the sediment surface (e.g., infauna [subsurface] or epifauna [above the sediment]); by size (e.g., meiofauna or macroinfauna), depending on the sieve size on which they are captured or pass through. Benthic organisms above the sediment surface may be sedentary, such as deep-sea corals (epifauna); stationary benthos below the sediment surface (macroinfauna, meiofauna); or mobile benthic organisms dependent on the bottom (demersal), such as shrimp or crabs with limited ability to move up in the water column or horizontally. Benthic organisms, because of their limited mobility, are monitored in response to pollutants, including petroleum hydrocarbons. The typical response of a benthic community to a continuous stress is a change from a deep-burrowing, larger, and more diverse fauna (not polluted) to high abundance of small organisms that are usually opportunistic surface deposit feeders in response to organic loading in polluted conditions (see Figure 6.31). The nematode:harpacticoid copepod ratio for meiofauna increases with concentration of contaminants in either chronic or acute oiling. Within a given habitat certain species of nematodes are typically most tolerant to stress variables, whereas crustacean meiofauna often are least tolerant (Wetzel et al., 2001). The petroleum inputs may be chronic (drilling or production activities from a platform) or acute (from an oil spill).

The benthos in this section are restricted to open-water habitats, such as estuarine, continental shelf and slope, and deep-sea sediments. Other benthos associated with shoreline habitats will be discussed in those sections (see Section 6.5.3 and others), as relevant.

6.5.5.1 Continental Shelf Soft Sediments

Continental shelf benthos can be affected by the sinking of oil or exposure to production fluids. An example of an oil spill negatively impacting benthic communities is the *Amoco Cadiz* spill of 1978 in the Bay of Moraix in the western end of the English Channel. A benthic study of a biologically diverse macroinfaunal community was dominated in biomass by *Arba alba* (bivalve) but the highest abundance among 25 species was composed of three species of *Ampelisca* amphipods (Dauvin, 1998). The same species of *Ampelisca* were absent from the community immediately after the oil spill for at least 2 years. These crustaceans lack pelagic larvae, and recruitment to prior population levels would take longer than for the polychaete populations that broadcast numerous larvae to the water column. The Bay of Moraix benthic community was repopulated by one species of *Ampelisca* after 3 years, but not all three of the originally dominant *Ampelisca* spp.

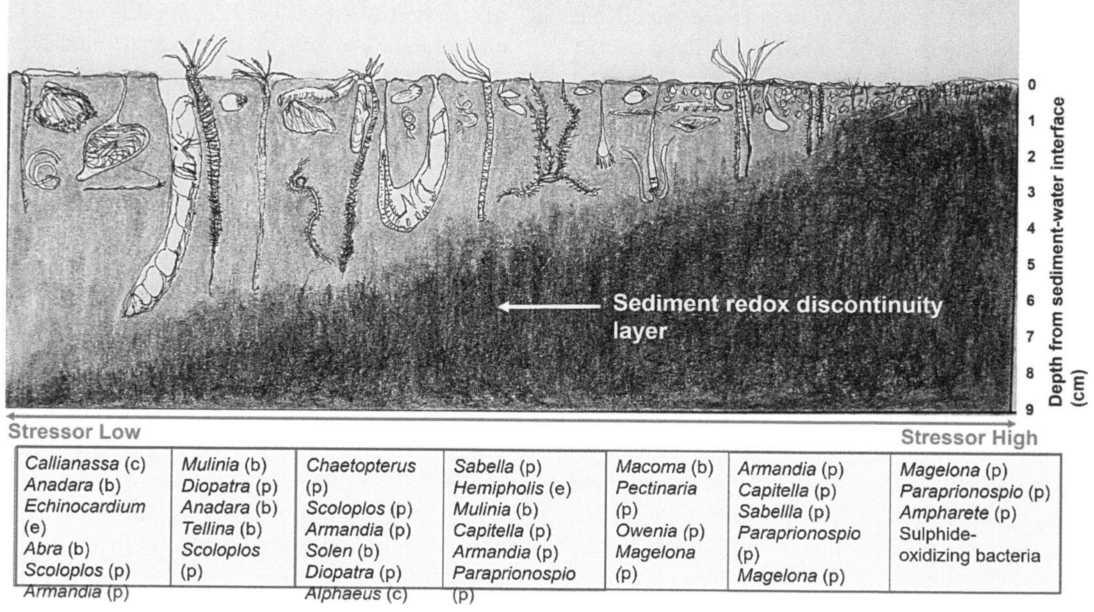

Callianassa (c) *Anadara* (b) *Echinocardium* (e) *Abra* (b) *Scoloplos* (p) *Armandia* (p)	*Mulinia* (b) *Diopatra* (p) *Anadara* (b) *Tellina* (b) *Scoloplos* (p)	*Chaetopterus* (p) *Scoloplos* (p) *Armandia* (p) *Solen* (b) *Diopatra* (p) *Alphaeus* (c)	*Sabella* (p) *Hemipholis* (e) *Mulinia* (b) *Capitella* (p) *Armandia* (p) *Paraprionospio* (p)	*Macoma* (b) *Pectinaria* (p) *Owenia* (p) *Magelona* (p)	*Armandia* (p) *Capitella* (p) *Sabellla* (p) *Paraprionospio* (p) *Magelona* (p)	*Magelona* (p) *Paraprionospio* (p) *Ampharete* (p) Sulphide-oxidizing bacteria

FIGURE 6.31 Changes in benthic macroinfauna (by genera) and sediment structure including the redox discontinuity layer, from non-polluted waters on the left, to high pollution on the right. Letters are (b) bivalve, (c) crustacean, (e) echinoderm, (p) polychaete.
SOURCE: N.N. Rabalais, modified from Pearson and Rosenberg (1978).

until 13 years after the oil spill through the total period of the study (18 years). At the same time, similar proportions of other infauna returned to similar pre-spill proportions.

Chronic exposure of benthic communities to petroleum hydrocarbons was examined in meiofaunal and macroinfaunal communities around three gas platforms on the continental shelf (29–157 m water depths) in the Gulf of Mexico associated with long-term production (Montagna and Harper, 1996). Effects only extended to the local area 100 m from the platforms. Total polychaete and nonselective deposit-feeding nematode density increased near platforms, which could be a response to organic enrichment. Amphipod and harpacticoid profusion and diversity and harpacticoid reproductive success declined near platforms, consistent with other studies of these organisms having lower recruitment because of the lack of pelagic larvae.

6.5.5.2 Deep-Sea Soft Sediments

Meiofauna (size > 63 but < 300 μm) and macrofauna (size retained on a 300 μm sieve) were used to examine the effects of the DWH oil spill of 2010 (Montagna et al., 2013). Ryerson et al. (2012) estimated that up to 35% of the hydrocarbons were trapped and transported in continuous deep sea plumes. Direct sinking of oil from the deepwater plume transported it to deepwater sediments, adsorption of small oil droplets onto particles in marine snow, among others; and sinking of oil-mud complexes resulting from the injection of drilling muds during top-kill operations (reviewed by Mongagna et al., 2013). Meiofaunal and macrofaunal samples were collected 2–3 months after the well was capped in water depths ranging from 76 to 2,767 m along a gradient of suspected oil contamination. The benthic communities experienced severe and moderate damage in 58 samples within an area of 148 km^2. Impacts included low diversity, low evenness, and low taxonomic richness, correlated with high levels of total petroleum hydrocarbons (TPHs) and polycyclic aromatic hydrocarbons (PAHs). Additionally, barium levels near the wellhead were very high. High nematode-to-harpacticoid copepod ratios corroborated the severe disturbance of meiofauna communities. For macroinfauna, the impacts were loss of biodiversity and low abundance of amphipods.

An additional 58 station samples were analyzed to enhance the resolution of the original Montagna et al. (2013) assessment and determine if impacts occurred further afield (Reuscher et al., 2020). These samples indicated that an area covering about 24 km^2, extending 3 km in all directions from the wellhead, displayed the most severe relative reduction of faunal abundance and diversity. Moderate impacts were observed up to 17 km toward the southwest and 8.5 km toward the northeast of the wellhead. The samples also correlated benthic effects to total petroleum hydrocarbons, polycyclic aromatic hydrocarbons, barium concentrations, and distance to the wellhead. Hydrocarbon seeps located 100 km to the east and 240 km to the southwest of the Macondo wellhead were not implicated in the effects. The work of Reuscher et al. (2020) calculated the affected area to be 78% higher than original estimates, covering approximately 263 km^2 around the wellhead. Adding new sampling stations extended the benthic footprint map to about twice as large as Montagna et al. (2013) originally estimated and improved the resolution of the spatial interpolation.

Initial impacts to benthic infauna were greater in spatial extent than defined by resource assessment sampling. This sentiment was echoed in the Trustee Council's Final Programmatic Restoration Plan (2016).

In the future, understanding of impacts on benthic infauna could be improved by increasing the spatial and temporal extent of sampling when designing assessment studies. Such an expansion is applicable to many other assessment indicators for impacts of and recovery from oil spills.

There was one sample in 76-m water depth designated as "moderately affected" by the DWH oil spill among all other sites in >300-m water depth for the deep-water study of Montagna et al. (2013), but this categorization was inconsistent for less contamination and fewer impacts on the fauna in stations in closest distance and in a northwest-to-southeast transect from the wellhead. This station was closest to the birdfoot delta and subject to high sediment and organic loading that could be detrimental to the formation of a typical continental shelf area. Another station in the extended analyses of samples by Reuscher et al. (2020), located 12 km offshore of Grand Isle in the Mississippi Bight, was characterized by faunal indicators that would indicate a highly stressed faunal community but no sign of hydrocarbon contamination. The station had the highest nematode-to-harpacticoid copepod ratio and low values for macrofauna and meiofauna diversity and evenness among the additional stations. The disturbance of the infauna was likely caused by hypoxic bottom-water conditions, which are common along the Louisiana coast during summer months (Rabalais and Turner, 2019). Hypoxia causes local extinction of sensitive organisms, while tolerant taxa may thrive (Baustian and Rabalais, 2009). These two atypical continental shelf stations amid a study of deep-water benthos exposed to DWH oil spill contaminants emphasizes the need to know the potential multiple stressors that may be affecting a habitat (see Section 6.5.5.1).

The above-identified benthic work for the DWH oil spill was coupled with other data in a synthesis of benthic faunal impacts (Schwing et al., 2020b) including microbes, foraminifera, macrofauna, meiofauna, megafauna (invertebrates), corals (see Section 6.5.4.2 on Mesophotic and Deep-Sea Cold-Water Corals), and demersal fishes. Where there were adequate baseline data from before the oil spill, compilations were able to identify impacts of shifts in community structure and changes in diversity and abundance. Time to return to a pre-DWH oil spill condition was (1) within 2 years for the microbial community, (2) within 5 years following a decrease in density and diversity for foraminiferans, (3) not within 4 years for meiofaunal taxa richness in the impacted area, (4) also not for decreased macrofaunal taxa richness as of 4 years post-spill, (5) inconclusive for megafauna due to lack of suitable data for comparison, and (6) a progression of some recovery for deep-sea corals after 7 years and within 10 years, but remaining predicted impact for as much as 50 for some species while other recoveries are predicted to extend to 100 years.

6.5.5.3 Hydrocarbon Seeps

Hydrocarbon and methane seeps are most abundant and most prolific in the central and western regions of the northern Gulf of Mexico in depths of 300 to 1,500 m (Garcia-Pineda et al., 2014), and also in the Bay of Campeche, southern Gulf of Mexico (MacDonald et al., 2015). Hydrocarbon seeps are also a feature off southern California, but at shallower depths (16–18 m) in the Santa Barbara Channel between Coal Oil Point and Goleta Point off Santa Barbara, California (Spies and Davis, 1979). The volume of inputs, composition, and fates of the hydrocarbons are covered in Sections 3.2 and 5.4.1.

Because of their slow seepage through the marine sediments before release, which allows solute-equilibrium with the subsurface water, liquid oil is not expected to contain significant fractions of light components. Furthermore, natural seeps have limited benthic footprint of impact because most of the seep oil is weathered and rises to the ocean surface in droplets when it releases from the sea floor (Sassen et al., 1999; MacDonald et al., 2002).

The nature of sediment organic matter enrichment owing to increased bacterial biomass of hydrocarbon-degrading bacteria is similar to that in shallow-water petroleum seeps. The organic enrichment is coupled with a response of increased meiofaunal and macrofaunal abundance and biomass compared to non-seep sediments (Montagna et al., 1987). Rather than a toxic and negative impact on benthic communities, there is a response of higher abundance of nematodes (meiofauna) but not harpacticoid copepods, which were similar in abundance in non-oiled seep areas. The macroinfauna, which are dominated by deposit-feeding polychaetes, were more abundant at the seep than at a nearby non-seep station (Spies and Davis, 1979). Mats of the sulfide-oxidizing bacteria *Beggiatoa* spp. are a common feature of oil seeps, as they are with deep-water seeps. In both situations, the seeped oil is not as toxic as a fresh oil because of the slow release from the formation, but supports an enriched organic sediment for higher abundances of organisms that feed on the bacteria.

Associated with Gulf of Mexico deep-water hydrocarbon seeps are low-temperature complex chemosynthetic communities (tube worms, methanotrophic mussels, clams, and various other fauna) that derive energy from reduced carbon, mainly methane, and bacterial H_2S (MacDonald et al., 1994; Sassen et al., 1999). Later research confirmed that a number of complex chemosynthetic communities were spatially associated with gas hydrates on the continental slope. At seeps below the photic zone (>200 m), where food sources are limited, the consistent supply of oil and gas supports dense aggregations of sessile invertebrates and associated fish and crustaceans (MacDonald et al., 1994). The oil and gas seeps and hydrate communities host chemoautotrophic symbionts that are unique biogenic communities with a high diversity of organisms, including bacteria consortia that support numerous, as yet unknown, biogeochemical processes. Metazoan invertebrates such as tubeworms, clams, and mussels host chemoautotrophic symbionts that are able to fix new carbon using H_2S and methane as energy sources (Fisher, 1990). These organisms irrigate seep sediments and further promote degradation of hydrocarbons (Cordes et al., 2016). The Gulf of Mexico deep-sea gas and hydrate seeps are not unique in the ocean; similar ecological adaptations occur worldwide (Dubilier et al., 2008). **As relatively understudied habitats, research is needed to increase understanding of these communities, especially in deeper sea locations, to identify novel species/biochemical pathways and chemosynthesis, and to identify bacteria that may be useful in oil spill response (i.e., oil degraders).**

6.5.6 Water Column

The effects of oil spills are obvious as surface-water oil slicks impinge on shorelines or as larger organisms such as marine mammals, sea turtles, diving ducks, and piscivorous birds encounter the oil at the sea surface. The oil in the water column below the surface slick, in contrast, affects different pelagic communities. The naturally dispersed oil below the surface can affect the 1 to 1.5 m of the water column below the surface (personal observations, diving operations, N.N. Rabalais, Louisiana State University) (see Figure 1 in Peterson et al., 2012), and may extend to depths of 25 m or more (Short and Harris, 1996). This dispersed oil mixture with water and finely dispersed droplets of weathered surface oil may affect organisms in the pelagic water column, in particular phytoplankton and zooplankton, in upper water column communities.

6.5.6.1 Bacterial Communities

The stimulation of oil-degrading bacteria was evident in the deep-water dispersed hydrocarbon plume from the DWH spill, with the peak at approximately 1,200 m water depth (Hazen et al., 2010). Cell densities within the plume were twice those detected outside the plume. The presence of these aerobic oil-degrading bacteria resulted in oxygen saturations of 59% within the plume compared to 67% outside the plume. These saturation levels were well above the critical saturation level of 30%, below which fishes, crabs, and shrimp are negatively affected in the coastal Gulf of Mexico oxygen-depleted waters (Rabalais and Turner, 2019).

Associated flocculants such as those detected within oil contaminated waters of the DWH oil spill (Achberger et al., 2021) eventually lead to the sedimentation of oil as marine oil snow (MOS; see Section 5.3.2.2). In an experiment to understand the role of aggregates in hydrocarbon degradation and transport, a MOS sedimentation event was produced using Gulf of Mexico coastal waters amended with oil or oil plus dispersant. Results showed smaller micrometer-scale (10- to 150-μm) microbial aggregate formation in addition to MOS. These microaggregates were most abundant in the oil-amended treatments and commonly associated with oil droplets, connecting their formation to the presence of oil. The maximum observations of the microaggregates overlapped with the maximum rates of biological hydrocarbon oxidation estimated by the mineralization of ^{14}C-labeled hexadecane and naphthalene. To clarify the prospect of microaggregates serving as hot spots for degradation, Achberger et al. (2021) categorized the free-living and aggregate hydrocarbon associated microbial collections using 16S rRNA gene sequencing. The study found the microaggregate population dominated by bacteria and enriched with supposed hydrocarbon-degrading taxa. Using catalyzed reporter deposition fluorescence in situ hybridization (CARD-FISH) (Pernthaler et al., 2002; Pernthaler and Pernthaler, 2007) for observation of these taxa confirmed higher amounts within microaggregates compared to the surrounding seawater (Achberger et al., 2021). Metagenomic sequencing of these bacteria-oil microaggregates (BOMAs) further revealed their community's ability to use various hydrocarbon compounds. These data lead to the fact that BOMAs are intrinsic features in the biological response to oil spills. They are also possibly important hot spots for hydrocarbon oxidation in the water column. These natural biological responses to oil in the environment from the DWH spill produced vast quantities of oil-associated marine snow (MOS). However, ambiguity remains about the forces controlling MOS formation and how it influences the environment.

There is detailed information on microbial communities in Chapter 5 (see Sections 5.2.8.2 and 5.3.2) concerning microbial community composition related to shifts that occur after oil spills and those associated with MOS. Furthermore, there is a conclusion in Section 5.4.5, Fates of Oil in Coastal Ecosystems: Monitoring PAH Profiles in Sediments and Bivalves, stating that

> [t]he development and application of 'omics has revolutionized microbial ecology and understanding of the microbes that respond to and biodegrade oil in the sea. . . . 'Omics techniques provide insight into the composition of microbial communities, their succession patterns, and both individual and composite biochemical activities which, together, influence the fate of oil in the sea.

Observations found part of the natural biological response to the DWH drilling oil spill resulted in vast quantities of oil-associated MOS forming throughout the water. However, the mechanisms controlling MOS formation and its impact on the environment remain largely unknown. Continued research is needed on the formation of MOS, influences on the processes of oil degradation and eventual hydrocarbon fates, as well as ecological consequences.

6.5.6.2 Phytoplankton Communities

Phytoplankton inhabit near-surface waters where they are most likely to be exposed to dispersed oil in the upper water column mixed layer. The conditions in these waters reduce the sunlight necessary for phytoplankton primary production, and the water may contain toxic hydrocarbons. Zooplankton vertically migrate, usually to take advantage of phytoplankton prey during the dark, so they are exposed to oil dispersed throughout the upper water column.

Studies of the effects of oil on phytoplankton often use single algal species lab cultures in laboratory microcosms (reviewed by Ozhan et al., 2014), and sometimes "natural" phytoplankton communities collected from the field (e.g., González et al., 2009). The field-collected phytoplankton studies examining multiple effects may be held at in situ environmental conditions, but the communities remain confined to microcosms within simulated environmental conditions or incubated within the environment from which the samples were collected (Ren et al., 2009; Gilde and Pinckney, 2012, respectively). Determining shifts in phytoplankton community composition, abundance, and biomass over time within an area that experiences an oil spill is complicated by lack of a suitable reference area, unknown exposure level and history, and other environmental factors (see Section 6.5.1).

Phytoplankton communities differ by season, physical variables, and nutrient availability (Parsons et al., 2021), which complicates the determination of the effects of an oil spill. The DWH oil spill and an existing long-term data set enabled a comparison of pre-, during-, and post-spill phytoplankton composition (Parsons et al., 2015). The study period encompassed spring and summer of 1990–2009, 2020, and 2011. The baseline data were collected as part of the Rabalais and Turner (2019) Gulf of Mexico hypoxia legacy data and represented monthly samples between 1989 and 2009 and monthly samples during the spill period of May–October 2010 (Parsons et al., 2015). The results of the CLUSTER and SIMPROF analyses (with PRIMER 7) indicated that the years 1994, 1996, 1998, 2001, 2003, and 2008 were not statistically different ($p > 0.05$) from 2010 in terms of the environmental parameters between the months of May and October, thus removing potential seasonal effects from consideration. Additionally, the CLUSTER and SIMPROF results on the baseline-averaged monthly phytoplankton data versus the 2010 monthly phytoplankton data indicated that the phytoplankton assemblage was different in 2010 compared to the baseline data and that the overall abundance was 22% lower following the DWH oil spill ($p < 0.05$). The cyanobacteria, autotrophic ciliates, cryptomonads, and chlorophytes accounted for the majority of the decrease. Diatoms and euglenophytes were more abundant in 2010, suggesting a possible stimulation from the oil or a relaxation in grazing pressure. Similarly, some individual phytoplankton species increased in abundance (e.g., small centric diatoms and *Cerataulina pelagica*), whereas others decreased (e.g., *Thalassionema nitzschioides* and *Mesodinium rubrum*) during the oil spill.

For comparison of phytoplankton communities potentially affected by the DWH oil spill, there were increases and decreases in the abundances of phytoplankton east and west of the Mississippi River delta (Quigg et al., 2021b). Small temporal and spatial scale variability of phytoplankton community dynamics precluded any inferences of petroleum exposure related to their composition. Phytoplankton studies to the east of the Mississippi River were limited and lacked good long-term data.

6.5.6.3 Zooplankton Communities

Most zooplankton do not occupy the upper water column during the day, in order to avoid their predators. The result is exposure in the surface layer at night and what might be encountered subsurface in day light. Daly et al. (2021) assessed the zooplankton community in the area from the DWH well head northeastward through the DeSoto Canyon to nearshore. Parameters included abundance, biomass, spatial distribution, species composition, and diversity indices in spring, summer, and winter, May 2010 to August 2014. SEAMAP (Southeast Area Monitoring and Assessment Program, National Marine Fisheries) samples collected between spring and summer 2005–2009 were analyzed as a baseline against which supplemental studies during the oil spill may be compared. The results of Daly et al. (2021) demonstrated that zooplankton community dynamics are strongly governed by environmental variability and riverine processes. The oil spill in spring 2010 did not significantly affect the Zooplankton abundances compared with those from spring 2011 and 2012. Over the period from 2005 to 2014, the summer 2010 zooplankton abundances were the highest observed and correlated with a high river discharge, high chlorophyll, and aggregation in eddies.

Carassou et al. (2014) examined zooplankton communities from the Alabama continental shelf (north central Gulf of Mexico) before and during (2005–2009) the DWH oil spill (for the months of May–August). They observed shifts in assemblage structure in May and June 2010, but these differences were no longer significant by July 2010. ANOSIM testing confirmed weakly significant, differences in mesozooplankton population composition during the oil spill years, when compared to historic years. Mesozooplankton assemblages were different during the oil spill at both sites when all months were combined together ($p < 0.2$).

Many taxa had higher densities during the oil spill year (e.g., calanoid and cyclopoid copepods, ostracods, bivalve larvae and cladocerans, and echinoderm larvae) but the differences were inconsistent among stations. Daly et al. (2021) cited environmental variables as important factors affecting the zooplankton communities in their study, but similar environmental factors were dismissed by Carassou et al. (2014), and the differences were attributed to the DWH oil spill. In neither case were data for potential exposure to petroleum hydrocarbons provided.

6.5.6.4 Kelp Beds

Kelp is a macroalga, attached to the seabed with gas-filled structures at the base of the blades to hold the kelp blades close to the water's surface. Thus, its oil exposure may be in the surface sheen or in the subsurface water column. Results are mixed concerning kelp communities exposed to an oil spill. The *World Prodigy* tanker released approximately 922 tons of No. 2 fuel oil on Brenton Reef, Rhode Island, and into surrounding coastal waters in 1989. Peckol et al. (1990) investigated the effects of oiling on the subtidal kelps *Laminaria saccharina* and *L. digitata*. Data on conditions of kelp at the same site, during pre-spill conditions (1984–1987) were compared with post-spill kelp condition, growth rates with depth, and pigment acclimation. The data indicated no evidence of detrimental effects by oiling on the kelps. There were no necrotic or bleached tissues on any kelps in an oiled cove. Both kelp species continued growth rates within the range of previous years' data and pigment acclimation was similar for all years. A significant brown tide in 1985 caused the lowest growth rates. This study and other data suggest that Narragansett Bay avoided harmful effects of the oil spill because little fuel oil mixed into the water column and therefore subtidal vegetation was spared.

A massive outflow of oil from an offshore drilling accident occurred near Santa Barbara, California, in January 1969. A part of the offshore-through-intertidal transitions of vegetation, algae, and invertebrate communities exposed to oil were offshore kelp communities. Some direct observations close in time to the spill (Foster et al., 1971) indicated that the kelp, consisting almost entirely of *Macrocystis angustifolia,* received the first dose of incoming oil. The floating fronds initially held large quantities of oil, especially during low tides, and the brown color of the beds turned black. Most of the oil in the offshore fringes of the kelp was dispersed by winds, currents, and tides, then moved shoreward. Oil did not adhere to healthy kelp fronds. Furthermore, invertebrate organisms associated with the oiled kelp beds did not differ from those in communities associated with non-oiled kelp.

6.5.6.5 Fish and Other Water-Column Inhabitants

Although widespread mortalities of fish, crustaceans, and other megafauna inhabiting the marine water column are infrequent, high winds during or immediately after spills that promote and entrain dispersion of oil droplets into seawater increase deleterious effects on these organisms. The 1996 *North Cape* spill provides an example of serious effects attributed to oil naturally dispersed into the air and water. Amid a winter storm near the coast of Rhode Island the barge *North Cape* discharged ~3,000 tonnes of No. 2 fuel oil. Storm winds above 100 km/h and breaking waves >5 m spread the oil along the coast and into inshore salt ponds, and dispersed the oil throughout the water column (Reddy and Quinn, 2001). Concentrations of 26 PAC (polycyclic aromatic compounds)[5] and of total petroleum hydrocarbons, existing as droplets of dispersed oil and dissolved compounds, in the water column reached 115 and 3,940 µg/L, respectively. These measurements are some of the highest concentrations of PAHs in the water column ever recorded after an oil spill (Reddy and Quinn, 2001), causing substantial mortality to aquatic organisms. Estimated deaths include 2,292 birds, 312,000 kg of lobsters, nearly 1 million kg of shellfish, and 116,000 kg of fish (NOAA et al., 1999).

Detection of photoenhanced toxicity effects in the open marine water column has not been clearly established. Studied of natural seawater polluted with oil from the North Cape showed a significant increase in the toxic effects of the oil to embryos of dwarf surf clams (*Mulinea lateralis*; Ho et al., 1999) after photoenhancement (Arfsten et al., 1996). However, photoenhanced toxicity effects of oil from the 2002 *Prestige* oil spill were not detected in embryogenesis bioassays involving mussels (*Mytilus galloprovincialis*) or sea urchins (*Paracentrotus lividus*), nor in the copepod *Acartia tonsa* or the fish *Cyprinodon variegatus* (Saco-Álvarez et al., 2008). Photoenhanced toxicity effects in the sea water column were also not reported after the DWH oil spill, but that may have been a consequence of lack of sampling effort.

Cardiotoxic effects appeared in Pacific herring (*Clupea pallasii*) embryos reared in subtidal cages near shorelines oiled by the 2007 *Cosco Busan* spill (Incardona et al., 2012). Though cardiotoxic effects probably affected developing embryos of fish and perhaps other organisms exposed to PACs dissolved from Macondo 252 oil following the DWH spill, clear evidence supporting this has not been presented.

6.5.6.6 Produced Water Discharges

Another type of dispersed petroleum hydrocarbon results from the disposal of produced waters, a by-product of oil and gas exploration and production with contamination by elevated salinity (usually), petroleum hydrocarbons, trace metals, and radionuclides (see Sections 3.4.1.2 and 5.2.5). Produced waters are usually discharged mid-water but sometimes onto the water surface or near the seabed (Neff et al., 2011). Produced waters are the primary source of hydrocarbons from these activities, with minor contributions from deck washing and drilling fluids and cuttings. The constituents of produced water and their relative proportions vary widely according to the petroleum reservoir (Neff et al., 2011). The consensus of several reviews (Holdway, 2002; Neff et al., 2011; International Association of Oil & Gas Producers, 2020) is that the produced water pollutants disperse quickly into the water column, and laboratory studies of toxicity and field surveys suggest that the overall risk of produced water discharge inducing adverse impacts in populations of pelagic organisms is low.

The Canadian Environmental Research Studies Funds supported monitoring of juvenile fish that were exposed to hydrocarbon discharges from three oil and gas production platforms on the Grand Banks and a reference area at least 2 km from the production platforms (LGL Limited and Ocean LTD, 2018). Four species of fish (American plaice, Atlantic cod, capelin, and sand lance) were collected in both bottom trawls and mid-water trawls at least 1 km from the discharges. Comparisons of ethoxyresorufin-O-deethylase (EROD) activity as indicators of increased enzymatic responses and metabolites in fish tissues and bile were inconsistent with relationship to a fish type, fish size, discharge platform or distance from it. There were no clear relationships of produced water discharges with metabolic indicators in the fish (LGL Limited and Ocean LTD, 2018). Potential reasons for this were the mobility of the fish collected at least 1 km from the discharge, and the uniqueness of produced water chemical constituents by platform.

Produced water also contains substantial concentrations of alkyl-phenols, some of which are estrogen-mimic endocrine disruptors (Boitsov et al., 2007; Meier et al., 2007). Discharge in hypersaline produced waters may transport these compounds to the seafloor. Adverse effects of these compounds appear to be limited to 1–2 km from the point of discharge (Bakke et al., 2013). Examination of organisms in situ that are exposed to produced water discharges include studies of kelp and deployed mussels. Giant kelp (*Macrocystis pyrifera*) recruitment near a produced water diffuser off the Santa Barbara, California, coast was affected only within 50 m of the outfall, most likely related to gametophyte survival (Reed et al., 1994). Osenberg et al. (1992) deployed mussels (*Mytilus edulis and M. californianus*) down-plume from the same discharge off Santa Barbara and found distance-from-source sublethal effects in shell growth and condition. Osenberg et al. (1992) noted that it was difficult to separate differences in mussel recruitment from planktonic larvae and local production of propagules.

Although not water column organisms, the produced water discharge affected benthic communities with nematodes being more abundant closer to the diffuser off Santa Barbara, but there were reduced abundances of most carnivorous groups, including nemerteans and several families of polychaetes (Osenberg et al., 1992). Similarly, Rabalais et al. (1991, 1992) documented adverse effects on benthic communities in the northern Gulf of Mexico, such as

[5] Although PACS and PAHs are sometimes used interchangeably in the literature, PAHs containing only carbon and hydrogen are a subset of PACs. See Section 2.1.3 for more details.

mortality, lower abundance, and dominance by a few species of opportunistic polychaetes. These effects were within 500 m of the discharge. Discharges closer to the seabed (in this case, at 19 m in a 20-m water column) contained higher concentrations of petroleum compounds than discharges in mid water, because the produced waters were entrained in the hypersaline plume at the sea bed and not dispersed.

Determining effects on pelagic communities is difficult, primarily because of unknown exposure, wide dispersal of oil hydrocarbons, multiple environmental factors in this water habitat, and the ephemeral nature of these communities. To fill this gap, in situ sensors could be developed for petroleum hydrocarbon detection and image analysis for plankton, as well as using autonomous underwater vehicles (AUVs) for determination of water column effects. However, these communities are ephemeral.

6.5.7 Ecosystem Effects

Oil in the Sea III (NRC, 2003) noted the difficulty of translating individual-level effects of an oil spill to the level of population effects. Extrapolation of population-level effects to the level of ecosystem effects is more difficult. Reasons for the difficulty include the high variability of temporal and spatial characteristics of the habitat, along with the variability in population structure and dynamics, environmental forcing factors, and unknown community structure and trophic interactions. Within the context of open ecosystems, the successful recruitment of individuals is essential for the recovery of populations and subsequent community interactions.

6.5.7.1 Trophic Transfer

Oil pollution may become incorporated into marine food webs through trophic transfer of toxic compounds via ingestion of PAHs from the primary producer level to higher organism levels. Evidence of trophic transfer is found in studies where the source of hydrocarbons contained in prey organisms was clearly linked to oil components ingested or absorbed by those organisms, and the transfer of those hydrocarbons from prey to their predators is also clear. These hydrocarbons are bioaccumulated when ingestion leads to incorporation into the somatic tissues of the organisms that consume them. Biomagnification occurs when organisms are unable to efficiently excrete accumulated contaminants, resulting in contaminant concentrations that increase in species occupying higher trophic levels. The enzymatic pathways of CYP1A induce metabolism of PAHs in vertebrates, but there is no biomagnification of these contaminants, as in growing-body burdens of mercury. Oil-contaminated plants and animals, whether through contact with or ingestion of oil, provide a route of additional contamination of species that consume them. Animals that consume oil-contaminated prey may incorporate various components of the ingested oil into their tissues. These oil components tend to migrate into the most lipid-rich tissues because of the higher solubility of oil components in natural oils compared with water. The persistence of oil components in tissues depends mainly on how fast the organism turns over its lipid reserves, and whether the organism is able to biochemically degrade the accumulated oil components.

Most invertebrates are cold-blooded and hence have relatively slow metabolic rates, and also have less well-developed biochemical pathways to degrade the more persistent and toxic oil components such as aromatic hydrocarbons, so half-lives of accumulated oil components may range from several days to months. However, fish and most other vertebrates, as well as some invertebrates, can produce enzymes that rapidly degrade aromatic hydrocarbons into more water-soluble products that are readily excreted. These pathways may be activated within an hour of initial exposure, and can rapidly degrade most ingested aromatic hydrocarbons within 1 day or so.

Vertebrates constitute most of the species that occupy the higher trophic levels of marine food webs, and their ability to rapidly degrade aromatic hydrocarbons precludes biomagnification of these compounds in marine food webs. Although organisms may readily bioaccumulate oil and oil-derived compounds through ingestion or absorption, these compounds rarely biomagnify in marine food webs. Finfish and most arthropods are able to metabolize PACs relatively rapidly (Meador et al., 1995), and consequently are considerably less likely to be detectably tainted by petroleum-derived compounds after a spill in comparison with filter- or suspension-feeding shellfish. Filter- and suspension-feeding shellfish can efficiently accumulate dispersed organic particles, including oil droplets, from seawater, and their depuration rate of these compounds is relatively slow. For example, blue mussels (*Mytilus edulis*) ingest food particles ranging from 1–35 μm (Strohmeier et al., 2012), and oysters (*Crassostrea virginica*) ingest particles as large as 400 μm (Tamburri and Zimmer-Faust, 1996). Once ingested petroleum compounds may remain detectable for months (Meador et al., 1995) following surface oil spills that generate widespread dispersions of oil microdroplets into the water column near the sea surface.

6.5.7.2 Community Effects

Damage inflicted by large oil spills on populations and habitats can perturb ecological communities, with consequences that are long term and possibly permanent. For example, effects may result from widespread initial mortalities of keystone species that subsequently alter populations of other species and communities that have strong links with them. Avoidance of persistent oil-contaminated habitats by organisms, aquatic and terrestrial, can alter community interactions, including permanent alteration following local extirpation of one or more aquatic or terrestrial species. Community-level effects remain among the least well-understood consequences of oil spills. Potential pitfalls for lack of evidence are identified in Section 6.5.1. This section examines community-level effects of major oil spills and makes appropriate suggestions of altered ecological processes. Conclusions may not be definitive and remain inconclusive, but may outline suitable rationale for an "effect" or no "effect" call.

6.5.7.3 Indirect Effects and Trophic Cascades

Community responses to strong environmental perturbations provide invaluable insights regarding ecological structure, linkages, and functioning, especially in marine ecosystems at large spatial scales that are not readily amenable to direct experimental manipulation and composed of species populations that are usually difficult to monitor. The appropriate standards of evidence for evaluating hypotheses in these cases are similar to those often applicable in geology, where ancient processes that are not directly observable must be inferred from available observational data in the present. In such cases, acceptance of proposed hypotheses depends on the ability of a candidate hypothesis to account for an extensive body of widely ranging qualitative and quantitative observations, along with the absence of any conclusively contradictory evidence.

Large marine oil spills have the potential to cause strong ecological perturbations if they result in widespread mortalities of major component species. These indirect effects have often been suspected after major spills, but have rarely been clearly documented. The most compelling examples of such indirect effects are associated with the DWH and *Exxon Valdez* oil spills, and even for these the available evidence remains merely suggestive. Nevertheless, these are summarized here in recognition that appropriate appreciation for the weight of this evidence provides guidance for more careful evaluation of similar effects following future oil spills or other perturbations of marine ecosystems. Substantial evidence suggests that the extensive mortalities of piscivorous seabirds following the DWH oil spill (Haney et al., 2014a,b) released juvenile Gulf menhaden from predation, triggering a trophic cascade response (see Figure 6.32). Gulf menhaden abundance and biomass reached record-breaking levels in 2011 and 2012, which were associated with the poorest body condition on record as well (Short et al., 2017, 2021), suggesting that the population had outstripped its available supply of phyto- and zooplankton food. As an important forage fish in the coastal waters that were contaminated by oil from the DWH spill, the poor body condition of Gulf menhaden in 2011 and 2012 made them less nutritious for the many species that consume them. This is an example of a community-level response that may have gone undetected in prior oil spills such as the *Exxon Valdez* oil spill.

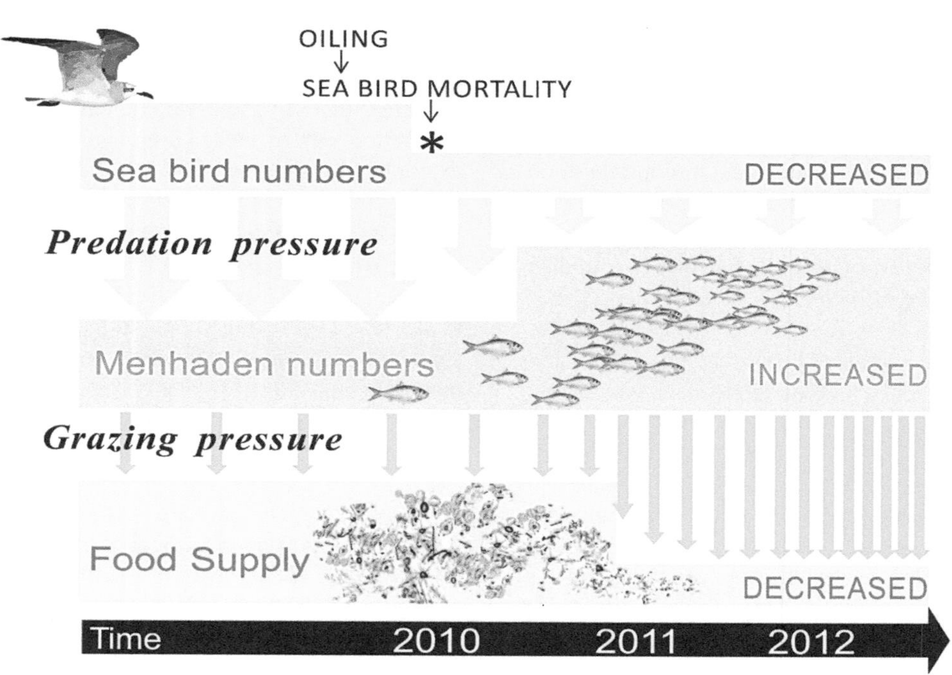

FIGURE 6.32 Conceptualized response of Gulf menhaden to reduced predation that resulted from mortalities of piscivorous seabirds from oiling by the DWH oil spill.
SOURCE: Short et al. (2017). CC BY.

Initial loss of cover by the biogenic habitat—the rockweed *Fucus gardneri*—following the *Exxon Valdez* oil spill triggered a cascade of indirect impacts that lasted for a decade or more. Studies that distinguished effects from natural variability at un-oiled locations, from shoreline treatment efforts, and from oil exposure (reviewed by Peterson, 2001) showed oil-associated reductions in *Fucus* and important predatory limpets and gastropods, increased space on rocks for blooms of ephemeral green algae in 1989 and 1990, and for an opportunistic barnacle (*Chthamalus dalli*) in 1991. Loss of the structure of the *Fucus* itself led to declines in associated invertebrates and inhibited recovery of the *Fucus*. *Fucus* plants that subsequently settled on tests of *C. dalli* became dislodged during storms because of the structural instability of the attachment of this opportunistic barnacle. After apparent recovery of *Fucus*, previously oiled shores exhibited mass rockweed mortality in 1994, which is a cyclic senility and resulting instability (see Box 6.5). The general sequence of succession on rocky intertidal shores is of rapid recovery of short general times of intertidal plants and animals (Menge, 1995). Expectations of rapid recovery based on short generation times of most intertidal plants and animals are not supported by the data. The general sequence on rocky intertidal shores after the *Exxon Valdez* spill was a delayed set of indirect effects over a decade or longer (Peterson et al., 2003).

Indirect interactions are not restricted to trophic cascades or to intertidal benthos. Interaction cascades defined broadly include loss of key individuals in complexly organized populations, which then suffer subsequently enhanced mortality or depressed reproduction. The most compelling example of a trophic cascade radically modifying a marine community comes from the Gulf of Alaska kelp ecosystem (Estes and Duggins, 1995). Sea otters control sea urchin populations, preventing them from overgrazing kelp and other macroalgae, and thereby retaining structural habitat for fishes and invertebrates. If sea otters were eliminated from an area by an oil spill, the otter-urchin-kelp cascade is put at risk. However, a ~50% reduction of sea otters was apparently not sufficient to induce this cascade following the *Exxon Valdez* oil spill (Dean et al., 2000).

6.5.7.4 Ecosystem-Level Effects

Media statements regarding the effect of DWH hydrocarbon on the 2010 continental shelf "dead zone" were dire because they extrapolated nominal reductions in deep waters (approximately 1,200 m) by bacterial respiration (Hazen et al., 2010). However, there was no indication that DWH oil residues influenced the seasonal occurrence of bottom-water hypoxia in 2010. There were indications of different carbon signatures near the Mississippi River, but these stations also demonstrated characteristics of high freshwater inputs onto the continental shelf (Hu et al., 2016). The combination of oil and dispersants mixed in the water column from the DWH oil spill combined with the annual development of oxygen-depleted bottom waters on the Louisiana continental shelf. It was unknown whether the oil and dispersants from the spill might affect the seasonal hypoxic area formation by either worsening the extent or severity. In May, measurements of the surface and bottom water hydrocarbons were higher than levels observed in June and July. The dissolved oxygen concentrations in bottom water were higher in May and June than in July. It was unknown the level of oil degradation in the water column or sediment. Statistical analysis of the progression of hypoxia development in May, June, and July 2010, and an analysis of conditions in July compared background levels collected over a 27-year period, indicated no difference in oxygen concentrations for May, June, or July 2010, with or without oil data included. The analysis also did not find any difference in July 2010 compared to the background years. Findings instead showed, throughout the dataset, that, the hypoxic area increased with higher river discharge, higher nitrate-N load, an easterly (westward) wind, and reduced wind speed. The analyses could not determine whether the oil spill affected, or did not affect, the size of the 2010 hypoxic zone, but there were signals that the 2010 hypoxia season was similar to the long-term record (Rabalais et al., 2018).

6.5.7.5 Ecosystem Services

The effects of an oil spill on ecosystem services is not an area of coverage for this report, but it is important to note that ecosystem services are curtailed with negative impacts on ecosystems or their components. Intertidal biogenic habitats, such as salt marshes and mangrove swamps, provide niches for high biodiversity. They also accumulate sediments and serve as a means of their cohesion against sediment resuspension and erosion. They serve as nursery habitat for larval, post-larval, and juvenile organisms in estuarine-coastal water networks. Salt marshes and mangroves also provide a defense against high tides and waves during tropical storms and hurricanes. Salt marshes filter nutrients and pollutants from the water column. Deep-sea corals and their associated invertebrate and fish communities are not only biodiversity hotspots but also unique among marine habitats. Intertidal and subtidal coral reefs also host high biodiversity. Ecosystem services can be converted to economic value. Should an ecosystem be negatively affected by an oil spill, the loss of ecosystem functions will convert to an economic loss for otherwise societal and environmental benefits. The National Research Council provided a report on ecosystem services related to the DWH oil spill and recommended several avenues of research that would more fully address this issue and future evaluation of effects (NRC, 2013).

Ecosystem-level effect conclusions remain elusive; the addition of longer-term observations and experiments that include higher organization-level components and trophic interactions could shed light on these effects.

6.6 EFFECTS IN ARCTIC ENVIRONMENTS

Decline by nearly one-third of the summertime minimum sea ice cover in the Arctic since the early 1980s (NASA)[6] has opened up more of the region's coastal waters to exploration for oil and gas, commercial fishing, trans-continental shipping, and commercial tourism. As described in Box 3.1, these activities bring greater risks of accidental oil discharges in a region with scant infrastructure, a challenging climate, highly variable weather that can turn treacherous with little warning, and often the presence of sea ice. These conditions also make response efforts to accidental oil discharges, as well as efforts to study the Arctic marine ecosystem and evaluate environmental damage from spills and other oil discharges, considerably more difficult than in more temperate waters. Consequently, marine ecosystem structure and functioning in the Arctic remain poorly understood, and the functioning of these ecosystems is changing rapidly in response to accelerating warming of the Arctic Ocean. The challenges to understanding Arctic marine ecosystems and the effects of oil on vulnerable ecosystems are briefly summarized in this section.

6.6.1 Marine Ecosystems in the North American Arctic

The annual marine production cycle in the North American Arctic is strongly modulated by ice. Complete ice cover during winter and most of spring results in nearly all of the solar radiation prior to the summer solstice impinging on ice which, with an albedo of 0.5 to 0.7, reflects more than half of it back into the atmosphere. Open sea water absorbs more than 90% of incident solar radiation, and open water is most widespread during summer. Primary production rates are consequently highest during summer, but considerable primary production occurs during spring as well despite the ice cover. This is for two reasons. First, most of the ice cover in the Arctic Ocean is relatively thin first-year ice, usually 1–2 m thick (Renner et al., 2013), and can transmit much of the incident photosynthetically active radiation (PAR) to the lower sea ice surface interface with sea water. The lower ice surface provides habitat that keeps attached marine algae exposed to PAR, and upwelled inorganic nutrients from the continental shelf break current in the Bering Sea are transported through the Bering Strait into the Chukchi Sea, fueling rapid growth of these epontic algal communities (Springer et al., 1996).

Second, and more importantly, as sea ice melts during the spring and especially summer retreating phase, the meltwater stabilizes the water column, allowing dispersal of the epontic algae into the mixed layer that often extends to a shallow seafloor where, along with other algal species, very rapid growth is supported by the presence of considerably higher PAR intensities present throughout most of the day, and the steady supply of inorganic nutrients. These rapid, intense algal blooms typically follow the retreating sea ice edge, and saturate the capacity of zooplanktonic grazers to consume them (see Figure 6.33).

Annual primary productivity in the western Arctic Ocean can be quite high, reaching several hundred grams carbon per m^2 of sea surface per year (Springer et al., 1996; Smith et al., 2017). Some measurements exceeded 800 g C/m^2-yr north of the Bering Strait (Sapozhnikov et al., 1993). Most of

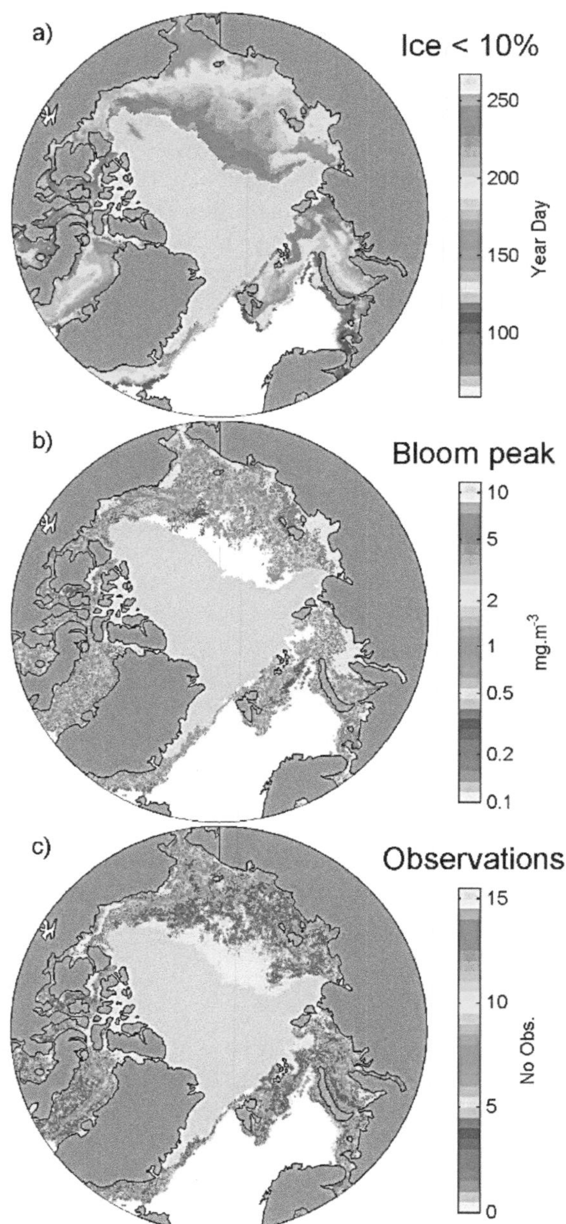

FIGURE 6.33 Correlation between peak algal bloom and location of sea ice retreat in the Arctic during 2007.
NOTE: MIZ = Marginal Ice Zone.
SOURCE: Perette et al. (2011).

[6] See https://climate.nasa.gov/vital-signs/arctic-sea-ice.

the algal production falls unconsumed to the seafloor, where it sustains high secondary production of the benthic invertebrate community, including dense accumulations of numerous species of crabs, starfish, and various marine worms (Smith et al., 2017).

Marine consumer species inhabiting the Arctic Ocean year-round must develop bioenergetic strategies to survive the fall, winter, and early spring months, when primary production is low or negligible. Fish, especially young-of-the-year, must accumulate sufficient lipid reserves during the brief season of abundant availability of phyto- and zooplankton prey to survive until the following spring, something relatively few species such as saffron cod and Arctic cod manage. Another strategy is to specialize in consuming live or dead animals in the benthic community, because this food supply is available throughout the year. Crabs, starfish, bivalve and other mollusks, and other species that can feed on living organisms or decaying detrital matter produced by the benthic community adopt this strategy. So do most species of marine mammals that reside in the Arctic Ocean year round, as well as several marine mammal species that migrate seasonally to the Arctic Ocean.

6.6.2 Arctic Marine Organisms Vulnerable to Oil Pollution

As elsewhere, seabirds and marine mammals are particularly vulnerable to adverse effects of oil pollution in the western Arctic Ocean. This being said, challenges associated with maintaining core body temperatures in the face of external oiling for these taxa become even more challenging than in temperate climates, even in the face of light oiling. Several species of marine mammals excavate and maintain breathing holes in first-year sea ice throughout the ice-cover seasons (Smith et al., 2010). Accidental releases of oil would tend to accumulate in these holes, exposing these mammals to oil through inhalation and possibly ingestion. They may also expose polar bears to oil, because these bears often wait for marine mammals to surface in these holes to attack them. Similarly, oil released in broken sea ice will be herded by the ice patches into the channels separating them, presenting a similar contact and inhalation hazard to marine mammals. Most seabirds in the Bering and Chukchi Seas follow the retreating sea ice edge northward in spring, feeding on fish and invertebrates associated with the high primary production associated with the ice edge, and concentrations of seabirds in large numbers occur along the northwest Alaska coast during summer (Smith et al., 2017). An accidental oil discharge near the retreating ice edge could cause widespread mortalities of seabirds. In the water column, acute toxicity tests indicate that aquatic organisms inhabiting Arctic waters are about as sensitive as comparable organisms living in more temperate waters, although adverse effects of exposure to oil may take longer to manifest in Arctic organisms.[7]

[7] See https://neba.arcticresponsetechnology.org.

An accidental subsurface oil release would tend to spread horizontally at the seawater–sea-ice interface (see Section 5.3.5), where it could contaminate the epontic communities associated with the ice, and also be drawn into the ice by capillary action into brine channels (Faksness and Brandvik, 2008). Brine channels form as sea ice accumulates through exclusion of salt by the ice crystals, which increases the concentration of salt in the receiving boundary layers of seawater. This lowers the freezing temperature of the seawater, which can create microchannels within ice floes.

Accidentally discharged oil that becomes encapsulated in growing sea ice during fall may be transported several km per day (Nansen, 1902; Kwok et al., 2013). If oil leaking from a future pipeline or other source in the Arctic Ocean during fall went undetected, it could reappear as a "mystery spill" 100 or more km distant from the point of initial release during the subsequent spring thaw.

6.7 OIL EFFECTS MODELING

6.7.1 Overview of Fate, Exposure, and Toxicity Models

Modeling efforts have been directed toward understanding the past, current, and future trajectories of oil location, fate, exposure, and impacts of oil to organisms. These tools (see Section 4.2.1.4) have been essential components in contingency planning and risk assessments and in oil-spill response decisions (NASEM, 2020). Oil spill models have been used to support oil spill response decisions as part of NEBA and its refinements (i.e., spill impact mitigation assessments [SIMAs]) (NASEM, 2020). For example, oil spill models have been used to estimate habitats exposed to oil concentrations above certain thresholds for certain periods of time in comparative risk assessments (CRAs; a SIMA-type of approach) that allow for tradeoff decisions on potentially affected resources (French-McCay et al., 2018b). They are also essential in NRDA and hindcasting efforts.

Since *Oil in the Sea III* (NRC, 2003), new models have been developed, and existing ones have been updated and refined based on new knowledge. For impacts to occur, organisms need to be exposed to oil, and field sampling cannot fully quantify—spatially or temporally—the details of concentration and hydrocarbon composition that would be required. Therefore, models have been directed at understanding the fate (e.g., location, concentration, specific chemical constituents, among other features) of oil; these were previously discussed in Chapters 3 and 4 and thus are only briefly mentioned in this chapter. Oil trajectory models provide information regarding where the oil may go, providing estimates of surface or subsurface oil exposure (Boufadel and Geng, 2014; Ji et al., 2020; Keramea et al., 2021; Nordtug and Hansen, 2021). Oil fate and exposure models employ multiple steps and provide estimates on the composition, concentration, and partitioning of oil constituents both temporally and spatially and can also include the distribution, movements, and behaviors of aquatic organisms to estimate their exposure to oil (see reviews by McCay et al., 2018a,b, 2021a).

Integrated trajectory, fate, and effects models also show the evolution of the spill, including oil component concentrations, which can be used for planning response options, response optimization in drills and aid in responding to oil spill events. Oil trajectory models have been expanded in recent years (NRDA) for use in quantifying the extent of oil impact (French-McCay, 2004, 2009) and in forecasting oil droplet distributions and oil constituents, and improved for use in Arctic locations (Boufadel and Geng, 2014; Nelson and Grubesic, 2018; Nordam et al., 2019; Ji et al., 2020; Keramea et al., 2021). These models have been used to identify and predict the resources of concern for impacts although specific impacts are not identified given that chemical toxicity depends on the concentration, specific type, and weathering state of the oil unique to each spill event.

Data derived from laboratory toxicity tests have been used in developing and also validating models to predict biological effects, using mechanistic studies of single hydrocarbons, which are then validated using both hydrocarbon mixture and whole oil exposure tests (e.g., French-McCay, 2002). Laboratory toxicity test data suitable for model inclusion (see Section 6.4 regarding laboratory toxicity test data limitations) have also been collected. Approaches used allow a comparison of species to identify those that may be more sensitive than others (e.g., see Section 6.7.2.5 on CAFE) or are able to predict the toxicity of oil and its constituents to species that have not yet been studied (e.g., see Section 6.7.2.6 on ICE). Numerous issues have been highlighted regarding the utility of toxicity tests, especially in terms of representing field conditions for dynamic complex mixtures (see Section 6.4). Indeed, a recommendation made in the recent oil spill dispersant report (NASEM, 2020) was that toxicity tests should not be developed to represent or replicate field-exposure conditions (which they cannot do), but rather for use to further develop and refine models so that integrated fate, exposure and effects models can support the decision-making process during an oil spill response (i.e., by comparing all of the options; NASEM, 2020).

The comparative advantages of specific exposure metrics (e.g., chemistry reporting) have also been discussed in detail in the National Academies oil spill dispersant report and will not be repeated here (see NASEM, 2020). Briefly, models use chemistry concentration inputs as a number of pseudo-components, hydrocarbon groups or specific individual hydrocarbons and assume equal additive toxicity effects, or they input specific hydrocarbons and weight their individual toxic effects using a toxicity unit (TU) approach. Ultimately, the type of exposure metric chosen (in addition to specific analytical approaches used) may under- or over-represent the toxicity reported, and this should be considered when the choice is made. Two models often used that use boiling cuts (soluble and insoluble divisions) to reflect around eight pseudo-components are SIMAP (Spill Impact MAPping; see French-McCay, 2004; French-McCay et al., 2018a,b, 2021a) and also OSCAR (Oil Spill Contingency and Response; Reed, 2004; Stephansen et al., 2021). The SIMAP model has been modified continually and incorporates exposure to oil droplets, the influence of temperature and light, and the movements of organisms to provide a biological effects model to which specific aquatic toxicity models are then applied (e.g., OilToxEx; French McCay, 2002; see Section 6.7.2).

Trajectory models have been used in quantifying impacts to shoreline habitats, birds, mammals and reptiles. Using satellite-derived surface oil distributions and direct observations of oiled turtles a spatio-temporally explicit model was developed during the NRDA process in the DWH oil spill to statistically estimate the number of turtles that may have been exposed to oil (Wallace et al., 2017). These turtle numbers were later used to provide estimates of the number of turtles affected by chemical exposure and toxicity mechanisms (Mitchelmore et al., 2017; see Section 6.7.2). Specific advances discussed in Chapters 3 and 4, that are pertinent to this chapter include transport, fate and exposure models that estimate the concentration and composition of oil components, which are particularly focused on dissolved phase concentration and composition estimations. Since *Oil in the Sea III*, models have also been used to refine estimations for particular oil components (e.g., photo-oxidation products, MOS), predicting droplet size/quantity and particulate phase components and including environmental covariables that may alter fate and exposure (see NASEM, 2020).

6.7.2 Models to Estimate Aquatic Toxicity to Individuals

Numerous approaches and specific models have been used to provide information at multiple scales and levels of complexity regarding the adverse effects that may result from oil exposure, including future projections of population losses. Some of these modeling efforts provide predictive estimations of toxicity and can include considerations of specific environmental conditions (co-variables such as temperature, light, and pressure) (French-McCay, 2002, 2004, 2009; McGrath and Di Toro, 2009; Bejarano and Mearns, 2015; Marzooghi and Di Toro, 2017; Carroll et al., 2018; French-McCay et al., 2018b; Gallaway et al., 2019; Bejarano and Wheeler, 2020; Colvin et al., 2020; Stephansen et al., 2021). Initial models used representative oil chemical compositions, but are now more sophisticated and include mixture-based approaches, such as summations of effects of constituents using a TU approach. The validation of many models has been performed using hindcasts after a spill event and the use of data derived from laboratory toxicity test studies. These models are briefly described here, with pros and cons of each approach highlighted using examples of their application in understanding the impacts of oil in the environment. As described later, these models have been combined with physical fate and biological effects models to quantify impacts (e.g., OilToxEx in SIMAP, PETROTOX).

Regarding aquatic organisms, models have been used to estimate oil entrainment and dissolution into the water column. These models predict the concentrations of oil components that have been a focus of toxicological impacts: namely, the more soluble lower- and intermediate-molecular weight monoaromatic hydrocarbons (MAHs) and the PAHs (e.g., OilToxEx; French-McCay, 2002). These are predictive models that have over time been modified to incorporate evolving knowledge. Additional models include the target lipid model (TLM; McGrath and Di Toro, 2009) and associated PETROTOX model (Redman et al., 2012, 2017). Since *Oil in the Sea III* was published, other components in oil that also contribute to toxicity (e.g., heterocyclics) have been identified, in addition to covariates (e.g., UV light and phototoxicity) that modify toxicity. Thus, models have been refined and further developed based on these advances (Marzooghi and Di Toro, 2017). Furthermore, models traditionally focused on acute toxicity whereas refinements of models have included chronic toxicity endpoints (e.g., McGrath et al., 2018). A third and more recent type of model developed for oil toxicity has been directed toward hazard assessments. These models, which use laboratory toxicity data, were developed to predict the impact of oil constituents on standard or non-standard species (e.g., Interspecies Correlation Estimation or ICE models) or to estimate concentrations that are protective for 95% of the organisms (e.g., preparing species sensitivity distributions [SSDs] using CAFE).

6.7.2.1 Narcosis Target Lipid Models

Target lipid models quantify toxicity based on a relationship between octanol-water partition coefficient (K_{ow}) and are applicable to dissolved oil components on the assumption that such components result in narcosis effects directly proportional to the extent of accumulation in the tissues. This also assumes that each chemical results in an additive effect (Di Toro et al., 2000). The methodology is intended to be applicable to 95% of species (i.e., HC5), particularly for early life stages. McGrath and Di Toro (2009) developed the concept for aromatic hydrocarbons, using TUs for hydrocarbon mixtures (for a description on the use of toxic units, see NASEM, 2020). Essentially, this increases the complexity, because although individual components are still additive, their toxicity is not equal and each component provides its own proportion into the toxicity calculation and the specific composition and concentration of each chemical results in an estimation of toxicity. This is very different from the approach of reporting toxicity based on concentrations of mixtures using units of TPH or TPAH (see NASEM, 2020). Briefly, TUs for mortality are calculated from each of the pseudo-component chemical constituents' dissolved exposure concentrations divided by the $L(E)C_{50}$, with each constituent considered additive. This provides an estimate of the mixture's LC_{50} (i.e., sum of TUs = 1) at a certain concentration. The $L(E)C_{50}$s for this model were calculated using quantitative structure activity relationship (QSAR) models (McGrath et al., 2018).

The original concept is that 95% of species would be protected using this modeling approach. With increased understanding of toxicity and exposure mechanisms, the model has been updated (McGrath et al., 2018). Based on toxicity work during the *T/V Prestige* oil spill, Barata et al.'s (2005) work with copepods agrees with the concept of octanol-water partitioning being successful as they found the toxic effects of components to be additive. An early additive toxicity model, OilToxEx, was developed by French-McCay (2002) and used the TLM and TU approach to predict oil toxicity based on the effects from the dissolved chemical composition. Additive toxicity models are simple in terms of adding the individual toxicities from a mixture and are incorporated into the SIMAP model (French-McCay, 2003). These models are generalized by adding a coefficient to scale the model with observations. These models have been updated (e.g., PETROTOX) from the first models developed (French-McCay, 2002).

Given that toxicity data typically are provided for temperate standard test species, both TLM and experimental approaches were recently compared to assess the protection for tropical species exposed to gas condensate oils (Negri et al., 2021). Although several of the tropical species were among the most sensitive to be included in the TLM database, the comparison of the experimental and modeled efforts demonstrated the utility of the model for tropical species. The TLM-modeled HC5 was found to be more conservative than the experimentally derived hazard concentration after testing eight diverse tropical taxa (i.e., 78 and 167 µg/L, respectively; Negri et al., 2021). The authors did note that further research should be conducted to validate this with additional oil types, covariables (e.g., UV light), and especially with other keystone species, including additional coral species.

6.7.2.2 PETROTOX Model

PETROTOX can predict the toxicity of any oil, providing that its detailed oil composition is known. Equilibrium calculations are made on the distribution of the hydrocarbons based on the physical properties such as octanol-water partitioning. The most recent input type is based on GC×GC-FID. The higher-resolution version has 16 hydrocarbon classes based on hydrocarbon number for input. The model was developed to predict the aquatic toxicity of water in contact with oil as described in Redman et al. (2012). It employs the TLM to predict individual oil component toxicities and the TU mixture model to predict the whole toxicity of the oil mixture. The acute and chronic HC5 critical body burdens used in the calculations are from McGrath et al. (2018). So, PETROTOX is used to predict oil toxicity but also is used to examine the utility of TPAH/TPH for oil constituent

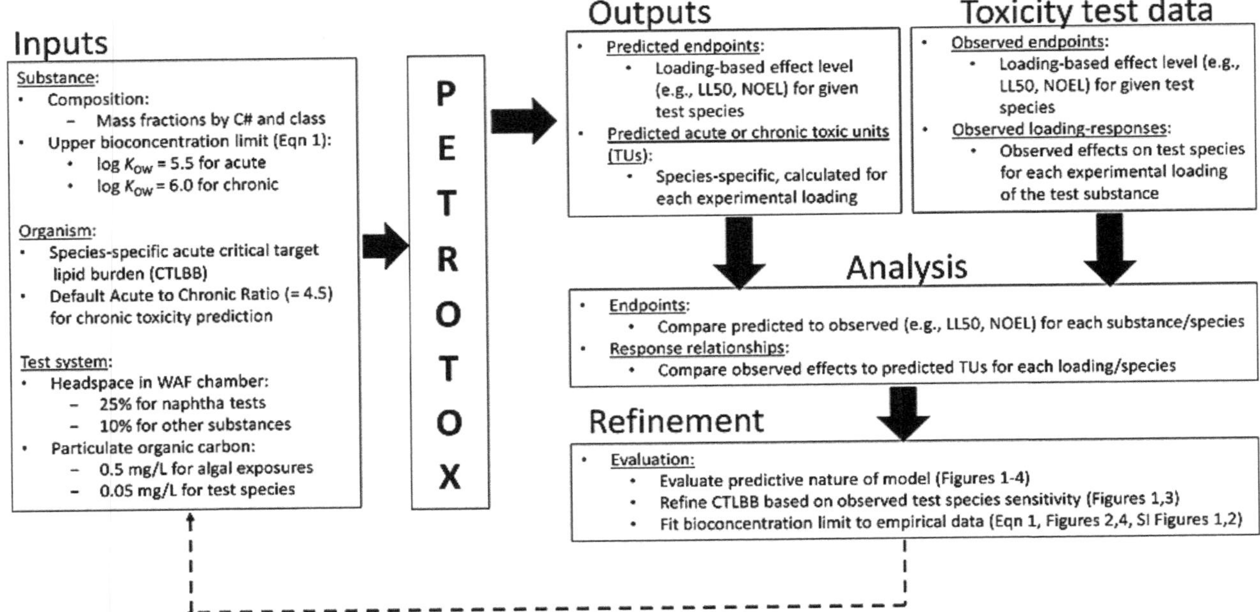

FIGURE 6.34 Flow chart describing analysis of toxicity data using PETROTOX.
NOTE: CTLBB = critical target lipid body burden; LL = lethal loading; NOEL = no observed effect loading; WAF = water-accommodated fraction.
SOURCE: Redman et al. (2017).

concentrations and evaluate how toxicities vary with different oil types, oil loading, and oil droplet concentrations (see NRC, 2020). The latest advancement in the PETROTOX model is a recalibration of two parameters, discussed in Redman et al. (2017) demonstrating the utility of the TLM-TU model in PETROTOX to reproduce observed chronic toxicity (see Figure 6.34).

The effect of UV in sunlight on oil can add oxygen molecules to the hydrocarbons as well as interact with polycyclic aromatic compounds (PACs). This can greatly increase toxicity by up to three orders of magnitude (see Barron, 2017b). A number of models of PAH phototoxicity have been developed (see Table 3.3 in NASEM, 2020). Marzooghi et al. (2017, 2018) recently developed the Phototoxic Target Lipid Model (PTLM) as found in the PETROTOX model (Marzooghi and Di Toro, 2018). This model calculates the ratio of the phototoxic LC_{50} (PLC_{50}) to the TLM LC_{50} as a function of PAH spectral absorbance and the spectral distribution of the incident light exposure (NASEM, 2020).

6.7.2.3 De Minimiz Risk Models

This model by Sellin Jeffries et al. (2013) used ultraviolet A (UVA) data to calculate the depth of PAH phototoxicity for young herring during the T/V *Exxon Valdez* oil spill. The results were a worst case estimate of 2-m depth, which correlated well with experimental data, and where 1% of the population would be located. The model compares well with data using median lethal times (LT_{50}).

6.7.2.4 The Dispersant and Chemically Dispersed Oil Toxicity Database (DTox)

Similar to the CAFE database, the DTox database provides a publicly accessible compilation of carefully selected available dispersant and chemically dispersed oil toxicity data (Bejarano et al., 2013). Data for inclusion had to meet a number of criteria (see details in Bejarano et al., 2013a), including a minimum requirement for method descriptions and chemical analysis and various quality assurance/quality control (QA/QC) evaluations. DTox is a user-friendly search tool and can provide the data in the form of SSD plots. It provides a centralized repository useful not just to the scientific community but also to oil spill responders in making decisions regarding the use (or not) of various response options (e.g., use of oil spill chemical dispersants and comparisons between specific ones). The database can also be used to compare the relative toxicity of chemically (CEWAF) versus physically dispersed oils (WAF) using a combined number of petroleum hydrocarbon products. Even despite the caveats discussed in Section 6.3 and NASEM (2020) regarding differential interpretations of toxicity based on the choice of analytical chemistry used, DTox database queries were consistent in demonstrating the lack of evidence that CEWAFs were more toxic than WAFs (NRC, 2005; Bejarano et al., 2013a). This was evidenced by comparing CEWAF versus WAF data for fish and crustacean species and also alternatively using all WAF and CEWAF data for a single species. The comparison showed that data were interspersed over the curve and that CEWAF data points did not cluster at the lower end of the curve indicative of higher toxicity. The dataset is also able to compare acute versus chronic endpoints, the toxicity for different species (e.g., temperate versus Arctic) and different oil types. For example, SSDs were similar irrespective of use of LC or EC50 data and showed that polar species demonstrated similar responses to temperate species (see Figure 6.35A); thus temperate species could be suitable surrogates for understudied Arctic species. The dataset also highlighted that Venezuelan crude oil was much more toxic than two Alaskan oils (i.e., Prudhoe Bay and Alaska North Slope) (Bejarano et al., 2013a; see Figure 6.35B).

As demonstrated, DTox allows for queries to investigate and compare the toxicity of different oils, oil preparation types, and species and can also be informative for response decisions regarding the use of dispersants. This DTox database was ultimately incorporated into a newly developed database, NOAA's CAFE model as described in Section 6.7.2.5.

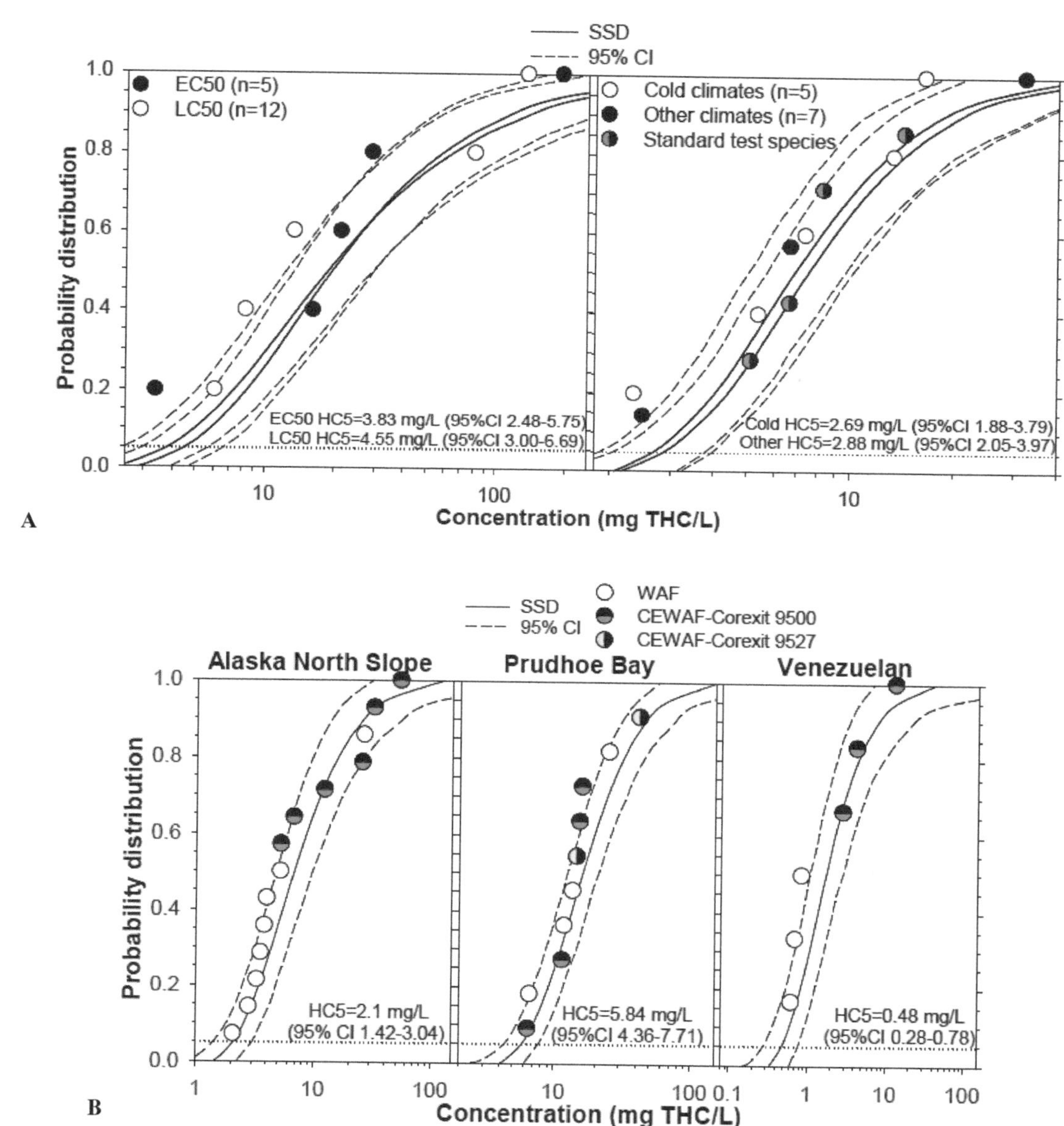

FIGURE 6.35 (A) Comparison of SSDs using EC or LC50 endpoints (left) and from species from cold or other climates (right). (B) Display of the relative aqueous toxicity of several oils, and relative position of CEWAF within oil type.
NOTE: 96-hour LC50/EC50 data for fish and crustacean from flow-through contact and studies reporting toxicity based on measured THC made up SSDs.
SOURCE: Bejarano et al. (2013a).

6.7.2.5 CAFE and Species-Sensitivity Distributions

In environmental risk assessments (ERAs), a number of approaches are used to determine the risk of chemical contaminants to resident species and estimate the acceptable concentration of a chemical in the environment: that is, when predicted no effect concentrations (PNECs) are lower than the predicted environmental concentration (PEC; see Box 6.10). Species sensitivity distribution curves (SSDs) are an important technique used in risk assessments; they estimate the first and fifth percentile hazard concentrations (HCs), which are the concentrations considered protective of 99% and 95% of the species, respectively. HC estimates are used as levels of concern protective to a broad array of aquatic species, as they inform of potential impacts and are used to derive PNECs because they are a community-level threshold making full use of the available data on the toxicity of a chemical to standard and non-standard toxicity test species and so relate concentration to ecological impacts (i.e., the proportion of species at risk). This is in contrast to the alternate (traditional) approach, in which hazard is predicted using concentration-effect data from the most sensitive single-species laboratory test data combined with an application (or uncertainty) factor (see Belanger et al., 2017). The application factor (AF) applied to SSDs is also much lower (i.e., 1–5) compared to those applied using the alternate traditional individual sensitive species approach (see Box 6.10). SSDs also reduce the need to study every species with every oil, although some sensitive taxa may still have to be investigated. Although the SSD approach is becoming the more common approach, because it estimates potential hazard to communities, SSDs do not address ecosystem processes or interspecific interactions (e.g., competition or predation). An additional advantage is that often the shape and form of the SSD can inform about some specific modes of action (e.g., narcosis) (Belanger et al., 2017).

The NOAA Chemical Aquatic Fate and Effects (CAFE) database, developed by NOAA's Emergency Response Division (ERD), includes fate and effect data compilations that provide a quantitative basis on which to assess hazard concentrations for hundreds of chemicals (see Bejarano and Mearns, 2015; Bejarano et al., 2016b, 2017; NOAA ERD, 2015; see Figure 6.36). This database was developed to facilitate access to fate and toxicity data following an oil spill and allows for SSD development to improve hazard estimates during oil spill response planning and activities. In CAFE SSDs summarize the aquatic toxicity data and provide estimates of the first and fifth percentile HCs.

Although there is a significant amount of toxicity data available regarding the impact of oil, its individual chemical constituents, and oil spill response chemicals (e.g., dispersants), often data are limited regarding the number of species, so SSDs cannot be developed (SSDs typically require between four to eight species; CAFE requires five [Bejarano et al., 2017]). To address this data gap without conducting an extensive number of laboratory toxicity tests, further modeling approaches have been developed to predict toxicity; for example, the Interspecies Correlation Estimation (ICE) models described in Section 6.7.2.6.

6.7.2.6 Interspecies Correlation Estimation Models

ICE models are a type of interspecies correlation model (ICM). ICE models extrapolate the known toxicity data from surrogate species to generate median lethal or effect concentration (L/EC50) toxicity predictions for one or several new species (Bejarano and Barron, 2014; Bejarano, 2019) The advantage of this method lies in leveraging older or broader toxicity results as a baseline for prediction of new LC_{50} curves for new or related species. Bejarano and Barron (2014) developed the first ICMs for the toxicity of oils and chemical dispersants (also see Bejarano and Wheeler, 2020). The most recent integration of ICE models into CAFE addresses, at least partially, utility for many chemicals for which aquatic toxicity data are limited. ICE predicted values enhance species diversity in CAFE and allow the derivation of SSDs, both improving the capabilities of this tool and ultimately increasing confidence in environmental assessments related to accidental chemical releases in aquatic environments. Also, broader access to ICE models available to date facilitate their use and further refinement. Numerous studies have now shown the utility and validity of this approach, by demonstrating that SSDs developed using ICE-predicted values have HC estimates similar (i.e., less than 3-fold difference) to those SSDs calculated using measured toxicity data from the most sensitive laboratory test species (Dyer et al., 2006, 2008; Awkerman et al., 2014; Bejarano and Barron, 2014; Sorgog and Kamo, 2019). They are often more conservative; in less than 10% of cases was toxicity underpredicted compared with measured values (Bejarano et al., 2017). The main advantage of the ICE approach is that it increases the number of species that can be included in an SSD (i.e., by an average of 34 species), which generally reduces the uncertainty around the HC values generated and reduces the need for additional toxicity tests. Therefore, the integration of ICE-predicted values into the CAFE database will increase the confidence in environmental risk assessments (Bejarano et al., 2017).

The use of ICE/CAFE also has demonstrated that we do have a significant amount of data regarding the toxicity of oil, its constituents, and chemical dispersants and that research efforts toward conducting further toxicity tests (especially acute and standard test species) may not be the most useful allocation of resources. However, ICE models could be developed even further with the inclusion of aquatic toxicity data directed toward sensitive taxa (Bejarano et al., 2017). A related recommendation made in the oil spill dispersant report (NASEM, 2000) is to rethink how toxicity tests are used, moving away from trying to replicate field exposure scenarios (which is not possible) to providing metrics that

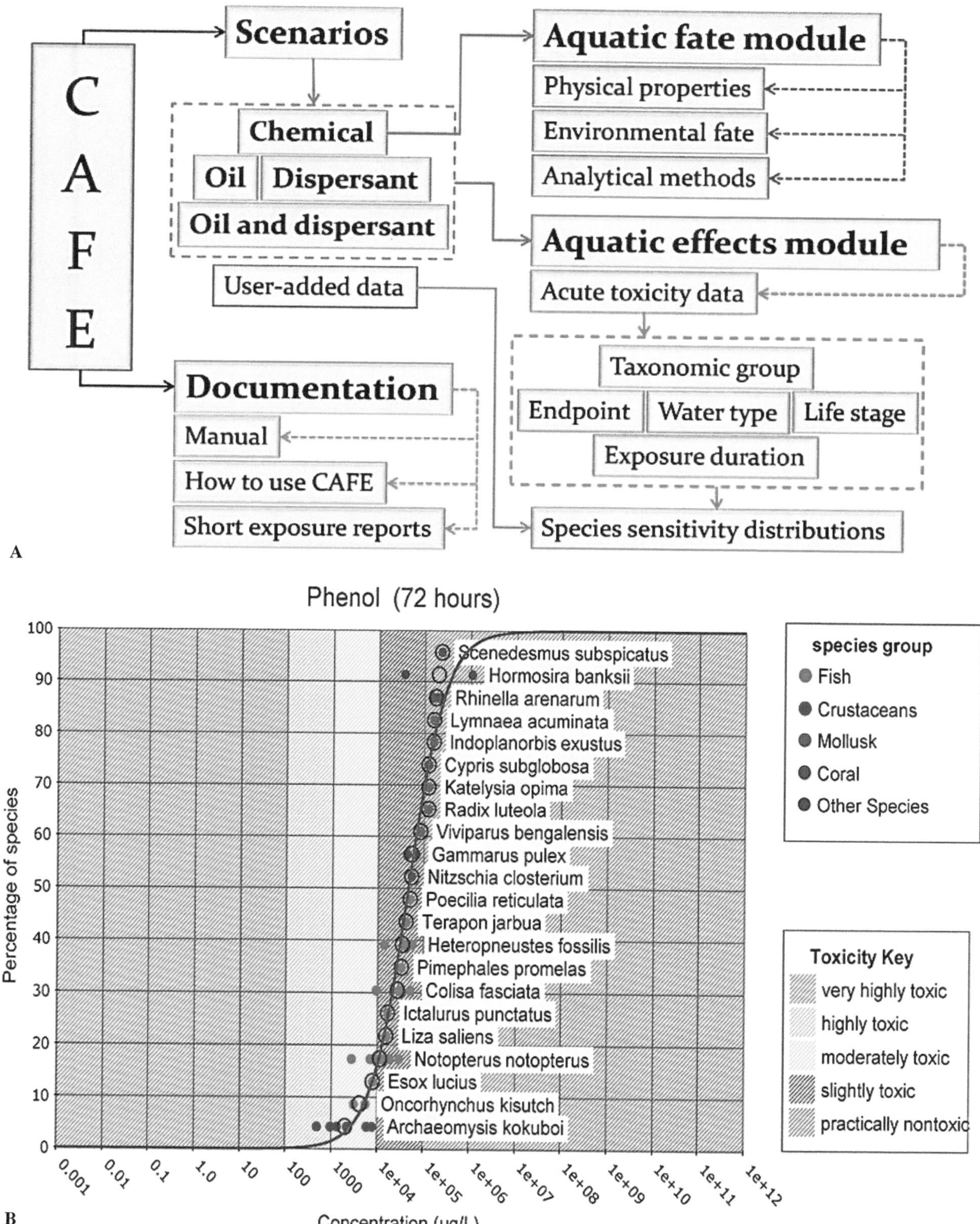

FIGURE 6.36 The CAFE database (from Bejarano et al., 2016). (A) Diagram listing the key characteristics of the Chemical Aquatic Fate and Effects database; (B) Examples of the plot of toxicity data organized as species sensitivity distributions (SSDs) within the effects module of the CAFE database.
SOURCE: Bejarano et al., 2016.

BOX 6.10
Approaches Used in Environmental Risk Assessments

Assessment of ecological risk requires knowledge of the concentration of the chemical in the environment using measured (i.e. monitoring) or modeled concentrations (exposure) coupled with information regarding the hazard (or toxicity) of the chemical using one of two approaches (see Figure 6.37).

Terminology Used:

LC50: lethal concentration in which 50% of the organisms die

EC50/10: concentration at which 50% or 10% effect observed compared to controls

NOEC: no observed effect concentration; highest concentration tested where there is no statistical difference from the control

PEC: predicted environmental concentration

PNEC: predicted no-effect concentration

RQ: risk quotient

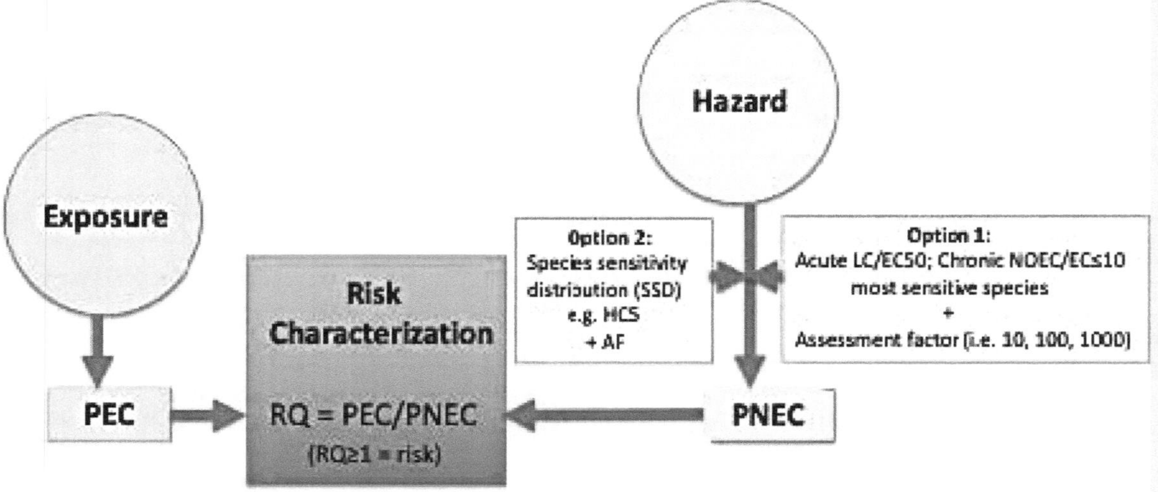

FIGURE 6.37 Overview of the steps involved in an ecological risk assessment.
NOTES: PEC = predicted environmental concentration derived from measured or modeling efforts; PNEC = predicted no effect concentration estimated from laboratory toxicity data (option 1) or derived from species sensitivity distributions (option 2).

can be used in fate, exposure, and mechanistic mixture-based oil toxicity determinations. Employing these integrated modeling approaches supports response decision-making by allowing predictions of exposure and toxicity using various spill response agents to be compared.

6.7.3 Limitations and Challenges of Modeling Approaches

The main weakness in modeling still revolves around gaps in our knowledge of the impacts of oil spills, although the models have been improved and verified using extensive datasets collected from the DWH incident. Limitations in toxicity models derived from laboratory data include lack of appropriate experimental design. Underpinning predictive models of impact are laboratory toxicity tests, the utility of which depend on how exposure solutions were made and reported and toxicological thresholds calculated. Unfortunately, even despite standard method guidelines outlined more than 20 years ago (CROSERF; Singer et al., 2000), the toxicological implications of many studies cannot be established and/or used in model development/validation. Recent reviews have addressed this issue in detail (e.g., Bejarano, 2018; Hodson et al., 2019; Murawski et al., 2020) including the recent report titled *Use of Dispersants in Marine Oil Spill Response* (NASEM, 2020). In this report a recommendation was made that toxicity tests should be used, not to attempt to reproduce field exposures, but to develop a consistent means of using toxicity metrics, such as HC_5 and LC_{50} for toxicity models used with fate and transport models to compare the exposure and toxicity of various response options, including dispersants (NASEM, 2020).

The toxicity models described earlier are also based on acute toxicity and a narcotic mode of action, although (as this chapter highlights) there are many other mechanisms of action and sublethal effects that have implications at the individual, population, and ecosystem levels as a result of both acute (not lethal and also including delayed responses) and chronic longer-term exposures. Accounting for all of these potential effects in a model would be extremely challenging, so current approaches to estimate chronic longer-term impacts have applied the acute toxicity data to predict chronic toxicity by calculating acute to chronic ratios (ACRs). An additional limitation of the narcosis models is that they represent a specific group of chemicals, and new research has highlighted the importance of other components in driving acute and chronic toxicity in aquatic organisms. Further research is needed to identify and characterize other toxic components and degradation/metabolite products (e.g., photo-oxidized constituents) (see Hodson et al., 2019).

Although new models have been developed and existing ones refined, particularly to include the influence of oil droplets and also co-variables that modify exposure and/or toxicity (such as, temperature, UV light, and pressure), there are still a number of data gaps for improvement and refinements of the existing models, including validation with tropical and other non-standard test species (Negri et al., 2021) and with new oil types (e.g., very low sulfur fuel oils, diluted bitumen; NASEM, 2016a; Ruberg et al., 2021). One modifying toxicity factor investigated recently as a result of the DWH spill has been that of pressure. Paquin et al. (2018) accounted for pressure in a modified target lipid model and demonstrated that increased pressure resulted in lower toxicity, although further toxicity data are needed to validate this. Exposure time is also a critical variable component, as typically toxicity tests follow standard guidelines of typically 48–96 hours for acute tests, whereas most acute oil spill exposure scenarios may last only a few hours. This potential disconnect could be easily addressed by reporting additional times when conducting a toxicity test (e.g., providing time-to-death estimates and also LC_{50}s at multiple time points) or conducting tests of shorter duration.

As highlighted in Section 6.7.3, there are many models for assessing individual toxicity to organisms, but more research is needed directed toward studying the impacts of oil and response options on populations. The use of population models is common practice in conservation biology and a tool frequently used by resource managers for managing wildlife, but has yet to be broadly used within the context of oil spills. Indeed, the NRC (2013) urged the inclusion of population models for assessing pesticide risks to threatened and endangered species, which could also prove useful for setting a framework applicable to the implementation of these models in oil spill assessments. However, there are examples (mainly with fish) of the integration of dose-response toxicity data with species-specific and spatially explicit information resulting in models potentially useful for assessing population risks (e.g., Awkerman et al., 2016; Gallaway et al., 2017, 2019).

6.8 HUMAN HEALTH EFFECTS

The first two committees formed by the National Academy of Sciences to consider oil in the sea (NAS, 1975, 1985) included short sections in their reports discussing the potential for human health effects, primarily focusing on polycyclic aromatic hydrocarbons. The introductory chapter to *Oil in the Sea III* (NRC, 2003), in a section titled "Scope of the Present Study" describing how the goals of the study were determined, contains a parenthetical sentence: "It was agreed early on that inclusion of an examination of potential effects on human populations, while undoubtedly of interest, would overly complicate an already daunting task" (NRC, 2003).

Human health also was not part of the original conceptualization of the present *Oil in the Sea IV* study. However, discussion at the first committee meeting led to the recognition that human health, including community health, would fit within the formal charge. As a result, this section on the human health effects of oil spills was added.

The approach taken to the human health aspects of *Oil in the Sea IV* follows a dictum stated by Savitz and Engler (2010) in a perspective providing advice on the health research approach to the then-ongoing DWH oil spill work: "Studies should focus on the health effects of the oil spill rather than solely on the health effects of the oil."

The ability to do so in the current document is based largely on the marked increase in depth and breadth of the published information available from recent spills, including the DWH. Valuable information about the human health impact of oil spills has also been obtained from recent studies of the *Hebei Spirit* and *Prestige* oil spills, which led the South Korean and Spanish governments to support ongoing follow up studies. Extensive reviews of the available information, from these and other studies of oil spill health effects, are now available (Aguilera et al., 2010; Solomon and Janssen, 2010; Goldstein et al., 2011; Zock et al., 2011; Laffon et al., 2016; Rusiecki et al., 2018; Park et al., 2019; Partyka et al., 2021; Sandifer et al., 2021). Particularly notable have been studies evaluating the indirect health effects of oil spills on communities (e.g., mental and behavioral effects; see Section 6.8.8), as well as greater understanding of the linkages between human health and the well-being of other species. Evidence that human health and well-being are dependent on interaction with nature has grown, as has the linkage between healthy ecosystems and healthy humans (Rabinowitz and Conti, 2013; Sandifer and Walker, 2018; Bratman et al., 2019).

Humans are an integral part of marine ecosystems in much of the world. We interact with marine environments as workers, as nearby residents, and for recreation. Our complex interactions with other marine species include serving as direct and indirect predators and despoilers as well as stewards and protectors. Historically, many communities have developed close to the shore. Their health and well-being have depended on successful interaction of human communities with the marine environment—an interaction that can be put at risk by oil in the sea. The U.S. Census Bureau (2019) reported that 94.7 million Americans (almost 30% of the U.S. population) live in shoreline regions, that this number is growing rapidly, and that many of these regions are characterized by great ethnic diversity. Globally, the United Nations

Environmental Programme (UNEP) estimates that 40% of the world population lives within 100 km of the sea.[8]

Just as every oil spill has different effects on ecosystems, so they also differ greatly in their direct and indirect effects on humans. Our focus here is on the relatively recent DWH oil spill, for a number of reasons. Increasing occurrence of disasters has led to a larger research community interested in the human health impact of disasters and better able to respond with focused studies. The recurring disasters in the Gulf Coast area had led to baseline information which obviated some of the shortcomings of post-disaster studies. Perhaps of greatest importance was that following the DWH oil spill, much more funding was made available for health research, in part because of a rapid response by the National Institute of Environmental Health Sciences (NIEHS) and other U.S. governmental organizations and because of the willingness of BP to be responsive, thus ameliorating, however incompletely, many of the litigation related issues that have complicated the study of other oil spill responses in the United States and elsewhere (Shore et al., 1986; Picou, 2011; Ritchie et al., 2018; Park et al., 2019). Also helpful was the coordination of federal health research at the level of the Assistant Secretary of Health and Human Services (HHS) for Preparedness and Response, an office that had been formed in 2006 as a result of problems in responding to Hurricane Katrina. The NIEHS funded a variety of programs (NIEHS, n.d.). A joint intramural/extramural program, the Gulf Long-term Follow-up Study (GuLF) is performing a 10-year follow-up study of more than 30,000 response workers which led to 40 peer-reviewed publications by early 2021 (Kwok et al., 2017b; NIEHS, 2021a,b). A 5-year program of university–community partnerships was competitively funded from which a consortium developed that performed research responsive to community needs. Working with the National Library of Medicine, a National Institutes of Health Disaster Response Program was developed (DR2, n.d.). The National Toxicology Program formed a program to further the study of the toxicology of PAHs; and, working with other agencies, the NIEHS Worker Training Program expanded its on-site training program for cleanup workers (NIEHS, 2012).

Another important federally funded program was the DWH Oil Spill Coast Guard Cohort study. This study compared 8,696 responders with 44,823 non-responders, all of whom were members of the U.S. Coast Guard (Rusiecki et al., 2018). The study design allowed for comparison of responders with a similar population of non-responders, as well as providing a comparison between those responders who were oil-exposed and those who were not.

Also unique for a major oil spill, BP, the principal responsible party, was a relatively rapid source of direct and indirect hands-off funding that produced significant research efforts. The Gulf of Mexico Research Initiative (GoMRI), formed with a distinguished leadership group, received $500 million of BP funding for a 10-year period. One of its five research themes was public health, including behavioral and socioeconomic impacts; however, the public health portion of the research portfolio was not as pronounced as others, for many reasons (Eklund et al., 2019; Goldstein, 2020).

Another source of published research was indirectly funded by BP through a set-aside of $105 million of the $7.6 billion medical settlement to form the Gulf Region Health Outreach Program (GRHOP). Its goal was to improve local health in specified counties in four states most affected by the oil spill, with a particular focus on mental and behavioral health and on environmental health. Overseen by the court, GRHOP included an academic program primarily focused on environmental health, and four programs working on mental and behavioral health and social and community issues at state universities in the four affected states. The latter resulted in the Gulf Coast Resilience Coalition (Hansel et al., 2015). Each program worked with community organizations (Buckner et al., 2017). Although not formally a research program, GRHOP produced a large number of peer-reviewed publications, primarily focused on mental and behavioral health, including a special supplement to the *Journal of Public Health Management and Practice* (2017). That supplement included consideration of environmental justice issues as well as the training of community health workers (Hansel et al., 2017; Nicholls et al., 2017). GRHOP also was part of a groundbreaking effort in enterprise evaluation, an approach increasingly needed to gauge the success of broad multidisciplinary efforts (NIEHS, 2018; Sherman et al., 2019). More recently, the National Academies created the Gulf Research Program (GRP) with $500 million received through the U.S. Department of Justice from the criminal settlement of the DWH oil spill for activities related to the prevention of and response to future oil disasters. Human health research has already become a significant part of its broad portfolio. With its 30-year time frame (ending in 2043), and a mandate that extends to U.S. coastal waters beyond the Gulf Coast, the GRP has the opportunity for much needed longer-term integrated investigations of the impact of oil spills (NASEM, n.d.).

In this section we address the components of crude oil and their derivatives that may affect human health directly or indirectly, the pathways of human exposure, the direct and indirect health effects of concern, and the modulating effect of factors such as individual and community susceptibility and resilience; and we touch on the potential health impacts of the different cleanup responses to oil in the sea, such as burning of surface oil and the use of dispersants. Safety issues, including fire and explosions and the safety of response workers, are considered briefly. In keeping with the recommendations of a 2010 Institute of Medicine Letter Report titled *Research Priorities for Assessing Health Effects from the Gulf of Mexico Oil Spill*, particular attention will be paid to psychological and behavioral effects, exposure

[8] See https://www.unep.org/explore-topics/oceans-seas/what-we-do/working-regional-seas/coastal-zone-management.

assessment, and seafood safety (IOM, 2010a,b). We also focus on studies considering the potential linkage between the economic effects of an oil spill disaster and a disaster's mental and behavioral effects.

6.8.1 Components and Derivatives of Crude Oil of Known Importance to Human Health Effects

Crude oil is a mixture of hydrocarbons which vary according to source. As with other species, two of the major factors determining the human health effects of oil in the sea are the chemical and physical composition and properties of the oil, and the products of chemical and physical transformations of oil components within the environment (see Chapter 2). Following release into the sea, crude oil changes its physicochemical properties, with different components undergoing different pathways and fates. Lighter weight components, such as BTEX, tend to volatilize at rates depending on water and air temperature and on sea surface turbulence. Heavier crude oil components are particularly affected by the process known as "weathering." This leads to greater oil viscosity, which can allow the formation of "tar balls" that can be brought to the shore or sink to the sea bottom. Little information is available about the potential human health effects of weathered oil or other oil spill degradation products (Black et al., 2016; NASEM, 2020). Weathered crude oil has been evaluated in other species with findings that may be relevant to humans (see Chapter 6).

The major components known to have human health impacts are the monocyclic and polycyclic aromatic hydrocarbons, which are considered in more detail later in this section. Microbial degradation and photo-oxidation participate in changes in the chemical structure of oil products with different resultant degradation products (Garrett et al., 1998). Crude oils also vary in their composition of trace elements such as arsenic.

As noted in Section 5.2.5, there is increasing information about the photo-oxidation process and its products. Volatile hydrocarbons, along with oxides of nitrogen (NOx) and sunlight, underlie the photochemical process that leads to tropospheric ozone pollution. Middlebrook et al. (2012), in an extensive analysis of the air pollution resulting from the DWH oil spill, attributed ozone formation to NOx from oil spill response activities, including emissions from the many ships involved and the flaring of natural gas. Burning of oil at sea contributed to production of NOx as well as particulates (Middlebrook et al., 2012). Particulates and ozone are air pollutants of particular concern as causes of cardiovascular and lung disease which could potentially affect community members and cleanup workers. Coastal areas with drilling sites, refineries, and other major sources of hydrocarbons such as South Texas, appear to be particularly at risk for exceedances of the health-based ozone standard from offshore oil spills (Sanchez et al., 2008). Atmospheric photo-oxidation also produces potentially toxic carbonyl compounds such as acrolein and peroxyacyl nitrates (Middlebrook et al., 2012; Weitekamp et al., 2020). Impetus for further study of the potential human health effects of photo-oxidation also comes from their persistence in degradation products that may come in contact with humans. Black et al. (2016) noted that weathered oil oxidative by-products were present in beach sand 5 years after the DWH oil spill, and that the lack of toxicological data on such products (as well as dispersants) was a limitation of their risk assessment for children exposed to beach sands affected by the DWH oil spill.

An extensive literature exists on the role of free radicals and active species of oxygen in human disease, including cancer (Klaunig, 2018). While externally produced short-lived reactive species such as hydroxyl radicals, superoxide anion radical, and singlet oxygen all decay much too rapidly to enter the human body, photo-oxidation results in a variety of longer lived potentially active compounds, including peroxides, which once ingested or inhaled could conceivably have sufficiently long lifetimes in vivo to cause adverse effects. This could occur through their further release of active oxygen species or through the mutagenic action of carbonyl derivatives from peroxidized lipid leading to their attachment to DNA, including producing DNA-DNA or DNA-protein crosslinks (Mukai and Goldstein, 1976; Lu et al., 2010). The presence of oxidized forms of PAHs in air may be toxicologically important (Lammel et al., 2020). **Further study is warranted of the formation, persistence, and potential adverse human health effects of photodegradation products of oil in the sea.**

Various metals are present in trace amounts in crude oil. An association of these metals with endocrine and genotoxic effects among Prestige spill cleanup workers who remove the heavy fuel oil has been reported (Perez-Cadahia et al., 2008). Gohlke et al. (2011) discussed the evidence for bioaccumulation of metals in marine organisms following oil spills and criticized seafood safety determinations for failure to include analysis of metals. Black et al. (2016) noted that the levels of arsenic, vanadium, and barium, as well as four PAH components, exceeded guideline levels.

Black et al. (2016) followed up on their finds of guideline exceedances for metals and PAHs by performing a formal risk assessment for a child playing on the beach. They found overall low levels of risk, consistent with the findings of other investigators who performed risk assessments following oil spills. Two prior risk assessments related to beach activities following the Erika heavy fuel oil spill in France also reported low or limited risk depending on the scenario employed (Baars, 2002; Dor et al., 2003). Dor et al. (2003) suggested concerns about exposure to relatively high levels of PAHs for the scenario of a pregnant woman walking on a rocky shore from which it is difficult to remove oil. Another approach to estimating risk was taken by Afshar-Mujaher et al. (2019), who measured the air emissions from a tank containing crude oil with or without dispersants which was subjected to a horizontal wave simulator. They found airborne levels of VOCs that on an exploratory analysis were within ranges of concern depending on the estimated time of exposure.

6.8.2 Pathways of Human Exposure to Oil in the Sea

The DWH oil spill, and other spills affecting human populations, conform to the five elements of a complete exposure pathway: sources of contaminants; readily identifiable environmental media, including air, water, and soil; points of exposure for both response workers and the public; routes of exposure including inhalation, ingestion, and absorption through the skin; and a wide variety of receptor populations encompassing the diversity of the Gulf area and of the response workforce (Goldstein et al., 2011).

Similar to marine and shore biota, the major factor determining the extent of human exposure to oil in the sea and its derivatives is location in relation to human habitation and activities. Seafood gatherers, other maritime workers, and those employed in the offshore oil industry are more likely to be exposed at work. Again, similar to many marine species, the location of humans often depends on the stage of life. Relatively placid water areas with sandy beaches are where children may be found, as are their mothers, who are more likely to be pregnant with another child than are other women. Areas of high surf in which breaking waves might lead to more off-gassing of volatile components tend to attract teenagers (Afshar-Mohajer et al., 2018). Individuals on chartered fishing boats tend to be older and male. Beachcombers may encounter weathered oil on the beach. Children, with their curiosity and hand to mouth habits, may be particularly at risk from touching and ingesting tar balls. Local weather is important in determining both the direction of movement of the oil components in air or water, and the activities of humans that result in exposure. These all factor into decisions by authorities restricting public access to contaminated shorelines until cleanup activities have been completed.

Human exposure pathways will differ greatly for different crude oil components. Those living or working on or near to the shoreline will generally have relatively larger exposures to airborne volatiles, such as BTEX, particularly when oil releases are near the shore. The off-shore location of the DWH spill contributed to the lack of evidence of significant levels of benzene reaching shore. Factors affecting the extent of human exposure to BTEX include their chemical degradation rates, the amount of BTEX in the crude, the usage of subsea dispersant injection, the length of time for the crude to travel toward the shore, water temperature, wind direction, tidal direction, water turbulence, the proximity of humans to the shore, their activities (including breathing rate), and (if they are indoors) air exchanges and flow rates. Skin absorption of BTEX also occurs (Gorman et al., 2019). Theoretically, skin absorption could be enhanced by the presence of dispersants in the same mixture, although exposure to such a mixture is unlikely. BTEX components degrade relatively rapidly, with a half-life usually less than 1 week. Sunlight hastens degradation in air, and microbes hasten degradation in water or soil.

The major pathway for exposure to PAHs is through ingestion, either of seafood or, particularly in children, during hand-to-mouth activities. Inhalation of PAHs also occurs, as does direct skin contact, particularly among response workers and curious children. In humans and higher predators, active loss through metabolism occurs, with differing rates in different food species.

Idiosyncratic factors related to release of oil components into the sea and the vagaries of human activities can make a significant difference in the extent of exposure. For example, fishing is said to be relatively common around abandoned drilling sites at which greater volatile hydrocarbon (VHC) release might be anticipated.

Furthermore, response to oil spills may reduce the extent of volatilization, as occurs with dispersants, or may lead to the secondary production of PAHs and benzene by offshore burning (NASEM, 2020). Transdermal exposure also can occur. Aerosols of oil-containing particles have been suggested to have potential for inhalation or dermal exposure (Middlebrook et al., 2012; Ehrenhauser et al., 2014; Afshar-Mojafer et al., 2019).

6.8.3 Methodological Approaches to Estimating Human Exposure

The quantitative estimation of risk requires understanding of both the intrinsic hazard of the agent and the exposure dose. Dose is itself dependent on two factors: the pollutant level and the duration or frequency of exposure. Quantitative estimation of individual exposure levels is often challenging. Many of the oil in the sea scenarios provide at best an indirect estimate of actual exposure, usually through data from measuring sites in proximity to exposed individuals, or at an intermediary site between the source and the potential receptor. Such factors as wind or current speed and direction, and whether the individual was outdoors or using personal protective equipment (PPE) may also be considered. For inhaled agents, the respiration rate is an important determinant of personal exposure.

For most epidemiological studies of oil spills, estimation of whether the release of oil caused adverse health effects has depended on qualitative metrics of exposure, such as comparing those who worked in the spill response with those who did not, or time series studies comparing symptoms or findings in responders during and subsequent to exposure. Rarely is information available beforehand. Particularly problematic has been the use of self-reporting of exposure, which is subject to recall bias. *Recall bias* is the well-documented tendency for those with adverse consequences to give a positive response to questions about the extent of exposure to potential causes; for example, a mother with a child who is born with a congenital malformation is much more likely to remember a mild respiratory infection that occurred during the first trimester than the mother whose baby was born with no malformations (Thorpe et al., 2015; Crump, 2020).

Among the reasons for major advances in the science of environmental epidemiology has been the development of valid biological markers of exposure suitable for linking relatively low-level human exposures to adverse outcomes or to valid biomarkers of effect. For organic compounds that undergo metabolism, such as almost all crude oil components, the use of biological markers of exposure is limited by the length of time that the agent or its products persist in the human body. Although biological markers of exposure are available for crude oil components of particular interest, such as PAHs and benzene, most studies of the DWH oil spill began well after those components' expected persistence. For example, although there are some relatively long-lasting metabolites of benzene, the most sensitive and reproducible measurements of exposure to low levels of benzene in the workplace are best obtained at the end of the work shift (Loomis et al., 2017).

Techniques to improve measurement of individual exposure suitable for use in epidemiological studies have been developed, and do provide a better opportunity to accurately relate dose to effects. These include the use of personal monitors which can be worn by potentially exposed individuals. Personal monitors which are sensitive to 1 ppb are available for benzene (as a comparison, the Occupational Safety and Health Administration [OSHA] standard is 1,000 ppb as an 8-hr average), and for total hydrocarbons. Notable in response to the DWH spill was the development of a sophisticated Job Exposure Matrix (JEM), which allowed estimation of exposure levels for each specific task (Stewart et al., 2018). JEMs also have been developed for offshore petroleum workers (Stenehjem et al., 2021). JEMs are refinements of classic occupational epidemiology in which assignment of workers to different exposure levels, qualitatively or quantitatively, allows assessment of whether the outcome of concern followed expected dose-response relationships, where those more highly exposed to the agent of concern had a higher risk of an adverse outcome.

The pathways by which a disaster, such as a major oil spill, produces social and behavioral effects in communities and community members are discussed in Section 6.8.8.

6.8.4 Human Susceptibility Factors Related to the Potential Toxicity of Oil in the Sea

Human susceptibility, also known as vulnerability, plays a major role in health. Individual, family, and community choices related to potential exposure constitutes the major susceptibility factor related to oil in the sea, including location of home, work, and recreation. Because of the centrality of susceptibility factors to human medicine, and the generally greater emphasis given to protecting and ministering to the health of individuals, much is generally known about what potentially might make different humans inherently more or less susceptible to chemicals. As many of the chemical components of particular interest require metabolism for their toxicity, enzyme polymorphisms or other causes of individual variation in metabolism can be important. (See discussion of benzene later in chapter.)

Resilience can be defined as the opposite of vulnerability. Differences in vulnerability are mediated by different levels of individual and community resilience in responding to disasters. Sandifer and Walker (2018) have extensively reviewed the relation between disaster resilience and stress-related health impacts. They and others have made multiple recommendations to improve individual and community resilience (Morris et al., 2013; Buckner et al., 2017; Nicholls et al., 2017; Sandifer and Walker, 2018; Fuchs et al., 2021). Evidence that repetitive disasters, as has occurred on the Gulf Coast, create a cumulative effect on mental and behavioral health, has been among the factors leading to efforts to improve individual and community resilience (Hansel et al., 2015; Harville et al., 2018; Lowe et al., 2019a,b)

As discussed in more detail in Section 6.8.8.2, economic well-being is a major factor in resilience, particularly in relation to mental and behavioral health of individuals and social disruption of communities. Poverty, and other issues potentially reflecting environmental injustice, limits choices in responding to the economic losses caused by the loss of jobs directly or indirectly due to the closure of seafood gathering. Functional disruption of a community and the loss of social capital means there is less of a safety net for individual community members.

6.8.5 Human Health Effects Potentially Due to Oil in the Sea: Direct Toxicity of Crude Oil and Its Degradation Products

6.8.5.1 Acute Effects

Acute effects of crude oil include skin and eye irritation, dizziness and other neurotoxic effects, and effects on the respiratory tract including throat irritation and cough. These have been reported in response workers (e.g., Meo et al., 2009; Gwack et al., 2012; Ha et al., 2012; Rusiecki et al., 2018) as well as in community members when the spills were near shore (e.g., the *Tasman Spirit* oil spill in the port area of Karachi [Janjua et al., 2006], the *Hebei Spirit* oil spill in Korea [Park et al., 2019] and the MV *Braer* oil spill near the Shetland Islands (Campbell et al., 1993, 1994). Skin and eye irritation and respiratory difficulties are generally ascribed to the complex crude oil mixture and its immediate derivatives, with inhalation effects occurring particularly with freshly spilled oil. More recently, the fouling of Israel's beaches by a tanker apparently intentionally releasing oil reportedly led to the initial response workers suffering respiratory difficulties that required supplemental oxygen (Rasgon, 2021), and a refinery oil spill affecting beaches in Peru was said to cause early responders to pass out from breathing the fumes (Taj, 2022). The DWH oil spill Coast Guard Cohort Study noted an increase in gastrointestinal and genitourinary effects (Rusiecki, 2018), perhaps related to the heat stress. An increase in deaths due to heart attacks has been reported in relation to some disasters, usually ascribed to stress (Kario et al., 2003; Yousuf et al., 2020). Mice exposed to particulate matter obtained from the burning of surface oil during the DWH disaster have developed pulmonary inflammation as well as an alteration in immune response (Jaligama et al., 2015).

6.8.5.2 Cancer

The known human carcinogens present in crude oil are benzene and the class of compounds known as PAHs. These are discussed separately later in the chapter. In this section we consider studies of cancer incidence or genotoxicity related to crude oil in aggregate rather than to the individual components.

One of the largest prospective cancer studies of workers exposed to crude oil directly related to oil in the sea is the Norwegian Offshore Petroleum Workers (NOPW) cohort. It began in 1998 based on questionnaires returned by 27,917 workers, about 10% of whom were females (Stenehjem et al., 2021). Excluded from their cohort were those who were unlikely to have had offshore exposure to petroleum. The country-wide Norwegian Registry of Employers and Employees was utilized to cross-check data and compare findings. Exposure was estimated based on a series of job-exposure matrices. Data collection about workplace activities was done only at the beginning of the study, thereby limiting the use of the data in estimating the effects of subsequent exposures. The major finding has been a statistically significant increase in the standardized incident ratios (SIRs) of all cancers in males, SIR 1.07 (95% confidence interval [CI] 1.04–1.11) and females SIR 1.13 (95% CI 1.01–1.36). Not surprisingly, statistically significant increases in hematological and skin cancer types were noted. For skin cancer, the extent of exposure to sun or artificial tanning equipment are potential confounders (Stenehjem et al., 2017). Unexpected was an increase in breast cancer both in males and females, although in females it was not statistically significant. A study of cancer incidence in communities exposed to the *Hebei Spirit* oil spill 7 years before and 7 years after the spill reported an increase in prostate cancer (Choi et al., 2018).

Studies have evaluated indicators of genotoxicity in those exposed during oil spills. Following the MV *Braer* incident, tests used to assess genotoxicity and mutagenesis were both negative (Cole et al., 1997). However, subsequently a number of studies have been suggestive of genotoxic effects (see Aguilera et al., 2010, for an extensive review of these studies). Using different genotoxicity assays, two different studies after the *Prestige* oil spill reported findings indicative of DNA damage (Rodriguez-Trigo et al., 2010). In the long-term follow-up of the *Prestige* oil spill volunteer cohort, persistent DNA damage was observed after 2 years but not after 7 years following exposure (Laffon et al., 2016). In a separate study of local fishermen who participated in the *Prestige* oil spill cleanup and were assumed to be more heavily exposed, persistent chromosomal damage continued to be observed 6 years after the initial exposure. Transient decreases in DNA repair activity were also noted. The authors report on chromosome bands that appear to be prone to breakage and that are said to be associated with hematological cancers. Genotoxicity and some alteration in DNA repair activity were noted in rats inhaling fuel oil similar to that released from the *Prestige* (Valdiglesias et al., 2012).

Two urinary measures of oxidative stress in heavily exposed community members, including 8-hydroxydeoxyguanosine, were reported to be increased as late as 6 years following the *Hebei Spirit* crude oil spill (Kim et al., 2017), although the mechanism for this delayed response is not fully apparent.

6.8.5.3 Other Longer-Term Effects

Reports of longer-term respiratory tract effects have been noted by a number of investigators Community studies include the finding of asthma, respiratory tract allergies, and deficits in pulmonary function in children after the *Hebei Spirit* oil spill (Park et al., 2019). One year after the *Tasman Spirit* oil spill a decline in pulmonary function was observed in healthy response workers, though this improved over time (Meo et al., 2008).

Intensive studies after the *Prestige* oil spill included a long-term study by the Spanish Society of Respiratory Medicine of a cohort of fishermen who were involved in the cleanup efforts and who for the most part did not use PPE during the initial response period (Zock, 2011). At a little more than 1 year after the oil spill, these relatively heavily exposed workers were found to have a higher prevalence of upper and lower respiratory tract symptoms (Zock et al., 2007). A subsequent study at 2 years post-spill found similar symptomatic differences from a control group of lesser-exposed fishermen but no significant difference in pulmonary function as measured by spirometry was observed (Rodriguez-Trigo et al., 2010). The latter study also found increases in levels of biomarkers consistent with respiratory tract damage and oxidative stress. A subsequent study of a sample of this cohort 6 years following the *Prestige* oil spill showed no clear differences between the exposed and control groups (Zock et al., 2012). As pointed out by the authors, the study was limited to non-smokers and may have missed an effect in those whose lungs were compromised by smoking.

In the NIEHS GuLF study, suggestion of a reduction in lung function among those involved with handling oily plants or dead animals was observed in participants tested from 1–3 years following the spill, but lung function was unrelated to the extent of estimated THC exposure. It was suggested that lung function was affected by the stressful nature of the cleanup job (Gam et al., 2018). However, lung function was not affected by working in those jobs for which Kwok et al. (2017a) had found the highest levels of mental and behavioral effects. A return to normal lung function was seen during the 4–6-year time frame with greatest improvement in those with the highest exposure (Lawrence et al., 2020).

Longer-term cardiovascular effects have also been suggested. The researchers conducting the NIEHS GuLF study follow-up of 24,375 workers have published two studies evaluating heart attacks (Strelitz et al., 2018, 2019). During the first 3 years of follow-up, the authors noted a suggestive association of heart attacks with individual exposures to total hydrocarbons (THCs) while working as responders (Strelitz et al., 2018). Their subsequent study extended the follow-up period to 5 years and reported that when maximum THC levels were above 3.00 ppm, there was a statistically significant 1.8-fold increase in heart attack risk as compared to those with a maximum THC level of less than 0.30 ppm. They point out that particulate air pollution has been associated with cardiovascular disease but they do not appear to have considered this as a possible confounder, although evidence of increased particulate levels was reported by Middlebrook et al. (2012), who noted that burning of oil was a significant cause. This was recently confirmed by Pratt et al. (2020) of the GuLF program, who estimated that major exceedances of the particulate air standard would result from in situ burning. Strelitz et al. (2018, 2019) also do not review whether there was an increase in heart attack rates in the many previous studies of workers in the petroleum industry or other industries whose THC exposure levels were presumably far higher than those of the DWH response workers. However, in comparing the two studies by Strelitz et al. (2018, 2019), one finds that in the earlier publication the authors consider that their preliminary findings of an increased risk of heart attacks may be due to psychosocial stress and anxiety associated with those response jobs that had higher exposure to THCs. In the second study, although the authors expertly adjusted for many possible confounders, they did not consider psychosocial stress factors. Furthermore, current understanding of the toxicological mechanism of action of THC components would not readily explain the observed higher association with maximum THC levels than median THC levels, particularly as there does not appear to be an attenuation of the effect during the 5-year follow up. The authors do cite a study associating benzene exposure with cardiac risk among 210 clinic patients with cardiovascular risk (Abplanalp et al., 2017). However, this study has only a single measurement in each subject of a urinary metabolite were by itself is inadequate to determine long term benzene effects even if the metabolite was not also produced by agents other than benzene. Nevertheless, whether THCs are a causative factor, or a proxy for another cause related to responding to oil spills, these observations deserve careful follow-up.

In view of the large number of workers, and the extent of long-term occupational surveillance of workers directly exposed to relatively high levels of crude oil in the past, it would seem unlikely that unrecognized long-term hazards exist. But such workforces were composed primarily of relatively healthy males. Toxicological evidence suggests the possibility of reproductive and developmental toxicity from crude oil components (Feuston, 1997; Diamante, 2017), particularly PAHs. Epidemiological studies of the relation of PAHs to birth defects have not been conclusive. For example, case control studies using the National Birth Defects Prevention Database to investigate neural or cardiac birth defects in babies whose mothers had been occupationally exposed to PAHs found some evidence of an association of PAH exposure with spina bifida but not other neural tube defects, and not with congenital heart disease (Langlois et al., 2012; Lupo et al., 2012). Though not a consistent finding, studies suggest that disasters affecting pregnant women may have a behavioral impact on their infants (Zhang et al., 2018).

Further evaluation of potential reproductive and developmental toxicity of crude oil mixtures and components is warranted. The recent finding in Norwegian offshore workers of an increase in breast cancer in both sexes (although not statistically significant in females), if confirmed, could be related to an endocrine effect (Stenehjem et al., 2021).

6.8.6 Toxicity of Crude Oil Components of Particular Concern

6.8.6.1 Polycyclic Aromatic Hydrocarbons

PAHs are well-studied known human carcinogens that are present naturally in crude oil in varying total amounts, and with varying distributions of the individual PAH components. A major source of human exposure is through inhalation of products of hydrocarbon combustion, although the resultant mixture of pyrogenic PAHs differs somewhat from the petrogenic PAH mixtures found in crude oil (Straif et al., 2005). PAHs also are formed in nature from terpenes and other related ring structures.

PAHs are likely to be responsible for one of the earliest recognized occupational causes of human cancer, that of scrotal cancer observed in chimney sweeps and reported by Sir Percival Pott in 1775. Although the individual PAH compounds differ widely in their toxicity and carcinogenicity, when present as a mixture, such as in soot or cigarette smoke, the overall mixture is carcinogenic. Current evidence indicates a direct relation between PAHs and cancer of the skin, bladder and lung, with lung cancer primarily due to inhalation of PAHs in cigarette smoke. Benzo(a)pyrene, a common

PAH, was the first thoroughly studied carcinogen in laboratory animals. As with other PAHs, it acts primarily through metabolic activation leading to products that bind to DNA (Mumtaz et al., 1996).

A major direct and indirect impact of oil in the sea is the uptake of PAHs into seafood which is then eaten by humans. PAHs are relatively persistent in the environment as compared to volatile hydrocarbon components. As discussed in Chapter 2 of this report, the different chemical structures result in different physical properties that affect their liquid/vapor phase distribution and environmental persistence, and thus the extent of human exposure. Furthermore, the PAHs vary greatly in the extent of their mutagenicity and presumed human carcinogenicity, a fact that has complicated the risk assessment for determining levels of PAHs in seafood pertinent to recommendations concerning reopening of seafood gathering (see discussion in Section 6.8.11).

6.8.6.2 Benzene and Alkyl Benzenes

BTEX are ubiquitous components of our petrochemical era. Benzene is a well-studied known chemical carcinogen, as demonstrated in epidemiological evaluation of large workforces relatively closely and followed for many decades in many different countries. All of the BTEX components, in common with most other volatile hydrocarbons, have properties similar to anesthetics at high concentrations. Measurements of BTEX during the DWH oil spill were variously reported. Huynh et al. (2022) estimated BTEX levels on supporting vessels as being in the part per billion (ppb) range. In contrast, the on-scene coordinator reported that levels of benzene above 200 ppm (i.e., 200,000 ppb) were observed on vessels drilling relief wells, and that on these vessels the overall levels of volatile hydrocarbons raised concerns about fire (Tavares, 2011).

Central Nervous System Effects of Benzene and Alkyl Benzenes

At petrochemical industry workplaces, a feared event is buildup of gas concentrations from BTEX and other petroleum hydrocarbons to lethal levels in enclosed spaces. At sufficient concentrations BTEX compounds all have effects on the central nervous system. The effects are believed to be additive (Wilbur and Bosch, 2004). At lower levels, drowsiness and lack of coordination can contribute to safety incidents. This concern is supported by a U.S. EPA study in rats equating relatively low levels of toluene with the effects of ethanol on coordination (Benignus et al., 2011). The adverse central nervous system effects of toluene are among the most thoroughly studied of the alkyl benzenes, due in part to toluene being intentionally inhaled at high concentrations for intoxication purposes (e.g., glue sniffing).

n-hexane is also a known neurotoxin. It causes peripheral nerve damage through a metabolite that cross-links nervous tissue (Spencer, 2020). As with other aliphatic components of petroleum, it also contributes to the anesthetic-like effects of crude oil fractions.

The U.S. Coast Guard study found an association between exposure to crude oil and headaches, lightheadedness, difficulty concentrating, numbness/tingling sensation, blurred vision, and memory loss/confusion (Krishnamurthy et al., 2019). Similar results were found with inhalation or skin exposure. For all neurological symptoms, the results were of greater magnitude for those reported to also have dispersant exposure than for those exposed to oil alone, perhaps indicative of recall bias.

In the GuLF study, longer-term evaluation of neurobehavioral effects in response workers 4–6 years following the DWH event found modest decreases in neurobehavioral function, especially attention, memory, and executive function (Quist et al., 2019). It is unclear how these longer-term effects would occur. While present understanding of the mechanism of action of anesthetics appears to be moving from older concepts of lipid solubility to more specific effects on membrane function, neither appears to be a mechanism for longer-term effects.

Werder et al. (2019) found a relationship between blood levels of individual BTEX components, particularly benzene and to some extent toluene, and the results of a questionnaire about potential central nervous system or peripheral nervous system self-reported symptoms (e.g., dizziness, stumbling). The study population was 690 Gulf area residents who had previously been exposed to the DWH oil spill as residents or responders. However, the investigators point out that the blood levels of BTEX generally reflect exposure within the past 24 hours, and the DWH exposure occurred years previously. Only one blood level was obtained, and the blood BTEX levels were not different from those found in the National Health and Nutrition Examination Survey—which used the same laboratory to measure blood benzene as did this study—so presumably the blood benzene levels were representative of the general population. Nor was there information about when during the day the blood was drawn beyond it usually being the morning to facilitate overnight shipping of samples to the lab (Engel et al., 2017), nor the length of time before the blood draw that the individual had been at home where benzene levels are generally much higher than outdoor levels. While cigarette smoking and environmental tobacco smoke were considered in the analysis, no questions appear to have been asked about other household sources of benzene. As just one example, the World Health Organization's review of indoor benzene exposures states that 40–60% of indoor benzene may be attributable to an attached garage (Harrison et al., 2010). Finally, as with many of the other DWH-related studies attributing effects to relatively low BTEX levels, the authors have not provided a rationale for why this outcome was not recognized in previous comprehensive studies of workers exposed to perhaps three orders of magnitude higher levels of benzene or toluene.

Hematological Effects of Benzene

Benzene is the only hematotoxic BTEX component. At high doses one or more of its phase 1 metabolites destroys hematological stem cells and can cause death due to the failure to produce red blood cells, white blood cells, and platelets. Lesser loss of stem cells leads to pancytopenia, a decrease in all of the blood cell types. Subtle changes in the number of these cells in the circulation have been reported in Chinese workers chronically exposed to levels as low as 1 ppm (Lan et al., 2004). In comparison, the OSHA benzene standard is 1 ppm for an 8-hr time-weighted average (TWA); the exposure level recommended by the National Institute for Occupational Safety and Health (NIOSH) is 0.1 ppm for a 10-hr TWA; outdoor levels in the United States are usually about 1 ppb; and indoor levels can be as high as about 10–12 ppb depending on indoor sources, such as the extent of cigarette smoking, the presence of an attached garage or other benzene source, and indoor/outdoor air exchange.

Benzene itself does not directly affect blood cells. It causes hematological effects through its active metabolites that produce mutations in stem cells. These mutations may cause hematological neoplasms, including leukemia and lymphoma.

Contrary to the appropriate concerns that mixtures of chemicals may enhance the toxicity of any one component, the evidence indicates that relatively high concentrations of toluene actually protect against benzene hematotoxicity. This is not surprising in that both undergo oxidative metabolism by Cytochrome P450 2E1 which in the case of benzene, but not toluene, produces hematotoxic metabolites. Usually about half of inhaled benzene is exhaled harmlessly. If there is sufficient toluene to competitively inhibit benzene metabolism such that more is exhaled unmetabolized, this would decrease benzene toxicity. Note that, except in heavily exposed workers, the relatively lower levels of benzene and toluene inhaled by most of those exposed to oil in the sea makes an inhibitory interaction unlikely, as sufficient metabolic machinery appears to exist to metabolize both. Similarly, variations in the rate of benzene metabolism, including due to genetic polymorphisms, affect susceptibility to hematological effects (Rothman et al., 1997; Hosgood et al., 2009).

Many of these same issues about the utility of a single blood draw to measure a rapidly changing parameter such as blood benzene, and potential confounding by indoor benzene sources other than smoking and environmental tobacco smoke, are pertinent to the GuLF-related study of Doherty et al. (2017). The investigators report an inverse relation between blood benzene levels and a number of red blood cell parameters such as hemoglobin. Particularly problematic, as recognized by the investigators, is that benzene's well-studied impact on circulating red blood cells is through benzene's attack on nucleated stem cells. Circulating red blood cells obtained in a blood draw have lost their nucleus, and in normal individuals will survive 120 days after leaving the bone marrow. In essence, the investigators have compared endpoints that integrate what has happened in the bone marrow over the past 4 months with the level of benzene obtained from an evanescent exposure measure which primarily reflects about 24 hours of benzene exposure, and that itself may vary greatly depending on such factors as whether a window happened to be open. Furthermore, the effect estimates for toluene were in the same range or even higher than for benzene, despite toluene not having any known hematological effects at exposure levels three or four orders of magnitude higher.

Hematologic effects were also reported by Park et al. (2019) in *Hebei Spirit* oil spill residents and responders at two time frames following the oil spill. In this case, though, the hematocrit went up from an average of 39.3% to 39.5%, a miniscule change that was said to be statistically significant. This is well within the range of instrument variability due to slight recalibration. The white blood count also increased, which is contrary to the usual effect of benzene.

Other Effects of Alkyl Benzenes

In addition to BTEX there are many other alkyl benzenes present in crude oil. One of the few that has been studied in some detail is cumene (isopropyl benzene) which is usually present in crude oil at concentrations of 0.1–1.0%. Based on genotoxicity and cancer studies in laboratory animals, cumene is classified by the National Toxicology Program as "Reasonably Anticipated to Be a Human Carcinogen."

The xylene isomers are also irritants, particularly of the eyes and upper respiratory tract, as to a lesser extent are toluene and ethylbenzene.

A number of recent studies have associated relatively low levels of benzene or BTEX with asthma or with endocrine disruption (Bolden et al., 2015). In the absence of toxicological mechanistic support, and the previous lack of such observations in relatively heavily exposed workforces, these observations require further rigorous validation (Bolden and Kwiatkowski, 2016; Lynch, 2016).

6.8.7 Dispersants

Dispersants are used to prevent or reduce oil from subsea releases from reaching the surface and forming slicks, or to disperse surface slicks already formed to prevent their direct contact with birds and marine mammals, including humans (e.g., response workers on vessels). Dispersing an offshore slick also prevents oil from contaminating shoreline where it could cause exposure to response workers and to the general public. Intended operational best practices lessen the likelihood of direct contact of the general public with dispersants or dispersed oil. Dispersant application procedures also are designed to limit the opportunity for response workers to interact with dispersants directly. See Chapters 4 and 5 for detailed discussion on dispersant use and fate of dispersed oil.

In 2020, a National Academies committee fully reviewed the issues associated with the unprecedented amount of dispersants used in response to the DWH oil spill (NASEM, 2020). Both the U.S. Coast Guard and NIEHS studies reported that response workers who believed they were exposed to dispersants had higher levels of reported adverse respiratory effects (McGowan et al., 2017; Alexander et al., 2018). However, in the absence of biological markers and more specific information about when and where dispersants were used in relation to much of the workforce, it was not possible to verify personal reports of dispersant exposure. The paucity of information and the lack of a biological marker made it not possible to add dispersant exposure to the JEM. Accordingly, recall bias likely was a strong potential confounder in both studies. Also, no distinction was made between the personal protective equipment used whether or not dispersant was present, thus making it harder for the individual responder to ascertain whether dispersant was present (NASEM, 2020).

The well-publicized debate about the use of an unprecedented amount of dispersants added to the concern of families affected by job loss related to the oil spill (Osofsky and Osofsky, 2021). Also adding to public concern was the initial failure to release total information about the chemical composition of the dispersant, leading to publicity about a secret ingredient (Goldstein, 2020). Lack of transparency contributes to what is called the "social amplification of risk," which itself is related to the trust of the information source (Kasperson et al., 1988). Avoidance of unnecessary secrecy about dispersants was recommended by the National Academies report, *The Use of Dispersants in Marine Oil Spill Response* (NASEM, 2020).

6.8.8 Human Health Effects Due to Oil Spilled in the Sea: Mental and Behavioral Effects

6.8.8.1 Evidence of Disasters as a Cause of Mental and Behavioral Health Effects

An extensive body of literature unequivocally documents mental and behavioral effects in those directly involved in disasters, such as military experiences, the London King's Cross tube fire (Rosser et al., 1991); the Buffalo Creek, West Virginia, dam failure (Titchener and Kapp, 1976), and the World Trade Center disaster (Diab et al., 2020). The failure to fully consider the importance of mental and behavioral effects in disasters is believed to have exerted a major toll in lack of preparedness and responsiveness. For example, Murthy (2014) argued that the relative lack of consideration of community mental and behavioral effects following the Bhopal disaster, in which more than 3,000 residents died, left a wide range of unaddressed adverse effects that could have been ameliorated by psychological support for community members. Natural disasters, such as hurricanes, are believed to produce mental and behavioral effects in more individuals than are affected by the physical force of the hurricane (Espinel et al., 2019).

McCoy and Salerno, in summarizing an Institute of Medicine workshop on "Assessing the Effects of the Gulf of Mexico Oil Spill on Human Health," held 2 months after the DWH explosion, noted the need for evaluation of psychological and socioeconomic health. This has been more than confirmed in the many studies that followed. The three most common findings have been relatively high levels of post-traumatic stress disorder (PTSD), depression, and anxiety (IOM, 2010a,b).

Further indication of the heightened prominence of mental and behavioral health in disaster response was a "Perspective" article in the *New England Journal of Medicine* titled "Moving Mental Health into the Disaster-Preparedness Spotlight" (Yun et al., 2010). It appeared in September 2010, soon after the DWH oil source was finally capped, and was co-authored by Nicole Lurie, the Assistant Secretary for Preparedness and Response of HHS. The Perspective began with the statement: "As the Deepwater Horizon oil disaster enters its next phase, consensus is emerging that among its most profound immediate health effects are those on the emotional and psychosocial health of Gulf Coast communities."

The authors added: "If one bright spot emerges from this catastrophe, it will be incorporation of mental health-related emergency response into the core competencies for disaster preparedness" (Yun et al., 2010).

The evidence gathered from studies of the DWH and other oil spills fully supports this emerging consensus. **Of particular pertinence is the growing evidence that major oil spills produce mental and behavioral effects unrelated to direct contact of community members with crude oil or its components.** This is distinct from the earlier reported direct effects on mental and behavioral health such as following the 1990 explosion on the Piper Alpha oil rig in the North Sea, in which there were 125 deaths (Alexander, 1991). Psychiatric intervention was needed for survivors on the oil rig, as well as for police officers handling the charred bodies. While not stated, presumably interventions would also have been helpful for family members and close friends of those who were directly affected.

The committee does not at all belittle the impact of the DWH disaster on affected workers and their families, colleagues, and friends. For example, the psychological impact on the survivors of the Piper Alpha explosion were still apparent 10 years after the event (Hull et al., 2002). Rather, we wish to extend this review of health effects due to oil in the sea to include the broader effects of oil spills on the psychosocial health of community members.

Studies of Alaskan coastal communities affected by the *Exxon Valdez* oil spill provided significant evidence that community disruption impacted the mental and behavioral effects of individual community members (Picou et al., 1992; Palinkas et al., 1993; Gill and Picou, 1998; Arata et al., 2000). Comparison of the impact of the *Exxon Valdez* oil spill on Native Alaskan communities with others involved in the oil spill reported differences in the extent and type of the mental and behavioral impacts (Palinkas et al., 2004; see also Section 6.6.6).

Building on these earlier studies of the *Exxon Valdez* and other disasters, more recent work, and particularly studies of the impact of the DWH oil spill, have provided the evidence base to strongly confirm the relation between a major oil spill and the disruption and mental and behavioral factors.

Before the DWH oil spill, the mental and behavioral impact of oil spills on community members was usually determined in one or at most a few studies. The extent of this literature now has been dwarfed by numerous studies of different investigators using different approaches on different populations which show mental and behavioral effects in response workers and community members following the DWH spill. Comparing the studies is difficult as they usually depend on questionnaires of different length and different levels of standardization. However, as reviewed here, the published evidence relatively consistently finds that populations affected by the DWH oil spill show an increase in mental and behavioral health effects.

Just prior to the DWH oil spill, Harville and her colleagues (2010) had reviewed the impact of disasters on perinatal health, exemplifying how the greater extent of disaster research prepared investigators to evaluate the health effects of the DWH incident. Similarly, studies of children and adolescents following the DWH oil spill were able to use the baseline of prior studies of children of the same age group, often in the same group (Osofsky et al., 2014).

A number of studies have shown additive effects of exposure to multiple disasters, including exposure to both Hurricane Katrina and the DWH spill (Osofsky et al., 2011). Harville et al. (2018), in a long-term study of 1,366 women of childbearing age, specifically looked for interactive effects of exposure to up to four Gulf Coast hurricanes and the DWH spill. Their findings were most consistent with a cumulative effect for PTSD and depression. When the experiences were particularly severe, some evidence was observed for sensitization, in which exposure to one disaster would increase the extent of effects from a second disaster. But there was no evidence in this population of habituation, in which exposure to one disaster might lessen the impact of a second disaster, as was seen for PTSD in a study of Gulf residents affected by repetitive natural disasters (Hu et al., 2021).

As a result of the oil spill, particularly in tandem with the effects of Hurricane Katrina and other natural disasters, teenagers seemed to be at particular risk for mental and behavioral effects and such outcomes as substance abuse (Fuchs et al., 2021). The authors suggest that psychological resilience protects the individual teenager. Ha et al. (2013) noted the sensitivity of children to depression following the *Hebei Spirit* oil spill.

Taking into account the importance of developmental factors in children and adolescents in their response to disaster and their resilience has been stressed (Osofsky and Osofsky, 2021).

The importance of a baseline for the study of the mental health of disasters was illustrated when two floods occurred during an ongoing longitudinal study of older adult mental health in a relatively impoverished area of eastern Kentucky in 1981 and 1984 (Phifer and Norris, 1989). Personal loss was found to be particularly associated with shorter-term increases in mental and behavioral effects; community disruption led to longer-term effects, and the intensity of the flood was associated with greater effects.

6.8.8.2 Evidence for the Role of Economic Impacts on the Mental and Behavioral Effects and Social Effects of Oil Spills

Communities in the area of the Gulf of Mexico benefit substantially from the economic value of the Gulf ecosystem. Shepard et al. (2013) have estimated that the 2010 revenues from provisioning ecosystem goods and services from the five states bordering the Gulf of Mexico amounted to more than $2 trillion.

As described in Section 6.8.8.1, the evidence now strongly supports the contention that virtually any major disaster, including an oil spill, causes mental and behavioral effects. However, the *Oil in the Sea IV* committee has also explored a slightly different question: whether there is evidence that those who lose income as a result of an oil spill are more likely to have mental and behavioral effects than those who do not. If so, then oil spill prevention and response efforts should consider not only community economic health but also what would appear to be inevitable community mental and behavioral health effects due to the oil spill causing economic loss. In making such a determination, associated factors such as resilience and community social capital would have to be taken into account (Clay and Abramson, 2021).

A related question is whether the duration of economic impact is also a factor in the extent of mental and behavioral effects. The duration of closure of seafood gathering is not only an economic issue but also one related to human mental and behavioral health. This issue was considered by the National Academies' Committee on the Use of Dispersants in Marine Oil Response (NASEM, 2020). As summarized here, the DWH oil spill substantially added to the evidence base that **both the extent of economic loss and the duration of time until reopening of seafood gathering are significant factors in the causation of human mental and behavioral health effects.**

Evidence that spill-related income loss appears to cause greater psychological distress than the presence of shoreline oil comes from studies comparing two fishing communities, one directly affected by the DWH oil spill (Baldwin County, Alabama), and the other (Franklin County, Florida) having economic consequences from loss of fisheries and tourism, but no fouling of its shoreline. Grattan et al. (2011) found no difference in the measures of psychological distress between residents of these two communities, but within both communities those who had spill-related income loss had much more psychological distress than those with stable incomes. This finding persisted in a follow-up study 1 year after the oil spill (Morris et al., 2013). The authors conclude that spill-related income loss may have greater psychological effects than the presence of shoreline oil.

Choi et al. (2021), in their ongoing study of local residents 9 years following the *Hebei Spirit* oil spill, observed persistent effects that were more common among those with lower income (Choi et al., 2021). Loss of income from the oil spill also appeared to be a factor. In the Hu et al. (2021) study cited earlier, the severity of property or crop loss was associated with anxiety, depression, and PTSD.

Using the lens of the relative weight of psychological risk factors associated with disasters, shorter term economic loss figured heavily in a tabulation of those risk factors that were "Present and Prominent," as compared to "Absent or Minimal" (Shultz et al., 2015). The authors point out that the psychological impact of the DWH oil spill was less than many feared, despite the multiple-impact effect of Hurricane Katrina, which they attribute to coastal residents generally not experiencing significant injury or mortality, disruption of vital services, or population displacements, as well as to the relatively rapid infusion of economic support.

Lowe et al. (2015) noted that workers enrolled in the GuLF program were relatively well paid but yet had an increased incidence of major depression, PTSD, and generalized anxiety in which a greater level of physical effects played a major role. Even within this group of relatively well-paid workers, they found evidence that greater income mitigated mental depression and generalized anxiety disorder. Lowe et al. (2015) also analyzed the factors that led members of this cohort to seek mental health support.

Osofsky and Osofsky (2021) considered the lessons from both Hurricane Katrina and the DWH oil spill, and noted the prolonged mental health impact on those having fewer resources and who received less help to recover. Other studies include those of Drescher et al. (2014), who, in a study of 1,119 coastal Mississippi residents receiving mental and behavioral health services, reported that the chronic problems of life affecting Gulf coast communities were significantly associated with psychological distress, with those having income below the poverty line reporting higher levels of distress related to the DWH oil spill. Similarly, in a telephone survey of 812 adults, those with spill-related economic impact had more symptoms consistent with depression and PTSD (Shenesy and Langhinrichsen-Rohling, 2015). The authors also found an association between an individual's perceived resilience and lower levels of psychological symptoms. Gould et al. (2015) attributed the rapid distribution of funds from various sources, including the BP tort settlement, for the findings of relatively minor impact on substance abuse in two federal surveys. Similarly, the unusually rapid distribution of funding support to local residents in Spanish communities affected by the *Prestige* oil spill was posited to play a role in the lack of a statistically significant effect on measures of health quality of life, including mental health (Carrasco et al., 2007).

As part of the Women and Their Children Health Study (WaTCH), which followed a cohort of women from southern Louisiana, Rung et al. (2017) reported that the impact of economic loss on causing depression was largely mediated by the concomitant loss of social capital (Rung et al., 2017). Gaston et al. (2016), in a study based on the same WaTCH cohort, reported that the three stressors of economic problems, physical/environmental exposure, and signs of neighborhood decline were independent predictors of depressive symptoms. Evaluating PTSD symptoms, Nugent et al. (2019) found that wealthier women were more likely to be in their low-symptom category.

Johnson et al. (2020) recently pointed out the semantic ambiguity involved in many of the approaches to understanding why natural disasters have different impacts on different communities. Using factor analysis, the authors have attempted to tease out the commonalities among the 130 variables used in various indices that have been developed to measure community vulnerability and resilience. They report that 50 of these variables fit within five factors: wealth, poverty, agencies per capita, elderly populations, and non-English speaking populations. Two of these five factors, wealth and poverty, reflect the role of economics in the vulnerability and resilience of communities to natural disasters. A similar analysis for type-specific technological disasters, such as oil spills, could be helpful.

Economic factors also play a major role in social vulnerability to environmental disasters. In assessing the dimensions of community social vulnerability, Cutter et al. (2003) found that personal wealth ranked first in explaining the observed variance. The association of community attachment with health impacts and with employment in seafood gathering or the oil industry was studied in a repeated telephone survey of Louisiana coastal households (Cope et al., 2013). The first survey was done in June 2010 while the oil was still flowing, and subsequent studies were in June 2010 and April 2011. Major findings were that both mental and physical impacts were highest initially but persisted after the oil was contained; that higher levels of community attachment were associated with lower levels of negative mental and physical health impacts; and that fishing households had significantly higher adverse mental health impacts than others. The greater impact on seafood gatherers than other community members had previously been noted following the *Exxon Valdez* oil spill (Picou et al., 1992; Picou and Gill, 1997; Arata et al., 2000).

The economic impact of an oil spill disaster similar to that of the DWH would seem to be dependent on the duration of closure of seafood gathering and other maritime-related activities. Berren et al. (1980) listed the duration of a disaster as one of five factors in their suggested typology of disasters. Evidence that the duration of a disaster is accepted as a factor in the extent of mental and behavioral effects includes a study of Dutch survivors of an Asian tsunami which found that duration of threat to life was a major determinant of ongoing physiological and psychological health problems (Marres et al., 2011); and a study by Caramello et al. (2019), who reported less mental and behavioral impact in responders to a mass casualty event in Turin, Italy, which they ascribed in part to its brief duration. Lowe et al. (2015), in their study of GuLF cleanup workers, found a complex relation of duration of work to greater physical effects also leading to more mental and behavioral health problems, which was further complicated by greater income appearing to lead to lesser mental and behavioral problems. Choi et al. (2021), in their follow up study 9 years after the *Hebei Spirit* oil spill, found that duration of cleanup work was related to the extent of persistent mental health effects. Continued studies of this cohort will be helpful.

The many studies following the DWH oil spill amply confirm that a major impact of oil spills affecting shore communities is on their mental and behavioral health, and that economic impacts of the oil spill play a significant role, as does the vulnerability and resilience of the affected communities. One potential ameliorating response is to provide greater prominence to preventing mental and behavioral effects in the decision process responding to oil spills (Goldstein, 2020). The Incident Command System (ICS) currently plays a major role in responding to oil spills (refer to Section 4.2). The ICS, under the direction of the U.S. Coast Guard for offshore spills and the U.S. EPA for inland spills within the United States, has been a positive development in disaster response. Its strategic priorities initially focus on human health, which is interpreted as preventing acute safety hazards such as explosion and fire. This is followed by responding to the need to protect ecosystems, which is assisted by a relatively formal science-based evaluation (see Chapter 4 for consideration of NEBA and similar ecological approaches of value for ICS decision-making).

The relatively low priority given to health concerns beyond immediate safety considerations is apparent from a review of the post-DWH spill On Scene Coordinator report (Tavares, 2011). Human health issues are subsumed under the heading of "Safety." The on-scene coordinator (OSC) report contains an organizational chart for the ICS Houston Headquarters (see Figure 3.1, Organizational Chart Task Forces). It shows four operating branches. The first three are Procedures, Source Control, and Dispersant. The fourth operating branch has a catch-all title of "Safety, Human Resources, Information, Legal, Security." Chapter 4 of the OSC report is on safety. It is positive about the relatively good safety record related to the health of response workers, and goes into detail about seafood safety and the decision process concerning closures of seafood gathering. Mental health concerns, which led to the specific request for HHS involvement, are described as follows:

> Although not covered in the National Contingency Plan, the combined effects of the spill on a population that had only five years earlier endured Hurricane Katrina raised concerns about impacts on the mental health of the people living near oil-impacted areas. (Tavares, 2011)

Using the find function on this 244-page report reveals three other instances in which human health is mentioned. Two are related to the handling of oiled wastes, and the third concerns a report on potential post-cleanup risks to humans and ecosystems due to residual oil. In contrast, "turtle" elicits 149 entries and "bird" 71 entries.

Also instructive is a review of the U.S. EPA's *Handbook on Area Contingency Planning* (2018), which includes detailed consideration of ecosystems, including endangered species, historic preservation, tradeoffs using NEBA, and collaborative efforts with the U.S. Fish and Wildlife Service, NOAA, and other agencies—but not HHS. The *Handbook* does provide an overview of EPA's legal authority to consider public health. However, it contains no planning tools to identify population health risks, vulnerabilities, evacuation routes, or other standard public health approaches to minimizing disaster risks.

In view of the increasing evidence of mental and behavioral effects, additional focus on the role of the ICS in preventing mental and behavioral effects of an oil spill would seem to be deserved. However, **achieving the level of information needed to inform ICS decisions would require mental and behavioral health scientists to focus on further developing and validating standardized measures of community mental and behavioral health equivalent to those now provided for decisions about ecosystems. Approaches to measure and improve community resilience are also needed to include human health within National and Area Contingency Planning.**

6.8.9 Dinoflagellate Toxin Poisoning

Dinoflagellates are a class of primarily unicellular organisms which when present in seafood (particularly shellfish), produce toxins that are responsible for a variety of effects including neurotoxicity, diarrheal diseases and ciguatera fish poisoning. In the United States, deaths are uncommonly reported but may be underestimated due to physicians' lack of familiarity with dinoflagellate toxicity. The linkage of dinoflagellate blooms to oil in the sea has been controversial, although recent studies have suggested that toxicity of oil to dinoflagellates with or without dispersants is less than to their main grazers (ciliates and heterotrophic). The resulting higher abundance of dinoflagellates may result more from the release from predation than from stimulation by the oil. The disruption of the bottom-up limiting processes may allow blooms of dinoflagellates, including harmful ones, to develop (Almeda et al., 2018). In another study by Park et al. (2020), the combination of bacterial communities exposed to oil-contaminated sediment from the Texas City, Texas "Y" oil spill stimulated the growth of the dinoflagellate *Prorocentrum texanum*. Furthermore, when isolates of oil-degrading bacteria were co-cultured with the dinoflagellates, dinoflagellate growth was stimulated by unidentified bacterially generated substances.

6.8.10 Other Effects on Human Health: Workers Health and Safety

Much of the earlier information about the health effects of oil spills has come from study of cleanup workers, both professionals and citizen volunteers. This is reflected in the discussion earlier in this section. As described in the various health-based sections, much of the information about effects of oil in the sea comes from studies of response workers. This includes mental and behavioral health problems, so it is not surprising that such findings were also noted in DWH responders (Fullerton et al., 2004; Goldmann and Galea, 2014; Lowe et al., 2016, 2017; Quist et al., 2019).

Worker safety issues and the safety culture in the oil industry are entwined in the subject of oil in the sea. Failure to adhere to worker safety rules arguably was a significant causative factor in the *Exxon Valdez* oil spill and a contributor to the system failure that led to the DWH oil spill. Eleven rig workers died in the latter (NRC, 2012). Many of the worker safety issues are an amalgam of those related to the sea and those common to the petroleum industry on shore, including drilling for and transporting flammable and explosive materials. Response activities can be more intense or at variance from standard industry activities (e.g., dealing with the carcasses of oiled birds or mammals, or spending much more time cleaning ships and gear). Response workers in particularly stressful situations and those exposed to high amounts of total hydrocarbons have an increased risk of mental health effects such as PTSD or depression (Kwok et al., 2017b). Performing research related to oil spills and seeps can lead to risks specific to deep sea diving, although humans are largely being replaced by controlled submersibles.

In recent decades, coastal fouling from major oil spills has led to complex response operations by petroleum company employees and governmental organizations such as the U.S. Coast Guard. In addition, there has often been a major influx of volunteers and of paid workers to help clean the shore and the oiled birds and mammals. Significant concerns about the well-being of these cleanup workers have included standard safety issues, mental health concerns, an overlay of local climate issues such as sunburn and dehydration on Gulf beaches, and a concern that the new responders are not sufficiently trained or have unreported pre-existing medical conditions that might add to their risks.

6.8.11 Seafood Safety

The complexities of seafood safety determinations after an oil spill have been relatively well explained for public understanding by NOAA and in a video by the Sea Grant Program (NOAA, 2021d) as well as in the *Report of the On Scene Coordinator* (Tavares, 2011). Many steps are taken to avoid contaminated seafood reaching the public. Seafood gathering is not allowed if there is visual oil on the water. Once the sheen of oil is no longer visible, the next step is sensory testing in which trained observers test gathered or cooked seafood for the smell of oil. Finally, a chemical determination is made to ensure that seafood PAH levels are within allowable standards. However, despite these lucid explanations, Simon-Friedt et al. (2016), in a 3-year survey of residents of Southeastern Louisiana, found that almost 50% reported not having sufficient information to determine whether they should eat local seafood. **Further transparency about the seafood safety risk determination would be helpful.**

As discussed earlier in this report, there are thousands of PAHs whose different chemical structures result in different physical properties that affect their liquid/vapor phase distribution and environmental persistence, and thus the extent of seafood and human exposure (Allan et al., 2012; see also more detailed discussion of PAH chemistry, toxicology, and analysis in Section 2.1.3, including Figure 2.8; Section 2.1.7.7; and Section 3.3.2). Factors that influence the extent of seafood contamination following an oil spill include the quantity and composition of the oil; its proximity to seafood gathering areas; weather and other seasonal factors; and the specifics of the ecosystem that affect uptake into seafood, including the potential for bioaccumulation and depuration. For example, vertebrates can catabolically excrete accumulated oil contaminants much more rapidly than most invertebrates.

PAHs vary greatly in the extent of their bioavailability, persistence, mutagenicity, and presumed human carcinogenicity. As has been done with other toxicologically relevant large classes of compounds having similar properties, such as polychlorinated biphenyls (PCBs), the allowable risk-based standard for seafood PAH levels is developed by first assigning an equivalent risk level for each of the PAHs whose toxicity and usual relative concentration are most likely to cause adverse effects. Based on the measured amounts of these selected PAHs, multiplied by the equivalency factor for each, a total PAH risk is determined which is used as the basis for decisions about seafood safety.

Although this is a useful approach, it inherently requires updating to accommodate additional information that becomes available about the toxicity of different PAHs alone and in a mixture. For example, Farrington (2020) reviewed information on the PAH compositions in (petrogenic) petroleum compared to pyrogenic sources that had been obtained during the several decades since the adoption of the specific PAHs included in the U.S. EPA's Priority Pollutant listing. The author advocated reevaluating human health risk protocols for seafood because the level of alkylated compounds and their potential toxicity may have been underestimated. Similarly, as offshore burning is a potential response to oil released to the sea, an issue particularly relevant to oil spills is the potential toxicological differences between petrogenic (petroleum) PAHs, which would be decreased by burning the oil, and pyrogenic PAHs, which are formed when oil is burned.

Gohlke et al. (2011) provide an overview of the various federal and state agencies involved and their different jurisdictions (see also Yender et al., 2002; NASEM, 2020). Gohlke et al. (2011) noted inconsistencies in the risk methodology used to close or reopen seafood gathering in the past. Further studies by this group include finding higher seafood consumption among children living close to the Gulf shore as compared to those living farther inland, although in all cases the intake by children of PAHs, metals, or a dispersant component were below the level of concern (Sathiakumar et al., 2017); and that testing of fish caught by commercial fishing folk found detectable levels of PAH in only 2 of 92 fish, both of which were far below the level of concern (Fitzgerald and Gohlke, 2014). Both Ylitalo et al. (2012) and Gohlke et al. (2011) reported that all of the officially measured PAH levels in seafood were below the level of concern. These were fish that were tested beginning after there was no longer a sheen in the water and the trained observers could detect no smell. Similarly, Wickliffe et al. (2018) reported that consumption of fish and shrimp from Southeast Louisiana posed no unacceptable lifetime cancer risk. (See also NASEM [2020] for discussion of seafood safety issues after the DWH oil spill.)

The uncertainties in the PAH risk methodology are evident from a critical published review of the U.S. Food and Drug Administration's (FDA's) approach to estimating the risk of PAHs in seafood which particularly emphasized concerns about pregnant women and children (Rotkin-Ellman et al., 2012). The response by Robert Dickey of the FDA (Dickey, 2012), argued that the conservative approach used by FDA likely overestimated actual risk, including by assuming a 5-year consumption period of the contaminated seafood. He also contrasted the 1/100,000 risk level for PAHs used to determine seafood safety with the much higher PAH background levels in the general U.S. population and the value of fish consumption to human health. This was responded to by two of the original authors who again highlighted the uncertainties in the risk approaches used for vulnerable populations (Rotkin-Ellman and Solomon, 2012).

Given that the economic and social consequences of the inability to gather seafood are a significant cause of the mental and behavioral health effects and community disruption caused by major oil spills (see preceding discussion), much more information is needed to narrow these risk-related scientific uncertainties. **The tradeoff between the public health risks of PAHs in seafood and the public health impact of seafood closures on community well-being also warrants attention by policy makers.**

6.8.12 Vulnerability of Humans to Oil Pollution in the Arctic

Indigenous cultures who have lived along the Arctic coast of Alaska and western Canada for many thousands of years depend heavily on stable ecosystems for their food security through subsistence harvests from the sea. Because participation in subsistence harvests are essential for the intergenerational transmission and maintenance of their cultures, members of these communities are uniquely vulnerable to adverse effects of accidental oil discharges. Members of these communities harvest, on average, more than 300 kg of marine mammals per person per year (based on data from the Alaska Department of Fish and Game), providing their primary source of calories as well as protein. Many marine mammals can be harvested throughout the year, and it is their abundance and availability that make life, and these ancient cultures, possible. These community sites have been occupied for as long as 6,000 years or more, and are the longest continuously occupied sites in North America (Giddings, 1962). This long and intimate relation between these subsistence-based cultures and the marine mammals (as well as all the other resources) that they harvest has led to development of extensive and intricate methods for making use of nearly every part of the animals and plants harvested.

Subsistence harvesting in Indigenous communities of the Alaskan and western Canadian Arctic also provides the foundation for maintenance and transmission of their cultures. Subsistence harvests are community efforts, with members of both sexes engaged in various aspects of harvesting, processing, storing, and utilizing harvested plants and animals from an early age. These activities provide the means through which art, technology, hunting skills and techniques, and cultural history are transmitted across generations. Consequently, any disruption of these subsistence food-gathering activities poses a serious threat to the integrity and maintenance of these cultures.

Following the 1989 *Exxon Valdez* oil spill, members of Indigenous communities in the spill-affected region became deeply concerned about the safety and continued availability of their subsistence food supplies. As noted by Fall (1999), after personal observations of erratic effects of oil on fish, wildlife and their habitats,

> [T]he following questions became very important for the people of the villages: Are our subsistence foods still safe to eat? If some beaches, waters, and animals were oiled and other, seemingly unoiled animals are inexplicably dying, are any resources safe to use?

After hearing assurances from health officials that absence of an oily taste or smell probably indicated that consumption of subsistence-gathered foods was safe, Fall (1999) noted that "villagers responded with skepticism and disbelief" and quoted one villager as saying "We saw too much oil, and we didn't want nothing to do with [the fish]. . . . We don't want to eat them until we find out what's really going on." Consequently, "subsistence harvests in many villages virtually ceased soon after the spill," and had not fully recovered 5 years later.

Indigenous communities had no cultural experience that deals with the effects of oil spills, in contrast with firmly rooted cultural approaches to dealing with natural disasters. This, along with overwhelming numbers of strangers having little familiarity with Native culture or institutions implementing oil spill response and remediation over which Natives had little or no influence, led to profoundly adverse psychological and social effects (Fall et al., 1999), exacerbated by the fact that oil spills are technological and not natural disasters (Picou et al., 1997).

Once interrupted, re-establishing traditional subsistence lifestyles can be difficult or impossible. Members of younger generations may abandon their villages and turn to cash economies in urban settings in search of food and other necessities. If widespread and sustained, this can seriously jeopardize the vitality of the Native cultures affected (Gill and Picou, 1997).

6.9 CONCLUSIONS AND RESEARCH NEEDS

The following conclusions and research needs are aimed at filling in important research gaps in understanding the effects of oil in the marine environment so that the environment may be better protected in the future.

Conclusion—Evolving Baseline Knowledge and Data: There are numerous and diverse sources of data and knowledge that can be reviewed and assembled into evolving baseline knowledge and data for marine ecosystems in North American marine waters to inform response, damage assessment, and fates and effects of oil spills.

Conclusion—Long-Term Effects: The committee noted increasing evidence suggesting significant longer-term effects on multiple aquatic and shoreline species and communities, including humans, than previously estimated. Short-term assessments do not cover enough time or provide opportunities for holistic studies to recognize changes in environmental conditions or resources in current time frames, particularly in relation to potentially changing baseline conditions.

Conclusion—Rapid Scientific Response, Communication, and Coordination: During and after an oil spill, there is often an extended time before scientists become engaged in field data collection resulting in missed opportunities to assess critical issues, including establishing environmental baselines and determining initial impacts. Rapid decision-making, funding, and deployment of scientists into the field provided critical information in both the *Exxon Valdez* and DWH oil spills, though there is still room for improvement.

Conclusion—Protection of Key Foodweb Components and Endangered Species: Mounting evidence suggests that widespread adverse effects on species that are endangered or are major components of marine food webs, such as seabirds and marine mammals, may have substantial repercussions on other species, operating through strong trophic linkages or cascades. Furthermore, adverse health effects on marine mammals and sea turtles exposed at the air-sea interface, primarily on those with rapid gas exchange and high air flow (cetaceans) or prolonged exposure (sea otters) manifest in short- and long-term morbidity and mortality. Lack of avoidance of oil, especially in calm waters, leads to significant risk in these species.

Conclusion—Natural Seeps: Not all effects of oil in the sea are negative. Natural seeps contain areas of (often unique) biological communities that have evolved to use oil as an energy and nutrient source. The oil and gas seeps and hydrate communities host chemoautotrophic symbionts that are unique biogenic communities, particularly in deep-sea locations with a high diversity of organisms, including bacteria consortia (including oil degraders) that support numerous, as yet unknown, biogeochemical processes (including oil degradation), particularly in deep-sea locations.

Conclusion—Marine Oil Snow: Observations found part of the natural biological response to the DWH drilling oil spill resulted in vast quantities of oil-associated marine oil snow (MOS) forming throughout the water. However, the mechanisms controlling MOS formation and its impact on the environment remain largely unknown.

Conclusion—Shoreline Oiling Characterization: The SCAT program supports response efforts by conducting rapid assessment of shoreline oiling, recommending shoreline treatment techniques, and identifying cleanup endpoints. An ancillary sampling program focused on gathering more detailed information on hydrocarbon composition, concentration, and distribution over time would complement SCAT-collected data and provide valuable information for the environmental impact analysis and monitoring of restoration progress.

Conclusion—Behavioral Effects of Oil: Recent research has demonstrated complex effects of oil on olfaction in fish that altered behavioral responses to oil exposure. These findings suggest that these and possibly other effects of oil on organismal behavior may reveal adverse effects of oil that are currently unrecognized. Further research focused on behavioral responses to oil exposure is needed.

Conclusion—Toxicity Studies and Models: Laboratory toxicity studies focus extensively on water column and acute toxicity effects of relatively fresh oil, for which a large amount of data is already available, and models have been developed that can predict the acute toxicity to additional non-standard test species. Limited data are available for chronic, delayed and multi-generational impacts in addition to field related effects and the influence of multiple stressors and environmental covariables.

Despite previous efforts (i.e., CROSERF) to standardize laboratory toxicity testing approaches, many laboratory toxicity studies are still conducted/reported in such a way that results cannot be used in models or implications of the findings understood and many may under- or over-represent toxicity. Furthermore, new knowledge and/or technologies have been developed since the original standard methods were published.

Inaccurate predictions of toxicity can occur based on how exposure media are prepared and characterized, and the choice of the concentration metric used. Depending on the specific species/life-stage and associated exposure pathways, oil mixtures containing droplets (i.e., high-energy preparations or those containing dispersants) will result in under- or over-estimations of toxicity if oil partitioning (dissolved/particulate) and/or combined chemical metrics (e.g., total petroleum hydrocarbons or total polycyclic aromatic hydrocarbons) are used.

Models of fate, exposure, and effects have been developed and refined and are useful for informing response options and for predicting effects for NRDA. Models have been refined to be inclusive of new information regarding the role of environmental variables, including temperature, light and pressure and the partitioning of oil constituents; thus, they now include oil droplets in addition to dissolved phase components, although many models require validation by laboratory and/or field data.

There are understudied exposure pathways that may be critical for the health of some marine species, for example, exposure to volatile oil constituents in air-breathing marine mammals and turtles based on studies of DWH-affected dolphins. Furthermore, the role of understudied and novel toxicological mechanisms has been highlighted in recent studies, including toxicity and toxic mechanisms of photo-oxidation products, impacts to the adrenal and immune systems, and the importance of oil inducing microbiome changes and ultimate health consequences. In some instances, no appropriate laboratory models exist to better understand these pathways (e.g., effects on HPA axis for dolphins). The development of various 'omics techniques have elucidated the potential role and impact of oil on many new toxicological pathways, although many require further development and investigation to resolve consequences at the individual and ultimately population levels.

Conclusion—Seafood Safety: Standard U.S. Food and Drug Administration designations of seafood safety do not adequately describe potential toxicity effects on humans. The basis for re-opening fishery resources is in part related to the risk assessment of the U.S. EPA priority-list PAHs established several decades ago. There are additional PAH and reaction product analytes that may be important to consider. Further evaluations are needed of these and the originally listed PAHs, including the implications of this information to the mechanism of carcinogenesis, to understand whether the current risk is over- or under-estimated.

Indigenous communities along coastlines rely heavily on subsistence harvests of marine food sources. Following a major oil exposure, members of these communities will have concerns regarding additional contamination of these food sources by the discharged oil.

The major economic impact of the DWH oil spill, and others, has been the closing of seafood gathering based on protection of human health. There is a need for improved understanding of existing uncertainties in the risk basis for reopening of seafood gathering in ocean areas that had been oiled and for transparent communication of this information. New information available about the chemistry and toxicology of the mixtures of PAHs, as well as increasing recognition of the impact of the closing of seafood gathering on community health effects, indicate a need to reassess risk to humans related to ingesting seafood following an oil spill. (The committee notes that further research to narrow uncertainties could provide the scientific basis to make the criteria more or less stringent.)

Conclusion—Photo-Oxidation: UV light has been shown to influence (often increase) the toxicity of oil constituents, especially to certain life stages and species in the photic zone. This includes photoenhanced toxicity mechanisms and the production of photo-oxidation products.

Conclusion—Arctic Industrialization: The Arctic is an extraordinarily challenging and difficult place to live and work, which increases risks of industrial accidents associated with oil and gas exploration and production, and commercial fishing, tourism cruises, and container ships. Within the circumboreal continental shelves, the region with the greatest potential for oil and gas exploration and development includes the Alaskan and western Canadian Arctic coasts, placing this region at the highest risk of a major oil spill should such development progress. If a spill were to occur, the likelihood of effective oil discharge response and remediation efforts beyond "natural recovery" is low given the scant availability of support infrastructure and the high frequency of inclement weather conditions.

Conclusion—Risk to Coastal Communities: North American coastal communities have unique social and cultural characteristics which differentiate the effect of an oil spill and can make these communities particularly vulnerable to natural disasters and the effects of oil spills, as seen following the *Exxon Valdez* and DWH oil spills.

Conclusion—Indirect Human Health Effects: The ICS has proven to be a very effective approach to decision-making following an oil spill. The ICS ranks human health as its highest priority, but human health is primarily conceptualized as safety from direct oil spill effects (including, but not limited to, inhalation, contact, and the risk of fire and explosion). As has been seen as a consequence of the DWH and other oil spills, oil spills have the potential to indirectly affect human health; studies of the DWH oil spill have confirmed and further demonstrated the major impact of oil spills on community mental and behavioral health, as well as the role of economic impacts from oil spills.

RESEARCH NEEDS

Continued research to better understand and model effects of oil on the marine ecosystem, including humans, is encouraged. More specifically, the research included in Table 6.3 would continue to advance this important component of oil spill science.

TABLE 6.3 Research Recommended to Advance Understanding of the Effects of Oil in the Sea on the Marine Environment

6.1	**Natural Seeps:** As relatively understudied habitats, research is needed to increase our understanding of unique chemosynthetic communities near natural seeps, especially deeper sea locations, to identify novel species/biochemical pathways and chemosynthesis, and to identify bacteria that may be useful in oil spill response (i.e., oil degraders), and to understand how organisms and communities respond to the presence of oil.
6.2	**Marine Oil Snow:** Continuing research on the formation of marine oil snow (MOS) is needed with respect to influences on processes of oil degradation and eventual hydrocarbon fates, such as flux through the water column, interactions with water column organisms, and short- and long-term deep-water biogenic communities.
6.3	**Assessment Techniques:** Potential research areas would be the development of sensors for petroleum hydrocarbons and image analysis for plankton in situ, and autonomous underwater vehicles (AUVs) for determination of water column effects.
6.4	**Marsh Ecosystem Health:** The protocols of natural resource damage assessment for marshes following exposure to an oil spill should incorporate additional measures of the health of marshes, including their structural integrity. Research into other measures or development of technological advances (e.g., portable photosynthesis systems for gas exchange and chlorophyll fluorescence measurements in plants) may also generate more universal representative indicators. The longer-term study with multiple integrated features of the salt marsh ecosystem points to the need for these types of studies to integrate the multiple aspects of contamination, salt marsh ecology, trophic structure, predator-prey dynamics, understanding of microhabitats within the marsh, and an integrated approach from genetics and enzymatic responses to ecosystem-level effects.
6.5	**Marine Vertebrates:** Studies focused on better estimation of mortalities of seabirds, marine mammals, and sea turtles during future spills of opportunity are needed. These studies should include sampling methods that permit estimation of statistical confidence intervals in addition to point estimates of, for example, numbers of animals killed.
6.6	**Corals:** The numerous environmental and coral health co-stressors should be studied together with additional investigations of species and life stage sensitivities to better understand the impact of oil on these important coral reef ecosystems. Studies of mesophotic and deep-sea corals highlight the need for prior baseline studies of the health of benthic ecosystems, together with long-term follow-up studies on recovery, delayed mortality, and continued declines in health in these species, particularly given their slow growth and lower recruitment compared to other marine species and hence potential for a protracted recovery period.
6.7	**Ecosystem-Level Effects:** Better understanding of trophic structure in marine systems could be accomplished with experimental design that incorporates populations, the community trophic interactions, multiple stressors, and interrelationships that could anticipate indirect or cascading effects of an oil spill. Field studies that incorporate all these features may not be able to reproduce the complexity of a marine ecosystem, but models could provide a basis for further exploration. The addition of longer-term observations and experiments that include higher organization level components and trophic interactions should be funded by the appropriate agencies and responsible parties. To these ends, appropriate agencies should encourage and support efforts to develop ecological atlases of marine resources that identify especially important ecological areas and habitats of threatened or endangered species, similar to those developed by Audubon, Oceana, and other environmental nongovernmental organizations (NGOs) for the Alaskan Arctic, which may serve to extend to offshore waters the Environmental Sensitivity Index approach currently used for shorelines.
6.8	**Shoreline Oiling Characterization:** A comparison of Shoreline Cleanup and Assessment Technique (SCAT) oiling categories with chemical characterization of hydrocarbons would benefit subsequent field studies and comparison of effects results based primarily on the four-tier SCAT categories. Also, the use of statistically-based sampling methods to estimate upper and lower bounds of the extent and intensity of shoreline oiling, including estimates of subsurface oil, would permit meaningful comparisons among sampling periods to estimate temporal trends of oil persistence.
6.9	**Photo-Oxidation:** The toxicity of photo-oxidized products of compounds that dissolve from oil into seawater should be evaluated to determine how much they contribute to acute and chronic toxicity effects on test organisms. Further studies are needed to characterize the array of photo-oxidation products produced from various oils and assess their persistence, bioaccumulation, and toxicity to standard toxicity test organisms. Additional organisms and numerous life stages should be included in these studies. Further research on UV radiation, elevated temperature, and decreased pH is needed to fully assess the effects of these co-factors and changing climate conditions on toxicity and the impact of oil pollution in exposed ecosystems.
6.10	**Behavioral Effects of Oil:** As a relatively understudied effect, research is needed to more fully evaluate the effects of exposure to oil on organismal behavioral responses.

TABLE 6.3 Continued

6.11 **Toxicity Studies and Models:**

1. Redesign the toxicity tests needed to provide data useful for model verification and in the decision making process (i.e., response tradeoffs). Enable further integration among laboratory studies and models useful for oil spill responders. Toxicity studies should focus on chronic, sublethal impacts, particularly of novel mechanisms of action that have health and population-level consequences (e.g., immune responses). The influence of environmental co-variables (e.g., light/photochemical reactions and pressure) should be further evaluated and included in models.
2. Revisit and update standard procedures for the preparation and characterization of toxicity test media, and the conduct of toxicity tests, including control of additional covariables and how they are reported (e.g., updating the chemical response to oil spills: Ecological Research Forum (CROSERF) protocols). Descriptions of the types of chemical characterization should be required (e.g., regarding chemical constituent resolution and partitioning [i.e. dissolved and droplet concentrations and droplet size/volume characterization]). The minimum data reporting should be identified and required.
3. Study test designs should clearly identify the questions being asked, the ecological relevance, and the species/life stages being used, and employ appropriate chemical analyses (e.g., filtration of exposure media and specific analytes measured) in addition to reporting toxicological endpoints appropriate for the study, including the use of toxic units (TUs) if appropriate.
4. Studies should be directed toward investigation of novel exposure pathways (e.g., inhalation in marine mammals, seabirds, and turtles; the influence of maternal transfer pathways) and new focused mechanisms of toxicity (e.g., adrenal function, immune responses, and the influence and importance of the microbiome).
5. Studies on the effects of surface oil ingested by fish for cardiotoxic effects of PACs from surface oil and on fish embryos should be supported in laboratory studies and during future spills of opportunity.
6. Models (e.g. OilTox, Petrotox) should continue to be refined based on inputs (e.g., role of droplets, additional chemical constituents) and influence of environmental variables (e.g., photo-oxidation, UV light, temperature, pressure). Models should also continue to be validated, particularly involving national or global inter-laboratory efforts or in-field trials (if applicable).

6.12 **Seafood Safety:**

1. Update the list of analytes included in risk assessment for seafood safety considerations and refine the current approaches used (e.g., benzo(a) pyrene equivalents).
2. The governmental agencies involved in responding to an oil spill should make a concerted effort to thoroughly reevaluate the health risk basis for the closing and reopening of seafood gathering following an oil spill, including an in-depth analysis of different types of PAHs likely to be present and their relative contribution to health risk. Governmental agencies should sponsor an ongoing integrated program supporting a broad range of risk-related research relevant to decisions about closing and reopening seafood-gathering areas following oil contamination.
3. Establish current polycyclic aromatic hydrocarbon contaminant concentrations in edible tissues of marine mammals consumed by Indigenous communities in the Arctic.

6.13 **Coastal Community Response:** Programs to improve community resilience in response to future oil spills must be tailored to the individual communities at risk. Support is needed for social science research initiatives that will incorporate understanding of the broad range of local factors as a basis for preparation for the next oil spill or other disasters. Existing and new findings should be incorporated by EPA and other agencies into disaster planning (e.g., the EPA *Handbook on Area Contingency Planning*, which at present does not consider community health).

6.14 **Follow up of Epidemiological Studies:** Studies of response workers have noted a longer-term association between cardiovascular and central nervous system effects at levels of petroleum hydrocarbons exposure that are orders of magnitude below currently allowable worker exposures. This association seems unlikely to be causative in the absence of such findings in well-studied workforces exposed to the higher levels for longer periods of time, but should not be discounted without further rigorous follow-up.

6.15 **Maternal and Child Health:** The relative absence of female workers in the petroleum industry in the past has limited the availability of information about potential maternal and child health impacts. Longer-term follow-up of cohorts of women pregnant during the DWH oil spill and of community children should be performed. An increase in the limited amount of toxicological information on the potential reproductive and developmental effects of crude oil derivatives possibly reaching shore communities would also be useful.

7

Recommendations

Chapter 7 includes the committee's recommendations to improve oil spill science based on the conclusions drawn in previous chapters. Recommendations are targeted toward improving quantification of oil entering the sea, preventing oil from entering the sea, minimizing effects of oil that pollutes the sea, as well as recommendations for the data, framework, and research needed to better understand the fate and effects of oil in the sea and to minimize those effects on humans and on the marine environment. Recommendations are directed toward government, industry, and research communities with interest in North American waters. Recommendations targeted toward government agencies include relevant agencies at national, state, and local levels in the United States, in Canada, and in Mexico. National agencies include, but are not limited to:

Canadian Agencies:

- Canada Energy Regulator (CER)
- Canada–Newfoundland and Labrador Offshore Petroleum Board
- Canada–Nova Scotia Offshore Petroleum Board
- Department of Fisheries and Oceans, Canada (DFO)
- Environment and Climate Change Canada (ECCC)
- National Research Council
- Natural Resources Canada
- Transport Canada

Mexican Agencies:

- CNH Comisión Nacional de Hidrocarburos (National Commision for Hydrocarbons)
- Instituto Mexicano del Petróleo (National Institute of Oil Studies)
- Mexican Navy (Marine Environment Protection Division) (PROMAM)
- Ministry of Communications and Transport
- Ministry of Environment, Natural Resources and Fisheries (SEMARNAT)
    - Federal Attorney for the Protection of the Environment (PROFEPA)
    - National Agency for Safety of Energy and Environment (ASEA)
- Ministry of National Defense
- Ministry of the Energy (Secretaría de Energía) (SENER)

U.S. Agencies:

- Interagency Cooperating Committee on Oil Pollution Research (ICCOPR)
- National and Regional Response Teams (NRTs, RRTs)
- National Oceanic and Atmospheric Administration (NOAA)
- Pipeline and Hazardous Materials Safety Administration (PHMSA)
- U.S. Arctic Research Commission (USARC)
- U.S. Coast Guard (USCG)
- U.S. Department of Energy (DOE)
- U.S. Department of Health and Human Services (HHS)
- U.S. Department of the Interior (DOI)
    - Bureau of Ocean Energy Management (BOEM)
    - Bureau of Safety and Environmental Enforcement (BSEE)
    - U.S. Fish and Wildlife Service (USFW)
    - U.S. Geological Survey (USGS)
- U.S. Environmental Protection Agency (U.S. EPA)

State and local agencies are too numerous to list but include applicable entities from the state and province level to the coastal community level of government, including tribal and First Nations government.

The preceding list of government agencies underscores the breadth and depth of government responsibilities as well as the diversity of roles served by different agencies for understanding, managing, and mitigating oil in the sea. Some of these organizations have undergone recent reorganizations, and many are continually adapting their scope and responsibilities to address new charges and government leadership. Moreover, it is unlikely that the responsibilities of each agency within one nation are fully understood by those in other North American agencies. Neither oil nor its environmental effects are necessarily contained in the jurisdiction in which the oil was spilled, requiring further understanding and coordination between North American agencies.

> **Recommendation—Defining Roles and Responsibilities:** In light of the complexity of agency responsibilities for oil in the sea and recent reorganizations of many of these agencies, the committee recommends the roles and responsibilities of various authorities involved in overseeing the oil and gas industry and marine environmental protection should be clarified in order to make specific recommendations to these authorities. Specific to the United States, the committee recommends that the Interagency Cooperating Committee on Oil Pollution Research examine the recommendations, assign responsibility, and form a working group to further explore potential funding mechanisms such as the Oil Spill Liability Trust Fund.

In addition to the complexity of agency responsibility for oil in the sea, industrial organizations, nongovernmental organizations, and private foundations also have roles and responsibilities in addressing or supporting many of the following recommendations.

7.1 QUANTIFICATION OF OIL IN THE SEA

Recommendation—Improve Quantification of Oil Inputs into the Sea: In recognition of decades of inaction on past *Oil in the Sea* report recommendations to measure natural and anthropogenic oil inputs to inform mitigation of the effects on the marine environment, the committee recommends that an independent group report on measures and responsibilities of North American agencies to acquire appropriate data to better achieve quantification of oil inputs to the sea. The following actions should be taken to improve quantification of oil inputs to the marine environment:

- Federal agencies should work with industry and academia to use existing techniques, along with exploration of new technologies, for identification and quantification of inputs from natural seeps including less recently studied areas such as the North American Pacific margin, North American Arctic margin, and newly discovered seeps in the North American Atlantic margin. Priority should be given to areas with offshore energy exploration and production and along marine shipping routes to better assess background levels of oil in the sea and inform damage assessment.
- Federal agencies should work with state and local authorities to undertake regular monitoring of oil inputs from land-based sources of water (runoff, rivers, harbors, and direct ocean sewage discharge) to determine oil inputs into marine environments.
- Federal agencies should support research to refine estimates of land-based inputs of oil that are transported via the atmosphere to the sea (including fuel jettison) by (1) expanding the geographic and temporal coverage of data collection, and (2) refining understanding of the source of hydrocarbons measured in marine atmospheres and surface seawater.
- Relevant agencies should work with industry to gain a better understanding of composition and concentrations in produced water (from offshore exploration and production activities) released into the marine environment and implement practices to reduce potential environmental impacts of these discharges.
- A study should be conducted on the level of MARPOL Annex I compliance to establish a baseline to monitor changes in compliance.

7.2 PREVENTION

Recommendation—Prevention of Anthropogenic Oil Inputs into the Sea: During the transition to more renewable energy sources, the following recommendations should be acted upon by industry and by federal and state agencies to prevent future spills in North American waters along with their actual and perceived effects:

- Consistent with previous reports, government and industry should continue their efforts to develop and implement technologies and best practices to prevent and reduce the magnitude of accidental spills from onshore and offshore pipelines and facilities and marine transportation. The committee recognizes the risk of complacency following periods of reduced spillage and advises government and industry maintain vigilance.
- Government should review whether technical recommendations arising from the extensive investigations in the aftermath of the *Deepwater Horizon* incident regarding blowout preventers, as well as operational issues and safety culture, have been implemented, and identify those that have not yet but could be implemented.
- Government should conduct a comprehensive review of the integrity of coastal onshore and offshore energy infrastructure to determine if it can withstand increased

frequency and intensity of extreme weather events and other natural hazards, including:
  - Review and update of design criteria for extreme events in light of new data.
  - Assessment of modifications to existing structures needed to prevent or mitigate damage or spillage of oil resulting from extreme events.
  - Development of response plans and corresponding response capabilities to reduce and mitigate spills in case of damage due to extreme weather events.
- To mitigate potential spills from aging infrastructure, appropriate agencies should take inventory of the existing remnants of oil storage, transport, or production activities that still need to be identified. Salvage or capping of these facilities should be prioritized based on the potential impact if the infrastructure fails.
- In order to maintain response readiness, government should assess the economic and environmental impacts of changes in marine vessel transportation on pollution risk, such as introduction of new fuel types, increased vessel size, new types of cargoes (LNG, other gases, biofuels, and diluted bitumen), and new traffic patterns and offshore infrastructure.

7.3 MINIMIZING EFFECTS

Recommendation—Response Toolbox: Regulatory mechanisms should be introduced to encourage evaluation, permitting, and deployment of new advanced response techniques when they become available. The use of these techniques during actual emergency response should be guided by the specific scenario to ensure they add value and maximize health, safety, environmental, cultural, and socioeconomic protection.

Recommendation—Integration of Science with Response: Appropriate responsible agencies should plan for effective ways for rapid scientific response to oil spills to enable scientists to mobilize to the field quickly and in concert with response operations. This should involve rapid communication and approvals between all parties to define the operationally relevant science direction and gather relevant information for future decision making with respect to minimizing the effects of the spill.

Recommendation—New Products and Fuel Types: Government should fund research needed to study the composition, toxicity, and behavior of new types of marine fuels (e.g., low-sulfur fuel oil, very low sulfur fuel oil, biofuels) and petroleum products (e.g., diluted bitumen) so that fate and effects of these products can be understood and response operations can be planned and executed most effectively to reduce impacts.

Recommendation—Human Health Effects: The governmental agencies involved in responding to an oil spill should upgrade the priority and attention given to individual and community mental and behavioral effects and community social-economic disruptions in Incident Command Structure (ICS) decision-making and response processes. The inclusion of community-based human health assessment and mitigation measures into the ICS is needed to provide a more holistic approach regarding both human and ecosystem health.

Recommendation—Protection of Key Food Web Components and Endangered Species: Large oil spills contaminating productive marine ecosystems and shorelines may inflict mass mortalities on vulnerable species such as seabirds, marine mammals, and shoreline biota and disrupt the ecology in that location. Following a spill, appropriate environmental specialists (e.g., an environmental technical specialist in the incident command structure) should promptly identify these species, and ecological linkages should be promptly identified and their abundances in the affected region should be monitored, to enable detection of these indirect effects on populations at the community level and ensure their protection.

7.4 DATA TO ADVANCE THE SCIENCE

Recommendation—Baseline Knowledge and Data: There is a need to review how pertinent knowledge and data from numerous sources are most effectively assembled, made available, and archived, given the advances and gaps in understanding noted in this report.

- The review should assess what is needed for baseline knowledge and data with recognition that both natural and anthropogenic influences (other than inputs of oil) result in baselines that are dynamic in space and time.
- Funding should be established for appropriate baseline data acquisition and curation in locations of particular interest, such as coastal areas, areas with offshore energy exploration and production, and marine transportation routes.
- Data collections would include aspects such as physical oceanography, biogeochemical processes, contaminant source surveys, critical species (e.g., endangered, abundant, vulnerable, or of commercial importance) and marine biodiversity, and pertinent metrics of human health and well-being.
- As a corollary, guidelines should be developed for collecting and analyzing baseline data immediately after and in the midst of a spill from neighboring, unaffected (control) areas, where possible.
- For example, U.S. Interagency Cooperating Committee on Oil Pollution Research member agencies, in cooperation with relevant agencies from Canada and Mexico and other interested parties, should convene a series of regional workshops and studies to inform the most efficient process of defining and assembling the evolving baseline knowledge and data. These workshops should be inclusive in gathering relevant

knowledge from stakeholders such as indigenous peoples; diverse rural, suburban, and urban coastal communities; government agencies; business and industry; nonprofit groups; and the academic sector. The assembled knowledge and data should be stored in a useful format to support oil spill response, oil spill damage assessment, and assessment of fates and effects of oils spills and other oil inputs.

Recommendation—Data Management and Interdisciplinary Research: Given the enormous datasets being generated through advanced chemical analyses, 'omics techniques, geoscience surveys (among others), and especially field and laboratory studies pursuant to oil spills, the committee recommends the formation of a free central, universally accessible, and curated repository of information pertinent to oil in the sea. Optimum use of such archives will require development of data analytics, data quality control, and reporting standards for associated metadata to enable integration and interpretation by, and training of, interdisciplinary teams.

Recommendation—Arctic Studies: In agreement with previous reports, there should be a concerted effort to gather information about the fate of oil in Arctic marine ecosystems, with and without ice cover, in advance of further development of this region. This would include baseline surveys (geophysical and biological), efficacy of response and mitigation options, data acquisition on natural attenuation and active remediation strategies including biodegradation kinetics at low temperature, and effects on higher organisms, populations, and ecosystems in Arctic waters and on shorelines.

Recommendation—Monitoring Polycyclic Aromatic Hydrocarbon Profiles in Sediments and Bivalves: The National Oceanic and Atmospheric Administration (NOAA) should reinstitute the nationwide NOAA National Status and Trends Mussel Watch Program involving nationwide sampling at the suite of stations previously sampled annually for more than two decades. These samples should be analyzed for an expanded suite of polycyclic aromatic hydrocarbon analytes and selected petroleum biomarker compounds to provide a better assessment of the status and trends of petrogenic hydrocarbon contamination of coastal waters. Instituting this simultaneously with a collaborative program with Canada could be beneficial to both countries in assessing the status and trends of petrogenic hydrocarbon contamination in coastal waters. The human health implications of these findings to seafood safety determinations should be evaluated.

7.5 FRAMEWORK TO ADVANCE THE RESEARCH

Recommendation—Long-Term Funding for Inputs, Fates, and Effects of Oil in the Sea: As recommended in *Oil in the Sea III*, there remains a need for long-term, sustained funding focused on oil in the sea to support multi-disciplinary research projects that address current knowledge gaps, including those listed as Research Needs throughout this report. Research is needed to address new regulatory requirements and to improve response capabilities. The application of new data and technologies to advance interdisciplinary knowledge of fates and effects of oil in the sea will require a longer funding commitment than in typical.

Recommendation—Field Studies: As recommended in previous studies, controlled in situ field trials using real oils should be planned, permitted, and funded to incorporate multi-disciplinary research focused on important processes as well as response techniques that do not accurately scale from in vitro or ex situ experiments to in situ conditions. Additionally, funding and systemic mechanisms should be set in place by appropriate agencies to enable rapid deployment of qualified scientific personnel during actual oil spill events (i.e., spills of opportunity) to conduct appropriate, time-critical research in situ, outside the Natural Resource Damage Assessment process, while having minimal or no interference with spill response activities.

Recommendation—Employing Advanced Analytical Techniques: To answer questions asked for appropriate use in forensic analyses of spilled oils and other inputs and in damage assessment and response activities, relevant agencies and research communities should:

- Incorporate recent advances in analytical chemistry techniques in standardized protocols. Continued expansion of the use of these techniques in research efforts connected with inputs, fates, and effect studies is strongly encouraged. The preceding should be accompanied by rigorous quality control and quality assurances for sampling and analyses underpinned by preparation, calibration, and maintenance of a suite of existing and expanded appropriate Standard Reference Materials by the U.S. National Institute of Standards and Technology and comparable agencies in Canada and Mexico.
- Expand the current libraries of oil characteristics with more detailed analyses and reference oils.
- Utilize and continue to develop and validate 'omic technologies, and their rapid advancements, targeted at individuals (microbes to higher organisms) through environmental samples encompassing population and community compositions to provide information on oil fate and short- and long-term direct and indirect effects of oil exposure.

7.6 RESEARCH NEEDS TO ADVANCE THE SCIENCE

Chapters 4, 5, and 6 include detailed lists of specific research needed to advance understanding of fates of oil in the sea, of effects of oil on the marine environment, and of minimizing those effects through improved response. Tables 4.5, 5.4, and 6.3 include the committee's recommendations on research needed to fill important gaps and to continue to progress oil spill science.

References

Abbasnezhad, H., M. Gray, and J. M. Foght. 2011. Influence of adhesion on aerobic biodegradation and bioremediation of liquid hydrocarbons. *Applied Microbiology and Biotechnology* 92(4):653–675.

Abdulredha, M. M., H. S. Aslina, and C. A. Luqman. 2018. Overview on petroleum emulsions, formation, influence and demulsification treatment techniques. *Arabian Journal of Chemistry* 13(1):3403–3428.

Able, K. W., P. C. López-Duarte, F. J. Fodrie, O. P. Jensen, C. W. Martin, B. J. Roberts, J. Valenti, K. O'Connor, and S. C. Halbert. 2015. Fish assemblages in Louisiana salt marshes: Effects of the Macondo oil spill. *Estuaries and Coasts* 38(5):1385–1398.

Abplanalp, W., N. DeJarnett, D. W. Riggs, D. J. Conklin, J. P. McCracken, S. Srivastava, Z. Xie, S. Rai, A. Bhatnagar, and T. E. O'Toole. 2017. Benzene exposure is associated with cardiovascular disease risk. *PloS ONE* 12(9):e0183602.

ABS (American Bureau of Shipping). 2020. *Pathways to sustainable shipping: Setting the course to low carbon shipping.* Spring, TX: American Bureau of Shipping.

Abu Laban, N., A. Dao, K. Semple, and J. Foght. 2015. Biodegradation of C 7 and C 8 iso-alkanes under methanogenic conditions. *Environmental Microbiology* 17(12):4898–4915.

Achberger, A. M., S. M. Doyle, M. I. Mills, C. P. Holmes, A. Quigg, and J. B. Sylvan. 2021. Bacteria-oil microaggregates are an important mechanism for hydrocarbon degradation in the marine water column. *mSystems* 6(5):e01105–e01121.

ACS (Alaska Clean Seas). 2015. *Technical manual. Volume 1. Tactics descriptions.* Prudhoe Bay, AK: Alaska Clean Seas.

Adamovsky, O., A. N. Buerger, A. M. Wormington, N. Ector, R. J. Griffitt, J. H. Bisesi, Jr., and C. J. Martyniuk. 2018. The gut microbiome and aquatic toxicology: An emerging concept for environmental health. *Environmental Toxicology and Chemistry* 37(11):2758–2775.

Adams, E. E., and P. S. A. Socolofsky. 2005. *Review of deep oil spill modeling activity supported by the deep spill jip and offshore operator's committee. Rep., final report.* Washington, DC: Minerals Management Service.

Adams, J. E., and Canadian Science Advisory Secretariat. 2017. *Review of methods for measuring the toxicity to aquatic organisms of the water accommodated fraction (WAF) and chemically-enhanced water accommodated fraction (CEWAF) of petroleum.* Ottawa, Ontario, Canada: Canadian Science Advisory Secretariat.

Adegboye, M. A., W.-K. Fung, and A. Karnik. 2019. Recent advances in pipeline monitoring and oil leakage detection technologies: Principles and approaches. *Sensors* 19(11):2548.

Aeppli, C., C. A. Carmichael, R. K. Nelson, K. L. Lemkau, W. M. Graham, M. C. Redmond, D. L. Valentine, and C. M. Reddy. 2012. Oil weathering after the *Deepwater Horizon* disaster led to the formation of oxygenated residues. *Environmental Science & Technology* 46(16):8799–8807. https://doi.org/10.1021/es3015138.

Aeppli, C., R. K. Nelson, J. R. Radovic, C. A. Carmichael, D. L. Valentine, and C. M. Reddy. 2014. Recalcitrance and degradation of petroleum biomarkers upon abiotic and biotic natural weathering of *Deepwater Horizon* oil. *Environmental Science & Technology* 48(12):6726–6734.

Afshar-Mohajer, N., C. Li, A. M. Rule, J. Katz, and K. Koehler. 2018. A laboratory study of particulate and gaseous emissions from crude oil and crude oil-dispersant contaminated seawater due to breaking waves. *Atmospheric Environment* 179:177–186.

Afshar-Mohajer, N., M. A. Fox, and K. Koehler. 2019. The human health risk estimation of inhaled oil spill emissions with and without adding dispersant. *Science of the Total Environment* 654:924–932.

Afshar-Mohajer, N., A. Lam, L. Dora, J. Katz, A. M. Rule, and K. Koehler. 2020. Impact of dispersant on crude oil content of airborne fine particulate matter emitted from seawater after an oil spill. *Chemosphere* 256:127063. https://doi.org/10.1016/j.chemosphere.2020.127063.

Agrawal, H., W. A. Welch, J. W. Miller, and D. R. Cocker. 2008. Emission measurements from a crude oil tanker at sea. *Environmental Science & Technology* 42(19):7098–7103.

Agrawal, H., W. A. Welch, S. Henningsen, J. W. Miller, and D. R. Cocker III. 2010. Emissions from main propulsion engine on container ship at sea. *Journal of Geophysical Research: Atmospheres* 115(D23).

Aguilera, F., J. Mendez, E. Pasaro, and B. Laffon. 2010. Review on the effects of exposure to spilled oils on human health. *Journal of Applied Toxicology* 30(4):291–301.

Ainsworth, C., E. Chassignet, D. French-McCay, C. Beegle-Krause, I. Berenshtein, J. Englehardt, T. Fiddaman, H. Huang, M. Huettel, and D. Justic. 2021. Ten years of modeling the *Deepwater Horizon* oil spill. *Environmental Modelling & Software* 142:105070.

Aiyer, A., and C. Meneveau. 2020. Coupled population balance and large eddy simulation model for polydisperse droplet evolution in a turbulent round jet. *Physical Review Fluids* 5(11):114305.

Aiyer, A., D. Yang, M. Chamecki, and C. Meneveau. 2019. A population balance model for large eddy simulation of polydisperse droplet evolution. *Journal of Fluid Mechanics* 878:700–739.

Aksenov, A. A., L. Yeates, A. Pasamontes, C. Siebe, Y. Zrodnikov, J. Simmons, M. M. McCartney, J.-P. Deplanque, R. S. Wells, and C. E. Davis. 2014. Metabolite content profiling of bottlenose dolphin exhaled breath. *Analytical Chemistry* 86(21):10616–10624.

Albers, P. H. 1995. *Oil, biological communities and contingency planning.* Newark, DE: Tri-State Bird Rescue & Research Inc.

Alexander, D. A. 1991. Psychiatric intervention after the Piper Alpha disaster. *Journal of the Royal Society of Medicine* 84(1):8–11.

Alexander, M. 1985. Biodegradation of organic chemicals. *Environmental Science & Technology* 19(2):106–111.

Alexander, M., L. S. Engel, N. Olaiya, L. Wang, J. Barrett, L. Weems, E. G. Schwartz, and J. A. Rusiecki. 2018. The *Deepwater Horizon* oil spill coast guard cohort study: A cross-sectional study of acute respiratory health symptoms. *Environmental Research* 162:196–202.

Allan, S. E., B. W. Smith, and K. A. Anderson. 2012. Impact of the *Deepwater Horizon* oil spill on bioavailable polycyclic aromatic hydrocarbons in Gulf of Mexico coastal waters. *Environmental Science & Technology* 46(4):2033–2039.

Alldredge, A. L., J. J. Cole, and D. A. Caron. 1986. Production of heterotrophic bacteria inhabiting macroscopic organic aggregates (marine snow) from surface waters 1. *Limnology and Oceanography* 31(1):68–78.

Allen, A. 1999. New tools and techniques for controlled in-situ burning. *Proceedings Arctic and Marine Oilspill Program Technical Seminar* 22(2):613–628.

Allen, A. A., D. Jaeger, N. J. Mabile, and D. Costanzo. 2011. The use of controlled burning during the Gulf of Mexico *Deepwater Horizon* MC-252 oil spill response. *International Oil Spill Conference Proceedings* 2011(1):abs194.

Allen, J., and B. Walsh. 2008. Enhanced oil spill surveillance, detection and monitoring through the applied technology of unmanned air systems. *International Oil Spill Conference Proceedings* 2008(1):113–120.

Almeda, R., S. Cosgrove, and E. J. Buskey. 2018. Oil spills and dispersants can cause the initiation of potentially harmful dinoflagellate blooms ("red tides"). *Environmental Science & Technology* 52(10):5718–5724.

Aman, Z. M., C. B. Paris, E. F. May, M. L. Johns, and D. Lindo-Atichati. 2015. High-pressure visual experimental studies of oil-in-water dispersion droplet size. *Chemical Engineering Science* 127:392–400.

AMAP (Arctic Monitoring and Assessment Programme). 2010. *Assessment 2007: Oil and gas activities in the Arctic-effects and potential effects. Volume 2.* Oslo, Norway: Arctic Monitoring and Assessment Programme.

Amec Foster Wheeler. 2019. *Hibernia platform (year 10) and Hibernia southern extension (year 3) environmental effects monitoring program (2016): Volume I: Interpretation.* London, UK: Amec Foster Wheeler.

Amos, A. F., S. C. Rabalais, and R. S. Scalan. 1983. Oil-field callianassa burrows on a Texas barrier-island beach. *Journal of Sedimentary Research* 53(2):411–416.

Anderson, J., J. Neff, B. Cox, H. Tatem, and G. Hightower. 1974. Characteristics of dispersions and water-soluble extracts of crude and refined oils and their toxicity to estuarine crustaceans and fish. *Marine Biology* 27(1):75–88.

Anderson, K., G. Bhatnagar, D. Crosby, G. Hatton, P. Manfield, A. Kuzmicki, N. Fenwick, J. Pontaza, M. Wicks, and S. Socolofsky. 2012. Hydrates in the ocean beneath, around, and above production equipment. *Energy & Fuels* 26(7):4167–4176.

Andrady, A. L. 2011. Microplastics in the marine environment. *Marine Pollution Bulletin* 62(8):1596–1605.

Andreussi, H., and G. De Ghetto. 2013. Cube: A new technology for the containment of subsea blowouts. *Journal of Petroleum Technology* 65(1):32–35.

Anon. 2014. Eighty years of redfield. *Nature Geoscience* 7:849.

ANSI (American National Standards Institute) and API (American Petroleum Institute). 2015. *Pipeline safety management systems.* Recommended Practice 1173. Washington, DC: American Petroleum Institute.

Apeti, D., D. Whitall, G. Lauenstein, T. McTigue, K. Kimbrough, A. Jacob, and A. Mason. 2013. *Assessing the impacts of the* Deepwater Horizon *oil spill: The National Status and Trends Program response. A summary report of coastal contamination.* Silver Spring, MD: Center for Coastal Monitoring and Assessment, National Centers for Coastal Ocean Science.

API (American Petroleum Institute). 2006. *Recommended practice for well control operations.* Recommended Practice 59, 2nd edition. Washington, DC: American Petroleum Institute.

API. 2010. *Public awareness programs for pipeline operators.* Recommended Practice 1162, 2nd edition. Washington, DC: American Petroleum Institute.

API. 2012. *Computational pipeline monitoring for liquids.* Recommended Practice 1130, 1st edition. Washington, DC: American Petroleum Institute.

API. 2013a. *An Evaluation of the Alternative Response Technology Evaluation System (ARTES): Based on the Deepwater Horizon Experience.* Technical Report 1142. https://www.oilspillprevention.org/~/media/Oil-Spill-Prevention/spillprevention/r-and-d/alternative-response-technologies/1142.pdf.

API. 2013b. *Shoreline protection on sand beaches.* Technical Report 1150-1. http://www.oilspillprevention.org/~/media/Oil-Spill-Prevention/spillprevention/r-and-d/shoreline-protection/1150-1-shoreline-protection-report.pdf.

API. 2014. *Biodegradation and bioremediation of oiled beaches: A primer for planners and managers.* API Technical Report 1147. http://www.oilspillprevention.org/~/media/Oil-Spill-Prevention/spillprevention/r-and-d/shoreline-protection/biodegradation-bioremediation-on-sand-be.pdf.

API. 2015a. *Aerial and vessel dispersant preparedness and operations guide.* Technical Report 1148. Washington, DC: American Petroleum Institute.

API. 2015b. *Field operations guide for in-situ burning of on-water oil spills.* Technical Report 1252. Washington, DC: American Petroleum Institute.

API. 2015c. *Pipeline leak detection-program management.* Recommended Practice 1175, 1st edition. Washington, DC: American Petroleum Institute.

API. 2016a. *Canine oil detection (K9-SCAT) guidelines.* Technical Report 1149-4. Washington, DC: American Petroleum Institute. http://www.oilspillprevention.org/~/media/Oil-Spill-Prevention/spillprevention/r-and-d/shoreline-protection/canine-oil-detection-field-trials-report.pdf.

API. 2016b. *Tidal Inlet Protection Strategies (TIPS) field guide.* Technical Report 1153-2. Washington, DC: American Petroleum Institute. http://www.oilspillprevention.org/-/media/Oil-Spill-Prevention/spillprevention/r-and-d/shoreline-protection/tips-field-guide-final.pdf.

API. 2016c. *Selection and training guidelines for in situ burning personnel.* Technical Report 1253. Washington, DC: American Petroleum Institute.

API. 2016d. *Sunken oil detection and recovery.* Technical Report 1154-1. Washington, DC: American Petroleum Institute.

API. 2018a. *In-situ burning guidance for safety officers and safety and health professionals.* Technical Report 1254. Washington, DC: American Petroleum Institute.

API. 2018b. *Well control equipment systems for drilling wells.* Standard 53, 5th edition. Washington, DC: American Petroleum Institute.

API. 2019a. *Safety and environmental management system for offshore operations and assets.* Recommended Practice 75, 4th edition. Washington, DC: American Petroleum Institute.

API. 2019b. *Pipeline safety excellence performance report and 2020–2022 strategic plan.* Washington, DC: American Petroleum Institute.

API. 2019c. *Managing system integrity for hazardous liquid pipelines.* Recommended Practice 1160, 3rd edition. Washington, DC: American Petroleum Institute.

API. 2020. *Industry recommended subsea dispersant monitoring plan.* 2nd edition. Technical Report 1152. Washington, DC: American Petroleum Institute.

API. 2022. *Understanding crude oil and product markets.* Prepared by The Brattle Group. Washington, DC: American Petroleum Institute. https://www.api.org/-/media/Files/Oil-and-Natural-Gas/Crude-Oil-Product-Markets/Crude-Oil-Primer/Understanding-Crude-Oil-and-Product-Markets-Primer-High.pdf.

Arata, C. M., J. S. Picou, G. D. Johnson, and T. S. McNally. 2000. Coping with technological disaster: An application of the conservation of resources model to the *Exxon Valdez* oil spill. *Journal of Traumatic Stress* 13(1):23–39.

Araújo, K. C., M. C. Barreto, A. S. Siqueira, A. C. P. Freitas, L. G. Oliveira, M. E. P. A. Bastos, M. E. P. Rocha, L. A. Silva, and W. D. Fragoso. 2021. Oil spill in northeastern Brazil: Application of fluorescence spectroscopy and PARAFAC in the analysis of oil-related compounds. *Chemosphere* 267:129154. https://doi.org/10.1016/j.chemosphere.2020.129154.

Arfsten, D. P., D. J. Schaeffer, and D. C. Mulveny. 1996. The effects of near ultraviolet radiation on the toxic effects of polycyclic aromatic hydrocarbons in animals and plants: A review. *Ecotoxicology and Environmental Safety* 33(1):1–24. https://www.sciencedirect.com/science/article/pii/S0147651396900019.

Arnosti, C., K. Ziervogel, T. Yang, and A. Teske. 2016. Oil-derived marine aggregates—hot spots of polysaccharide degradation by specialized bacterial communities. *Deep Sea Research Part II: Topical Studies in Oceanography* 129:179–186.

ASTM. 2014. *Standard guide for in-situ burning of oil spills on water: Environmental and operational considerations.* West Conshohocken, PA: American Society for Testing and Materials.

Atlas, R. M. 1995. Petroleum biodegradation and oil spill bioremediation. *Marine Pollution Bulletin* 31(4):178–182. https://www.sciencedirect.com/science/article/pii/0025326X95001132.

Atlas, R. M., D. M. Stoeckel, S. A. Faith, A. Minard-Smith, J. R. Thorn, and M. J. Benotti. 2015. Oil biodegradation and oil-degrading microbial populations in marsh sediments impacted by oil from the *Deepwater Horizon* well blowout. *Environmental Science & Technology* 49(14):8356–8366.

ATSDR (Agency for Toxic Substances and Disease Registry). 1995. *Toxicological Profile for Polycyclic Aromatic Hydrocarbons (PAHs).* Atlanta, GA: American Society for Testing and Materials. https://www.atsdr.cdc.gov/toxprofiles/tp69.

Ault, A. P., C. J. Gaston, Y. Wang, G. Dominguez, M. H. Thiemens, and K. A. Prather. 2010. Characterization of the single particle mixing state of individual ship plume events measured at the port of Los Angeles. *Environmental Science & Technology* 44(6):1954–1961.

Aurand, D., and G. Coelho. 2005. *Cooperative aquatic toxicity testing of dispersed oil and the chemical response to oil spills: Ecological effects research forum (CROSERF).* Inc. Lusby, MD. Tech. Report 07-03.

Aurand, D., L. Walko, and R. Pond. 2000. *Developing consensus ecological risk assessments: Environmental protection in oil spill response planning a guidebook.* Washington, DC: United States Coast Guard.

Aurell, J., and B. K. Gullett. 2010. Aerostat sampling of PCDD/PCDF emissions from the Gulf oil spill in situ burns. *Environmental Science & Technology* 44(24):9431–9437. https://doi.org/10.1021/es103554y.

Ausuri, J., G. A. Vitale, D. Coppola, F. Palma Esposito, C. Buonocore, and D. de Pascale. 2021. Assessment of the degradation potential and genomic insights towards phenanthrene by dietzia psychralcaliphila ji1d. *Microorganisms* 9(6):1327.

Avigan, J., and M. Blumer. 1968. On the origin of pristane in marine organisms. *Journal of Lipid Research* 9(3):350–352.

Awkerman, J. A., B. Hemmer, A. Almario, C. Lilavois, M. G. Barron, and S. Raimondo. 2016. Spatially explicit assessment of estuarine fish after *Deepwater Horizon* oil spill: Trade-off in complexity and parsimony. *Ecological Applications* 26(6):1708–1720.

Azam, F., and F. Malfatti. 2007. Microbial structuring of marine ecosystems. *Nature Reviews Microbiology* 5(10):782–791.

Baars, B.-J. 2002. The wreckage of the oil tanker "Erika"—human health risk assessment of beach cleaning, sunbathing and swimming. *Toxicology Letters* 128(1–3):55–68.

Bacosa, H. P., D. L. Erdner, and Z. Liu. 2015. Differentiating the roles of photooxidation and biodegradation in the weathering of light Louisiana sweet crude oil in surface water from the *Deepwater Horizon* site. *Marine Pollution Bulletin* 95(1):265–272.

Bælum, J., S. Borglin, R. Chakraborty, J. L. Fortney, R. Lamendella, O. U. Mason, M. Auer, M. Zemla, M. Bill, and M. E. Conrad. 2012. Deep-sea bacteria enriched by oil and dispersant from the *Deepwater Horizon* spill. *Environmental Microbiology* 14(9):2405–2416.

Bagby, S. C., C. M. Reddy, C. Aeppli, G. B. Fisher, and D. L. Valentine. 2016. Persistence and biodegradation of oil at the ocean floor following *Deepwater Horizon*. *Proceedings of the National Academy of Sciences* 114(1):E9–E18.

Bagi, A., D. M. Pampanin, A. Lanzén, T. Bilstad, and R. Kommedal. 2014. Naphthalene biodegradation in temperate and Arctic marine microcosms. *Biodegradation* 25(1):111–125.

Bak, R. 1987. Effects of chronic oil pollution on a Caribbean coral reef. *Marine Pollution Bulletin* 18(10):534–539.

Baker, J. M. 1983. Impact of oil pollution on living resources. *The Environmentalist* 3(Suppl 4):1-48.

Bakke, T., J. Klungsøyr, and S. Sanni. 2013. Environmental impacts of produced water and drilling waste discharges from the Norwegian offshore petroleum industry. *Marine Environmental Research* 92:154–169.

Balseiro, A., A. Espí, I. Márquez, V. Pérez, M. C. Ferreras, J. F. Marín, and J. M. Prieto. 2005. Pathological features in marine birds affected by the prestige's oil spill in the north of Spain. *Journal of Wildlife Diseases* 41(2):371–378. https://doi.org/10.7589/0090-3558-41.2.371.

Ballou, T. G., S. C. Hess, C. D. Getter, A. Knap, R. Dodge, and T. Sleeter. 1987. Final results of the API tropics oil spill and dispersant use experiments in Panama. *International Oil Spill Conference Proceedings* 1987(1):634B.

Bar Dolev, M., R. Bernheim, S. Guo, P. L. Davies, and I. Braslavsky. 2016. Putting life on ice: Bacteria that bind to frozen water. *Journal of the Royal Society Interface* 13(121):20160210.

Barata, C., A. Calbet, E. Saiz, L. Ortiz, and J. M. Bayona. 2005. Predicting single and mixture toxicity of petrogenic polycyclic aromatic hydrocarbons to the copepod oithona davisae. *Environmental Toxicology and Chemistry* 24(11):2992–2999.

Barbato, M., and A. Scoma. 2020. Mild hydrostatic-pressure (15 MPA) affects the assembly, but not the growth, of oil-degrading coastal microbial communities tested under limiting conditions (5°C, no added nutrients). *FEMS Microbiology Ecology* 96(11):fiaa160.

Barnea, N. 1995. *Health and safety aspects of in-situ burning of oil.* https://response.restoration.noaa.gov/sites/default/files/health-safety-ISB.pdf.

Barone, M., A. Campanile, F. Caprio, and E. Fasano. 2007. The impact of new marpol regulations on bulker design: A case study. *Proceedings of the 2nd International Conference on Marine Research and Transportation.*

Barron, M. G. 2007. Sediment-associated phototoxicity to aquatic organisms. *Human and Ecological Risk Assessment* 13(2):317–321.

Barron, M. G. 2017. Photoenhanced toxicity of petroleum to aquatic invertebrates and fish. *Archives of Environmental Contamination and Toxicology* 73(1):40–46.

Barron, M. G., and L. Ka'aihue. 2001. Potential for photoenhanced toxicity of spilled oil in Prince William Sound and Gulf of Alaska waters. *Marine Pollution Bulletin* 43(1–6):86–92.

Barron, M. G., and L. Ka'aihue. 2003. Critical evaluation of *CROSERF* test methods for oil dispersant toxicity testing under subarctic conditions. *Marine Pollution Bulletin* 46(9):1191–1199.

Barron, M. G., R. Heintz, and S. D. Rice. 2004. Relative potency of PAHs and heterocycles as aryl hydrocarbon receptor agonists in fish. *Marine Environmental Research* 58(2–5):95–100.

Barros, E. V., P. R. Filgueiras, V. Lacerda, R. P. Rodgers, and W. Romão. 2022. Characterization of naphthenic acids in crude oil samples—a literature review. *Fuel* 319:123775. https://doi.org/10.1016/j.fuel.2022.123775.

Batchelor, G. 1951. Pressure fluctuations in isotropic turbulence. *Mathematical Proceedings of the Cambridge Philosophical Society* 47(2):359–374.

Bauck, L., S. Orosz, and G. Dorrestein. 1997. Avian dermatology. *Avian medicine and surgery*, p. 549. Philadelphia, PA: W.B. Saunders.

Baustian, M. M., and N. N. Rabalais. 2009. Seasonal composition of benthic macroinfauna exposed to hypoxia in the northern Gulf of Mexico. *Estuaries and Coasts* 32(5):975–983.

Bayha, K. M., N. Ortell, C. N. Ryan, K. J. Griffitt, M. Krasnec, J. Sena, T. Ramaraj, R. Takeshita, G. D. Mayer, and F. Schilkey. 2017. Crude oil impairs immune function and increases susceptibility to pathogenic bacteria in southern flounder. *PloS ONE* 12(5):e0176559.

Beazley, M. J., R. J. Martinez, S. Rajan, J. Powell, Y. M. Piceno, L. M. Tom, G. L. Andersen, T. C. Hazen, J. D. Van Nostrand, and J. Zhou. 2012. Microbial community analysis of a coastal salt marsh affected by the *Deepwater Horizon* oil spill. *PloS ONE* 7(7):e41305.

Beckett, K., R. Aulerich, L. Duffy, J. Patterson, and S. Bursian. 2002. Effects of dietary exposure to environmentally relevant concentrations of weathered Prudhoe Bay crude oil in ranch-raised mink (Mustela vison). *Bulletin of Environmental Contamination and Toxicology* 69(4):593–600.

Bejarano, A. C. 2018. Critical review and analysis of aquatic toxicity data on oil spill dispersants. *Environmental Toxicology and Chemistry* 37(12):2989–3001.

Bejarano, A. C., and M. G. Barron. 2014. Development and practical application of petroleum and dispersant interspecies correlation models for aquatic species. *Environmental Science & Technology* 48(8):4564–4572.

Bejarano, A. C., and M. G. Barron. 2016. Aqueous and tissue residue-based interspecies correlation estimation models provide conservative hazard estimates for aromatic compounds. *Environmental Toxicology and Chemistry* 35(1):56–64.

Bejarano, A. C., and A. J. Mearns. 2015. Improving environmental assessments by integrating species sensitivity distributions into environmental modeling: Examples with two hypothetical oil spills. *Marine Pollution Bulletin* 93(1–2):172–182.

Bejarano, A. C., and J. Michel. 2016. Oil spills and their impacts on sand beach invertebrate communities: A literature review. *Environmental Pollution* 218:709–722.

Bejarano, A. C., and J. R. Wheeler. 2020. Scientific basis for expanding the use of interspecies correlation estimation models. *Integrated Environmental Assessment and Management* 16(4):528–530.

Bejarano, A. C., E. Levine, and A. J. Mearns. 2013. Effectiveness and potential ecological effects of offshore surface dispersant use during the *Deepwater Horizon* oil spill: A retrospective analysis of monitoring data. *Environmental Monitoring and Assessment* 185(12):10281–10295.

Bejarano, A. C., J. R. Clark, and G. M. Coelho. 2014. Issues and challenges with oil toxicity data and implications for their use in decision making: A quantitative review. *Environmental Toxicology and Chemistry* 33(4):732–742.

Bejarano, A. C., J. K. Farr, P. Jenne, V. Chu, and A. Hielscher. 2016. The chemical aquatic fate and effects database (CAFE), a tool that supports assessments of chemical spills in aquatic environments. *Environmental Toxicology and Chemistry* 35(6):1576–1586.

Bejarano, A. C., S. Raimondo, and M. G. Barron. 2017. Framework for optimizing selection of interspecies correlation estimation models to address species diversity and toxicity gaps in an aquatic database. *Environmental Science & Technology* 51(14):8158–8165.

Beland, M., T. W. Biggs, D. A. Roberts, S. H. Peterson, R. F. Kokaly, and S. Piazza. 2017. Oiling accelerates loss of salt marshes, southeastern Louisiana. *PloS ONE* 12(8):e0181197.

Belanger, S., M. Barron, P. Craig, S. Dyer, M. Galay-Burgos, M. Hamer, S. Marshall, L. Posthuma, S. Raimondo, and P. Whitehouse. 2017. Future needs and recommendations in the development of species sensitivity distributions: Estimating toxicity thresholds for aquatic ecological communities and assessing impacts of chemical exposures. *Integrated Environmental Assessment and Management* 13(4):664–674.

Belore, R., A. Lewis, A. Guarino, and J. Mullin. 2008. Dispersant effectiveness testing on viscous, us outer continental shelf crude oils and water-in-oil emulsions at Ohmsett. *International Oil Spill Conference Proceedings* 2008(1):823–828.

Belore, R. C., K. Trudel, J. V. Mullin, and A. Guarino. 2009. Large-scale cold water dispersant effectiveness experiments with Alaskan crude oils and COREXIT 9500 and 9527 dispersants. *Marine Pollution Bulletin* 58(1):118–128.

Belt, S. T., L. Smik, D. Köseoğlu, J. Knies, and K. Husum. 2019. A novel biomarker-based proxy for the spring phytoplankton bloom in Arctic and sub-Arctic settings—HBI T_{25}. *Earth and Planetary Science Letters* 523:115703.

Bender, M. A., T. R. Knutson, R. E. Tuleya, J. J. Sirutis, G. A. Vecchi, S. T. Garner, and I. M. Held. 2010. Modeled impact of anthropogenic warming on the frequency of intense Atlantic hurricanes. *Science* 327(5964):454–458.

Benigni, P., J. D. DeBord, C. J. Thompson, P. Gardinali, and F. Fernandez-Lima. 2016. Increasing polyaromatic hydrocarbon (PAH) molecular coverage during fossil oil analysis by combining gas chromatography and atmospheric-pressure laser ionization fourier transform ion cyclotron resonance mass spectrometry (FT-ICR MS). *Energy & Fuels* 30(1):196–203.

Benignus, V. A., P. J. Bushnell, and W. K. Boyes. 2011. Estimated rate of fatal automobile accidents attributable to acute solvent exposure at low inhaled concentrations. *Risk Analysis* 31(12):1935–1948.

BenKinney, M., J. Brown, S. Mudge, M. Russell, A. Nevin, and C. Huber. 2011. Monitoring effects of aerial dispersant application during the MC252 *Deepwater Horizon* incident. *International Oil Spill Conference Proceedings* 2011(1):abs368.

Bergeon Burns, C. M., J. A. Olin, S. Woltmann, P. C. Stouffer, and S. S. Taylor. 2014. Effects of oil on terrestrial vertebrates: Predicting impacts of the Macondo blowout. *Bioscience* 64(9):820–828.

Bernabeu, A., S. Fernández-Fernández, F. Bouchette, D. Rey, A. Arcos, J. Bayona, and J. Albaiges. 2013. Recurrent arrival of oil to Galician coast: The final step of the prestige deep oil spill. *Journal of Hazardous Materials* 250:82–90.

Bernier, C., S. Kameshwar, and J. Padgett. 2017. Performance of oil infrastructure during Hurricane Harvey. *AGU Fall Meeting Abstracts*. https://www.researchgate.net/publication/340048926_Performance_of_oil_infrastructure_during_Hurricane_Harvey.

Berren, M. R., A. Beigel, and S. Ghertner. 1980. A typology for the classification of disasters. *Community Mental Health Journal* 16(2):103–111.

Betha, R., L. M. Russell, K. J. Sanchez, J. Liu, D. J. Price, M. A. Lamjiri, C.-L. Chen, X. M. Kuang, G. O. da Rocha, and S. E. Paulson. 2017. Lower no x but higher particle and black carbon emissions from renewable diesel compared to ultra low sulfur diesel in at-sea operations of a research vessel. *Aerosol Science and Technology* 51(2):123–134.

Bianchi, T. S., C. Osburn, M. R. Shields, S. Yvon-Lewis, J. Young, L. Guo, and Z. Zhou. 2014. *Deepwater Horizon* oil in Gulf of Mexico waters after 2 years: Transformation into the dissolved organic matter pool. *Environmental Science & Technology* 48(16):9288–9297. https://doi.org/10.1021/es501547b.

Björklund, K. 2010. Substance flow analyses of phthalates and nonylphenols in stormwater. *Water Science and Technology* 62(5):1154–1160.

Black, J. C., J. N. Welday, B. Buckley, A. Ferguson, P. L. Gurian, K. D. Mena, I. Yang, E. McCandlish, and H. M. Solo-Gabriele. 2016. Risk assessment for children exposed to beach sands impacted by oil spill chemicals. *International Journal of Environmental Research and Public Health* 13(9):853.

Blanchard, D., and A. Woodcock. 1957. Bubble formation and modification in the sea and its meteorological significance. *Tellus* 9(2):145–158.

Blenkinsopp, S. A., G. Sergy, K. Doe, G. Wohlgeschaffen, K. Li, and M. Fingas. 1996. Toxicity of the weathered crude oil used at the Newfoundland Offshore Burn Experiment (NOBE) and the resultant burn residue. *Spill Science & Technology Bulletin* 3(4):277–280. https://doi.org/10.1016/S1353-2561(97)00028-5.

Blenkinsopp, S., G. Sergy, K. Doe, G. Wohlgeschaffen, K. Li, and M. Fingas. 1997. *Evaluation of the toxicity of the weathered crude oil used at the Newfoundland offshore burn experiment (NOBE) and the resultant burn residue*, pp. 677–684. Ottawa, Ontario, Canada: Emergencies Science Division, Environment Canada.

Blumer, M., M. M. Mullin, and D. W. Thomas. 1963. Pristane in zooplankton. *Science* 140(3570):974.

Blumer, M., R. Guillard, and T. Chase. 1971. Hydrocarbons of marine phytoplankton. *Marine Biology* 8(3):183–189.

Blumer, M., M. Ehrhardt, and J. Jones. 1973. The environmental fate of stranded crude oil. *Deep Sea Research and Oceanographic Abstracts* 20(3):239–259. https://doi.org/10.1016/0011-7471(73)90014-4.

Boccadoro, C., A. Krolicka, J. Receveur, C. Aeppli, and S. Le Floch. 2018. Microbial community response and migration of petroleum compounds during a sea-ice oil spill experiment in Svalbard. *Marine Environmental Research* 142:214–233.

Bociu, I., B. Shin, W. B. Wells, J. E. Kostka, K. T. Konstantinidis, and M. Huettel. 2019. Decomposition of sediment-oil-agglomerates in a Gulf of Mexico sandy beach. *Scientific Reports* 9(1):1–13.

Bock, M., H. Robinson, R. Wenning, D. French-McCay, J. Rowe, and A. H. Walker. 2018. Comparative risk assessment of oil spill response options for a deepwater oil well blowout: Part II. Relative risk methodology. *Marine Pollution Bulletin* 133:984–1000.

Bodkin, J. L., and M. S. Udevitz. 1994. An intersection model for estimating sea otter. In *Marine mammals and the Exxon Valdez*, T. R. Loughlin (ed.), p. 81. Cambridge, MA: Academic Press.

Bodkin, J. L., B. E. Ballachey, H. A. Coletti, G. G. Esslinger, K. A. Kloecker, S. D. Rice, J. A. Reed, and D. H. Monson. 2012. Long-term effects of the "Exxon Valdez" oil spill: Sea otter foraging in the intertidal as a pathway of exposure to lingering oil. *Marine Ecology Progress Series* 447:273–287. https://www.int-res.com/abstracts/meps/v447/p273–287.

Boehm, P. D., and D. L. Fiest. 1982. Subsurface distributions of petroleum from an offshore well blowout. The Ixtoc I blowout, Bay of Campeche. *Environmental Science & Technology* 16(2):67–74. https://doi.org/10.1021/es00096a003.

Boehm, P. D., W. Steinhauer, A. Requejo, D. Cobb, S. Duffy, and J. Brown. 1985. Comparative fate of chemically dispersed and untreated oil in the Arctic: Baffin Island oil spill studies 1980–1983. *International Oil Spill Conference Proceedings* 1985(1):561–569.

Boehm, P., M. Steinhauer, D. Green, B. Fowler, B. Humphrey, D. Fiest, and W. Cretney. 1987. Comparative fate of chemically dispersed and beached crude oil in subtidal sediments of the Arctic nearshore. *Arctic* 40(Suppl 1):133–148.

Boehm, P. D., D. S. Page, J. S. Brown, J. M. Neff, and E. Gundlach. 2014. Long-term fate and persistence of oil from the *Exxon Valdez* oil spill: Lessons learned or history repeated? *International Oil Spill Conference Proceedings* 2014(1):63–79.

Boese, B. L., J. O. Lamberson, R. C. Swartz, and R. J. Ozretich. 1997. Photoinduced toxicity of fluoranthene to seven marine benthic crustaceans. *Archives of Environmental Contamination and Toxicology* 32(4):389–393.

Boglaienko, D., and B. Tansel. 2018. Classification of oil-particle interactions in aqueous environments: Aggregate types depending on state of oil and particle characteristics. *Marine Pollution Bulletin* 133:693–700.

Boitsov, S., S. A. Mjøs, and S. Meier. 2007. Identification of estrogen-like alkylphenols in produced water from offshore oil installations. *Marine Environmental Research* 64(5):651–665.

Bolden, A. L., and C. F. Kwiatkowski. 2016. Response to comment on "New look at BTEX: Are ambient levels a problem?" *Environmental Science & Technology* 50(2):1072–1073.

Bolden, A. L., C. F. Kwiatkowski, and T. Colborn. 2015. New look at BTEX: Are ambient levels a problem? *Environmental Science & Technology* 49(9):5261–5276.

Bolina, I. C. A., R. A. B. Gomes, and A. A. Mendes. 2021. Biolubricant production from several oleaginous feedstocks using lipases as catalysts: Current scenario and future perspectives. *BioEnergy Research* 14(4):1039–1057. https://doi.org/10.1007/s12155-020-10242-4.

Bomboi, M., and A. Hernandez. 1991. Hydrocarbons in urban runoff: Their contribution to the wastewaters. *Water Research* 25(5):557–565.

Bonisoli-Alquati, A., W. Xu, P. Stouffer, and S. Taylor. 2020. Transcriptome analysis indicates a broad range of toxic effects of *Deepwater Horizon* oil on seaside sparrows. *Science of the Total Environment* 720:137583.

Bonn Agreement. 2012. *Bonn agreement aerial operations handbook*. www.bonnagreement.org.

Boothroyd, I., S. Almond, S. Qassim, F. Worrall, and R. Davies. 2016. Fugitive emissions of methane from abandoned, decommissioned oil and gas wells. *Science of the Total Environment* 547:461–469.

Bostic, J. T., C. Aeppli, R. F. Swarthout, C. M. Reddy, and L. A. Ziolkowski. 2018. Ongoing biodegradation of *Deepwater Horizon* oil in beach sands: Insights from tracing petroleum carbon into microbial biomass. *Marine Pollution Bulletin* 126:130–136.

Boucher, G. 1985. Long-term monitoring of meiofauna densities after the *Amoco Cadiz* oil spill. *Marine Pollution Bulletin* 16(8):328–333.

Boufadel, M. C., and X. Geng. 2014. A new paradigm in oil spill modeling for decision making? *Environmental Research Letters* 9(8):081001.

Boufadel, M. C., F. Gao, L. Zhao, T. Özgökmen, R. Miller, T. King, B. Robinson, K. Lee, and I. Leifer. 2018. Was the *Deepwater Horizon* well discharge churn flow? Implications on the estimation of the oil discharge and droplet size distribution. *Geophysical Research Letters* 45(5):2396–2403.

Boufadel, M. C., S. Socolofsky, J. Katz, D. Yang, C. Daskiran, and W. Dewar. 2020. A review on multiphase underwater jets and plumes: Droplets, hydrodynamics, and chemistry. *Reviews of Geophysics* 58(3):e2020RG000703.

Bowman, J. S. 2015. The relationship between sea ice bacterial community structure and biogeochemistry: A synthesis of current knowledge and known unknowns. *Elementa: Science of the Anthropocene* 3:000072. https://doi.org/10.12952/journal.elementa.000072.

Bowman, J. S., and J. W. Deming. 2014. Alkane hydroxylase genes in psychrophile genomes and the potential for cold active catalysis. *BMC Genomics* 15(1):1–11.

Boxall, J. A., C. A. Koh, E. D. Sloan, A. K. Sum, and D. T. Wu. 2012. Droplet size scaling of water-in-oil emulsions under turbulent flow. *Langmuir* 28(1):104–110.

BP. 2020. *Statistical review of world energy*. 69th edition. https://www.bp.com/content/dam/bp/business-sites/en/global/corporate/pdfs/energy-economics/statistical-review/bp-stats-review-2020-full-report.pdf.

BP. 2021. *Statistical review of world energy*. 70th edition. https://www.bp.com/content/dam/bp/business-sites/en/global/corporate/pdfs/energy-economics/statistical-review/bp-stats-review-2021-full-report.pdf.

Bracco, A., C. B. Paris, A. J. Esbaugh, K. Frasier, S. Joye, G. Liu, K. Polzin, and A. C. Vaz. 2020. Transport, fate and impacts of the deep plume of petroleum hydrocarbons formed during the Macondo blowout. *Frontiers in Marine Science* 7:764.

Brakstad, O. G., I. Nonstad, L.-G. Faksness, and P. J. Brandvik. 2008. Responses of microbial communities in Arctic sea ice after contamination by crude petroleum oil. *Microbial Ecology* 55(3):540–552.

Brakstad, O. G., T. Nordtug, and M. Throne-Holst. 2015a. Biodegradation of dispersed Macondo oil in seawater at low temperature and different oil droplet sizes. *Marine Pollution Bulletin* 93(1–2):144–152.

Brakstad, O. G., M. Throne-Holst, R. Netzer, D. M. Stoeckel, and R. M. Atlas. 2015b. Microbial communities related to biodegradation of dispersed Macondo oil at low seawater temperature with Norwegian coastal seawater. *Microbial Biotechnology* 8(6):989–998.

Brakstad, O. G., E. J. Davies, D. Ribicic, A. Winkler, U. Brönner, and R. Netzer. 2018a. Biodegradation of dispersed oil in natural seawaters from western Greenland and a Norwegian fjord. *Polar Biology* 41(12):2435–2450.

Brakstad, O. G., U. Farooq, D. Ribicic, and R. Netzer. 2018b. Dispersibility and biotransformation of oils with different properties in seawater. *Chemosphere* 191:44–53.

Brakstad, O. G., A. Lewis, and C. Beegle-Krause. 2018c. A critical review of marine snow in the context of oil spills and oil spill dispersant treatment with focus on the *Deepwater Horizon* oil spill. *Marine Pollution Bulletin* 135:346–356.

Brandvik, J., K. R. Sørheim, I. Singsaas, M. Reed, J. Industr, G. H. Lille, E. Garpestad, K. Bakke, U.-E. Moltu, and A. Myrvold. 2006. *Short state-of-the-art report on oil spills in ice-infested waters*. Trondheim, Norway: Marine Environmental Technology, SINTEF Materials and Chemistry.

Brandvik, P. J., and L.-G. Faksness. 2009. Weathering processes in Arctic oil spills: Meso-scale experiments with different ice conditions. *Cold Regions Science and Technology* 55(1):160–166.

Brandvik, P., O. Knudsen, M. Moldestad, and P. Daling. 1995. Laboratory testing of dispersants under Arctic conditions. In *The use of chemicals in oil spill response*. West Conshohocken, PA: ASTM International.

Brandvik, P., L.-G. Faksness, D. Dickins, and J. Bradford. 2008. Weathering of oil spills under Arctic conditions: Field experiments with different ice conditions followed by in-situ burning. In *Oil spill response: A global perspective,* pp. 63–64. Dordrecht, Netherlands: Springer.

Brandvik, P., J. M. Resby, P. Daling, F. Leirvik, and J. Fritt-Rasmussen. 2010. Meso-scale weathering of oil as a function of ice conditions. In *Oil properties, dispersibility and in situ burnability of weathered oil as a function of time. SINTEF A* 15563. Trondheim, Norway: SINTEF Materials and Chemistry.

Brandvik, P. J., Ø. Johansen, F. Leirvik, U. Farooq, and P. S. Daling. 2013. Droplet breakup in subsurface oil releases—part 1: Experimental study of droplet breakup and effectiveness of dispersant injection. *Marine Pollution Bulletin* 73(1):319–326.

Brandvik, P. J., Ø. Johansen, U. Farooq, E. Davies, D. Krause, and L. F. 2015. *Sub-surface oil releases—experimental study of droplet size distributions phase-II: A scaled experimental approach using the SINTEF Tower Basin.* Trondheim, Norway. http://www.oilspillprevention.org/-/media/Oil-Spill-Prevention/spillprevention/r-and-d/dispersants/sintef-api-d3-phase-ii-final-report-a268.pdf.

Brandvik, P. J., Ø. Johansen, F. Leirvik, D. F. Krause, and P. S. Daling. 2018. Subsea dispersants injection (SSDI), effectiveness of different dispersant injection techniques: An experimental approach. *Marine Pollution Bulletin* 136:385–393.

Brandvik, P. J., C. Storey, E. J. Davies, and Ø. Johansen. 2019. Combined releases of oil and gas under pressure: The influence of live oil and natural gas on initial oil droplet formation. *Marine Pollution Bulletin* 140:485–492.

Brandvik, P. J., E. Davies, F. Leirvik, Ø. Johansen, and R. Belore. 2021. Large-scale basin testing to simulate realistic oil droplet distributions from subsea release of oil and the effect of subsea dispersant injection. *Marine Pollution Bulletin* 163:111934.

Brannon, E. L., K. C. Collins, L. L. Moulton, and K. R. Parker. 2001. Resolving allegations of oil damage to incubating pink salmon eggs in Prince William Sound. *Canadian Journal of Fisheries and Aquatic Sciences* 58(6):1070–1076.

Bratman, G. N., C. B. Anderson, M. G. Berman, B. Cochran, S. De Vries, J. Flanders, C. Folke, H. Frumkin, J. J. Gross, and T. Hartig. 2019. Nature and mental health: An ecosystem service perspective. *Science Advances* 5(7):eaax0903.

Brekke, C., and A. H. Solberg. 2005. Oil spill detection by satellite remote sensing. *Remote Sensing of Environment* 95(1):1–13.

Brette, F., B. Machado, C. Cros, J. P. Incardona, N. L. Scholz, and B. A. Block. 2014. Crude oil impairs cardiac excitation-contraction coupling in fish. *Science* 343(6172):772–776.

Brette, F., H. A. Shiels, G. L. Galli, C. Cros, J. P. Incardona, N. L. Scholz, and B. A. Block. 2017. A novel cardiotoxic mechanism for a pervasive global pollutant. *Scientific Reports* 7(1):1–9.

Briggs, K. T., S. H. Yoshida, and M. E. Gershwin. 1996. The influence of petrochemicals and stress on the immune system of seabirds. *Regulatory Toxicology and Pharmacology* 23(2):145–155.

Briggs, K. T., M. E. Gershwin, and D. W. Anderson. 1997. Consequences of petrochemical ingestion and stress on the immune system of seabirds. *ICES Journal of Marine Science* 54(4):718–725.

Brinkmann, W. 1985. Urban storm water pollutants: Sources and loadings. *GeoJournal* 11(3):277–283.

Brooks, G. R., R. A. Larson, P. T. Schwing, I. Romero, C. Moore, G.-J. Reichart, T. Jilbert, J. P. Chanton, D. W. Hastings, and W. A. Overholt. 2015. Sedimentation pulse in the NE Gulf of Mexico following the 2010 DWH blowout. *PloS ONE* 10(7):e0132341.

Brown, D. M., L. Camenzuli, A. D. Redman, C. Hughes, N. Wang, E. Vaiopoulou, D. Saunders, A. Villalobos, and S. Linington. 2020. Is the Arrhenius-correction of biodegradation rates, as recommended through reach guidance, fit for environmentally relevant conditions? An example from petroleum biodegradation in environmental systems. *Science of the Total Environment* 732:139293.

Brown, H. M., and R. H. Goodman. 1996. The use of dispersants in broken ice. *Proceedings of the 19th Arctic and Marine Oilspill Program Technical Seminar*, pp. 453–460. Ottawa, Ontario, Canada: Emergencies Science Division, Environment Canada.

Brown, R. E., J. D. Brain, and N. Wang. 1997. The avian respiratory system: A unique model for studies of respiratory toxicosis and for monitoring air quality. *Environmental Health Perspectives* 105(2):188–200.

Brown-Peterson, N. J., M. O. Krasnec, C. R. Lay, J. M. Morris, and R. J. Griffitt. 2017. Responses of juvenile southern flounder exposed to *Deepwater Horizon* oil-contaminated sediments. *Environmental Toxicology and Chemistry* 36(4):1067–1076.

Browne, M. A., T. S. Galloway, and R. C. Thompson. 2010. Spatial patterns of plastic debris along estuarine shorelines. *Environmental Science & Technology* 44(9):3404–3409.

Browse, J., K. Carslaw, A. Schmidt, and J. Corbett. 2013. Impact of future Arctic shipping on high-latitude black carbon deposition. *Geophysical Research Letters* 40(16):4459–4463.

Bruner, R. 1984. *Pathologic findings in laboratory animals exposed to hydrocarbon fuels of military interest.* Bethesda, MD: Naval Medical Research Center.

Bryant, W. L., R. Camilli, G. B. Fisher, E. B. Overton, C. M. Reddy, D. Reible, R. F. Swarthout, and D. L. Valentine. 2020. Harnessing a decade of data to inform future decisions: Insights into the ongoing hydrocarbon release at Taylor Energy's Mississippi Canyon block 20 (MC20) site. *Marine Pollution Bulletin* 155:111056.

BTS (Bureau of Transportation Statistcs). 2019. *Table 4-41: Federal exhaust emissions standards for marine spark-ignition engines and vehicles.* Washington, DC: Bureau of Transportation Statistics.

Buchholz, K., A. Krieger, J. Rowe, D. Etkin, D. F. McCay, M. S. Gearon, M. Grennan, and J. Turner. 2016. *Worst case discharge analysis (volume I): Oil Spill Response Plan (OSRP) equipment capabilities review.* BPA No. E14PB00072.

Buckner, A. V., B. D. Goldstein, and L. M. Beitsch. 2017. Building resilience among disadvantaged communities: Gulf region health outreach program overview. *Journal of Public Health Management and Practice* 23:S1–S4.

Bue, B., S. Sharr, S. Moffitt, and A. K. Craig. 1996. Effects of the *Exxon Valdez* oil spill on pink salmon embryos and preemergent fry. *American Fisheries Society Symposium.*

Buffaloe, G., D. Lack, E. Williams, D. Coffman, K. Hayden, B. Lerner, S.-M. Li, I. Nuaaman, P. Massoli, and T. Onasch. 2014. Black carbon emissions from in-use ships: A California regional assessment. *Atmospheric Chemistry and Physics* 14(4):1881–1896.

Bugden, J. B., C. W. Yeung, P. E. Kepkay, and K. Lee. 2008. Application of ultraviolet fluorometry and excitation-emission matrix spectroscopy (EEMS) to fingerprint oil and chemically dispersed oil in seawater. *Marine Pollution Bulletin* 56(4):677–685. https://doi.org/10.1016/j.marpolbul.2007.12.022.

Buist, I., and D. Dickins. 1987. Experimental spills of crude oil in pack ice. *International Oil Spill Conference Proceedings* 1987(1):373–381.

Buist, I., R. Belore, A. Guarino, D. Hackenberg, D. Dickins, and Z. Wang. 2008. *Empirical weathering properties of oil in ice and snow.* Project Number 1435-01-04-RP-34501 OCS Study MMS 2008-033.

Buist, I., S. Potter, and S. E. Sørstrøm. 2010. Barents Sea field test of herder to thicken oil for in situ burning in drift ice. *Proceedings of the 33rd Arctic and Marine Oilspill Program Technical Seminar*, pp. 725–742. Ottawa, Ontario, Canada: Emergencies Science Division, Environment Canada.

Buist, I., S. Potter, B. Trudel, S. Shelnutt, A. Walker, D. Scholz, P. Brandvik, J. Fritt-Rasmussen, A. Allen, and P. Smith. 2013a. In situ burning in ice-affected waters: State of knowledge report, final report 7.1.1. *Arctic Response Technology* 293.

Buist, I., S. Potter, S. Trudel, A. Walker, D. Scholz, P. Brandvik, J. Fritt Rasmussen, A. Allen, and P. Smith. 2013b. In-situ burning in ice-affected waters: A technology summary and lessons from key experiments. In *Final report 7.1.2, Arctic response technology-oil spill preparedness.*

Buist, I., S. Potter, and P. Lane. 2016. Historical review and state of the art for oil slick ignition for ISB. *Final Report for the Arctic Oil Spill Response Technology-Joint Industry Programme* 36.

Buist, I., D. Cooper, K. Trudel, J. Fritt–Rasmussen, S. Wegeberg, K. Gustavson, P. Lassen, W. R. Alva, G. Jomaas, and L. Zabilansky. 2017. Research investigations into herder fate, effects and windows-of-opportunity. *Arctic Response Technology*. https://doi.org/10.13140/RG.2.2.36120.70403.

Bullock, R. J., R. A. Perkins, and S. Aggarwal. 2019. In-situ burning with chemical herders for Arctic oil spill response: Meta-analysis and review. *Science of the Total Environment* 675:705–716.

Bunch, J. 1987. Effects of petroleum releases on bacterial numbers and microheterotrophic activity in the water and sediment of an Arctic marine ecosystem. *Arctic* 40(Suppl 1):172–183.

Burd, A. B., J. P. Chanton, K. L. Daly, S. Gilbert, U. Passow, and A. Quigg. 2020. The science behind marine-oil snow and MOSSFA: Past, present, and future. *Progress in Oceanography* 102398. https://www.ncbi.nlm.nih.gov/pubmed/21366214.

Burgin, S., and N. Hardiman. 2011. The direct physical, chemical and biotic impacts on Australian coastal waters due to recreational boating. *Biodiversity and Conservation* 20(4):683–701. https://doi.org/10.1007/s10531-011-0003-6.

Burns, J. G., and R. C. Ydenberg. 2002. The effects of wing loading and gender on the escape flights of least sandpipers (Calidris minutilla) and western sandpipers (Calidris mauri). *Behavioral Ecology and Sociobiology* 52(2):128–136. https://doi.org/10.1007/s00265-002-0494-y.

Burns, K. A., S. D. Garrity, and S. C. Levings. 1993. How many years until mangrove ecosystems recover from catastrophic oil spills? *Marine Pollution Bulletin* 26(5):239–248.

Burridge, L., M. Boudreau, M. Lyons, S. Courtenay, and K. Lee. 2011. Effects of Hibernia production water on the survival, growth and biochemistry of juvenile Atlantic cod (Gadus morhua) and northern mummichog (Fundulus heteroclitus macrolepidotus). In *Produced water*, pp. 329–344. Dordrecht, Netherlands: Springer.

Bursian, S. J., K. M. Dean, K. E. Harr, L. Kennedy, J. E. Link, I. Maggini, C. Pritsos, K. L. Pritsos, R. E. Schmidt, and C. G. Guglielmo. 2017. Effect of oral exposure to artificially weathered *Deepwater Horizon* crude oil on blood chemistries, hepatic antioxidant enzyme activities, organ weights and histopathology in western sandpipers (Calidris mauri). *Ecotoxicology and Environmental Safety* 146:91–97. https://doi.org/10.1016/j.ecoenv.2017.03.045.

Burwood, R., and G. Speers. 1974. Photo-oxidation as a factor in the environmental dispersal of crude oil. *Estuarine and Coastal Marine Science* 2(2):117–135.

Butler, R. G., and P. Lukasiewicz. 1979. A field study of the effect of crude oil on Herring Gull (larus argentatus) chick growth. *The Auk* 96(4):809–812.

Byrne, R., P. D. Panetta, H. Du, and A. Podolski. 2018. *Determining the operability limits of chemical herders*. Washington, DC: Bureau of Safety and Environmental Enforcement, U.S. Department of the Interior.

Byrnes, M. R., R. A. Davis, Jr., M. C. Kennicutt II, R. T. Kneib, I. A. Mendelssohn, G. T. Rowe, J. W. Tunnell, Jr., B. A. Vittor, and C. H. Ward. 2017. *Habitats and biota of the Gulf of Mexico: Before the* Deepwater Horizon *oil spill. Volume 1: Water quality, sediments, sediment contaminants, oil and gas seeps, coastal habitats, offshore plankton and benthos, and shellfish.*

Bytingsvik, J., T. F. Parkerton, J. Guyomarch, L. Tassara, S. LeFloch, W. R. Arnold, S. M. Brander, A. Volety, and L. Camus. 2020. The sensitivity of the deepsea species Northern Shrimp (pandalus borealis) and the cold-water coral (lophelia pertusa) to oil-associated aromatic compounds, dispersant, and Alaskan North Slope crude oil. *Marine Pollution Bulletin* 156:111202.

C-NLOPB (Canada-Newfoundland and Labrador Offshore Petroleum Board). 2018. *Special oversight measures for Deepwater and critical wells in harsh environments*. https://www.cnlopb.ca/wp-content/uploads/sr/specovermeas.pdf.

Cabrerizo, A., C. Galbán-Malagón, S. Del Vento, and J. Dachs. 2014. Sources and fate of polycyclic aromatic hydrocarbons in the Antarctic and southern ocean atmosphere. *Global Biogeochemical Cycles* 28(12):1424–1436. https://doi.org/10.1002/2014GB004910.

Caia, A., A. G. Di Lullo, G. De Ghetto, and A. Guadagnini. 2018. Probabilistic analysis of risk and mitigation of deepwater well blowouts and oil spills. *Stochastic Environmental Research and Risk Assessment* 32(9):2647–2666.

Calabrese, R. V., T. Chang, and P. Dang. 1986. Drop breakup in turbulent stirred-tank contactors. Part I: Effect of dispersed-phase viscosity. *AIChE Journal* 32(4):657–666.

Callaghan, A. V. 2013. Enzymes involved in the anaerobic oxidation of n-alkanes: From methane to long-chain paraffins. *Frontiers in Microbiology* 4:89.

Camacho, M., P. Calabuig, O. P. Luzardo, L. D. Boada, M. Zumbado, and J. Orós. 2013. Crude oil as a stranding cause among loggerhead sea turtles (caretta caretta) in the Canary Islands, Spain (1998–2011). *Journal of Wildlife Diseases* 49(3):637–640.

Camarillo, M. K., and W. T. Stringfellow. 2018. Biological treatment of oil and gas produced water: A review and meta-analysis. *Clean Technologies and Environmental Policy* 20(6):1127–1146.

Camilli, R., C. M. Reddy, D. R. Yoerger, B. A. Van Mooy, M. V. Jakuba, J. C. Kinsey, C. P. McIntyre, S. P. Sylva, and J. V. Maloney. 2010. Tracking hydrocarbon plume transport and biodegradation at *Deepwater Horizon*. *Science* 330(6001):201–204.

Campbell, D., D. Cox, J. Crum, K. Foster, P. Christie, and D. Brewster. 1993. Initial effects of the grounding of the tanker braer on health in Shetland. The Shetland health study group. *BMJ* 307(6914):1251–1255.

Campbell, D., D. Cox, J. Crum, K. Foster, and A. Riley. 1994. Later effects of grounding of tanker braer on health in Shetland. *BMJ* 309(6957):773–774.

Campo, P., A. D. Venosa, and M. T. Suidan. 2013. Biodegradability of COREXIT 9500 and dispersed south Louisiana crude oil at 5 and 25 C. *Environmental Science & Technology* 47(4):1960–1967.

Cao, X., and M. Tarr, 2017. Aldehyde and ketone photoproducts from solar-irradiated crude oil–seawater systems determined by electrospray ionization–tandem mass spectrometry. *Environmental Science & Technology* 51(20):11858–11866. https://pubs.acs.org/doi/10.1021/acs.est.7b01991.

Cappa, C., E. Williams, D. Lack, G. Buffaloe, D. Coffman, K. Hayden, S. Herndon, B. Lerner, S.-M. Li, and P. Massoli. 2014. A case study into the measurement of ship emissions from plume intercepts of the NOAA ship Miller Freeman. *Atmospheric Chemistry and Physics* 14(3):1337–1352.

Caramello, V., L. Bertuzzi, F. Ricceri, U. Albert, G. Maina, A. Boccuzzi, F. Della Corte, and M. C. Schreiber. 2019. The mass casualty incident in Turin, 2017: A case study of disaster responders' mental health in an Italian level I hospital. *Disaster Medicine and Public Health Preparedness* 13(5–6):880–888.

Carassou, L., F. Hernandez, and W. Graham. 2014. Change and recovery of coastal mesozooplankton community structure during the *Deepwater Horizon* oil spill. *Environmental Research Letters* 9(12):124003.

Carey, J. H., E. D. Ongley, and E. Nagy. 1990. Hydrocarbon transport in the Mackenzie River, Canada. *Science of the Total Environment* 97:69–88.

Carls, M., L. Holland, M. Larsen, J. Lum, D. Mortensen, S. Wang, and A. Wertheimer. 1996. Growth, feeding, and survival of pink salmon fry exposed to food contaminated with crude oil. *American Fisheries Society Symposium*.

Carls, M. G., L. Holland, M. Larsen, T. K. Collier, N. L. Scholz, and J. P. Incardona. 2008. Fish embryos are damaged by dissolved PAHs, not oil particles. *Aquatic Toxicology* 88(2):121–127.

Carmen Guzmán Martínez, M. D., P. R. Romero, and A. T. Banaszak. 2007. Photoinduced toxicity of the polycyclic aromatic hydrocarbon, fluoranthene, on the coral, Porites divaricata. *Journal of Environmental Science and Health, Part A* 42(10):1495–1502.

Carrasco, J. M., B. Pérez-Gómez, M. J. García-Mendizábal, V. Lope, N. Aragonés, M. J. Forjaz, P. Guallar-Castillón, G. López-Abente, F. Rodríguez-Artalejo, and M. Pollán. 2007. Health-related quality of life and mental health in the medium-term aftermath of the prestige oil spill in Galiza (Spain): A cross-sectional study. *BMC Public Health* 7(1):1–12.

Carroll, J., F. Vikebø, D. Howell, O. J. Broch, R. Nepstad, S. Augustine, G. M. Skeie, R. Bast, and J. Juselius. 2018. Assessing impacts of simulated oil spills on the northeast Arctic cod fishery. *Marine Pollution Bulletin* 126:63–73.

Casal, P., A. Cabrerizo, M. Vila-Costa, M. Pizarro, B. Jiménez, and J. Dachs. 2018. Pivotal role of snow deposition and melting driving fluxes of polycyclic aromatic hydrocarbons at coastal Livingston Island (Antarctica). *Environmental Science & Technology* 52(21):12327–12337. https://doi.org/10.1021/acs.est.8b03640.

Castro-Jiménez, J., N. Berrojalbiz, J. Wollgast, and J. Dachs. 2012. Polycyclic aromatic hydrocarbons (PAHs) in the Mediterranean Sea: Atmospheric occurrence, deposition and decoupling with settling fluxes in the water column. *Environmental Pollution* 166:40–47.

Cedre, F. 2005. *Using dispersants to treat oil slicks at sea: Airborne and shipborne treatment response manual*. Brest, France: Centre for Documentation, Research and Experimentation on Accidental Water Pollution.

Celo, V., E. Dabek-Zlotorzynska, and M. McCurdy. 2015. Chemical characterization of exhaust emissions from selected Canadian marine vessels: The case of trace metals and lanthanoids. *Environmental Science & Technology* 49(8):5220–5226.

Chacón-Patiño, M. L., J. Nelson, E. Rogel, K. Hench, L. Poirier, F. Lopez-Linares, and C. Ovalles. 2021. Vanadium and nickel distributions in Pentane: In-between C5-C7 Asphaltenes, and heptane asphaltenes of heavy crude oils. *Fuel* 292:120259. https://doi.org/10.1016/j.fuel.2021.120259.

Chamberlain, S. D., P. G. Wells, and B. H. MacDonald. 2018. The Gulfwatch contaminants monitoring program in the Gulf of Maine: Are its data being used for ocean protection, with special reference to Nova Scotia, Canada? *Marine Pollution Bulletin* 127:781–787. https://doi.org/10.1016/j.marpolbul.2017.09.050.

Chamecki, M., T. Chor, D. Yang, and C. Meneveau. 2019. Material transport in the ocean mixed layer: Recent developments enabled by large eddy simulations. *Reviews of Geophysics* 57(4):1338–1371.

Chanton, J., J. Cherrier, R. Wilson, J. Sarkodee-Adoo, S. Bosman, A. Mickle, and W. Graham. 2012. Radiocarbon evidence that carbon from the *Deepwater Horizon* spill entered the planktonic food web of the Gulf of Mexico. *Environmental Research Letters* 7(4):045303.

Chen, A., X. Wu, S. L. M. Simonich, H. Kang, and Z. Xie. 2021. Volatilization of polycyclic aromatic hydrocarbons (PAHs) over the North Pacific and adjacent Arctic Ocean: The impact of offshore oil drilling. *Environmental Pollution* 268:115963.

Chen, B., D. Yang, C. Meneveau, and M. Chamecki. 2018. Numerical study of the effects of chemical dispersant on oil transport from an idealized underwater blowout. *Physical Review Fluids* 3(8):083801.

Chen, H., A. Hou, Y. E. Corilo, Q. Lin, J. Lu, I. A. Mendelssohn, R. Zhang, R. P. Rodgers, and A. M. McKenna. 2016. 4 years after the *Deepwater Horizon* spill: Molecular transformation of Macondo well oil in Louisiana salt marsh sediments revealed by FT-ICR mass spectrometry. *Environmental Science & Technology* 50(17):9061–9069.

Chen, J., W. Zhang, Z. Wan, S. Li, T. Huang, and Y. Fei. 2019. Oil spills from global tankers: Status review and future governance. *Journal of Cleaner Production* 227:20–32. https://doi.org/10.1016/j.jclepro.2019.04.020.

Chen, J. L., K. C. Au, Y. S. Wong, and N. F. Y. Tam. 2010. Using orthogonal design to determine optimal conditions for biodegradation of phenanthrene in mangrove sediment slurry. *Journal of Hazardous Materials* 176(1–3):666–671.

Chen, S.-C., N. Musat, O. J. Lechtenfeld, H. Paschke, M. Schmidt, N. Said, D. Popp, F. Calabrese, H. Stryhanyuk, and U. Jaekel. 2019. Anaerobic oxidation of ethane by archaea from a marine hydrocarbon seep. *Nature* 568(7750):108–111.

Chernikova, T. N., R. Bargiela, S. V. Toshchakov, V. Shivaraman, E. A. Lunev, M. M. Yakimov, D. N. Thomas, and P. N. Golyshin. 2020. Hydrocarbon-degrading bacteria alcanivorax and marinobacter associated with microalgae pavlova lutheri and nannochloropsis oculata. *Frontiers in Microbiology* 11:2650.

Cho, H. J., P. Biber, K. M. Darnell, and K. Dunton. 2017. Seasonal and annual dynamics in seagrass beds of the grand bay national estuarine research reserve, Mississippi. *Southeastern Geographer* 57(3):246–272.

Choi, J., and J. T. Oris. 2003. Assessment of the toxicity of anthracene photo-modification products using the Topminnow (poeciliopsis lucida) hepatoma cell line (PLHC-1). *Aquatic Toxicology* 65(3):243–251.

Choi, K.-H., M.-S. Park, M. Ha, J.-I. Hur, and H.-K. Cheong. 2018. Cancer incidence trend in the Hebei Spirit oil spill area, from 1999 to 2014: An ecological study. *International Journal of Environmental Research and Public Health* 15(5):1006.

Choi, K.-H., M.-S. Park, M. H. Lim, J.-I. Hur, S. R. Noh, W.-C. Jeong, H.-K. Cheong, and M. Ha. 2021. Who has sustained psychological symptoms nine years after the Hebei Spirit oil spill? The health effect research on Hebei Spirit oil spill (HEROS) study. *Journal of Environmental Management* 294:112936.

Chor, T., D. Yang, C. Meneveau, and M. Chamecki. 2018. A turbulence velocity scale for predicting the fate of buoyant materials in the oceanic mixed layer. *Geophysical Research Letters* 45(21):11817–11826.

Chrisseufert. 2018. *Nauset Beach.* https://www.atlasobscura.com/places/nauset-beach.

Chua, E. J., W. Savidge, R. T. Short, A. M. Cardenas-Valencia, and R. W. Fulweiler. 2016. A review of the emerging field of underwater mass spectrometry. *Frontiers in Marine Science* 3:209.

Claireaux, G., P. Quéau, S. Marras, S. Le Floch, A. P. Farrell, A. Nicolas-Kopec, P. Lemaire, and P. Domenici. 2018. Avoidance threshold to oil water-soluble fraction by a juvenile marine teleost fish. *Environmental Toxicology and Chemistry* 37(3):854–859.

Clark, C. E., and J. A. Veil. 2009. *Produced water volumes and management practices in the United States.* Argonne, IL: Argonne National Laboratory.

Clark, R. B., and R. Kennedy. 1968. *Rehabilitation of oiled seabirds.* Newcastle upon Tyne, UK: Department of Zoology, University of Newcastle upon Tyne.

Clark, Jr., R. C., and M. Blumer. 1967. Distribution of n-paraffins in marine organisms and sediment 1. *Limnology and Oceanography* 12(1):79–87.

Clay, L. A., and D. M. Abramson. 2021. Bowling together: Community social institutions protective against poor child mental health. *Environmental Justice* 14(3):206–215.

Clayton, J. R., J. R. Payne, J. S. Farlow, and C. Sarwar. 1993. *Oil spill dispersants: Mechanisms of action and laboratory tests.* Boca Raton, FL: CRC Press.

Clemente, J. S., and P. M. Fedorak. 2005. A review of the occurrence, analyses, toxicity, and biodegradation of naphthenic acids. *Chemosphere* 60(5):585–600.

Clift, R., J. R. Grace, and M. E. Weber. 1978. *Bubbles, drops, and particles.* Mineola, NY: Dover Publications, Inc.

Coelho, G., D. Aurand, L. Essex, A. Parkin, and L. Robinson. 2011. *Monitoring subsurface dispersant injection during the MC252 incident*, Volume 1. EM&A Report Number 11-05. Lusby, MD. 35 pp.

Cole, J., D. M. Beare, A. P. Waugh, E. Capulas, K. E. Aldridge, C. F. Arlett, M. H. Green, J. E. Crum, D. Cox, and R. C. Garner. 1997. Biomonitoring of possible human exposure to environmental genotoxic chemicals: Lessons from a study following the wreck of the oil tanker braer. *Environmental and Molecular Mutagenesis* 30(2):97–111.

Collin, T. D., R. Cunningham, M. Q. Asghar, R. Villa, J. MacAdam, and B. Jefferson. 2020. Assessing the potential of enhanced primary clarification to manage fats, oils and grease (FOG) at wastewater treatment works. *Science of the Total Environment* 728:138415.

Collins, A. W., V. Elango, D. Curtis, M. Rodrigue, and J. H. Pardue. 2020. Biogeochemical controls on biodegradation of buried oil along a coastal headland beach. *Marine Pollution Bulletin* 154:111051.

Colvin, K. A., C. Lewis, and T. S. Galloway. 2020. Current issues confounding the rapid toxicological assessment of oil spills. *Chemosphere* 245:125585.

Comet Program. 2014. *Introduction to observing oil from helicopters and planes*. https://www.meted.ucar.edu/training_module.php?id=1044.

Conlon, K. J., and C. A. Journey. 2008. *Evaluation of four structural best management practices for highway runoff in Beaufort and Colleton Counties, South Carolina, 2005–2006*. http://pubs.er.usgs.gov/publication/sir20085150.

Connolly, R. M., F. N. Connolly, and M. A. Hayes. 2020. Oil spill from the era: Mangroves taking eons to recover. *Marine Pollution Bulletin* 153:110965.

Conroy, J. W. H., H. Kruuk, and S. E. George. 1994. *Otters in Shetland: Monitoring the impact of the Braer oil spill: Final report*. Aberdeenshire, Scotland: Banchory Research Station.

Cooper, Z. S., J. Z. Rapp, S. D. Carpenter, G. Iwahana, H. Eicken, and J. W. Deming. 2019. Distinctive microbial communities in subzero hypersaline brines from Arctic coastal sea ice and rarely sampled cryopegs. *FEMS Microbiology Ecology* 95(12):fiz166.

Cope, M. R., T. Slack, T. C. Blanchard, and M. R. Lee. 2013. Does time heal all wounds? Community attachment, natural resource employment, and health impacts in the wake of the BP *Deepwater Horizon* disaster. *Social Science Research* 42(3):872–881.

Corbett, J. J., and H. W. Koehler. 2003. Updated emissions from ocean shipping. *Journal of Geophysical Research: Atmospheres* 108(D20). http://dx.doi.org/10.1029/2003JD003751.

Corbin, J. C., S. M. Pieber, H. Czech, M. Zanatta, G. Jakobi, D. Massabò, J. Orasche, I. El Haddad, A. A. Mensah, and B. Stengel. 2018. Brown and black carbon emitted by a marine engine operated on heavy fuel oil and distillate fuels: Optical properties, size distributions, and emission factors. *Journal of Geophysical Research: Atmospheres* 123(11):6175–6195.

Cordes, E. E., D. O. Jones, T. A. Schlacher, D. J. Amon, A. F. Bernardino, S. Brooke, R. Carney, D. M. DeLeo, K. M. Dunlop, and E. G. Escobar-Briones. 2016. Environmental impacts of the deep-water oil and gas industry: A review to guide management strategies. *Frontiers in Environmental Science* 4:58.

Cormack, D., and J. Nichols. 1977. The concentrations of oil in sea water resulting from natural and chemically induced dispersion of oil slicks. *International Oil Spill Conference Proceedings* 1977(1):381–385.

Costa, D. P., and G. L. Kooyman. 1982. Oxygen consumption, thermoregulation, and the effect of fur oiling and washing on the sea otter, Enhydra lutris. *Canadian Journal of Zoology* 60(11):2761–2767. https://doi.org/10.1139/z82-354.

Couillard, C. M., and F. A. Leighton. 1990. The toxicopathology of Prudhoe Bay crude oil in chicken embryos. *Fundamental and Applied Toxicology* 14(1):30–39. https://doi.org/10.1016/0272-0590(90)90228-C.

Couvillion, B. R., J. A. Barras, G. D. Steyer, W. Sleavin, M. Fischer, H. Beck, N. Trahan, B. Griffin, and D. Heckman. 2011. *Land area change in coastal Louisiana from 1932 to 2010*. Washington, DC: United States Geological Survey.

Crabb, M., P. Wright, O. Humphrey, G. Johnson, S. Rush, H. van Rein, and H. Hinchen. 2019. *Unmanned aerial vehicles for use in marine benthic monitoring*. Marine Monitoring Platform Guidelines No. 3. Peterborough, England: JNCC (Joint Nature Conservation Committee).

Craig, J. K., B. J. Burke, L. B. Crowder, and J. A. Rice. 2006. Prey growth and size-dependent predation in juvenile estuarine fishes: experimental and model analyses. *Ecology* 87(9):2366–2377. https://doi.org/10.1890/0012-9658(2006)87[2366:PGASPI]2.0.CO;2.

Cramm, M. A., B. de Moura Neves, C. C. Manning, T. B. Oldenburg, P. Archambault, A. Chakraborty, A. Cyr-Parent, E. N. Edinger, A. Jaggi, and A. Mort. 2021. Characterization of marine microbial communities around an Arctic seabed hydrocarbon seep at Scott Inlet, Baffin Bay. *Science of the Total Environment* 762:143961.

Crane, J. 1975. *Fiddler crabs of the world: Ocypodidae: Genus uca*. Vol. 1276. Princeton, NJ: Princeton University Press.

Crimmins, B. S., R. R. Dickerson, B. G. Doddridge, and J. E. Baker. 2004. Particulate polycyclic aromatic hydrocarbons in the Atlantic and Indian Ocean atmospheres during the Indian Ocean experiment and aerosols99: Continental sources to the marine atmosphere. *Journal of Geophysical Research: Atmospheres* 109(D5). https://doi.org/10.1029/2003JD004192.

Crump, K. 2020. The potential effects of recall bias and selection bias on the epidemiological evidence for the carcinogenicity of glyphosate. *Risk Analysis* 40(4):696–704.

Cruz-Martinez, L., J. E. Smits, and K. Fernie. 2015. Stress response, biotransformation effort, and immunotoxicity in captive birds exposed to inhaled benzene, toluene, nitrogen dioxide, and sulfur dioxide. *Ecotoxicology and Environmental Safety* 112:223–230.

CTEH. 2019. *Comparison of emissions from burning of petroleum, petroleum-derived fuels, and common vegetative fuels*. https://www.oilspillprevention.org/-/media/Oil-Spill-Prevention/spillprevention/r-and-d/in-situ-burning/isb-emissions-comparison-2020-final.pdf.

Cui, F., M. C. Boufadel, X. Geng, F. Gao, L. Zhao, T. King, and K. Lee. 2018. Oil droplets transport under a deep-water plunging breaker: Impact of droplet inertia. *Journal of Geophysical Research: Oceans* 123(12):9082–9100.

Cui, F., C. Daskiran, T. King, B. Robinson, K. Lee, J. Katz, and M. C. Boufadel. 2020a. Modeling oil dispersion under breaking waves. Part I: Wave hydrodynamics. *Environmental Fluid Mechanics* 20(6):1527–1551.

Cui, F., L. Zhao, C. Daskiran, T. King, K. Lee, J. Katz, and M. C. Boufadel. 2020b. Modeling oil dispersion under breaking waves. Part II: Coupling lagrangian particle tracking with population balance model. *Environmental Fluid Mechanics* 20(6):1553–1578.

Cui, F., X. Geng, B. Robinson, T. King, K. Lee, and M. C. Boufadel. 2020c. Oil droplet dispersion under a deep-water plunging breaker: Experimental measurement and numerical modeling. *Journal of Marine Science and Engineering* 8(4):230.

Culbertson, J. B., I. Valiela, E. E. Peacock, C. M. Reddy, A. Carter, and R. VanderKruik. 2007. Long-term biological effects of petroleum residues on fiddler crabs in salt marshes. *Marine Pollution Bulletin* 54(7):955–962.

Culbertson, J. B., I. Valiela, M. Pickart, E. E. Peacock, and C. M. Reddy. 2008a. Long-term consequences of residual petroleum on salt marsh grass. *Journal of Applied Ecology* 45(4):1284–1292.

Culbertson, J. B., I. Valiela, Y. S. Olsen, and C. M. Reddy. 2008b. Effect of field exposure to 38-year-old residual petroleum hydrocarbons on growth, condition index, and filtration rate of the ribbed mussel, geukensia demissa. *Environmental Pollution* 154(2):312–319.

Culliton, T. J., M. A. Warren, T. R. Goodspeed, D. G. Remer, C. M. Blackwell, and J. J. McDonough III. 1990. Fifty Years of Population Change Along the Nation's Coasts: 1960-2010. *Coastal Trends Series 2nd Report*, Rockville, MD, National Oceanic and Atmospheric Administration.

Cunningham, F., K. Dean, K. Hanson–Dorr, K. Harr, K. Healy, K. Horak, J. Link, S. Shriner, S. Bursian, and B. Dorr. 2017. Development of methods for avian oil toxicity studies using the double crested cormorant (phalacrocorax auritus). *Ecotoxicology and Environmental Safety* 141:199–208. https://doi.org/10.1016/j.ecoenv.2017.03.025.

Curtis, D., V. Elango, A. W. Collins, M. Rodrigue, and J. H. Pardue. 2018. Transport of crude oil and associated microbial populations by washover events on coastal headland beaches. *Marine Pollution Bulletin* 130:229–239.

Cutter, S. L., B. J. Boruff, and W. L. Shirley. 2003. Social vulnerability to environmental hazards. *Social Science Quarterly* 84(2):242–261.

Dachs, J., R. Lohmann, W. A. Ockenden, L. Méjanelle, S. J. Eisenreich, and K. C. Jones. 2002. Oceanic biogeochemical controls on global dynamics of persistent organic pollutants. *Environmental Science & Technology* 36(20):4229–4237.

Dal Ferro, B., and M. Smith. 2007. Global onshore and offshore water production. *Oil & Gas Review OTC Edition*.

Daling, P. S., and G. Indrebo. 1996. Recent improvements in optimizing use of dispersants as a cost–effective oil spill countermeasure technique. SPE Health, Safety and Environment in Oil and Gas Exploration and Production Conference. https://doi.org/10.2118/36072-MS.

Daling, P. S., and T. Strøm. 1999. Weathering of oils at sea: Model/field data comparisons. *Spill Science & Technology Bulletin* 5(1):63–74.

Daling, P. S., P. J. Brandvik, D. Mackay, and O. Johansen. 1990. Characterization of crude oils for environmental purposes. *Oil and Chemical Pollution* 7(3):199–224.

Daling, P. S., M. Ø. Moldestad, Ø. Johansen, A. Lewis, and J. Rødal. 2003. Norwegian testing of emulsion properties at sea—the importance of oil type and release conditions. *Spill Science & Technology Bulletin* 8(2):123–136.

Daling, P., P. Brandvik, F. Leirvik, A. Holumsnes, and C. Rasmussen. 2010. Development and field testing of a flexible system for application of dispersants on oil spills in ice. *Proceedings of the 33rd Arctic and Marine Oilspill Program Technical Seminar (AMOP) Technical Seminar, June 7–9, Halifax, Nova Scotia.*

Daling, P. S., F. Leirvik, I. K. Almås, P. J. Brandvik, B. H. Hansen, A. Lewis, and M. Reed. 2014. Surface weathering and dispersibility of MC252 crude oil. *Marine Pollution Bulletin* 87(1–2):300–310.

Daly, K. L., U. Passow, J. Chanton, and D. Hollander. 2016. Assessing the impacts of oil-associated marine snow formation and sedimentation during and after the *Deepwater Horizon* oil spill. *Anthropocene* 13:18–33.

Daly, K. L., A. Remsen, D. M. Outram, H. Broadbent, K. Kramer, and K. Dubickas. 2021. Resilience of the zooplankton community in the northeast Gulf of Mexico during and after the *Deepwater Horizon* oil spill. *Marine Pollution Bulletin* 163:111882.

Dang, H., and C. R. Lovell. 2016. Microbial surface colonization and biofilm development in marine environments. *Microbiology and Molecular Biology Reviews* 80(1):91–138.

Dang, N. P., C. Petrich, M. O'Sadnick, and L. Toske. 2020. Biotransformation of chemically dispersed diesel at sub-zero temperatures using artificial brines. *Environmental Technology* 42(17):2624–2630.

Dannreuther, N. M., D. Halpern, J. Rullkötter, and D. Yoerger. 2021. Technological developments since the *Deepwater Horizon* oil spill. *Oceanography* 34(1):192–211.

Danovaro, R., P. V. Snelgrove, and P. Tyler. 2014. Challenging the paradigms of deep-sea ecology. *Trends in Ecology & Evolution* 29(8):465–475.

Danovaro, R., M. Molari, C. Corinaldesi, and A. Dell'Anno. 2016. Macroecological drivers of archaea and bacteria in benthic deep-sea ecosystems. *Science Advances* 2(4):e1500961.

Das, P., X.-P. Yang, and L. Z. Ma. 2014. Analysis of biosurfactants from industrially viable pseudomonas strain isolated from crude oil suggests how rhamnolipids congeners affect emulsification property and antimicrobial activity. *Frontiers in Microbiology* 5:696.

D'Asaro, E. A., A. Y. Shcherbina, J. M. Klymak, J. Molemaker, G. Novelli, C. M. Guigand, A. C. Haza, B. K. Haus, E. H. Ryan, G. A. Jacobs, H. S. Huntley, N. J. M. Laxague, S. Chen, F. Judt, J. C. McWilliams, R. Barkan, A. D. Kirwan, Jr., A. C. Poje, and T. M. Özgökmen. 2018. Ocean convergence and the dispersion of flotsam. *Proceedings of the National Academy of Sciences* 115(6):1162–1167. https://doi.org/10.1073/pnas.1718453115.

D'Asaro, E. A., D. F. Carlson, M. Chamecki, R. R. Harcourt, B. K. Haus, B. Fox-Kemper, M. J. Molemaker, A. C. Poje, and D. Yang. 2020. Advances in observing and understanding small-scale open ocean circulation during the Gulf of Mexico research initiative era. *Frontiers in Marine Science* 7:349.

Daskiran, C., F. Cui, M. C. Boufadel, L. Zhao, S. A. Socolofsky, T. Ozgokmen, B. Robinson, and T. King. 2020. Hydrodynamics and dilution of an oil jet in crossflow: The role of small-scale motions from laboratory experiment and large eddy simulations. *International Journal of Heat and Fluid Flow* 85:108634.

Dauvin, J. C. 1998. The fine sand Abra Alba community of the Bay of Morlaix twenty years after the *Amoco Cadiz* oil spill. *Marine Pollution Bulletin* 36(9):669–676.

Davies, E. J., D. A. Dunnebier, Ø. Johansen, S. Masutani, I. Nagamine, and P. J. Brandvik. 2019. Shedding from chemically-treated oil droplets rising in seawater. *Marine Pollution Bulletin* 143:256–263.

Davis, R., T. Williams, J. Thomas, R. Kastelein, and L. Cornell. 1988. The effects of oil contamination and cleaning on sea otters (Enhydra lutris). II. Metabolism, thermoregulation, and behavior. *Canadian Journal of Zoology* 66(12):2782–2790.

Daykin, M., A. Tang, G. Sergy, and D. Aurand. June 1994. *Aquatic toxicity resulting from in situ burning of oil-on-water. Arctic and Marine Oilspill Program Technical Seminar.* Environment Canada.

De Beukelaer, S., I. MacDonald, N. Guinnasso, and J. Murray. 2003. Distinct side-scan sonar, radarsat SAR, and acoustic profiler signatures of gas and oil seeps on the Gulf of Mexico slope. *Geo-Marine Letters* 23(3):177–186.

De Gouw, J., A. Middlebrook, C. Warneke, R. Ahmadov, E. L. Atlas, R. Bahreini, D. Blake, C. Brock, J. Brioude, and D. Fahey. 2011. Organic aerosol formation downwind from the *Deepwater Horizon* oil spill. *Science* 331(6022):1295–1299.

De Padova, D., M. Mossa, M. Adamo, G. De Carolis, and G. Pasquariello. 2017. Synergistic use of an oil drift model and remote sensing observations for oil spill monitoring. *Environmental Science and Pollution Research* 24(6):5530–5543.

de Sousa, A. G. G., M. P. Tomasino, P. Duarte, M. Fernández-Méndez, P. Assmy, H. Ribeiro, J. Surkont, R. B. Leite, J. B. Pereira-Leal, and L. Torgo. 2019. Diversity and composition of pelagic prokaryotic and protist communities in a thin Arctic sea-ice regime. *Microbial Ecology* 78(2):388–408.

Dean, K. M., D. Cacela, M. W. Carney, F. Cunningham, C. Ellis, A. Gerson, C. Guglielmo, K. C. Hanson-Dorr, K. Harr, and K. A. Healy. 2017. Testing of an oral dosing technique for double-crested cormorants, Phalacocorax auritus, laughing gulls, Leucophaeus atricilla, homing pigeons, Columba livia, and western sandpipers, Calidris mauri, with artificially weather MC252 oil. *Ecotoxicology and Environmental Safety* 146:11–18.

Dean, T. A., J. L. Bodkin, S. C. Jewett, D. H. Monson, and D. Jung. 2000. Changes in sea urchins and kelp following a reduction in sea otter density as a result of the *Exxon Valdez* oil spill. *Marine Ecology Progress Series* 199:281–291.

Deepwater Horizon Natural Resource Damage Assessment Trustees. 2016. Deepwater Horizon *oil spill: Final programmatic damage assessment and restoration plan and final programmatic environmental impact statement.* http://www.gulfspillrestoration.noaa.gov/restoration-planning/gulf-plan.

Deis, D. R., J. W. Fleeger, S. M. Bourgoin, I. A. Mendelssohn, Q. Lin, and A. Hou. 2017. Shoreline oiling effects and recovery of salt marsh macroinvertebrates from the *Deepwater Horizon* oil spill. *PeerJ* 5:e3680.

Deis, D. R., J. W. Fleeger, D. S. Johnson, I. A. Mendelssohn, Q. Lin, S. A. Graham, S. Zengel, and A. Hou. 2020. Recovery of the salt marsh Periwinkle (littoraria irrorata) 9 years after the *Deepwater Horizon* oil spill: Size matters. *Marine Pollution Bulletin* 160:111581.

DeLeo, D. M., D. V. Ruiz-Ramos, I. B. Baums, and E. E. Cordes. 2016. Response of deep-water corals to oil and chemical dispersant exposure. *Deep Sea Research Part II: Topical Studies in Oceanography* 129:137–147.

DeLeo, D. M., S. Herrera, S. D. Lengyel, A. M. Quattrini, R. J. Kulathinal, and E. E. Cordes. 2018. Gene expression profiling reveals deep-sea coral response to the *Deepwater Horizon* oil spill. *Molecular Ecology* 27(20):4066–4077.

DeLeo, D. M., A. Glazier, S. Herrera, A. Barkman, and E. E. Cordes. 2021. Transcriptomic responses of deep-sea corals experimentally exposed to crude oil and dispersant. *Frontiers in Marine Science* 8:308.

Delvigne, G. A. L., and C. E. Sweeney. 1988. Natural dispersion of oil. *Oil and Chemical Pollution* 4(4):281–310.

Deming, J.W. and Eric Collins, R. 2017. Sea ice as a habitat for Bacteria, Archaea and viruses. In *Sea ice*, D. N. Thomas (ed.). pp. 326–351. https://doi.org/10.1002/9781118778371.ch13.

Den Hartog, C., and R. Jacobs. 1980. Effects of the "Amoco Cadiz" oil spill on an eelgrass community at Roscoff (France) with special reference to the mobile benthic fauna. *Helgoländer Meeresuntersuchungen* 33(1):182–191.

Deng, S., B. Wang, W. Zhang, S. Su, H. Dong, I. M. Banat, S. Sun, J. Guo, W. Liu, and L. Wang. 2021. Elucidate microbial characteristics in a full-scale treatment plant for offshore oil produced wastewater. *PloS ONE* 16(8):e0255836.

Department of Resources Energy and Tourism Australia. 2011. *Final Government Response to the Report of the Montara Commission of Inquiry.* Commonwealth of Australia Canberra, Australia. https://www.industry.gov.au/sites/default/files/2018-11/montara-commission-of-inquiry-report-final-government-response-may-2011.pdf.

Deppe, U., H.-H. Richnow, W. Michaelis, and G. Antranikian. 2005. Degradation of crude oil by an Arctic microbial consortium. *Extremophiles* 9(6):461–470.

Desmond, D. S., D. Saltymakova, T. D. Neusitzer, N. Firoozy, D. Isleifson, D. G. Barber, and G. A. Stern. 2019. Oil behavior in sea ice: Changes in chemical composition and resultant effect on sea ice dielectrics. *Marine Pollution Bulletin* 142:216–233.

Desmond, D. S., D. Saltymakova, A. Smith, T. Wolfe, N. Snyder, K. Polcwiartek, M. Bautista, M. Lemes, C. R. Hubert, and D. G. Barber. 2021a. Photooxidation and biodegradation potential of a light crude oil in first-year sea ice. *Marine Pollution Bulletin* 165:112154.

Desmond, D. S., O. Crabeck, M. Lemes, M. L. Harasyn, A. Mansoori, D. Saltymakova, M. C. Fuller, S. Rysgaard, D. G. Barber, and D. Isleifson. 2021b. Investigation into the geometry and distribution of oil inclusions in sea ice using non-destructive x-ray microtomography and its implications for remote sensing and mitigation potential. *Marine Pollution Bulletin* 173:112996.

Dewangan, N. K., and J. C. Conrad. 2020. Bacterial motility enhances adhesion to oil droplets. *Soft Matter* 16(35):8237–8244.

DFO (Department of Fisheries and Oceans Canada). 2021. *State of Knowledge on Chemical Dispersants for Canadian Marine Oil Spills.* DFO Canada Science Advisory Secretariat Science Advisory Report 2021/051. https://www.dfo-mpo.gc.ca/csas-sccs/Publications/SAR-AS/2021/2021_051-eng.html.

Dhima, A., J.-C. de Hemptinne, and J. Jose. 1999. Solubility of hydrocarbons and CO2 mixtures in water under high pressure. *Industrial & Engineering Chemistry Research* 38(8):3144–3161.

D'Hondt, S., F. Inagaki, C. A. Zarikian, L. J. Abrams, N. Dubois, T. Engelhardt, H. Evans, T. Ferdelman, B. Gribsholt, R. N. Harris, Bryce W. Hoppie, J.-H. Hyun, J. Kallmeyer, J. Kim, J. E. Lynch, Claire C. McKinley, S. Mitsunobu, Y. Morono, R. W. Murray, R. Pockalny, J. Sauvage, T. Shimono, F. Shiraishi, D. C. Smith, Christopher E. Smith-Duque, A. J. Spivack, B. O. Steinsbu, Y. Suzuki, M. Szpak, L. Toffin, G. Uramoto, Y. T. Yamaguchi, G.-l. Zhang, X.-H. Zhang, and W. Ziebis. 2015. Presence of oxygen and aerobic communities from sea floor to basement in deep-sea sediments. *Nature Geoscience* 8(4):299–304. https://doi.org/10.1038/ngeo2387.

DHS (U.S. Department of Homeland Security). 2008. *System assessment and validation for emergency responders (SAVER): Common operating picture for emergency responders.* https://www.dhs.gov/sites/default/files/publications/CommonOpER_HLT_0908-508.pdf.

Di Toro, D. M., J. A. McGrath, and D. J. Hansen. 2000. Technical basis for narcotic chemicals and polycyclic aromatic hydrocarbon criteria. I. Water and tissue. *Environmental Toxicology and Chemistry* 19(8):1951–1970.

Diab, O., J. DePierro, L. Cancelmo, J. Schaffer, C. Schechter, C. R. Dasaro, A. Todd, M. Crane, I. Udasin, and D. Harrison. 2020. Mental healthcare needs in world trade center responders: Results from a large, population-based health monitoring cohort. *Administration and Policy in Mental Health and Mental Health Services Research* 47(3):427–434.

Diamante, G., N. Menjivar-Cervantes, E. G. Xu, D. C. Volz, A. C. D. Bainy, and D. Schlenk. 2017. Developmental toxicity of hydroxylated chrysene metabolites in zebrafish embryos. *Aquatic Toxicology* 189:77–86.

Diamond, S. A. 2003. Photoactivated toxicity in aquatic environments. In *UV effects in aquatic organisms and ecosystems*, E. W. Helbling and H. Zagarese (eds.), pp. 219–250. Cambridge, UK: The Royal Society of Chemistry.

Díaz-Jaramillo, M., M. S. Islas, and M. Gonzalez. 2021. Spatial distribution patterns and identification of microplastics on intertidal sediments from urban and semi-natural SW Atlantic estuaries. *Environmental Pollution* 273:116398.

Dickey, R. 2012. FDA risk assessment of seafood contamination after the BP oil spill. *Environmental Health Perspectives* 120:A54–A55. https://doi.org/10.1289/ehp.1104539.

Dickey, R. and W. Dickhoff. 2011. *Dispersants and Seafood Safety Assessment of the potential impact of Corexit® oil dispersants on seafood safety.* A White Paper for the Coastal Response Research Center. Dispersant Initiative and Workshop "The Future of Dispersant Use in Spill Response." https://scholars.unh.edu/crrc/10.

Dickins, D. 2011. Behavior of oil spills in ice and implications for Arctic spill response. *Proceedings of the Arctic Technology Conference.* https://doi.org/10.4043/22126-MS.

Dickins, D. F., and I. A. Buist. 1981. *Oil and gas under sea ice study: Vols. I&2.* (Prepared by Dome Petroleum Ltd. for COOSRA, Report CV-1, Calgary, AB, Canada [also published in Proceedings of 1981 International Oil Spill Conference, Atlanta, GA]).

Dickins, D. F., D. E. Thornton, and W. J. Cretney. 1987. Design and operation of oil discharge systems and characteristics of oil used in the Baffin Island oil spill project. *Arctic* 40(Suppl 1):100–108.

Diercks, A. R., I. C. Romero, R. A. Larson, P. Schwing, A. Harris, S. Bosman, J. P. Chanton, and G. Brooks. 2021. Resuspension, redistribution, and deposition of oil-residues to offshore depocenters after the *Deepwater Horizon* oil spill. *Frontiers in Marine Science* 8. https://doi.org/10.3389/fmars.2021.630183.

Ding, X., X.-M. Wang, Z.-Q. Xie, C.-H. Xiang, B.-X. Mai, L.-G. Sun, M. Zheng, G.-Y. Sheng, J.-M. Fu, and U. Poeschl. 2007. Atmospheric polycyclic aromatic hydrocarbons observed over the North Pacific Ocean and the Arctic area: Spatial distribution and source identification. *Atmospheric Environment* 41(10):2061–2072.

Dissanayake, A. L., J. Gros, and S. A. Socolofsky. 2018. Integral models for bubble, droplet, and multiphase plume dynamics in stratification and crossflow. *Environmental Fluid Mechanics* 18(5):1167–1202.

DNV (Det Norske Veritas). 1999. *Risk Assessment of Pollution from Oil and HNS Spills in Australian Ports and Waters.* Det Norske Project 9330-3972, December 1999.

DNV. 2016. *Offshore leak detection. Recommended practice.* DNVGL-RP-F302. Edition April 2016. https://www.dnv.com/oilgas/download/dnv-rp-f302-offshore-leak-detection.html.

Dodge, R. E., B. J. Baca, A. Knap, S. Snedaker, and T. Sleeter. 1995. The effects of oil and chemically dispersed oil in tropical ecosystems: 10 years of monitoring experimental sites.

DOE (U.S. Department of Energy). 2005. *Liquefied natural gas: Understanding the basic facts.* energy.gov/sites/prod/files/2013/04/10LNG_primerudp.pdf.

Doherty, B. T., R. K. Kwok, M. D. Curry, C. Ekenga, D. Chambers, D. P. Sandler, and L. S. Engel. 2017. Associations between blood BTEXs concentrations and hematologic parameters among adult residents of the US Gulf States. *Environmental Research* 156:579–587.

Dokken, Q. R. 2011. IXTOC I versus Macondo well blowout: Anatomy of an oil spill event then and now. *International Oil Spill Conference Proceedings* 2011(1):420.

Donahue, N. M., A. Robinson, C. Stanier, and S. Pandis. 2006. Coupled partitioning, dilution, and chemical aging of semivolatile organics. *Environmental Science & Technology* 40(8):2635–2643.

Donahue, N. M., S. A. Epstein, S. N. Pandis, and A. L. Robinson. 2011. A two-dimensional volatility basis set: part 1. Organic-aerosol mixing thermodynamics. *Atmospheric Chemistry and Physics* 11:3303–3318.

Donahue, N. M., J. Kroll, S. N. Pandis, and A. L. Robinson. 2012. A two-dimensional volatility basis set: Part 2. Diagnostics of organic-aerosol evolution. *Atmospheric Chemistry and Physics* 12(2):615–634.

Dong, C., X. Bai, H. Sheng, L. Jiao, H. Zhou, and Z. Shao. 2015. Distribution of PAHs and the PAH-degrading bacteria in the deep-sea sediments of the high-latitude Arctic Ocean. *Biogeosciences* 12(7):2163–2177.

Dong, X., C. Greening, J. E. Rattray, A. Chakraborty, M. Chuvochina, D. Mayumi, J. Dolfing, C. Li, J. M. Brooks, and B. B. Bernard. 2019. Metabolic potential of uncultured bacteria and archaea associated with petroleum seepage in deep-sea sediments. *Nature Communications* 10(1):1–12.

Dong, X., J. E. Rattray, D. C. Campbell, J. Webb, A. Chakraborty, O. Adebayo, S. Matthews, C. Li, M. Fowler, and N. M. Morrison. 2020. Thermogenic hydrocarbon biodegradation by diverse depth-stratified microbial populations at a scotian basin cold seep. *Nature Communications* 11(1):1–14.

Dor, F., R. Bonnard, C. Gourier-Fréry, A. Cicolella, R. Dujardin, and D. Zmirou. 2003. Health risk assessment after decontamination of the beaches polluted by the wrecked Erika tanker. *Risk Analysis* 23(6):1199–1208.

Dorobantu, L. S., A. K. Yeung, J. M. Foght, and M. R. Gray. 2004. Stabilization of oil-water emulsions by hydrophobic bacteria. *Applied and Environmental Microbiology* 70(10):6333–6336.

Dorobantu, L. S., S. Bhattacharjee, J. M. Foght, and M. R. Gray. 2008. Atomic force microscopy measurement of heterogeneity in bacterial surface hydrophobicity. *Langmuir* 24(9):4944–4951.

Doyle, S. M., E. A. Whitaker, V. De Pascuale, T. L. Wade, A. H. Knap, P. H. Santschi, A. Quigg, and J. B. Sylvan. 2018. Rapid formation of microbe-oil aggregates and changes in community composition in coastal surface water following exposure to oil and the dispersant COREXIT. *Frontiers in Microbiology* 9:689.

Drescher, C. F., S. E. Schulenberg, and C. V. Smith. 2014. The *Deepwater Horizon* oil spill and the Mississippi gulf coast: Mental health in the context of a technological disaster. *American Journal of Orthopsychiatry* 84(2):142.

Dris, R., J. Gasperi, M. Saad, C. Mirande, and B. Tassin. 2016. Synthetic fibers in atmospheric fallout: A source of microplastics in the environment? *Marine Pollution Bulletin* 104(1–2):290–293.

Drozd, G. T., D. R. Worton, C. Aeppli, C. M. Reddy, H. Zhang, E. Variano, and A. H. Goldstein. 2015. Modeling comprehensive chemical composition of weathered oil following a marine spill to predict ozone and potential secondary aerosol formation and constrain transport pathways. *Journal of Geophysical Research: Oceans* 120(11):7300–7315.

Du, M., and J. D. Kessler. 2012. Assessment of the spatial and temporal variability of bulk hydrocarbon respiration following the *Deepwater Horizon* oil spill. *Environmental Science & Technology* 46(19):10499–10507. https://doi.org/10.1021/es301363k.

Dubilier, N., C. Bergin, and C. Lott. 2008. Symbiotic diversity in marine animals: The art of harnessing chemosynthesis. *Nature Reviews Microbiology* 6(10):725–740.

Dubinsky, E. A., M. E. Conrad, R. Chakraborty, M. Bill, S. E. Borglin, J. T. Hollibaugh, O. U. Mason, Y. M. Piceno, F. C. Reid, and W. T. Stringfellow. 2013. Succession of hydrocarbon-degrading bacteria in the aftermath of the *Deepwater Horizon* oil spill in the Gulf of Mexico. *Environmental Science & Technology* 47(19):10860–10867.

Duke, N. C., Z. S. Pinzón, and M. C. Prada T. 1997. Large-scale damage to mangrove forests following two large oil spills in panama 1. *Biotropica* 29(1):2–14.

Duncker, H.-R. 1974. Structure of the avian respiratory tract. *Respiration Physiology* 22(1–2):1–19.

Duperron, S., S. Halary, A. Gallet, and B. Marie. 2020. Microbiome-Aware Ecotoxicology of Organisms: Relevance, Pitfalls, and Challenges. *Frontiers in Public Health* 8. https://doi.org/10.3389/fpubh.2020.00407.

Dutta, T. K., and S. Harayama. 2001. Biodegradation of *n*-Alkylcycloalkanes and *n*-Alkylbenzenes via new pathways in *Alcanivorax* sp. Strain MBIC 4326. *Applied and Environmental Microbiology* 67(4):1970–1974.

DWH (Deepwater Horizon) Trustees. 2015. Deepwater Horizon *oil spill: Draft programmatic damage assessment and restoration plan and draft programmatic environmental impact statement.* Silver Spring, MD: National Marine Fisheries Service, National Oceanic and Atmospheric Administration. www.gulfspillrestoration.noaa.gov/restoration-planning/gulf–plan.

Dyer, S. D., D. J. Versteeg, S. E. Belanger, J. G. Chaney, and F. L. Mayer. 2006. Interspecies correlation estimates predict protective environmental concentrations. *Environmental Science & Technology* 40(9):3102–3111. https://doi.org/10.1021/es051738p.

Dyer, S. D., D. J. Versteeg, S. E. Belanger, J. G. Chaney, S. Raimondo, and M. G. Barron. 2008. Comparison of species sensitivity distributions derived from interspecies correlation models to distributions used to derive water quality criteria. *Environmental Science & Technology* 42(8):3076–3083. https://doi.org/10.1021/es702302e.

ECCC (Environment and Climate Change Canada). 2018. *Shoreline cleanup assessment technique (SCAT) manual, third edition.* Ottawa, Ontario: Owens Coastal Consultants. Prepared and provided by Triox Environmental Emergencies, Environmental Mapping Ltd.

Echols, M. 2006. Evaluating and treating the kidneys. *Clinical Avian Medicine* 2:451–492.

Eganhouse, R. P., and I. R. Kaplan. 1982. Extractable organic matter in municipal wastewaters: 1. Petroleum hydrocarbons: temporal variations and mass emission rates to the oceans. *Environmental Science and Technology* 16:180–186.

Egres, A. G., C. C. Martins, V. M. de Oliveira, and P. da Cunha Lana. 2012. Effects of an experimental in situ diesel oil spill on the benthic community of unvegetated tidal flats in a subtropical estuary (Paranaguá Bay, Brazil). *Marine Pollution Bulletin* 64(12):2681–2691.

Ehrenhauser, F. S., P. Avij, X. Shu, V. Dugas, I. Woodson, T. Liyana–Arachchi, Z. Zhang, F. R. Hung, and K. T. Valsaraj. 2014. Bubble bursting as an aerosol generation mechanism during an oil spill in the deep-sea environment: Laboratory experimental demonstration of the transport pathway. *Environmental Science: Processes & Impacts* 16(1):65–73.

EIA (U.S. Energy Information Administration). 2019. *Annual energy outlook 2019.* https://www.eia.gov/pressroom/presentations/Capuano_01242019.pdf.

EIA. 2021a. *Annual energy outlook 2021.* https://www.eia.gove/outlooks/ieo.

EIA. 2021b. *Movements of crude oil and selected products by rail.* https://www.eia.gov/dnav/pet/PET_MOVE_RAILNA_A_EPC0_RAIL_MBBL_M.htm.

EIA. 2022. *April 2022 monthly energy review.* https://www.eia.gov/totalenergy/data/monthly/archive/00352204.pdf.

Eklund, R. L., L. C. Knapp, P. A. Sandifer, and R. C. Colwell. 2019. Oil spills and human health: Contributions of the Gulf of Mexico research initiative. *Geohealth* 3(12):391–406. https://doi.org/10.1029/2019GH000217.

Elango, V., M. Urbano, K. R. Lemelle, and J. H. Pardue. 2014. Biodegradation of MC252 oil in oil: Sand aggregates in a coastal headland beach environment. *Frontiers in Microbiology* 5:161.

Elliott, J. E., and D. DeVilbiss. 2014. Advancements in underwater oil detection and recovery techniques. *International Oil Spill Conference Proceedings* 2014(1):2037–2052.

Ellis, J. B. 1986. Pollutional aspects of urban runoff. In *Urban Runoff Pollution*, pp. 1–38. Dordrecht, Netherlands: Springer.

Elsner, J. B., J. P. Kossin, and T. H. Jagger. 2008. The increasing intensity of the strongest tropical cyclones. *Nature* 455(7209):92–95.

Elyamine, A. M., J. Kan, S. Meng, P. Tao, H. Wang, and Z. Hu. 2021. Aerobic and anaerobic bacterial and fungal degradation of pyrene: Mechanism pathway including biochemical reaction and catabolic genes. *International Journal of Molecular Sciences* 22(15):8202.

Endresen, Ø., E. Sørgård, J. K. Sundet, S. B. Dalsøren, I. S. Isaksen, T. F. Berglen, and G. Gravir. 2003. Emission from international sea transportation and environmental impact. *Journal of Geophysical Research: Atmospheres* 108(D17):4560.

Energo Engineering. 2006. *Assessment of fixed offshore platform performance in hurricanes Andrew, Lili, and Ivan.* Prepared by Energo Engineering, Inc., Houston, Texas, for the U.S. Department of the Interior, Minerals Management Service, Herndon, Virginia.

Energo Engineering. 2010. *Assessment of damage and failure mechanisms for offshore structures and pipelines in Hurricanes Gustav and Ike.* Prepared by Energo Engineering, Inc., Houston, Texas, for the U.S. Department of the Interior, Minerals Management Service, Herndon, Virginia.

Engel, L. S., R. K. Kwok, A. K. Miller, A. Blair, M. D. Curry, J. A. McGrath, and D. P. Sandler. 2017. The gulf long-term follow-up study (Gulf study): Biospecimen collection at enrollment. *Journal of Toxicology and Environmental Health, Part A* 80(4):218–229.

Engelhardt, F. R. 1982. Hydrocarbon metabolism and cortisol balance in oil-exposed ringed seals, Phoca hispida. *Comparative Biochemistry and Physiology Part C: Comparative Pharmacology* 72(1):133–136.

Engelhardt, F. R. 1983. Petroleum effects on marine mammals. *Aquatic Toxicology* 4(3):199–217.

EPA (U.S. Environmental Protection Agency). 2016. *FOSC desk report for the Enbridge Line 6b oil spill.* Marshall, Michigan. https://www.epa.gov/enbridge-spill-michigan/fosc-desk-report-enbridge-oil-spill.

EPA. 2018. *Area contingency planning (ACP) handbook.* https://www.epa.gov/oil-spills-prevention-and-preparedness-regulations/area-contingency-planning-handbook.

EPA. n.d.a. *National contingency plan product schedule toxicity and effectiveness summaries.* https://www.epa.gov/emergency-response/national-contingency-plan-product-schedule-toxicity-and-effectiveness-summaries.

EPA. n.d.b. *Revision to the National Oil and Hazardous Substances Pollution Contingency Plan—Subpart J monitoring requirements.* https://www.epa.gov/emergency-response/revisions-national-oil-and-hazardous-substances-pollution-contingency-plan-0.

EPA. n.d.c. *Technical overview of volatile organic compounds.* https://www.epa.gov/indoor-air-quality-iaq/technical-overview-volatile-organic-compounds.

EPPR (Emergency Prevention, Preparedness, and Response). 2015. *Guide to oil spill response in snow and ice conditiond in the Arctic.* https://oaarchive.arctic-council.org/handle/11374/403.

Epstein, N., R. Bak, and B. Rinkevich. 2000. Toxicity of third generation dispersants and dispersed Egyptian crude oil on Red Sea coral larvae. *Marine Pollution Bulletin* 40(6):497–503.

Eriksen, M., L. C. Lebreton, H. S. Carson, M. Thiel, C. J. Moore, J. C. Borerro, F. Galgani, P. G. Ryan, and J. Reisser. 2014. Plastic pollution in the world's oceans: More than 5 trillion plastic pieces weighing over 250,000 tons afloat at sea. *PloS ONE* 9(12):e111913.

Erni-Cassola, G., V. Zadjelovic, M. I. Gibson, and J. A. Christie-Oleza. 2019. Distribution of plastic polymer types in the marine environment: A meta-analysis. *Journal of Hazardous Materials* 369:691–698.

ESGOSS (Ecological Steering Group on the Oil Spill in Shetland). 1994. *The environmental impact of the wreck of the braer.* Report of the Ecological Steering Group on the Oil Spill in Shetland (ESGOSS). Edinburgh, UK: Scottish Office.

Esler, D., K. A. Trust, B. E. Ballachey, S. A. Iverson, T. L. Lewis, D. J. Rizzolo, D. M. Mulcahy, A. K. Miles, B. R. Woodin, and J. J. Stegeman. 2010. Cytochrome P4501A biomarker indication of oil exposure in harlequin ducks up to 20 years after the *Exxon Valdez* oil spill. *Environmental Toxicology and Chemistry* 29(5):1138–1145.

Esler, D., B. E. Ballachey, C. Matkin, D. Cushing, R. Kaler, J. Bodkin, D. Monson, G. Esslinger, and K. Kloecker. 2018. Timelines and mechanisms of wildlife population recovery following the *Exxon Valdez* oil spill. *Deep Sea Research Part II: Topical Studies in Oceanography* 147:36–42.

Espinel, Z., J. P. Kossin, S. Galea, A. S. Richardson, and J. M. Shultz. 2019. Forecast: Increasing mental health consequences from Atlantic hurricanes throughout the 21st century. *Psychiatric Services* 70(12):1165–1167.

Estes, J. A. 1998. Concerns about rehabilitation of oiled wildlife. *Conservation Biology* 12(5):1156–1157.

Estes, J. A., and D. O. Duggins. 1995. Sea otters and kelp forests in Alaska: Generality and variation in a community ecological paradigm. *Ecological Monographs* 65(1):75–100.

Etkin, D. 2014. Risk of crude and bitumen pipeline spills in the United States: 35 analyses of historical data and case studies (1968–2012). *Arctic and Marine Oilspill Program Proceedings* 36:37.

Etkin, D., and K. Michel. 2003. Bio-economic modeling for oil spills from tanker/freighter groundings on rock pinnacles in San Francisco bay, vols. III and IV, spill response reports—shag rock and blossom rock, final report. Contract DACW07-01-C-0018. Sacramento District, Sacramento, CA: U.S. Army Corps of Engineers. P. 314.

Etkin, D., D. French-McCay, and J. Michel. 2007. *Review of the state-of-the-art on modeling interactions between spilled oil and shorelines for the development of algorithms for oil spill risk analysis modeling.* Cortlandt Manor, NY: Minerals Management Services & Environmental Research Consulting, U.S. Department of the Interior.

Etkin, D. S. 2006. *Trends in oil spills from large vessels in the US and California with implications for anticipated oil spill prevention and mitigation based on the Washington oil transfer rule.* Contract No. C040018. Olympia, WA. Prepared by Environmental Research Consulting for the Washington Department of Ecology.

Etkin, D. S. 2009. *Analysis of us oil spillage.* Prepared by Environmental Research Consulting for the American Petroleum Institute, Washington, DC.

Etkin, D. S. 2010a. Forty-year analysis of U.S. oil spillage rates. *Proceedings of the 33rd Arctic & Marine Oilspill Program Technical Seminar on Environmental Contamination and Response*, pp. 505–528.

Etkin, D. S. 2010b. Worldwide analysis of in-port vessel operational lubricant discharges and leakages. *Proceedings of the 33rd Arctic & Marine Oilspill Program Technical Seminar on Environmental Contamination and Response*, pp. 529–554.

Etkin, D. S. 2015. Offshore well blowout probability model. *Proceedings of the 38th Arctic & Marine Oilspill Program Technical Seminar on Environmental Contamination and Response*, pp. 169–192.

Etkin, D. S. 2017a. Analysis of US crude-by-rail oil spillage and potential future trends. *Proceedings of the 40th Arctic & Marine Oilspill Program Technical Seminar on Environmental Contamination and Response.*

Etkin, D. S. 2017b. Historical analysis of US pipeline spills and implications for contingency planning. *Proceedings of the 40th Arctic & Marine Oilspill Program Technical Seminar on Environmental Contamination and Response.*

Etkin, D. S. 2019. Developments in risk assessments for potentially-polluting sunken vessels. *Proceedings of the 42nd Arctic & Marine Oilspill Program Technical Seminar on Environmental Contamination and Response.*

Etkin, D. S., and J. Neel. 2001. Investing in spill prevention: Has it reduced vessel spills and accidents in Washington state? *International Oil Spill Conference Proceedings* 2001(1):47–56.

Etkin, D. S., J. Joeckel, A.-H. Walker, D. Scholz, D. L. Hatzenbuhler, E. J. Lyman, and R. G. Patton. 2015. New risks from crude-by-rail transportation. *Proceedings of the 38th Arctic & Marine Oilspill Program Technical Seminar on Environmental Contamination and Response.*

Etnoyer, P. J., and S. D. Cairns. 2017. Deep-sea coral taxa in the US Gulf of Mexico: Depth and geographical distribution. https://repository.si.edu/handle/10088/34996.

Etnoyer, P. J., L. N. Wickes, M. Silva, J. Dubick, L. Balthis, E. Salgado, and I. R. MacDonald. 2016. Decline in condition of gorgonian octocorals on mesophotic reefs in the northern Gulf of Mexico: Before and after the *Deepwater Horizon* oil spill. *Coral Reefs* 35(1):77–90.

Evans, M., J. Liu, H. Bacosa, B. E. Rosenheim, and Z. Liu. 2017. Petroleum hydrocarbon persistence following the *Deepwater Horizon* oil spill as a function of shoreline energy. *Marine Pollution Bulletin* 115(1–2):47–56.

Everaert, G., A. Ruus, D. Ø. Hjermann, K. Borgå, N. Green, S. Boitsov, H. Jensen, and A. Poste. 2017. Additive models reveal sources of metals and organic pollutants in Norwegian marine sediments. *Environmental Science & Technology* 51(21):12764–12773.

Exxon-Mobil Research and Engineering Company. 2014. *Oil spill response field manual.* https://corporate.exxonmobil.com/-/media/Global/Files/risk-management-and-safety/Oil-Spill-Response-Field-Manual_2014.

Eyring, V., H. Köhler, J. Van Aardenne, and A. Lauer. 2005. Emissions from international shipping: 1. The last 50 years. *Journal of Geophysical Research: Atmospheres* 110(D17).

Fairbrother, A., J. Smits, and K. A. Grasman. 2004. Avian immunotoxicology. *Journal of Toxicology and Environmental Health, Part B* 7(2):105–137. https://doi.org/10.1080/10937400490258873.

Fakhru'l-Razi, A., A. Pendashteh, L. C. Abdullah, D. R. A. Biak, S. S. Madaeni, and Z. Z. Abidin. 2009. Review of technologies for oil and gas produced water treatment. *Journal of Hazardous Materials* 170(2–3):530–551.

Faksness, L.-G., and P. J. Brandvik. 2008a. Distribution of water soluble components from Arctic marine oil spills—a combined laboratory and field study. *Cold Regions Science and Technology* 54(2):97–105.

Faksness, L.-G., and P. J. Brandvik. 2008b. Distribution of water soluble components from oil encapsulated in Arctic sea ice: Summary of three field seasons. *Cold Regions Science and Technology* 54(2):106–114.

Faksness, L.-G., P. G. Grini, and P. S. Daling. 2004. Partitioning of semi-soluble organic compounds between the water phase and oil droplets in produced water. *Marine Pollution Bulletin* 48(7–8):731–742.

Faksness, L.-G., P. Daling, D. Altin, H. Dolva, B. Fosbæk, and R. Bergstrøm. 2015. Relative bioavailability and toxicity of fuel oils leaking from World War II shipwrecks. *Marine Pollution Bulletin* 94(1–2):123–130.

Faksness, L.-G., R. C. Belore, J. McCourt, M. Johnsen, T.-A. Pettersen, and P. S. Daling. 2017. Effectiveness of chemical dispersants used in broken ice conditions. *International Oil Spill Conference Proceedings* 2017(1):1543–1558.

Fall, J. A. 1999. *Subsistence.* Exxon Valdez *Oil Spill Trustee Council.* https://evostc.state.ak.us/status-of-restoration/subsistence.

Fallon, J. A., E. P. Smith, N. Schoch, J. D. Paruk, E. A. Adams, D. C. Evers, P. G. Jodice, C. Perkins, S. Schulte, and W. A. Hopkins. 2018. Hematological indices of injury to lightly oiled birds from the *Deepwater Horizon* oil spill. *Environmental Toxicology and Chemistry* 37(2):451–461.

Fallon, J. A., E. P. Smith, N. Schoch, J. D. Paruk, E. M. Adams, D. C. Evers, P. G. Jodice, M. Perkins, D. E. Meattey, and W. A. Hopkins. 2020. Ultraviolet-assisted oiling assessment improves detection of oiled birds experiencing clinical signs of hemolytic anemia after exposure to the *Deepwater Horizon* oil spill. *Ecotoxicology* 29(9):1399–1408.

Fam, S., M. K. Stenstrom, and G. Silverman. 1987. Hydrocarbons in runoff. *Journal of Environmental Engineering* 113(5):1032–1046.

Farrington, J. W. 2020. Need to update human health risk assessment protocols for polycyclic aromatic hydrocarbons in seafood after oil spills. *Marine Pollution Bulletin* 150:110744. https://doi.org/10.1016/j.marpolbul.2019.110744.

Farrington, J. W., and J. G. Quinn. 2015. "Unresolved complex mixture" (USM): A brief history of the term and moving beyond it. *Marine Pollution Bulletin* 96(1–2):29–31.

Farrington, J. W., A. C. Davis, N. M. Frew, and A. Knap. 1988. ICES/IOC intercomparison exercise on the determination of petroleum hydrocarbons in biological tissues (mussel homogenate). *Marine Pollution Bulletin* 19(8):372–380.

Farrington, J. W., B. W. Tripp, S. Tanabe, A. Subramanian, J. L. Sericano, T. L. Wade, and A. H. Knap. 2016. Edward d. Goldberg's proposal of "the mussel watch": Reflections after 40 years. *Marine Pollution Bulletin* 110(1):501–510.

FDA (U.S. Food and Drug Administration). 2011. Gulf of Mexico Oil Spill Update FDA's Role in Seafood Safety. Food Safety During Emergencies. https://www.fda.gov/food/food-safety-during-emergencies/gulf-mexico-oil-spill.

Federal Interagency Solutions Group. 2010. *Oil budget calculator:* Deepwater Horizon*: A report to the National Incident Command.* https://www.restorethegulf.gov/sites/default/files/documents/pdf/OilBudgetCalc_Full_HQ-Print_111110.pdf.

Ferguson, R. M., E. Gontikaki, J. A. Anderson, and U. Witte. 2017. The variable influence of dispersant on degradation of oil hydrocarbons in subarctic deep-sea sediments at low temperatures (0–5°C). *Scientific Reports* 7(1):1–13.

Fernández-Carrera, A., K. Rogers, S. Weber, J. Chanton, and J. Montoya. 2016. Deep water horizon oil and methane carbon entered the food web in the Gulf of Mexico. *Limnology and Oceanography* 61(S1):S387–S400.

Fernández-Gómez, B., B. Díez, M. F. Polz, J. I. Arroyo, F. D. Alfaro, G. Marchandon, C. Sanhueza, L. Farías, N. Trefault, and P. A. Marquet. 2019. Bacterial community structure in a sympagic habitat expanding with global warming: Brackish ice brine at 85–90°N. *The ISME Journal* 13(2):316–333.

Fernando, H. J. 2013a. *Handbook of environmental fluid dynamics, volume one.* Boca Raton, London, New York: CRC Press, Taylor and Francis Group.

Fernando, H. J. 2013b. *Handbook of environmental fluid dynamics, volume two.* Boca Raton, London, New York: CRC Press, Taylor and Francis Group.

Feuston, M. H., C. E. Hamilton, C. A. Schreiner, and C. R. Mackerer. 1997. Developmental toxicity of dermally applied crude oils in rats. *Journal of Toxicology and Environmental Health* 52(1):79–93.

Findlay, A. 2003. Practical considerations in the recovery of oil from sunken and abandoned vessels. *International Oil Spill Conference.* 2003(1). https://doi.org/10.7901/2169-3358-2003-1-161.

Fingas, M. 2010. Soot production from in-situ oil fires: Literature review and calculation of values from experimental spills. *Proceedings of the 33rd AMOP Technical Seminar on Environmental Contamination and Response Environment Canada*, pp. 1017–1054. Ottawa, Ontario.

Fingas, M. 2011a. Models for water-in-oil emulsion formation. In *Oil Spill Science and Technology*, pp. 243–273. Amsterdam, Netherlands: Elsevier.

Fingas, M. 2011b. Oil spill dispersants: A technical summary. In *Oil Spill Science and Technology*, pp. 435–582. Amsterdam, Netherlands: Elsevier.

Fingas, M. 2015. Review of the properties and behaviour of diluted bitumens. *Proceedings of the 38th Arctic and Marine Oilspill Program Technical Seminar on Environmental Contamination and Response.* Ottawa, Ontario, Canada: Environment Canada.

Fingas, M. 2017. The fate of PAHs resulting from in-situ oil burns. *International Oil Spill Conference Proceedings* 2017(1):1041–1056.

Fingas, M. 2018. *In-situ burning for oil spill countermeasures.* M. Fingas (ed.). Boca Raton, FL: CRC Press. https://doi.org/10.1201/9780429506376.

Fingas, M. 2021. Visual appearance of oil on the sea. *Journal of Marine Science and Engineering* 9(1):97.

Fingas, M., and J. Banta. 2017. A review of literature related to oil spill dispersants. *Prince William Sound Regional Citizens' Advisory Council.* http://www.pwsrcac.org/wpcontent/uploads/filebase/programs/environmental_monitorin/dispersants/review_of_osd_literature.pdf.

Fingas, M., and C. Brown. 2014. Review of oil spill remote sensing. *Marine Pollution Bulletin* 83(1):9–23. https://doi.org/10.1016/j.marpolbul.2014.03.059.

Fingas, M., and C. E. Brown. 2018. A review of oil spill remote sensing. *Sensors* 18(1):91.

Fingas, M., and B. Fieldhouse. 2009. Studies on crude oil and petroleum product emulsions: Water resolution and rheology. *Colloids and Surfaces A: Physicochemical and Engineering Aspects* 333(1–3):67–81.

Fingas, M., and B. Fieldhouse. 2011. Review of solidifiers. In *Oil spill science and technology*, pp. 713–733. Amsterdam, Netherlands: Elsevier.

Fingas, M., and B. Fieldhouse. 2014. Water-in-oil emulsions: Formation and prediction. In *Handbook of oil spill science and technology*, M. Fingas (ed.), p. 225. New York: John Wiley and Sons Inc.

Fingas, M., and B. Fieldhouse. 2015. Water-in-oil emulsions: Formation and predictions. In *Handbook of oil spill science and technology*, pp. 460–510. M. Fingas (ed.). New York: John Wiley & Sons.

Fingas, M., and B. Hollebone. 2003. Review of behaviour of oil in freezing environments. *Marine Pollution Bulletin* 47(9–12):333–340.

Fingas, M., C. Brown, and L. Gamble. 1999. The visibility and detectability of oil slicks and oil discharges on water. Arctic and Marine Oilspill Program Technical Seminar. Ottawa, Ontario, Canada: Environment Canada.

Fingas, M. F. 1999. The evaporation of oil spills: Development and implementation of new prediction methodology. *International Oil Spill Conference Proceedings* 1999(1):281–287.

Fingas, M. F., G. Halley, F. Ackerman, R. Nelson, M. Bissonnette, N. Laroche, Z. Wang, P. Lambert, K. Li, and P. Jokuty. 1995. The Newfoundland offshore burn experiment-NOBE. *International Oil Spill Conference Proceedings* 1995(1):123–132.

Fingas, M. F., P. Lambert, K. Li, Z. Wang, F. Ackerman, S. Whiticar, M. Goldthorp, S. Schutz, M. Morganti, R. Nadeau, P. Campagna, and R. Hiltabrand. 2001. Studies of emissions from oil fires. *International Oil Spill Conference Proceedings* 2001(1):539–544. https://doi.org/10.7901/2169-3358-2001-1-539.

Firoozy, N., T. Neusitzer, D. S. Desmond, T. Tiede, M. J. L. Lemes, J. Landy, P. Mojabi, S. Rysgaard, G. Stern, and D. G. Barber. 2017. An electromagnetic detection case study on crude oil injection in a young sea ice environment. *IEEE Transactions on Geoscience and Remote Sensing* 55(8):4465–4475. https://doi.org/10.1109/TGRS.2017.2692734.

Fischer, H. B., E. J. List, R. C. Y. Koh, J. Imberger, and N. H. Brooks 1979. *Mixing in inland and coastal waters.* New York: Academic Press.

Fisher, C. R. 1990. Chemoautotrophic and methanotrophic symbioses in marine invertebrates. *Reviews in Aquatic Sciences* 2:399–436.

Fitch, W., K. Kirby, J. Dragna, D. Kuchler, D. Haycraft, R. Godfrey, J. Langan, B. Fields, H. Karis, and M. Regan. 2013. BP and Anadarko's phase 2 pre-trial memorandum quantification segment. In *Document submitted in the US district court for the eastern district of Louisiana mdl no. 2179 section J. In re: Oil spill by the oil rig "Deepwater Horizon" in the Gulf of Mexico, on April 20, 2010.* P. 14.

Fitzgerald, T. P., and J. M. Gohlke. 2014. Contaminant levels in Gulf of Mexico reef fish after the *Deepwater Horizon* oil spill as measured by a fishermen-led testing program. *Environmental Science & Technology* 48(3):1993–2000.

Fleeger, J., K. Carman, M. Riggio, I. Mendelssohn, Q. Lin, A. Hou, D. Deis, and S. Zengel. 2015. Recovery of salt marsh benthic microalgae and meiofauna following the *Deepwater Horizon* oil spill linked to recovery of spartina alterniflora. *Marine Ecology Progress Series* 536:39–54.

Fleeger, J., D. Johnson, S. Zengel, I. Mendelssohn, D. Deis, S. Graham, Q. Lin, M. Christman, M. Riggio, and M. Pant. 2020. Macroinfauna responses and recovery trajectories after an oil spill differ from those following saltmarsh restoration. *Marine Environmental Research* 155:104881.

Fleming Environmental and Hyde Marine. 2003. *Final report—the Joint Viscous Oil Pumping System (JVOPS) workshop.* A united effort by the U.S. Coast Guard, Canadian Coast Guard, U.S. Navy, and Response Community to promote and improve heavy viscous oil marine lightering response.

Flint, R. W., and N. N. Rabalais. 1981. *Environmental studies of a marine ecosystem: South Texas outer continental shelf.* Austin, TX: University of Texas Press.

Flombaum, P., J. L. Gallegos, R. A. Gordillo, J. Rincón, L. L. Zabala, N. Jiao, D. M. Karl, W. K. Li, M. W. Lomas, and D. Veneziano. 2013. Present and future global distributions of the marine cyanobacteria prochlorococcus and synechococcus. *Proceedings of the National Academy of Sciences* 110(24):9824–9829.

Fodrie, F. J., and K. L. Heck, Jr. 2011. Response of coastal fishes to the Gulf of Mexico oil disaster. *PloS ONE* 6(7):e21609.

Fodrie, F. J., K. W. Able, F. Galvez, K. L. Heck, Jr., O. P. Jensen, P. C. López-Duarte, C. W. Martin, R. E. Turner, and A. Whitehead. 2014. Integrating organismal and population responses of estuarine fishes in Macondo spill research. *Bioscience* 64(9):778–788.

Foght, J. 2008. Anaerobic biodegradation of aromatic hydrocarbons: Pathways and prospects. *Microbial Physiology* 15(2–3):93–120.

Foght, J. 2018. Nitrogen fixation and hydrocarbon-oxidizing bacteria. Cellular Ecophysiology of Microbe: Hydrocarbon and lipid interactions. In *Handbook of hydrocarbon and lipid microbiology*, pp. 431–448. Dordrecht, Netherlands: Springer.

Foght, J., P. Fedorak, and D. Westlake. 1990. Mineralization of [14C] hexadecane and [14C] phenanthrene in crude oil: Specificity among bacterial isolates. *Canadian Journal of Microbiology* 36(3):169–175.

Fonseca, M., G. A. Piniak, and N. Cosentino-Manning. 2017. Susceptibility of seagrass to oil spills: A case study with eelgrass, Zostera Marina in San Francisco bay, USA. *Marine Pollution Bulletin* 115(1–2):29–38.

Ford, R. G., M. L. Bonnell, D. H. Varoujean, G. W. Page, H. R. Carter, B. E. Sharp, D. Heinemann, and J. L. Casey. 1996. *Total direct mortality of seabirds from the* Exxon Valdez *oil spill.* Bethesda, MD: Environmental Science.

Forth, H. P., C. L. Mitchelmore, J. M. Morris, C. R. Lay, and J. Lipton. 2017a. Characterization of dissolved and particulate phases of water accommodated fractions used to conduct aquatic toxicity testing in support of the *Deepwater Horizon* natural resource damage assessment. *Environmental Toxicology and Chemistry* 36(6):1460–1472.

Forth, H. P., C. L. Mitchelmore, J. M. Morris, and J. Lipton. 2017b. Characterization of oil and water accommodated fractions used to conduct aquatic toxicity testing in support of the *Deepwater Horizon* oil spill natural resource damage assessment. *Environmental Toxicology and Chemistry* 36(6):1450–1459.

Foster, M., M. Neushul, and R. Zingmark. 1971. The Santa Barbara oil spill part 2: Initial effects on intertidal and kelp bed organisms. *Environmental Pollution (1970)* 2(2):115–134.

Fraga, B., and T. Stoesser. 2016. Influence of bubble size, diffuser width, and flow rate on the integral behavior of bubble plumes. *Journal of Geophysical Research: Oceans* 121(6):3887–3904.

Fraga, B., T. Stoesser, C. C. Lai, and S. A. Socolofsky. 2016. A LES-based Eulerian–Lagrangian approach to predict the dynamics of bubble plumes. *Ocean Modelling* 97:27–36.

Franks, N., and W. Lieb. 1994. Molecular and cellular mechanisms of general anaesthesia. *Nature* 367(6464):607–614.

Fravel, V. A., W. Van Bonn, F. Gulland, C. Rios, A. Fahlman, J. L. Graham, and P. J. Havel. 2016. Intraperitoneal dextrose administration as an alternative emergency treatment for hypoglycemic yearling California sea lions (zalophus californianus). *Journal of Zoo and Wildlife Medicine* 47(1):76–82. http://www.jstor.org/stable/24773869.

Freeman, D. H., and C. P. Ward. 2022. Sunlight-driven dissolution is a major fate of oil at sea. *Science Advances* 8(7):eabl7605. https://doi.org/10.1126/sciadv.abl7605.

French-McCay, D. 2009. State-of-the-art and research needs for oil spill impact assessment modeling. *Proceedings of the 32nd Arctic and Marine Oilspill Program Technical Seminar on Environmental Contamination and Response.* Volume 2. Ottawa, Ontario, Canada: Emergencies Science Division, Environment Canada.

French-McCay, D. 2016. Potential effects thresholds for oil spill risk assessments. *Proceedings of the 39th Arctic and Marine Oilspill Program Technical Seminar on Environmental Contamination and Response.* Ottawa, Ontario, Canada: Emergencies Science Division, Environment Canada.

French-McCay, D., D. Reich, J. Michel, D. Etkin, L. Symons, D. Helton, and J. Wagner. 2012. Oil spill consequence analyses of potentially polluting shipwrecks. *Proceedings of the 35th Arctic & Marine Oilspill Program Technical Seminar*, pp. 751–774.

French-McCay, D., M. C. McManus, R. Balouskus, J. Rowe, M. Schroeder, A. Morandi, E. Bohaboy, and E. Graham. 2015. *Technical reports for the* Deepwater Horizon *water column injury assessment. Wc_tr.14: Modeling oil fate and exposure concentrations in the deepwater plume and risinig oil resulting from the* Deepwater Horizon *oil spill.* South Kingstown, RI.

French-McCay, D., D. Crowley, J. J. Rowe, M. Bock, H. Robinson, R. Wenning, A. H. Walker, J. Joeckel, T. J. Nedwed, and T. F. Parkerton. 2018b. Comparative risk assessment of spill response options for a deepwater oil well blowout: Part 1. Oil spill modeling. *Marine Pollution Bulletin* 133:1001–1015.

French-McCay, D. P. 2002. Development and application of an oil toxicity and exposure model, OilToxEx. *Environmental Toxicology and Chemistry* 21(10):2080–2094.

French-McCay, D. P. 2004. Oil spill impact modeling: Development and validation. *Environmental Toxicology and Chemistry* 23(10):2441–2456.

French-McCay, D. P., M. Horn, Z. Li, K. Jayko, M. L. Spaulding, D. Crowley, and D. Mendelsohn. 2018a. Modeling distribution, fate, and concentrations of *Deepwater Horizon* oil in subsurface waters of the Gulf of Mexico. In *Oil spill environmental forensics case studies*, pp. 683–735. Amsterdam, Netherlands: Elsevier.

French-McCay, D. P., K. Jayko, Z. Li, M. L. Spaulding, D. Crowley, D. Mendelsohn, M. Horn, T. Isaji, Y. H. Kim, and J. Fontenault. 2021a. Oil fate and mass balance for the *Deepwater Horizon* oil spill. *Marine Pollution Bulletin* 171:112681.

French-McCay, D. P., H. J. Robinson, M. L. Spaulding, Z. Li, M. Horn, M. D. Gloekler, Y. H. Kim, D. Crowley, and D. Mendelsohn. 2021b. Validation of oil fate and mass balance for the *Deepwater Horizon* oil spill: Evaluation of water column partitioning. *Marine Pollution Bulletin* 173:113064.

Fritt-Rasmussen, J., B. E. Ascanius, P. J. Brandvik, A. Villumsen, and E. H. Stenby. 2012. Composition of in situ burn residue as a function of weathering conditions. *Marine Pollution Bulletin* 67(1–2):75–81.

Frometa, J., M. E. DeLorenzo, E. C. Pisarski, and P. J. Etnoyer. 2017. Toxicity of oil and dispersant on the deep water gorgonian octocoral swiftia exserta, with implications for the effects of the *Deepwater Horizon* oil spill. *Marine Pollution Bulletin* 122(1–2):91–99.

Fry, D. M., and L. J. Lowenstine. 1985. Pathology of common murres and Cassin's auklets exposed to oil. *Archives of Environmental Contamination and Toxicology* 14(6):725–737. https://doi.org/10.1007/BF01055780.

Frysinger, G. S., and R. B. Gaines. 1999. Comprehensive two-dimensional gas chromatography with mass spectrometric detection (GC× GC/MS) applied to the analysis of petroleum. *Journal of High Resolution Chromatography* 22(5):251–255.

Frysinger, G. S., and R. B. Gaines. 2001. Separation and identification of petroleum biomarkers by comprehensive two-dimensional gas chromatography. *Journal of Separation Science* 24(2):87–96.

Frysinger, G. S., R. B. Gaines, and E. B. Ledford, Jr. 1999. Quantitative determination of BTEX and total aromatic compounds in gasoline by comprehensive two-dimensional gas chromatography (GC×GC). *Journal of High Resolution Chromatography* 22(4):195–200.

Frysinger, G. S., R. B. Gaines, L. Xu, and C. M. Reddy. 2003. Resolving the unresolved complex mixture in petroleum-contaminated sediments. *Environmental Science & Technology* 37(8):1653–1662.

Fu, J., Y. Gong, X. Zhao, S. E. O'Reilly, and D. Zhao. 2014. Effects of oil and dispersant on formation of marine oil snow and transport of oil hydrocarbons. *Environmental Science & Technology* 48(24):14392–14399. https://doi.org/10.1021/es5042157.

Fuchs, R., M. Glaude, T. Hansel, J. Osofsky, and H. Osofsky. 2021. Adolescent risk substance use behavior, posttraumatic stress, depression, and resilience: Innovative considerations for disaster recovery. *Substance Abuse* 42(3):358–365.

Fuller, C., and J. S. Bonner. 2001. Comparative toxicity of oil, dispersant, and dispersed oil to Texas marine species. *International Oil Spill Conference Proceedings* 2001(2):1243–1248.

Fullerton, C. S., R. J. Ursano, and L. Wang. 2004. Acute stress disorder, posttraumatic stress disorder, and depression in disaster or rescue workers. *American Journal of Psychiatry* 161(8):1370–1376.

Gaines, R. B., G. S. Frysinger, M. S. Hendrick-Smith, and J. D. Stuart. 1999. Oil spill source identification by comprehensive two-dimensional gas chromatography. *Environmental Science & Technology* 33(12): 2106–2112.

Gallaway, B. J., W. J. Konkel, and B. L. Norcross. 2017. Some thoughts on estimating change to Arctic cod populations from hypothetical oil spills in the eastern Alaska Beaufort Sea. *Arctic Science* 3(4):716–729.

Gallaway, B. J., W. J. Konkel, and J. G. Cole. 2019. The effects of modeled dispersed and undispersed hypothetical oil spills on red snapper, lutjanus campechanus, stocks in the Gulf of Mexico. In *Red snapper biology in a changing world*, pp. 123–139. Boca Raton, FL: CRC Press.

Gam, K. B., L. S. Engel, R. K. Kwok, M. D. Curry, P. A. Stewart, M. R. Stenzel, J. A. McGrath, W. B. Jackson II, M. Y. Lichtveld, and D. P. Sandler. 2018. Association between *Deepwater Horizon* oil spill response and cleanup work experiences and lung function. *Environment International* 121:695–702.

GAO (U.S. Government Accounting Office). 2006. *Coast Guard: Observations on the preparation, response, and recovery missions related to Hurricane Katrina.* Report to Congressional Committees. https://www.gao.gov/assets/gao-06-903.pdf.

GAO (U.S. Government Accountability Organization). 2020. *Coast Guard: Improved analysis of vessel response plan use could help mitigate marine pollution risk.* Report to Congressional Committees. https://www.gao.gov/assets/gao-20-554.pdf.

GAO. 2021. *Additional information is needed to better understand the environmental tradeoffs of using chemical dispersants.* https://www.gao.gov/assets/gao-22-104153.pdf.

Gao, S., L. Smik, M. Kulikovskiy, N. Shkurina, E. Gusev, N. Pedentchouk, T. Mock, and S. T. Belt. 2020. A novel tri-unsaturated highly branched isoprenoid (HBI) alkene from the marine diatom Navicula salinicola. *Organic Geochemistry* 146:104050.

Garcia, M., E. Campos, A. Marsal, and I. Ribosa. 2009. Biodegradability and toxicity of sulphonate-based surfactants in aerobic and anaerobic aquatic environments. *Water Research* 43(2):295–302.

Garcia-Pineda, O., B. Zimmer, M. Howard, W. Pichel, X. Li, and I. R. MacDonald. 2009. Using SAR images to delineate ocean oil slicks with a texture-classifying neural network algorithm (TCNNA). *Canadian Journal of Remote Sensing* 35(5):411–421.

Garcia-Pineda, O., I. MacDonald, B. Zimmer, B. Shedd, and H. Roberts. 2010. Remote-sensing evaluation of geophysical anomaly sites in the outer continental slope, northern Gulf of Mexico. *Deep Sea Research Part II: Topical Studies in Oceanography* 57(21–23):1859–1869.

Garcia-Pineda, O., I. MacDonald, and W. Shedd. 2014. Analysis of Oil-Volume Fluxes of Hydrocarbon-Seep Formations on the Green Canyon and Mississippi Canyon: A Study with 3d-Seismic Attributes in Combination with Satellite and Acoustic Data. *SPE Reservoir Evaluation & Engineering* 17(4):430–435.

Garcia-Pineda, O., G. Staples, C. E. Jones, C. Hu, B. Holt, V. Kourafalou, G. Graettinger, L. DiPinto, E. Ramirez, and D. Streett. 2020. Classification of oil spill by thicknesses using multiple remote sensors. *Remote Sensing of Environment* 236:111421.

Garneau, M.-È., C. Michel, G. Meisterhans, N. Fortin, T. L. King, C. W. Greer, and K. Lee. 2016. Hydrocarbon biodegradation by Arctic sea-ice and sub-ice microbial communities during microcosm experiments, Northwest Passage (Nunavut, Canada). *FEMS Microbiology Ecology* 92(10):fiw130.

Garrett, R. M., I. J. Pickering, C. E. Haith, and R. C. Prince. 1998. Photooxidation of crude oils. *Environmental Science & Technology* 32(23):3719–3723.

Garrett, R. M., C. C. Guénette, C. E. Haith, and R. C. Prince. 2000. Pyrogenic polycyclic aromatic hydrocarbons in oil burn residues. *Environmental Science & Technology* 34(10):1934–1937.

Gaston, C. J., P. K. Quinn, T. S. Bates, J. B. Gilman, D. M. Bon, W. C. Kuster, and K. A. Prather. 2013. The impact of shipping, agricultural, and urban emissions on single particle chemistry observed aboard the r/v Atlantis during calnex. *Journal of Geophysical Research: Atmospheres* 118(10):5003–5017.

Gaston, S., N. Nugent, E. S. Peters, T. F. Ferguson, E. J. Trapido, W. T. Robinson, and A. L. Rung. 2016. Exploring heterogeneity and correlates of depressive symptoms in the women and their children's health (WATCH) study. *Journal of Affective Disorders* 205:190–199.

Gauthier, M. J., B. Lafay, R. Christen, L. Fernandez, M. Acquaviva, P. Bonin, and J.-C. Bertrand. 1992. Marinobacter hydrocarbonoclasticus gen. Nov., sp. Nov., a new, extremely halotolerant, hydrocarbon-degrading marine bacterium. *International Journal of Systematic and Evolutionary Microbiology* 42(4):568–576.

Gautier, D. L., K. J. Bird, R. R. Charpentier, A. Grantz, D. W. Houseknecht, T. R. Klett, T. E. Moore, J. K. Pitman, C. J. Schenk, J. H. Schuenemeyer, K. Sorensen, M. E. Tennyson, Z. C. Valin, and C. J. Wandrey. 2009. Assessment of undiscovered oil and gas in the Arctic. *Science* 324(5931):1175–1179. https://doi.org/10.1126/science.1169467.

Geiger, D., C. Northcott, D. Call, and L. Brooke. 1985. *Acute toxicities of organic chemicals to fathead minnows (Pimephales promelas).* Superior, WI: Centre for Lake Superior Environmental Studies, University of Wisconsin.

Geng, X., and M. C. Boufadel. 2015. Numerical study of solute transport in shallow beach aquifers subjected to waves and tides. *Journal of Geophysical Research: Oceans* 120(2):1409–1428.

Geng, X., C. A. Khalil, R. C. Prince, K. Lee, C. An, and M. C. Boufadel. 2021. Hypersaline pore water in Gulf of Mexico beaches prevented efficient biodegradation of *Deepwater Horizon* beached oil. *Environmental Science & Technology* 55(20):13792–13801.

Genuino, H. C., D. T. Horvath, C. K. King'ondu, G. E. Hoag, J. B. Collins, and S. L. Suib. 2012. Effects of visible and UV light on the characteristics and properties of crude oil-in-water (o/w) emulsions. *Photochemical & Photobiological Sciences* 11(4):692–702.

Geraci, J., D. St. Aubin, and R. Reisman. 1983. Bottlenose dolphins, Tursiops truncatus, can detect oil. *Canadian Journal of Fisheries and Aquatic Sciences* 40(9):1516–1522.

Geraci, J. R. 1990. Physiologic and toxic effects on cetaceans. In *Sea mammals and oil: Confronting the risks*, pp. 167–197. Cambridge, MA: Academic Press.

Geraci, J. R., and D. St Aubin. 1982. *Study of the effects of oil on cetaceans.* Prepared for the Bureau of Land Management, U.S. Department of the Interior, Washington, DC.

GESAMP. 2007. Estimates of oil entering the marine environment from sea-based activities. *GESAMP Reports and Studies.* London, UK: International Maritime Organization.

Gesser, H. D., T. A. Wildman, and Y. B. Tewari. 1977. Photooxidation of n–hexadecane sensitized by xanthone. *Environmental Science & Technology* 11(6):605–608.

Getter, C. D., G. I. Scott, and J. Michel. 1981. The effects of oil spills on mangrove forests: A comparison of five oil spill sites in the Gulf of Mexico and the Caribbean Sea. *International Oil Spill Conference Proceedings* 1981(1):535–540.

Getter, C. D., T. G. Ballou, and C. B. Koons. 1985. Effects of dispersed oil on mangroves synthesis of a seven-year study. *Marine Pollution Bulletin* 16(8):318–324.

Ghosal, D., S. Ghosh, T. K. Dutta, and Y. Ahn. 2016. Corrigendum: Current state of knowledge in microbial degradation of polycyclic aromatic hydrocarbons (PAHs): A review. *Frontiers in Microbiology* 7:1837.

Gibeaut, J. 2016. Enabling data sharing through the Gulf of Mexico Research Initiative Information and Data Cooperative (GRIIDC). *Oceanography* 29(3):33–37.

Giddings, J. L. 1962. Onion portage and other flint sites of the Kobuk River. *Arctic Anthropology* 1(1):6–27. http://www.jstor.org/stable/40315537.

Gieg, L. M., T. R. Jack, and J. M. Foght. 2011. Biological souring and mitigation in oil reservoirs. *Applied Microbiology and Biotechnology* 92(2):263–282.

Gieg, L. M., S. J. Fowler, and C. Berdugo-Clavijo. 2014. Syntrophic biodegradation of hydrocarbon contaminants. *Current Opinion in Biotechnology* 27:21–29.

Gilbert, T. 2001. Report of the strategic environmental assessment, USS Mississinewa oil spill, Ulithi Lagoon, Yap state. Micronesia. Pub. September 18.

Gilbert, T., S. Nawadra, A. Tafileichig, and L. Yinug. 2003. Response to an oil spill from a sunken WWII oil tanker in Yap state, Micronesia. *International Oil Spill Conference Proceedings* 2003(1):175–182.

Gilde, K., and J. L. Pinckney. 2012. Sublethal effects of crude oil on the community structure of estuarine phytoplankton. *Estuaries and Coasts* 35(3):853–861.

Gill, D. A., and J. S. Picou. 1997. The day the water died: Cultural impacts of the *Exxon Valdez* oil spill. In *The* Exxon Valdez *disaster: Readings on a modern social problem*, J. S. Picou, D. A. Gill and M. Cohen (eds.), pp. 167–187. Dubuque, IA: Kendall/Hunt Publishing Company.

Gill, D. A., and J. S. Picou. 1998. Technological disaster and chronic community stress. *Society & Natural Resources* 11(8):795–815.

Gillham, D. M. 1978. *Peaty topsoil, organic clay and pale estuarine clay cliff. Peat fragment on beach. Rumney sea wall. Cley-flecked ferrous blue and ferric orange.* https://www.flickr.com/photos/marygillhamarchiveproject/34419847801/in/photostream.

Giovannoni, S. J. 2017. SAR11 bacteria: The most abundant plankton in the oceans. *Annual Review of Marine Science* 9:231–255.

Girard, F., K. Shea, and C. R. Fisher. 2018. Projecting the recovery of a long-lived deep-sea coral species after the *Deepwater Horizon* oil spill using state-structured models. *Journal of Applied Ecology* 55(4):1812–1822.

Girard, F., R. Cruz, O. Glickman, T. Harpster, C. R. Fisher, and L. Thomsen. 2019. In situ growth of deep-sea octocorals after the *Deepwater Horizon* oil spill. *Elementa: Science of the Anthropocene* 7(12).

Girin, M. 2004. European experience in response to potentially polluting shipwrecks. *Marine Technology Society Journal* 38(3):21–25.

Gittel, A., J. Donhauser, H. Røy, P. R. Girguis, B. B. Jørgensen, and K. U. Kjeldsen. 2015. Ubiquitous presence and novel diversity of anaerobic alkane degraders in cold marine sediments. *Frontiers in Microbiology* 6:1414.

Gocht, T., O. Klemm, and P. Grathwohl. 2007. Long-term atmospheric bulk deposition of polycyclic aromatic hydrocarbons (PAHs) in rural areas of southern Germany. *Atmospheric Environment* 41(6):1315–1327. https://doi.org/10.1016/j.atmosenv.2006.09.036.

Godfrin, M. P., M. Sihlabela, A. Bose, and A. Tripathi. 2018. Behavior of marine bacteria in clean environment and oil spill conditions. *Langmuir* 34(30):9047–9053.

Gofstein, T. R., M. Perkins, J. Field, and M. B. Leigh. 2020. The interactive effects of crude oil and COREXIT 9500 on their biodegradation in Arctic seawater. *Applied and Environmental Microbiology* 86(21):e01194–e01120.

Gohlke, J. M., D. Doke, M. Tipre, M. Leader, and T. Fitzgerald. 2011. A review of seafood safety after the *Deepwater Horizon* blowout. *Environmental Health Perspectives* 119(8):1062–1069.

Goldberg, E. D. 1975. The mussel watch—a first step in global marine monitoring. *Marine Pollution Bulletin* 6:111.

Goldmann, E., and S. Galea. 2014. Mental health consequences of disasters. *Annual Review of Public Health* 35:169–183.

Goldstein, B. D. 2020. Broadening the mandate of the incident command system to address community mental and behavioral health effects as part of the federal response to disasters. *Current Environmental Health Reports* 1–10.

Goldstein, B. D., H. J. Osofsky, and M. Y. Lichtveld. 2011. The Gulf oil spill. *New England Journal of Medicine* 364(14):1334–1348.

Golightly, R. T., S. H. Newman, E. N. Craig, H. R. Carter, and J. A. Mazet. 2002. Survival and behavior of western gulls following exposure to oil and rehabilitation. *Wildlife Society Bulletin* 539–546.

Golyshin, P. N., T. N. Chernikova, W.-R. Abraham, H. Lünsdorf, K. N. Timmis, and M. M. Yakimov. 2002. Oleiphilaceae fam. Nov., to include oleiphilus messinensis gen. Nov., sp. Nov., a novel marine bacterium that obligately utilizes hydrocarbons. *International Journal of Systematic and Evolutionary Microbiology* 52(3):901–911.

GoMRI (Gulf of Mexico Research Initiative). *GoMRI database.* https://data.gulfresearchinitiative.org.

González, J., F. Figueiras, M. Aranguren-Gassis, B. Crespo, E. Fernández, X. A. G. Morán, and M. Nieto-Cid. 2009. Effect of a simulated oil spill on natural assemblages of marine phytoplankton enclosed in microcosms. *Estuarine, Coastal and Shelf Science* 83(3):265–276.

González-Gaya, B., M.-C. Fernández-Pinos, L. Morales, L. Méjanelle, E. Abad, B. Piña, C. M. Duarte, B. Jiménez, and J. Dachs. 2016. High atmosphere-ocean exchange of semivolatile aromatic hydrocarbons. *Nature Geoscience* 9(6):438–442. https://doi.org/10.1038/ngeo2714.

González-Penagos, C. E., J. A. Zamora-Briseño, D. Cerqueda-García, M. Améndola-Pimenta, J. A. Pérez-Vega, E. Hernández-Nuñez, and R. Rodríguez-Canul. 2020. Alterations in the gut microbiota of zebrafish (danio rerio) in response to water-soluble crude oil components and its mixture with a chemical dispersant. *Frontiers in Public Health* 8:705.

Goodbody-Gringley, G., D. L. Wetzel, D. Gillon, E. Pulster, A. Miller, and K. B. Ritchie. 2013. Toxicity of *Deepwater Horizon* source oil and the chemical dispersant, COREXIT® 9500, to coral larvae. *PloS ONE* 8(1):e45574.

Gopalan, B., and J. Katz. 2009. Break up of viscous crude oil droplets mixed with dispersants in locally isotropic turbulence. *Fluids Engineering Division Summer Meeting*. 2.

Gopalan, B., and J. Katz. 2010. Turbulent shearing of crude oil mixed with dispersants generates long microthreads and microdroplets. *Physical Review Letters* 104(5):054501.

Gordon, Jr., D. C., J. Dale, and P. D. Keizer. 1978. Importance of sediment working by the deposit-feeding polychaete arenicola marina on the weathering rate of sediment-bound oil. *Journal of the Fisheries Research Board of Canada* 35(5):591–603. https://doi.org/10.1139/f78-105.

Górecki, T. 2021. GC×GC: The road not taken? *Analytical Scientist*. https://theanalyticalscientist.com/techniques-tools/gcxgc-the-road-not-taken-tadeusz-gorecki.

Gorman Ng, M., J. W. Cherrie, A. Sleeuwenhoek, M. Stenzel, R. K. Kwok, L. S. Engel, J. M. Cavallari, A. Blair, D. P. Sandler, and P. Stewart. 2019. GuLF dream: A model to estimate dermal exposure among oil spill response and clean-up workers. *Annals of Work Exposures and Health*. https://doi.org/10.1093/annweh/wxz037.

Gould, D. W., J. L. Teich, M. R. Pemberton, C. Pierannunzi, and S. Larson. 2015. Behavioral health in the Gulf Coast region following the *Deepwater Horizon* oil spill: Findings from two federal surveys. *The Journal of Behavioral Health Services & Research* 42(1):6–22.

Grace, J., T. Wairegi, and J. Brophy. 1978. Break-up of drops and bubbles in stagnant media. *The Canadian Journal of Chemical Engineering* 56(1):3–8.

Grattan, L. M., S. Roberts, W. T. Mahan, Jr., P. K. McLaughlin, W. S. Otwell, and J. G. Morris, Jr. 2011. The early psychological impacts of the *Deepwater Horizon* oil spill on Florida and Alabama communities. *Environmental Health Perspectives* 119(6):838–843.

Gray, D. L., B. L. Hutchinson, D. S. Etkin, K. Michel, and M. Grabowski. 2005. *Study of tug escorts in Puget Sound*. Prepared by The Glosten Associates, Herbert Engineering, Environmental Research Consulting, and M. Grabowski for the Washington Department of Ecology, Olympia, Washington.

Gray, M. R. 2021. Whatsoever things are true: Hypothesis, artefact, and bias in chemical engineering research. *The Canadian Journal of Chemical Engineering* 99(10):2055–2068.

Green, H. S., S. A. Fuller, A. W. Meyer, P. S. Joyce, C. Aeppli, R. K. Nelson, R. F. Swarthout, D. L. Valentine, H. K. White, and C. M. Reddy. 2018. Pelagic tar balls collected in the North Atlantic Ocean and Caribbean Sea from 1988 to 2016 have natural and anthropogenic origins. *Marine Pollution Bulletin* 137:352–359.

Greene Economics. 2021. *Energy outlook review supporting Oil in the Sea IV.* Prepared by Greene Economics LLC, Ridgefield, Washington, for the National Academies of Sciences, Engineering, and Medicine.

Gregson, B. H., G. Metodieva, M. V. Metodiev, P. N. Golyshin, and B. A. McKew. 2018. Differential protein expression during growth on medium versus long-chain alkanes in the obligate marine hydrocarbon-degrading bacterium thalassolituus oleivorans MIL-1. *Frontiers in Microbiology* 9:3130.

Gregson, B. H., G. Metodieva, M. V. Metodiev, and B. A. McKew. 2019. Differential protein expression during growth on linear versus branched alkanes in the obligate marine hydrocarbon-degrading bacterium alcanivorax borkumensis SK2[t]. *Environmental Microbiology* 21(7):2347–2359.

Gregson, B. H., G. Metodieva, M. V. Metodiev, P. N. Golyshin, and B. A. McKew. 2020. Protein expression in the obligate hydrocarbon-degrading psychrophile oleispira Antarctica RB-8 during alkane degradation and cold tolerance. *Environmental Microbiology* 22(5):1870–1883.

Gregson, B. H., B. A. McKew, R. D. Holland, T. J. Nedwed, R. C. Prince, and T. J. McGenity. 2021. Marine oil snow, a microbial perspective. *Frontiers in Marine Science* 8:11.

Gros, J., C. M. Reddy, C. Aeppli, R. K. Nelson, C. A. Carmichael, and J. S. Arey. 2014. Resolving biodegradation patterns of persistent saturated hydrocarbons in weathered oil samples from the *Deepwater Horizon* disaster. *Environmental Science & Technology* 48(3):1628–1637.

Gros, J., C. M. Reddy, R. K. Nelson, S. A. Socolofsky, and J. S. Arey. 2016. Simulating gas-liquid-water partitioning and fluid properties of petroleum under pressure: Implications for deep-sea blowouts. *Environmental Science & Technology* 50(14):7397–7408.

Gros, J., S. A. Socolofsky, A. L. Dissanayake, I. Jun, L. Zhao, M. C. Boufadel, C. M. Reddy, and J. S. Arey. 2017. Petroleum dynamics in the sea and influence of subsea dispersant injection during *Deepwater Horizon*. *Proceedings of the National Academy of Sciences* 114(38):10065–10070.

Gros, J., A. L. Dissanayake, M. M. Daniels, C. H. Barker, W. Lehr, and S. A. Socolofsky. 2018. Oil spill modeling in deep waters: Estimation of pseudo-component properties for cubic equations of state from distillation data. *Marine Pollution Bulletin* 137:627–637.

Gros, J., J. S. Arey, S. A. Socolofsky, and A. L. Dissanayake. 2020. Dynamics of live oil droplets and natural gas bubbles in deep water. *Environmental Science & Technology* 54(19):11865–11875.

Grosell, M., R. J. Griffitt, T. A. Sherwood, and D. L. Wetzel. 2020. Digging deeper than LC/EC50: Nontraditional endpoints and non-model species in oil spill toxicology. In *Deep oil spills*, pp. 497–514. Dordrecht, Netherlands: Springer.

Grossart, H.-P., S. Van den Wyngaert, M. Kagami, C. Wurzbacher, M. Cunliffe, and K. Rojas-Jimenez. 2019. Fungi in aquatic ecosystems. *Nature Reviews Microbiology* 17(6):339–354.

Grossart, H. P., R. Massana, K. D. McMahon, and D. A. Walsh. 2020. Linking metagenomics to aquatic microbial ecology and biogeochemical cycles. *Limnology and Oceanography* 65:S2–S20.

Gründger, F., V. Carrier, M. M. Svenning, G. Panieri, T. R. Vonnahme, S. Klasek, and H. Niemann. 2019. Methane-fuelled biofilms predominantly composed of methanotrophic ANME-1 in Arctic gas hydrate-related sediments. *Scientific Reports* 9(1):1–10.

Gu, Y., J. C. Goez, M. Guajardo, and S. W. Wallace. 2021. Autonomous vessels: State of the art and potential opportunities in logistics. *International Transactions in Operational Research* 28(4):1706–1739.

Guena, A. 2012. *Custom-made spill response barriers. Operational guide.* Cedre. http://wwz.cedre.fr/cedre_en/content/download/1767/139998/file/extract-custom-made-barriers.pdf.

Guenette, C., P. Sveum, I. Buist, T. Aunaas, and L. Godal. 1994. *In-situ burning of water-in-oil emulsions*. Norway.

Guengerich, F. P. 2008. Cytochrome P450 and chemical toxicology. *Chemical Research in Toxicology* 21(1):70–83.

Gulec, I., and D. A. Holdway. 1999. The toxicity of laboratory burned oil to the amphipod *Allorchestes compressa* and the snail *Polinices conicus*. *Spill Science & Technology Bulletin* 5(2):135–139. https://doi.org/10.1016/S1353-2561(98)00025-5.

Guo, S., C. A. Stevens, T. D. Vance, L. L. Olijve, L. A. Graham, R. L. Campbell, S. R. Yazdi, C. Escobedo, M. Bar-Dolev, and V. Yashunsky. 2017. Structure of a 1.5-Mda adhesin that binds its Antarctic bacterium to diatoms and ice. *Science Advances* 3(8):e1701440.

Gustafsson, O., C. M. Long, J. Macfarlane, and P. M. Gschwend. 2001. Fate of linear alkylbenzenes released to the coastal environment near Boston Harbor. *Environmental Science and Technology* 35(10):2040–2048. https://doi.org/10.1021/es000188m.

Gustafsson, Ö., M. Kruså, Z. Zencak, R. J. Sheesley, L. Granat, E. Engström, P. S. Praveen, P. S. P. Rao, C. Leck, and H. Rodhe. 2009. Brown clouds over south Asia: Biomass or fossil fuel combustion? *Science* 323(5913):495–498. https://doi.org/10.1126/science.1164857.

Gustavson, K., S. V. Hansson, F. M. van Beest, J. Fritt-Rasmussen, P. Lassen, O. Geertz-Hansen, and S. Wegeberg. 2020. Natural removal of crude and heavy fuel oil on rocky shorelines in Arctic climate regimes. *Water, Air, & Soil Pollution* 231(9):1–15.

Gustitus, S. A., and T. P. Clement. 2017. Formation, fate, and impacts of microscopic and macroscopic oil-sediment residues in nearshore marine environments: A critical review. *Reviews of Geophysics* 55(4):1130–1157.

Gutierrez, T., D. Berry, T. Yang, S. Mishamandani, L. McKay, A. Teske, and M. D. Aitken. 2013. Role of bacterial exopolysaccharides (EPS) in the fate of the oil released during the *Deepwater Horizon* oil spill. *PloS ONE* 8(6):e67717.

Gutierrez, T., G. Morris, D. Ellis, B. Bowler, M. Jones, K. Salek, B. Mulloy, and A. Teske. 2018. Hydrocarbon-degradation and MOS-formation capabilities of the dominant bacteria enriched in sea surface oil slicks during the *Deepwater Horizon* oil spill. *Marine Pollution Bulletin* 135:205–215.

Gutteridge, J. M. C., and B. Halliwell. 2010. Antioxidants: Molecules, medicines, and myths. *Biochemical and Biophysical Research Communications* 393(4):561–564. https://doi.org/10.1016/j.bbrc.2010.02.071.

Guzman, H. M., S. Kaiser, and E. Weil. 2020. Assessing the long-term effects of a catastrophic oil spill on subtidal coral reef communities off the Caribbean coast of Panama (1985–2017). *Marine Biodiversity* 50(3):28. https://doi.org/10.1007/s12526-020-01057-9.

Gwack, J., J. H. Lee, Y. A. Kang, K.-J. Chang, M. S. Lee, and J. Y. Hong. 2012. Acute health effects among military personnel participating in the cleanup of the Hebei Spirit oil spill, 2007, in Taean County, Korea. *Osong Public Health and Research Perspectives* 3(4):206–212.

GWPC (Ground Water Protection Council). 2019. *Produced water report—regulations, current practices, and research needs.* Prepared by the Ground Water Protection Council.

Ha, M., H. Kwon, H.-K. Cheong, S. Lim, S. J. Yoo, E.-J. Kim, S. G. Park, J. Lee, and B. C. Chung. 2012. Urinary metabolites before and after cleanup and subjective symptoms in volunteer participants in cleanup of the *Hebei Spirit* oil spill. *Science of the Total Environment* 429:167–173. https://doi.org/10.1016/j.scitotenv.2012.04.036.

Ha, M., W.-C. Jeong, M. Lim, H. Kwon, Y. Choi, S.-J. Yoo, S. R. Noh, and H.-K. Cheong. 2013. Children's mental health in the area affected by the Hebei Spirit oil spill accident. *Environmental Health and Toxicology* 28.

Hale, R. C., M. E. Seeley, M. J. La Guardia, L. Mai, and E. Y. Zeng. 2020. A global perspective on microplastics. *Journal of Geophysical Research: Oceans* 125(1):e2018JC014719.

Hall, R. J., A. A. Belisle, and L. Sileo. 1983. Residues of petroleum hydrocarbons in tissues of sea turtles exposed to the Ixtoc I oil spill. *Journal of Wildlife Diseases* 19(2):106–109. https://doi.org/10.7589/0090-3558-19.2.106.

Haller, G. 2015. Lagrangian coherent Structures. *Annual Review of Fluid Mechanics* 47(1):137–162. https://doi.org/10.1146/annurev-fluid-010313-141322.

Hampton, S., R. G. Ford, H. R. Carter, C. Abraham, and D. Humple. 2003. Chronic oiling and seabird mortality from the sunken vessel SS Jacob Luckenbach in central California. *Marine Ornithology* 31:35–41.

Haney, J. C., H. J. Geiger, and J. W. Short. 2014a. Bird mortality from the *Deepwater Horizon* oil spill. I. Exposure probability in the offshore Gulf of Mexico. *Marine Ecology Progress Series* 513:225–237. https://www.int-res.com/abstracts/meps/v513/p225-237.

Haney, J. C., H. J. Geiger, and J. W. Short. 2014b. Bird mortality from the *Deepwater Horizon* oil spill. II. Carcass sampling and exposure probability in the coastal Gulf of Mexico. *Marine Ecology Progress Series* 513:239–252. https://www.int-res.com/abstracts/meps/v513/p239-252.

Hansel, T. C., H. J. Osofsky, J. Langhinrichsen-Rohling, A. Speier, T. Rehner, J. D. Osofsky, and G. Rohrer. 2015. Gulf Coast resilience coalition: An evolved collaborative built on shared disaster experiences, response, and future preparedness. *Disaster Medicine and Public Health Preparedness* 9(6):657–665.

Hansel, T. C., H. Osofsky, E. Baumgartner, S. Bradberry, L. Brown, K. Kirkland, J. Langhinrichsen-Rohling, J. Osofsky, A. H. Speier, and B. D. Goldstein. 2017. Social and environmental justice as a lens to approach the distribution of $105 million of directed funding in response to the *Deepwater Horizon* oil disaster. *Environmental Justice* 10(4):119–127.

Hansen, B. H., L. Sørensen, P. A. Carvalho, S. Meier, A. M. Booth, D. Altin, J. Farkas, and T. Nordtug. 2018. Adhesion of mechanically and chemically dispersed crude oil droplets to eggs of Atlantic cod (gadus morhua) and haddock (melanogrammus aeglefinus). *Science of the Total Environment* 640:138–143.

Hansen, K. 2000. *Boom vane field tests.* Report prepared for the Marine Safety and Environmental Protection Systems, United States Coast Guard, Washington, DC.

Haritash, A., and C. Kaushik. 2009. Biodegradation aspects of polycyclic aromatic hydrocarbons (PAHs): A review. *Journal of Hazardous Materials* 169(1–3):1–15.

Harman, C., K. Langford, R. C. Sundt, and S. Brooks. 2014. Measurement of naphthenic acids in the receiving waters around an offshore oil platform by passive sampling. *Environmental Toxicology and Chemistry* 33(9):1946–1949.

Harr, K., M. Rishniw, T. Rupp, D. Cacela, K. Dean, B. Dorr, K. Hanson-Dorr, K. Healy, K. Horak, and J. Link. 2017a. Dermal exposure to weathered MC252 crude oil results in echocardiographically identifiable systolic myocardial dysfunction in double-crested cormorants (Phalacrocorax auritus). *Ecotoxicology and Environmental Safety* 146:76–82.

Harriman, B. H., P. Zito, D. C. Podgorski, M. A. Tarr, and J. M. Suflita. 2017. Impact of photooxidation and biodegradation on the fate of oil spilled during the *Deepwater Horizon* incident: Advanced stages of weathering. *Environmental Science & Technology* 51(13):7412–7421.

Harris, C. 1997. The Sea Empress incident: Overview and response at sea. *International Oil Spill Conference Proceedings* 1997(1):177–184.

Harrison, R., J. M. D. Saborit, F. Dor, R. Henderson, D. Penney, V. Benignus, S. Kephalopoulos, D. Kotzias, M. Kleinman, and A. Verrier. 2010. *WHO guidelines for indoor air quality: Selected pollutants.* Geneva, Switzerland: World Health Organization.

Harrison, R. J., K. Retzer, M. J. Kosnett, M. Hodgson, T. Jordan, S. Ridl, and M. Kiefer. 2016. Sudden deaths among oil and gas extraction workers resulting from oxygen deficiency and inhalation of hydrocarbon gases and vapors—United States, January 2010–March 2015. *Morbidity and Mortality Weekly Report* 65(1):6–9.

Hart, M. E., A. Perez-Umphrey, P. C. Stouffer, C. B. Burns, A. Bonisoli-Alquati, S. S. Taylor, and S. Woltmann. 2021. Nest survival of Seaside Sparrows (ammospiza maritima) in the wake of the *Deepwater Horizon* oil spill. *PloS ONE* 16(10):e0259022.

Hartmann, A. C., S. A. Sandin, V. F. Chamberland, K. L. Marhaver, J. M. de Goeij, and M. J. Vermeij. 2015. Crude oil contamination interrupts settlement of coral larvae after direct exposure ends. *Marine Ecology Progress Series* 536:163–173.

Hartung, R. 1963. *Ingestion of oil by waterfowl.* Ph.D. thesis. Ann Arbor, MI: University of Michigan.

Hartung, R. 1967. Energy metabolism in oil-covered ducks. *The Journal of Wildlife Management* 31(4):798–804. https://doi.org/10.2307/3797987.

Hartung, R., and G. S. Hunt. 1966. Toxicity of some oils to waterfowl. *Journal of Wildlife Management* 30:564–570. https://doi.org/10.2307/3798748.

Harvey, R. G. 1991. *Polycyclic aromatic hydrocarbons: Chemistry and carcinogenicity.* CUP Archive.

Harville, E., X. Xiong, and P. Buekens. 2010. Disasters and perinatal health: A systematic review. *Obstetrical & Gynecological Survey* 65(11):713.

Harville, E. W., A. Shankar, C. Dunkel Schetter, and M. Lichtveld. 2018. Cumulative effects of the Gulf oil spill and other disasters on mental health among reproductive-aged women: The Gulf resilience on women's health study. *Psychological Trauma: Theory, Research, Practice, and Policy* 10(5):533.

Hassanpouryouzband, A., E. Joonaki, M. V. Farahani, S. Takeya, C. Ruppel, J. Yang, N. J. English, J. M. Schicks, K. Edlmann, and H. Mehrabian. 2020. Gas hydrates in sustainable chemistry. *Chemical Society Reviews* 49(15):5225–5309.

Hassanshahian, M., G. Emtiazi, G. Caruso, and S. Cappello. 2014. Bioremediation (bioaugmentation/biostimulation) trials of oil polluted seawater: A mesocosm simulation study. *Marine Environmental Research* 95:28–38. https://doi.org/10.1016/j.marenvres.2013.12.010.

Hastings, D. W., T. Bartlett, G. R. Brooks, R. A. Larson, K. A. Quinn, D. Razionale, P. T. Schwing, L. H. P. Bernal, A. C. Ruiz-Fernández, and J.-A. Sánchez-Cabeza. 2020. Changes in redox conditions of surface sediments following the *Deepwater Horizon* and Ixtoc 1 events. In *Deep oil spills*, pp. 269–284. Dordrecht, Netherlands: Springer.

Hawkins, S. J., A. J. Evans, J. Moore, M. Whittington, K. Pack, L. B. Firth, L. C. Adams, P. J. Moore, P. Masterson-Algar, and N. Mieszkowska. 2017. From the Torrey Canyon to today: A 50-year retrospective of recovery from the oil spill and interaction with climate-driven fluctuations on Cornish rocky shores. *International Oil Spill Conference Proceedings* 2017(1):74–103.

Hayworth, J., T. Clement, and J. Valentine. 2011. *Deepwater Horizon* oil spill impacts on Alabama beaches. *Hydrology and Earth System Sciences* 15(12):3639–3649.

Hayworth, J. S., T. P. Clement, G. F. John, and F. Yin. 2015. Fate of *Deepwater Horizon* oil in Alabama's beach system: Understanding physical evolution processes based on observational data. *Marine Pollution Bulletin* 90(1–2):95–105.

Hazelwood, R. L. 1986. Carbohydrate metabolism. In *Avian physiology*, P. D. Sturkie (ed.), pp. 303–325. New York: Springer New York.

Hazen, T. C., E. A. Dubinsky, T. Z. DeSantis, G. L. Andersen, Y. M. Piceno, N. Singh, J. K. Jansson, A. Probst, S. E. Borglin, J. L. Fortney, W. T. Stringfellow, M. Bill, M. S. Conrad, L. M. Tom, K. L. Chavarria, T. R. Alusi, R. Lamendella, D. C. Joyner, C. Spier, J. Baelum, M. Auer, M. L. Zemla, R. Chakraboty, E. L. Sonnenthal, P. D'haeseller, H.-Y. N. Holman, S. Osman, L. Zhenmei, J. D. Van Nostrand, Y. Deng, J. Zhou, and O. U. Mason. 2010. Deep-sea oil plume enriches indigenous oil-degrading bacteria. *Science* 330(6001):204–208.

Hazen, T. C., R. C. Prince, and N. Mahmoudi. 2016. *Marine oil biodegradation*. Washington, DC: ACS Publications.

Head, I. M., D. M. Jones, and W. F. Röling. 2006. Marine microorganisms make a meal of oil. *Nature Reviews Microbiology* 4(3):173–182.

Helm, R. C., H. R. Carter, R. G. Ford, D. M. Fry, R. L. Moreno, C. Sanpera, and F. S. Tseng. 2014. Overview of efforts to document and reduce impacts of oil spills on seabirds. In *Handbook of oil spill science and technology*, pp. 431–453. New York: John Wiley & Sons.

Helton, D. 2021. Historic dispersant use. Poster at the International Oil Spill Conference.

Hemmer, M. J., M. G. Barron, and R. M. Greene. 2011. Comparative toxicity of eight oil dispersants, Louisiana sweet crude oil (LSC), and chemically dispersed LSC to two aquatic test species. *Environmental Toxicology and Chemistry* 30(10):2244–2252.

Henderson, S., S. Grigson, P. Johnson, and B. Roddie. 1999. Potential impact of production chemicals on the toxicity of produced water discharges from North Sea oil platforms. *Marine Pollution Bulletin* 38(12):1141–1151.

Henkel, L. A., and M. H. Ziccardi. 2018. Life and death: How should we respond to oiled wildlife? *Journal of Fish and Wildlife Management* 9(1):296–301.

Henkel, L. A., H. Nevins, M. Martin, S. Sugarman, J. T. Harvey, and M. H. Ziccardi. 2014. Chronic oiling of marine birds in California by natural petroleum seeps, shipwrecks, and other sources. *Marine Pollution Bulletin* 79(1–2):155–163.

Henry, I. A., R. Netzer, E. J. Davies, and O. G. Brakstad. 2020. Formation and fate of oil-related aggregates (ORAs) in seawater at different temperatures. *Marine Pollution Bulletin* 159:111483.

Herbert Engineering Corp. and Designers & Planners Inc. 2003. *Evaluation of accidental oil spills from bunker tanks (Phase 1)*. Prepared for the United States Coast Guard Ship Structure Committee and Research, Washington, DC, and the United States Coast Guard Development Center, Groton, Connecticut.

Hernández-López, E., M. Ayala, and R. Vazquez-Duhalt. 2015. Microbial and enzymatic biotransformations of asphaltenes. *Petroleum Science and Technology* 33(9):1017–1029.

Hester, M., and J. Willis. 2011. *National Oceanic and Atmospheric Administration, sampling and monitoring plan for the assessment of MC252 oil impacts to coastal wetland vegetation in the Gulf of Mexico, August 4, 2011.*

Hewitt, C. N., and M. B. Rashed. 1990. An integrated budget for selected pollutants for a major rural highway. *Science of the Total Environment* 93:375–384.

Hinze, J. O. 1955. Fundamentals of the hydrodynamic mechanism of splitting in dispersion processes. *AIChE Journal* 1(3):289–295.

Hites, R. A. 2016. Development of gas chromatographic mass spectrometry. *Analytical Chemistry* 88(14):6955–6961.

Ho, K. A., L. Patton, J. S. Latimer, R. J. Pruell, M. Pelletier, R. McKinney, and S. Jayaraman. 1999. The chemistry and toxicity of sediment affected by oil from the North Cape spilled into Rhode Island sound. *Marine Pollution Bulletin* 38:275–282. https://doi.org/10:1016/j.scitotenv.2008.01.045.

Ho, M., S. El-Borgi, D. Patil, and G. Song. 2019. Inspection and monitoring systems subsea pipelines: A review paper. *Structural Health Monitoring* 19(2):606–645. https://doi.org/10.1177/1475921719837718.

Hochleithner, M., C. Hochleithner, and L. Harrison. 2006. Evaluating and treating the liver. *Clinical Avian Medicine* 1:441–450.

Hodson, P. V. 2017. The toxicity to fish embryos of PAH in crude and refined oils. *Archives of Environmental Contamination and Toxicology* 73(1):12–18.

Hodson, P. V., J. Adams, and R. S. Brown. 2019. Oil toxicity test methods must be improved. *Environmental Toxicology and Chemistry* 38(2):302–311.

Hodzic, A., J. L. Jimenez, S. Madronich, M. Canagaratna, P. F. DeCarlo, L. Kleinman, and J. Fast. 2010. Modeling organic aerosols in a megacity: Potential contribution of semi-volatile and intermediate volatility primary organic compounds to secondary organic aerosol formation. *Atmospheric Chemistry and Physics* 10(12):5491–5514.

Hoff, R. Z. 1996. Responding to oil spills in marshes: The fine line between help and hindrance. In *Symposium proceedings: Gulf of Mexico and Caribbean Oil spills in coastal ecosystems: Assessing effects, natural recovery, and progress in remediation research*, C. E. Proffitt and P. F. Roscigno (eds.). OCS Study MMS 95-0063. New Orleans, LA: Minerals Management Service, U.S. Department of the Interior.

Hoffman, E. J., and J. G. Quinn. 1987a. Chronic hydrocarbon discharges into aquatic environments: I- Municipal treatment facilities. Pp. 97–113 in Oil in Freshwater: Chemistry, Biology, Countermeasure Technology, J. H. Vandermeulen and S. E. Hrudey (eds.). *Proceedings of the Symposium on Oil Pollution in Freshwater*. Alberta, Canada: Pergamon Press.

Hoffman, E. J., and J. G. Quinn. 1987b. Chronic hydrocarbon discharges into aquatic environments: II-Urban runoff and combined sewer overflows. In Oil in freshwater: chemistry, biology, countermeasure technology, J. H. Vandermeulen and S. E. Hrudey (eds.), pp. 114–137. *Proceedings of the Symposium on Oil Pollution in Freshwater*. Alberta, Canada: Pergamon Press.

Holand, P. 2006. *Blowout and well release characteristics and frequencies.* SINTEF Report STF50 F06112. Trondheim, Norway: SINTEF Technology and Society.

Holand, P. 2013. *Blowout and well release characteristics and frequencies.* Trondheim, Norway: SINTEF Technology and Society.

Holand, P. 2016. *Loss of well control occurrence and size estimators, phase I and II.* Report #ES201471/2. Report prepared for the Bureau of Safety and Environmental Enforcement, Washington, DC.

Holdway, D. A. 2002. The acute and chronic effects of wastes associated with offshore oil and gas production on temperate and tropical marine ecological processes. *Marine Pollution Bulletin* 44(3):185–203.

Holmes, W., and J. Cronshaw. 1977 Biological effects of petroleum on marine birds. In *The effects of petroleum on Arctic and sub-Arctic marine ecosystems*, D. C. Malins (ed.), pp. 359–398. New York: Academic Press.

Holmes, W. N., J. Cronshaw, and J. Gorsline. 1978. Some effects of ingested petroleum on seawater-adapted ducks (Anas platyrhynchos). *Environmental Research* 17(2):177–190. https://doi.org/10.1016/0013-9351(78)90020-8.

Holmes, W., J. Gorsline, and J. Cronshaw. 1979. Effects of mild cold stress on the survival of seawater-adapted mallard ducks (Anas platyrhynchos) maintained on food contaminated with petroleum. *Environmental Research* 20(2):425–444.

Homan, A. C., and T. Steiner. 2008. OPA 90's impact at reducing oil spills. *Marine Policy* 32(4):711–718.

Horak, K. E., S. J. Bursian, C. Ellis, K. Dean, J. Link, K. C. Hanson-Dorr, F. Cunningham, K. Harr, C. Pritsos, and K. L. Pritsos. 2017. Toxic effects of orally ingested oil from the *Deepwater Horizon* spill on laughing gulls. *Ecotoxicology and Environmental Safety* 146:83–90.

Hosgood, H. D., L. Zhang, M. Shen, S. I. Berndt, R. Vermeulen, G. Li, S. Yin, M. Yeager, J. Yuenger, and N. Rothman. 2009. Association between genetic variants in VEGF, ERCC3 and occupational benzene haematotoxicity. *Occupational and Environmental Medicine* 66(12):848–853.

Hostettler, F. D., R. J. Rosenbauer, and K. A. Kvenvolden. 1999. PAH refractory index as a source discriminant of hydrocarbon input from crude oil and coal in Prince William Sound, Alaska. *Proceedings of the 1998 ACS National Meeting on the Biogeochemistry of Polycyclic Aromatic Hydrocarbons: Sources, Interactions, Biodegradation, Toxicity and Analytical Developments, Dallas, Texas* 30(8.2):873–879.

Hou, Y., Y. Li, B. Liu, Y. Liu, and T. Wang. 2018. Design and implementation of a coastal-mounted sensor for oil film detection on seawater. *Sensors* 18(1). https://doi.org/10.3390/s18010070.

Houghton, J. P., D. C. Lees, W. B. Driskell, S. C. Lindstrom, and A. J. Mearns. 1996. Recovery of Prince William Sound intertidal epibiota from *Exxon Valdez* oiling and shoreline treatments, 1989 through 1992. In *Proceedings of the Exxon Valdez Oil Spill Symposium*, S. D. Rice, R. B. Spies, D. A. Wolfe, and B. A. Wright (eds.), pp. 379–411. Bethesda, MD: American Fisheries Society.

Howell, E. A., S. J. Bograd, C. Morishige, M. P. Seki, and J. J. Polovina. 2012. On north Pacific circulation and associated marine debris concentration. *Marine Pollution Bulletin* 65(1–3):16–22.

Hsing, P.-Y., B. Fu, E. A. Larcom, S. P. Berlet, T. M. Shank, A. F. Govindarajan, A. J. Lukasiewicz, P. M. Dixon, C. R. Fisher, and L. Thomsen. 2013. Evidence of lasting impact of the *Deepwater Horizon* oil spill on a deep Gulf of Mexico coral community oil spill impact on deep coral community. *Elementa: Science of the Anthropocene* 1(000012).

HSPD 5 (Homeland Security Presidential Directive 5). 2003. *Homeland Security Presidential Directive 5*. https://www.dhs.gov/publication/homeland-security-presidential-directive-5.

Hu, M. D., K. G. Lawrence, M. Gall, C. T. Emrich, M. R. Bodkin, W. B. Jackson, N. MacNell, R. K. Kwok, L. S. Engel, and D. P. Sandler. 2021. Natural hazards and mental health among US Gulf Coast residents. *Journal of Exposure Science & Environmental Epidemiology* 31:842–851.

Hu, P., E. A. Dubinsky, A. J. Probst, J. Wang, C. M. Sieber, L. M. Tom, P. R. Gardinali, J. F. Banfield, R. M. Atlas, and G. L. Andersen. 2017. Simulation of *Deepwater Horizon* oil plume reveals substrate specialization within a complex community of hydrocarbon degraders. *Proceedings of the National Academy of Sciences* 114(28):7432–7437.

Hu, X., W.-J. Cai, N. N. Rabalais, and J. Xue. 2016. Coupled oxygen and dissolved inorganic carbon dynamics in coastal ocean and its use as a potential indicator for detecting water column oil degradation. *Deep Sea Research Part II: Topical Studies in Oceanography* 129:311–318.

Hu, X., Q. Li, W.-J. Huang, B. Chen, W.-J. Cai, N. N. Rabalais, and R. E. Turner. 2017. Effects of eutrophication and benthic respiration on water column carbonate chemistry in a traditional hypoxic zone in the northern Gulf of Mexico. *Marine Chemistry* 194:33–42.

Hua, Y., F. S. Mirnaghi, Z. Yang, B. P. Hollebone, and C. E. Brown. 2018. Effect of evaporative weathering and oil-sediment interactions on the fate and behavior of diluted bitumen in marine environments. Part 1. Spill-related properties, oil buoyancy, and oil-particulate aggregates characterization. *Chemosphere* 191:1038–1047.

Huber, M., A. Welker, and B. Helmreich. 2016. Critical review of heavy metal pollution of traffic area runoff: Occurrence, influencing factors, and partitioning. *Science of the Total Environment* 541:895–919.

Huettel, M., W. A. Overholt, J. E. Kostka, C. Hagan, J. Kaba, W. B. Wells, and S. Dudley. 2018. Degradation of *Deepwater Horizon* oil buried in a Florida beach influenced by tidal pumping. *Marine Pollution Bulletin* 126:488–500.

Hull, A. M., D. A. Alexander, and S. Klein. 2002. Survivors of the piper alpha oil platform disaster: Long-term follow-up study. *The British Journal of Psychiatry* 181(5):433–438.

Hunt, J. M. 1996. *Petroleum geochemistry and geology.* New York: W.H. Freeman and Company.

Husky Energy. 2019. *White rose environmental effects monitoring program.*

Husseneder, C., J. R. Donaldson, and L. D. Foil. 2016. Impact of the 2010 *Deepwater Horizon* oil spill on population size and genetic structure of horse flies in Louisiana marshes. *Scientific Reports* 6(1):1–11.

Husseneder, C., J.-S. Park, and L. D. Foil. 2018. Recovery of horse fly populations in Louisiana marshes following the *Deepwater Horizon* oil spill. *Scientific Reports* 8(1):1–10.

Huynh, T. B., C. P. Groth, G. Ramachandran, S. Banerjee, M. Stenzel, A. Blair, D. P. Sandler, L. S. Engel, R. K. Kwok, and P. A. Stewart. 2022. Estimates of inhalation exposures to oil-related components on the supporting vessels during the *Deepwater Horizon* oil spill. *Annals of Work Exposures and Health* 66(Suppl 1):i111–i123.

Hvidbak, F. 2001. *The development and test of techniques for emergency transfer of extreme viscosity oil.* Canada.

Hvitved-Jacobson, T., and Y. Yenisei. 1991. Highway runoff quality, environmental impacts and control. In *Studies in Environmental Science*, pp. 165–208. Amsterdam, Netherlands: Elsevier.

ICAO (International Civil Aviation Organization). 2016. *Doc 444: Procedures for air navigation services—air traffic management.* https://ops.group/blog/wp-content/uploads/2017/03/ICAO-Doc4444-Pans-Atm-16thEdition-2016-OPSGROUP.pdf.

ICCOPR (Interagency Coordinating Committee on Oil Pollution Research). 2015. *Oil pollution research and technology plan.* U.S. Coast Guard. https://www.dco.uscg.mil/Portals/9/CG-5R/ICCOPR/Files/2015-2021%20Research%20and%20Technology%20Plan.pdf?ver=2018-01-05-133442-250.

IEA (International Energy Agency). 2021a. *Key world energy statistics 2021.* https://www.iea.org/reports/key-world-energy-statistics-2021/supply.

IEA. 2021b. *Net zero by 2050: A roadmap for the global energy sector.* Paris, France: International Energy Agency.

Igunnu, E. T., and G. Z. Chen. 2012. Produced water treatment technologies. *International Journal of Low-Carbon Technologies* 9(3):157–177.

Imperial Oil Resources Ventures Limited. 2013. *Beaufort Sea exploration joint venture drilling program: Project description.* Submitted to Environmental Impact Screening Committee, Calgary, Alberta, Canada.

Incardona, J. P. 2017. Molecular mechanisms of crude oil developmental toxicity in fish. *Archives of Environmental Contamination and Toxicology* 73(1):19–32.

Incardona, J. P., T. K. Collier, and N. L. Scholz. 2004. Defects in cardiac function precede morphological abnormalities in fish embryos exposed to polycyclic aromatic hydrocarbons. *Toxicology and Applied Pharmacology* 196(2):191–205.

Incardona, J. P., C. A. Vines, B. F. Anulacion, D. H. Baldwin, H. L. Day, B. L. French, J. S. Labenia, T. L. Linbo, M. S. Myers, O. P. Olson, C. A. Sloan, S. Sol, F. J. Griffin, K. Menard, S. G. Morgan, J. E. West, T. K. Collier, G. M. Ylitalo, G. N. Cherr, and N. L. Scholz. 2012. Unexpectedly high mortality in Pacific herring embryos exposed to the 2007 *Cosco Busan* oil spill in San Francisco bay. *Proceedings of the National Academy of Sciences* 109(2):E51–E58. https://doi.org/10.1073/pnas.1108884109.

Incardona, J. P., T. L. Swarts, R. C. Edmunds, T. L. Linbo, A. Aquilina-Beck, C. A. Sloan, L. D. Gardner, B. A. Block, and N. L. Scholz. 2013. *Exxon Valdez* to *Deepwater Horizon*: Comparable toxicity of both crude oils to fish early life stages. *Aquatic Toxicology* 142:303–316.

Incardona, J. P., L. D. Gardner, T. L. Linbo, T. L. Brown, A. J. Esbaugh, E. M. Mager, J. D. Stieglitz, B. L. French, J. S. Labenia, and C. A. Laetz. 2014. *Deepwater Horizon* crude oil impacts the developing hearts of large predatory pelagic fish. *Proceedings of the National Academy of Sciences* 111(15):E1510–E1518.

Incardona, J. P., M. G. Carls, L. Holland, T. L. Linbo, D. H. Baldwin, M. S. Myers, K. A. Peck, M. Tagal, S. D. Rice, and N. L. Scholz. 2015. Very low embryonic crude oil exposures cause lasting cardiac defects in salmon and herring. *Scientific Reports* 5(1):1–13.

Incardona, J. P., T. L. Linbo, B. L. French, J. Cameron, K. A. Peck, C. A. Laetz, M. B. Hicks, G. Hutchinson, S. E. Allan, and D. T. Boyd. 2021. Low-level embryonic crude oil exposure disrupts ventricular ballooning and subsequent trabeculation in Pacific herring. *Aquatic Toxicology* 235:105810.

Ingraffea, A. R., M. T. Wells, R. L. Santoro, and S. B. Shonkoff. 2014. Assessment and risk analysis of casing and cement impairment in oil and gas wells in Pennsylvania, 2000–2012. *Proceedings of the National Academy of Sciences* 111(30):10955–10960.

IOGP (International Association of Oil & Gas Producers). 2009. *Guidelines for Waste Management with Special Focus on Areas with Limited Infrastructure.* OGP Publications Report 413 version 1.1.

IOGP. 2020. *Risk based assessment of offshore produced water discharges, version 1.0.* Houston, Texas. www.iogp.org.

IOGP and IPIECA. 2012. *Sensitivity mapping for oil spill response. Report number 477.* https://www.IPIECA.org/resources/good-practice/sensitivity-mapping-for-oil-spill-response.

IOGP and IPIECA. 2019. *Source control emergency response planning guide for subsea wells.* Report 594.

IOM (Institute of Medicine). 2010a. *Research priorities for assessing health effects from the Gulf of Mexico oil spill: A letter report.* Washington, DC: The National Academies Press.

IOM. 2010b. *Assessing the effects of the Gulf of Mexico oil spill on human health: A summary of the June 2010 workshop.* Washington, DC: The National Academies Press.

IPCC (Intergovernmental Panel on Climate Change). 2021. Climate Change 2021: The Physical Science Basis. *Contribution of Working Group I to the Sixth Assessment Report of the Intergovernmental Panel on Climate.* Cambridge University Press, Cambridge, UK, and New York, NY. https://doi.org/10.1017/9781009157896.

IPIECA. 2014. *Wildlife response preparedness.* https://www.IPIECA.org/resources/good-practice/wildlife-response-preparedness.

IPIECA. 2015a. *Dispersants: Surface application. Good practice guidelines for incident management and emergency response personnel.* IOGP Report 533.

IPIECA. 2015b. *Response strategy development using net environmental benefit analysis (NEBA).*

IPIECA. 2016. *Aerial observation of oil spills at sea. Good practice guide for incident management and emergency response personnel.* https://www.IPIECA.org/resources/good-practice/aerial-observation-of-oil-spills-at-sea.

IPIECA. 2017. *Key principles for the protection, care and rehabilitation of oiled wildlife.* https://www.ipieca.org/resources/awareness-briefing/key-principles-for-the-protection-care-and-rehabilitation-of-oiled-wildlife.

IPIECA. 2020. *Good practice guidelines for incident management and emergency responses personnel. Oil spill monitoring and sampling.* International Association of Oil & Gas Producers. IPIECA-IOG Report 639.

IPIECA. 2021. *Oil spill surveillance planning guidance. A technical support document to accompany the IPIECA–IOGP guidance on contingency planning, aerial observation, remote sensing and the incident management system.*

IPIECA and IOGP (International Association of Oil & Gas Producers). 2018. *Guidelines on implementing spill impact mitigation assessment (SIMA).* https://www.ipieca.org/resources/awareness-briefing/guidelines-on-implementing-spill-impact-mitigation-assessment-sima.

ITOPF (International Tanker Owners Pollution Federation). 2021. *Oil tanker spill statistics 2020.* London, UK: International Association of Oil and Gas Producers.

Izon, D., E. Danenberger, and M. Mayes. 2007. Absence of fatalities in blowouts encouraging in MMS study of OCS incidents 1992–2006. *Drilling Contractor* 63(4):84–89.

Jaekel, U., J. Zedelius, H. Wilkes, and F. Musat. 2015. Anaerobic degradation of cyclohexane by sulfate-reducing bacteria from hydrocarbon-contaminated marine sediments. *Frontiers in Microbiology* 6:116.

Jaligama, S., Z. Chen, J. Saravia, N. Yadav, S. M. Lomnicki, T. R. Dugas, and S. A. Cormier. 2015. Exposure to *Deepwater Horizon* crude oil burnoff particulate matter induces pulmonary inflammation and alters adaptive immune response. *Environmental Science & Technology* 49(14):8769–8776.

Jameel, A. G. A., Y. Han, O. Brignoli, S. Telalović, A. M. Elbaz, H. G. Im, W. L. Roberts, and S. M. Sarathy. 2017. Heavy fuel oil pyrolysis and combustion: Kinetics and evolved gases investigated by TGA-FTIR. *Journal of Analytical and Applied Pyrolysis* 127:183–195.

Janjua, N. Z., P. M. Kasi, H. Nawaz, S. Z. Farooqui, U. B. Khuwaja, S. N. Jafri, S. A. Lutfi, M. M. Kadir, and N. Sathiakumar. 2006. Acute health effects of the Tasman Spirit oil spill on residents of Karachi, Pakistan. *BMC Public Health* 6(1):1–11.

Jaouen-Madoulet, A., A. Abarnou, A. M. Le Guellec, V. Loizeau, and F. Leboulenger. 2000. Validation of an analytical procedure for polychlorinated biphenyls, coplanar polychlorinated biphenyls and polycyclic aromatic hydrocarbons in environmental samples. *Journal of Chromatography A* 886(1):153–173. https://doi.org/10.1016/S0021-9673(00)00422-2.

Jensen, H. V., S. L. Gyltnes, and S. Andersen. 2012. *Norwegian oil spill response technology development.* Norwegian Clean Seas Association for Operating Companies.

Jensen, O. P., C. W. Martin, K. L. Oken, F. J. Fodrie, P. C. López-Duarte, K. W. Able, and B. J. Roberts. 2019. Simultaneous estimation of dispersal and survival of the Gulf killifish fundulus grandis from a batch-tagging experiment. *Marine Ecology Progress Series* 624:183–194.

Jenssen, B. M. 1994. Effects of oil pollution, chemically treated oil, and cleaning on thermal balance of birds. *Environmental Pollution* 86(2):207–215.

Jenssen, B. M., and M. Ekker. 1988. A method for evaluating the cleaning of oiled seabirds. *Wildlife Society Bulletin (1973–2006)* 16(2):213–215. http://www.jstor.org/stable/3782193.

Jenssen, B. M., and M. Ekker. 1991. Effects of plumage contamination with crude oil dispersant mixtures on thermoregulation in common eiders and mallards. *Archives of Environmental Contamination and Toxicology* 20(3):398–403. https://doi.org/10.1007/BF01064410.

Jenssen, B. M., M. Ekker, and K. Zahlsen. 1990. Effects of ingested crude oil on thyroid hormones and on the mixed function oxidase system in ducks. *Comparative Biochemistry and Physiology Part C: Comparative Pharmacology and Toxicology* 95(2):213–216.

Jernelöv, A. 2010. The threats from oil spills: Now, then, and in the future. *Ambio* 39(5):353–366.

Jernelöv, A., and O. Lindén. 1981. Ixtoc I: A case study of the world's largest oil spill. *Ambio* 299–306.

Jessup, D. A. 1998. Diversity-rehabilitation of oiled wildlife. *Conservation Biology* 12(5):1153–1155.

Jessup, D., and F. Leighton. 1996. Oil pollution and petroleum toxicity to wildlife. In *Non-infectious diseases of wildlife*, pp. 141–157. Ames, IA: Iowa State University Press.

Ji, C., C. J. Beegle-Krause, and J. D. Englehardt. 2020. Formation, detection, and modeling of submerged oil: A review. *Journal of Marine Science and Engineering* 8(9):642.

Ji, Z.-G., W. R. Johnson, and G. L. Wikel. 2014. Statistics of extremes in oil spill risk analysis. *Environmental Science & Technology* 48(17):10505–10510.

Jiang, Y., J. Yang, S. Gagné, T. W. Chan, K. Thomson, E. Fofie, R. A. Cary, D. Rutherford, B. Comer, and J. Swanson. 2018. Sources of variance in BC mass measurements from a small marine engine: Influence of the instruments, fuels and loads. *Atmospheric Environment* 182:128–137.

Jiménez, S., M. Micó, M. Arnaldos, F. Medina, and S. Contreras. 2018. State of the art of produced water treatment. *Chemosphere* 192:186–208.

Jindrova, E., M. Chocova, K. Demnerova, and V. Brenner. 2002. Bacterial aerobic degradation of benzene, toluene, ethylbenzene and xylene. *Folia Microbiologica* 47(2):83–93.

Johannessen, S. C., C. W. Greer, C. G. Hannah, T. L. King, K. Lee, R. Pawlowicz, and C. A. Wright. 2020. Fate of diluted bitumen spilled in the coastal waters of British Columbia, Canada. *Marine Pollution Bulletin* 150:110691.

Johansen, C., A. C. Todd, and I. R. MacDonald. 2017. Time series video analysis of bubble release processes at natural hydrocarbon seeps in the northern Gulf of Mexico. *Marine and Petroleum Geology* 82:21–34.

Johansen, Ø., H. Rye, and C. Cooper. 2003. Deepspill—field study of a simulated oil and gas blowout in deep water. *Spill Science & Technology Bulletin* 8(5–6):433–443.

Johansen, Ø., P. J. Brandvik, and U. Farooq. 2013. Droplet breakup in subsea oil releases—part 2: Predictions of droplet size distributions with and without injection of chemical dispersants. *Marine Pollution Bulletin* 73(1):327–335.

Johansen, Ø., M. Reed, and N. R. Bodsberg. 2015. Natural dispersion revisited. *Marine Pollution Bulletin* 93(1–2):20–26.

Johansson, K., M. Head-Gordon, P. Schrader, K. Wilson, and H. Michelsen. 2018. Resonance-stabilized hydrocarbon-radical chain reactions may explain soot inception and growth. *Science* 361(6406):997–1000.

John, G. F., Y. Han, and T. P. Clement. 2016. Weathering patterns of polycyclic aromatic hydrocarbons contained in submerged *Deepwater Horizon* oil spill residues when re-exposed to sunlight. *Science of the Total Environment* 573:189–202. https://doi.org/10.1016/j.scitotenv.2016.08.059.

Johnsen, S., T. I. Røe Utvik, E. Garland, B. de Vals, and J. Campbell. 2004. *Environmental fate and effect of contaminants in produced water. SPE International Conference on Health, Safety, and Environment in Oil and Gas Exploration and Production*. Calgary, Alberta, Canada, March 2004.

Johnson, A., F. Besik, and A. Hamielec. 1969. Mass transfer from a single rising bubble. *The Canadian Journal of Chemical Engineering* 47(6):559–564.

Johnson, P. M., C. E. Brady, C. Philip, H. Baroud, J. V. Camp, and M. Abkowitz. 2020. A factor analysis approach toward reconciling community vulnerability and resilience indices for natural hazards. *Risk Analysis* 40(9):1795–1810.

Jones, D. O., A. R. Gates, V. A. Huvenne, A. B. Phillips, and B. J. Bett. 2019. Autonomous marine environmental monitoring: Application in decommissioned oil fields. *Science of the Total Environment* 668:835–853.

Jones, R. 1997. *A simplified pseudo-component oil evaporation model.* Canada.

Jørgensen, B. B. 2012. Shrinking majority of the deep biosphere. *Proceedings of the National Academy of Sciences* 109(40):15976–15977.

Joye, S., and J. E. Kostka. 2020. *Microbial genomics of the global ocean system.* Report on an American Academy of Microbiology, American Geophysical Union, and Gulf of Mexico Research Initiative Colloquium held on 9 and 10 April 2019.

Joye, S. B. 2020. The geology and biogeochemistry of hydrocarbon seeps. *Annual Review of Earth and Planetary Sciences* 48:205–231.

Judson, R. S., M. T. Martin, D. M. Reif, K. A. Houck, T. B. Knudsen, D. M. Rotroff, M. Xia, S. Sakamuru, R. Huang, and P. Shinn. 2010. Analysis of eight oil spill dispersants using rapid, in vitro tests for endocrine and other biological activity. *Environmental Science & Technology* 44(15):5979–5985.

Judy, C. R., S. A. Graham, Q. Lin, A. Hou, and I. A. Mendelssohn. 2014. Impacts of Macondo oil from *Deepwater Horizon* spill on the growth response of the common reed phragmites australis: A mesocosm study. *Marine Pollution Bulletin* 79(1–2):69–76.

Jung, J.-H., C. E. Hicken, D. Boyd, B. F. Anulacion, M. G. Carls, W. J. Shim, and J. P. Incardona. 2013. Geologically distinct crude oils cause a common cardiotoxicity syndrome in developing zebrafish. *Chemosphere* 91(8):1146–1155.

Jung, J.-H., M. Kim, U. H. Yim, S. Y. Ha, W. J. Shim, Y. S. Chae, H. Kim, J. P. Incardona, T. L. Linbo, and J.-H. Kwon. 2015. Differential toxicokinetics determines the sensitivity of two marine embryonic fish exposed to Iranian heavy crude oil. *Environmental Science & Technology* 49(22):13639–13648.

Jung, J.-H., J. Ko, E.-H. Lee, K.-M. Choi, M. Kim, U. H. Yim, J.-S. Lee, and W. J. Shim. 2017. RNA seq- and DEG-based comparison of developmental toxicity in fish embryos of two species exposed to Iranian heavy crude oil. *Comparative Biochemistry and Physiology Part C: Toxicology & Pharmacology* 196:1–10.

Junge, K., H. Eicken, and J. W. Deming. 2004. Bacterial activity at −2 to −20°C in Arctic wintertime sea ice. *Applied and Environmental Microbiology* 70(1):550–557.

Kaiser, M. J. 2017. FERC pipeline decommissioning cost in the US Gulf of Mexico, 1995–2015. *Marine Policy* 82:167–180.

Kaiser, M. J., and S. Narra. 2019. US Gulf of Mexico pipeline activity statistics, trends and correlations. *Ships and Offshore Structures* 14(1):1–22.

Kalimov, N., P. M. Taylor, Z. Kulekeyev, G. Nurtayeva. 2021. Test of herder and controlled burning of spilled oil in Kazakhstan. *International Oil Spill Conference Proceedings* 2021(1):686904.

Kamalanathan, M., C. Xu, K. Schwehr, L. Bretherton, M. Beaver, S. M. Doyle, J. Genzer, J. Hillhouse, J. B. Sylvan, and P. Santschi. 2018. Extracellular enzyme activity profile in a chemically enhanced water accommodated fraction of surrogate oil: Toward understanding microbial activities after the *Deepwater Horizon* oil spill. *Frontiers in Microbiology* 9:798.

Kanaly, R. A., and S. Harayama. 2000. Biodegradation of high-molecular-weight polycyclic aromatic hydrocarbons by bacteria. *Journal of Bacteriology* 182(8):2059–2067.

Kanaly, R. A., and S. Harayama. 2010. Advances in the field of high-molecular-weight polycyclic aromatic hydrocarbon biodegradation by bacteria. *Microbial Biotechnology* 3(2):136–164.

Kang, N.-Y., and J. B. Elsner. 2015. Trade-off between intensity and frequency of global tropical cyclones. *Nature Climate Change* 5(7):661–664.

Kang, Z., A. Yeung, J. M. Foght, and M. R. Gray. 2008a. Mechanical properties of hexadecane–water interfaces with adsorbed hydrophobic bacteria. *Colloids and Surfaces B: Biointerfaces* 62(2):273–279.

Kang, Z., A. Yeung, J. M. Foght, and M. R. Gray. 2008b. Hydrophobic bacteria at the hexadecane–water interface: Examination of micrometer-scale interfacial properties. *Colloids and Surfaces B: Biointerfaces* 67(1):59–66.

Kanhai, L. D. K., C. Johansson, J. Frias, K. Gardfeldt, R. C. Thompson, and I. O'Connor. 2019. Deep sea sediments of the Arctic central basin: A potential sink for microplastics. *Deep Sea Research Part I: Oceanographic Research Papers* 145:137–142.

Karinen, J. F. 1993. *Hydrocarbons in intertidal sediments and mussels from Prince William Sound, Alaska, 1977–1980: Characterization and probable sources*. U.S. Department of Commerce, NOAA Technical Memorandum. NMFS-AFSC-9, 70.

Kario, K., S. M. Bruce, and G. P. Thomas. 2003. Disasters and the heart: A review of the effects of earthquake-induced stress on cardiovascular disease. *Hypertension Research* 26(5):355–367.

Karp, A. T., A. I. Holman, P. Hopper, K. Grice, and K. H. Freeman. 2020. Fire distinguishers: Refined interpretations of polycyclic aromatic hydrocarbons for paleo-applications. *Geochimica et Cosmochimica Acta* 289:93–113.

Karthikeyan, S., L. M. Rodriguez-R, P. Heritier-Robbins, M. Kim, W. A. Overholt, J. C. Gaby, J. K. Hatt, J. C. Spain, R. Rosselló-Móra, and M. Huettel. 2019. "Candidatus macondimonas diazotrophica": A novel gammaproteobacterial genus dominating crude-oil-contaminated coastal sediments. *The ISME Journal* 13(8):2129–2134.

Karthikeyan, S., L. M. Rodriguez-R, P. Heritier-Robbins, J. K. Hatt, M. Huettel, J. E. Kostka, and K. T. Konstantinidis. 2020. Genome repository of oil systems: An interactive and searchable database that expands the catalogued diversity of crude oil-associated microbes. *Environmental Microbiology* 22(6):2094–2106.

Kasai, Y., H. Kishira, T. Sasaki, K. Syutsubo, K. Watanabe, and S. Harayama. 2002. Predominant growth of alcanivorax strains in oil-contaminated and nutrient-supplemented sea water. *Environmental Microbiology* 4(3):141–147.

Kasperson, R. E., O. Renn, P. Slovic, H. S. Brown, J. Emel, R. Goble, J. X. Kasperson, and S. Ratick. 1988. The social amplification of risk: A conceptual framework. *Risk Analysis* 8(2):177–187.

Ke, H., M. Chen, M. Liu, M. Chen, M. Duan, P. Huang, J. Hong, Y. Lin, S. Cheng, and X. Wang. 2017. Fate of polycyclic aromatic hydrocarbons from the north Pacific to the Arctic: Field measurements and fugacity model simulation. *Chemosphere* 184:916–923.

Kegler, P., G. Baum, L. F. Indriana, C. Wild, and A. Kunzmann. 2015. Physiological response of the hard coral pocillopora verrucosa from Lombok, Indonesia, to two common pollutants in combination with high temperature. *PloS ONE* 10(11):e0142744.

Keith, V. F. 1993. Double hull oil tankers—how effective are they? *International Oil Spill Conference Proceedings* 1993(1):745–751.

Kellogg, C. T., J. W. McClelland, K. H. Dunton, and B. C. Crump. 2019. Strong seasonality in Arctic estuarine microbial food webs. *Frontiers in Microbiology* 10:2628.

Kemball-Cook, S., A. Bar-Ilan, J. Grant, L. Parker, J. Jung, W. Santamaria, and G. Yarwood. n.d. *Development of an emission inventory for natural gas exploration and production in the haynesville shale and evaluation of ozone impacts.* https://gaftp.epa.gov/Air/nei/ei_conference/EI19/session2/kemball_cook.pdf.

Kenworthy, W. J., N. Cosentino-Manning, L. Handley, M. Wild, and S. Rouhani. 2017. Seagrass response following exposure to *Deepwater Horizon* oil in the Chandeleur Islands, Louisiana (USA). *Marine Ecology Progress Series* 576:145–161.

Kenyon, K. W. 1969. *The sea otter in the eastern Pacific Ocean.* Vol. 68. U.S. Bureau of Sport Fisheries and Wildlife.

Keramea, P., K. Spanoudaki, G. Zodiatis, G. Gikas, and G. Sylaios. 2021. Oil spill modeling: A critical review on current trends, perspectives, and challenges. *Journal of Marine Science and Engineering* 9(2):181.

Kessler, J. D., and T. Weber. 2021. *Report on seep methane inputs into North American coastal waters.* Prepared for the National Academies Committee on Oil in the Sea IV: Sources, Fates and Effects, Washington, DC.

Kessler, J. D., D. L. Valentine, M. C. Redmond, M. Du, E. W. Chan, S. D. Mendes, E. W. Quiroz, C. J. Villanueva, S. S. Shusta, and L. M. Werra. 2011. A persistent oxygen anomaly reveals the fate of spilled methane in the deep Gulf of Mexico. *Science* 331(6015):312–315.

Keyte, I. J., R. M. Harrison, and G. Lammel. 2013. Chemical reactivity and long-range transport potential of polycyclic aromatic hydrocarbons—a review. *Chemical Society Reviews* 42(24):9333–9391. https://doi.org/10.1039/C3CS60147A.

Khan, M. Y., M. Giordano, J. Gutierrez, W. A. Welch, A. Asa-Awuku, J. W. Miller, and D. R. Cocker III. 2012. Benefits of two mitigation strategies for container vessels: Cleaner engines and cleaner fuels. *Environmental Science & Technology* 46(9):5049–5056.

Khan, R., and K. Nag. 1993. Estimation of hemosiderosis in seabirds and fish exposed to petroleum. *Bulletin of Environmental Contamination and Toxicology (United States)* 50(1):125–131.

Khatib, Z., and P. Verbeek. 2003. Water to value-produced water management for sustainable field development of mature and green fields. *SPE International Conference on Health, Safety and Environment in Oil and Gas Exploration and Production.* San Antonio, Texas.

Khot, V., J. Zorz, D. A. Gittins, A. Chakraborty, E. Bell, M. A. Bautista, A. J. Paquette, A. K. Hawley, B. Novotnik, and C. R. Hubert. 2022. CANT–HYD: A curated database of phylogeny-derived hidden Markov models for annotation of marker genes involved in hydrocarbon degradation. *Frontiers in Microbiology* 12:764058. https://doi.org/10.3389/fmicb.2021.764058.

Kim, A.-R., and Y.-J. Seo. 2019. The reduction of sox emissions in the shipping industry: The case of Korean companies. *Marine Policy* 100:98–106.

Kim, J.-A., S. R. Noh, H.-K. Cheong, M. Ha, S.-Y. Eom, H. Kim, M.-S. Park, Y. Chu, S.-H. Lee, and K. Choi. 2017. Urinary oxidative stress biomarkers among local residents measured 6 years after the *Hebei Spirit* oil spill. *Science of the Total Environment* 580:946–952.

Kimbrough, K. L., G. Lauenstein, J. Christensen, and D. Apeti. 2008. *An assessment of two decades of contaminant monitoring in the nation's coastal zone.* NOAA Technical Memorandum NOS NCCOS 74. Silver Spring, MD: National Centers for Coastal Ocean Science.

Kimes, N. E., A. V. Callaghan, J. M. Suflita, and P. J. Morris. 2014. Microbial transformation of the *Deepwater Horizon* oil spill—past, present, and future perspectives. *Frontiers in Microbiology* 5:603.

King, B. S., and J. D. Gibbins. 2011. *Health hazard evaluation of* Deepwater Horizon *response workers.* Health Hazard Evaluation Report HETA 2010-0115 and 2010-0129-3138. https://www.cdc.gov/niosh/hhe/reports/pdfs/2010-0115-0129-3138.pdf.

King, M. B. 1969. *Phase equilibrium in mixtures.* New York: Elsevier.

King, T. L., B. Robinson, M. Boufadel, and K. Lee. 2014. Flume tank studies to elucidate the fate and behavior of diluted bitumen spilled at sea. *Marine Pollution Bulletin* 83(1):32–37.

King, T. L., B. J. Robinson, S. A. Ryan, J. Mason, P. Thamer, G. Wohlgeschaffen, and K. Lee. 2016. *Oil sheen formation and fate in an open sea environment.* Canadian Technical Report of Fisheries and Aquatic Sciences 3178:31.

Kirkwood, K. M., P. Chernik, J. M. Foght, and M. R. Gray. 2008. Aerobic biotransformation of decalin (decahydronaphthalene) by rhodococcus spp. *Biodegradation* 19(6):785–794.

Klaunig, J. E. 2018. Oxidative stress and cancer. *Current Pharmaceutical Design* 24(40):4771–4778.

Kleindienst, S., F.-A. Herbst, M. Stagars, F. Von Netzer, M. Von Bergen, J. Seifert, J. Peplies, R. Amann, F. Musat, and T. Lueders. 2014. Diverse sulfate-reducing bacteria of the desulfosarcina/desulfococcus clade are the key alkane degraders at marine seeps. *The ISME Journal* 8(10):2029–2044.

Kleindienst, S., M. Seidel, K. Ziervogel, S. Grim, K. Loftis, S. Harrison, S. Y. Malkin, M. J. Perkins, J. Field, and M. L. Sogin. 2015. Chemical dispersants can suppress the activity of natural oil-degrading microorganisms. *Proceedings of the National Academy of Sciences* 112(48):14900–14905.

Kleindienst, S., S. Grim, M. Sogin, A. Bracco, M. Crespo-Medina, and S. B. Joye. 2016. Diverse, rare microbial taxa responded to the *Deepwater Horizon* deep-sea hydrocarbon plume. *The ISME Journal* 10(2):400–415.

Kleinman, M. T., G. R. Mueller, E. Stevenson, R. Alvarez, A. J. Marchese, and D. Allen. 2016. Emissions from oil and gas operations in the United States and their air quality implications. *Journal of the Air & Waste Management Association* 66(12):1165–1170.

Klekowski, E. J., J. E. Corredor, J. M. Morell, and C. A. Del Castillo. 1994. Petroleum pollution and mutation in mangroves. *Marine Pollution Bulletin* 28(3):166–169. https://doi.org/10.1016/0025-326X(94)90393-X.

Klimaszewska, Ż. P., Z. Polkowsla, and J. Namieśnik. 2007. Influence of mobile sources on pollution of runoff waters from roads with high traffic intensity. *Polish Journal of Environmental Studies* 16(6):889–897. http://www.pjoes.com/Influence-of-Mobile-Sources-on-Pollution-of-r-nRunoff-Waters-from-Roads-with-High,88063,0,2.html.

Knap, A.H. 1992. Biological impacts of oil pollution: coral reefs. London, UK: IPIECA.

Kniemeyer, O., F. Musat, S. M. Sievert, K. Knittel, H. Wilkes, M. Blumenberg, W. Michaelis, A. Classen, C. Bolm, and S. B. Joye. 2007. Anaerobic oxidation of short-chain hydrocarbons by marine sulphate-reducing bacteria. *Nature* 449(7164):898–901.

Knightes, C. D., and C. A. Peters. 2006. Multisubstrate biodegradation kinetics for binary and complex mixtures of polycyclic aromatic hydrocarbons. *Environmental Toxicology and Chemistry* 25(7):1746–1756.

Knittel, K., and A. Boetius. 2009. Anaerobic oxidation of methane: Progress with an unknown process. *Annual Review of Microbiology* 63:311–334.

Knittel, K., T. Lösekann, A. Boetius, R. Kort, and R. Amann. 2005. Diversity and distribution of methanotrophic archaea at cold seeps. *Applied and Environmental Microbiology* 71(1):467–479.

Knutson, T. R., J. L. McBride, J. Chan, K. Emanuel, G. Holland, C. Landsea, I. Held, J. P. Kossin, A. Srivastava, and M. Sugi. 2010. Tropical cyclones and climate change. *Nature Geoscience* 3(3):157–163.

Koh, C. A., E. D. Sloan, A. K. Sum, and D. T. Wu. 2011. Fundamentals and applications of gas hydrates. *Annual Review of Chemical and Biomolecular Engineering* 2:237–257.

Koseki, H. 1993. *Thermography research for radiation measurement on an oil spill fire.* (EC/TDTS—94-02286-Vol1-2). Canada.

Kostka, J.E., O. Prakash, W. Overholt, S. Green, G. Freyer, A. Canion, J. Delgardio, N. Norton, T.C. Hazen, and M. Huettel. 2011. Hydrocarbon-degrading bacteria and the bacterial community response in Gulf of Mexico beach sands impacted by the *Deepwater Horizon* oil spill. *Applied and Environmental Microbiology* 77:7962–7974.

Krick, T., M. Forstater, P. Monaghan, and M. Sillanpaa. 2005. *The stakeholder engagement manual. Volume 2. The practitioner's handbook on stakeholder engagement.* AccountAbility, United Nations Environment Programme and Stakeholder Research Associates. http://www.mas-business.com/docs/English%20Stakeholder%20Engagement%20Handbook.pdf.

Krishnamurthy, J., L. S. Engel, L. Wang, E. G. Schwartz, K. Christenbury, B. Kondrup, J. Barrett, and J. A. Rusiecki. 2019. Neurological symptoms associated with oil spill response exposures: Results from the *Deepwater Horizon* oil spill coast guard cohort study. *Environment International* 131:104963.

Kristensen, M., A. R. Johnsen, and J. H. Christensen. 2015. Marine biodegradation of crude oil in temperate and Arctic water samples. *Journal of Hazardous Materials* 300:75–83.

Kristensen, M., A. R. Johnsen, and J. H. Christensen. 2021. Super-complex mixtures of aliphatic- and aromatic acids may be common degradation products after marine oil spills: A lab-study of microbial oil degradation in a warm, pre-exposed marine environment. *Environmental Pollution* 285:117264.

Krolicka, A., C. Boccadoro, M. M. Nilsen, E. Demir-Hilton, J. Birch, C. Preston, C. Scholin, and T. Baussant. 2019. Identification of microbial key-indicators of oil contamination at sea through tracking of oil biotransformation: An Arctic field and laboratory study. *Science of the Total Environment* 696:133715.

Kroon, F. J., K. L. Berry, D. L. Brinkman, R. Kookana, F. D. Leusch, S. D. Melvin, P. A. Neale, A. P. Negri, M. Puotinen, and J. J. Tsang. 2020. Sources, presence and potential effects of contaminants of emerging concern in the marine environments of the Great Barrier Reef and Torres Strait, Australia. *Science of the Total Environment* 719:135140.

Kruge, M. A., J. L. R. Gallego, A. Lara-Gonzalo, and N. Esquinas. 2018. Environmental forensics study of crude oil and petroleum product spills in coastal and oilfield settings: Combined insights from conventional GC-MS, thermodesorption-GC-MS, and pyrolysis-GC-MS. In *Oil spill environmental forensics case studies*, pp. 131–155. Amsterdam, Netherlands: Elsevier.

Kube, M., T. N. Chernikova, Y. Al-Ramahi, A. Beloqui, N. Lopez-Cortez, M.-E. Guazzaroni, H. J. Heipieper, S. Klages, O. R. Kotsyurbenko, and I. Langer. 2013. Genome sequence and functional genomic analysis of the oil-degrading bacterium oleispira Antarctica. *Nature Communications* 4(1):1–11.

Kujawinski, E. B., M. C. Kido Soule, D. L. Valentine, A. K. Boysen, K. Longnecker, and M. C. Redmond. 2011. Fate of dispersants associated with the *Deepwater Horizon* oil spill. *Environmental Science & Technology* 45(4):1298–1306.

Kvenvolden, K., and C. Cooper. 2003. Natural seepage of crude oil into the marine environment. *Geo-Marine Letters* 23(3):140–146.

Kwok, R., G. Spreen, and S. Pang. 2013. Arctic Sea Ice Circulation and Drift Speed: Decadal Trends and Ocean Currents. *Journal of Geophysical Research: Oceans* 118(5):2408–2425.

Kwok, R. K., L. S. Engel, A. K. Miller, A. Blair, M. D. Curry, W. B. Jackson, P. A. Stewart, M. R. Stenzel, L. S. Birnbaum, and D. P. Sandler. 2017a. The Gulf study: A prospective study of persons involved in the *Deepwater Horizon* oil spill response and clean-up. *Environmental Health Perspectives* 125(4):570–578.

Kwok, R. K., J. A. McGrath, S. R. Lowe, L. S. Engel, W. B. Jackson 2nd, M. D. Curry, J. Payne, S. Galea, and D. P. Sandler. 2017b. Mental health indicators associated with oil spill response and clean-up: Cross-sectional analysis of the gulf study cohort. *The Lancet Public Health* 2(12):e560–e567.

Lacaze, J. C., and O. Villedon de Naide. 1976. Influence of illumination on phytotoxicity of crude oil. *Marine Pollution Bulletin* 7:73–76.

Lack, D., and J. Corbett. 2012. Black carbon from ships: A review of the effects of ship speed, fuel quality and exhaust gas scrubbing. *Atmospheric Chemistry and Physics* 12(9):3985–4000.

Laczi, K., Á. Erdeiné Kis, Á. Szilágyi, N. Bounedjoum, A. Bodor, G. E. Vincze, T. Kovács, G. Rákhely, and K. Perei. 2020. New frontiers of anaerobic hydrocarbon biodegradation in the multi-omics era. *Frontiers in Microbiology* 11:2886.

Laffon, B., E. Pásaro, and V. Valdiglesias. 2016. Effects of exposure to oil spills on human health: Updated review. *Journal of Toxicology and Environmental Health, Part B* 19(3–4):105–128.

Lai, C. C., and S. A. Socolofsky. 2019. The turbulent kinetic energy budget in a bubble plume. *Journal of Fluid Mechanics* 865:993–1041.

Lambert, R. A., and E. A. Variano. 2016. Collision of oil droplets with marine aggregates: Effect of droplet size. *Journal of Geophysical Research: Oceans* 121(5):3250–3260.

Lamberts, R. F., J. H. Christensen, P. Mayer, O. Andersen, and A. R. Johnsen. 2008. Isomer-specific biodegradation of methylphenanthrenes by soil bacteria. *Environmental Science & Technology* 42(13):4790–4796.

Lammel, G., F. X. Meixner, B. Vrana, C. I. Efstathiou, J. Kohoutek, P. Kukučka, M. D. Mulder, P. Přibylová, R. Prokeš, and T. P. Rusina. 2016. Bidirectional air-sea exchange and accumulation of POPs (PAHs, PCBs, OCPs and PBDEs) in the nocturnal marine boundary layer. *Atmospheric Chemistry and Physics* 16(10):6381–6393.

Lammel, G., Z. Kitanovski, P. Kukučka, J. í. Novák, A. M. Arangio, G. P. Codling, A. Filippi, J. Hovorka, J. Kuta, and C. Leoni. 2020. Oxygenated and nitrated polycyclic aromatic hydrocarbons in ambient air-levels, phase partitioning, mass size distributions, and inhalation bioaccessibility. *Environmental Science & Technology* 54(5):2615–2625.

Lan, Q., L. Zhang, G. Li, R. Vermeulen, R. S. Weinberg, M. Dosemeci, S. M. Rappaport, M. Shen, B. P. Alter, and Y. Wu. 2004. Hematotoxicity in workers exposed to low levels of benzene. *Science* 306(5702):1774–1776.

Landquist, H., I. M. Hassellöv, L. Rosén, J. F. Lindgren, and I. Dahllöf. 2013. Evaluating the needs of risk assessment methods of potentially polluting shipwrecks. *Journal of Environmental Management* 119: 85–92. https://doi.org/10.1016/j.jenvman.2012.12.036.

Landquist, H., L. Rosén, A. Lindhe, T. Norberg, I. M. Hassellöv, J. F. Lindgren, and I. Dahllöf. 2014. A fault tree model to assess probability of contaminant discharge from shipwrecks. *Marine Pollution Bulletin* 88(1–2):239–248. https://doi.org/10.1016/j.marpolbul.2014.08.037.

Landquist, H., L. Rosén, A. Lindhe, T. Norberg, and I. M. Hassellöv. 2017. Bayesian updating in a fault tree model for shipwreck risk assessment. *Science of the Total Environment* 590–591:80–91. https://doi.org/10.1016/j.scitotenv.2017.03.033.

Landrum, P., J. Giesy, J. Oris, and P. Allred. 1987. Photoinduced toxicity of polycyclic aromatic hydrocarbons to aquatic organisms. In *Oil in freshwater: Chemistry, biology, countermeasure technology*, pp. 304–318. Amsterdam, Netherlands: Elsevier.

Langenhoff, A. A., S. Rahsepar, J. S. van Eenennaam, J. R. Radović, T. B. Oldenburg, E. Foekema, and A. J. Murk. 2020. Effect of marine snow on microbial oil degradation. In *Deep oil spills*, pp. 301–311. Dordrecht, Netherlands: Springer.

Langlois, P. H., A. T. Hoyt, P. J. Lupo, C. C. Lawson, M. A. Waters, T. A. Desrosiers, G. M. Shaw, P. A. Romitti, E. J. Lammer, and National Birth Defects Prevention Study. 2012. Maternal occupational exposure to polycyclic aromatic hydrocarbons and risk of neural tube defect-affected pregnancies. *Birth Defects Research Part A: Clinical and Molecular Teratology* 94(9):693–700.

Lanksbury, J., B. Lubliner, M. Langness, and J. West. 2017. *Stormwater action monitoring 2015/16 mussel monitoring survey.* WDFW report number FPT 17-06. Prepared by the Washington Department of Fish and Wildlife, Olympia, Washington, for stormwater action monitoring.

Larson, R. A., G. R. Brooks, P. T. Schwing, C. W. Holmes, S. R. Carter, and D. J. Hollander. 2018. High-resolution investigation of event driven sedimentation: Northeastern Gulf of Mexico. *Anthropocene* 24:40–50. https://doi.org/10.1016/j.ancene.2018.11.002.

Latimer, J., and J. Quinn. 1998. Aliphatic petroleum and biogenic hydrocarbons entering Narragansett Bay from tributaries under dry weather conditions. *Estuaries* 21(1):91–107.

Latimer, J., E. Hoffman, G. Hoffman, J. L. Fasching, and J. G. Quinn. 1990. Sources of petroleum hydrocarbons in urban runoff. *Water, Air, and Soil Pollution* 52:1–21.

Lattin, C. R., H. M. Ngai, and L. M. Romero. 2014. Evaluating the stress response as a bioindicator of sub-lethal effects of crude oil exposure in wild house sparrows (Passer domesticus). *PloS ONE* 9(7):e102106.

Lauenstein, G., and A. Y. Cantillo. 1998. Sampling and analytical methods of the national status and trends program mussel watch project: 1993–1996 update. NOAA Technical Memorandum NOS ORCA 1430. Silver Spring: MD: Office of Ocean Resources Conservation and Assessment, Coastal Monitoring and Bioeffects Assessment Division, National Ocean Service, National Oceanic and Atmospheric Administration.

Laurel, B. J., L. A. Copeman, P. Iseri, M. L. Spencer, G. Hutchinson, T. Nordtug, C. E. Donald, S. Meier, S. E. Allan, D. T. Boyd, G. M. Ylitalo, J. R. Cameron, B. L. French, T. L. Linbo, N. L. Scholz, J. P. Incardona. 2019. Embryonic Crude Oil Exposure Impairs Growth and Lipid Allocation in a Keystone Arctic Forage Fish. *iScience* 19:1101–1113.

Lauritano, C., C. Rizzo, A. Lo Giudice, and M. Saggiomo. 2020. Physiological and molecular responses to main environmental stressors of microalgae and bacteria in polar marine environments. *Microorganisms* 8(12):1957.

Lawrence, K. G., A. P. Keil, S. Garantziotis, D. M. Umbach, P. A. Stewart, M. R. Stenzel, J. A. McGrath, W. B. Jackson, R. K. Kwok, and M. D. Curry. 2020. Lung function in oil spill responders 4–6 years after the *Deepwater Horizon* disaster. *Journal of Toxicology and Environmental Health, Part A* 83(6):233–248.

Laxague, N. J., T. M. Özgökmen, B. K. Haus, G. Novelli, A. Shcherbina, P. Sutherland, C. M. Guigand, B. Lund, S. Mehta, and M. Alday. 2018. Observations of near-surface current shear help describe oceanic oil and plastic transport. *Geophysical Research Letters* 45(1):245–249.

Lea-Smith, D. J., S. J. Biller, M. P. Davey, C. A. Cotton, B. M. P. Sepulveda, A. V. Turchyn, D. J. Scanlan, A. G. Smith, S. W. Chisholm, and C. J. Howe. 2015. Contribution of cyanobacterial alkane production to the ocean hydrocarbon cycle. *Proceedings of the National Academy of Sciences* 112(44):13591–13596.

Lea-Smith, D. J., M. L. Ortiz-Suarez, T. Lenn, D. J. Nürnberg, L. L. Baers, M. P. Davey, L. Parolini, R. G. Huber, C. A. R. Cotton, G. Mastroianni, P. Bombelli, P. Ungerer, T. J. Stevens, A. G. Smith, P. J. Bond, C. W. Mullineaux, and C. J. Howe. 2016. Hydrocarbons are essential for optimal cell size, division, and growth of cyanobacteria. *Plant Physiology* 172(3):1928–1940. https://doi.org/10.1104/pp.16.01205.

Lee, C.-H., C.-G. Sung, S.-K. Kang, S.-D. Moon, J.-H. Lee, and J.-H. Lee. 2013. Effects of ultraviolet radiation on the toxicity of water-accommodated fraction and chemically enhanced water-accommodated fraction of *Hebei Spirit* crude oil to the embryonic development of the manila clam, ruditapes philippinarum. *The Korean Journal of Malacology* 29(1):23–32.

Lee, J. H.-W., V. Chu, and V. H. Chu. 2003. *Turbulent jets and plumes: A lagrangian approach.* Vol. 1. Berlin, Germany: Springer Science & Business Media.

Lee, K. 2021. The multi-partner research initiative: A scientific research network to support decision making in oil spill response. *Presentation at the International Oil Spill Conference*, May 12, 2021.

Lee, K., and P. Stoffyn-Egli. 2001. Characterization of oil-mineral aggregates. *International Oil Spill Conference Proceedings* 2001:991–996.

Lee, K., T. Lunel, P. Wood, R. Swannell, and P. Stoffyn-Egli. 1997. Shoreline cleanup by acceleration of clay-oil flocculation processes. *International Oil Spill Conference Proceedings* 1997(1):235–240.

Lee, K., R. Prince, C. Greer, K. Doe, J. Wilson, S. Cobanli, G. Wohlgeschaffen, D. Alroumi, T. King, and G. Tremblay. 2003a. Composition and toxicity of residual bunker C fuel oil in intertidal sediments after 30 years. *Spill Science & Technology Bulletin* 8(2):187–199.

Lee, K., P. Stoffyn-Egli, G. H. Tremblay, E. H. Owens, G. A. Sergy, C. C. Guénette, and R. C. Prince. 2003b. Oil-mineral aggregate formation on oiled beaches: Natural attenuation and sediment relocation. *Spill Science & Technology Bulletin* 8(3):285–296.

Lee, K., K. Azetsu-Scott, S. Cobanli, J. Dalziel, S. Niven, G. Wohlgeschaffen, and P. Yeats. 2005. Overview of potential impacts from produced water discharges in Atlantic Canada. In *Offshore oil and gas environmental effects monitoring: Approaches and technologies*. Columbus, OH: Battelle Press.

Lee, K., M. Boudreau, J. Bugden, L. Burridge, S. Cobanli, S. Courtenay, S. Grenon, B. Hollebone, P. Kepkay, and Z. Li. 2011a. *State of knowledge review of fate and effect of oil in the Arctic marine environment.* A report prepared for the National Energy Board of Canada.

Lee, K., Z. Li, P. Kepkay, and S. Ryan. 2011b. Time-series monitoring the subsurface oil plume released from *Deepwater Horizon* MC252 in the Gulf of Mexico. *International Oil Spill Conference Proceedings.* https://www.researchgate.net/profile/Kenneth-Lee-24/publication/271232223_Time-series_Monitoring_the_Subsurface_Oil_Plume_released_from_Deepwater_Horizon_MC252_in_the_Gulf_of_Mexico/links/5e1b8a4a92851c8364c8d800/Time-series-Monitoring-the-Subsurface-Oil-Plume-released-from-Deepwater-Horizon-MC252-in-the-Gulf-of-Mexico.pdf.

Lee, K., T. Nedwed, R. C. Prince, and D. Palandro. 2013. Lab tests on the biodegradation of chemically dispersed oil should consider the rapid dilution that occurs at sea. *Marine Pollution Bulletin* 73(1):314–318.

Lee, K., M. Boufadel, B. Chen, J. Foght, P. Hodson, S. Swanson, and A. Venosa. 2015. *Expert panel report on the behaviour and environmental impacts of crude oil released into aqueous environments.* Ottawa, Ontario, Canada: Royal Society of Canada.

Lee, L.-H., and H.-J. Lin. 2013. Effects of an oil spill on benthic community production and respiration on subtropical intertidal sandflats. *Marine Pollution Bulletin* 73(1):291–299.

Lee, R. F. 2003. Photo-oxidation and photo-toxicity of crude and refined oils. *Spill Science & Technology Bulletin* 8(2):157–162.

Lee, Y.-Z., F. Leighton, D. Peakall, R. Norstrom, P. O'Brien, J. Payne, and A. Rahimtula. 1985. Effects of ingestion of Hibernia and Prudhoe Bay crude oils on hepatic and renal mixed function oxidase in nestling herring gulls (Larus argentatus). *Environmental Research* 36(1):248–255.

LeGore, S., D. S. Marszalek, L. J. Danek, M. S. Tomlinson, J. E. Hofmann, and J. E. Cuddeback. 1989. Effect of chemically dispersed oil on Arabian Gulf corals: A field experiment. *International Oil Spill Conference Proceedings* 1989(1):375–380

Lehr, B., S. Bristol, and A. Possolo. 2010. Deepwater Horizon *oil budget calculator: A report to the national incident command.* Federal Interagency Solutions Group, Oil Budget Calculator Science and Engineering Team: National Incident Command.

Lehr, W., R. Jones, M. Evans, D. Simecek-Beatty, and R. Overstreet. 2002. Revisions of the ADIOS oil spill model. *Environmental Modelling & Software* 17(2):189–197.

Leifer, I., W. J. Lehr, D. Simecek-Beatty, E. Bradley, R. Clark, P. Dennison, Y. Hu, S. Matheson, C. E. Jones, and B. Holt. 2012. State of the art satellite and airborne marine oil spill remote sensing: Application to the BP *Deepwater Horizon* oil spill. *Remote Sensing of Environment* 124:185–209.

Leighton, F. 1985. Morphological lesions in red blood cells from herring gulls and Atlantic puffins ingesting Prudhoe Bay crude oil. *Veterinary Pathology* 22(4):393–402.

Leighton, F. 1986. Clinical, gross, and histological findings in herring gulls and Atlantic puffins that ingested Prudhoe Bay crude oil. *Veterinary Pathology* 23(3):254–263.

Leighton, F. A. 1993. The toxicity of petroleum oils to birds. *Environmental Reviews* 1(2):92–103. https://doi.org/10.1139/a93-008.

Lemcke, S., J. Holding, E. F. Møller, J. Thyrring, K. Gustavson, T. Juul-Pedersen, and M. K. Sejr. 2019. Acute oil exposure reduces physiological process rates in Arctic phyto- and zooplankton. *Ecotoxicology* 28(1):26–36.

Lemkau, K. L., E. E. Peacock, R. K. Nelson, G. T. Ventura, J. L. Kovecses, and C. M. Reddy. 2010. The m/v Cosco Busan spill: Source identification and short-term fate. *Marine Pollution Bulletin* 60(11):2123–2129. https://doi.org/10.1016/j.marpolbul.2010.09.001.

Leonte, M., B. Wang, S. Socolofsky, S. Mau, J. Breier, and J. Kessler. 2018. Using carbon isotope fractionation to constrain the extent of methane dissolution into the water column surrounding a natural hydrocarbon gas seep in the northern Gulf of Mexico. *Geochemistry, Geophysics, Geosystems* 19(11):4459–4475.

Lessard, R. R., and G. DeMarco. 2000. The significance of oil spill dispersants. *Spill Science & Technology Bulletin* 6(1):59–68.

Levine, J., I. Haljasmaa, R. Lynn, F. Shaffer, and R. Warzinski. 2015. *Detection of hydrates on gas bubbles during a subsea oil/gas leak.* Pittsburgh, PA, Morgantown, WV: National Energy Technology Laboratory.

Lewis, A. 2013. Dispersant testing in realistic conditions. *Final Report 2.1 for the Arctic Oil Spill Response Technology-Joint Industry Programme* 33.

Lewis, A., and P. Daling. 2007. A review of studies of oil spill dispersant effectiveness in Arctic conditions. SINTEF Report STF80MKF07095. *Oil in Ice JIP Report* (11).

Lewis, A., and R. C. Prince. 2018. Integrating dispersants in oil spill response in Arctic and other icy environments. *Environmental Science & Technology* 52(11):6098–6112.

Lewis, A., P. S. Daling, T. Strøm-Kristiansen, A. B. Nordvik, and R. J. Fiocco. 1995. Weathering and chemical dispersion of oil at sea. *International Oil Spill Conference Proceedings* 1995(1):157–164.

LGL Limited and Oceans Ltd. 2018. *Exposure of juveniles of four marine fish species to hydrocarbon discharges from oil and gas production platforms on the Grand Banks.* Prepared by LGL Limited, Environmental Research Associates, and Oceans Ltd. for the Environmental Studies Research Fund.

Li, C., J. Miller, J. Wang, S. Koley, and J. Katz. 2017. Size distribution and dispersion of droplets generated by impingement of breaking waves on oil slicks. *Journal of Geophysical Research: Oceans* 122(10):7938–7957.

Li, H., and M. C. Boufadel. 2010. Long-term persistence of oil from the *Exxon Valdez* spill in two-layer beaches. *Nature Geoscience* 3(2): 96–99.

Li, M., Z. Zhao, Y. Pandya, G. V. Iungo, and D. Yang. 2019. Large-eddy simulations of oil droplet aerosol transport in the marine atmospheric boundary layer. *Atmosphere* 10(8):459.

Li, Y., H. Zhang, and C. Tang. 2020. A review of possible pathways of marine microplastics transport in the ocean. *Anthropocene Coasts* 3(1):6–13.

Li, Z., P. Kepkay, K. Lee, T. King, M. C. Boufadel, and A. D. Venosa. 2007. Effects of chemical dispersants and mineral fines on crude oil dispersion in a wave tank under breaking waves. *Marine Pollution Bulletin* 54(7):983–993.

Li, Z., K. Lee, T. King, M. C. Boufadel, and A. D. Venosa. 2008. Assessment of chemical dispersant effectiveness in a wave tank under regular non-breaking and breaking wave conditions. *Marine Pollution Bulletin* 56(5):903–912.

Li, Z., K. Lee, T. King, M. C. Boufadel, and A. D. Venosa. 2009a. Evaluating chemical dispersant efficacy in an experimental wave tank: 2-significant factors determining in situ oil droplet size distribution. *Environmental Engineering Science* 26(9):1407–1418.

Li, Z., K. Lee, T. King, M. C. Boufadel, and A. D. Venosa. 2009b. Evaluating crude oil chemical dispersion efficacy in a flow-through wave tank under regular non-breaking wave and breaking wave conditions. *Marine Pollution Bulletin* 58(5):735–744.

Li, Z., K. Lee, T. King, P. Kepkay, M. C. Boufadel, and A. D. Venosa. 2009c. Evaluating chemical dispersant efficacy in an experimental wave tank: 1, dispersant effectiveness as a function of energy dissipation rate. *Environmental Engineering Science* 26(6):1139–1148.

Li, Z., K. Lee, T. King, M. C. Boufadel, and A. D. Venosa. 2010. Effects of temperature and wave conditions on chemical dispersion efficacy of heavy fuel oil in an experimental flow-through wave tank. *Marine Pollution Bulletin* 60(9):1550–1559.

Li, Z., M. Spaulding, D. F. McCay, D. Crowley, and J. R. Payne. 2017. Development of a unified oil droplet size distribution model with application to surface breaking waves and subsea blowout releases considering dispersant effects. *Marine Pollution Bulletin* 114(1): 247–257.

Liang, Y., Y. Ning, L. Liao, and B. Yuan. 2018. Special focus on produced water in oil and gas fields: Origin, management, and reinjection practice. In *Formation damage during improved oil recovery*, pp. 515–586. Amsterdam, Netherlands: Elsevier.

Lima, A. L. C., J. W. Farrington, and C. M. Reddy. 2005. Combustion-derived polycyclic aromatic hydrocarbons in the environment—a review. *Environmental Forensics* 6(2):109–131.

Lima-Neto, I. E., D. Z. Zhu, and N. Rajaratnam. 2008. Bubbly jets in stagnant water. *International Journal of Multiphase Flow* 34(12):1130–1141.

Lin, C.-Y. 2013. Strategies for promoting biodiesel use in marine vessels. *Marine Policy* 40:84–90.

Lin, H., G. D. Morandi, R. S. Brown, V. Snieckus, T. Rantanen, K. B. Jørgensen, and P. V. Hodson. 2015. Quantitative structure-activity relationships for chronic toxicity of alkyl-chrysenes and alkyl-benz [a] anthracenes to Japanese Medaka embryos (oryzias latipes). *Aquatic Toxicology* 159:109–118.

Lin, Q., and I. A. Mendelssohn. 2012. Impacts and recovery of the *Deepwater Horizon* oil spill on vegetation structure and function of coastal salt marshes in the northern Gulf of Mexico. *Environmental Science & Technology* 46(7):3737–3743.

Lin, Q., I. A. Mendelssohn, S. A. Graham, A. Hou, J. W. Fleeger, and D. R. Deis. 2016. Response of salt marshes to oiling from the *Deepwater Horizon* spill: Implications for plant growth, soil surface-erosion, and shoreline stability. *Science of the Total Environment* 557:369–377.

Lipscomb, T., R. Harris, R. Moeller, J. Pletcher, R. Haebler, and B. E. Ballachey. 1993. Histopathologic lesions in sea otters exposed to crude oil. *Veterinary Pathology* 30(1):1–11.

Litman, E., S. Emsbo-Mattingly, and W. Wong. 2018. Critical review of an interlaboratory forensic dataset: Effects on data interpretation in oil spill studies. *Oil Spill Environmental Forensics Case Studies* 1–23.

Liu, A., N. Hong, P. Zhu, and Y. Guan. 2018. Understanding benzene series (BTEX) pollutant load characteristics in the urban environment. *Science of the Total Environment* 619–620:938–945. https://doi.org/10.1016/j.scitotenv.2017.11.184.

Liu, F., K. B. Olesen, A. R. Borregaard, and J. Vollertsen. 2019. Microplastics in urban and highway stormwater retention ponds. *Science of the Total Environment* 671:992–1000.

Liu, J., H. P. Bacosa, and Z. Liu. 2017. Potential environmental factors affecting oil-degrading bacterial populations in deep and surface waters of the northern Gulf of Mexico. *Frontiers in Microbiology* 7:2131.

Liu, J., Y. Zheng, H. Lin, X. Wang, M. Li, Y. Liu, M. Yu, M. Zhao, N. Pedentchouk, and D. J. Lea-Smith. 2019. Proliferation of hydrocarbon-degrading microbes at the bottom of the Mariana trench. *Microbiome* 7(1):1–13.

Liu, Y., and E. B. Kujawinski. 2015. Chemical Composition and Potential Environmental Impacts of Water-Soluble Polar Crude Oil Components Inferred from ESI FT-ICR MS. *PLoS ONE* 10(9):e0136376. https://doi.org/10.1371/journal.pone.0136376. https://doi.org/10.1371/journal.pone.0136376.

Liu, Y., H. Lu, Y. Li, H. Xu, Z. Pan, P. Dai, H. Wang, and Q. Yang. 2021. A review of treatment technologies for produced water in offshore oil and gas fields. *Science of the Total Environment* Volume 775(2021):145485.

Liu, Z., and J. B. Phillips. 1991. Comprehensive two-dimensional gas chromatography using an on-column thermal modulator interface. *Journal of Chromatographic Science* 29(6):227–231. https://doi.org/10.1093/chromsci/29.6.227.

Liu, Z., and J. Liu. 2013. Evaluating bacterial community structures in oil collected from the sea surface and sediment in the northern Gulf of Mexico after the *Deepwater Horizon* oil spill. *MicrobiologyOpen* 2(3):492–504.

Liu, Z., J. Liu, Q. Zhu, and W. Wu. 2012. The weathering of oil after the *Deepwater Horizon* oil spill: Insights from the chemical composition of the oil from the sea surface, salt marshes and sediments. *Environmental Research Letters* 7(3):035302.

Lo, N. C.-h., L. D. Jacobson, and J. L. Squire. 1992. Indices of relative abundance from fish spotter data based on delta-lognornial models. *Canadian Journal of Fisheries and Aquatic Sciences* 49(12):2515–2526.

Lofthus, S., R. Netzer, A. S. Lewin, T. M. Heggeset, T. Haugen, and O. G. Brakstad. 2018. Biodegradation of n-alkanes on oil-seawater interfaces at different temperatures and microbial communities associated with the degradation. *Biodegradation* 29(2):141–157.

Lofthus, S., I. Bakke, J. Tremblay, C. W. Greer, and O. G. Brakstad. 2020. Biodegradation of weathered crude oil in seawater with frazil ice. *Marine Pollution Bulletin* 154:111090.

Lohmann, R., and I. M. Belkin. 2014. Organic pollutants and ocean fronts across the Atlantic Ocean: A review. *Progress in Oceanography* 128:172–184.

Lohmann, R., M. Dapsis, E. J. Morgan, V. Dekany, and P. J. Luey. 2011. Determining air-water exchange, spatial and temporal trends of freely dissolved PAHs in an urban estuary using passive polyethylene samplers. *Environmental Science & Technology* 45(7):2655–2662. https://doi.org/10.1021/es1025883.

Lohmann, R., J. Klanova, P. Pribylova, H. Liskova, S. Yonis, and K. Bollinger. 2013. PAHs on a west-to-east transect across the tropical Atlantic Ocean. *Environmental Science & Technology* 47(6):2570–2578. https://doi.org/10.1021/es304764e.

Loomis, D., K. Z. Guyton, Y. Grosse, F. El Ghissassi, V. Bouvard, L. Benbrahim-Tallaa, N. Guha, N. Vilahur, H. Mattock, and K. Straif. 2017. Carcinogenicity of benzene. *The Lancet Oncology* 18(12):1574–1575.

Louvado, A., N. Gomes, M. M. Simões, A. Almeida, D. F. Cleary, and A. Cunha. 2015. Polycyclic aromatic hydrocarbons in deep sea sediments: Microbe-pollutant interactions in a remote environment. *Science of the Total Environment* 526:312–328.

Love, C. R., E. C. Arrington, K. M. Gosselin, C. M. Reddy, B. A. Van Mooy, R. K. Nelson, and D. L. Valentine. 2021. Microbial production and consumption of hydrocarbons in the global ocean. *Nature Microbiology* 6(4):489–498.

Lowe, S. R., R. K. Kwok, J. Payne, L. S. Engel, S. Galea, and D. P. Sandler. 2015. Mental health service use by cleanup workers in the aftermath of the *Deepwater Horizon* oil spill. *Social Science & Medicine* 130:125–134.

Lowe, S. R., R. K. Kwok, J. Payne, L. S. Engel, S. Galea, and D. P. Sandler. 2016. Why does disaster recovery work influence mental health? Pathways through physical health and household income. *American Journal of Community Psychology* 58(3–4):354–364.

Lowe, S. R., J. L. Bonumwezi, Z. Valdespino-Hayden, and S. Galea. 2019a. Posttraumatic stress and depression in the aftermath of environmental disasters: A review of quantitative studies published in 2018. *Current Environmental Health Reports* 6(4):344–360.

Lowe, S. R., J. A. McGrath, M. N. Young, R. K. Kwok, L. S. Engel, S. Galea, and D. P. Sandler. 2019b. Cumulative disaster exposure and mental and physical health symptoms among a large sample of gulf coast residents. *Journal of Traumatic Stress* 32(2):196–205.

Loya, Y., and B. Rinkevich. 1979. Abortion effect in corals induced by oil pollution. *Marine Ecology Progress Series* 1(1):77–80.

Lozada, M., M. S. Marcos, M. G. Commendatore, M. N. Gil, and H. M. Dionisi. 2014. The bacterial community structure of hydrocarbon-polluted marine environments as the basis for the definition of an ecological index of hydrocarbon exposure. *Microbes and Environments* 29(3):269–276.

Lu, K., L. B. Collins, H. Ru, E. Bermudez, and J. A. Swenberg. 2010. Distribution of DNA adducts caused by inhaled formaldehyde is consistent with induction of nasal carcinoma but not leukemia. *Toxicological Sciences* 116(2):441–451.

Lu, Y., J. Shi, C. Hu, M. Zhang, S. Sun, and Y. Liu. 2020. Optical interpretation of oil emulsions in the ocean—part II: Applications to multi-band coarse-resolution imagery. *Remote Sensing of Environment* 242:111778.

Luckenbach Trustee Council. 2006. *SS Jacob Luckenbach and associated mystery oil spills*. California Department of Fish and Game, National Oceanic and Atmospheric Administration, U.S. Fish & Wildlife Service, National Park Service.

Lunel, T. June 1994. Dispersion of a large experimental slick by aerial application of dispersant. *Arctic and Marine Oilspill Program Technical Seminar.* Canada.

Luo, Q., D. Hou, D. Jiang, and W. Chen. 2021. Bioremediation of marine oil spills by immobilized oil-degrading bacteria and nutrition emulsion. *Biodegradation* 32(2):165–177.

Lupo, P. J., E. Symanski, P. H. Langlois, C. C. Lawson, S. Malik, S. M. Gilboa, L. J. Lee, A. Agopian, T. A. Desrosiers, and M. A. Waters. 2012. Maternal occupational exposure to polycyclic aromatic hydrocarbons and congenital heart defects among offspring in the national birth defects prevention study. *Birth Defects Research Part A: Clinical and Molecular Teratology* 94(11):875–881.

Luquet, P., J. P. Cravedi, G. Choubert, J. Tulliez, and G. Bories. 1983. Long-term ingestion by rainbow trout of saturated hydrocarbons: Effects of n-paraffins, pristane and dodecylcyclohexane on growth, feed intake, lipid digestibility and canthaxanthin deposition. *Aquaculture* 34(1–2):15–25.

Luquet, P., J. P. Cravedi, J. Tulliez, and G. Bories. 1984. Growth reduction in trout induced by naphthenic and isoprenoid hydrocarbons (dodecylcyclohexane and pristane). *Ecotoxicology and Environmental Safety* 8(3):219–226.

Lynch, A. M. 2016. Comment on "New look at BTEX: Are ambient levels a problem?" *Environmental Science & Technology* 50(2):1070–1071.

Lyu, L.-N., H. Ding, Z. Cui, and D. L. Valentine. 2018. The wax-liquid transition modulates hydrocarbon respiration rates in Alcanivorax borkumensis SK2. *Environmental Science & Technology Letters* 5(5):277–282. https://doi.org/10.1021/acs.estlett.8b00143.

Ma, W.-L., L.-Y. Liu, H. Qi, Z.-F. Zhang, W.-W. Song, J.-M. Shen, Z.-L. Chen, N.-Q. Ren, J. Grabuski, and Y.-F. Li. 2013. Polycyclic aromatic hydrocarbons in water, sediment and soil of the Songhua River basin, China. *Environmental Monitoring and Assessment* 185(10):8399–8409. https://doi.org/10.1007/s10661-013-3182-7.

Ma, Y., Z. Xie, H. Yang, A. Möller, C. Halsall, M. Cai, R. Sturm, and R. Ebinghaus. 2013. Deposition of polycyclic aromatic hydrocarbons in the North Pacific and the Arctic. *Journal of Geophysical Research: Atmospheres* 118(11):5822–5829. https://doi.org/10.1002/jgrd.50473.

Mabile, N. 2010. *Fire boom performance evaluation: Controlled burning during the* Deepwater Horizon *spill operational period.* Houston, TX: BP Americas.

Mabile, N. J. 2013. Considerations for the application of controlled in-situ burning. *Oil and Gas Facilities* 2(2):72–84.

MacDonald, I., N. Guinasso, Jr., S. Ackleson, J. Amos, R. Duckworth, R. Sassen, and J. Brooks. 1993. Natural oil slicks in the Gulf of Mexico visible from space. *Journal of Geophysical Research: Oceans* 98(C9):16351–16364.

MacDonald, I., N. Guinasso, Jr., R. Sassen, J. Brooks, L. Lee, and K. Scott. 1994. Gas hydrate that breaches the sea floor on the continental slope of the Gulf of Mexico. *Geology* 22(8):699–702.

MacDonald, I., I. Leifer, R. Sassen, P. Stine, R. Mitchell, and N. Guinasso, Jr. 2002. Transfer of hydrocarbons from natural seeps to the water column and atmosphere. *Geofluids* 2(2):95–107.

MacDonald, I. R., O. Garcia-Pineda, A. Beet, S. Daneshgar Asl, L. Feng, G. Graettinger, D. French-McCay, J. Holmes, C. Hu, F. Huffer, I. Leifer, F. Muller-Karger, A. Solow, M. Silva, and G. Swayze. 2015. Natural and unnatural oil slicks in the Gulf of Mexico. *Journal of Geophysical Research: Oceans* 120(12):8364–8380. https://doi.org/10.1002/2015JC011062.

MacFadyen, A., E. Wei, C. Warren, C. Henry, and G. Watabayashi. 2014. Utilization of the northern gulf operational forecast system to predict trajectories of surface oil from a persistent source offshore of the Mississippi river delta. *International Oil Spill Conference Proceedings* 2014(1):531–543. https://doi.org/10.7901/2169-3358-2014.1.531.

Macías-Zamora, J. V., and N. Ramírez-Alvarez. 2004. Tracing sewage pollution using linear alkylbenzenes (LABs) in surface sediments at the south end of the Southern California Bight. *Environmental Pollution* 130(2):229–238. https://doi.org/10.1016/j.envpol.2003.12.004.

Mackay, D. 1977. *Mathematical model of the behavior of oil spills on water with natural and chemical dispersion*. Report EPS-3-EC-77-19. Ottawa, Ontario, Canada: Environment Canada.

Mackay, D., and P. J. Leinonen. 1977. Mathematical model of the behavior of oil spills on water with natural and chemical dispersion. Prepared for Fisheries and Environment Canada. *Economic and Technical Review Report EPS-3-EC-77-19*.

Mackay, D., and R. S. Matsugu. 1973. Evaporation rates of liquid hydrocarbon spills on land and water. *The Canadian Journal of Chemical Engineering* 51(4):434–439.

Mackay, D., and W. Zagorski. 1982. *Studies of water-in-oil emulsions*. Ottawa, Ontario, Canada: Environmental Protection Service, Environmental Impact Control Directorate, Environmental Emergency Branch, Research and Development Division, Environment Canada.

Mackay, D., W. Shiu, and A. Wolkoff. 1975. *Gas chromatographic determination of low concentrations of hydrocarbons in water by vapor phase extraction*. West Conshohocken, PA: ASTM International.

Mackay, D., J. A. Arnot, F. Wania, and R. E. Bailey. 2011. Chemical activity as an integrating concept in environmental assessment and management of contaminants. *Integrated Environmental Assessment and Management* 7(2):248–255.

Maggini, I., L. V. Kennedy, A. Macmillan, K. H. Elliott, K. Dean, and C. G. Guglielmo. 2017. Light oiling of feathers increases flight energy expenditure in a migratory shorebird. *Journal of Experimental Biology* 220(13):2372–2379. https://doi.org/10.1242/jeb.158220.

Malins, D. C. 1977. *Effects of petroleum on Arctic and subarctic marine environments and organisms*. Cambridge: MA: Academic Press.

Malone, K., S. Pesch, M. Schlüter, and D. Krause. 2018. Oil droplet size distributions in deep-sea blowouts: Influence of pressure and dissolved gases. *Environmental Science & Technology* 52(11):6326–6333.

Mandalakis, M., Ö. Gustafsson, T. Alsberg, A.-L. Egebäck, C. M. Reddy, L. Xu, J. Klanova, I. Holoubek, and E. G. Stephanou. 2005. Contribution of biomass burning to atmospheric polycyclic aromatic hydrocarbons at three European background sites. *Environmental Science & Technology* 39(9):2976–2982. https://doi.org/10.1021/es048184v.

Margesin, R., and F. Schinner. 2001. Biodegradation and bioremediation of hydrocarbons in extreme environments. *Applied Microbiology and Biotechnology* 56(5):650–663.

Marietou, A., R. Chastain, F. Beulig, A. Scoma, T. C. Hazen, and D. H. Bartlett. 2018. The effect of hydrostatic pressure on enrichments of hydrocarbon degrading microbes from the Gulf of Mexico following the *Deepwater Horizon* oil spill. *Frontiers in Microbiology* 9:808.

Markiewicz, A., K. Björklund, E. Eriksson, Y. Kalmykova, A.-M. Strömvall, and A. Siopi. 2017. Emissions of organic pollutants from traffic and roads: Priority pollutants selection and substance flow analysis. *Science of the Total Environment* 580:1162–1174.

Marres, G. M., L. P. Leenen, J. de Vries, P. G. Mulder, and E. Vermetten. 2011. Disaster-related injury and predictors of health complaints after exposure to a natural disaster: An online survey. *BMJ Open* 1(2):e000248.

Marshall, A. G., C. L. Hendrickson, and G. S. Jackson. 1998. Fourier transform ion cyclotron resonance mass spectrometry: A primer. *Mass Spectrometry Reviews* 17(1):1–35.

Martinelli, F., and D. Cormack. 1979. *Investigation of the effects of oil viscosity and water-in-oil emulsion formation on dispersant efficiency*. Stevenage, UK: Warren Spring Lab.

Martínez, M. D., P. R. Romero, and A. T. Banaszak. 2007. Photoinduced toxicity of the polycyclic aromatic hydrocarbon, fluoranthene, on the coral, porites divaricata. *Journal of Environmental Science and Health, Part A* 42(10):1495–1502.

Martínez, M. L., R. A. Feagin, K. M. Yeager, J. Day, R. Costanza, J. A. Harris, R. J. Hobbs, J. López-Portillo, I. J. Walker, E. Higgs, P. Moreno-Casasola, J. Sheinbaum, and A. Yáñez-Arancibia. 2012. Artificial modifications of the coast in response to the *Deepwater Horizon* oil spill: Quick solutions or long-term liabilities? *Frontiers in Ecology and the Environment* 10(1):44–49. https://doi.org/10.1890/100151.

Martins, C., J. Ferreira, S. Taniguchi, M. Mahiques, M. Bicego, and R. Montone. 2008. Spatial distribution of sedimentary linear alkylbenzenes and faecal steroids of Santos Bay and adjoining continental shelf, SW Atlantic, Brazil: Origin and fate of sewage contamination in the shallow coastal environment. *Marine Pollution Bulletin* 56:1359–1363. 10.1016/j.marpolbul.2008.04.011.

Marzooghi, S., and D. M. Di Toro. 2017. A critical review of polycyclic aromatic hydrocarbon phototoxicity models. *Environmental Toxicology and Chemistry* 36(5):1138–1148.

Marzooghi, S., B. E. Finch, W. A. Stubblefield, O. Dmitrenko, S. L. Neal, and D. M. Di Toro. 2017. Phototoxic target lipid model of single polycyclic aromatic hydrocarbons. *Environmental Toxicology and Chemistry* 36(4):926–937.

Marzooghi, S., B. E. Finch, W. A. Stubblefield, and D. M. Di Toro. 2018. Predicting phototoxicity of alkylated PAHs, mixtures of PAHs, and water accommodated fractions (WAFs) of neat and weathered petroleum with the phototoxic target lipid model. *Environmental Toxicology and Chemistry* 37(8):2165–2174.

Mason, J., A. C. Ortmann, S. E. Cobanli, G. Wohlgeschaffen, P. Thamer, C. McIntyre, and T. L. King. 2019. Inorganic nutrients have a significant, but minimal, impact on a coastal microbial community's response to fresh diluted bitumen. *Marine Pollution Bulletin* 139:381–389.

Mason, O. U., T. C. Hazen, S. Borglin, P. S. Chain, E. A. Dubinsky, J. L. Fortney, J. Han, H.-Y. N. Holman, J. Hultman, and R. Lamendella. 2012. Metagenome, metatranscriptome and single-cell sequencing reveal microbial response to *Deepwater Horizon* oil spill. *The ISME Journal* 6(9):1715–1727.

Mason, O. U., J. Han, T. Woyke, and J. K. Jansson. 2014a. Single-cell genomics reveals features of a colwellia species that was dominant during the *Deepwater Horizon* oil spill. *Frontiers in Microbiology* 5:332.

Mason, O. U., N. M. Scott, A. Gonzalez, A. Robbins-Pianka, J. Bælum, J. Kimbrel, N. J. Bouskill, E. Prestat, S. Borglin, and D. C. Joyner. 2014b. Metagenomics reveals sediment microbial community response to *Deepwater Horizon* oil spill. *The ISME Journal* 8(7):1464–1475.

Massey, J. G. 2006. Summary of an oiled bird response. *Journal of Exotic Pet Medicine* 15(1):33–39.

Masutani, S., and E. Adams. 2000. Experimental study of multiphase plumes with application to deep ocean oil spills, final report. *Contract* (2000):1435–1401.

Mathewson, P. D., K. C. Hanson-Dorr, W. P. Porter, S. J. Bursian, K. M. Dean, K. Healy, K. Horak, J. E. Link, K. Harr, and B. S. Dorr. 2018. Experimental and modeled thermoregulatory costs of repeated sublethal oil exposure in the double-crested cormorant, Phalacrocorax auritus. *Marine Pollution Bulletin* 135:216–223.

Maunder, M. N., and A. E. Punt. 2004. Standardizing catch and effort data: A review of recent approaches. *Fisheries Research* 70(2–3):141–159.

Maximenko, N., J. Hafner, M. Kamachi, and A. MacFadyen. 2018. Numerical simulations of debris drift from the great japan tsunami of 2011 and their verification with observational reports. *Marine Pollution Bulletin* 132:5–25.

May, L. A., A. R. Burnett, C. V. Miller, E. Pisarski, L. F. Webster, Z. J. Moffitt, P. Pennington, E. Wirth, G. Baker, and R. Ricker. 2020. Effect of Louisiana sweet crude oil on a Pacific coral, pocillopora damicornis. *Aquatic Toxicology* 222:105454.

May, W. E. 1980. Chapter 7: The solubility behavior of polycyclic aromatic hydrocarbons in aqueous systems. In *Petroleum in the marine environment, advances in chemistry series*, L. Petrakis and F. T. Weiss (eds.), pp. 143–192. Washington, DC: American Chemical Society.

Mazet, J. A. K., S. H. Newman, K. V. K. Gilardi, F. S. Tseng, J. B. Holcomb, D. A. Jessup, and M. H. Ziccardi. 2002. Advances in oiled bird emergency medicine and management. *Journal of Avian Medicine and Surgery* 16(2):144, 146–149. https://doi.org/10.1647/1082-6742(2002)016[0146:AIOBEM]2.0.CO;2.

Mazet, J. K., I. A. Gardner, D. A. Jessup, L. J. Lowenstine, and W. M. Boyce. 2000. Evaluation of changes in hematologic and clinical biochemical values after exposure to petroleum products in mink (Mustela vison) as a model for assessment of sea otters (Enhydra lutris). *American Journal of Veterinary Research* 61(10):1197–1203.

Mbadinga, S. M., L.-Y. Wang, L. Zhou, J.-F. Liu, J.-D. Gu, and B.-Z. Mu. 2011. Microbial communities involved in anaerobic degradation of alkanes. *International Biodeterioration & Biodegradation* 65(1):1–13.

McAlexander, B. L. 2014. A suggestion to assess spilled hydrocarbons as a greenhouse gas source. *Environmental Impact Assessment Review* 49:57–58.

McAuliffe, C. D. 1989. The weathering of volatile hydrocarbons from crude oil slicks on water. *International Oil Spill Conference Proceedings* 1989(1):357–363.

McAuliffe, C. D., J. C. Johnson, S. H. Greene, G. P. Canevari, and T. D. Searl. 1980. Dispersion and weathering of chemically treated crude oils on the ocean. *Environmental Science & Technology* 14(12):1509–1518.

McAuliffe, C. D., B. L. Steelman, W. R. Leek, D. E. Fitzgerald, J. P. Ray, and C. D. Barker. 1981. The 1979 southern California dispersant treated research oil spills. *International Oil Spill Conference Proceedings* 1981(1):269–282.

McCain, Jr., W. D. 1990. *The properties of petroleum fluids, 2nd edition*. Tulsa, OK: PennWell Publishing Company.

McCall, B. D., and S. C. Pennings. 2012. Disturbance and recovery of salt marsh arthropod communities following BP *Deepwater Horizon* oil spill. *PloS ONE* 7(3):e32735.

McCarty, L. S., and D. Mackay. 1993. Enhancing ecotoxicological modeling and assessment. Body residues and modes of toxic action. *Environmental Science & Technology* 27(9):1718–1728.

McCay, D., J. Rowe, W. Nordhausen, and J. Payne. 2006. Modelling potential impacts of effective dispersant use on aquatic biota. *Proceedings of the 29th Arctic and Marine Oilspill Program Technical Seminar*, pp. 855–878. Ottawa, Ontario, Canada: Environment Canada.

McCay, D. F. 2003. Development and application of damage assessment modeling: Example assessment for the North Cape oil spill. *Marine Pollution Bulletin* 47(9–12):341–359. https://doi.org/10.1016/S0025-326X(03)00208-X.

McCay, D. F., and J. R. Payne. 2001. *Model of oil fate and water concentrations with and without application of dispersants*. Arctic and Marine Oilspill Program Technical Seminar. Ottawa, Ontario, Canada: Environment Canada.

McCay, D. F., and J. J. Rowe. 2004. *Evaluation of bird impacts in historical oil spill cases using the SIMAP oil spill model*. Arctic and Marine Oilspill Program Technical Seminar. Ottawa, Ontario, Canada: Environment Canada.

McClenachan, G., R. E. Turner, and A. W. Tweel. 2013. Effects of oil on the rate and trajectory of Louisiana marsh shoreline erosion. *Environmental Research Letters* 8(4):044030.

McDonald, T. L., B. A. Schroeder, B. A. Stacy, B. P. Wallace, L. A. Starcevich, J. Gorham, M. C. Tumlin, D. Cacela, M. Rissing, and D. B. McLamb. 2017. Density and exposure of surface-pelagic juvenile sea turtles to *Deepwater Horizon* oil. *Endangered Species Research* 33:69–82.

McEwan, E. H., and A. F. C. Koelink. 1973. The heat production of oiled mallards and scaup. *Canadian Journal of Zoology* 51(1):27–31. https://doi.org/10.1139/z73-005.

McFarlin, K. M., R. C. Prince, R. Perkins, and M. B. Leigh. 2014. Biodegradation of dispersed oil in Arctic seawater at −1°C. *PloS ONE* 9(1):e84297.

McFarlin, K., M. Perkins, J. Field, and M. Leigh. 2018. Biodegradation of crude oil and COREXIT 9500 in Arctic seawater. *Frontiers in Microbiology* 9:1788.

McGenity, T. J. 2018. *Taxonomy, genomics and ecophysiology of hydrocarbon-degrading microbes*. Dordrecht, Netherlands: Springer.

McGenity, T. J., B. D. Folwell, B. A. McKew, and G. O. Sanni. 2012. Marine crude-oil biodegradation: A central role for interspecies interactions. *Aquatic Biosystems* 8(1):1–19.

McGenity, T. J., A. T. Crombie, and J. C. Murrell. 2018. Microbial cycling of isoprene, the most abundantly produced biological volatile organic compound on earth. *The ISME Journal* 12(4):931–941.

McGinnis, D. F., J. Greinert, Y. Artemov, S. Beaubien, and A. Wüest. 2006. Fate of rising methane bubbles in stratified waters: How much methane reaches the atmosphere? *Journal of Geophysical Research: Oceans* 111(C9).

McGowan, C. J., R. K. Kwok, L. S. Engel, M. R. Stenzel, P. A. Stewart, and D. P. Sandler. 2017. Respiratory, dermal, and eye irritation symptoms associated with COREXIT™ EC9527A/EC9500A following the *Deepwater Horizon* oil spill: Findings from the Gulf study. *Environmental Health Perspectives* 125(9):097015.

McGrath, J. A., and D. M. Di Toro. 2009. Validation of the target lipid model for toxicity assessment of residual petroleum constituents: Monocyclic and polycyclic aromatic hydrocarbons. *Environmental Toxicology and Chemistry* 28(6):1130–1148.

McGrath, J. A., C. J. Fanelli, D. M. Di Toro, T. F. Parkerton, A. D. Redman, M. L. Paumen, M. Comber, C. V. Eadsforth, and K. den Haan. 2018. Re-evaluation of target lipid model-derived HC5 predictions for hydrocarbons. *Environmental Toxicology and Chemistry* 37(6):1579–1593.

McGrattan, K. B., A. D. Putorti, W. H. Twilley, and D. D. Evans. 1993. *Smoke plume trajectory of in situ burning of crude oil in Alaska*. Gaithersburg, MD: National Institute of Standards and Technology, U.S. Department of Commerce.

McGrattan, K. B., H. R. Baum, and R. G. Rehm. 1995. Numerical simulation of smoke plumes from large oil fires. *Atmospheric Environment* 30(24):4125–4136. https://doi.org/10.1016/1352-2310(96)00151-3.

McIntyre, J. K., R. C. Edmunds, M. G. Redig, E. M. Mudrock, J. W. Davis, J. P. Incardona, J. D. Stark, and N. L. Scholz. 2016. Confirmation of stormwater bioretention treatment effectiveness using molecular indicators of cardiovascular toxicity in developing fish. *Environmental Science & Technology* 50(3):1561–1569.

McKew, B. A., F. Coulon, M. M. Yakimov, R. Denaro, M. Genovese, C. J. Smith, A. M. Osborn, K. N. Timmis, and T. J. McGenity. 2007. Efficacy of intervention strategies for bioremediation of crude oil in marine systems and effects on indigenous hydrocarbonoclastic bacteria. *Environmental Microbiology* 9(6):1562–1571. https://doi.org/10.1111/j.1462–2920.2007.01277.x.

McKim, J. M., S. P. Bradbury, and G. J. Niemi. 1987. Fish acute toxicity syndromes and their use in the qsar approach to hazard assessment. *Environmental Health Perspectives* 71:171–186.

McLeod, W., and D. McLeod. 1972. *Measures to combat offshore artie oil spills*. Offshore Technology Conference, Houston, Texas. https://doi.org/10.4043/1523-MS.

McNutt, M. K., R. Camilli, T. J. Crone, G. D. Guthrie, P. A. Hsieh, T. B. Ryerson, O. Savas, and F. Shaffer. 2011. Review of flow rate estimates of the *Deepwater Horizon* oil spill. *Proceedings of the National Academy of Sciences* 109(50):20260–20267.

McNutt, M., K. Lehnert, B. Hanson, B. A. Nosek, A. M. Ellison, and J. L. King. 2016. Liberating field science samples and data. *Science* 351(6277):1024–1026.

Meador, J., J. Stein, W. Reichert, and U. Varanasi. 1995. Bioaccumulation of polycyclic aromatic hydrocarbons by marine organisms. *Reviews of Environmental Contamination and Toxicology* 143:79–165.

Mearns, A., D. Janka, R. Campbell, S. Pegau, K. McLaughlin, M. Lindeberg, P. Eiting, and G. Shigenaka. 2017. Twenty-six years after the *Exxon Valdez* oil spill: Volunteers continue monitor long-term variability of intertidal biology in western Prince William Sound. *International Oil Spill Conference Proceedings* 2017(1):2017340.

Mearns, A. J. 1997. Cleaning oiled shores: Putting bioremediation to the test. *Spill Science & Technology Bulletin* 4(4):209–217.

Mearns, A. J., and Evans, M. B. 2007. *Using consensus ecological risk assessment (CERA) to evaluate oil spill response options do more skimmers equal better response?* Seattle, WA: Emergency Response Division, Office of Response and Restoration, National Oceanic and Atmospheric Administration. https://response.restoration.noaa.gov/sites/default/files/CERA_poster_SETAC_2007.pdf.

Meier, S., T.E. Andersen, B. Norberg, A. Thorsen, G.L. Taranger, O.S. Kjesbu, R. Dale, H.C. Morton, J. Klungsoyr, A. Svardal. 2007. Effects of alkylphenols on the reproductive system of Atlantic cod (Gadus morhua). *Aquatic Toxicology* 81:207–218.

Meiners, K. M., and C. Michel. 2017. Dynamics of nutrients, dissolved organic matter and exopolymers. *Sea Ice, Third Edition*.

Menard, H. W., and S. M. Smith. 1966. Hypsometry of ocean basin provinces. *Journal of Geophysical Research (1896–1977)* 71(18):4305–4325. https://doi.org/10.1029/JZ071i018p04305.

Menge, B. A. 1995. Indirect effects in marine rocky intertidal interaction webs: Patterns and importance. *Ecological Monographs* 65(1):21–74.

Meo, S. A., A. M. Al-Drees, I. M. Meo, M. M. Al-Saadi, and M. A. Azeem. 2008. Lung function in subjects exposed to crude oil spill into sea water. *Marine Pollution Bulletin* 56(1):88–94.

Meo, S. A., A. M. Al-Drees, S. Rasheed, I. M. Meo, M. M. Khan, M. M. Al-Saadi, and J. R. Alkandari. 2009. Effect of duration of exposure to polluted air environment on lung function in subjects exposed to crude oil spill into sea water. *International Journal of Occupational Medicine and Environmental Health* 22(1):35–41.

Merlin, F., and P. Le Guerroue. 2009. *Use of sorbents for spill response*. Cedre. https://wwz.cedre.fr/en/content/download/1776/140008/file/extract-sorbents.pdf.

Messina, E., R. Denaro, F. Crisafi, F. Smedile, S. Cappello, M. Genovese, L. Genovese, L. Giuliano, D. Russo, and M. Ferrer. 2016. Genome sequence of obligate marine polycyclic aromatic hydrocarbons-degrading bacterium *Cycloclasticus* sp. 78-ME, isolated from petroleum deposits of the sunken tanker *Amoco Milford Haven*, Mediterranean Sea. *Marine Genomics* 25:11–13.

Meyer, P. 2014. Testing of oil recovery skimmers in ice at Ohmsett, the national oil spill response research & renewable energy test facility. *International Oil Spill Conference Proceedings* 2014(1):618–633.

Meyer, P., B. Schmidt, D. DeVitis, and J-E. Delgado. 2012. *High capacity advancing oil recovery system performance testing at Ohmsett for the Wendy Schmidt oil cleanup X challenge*. Interspill Conference. https://www.interspill.org/wp-content/uploads/2021/11/William-Schmidt-High-Capacity-Advancing-Oil-Recovery-System-Performance-Testing-at-Ohmsett-for-the-Wendy-Schmidt-Oil-Cleanup-X-CHALLENGE.pdf.

Michalis, V. K., O. A. Moultos, I. N. Tsimpanogiannis, and I. G. Economou. 2016. Molecular dynamics simulations of the diffusion coefficients of light n-alkanes in water over a wide range of temperature and pressure. *Fluid Phase Equilibria* 407:236–242.

Michel, J., and P. Bambach. 2020. *A response guide for sunken oil mats (SOMs): Formation, behavior, detection and recovery*. Silver Spring, MD: National Ocean Service, National Oceanic and Atmospheric Administration.

Michel, J., and M. Fingas. 2016. Oil spills: Causes, consequences, prevention, and countermeasures. In *Fossil fuels: Current status and future directions*, G. M. Crawley (ed.), pp. 159–201. Hackensack, NJ: World Scientific.

Michel, J., and M. Ploen. 2017. Options for minimizing environmental impacts of inland spill response: New guide from the American Petroleum Institute. *International Oil Spill Conference Proceedings* 2017(1):1770–1783.

Michel, J., and N. Rutherford. 2014. Impacts, recovery rates, and treatment options for spilled oil in marshes. *Marine Pollution Bulletin* 82(1–2):19–25.

Michel, J., M. O. Hayes, and P. J. Brown. 1978. Application of an oil spill vulnerability index to the shoreline of lower Cook Inlet, Alaska. *Environmental Geology* 2(2):107–117.

Michel, J., F. Csulak, D. French, and M. Sperduto. 1997. Natural resource impacts from the North Cape oil spill. *International Oil Spill Conference Proceedings* 1997(1):841–850.

Michel, J., T. Gilbert, D. Etkin, R. Urban, J. Waldron, and C. T. Blocksidge. 2005. *Potentially polluting wrecks in marine waters*. Annals of the 2005 International Oil Spill Conference, Maio. https://portal.helcom.fi/meetings/SUBMERGED%205-2016-377/Related%20Information/Potentially%20Polluting%20Wrecks%20in%20Marine%20Water_Michel_etal_2005.pdf.

Michel, J., E. H. Owens, S. Zengel, A. Graham, Z. Nixon, T. Allard, W. Holton, P. D. Reimer, A. Lamarche, and M. White. 2013. Extent and degree of shoreline oiling: *Deepwater Horizon* oil spill, Gulf of Mexico, USA. *PloS ONE* 8(6):e65087.

Michel, J., S. Fegley, J. Dahlin, and C. Wood. 2017. Oil spill response-related injuries on sand beaches: When shoreline treatment extends the impacts beyond the oil. *Marine Ecology Progress Series* 576. https://doi.org/10.3354/meps11917.

Michel, K., C. Moore, and R. Tagg. 1996. A simplified methodology for evaluating alternative tanker configurations. *Journal of Marine Science and Technology* 1(4):209–219.

Michel, K., and T. S. Winslow. 2000. Cargo ship bunker tanks: Designing to mitigate oil spillage. *Marine Technology and SNAME News* 37(4):191–199.

Middlebrook, A. M., D. M. Murphy, R. Ahmadov, E. L. Atlas, R. Bahreini, D. R. Blake, J. Brioude, J. A. De Gouw, F. C. Fehsenfeld, and G. J. Frost. 2012. Air quality implications of the *Deepwater Horizon* oil spill. *Proceedings of the National Academy of Sciences* 109(50):20280–20285.

Miles, J. A. 1989. *Illustrated glossary of petroleum geochemistry*. New York: Oxford University Press.

Milgram, J. 1983. Mean flow in round bubble plumes. *Journal of Fluid Mechanics* 133:345–376.

Miller, S., and J. Kotula. 2014. Alaska's approach to determining oil recovery rates and efficiencies. *International Oil Spill Proceedings* 2014(1):1749–1758.

Mills, C. G., and K. S. McNeal. 2014. Salt marsh sediment biogeochemical response to the BP *Deepwater Horizon* blowout. *Journal of Environmental Quality* 43(5):1813–1819.

Mitchelmore, C., C. Bishop, and T. Collier. 2017. Toxicological estimation of mortality of oceanic sea turtles oiled during the *Deepwater Horizon* oil spill. *Endangered Species Research* 33:39–50.

Mitchelmore, C. L., R. J. Griffitt, G. M. Coelho, and D. L. Wetzel. 2020a. Modernizing protocols for aquatic toxicity testing of oil and dispersant. In *Scenarios and responses to future deep oil spills*, pp. 239–252. Dordrecht, Netherlands: Springer.

Mitchelmore, C. L., A. C. Bejarano, and D. L. Wetzel. 2020b. A synthesis of DWH oil: Chemical dispersant and chemically dispersed oil aquatic standard laboratory acute and chronic toxicity studies. In *Deep oil spills*, pp. 480–496. Dordrecht, Netherlands: Springer.

Mitchelmore, C. L., E. E. Burns, A. Conway, A. Heyes, and I. A. Davies. 2021. A Critical Review of Organic Ultraviolet Filter Exposure, Hazard, and Risk to Corals. *Environmental Toxicology and Chemistry* 40(4):967–988.

Miyata, N., K. Iwahori, J. M. Foght, and M. R. Gray. 2004. Saturable, energy-dependent uptake of phenanthrene in aqueous phase by mycobacterium sp. Strain RJGII-135. *Applied and Environmental Microbiology* 70(1):363–369.

Mobarak, H., E. N. Mohamad, H. H. Masjuki, M. Kalam, K. Al Mahmud, M. Habibullah, and A. Ashraful. 2014. The prospects of biolubricants as alternatives in automotive applications. *Renewable and Sustainable Energy Reviews* 33:34–43.

Mohr, F. C., B. Lasley, and S. Bursian. 2010. Fuel oil–induced adrenal hypertrophy in ranch mink (mustela vison): Effects of sex, fuel oil weathering, and response to adrenocorticotropic hormone. *Journal of Wildlife Diseases* 46(1):103–110.

Moilleron, R., A. Gonzalez, G. Chebbo, and D. R. Thévenot. 2002. Determination of aliphatic hydrocarbons in urban runoff samples from the "le Marais" experimental catchment in Paris centre. *Water Research* 36(5):1275–1285.

Moldanová, J., E. Fridell, O. Popovicheva, B. Demirdjian, V. Tishkova, A. Faccinetto, and C. Focsa. 2009. Characterisation of particulate matter and gaseous emissions from a large ship diesel engine. *Atmospheric Environment* 43(16):2632–2641.

Moldanová, J., E. Fridell, H. Winnes, S. Holmin-Fridell, J. Boman, A. Jedynska, V. Tishkova, B. Demirdjian, S. Joulie, and H. Bladt. 2013. Physical and chemical characterisation of PM emissions from two ships operating in European emission control areas. *Atmospheric Measurement Techniques* 6(12):3577–3596.

Monaghan, J., L. C. Richards, G. W. Vandergrift, L. J. Hounjet, S. R. Stoyanov, C. G. Gill, and E. T. Krogh. 2021. Direct mass spectrometric analysis of naphthenic acids and polycyclic aromatic hydrocarbons in waters impacted by diluted bitumen and conventional crude oil. *Science of the Total Environment* 765:144206.

Monahan, E. C., C. W. Fairall, K. L. Davidson, and P. J. Boyle. 1983. Observed inter-relations between 10m winds, ocean whitecaps and marine aerosols. *Quarterly Journal of the Royal Meteorological Society* 109(460):379–392.

Montagna, P. A., and D. E. Harper, Jr. 1996. Benthic infaunal long-term response to offshore production platforms in the Gulf of Mexico. *Canadian Journal of Fisheries and Aquatic Sciences* 53(11): 2567–2588.

Montagna, P. A., J. E. Bauer, J. Toal, D. Hardin, and R. B. Spies. 1987. Temporal variability and the relationship between benthic meiofaunal and microbial populations of a natural coastal petroleum seep. *Journal of Marine Research* 45(3):761–789.

Montagna, P. A., J. G. Baguley, C. Cooksey, I. Hartwell, L. J. Hyde, J. L. Hyland, R. D. Kalke, L. M. Kracker, M. Reuscher, and A. C. Rhodes. 2013. Deep-sea benthic footprint of the *Deepwater Horizon* blowout. *PloS ONE* 8(8):e70540.

Montilla, L. M., R. Ramos, E. García, and A. Cróquer. 2016. Caribbean yellow band disease compromises the activity of catalase and glutathione S-transferase in the reef-building coral Orbicella faveolata exposed to anthracene. *Diseases of Aquatic Organisms* 119(2):153–161.

Moody, R. M., J. Cebrian, and K. L. Heck, Jr. 2013. Interannual recruitment dynamics for resident and transient marsh species: Evidence for a lack of impact by the Macondo oil spill. *PloS ONE* 8(3):e58376.

Morris, Jr., J. G., L. M. Grattan, B. M. Mayer, and J. K. Blackburn. 2013. Psychological responses and resilience of people and communities impacted by the *Deepwater Horizon* oil spill. *Transactions of the American Clinical and Climatological Association* 124:191.

MOSSFA (Marine Oil Snow Sedimentation and Flocculent Accumulation). 2013. *MOSSFA workshop report.* Tallahassee, FL: Florida State University.

Moyer, C. L., and R. Y. Morita. 2007. Psychrophiles and psychrotrophs. *Encyclopedia of Life Sciences* 1(6).

Mukai, F. H., and B. D. Goldstein. 1976. Mutagenicity of malonaldehyde, a decomposition product of peroxidized polyunsaturated fatty acids. *Science* 191(4229):868–869.

Mulabagal, V., F. Yin, G. John, J. Hayworth, and T. Clement. 2013. Chemical fingerprinting of petroleum biomarkers in *Deepwater Horizon* oil spill samples collected from Alabama shoreline. *Marine Pollution Bulletin* 70(1–2):147–154.

Müller, A., H. Österlund, J. Marsalek, and M. Viklander. 2020. The pollution conveyed by urban runoff: A review of sources. *Science of the Total Environment* 709:136125.

Mumtaz, M., J. George, K. Gold, W. Cibulas, and C. DeRosa. 1996. Atsdr evaluation of health effects of chemicals. IV. Polycyclic aromatic hydrocarbons (PAHs): Understanding a complex problem. *Toxicology and Industrial Health* 12(6):742–971.

Murawski, S. A., W. T. Hogarth, E. B. Peebles, and L. Barbeiri. 2014. Prevalence of external skin lesions and polycyclic aromatic hydrocarbon concentrations in Gulf of Mexico fishes, post-*Deepwater Horizon*. *Transactions of the American Fisheries Society* 143(4):1084–1097. https://doi.org/10.1080/00028487.2014.911205.

Murawski, S. A., J. P. Kilborn, A. C. Bejarano, D. Chagaris, D. Donaldson, F. J. Hernandez, T. C. MacDonald, C. Newton, E. Peebles, and K. L. Robinson. 2021. A synthesis of *Deepwater Horizon* impacts on coastal and nearshore living marine resources. *Frontiers in Marine Science* 7. https://doi.org/10.3389/fmars.2020.594862.

Murphy, D., B. Gemmell, L. Vaccari, C. Li, H. Bacosa, M. Evans, C. Gemmell, T. Harvey, M. Jalali, and T. H. R. Niepa. 2016. An in-depth survey of the oil spill literature since 1968: Long-term trends and changes since *Deepwater Horizon*. *Marine Pollution Bulletin* 113:371–379. https://doi.org/10.7266/N7SN06Z4.

Murphy, D. W., C. Li, V. d'Albignac, D. Morra, and J. Katz. 2015. Splash behaviour and oily marine aerosol production by raindrops impacting oil slicks. *Journal of Fluid Mechanics* 780:536–577.

Murphy, D. W., X. Xue, K. Sampath, and J. Katz. 2016. Crude oil jets in crossflow: Effects of dispersant concentration on plume behavior. *Journal of Geophysical Research: Oceans* 121(6):4264–4281.

Murphy, S. M., M. A. Bautista, M. A. Cramm, and C. R. Hubert. 2021. Diesel and crude oil biodegradation by cold-adapted microbial communities in the Labrador Sea. *Applied and Environmental Microbiology* 87(20):e00800–e00821.

Murray, J. A., C. M. Reddy, L. C. Sander, and S. A. Wise. 2015. *Gulf of Mexico research iniutiative 2014–2015 hydrocarbon intercalibration experiment: Description and results for SRM 2779 Gulf of Mexico crude oil and candidate SRM 2777 weathered Gulf of Mexico crude oil.* NISTIR 8123. Gaithersburg, MD: National Institute of Standards and Technology, U.S. Department of Commerce.

Murray, S. P. 1998. *An observational study of the Mississippi-Atchafalaya coastal plume*. New Orleans, LA: U.S. Department of the Interior.

Murru, C., R. Badía-Laíño, and M. E. Díaz-García. 2021. Oxidative stability of vegetal oil-based lubricants. *ACS Sustainable Chemistry & Engineering* 9(4):1459–1476. https://doi.org/10.1021/acssuschemeng.0c06988.

Murthy, R. S. 2014. Mental health of survivors of 1984 Bhopal disaster: A continuing challenge. *Industrial Psychiatry Journal* 23(2):86.

Muschak, W. 1990. Pollution of street run-off by traffic and local conditions. *Science of the Total Environment (Netherlands)* 93(1).

Muskat, J. 2021. Emergency oil spill response: Data collection automation, and dissemination. *International Oil Spill Conference Proceedings* 2021(1):680795.

NAE (National Academy of Engineering) and NRC (National Research Council). 2012. *Macondo well* Deepwater Horizon *blowout: Lessons for improving offshore drilling safety*. Washington, DC: The National Academies Press.

Nansen, F. 1902. The oceanography of the North Polar Basin: The Norwegian North Polar expedition 1893–1896. *Nature 67, 97–98 (1902).* https://doi.org/10.1038/067097a0.

NASA (National Aeronautics and Space Administration). 2010. *NASA images show oil's invasion along Louisiana coast.* https://www.nasa.gov/topics/earth/features/oil20100602.html.

NASEM (National Academies of Sciences, Engineering, and Medicine). 2016a. *Spills of diluted bitumen from pipelines: A comparative study of environmental fate, effects, and response.* Washington, DC: The National Academies Press.

NASEM. 2016b. *Attribution of extreme weather events in the context of climate change.* Washington, DC: The National Academies Press.

NASEM. 2018. *The human factors of process safety and worker empowerment in the offshore oil industry: Proceedings of a workshop.* Washington, DC: The National Academies Press.

NASEM. 2019. *Reproducibility and replicability in science.* Washington, DC: The National Academies Press.

NASEM. 2020. *The use of dispersants in marine oil spill response.* Washington, DC: The National Academies Press.

NASEM. 2021. *Reckoning with the U.S. role in global ocean plastic waste.* Washington, DC: The National Academies Press.

NASEM. n.d. *Gulf Research Program.* https://www.nationalacademies.org/gulf/gulf-research-program.

Nation Master. 2014. https://www.nationmaster.com/country-info/stats/Transport/Road/Motor-vehicles-per-1000-people.

National Commission on the BP Deepwater Horizon Oil Spill and Offshore Drilling. 2011a. *Deep water: The Gulf oil disaster and the future of offshore drilling: Report to the president.*

National Commission on the BP Deepwater Horizon Oil Spill and Offshore Drilling. 2011b. *Macondo: The Gulf oil disaster: Chief Counsel's report.*

National Commission on the BP Deepwater Horizon Oil Spill and Offshore Drilling. 2011c. *Deep water: The Gulf oil disaster and the future of offshore drilling: Recommendations.*

Natter, M., J. Keevan, Y. Wang, A. R. Keimowitz, B. C. Okeke, A. Son, and M.-K. Lee. 2012. Level and degradation of *Deepwater Horizon* spilled oil in coastal marsh sediments and pore-water. *Environmental Science & Technology* 46(11):5744–5755.

NEB (National Energy Board). 2010. *Offshore waste treatment guidelines.* Canada-Newfoundland and Labrador Offshore Petroleum Board. https://www.cer-rec.gc.ca/en/about/acts-regulations/other-acts/offshore-waste-treatment-guidelines/2010ffshrwstgd-eng.pdf.

Nedwed, T., and T. Coolbaugh. 2008. Do basins and beakers negatively bias dispersant-effectiveness tests? *International Oil Spill Conference Proceedings* 2008(1):835–841.

Nedwed, T., J. L. M. Resby, and J. Guyomarch. 2006. Dispersant effectiveness after extended low-energy soak times. *Proceedings from Interspill.* https://www.interspill.org/wp-content/uploads/2021/11/science_lowenergy_doc.pdf.

Nedwed, T., R. Belore, W. Spring, and D. Blanchet. 2007. *Basin-scale testing of asd icebreaker enhanced chemical dispersion of oil spills.* Canada: Environment Canada.

Nedwed, T., G. P. Canevari, J. R. Clark, and R. Belore. 2008. New dispersant delivered as a gel. *International Oil Spill Conference Proceedings* 2008(1):121–125.

Nedwed, T., V. Broje, and S. Le Floch. 2017. *The effectiveness of dispersants on oil encapsulated in ice for extended periods.* Arctic and Marine Oil Spill Program.

Nedwed, T., S. Pegau, and K. Stone. 2021. Recent development on herder commercialization. *International Oil Spill Conference Proceedings* 2021(1). https://doi.org/10.7901/2169-3358-2021.1.687208.

Neff, J., K. Lee, and E. M. DeBlois. 2011. Produced water: Overview of composition, fates, and effects. *Produced Water* 3–54.

Neff, J. M. 2002. *Bioaccumulation in marine organisms: Effect of contaminants from oil well produced water.* Amsterdam, Netherlands: Elsevier.

Neff, J. M., and W. A. Stubblefield. 1995. Chemical and toxicological evaluation of water quality following the *Exxon Valdez* oil spill. In Exxon Valdez *oil spill: Fate and effects in Alaskan waters.* West Conshohocken, PA: ASTM International.

Neff, J. M., N. N. Rabalais, and D. F. Boesch. 1987. Offshore oil and gas activities potentially causing long-term effects. In *Long-term environmental effects of oil and gas development*, D. F. Boesch and N. N. Rabalais (eds.), pp. 149–174. London, UK: Elsevier Applied Science Publishers.

Negri, A. P., and A. J. Heyward. 2000. Inhibition of fertilization and larval metamorphosis of the coral acropora millepora (Ehrenberg, 1834) by petroleum products. *Marine Pollution Bulletin* 41(7–12):420–427.

Negri, A. P., D. L. Brinkman, F. Flores, E. S. Botté, R. J. Jones, and N. S. Webster. 2016. Acute ecotoxicology of natural oil and gas condensate to coral reef larvae. *Scientific Reports* 6(1):21153. https://doi.org/10.1038/srep21153.

Negri, A. P., H. M. Luter, R. Fisher, D. L. Brinkman, and P. Irving. 2018. Comparative toxicity of five dispersants to coral larvae. *Scientific Reports* 8(1):3043. https://doi.org/10.1038/s41598-018-20709-2.

Negri, A. P., D. L. Brinkman, F. Flores, J. van Dam, H. M. Luter, M. C. Thomas, R. Fisher, L. S. Stapp, P. Kurtenbach, and A. Severati. 2021. Derivation of toxicity thresholds for gas condensate oils protective of tropical species using experimental and modelling approaches. *Marine Pollution Bulletin* 172:112899.

Nelson, J. R., and T. H. Grubesic. 2018. Oil spill modeling: Risk, spatial vulnerability, and impact assessment. *Progress in Physical Geography: Earth and Environment* 42(1):112–127.

Nelson, R. K., C. Aeppli, J. S. Arey, H. Chen, A. H. de Oliveira, C. Eiserbeck, G. S. Frysinger, R. B. Gaines, K. Grice, and J. Gros. 2016. Applications of comprehensive two-dimensional gas chromatography (GC×GC) in studying the source, transport, and fate of petroleum hydrocarbons in the environment. In *Standard handbook oil spill environmental forensics*, pp. 399–448. Amsterdam, Netherlands: Elsevier.

Nepstad, R., B. H. Hansen, and J. Skancke. 2021. North Sea produced water PAH exposure and uptake in early life stages of Atlantic cod. *Marine Environmental Research* 163:105203.

Newman, S. H., D. W. Anderson, M. H. Ziccardi, J. G. Trupkiewicz, F. S. Tseng, M. M. Christopher, and J. G. Zinkl. 2000. An experimental soft-release of oil-spill rehabilitated American Coots (fulica Americana): II. Effects on health and blood parameters. *Environmental Pollution* 107(3):295–304. https://doi.org/10.1016/S0269-7491(99)00171-2.

Nguyen, U. T., S. A. Lincoln, A. G. Valladares Juárez, M. Schedler, J. L. Macalady, R. Müller, and K. H. Freeman. 2018. The influence of pressure on crude oil biodegradation in shallow and deep Gulf of Mexico sediments. *PloS ONE* 13(7):e0199784.

Nicholls, K., S. J. Picou, and S. C. McCord. 2017. Training community health workers to enhance disaster resilience. *Journal of Public Health Management and Practice* 23:S78–S84.

NIEHS (National Institute of Environmental Health Sciences). 2012. *Improving safety and health training for disaster cleanup workers: Lessons learned from the 2010* Deepwater Horizon *oil spill.* https://www.niehs.nih.gov/news/events/pastmtg/hazmat/assets/2011/wtp_workshop_report_spring_2011_training_disaster_cleanup_workers_508.pdf.

NIEHS. 2021a. *Disaster research response (DR2) program.* https://www.niehs.nih.gov/research/programs/disaster/index.cfm.

NIEHS. 2021b. *Gulf oil spill response efforts.* https://www.niehs.nih.gov/research/programs/gulfspill/index.cfm.

NIEHS Environmental Factor. 2018. *Lichtveld discusses enterprise evaluation.* https://factor.niehs.nih.gov/2018/3/science-highlights/lichtveld/index.htm.

Niles, S. F., M. L. Chacón-Patiño, H. Chen, A. M. McKenna, G. T. Blakney, R. P. Rodgers, and A. G. Marshall. 2019. Molecular-level characterization of oil-soluble ketone/aldehyde photo-oxidation products by Fourier transform ion cyclotron resonance mass spectrometry reveals similarity between microcosm and field samples. *Environmental Science & Technology* 53(12):6887–6894.

Niles, S. F., M. L. Chacón-Patiño, A. G. Marshall, and R. P. Rodgers. 2020. Molecular composition of photooxidation products derived from sulfur-containing compounds isolated from petroleum samples. *Energy & Fuels* 34(11):14493–14504.

NIOSH (National Institute for Occupational Safety and Health). 2010. *Interim report 1: Health hazard evaluation of* Deepwater Horizon *response workers.* Letter from A. Tepper to F. Tremmel. http://www.cdc.gov/niosh/topics/oilspillresponse/gulfspillhhe.html.

Nissanka, I. D., and P. D. Yapa. 2016. Calculation of oil droplet size distribution in an underwater oil well blowout. *Journal of Hydraulic Research* 54(3):307–320.

NIST (National Institute of Standards and Technology). 2021. *PAH analyses*. https://www.nist.gov/fusion-search?s=PAH+analyses.

Nixon, Z., and J. Michel. 2018. A review of distribution and quantity of lingering subsurface oil from the *Exxon Valdez* oil spill. *Deep Sea Research Part II: Topical Studies in Oceanography* 147:20–26. https://doi.org/10.1016/j.dsr2.2017.07.009.

Nixon, Z., S. Zengel, M. Baker, M. Steinhoff, G. Fricano, S. Rouhani, and J. Michel. 2016. Shoreline oiling from the *Deepwater Horizon* oil spill. *Marine Pollution Bulletin* 107(1):170–178.

Nizzetto, L., R. Lohmann, R. Gioia, A. Jahnke, C. Temme, J. Dachs, P. Herckes, A. D. Guardo, and K. C. Jones. 2008. PAHs in air and seawater along a north-south Atlantic transect: Trends, processes and possible sources. *Environmental Science & Technology* 42(5):1580–1585.

NMMA (National Marine Manufacturers Association). 2019. *U.S. recreational boating industry sees seventh consecutive year of growth in 2018, expects additional increase in 2019*. https://www.nmma.org/press/article/22428.

NOAA (National Oceanic and Atmospheric Administration). 1996. Aerial observations of oil at sea. Hazmat report 96-7, April 1996. Office of Ocean Resources, National Oceanic and Atmospheric Administration.

NOAA. 2004. *M/V Athos I; Delaware River, NJ/PA*. NOAA Incident News. https://incidentnews.noaa.gov/incident/1236.

NOAA. 2010 (revised 2017). *Characteristic coastal habitats, choosing spill response alternatives*. Seattle, WA: Office of Response and Restoration, Emergency Response Division, National Ocean Service, National Oceanic and Atmospheric Administration, U.S. Department of Commerce.

NOAA. 2013a. *Characteristics of response strategies: A guide for spill response planning in marine environments*. https://response.restoration.noaa.gov/oil-and-chemical-spills/oil-spills/resources/characteristics-response-strategies.html.

NOAA. 2013b. *Shoreline assessment manual. 4th edition*. Seattle, WA: Office of Response and Restoration, Emergency Response Division, National Oceanic and Atmospheric Administration, U.S. Department of Commerce.

NOAA. 2014a. *Environmental response management application (ERMA)*. https://response.restoration.noaa.gov/erma.

NOAA. 2014b. *Oil spills in mangroves. Planning and response considerations*. Washington, DC: Office of Response and Restoration, National Ocean Service, National Oceanic and Atmospheric Administration, U.S. Department of Commerce.

NOAA. 2015. *Pinniped and cetacean oil spill response guidelines*. https://www.fisheries.noaa.gov/resource/document/pinniped-and-cetacean-oil-spill-response-guidelines.

NOAA. 2016. *Open water oil identification job aid (NOAA-code) for aerial observation*. Seattle, WA: National Oceanic and Atmospheric Administration, U.S. Department of Commerce.

NOAA. 2017. *Characteristic coastal habitats: Choosing spill response alternatives*. https://response.restoration.noaa.gov/oil-and-chemical-spills/oil-spills/resources/characteristic-coastal-habitats.html.

NOAA. 2019a. *Environmental Sensitivity Index Guidelines, version 4.0*. U.S. Department of Commerce National Oceanic and Atmospheric Administration: Washington, DC.

NOAA. 2019b. *ARTES: During the chaos of oil spills, seeking a system to test potential solutions*. https://response.restoration.noaa.gov/about/media/during-chaos-oil-spills-seeking-system-test-potential-solutions.html.

NOAA. 2021. *Oil types*. https://response.restoration.noaa.gov/oil-and-chemical-spills/oil-spills/oil-types.html.

NOAA. 2021a. *Office of Response and Restoration: Glossary of terms*. https://response.restoration.noaa.gov/oil-and-chemical-spills/glossary-terms.html.

NOAA. 2021b. *Office of Response and Restoration: What happens to dispersed oil?* https://response.restoration.noaa.gov/oil-and-chemical-spills/oil-spills/resources/8-what-happens-dispersed-oil.html.

NOAA. 2021c. *Office of Response and Restoration: GNOME suite for oil modeling*. https://response.restoration.noaa.gov/gnome.

NOAA. 2021d. *Seafood safety after an oil spill*. https://response.restoration.noaa.gov/oil-and-chemical-spills/oil-spills/resources/seafood-safety-after-oil-spill.html.

NOAA. n.d. *U.S. NOAA diver database*. Edited by the National Oceanic and Atmospheric Administration.

NOAA and API (American Petroleum Institute). 2013. *Oil spills in marshes: Planning and response considerations*. Office of Response and Restoration, National Oceanic and Atmospheric Administration, U.S. Department of Commerce.

NOAA ERD (Emergency Response Division). 2015. *The chemical aquatic fate and effects (CAFE) database*. Seattle, WA: Office of Response and Restoration, Emergency Response Division, National Oceanic and Atmospheric Administration, U.S. Department of Commerce.

NOAA, Rhode Island Department of Environmental Management, U.S. Department of the Interior, and U.S. Fish & Wildlife Service. 1999. *Restoration plan and environmental assessment for the January, 19, 1996* North Cape *oil spill*. http://www.dem.ri.gov/pubs/damage/rptchooz.htm.

Noirungsee, N., S. Hackbusch, J. Viamonte, P. Bubenheim, A. Liese, and R. Müller. 2020. Influence of oil, dispersant, and pressure on microbial communities from the Gulf of Mexico. *Scientific Reports* 10(1):1–9.

Noor, C. M., M. Noor, and R. Mamat. 2018. Biodiesel as alternative fuel for marine diesel engine applications: A review. *Renewable and Sustainable Energy Reviews* 94:127–142.

Norcor. 1975. *The interaction of crude oil with Arctic sea ice*. Beaufort Sea project Technical Report No. 27. Victoria, British Columbia, Canada: Canadian Department of Environment.

Nordam, T., C. Beegle-Krause, J. Skancke, R. Nepstad, and M. Reed. 2019. Improving oil spill trajectory modelling in the Arctic. *Marine Pollution Bulletin* 140:65–74.

Nordam, T., S. Lofthus, and O. G. Brakstad. 2020. Modelling biodegradation of crude oil components at low temperatures. *Chemosphere* 254: 126836.

Nordborg, F. M., F. Flores, D. L. Brinkman, S. Agustí, and A. P. Negri. 2018. Phototoxic effects of two common marine fuels on the settlement success of the coral Acropora tenuis. *Scientific Reports* 8(1):8635. https://doi.org/10.1038/s41598-018-26972-7.

Nordborg, F. M., R. J. Jones, M. Oelgemöller, and A. P. Negri. 2020. The effects of ultraviolet radiation and climate on oil toxicity to coral reef organisms—a review. *Science of the Total Environment* 720:137486. https://doi.org/10.1016/j.scitotenv.2020.137486.

Nordborg, F. M., D. L. Brinkman, G. F. Ricardo, S. Agustí, and A. P. Negri. 2021. Comparative sensitivity of the early life stages of a coral to heavy fuel oil and UV radiation. *Science of the Total Environment* 781:146676. https://doi.org/10.1016/j.scitotenv.2021.146676.

Nordtug, T., and B. H. Hansen. 2021. Assessment of oil toxicity in water. In *Marine hydrocarbon spill assessments*, pp. 199–220. Amsterdam, Netherlands: Elsevier.

NPS (National Park Service). 2017. *Wildland fire: Incident command system*. https://www.nps.gov/articles/wildland-fire-incident-command-system.htm.

NRC (National Research Council). 1975. *Petroleum in the marine environment*. Washington, DC: National Academy Press.

NRC. 1985. *Oil in the sea: Inputs, fates, and effects*. Vol. 1. Washington, DC: National Academy Press.

NRC. 1991. *Tanker spills: Prevention by design*. Washington, DC: National Academy Press.

NRC. 1993. *Assessment of the U.S. outer continental shelf environmental studies program: III. Social and economic studies*. Washington, DC: National Academy Press.

NRC. 1994. *Minding the helm: Marine navigation and piloting*. Washington, DC: National Academy Press.

NRC. 1998. *Double-hull tanker legislation: An assessment of the Oil Pollution Act of 1990*. Washington, DC: National Academy Press.

NRC. 2001. *Environmental performance of tanker designs in collision and grounding: Method for comparision—Special Report 259*. Washington, DC: National Academy Press.

NRC. 2003. *Oil in the sea III: Inputs, fates, and effects*. Washington, DC: The National Academies Press.

NRC. 2005. *Oil spill dispersants; efficacy and effects*.Washington, DC: The National Academies Press.

NRC. 2013. *An ecosystem services approach to assessing the impacts of the* Deepwater Horizon *oil spill in the Gulf of Mexico*, pp. 71–102. Washington, DC: The National Academies Press.

NRC. 2014. *Responding to oil spills in the U.S. Arctic marine environment*. Washington, DC: The National Academies Press.

NRT (National Response Team). 2013. *Environmental monitoring for atypical dispersant operations*. https://www.nrt.org/sites/2/files/NRT_Atypical_Dispersant_Guidance_Final_5-30-2013.pdf.

NRT. 2021. *National Oil and Hazardous Substances Pollution Contingency Plan: Monitoring requirements for use of dispersants and other chemicals*. https://www.federalregister.gov/documents/2021/07/27/2021-15122/national-oil-and-hazardous-substances-pollution-contingency-plan-monitoring-requirements-for-use-of.

Nugent, N., S. A. Gaston, J. Perry, A. L. Rung, E. J. Trapido, and E. S. Peters. 2019. PTSD symptom profiles among Louisiana women affected by the 2010 *Deepwater Horizon* oil spill: A latent profile analysis. *Journal of Affective Disorders* 250:289–297.

Nwipie, G. N., A. I. Hart, N. Zabbey, K. Sam, G. Prpich, and P. E. Kika. 2019. Recovery of infauna macrobenthic invertebrates in oil-polluted tropical soft-bottom tidal flats: 7 years post spill. *Environmental Science and Pollution Research* 26(22):22407–22420.

OCIMF (Oil Companies International Marine Forum). 2019. *Volatile organic compound emissions from cargo systems on oil tankers*. https://www.ocimf.org/document-libary/60-volatile-organic-compound-emissions-from-cargo-systems-on-oil-tankers/file.

O'Clair, C. E., J. W. Short, and S. D. Rice (eds.). 1996. *Contamination of intertidal and subtidal sediments by oil from the* Exxon Valdez *in Prince William Sound*, S. D. Rice, R. B. Spies, D. A. Wolfe and B. A. Wright (eds.). Vol. S, *American fisheries society symposium series*. Bethesda, MD: American Fisheries Society.

O'Dowd, C. D., and G. De Leeuw. 2007. Marine aerosol production: A review of the current knowledge. *Philosophical Transactions of the Royal Society A: Mathematical, Physical and Engineering Sciences* 365(1856):1753–1774.

Oggier, M., H. Eicken, J. Wilkinson, C. Petrich, and M. O'Sadnick. 2020. Crude oil migration in sea-ice: Laboratory studies of constraints on oil mobilization and seasonal evolution. *Cold Regions Science and Technology* 174:102924.

O'Hara, P. D., and L. A. Morandin. 2010. Effects of sheens associated with offshore oil and gas development on the feather microstructure of pelagic seabirds. *Marine Pollution Bulletin* 60(5):672–678. https://www.sciencedirect.com/science/article/pii/S0025326X09005104.

Oil Budget Calculator Science and Engineering Team (U.S.). 2010. *Oil budget calculator:* Deepwater Horizon. Federal Interagency Solutions Group, Oil Budget Calculator Science and Engineering Team.

Oka, N., and M. Okuyama. 2000. Nutritional status of dead oiled rhinoceros auklets (Cerorhinca monocerata) in the Southern Japan Sea. *Marine Pollution Bulletin* 40(4):340–347. https://doi.org/10.1016/S0025-326X(99)00223-4.

Okubo, A. 1972. *Some speculations on oceanic diffusion diagrams*. https://doi.org/10.2172/4626706.

Oliveira, G., F. Khan, and L. James. 2019. Ecological risk assessment of oil spills in ice-covered waters: A surface slick model coupled with a food-web bioaccumulation model. *Integrated Environmental Assessment and Management* 16(5):729–744.

Oliveira, L. G., K. C. Araújo, M. C. Barreto, M. E. P. A. Bastos, S. G. Lemos, and W. D. Fragoso. 2021. Applications of chemometrics in oil spill studies. *Microchemical Journal* 166:106216. https://doi.org/10.1016/j.microc.2021.106216.

Oliveira, M. B., V. L. Oliveira, J. A. Coutinho, and A. J. Queimada. 2009. Thermodynamic modeling of the aqueous solubility of PAHs. *Industrial & Engineering Chemistry Research* 48(11):5530–5536.

Olsen, J. E., D. F. Krause, E. J. Davies, and P. Skjetne. 2019. Observations of rising methane bubbles in Trondheim Fjord and its implications to gas dissolution. *Journal of Geophysical Research: Oceans* 124(3):1399–1409.

Olsgard, M. L., G. R. Bortolotti, B. R. Trask, and J. E. Smits. 2008. Effects of inhalation exposure to a binary mixture of benzene and toluene on vitamin a status and humoral and cell-mediated immunity in wild and captive American kestrels. *Journal of Toxicology and Environmental Health, Part A* 71(16):1100–1108.

Olson, M. C., J. L. Iverson, E. T. Furlong, and M. P. Schroeder. 2004. *Methods of analysis by the U.S. Geological Survey national water quality laboratory determination of polycyclic aromatic hydrocarbons compounds in sediment by gas chromatography*. Water Resources Investigations Report 03-4318. Denver, CO: U.S. Geological Survey Information Services.

Omarova, M., L. T. Swientoniewski, I. K. Mkam Tsengam, D. A. Blake, V. John, A. McCormick, G. D. Bothun, S. R. Raghavan, and A. Bose. 2019. Biofilm formation by hydrocarbon-degrading marine bacteria and its effects on oil dispersion. *ACS Sustainable Chemistry & Engineering* 7(17):14490–14499.

Opedal, N. v. d. T., G. Sørland, and J. Sjöblom. 2009. Methods for droplet size distribution determination of water-in-oil emulsions using low-field NMR. *The Open-Access Journal for the Basic Principles of Diffusion Theory, Experiment and Application* 9(7):1–29.

Orcutt, B. N., J. B. Sylvan, N. J. Knab, and K. J. Edwards. 2011. Microbial ecology of the dark ocean above, at, and below the seafloor. *Microbiology and Molecular Biology Reviews* 75(2):361–422.

Orosz, S. 2013. Critical care nutrition for exotic animals. *Journal of Exotic Pet Medicine* 22:163–177.

Orsi, W. D. 2018. Ecology and evolution of seafloor and subseafloor microbial communities. *Nature Reviews Microbiology* 16(11):671–683.

Ortmann, A. C., S. E. Cobanli, G. Wohlgeschaffen, J. MacDonald, A. Gladwell, A. Davis, B. Robinson, J. Mason, and T. L. King. 2020. Measuring the fate of different diluted bitumen products in coastal surface waters. *Marine Pollution Bulletin* 153:111003.

OSAT-2 (Operational Science Advisory Team). 2011. *Summary reports for fate and effects of remnant oil in the beach environment*. Operational Science Advisory Team (OSAT-2) Gulf Coast Incident Management Team.

Osenberg, C., R. Schmitt, S. Holbrook, and D. Canestro. 1992. Spatial scale of ecological effects associated with an open coast discharge of produced water. In *Produced water*, pp. 387–402. Dordrecht, Netherlands: Springer.

Osofsky, H. J., J. D. Osofsky, and T. C. Hansel. 2011. *Deepwater Horizon* oil spill: Mental health effects on residents in heavily affected areas. *Disaster Medicine and Public Health Preparedness* 5(4):280–286.

Osofsky, H. J., J. D. Osofsky, J. H. Wells, and C. Weems. 2014. Integrated care: Meeting mental health needs after the Gulf oil spill. *Psychiatric Services* 65(3):280–283.

Osofsky, J. D., and H. J. Osofsky. 2021. Hurricane Katrina and the Gulf oil spill: Lessons learned about short-term and long-term effects. *International Journal of Psychology* 56(1):56–63.

Ou, X., and K. S. Ramos. 1992. Proliferative responses of quail aortic smooth muscle cells to benzo [a] pyrene: Implications in PAH-induced atherogenesis. *Toxicology* 74(2–3):243–258.

Overholt, W. A., P. Schwing, K. M. Raz, D. Hastings, D. J. Hollander, and J. E. Kostka. 2019. The core seafloor microbiome in the Gulf of Mexico is remarkably consistent and shows evidence of recovery from disturbance caused by major oil spills. *Environmental Microbiology* 21(11):4316–4329.

Overstreet, R., and J. A. Galt. 1995. *Physical processes affecting the movement and spreading of oils in inland waters*. National Oceanic and Atmospheric Administration (NOAA)/Hazardous Materials: Silver Spring, Maryland.

Overstreet, R., A. Lewandowski, W. Lehr, R. Jones, D. Simecek-Beatty, and D. Calhoun. 1995. Sensitivity analysis in oil spill models: Case study using ADIOS. *International Oil Spill Conference Proceedings* 1995(1):898–900.

Overton, E. G., J. L. Laseter, W. Mascarella, C. Raschke, I. Nuiry, and J. W. Farrington. 1980. Photo-chemical oxidation of IXTOC-I oil. In *Proceedings of Symposium on Preliminary Results from the September, 1979 Researcher/Pierce IXTOC-I Cruise*. Washington, DC: National Oceanic and Atmospheric Administration, U.S. Department of Commerce. https://archive.org/details/proceedingsofsym00unit.

Owens, C., and R. Belore. 2004. Dispersant effectiveness testing in cold water and brash ice. *Proceedings of the 27th Arctic and Marine Oilspill Program Technical Seminar.* Ottawa, Ontario, Canada: S.L. Ross Environmental Research Ltd.

Owens, E., C. Haselwimmer, and D. Palandro. 2018. Recent lessons learned from UAS field activities for shoreline oiling survey applications. In *Proceedings from Interspill*. https://www.interspill.org/wp-content/uploads/2021/11/Recent-Lessons-Learned-E-Owens-Owens-Coastal-Consultants.pdf.

Owens, E. H., and P. C. Bunker. 2021. Chapter 20: Canine detection teams to support oil spill response surveys. In *Canines: The original biosensors*, L. E. Degreef (ed.), pp. 697–748. Boca Raton, FL: Pan Stanford Publishing, CRC Press.

Owens, E. H., and K. Lee. 2003. Interaction of oil and mineral fines on shorelines: Review and assessment. *Marine Pollution Bulletin* 47(9–12):397–405.

Owens, E. H., and W. Robson. 1987. Experimental design and the retention of oil on Arctic test beaches. *Arctic* 230–243.

Owens, E. H., J. R. Harper, W. Robson, and P. D. Boehm. 1987. Fate and persistence of crude oil stranded on a sheltered beach. *Arctic* 109–123.

Owens, E. H., E. Taylor, and B. Humphrey. 2008. The persistence and character of stranded oil on coarse-sediment beaches. *Marine Pollution Bulletin* 56(1):14–26.

Owens, E. H., P. Bunker, H. C. Dubach, and E. Taylor. 2021. Field trials using canines to detect deep subsurface weathered and heavy oils. *43rd Arctic and Marine Oilspill Program Technical Seminar on Environmental Contamination and Response Environment and Climate Change.* Canada.

Özgökmen, T. M., E. P. Chassignet, C. N. Dawson, D. Dukhovskoy, G. Jacobs, J. Ledwell, O. Garcia-Pineda, I. R. MacDonald, S. L. Morey, and M. J. Olascoaga. 2016. Over what area did the oil and gas spread during the 2010 *Deepwater Horizon* oil spill? *Oceanography* 29(3):96–107.

Özgökmen, T. M., M. Boufadel, D. F. Carlson, C. Cousin, C. Guigand, B. K. Haus, J. Horstmann, B. Lund, J. Molemaker, and G. Novelli. 2018. Technological advances for ocean surface measurements by the Consortium for Advanced Research on Transport of Hydrocarbons in the Environment (CARTHE). *Marine Technology Society Journal* 52(6):71–76. https://doi.org/10.4031/MTSJ.52.6.11.

Ozhan, K., and S. Bargu. 2014. Distinct responses of Gulf of Mexico phytoplankton communities to crude oil and the dispersant COREXIT® EC9500A under different nutrient regimes. *Ecotoxicology* 23(3):370–384.

Ozhan, K., M. L. Parsons, and S. Bargu. 2014. How were phytoplankton affected by the *Deepwater Horizon* oil spill? *Bioscience* 64(9):829–836.

Palinkas, L. A., J. S. Petterson, J. Russell, and M. A. Downs. 1993. Community patterns of psychiatric disorders after the *Exxon Valdez* oil spill. *The American Journal of Psychiatry* 150(10):1517–1523.

Palinkas, L. A., J. S. Petterson, J. C. Russell, and M. A. Downs. 2004. Ethnic differences in symptoms of post-traumatic stress after the *Exxon Valdez* oil spill. *Prehospital and Disaster Medicine* 19(1):102–112.

Pallen, M. J., A. Telatin, and A. Oren. 2020. The next million names for archaea and bacteria. *Trends in Microbiology* 29(4):P289–P298.

Panetta, P. D., and S. Potter. 2016. *TRL definitions for oil spill response technologies and equipment*. Final report prepared for the Bureau of Safety and Environmental Enforcement, U.S. Department of the Interior.

Papanikolaou, A., and E. Eliopoulou. 2008. Impact of ship age on tanker accidents. *2nd International Symposium on Ship Operations, Management and Economics*.

Paquin, P. R., J. McGrath, C. J. Fanelli, and D. M. Di Toro. 2018. The aquatic hazard of hydrocarbon liquids and gases and the modulating role of pressure on dissolved gas and oil toxicity. *Marine Pollution Bulletin* 133:930–942.

Pardue, J. H., K. R. Lemelle, M. Urbano, and V. Elango. 2014. Distribution and biodegradation potential of buried oil on a coastal headland beach. *International Oil Spill Conference Proceedings* 2014(1):1073–1086.

Paris, C. B., M. L. Hénaff, Z. M. Aman, A. Subramaniam, J. Helgers, D.-P. Wang, V. H. Kourafalou, and A. Srinivasan. 2012. Evolution of the Macondo well blowout: Simulating the effects of the circulation and synthetic dispersants on the subsea oil transport. *Environmental Science & Technology* 46(24):13293–13302.

Park, B. S., D. L. Erdner, H. P. Bacosa, Z. Liu, and E. J. Buskey. 2020. Potential effects of bacterial communities on the formation of blooms of the harmful dinoflagellate *Prorocentrum* after the 2014 Texas City "Y" oil spill (USA). *Harmful Algae* 95:101802.

Park, M. S., K.-H. Choi, S.-H. Lee, J.-I. Hur, S. R. Noh, W.-C. Jeong, H.-K. Cheong, and M. Ha. 2019. Health effect research on Hebei Spirit oil spill (HEROS) in Korea: A cohort profile. *BMJ Open* 9(8):e026740.

Parker, A. M., I. Ferrer, E. M. Thurman, F. L. Rosario-Ortiz, and K. G. Linden. 2014. Determination of COREXIT components used in the *Deepwater Horizon* cleanup by liquid chromatography-ion trap mass spectrometry. *Analytical Methods* 6(15):5498–5502.

Parker, C. A., M. Freegarde, and G. C. Hatchard. 1971. The effect of some chemical and biological factors on the degredation of crude oil at sea. In *Water pollution by oil*, P. Hepple (ed.), pp. 237–244. London, UK: Institute of Petroleum.

Parkerton, T. F., D. J. Letinski, E. J. Febbo, J. D. Butler, C. A. Sutherland, G. E. Bragin, B. M. Hedgpeth, B. A. Kelley, A. D. Redman, and P. Mayer. 2021. Assessing toxicity of hydrophobic aliphatic and monoaromatic hydrocarbons at the solubility limit using novel dosing methods. *Chemosphere* 265:129174.

Parsons, M., W. Morrison, N. Rabalais, R. E. Turner, and K. Tyre. 2015. Phytoplankton and the Macondo oil spill: A comparison of the 2010 phytoplankton assemblage to baseline conditions on the Louisiana shelf. *Environmental Pollution* 207:152–160.

Parsons, M. L., A. L. Brandt, R. E. Turner, W. L. Morrison, and N. N. Ralabais. 2021. Characterization of common phytoplankton on the Louisiana shelf. *Marine Pollution Bulletin* 168:112458.

Pärt, S., H. Kankaanpää, J.-V. Björkqvist, and R. Uiboupin. 2021. Oil spill detection using fluorometric sensors: Laboratory validation and implementation to a ferrybox and a moored smartbuoy. *Frontiers in Marine Science* 8. https://doi.org/10.3389/fmars.2021.778136.

Partington, K. 2014. *An assessment of surface surveillance capabilities for oil spill response using satellite remote sensing*. https://www.ipieca.org/resources/awareness-briefing/an-assessment-of-surface-surveillance-capabilities-for-oil-spill-response-using-satellite-remotesensing.

Partyka, M., D. Bailey, E. Maung-Douglass, S. Sempier, T. Skelton, and M. Wilson. 2021. *Understanding the human health and socioeconomic impacts from the* Deepwater Horizon *oil spill*. GOMSG-G-21-002.

Passow, U. 2016. Formation of rapidly-sinking, oil-associated marine snow. *Deep Sea Research Part II: Topical Studies in Oceanography* 129:232–240.

Passow, U., and E. B. Overton. 2021. The complexity of spills: The fate of the *Deepwater Horizon* oil. *Annual Review of Marine Science* 13:109–136. https://doi.org/10.1146/annurev-marine-032320-095153.

Passow, U., and S. A. Stout. 2020. Character and sedimentation of "lingering" Macondo oil to the deep-sea after the *Deepwater Horizon* oil spill. *Marine Chemistry* 218:103733.

Passow, U., and K. Ziervogel. 2016. Marine snow sedimented oil released during the *Deepwater Horizon* spill. *Oceanography* 29(3):118–125. https://doi.org/10.5670/oceanog.2016.76.

Passow, U., K. Ziervogel, V. Asper, and A. Diercks. 2012. Marine snow formation in the aftermath of the *Deepwater Horizon* oil spill in the Gulf of Mexico. *Environmental Research Letters* 7(3):035301.

Passow, U., J. Sweet, and A. Quigg. 2017. How the dispersant COREXIT impacts the formation of sinking marine oil snow. *Marine Pollution Bulletin* 125(1–2):139–145.

Payne, J., and W. Driskell. 2015. *Deepwater Horizon* oil spill NRDA offshore adaptive sampling strategies and field observations. In *PECI technical report to the trustees in support of the PDARP*.

Payne, J. R., and W. B. Driskell. 2003. The importance of distinguishing dissolved—versus oil—droplet phases in assessing the fate, transport, and toxic effects of marine oil pollution. *International Oil Spill Conference*. April 2003. Vancouver, Canada.

Payne, J. R., and C. R. Phillips. 1985. Photochemistry of petroleum in water. *Environmental Science & Technology* 19(7):569–579.

Peacock, E. E., R. K. Nelson, A. R. Solow, J. D. Warren, J. L. Baker, and C. M. Reddy. 2005. The West Falmouth oil spill: ~100 kg of oil found to persist decades later. *Environmental Forensics* 6(3):273–281.

Peakall, D. B., D. J. Hallett, J. R. Bend, G. L. Foureman, and D. S. Miller. 1982. Toxicity of Prudhoe Bay crude oil and its aromatic fractions to nestling herring gulls. *Environmental Research* 27(1):206–215. https://doi.org/10.1016/0013-9351(82)90071-8.

Peakall, D. B., D. S. Miller, and W. B. Kinter. 1983. Toxicity of crude oils and their fractions to nestling herring gulls—1. Physiological and biochemical effects. *Marine Environmental Research* 8(2):63–71. https://doi.org/10.1016/0141-1136(83)90027-2.

Peakall, D., R. Norstrom, D. Jeffrey, and F. Leighton. 1989. Induction of hepatic mixed function oxidases in the herring gull (Larus argentatus) by Prudhoe Bay crude oil and its fractions. *Comparative Biochemistry and Physiology Part C: Comparative Pharmacology* 94(2):461–463.

Pearson, T. H., and R. Rosenberg. 1978. Macrobenthic succession in relation to organic enrichment and pollution of the marine environment. *Oceanography and Marine Biology: An Annual Review* 16:229–311.

Peckol, P., S. Levings, and S. Garrity. 1990. Kelp response following the world prodigy oil spill. *Marine Pollution Bulletin* 21(10):473–476.

Pegoraro, C. N., S. L. Quiroga, H. A. Montejano, G. N. Rimondino, G. A. Argüello, and M. S. Chiappero. 2020. Assessing polycyclic aromatic hydrocarbons in the marine atmosphere on a transect across the southwest Atlantic Ocean. *Atmospheric Pollution Research* 11(7):1035–1041.

Pei, X., A. G. Abdul Jameel, C. Chen, I. A. AlGhamdi, K. AlAhmadi, E. AlBarakati, S. Saxena, and W. L. Roberts. 2021. Swirling flame combustion of heavy fuel oil: Effect of fuel sulfur content. *Journal of Energy Resources Technology* 143(8):082103.

Pelletier, M. C., R. M. Burgess, K. T. Ho, A. Kuhn, R. A. McKinney, and S. A. Ryba. 1997. Phototoxicity of individual polycyclic aromatic hydrocarbons and petroleum to marine invertebrate larvae and juveniles. *Environmental Toxicology and Chemistry* 16(10):2190–2199.

Péneloux, A., E. Rauzy, and R. Fréze. 1982. A consistent correction for Redlich-Kwong-Soave volumes. *Fluid Phase Equilibria* 8(1):7–23.

Peng, C., S. Xiao, and D. Yang. 2020. Large-eddy simulation model for the effect of gas bubble dissolution on the dynamics of hydrocarbon plume from deep-water blowout. *Journal of Geophysical Research: Oceans* 125(6):e2019JC016037.

Peng, D.-Y., and D. B. Robinson. 1976. A new two-constant equation of state. *Industrial & Engineering Chemistry Fundamentals* 15(1):59–64.

Peng, X., Y. Wen, L. Wu, C. Xiao, C. Zhou, and D. Han. 2020. A sampling method for calculating regional ship emission inventories. *Transportation Research Part D: Transport and Environment* 89:102617.

Pennings, S. C., B. D. McCall, and L. Hooper-Bui. 2014. Effects of oil spills on terrestrial arthropods in coastal wetlands. *Bioscience* 64(9):789–795.

Pepi, M., A. Cesàro, G. Liut, and F. Baldi. 2005. An Antarctic psychrotrophic bacterium Halomonas sp. ANT-3b, growing on n-hexadecane, produces a new emulsyfying glycolipid. *FEMS Microbiology Ecology* 53(1):157–166.

Péquin, B., Q. Cai, K. Lee, and C. W. Greer. 2022. Natural attenuation of oil in marine environments: A review. *Marine Pollution Bulletin* 176:113464. https://doi.org/10.1016/j.marpolbul.2022.113464.

Perez, C. R., J. K. Moye, D. Cacela, K. M. Dean, and C. A. Pritsos. 2017. Low level exposure to crude oil impacts avian flight performance: The *Deepwater Horizon* oil spill effect on migratory birds. *Ecotoxicology and Environmental Safety* 146:98–103.

Pérez-Cadahía, B., B. Laffon, M. Porta, A. Lafuente, T. Cabaleiro, T. López, A. Caride, J. Pumarega, A. Romero, and E. Pásaro. 2008. Relationship between blood concentrations of heavy metals and cytogenetic and endocrine parameters among subjects involved in cleaning coastal areas affected by the "prestige" tanker oil spill. *Chemosphere* 71(3):447–455.

Pernthaler, A., and J. Pernthaler. 2007. Fluorescence in situ hybridization for the identification of environmental microbes. In *Protocols for nucleic acid analysis by nonradioactive probes*, pp. 153–164. Dordrecht, Netherlands: Springer.

Pernthaler, A., J. Pernthaler, and R. Amann. 2002. Fluorescence in situ hybridization and catalyzed reporter deposition for the identification of marine bacteria. *Applied and Environmental Microbiology* 68(6):3094–3101.

Perraud, V., J. R. Horne, A. S. Martinez, J. Kalinowski, S. Meinardi, M. L. Dawson, L. M. Wingen, D. Dabdub, D. R. Blake, and R. B. Gerber. 2015. The future of airborne sulfur-containing particles in the absence of fossil fuel sulfur dioxide emissions. *Proceedings of the National Academy of Sciences* 112(44):13514–13519.

Perrette, M., A. Yool, G. D. Quartly, and E. E. Popova. 2011. Near-ubiquity of ice-edge blooms in the Arctic. *Biogeosciences* 8(2):515–524.

Perring, A., J. Schwarz, J. Spackman, R. Bahreini, J. De Gouw, R. Gao, J. Holloway, D. Lack, J. Langridge, and J. Peischl. 2011. Characteristics of black carbon aerosol from a surface oil burn during the *Deepwater Horizon* oil spill. *Geophysical Research Letters* 38(17):L17809.

Pesch, S., P. Jaeger, A. Jaggi, K. Malone, M. Hoffmann, D. Krause, T. B. Oldenburg, and M. Schlüter. 2018. Rise velocity of live-oil droplets in deep-sea oil spills. *Environmental Engineering Science* 35(4):289–299.

Peters, A., and A. Siuda. 2014. A review of observations of floating tar in the Sargasso Sea. *Oceanography* 27(1):217–221. https://doi.org/10.5670/oceanog.2014.25.

Peters, K. E., and M. G. Fowler. 2002. Applications of petroleum geochemistry to exploration and reservoir management. *Organic Geochemistry* 33(1):5–36.

Peters, K. E., C. C. Walters, and J. M. Moldowan. 2005. *The biomarker guide. Biomarkers and isotopes in the environment and human history.* Vol. 2. Cambridge, UK: Cambridge University Press.

Petersen, J., D. Nelson, T. Marcella, J. Michel, M. Atkinson, M. White, C. Boring, L. Szathmary, and J. Weaver. 2019. *Environmental sensitivity index guidelines, version 4.0.* NOAA technical memorandum NOS OR&R 52.

Peterson, C. H., L. L. McDonald, R. H. Green, and W. P. Erickson. 2001. Sampling design begets conclusions: The statistical basis for detection of injury to and recovery of shoreline communities after the *Exxon Valdez* oil spill. *Marine Ecology Progress Series* 210:255–283.

Peterson, C. H., S. D. Rice, J. W. Short, D. Esler, J. L. Bodkin, B. E. Ballachey, and D. B. Irons. 2003. Long-term ecosystem response to the *Exxon Valdez* oil spill. *Science* 302(5653):2082–2086.

Peterson, C. H., S. S. Anderson, G. N. Cherr, R. F. Ambrose, S. Anghera, S. Bay, M. Blum, R. Condon, T. A. Dean, and M. Graham. 2012. A tale of two spills: Novel science and policy implications of an emerging new oil spill model. *Bioscience* 62(5):461–469.

Petrich, C., J. Karlsson, and H. Eicken. 2013. Porosity of growing sea ice and potential for oil entrainment. *Cold Regions Science and Technology* 87:27–32.

Phifer, J. F., and F. H. Norris. 1989. Psychological symptoms in older adults following natural disaster: Nature, timing, duration, and course. *Journal of Gerontology* 44(6):S207–S212.

Philp, R. P. 2018. Fifty years of petroleum geochemistry: A valuable asset in oil spill environmental forensics. In *Oil spill environmental forensics case studies*, pp. 25–48. Amsterdam, Netherlands: Elsevier.

Picou, J. S. 2011. When the solution becomes the problem: The impacts of adversarial litigation on survivors of the *Exxon Valdez* oil spill. *University of St. Thomas Law Journal* 7:68.

Picou, J. S., D. A. Gill, C. L. Dyer, and E. W. Curry. 1992. Disruption and stress in an Alaskan fishing community: Initial and continuing impacts of the *Exxon Valdez* oil spill. *Industrial Crisis Quarterly* 6(3):235–257.

Picou, J. S., D. A. Gill, and M. J. Cohen. 1997. Technological Disasters and Social Policy. In *The* Exxon Valdez *disaster: Readings on a modern social problem*, p. 309.

Pitt, R., R. Field, M. Lalor, and M. Brown. 1995. Urban stormwater toxic pollutants: Assessment, sources, and treatability. *Water Environment Research* 67(3):260–275.

Pitt, R., A. Maestre, and J. Clary. 2018. *The National Stormwater Quality Database (NSQD), version 4.02*. Department of Civil and Environmental Engineering.

Plant, N. G., J. W. Long, P. S. Dalyander, D. M. Thompson, and E. A. Raabe. 2013. *Application of a hydrodynamic and sediment transport model for guidance of response efforts related to the* Deepwater Horizon *oil spill in the northern Gulf of Mexico along the coast of Alabama and Florida*. Washington, DC: United States Geological Survey.

Poje Andrew, C., M. Özgökmen Tamay, L. Lipphardt Bruce, K. Haus Brian, H. Ryan Edward, C. Haza Angelique, A. Jacobs Gregg, A. J. H. M. Reniers, J. Olascoaga Maria, G. Novelli, A. Griffa, J. Beron-Vera Francisco, S. Chen Shuyi, E. Coelho, J. Hogan Patrick, D. Kirwan Albert, S. Huntley Helga, and J. Mariano Arthur. 2014. Submesoscale dispersion in the vicinity of the *Deepwater Horizon* spill. *Proceedings of the National Academy of Sciences* 111(35):12693–12698. https://doi.org/10.1073/pnas.1402452111.

Poling, B. E., J. M. Prausnitz, and J. P. O'Connell. 2001. *Properties of gases and liquids*. New York: McGraw-Hill Education.

Pontes, J., A. P. Mucha, H. Santos, I. Reis, A. Bordalo, M. C. Basto, A. Bernabeu, and C. M. R. Almeida. 2013. Potential of bioremediation for buried oil removal in beaches after an oil spill. *Marine Pollution Bulletin* 76(1):258–265. https://doi.org/10.1016/j.marpolbul.2013.08.029.

Pope, S. B., and S. B. Pope. 2000. *Turbulent flows*. Cambridge, UK: Cambridge University Press.

Popovicheva, O., E. Kireeva, N. Shonija, N. Zubareva, N. Persiantseva, V. Tishkova, B. Demirdjian, J. Moldanova, and V. Mogilnikov. 2009. Ship particulate pollutants: Characterization in terms of environmental implication. *Journal of Environmental Monitoring* 11(11):2077–2086.

Porter, A., B. P. Lyons, T. S. Galloway, and C. Lewis. 2018. Role of Marine Snow in Microplastic Fate and Bioavailability. *Environmental Science & Technology* 52(12):7111–7119. https://doi.org/10.1021/acs.est.8b01000.

Portnoy, D. S., A. T. Fields, J. B. Greer, and D. Schlenk. 2020. Genetics and oil: Transcriptomics, epigenetics, and population genomics as tools to understand animal responses to exposure across different time scales. In *Deep oil spills*, pp. 515–532.

Potter, S., and I. Buist. 2010. In-situ burning in Arctic and ice-covered waters: Tests of fire-resistant boom in low concentrations of drift ice. *Proceedings of the 33rd Arctic and Marine Oilspill Program Technical Seminar*, pp. 743–754. Halifax, Nova Scotia, Canada.

Potter, S., I. Buist, K. Trudel, D. Dickins, and E. Owens. 2012. Spill response in the Arctic offshore. *Shell Exploration and Production Services* 463–467.

Powers, S. P., F. J. Hernandez, R. H. Condon, J. M. Drymon, and C. M. Free. 2013. Novel pathways for injury from offshore oil spills: Direct, sublethal and indirect effects of the *Deepwater Horizon* oil spill on pelagic sargassum communities. *PloS ONE* 8(9):e74802.

Prascal, B., M. Ziska, and J. Williams. 2014. A field evaluation of unmanned aircraft systems for oil spill response. *International Oil Spill Conference Proceedings* 2014(1):373–387.

Pratt, G. C., M. R. Stenzel, R. K. Kwok, C. P. Groth, S. Banerjee, S. F. Arnold, L. S. Engel, D. P. Sandler, and P. A. Stewart. 2020. Modeled air pollution from in situ burning and flaring of oil and gas released following the *Deepwater Horizon* disaster. *Annals of Work Exposures and Health* 66(Suppl 1):i172–i187.

Price, D. J., C.-L. Chen, L. M. Russell, M. A. Lamjiri, R. Betha, K. Sanchez, J. Liu, A. K. Lee, and D. R. Cocker. 2017. More unsaturated, cooking-type hydrocarbon-like organic aerosol particle emissions from renewable diesel compared to ultra low sulfur diesel in at-sea operations of a research vessel. *Aerosol Science and Technology* 51(2):135–146.

Price, E. R., and E. M. Mager. 2020. The effects of exposure to crude oil or PAHs on fish swim bladder development and function. *Comparative Biochemistry and Physiology Part C: Toxicology & Pharmacology* 238(December 2020):108853.

Prince, R., T. Amande, and T. McGenity. 2018. Taxonomy, genomics and ecophysiology of hydrocarbon-degrading microbes. *Handbook of hydrocarbon and lipid microbiology*. Dordrecht, Netherlands: Springer.

Prince, R. C. 1993. Petroleum spill bioremediation in marine environments. *Critical Reviews in Microbiology* 19(4):217–240. https://doi.org/10.3109/10408419309113530.

Prince, R. C. 2015. Oil spill dispersants: Boon or bane? *Environmental Science & Technology* 49(11):6376–6384.

Prince, R. C. 2017. Biostimulation of marine crude oil spills using dispersants. In *Hydrocarbon and lipid microbiology protocols*, pp. 95–104. Dordrecht, Netherlands: Springer.

Prince, R. C., and J. D. Butler. 2014. A protocol for assessing the effectiveness of oil spill dispersants in stimulating the biodegradation of oil. *Environmental Science and Pollution Research* 21(16):9506–9510.

Prince, R. C., and C. C. Walters. 2007. Chapter 11: Biodegradation of oil hydrocarbons and its implications for source identification. In *Oil spill environmental forensics: Fingerprinting and source identification*, Z. Wang and S. Stout (eds.). Amsterdam, Netherlands: Elsevier.

Prince, R. C., E. H. Owens, and G. A. Sergy. 2002. Weathering of an Arctic oil spill over 20 years: The bios experiment revisited. *Marine Pollution Bulletin* 44(11):1236–1242.

Prince, R. C., K. M. McFarlin, J. D. Butler, E. J. Febbo, F. C. Wang, and T. J. Nedwed. 2013. The primary biodegradation of dispersed crude oil in the sea. *Chemosphere* 90(2):521–526.

Prince, R. C., G. W. Nash, and S. J. Hill. 2016a. The biodegradation of crude oil in the deep ocean. *Marine Pollution Bulletin* 111(1–2):354–357.

Prince, R. C., J. D. Butler, G. E. Bragin, T. F. Parkerton, A. D. Redman, B. A. Kelley, and D. J. Letinski. 2016b. Preparing the hydrocarbon/crude oil. In *Hydrocarbon and lipid microbiology protocols*, pp. 15–32. Dordrecht, Netherlands: Springer.

Prince, R. C., J. D. Butler, and A. D. Redman. 2017. The rate of crude oil biodegradation in the sea. *Environmental Science & Technology* 51(3):1278–1284.

Prince, R. C., T. J. Amande, and T. J. McGenity. 2018. Prokaryotic hydrocarbon degraders. In *Taxonomy, genomics and ecophysiology of hydrocarbon-degrading microbes*, T. J. McGenity (ed.). Dordrecht, Netherlands: Springer.

Puestow, T., L. Parsons, I. Zakharov, N. Cater, P. Bobby, M. Fuglem, G. Parr, A. Jayasiri, S. Warren, and G. Warbanski. 2013. *Oil spill detection and mapping in low visibility and ice: Surface remote sensing*. London, UK: Arctic Oil Spill Response Technology Joint Industry Programme.

Quigg, A., U. Passow, W. C. Chin, C. Xu, S. Doyle, L. Bretherton, M. Kamalanathan, A. K. Williams, J. B. Sylvan, and Z. V. Finkel. 2016. The role of microbial exopolymers in determining the fate of oil and chemical dispersants in the ocean. *Limnology and Oceanography Letters* 1(1):3–26.

Quigg, A., U. Passow, K. L. Daly, A. Burd, D. J. Hollander, P. T. Schwing, and K. Lee. 2020. Marine Oil Snow Sedimentation and Flocculent Accumulation (MOSSFA) events: Learning from the past to predict the future. In *Deep oil spills: Facts, fate, and effects*, S. A. Murawski, C. H. Ainsworth, S. Gilbert, D. J. Hollander, C. B. Paris, M. Schlüter and D. L. Wetzel (eds.), pp. 196–220. Cham, Germany: Springer International Publishing.

Quigg, A., J. W. Farrington, S. Gilbert, S. A. Murawski, and V. T. John. 2021a. A decade of GoMRI dispersant science. *Oceanography* 34(1):98–111.

Quigg, A., M. Parsons, S. Bargu, K. Ozhan, K. L. Daly, S. Chakraborty, M. Kamalanathan, D. Erdner, S. Cosgrove, and E. J. Buskey. 2021b. Marine phytoplankton responses to oil and dispersant exposures: Knowledge gained since the *Deepwater Horizon* oil spill. *Marine Pollution Bulletin* 164:112074.

Quist, A. J., D. S. Rohlman, R. K. Kwok, P. A. Stewart, M. R. Stenzel, A. Blair, A. K. Miller, M. D. Curry, D. P. Sandler, and L. S. Engel. 2019. *Deepwater Horizon* oil spill exposures and neurobehavioral function in gulf study participants. *Environmental Research* 179:108834.

Rabalais, N., B. McKee, D. Reed, and J. Means. 1992. Fate and effects of produced water discharges in coastal Louisiana, Gulf of Mexico, USA. In *Produced water*, pp. 355–369. Dordrecht, Netherlands: Springer.

Rabalais, N. N. 1991. *Fate and effects of nearshore discharges of OCS produced waters: Technical report.* Vol. 2. Washington, DC: Minerals Management Service, Gulf of Mexico, U.S. Department of the Interior.

Rabalais, N. N., and R. E. Turner. 2019. Gulf of Mexico hypoxia: Past, present, and future. *Limnology and Oceanography Bulletin* 28(4):117–124.

Rabalais, N. N., L. M. Smith, and R. E. Turner. 2018. The *Deepwater Horizon* oil spill and Gulf of Mexico shelf hypoxia. *Continental Shelf Research* 152:98–107.

Rabinowitz, P., and L. Conti. 2013. Links among human health, animal health, and ecosystem health. *Annual Review of Public Health* 34:189–204.

Radović, J. R., T. B. Oldenburg, and S. R. Larter. 2018. Environmental assessment of spills related to oil exploitation in Canada's oil sands region. In *Oil spill environmental forensics case studies*, pp. 401–417. Amsterdam, Netherlands: Elsevier.

Ramachandran, S., C. Khan, P. Hodson, K. Lee, and T. King. 2004. *Role of droplets in promoting uptake of PAHs by fish exposed to chemically dispersed crude oil.* Arctic and Marine Oilspill Program Technical Seminar.

Rangoonwala, A., C. E. Jones, and E. Ramsey III. 2016. Wetland shoreline recession in the Mississippi river delta from petroleum oiling and cyclonic storms. *Geophysical Research Letters* 43(22):11652–11660.

Rasgon, A. 2021. *Israel's beaches are littered with tar after mysterious oil spill. The New York Times.* https://www.nytimes.com/2021/02/23/world/middleeast/israel-oil-spill-mystery.html.

Rattner, B. A., and W. C. Eastin. 1981. Plasma corticosterone and thyroxine concentrations during chronic ingestion of crude oil in mallard ducks (anas platyrhynchos). *Comparative Biochemistry and Physiology Part C: Comparative Pharmacology* 68(2):103–107. https://doi.org/10.1016/0306-4492(81)90002-2.

Rattner, B. A., V. P. Eroschenko, G. A. Fox, D. M. Fry, and J. Gorsline. 1984. Avian endocrine responses to environmental pollutants. *Journal of Experimental Zoology* 232(3):683–689.

Rauch, S., H. F. Hemond, C. Barbante, M. Owari, G. M. Morrison, B. Peucker-Ehrenbrink, and U. Wass. 2005. Importance of automobile exhaust catalyst emissions for the deposition of platinum, palladium, and rhodium in the northern hemisphere. *Environmental Science & Technology* 39(21):8156–8162.

Rawson, C., K. Crake, and A. Brown. 1998. Assessing the environmental performance of tankers in accidental grounding and collision. *SNAME Transactions* 106:41–58.

Ray, P. Z., and M. A. Tarr. 2014a. Solar production of singlet oxygen from crude oil films on water. *Journal of Photochemistry and Photobiology A: Chemistry* 286:22–28. https://doi.org/10.1016/j.jphotochem.2014.04.016.

Ray, P. Z., and M. A. Tarr. 2014b. Petroleum films exposed to sunlight produce hydroxyl radical. *Chemosphere* 103:220–227. https://doi.org/10.1016/j.chemosphere.2013.12.005.

Rebar, A., T. Lipscomb, R. Harris, and B. E. Ballachey. 1995. Clinical and clinical laboratory correlates in sea otters dying unexpectedly in rehabilitation centers following the *Exxon Valdez* oil spill. *Veterinary Pathology* 32(4):346–350.

Reddy, C., J. Arey, J. Seewald, S. Sylva, K. Lemkau, R. Nelson, C. Carmichael, C. McIntyre, J. Fenwick, and G. Ventura. 2012. Science applications in the *Deepwater Horizon* oil spill special feature: Composition and fate of gas and oil released to the water column during the *Deepwater Horizon* oil spill. *Proceedings of the National Academy of Sciences* 109:20229–20234.

Reddy, C. M., and J. G. Quinn. 2001. The North Cape oil spill: Hydrocarbons in Rhode Island coastal waters and Point Judith Pond. *Marine Environmental Research* 52(5):445–461. https://doi.org/10.1016/s0141-1136(01)00100-3.

Reddy, C. M., T. I. Eglinton, A. Hounshell, H. K. White, L. Xu, R. B. Gaines, and G. S. Frysinger. 2002. The west Falmouth oil spill after thirty years: The persistence of petroleum hydrocarbons in marsh sediments. *Environmental Science & Technology* 36(22):4754–4760.

Reddy, C. M., R. K. Nelson, S. P. Sylva, L. Xu, E. A. Peacock, B. Raghuraman, and O. C. Mullins. 2007. Identification and quantification of alkene-based drilling fluids in crude oils by comprehensive two-dimensional gas chromatography with flame ionization detection. *Journal of Chromatography A* 1148(1):100–107. https://doi.org/10.1016/j.chroma.2007.03.001.

Redman, A. D., and T. F. Parkerton. 2015. Guidance for improving comparability and relevance of oil toxicity tests. *Marine Pollution Bulletin* 98(1–2):156–170.

Redman, A. D., T. F. Parkerton, J. A. McGrath, and D. M. Di Toro. 2012. Petrotox: An aquatic toxicity model for petroleum substances. *Environmental Toxicology and Chemistry* 31(11):2498–2506.

Redman, A. D., J. D. Butler, D. J. Letinski, and T. F. Parkerton. 2017. Investigating the role of dissolved and droplet oil in aquatic toxicity using dispersed and passive dosing systems. *Environmental Toxicology and Chemistry* 36(4):1020–1028.

Redmond, M. C., and D. L. Valentine. 2012. Natural gas and temperature structured a microbial community response to the *Deepwater Horizon* oil spill. *Proceedings of the National Academy of Sciences* 109(50):20292–20297.

Reeburgh, W. S. 2007. Oceanic methane biogeochemistry. *Chemical Reviews* 107(2):486–513. https://doi.org/10.1021/cr050362v.

Reed, D., R. Lewis, and M. Anghera. 1994. Effects of an open-coast oil-production outfall on patterns of giant kelp (macrocystis pyrifera) recruitment. *Marine Biology* 120(1):25–31.

Reed, M., and H. Rye. 2011. The dream model and the environmental impact factor: Decision support for environmental risk management. In *Produced water*, pp. 189–203. Dordrecht, Netherlands: Springer.

Reed, M., P. Daling, A. Lewis, M. K. Ditlevsen, B. Brørs, J. Clark, and D. Aurand. 2004. Modelling of dispersant application to oil spills in shallow coastal waters. *Environmental Modelling & Software* 19(7–8):681–690.

Regnier, Z. R., and B. F. Scott. 1975. Evaporation rates of oil components. *Environmental Science & Technology* 9(5):469–472.

Rehder, G., P. W. Brewer, E. T. Peltzer, and G. Friederich. 2002. Enhanced lifetime of methane bubble streams within the deep ocean. *Geophysical Research Letters* 29(15):21-1–21-4.

Rehder, G., I. Leifer, P. G. Brewer, G. Friederich, and E. T. Peltzer. 2009. Controls on methane bubble dissolution inside and outside the hydrate stability field from open ocean field experiments and numerical modeling. *Marine Chemistry* 114(1–2):19–30.

Ren, L., N. N. Rabalais, R. E. Turner, W. Morrison, and W. Mendenhall. 2009. Nutrient limitation on phytoplankton growth in the upper barataria basin, Louisiana: Microcosm bioassays. *Estuaries and Coasts* 32(5):958–974.

Renegar, D. A., and N. R. Turner. 2021. Species sensitivity assessment of five Atlantic scleractinian coral species to 1-methylnaphthalene. *Scientific Reports* 11(1):1–17.

Renegar, D. A., P. Schuler, N. Turner, R. Dodge, B. Riegl, A. Knap, G. Bera, R. Jézéquel, and B. Benggio. 2017. Tropics field study (Panama), 32-year site visit: Observations and conclusions for near shore dispersant use neba and tradeoffs. *International Oil Spill Conference Proceedings* 2017(1):3030–3050.

Renegar, D. A., P. A. Schuler, A. H. Knap, and R. A. Dodge. 2022. TRopical Oil Pollution Investigations in Coastal Systems [TROPICS]: A synopsis of impacts and recovery. *Marine Pollution Bulletin* 181:113880.

Renner, A. H. H., M. Dumont, J. Beckers, S. Gerland, and C. Haas. 2013. Improved characterisation of sea ice using simultaneous aerial photography and sea ice thickness measurements. *Cold Regions Science and Technology* 92:37–47. https://doi.org/10.1016/j.coldregions.2013.03.009.

Reuscher, M. G., J. G. Baguley, and P. A. Montagna. 2020. The expanded footprint of the *Deepwater Horizon* oil spill in the Gulf of Mexico deep-sea benthos. *PloS ONE* 15(6):e0235167.

Revsbech Niels, P., B. Jørgensen Bo, and T. H. Blackburn. 1980. Oxygen in the sea bottom measured with a microelectrode. *Science* 207(4437):1355–1356. https://doi.org/10.1126/science.207.4437.1355.

Reynolds, R. R., and R. D. Kiker. 2003. *Produced water and associated issues*. Oklahoma Geological Survey. http://www.ogs.ou.edu/PTTC/pwm/produced_water.pdf.

Rice, S. D. 1973. Toxicity and avoidance tests with Prudhoe Bay oil and pink salmon fry. *International Oil Spill Conference Proceedings* 1973(1):667–670.

Rice, S. D., J. W. Short, and J. F. Karinen. 1977. Comparative oil toxicity and comparative animal sensitivity. In *Fate and effects of petroleum hydrocarbons in marine ecosystems and organisms*, pp. 78–94.

Rice, S. D., R. E. Thomas, M. G. Carls, R. A. Heintz, A. C. Wertheimer, M. L. Murphy, J. W. Short, and A. Moles. 2001. Impacts to pink salmon following the *Exxon Valdez* oil spill: Persistence, toxicity, sensitivity, and controversy. *Reviews in Fisheries Science* 9(3):165–211.

Richards, T. A., M. D. Jones, G. Leonard, and D. Bass. 2012. Marine fungi: Their ecology and molecular diversity. *Annual Review of Marine Science* 4:495–522.

Rinkevich, B. 1977. Harmful effects of chronic oil pollution on a Red Sea scleractinian coral population. In *Proceedings of the 3rd International Coral Reef Symposium*, pp. 585–591.

Risk, M. J., J. M. Heikoop, M. G. Snow, and R. Beukens. 2002. Lifespans and growth patterns of two deep-sea corals: Primnoa resedaeformis and Desmophyllum cristagalli. *Hydrobiologia* 471(1):125–131. https://doi.org/10.1023/A:1016557405185.

Ritchie, L. A., D. A. Gill, and M. A. Long. 2018. Mitigating litigating: An examination of psychosocial impacts of compensation processes associated with the 2010 BP *Deepwater Horizon* oil spill. *Risk Analysis* 38(8):1656–1671.

Røberg, S., J. I. Østerhus, and B. Landfald. 2011. Dynamics of bacterial community exposed to hydrocarbons and oleophilic fertilizer in high-Arctic intertidal beach. *Polar Biology* 34(10):1455–1465.

Robinson, A. L., N. M. Donahue, M. K. Shrivastava, E. A. Weitkamp, A. M. Sage, A. P. Grieshop, T. E. Lane, J. R. Pierce, and S. N. Pandis. 2007. Rethinking organic aerosols: Semivolatile emissions and photochemical aging. *Science* 315(5816):1259–1262.

Rocke, T., T. Yuill, and R. Hinsdill. 1984. Oil and related toxicant effects on mallard immune defenses. *Environmental Research* 33(2):343–352.

Rodgers, R. P., M. L. Chacón-Patiño, S. F. Niles, H. Chen, A. M. McKenna, P. Zito, M. A. Tarr, and A. G. Marshall. 2021. High resolution mass spectrometry advances in oil spill analysis. *International Oil Spill Conference Proceedings* 2021(1):685482.

Rodriguez, L. M., W. A. Overholt, C. Hagan, M. Huettel, J. E. Kostka, and K. T. Konstantinidis. 2015. Microbial community successional patterns in beach sands impacted by the *Deepwater Horizon* oil spill. *The ISME Journal* 9(9):1928–1940.

Rodríguez-Trigo, G., J.-P. Zock, F. Pozo-Rodríguez, F. P. Gómez, G. Monyarch, L. Bouso, M. D. Coll, H. Verea, J. M. Antó, and C. Fuster. 2010. Health changes in fishermen 2 years after clean-up of the prestige oil spill. *Annals of Internal Medicine* 153(8):489–498.

Rojo, F. 2009. Degradation of alkanes by bacteria. *Environmental Microbiology* 11(10):2477–2490.

Rolland, R. M. 2000. A review of chemically-induced alterations in thyroid and vitamin A status from field studies of wildlife and fish. *Journal of Wildlife Diseases* 36(4):615–635.

Roman-Hubers, A. T., T. J. McDonald, E. S. Baker, W. A. Chiu, and I. Rusyn. 2021. A comparative analysis of analytical techniques for rapid oil spill identification. *Environmental Toxicology and Chemistry* 40(4):1034–1049.

Römer, M., H. Sahling, T. Pape, G. Bohrmann, and V. Spieß. 2012. Quantification of gas bubble emissions from submarine hydrocarbon seeps at the makran continental margin (offshore Pakistan). *Journal of Geophysical Research: Oceans* 117(C10):15.

Römer, M., C.-W. Hsu, M. Loher, I. R. MacDonald, C. dos Santos Ferreira, T. Pape, S. Mau, G. Bohrmann, and H. Sahling. 2019. Amount and fate of gas and oil discharged at 3400 m water depth from a natural seep site in the southern Gulf of Mexico. *Frontiers in Marine Science* 6:700.

Romero, I. C., T. Sutton, B. Carr, E. Quintana-Rizzo, S. W. Ross, D. J. Hollander, and J. J. Torres. 2018. Decadal assessment of polycyclic aromatic hydrocarbons in mesopelagic fishes from the Gulf of Mexico reveals exposure to oil-derived sources. *Environmental Science & Technology* 52(19):10985–10996. https://doi.org/10.1021/acs.est.8b02243.

Romero, I. C., J. P. Chanton, B. E. Roseheim, J. R. Radović, P. T. Schwing, D. J. Hollander, S. R. Larter, and T. B. Oldenburg. 2020. Long-term preservation of oil spill events in sediments: The case for the *Deepwater Horizon* oil spill in the northern Gulf of Mexico. In *Deep oil spills*, pp. 285–300. Dordrecht, Netherlands: Springer.

Romero, I. C., J. P. Chanton, G. R. Brooks, S. Bosman, R. A. Larson, A. Harris, P. Schwing, and A. Diercks. 2021. Molecular markers of biogenic and oil-derived hydrocarbons in deep-sea sediments following the *Deepwater Horizon* spill. *Frontiers in Marine Science* 8. https://doi.org/10.3389/fmars.2021.637970.

Ron, E. Z., and E. Rosenberg. 2002. Biosurfactants and oil bioremediation. *Current Opinion in Biotechnology* 13(3):249–252.

Ross, J., D. Hollander, S. Saupe, A. B. Burd, S. Gilbert, and A. Quigg. 2021. Integrating marine oil snow and MOSSFA into oil spill response and damage assessment. *Marine Pollution Bulletin* 165:112025. https://doi.org/10.1016/j.marpolbul.2021.112025.

Rosser, R., S. Dewar, and J. Thompson. 1991. Psychological aftermath of the King's Cross fire. *Journal of the Royal Society of Medicine* 84(1):4–8.

Rothman, N., M. T. Smith, R. B. Hayes, R. D. Traver, B.-a. Hoener, S. Campleman, G.-L. Li, M. Dosemeci, M. Linet, and L. Zhang. 1997. Benzene poisoning, a risk factor for hematological malignancy, is associated with the NQO1 609C→ T mutation and rapid fractional excretion of chlorzoxazone. *Cancer Research* 57(14):2839–2842.

Rotkin-Ellman, M., Karen K. Wong, and Gina M. Solomon. 2012. Seafood contamination after the BP Gulf oil spill and risks to vulnerable populations: A critique of the FDA risk assessment. *Environmental Health Perspectives* 120(2):157–161. https://doi.org/10.1289/ehp.1103695.

Ruberg, E. J., J. E. Elliott, and T. D. Williams. 2021. Review of petroleum toxicity and identifying common endpoints for future research on diluted bitumen toxicity in marine mammals. *Ecotoxicology* 30(4):537–551. https://doi.org/10.1007/s10646-021-02373-x.

Rubin-Blum, M., C. P. Antony, C. Borowski, L. Sayavedra, T. Pape, H. Sahling, G. Bohrmann, M. Kleiner, M. C. Redmond, and D. L. Valentine. 2017. Short-chain alkanes fuel mussel and sponge cycloclasticus symbionts from deep-sea gas and oil seeps. *Nature Microbiology* 2(8):1–11.

Rughöft, S., A. L. Vogel, S. B. Joye, T. Gutierrez, and S. Kleindienst. 2020. Starvation-dependent inhibition of the hydrocarbon degrader Marinobacter sp. Tt1 by a chemical dispersant. *Journal of Marine Science and Engineering* 8(11):925.

Ruiz-Ramos, D. V., C. R. Fisher, and I. B. Baums. 2017. Stress response of the black coral Leiopathes glaberrima when exposed to sub-lethal amounts of crude oil and dispersant. *Elementa: Science of the Anthropocene* 5. https://doi.org/10.1525/elementa.261.

Rullkötter, J., and J. W. Farrington. 2021. What was released? Assessing the physical properties and chemical composition of petroleum and products of burned oil. *Oceanography* 34(1):44–57. https://www.jstor.org/stable/27020060.

Rung, A. L., S. Gaston, W. T. Robinson, E. J. Trapido, and E. S. Peters. 2017. Untangling the disaster-depression knot: The role of social ties after *Deepwater Horizon*. *Social Science & Medicine* 177:19–26.

Ruppel, C. D., and J. Kessler. 2017. The interaction of climate change and methane hydrates: Climatehydrates interactions. *Reviews of Geophysics* 55(1):126–168.

Rusiecki, J., M. Alexander, E. G. Schwartz, L. Wang, L. Weems, J. Barrett, K. Christenbury, D. Johndrow, R. H. Funk, and L. S. Engel. 2018. The *Deepwater Horizon* oil spill coast guard cohort study. *Occupational and Environmental Medicine* 75(3):165–175.

Ryerson, T., K. Aikin, W. Angevine, E. Atlas, D. Blake, C. Brock, F. Fehsenfeld, R. S. Gao, J. De Gouw, and D. Fahey. 2011. Atmospheric emissions from the *Deepwater Horizon* spill constrain air-water partitioning, hydrocarbon fate, and leak rate. *Geophysical Research Letters* 38(7):L07803.

Ryerson, T. B., R. Camilli, J. D. Kessler, E. B. Kujawinski, C. M. Reddy, D. L. Valentine, E. Atlas, D. R. Blake, J. De Gouw, and S. Meinardi. 2012. Chemical data quantify *Deepwater Horizon* hydrocarbon flow rate and environmental distribution. *Proceedings of the National Academy of Sciences* 109(50):20246–20253.

Saco-Alvarez, L., J. Bellas, O. Nieto, J. M. Bayona, J. Albaigés, and R. Beiras. 2008. Toxicity and phototoxicity of water-accommodated fraction obtained from prestige fuel oil and marine fuel oil evaluated by marine bioassays. *Science of the Total Environment* 394(2–3):275–282. https://doi.org/10.1016/j.scitotenv.2008.01.045.

Saltymakova, D., D. S. Desmond, D. Isleifson, N. Firoozy, T. D. Neusitzer, Z. Xu, M. Lemes, D. G. Barber, and G. A. Stern. 2020. Effect of dissolution, evaporation, and photooxidation on crude oil chemical composition, dielectric properties and its radar signature in the Arctic environment. *Marine Pollution Bulletin* 151:110629.

Sampath, K., N. Afshar-Mohajer, L. D. Chandrala, W. S. Heo, J. Gilbert, D. Austin, K. Koehler, and J. Katz. 2019. Aerosolization of crude oil-dispersant slicks due to bubble bursting. *Journal of Geophysical Research: Atmospheres* 124(10):5555–5578.

Sanchez, M., S. Karnae, and K. John. 2008. Source characterization of volatile organic compounds affecting the air quality in a coastal urban area of South Texas. *International Journal of Environmental Research and Public Health* 5(3):130–138.

Sandifer, P. A., and A. H. Walker. 2018. Enhancing disaster resilience by reducing stress-associated health impacts. *Frontiers in Public Health* 6:373.

Sandifer, P. A., A. Ferguson, M. L. Finucane, M. Partyka, H. M. Solo-Gabriele, A. H. Walker, K. Wowk, R. Caffey, and D. Yoskowitz. 2021. Human health and socioeconomic effects of the *Deepwater Horizon* oil spill in the Gulf of Mexico. *Oceanography* 34(1):174–191.

Sandkvist, J. n.d. *Oil spill recovery in ice*.

Sandoval, K., Y. Ding, and P. Gardinali. 2017. Characterization and environmental relevance of oil water preparations of fresh and weathered MC-252 Macondo oils used in toxicology testing. *Science of the Total Environment* 576:118–128.

Sanli, K., J. Bengtsson-Palme, R. H. Nilsson, E. Kristiansson, M. Alm Rosenblad, H. Blanck, and K. M. Eriksson. 2015. Metagenomic sequencing of marine periphyton: Taxonomic and functional insights into biofilm communities. *Frontiers in Microbiology* 6:1192.

Sapozhnikov, V., N. Arzhanova, and V. Zubarevitch. 1993. Estimation of primary production in the ecosystem of western Bering Sea. *Russian Journal of Aquatic Ecology* 2(1):23–34.

Sapozhnikov, V. V., A. Gruzevich, V. Zubarevich, N. Arzhanova, N. V. Mordasova, I. Nalyotova, N. I. Torgunova, Y. Mikhailovskiy, and I. Smolyar. 2001. Hydrochemical atlas of the sea of Okhotsk 2001. *International Ocean Atlas Series, Laboratory of the Marine Ecology, Russian Federal Research Institute of Fisheries and Oceanography* 3. http://www.nodc.noaa.gov/OC5/okhotsk/start_ok.html.

Sassen, R., S. Joye, S. T. Sweet, D. A. DeFreitas, A. V. Milkov, and I. R. MacDonald. 1999. Thermogenic gas hydrates and hydrocarbon gases in complex chemosynthetic communities, Gulf of Mexico continental slope. *Organic Geochemistry* 30(7):485–497.

Sathiakumar, N., M. Tipre, A. Turner-Henson, L. Chen, M. Leader, and J. Gohlke. 2017. Post-*Deepwater Horizon* blowout seafood consumption patterns and community-specific levels of concern for selected chemicals among children in Mobile County, Alabama. *International Journal of Hygiene and Environmental Health* 220(1):1–7. https://doi.org/10.1016/j.ijheh.2016.08.003.

Savitz, D. A., and L. S. Engel. 2010. Lessons for study of the health effects of oil spills. *Annals of Internal Medicine* 153(8):540–541.

Schaefer, J., N. Frazier, and J. Barr. 2016. Dynamics of near-coastal fish assemblages following the *Deepwater Horizon* oil spill in the northern Gulf of Mexico. *Transactions of the American Fisheries Society* 145(1):108–119.

Scarlett, A. G., R. K. Nelson, M. M. Gagnon, A. I. Holman, C. M. Reddy, P. A. Sutton, and K. Grice. 2021. Mv Wakashio grounding incident in Mauritius 2020: The world's first major spillage of very low sulfur fuel oil. *Marine Pollution Bulletin* 171:112917.

Schaum, J., M. Cohen, S. Perry, R. Artz, R. Draxler, J. B. Frithsen, D. Heist, M. Lorber, and L. Phillips. 2010. Screening level assessment of risks due to dioxin emissions from burning oil from the BP *Deepwater Horizon* Gulf of Mexico spill. *Environmental Science & Technology* 44(24):9383–9389. https://doi.org/10.1021/es103559w.

Schedler, M., R. Hiessl, A. G. V. Juárez, G. Gust, and R. Müller. 2014. Effect of high pressure on hydrocarbon-degrading bacteria. *AMB Express* 4(1):1–7.

Scheibye, K., J. H. Christensen, and A. R. Johnsen. 2017. Biodegradation of crude oil in Arctic subsurface water from the Disko Bay (Greenland) is limited. *Environmental Pollution* 223:73–80.

Schlenker, L. S., M. J. Welch, E. M. Mager, J. D. Stieglitz, D. D. Benetti, P. L. Munday, and M. Grosell. 2019. Exposure to crude oil from the *Deepwater Horizon* oil spill impairs oil avoidance behavior without affecting olfactory physiology in juvenile mahi-mahi (Coryphaena hippurus). *Environmental Science & Technology* 53(23):14001–14009.

Schmale, O., I. Leifer, J. S. v. Deimling, C. Stolle, S. Krause, K. Kießlich, A. Frahm, and T. Treude. 2015. Bubble transport mechanism: Indications for a gas bubble-mediated inoculation of benthic methanotrophs into the water column. *Continental Shelf Research* 103:70–78.

Schmidt-Etkin, D., and T. J. Nedwed. 2021. Effectiveness of mechanical recovery for large offshore oil spills. *Marine Pollution Bulletin* 163:111848.

Scholz, D. K., S. R. Warren, Jr., A. H. Walker, and J. Michel. 2004. *Risk communication for in-situ burning: The fate of burned oil.* Washington, DC: American Petroleum Institute.

Schreiber, L., N. Fortin, J. Tremblay, J. Wasserscheid, M. Elias, J. Mason, S. Sanschagrin, S. Cobanli, T. King, and K. Lee. 2019. Potential for microbially mediated natural attenuation of diluted bitumen on the coast of British Columbia (Canada). *Applied and Environmental Microbiology* 85(10):e00086-19.

Schreiber, L., N. Fortin, J. Tremblay, J. Wasserscheid, S. Sanschagrin, J. Mason, C. A. Wright, D. Spear, S. C. Johannessen, and B. Robinson. 2021. In situ microcosms deployed at the coast of British Columbia (Canada) to study dilbit weathering and associated microbial communities under marine conditions. *FEMS Microbiology Ecology* 97(7):fiab082.

Schuler, B., Y. Zhang, F. Liu, A. E. Pomerantz, A. B. Andrews, L. Gross, V. Pauchard, S. Banerjee, and O. C. Mullins. 2020. Overview of asphaltene nanostructures and thermodynamic applications. *Energy & Fuels* 34(12):15082–15105.

Schwacke, L. H., C. R. Smith, F. I. Townsend, R. S. Wells, L. B. Hart, B. C. Balmer, T. K. Collier, S. De Guise, M. M. Fry, and L. J. Guillette, Jr. 2014. Health of common bottlenose dolphins (Tursiops truncatus) in Barataria Bay, Louisiana, following the *Deepwater Horizon* oil spill. *Environmental Science & Technology* 48(1):93–103.

Schwartz, J. A., B. M. Aldridge, B. L. Lasley, P. W. Snyder, J. L. Stott, and F. C. Mohr. 2004a. Chronic fuel oil toxicity in American mink (Mustela vison): Systemic and hematological effects of ingestion of a low-concentration of bunker C fuel oil. *Toxicology and Applied Pharmacology* 200(2):146–158.

Schwartz, J. A., B. M. Aldridge, J. L. Stott, and F. C. Mohr. 2004b. Immunophenotypic and functional effects of bunker C fuel oil on the immune system of American mink (Mustela vison). *Veterinary Immunology and Immunopathology* 101(3–4):179–190.

Schwarz, J. R., J. D. Walker, and R. R. Colwell. 1975. Deep-sea bacteria: Growth and utilization of n-hexadecane at in situ temperature and pressure. *Canadian Journal of Microbiology* 21(5):682–687. https://doi.org/10.1139/m75-098.

Schwarzenbach, R. P., P. M. Gschwend, and D. M. Imboden. 2003. *Environmental organic chemistry, 2nd edition*. Hoboken, NJ: John Wiley & Sons.

Schwarzenbach, R. P., P. M. Gschwend, and D. M. Imboden. 2016. *Environmental organic chemistry, 3rd edition*. Hoboken, NJ: John Wiley & Sons.

Schwing, P. T., D. J. Hollander, G. R. Brooks, R. A. Larson, D. W. Hastings, J. P. Chanton, S. A. Lincoln, J. R. Radović, and A. Langenhoff. 2020a. Chapter 13: The sedimentary record of MOSSFA events in the Gulf of Mexico: A comparison of the *Deepwater Horizon* (2010) and Ixtoc 1 (1979) oil spills. In *Deep oil spills*, pp. 221–234. Dordrecht, Netherlands: Springer.

Schwing, P. T., P. A. Montagna, S. B. Joye, C. B. Paris, E. E. Cordes, C. R. McClain, J. P. Kilborn, and S. A. Murawski. 2020b. A synthesis of deep benthic faunal impacts and resilience following the *Deepwater Horizon* oil spill. *Frontiers in Marine Science* 7. https://doi.org/10.3389/fmars.2020.560012.

Scoma, A., M. Barbato, E. Hernandez-Sanabria, F. Mapelli, D. Daffonchio, S. Borin, and N. Boon. 2016a. Microbial oil-degradation under mild hydrostatic pressure (10 MPa): Which pathways are impacted in piezosensitive hydrocarbonoclastic bacteria? *Scientific Reports* 6(1):1–14.

Scoma, A., M. M. Yakimov, and N. Boon. 2016b. Challenging oil bioremediation at deep-sea hydrostatic pressure. *Frontiers in Microbiology* 7: 1203.

Scott, D. E., M. Schulze, J. M. Stryker, and R. R. Tykwinski. 2021. Deciphering structure and aggregation in asphaltenes: Hypothesis-driven design and development of synthetic model compounds. *Chemical Society Reviews* 50(16):9202–9239.

Scott, N. M., M. Hess, N. J. Bouskill, O. U. Mason, J. K. Jansson, and J. A. Gilbert. 2014. The microbial nitrogen cycling potential is impacted by polyaromatic hydrocarbon pollution of marine sediments. *Frontiers in Microbiology* 5:108.

Seeley, M. E., Q. Wang, H. Bacosa, B. E. Rosenheim, and Z. Liu. 2018. Environmental petroleum pollution analysis using ramped pyrolysis-gas chromatography-mass spectrometry. *Organic Geochemistry* 124:180–189.

Sell, D., L. Conway, T. Clark, G. B. Picken, J. M. Baker, G. M. Dunnet, A. D. McIntyre, and R. B. Clark. 1995. Scientific criteria to optimize oil spill cleanup. *International Oil Spill Conference Proceedings* 1995(1):595–610.

Sellin Jeffries, M. K., C. Claytor, W. Stubblefield, W. H. Pearson, and J. T. Oris. 2013. Quantitative risk model for polycyclic aromatic hydrocarbon photoinduced toxicity in Pacific herring following the *Exxon Valdez* oil spill. *Environmental Science & Technology* 47(10):5450–5458.

Seol, D.-G., T. Bhaumik, C. Bergmann, and S. A. Socolofsky. 2007. Particle image velocimetry measurements of the mean flow characteristics in a bubble plume. *Journal of Engineering Mechanics* 133(6):665–676.

Seol, D.-G., D. B. Bryant, and S. A. Socolofsky. 2009. Measurement of behavioral properties of entrained ambient water in a stratified bubble plume. *Journal of Hydraulic Engineering* 135(11):983–988.

Sergy, G. A., and P. J. Blackall. 1987. Design and conclusions of the Baffin Island oil spill project. *Arctic* 1–9.

Shapiro, K., S. Khanna, and S. L. Ustin. 2016. Vegetation impact and recovery from oil-induced stress on three ecologically distinct wetland sites in the Gulf of Mexico. *Journal of Marine Science and Engineering* 4(2):33.

Sharp, B. E. 1996. Post-release survival of oiled, cleaned seabirds in North America. *Ibis* 138(2):222–228.

Shaw, S. L., B. Gantt, and N. Meskhidze. 2010. Production and emissions of marine isoprene and monoterpenes: A review. *Advances in Meteorology*. https://doi.org/10.1155/2010/408696.

Shenesey, J. W., and J. Langhinrichsen-Rohling. 2015. Perceived resilience: Examining impacts of the *Deepwater Horizon* oil spill one-year post-spill. *Psychological Trauma: Theory, Research, Practice, and Policy* 7(3):252.

Shepard, A. N., J. F. Valentine, C. F. D'Elia, D. Yoskowitz, and D. E. Dismukes. 2013. Economic impact of Gulf of Mexico ecosystem goods and services and integration into restoration decision-making. *Gulf of Mexico Science* 31(1):10–27.

Sherblom, P. M., and R. P. Eganhouse. 1991. Bioaccumulation of molecular markers for municipal wastes by mytilus edulis. *Organic Substances and Sediments in Water* 3:139–158.

Sherman, M., H. Covert, L. Brown, J. Langhinrichsen-Rohling, T. Hansel, T. Rehner, A. Buckner, and M. Lichtveld. 2019. Practice full report: Enterprise evaluation: A new opportunity for public health policy. *Journal of Public Health Management and Practice* 25(5):479.

Shigenaka, G. 2001. *Toxicity of oil to reef-building corals: A spill response perspective*. Seattle, WA: National Ocean Service, National Oceanic and Atmospheric Administration.

Shigenaka, G., and S. Milton. 2003. *Oil and sea turtles: Biology, planning, and response.* Seattle, WA: National Ocean Service, National Oceanic and Atmospheric Administration.

Shiller, A. M., and D. Joung. 2012. Nutrient depletion as a proxy for microbial growth in *Deepwater Horizon* subsurface oil/gas plumes. *Environmental Research Letters* 7(4):045301.

Shiller, A. M., E. W. Chan, D. Joung, M. Redmond, and J. D. Kessler. 2017. Light rare Earth element depletion during *Deepwater Horizon* blowout methanotrophy. *Scientific Reports* 7(1):1–9.

Shin, B., I. Bociu, M. Kolton, M. Huettel, and J. E. Kostka. 2019a. Succession of microbial populations and nitrogen-fixation associated with the biodegradation of sediment-oil-agglomerates buried in a Florida sandy beach. *Scientific Reports* 9(1):1–11.

Shin, B., M. Kim, K. Zengler, K.-J. Chin, W. A. Overholt, L. M. Gieg, K. T. Konstantinidis, and J. E. Kostka. 2019b. Anaerobic degradation of hexadecane and phenanthrene coupled to sulfate reduction by enriched consortia from northern Gulf of Mexico seafloor sediment. *Scientific Reports* 9(1):1–13.

Shiu, W. Y., A. Maijanen, A. L. Ng, and D. Mackay. 1988. Preparation of aqueous solutions of sparingly soluble organic substances: II. Multicomponent systems-hydrocarbon mixtures and petroleum products. *Environmental Toxicology and Chemistry* 7(2):125–137.

Shore, J. H., E. L. Tatum, and W. M. Vollmer. 1986. Evaluation of mental effects of disaster, Mount St. Helens eruption. *American Journal of Public Health* 76(Suppl):76–83.

Short, A. D. 2012. Coastal processes and beaches. *Nature Education Knowledge* 3(10):15.

Short, J. W. 2005. *Seasonal variability of pristane in mussels (mytilus trossulus) in Prince William Sound, Alaska.* Fairbanks, AK: University of Alaska Fairbanks.

Short, J. W., and P. M. Harris. 1996. Petroleum hydrocarbons in caged mussels deployed in Prince William Sound after the *Exxon Valdez* oil spill. *American Fisheries Society Symposium.*

Short, J. W., M. R. Lindeberg, P. M. Harris, J. M. Maselko, J. J. Pella, and S. D. Rice. 2004. Estimate of oil persisting on the beaches of Prince William Sound 12 years after the *Exxon Valdez* oil spill. *Environmental Science & Technology* 38(1):19–25.

Short, J. W., J. M. Maselko, M. R. Lindeberg, P. M. Harris, and S. D. Rice. 2006. Vertical distribution and probability of encountering intertidal *Exxon Valdez* oil on shorelines of three embayments within Prince William Sound, Alaska. *Environmental Science & Technology* 40(12):3723–3729.

Short, J. W., G. V. Irvine, D. H. Mann, J. M. Maselko, J. J. Pella, M. R. Lindeberg, J. R. Payne, W. B. Driskell, and S. D. Rice. 2007. Slightly weathered *Exxon Valdez* oil persists in Gulf of Alaska beach sediments after 16 years. *Environmental Science & Technology* 41(4): 1245–1250.

Short, J. W., H. J. Geiger, J. C. Haney, C. M. Voss, M. L. Vozzo, V. Guillory, and C. H. Peterson. 2017. Anomalously high recruitment of the 2010 Gulf Menhaden (brevoortia patronus) year class: Evidence of indirect effects from the *Deepwater Horizon* blowout in the Gulf of Mexico. *Archives of Environmental Contamination and Toxicology* 73(1):76–92.

Short, J. W., C. M. Voss, M. L. Vozzo, V. Guillory, H. J. Geiger, J. C. Haney, and C. H. Peterson. 2021. Evidence for ecosystem-level trophic cascade effects involving Gulf Menhaden (brevoortia patronus) triggered by the *Deepwater Horizon* blowout. *Journal of Marine Science and Engineering* 9(2):190.

Shrivastava, M., R. C. Easter, X. Liu, A. Zelenyuk, B. Singh, K. Zhang, P. L. Ma, D. Chand, S. Ghan, and J. L. Jimenez. 2015. Global transformation and fate of SOA: Implications of low-volatility SOA and gas-phase fragmentation reactions. *Journal of Geophysical Research: Atmospheres* 120(9):4169–4195.

Shtratnikova, V. Y., I. Belalov, A. S. Kasianov, M. I. Schelkunov, D. L. Maria, A. D. Novikov, A. A. Shatalov, T. V. Gerasimova, A. S. Yanenko, and V. J. Makeev. 2018. The complete genome of the oil emulsifying strain thalassolituus oleivorans K-188 from the Barents Sea. *Marine Genomics* 37:18–20.

Shultz, J. M., L. Walsh, D. R. Garfin, F. E. Wilson, and Y. Neria. 2015. The 2010 *Deepwater Horizon* oil spill: The trauma signature of an ecological disaster. *The Journal of Behavioral Health Services & Research* 42(1):58–76.

Sicot, G., M. Lennon, V. Miegebielle, and D. Dubucq. 2015. Estimation of the thickness and emulsion rate of oil spilled at sea using hyperspectral remote sensing imagery in the SWIR domain. *The International Archives of Photogrammetry, Remote Sensing and Spatial Information Sciences* 40(3):445.

Siddique, T., M. F. Mohamad Shahimin, S. Zamir, K. Semple, C. Li, and J. M. Foght. 2015. Long-term incubation reveals methanogenic biodegradation of C5 and C6 iso-alkanes in oil sands tailings. *Environmental Science & Technology* 49(24):14732–14739.

Sievwright, K. A. 2014. *Post-release survival and productivity of oiled little blue penguins (Eudyptula minor) rehabilitated after the 2011 C/V Rena oil spill.* Palmerston North, New Zealand: Master of Science, Conservation Biology, Massey University.

Silliman, B. R., J. van de Koppel, M. W. McCoy, J. Diller, G. N. Kasozi, K. Earl, P. N. Adams, and A. R. Zimmerman. 2012. Degradation and resilience in Louisiana salt marshes after the BP-*Deepwater Horizon* oil spill. *Proceedings of the National Academy of Sciences* 109(28):11234–11239.

Silva, M., P. J. Etnoyer, and I. R. MacDonald. 2016. Coral injuries observed at mesophotic reefs after the *Deepwater Horizon* oil discharge. *Deep Sea Research Part II: Topical Studies in Oceanography* 129:96–107.

Simon-Friedt, B. R., J. L. Howard, M. J. Wilson, D. Gauthe, D. Bogen, D. Nguyen, E. Frahm, and J. K. Wickliffe. 2016. Louisiana residents' self-reported lack of information following the *Deepwater Horizon* oil spill: Effects on seafood consumption and risk perception. *Journal of Environmental Management* 180:526–537.

Singer, M., D. Aurand, G. Bragin, J. Clark, G. Coelho, M. Sowby, and R. Tjeerdema. 2000. Standardization of the preparation and quantitation of water-accommodated fractions of petroleum for toxicity testing. *Marine Pollution Bulletin* 40(11):1007–1016.

Singh, A. K., A. Sherry, N. D. Gray, D. M. Jones, B. F. Bowler, and I. M. Head. 2014. Kinetic parameters for nutrient enhanced crude oil biodegradation in intertidal marine sediments. *Frontiers in Microbiology* 5:160.

Singh, R., M. S. Guzman, and A. Bose. 2017. Anaerobic oxidation of ethane, propane, and butane by marine microbes: A mini review. *Frontiers in Microbiology* 8:2056.

Skaare, B. B., H. Wilkes, A. Vieth, E. Rein, and T. Barth. 2007. Alteration of crude oils from the troll area by biodegradation: Analysis of oil and water samples. *Organic Geochemistry* 38(11):1865–1883.

Skarke, A., C. Ruppel, M. Kodis, D. Brothers, and E. Lobecker. 2014. Widespread methane leakage from the sea floor on the northern US Atlantic margin. *Nature Geoscience* 7(9):657–661.

S.L. Ross Environmental Research Ltd. 2011. *Comparison of large-scale (Ohmsett) and small-scale dispersant effectiveness test results.* Final report for the Regulation and Enforcement, Bureau of Ocean Energy Management, U.S. Department of the Interior, Herndon, VA. https://www.bsee.gov/sites/bsee.gov/files/osrr-oil-spill-response-research/638ab.pdf.

Slater, G. F., H. K. White, T. I. Eglinton, and C. M. Reddy. 2005. Determination of microbial carbon sources in petroleum contaminated sediments using molecular 14C analysis. *Environmental Science & Technology* 39(8):2552–2558.

Sloan, E. D., and C. A. Koh. 2008. *Clathrate hydrates of natural gases, 3rd edition*, p. 119. Boca Raton, FL: CRC Press.

SMART (Special Monitoring of Applied Response Technologies). 2006. *The SMART protocol, updated in August 2006.*

Smith, C. R., T. K. Rowles, L. B. Hart, F. I. Townsend, R. S. Wells, E. S. Zolman, B. C. Balmer, B. Quigley, M. Ivančić, and W. McKercher. 2017. Slow recovery of Barataria Bay dolphin health following the *Deepwater Horizon* oil spill (2013–2014), with evidence of persistent lung disease and impaired stress response. *Endangered Species Research* 33:127–142.

Smith, M. A., Q. Smith, J. Morse, A. Baldivieso, and D. Tosa. 2010. *Arctic Marine Synthesis: Atlas of the Chukchi and Beaufort Seas, 1st edition.* Anchorage, AK: Audubon Alaska and Oceana.

Smith, M. A., M. S. Goldman, E. J. Knight, and J. J. Warrenchuk. 2017. *Ecological atlas of the Bering Chukchi, and Beaufort seas, 2nd edition.* Anchorage, AK: Audubon Alaska.

Smith, S. J., J. v. Aardenne, Z. Klimont, R. J. Andres, A. Volke, and S. Delgado Arias. 2011. Anthropogenic sulfur dioxide emissions: 1850–2005. *Atmospheric Chemistry and Physics* 11(3):1101–1116.

Smultea, M. A., and B. Würsig. 1995. Behavioral reactions of bottlenose dolphins to the *Mega Borg* oil spill, Gulf of Mexico 1990. *Aquatic Mammals* 21(3):171–181.

Socolofsky, S., and E. Adams. 2002. Multi-phase plumes in uniform and stratified crossflow. *Journal of Hydraulic Research* 40(6):661–672.

Socolofsky, S. A., and E. E. Adams. 2003. Liquid volume fluxes in stratified multiphase plumes. *Journal of Hydraulic Engineering* 129(11): 905–914.

Socolofsky, S. A., and E. E. Adams. 2005. Role of slip velocity in the behavior of stratified multiphase plumes. *Journal of Hydraulic Engineering* 131(4):273–282.

Socolofsky, S. A., T. Bhaumik, and D.-G. Seol. 2008. Double-plume integral models for near-field mixing in multiphase plumes. *Journal of Hydraulic Engineering* 134(6):772–783.

Socolofsky, S. A., E. E. Adams, and C. R. Sherwood. 2011. Formation dynamics of subsurface hydrocarbon intrusions following the *Deepwater Horizon* blowout. *Geophysical Research Letters* 38(9). https://doi.org/10.1029/2011GL047174.

Socolofsky, S. A., E. E. Adams, M. C. Boufadel, Z. M. Aman, Ø. Johansen, W. J. Konkel, D. Lindo, M. N. Madsen, E. W. North, and C. B. Paris. 2015. Intercomparison of oil spill prediction models for accidental blowout scenarios with and without subsea chemical dispersant injection. *Marine Pollution Bulletin* 96(1–2):110–126.

Socolofsky, S. A., E. E. Adams, C. B. Paris, and D. Yang. 2016. How do oil, gas, and water interact near a subsea blowout? *Oceanography* 29(3):64–75.

Socolofsky, S. A., J. Gros, E. North, M. C. Boufadel, T. F. Parkerton, and E. E. Adams. 2019. The treatment of biodegradation in models of sub-surface oil spills: A review and sensitivity study. *Marine Pollution Bulletin* 143:204–219.

Solomon, G. M., and S. Janssen. 2010. Health effects of the Gulf oil spill. *JAMA* 304(10):1118–1119.

Song, I.-C., E.-J. Jeon, S. Kim, S.-J. Hwang, and J.-M. Seo. 2021. Oil spill fingerprint of low sulfur fuel oil in South Korea. *Marine Pollution Bulletin* 171:112721.

Sørensen, J., B. B. Jørgensen, and N. P. Revsbech. 1979. A comparison of oxygen, nitrate, and sulfate respiration in coastal marine sediments. *Microbial Ecology* 5(2):105–115. http://www.jstor.org/stable/4250563.

Sørensen, L., E. Sørhus, T. Nordtug, J. P. Incardona, T. L. Linbo, L. Giovanetti, Ø. Karlsen, and S. Meier. 2017. Oil droplet fouling and differential toxicokinetics of polycyclic aromatic hydrocarbons in embryos of Atlantic haddock and cod. *PloS ONE* 12(7):e0180048.

Sorgog, K., and M. Kamo. 2019. Quantifying the precision of ecological risk: Conventional assessment factor method vs. species sensitivity distribution method. *Ecotoxicology and Environmental Safety* 183:109494.

Sørheim, K. R., P. S. Daling, D. Cooper, I. Bust, L. G. Faksness, D. Altin, T.-A. Pettersen, and O. M. Bakken. 2020. Characterization of low sulfur fuel oils (LSFO)—a new generation of marine fuel oils—OC2020 A-050. SINTEF Ocean AS.

Sørhus, E., R. B. Edvardsen, Ø. Karlsen, T. Nordtug, T. van der Meeren, A. Thorsen, C. Harman, S. Jentoft, and S. Meier. 2015. Unexpected interaction with dispersed crude oil droplets drives severe toxicity in Atlantic haddock embryos. *PloS ONE* 10(4):e0124376.

Sørhus, E., J. P. Incardona, Ø. Karlsen, T. Linbo, L. Sørensen, T. Nordtug, T. van der Meeren, A. Thorsen, M. Thorbjørnsen, and S. Jentoft. 2016. Crude oil exposures reveal roles for intracellular calcium cycling in haddock craniofacial and cardiac development. *Scientific Reports* 6(1):1–21.

Sörme, L., B. Bergbäck, and U. Lohm. 2001. Goods in the anthroposphere as a metal emission source a case study of Stockholm, Sweden. *Water, Air and Soil Pollution: Focus* 1(3):213–227.

Sørstrøm, S. E. 2009. *Full scale field experiment 2009 cruise report.* Trondheim, Norway: SINTEF Materials and Chemistry.

Sørstrøm, S. E., P. J. Brandvik, I. Buist, P. Daling, D. Dickins, L.-G. Faksness, S. Potter, J. Fritt-Rasmussen, and I. Singsaas. 2010. *Joint industry program on oil spill contingency for Arctic and ice-covered waters: Summary report.*

Southward, A. J., and E. C. Southward. 1978. Recolonization of rocky shores in Cornwall after use of toxic dispersants to clean up the Torrey Canyon Spill. *Journal of the Fisheries Research Board of Canada* 35(5):682–706. https://doi.org/10.1139/f78-120.

Spaulding, M. L. 1988. A state-of-the-art review of oil spill trajectory and fate modeling. *Oil and Chemical Pollution* 4(1):39–55.

Spaulding, M. L. 2017. State of the art review and future directions in oil spill modeling. *Marine Pollution Bulletin* 115(1–2):7–19.

Spehar, R., S. Poucher, L. Brooke, D. Hansen, D. Champlin, and D. Cox. 1999. Comparative toxicity of fluoranthene to freshwater and saltwater species under fluorescent and ultraviolet light. *Archives of Environmental Contamination and Toxicology* 37(4):496–502.

Speight, J. G. 2014. *The chemistry and technology of petroleum, 5th edition.* Boca Raton, FL: CRC Press.

Spencer, P. S. 2020. Neuroprotein targets of γ-diketone metabolites of aliphatic and aromatic solvents that induce central-peripheral axonopathy. *Toxicologic Pathology* 48(3):411–421.

Spies, R. B., and P. Davis. 1979. The infaunal benthos of a natural oil seep in the Santa Barbara channel. *Marine Biology* 50(3):227–237.

Spring, W., T. Nedwed, and R. Belore. 2006. *Icebreaker enhanced chemical dispersion of oil spills.*

Springer, A. M., C. P. McRoy, and M. V. Flint. 1996. The Bearing Sea Green Belt: shelf-edge processes and ecosystem production. *Fisheries Oceanography* 5(3–4):205–223.

Stacy, B. 2012. *Summary of findings for sea turtles documented by directed captures, stranding response, and incidental captures under response operations during the BP DWH MC252 oil spill.* DWH Sea Turtles NRDA Technical Working Group Report. Seattle, WA: Assessment and Restoration Division, National Oceanic and Atmospheric Administration.

Stacy, N., C. Field, L. Staggs, R. MacLean, B. Stacy, J. Keene, D. Cacela, C. Pelton, C. Cray, and M. Kelley. 2017. Clinicopathological findings in sea turtles assessed during the *Deepwater Horizon* oil spill response. *Endangered Species Research* 33:25–37.

Stantec. 2018. *Terra nova 2017 environmental effects monitoring program.* https://www.cnlopb.ca/wp-content/uploads/eem/2017tneem.pdf.

Statistica Research Department. 2021. *Capacity of oil tankers in seaborne trade from 1980 to 2020.* https://www.statista.com/statistics/267605/capacity-of-oil-tankers-in-the-world-maritime-trade-since-1980.

Statistics Canada. 2016. *The changing landscape of Canadian metropolitan areas.* Human Activity and the Environment 2015, Minister of Industry, 2016. https://www150.statcan.gc.ca/n1/pub/16-201-x/16-201-x2016000-eng.pdf.

Steen, A., and A. Findlay. 2008. Frequency of dispersant use worldwide. *International Oil Spill Conference Proceedings* 2008(1):645–649. https://doi.org/10.7901/2169-3358-2008-1-645.

Stenehjem, J. S., T. E. Robsahm, M. Bråtveit, S. O. Samuelsen, J. Kirkeleit, and T. K. Grimsrud. 2017. Ultraviolet radiation and skin cancer risk in offshore workers. *Occupational Medicine* 67(7):569–573.

Stenehjem, J. S., R. Babigumira, H. D. Hosgood, M. B. Veierød, S. O. Samuelsen, M. Bråtveit, J. Kirkeleit, N. Rothman, Q. Lan, and D. T. Silverman. 2021. Cohort profile: Norwegian offshore petroleum workers (NOPW) cohort. *International Journal of Epidemiology* 50(2):398–399.

Stephansen, C., A. Bjørgesæter, O. W. Brude, U. Brönner, T. W. Rogstad, G. Kjeilen-Eilertsen, J.-M. Libre, and C. Collin-Hansen. 2021. *Assessing environmental risk of oil spills with era acute: A new methodology.* Dordrecht, Netherlands: Springer Nature.

Stephenson, R., and C. A. Andrews. 1997. The effect of water surface tension on feather wettability in aquatic birds. *Canadian Journal of Zoology* 75(2):288–294. https://doi.org/10.1139/z97-036.

Stewart, P. A., M. R. Stenzel, G. Ramachandran, S. Banerjee, T. B. Huynh, C. P. Groth, R. K. Kwok, A. Blair, L. S. Engel, and D. P. Sandler. 2018. Development of a total hydrocarbon ordinal job-exposure matrix for workers responding to the *Deepwater Horizon* disaster: The Gulf study. *Journal of Exposure Science & Environmental Epidemiology* 28(3):223–230.

Stiver, W., and D. Mackay. 1984. Evaporation rate of spills of hydrocarbons and petroleum mixtures. *Environmental Science & Technology* 18(11):834–840.

Stoffyn-Egli, P., and K. Lee. 2002. Formation and characterization of oil-mineral aggregates. *Spill Science & Technology Bulletin* 8(1):31–44.

Stout, A., S. Rouhani, B. Liu, and J. Oehrig. 2015. *Spatial extent ("footprint") and volume of Macondo oil found on the deep-sea floor following the* Deepwater Horizon *oil spill.* Administrative Record, DWH-AR0260244. Deepwater Horizon Response & Restoration, U.S. Department of the Interior.

Stout, S., and Z. Wang. 2016. *Standard handbook oil spill environmental forensics: Fingerprinting and source identification.* Cambridge, MA: Academic Press.

Stout, S., and Z. Wang. 2018. *Oil spill environmental forensics case studies.* Oxford, UK: Butterworth-Heinemann.

Stout, S. A., and E. R. Litman. 2022. Quantification of synthetic-based drilling mud olefins in crude oil and oiled sediment by liquid column silver nitrate and gas chromatography. *Environmental Forensics* 1–13. https://doi.org/10.1080/15275922.2022.2047834.

Stout, S. A., and J. R. Payne. 2016. Macondo oil in deep-sea sediments: Part 1: Sub-sea weathering of oil deposited on the seafloor. *Marine Pollution Bulletin* 111(1–2):365–380.

Stout, S. A., and J. R. Payne. 2017. Footprint, weathering, and persistence of synthetic-base drilling mud olefins in deep-sea sediments following the *Deepwater Horizon* disaster. *Marine Pollution Bulletin* 118(1):328–340. https://doi.org/10.1016/j.marpolbul.2017.03.013.

Stoyanovich, S. S., Z. Yang, M. Hanson, B. P. Hollebone, D. M. Orihel, V. Palace, J. L. Rodriguez-Gil, R. Faragher, F. S. Mirnaghi, and K. Shah. 2019. Simulating a spill of diluted bitumen: Environmental weathering and submergence in a model freshwater system. *Environmental Toxicology and Chemistry* 38(12):2621–2628.

Straif, K., R. Baan, Y. Grosse, B. Secretan, F. El Ghissassi, V. Cogliano, D. Drewski, T. Partanen, K. Vähäkangas, and I. Stücker. 2005. Carcinogenicity of polycyclic aromatic hydrocarbons. *The Lancet Oncology* 6(12):931–932.

Strang, C. 2013. *Lessons from travels: Upland vs. lowland tundra.* https://natureinquiries.wordpress.com/tag/lowland-tundra.

Streets, D. G., S. K. Guttikunda, and G. R. Carmichael. 2000. The growing contribution of sulfur emissions from ships in Asian waters, 1988–1995. *Atmospheric Environment* 34(26):4425–4439.

Streibel, T., J. Schnelle-Kreis, H. Czech, H. Harndorf, G. Jakobi, J. Jokiniemi, E. Karg, J. Lintelmann, G. Matuschek, and B. Michalke. 2017. Aerosol emissions of a ship diesel engine operated with diesel fuel or heavy fuel oil. *Environmental Science and Pollution Research* 24(12):10976–10991.

Strelitz, J., L. S. Engel, R. K. Kwok, A. K. Miller, A. Blair, and D. P. Sandler. 2018. *Deepwater Horizon* oil spill exposures and nonfatal myocardial infarction in the gulf study. *Environmental Health* 17(1):1–12.

Strelitz, J., D. P. Sandler, A. P. Keil, D. B. Richardson, G. Heiss, M. D. Gammon, R. K. Kwok, P. A. Stewart, M. R. Stenzel, and L. S. Engel. 2019. Exposure to total hydrocarbons during cleanup of the *Deepwater Horizon* oil spill and risk of heart attack across 5 years of follow-up. *American Journal of Epidemiology* 188(5):917–927.

Strohmeier, T., Ø. Strand, M. Alunno-Bruscia, A. Duinker, and P. J. Cranford. 2012. Variability in particle retention efficiency by the mussel Mytilus edulis. *Journal of Experimental Marine Biology and Ecology* 412: 96–102. https://doi.org/10.1016/j.jembe.2011.11.006.

Strøm-Kristiansen, T., A. Lewis, P. S. Daling, J. N. Hokstad, and I. Singsaas. 1997. Weathering and dispersion of naphthenic, asphaltenic, and waxy crude oils. *International Oil Spill Conference Proceedings* 1997(1):631–636.

Suja, L. D., X. Chen, S. Summers, D. M. Paterson, and T. Gutierrez. 2019. Chemical dispersant enhances microbial exopolymer (EPS) production and formation of marine oil/dispersant snow in surface waters of the subarctic northeast Atlantic. *Frontiers in Microbiology* 10:553.

Sun, X., and J. E. Kostka. 2019. Hydrocarbon-degrading microbial communities are site specific, and their activity is limited by synergies in temperature and nutrient availability in surface ocean waters. *Applied and Environmental Microbiology* 85(15):e00443-19.

Sun, X., L. Chu, E. Mercando, I. Romero, D. Hollander, and J. E. Kostka. 2019. Dispersant enhances hydrocarbon degradation and alters the structure of metabolically active microbial communities in shallow seawater from the northeastern Gulf of Mexico. *Frontiers in Microbiology* 10:2387.

Sunagawa, S., L. P. Coelho, S. Chaffron, J. R. Kultima, K. Labadie, G. Salazar, B. Djahanschiri, G. Zeller, D. R. Mende, and A. Alberti. 2015. Structure and function of the global ocean microbiome. *Science* 348(6237):1261359.

Suttle, C. A. 2007. Marine viruses-major players in the global ecosystem. *Nature Reviews Microbiology* 5(10):801–812.

Svanberg, M., J. Ellis, J. Lundgren, and I. Landälv. 2018. Renewable methanol as a fuel for the shipping industry. *Renewable and Sustainable Energy Reviews* 94:1217–1228.

Symons, L., J. Michel, J. Delgado, D. Reich, D. F. McCay, D. S. Etkin, and D. Helton. 2013. The remediation of underwater legacy environmental threats (RULET) risk assessment for potentially polluting shipwrecks in US waters. *International Oil Spill Conference Proceedings* 2014(1):783–793.

Szulejko, J. E., and T. Solouki. 2002. Potential analytical applications of interfacing a GC to an FT-ICR MS: Fingerprinting complex sample matrixes. *Analytical Chemistry* 74(14):3434–3442. https://doi.org/10.1021/ac011192z.

Taj, M. 2022. *Its beaches mired in crude oil, Peru vows to make refinery pay. The New York Times.* https://www.newsbreak.com/news/2505414405204/its-beaches-mired-in-crude-oil-peru-vows-to-make-refinery-pay.

Takada, H., and R. Ishiwatari. 1990. Biodegradation experiments of linear alkylbenzenes (LABs): Isomeric composition of C12 LABs as an indicator of the degree of lab degradation in the aquatic environment. *Environmental Science & Technology* 24(1):86–91.

Takada, H., R. Ishiwatari, and N. Ogura. 1992. Distribution of linear alkylbenzenes (LABs) and linear alkylbenzenesulphonates (LAS) in Tokyo bay sediments. *Estuarine, Coastal and Shelf Science* 35(2):141–156.

Takada, H., J. W. Farrington, M. H. Bothner, C. G. Johnson, and B. W. Tripp. 1994. Transport of sludge-derived organic pollutants to deep-sea sediments at deep water dump site 106. *Environmental Science & Technology* 28(6):1062–1072.

Takeshita, R., S. J. Bursian, K. M. Colegrove, T. K. Collier, K. Deak, K. M. Dean, S. De Guise, L. M. DiPinto, C. J. Elferink, and A. J. Esbaugh. 2021. A review of the toxicology of oil in vertebrates: What we have learned following the *Deepwater Horizon* oil spill. *Journal of Toxicology and Environmental Health, Part B* 24(8):355–394.

Tamburri, M. N., and R. K. Zimmer-Faust. 1996. Suspension feeding: Basic mechanisms controlling recognition and ingestion of larvae. *Limnology and Oceanography* 41(6):1188–1197. https://doi.org/10.4319/lo.1996.41.6.1188.

Tansel, B. 2018. Morphology, composition and aggregation mechanisms of soft bioflocs in marine snow and activated sludge: A comparative review. *Journal of Environmental Management* 205:231–243.

Tansel, B., A. Arreaza, D. Z. Tansel, and M. Lee. 2015. Decrease in osmotically driven water flux and transport through mangrove roots after oil spills in the presence and absence of dispersants. *Marine Pollution Bulletin* 98(1–2):34–39.

Tao, L., D. Fairley, M. J. Kleeman, and R. A. Harley. 2013. Effects of switching to lower sulfur marine fuel oil on air quality in the San Francisco bay area. *Environmental Science & Technology* 47(18):10171–10178.

Tarasoff, F. J. 1972. Anatomical observations on the river otter, sea otter and harp seal with reference to those structures that are of known significance in thermal regulation and diving. Doctor of Philosophy, Biology, McGill University.

Tarnecki, A. M., C. Miller, T. A. Sherwood, R. J. Griffitt, R. W. Schloesser, and D. L. Wetzel. 2022. Dispersed crude oil induces dysbiosis in the red snapper lutjanus campechanus external microbiota. *Microbiology Spectrum* 10(1):e00587-21.

Tarpley, J., J. Michel, S. Zengel, N. Rutherford, C. Childs, and F. Csulak. 2014. Best practices for shoreline cleanup and assessment technique (SCAT) from recent incidents. *International Oil Spill Conference Proceedings* 2014(1):1281–1297.

Tavares, P. 2011. *On scene coordinator report:* Deepwater Horizon *oil spill.* Washington, DC: United States Coast Guard, U.S. Department of Homeland Security.

Taylor, E., and D. Reimer. 2008. Oil persistence on beaches in Prince William Sound—a review of SCAT surveys conducted from 1989 to 2002. *Marine Pollution Bulletin* 56(3):458–474.

Te, F. T. 1991. Effects of two petroleum products on pocillopora damicornis planulae. *Pacific Science* 45(3):290–298.

Teal, J., J. Farrington, K. Burns, J. Stegeman, B. Tripp, B. Woodin, and C. Phinney. 1992. The west Falmouth oil spill after 20 years: Fate of fuel oil compounds and effects on animals. *Marine Pollution Bulletin* 24(12):607–614.

Tennekes, H., J. L. Lumley, and J. L. Lumley. 1972. *A First Course in Turbulence.* Cambrigde, MA: MIT Press.

Thessen, A. E., and E. W. North. 2017. Calculating in situ degradation rates of hydrocarbon compounds in deep waters of the Gulf of Mexico. *Marine Pollution Bulletin* 122(1–2):77–84.

Thompson, H. F., S. Summers, R. Yuecel, and T. Gutierrez. 2020. Hydrocarbon-degrading bacteria found tightly associated with the 50–70 μm cell-size population of eukaryotic phytoplankton in surface waters of a northeast Atlantic region. *Microorganisms* 8(12):1955.

Thorpe, L. E., S. Assari, S. Deppen, S. Glied, N. Lurie, M. P. Mauer, V. M. Mays, and E. Trapido. 2015. The role of epidemiology in disaster response policy development. *Annals of Epidemiology* 25(5): 377–386.

Thrift-Viveros, D. L., R. Jones, and M. Boufadel. 2015. Development of a new oil biodegradation algorithm for NOAA's oil spill modelling suite (GNOME/ADIOS). *Proceedings of the 38th Arctic and Marine Oilspill Program Technical Seminar.* Vancouver, British Columbia, Canada.

Tissot, B. P., and D. H. Welte. 1984. From kerogen to petroleum. In *Petroleum formation and occurrence*, pp. 160–198. Dordrecht, Netherlands: Springer.

Titchener, J. L., and F. T. Kapp. 1976. Family and character change at Buffalo Creek. *The American Journal of Psychiatry* 133(3):295–299.

Tomco, P. L., K. N. Duddleston, A. Driskill, J. J. Hatton, K. Grond, T. Wrenn, M. A. Tarr, D. C. Podgorski, and P. Zito. 2022. Dissolved organic matter production from herder application and in-situ burning of crude oil at high latitudes: Bioavailable molecular composition patterns and microbial community diversity effects. *Journal of Hazardous Materials* 424:127598. https://doi.org/10.1016/j.jhazmat.2021.127598.

Torlapati, J., and M. C. Boufadel. 2014. Evaluation of the biodegradation of Alaska North Slope oil in microcosms using the biodegradation model biob. *Frontiers in Microbiology* 5:212.

Transport Canada. 2014. *Marine oil spill preparedness and response regime: Report to parliament 2006–2011.* https://tc.canada.ca/en/marine-transportation/marine-safety/marine-oil-spill-preparedness-response-regime-tp-14539-0.

Tripp, B. W., J. W. Farrington, and J. M. Teal. 1981. Unburned coal as a source of hydrocarbons in surface sediments. *Marine Pollution Bulletin* 12(4):122–126.

Troisi, G., L. Borjesson, S. Bexton, and I. Robinson. 2007. Biomarkers of polycyclic aromatic hydrocarbon (PAH)-associated hemolytic anemia in oiled wildlife. *Environmental Research* 105(3):324–329.

Troisi, G. M. 2013. T-cell responses in oiled guillemots and swans in a rehabilitation setting. *Archives of Environmental Contamination and Toxicology* 65(1):142–148.

Troisi, G. M., S. Bexton, and I. Robinson. 2006. Polyaromatic hydrocarbon and PAH metabolite burdens in oiled common guillemots (Uria aalge) stranded on the east coast of England (2001–2002). *Environmental Science & Technology* 40(24):7938–7943.

Trudel, K., R. Belore, M. VanHaverbeke, and J. Mullin. 2009. Updating the US SMART dispersant efficacy monitoring protocol. *Proceedings of the 32nd Arctic and Marine Oilspill Program Technical Seminar*. Ottawa, Ontario, Canada: Environment Canada.

Trudnowska, E., L. Lacour, M. Ardyna, A. Rogge, J. O. Irisson, A. M. Waite, M. Babin, and L. Stemmann. 2021. Marine snow morphology illuminates the evolution of phytoplankton blooms and determines their subsequent vertical export. *Nature Communications* 12(1):2816. https://doi.org/10.1038/s41467-021-22994-4.

Trust, K. A., D. Esler, B. R. Woodin, and J. J. Stegeman. 2000. Cytochrome P450 1A induction in sea ducks inhabiting nearshore areas of Prince William Sound, Alaska. *Marine Pollution Bulletin* 40(5):397–403.

TSB (Transportation Safety Board of Canada). 2014. *Runaway and main-track derailment Montreal, Maine & Atlantic railway freight train MMA-002 mile 0.23, Sherbrooke subdivision, Lac-Mégantic, Quebec, 06 Jul 2013.* Minister of Public Works and Government Services Canada.

Tseng, F. S. 1999. Considerations in care for birds affected by oil spills. *Seminars in Avian and Exotic Pet Medicine* 8(1):21–31.

Tseng, F. S., and M. Ziccardi. 2019. Care of oiled wildlife. In *Medical management of wildlife species*, S. M. Hernandez, H. W. Barron, E. A. Miller, R. F. Aguilar and M. J. Yabsley (eds.).

Turcotte, D., P. Akhtar, M. Bowerman, Y. Kiparissis, R. S. Brown, and P. V. Hodson. 2011. Measuring the toxicity of alkyl-phenanthrenes to early life stages of medaka (oryzias latipes) using partition-controlled delivery. *Environmental Toxicology and Chemistry* 30(2):487–495.

Turner, J. 1986. Turbulent entrainment: The development of the entrainment assumption, and its application to geophysical flows. *Journal of Fluid Mechanics* 173:431–471.

Turner, J. T. 2015. Zooplankton fecal pellets, marine snow, phytodetritus and the ocean's biological pump. *Progress in Oceanography* 130:205–248. https://doi.org/10.1016/j.pocean.2014.08.005.

Turner, N. R., and D. A. Renegar. 2017. Petroleum hydrocarbon toxicity to corals: A review. *Marine Pollution Bulletin* 119(2):1–16.

Turner, R. E. 2011. Beneath the salt marsh canopy: Loss of soil strength with increasing nutrient loads. *Estuaries and Coasts* 34(5):1084–1093.

Turner, R. E., E. B. Overton, B. M. Meyer, M. S. Miles, G. McClenachan, L. Hooper-Bui, A. S. Engel, E. M. Swenson, J. M. Lee, and C. S. Milan. 2014. Distribution and recovery trajectory of Macondo (Mississippi Canyon 252) oil in Louisiana coastal wetlands. *Marine Pollution Bulletin* 87(1–2):57–67.

Turner, R. E., G. McClenachan, and A. W. Tweel. 2016. Islands in the oil: Quantifying salt marsh shoreline erosion after the *Deepwater Horizon* oiling. *Marine Pollution Bulletin* 110(1):316–323.

Turner, R. E., N. N. Rabalais, and D. Justić. 2017. Trends in summer bottom-water temperatures on the northern Gulf of Mexico continental shelf from 1985 to 2015. *PloS ONE* 12(9):e0184350.

Turner, R. E., N. N. Rabalais, E. B. Overton, B. M. Meyer, G. McClenachan, E. M. Swenson, M. Besonen, M. L. Parsons, and J. Zingre. 2019. Oiling of the continental shelf and coastal marshes over eight years after the 2010 *Deepwater Horizon* oil spill. *Environmental Pollution* 252:1367–1376.

Tuttle, S. G., K. M. Hinnant, T. N. Loegel, B. T. Fisher, A. D. Tuesta, and M. Weismiller. 2017. *Development of a low-emission spray combustor for emulsified crude oil.* Washington, DC: Navy Technology Center for Safety and Survivability.

UNCTAD (United Nations Conference on Trade and Development). 2020. *Review of maritime tansport 2020: Shipping in times of the COVID-19 pandemic.* https://unctad.org/system/files/official-document/rmt2020_en.pdf.

United States. 1983. The Endangered Species Act as amended by Public Law 97-304 (the Endangered Species Act Amendments of 1982). Washington, DC: U.S. Government Printing Office.

U.S. Census Bureau. 2019. *About 60.2m live in areas most vulnerable to hurricanes.* https://www.census.gov/library/stories/2019/07/millions-of-americans-live-coastline-regions.html.

U.S. Congress, Senate Committee on Environment. 1983. *The Endangered Species Act as Amended by Public Law 97-304 (the Endangered Species Act Amendments of 1982).* Vol. 98. Washington, DC: U.S. Government Printing Office.

U.S. Department of Commerce. 2014. *Oil spills in mangroves: Planning and response considerations.* Seattle, WA: Office of Response and Restoration, National Ocean Service, National Oceanic and Atmospheric Administration.

U.S. Department of Commerce: National Oceanic and Atmospheric Administration, State of New Jersey: Department of Environmental Protection, Commonwealth of Pennsylvania: Department of Conservation and Natural Resources, Commonwealth of Pennslvania: Department of Environmental Protection, Commonwealth of Pennslvania: Fish and Boat Commission, Commonwealth of Pennslvania: Game Commission, State of Delaware: Department of Natural Resources and Environmental Control, and U.S. Department of the Interior: U.S. Fish & Wildlife Service. 2006. *Final preassessment data report M/T Athos I oil spill, Delaware River.*

U.S. District Court for the Eastern District of Louisiana. 2015. *United States of America v. BP Exploration & Production, Inc., et al.*

USCG (United States Coast Guard). 2003. *Oil spill response offshore. In-situ burn operations manual.* Report no. CG-D-06-03.

USCSHIB. 2016. *Investigative report executive summary: Drilling rig explosion and fire at the Macondo well.*

Utvik, T. I. R. 1999. Chemical characterisation of produced water from four offshore oil production platforms in the North Sea. *Chemosphere* 39(15):2593–2606.

Valderrama, J. O. 2003. The state of the cubic equations of state. *Industrial & Engineering Chemistry Research* 42(8):1603–1618.

Valdiglesias, V., G. Kiliç, C. Costa, Ó. Amor-Carro, L. Mariñas-Pardo, D. Ramos-Barbón, J. Méndez, E. Pásaro, and B. Laffon. 2012. In vivo genotoxicity assessment in rats exposed to prestige-like oil by inhalation. *Journal of Toxicology and Environmental Health, Part A* 75(13–15):756–764.

Valentine, D. L., and C. M. Reddy. 2015. Latent hydrocarbons from cyanobacteria. *Proceedings of the National Academy of Sciences* 112(44):13434–13435.

Valentine, D. L., J. D. Kessler, M. C. Redmond, S. D. Mendes, M. B. Heintz, C. Farwell, L. Hu, F. S. Kinnaman, S. Yvon-Lewis, and M. Du. 2010. Propane respiration jump-starts microbial response to a deep oil spill. *Science* 330(6001):208–211.

Valentine, D. L., I. Mezić, S. Maćešić, N. Črnjarić-Žic, S. Ivić, J. Hogan Patrick, A. Fonoberov Vladimir, and S. Loire. 2012. Dynamic autoinoculation and the microbial ecology of a deep water hydrocarbon irruption. *Proceedings of the National Academy of Sciences* 109(50): 20286–20291. https://doi.org/10.1073/pnas.1108820109.

Valentine, D. L., G. B. Fisher, S. C. Bagby, R. K. Nelson, C. M. Reddy, S. P. Sylva, and M. A. Woo. 2014. Fallout plume of submerged oil from *Deepwater Horizon*. *Proceedings of the National Academy of Sciences* 111(45):15906–15911. https://doi.org/10.1073/pnas.1414873111.

van Dam, J. W., A. P. Negri, S. Uthicke, and J. F. Mueller. 2011. Chemical pollution on coral reefs: Exposure and ecological effects. In *Ecological impacts of toxic chemicals*, pp. 187–211. Sharjah, United Arab Emirates: Bentham Science Publishers Ltd.

Van Hamme, J. D., A. Singh, and O. P. Ward. 2003. Recent advances in petroleum microbiology. *Microbiology and Molecular Biology Reviews* 67(4):503–549.

Van Landuyt, J., L. Cimmino, C. Dumolin, I. Chatzigiannidou, F. Taveirne, V. Mattelin, Y. Zhang, P. Vandamme, A. Scoma, and A. Williamson. 2020. Microbial enrichment, functional characterization and isolation from a cold seep yield piezotolerant obligate hydrocarbon degraders. *FEMS Microbiology Ecology* 96(9):fiaa097.

Van Vleet, E. S., and J. G. Quinn. 1978. Contribution of chronic petroleum inputs to Narragansett Bay and Rhode Island Sound sediments. *Journal of the Fisheries Research Board of Canada* 35:536–543.

Vasconcelos, J. M., S. P. Orvalho, and S. S. Alves. 2002. Gas-liquid mass transfer to single bubbles: Effect of surface contamination. *The AIChE Journal* 48(6):1145–1154.

Veil, J. 2015. *US produced water volumes and management practices in 2012*. Oklahoma City, OK: Ground Water Protection Council.

Veil, J. 2020. *U.S. produced water volumes and management practices in 2017*. Report for the Ground Water Research and Education Foundation.

Veil, J. A. 2011. Produced water management options and technologies. In *Produced water*, pp. 537–571. Dordrecht, Netherlands: Springer.

Veil, J. A., M. G. Puder, D. Elcock, and R. J. Redweik, Jr. 2004. *A white paper describing produced water from production of crude oil, natural gas, and coal bed methane*. Argonne, IL: Argonne National Laboratory.

Veil, J. A., T. A. Kimmell, and A. C. Rechner. 2005. *Characteristics of produced water discharged to the Gulf of Mexico hypoxiczone*. Argonne, IL: Argonne National Laboratory.

Veith, G. D., and S. J. Broderius. 1990. Rules for distinguishing toxicants that cause type I and type II narcosis syndromes. *Environmental Health Perspectives* 87:207–211.

Veith, G. D., D. J. Call, and L. Brooke. 1983. Structure-toxicity relationships for the fathead minnow, pimephales promelas: Narcotic industrial chemicals. *Canadian Journal of Fisheries and Aquatic Sciences* 40(6):743–748.

Venn-Watson, S., K. M. Colegrove, J. Litz, M. Kinsel, K. Terio, J. Saliki, S. Fire, R. Carmichael, C. Chevis, and W. Hatchett. 2015. Adrenal gland and lung lesions in Gulf of Mexico common bottlenose dolphins (Tursiops truncatus) found dead following the *Deepwater Horizon* oil spill. *PloS ONE* 10(5):e0126538.

Vergeynst, L., S. Wegeberg, J. Aamand, P. Lassen, U. Gosewinkel, J. Fritt-Rasmussen, K. Gustavson, and A. Mosbech. 2018. Biodegradation of marine oil spills in the Arctic with a Greenland perspective. *Science of the Total Environment* 626:1243–1258.

Vergeynst, L., C. W. Greer, A. Mosbech, K. Gustavson, L. Meire, K. G. Poulsen, and J. H. Christensen. 2019a. Biodegradation, photo-oxidation, and dissolution of petroleum compounds in an Arctic fjord during summer. *Environmental Science & Technology* 53(21):12197–12206.

Vergeynst, L., J. H. Christensen, K. U. Kjeldsen, L. Meire, W. Boone, L. M. Malmquist, and S. Rysgaard. 2019b. In situ biodegradation, photooxidation and dissolution of petroleum compounds in Arctic seawater and sea ice. *Water Research* 148:459–468.

Vermeer, K., and G. G. Anweiler. 1975. Oil threat to aquatic birds along the Yukon coast. *The Wilson Bulletin* 467–480.

Vermeer, K., and R. Vermeer. 1975. Oil threat to birds on the Canadian west coast. *Canadian Field-Naturalist* 89:278–298.

Vianello, A., A. Boldrin, P. Guerriero, V. Moschino, R. Rella, A. Sturaro, and L. Da Ros. 2013. Microplastic particles in sediments of lagoon of Venice, Italy: First observations on occurrence, spatial patterns and identification. *Estuarine, Coastal and Shelf Science* 130:54–61.

Vigneron, A., P. Cruaud, F. Ducellier, I. M. Head, and N. Tsesmetzis. 2021. Syntrophic hydrocarbon degradation in a decommissioned off-shore subsea oil storage structure. *Microorganisms* 9(2):356.

Vogt, I. 1995. Coral reefs in Saudi Arabia: 3.5 years after the Gulf War oil spill. *Coral Reefs* 14(4):271–273.

Vonk, S. M., D. J. Hollander, and A. J. Murk. 2015. Was the extreme and wide-spread marine oil-snow sedimentation and flocculent accumulation (MOSSFA) event during the *Deepwater Horizon* blow-out unique? *Marine Pollution Bulletin* 100(1):5–12.

Wade, T., E. Atlas, J. Brooks, M. Kennicutt, R. Fox, J. Sericano, B. Garcia-Romero, and D. DeFreitas. 1988. NOAA Gulf of Mexico status and trends program: Trace organic contaminant distribution in sediments and oysters. *Estuaries* 11(3):171–179.

Wade, T. L., S. T. Sweet, J. L. Sericano, N. L. Guinasso, A.-R. Diercks, R. C. Highsmith, V. L. Asper, D. Joung, A. M. Shiller, and S. E. Lohrenz. 2011. Analyses of water samples from the *Deepwater Horizon* oil spill: Documentation of the subsurface plume. *Monitoring and Modeling the Deepwater Horizon Oil Spill: A Record-Breaking Enterprise* 195: 77–82.

Wade, T. L., J. L. Sericano, S. T. Sweet, A. H. Knap, and N. L. Guinasso, Jr. 2016. Spatial and temporal distribution of water column total polycyclic aromatic hydrocarbons (PAH) and total petroleum hydrocarbons (TPH) from the *Deepwater Horizon* (Macondo) incident. *Marine Pollution Bulletin* 103(1–2):286–293.

Wakeham, S. G., and E. A. Canuel. 2016. Biogenic polycyclic aromatic hydrocarbons in sediments of the San Joaquin River in California (USA), and current paradigms on their formation. *Environmental Science and Pollution Research* 23(11):10426–10442. https://doi.org/10.1007/s11356-015-5402-x.

Walker, A. H., R. Pavia, A. Bostrom, T. M. Leschine, and K. Starbird. 2015. Communication practices for oil spills: Stakeholder engagement during preparedness and response. *Human and Ecological Risk Assessment* 21(3):667–690.

Walker, A. H., D. Scholz, M. McPeek, D. French-McCay, J. Rowe, M. Bock, H. Robinson, and R. Wenning. 2018. Comparative risk assessment of spill response options for a deepwater oil well blowout: Part III. Stakeholder engagement. *Marine Pollution Bulletin* 133:970–983.

Wallace, B. P., B. A. Stacy, M. Rissing, D. Cacela, L. P. Garrison, G. D. Graettinger, J. V. Holmes, T. McDonald, D. McLamb, and B. Schroeder. 2017. Estimating sea turtle exposures to *Deepwater Horizon* oil. *Endangered Species Research* 33:51–67.

Walter, J. M., A. Bagi, and D. M. Pampanin. 2019. Insights into the potential of the Atlantic cod gut microbiome as biomarker of oil contamination in the marine environment. *Microorganisms* 7(7):209.

Walton, W. D., W. H. Twilley, J. A. McElroy, D. D. Evans, and E. J. Tennyson. 1995. Smoke measurements using a tethered miniblimp at the Newfoundland offshore oil burn experiment. In *Arctic and Marine Oilspill Program Technical Seminar, 17th proceedings*, pp. 1083–1098. Vancouver, British Columbia, and Ottawa, Ontario, Canada: Environment Canada.

Walton, W. D., K. B. McGrattan, and J. V. Mullin. 2003. ALOFT-PC: A smoke plume trajectory model for personal computers. *International Symposium on Oilfield Chemistry*. https://doi.org/10.2523/106178-MS.

Wammer, K. H., and C. A. Peters. 2005. Polycyclic aromatic hydrocarbon biodegradation rates: A structure-based study. *Environmental Science & Technology* 39(8):2571–2578.

Wang, B., and S. A. Socolofsky. 2015. A deep-sea, high-speed, stereoscopic imaging system for in situ measurement of natural seep bubble and droplet characteristics. *Deep Sea Research Part I: Oceanographic Research Papers* 104:134–148.

Wang, B., S. A. Socolofsky, J. A. Breier, and J. S. Seewald. 2016. Observations of bubbles in natural seep flares at MC 118 and GC 600 using in situ quantitative imaging. *Journal of Geophysical Research: Oceans* 121(4):2203–2230.

Wang, B., S. A. Socolofsky, C. C. Lai, E. E. Adams, and M. C. Boufadel. 2018. Behavior and dynamics of bubble breakup in gas pipeline leaks and accidental subsea oil well blowouts. *Marine Pollution Bulletin* 131:72–86.

Wang, B., C. C. Lai, and S. A. Socolofsky. 2019. Mean velocity, spreading and entrainment characteristics of weak bubble plumes in unstratified and stationary water. *Journal of Fluid Mechanics* 874:102–130.

Wang, B., I. Jun, S. A. Socolofsky, S. F. DiMarco, and J. D. Kessler. 2020a. Dynamics of gas bubbles from a submarine hydrocarbon seep within the hydrate stability zone. *Geophysical Research Letters* 47(18):e2020GL089256.

Wang, C., and R. V. Calabrese. 1986. Drop breakup in turbulent stirred-tank contactors. Part II: Relative influence of viscosity and interfacial tension. *The AIChE Journal* 32(4):667–676.

Wang, D. W., D. A. Mitchell, W. J. Teague, E. Jarosz, and M. S. Hulbert. 2005. Extreme waves under Hurricane Ivan. *Science* 309(5736): 896–896.

Wang, F., M. Pućko, and G. Stern. 2017. Chapter 19: Transport and transformation of contaminants in sea ice. In *Sea ice*, pp. 472–491. Hoboken, NJ: Wiley-Blackwell.

Wang, J., K. Sandoval, Y. Ding, D. Stoeckel, A. Minard-Smith, G. Andersen, E. A. Dubinsky, R. Atlas, and P. Gardinali. 2016. Biodegradation of dispersed Macondo crude oil by indigenous Gulf of Mexico microbial communities. *Science of the Total Environment* 557:453–468.

Wang, P., and T. M. Roberts. 2013. Distribution of surficial and buried oil contaminants across sandy beaches along NW Florida and Alabama coasts following the *Deepwater Horizon* oil spill in 2010. *Journal of Coastal Research* 29(6a):144–155.

Wang, Q., B. Leonce, M. E. Seeley, N. F. Adegboyega, K. Lu, W. C. Hockaday, and Z. Liu. 2020. Elucidating the formation pathway of photo-generated asphaltenes from light Louisiana sweet crude oil after exposure to natural sunlight in the Gulf of Mexico. *Organic Geochemistry* 150:104126. https://doi.org/10.1016/j.orggeochem.2020.104126.

Wang, W. S., and Z. Shao. 2013. Enzymes and genes involved in aerobic alkane degradation. *Frontiers in Microbiology* 4:116.

Wang, Y., J. Chen, F. Li, H. Qin, X. Qiao, and C. Hao. 2009. Modeling photoinduced toxicity of PAHs based on DFT-calculated descriptors. *Chemosphere* 76(7):999–1005.

Wang, Z. 2003. Fate and identification of spilled oils and petroleum products in the environment by GC-MS and GC-FID. *Energy Sources* 25(6):491–508. https://doi.org/10.1080/00908310390195570.

Wang, Z., M. Fingas, and G. Sergy. 1995. Chemical characterization of crude oil residues from an Arctic beach by GC/MS and GC/FID. *Environmental Science & Technology* 29(10):2622–2631.

Wang, Z., M. Fingas, S. Blenkinsopp, G. Sergy, M. Landriault, L. Sigouin, J. Foght, K. Semple, and D. Westlake. 1998. Comparison of oil composition changes due to biodegradation and physical weathering in different oils. *Journal of Chromatography A* 809(1–2):89–107.

Wang, Z., G. Na, X. Ma, X. Fang, L. Ge, H. Gao, and Z. Yao. 2013. Occurrence and gas/particle partitioning of PAHs in the atmosphere from the north Pacific to the Arctic Ocean. *Atmospheric Environment* 77:640–646.

Wang, Z., C. An, K. Lee, E. Owens, Z. Chen, M. Boufadel, E. Taylor, and Q. Feng. 2020. Factors influencing the fate of oil spilled on shorelines: A review. *Environmental Chemistry Letters* 19(2):1611–1628.

Wania, F., and D. Mackay. 1996. Peer reviewed: Tracking the distribution of persistent organic pollutants. *Environmental Science & Technology* 30(9):390A–396A.

Ward, C. P., and E. B. Overton. 2020. How the 2010 *Deepwater Horizon* spill reshaped our understanding of crude oil photochemical weathering at sea: A past, present, and future perspective. *Environmental Science: Processes & Impacts* 22(5):1125–1138. https://doi.org/10.1039/d0em00027b.

Ward, C. P., C. J. Armstrong, R. N. Conmy, D. P. French-McCay, and C. M. Reddy. 2018a. Photochemical oxidation reduced the efficacy of aerial dispersants applied in response to the *Deepwater Horizon* spill. *Environmental Science & Technology Letters* 5(5):226–231. https://doi.org/10.1021/acs.estlett.8b00084.

Ward, C. P., C. M. Sharpless, D. L. Valentine, D. P. French-McCay, C. Aeppli, H. K. White, R. P. Rodgers, K. M. Gosselin, R. K. Nelson, and C. M. Reddy. 2018b. Partial photochemical oxidation was a dominant fate of *Deepwater Horizon* surface oil. *Environmental Science & Technology* 52(4):1797–1805.

Ward, G. A., B. Baca, W. Cyriacks, R. E. Dodge, and A. Knap. 2003. Continuing long-term studies of the tropics Panama oil and dispersed oil spill sites. *International Oil Spill Conference Proceedings* 2003(1):259–267.

Warnock, A. M., S. C. Hagen, and D. L. Passeri. 2015. Marine tar residues: A review. *Water, Air, & Soil Pollution* 226(3):1–24.

Warzinski, R. P., F. Shaffer, R. Lynn, I. Haljasmaa, M. Schellhaas, B. J. Anderson, S. Velaga, I. Leifer, and J. Levine 2014a. *The role of gas hydrates during the release and transport of well fluids in the deep ocean.* Final report. Pittsburgh, PA: National Energy Technology Laboratory, U.S. Department of Energy.

Warzinski, R. P., R. Lynn, I. Haljasmaa, I. Leifer, F. Shaffer, B. J. Anderson, and J. S. Levine. 2014b. Dynamic morphology of gas hydrate on a methane bubble in water: Observations and new insights for hydrate film models. *Geophysical Research Letters* 41(19):6841–6847.

Watkins, R., A. Allen, and B. Ellis. 2016. *Remote sensing guide to oil spill detection in ice-covered waters.* Arctic response technology: Oil spill preparedness. https://crrc.unh.edu/sites/crrc.unh.edu/files/outreach/remote-sensing-guide-to-oil-spill-detection-in-ice-covered-waters.pdf.

Watling, L., S. C. France, E. Pante, and A. Simpson. 2011. Biology of deep-water octocorals. *Advances in Marine Biology* 60:41–122.

Weber, T. C., A. De Robertis, S. F. Greenaway, S. Smith, L. Mayer, and G. Rice. 2012. Estimating oil concentration and flow rate with calibrated vessel-mounted acoustic echo sounders. *Proceedings of the National Academy of Sciences* 109(50):20240–20245.

Weber, T. C., L. Mayer, K. Jerram, J. Beaudoin, Y. Rzhanov, and D. Lovalvo. 2014. Acoustic estimates of methane gas flux from the seabed in a 6000 km2 region in the northern Gulf of Mexico. *Geochemistry, Geophysics, Geosystems* 15(5):1911–1925.

Webster, P. J., G. J. Holland, J. A. Curry, and H.-R. Chang. 2005. Changes in tropical cyclone number, duration, and intensity in a warming environment. *Science* 309(5742):1844–1846.

Weiman, S., S. B. Joye, J. E. Kostka, K. M. Halanych, and R. R. Colwell. 2021. GoMRI insights into microbial genomics and hydrocarbon bioremediation response in marine ecosystems. *Oceanography* 34(1):124–135.

Weir, C. J. 2006. The molecular mechanisms of general anaesthesia: Dissecting the GABAA receptor. *Continuing Education in Anaesthesia, Critical Care and Pain* 6(2):49–53.

Weise, A., C. Nalewajko, and K. Lee. 1999. Oil-mineral fine interactions facilitate oil biodegradation in seawater. *Environmental Technology* 20(8):811–824.

Weitekamp, C. A., T. Stevens, M. J. Stewart, P. Bhave, and M. I. Gilmour. 2020. Health effects from freshly emitted versus oxidatively or photochemically aged air pollutants. *Science of the Total Environment* 704:135772.

Werder, E. J., L. S. Engel, A. Blair, R. K. Kwok, J. A. McGrath, and D. P. Sandler. 2019. Blood BTEX levels and neurologic symptoms in Gulf states residents. *Environmental Research* 175:100–107.

Westerlund, C., and M. Viklander. 2006. Particles and associated metals in road runoff during snowmelt and rainfall. *Science of the Total Environment* 362(1–3):143–156.

Wetzel, M. A., J. W. Fleeger, and S. P. Powers. 2001. Effects of hypoxia and anoxia on meiofauna: A review with new data from the Gulf of Mexico. *Coastal Hypoxia: Consequences for Living Resources and Ecosystems* 58:165–184.

White, H. K., C. M. Reddy, and T. I. Eglinton. 2005a. Isotopic constraints on the fate of petroleum residues sequestered in salt marsh sediments. *Environmental Science & Technology* 39(8):2545–2551.

White, H. K., L. Xu, T. I. Eglinton, and C. M. Reddy. 2005b. Abundance, composition, and vertical transport of PAHs in marsh sediments. *Environmental Science & Technology* 39(21):8273–8280.

White, H. K., P.-Y. Hsing, W. Cho, T. M. Shank, E. E. Cordes, A. M. Quattrini, R. K. Nelson, R. Camilli, A. W. Demopoulos, and C. R. German. 2012. Impact of the *Deepwater Horizon* oil spill on a deep-water coral community in the Gulf of Mexico. *Proceedings of the National Academy of Sciences* 109(50):20303–20308.

White, H. K., S. L. Lyons, S. J. Harrison, D. M. Findley, Y. Liu, and E. B. Kujawinski. 2014. Long-term persistence of dispersants following the *Deepwater Horizon* oil spill. *Environmental Science & Technology Letters* 1(7):295–299.

White, H. K., C. H. Wang, P. L. Williams, D. M. Findley, A. M. Thurston, R. L. Simister, C. Aeppli, R. K. Nelson, and C. M. Reddy. 2016. Long-term weathering and continued oxidation of oil residues from the *Deepwater Horizon* spill. *Marine Pollution Bulletin* 113(1–2):380–386.

Whitehead, A. 2013. Interactions between oil-spill pollutants and natural stressors can compound ecotoxicological effects. *Integrative and Comparative Biology* 53(4):635–647. https://doi.org/10.1093/icb/ict080.

Whitehead, A., B. Dubansky, C. Bodinier, T. I. Garcia, S. Miles, C. Pilley, V. Raghunathan, J. L. Roach, N. Walker, and R. B. Walter. 2012. Genomic and physiological footprint of the *Deepwater Horizon* oil spill on resident marsh fishes. *Proceedings of the National Academy of Sciences* 109(50):20298–20302.

Whitehouse, B. G. 1984. The effects of temperature and salinity on the aqueous solubility of polynuclear aromatic hydrocarbons. *Marine Chemistry* 14(4):319–332.

Whitman, W. B., D. C. Coleman, and W. J. Wiebe. 1998. Prokaryotes: The unseen majority. *Proceedings of the National Academy of Sciences* 95(12):6578–6583.

Whittington, P. 1999. The contribution made by cleaning oiled African penguins spheniscus demersus to population dynamics and conservation of the species. *Marine Ornithology* 27:177–180.

Wickliffe, J. K., B. Simon-Friedt, J. L. Howard, E. Frahm, B. Meyer, M. J. Wilson, D. Pangeni, and E. B. Overton. 2018. Consumption of fish and shrimp from southeast Louisiana poses no unacceptable lifetime cancer risks attributable to high-priority polycyclic aromatic hydrocarbons. *Risk Analysis* 38(9):1944–1961.

Wiegman, S., P. L. van Vlaardingen, E. A. Bleeker, P. de Voogt, and M. H. Kraak. 2001. Phototoxicity of azaarene isomers to the marine flagellate dunaliella tertiolecta. *Environmental Toxicology and Chemistry* 20(7):1544–1550.

Wiese, F. K., and G. J. Robertson. 2004. Assessing seabird mortality from chronic oil discharges at sea. *Journal of Wildlife Management* 68(3):627–638.

Wijesekera, H. W., D. W. Wang, W. J. Teague, and E. Jarosz. 2010. High seafloor stress induced by extreme hurricane waves. *Geophysical Research Letters* 37(11): L11604.

Wikelski, M., V. Wong, B. Chevalier, N. Rattenborg, and H. L. Snell. 2002. Marine iguanas die from trace oil pollution. *Nature* 417(6889):607–608. https://doi.org/10.1038/417607a.

Wilbur, S., and S. Bosch. 2004. *Interaction profile for benzene, toluene, ethylbenzene, and xylenes (BTEX)*. Atlanta, GA: Agency for Toxic Substances and Disease Registry.

Wilkinson, J., T. Boyd, B. Hagen, T. Maksym, W. Pegau, C. Roman, H. Singh, and L. Zabilansky. 2014. Detection and quantification of oil under sea ice: The view from below. *Cold Regions Science and Technology* 109. https://doi.org/10.1016/j.coldregions.2014.08.004.

Wilkman, G., E. Ritari, A. Uuskallio, and M. Niini. 2014. Technological development in oil recovery in ice conditions. *OTC Arctic Technology Conference*. Houston, Texas. February 2014.

Williams, T. D., D. D. Allen, J. M. Groff, and R. L. Glass. 1992. An analysis of California sea otter (Enhydra lutris) pelage and integument. *Marine Mammal Science* 8(1):1–18.

Williams, W. C. 2020. Leakage during decommissioning and from decommissioned wells. *National Academies of Sciences, Engineering, and Medicine Oil in the Sea IV open meeting.*

Willis, A. M., and J. T. Oris. 2014. Acute photo-induced toxicity and toxicokinetics of single compounds and mixtures of polycyclic aromatic hydrocarbons in zebrafish. *Environmental Toxicology and Chemistry* 33(9):2028–2037.

Willis, B. M., J. E. Strutt, and R. D. Eden. 2019. Long-term well plug integrity assurance—a probabilistic approach. *Offshore Technology Conference*. Houston, Texas. May 2019. OTC-29259-MS.

Wirth, M. A., U. Passow, J. Jeschek, I. Hand, and D. E. Schulz-Bull. 2018. Partitioning of oil compounds into marine oil snow: Insights into prevailing mechanisms and dispersant effects. *Marine Chemistry* 206:62–73.

Wise, S. A., R. P. Rodgers, C. M. Reddy, R. K. Nelson, E. B. Kujawinski, T. L. Wade, A. D. Campiglia, and Z. Liu. 2022. Advances in chemical analysis of oil spills since the *Deepwater Horizon* disaster. *Critical Reviews in Analytical Chemistry* 1–60. https://doi.org/10.1080/10408347.2022.2039093.

Wnek, S. M., C. L. Kuhlman, J. A. Harrill, P. A. Nony, G. C. Millner, and J. A. Kind. 2018. Forensic aspects of airborne constituents following releases of crude oil into the environment. In *Oil spill environmental forensics case studies*, pp. 87–115. Amsterdam, Netherlands: Elsevier.

Wolfe, D., J. Michel, M. Hameedi, J. Payne, J. Galt, G. Watabayashi, J. Braddock, J. Short, C. O'Claire, and S. Rice. 1994. The fate of the oil spilled from the *Exxon Valdez*. *Environmental Science & Technology* 28(13):560A–568A.

Word, J. Q., J. R. Clark, and L. S. Word. 2015. Comparison of the acute toxicity of COREXIT 9500 and household cleaning products. *Human and Ecological Risk Assessment* 21(3):707–725.

World Bank. 2020. *Population, total.* United Nations Population Division. World Population Prospects. https://data.worldbank.org/indicator/SP.POP.TOTL.

Wozniak, A. S., P. M. Prem, W. Obeid, D. C. Waggoner, A. Quigg, C. Xu, P. H. Santschi, K. A. Schwehr, and P. G. Hatcher. 2019. Rapid degradation of oil in mesocosm simulations of marine oil snow events. *Environmental Science & Technology* 53(7):3441–3450.

Wynn, J., M. Williamson, and J. Frank. 2017. Sequestered oil pollution mapping, and tracking active oil breakouts in sensitive rivers, bays, and estuaries. *International Oil Spill Conference Proceedings* 2017(1):879–891. https://doi.org/10.7901/2169-3358-2017.1.879.

Wynn, J. C., and J. A. Fleming. 2012. Seawater capacitance—a promising proxy for mapping and characterizing drifting hydrocarbon plumes in the deep ocean. *Ocean Science* 8(6):1099–1104. https://doi.org/10.5194/os-8-1099-2012.

Xia, Y., and M. C. Boufadel. 2011. Beach geomorphic factors for the persistence of subsurface oil from the *Exxon Valdez* spill in Alaska. *Environmental Monitoring and Assessment* 183(1):5–21.

Xiao, S., and D. Yang. 2020. Effect of oil plumes on upper-ocean radiative transfer—a numerical study. *Ocean Modelling* 145:101522.

Xie, H., P. D. Yapa, and K. Nakata. 2007. Modeling emulsification after an oil spill in the sea. *Journal of Marine Systems* 68(3–4):489–506.

Xu, Y., Y.-L. Zhang, J. Li, R. Gioia, G. Zhang, X.-D. Li, B. Spiro, R. S. Bhatia, and K. C. Jones. 2012. The spatial distribution and potential sources of polycyclic aromatic hydrocarbons (PAHs) over the Asian marginal seas and the Indian and Atlantic Oceans. *Journal of Geophysical Research: Atmospheres* 117(D7). https://doi.org/10.1029/2011JD016585.

Xue, X., and J. Katz. 2019. Formation of compound droplets during fragmentation of turbulent buoyant oil jet in water. *Journal of Fluid Mechanics* 878:98–112.

Xue, X., L. D. Chandrala, and J. Katz. 2021. Flow structure and turbulence in the near field of an immiscible buoyant oil jet. *Physical Review Fluids* 6(2):024301.

Yakimov, M. M., K. N. Timmis, and P. N. Golyshin. 2007. Obligate oil-degrading marine bacteria. *Current Opinion in Biotechnology* 18(3):257–266.

Yang, C., Z. Wang, Y. Liu, Z. Yang, Y. Li, K. Shah, G. Zhang, M. Landriault, B. Hollebone, and C. Brown. 2013. Aromatic steroids in crude oils and petroleum products and their applications in forensic oil spill identification. *Environmental Forensics* 14(4):278–293.

Yang, D., M. Chamecki, and C. Meneveau. 2014. Inhibition of oil plume dilution in Langmuir ocean circulation. *Geophysical Research Letters* 41(5):1632–1638.

Yang, D., B. Chen, M. Chamecki, and C. Meneveau. 2015. Oil plumes and dispersion in Langmuir, upper-ocean turbulence: Large-eddy simulations and k-profile parameterization. *Journal of Geophysical Research: Oceans* 120(7):4729–4759.

Yang, D., B. Chen, S. A. Socolofsky, M. Chamecki, and C. Meneveau. 2016. Large-eddy simulation and parameterization of buoyant plume dynamics in stratified flow. *Journal of Fluid Mechanics* 794:798–833.

Yang, F., P. Tchoukov, H. Dettman, R. B. Teklebrhan, L. Liu, T. Dabros, J. Czarnecki, J. Masliyah, and Z. Xu. 2015. Asphaltene subfractions responsible for stabilizing water-in-crude oil emulsions. Part 2: Molecular representations and molecular dynamics simulations. *Energy & Fuels* 29(8):4783–4794. https://doi.org/10.1021/acs.energyfuels.5b00657.

Yang, T., K. Speare, L. McKay, B. J. MacGregor, S. B. Joye, and A. Teske. 2016. Distinct bacterial communities in surficial seafloor sediments following the 2010 *Deepwater Horizon* blowout. *Frontiers in Microbiology* 7:1384.

Yender, R., J. M. Michel, and C. Lord. 2002. *Managing seafood safety after an oil spill.* Washington, DC: National Oceanic and Atmospheric Administration, U.S. Department of Commerce.

Yender, R., K. Stanzel, and A. Lloyd. 2008. Impacts and response challenges of the tanker solar 1 oil spill, guimaras, Philippines: Observations of international advisors. *International Oil Spill Conference Proceedings* 2008(1):77–81.

Yergeau, E., C. Michel, J. Tremblay, A. Niemi, T. L. King, J. Wyglinski, K. Lee, and C. W. Greer. 2017. Metagenomic survey of the taxonomic and functional microbial communities of seawater and sea ice from the Canadian Arctic. *Scientific Reports* 7(1):1–10.

Yergin, D. 2008. *The prize: The epic quest for oil, money & power.* New York: Simon and Schuster.

Yeung, C. W., K. Lee, and C. W. Greer. 2011. Microbial community characterization of produced water from the Hibernia oil production platform. In *Produced water*, pp. 345–352. Dordrecht, Netherlands: Springer.

Yim, U. H., M. Kim, S. Y. Ha, S. Kim, and W. J. Shim. 2012. Oil spill environmental forensics: The *Hebei Spirit* oil spill case. *Environmental Science & Technology* 46(12):6431–6437. https://doi.org/10.1021/es3004156.

Yin, F., G. F. John, J. S. Hayworth, and T. P. Clement. 2015. Long-term monitoring data to describe the fate of polycyclic aromatic hydrocarbons in *Deepwater Horizon* oil submerged off Alabama's beaches. *Science of the Total Environment* 508:46–56.

Yip, T. L., W. K. Talley, and D. Jin. 2011a. Determinants of vessel-accident bunker spillage. *Transportation Research Part D Transport and Environment* 17(8):605–609.

Yip, T. L., W. K. Talley, and D. Jin. 2011b. The effectiveness of double hulls in reducing vessel-accident oil spillage. *Marine Pollution Bulletin* 62(11):2427–2432.

Ylitalo, G. M., M. M. Krahn, W. W. Dickhoff, J. E. Stein, C. C. Walker, C. L. Lassitter, E. S. Garrett, L. L. Desfosse, K. M. Mitchell, and B. T. Noble. 2012. Federal seafood safety response to the *Deepwater Horizon* oil spill. *Proceedings of the National Academy of Sciences* 109(50):20274–20279.

Ylitalo, G. M., T. K. Collier, B. F. Anulacion, K. Juaire, R. H. Boyer, D. A. da Silva, J. L. Keene, and B. A. Stacy. 2017. Determining oil and dispersant exposure in sea turtles from the northern Gulf of Mexico resulting from the *Deepwater Horizon* oil spill. *Endangered Species Research* 33:9–24.

Young, W. 1994. What are vessel traffic services and what can they really do? *Navigation* 41(1):31–56.

Young, W. 1995. Marine traffic regulation in the United States. *Navigation* 42(1):259–286.

Yousuf, T., A. Nakhle, H. Rawal, D. Harrison, R. Maini, and A. Irimpen. 2020. Natural disasters and acute myocardial infarction. *Progress in Cardiovascular Diseases* 63(4):510–517.

Yu, F., C. Yang, Z. Zhu, X. Bai, and J. Ma. 2019. Adsorption behavior of organic pollutants and metals on micro/nanoplastics in the aquatic environment. *Science of the Total Environment* 694:133643.

Yun, K., N. Lurie, and P. S. Hyde. 2010. Moving mental health into the disaster-preparedness spotlight. *New England Journal of Medicine* 363(13):1193–1195.

Yunker, M. B., L. R. Snowdon, R. W. Macdonald, J. N. Smith, M. G. Fowler, D. N. Skibo, F. A. McLaughlin, A. I. Danyushevskaya, V. I. Petrova, and G. I. Ivanov. 1996. Polycyclic aromatic hydrocarbon composition and potential sources for sediment samples from the Beaufort and Barents Seas. *Environmental Science & Technology* 30(4):1310–1320. https://doi.org/10.1021/es950523k.

Yunker, M. B., S. M. Backus, E. Graf Pannatier, D. S. Jeffries, and R. W. Macdonald. 2002. Sources and significance of alkane and PAH hydrocarbons in Canadian Arctic rivers. *Estuarine, Coastal and Shelf Science* 55(1):1–31. https://doi.org/10.1006/ecss.2001.0880.

Zabbey, N., and H. Uyi. 2014. Community responses of intertidal soft-bottom macrozoobenthos to oil pollution in a tropical mangrove ecosystem, Niger delta, Nigeria. *Marine Pollution Bulletin* 82(1–2): 167–174.

Zafirakou, A. 2019. Oil spill dispersion forecasting models. In *Monitoring of marine pollution*. https://doi.org/10.5772/intechopen.81764.

Zencak, Z., J. Klanova, I. Holoubek, and Ö. Gustafsson. 2007. Source apportionment of atmospheric PAHs in the western Balkans by natural abundance radiocarbon analysis. *Environmental Science & Technology* 41(11):3850–3855. https://doi.org/10.1021/es0628957.

Zeng, E., and C. Vista. 1997. Organic pollutants in the coastal environment off San Diego, California. 1. Source identification and assessment by compositional indices of polycyclic aromatic hydrocarbons. *Environmental Toxicology and Chemistry* 16(2):179-188.

Zengel, S., B. M. Bernik, N. Rutherford, Z. Nixon, and J. Michel. 2015. Heavily oiled salt marsh following the *Deepwater Horizon* oil spill, ecological comparisons of shoreline cleanup treatments and recovery. *PloS ONE* 10(7):e0132324.

Zengel, S., C. L. Montague, S. C. Pennings, S. P. Powers, M. Steinhoff, G. Fricano, C. Schlemme, M. Zhang, J. Oehrig, and Z. Nixon. 2016a. Impacts of the *Deepwater Horizon* oil spill on salt marsh periwinkles (littoraria irrorata). *Environmental Science & Technology* 50(2):643–652.

Zengel, S., S. C. Pennings, B. Silliman, C. Montague, J. Weaver, D. R. Deis, M. O. Krasnec, N. Rutherford, and Z. Nixon. 2016b. *Deepwater Horizon* oil spill impacts on salt marsh fiddler crabs (uca spp.). *Estuaries and Coasts* 39(4):1154–1163.

Zengel, S., J. Weaver, I. A. Mendelssohn, S. A. Graham, Q. Lin, M. W. Hester, J. M. Willis, B. R. Silliman, J. W. Fleeger, and G. McClenachan. 2021. Meta-analysis of salt marsh vegetation impacts and recovery: A synthesis following the *Deepwater Horizon* oil spill. *Ecological Applications* e02489.

Zengler, K., and L. S. Zaramela. 2018. The social network of microorganisms—how auxotrophies shape complex communities. *Nature Reviews Microbiology* 16(6):383–390.

Zhang, Q., Z. Wan, B. Hemmings, and F. Abbasov. 2019. Reducing black carbon emissions from Arctic shipping: Solutions and policy implications. *Journal of Cleaner Production* 241:118261.

Zhang, W., K. Rajendran, J. Ham, J. Finik, J. Buthmann, K. Davey, P. M. Pehme, K. Dana, A. Pritchett, and H. Laws. 2018. Prenatal exposure to disaster-related traumatic stress and developmental trajectories of temperament in early childhood: Superstorm Sandy pregnancy study. *Journal of Affective Disorders* 234:335–345.

Zhang, Y., and S. Tao. 2009. Global atmospheric emission inventory of polycyclic aromatic hydrocarbons (PAHs) for 2004. *Atmospheric Environment* 43(4):812–819. https://doi.org/10.1016/j.atmosenv.2008.10.050.

Zhang, Y., R. S. McEwen, J. P. Ryan, J. G. Bellingham, H. Thomas, C. H. Thompson, and E. Rienecker. 2011. A peak-capture algorithm used on an autonomous underwater vehicle in the 2010 Gulf of Mexico oil spill response scientific survey. *Journal of Field Robotics* 28(4):484–496.

Zhang, Y., C. Wang, L. Huang, R. Chen, Y. Chen, and Z. Zuo. 2012. Low-level pyrene exposure causes cardiac toxicity in zebrafish (danio rerio) embryos. *Aquatic Toxicology* 114:119–124.

Zhao, L., M. C. Boufadel, S. A. Socolofsky, E. Adams, T. King, and K. Lee. 2014a. Evolution of droplets in subsea oil and gas blowouts: Development and validation of the numerical model VDROP-J. *Marine Pollution Bulletin* 83(1):58–69.

Zhao, L., J. Torlapati, M. C. Boufadel, T. King, B. Robinson, and K. Lee. 2014b. VDROP: A comprehensive model for droplet formation of oils and gases in liquids—incorporation of the interfacial tension and droplet viscosity. *Chemical Engineering Journal* 253:93–106.

Zhao, L., M. C. Boufadel, E. Adams, S. A. Socolofsky, T. King, K. Lee, and T. Nedwed. 2015. Simulation of scenarios of oil droplet formation from the *Deepwater Horizon* blowout. *Marine Pollution Bulletin* 101(1):304–319.

Zhao, L., M. C. Boufadel, K. Lee, T. King, N. Loney, and X. Geng. 2016. Evolution of bubble size distribution from gas blowout in shallow water. *Journal of Geophysical Research: Oceans* 121(3):1573–1599.

Zhao, L., M. C. Boufadel, T. King, B. Robinson, F. Gao, S. A. Socolofsky, and K. Lee. 2017. Droplet and bubble formation of combined oil and gas releases in subsea blowouts. *Marine Pollution Bulletin* 120(1–2):203–216.

Zhao, L., D. A. Mitchell, R. C. Prince, A. H. Walker, J. S. Arey, and T. J. Nedwed. 2021. *Deepwater Horizon* 2010: Subsea dispersants protected responders from VOC exposure. *Marine Pollution Bulletin* 173:113034.

Zhao, S., J. E. Ward, M. Danley, and T. J. Mincer. 2018. Field-based evidence for microplastic in marine aggregates and mussels: Implications for trophic transfer. *Environmental Science & Technology* 52(19):11038–11048. https://doi.org/10.1021/acs.est.8b03467.

Zheng, H., M. Cai, W. Zhao, M. Khairy, M. Chen, H. Deng, and R. Lohmann. 2021. Net volatilization of PAHs from the North Pacific to the Arctic Ocean observed by passive sampling. *Environmental Pollution* 276:116728.

Zheng, J., B. Chen, W. Thanyamanta, K. Hawboldt, B. Zhang, and B. Liu. 2016. Offshore produced water management: A review of current practice and challenges in harsh/Arctic environments. *Marine Pollution Bulletin* 104(1–2):7–19.

Zheng, L., and P. D. Yapa. 2000. Buoyant velocity of spherical and nonspherical bubbles/droplets. *Journal of Hydraulic Engineering* 126(11):852–854.

Zheng, Z., X. Tang, A. Asa-Awuku, and H. S. Jung. 2010. Characterization of a method for aerosol generation from heavy fuel oil (HFO) as an alternative to emissions from ship diesel engines. *Journal of Aerosol Science* 41(12):1143–1151.

Zhong, Z.-P., J. Z. Rapp, J. M. Wainaina, N. E. Solonenko, H. Maughan, S. D. Carpenter, Z. S. Cooper, H. B. Jang, B. Bolduc, and J. W. Deming. 2020. Viral ecogenomics of Arctic cryopeg brine and sea ice. *Msystems* 5(3):e00246-20.

Zhou, X., X. Xing, J. Hou, and J. Liu. 2017. Quantitative proteomics analysis of proteins involved in alkane uptake comparing the profiling of pseudomonas aeruginosa SJTD-1 in response to n-octadecane and n-hexadecane. *PloS ONE* 12(6):e0179842.

Zhou, Y., N. Pavlenko, D. Rutherford, L. Osipova, and B. Comer. 2020. The potential of liquid biofuels in reducing ship emissions. *International Council on Clean Transportation*. Working Paper 2020–2021.

Zhou, Z., C.-j. Zhang, P.-f. Liu, L. Fu, R. Laso-Pérez, L. Yang, L.-p. Bai, J. Li, M. Yang, J.-z. Lin, W.-d. Wang, G. Wegener, M. Li, and L. Cheng. 2022. Non-syntrophic methanogenic hydrocarbon degradation by an archaeal species. *Nature* 601(7892):257–262. https://doi.org/10.1038/s41586-021-04235-2.

Zhu, L., X. Zhao, L. Lai, J. Wang, L. Jiang, J. Ding, N. Liu, Y. Yu, J. Li, and N. Xiao. 2013. Soil TPH concentration estimation using vegetation indices in an oil polluted area of eastern china. *PloS ONE* 8(1): e54028.

Zhu, X., A. D. Venosa, M. T. Suidan, and K. Lee. 2001. *Guidelines for the bioremediation of marine shorelines and freshwater wetlands*. Washington, DC: U.S. Environmental Protection Agency.

Ziccardi, M. H. 2021. Creating a legacy of oiled wildlife response preparedness through the post-Macondo oil spill response-joint industry project. *International Oil Spill Conference Proceedings* 2021(1):687447.

Ziccardi, M. H., S. M. Wilkin, T. K. Rowles, and S. Johnson. 2015. *Pinniped and cetacean oil spill response guidelines*. https://repository.library.noaa.gov/view/noaa/10479.

Zick, A. 2013. Equation-of-state fluid characterization and analysis of the Macondo reservoir fluids. In *expert report prepared on behalf of the United States*, p. 175.

Zieman, J. C., R. Orth, R. C. Phillips, G. Thayer, and A. Thorhaug. 1984. Effects of oil on seagrass ecosystems. In *Restoration of habitats impacted by oil spills*, J. Cairns, Jr., and A. L. Buikema, Jr. (eds.), pp. 37–64, 5 Fig, 1 Tab, 82 Ref. Boston, MA: Butterworth.

Zimmermann, L. A., M. G. Feldman, D. S. Benoit, M. J. Carron, N. M. Dannreuther, K. H. Fillingham, J. C. Gibeaut, J. L. Petitt, J. B. Ritchie, and R. R. Rossi. 2021. From disaster to understanding. *Oceanography* 34(1):16–29.

Zito, P., D. C. Podgorski, T. Bartges, F. o. Guillemette, J. A. Roebuck, Jr., R. G. Spencer, R. P. Rodgers, and M. A. Tarr. 2020. Sunlight-induced molecular progression of oil into oxidized oil soluble.

Zock, J.-P., G. Rodríguez-Trigo, F. Pozo-Rodríguez, J. A. Barberà, L. Bouso, Y. Torralba, J. M. Antó, F. P. Gómez, C. Fuster, and H. Verea. 2007. Prolonged respiratory symptoms in clean-up workers of the prestige oil spill. *American Journal of Respiratory and Critical Care Medicine* 176(6):610–616.

Zock, J.-P., G. Rodríguez-Trigo, F. Pozo-Rodríguez, and J. Albert Barbera. 2011. Health effects of oil spills: Lessons from the prestige. *American Journal of Respiratory and Critical Care Medicine* 184(10):1094–1096.

Zock, J.-P., G. Rodríguez-Trigo, E. Rodríguez-Rodríguez, A. Espinosa, F. Pozo-Rodríguez, F. Gómez, C. Fuster, G. Castaño-Vinyals, J. M. Antó, and J. A. Barberà. 2012. Persistent respiratory symptoms in clean-up workers 5 years after the prestige oil spill. *Occupational and Environmental Medicine* 69(7):508–513.

Acronyms and Abbreviations

‰	parts per thousand or g/kg (used for salinity)
μg/g	microgram per gram
μg/L	microgram per liter
μm	micrometers
96-h LC_{50}	96-hour lethal concentration for 50% of the test organisms
ABS	Acrylonitrile butadiene styrene
ACR	acute-to-chronic ratio
ACS	American Chemical Society
ADCP	acoustic Doppler current profiler
ADEC	Alaska Department of Environmental Conservation
ADIOS	Automated Data Inquiry for Oil Spills
AF	application factor
AhR	aryl hydrocarbon receptor
AI	artificial intelligence
AIP	AhR-interacting protein
AIS	Automatic Identification System
AMOP	Arctic and Marine Oilspill Program
AMRA	Arctic marine risk assessment
ANS	Alaska North Slope crude oil
ANSI	American National Standards Institute
API	American Petroleum Institute
Arnt	AhR nuclear translocator protein
ARTES	Alternative Response Tool Evaluation System
ASEA	National Agency for Safety of Energy and Environment
ASTM	American Society for Testing and Materials
AU	assessment unit
AUV	autonomous underwater vehicle
AVIRIS	airborne visible/infrared imaging spectrometer
BAF	bioaccumulation factor
BaP	benzo(a)pyrene, a 5-ringed PAH
BAT	best available technology economically achievable
BAU	business as usual
bbl	barrel or barrels
bbl/d	barrels per day
BC	British Columbia

BCF	bioconcentration factor
BCT	best conventional pollutant control technology
BIOS	Baffin Island Oil Spill
BOA	bacteria-oil aggregate
BOD	biological oxygen demand
BOE	barrel of oil equivalents
BOEM	U.S. Bureau of Ocean Energy Management
BOMA	bacteria-oil microaggregates
BOP	blowout preventer
BOPD	barrels of oil per day
BP	British Petroleum
BSEE	U.S. Bureau of Safety and Environmental Enforcement
BTEX	benzene, toluene, ethylbenzene, and xylene isomers
°C	degrees Celsius (temperature)
C	carbon
C-NLOPB	Canada-Newfoundland and Labrador Offshore Petroleum Board
C:N:P	Redfield ratio; carbon:nitrogen:phosphorus
Ca^{2+}	calcium ion
CAFE	Chemical Aquatic Fate and Effects
CAPP	Canadian Association of Petroleum Producers
CARD-FISH	catalyzed reporter deposition fluorescence in situ hybridization
casq1, casq2	SR calcium-buffering proteins
CBBO	*Cosco Busan* bunker oil
CDOM	colored (or chromophoric) dissolved organic matter
CER	Canada Energy Regulator
CERA	Consensus Ecological Risk Assessment
CEWAF	chemically enhanced water-accommodated fraction of oil
CFD	computational fluid dynamics
CFR	Code of Federal Regulations
cm	centimeter
cm/s	centimeter per second
CMO	Churchill Marine Observatory (Manitoba, Canada)
CO_2	carbon dioxide
COD	chemical oxygen demand
COP	common operating procedure
COS	Center for Offshore Safety
COTP	Captain of the Port
COW	crude oil washing
CRA	comparative risk assessment
CROSERF	Chemical Response to Oil Spills Ecological Research Forum
CRRC	Coastal Response Research Center
CTD	conductivity, temperature, and depth
CYP1A	cytochrome P450-1A
d	day or days
D	diameter
DFO	Department of Fisheries and Oceans (Canada)
DHS	U.S. Department of Homeland Security
DIVER	Data Integration Visualization Exploration and Reporting (developed by NOAA)
DNA	deoxyribonucleic acid
DNAPL	dense nonaqueous phase liquids
DNS	dynamic numerical simulation
DO	dissolved oxygen
DOE	U.S. Department of Energy

DOI	U.S. Department of the Interior
DOM	dissolved organic matter
DOR	dispersant:oil ratio
DOSS	dioctyl sodium sulfosuccinate
DOT	U.S. Department of Transportation
DWH	*Deepwater Horizon* oil spill
DWOP	Drill the Well on Paper
DWT	dead weight tonnes
E&P	exploration and production
EAL	environmentally acceptable lubricant
EBSA	ecologically and biologically significant area
EC	excitation–contraction
EC_{50}	median effective concentration
ECA	emission control area
ECCC	Environment and Climate Change Canada
EDCF	effects-driven chemical fractionation
EEM	environmental effects monitoring
EEZ	Exclusive Economic Zone
EHSS	ethylhexyl sulfosuccinate
EIA	environmental impact assessment
EIA	U.S. Energy Information Administration
EMSA	European Maritime Safety Agency
EOS	equation of state (thermodynamics)
EPA	U.S. Environmental Protection Agency
EPS	extracellular polymeric substance
ERA	ecological risk assessment
ERG	ether-a-go-go related gene
ERMA®	Environmental Response Management Application
EROD	ethoxyresorufin-O-deethylase
ESEM	environmental scanning electron microscopy
ESI	Environmental Sensitivity Index
ETS	environmental tobacco smoke
ETS	erythroblast transformation-specific [-related gene (ERG)]
EVOS	*Exxon Valdez* oil spill
EXDET	Exxon Dispersant Effectiveness Test
FAA	U.S. Federal Aviation Administration
FAC	fluorescent aromatic compound
FDA	U.S. Food and Drug Administration
FID	flame ionization detector
FOG	fat, oils, and greases
FT-ICR-MS	Fourier transform ion cyclotron resonance mass spectrometer
FW	freshwater
g	gram
g/acre	gram per acre
g/cm^3	gram per centimeter cubed
GC	gas chromatography
GC×GC	two-dimensional gas chromatography
GC×GC-MS	two-dimensional gas chromatography with mass spectrometry
GC-FID	gas chromatography with flame ionization detector
GC-MS	gas chromatography with mass spectrometry
GFF	glass fiber water
GIS	geographic information system

GoMRI	Gulf of Mexico Research Initiative
GOR	gas-to-oil ratio
GPR	ground penetration radar
GRHOP	Gulf Region Health Outreach Program
GRIIDC	Gulf of Mexico Research Initiative Information and Data Cooperative
GROS	genome repository of oil systems
GRP	Gulf Research Program
GSI	Gonad Somatic Index
GSRN	Global Subsea Response Network
GST	glutathione transferase
GT	gross tonnes
GTH	glutathione
H	hydrogen
H_2S	hydrogen sulfide
ha	hectare
HC	hazard concentration
HC	hydrocarbon
HDPE	high-density polyethylene
HECEWAF	high energy chemically enhanced water-accommodated fraction
HEWAF	high-energy water-accommodated fraction (of oil)
HFO	heavy fuel oil (e.g., Bunker C)
HGL	hydrogen gas liquids
High Mag FT-ICR-MS	High Magnetic Field Fourier Transform–Ion Cyclotron Resonance–Mass Spectrometry or Ultra High Resolution Mass Spectrometry
HLB	hydrophilic-lipophilic balance
HMW	high molecular weight
HO_2	hydroperoxy radical
HP/HT	high pressure/high temperature
HPLC	high-performance liquid chromatography
HPWT	high-pressure water tunnel
hr	hour
HSFO	high sulfur fuel oil
hv	light
IBP	ice-binding protein
ICAO	International Civil Aviation Organization
ICCOPR	Interagency Coordinating Committee on Oil Pollution Research
ICE	Interspecies Correlation Estimation
ICES	International Council for the Exploration of the Seas
ICS	Incident Command System
IEA	International Energy Agency
IFO	intermediate fuel oil (e.g., IFO 180)
IM	interfacial material
IMH	Incident Management Handbook
IMO	International Maritime Organization
IMS-MS	ion mobility spectrometry-mass spectrometry
IOC	International Oceanographic Commission
IOGP	International Oil & Gas Producers
IOSC	International Oil Spill Conference
IP	intraperitoneal
IP	induced polarization
IPCC	Intergovernmental Panel on Climate Change
IPIECA	International Petroleum Industry Environmental Conservation Association
IR	infrared

ISB	in situ burning (of oil on water)
ISO	International Organization for Standardization
ITAC	Industry Technical Advisory Committee
ITOPF	International Tanker Owners Pollution Federation
JEM	job exposure matrix
K^+	potassium ion
K_{ow}	octanol-water partition coefficient
kg/m^3	kilogram per cubic meter (measure of density)
kHz	a unit of frequency equal to one thousand cycles per second
km	kilometers
km^2	kilometers squared
$km^2\ y^{-1}$	kilometers squared per year
kPa	kilopascal; unit of pressure equivalent to one newton per square meter
L	liter
l/ha	liter per hectare
LAB	linear alkylbenzenes
LAS	linear alkylbenzenesulfonates
LC_{50}	median lethal concentration
LDPE	low-density polyethylene
LES	large eddy simulation (an approach to modeling turbulent flow)
LEWAF	low energy water accommodated fractions
LH	lipid
LISST	laser in situ scattering and transmissivity
LL	lethal loading
LMW	low molecular weight
LNAPL	light non-aqueous phase liquid
LNG	liquefied natural gas
LOC	levels of concern
LOEC	lowest observable effect concentration
LOET	lowest observable effects time
LOOH	peroxidation
LOWC	loss of well control
LSC	Louisiana sweet crude
LSFO	low-sulfur fuel oil
LTCC	L-type calcium channel
m	meters
m^3	meters cubed
m/s	meters per second
M-BACI	multivariate before and after/control and impact
MAG	metagenome-assembled genome
MARPOL	International Convention for the Prevention of Pollution from Ships
MARSIS	Marine Safety Information System
MAS	marine autonomous system
MC-20	Mississippi Canyon, Block 20
MC-252	Macondo crude oil spilled in *Deepwater Horizon* event
MCF	1,000 cubic feet of gas
MCFD	1 million standard cubic feet per day
MERIS	Medium Resolution Imaging Spectrometer
MEWAF	mid energy water accommodated fraction
MG	million gallons
mg/cm^3	milligrams per cubic centimeter

mg L^{-1}	milligrams per 1 liter solution
MISLE	Marine Information for Safety and Law Enforcement
MISR	Multi-angle Imaging Spectroradiometer
MIZ	Marginal Ice Zone
mL	milliliter
mm	millimeter
MMCF	1 million cubic feet of gas
MMS	Minerals Management Service
MNA	monitored natural attenuation
MoA	mode of action
MOA	mechanism of action
MODIS	Moderate Resolution Imaging Spectroradiometer
MODU	mobile offshore drilling unit
MOS	marine oil snow
MOSSFA	marine oil snow sedimentation and flocculent accumulation
MP	microplastic
MPa	megapascal; one million pascals (Pa, unit of pressure)
MPRI	Multi-partner Research Inititiative
MPSR	Marine Pollution Surveillance Report (NOAA)
MS	mass spectrometry
MSO	marine safety office
MT	metric tonnes
MT/yr	metric tonnes per year
N	nitrogen
NA	naphthenic acids
NA	not applicable
NA^+	sodium ion
NAE	U.S. National Academy of Engineering
NAPL	non-aqueous phase liquid
NAS	U.S. National Academy of Sciences
NASF	non-accidental structural failures
ND	not determined
NEB	National Energy Board (Canada)
NEBA	net environmental benefits analysis
NESDIS	National Environmental Satellite, Data, and Information Service
ng/g	nanogram per gram
NGO	nongovernmental organization
NGS	next generation sequencing (of nucleic acids)
NHANES	U.S. National Health and Nutrition Examination Survey
Ni	nickel
NIEHS	U.S. National Institute of Environmental Health Sciences
NIOSH	U.S. National Institute for Occupational Safety and Health
NIST	U.S. National Institute of Standards and Technology
nm	nanometer
NMR	nuclear magnetic resonance
NMVOC	non-methane volatile organic compound
No.	number
NOAA	U.S. National Oceanic and Atmospheric Administration
NOBE	Newfoundland Offshore Burn Experiment
NOEC	no observable effect concentration
NOEL	no observed effect loading
NOIA	U.S. National Ocean Industries Association
NOPW	Norwegian Offshore Petroleum Workers
NOx	nitrogen oxide

NPDES	Clean Water Act National Pollutant Discharge Elimination System
NRC	National Research Council (in both Canada and the United States)
NRDA	Natural Resource Damage Assessment
NRT	National Response Team
NS&T	National Status and Trends Program (NOAA)
NTSB	U.S. National Transportation Safety Board
NWQMC	U.S. National Water Quality Monitoring Council
NZ2050	net zero by 2050
O	other
O	oxygen
O_2	molecular oxygen
o/w	oil-in-water emulsion
O&G	oil and gas
OCIMF	Oil Companies International Marine Forum
OCS	Outer Continental Shelf
OECD	Organisation for Economic Co-operation and Development
OH	hydroxyl radical
OHCB	obligate hydrocarbonoclastic bacteria
OITS	Oil in the Sea
OMA	oil-mineral aggregate
ONR	U.S. Office of Naval Research
OPA	oil-particle aggregate
OPA 90	Oil Pollution Act of 1990
OPRC	International Convention on Oil Pollution Preparedness, Response and Co-operation
OPS	Office of Pipeline Safety (PHMSA)
OPT/LFT	optical and laser flourosensor
OR&R	Office of Response and Restoration (NOAA)
ORA	oil-related aggregate
OSA	oil-sediment aggregate
OSCAR	Oil Spill Contingency and Response
OSHA	U.S. Occupational Safety and Health Administration
OSIM	Oil in Sea Ice Mesocosm facility at Churchill Marine Observatory, Hudson Bay
OSLTF	Oil Spill Liability Trust Fund
OSPRI	Oil Spill Preparedness Regional Initiative
OSRO	oil spill removal organization
OSRV	oil spill response vessel
OWD	oil-water dispersion
P	phosphorus
P&A	plugged and abandoned
PAC	polycyclic aromatic compound
PAH	polycyclic aromatic hydrocarbon
PAR	photosynthetically active radiation
PARAFAC	parallel factor analysis
PBCO	Prudhoe Bay Crude Oil
PCB	polychlorinated biphenyls
PCR	polymerase chain reaction
PEC	predicted environmental concentration
PET or PETE	polyethylene terephthalate
PETROTOX	model that can predict the toxicity of any oil
PHE	phenanthrene
PHMSA	U.S. Pipeline and Hazardous Materials Safety Administration
PIV	particle image velocimetry
PLIF	planar laser induced fluorescence

PM	particulate matter
PNEC	predicted no effect concentrations
POC	particulate organic carbon
POM	particulate organic matter
PP	polypropylene
ppb	parts-per-billion
PPE	personal protective equipment
ppm	parts-per-million
ppt	parts-per-thousand
PR	Peng-Robinson
PROFEPA	Federal Attorney for the Protection of the Environment
PROMAM	Mexican Navy's Marine Environment Protection Division
PS	polystyrene
PSU	practical salinity unit; a unit based on the properties of sea water conductivity equivalent to per-thousand (o/00) or g/kg
PTLM	Phototoxic Target Lipid Model
PTSD	posttraumatic stress disorder
PVC	polyvinyl chloride
PWS	Prince William Sound
QA/QC	quality assurance/quality control
Quad BTU	quadrillion British thermal unit
R	*n*-hexadecane free radical
RANS	Reynolds-averaged Navier-Stokes
RBCC	red blood cell count
RH	*n*-hexadecane
RNA	ribonucleic acid
RO	response organization
RO	oxygenated *n*-hexadecane radical
RO_2	*n*-hexadecane peroxy radical
RO_2H	*n*-hexadecanoic acid
ROH	*n*-hexadecanol
ROS	reactive oxygen species
ROSV	remotely operated surface vehicle
ROV	remotely operated vehicle
RP	responsible party
RPAS	remotely piloted aircraft system
RQ	risk quotient
RRI	Response Resource Inventory
RRT	Regional Response Team
RTG-2	rainbow trout gill (cell line)
RULET	Remediation of Underwater Legacy Environmental Threats project
RyR	ryanodine receptor
S	sulfur
SAB	Satellite Analysis Branch
SAG	single-cell amplified genome
SAR	synthetic aperture radar
SARA	saturates, aromatics, resins, and asphaltenes
SCAT	Shoreline Cleanup and Assessment Technique
SE	standard error of mean
SEMS	safety and environmental management system
SENER	Ministry of the Energy (Secretaría de Energía)
SERMARNAT	Ministry of Environment, Natural Resources and Fisheries

SHE-WAF	super high energy water accommodated fractions
SIR	standardized incidence ratio
SIRE	Ship Inspection Report Program
SIMA	spill impact mitigation assessment
SIMAP	Spill Impact MAPping
SLAR	side-looking airborne radar
SMART	Special Monitoring of Applied Response Technologies
SMFF	salvage and marine firefighting
SMS	safety management system
SOA	secondary organic aerosol
SOA	sediment-oil aggregate
SOD	superoxide dismutase
SOM	sediment-oil mat
SOM	submerged oil mat
SONS	spills of national significance
SOPEP	shipboard emergency plan
SOx	sulfur oxide
SPE	Society of Petroleum Engineers
SR	sarcoplasmic reticulum
SRB	sulfate-reducing bacteria
SRB	surface residual ball
SRM	submerged residual oil mat
SRP	sulfate-reducing prokaryotes
SRP	surface residue patty
SSD	species sensitivity distribution
SSDI	subsea dispersant injection
SW	saltwater
SWA	surface washing agent
t	tonne
t/d	tonnes per day
TCCA	Transport Canada Civil Aviation
TEP	transparent exopolymer particle
Tg	teragrams
TGLO	Texas General Land Office
THC	total hydrocarbon (sum of concentrations of all hydrocarbon measured)
TLM	target lipid model
TPAH	total polycyclic aromatic hydrocarbon
TPAH50	50 parent and/or more total polycyclic aromatic hydrocarbon
TPH	total petroleum hydrocarbons
TROPICS	TRopical Oil Pollution Investigations in Coastal Systems
TSS	total suspended solid
TU	toxic unit
TWA	time-weighted average
UAV	unmanned aerial vehicle
UCM	unresolved complex mixture
ULSFO	ultra-low sulfur fuel oils
UME	unusual mortality event
UNEP	United Nations Environment Programme
U.S. EPA	U.S. Environmental Protection Agency
USARC	U.S. Arctic Research Commission
USCG	United States Coast Guard
USDA	U.S. Department of Agriculture
USFW	U.S. Fish & Wildlife Service

USGS	U.S. Geological Survey
UUV	unmanned underwater vehicle
UV	ultraviolet
UV/IR	ultraviolet infrared [sensors]
UVE	UV epifluorescence microscopy
UVR	ultraviolet radiation
v/v	volume to volume (dilution)
VEC	valuable ecosystem component
VECS	vapor emission control system
VGP	Vessel General Permit
VHC	volatile hydrocarbon
VHS	viral hemorrhagic septicemia
VIDA	Vessel Incidental Discharge Act
VLSFO	very low sulfur fuel oil
VOC	volatile organic compound
VRP	vessel response plan
VTS	vessel traffic service
WAF	water-accommodated fraction of oil
We*	Weber number
WET	whole effluent toxicity
WGR	water-to-gas ratio
WOR	water-to-oil ratio
WQP	Water Quality Portal
WSF	water-soluble fraction of oil
X	xanthone
X*	xanthone triplet
XH	hydrated xanthone radical
XRE	xenobiotic response elements
ZFE	zone-of-flow-establishment

Glossary

Abiotic: non-biological (e.g., physical or geochemical).

Absorbent: a material that picks up and retains a liquid distributed throughout its molecular structure causing the solid to swell (50% or more). The absorbent is at least 70% insoluble in excess fluids.

Acute-to-chronic ratio: the ratio of the acute and chronic toxicity values for a given compound, usually the average of the ratios for a variety of species; used to estimate the chronic toxicity of a compound or a mixture of compounds, from the measured or modeled acute toxicity when no chronic toxicity data are available.

Adsorbent: an insoluble material that is coated by a liquid on its surface including pores and capillaries without the solid swelling more than 50% in excess fluid.

Adverse outcome pathways (AOPs): direct associations between specific molecular networks and deleterious effects to biological systems that can be used as signatures of exposure.

Allision: the striking of a vessel with a fixed or stationary object.

American Petroleum Institute (API) gravity: industry scale expressing the gravity or density of liquid petroleum products. The measuring scale is calibrated in terms of degrees API, it may be calculated using the following formula: Degrees API = (141.5/specific gravity at 15.6°C) – 131.5. The higher the API gravity, the lighter the compound. Light crudes generally exceed 38° API and heavy crudes are commonly labeled as all crudes with an API gravity of 22° or below. Intermediate crudes fall in the range of 22–38° API gravity. Light crudes yield more gasoline.

Anaerobic: metabolisms that occur independent of oxygen.

Annual exceedance probability: the chance or probability of a natural hazard event (usually a rainfall or flooding event) occurring annually, usually expressed as a percentage. Bigger rainfall events occur (are exceeded) less often and will therefore have a lesser annual probability.

Anoxic: depleted of oxygen.

Archaea: one of two prokaryote domains (microscopic single-celled organisms that have neither a distinct nucleus with a membrane nor other specialized organelles) that includes microorganisms that live in extreme environments.

Arctic: generally defined as the region above the Arctic Circle (i.e., above approximately 66° 34' N latitude, although other definitions are based on the tree line or on temperature data).

Aromatics: class of hydrocarbons having one or more five or six membered "fused" (clustered together) carbon atom planar rings. The single six carbon membered ring is named benzene. The multiple rings are made up of multiple six, and less often five, carbon atom planar rings, with single carbon to carbon (sigma) bonds between carbon atoms and a shared Pi electrons resonating between the carbon atoms in a "cloud" of electrons above and below the planar ring(s). Some aromatics are heteroatom aromatic compounds for which a carbon atom is replaced by a nitrogen, sulfur or oxygen atom. Multiple fused ring aromatic hydrocarbons are named polycyclic aromatic hydrocarbons and those with a nitrogen, sulfur, or oxygen atom are named polycyclic aromatic compounds (PACs). The abbreviation PAC is often used to mean all of the aromatic ring compounds. The benzene rings are often referred to as the "phenyl ring" when bonded to other aromatic ring structures or to other atoms in non-aromatic ring structures.

Aryl hydrocarbon receptor: a cellular protein that binds planar or plate-like polycyclic aromatic compounds having a molecular shape and dimensions that resemble 2,3,7,8 tetrachorodibenzo(p)dioxin.

Asphalt: a dark-brown-to-black cement-like material containing bitumens as the predominant constituent obtained by petroleum processing that is used primarily for road construction. It includes crude asphalt as well as finished products such as cements, fluxes, the asphalt content of emulsions (exclusive of water), and petroleum distillates blended with asphalt to make cutback asphalts.

Asphaltenes: class of petroleum compounds of high molecular weight and complexity; defined as the oil fraction that precipitates in low molecular weight n-alkanes (e.g., C5–C7) but is soluble in toluene.

Authigenic: a geological deposit that has formed, via geochemical processes, in the same site where it is currently located.

Bacteria: microscopic, single-celled organisms that lack a nucleus. One of the two kingdoms, along with the Archaea, that comprise the prokaryotes.

Ballast tank: tank that carries water for weight needed for maintaining ship stability or draft.

Barrel: term used as the standard measurement of volume for crude oil and large quantities of refined products in the petroleum industry. A unit of volume equal to 42 U.S. gallons (159 L), often abbreviated as bbl.

Bioaccumulation (or bioconcentration): the tendency of substances to accumulate in the body of organisms; the net uptake from their diet, respiration or transfer across skin and loss due to excretion or metabolism. The bioaccumulation factor or bioconcentration factor is the ratio of concentrations in tissue to concentrations in a source (i.e., water or diet).

Bioavailability or biological availability: the extent to which a compound can be assimilated by a living organism; also, the proportion of a chemical in an environmental compartment (e.g., water) that can be taken up by an organism.

Biodegradation: a natural process in which microbes enzymatically transform organic chemicals, such as oil, under aerobic or anaerobic conditions and typically resulting in cell growth; oil biodegradation usually requires nutrients, such as nitrogen and phosphorus; transformation may be complete, producing water, carbon dioxide and/or methane, or incomplete, producing partially oxidized chemicals.

Biofilm: any consortium of microbes that stick to each other (e.g., marine snow) and/or are attached to living or inanimate surfaces by biological macromolecules such as extracellular polymeric substances or transparent extracellular polymers.

Biofuel: a fuel derived from biomass, such as plant or animal material.

Bioinformatician: scientist who conducts research using an interdisciplinary approach to develop methods and software for analyzing and interpreting biological data, particularly using large, complex datasets.

Bioinformatics: a field of study at the intersection of molecular biology and computer science that enables acquisition, analysis, comparison and archiving of biological data, such as DNA and amino acid sequences.

Biomagnification: a food chain or food web phenomenon whereby a substance or element increases in concentration at successive trophic levels; occurs when a substance is persistent and is accumulated from the diet faster than it is lost due to excretion or metabolism.

Biomarkers: a term used in two different ways, depending on discipline. In petroleum chemistry, a biomarker is a relic chemical relating its presence to the original biological source (microbial, plant, or animal); biomarkers are usually poorly or non-biodegradable and so persist in the oil, enabling their use as internal standards in petroleum analysis. In environmental toxicology, a biomarker is a biochemical process, product, or cellular response that indicates the organism's exposure to a pollutant and/or the toxic effects of the pollutant.

Bioremediation: an environmental intervention strategy to enhance biodegradation of spilled oil (or other contaminants) ranging from no remedial action other than monitoring (monitored natural attenuation) to nutrient addition (biostimulation) to inoculation with competent microbial communities (bioaugmentation).

Bitumen: the heaviest class of petroleum, having high viscosity and density; widely considered to represent the residue from lighter oils that have undergone biodegradation over geological time.

BTEX: abbreviation for the mixture of benzene, toluene, ethylbenzene, and xylenes.

Bunker fuel: fuel oil used by ships for main propulsion or auxiliary power.

Chemically enhanced water-accommodated fraction of oil (CEWAF): a solution of hydrocarbons and a suspension of oil droplets created when a chemical dispersant is added to oil and water with stirring.

Chemotaxis: the movement of a motile cell or organism toward an attractant chemical or away from a repellent chemical, along the increasing or decreasing concentration gradient of that chemical.

Co-metabolism: simultaneous biodegradation of two compounds, in which the degradation of the second compound (the secondary substrate) depends on the presence of the first compound (the primary substrate).

Commingling: mixing of two petroleum products with similar specifications. Most branded gasoline firms require that their product not be commingled to preserve the integrity of the brand.

Conventional crude oil: commonly defined as liquid petroleum that flows in the reservoir and in pipelines and is recovered from traditional oil wells using established methods, including primary recovery and water flooding (e.g., condensates, light and medium crude oils), versus unconventional crude oils.

Crude oil: synonymous with petroleum; a naturally occurring and typically liquid complex mixture of thousands of different hydrocarbon and non-hydrocarbon molecules.

Cx: a molecule, such as a hydrocarbon, having x carbon atoms.

Cyanobacteria: a phylum of bacteria capable of photosynthesis; colloquially (but incorrectly) also called "blue-green algae."

CYP1A: a member of the cytochrome P450 family of proteins; an enzyme of vertebrates, including fish, birds, and mammals, that catalyzes the addition of oxygen to double bonds as a first step in the metabolism and excretion of polycyclic aromatic hydrocarbons.

cyp1a: the gene that codes for CYP1A proteins.

Dead oil: used to describe an oil at standard conditions in equilibrium with a gaseous headspace in a closed container. *Dead oil* has released most of the gaseous compounds dissolved in the oil but may retain all of its volatile constituents. See also *live oil*.

Dead weight tonnes: a measure of ships' carrying capacity (cargo, fuels, ballast water, provisions, passengers, and crew).

Demersal: can be pelagic but living on or associated with the bottom sediments of marine and freshwater aquatic systems. Typical organisms are fish, such as flounder, and invertebrates, such as shrimp.

Dilbit: bitumen diluted with a lighter petroleum class, such as condensate or naphtha, typically 70% bitumen and 30% diluent.

Dispersant: a chemical or mixture of chemicals applied, for example, to an oil spill to disperse the oil phase into small droplets in the water phase.

Dispersion: suspension of oil droplets in water accomplished by natural wind and wave action; production of biological materials (biosurfactants) and/or chemical dispersant formulations.

Distillate: No. 1 and No. 2 fuel oils, and No. 1 and No. 2 diesel fuels as well as some blends of gasoil. These are light fuel oils for transportation and heating oil, and include railroad engine fuel and diesel for agricultural machinery. Lower sulfur distillates also represent the preferable feedstock for power generation when utilities see natural gas curtailments or high natural gas prices.

Ebullition: the act or state of bubbling or boiling.

EC_{50}: the concentration of a substance that causes sublethal effects on the median or 50th percentile organism tested within a specified exposure time (e.g., 14-day EC_{50}).

Ecological risk assessment: process for analyzing and evaluating the possibility of adverse ecological effects caused by environmental pollutants.

Ecosystem: the interrelationships between all of the living things in an area and their non-living environment.

Effects-driven chemical fractionation (EDCF): the stepwise fractionation, toxicity testing, and chemical analysis of complex mixtures of compounds to isolate and identify the constituents responsible for the toxic effects of the whole mixture.

Emulsification: formation of water droplets in an oil matrix (water-in-oil) or conversely oil droplets in a water matrix (oil-in-water) achieved by the action of agitation, such as wind and wave activity; can be unstable, separating into oil and water phases again soon after formation, or stable for months or years (e.g., "chocolate mousse," a water-in-oil emulsion).

Environmental impact assessment: the process of measuring or estimating the environmental effects of pollutants, such as oil spills, relative to conditions at a reference site or to a time prior to a spill.

Epigenomics: the study of reversible heritable changes in gene function that occur without a change in the sequence of nuclear DNA (e.g., DNA methylation, histone modifications).

Eukaryote: a single-celled or multicellular organism (in the domain Eukarya) that has a nucleus and other intracellular organelles; includes plants, animals, fungi, and protists.

Evaporation: the physical loss of low molecular weight components of an oil to the atmosphere by volatilization.

Flame ionization detector (FID): used with analytical instruments like gas chromatographs to detect components of petroleum by combustion ionization, hence GC-FID.

Foraminifera: single-celled organisms (protists) with tests (shells). They may be pelagic or benthic and are abundant as fossils for the past 540 million years. The shells of benthic species are commonly divided into chambers that are added during growth and may be made of organic compounds, sand grains or other particles cemented together, or crystalline calcium carbonate.

Fossil fuel: a fuel source (such as oil, condensate, natural gas, natural gas liquids, or coal) formed in the earth from plant or animal remains.

Fractions: the different cuts of petroleum products that come off a distillation column contingent on their volatility or boiling range. Fractions are essentially crude cut points at different boiling ranges that produce the various finished products.

Fugitive emissions: emissions of gases or vapors from pressurized equipment, including pipelines, due to leakage or unintended or irregular releases of gases.

Gas chromatography (GC): an analytical method used to characterize petroleum components; GC is combined with different detection methods, hence GC-FID, GC-MS, etc.

Genomics: the study of the genomes (the full DNA sequence) of organisms, including taxonomic information, biochemical potential, evolutionary history, etc.

Gross tonnes: a measure of volume of enclosed spaces on ships used to categorize ships in regulations and commercial transactions.

Harpacticoids: small crustaceans (usually <0.3 mm) of the order Copepoda. They are generally benthic grazers and common members of the meiofauna.

Heteroatom: in petroleum, an atom such as nitrogen, sulfur, and/or oxygen that is part of a hydrocarbon skeleton, such as found in the resins fraction of crude oils.

High-energy water-accommodated fraction of oil (HEWAF): a solution of hydrocarbons and a suspension of oil droplets created when oil and water are mixed by high-energy agitation.

High molecular weight: relative term referring to the molecular mass of chemicals; in oil, asphaltenes would be typical of high molecular weight compounds.

High-performance liquid chromatography (HPLC): an analytical method for separating chemicals in solution.

Hydrocarbon: a chemical that is composed of only carbon and hydrogen; chemicals containing heteroatoms, such as nitrogen, sulfur, and/or oxygen, are not hydrocarbons, even though they may be petroleum constituents.

Hydrocarbonoclastic: term defining microorganisms that can metabolize or transform hydrocarbons (e.g., in crude oil or refined petroleum products).

Hydrogen sulfide (H_2S): a toxic, flammable, and corrosive gas sometimes associated with petroleum.

Hydrophilic-lipophilic balance (HLB): a measure of the properties of dispersants and surface-active agents at the interface of polar and non-polar liquids, such as oil and water. The higher the surfactant HLB value, the more hydrophilic (water-loving) it is.

Hydrophobic: having a tendency to repel water.

Hyporheic flow: the percolating flow of water through the sand, gravel, sediment, and other permeable soils under and beside the streambed.

Immiscible: a liquid that cannot be mixed with another liquid without separating from it.

Intermediate fuel oil (IFO): the heaviest commercial class of refined petroleum diluted with a lighter refined product, commonly burned in furnaces, boilers or ship engines (e.g., IFO 180).

Isomers: chemicals that have the same molecular formula (i.e., elemental composition) but different structures; may also have different properties, including water solubility, biodegradability, and toxicity.

Kinorhynch: a segmented, limbless animals <1 mm, with a body consisting of a head, neck, and a trunk of 11 segments. They are externally armored with a number of spines along the body, plus up to seven circles of spines around the head.

Kolmogorov scale: length scale of turbulent flow below which the effects of molecular viscosity are non-negligible. In three-dimensional turbulence, the Kolmogorov scale is $(\nu 2/\varepsilon)^{1/4}$, in which ν is the kinematic viscosity and ε is the energy dissipation rate per unit mass.

K_{ow}: the partition coefficient describing the equilibrium concentration ratio of a dissolved chemical in octanol versus in water, in a two-phase system at a specific temperature; used in the prediction of partitioning between dissolved and particulate or organic film or particle coating phase, prediction of bioaccumulation in organisms and in the related prediction of toxicity.

Labile: the property of being easily broken down (e.g., organic chemicals degraded by microbes; foodstuff metabolized by higher organisms).

Lacustrine: relating to lakes.

Lagrangian Coherent Structure (LCS): repelling, attracting, and shearing material surfaces that form the skeletons of Lagrangian particle dynamics. The theory of LCS seeks to reveal special surfaces of fluid trajectories that organize the rest of the flow into ordered patterns. Adapted from Haller (2015).

Langmuir cells: an alternating series of counter-rotating, helical vortex cells set up by the interaction of winds with waves. These cells resulting in alternating convergence and divergence lines on the sea surface that give the appearance of so-called wind rows.

LC_{50}: the concentration of a substance toxic to the median or 50th percentile organism tested within a specified exposure time (e.g., 96-hour LC_{50}).

Lipidomics: the study of the complete profile (lipidome) of lipids produced by an organism, such as those comprising cell membranes (and organelles in eukaryotes), and including identity, structure, enzymatic modification, and function.

Liquefied natural gas (LNG): a natural gas that has been cooled to a liquid form for transportation and/or storage purposes.

Liquefied petroleum gas (LPG): any of a group of hydrocarbon-based gases derived from crude oil refining or natural gas stream fractionation that are often liquified, through pressurization, for ease of transport. They include ethane, propane, normal butane, and isobutane. Uses of these fuels include home heating, industrial, automotive fuel, petrochemical feedstocks, and for drying purposes in farming.

Live oil: used to describe an oil whose composition would change if brought to equilibrium with standard conditions. Normally, *live oil* exists at pressures higher than standard pressure and contains various dissolved gases. See also *Dead oil*.

Long-chain alkanes: commonly considered to be alkanes having more than 16 carbons (>C16).

Low molecular weight: relative term referring to the molecular mass of chemicals; in oil, monoaromatics, and aliphatics up to C10 would be typical of these compounds.

Marine oil snow (MOS): an oil-particle aggregate (OPA) that forms when dispersed oil droplets, dispersed, or dissolved oil chemicals attach to, or interact (adsorb/absorb) with marine snow. Marine oil snow also forms when oil droplets, dispersed, or dissolved oil chemicals interact with organisms (mainly microbes and phytoplankton), causing exudation of polymeric organic material (exopolymers) that interact with the oil droplets and/or dispersed or dissolved oil chemicals. It is differentiated in one manner from other types of OPAs by higher proportions of microbes and organic material.

Marine snow: a descriptive term referring to aggregates with a general range in size from >0.5 mm to 10 cm that form mainly in upper layers of the ocean. Their composition can include organic and inorganic material from various sources (e.g., phytoplankton, microbes, plankton, fecal pellets, sand, soot, dust, sources) incorporated into a hydrated matrix of biopolymers. After initial formation, marine snow can have a range of densities that cause it to sink, rise, or be neutrally buoyant during its existence.

Mass spectrometry (MS): an analytical method used for detailed characterization of petroleum components, often in combination with GC, hence GC-MS.

MCF and MMCF: units of volume measurement for natural gas. An abbreviation that combines the Roman numeral M as a stand-in for 1,000, or MM for 1,000,000, with the term "cubic feet" (CF).

Mechanism of action (MOA): describes a functional or anatomical change at the molecular level.

Medium molecular weight (MMW): relative terms referring to the molecular mass of chemicals; in oil, 3- to 6-ringed polycyclic aromatic hydrocarbons and aliphatics up to C20 would be typical of MMW compounds.

Mesocosm: an experimental system designed to study the behavior of natural ecosystems under controlled conditions. Mesocosm experiments can help to link laboratory studies with observational field studies.

Metabolism: the biochemical processes occurring in a living entity (a single cell, an organ, or an entire organism) that maintain life.

Metabolomics: comprehensive chemical analysis of small molecules produced as intermediates or end products of biochemical reactions (i.e., metabolites) by an organism or community of organisms (i.e., an environmental sample) under optimal or stressed conditions.

Metagenomics: DNA sequence-based analysis of the collective genomes of (micro)biological assemblages from an environmental sample that provides information about community composition and/or function.

Metatranscriptomics: analysis of the function and activity of the collective suite of RNA transcripts extracted from a community of (micro)organisms to assess and profile which genes are being expressed in that environment at the sampling time.

Microbiota: the complete community of microorganisms present in a particular location or environmental habitat, including on or in higher organisms.

Microcosm: a small-scale, controlled experimental system intended to replicate all or parts of the conditions of an ecosystem in order to simulate and predict the behavior of natural ecosystems under specific conditions.

Mid-chain alkanes: commonly considered to be alkanes of six to 16 carbons in length (C6–C16).

Mineralization: complete oxidation of a compound (e.g., hydrocarbon) to carbon dioxide or methane, water, and biomass; may be accomplished by a single species of organism or by a community of microbes.

Mode of action (MoA): describes a functional or anatomical change at the cellular level, resulting from the exposure of a living organism to a substance.

Monitored natural attenuation: an environmental remediation strategy in which there is no intervention, but the site is monitored using various parameters.

Monoaromatics: aromatic hydrocarbons having only a single benzene ring; may also have one or more alkyl side chains.

MOS sedimentation and flocculent accumulation (MOSSFA): a process describing the sinking of oil chemicals through the water column and deposition to surface sediments through sedimentation and flocculent accumulation. First observed at some sampling and observational locations during and after the *Deepwater Horizon* oil spill, this process is thought to be a result of increases in density through the combined processes of aggregation and microbial degradation.

Naphthenic acids: a class of polar petroleum compounds that contributes to aquatic toxicity and to petroleum infrastructure corrosion.

Natural attenuation: remediation of a contaminated site by natural processes alone, without human intervention.

Natural dispersion: ability of oil to naturally disperse under ambient environmental conditions.

Natural gas: naturally occurring hydrocarbon gases found in porous rock formations. Its principal component is usually methane. Non-hydrocarbon gases such as carbon dioxide and hydrogen sulfide can sometimes be present in natural gas.

Neat oil: an oil that has not been diluted or altered by the addition of other components such as diluents or processing chemicals.

Nekton: Living organisms that are able to swim and move independently of currents, as opposed to most phytoplankton. Nekton are heterotrophic, have a large size range, and include marsh fish, squid, and sharks.

Obligate hydrocarbonoclastic bacteria: bacteria that have adapted to use hydrocarbons as their sole source of carbon and energy for growth and metabolism.

Oil-mineral aggregate (OMA): floc-containing oil adhering to mineral particle(s), which may float, sink, or resuspend in a water column; historically referred to as clay–oil flocculation.

Oil-particle aggregate (OPA): more general term than OMA, describing oil adhering to particles that may be inorganic (minerals) and/or organic, including microbial cells.

Olefin: a hydrocarbon containing one or more double bonds, such as an alkene or cycloalkene.

Oleophilic: having a strong affinity for oils.

Oligochaeta: a subclass of animals in the phylum Annelida, which is made up of many aquatic and terrestrial worms. Specifically, oligochaetes usually have few setae (chaetae) or "bristles" on their outer body surfaces and lack parapodia, unlike Polychaeta.

Omics: a collective term for (meta)genomics, (meta)transcriptomics, proteomics, lipidomics, and/or metabolomics.

Ostracod: a small crustacean, typically 1 mm in size, with a flattened body from side to side and protected by a bivalve-like, chitinous, or calcareous valve or "shell."

Oxidation: the loss of electrons by a molecule during a chemical reaction.

Partitioning: the diffusion of compounds between two immiscible liquid phases, including water and oil droplets and water and lipid membranes; may be described by the term K_{ow}, when at equilibrium.

Petroleum: synonymous with crude oil; a naturally occurring complex mixture of thousands of different hydrocarbon and non-hydrocarbon molecules.

Photochemical reaction: a chemical reaction initiated when a molecule absorbs light energy; a more general term than photo-oxidation.

Photo-enhanced toxicity: increased toxicity due to photo-oxidation in vivo.

Photo-oxidation: oxidation of chemicals due to the influence of photic energy, usually from UV light.

Phytoplankton: microscopic photosynthetic organisms that live in the upper regions of marine and freshwaters where light penetrates; includes Bacteria, eukaryotic protists, and microalgae.

Piezotolerant: capable of withstanding hydrostatic pressure (e.g., >10 MPa).

Pneumatophores: roots commonly found in mangrove species that grow in saline mud flats. The roots grow upward out of the mud and water to function as the site of oxygen intake for the submerged primary root system.

Polychaete: also known as bristle worms; a class of annelid worms, generally marine. Each body segment has a pair of fleshy protrusions called parapodia that bear many bristles, called setae.

Polycyclic aromatic hydrocarbons (PAHs): a subclass of aromatic hydrocarbons having two or more fused benzene rings; may also have one or more alkyl side chains, generating large suites of isomers; some are considered "priority pollutants" because of their toxicity and/or potential carcinogenicity.

Produced water: water produced in connection with oil and natural gas exploration and development activities.

Prokaryote: a division of life comprising the two prokaryotic domains of Bacteria and Archaea. All prokaryotes lack a nucleus, in contrast to eukaryotes.

Proteomics: study of the total complement of proteins (the proteome) produced by an organism under specific conditions, including identification, quantification, enzymatic modification, and inference of function.

Pseudo-component: a component having averaged thermodynamic properties that is used to represent a group of compounds in a petroleum mixture. Any number or type of compounds can be included within an arbitrary pseudo-component. Pseudo-components are used to reduce the thousands of individual molecules present in a crude oil mixture to a tractable number of quantifiable groups.

Psychrophiles: organisms (typically prokaryotes) that are capable of growth only at low temperatures (typically –20°C to +10°C) and perish at higher temperatures (e.g., above +15°C). They inhabit environments that are constantly cold, such as the deep-sea and polar ecosystems.

Psychrotolerant: property of cold-tolerant organisms (often prokaryotes) that are able to grow at cold temperatures (e.g., 4°C) but have optimum growth at higher temperatures (e.g., 20°–40°C).

Psychrotroph: cold-loving organisms (often prokaryotes) with growth temperatures slightly higher than psychrophiles (e.g., maximum growth temperature of ~+20°C).

Q10 rule: a generalized relationship of temperature sensitivity of (bio)chemical reactions over a range of 10°C (e.g., a measure of the change in enzyme rate over 10°C, within a tolerated range). Traditionally, a factor of 2 has been applied to estimate that biological activity (e.g., microbial growth) an organism decreases by 2 for a 10°C drop in ambient temperature, or increases 2-fold for a 10°C increase in temperature, to a minimum or maximum tolerated temperature.

Rare biosphere: a term describing organisms that persist in an environment but are present at low abundance (e.g., <1% of the total community) under a given condition. The rare biosphere cohort can change if environmental conditions change, and rare taxa can increase many-fold, "blooming" transiently or permanently.

Reactive oxygen species (ROS): a group of chemicals that can cause cellular damage due to their high reactivity and are released during reactions that add oxygen to double bonds. In turn, ROS can react with double bonds in lipids, proteins, and nucleic acids to change their structure and function.

Recalcitrant: a term describing a compound or molecule that persists in nature for a long time and resists biodegradation or chemical breakdown.

Redox discontinuity layer: the sediment transition boundary between oxic and anoxic (no oxygen) conditions.

Resins: a solubility class of poorly characterized, polar petroleum compounds in which each molecule contains one or more atoms of nitrogen, sulfur, and/or oxygen in a hydrocarbon skeleton.

Riparian: related to being situated on or dwelling on the bank of a river or other body of water, such as a lake or tidewater.

Riverine: relating to rivers.

Saturated hydrocarbons: class of hydrocarbons that may be straight-chain, branched-chain or cyclic, in which all carbon atoms have single bonds to either carbon or hydrogen.

Shale: a very fine-grained sedimentary rock that is formed by the consolidation of clay, mud, or silt and that usually has a finely stratified or laminated structure. Certain shale formations contain large amounts of oil and natural gas.

Shale oil: also known as "tight oil" (but not to be confused with "oil shale"); liquid petroleum that is produced from shale oil reservoirs, typically by hydraulic fracturing methods.

Short-chain alkanes: commonly considered to be alkanes less than six carbons in length (<C6).

Slick: a film or layer of oil floating on the water surface, of varying thickness.

Sorbent: an insoluble material or mixture of materials used to recover liquids through the mechanisms of absorption or adsorption, or both.

Sour crude: petroleum that has a >1% total sulfur content that may be present as hydrogen sulfide and/or as organic forms of sulfur in a hydrocarbon backbone.

Sour gas: natural gas or any other gas containing significant amounts of hydrogen sulfide.

Species sensitivity distributions (SSD): these compare the cumulative proportion of species (percentile) affected by a chemical to a toxicity end point measured for each species (e.g., 96-hour LC_{50}); the SSD model assumes that species sensitivity is randomly distributed.

Submerged oil: oil that is in the water column below the water surface, such as dispersed oil droplets and neutrally buoyant tar balls.

Surfactant: a surface-active chemical agent that reduces the interfacial tension between two liquids such as oil and water; surfactants are the active component of commercial oil dispersants and household detergents.

Sweet crude: petroleum with low total sulfur content, variously defined as <0.5% or <1% sulfur.

Sweet gas: natural gas that contains little or no hydrogen sulfide.

Sympagic: organisms that complete either their entire life cycle within the sea ice or spend at least part of their life cycle attached to the ice.

Synbit: bitumen diluted with synthetic crude oil, typically at 1:1 ratio.

Synthetic crude oil: a partially refined fraction of bitumen; may be used as a diluent to make dilbit for transport.

Syntrophy: a condition of two bacterial strains that rely on each other for growth by sharing resources that neither can utilize alone; a form of long-term stable cross-feeding.

Target lipid model: estimates the aqueous concentration of organic compounds, or mixtures of organic compounds, that cause toxicity by narcosis.

Taxon (plural, taxa): a coherent group of organisms considered to be evolutionarily related. In practice, a taxon can represent any named taxonomic level (e.g., class, family, species, etc. that may or may not have a formally recognized name and for taxonomic purposes may exist only as a discrete DNA sequence, such as a region of the 16S rRNA gene).

Total petroleum hydrocarbons (TPHs): the total mass of all hydrocarbons in an oil or environmental sample, including the volatile and extractable (non-volatile) hydrocarbons; may be further defined by stating the analytical method used (e.g., GC-detectable TPH or TPH-F [TPH measured by fluorescence], which vary in their rigor).

Total polycyclic aromatic hydrocarbons (TPAHs): including alkyl-PAHs and parent (unsubstituted) PAHs; the sum of all concentrations of PAHs measured by GC-MS or similar analytical method.

Toxic unit (TU): ratio between the concentration of a compound in water and a toxic end point (e.g., 96-hour LC_{50}).

Transcriptomics: study of the RNA transcripts in an organism (i.e., the expression of the complement of genes that are active under specific conditions).

Unresolved complex mixture (UCM): petroleum constituents that are not resolved by conventional GC and appear as a "hump" in the gas chromatogram; comprised of many compounds eluting from the gas chromatogram in an overlapping fashion (i.e., not resolved by gas chromatography).

Unusual mortality event (UME): term coined by the National Oceanic and Atmospheric Administration to describe a greater-than-usual rate of mortality of marine mammals.

Viscosity: a fluid's internal resistance to flow.

Volatile organic compounds (VOCs): chemicals having high vapor pressure at room temperature (and corresponding low boiling point) that therefore tend to evaporate or sublimate into the air (e.g., BTEX).

Water-accommodated fraction of oil (WAF): hydrocarbons that will partition from oil to water during gentle stirring or mixing; may contain droplets, in contrast to water-soluble fractions.

Water-soluble fraction of oil (WSF): aqueous solution of hydrocarbons that partition from oil; does not include droplet or particulate oil.

Weathering: a suite of processes that cause changes in spilled oil composition and properties. Weathering includes a variety of environmental processes such as spreading, evaporation, photo-oxidation, dissolution, emulsification, and biodegradation, among others.

Xenobiotic: a chemical that is foreign to an organism or ecological system.

Appendix A

North American Zone Descriptions

TABLE A.1 Descriptions of Coastal Zones as Defined by This Study

Zone	Description
A	Bounded by a line drawn northerly 0° true along the U.S./Canada border in the Arctic and a line bearing northerly 0° true from Lancaster Island in Canada (includes Arctic Canada west of Hudson Bay).
B	Bounded by a line bearing northerly 0° true from Lancaster Island in Canada and a line bearing easterly 90° true from the northern boundary of Newfoundland, Canada (includes Hudson Bay east to the Maritime Provinces in Canada).
C	Bounded by a line bearing easterly 90° true from the northern boundary of Newfoundland, Canada, and a line drawn easterly along the U.S./Canada border as follows: a line drawn 90° true from the shore along 44°11′12″ N to 67°16′46″ W; thence southwest to 42°53′14″ N, 67°44′35″ W; thence southeast to 42°31′08″ N, 67°28′05″ W; thence southeast to 40°27′05″ N, 65°41′59″ W.
D	Bounded by a line drawn easterly along the U.S./Canada border and a line bearing due east 90° true from the shore at the Virginia/North Carolina border (35°59.8′ N) to the offshore extent of the EEZ (includes U.S. Coast Guard MSOs Portland, Maine; Boston; Providence; Long Island Sound, Connecticut; New York; Philadelphia; Baltimore; and Hampton Roads, Virginia).
E	Bounded by a line bearing due east 90° true from the shore at the Virginia/North Carolina border (35°59.8′ N) to the offshore extent of the EEZ and by a line bearing 227° true drawn from 26°00′ N latitude and 81°30′ W longitude to the offshore extent of the EEZ (includes MSOs Charleston, Savannah, Jacksonville, and Miami).
F	Bounded by a line bearing 227° true drawn from 26°00′ N latitude and 81°30′ W longitude to the offshore extent of the EEZ and a line from the Mississippi coast at 89°10′ W bearing southeasterly to 29°10′ N, 88°00′ W; thence southerly bearing 180° true to the offshore extent of the EEZ (includes MSOs Tampa Bay and Mobile).
G	Bounded by a line from the Mississippi coast at 89°10′ W bearing southeasterly to 29°10′ N, 88°00′ W; thence southerly bearing 180° true to the offshore extent of the EEZ and a line drawn along the U.S./Mexico border in Texas bearing 90° true to the offshore extent of the EEZ (includes MSOs New Orleans and Morgan City, Louisiana; and Port Arthur, Houston, and Corpus Christi, Texas).
H	Includes Mexican offshore waters on the east coast of Mexico that are south of a line bearing 90° true from the U.S./Mexico border to the outermost extent of the EEZ and north of a line bearing 90° true from the Mexico/Belize border to the outermost extent of the EEZ.
I	Includes the areas of both the Commonwealth of Puerto Rico and the Territory of the U.S. Virgin Islands to the offshore extent of the EEZ (includes MSO San Juan).
J	Includes Mexican offshore waters on the west coast of Mexico that are south of a line bearing 270° true from the U.S./Mexico border to the outermost extent of the EEZ and north of a line bearing 270° true from the Mexico/Guatemala border to the outermost extent of the EEZ.
K	Includes a line along the U.S./Mexico border in California to the offshore extent of the EEZ and a line drawn from the intersection of 34°58′ N latitude and the California coastline bearing 229° true to the offshore extent of the EEZ (includes MSOs San Diego and Los Angeles/Long Beach).
L	Includes a line drawn from the intersection of 34°58′ N latitude and the California coastline bearing 229° true to the offshore extent of the EEZ and a line drawn along the 42°00′ N latitude from the shore to the offshore extent of the EEZ (includes MSO San Francisco).
M	Includes a line drawn along the 42°00′ N latitude from the shore to the offshore extent of the EEZ and a line drawn along the international boundary between the United States and Canada at the entrance to the Strait of Juan de Fuca to the offshore extent of the EEZ (includes MSOs Portland, Oregon; and Puget Sound, Washington).
N	Includes Hawaii, American Samoa, Johnston Atoll, Palmyra Atoll and Kingman Reef, Wake Island, Jarvis Island, Howland and Baker Islands, Guam, the Commonwealth of Northern Mariana Islands, and the Trust Territory of the Pacific Islands.
O	Includes a line drawn along the international boundary between the United States and Canada at the entrance to the strait of Juan de Fuca to the offshore extent of the EEZ (48°29′38.11″ N, 124°43′34.69″ W to 48°29′38.11″ N, 125°00′00″ to 48°04′00″ N, 126°10′35″ W) and a line along the international boundary bearing 270° true between the United States and Canada at the Dixon entrance to the offshore extent of the EEZ (includes all marine waters in western Canada).
P	Includes a line drawn along the international boundary bearing 270° true between the United States and Canada at the Dixon entrance to the offshore extent of the EEZ and a line drawn along the northern edge of the Aleutian Island chain to the offshore extent of the EEZ (includes MSO Southeast Alaska and a portion of MSO Western Alaska).
Q	Includes a line drawn along the northern edge of the Aleutian Island chain to the offshore extent of the EEZ and a line drawn northerly 0° true along the U.S./Canada border in the Arctic.

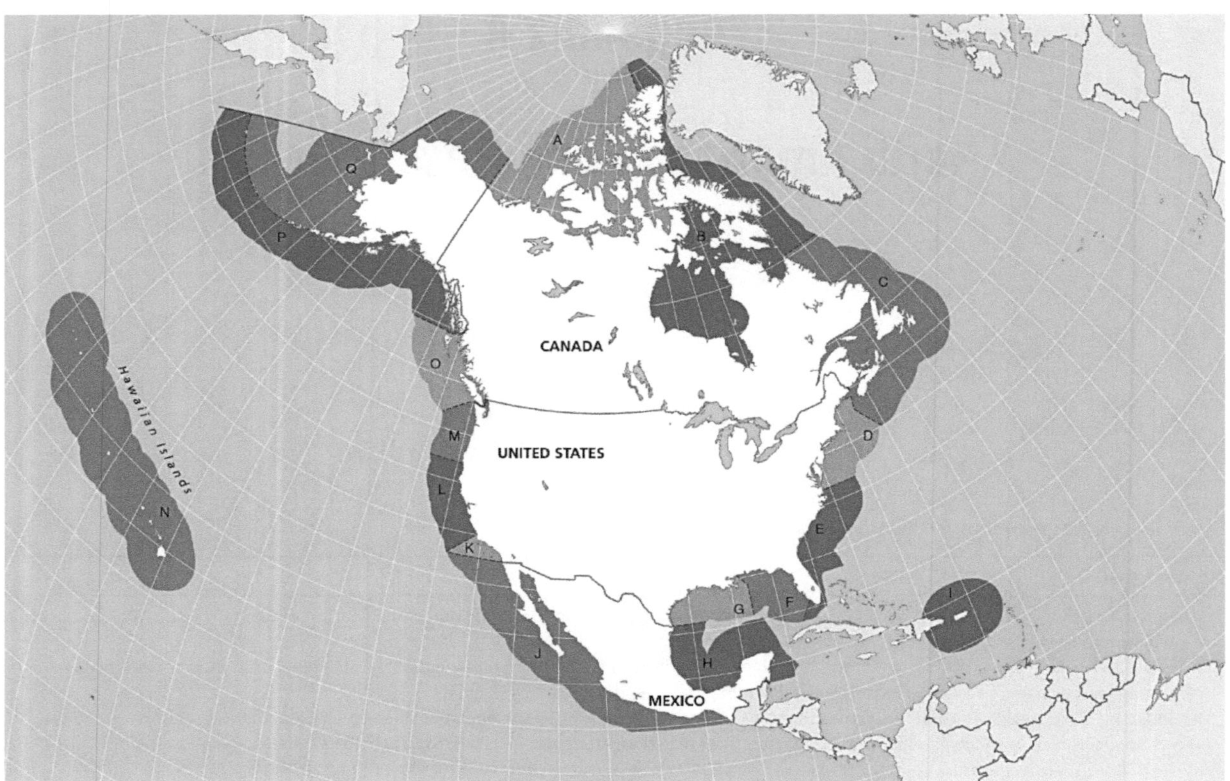

FIGURE A.1 Map showing geographic regions.

Appendix B

Energy Outlook Data Sources

The main sources for the predictions in the Greene Economics report (Greene, 2021), described in Chapters 1 and 3 in *Oil in the Sea IV* are shown in Tables B.1 to B.4.

TABLE B.1 Energy Consumption Data Sources for EOR Oil in the Sea (OITS) Database

Geography	1990–2020 (Actual)	2020–2050 (Forecasts)
Worldwide	U.S. Energy Information Administration (EIA), 2021 International Energy Statistics.	U.S. EIA, 2019. International Energy Outlook 2019 Table: World total energy consumption by region and fuel.
United States	Same as above	U.S. EIA, 2021. AEO2021_Release_Presentation.ppt, Slide 9 US Energy consumption by fuel AEO2021 Reference Case chart data.
Canada	Same as above	Canada Energy Regulator (CER), 2020. Canada's Energy Futures 2020 Supplement: Energy Demand.
Mexico	Same as above	Same as worldwide

TABLE B.2 Oil and Gas Production Data Sources for EOR OITS Database

Geography	1990–2020 (Actual)	2020–2050 (Forecasts)
United States	EIA, 2021. Crude Oil Production Worksheet Data 1, pet_crd_crpdn_adc_mbbl_a.xls, released May 28. AND EIA, 2021. Natural Gas Dry Production Worksheet Data 1, ng_prod_sum_a_epg0_fpd_mmcf_m.xls, released May 28.	International Energy Agency, 2020. World Energy Outlook 2020 (forecast through 2040). AND EIA, 2020. Annual Energy Outlook, released January 29.
Canada	EIA, 2021. International Energy Statistics.	CER. 2021. Canada's Energy Future 2020: Energy Supply and Demand Projections to 2050.
Mexico	Same as above	International Energy Agency, 2020. World Energy Outlook 2020 (forecast through 2040).

TABLE B.3 Offshore Oil and Gas Production Data Sources for EOR OITS Database

Geography	1990–2020 (Actual)	2020–2050 (Forecasts)
United States	EIA, 2021. Crude Oil Production Worksheet Data 1, pet_crd_crpdn_adc_mbbl_a.xls, released May 28. AND EIA, 2021. Bureau of Ocean Energy Management. Data Center, Annual Production for Entire Region.	DNV GL. 2018. Oil and Gas Forecast to 2050—Energy Transition Outlook 2018.
Canada	DNV GL. 2018. Oil and Gas Forecast to 2050—Energy Transition Outlook 2018.	Same as above
Mexico	EIA, 2016. Today in Energy, Offshore production nearly 30% of global crude oil output in 2015, October 25. (Data in charts 2005–2015 and anecdotal info.)	None found

TABLE B.4 Seaborne Transportation of Crude Oil and Petroleum Products Data Sources for EOR OITS Database

Geography	1990–2020 (Actual)	2020–2050 (Forecasts)
Worldwide	United Nations Conference on Trade and Development STAT (UNCTAD-STAT). 2021. World Seaborne Trade by Types of Cargo and by Group of Economies, Annual. United Nations Conference on Trade and Development (UNCTAD). 2020. Review of Maritime Transport—2020. Geneva. Şanlıer, Şengül. 2018. Should Chemical Tankers' Tank Cleaning Waters Be Banned from Discharging into the Sea?	American Bureau of Shipping (ABS). 2020. Pathways to Sustainable Shipping—Setting the Course to Low Carbon Shipping.
United States	FOR IMPORTS AND EXPORTS: U.S. Energy Information Administration (EIA). 2021. Petroleum and Other Liquids—Exports (1981–2020). U.S. Energy Information Administration (EIA). 2021. Petroleum and Other Liquids—Exports (1981–2020). FOR WITHIN THE UNITED STATES: EIA. 2021. Movements by Tanker and Barge between PAD Districts (1981–2020).	FOR IMPORTS AND EXPORTS: U.S. EIA. 2021. Annual Energy Outlook 2021—Table 11. Petroleum and Other Liquids Supply and Disposition (Reference Case) (2020–2050). FOR WITHIN THE UNITED STATES: (Elaborations on) U.S. EIA. 2021. Movements by Tanker and Barge between PAD Districts (1981–2020).

Appendix C

Estimating Land-Based Sources of Oil in the Sea

C.1 INTRODUCTION

Due to the scarcity of individual data samples for estimating land-based loads of oil to the sea, the loading estimates calculated in this analysis were based on loading per unit of urban land area. These calculations assume that most land-based runoff of oil is from urban areas. This approach was used for the United States and Canada and then extrapolated to estimate the other regions of the world. This appendix is an update of *Oil in the Sea III*, Appendix I.

C.1.1 Methodology and Sources of the Data

A review of The Water Quality Portal's STORET data indicated that oil and grease data had been collected for the major rivers in the United States; however, eight of those rivers had fewer than 10 observations. Only three rivers—Columbia, Mississippi, and Potomac—had petroleum hydrocarbon data.

Quantified estimates of oil and grease and petroleum hydrocarbon loadings were made for the United States and Canada. These estimates were made using unit loadings per urban land area. The annual loadings were calculated according to the coastal zones defined in this study, and the overall loadings for the United States and Canada were extrapolated to the world. For the calculations in the United States and Canada, the land-based sources were separated into two categories: inland basins and coastal basins. It was assumed that inland basins discharged into one of the following major river basins that outlet to the sea along the coast of the United States and Canada (coastal basins were assumed to discharge directly to the sea):

- Alabama–Tombigbee
- Apalachicola
- Altamaha
- Brazos
- Colorado (Texas)
- Columbia
- Copper (Alaska)
- Delaware
- Hudson
- James
- Mississippi
- Neuse
- Potomac
- Rio Grande
- Roanoke
- Sabine
- Sacramento
- San Joaquin
- Santee
- Saskatchewan
- Savannah
- St. Lawrence
- Susquehanna

C.1.2 Calculations for the Inland Rivers of the United States and Canada

The following methodology was used to estimate the loading of oil and grease to the sea from inland river basins in the United States and Canada:

1. Using the locations designated in *Oil in the Sea III* (see Table C.1), water quality data were requested from STORET. Using the Water Quality Portal, searches were made for all surface water quality data collected within these regions. Data with the Characteristic Group: Organics, Other and the following characteristics:
   - Oil and grease
   - Hydrocarbons, petroleum

2. Averages of all reported values from STORET were compiled for each river (see Table C.2) with the following assumptions (rivers not shown in Table C.2 did not have any usable oil and grease data):
   - Data were collected from 2000 onward
   - During calculations for each river, NA (not available) data were not included in the averages
3. An average annual load in tonne/yr was calculated for those rivers with reported oil and grease data by using the following formula:

   Equation C-1

   $L_i = c_i\, Q_i$,
   where L_i = average annual load for river i (tonne/yr),
   c_i = average oil and grease concentration for river i (mg/L),
   Q_i = average annual flow for river i (m^3/yr),
   tonne = 10^6 g.

   The average annual flow (per calendar year) was determined from U.S. Geological Survey (USGS) daily flow data available for each of the rivers at either the same station from *Oil in the Sea III* or the closest non-tidally influenced station to the collection site for the oil and grease samples (see Table C.3). The average annual flow of the major inland rivers is compared to the flows from 1980 to 1999, similar to the period of record in *Oil in the Sea III*, in order to assess the changes in river flow since the last report (see Table C-4).
4. Using data obtained from the State and Metropolitan Area Data Book (2010, 2020), unit loads per urban land area were calculated as follows:

   Equation C-2

   $$lai = \frac{Li(106 \text{ g/tonne})}{Aui}$$
   where lai = unit load per urban land area for river i (g/m^2yr),
   Aui = 2019 urban land area for river i (m^2)

   The 2019 urban land area in each river basin was determined by Table B-1 in *State and Metropolitan Area Data Book 2020*. Metropolitan areas in this table were partitioned in the major river basins identified in Table C.1, coastal areas, the Great Lakes, or areas not discharging to the coast of the United States or Canada (e.g., Great Salt Lake basin). Urban areas contributing to the Great Lakes fall within one of the counties defined by Indiana Business Research Center (2016). Metropolitan areas contributing urban runoff to the Great Lakes or areas not discharging to the coast of the United States or Canada were not included further in the analysis.
5. For the majority of the inland river basins, no usable oil and grease data were available in STORET. In addition, the number of non-NA observations for the Columbia, Delaware, James, Roanoke, Sacramento, San Joaquin, Susitna, and Susquehanna rivers were very small. It was therefore decided to use an alternative procedure based on the unit loads of oil and grease per urban land area and per capita calculated from Steps 1–4 to estimate the contributions of oil and grease from these other river basins. The procedure was as follows:
   a. The unit loads of oil and grease per urban land area calculated from Steps 1–4 were used for the other river basins with the following assumptions:
      - The Hudson, James, and Susquehanna rivers were assumed to have unit loads of oil and grease per urban land area of 12.34 g/m^2yr, the value calculated from four observations on the Delaware River (this small number of observations was deemed sufficient due to the consistency with the values of samples presented in *Oil in the Sea III*). The high unit loadings on the Delaware River are due to the highly industrialized nature of the waterway, and these three rivers are also very industrialized and in a similar geographic area.
      - It is assumed that the Alaskan rivers Copper and Susitna did not contribute significant loads of oil and grease to the ocean.
      - All other rivers for which the measured data were not adequate or were unavailable were assumed to have unit loads of oil and grease per urban land area of 0.15 g/m^2yr. This value was calculated from 404 non-NA observations on the Potomac. Rivers for which this value applied includes the Alabama–Tombigbee, Altamaha, Apalachicola, Brazos, Colorado (Texas), Columbia, Mississippi, Neuse, Rio Grande, Roanoke, Sabine, Sacramento, Santee, San Joaquin, Saskatchewan, Savannah, St. Lawrence, Trinity, and Yukon rivers.
   b. Using data obtained from the *State and Metropolitan Area Data Book* (2020) and Statistics Canada (2016), the annual loads per unit land area (Lai) were calculated as follows:

      Equation C-3

      $Lai = lai\ Aui$ (tonne/10^6 g)
      where lai was the unit load for river i as described in Step 5.a. The urban land area, Aui was calculated in the same manner as described in Step 4 for metropolitan areas

in the United States. For metropolitan areas in Canada, *Aui* was calculated using data from Statistics Canada (2016). The urban land area and population for each inland river was compared for the 1990s, 2000s, and 2010s to understand the comparative growth of urban land area and population within each river basin (see Table C.5).

C.1.3 Calculations for the Coastal Zones of the United States and Canada

For the United States, metropolitan areas in State and Metropolitan Area Data Book (2020) were classified as contributing to coastal basins if they fell within 1 of the 254 non–Great Lakes bordering coastal counties defined by Culliton et al. (1990). The individual coastal basin metropolitan areas were then aggregated into the appropriate coastal zones. The data for 2019 urban land area for metropolitan areas (*State and Metropolitan Area Data Book*, 2020) were then compiled for each coastal zone. Similarly, data from Statistics Canada (2016) for Canadian metropolitan areas along the coast were grouped into the appropriate coastal zones.

The annual load *Lai* was calculated for urban areas in each coastal zone *i* in the United States and Canada using Equation C-3. The unit load per urban land area for coastal zone *i*, *lai*, was 12.34 g/m^2yr for Coastal Zone D, and 0.15 g/m^2yr for all other coastal zones. The unit loads were set higher for Coastal Zone D because it is the coastal zone to which the Delaware River discharges.

The total oil and grease loading was determined by adding discharges from inland rivers and urban coastal areas to the appropriate coastal zones.

C.1.4 World Estimates of Oil and Grease

The data used for the calculations of oil and grease loading for North American were not available for other regions of the world. Therefore, a method to extrapolate the North American calculations to the rest of the world was used. It is widely thought that land-based contributions of oil and grease are due primarily to vehicle operation and maintenance (Fam et al., 1987; Hoffman and Quinn, 1987a,b; Latimer et al., 1990; Bomboi and Hernández, 1991; Zeng and Vista, 1997; Latimer and Quinn, 1998). Thus, oil and grease loading estimates for the world were based on the number of motor vehicles in different regions of the world as calculated by the motor vehicles per 1000 persons for each country and population of each country (World Bank, 2020). Oil and grease loading per vehicle was based on the calculations from *Oil in the Sea III*, 0.01573 tonne/vehicle yr. Redoing the original calculations with the new values found in this report produced a similar value of 0.01593 tonne/veh yr, so the value from the original report was used in order to compare more closely between the two reports.

Equation C-4

$$VEH_i = P_i veh_i n$$

where VEH_i = number of vehicles in country i,
P_i = population of country (World Bank, 2020),
veh_i = number of vehicles per capita in country i (Nation Master, 2014).

The number of vehicles in regions of the world was determined by applying Equation C-4 to country data and then compiling for each region. These numbers of vehicles were then multiplied by the loading per vehicle in North America calculated in *Oil in the Sea III* to obtain a world estimate of loading of oil and grease to the sea via land-based contributions. Because data on actual vehicle usage and maintenance in other countries were unavailable, it was assumed that the loadings of oil and grease per vehicle in North America were representative of oil and grease loadings per vehicle in other parts of the world. This assumption was considered reasonable because, while motor vehicles in other countries of the world are not as well maintained as vehicles in North America and therefore would likely contribute more oil and grease per vehicle while running, motor vehicles are less frequently used in other regions of the world.

C.1.5 Estimates of Petroleum Hydrocarbons and Polycyclic Aromatic Hydrocarbons

Within the STORET data, there were hydrocarbons and petroleum data for three rivers: Columbia, Mississippi, and Potomac (see Table C.2). In *Oil in the Sea III*, petroleum hydrocarbons loadings were estimated to be 1.5% of the oil and grease loadings. In contrast, the Potomac River data presents a proportion of petroleum hydrocarbons to oil and grease of about 90%. Without information to the contrary, the polycyclic aromatic hydrocarbon (PAH) proportion was considered to be 1% of the hydrocarbons and petroleum data as found in *Oil in the Sea III*.

C.2 RESULTS

The average annual loads of oil and grease discharged to the sea were calculated for those rivers with reported oil and grease data in STORET (see Table C.6). These total loads were then normalized to unit loads per urban land area. The final estimates of land-based contributions of oil and grease to the sea via all major inland river basins in the United States and Canada were then determined using the oil and grease data for the Delaware and Potomac rivers (see Table C.7) with urban land area data from *State and Metropolitan Area Data Book* (2020) and Statistics Canada (2016). About one-seventh of the estimated loading in North America was determined from actual measured data in STORET, with the remainder determined using the unit load approach.

The estimates of land-based contributions of oil and grease to the sea from both major inland rivers and coastal areas in the United States and Canada were totaled by coastal basin, based on the loads calculated in Table C.7 (see Table C.8). The total loading for North America (2.9 million tonne/yr; 4.5 million tonne/yr) was used to obtain a world estimate of land-based oil and grease loading (18.8 million tonne/yr; see Table C.10). The regional distribution of this loading shows that Europe, North America, and Asia contribute the majority of land-based oil and grease to the sea. The population and number of vehicle growth between the years 2000 and 2019 was also looked at for each region (see Table C.9).

Based on the calculations of *Oil in the Sea III*, a factor of 0.015 was applied to the total oil and grease loading to estimate the fraction of hydrocarbons in oil and grease. The estimated worldwide loading of hydrocarbons to the sea from land-based sources was 259,000 tonnes based on Table C.8 (see Table C.11). Based on *Oil in the Sea III*, a factor of 0.00015 was applied to the total oil and grease loading to estimate the fraction of PAH in oil and grease. The estimated worldwide loading of PAH to the sea from land-based sources was 2,584 tonne/yr based on Table C.8 (see Table C.11).

C.2.1 Discussion

The method used to estimate land-based oil and grease, hydrocarbon, and PAH contributions to the sea involved a large degree of uncertainty due to a number of factors, including (but not limited to):

- Lack of data; only 10 major rivers in the United States had oil and grease data in the U.S. Environmental Protection Agency's STORET database, and many of these consisted of very few observations.
- Estimating the proportion of petroleum-related hydrocarbons and PAH in oil and grease measurements; the data demonstrate a vastly different proportion of petroleum and hydrocarbon in oil and grease measurements in comparison to the original estimates in *Oil in the Sea III*.

Quantifying the uncertainty in the estimates presented in this analysis was not possible, but a reasonable estimate of the low and high ranges of the calculated oil and grease values was made (see Table C.12). The low estimate is the lowest unit load per urban land area, 0.15 for the Potomac River. The best estimate was either based on the calculation from available oil and grease data for the river (i.e., Columbia, Delaware, James, Potomac, Roanoke, Sacramento, San Joaquin, Savannah, Susitna, and Susquehanna) or, if those data were not available, an estimate of 1.25 from *Oil in the Sea III* was used. The estimate of 12.34 was the best estimate for Coastal D, Delaware, and Hudson. The high estimate is the highest unit load per urban land area, 15.88 for the Susquehanna River. Based on these estimates, the range of worldwide loadings of land-based sources of oil and grease to the sea was 7.4–82.5 million tonne/yr, with a best estimate of 21.1 million tonne/yr. For the vehicle-based calculations, the loading per vehicle is based on *Oil in the Sea III*. The low estimate for loading per vehicle is 0.007619, the best estimate is 0.01573, and the high estimate is 0.055799.

To estimate the range of total petroleum hydrocarbons, STORET data from the Columbia, Potomac, and Mississippi rivers was used. The low estimate was the lowest value for each region from Table C.8. The best estimate was built on the oil and grease and the hydrocarbon and petroleum STORET data, namely the average hydrocarbons and petroleum per urban land area and average oil and grease per urban land area (i.e., ~0.2). The high estimate is based on the proportion of Potomac oil and grease, and hydrocarbon and petroleum data (i.e., ~0.9). The best and high estimates were both proportions of the best estimates of oil and grease annual load (see Table C.12). The estimates of PAH followed suit, assuming PAH constitute 1% of total petroleum hydrocarbons (see Table C.13). The range of land-based petroleum hydrocarbon loading to the sea was 254,000–18,127,000 tonne/yr, with a best estimate of 4,028,000 tonne/yr. The range of PAH loading to the sea from land-based sources was 2,536–181,272 tonne/yr, with a best estimate of 40,281 tonne/yr.

C.2.2 Comparison of Estimates of Land-Based Loading with Other Estimates

The calculations of oil and grease loadings presented in this analysis were based on unit loadings per urban land area. Comparison calculations were also made based on unit loadings per capita urban population using 2019 urban populations in the United States obtained from the *State and Metropolitan Area Data Book* (2020) and 2016 urban populations in Canada from Statistics Canada (2016). These calculations resulted in oil and grease loadings of the same magnitude as calculations based on unit loadings per urban land area (see Table C.14).

The estimates of the land-based loadings of oil and grease were compared to global and regional oil consumption. According to BP Amoco (2020), North America consumed 1,029 million tonnes of oil in 2019. (The 2019 data was used for comparison, rather than 2020, due to the changes in global oil consumption due to the COVID-19 pandemic.) Assuming that all of the 5.8 million tonne/yr of oil and grease estimated in this study as returning to the sea from land-based sources were petroleum-derived, then only about 0.56% of consumed oil was returned to the sea from land-based sources. Furthermore, BP Amoco (2020) estimated that the North American annual consumption of oil was broken down as follows:

- Light Distillates (Gasoline, Naphtha): 519.3 million tonne/yr
- Middle Distillates (Diesel, Kerosene): 359.5 million tonne/yr
- Fuel Oil: 20.2 million tonne/yr
- Other: 312.6 million tonne/yr
- Total: 1211.6 million tonne/yr

Table C.15 shows comparisons of the computed land-based loads presented in the current study for North America and other regions with the BP Amoco (2020) data.

The best estimate of petroleum hydrocarbon loading from land-based sources was about 3 times as large as the best estimate from the National Research Council (1985), and 28 times as large as *Oil in the Sea III* (see Table C.16). These discrepancies indicate an increase in oil and grease loadings in the past two decades.

TABLE C.1 Regions Searched for Oil and Grease and Hydrocarbon Data from STORET

River	Latitude	Longitude	Radius (mi)
Alabama-Tombigbee	32°00'00", 30°00'00"	−87°15'00", −88°15'00"	[a]
Altamaha	32°31'30"	−81°15'45"	50
Apalachicola	[b]		
Brazos	29°34'56"	−95°45'27"	50
Colorado (Texas)	28°58'26"	−96°00'44"	30
Columbia	46°10'55"	−123°10'50"	50
Copper (Alaska)	61°00'00"	−144°45'00"	50
Delaware	39°30'03"	−75°34'07"	30
Hudson	41°43'18"	−73°56'28"	40
James	37°24'00"	−77°18'00"	50
Mississippi	29°16'26"	−89°21'00"	50
Neuse	35°06'33"	−77°01'59"	50
Potomac	38°55'46"	−77°07'02"	75
Rio Grande	25°52'35"	−97°27'15"	30
Roanoke	35°54'54"	−76°43'22"	70
Sabine	30°18'13"	−93°44'37"	50
Sacramento	37°30'00", 38°30'00"	−121°00'00", −123°00'00"	[a]
San Joaquin	37°30'00", 38°30'00"	−121°00'00", −123°00'00"	[a]
Santee	33°14'00"	−79°30'00"	40
Saskatchewan	[b]		
Savannah	32°31'30"	−81°15'45"	50
St. Lawrence	45°00'22"	−74°47'43"	50
Susitna	61°35'00"	−150°22'00"	40
Susquehanna	39°42'00"	−76°15'00"	50
Trinity	29°50'10"	−94°44'57"	30
Yukon	62°45'00"	−164°30'00"	30

[a] Rectangular polygons formed by the latitudinal and longitudinal coordinates were searched for these rivers
[b] No data were requested for the Appalachicola and Saskatchewan Rivers.

TABLE C.2 STORET Data Used to Calculate Average Oil and Grease; Hydrocarbons, Petroleum Concentration in Major Inland Rivers

River	Monitoring Location	Characteristic Name	Number of Observations (non-NA)	Date(s) of Observations	Average Concentration (mg/L)
Columbia	River/Stream at Superfund site	Hydrocarbons, Petroleum	249 (249)	4/27/04–11/30/05	1.35
		Oil and Grease	6 (6)	11/02/04–4/25/05	5.03
Delaware	River/Stream 2,000 yds up buoy	Oil and Grease	14 (4)	4/24/00–1/26/05	12.55
James	Logging runoff from land	Oil and Grease	6 (6)	4/19/01–4/28/15	6.83
Mississippi	Hurricane Rita/Urban Floodwater	Hydrocarbons, Petroleum	4 (2)	9/30/05–10/09/05	1.1
Potomac	Peter Pan Run Stream	Hydrocarbons, Petroleum	404 (404)	1/10/00–9/21/06	1.43
		Oil and Grease	413 (404)	1/10/00–9/21/06	1.55
Roanoke		Oil and Grease	2 (2)	5/10/01–8/06/01	5
Sacramento		Oil and Grease	20 (1)	1/24/12–3/26/14	0.68
San Joaquin	DELS	Oil and Grease	6 (6)	4/26/00	0
Savannah	Stream BMP intake	Oil and Grease	101 (82)	4/07/05–11/07/06	9.74
Susitna	Ship Creek, River/Stream	Oil and Grease	10 (5)	5/20/11–9/28/11	2.04
Susquehanna		Oil and Grease	39 (6)	2/15/001/26/05	11.9

TABLE C.3 Calculations of Average Annual Flows for Major Inland Rivers

River	Station Name	Period of Record Used	Average Annual Flow (m^3/yr)
Alabama–Tombigbee	02469761: Tombigbee River at Coffeeville L&D near Coffeeville, AL	2000–2012; 2014–2015; 2017–2019	24,596,158,257
Altamaha	02226000: Altamaha River at Doctortown, GA	2000–2020	9,894,382,025
Brazos	08114000: Brazos River at Richmond, TX	2000–2020	7,256,268,535
Colorado (Texas)	08162500: Colorado River near Bay City, TX	2000–2007	2,565,942,193
Columbia	14246900: Columbia River at Port Westward, near Quincy, OR	2000–2019	204,666,350,394
Copper (Alaska)	15214000: Copper River at Million Dollar Bridge near Cordova, AK	2010–2011; 2017–2019	59,627,302,886
Delaware	01463500: Delaware River at Trenton, NJ	2000–2019	12,109,284,340
Hudson	01335754: Hudson River above Lock 1 near Waterford, NY	2000–2019	8,257,334,855
James	02037500: James River Near Richmond, VA	2000–2020	6,636,808,160
Mississippi	07289000: Mississippi River at Vicksburg, MS	2000–2020	689,140,762,304
Neuse	02089500: Neuse River at Kinston, NC	2000–2015; 2017–2020	2,391,360,966
Potomac	01608500: South Branch Potomac River near Springfield, WV	2000–2020	1,274,215,458
Rio Grande	08375300: Rio Grande at Rio Grande Village, Big Bend NP, TX	2008–2019	682,154,193
Roanoke	02080500: Roanoke River at Roanoke Rapids, NC	2000–2020	7,115,455,412
Sabine	08030500: Sabine River near Ruliff, TX	2000–2020	7,064,129,254
Sacramento	11425500: Sacramento River at Verona, CA	2000–2020	15,926,970,319
San Joaquin	11303500: San Joaquin R near Vernalis, CA	2000–2020	5,720,145,091
Santee	02198500: Savannah River near Cylo, GA	2000–2020	3,107,890,290
Savannah	02171645: Rediv Canal at Santee River near St. Stephen, SC	2000–2020	8,292,980,383
Susitna	15292000: Susitna River at Gold Creek, AK	2002–2019	9,086,707,310
Susquehanna	01576000: Susquehanna River at Marietta, PA	2000–2020	36,942,927,072
Trinity	08066500: Trinity River at Romayor, TX	2000–2020	8,261,002,528
Yukon	15565447: Yukon River at Pilot Station, AK	2002–2019	210,866,737,379

TABLE C.4 Comparison and Change of Average Annual Flow

River	Station Name	Period of Record Used	Average Annual Flow (m^3/yr)	Percent Change
Alabama–Tombigbee	02469761: Tombigbee River at Coffeeville L&D near Coffeeville, AL	1980–1999	26,969,450,885	
		2000–2012; 2014–2015; 2017–2019	24,596,158,257	−8.80%
Altamaha	02226000: Altamaha River at Doctortown, GA	1980–1999	12,222,829,113	
		2000–2020	9,894,382,025	−19.05%
Brazos	08114000: Brazos River at Richmond, TX	1980–1999	7,083,041,689	
		2000–2020	7,256,268,535	2.45%
Colorado (Texas)	08162500: Colorado River near Bay City, TX	1980–1999	2,639,422,583	
		2000–2007	2,565,942,193	−2.78%
Columbia	14246900: Columbia River at Port Westward, near Quincy, OR	1992–1999	225,024,486,103	
		2000–2019	204,666,350,394	−9.05%
Copper (Alaska)	15214000: Copper River at Million Dollar Bridge near Cordova, AK	1989–1994	56,287,190,436	
		2010–2011; 2017–2019	59,627,302,886	5.93%
Delaware	01463500: Delaware River at Trenton, NJ	1980–1999	10,046,323,612	
		2000–2019	12,109,284,340	20.53%
Hudson	01335754: Hudson River above Lock 1 near Waterford, NY	1980–1999	7,065,137,067	
		2000–2019	8,257,334,855	16.87%
James	02037500: James River Near Richmond, VA	1980–1999	6,606,492,983	
		2000–2020	6,636,808,160	0.46%
Mississippi	07032000: Mississippi River at Memphis, TN	1980–1994	492,828,070,403	
	07289000: Mississippi River at Vicksburg, MS	2000–2020	689,140,762,304	39.83%
Neuse	02089500: Neuse River at Kinston, NC	1983–1999	2,643,275,872	
		2000–2015; 2017–2020	2,391,360,966	−9.53%
Potomac	01608500: South Branch Potomac River near Springfield, WV	1980–1999	1,327,513,867	
		2000–2020	1,274,215,458	−4.01%
Rio Grande	08361000: Rio Grande Below Elephant Butte Dam, NM	1980–1999	1,001,672,071	
	08375300: Rio Grande at Rio Grande Village, Big Bend NP, TX	2008–2019	682,154,193	−31.90%
Roanoke	02080500: Roanoke River at Roanoke Rapids, NC	1980–1999	7,469,397,536	
		2000–2020	7,115,455,412	−9.82%
Sabine	08030500: Sabine River near Ruliff, TX	1980–1999	7,890,714,278	
		2000–2020	7,064,129,254	−10.48%
Sacramento	11425500: Sacramento River at Verona, CA	1980–1999	18,506,011,951	
		2000–2020	15,926,970,319	−13.94%
San Joaquin	11303500: San Joaquin River near Vernalis, CA	1980–1999	4,962,946,910	
		2000–2020	3,107,890,290	−37.38%
Santee	02171645: Rediv Canal at Santee River near St. Stephen, SC	1987–1999	8,653,431,256	
		2000–2020	5,720,145,091	−33.90%
Savannah	02198500: Savannah River near Cylo, GA	1980–1985; 1987–1999	10,554,679,518	
		2000–2020	8,292,980,383	−21.43%
Susitna	15292000: Susitna River at Gold Creek, AK	1980–1986	8,904,929,565	
		2002–2019	9,086,707,310	2.04%
Susquehanna	01576000: Susquehanna River at Marietta, PA	1980–1999	32,899,408,125	
		2000–2020	36,942,927,072	12.29%
Trinity	08066500: Trinity River at Romayor, TX	1980–1999	8,461,251,087	
		2000–2020	8,261,002,528	−2.37%
Yukon	15565447: Yukon River at Pilot Station, AK	1980–1995	209,056,554,789	
		2002–2019	210,866,737,379	0.86%
Total		1980–1999	1,169,104,231,699	
		2000–2020	1,341,483,269,604	14.74%

TABLE C.5 Percent Change of Urban Land Area and Population in River Basins

River	Decade	Urban Land Area (m^2)	Percent Change	Population	Percent Change
Alabama–Tombigbee	1990	19,848,114,806		1,601,369	
	2000	35,059,114,917		2,270,148	
	2010	35,197,938,279	77%	2,339,409	46%
Altamaha	1990	5,498,803,735		454,600	
	2000	8,114,173,718		553,202	
	2010	8,482,211,028	54%	629,155	38%
Apalachicola	1990	21,708,503,258		4,016,893	
	2000	32,739,780,573		6,013,671	
	2010	34,848,289,886	61%	6,692,579	67%
Brazos	1990	14,384,534,909		987,859	
	2000	27,235,019,866		1,246,704	
	2010	31,468,355,415	119%	1,493,268	51%
Colorado (Texas)	1990	14,887,769,596		1,173,671	
	2000	17,605,703,109		1,762,165	
	2010	19,992,118,114	34%	2,349,110	100%
Columbia	1990	30,466,548,018		1,263,460	
	2000	101,947,370,591		2,316,164	
	2010	123,850,640,954	307%	2,953,885	134%
Delaware	1990	5,082,592,647		967,893	
	2000	6,004,369,412		1,211,805	
	2010	12,315,393,416	142%	3,419,661	253%
Hudson	1990	21,972,423,045		1,432,124	
	2000	23,837,991,474		1,587,549	
	2010	23,789,040,699	8%	1,598,036	12%
James	1990	7,686,825,682		354,043	
	2000	9,722,802,098		440,200	
	2010	9,751,305,197	27%	482,181	36%
Mississippi	1990	464,341,095,531		39,900,057	
	2000	654,764,272,610		48,226,561	
	2010	699,320,096,893	51%	53,628,488	34%
Neuse	1990	10,472,875,881		1,162,035	
	2000	11,483,230,239		1,692,198	
	2010	12,851,520,952	23%	2,158,283	86%
Potomac	1990	1,950,520,038		99,122	
	2000	6,960,852,018		339,811	
	2010	12,880,010,821	560%	793,235	700%
Rio Grande	1990	43,843,577,556		1,410,081	
	2000	52,220,376,067		1,689,829	
	2010	54,583,999,208	24%	1,912,049	36%
Roanoke	1990	4,830,068,808		337,136	
	2000	7,478,590,639		403,891	
	2010	7,461,755,716	54%	414,864	23%
Sabine	1990	6,964,737,000		374,973	
	2000	7,028,191,708		406,023	
	2010	9,326,547,148	34%	519,408	39%
Sacramento	1990	30,438,835,145		2,152,519	
	2000	33,061,198,097		2,873,315	
	2010	33,037,888,204	9%	3,159,043	47%
San Joaquin	1990	45,968,921,790		2,382,323	
	2000	49,560,458,289		3,062,479	
	2010	49,554,242,317	8%	3,366,051	41%
Santee	1990	26,807,930,828		3,183,877	
	2000	32,722,945,651		3,985,324	
	2010	40,737,922,825	52%	5,225,755	64%
Savannah	1990	6,342,621,858		457,228	
	2000	8,492,052,982		534,218	
	2010	9,015,748,576	42%	608,980	33%
Susquehanna	1990	27,400,261,106		2,788,354	
	2000	26,140,749,893		2,855,770	
	2010	27,676,612,837	1%	3,071,255	10%
Trinity	1990	23,581,064,654		4,683,013	
	2000	23,283,216,023		6,300,006	
	2010	22,465,556,779	25%	7,573,136	62%

TABLE C.6 Calculated Annual Load and Unit Load per Urban Land Area for Major Inland Rivers from STORET Data

River	Average Annual Flow (m^3/yr) 2000–2020	Average Concentration (mg/L)	Average Annual Load (tonne/yr)	Urban Land Area (m^2) (2019)	Unit Load per Urban Land Area (g/m^2/yr)
Columbia	204,666,350,394	5.03	1,029,471.74	123,850,640,954	8.31
Delaware	12,109,284,340	12.55	151,971.52	12,315,393,416	12.34
James	6,636,808,160	6.83	45,329.4	9,751,305,197	4.65
Potomac	1,274,215,458	1.55	1,975.03	12,880,010,821	0.15
Roanoke	7,115,455,412	5	35,577.28	7,461,755,716	4.77
Sacramento	15,926,970,319	0.68	10,830.34	33,037,888,204	0.33
San Joaquin	3,107,890,290	0	0	49,554,242,317	0
Savannah	8,292,980,383	9.74	80,773.63	9,015,748,576	8.96
Susitna	9,086,707,310	2.04	18,536.88	0	0
Susquehanna	36,942,927,072	11.9	439,620.83	27,676,612,837	15.88

TABLE C.7 Estimates of Land-Based Contributions of Oil and Grease to the Sea via Major Inland River Basins

River	Number of Non-NA Observations	Average Concentration of Oil and Grease, (mg/L)	Average Annual Flow (m^3/yr)	Urban Land Area in Watershed (m^2)	Annual Load (tonne/yr)	Unit Load per Urban Land Area (g/m^2/yr)
Calculated from STORET data						
Delaware	4	12.55	12.1×10^9	12.3×10^9	151,972	12.34
Potomac	404	1.55	1.3×10^9	12.9×10^9	1,975	0.15
Subtotal				25.2×10^9	153,947	
Calculated from alternative method						
Alabama–Tombigbee				35.2×10^9	5,280	0.15
Altamaha				8.5×10^9	1,275	0.15
Apalachicola				35.8×10^9	5,370	0.15
Brazos				31.5×10^9	4,725	0.15
Colorado (Texas)				20×10^9	3,000	0.15
Columbia				123.9×10^9	18,585	0.15
Copper (Alaska)				0	0	0
Hudson				23.8×10^9	293,692	12.34
James				9.8×10^9	120,932	12.34
Mississippi				699.3×10^9	104,895	0.15
Neuse				12.9×10^9	1,935	0.15
Rio Grande				54.6×10^9	8,190	0.15
Roanoke				7.5×10^9	1,125	0.15
Sabine				9.3×10^9	1,395	0.15
Sacramento				33×10^9	4,590	0.15
Santee				40.7×10^9	6,105	0.15
San Joaquin				49.6×10^9	7,440	0.15
Saskatchewan				2.1×10^9	315	0.15
Savannah				9×10^9	1,350	0.15
St. Lawrence				19.7×10^9	2,955	0.15
Susitna				0	0	0
Susquehanna				27.7×10^9	341,818	12.34
Trinity				22.5×10^9	3,375	0.15
Yukon				19×10^9	2,850	0.15
Subtotal				$1,295.4 \times 10^9$	941,197	
Average						2.01
Total				$1,320.6 \times 10^9$	1,095,144	

TABLE C.8 Estimate of Land-Based Oil and Grease to the Sea by Coastal Zone Based on Table C.7

Coastal Zone	Description	Urban Population in Watershed, *Pi*	Urban Land Area in Watershed, *Aui* (m^2)	Annual Load, *Lai* (tonne/yr)	z
A	No urban areas	0	0	0	0
B	Coastal	0	0	0	0
	Saskatchewan	3,009,130	2,113,000,000	317	0.15
	Subtotal	3,009,130	2,113,000,000	317	
C	Coastal	1,680,653	1,555,000,000	233	0.15
	St. Lawrence	6,278,758	19,676,169,378	2,951	0.15
	Coastal	7,959,411	21,231,169,378	3,184	
D	Coastal	51,808,560	139,545,968,840	1,721,998	12.34
	Delaware	3,419,661	12,315,393,416	151,967	12.34
	Hudson	1,598,036	23,789,040,699	293,556	12.34
	James	482,181	9,751,305,197	120,327	12.34
	Potomac	793,235	12,880,010,821	1,932	0.15
	Susquehanna	3,071,255	27,676,612,837	341,534	12.34
	Subtotal	61,172,928	225,958,331,810	2,631,314	
E	Coastal	17,810,577	102,074,021,009	15,311	0.15
	Altamaha	629,155	8,482,211,028	1,272	0.15
	Neuse	2,158,283	12,851,520,952	1,928	0.15
	Roanoke	414,864	7,461,755,716	1,119	0.15
	Santee	5,225,755	40,737,922,825	6,111	0.15
	Savannah	608,980	9,015,748,576	1,352	0.15
	Subtotal	26,847,614	180,623,180,106	27,093	
F	Coastal	8,252,789	52,675,177,978	7,901	0.15
	Alabama–Tombigbee	2,339,409	35,197,938,279	5,280	0.15
	Apalachicola	6,692,579	34,848,289,886	5,227	0.15
	Subtotal	17,284,777	122,721,406,143	18,408	
G	Coastal	12,773,639	97,880,830,275	14,682	0.15
	Brazos	1,493,268	31,468,355,415	4,720	0.15
	Colorado (Texas)	2,349,110	19,992,118,144	2,999	0.15
	Mississippi	53,628,488	699,320,096,893	104,898	0.15
	Rio Grande	1,912,049	54,583,999,208	8,188	0.15
	Sabine	519,408	9,326,547,148	1,399	0.15
	Trinity	7,573,136	22,465,556,779	3,370	0.15
	Subtotal	80,249,098	935,037,503,862	140,256	
I	No urban areas	0	0	0	0
K	Coastal	22,049,766	98,888,335,646	14,833	0.15
L	Coastal	9,239,070	46,772,595,098	7,016	0.15
	Sacramento	3,159,043	33,037,888,204	4,956	0.15
	San Joaquin	3,366,051	49,554,242,317	7,433	0.15
	Subtotal	15,764,164	129,364,725,619	19,405	
M	Coastal	8,542,083	72,757,945,705	10,914	0.15
	Columbia	2,953,885	123,850,640,954	18,578	0.15
	Subtotal	11,495,968	196,608,586,659	29,492	
N	Coastal	1,141,980	4,566,149,020	685	0.15
O	Coastal	3,011,719	1,367,000,000	205	0.15
P	Coastal	396,317	68,430,075,590	10,265	0.15
	Copper	0	0	0	0
	Susitna	0	0	0	0
	Subtotal	396,317	68,430,075,590	10,265	
Q	Coastal	0	0	0	0
	Yukon	96,849	18,984,612,773	2,848	0.15
	Subtotal	96,849	18,984,612,773	2,848	
	Total	250,479,721	$2,005.9 \times 10^9$	2,880,805	

TABLE C.9 Percent Change of Global Population and Number of Vehicles

Region	Year	Population	Population Percent Change	Number of Vehicles	Vehicle Percent Change
Africa	2000	778,484,000		15,569,680	
	2019	1,306,033,375	68%	44,902,492	188%
Europe	2000	729,406,000		196,939,620	
	2019	743,131,357	2%	346,027,255	76%
North America	2000	304,078,000		218,936,160	
	2019	365,889,132	20%	284,430,084	30%
Central America	2000	130,710,000		14,378,100	
	2019	217,693,617	67%	44,441,062	209%
South America	2000	331,889,000		29,870,010	
	2019	428,615,774	29%	84,113,596	182%
Asia	2000	3,588,877,000		107,666,310	
	2019	4,566,180,071	27%	370,533,424	244%
Oceania	2000	29,460,000		12,667,800	
	2019	42,461,759	44%	22,194,093	75%
Total	2000	5,892,904,000		596,027,680	
	2019	7,670,005,085	30%	1,196,642,006	101%

TABLE C.10 World Estimates of Land-Based Sources of Oil and Grease to the Sea

Region	Population	Motor Vehicles per Capita	Number of Vehicles	Loading per Vehicle	Loading (tonne/year)
Africa	1,306,033,375	0.03	44,902,492	0.01573	706,316
Europe	743,131,357	0.47	346,027,255	0.01573	5,443,009
North America	365,889,132	0.78	284,430,084	0.01573	4,474,085
Central America	217,693,617	0.204	44,441,062	0.01573	699,058
South America	428,615,774	0.196	84,113,596	0.01573	1,323,107
Asia	4,566,180,071	0.08	370,533,424	0.01573	5,828,491
Oceania	42,461,759	0.52	22,194,093	0.01573	349,113
Total	7,670,005,085		1,196,642,006	0.01573	18,823,179

TABLE C.11 Estimates of Worldwide Land-Based Contributions of Hydrocarbons and PAH to the Sea Based on Table C.8

World Region	Coastal Zone	Description	Hydrocarbon (tonne/year)	PAH (tonne/year)
North America	A	No urban area	0	0
	B	Coastal	0	0
		Saskatchewan	5	0
		Subtotal	5	0
	C	Coastal	3	0
		St. Lawrence	44	0
		Subtotal	47	0
	D	Coastal	25,830	258
		Delaware	2,280	23
		Hudson	4,403	44
		James	1,805	18
		Potomac	29	0
		Susquehanna	5,123	51
		Subtotal	39,470	394
	E	Coastal	230	2
		Altamaha	19	0
		Neuse	29	0
		Roanoke	17	0
		Santee	92	1
		Savannah	20	0
		Subtotal	407	3
	F	Coastal	119	1
		Alabama–Tombigbee	79	1
		Apalachicola	78	1
		Subtotal	276	3
	G	Coastal	220	2
		Brazos	71	1
		Colorado (TX)	45	0
		Mississippi	1,573	16
		Rio Grande	123	1
		Sabine	21	0
		Trinity	51	1
		Subtotal	2,104	21
	I	No urban areas	0	0
	K	Coastal	222	2
	L	Coastal	105	1
		Sacramento	74	1
		San Joaquin	111	1
		Subtotal	290	3
	M	Coastal	164	2
		Columbia	279	3
		Subtotal	443	5
	N	Coastal	10	0
	O	Coastal	3	0
	P	Coastal	154	2
		Copper	0	0
		Susitna	0	0
		Subtotal	154	2
	Q	Coastal	0	0
		Yukon	43	0
		Subtotal	43	0
	Subtotal		43,474	433
Africa			10,595	106
Europe			81,645	816
Central America			10,486	105
South America			19,847	198
Asia			87,427	874
Oceania			5,237	52
TOTAL			258,711	2,584

TABLE C.12 Ranges of Worldwide Land-Based Contributions of Oil and Grease to the Sea

World Region	Coastal Zone	Description	Unit Load Based on Urban Area (g/m²/yr)			Annual Load (tonne/yr)		
			Low	Best Estimate	High	Low	Best Estimate	High
North America	A	No urban area	0	0	0	0	0	0
	B	Coastal	0	0	0	0	0	0
		Saskatchewan	0.15	1.25	15.88	317	2,641	33,554
		Subtotal				317	2,641	33,554
	C	Coastal	0.15	1.25	15.88	233	1,944	24,693
		St. Lawrence	0.15	1.25	15.88	2,951	24,595	312,455
		Subtotal				3,184	26,539	337,148
	D	Coastal	0.15	12.34	15.88	20,932	1,721,998	2,215,990
		Delaware	12.34	12.34	12.34	151,967	151,967	151,967
		Hudson	0.15	12.34	15.88	3,568	293,556	377,769
		James	0.15	4.65	15.88	1,463	45,342	154,846
		Potomac	0.15	0.15	0.15	1,932	1,932	1,932
		Susquehanna	0.15	15.88	15.88	4,152	439,511	439,511
		Subtotal				184,014	2,654,306	3,342,015
	E	Coastal	0.15	1.25	15.88	15,311	127,593	1,620,935
		Altamaha	0.15	1.25	15.88	1,326	11,053	140,411
		Neuse	0.15	1.25	15.88	1,928	16,065	204,090
		Roanoke	0.15	4.77	15.88	1,119	35,594	118,497
		Santee	0.15	1.25	15.88	6,111	50,923	646,919
		Savannah	0.15	8.96	15.88	1,352	80,783	143,174
		Subtotal				27,147	322,011	2,874,026
	F	Coastal	0.15	1.25	15.88	7,901	65,844	836,479
		Ala-Tom	0.15	1.25	15.88	5,280	43,998	558,944
		Apalachicola	0.15	1.25	15.88	5,227	43,560	553,386
		Subtotal				18,408	153,402	1,948,859
	G	Coastal	0.15	1.25	15.88	14,682	122,351	1,554,350
		Brazos	0.15	1.25	15.88	4,720	39,335	499,712
		Colorado (TX)	0.15	1.25	15.88	2,999	24,990	317,473
		Mississippi	0.15	1.25	15.88	104,898	847,150	11,105,202
		Rio Grande	0.15	1.25	15.88	8,188	68,230	866,794
		Sabine	0.15	1.25	15.88	1,399	11,659	148,113
		Trinity	0.15	1.25	15.88	3,370	28,083	356,760
		Subtotal				140,256	1,141,798	14,848,404
	I	No urban areas	0	0	0	0	0	0
	K	Coastal	0.15	1.25	15.88	14,833	123,610	1,570,341
	L	Coastal	0.15	1.25	15.88	7,016	58,466	742,755
		Sacramento	0.15	0.33	15.88	4,956	10,903	524,643
		San Joaquin	0.15	1.25	15.88	7,433	61,943	786,918
		Subtotal				19,405	131,312	2,054,336
	M	Coastal	0.15	1.25	15.88	10,914	90,948	1,155,397
		Columbia	0.15	8.31	15.88	18,578	1,029,202	1,966,754
		Subtotal				29,492	1,120,150	3,122,151
	N	Coastal	0.15	1.25	15.88	685	5,708	72,508
	O	Coastal	0.15	1.25	15.88	205	1,709	21,708
	P	Coastal	0.15	1.25	15.88	10,265	85,538	1,086,668
		Copper	0	0	0	0	0	0
		Susitna	0	0	0	0	0	0
		Subtotal				10,265	85,538	1,086,668
	Q	Coastal	0	0	0	0	0	0
		Yukon	0.15	1.25	15.88	2,848	23,731	301,482
		Subtotal				2,848	23,731	301,482
	Subtotal					451,059	5,792,445	31,613,200
Africa						342,112	706,316	2,505,514
Europe						2,636,381	5,443,009	19,307,975
Central America						338,597	699,058	2,479,767
South America						640,862	1,323,107	4,693,455
Asia						2,823,094	5,828,491	20,675,395
Oceania						169,097	349,113	1,238,408
TOTAL						7,401,202	20,141,549	

TABLE C.13 Ranges of Worldwide Land-Based Contributions of Hydrocarbons and PAH to the Sea

			Hydrocarbons (tonne/yr)			PAH (tonne/yr)		
World Region	Coastal Zone	Description	Low	Best Estimate	High	Low	Best Estimate	High
North America	A	No urban area	0	0	0	0	0	0
	B	Coastal	0	0	0	0	0	0
		Saskatchewan	5	528	2,377	0	5	24
		Subtotal	5	528	2,377	0	5	24
	C	Coastal	3	389	1,750	0	4	18
		St. Lawrence	44	4,919	22,136	0	49	221
		Subtotal	47	5,308	23,886	0	53	239
	D	Coastal	25,579	344,400	1,549,798	256	3,444	15,498
		Delaware	2,257	30,393	136,770	23	304	1,368
		Hudson	4,403	58,711	264,200	44	587	2,642
		James	1,787	9,068	40,808	18	91	408
		Potomac	29	386	1,739	0	4	17
		Susquehanna	519	87,902	395,560	5	879	3,956
		Subtotal	34,574	530,860	2,388,875	346	5,309	23,889
	E	Coastal	230	25,519	114,834	2	255	1,148
		Altamaha	19	2,211	9,948	0	22	99
		Neuse	29	3,213	14,459	0	32	145
		Roanoke	17	7,119	32,035	0	71	320
		Santee	92	10,185	45,831	1	102	458
		Savannah	20	16,157	72,705	0	162	727
		Subtotal	407	64,414	289,812	3	644	2,897
	F	Coastal	88	13,169	59,260	1	132	593
		Ala-Tom	79	8,800	39,598	1	88	396
		Apalachicola	78	8,712	39,204	1	87	392
		Subtotal	245	30,681	138,062	3	307	1,381
	G	Coastal	220	24,470	110,116	2	245	1,101
		Brazos	71	7,867	35,402	1	79	354
		Colorado (TX)	45	4,998	22,491	0	50	225
		Mississippi	1,573	169,430	762,435	16	1,694	7,624
		Rio Grande	123	13,646	61,407	1	136	614
		Sabine	21	2,332	10,493	0	23	105
		Trinity	51	5,617	25,274	1	56	252
		Subtotal	2,104	228,360	1,027,618	21	2,283	10,275
	I	No urban areas	0	0	0	0	0	0
	K	Coastal	222	24,722	111,249	2	247	1,112
	L	Coastal	105	11,693	52,619	1	117	526
		Sacramento	74	2,181	9,813	1	22	98
		San Joaquin	111	12,389	55,749	1	124	557
		Subtotal	290	26,263	118,181	3	263	1,181
	M	Coastal	164	18,190	81,853	2	182	819
		Columbia	279	205,840	926,282	3	2,058	9,263
		Subtotal	443	224,030	1,008,135	5	2,240	10,082
	N	Coastal	10	1,142	5,137	0	11	51
	O	Coastal	3	342	1,538	0	3	15
	P	Coastal	154	17,108	76,984	2	171	770
		Copper	0	0	0	0	0	0
		Susitna	0	0	0	0	0	0
		Subtotal	154	17,108	76,984	2	171	770
	Q	Coastal	0	0	0	0	0	0
		Yukon	41	4,746	21,358	0	47	214
		Subtotal	41	4,746	21,358	0	47	214
	Subtotal		38,545	1,158,504	5,213,212	385	11,583	52,130
Africa			10,595	141,263	635,684	106	1,413	6,357
Europe			81,645	1,088,602	4,898,708	816	10,886	48,987
Central America			10,486	139,812	629,152	105	1,398	6,292
South America			19,847	264,621	1,190,796	198	2,646	11,908
Asia			87,427	1,165,698	5,245,642	874	11,657	52,456
Oceania			5,237	69,823	314,202	52	698	3,142
TOTAL			253,782	4,028,323	18,127,396	2,536	40,281	181,272

TABLE C.14 Comparison of Estimates of Worldwide of Land-Based Oil and Grease Based on Population and Area

World Region	Coastal Zone	Description	Annual Load Based on Population (tonne/yr)	Annual Load Based on Area (tonne/yr)
North America	A	No urban area	0	0
	B	Coastal	0	0
		Saskatchewan	7,329	317
		Subtotal	7,329	317
	C	Coastal	4,093	233
		St. Lawrence	15,293	2,951
		Subtotal	19,386	3,184
	D	Coastal	2,302,331	1,721,998
		Delaware	151,967	151,967
		Hudson	71,015	293,556
		James	45,342	45,342
		Potomac	1,932	1,932
		Susquehanna	439,511	439,511
		Subtotal	3,012,098	2,654,306
	E	Coastal	43,526	15,311
		Altamaha	1,532	1,272
		Neuse	5,257	1,928
		Roanoke	35,594	35,594
		Santee	12,728	6,111
		Savannah	80,783	80,783
		Subtotal	179,420	140,999
	F	Coastal	21,100	7,901
		Alabama–Tombigbee	5,698	5,280
		Apalachicola	16,300	5,227
		Subtotal	43,098	18,408
	G	Coastal	31,111	14,682
		Brazos	3,637	4,720
		Colorado (TX)	5,721	2,999
		Mississippi	130,617	104,898
		Rio Grande	4,657	8,188
		Sabine	1,265	1,399
		Trinity	18,445	3,370
		Subtotal	195,453	140,256
	I	No urban areas	0	0
	K	Coastal	53,704	14,833
	L	Coastal	22,502	7,016
		Sacramento	10,903	10,903
		San Joaquin	8,198	7,433
		Subtotal	41,603	25,352
	M	Coastal	20,805	10,914
		Columbia	1,029,202	1,029,202
		Subtotal	1,050,007	1,040,116
	N	Coastal	2,781	685
	O	Coastal	7,335	205
	P	Coastal	965	10,265
		Copper	0	0
		Susitna	0	0
		Subtotal	965	10,265
	Q	Coastal	0	0
		Yukon	236	2,848
		Subtotal	236	2,848
	Subtotal		4,613,415	4,051,774
Africa			1,027,411	706,316
Europe			7,917,425	5,443,009
Central America			1,016,853	699,058
South America			1,924,597	1,323,107
Asia			8,478,149	5,828,491
Oceania			507,822	349,113
TOTAL			25,485,672	18,400,868

TABLE C.15 Comparison of Oil Consumption with Estimated Oil and Grease Loading from Land-Based Sources to the Sea

Location	2019 Oil Consumption (million tonne/yr)	Oil and Grease Loading to the Sea from Land-Based Sources (million tonne/yr)	Ratio of Oil and Grease Loading to the Sea to Oil Consumption (percent)
North America	1,029	5.8	0.56
South and Central America	274.2	2	0.73
Europe	700	5.4	0.77
Africa	190.1	0.7	0.37
Asia	2,229.3	5.8	0.26
World	4,422.7	20.1	0.45

C.3 UNDERSTANDING LAND-BASED INPUTS OF FOSSIL FUEL HYDROCARBONS TRANSPORTED VIA THE ATMOSPHERE

To understand the inputs of fossil fuel hydrocarbons that are emitted from land and transported via the atmosphere, numerous sampling campaigns to measure petroleum hydrocarbons, specifically PAHs, in marine atmospheres and surface waters have been performed. This increase in data covers a broad geographic area and provides insights into the inputs and sources of PAH compounds to marine surface waters (see Table C.17).

The methods used in these studies are a combination of a high-volume sampler (air) or direct intake (water) passed through a filter to remove particulate matter, followed by a polyurethane foam (PUF; air) or resin (water and sometimes air) sampler to collect the compounds of interest (exception is Lohmann et al., 2011, and Zheng et al., 2021, which used passive polyethylene samplers). The PUF or resin is then extracted with an organic solvent and analyzed primarily via gas chromatography–mass spectrometry (GC-MS) in all but one instance (Nizzetto et al., 2008, uses high-performance liquid chromatography). Quality assurance and quality control including laboratory and field blanks, recoveries and analytical limits are reported in all studies described in the Table C.17. For González-Gaya et al. (2016), the quantification of the semivolatile aromatic-like compounds is more challenging as these compounds cannot be resolved by the gas chromatographic techniques used and are therefore not identified. Given this, the source of these compounds cannot be confirmed, and further examination is required to determine what these compounds are and what their source is. For example, there may be potential inputs from dissolved organic matter derived from biogenic sources. Quantification of the semivolatile aromatic-like compounds also requires further verification before this dataset could provide reliable estimates of inputs in the same way that the PAH measurements are.

TABLE C.16 Comparison of Petroleum Hydrocarbon Loading Estimates from Land-Based Sources from This Work and Other Studies

Reference	Comments	Hydrocarbon Loading (tonne/yr)		
		Low	Best Estimate	High
World estimates				
Baker (1983)	Petroleum hydrocarbons from municipal wastes, industrial waste, and runoff	700,000	1,400,000	2,800,000
National Research Council (1985)	World estimate of land-based sources	600,000	1,200,000	3,100,000
Van Vleet and Quinn (1978)	Petroleum hydrocarbons *from municipal wastes only* based on Rhode Island treatment plants	—	200,000	—
Oil in the Sea III (2003)	World estimate of land-based sources	6,800	141,000	5,000,000
This work	World estimate of land-based sources	254,000	4,000,000	18,000,000
Ratio (This work: *Oil in the Sea III*)		37	28	3.6
North American estimates				
Eganhouse and Kaplan (1982)	U.S. input of petroleum hydrocarbons based on mass emission rate for wastewater effluent in southern California	—	120,600	—
Oil in the Sea III (2003)	North American estimate of land-based sources	2,500	52,000	1,800,000
This work	North American estimate of land-based sources	38,500	1,158,500	5,213,200

TABLE C.17 Studies Examining Polycyclic Aromatic Hydrocarbons in Marine Surface Waters and Atmospheres

Location	Sample Type	Date	Proposed Sources	Reference
Atlantic and Indian Ocean	Marine aerosol ~15 m above the sea surface (81 samples)	1999	Predominantly fossil fuel with some biomass	Crimmins et al., 2004
South Atlantic	Air and water samples (14 samples)	2005	Uncombusted fuel, oil spills, gas flaring, ship transport, biomass burning	Nizzetto et al., 2008
Narragansett Bay, Atlantic Ocean	Air and water samples (72 of each)	2006	Fossil fuel combustion	Lohmann et al., 2011
Mediterranean and Black Sea	Gas and aerosol (66 samples) and 43 water samples (2–3 m depth)	2006 2007	Gas phase is pyrogenic, aerosol phase is mixture of pyrogenic and petrogenic	Castro-Jiménez et al., 2012
East and South China Seas, Bay of Bengal, Indian Ocean, Atlantic Ocean	Gaseous (60 samples, 9 PAH) and particle bound (44 samples, 15 PAH) from marine boundary layer	2008	Coal and coke from Mainland China, biomass burning in Africa and Southeast Asia	Xu et al., 2012
Southern Ocean	Gas (22 samples) and aerosol phase (30 samples)	2005 2008 2009	Long-range transport and local sources, specifics not provided	Cabrerizo et al., 2014
Tropical Atlantic Ocean	Water (57 samples) and air (47 samples)	2009	Traffic emissions and petroleum combustion products	Lohmann et al., 2013
North Pacific toward the Arctic Ocean	Boundary layer air (17 samples) and surface seawater (18 samples)	2010	Biomass or coal in air, mixture in seawater	Ma et al., 2013
Tropical and subtropical Atlantic, Pacific, Indian Oceans	Gas and aerosol (108 samples) and water samples (68)	2010	Mixed sources. SALC also calculated, but quantification more challenging	González-Gaya et al., 2016
North Pacific-Arctic Oceans	Atmospheric (32) and surface seawater (16)	2014	Combustion-derived via long-range transport, and sea ice melting and runoff	Zheng et al., 2021
Livingston Island, Antarctica	Air (52 samples), seawater (26 samples, 0.5–1 m depth)	2014 2015	Mixed sources (long-range and local)	Casal et al., 2018

Appendix D

Regional Values of Water-to-Oil Ratio for Calculating Inputs from Produced Water

Estimates of Produced Water for Oil and Gas Operations[a]

	Annual Production[b]			Average Rate of Produced Water (bbl) per Production[c]				Median Rate of Produced Water (bbl) per Production			
				Oil (per bbl)		Gas (per MCF)		Oil (per bbl)		Gas (per MCF)	
OITS IV Geographic Zone	Oil (bbl)	Gas (MCF)	Produced Water (bbl)	bbl Water per bbl Product	SD	bbl Water per MCF Product	SD	Median with All Values	Median Without Outliers	Median with All Values	Median Without Outliers
C: E Canada Offshore	104,005,970	1,615,773	61,037,083	2.2	3.3	0.15	0.14	1.02	0.8	0.03	0.02
F: E GOM Nearshore[d]	25,884	48,541,716	1,280,832	42.3	50.0	0.027	0.033	31.2	31.2	0.018	0.018
F: E GOM Offshore	34,991,045	78,403,905	11,811,098	18.6	45.0	0.40	0.91	1.16	0.80	0.031	0.030
G: W GOM Nearshore[e]	3,847,596	11,299,809	9,618,990[f]	No data	No data	No data	No data	No data	No data	No data	No data
G: W GOM Offshore	626,892,869	831,100,299	398,644,277	9.03	26.3	5.81	26.9	2.74	2.67	1.16	1.14
H: Mexican GOM[g] Offshore	620,500,000	73,000,000	1,551,250,000[h]	No data	No data	No data	No data	No data	No data	No data	No data
K: CA Pacific Nearshore[i]	1,496,500	393,470,000	16,242,500	10.9	10.3	0.04	0.11	13.16	17.94	0.01	0.01
K: CA Pacific Offshore[j]	4,383,418	2,702,654	42,233,459	8.8	5.2	18.8	14.9	9.38	9.00	10.82	10.38
P: South Alaska[k] Nearshore	3,253,102	23,610,069	35,371,703	18.52	51.96	0.013	0.033	2.19	2.19	0.0004	0.003
P: South Alaska Offshore	0	5,915,233	42,926	0	0	0.02	0.02	0	0	0.0047	0.0091
Q: North Alaska Nearshore	11,580,775	256,693,367	20,391,303	1.51	3.37	No data	No data	0.62	0.62	No data	No data
Total[l]	1,410,977,159	1,726,352,825	587,055,181								
Total[m]	1,410,977,159	1,726,352,825	2,147,924,171								

[a] Based on data available for 2020. Note that only nearshore (but not inland) and offshore oil and gas operations were included. Inland operations are outside of the scope of the OITS IV.

[b] U.S. offshore data based on publicly available data from BSEE (available at www.data.bsee.gov).

[c] U.S. estimates calculated based on publicly available data from BSEE (available at www.data.bsee.gov). Eastern Canada estimates for produced water for gas production based on value for exclusive gas production in Nova Scotia coupled with produced water data from the same area (data from Canada-Newfound and Labrador Offshore Petroleum Board and Canada-Nova Scotia Offshore Petroleum Board). Estimates for Alaska based on State of Alaska data on well gas and oil production and associated produced water. Calculations conducted for gas production wells and oil production wells. Oil production wells generally also produce some gas.

[d] Alabama nearshore data available at https://www.gsa.state.al.us/ogb/production.

[e] Louisiana data from Louisiana Department of Natural Resources Office of Conservation (data for 2020). Texas data from Texas Railroad Commission (data for 2020) available at http://webapps.rrc.texas.gov/PDQ/generalReportAction.do.a

[f] Estimated based on 2.5 bbl water per bbl oil as per Veil (2011).

[g] Mexican production data from U.S. Energy Information Administration (available at https://www.eia.gov/international/analysis/country/MEX).

[h] Estimated based on 2.5 bbl water per bbl oil as per Veil (2011).

[i] California nearshore (state waters) data for Platforms Eva, Emmy, and Esther from California State Lands Commission. The WOR ratios are relatively high because the production rates are now relatively low, totaling less than 1.5 million bbl per year for all three platforms.

[j] Data from BSEE Pacific. The WOR ratios are relatively high because the production rates are now relatively low, totaling less than 4.4 million bbl per year for 11 leases.

[k] Data for Cook Inlet offshore production from State of Alaska (available at http://aogweb.state.ak.us/DataMiner3/Forms/Production.aspx).

[l] Totals for produced water without Mexico and Western GOM nearshore because no data are available.

[m] Totals including estimates for produced water for Mexico and Western GOM nearshore based on an estimate of 2.5 bbl water per bbl oil as per Veil (2011).

NOTE: SD = standard deviation.

Appendix E

Common Shoreline Response Options

Oil spills in the environment that threaten a shoreline are met with a variety of response options for the containment, cleanup, or protection of sensitive resources and structures. This toolbox of response strategies is an ever-evolving list of options that has no singular "right" option for all spills. Each release of oil carries unique challenges and considerations. The oil type and its particular chemistry and characteristics, such as viscosity, emulsification, and environmental considerations, such as shoreline type and sensitivity of habitat, must be considered when choosing a response method. Appendix G describes shoreline types and expected oil behavior in detail; the more advanced responses: surface-washing agents, burning of oil in marshes, and bioremediation are described in Section 4.2.4. The following is an abbreviated description of some of the most common forms of shoreline response and cleanup methodologies.

E.1 BARRIERS, BERMS, AND TRENCHING

1. Barriers and berms are structures that are often built from materials at hand such as soils and sands that are purposely designed, built, and/or deployed to prevent migrating oil across it or to divert the direction of flow of the oil. Other materials and designs include but are not limited to sand bags, wood, timbers, metal, or fiberglass sheet-pile. Responders must seriously consider disruptions of water flow and destruction of vegetation in the construction or deployment process.
2. Underflow dams often share similar construction materials and designs as those of barriers and berms. An underflow dam utilizes a system of channels or pipes that divert water from below the water line, on the upstream side of the dam where the oil is collecting, to the downstream side, thereby stopping the oil migration for collection and allowing clean water to pass.
3. Trenching is a method of directing the movement or collection of oil by physical excavation of a pathway below the normal topographic elevation. As with any earth movement response options, care must be taken not to permanently disrupt normal water flow of an area or cause permanent vegetation or impacts to other habitats.

E.2 NATURAL ATTENUATION

Also known as response by natural recovery, this is one of the most often used response options available. Oil can be left in place to degrade naturally as a variety of processes immediately begin to remove, relocate, and degrade oil once it enters into the marine environment. A non-exhaustive list includes evaporation, dispersion, photo-oxidation, and microbial degradation (see Section 5.1). Components of crude oil are found in the environment and, assuming that they are accessible to the degrading processes and processors along with necessary nutrients, natural attenuation (NA) will occur, although residual oil may remain for years or decades. When this process of non-intervention is coupled with a monitoring and reporting system, it is called monitored natural attenuation (MNA).

One of the benefits to this method is that, unlike almost all other human responses to oil, NA and MNA are much less likely to inflict ancillary impacts on the environment. Whenever personnel actively enter a spill area, transferring the oil to a new vulnerable location is possible and having the responders themselves placed into the hazardous zone is unavoidable. By allowing for NA these possibilities are minimized. However, NA is a much slower process than many other responses, and this must be considered. During this residence time there is a higher potential of secondary contamination of nearby areas, further impacting the environment. Two methodologies used to augment the processes of NA are shoreline tilling and surfwashing, described below. It should be noted that the removal or redistribution of materials from a beach inherently carries an additional burden of potential habitat destruction and may accelerate erosion of the location.

E.2.1 Tilling and Aeration

Oil that has stranded on the shoreline is often subjected to burial by sand, shell, cobble, or other materials due to response activities, wave action, and tidal cycles. Once buried, this oil is often deprived of oxygen, which may slow the degradation processes. To bring the oil to the surface for the faster aerobic NA process to continue, substrate tilling is often done. This tilling brings oil into an area where it is no longer sequestered from oxygen and thus degrades much more quickly. The tilling process also tends to break up the oil into smaller particles, resulting in a larger surface area for natural processes to more efficiently proceed.

E.2.2 Surfwashing

Also called translocation, surfwashing is another form of response that not only allows for more efficient NA of oil but improves access to the oil for further physical cleanup measures. This response involves the manual or mechanical movement of oiled substrates such as sand, shell, or cobble that have been subjected to oiling back into the sea. This response measure utilizes the continuous force of wave action to remove and break up oil that has contaminated the shoreline materials. This oil can then be accessed by natural degraders and at times even recovered with other response technologies.

E.3 MANUAL OIL RECOVERY OR CLEANING

Manual oil recovery or cleaning of a shoreline is the technique of removing or remediating oil from a surface using hands and hand tools (rakes, shovels, scrapers, etc.) including cloth or sorbent materials, and placing the materials into containers for collection, recovery, removal, and possible disposal. These methods may be employed on all shoreline types. They are often labor intensive but also allow for a more precise cleanup of small or hard-to-reach surfaces. Manual operations may pose a risk to response personnel by placing the responders in direct or close contact with the oil and necessitating the need for the responders to access hazardous areas and positions where the oil made contact with the surface. Manual oil recovery is often used in conjunction with other cleanup methodologies such as surface-washing agents or mechanical cleanup.

E.4 MECHANICAL CLEANUP

Mechanical cleanup involves the use of light, medium, and heavy machinery, often not designed as a specific tool for oil spill response. Equipment such as road graders, maintainers, front end loaders, bulldozers, compact excavators, and lawn/garden equipment are just a few examples of equipment that is often brought into the response. This type of machinery has considerable pros and cons that need to be considered before use. Mechanical recovery has the potential to expedite the removal of large quantities of oil from an affected area. Oil stranded on a beach or buried in the sediment can be removed, allowing for an area to be reopened for public use or for species at risk to utilize without the hazard of secondary contamination. Oil is either scraped off the surface as efficiently as possible, or large amounts of sediments are collected for oil separation or removal. Regardless of the care taken, mechanical excavation of stranded oil additionally removes a massive amount of irreplaceable sediment that provides equally important areas of forage, habitat, and recreational opportunities. Another method involves the mechanical removal of gross amounts of contaminated sediment such as sand, shell, or cobble, which is then cleaned by methods such as incineration or cleaning baths and then returned to the original location. This methodology still has the propensity to harm biota within and utilizing the substrate.

1. *Specialized mechanical equipment.* Some mechanized equipment has been developed specific to oil spill response with the intent to remove the contamination while minimizing secondary detrimental impacts to the environment. These devices utilize methods such as sweeping, sifting, or vacuuming to remove the contamination, which is then recovered for proper disposal or reclamation. These pieces of equipment are heavy and, although less invasive than other mechanical equipment operations, still may disturb and impact local flora and fauna utilizing the area.
2. *Vacuum systems.* This method involves the use of small to large air vacuum systems, from small self-contained systems to large truck- and vessel-mounted systems, with or without external storage tank systems that can operate in a wet environment. Vacuum systems are used on floating oils that can be sucked off the surface of the water or substrate. The material collected may contain large quantities of water if skimmed on or near the water. A process of decanting can be utilized on site to separate the liquids, with oil being reclaimed and the water either returned to the environment, if permitted, or disposed of. Vacuum systems can be very efficient but are dependent on an operator's skill level.

E.5 SORBENTS

Sorbents are a method of oil removal utilizing an oleophilic material that is either absorbent or adsorbent to capture or clean oil from the water surface or a solid substrate for disposal or reclamation. Many types of sorbent materials are available, with many meant for specific applications. Oil sorbents are made of many types of materials including natural products such as cellulose from trees, plants, seeds, hair, and clays. Synthetic products are widely available and used and offer the ability to be both oleophilic as well as hydrophobic. Used materials can have the oils collected, squeezed out, and recovered, and the sorbent reused, but more often it is not efficient from the cost and logistics perspective; hence both the sorbents and oil are usually disposed of. A few of the methods are described below. It should be noted that the U.S. Environmental Protection Agency (U.S. EPA) requires that all sorbent materials used be removed after deployment and use.

1. *Flat pads and sheets.* These adsorbent products are often referred to as diapers, sorbent pads, or "sweep" when configured in an elongated fashion. These are some of the most widely used sorbent products within the industry. Sheets are made of either natural or synthetic material and are used extensively for routine maintenance and cleanup operations. Sheets that are synthetic are both oleophilic and hydrophobic, which makes them amenable to deployment directly into the water where the oil is taken up into the pad. Pads can be deployed in open water and in confined spaces or used directly on oil surfaces to remove the contamination. Sweep is sometimes deployed as a barrier around sheens to retard migration of the thin oil. Saturated sheets are then manually removed for disposal or recycling.
2. *Sausage boom.* This is made of the same variety of adsorbent materials as sorbent pads and sheets. This response option is manufactured in a cylindrical manner with varying diameters and lengths, often with attachment points for joining to create an elongated response option specific to the need. A sausage boom can be used as a response option where oil is actively corralled and collected by the boom and removed when saturated or as a defensive response at outfalls or around fueling vessels and oil transfers as a precautionary response.
3. *Snare.* Snare is also referred to as pom-poms, due to its resemblance to the hand-held tufts used by cheerleaders. This response option consists of varying lengths of thin strips of oleophilic, hydrophobic, usually synthetic material that can be deployed defensively or offensively. Snare is particularly good at picking up tacky oils that other sorbent products do not perform as well on. It can be used manually where it is directly administered to an oil product or left to collect oil that may migrate to its location.
4. *Loose sorbent.* Also known as bulk sorbents, this methodology involves the use of a particulate or granular absorbent or adsorbent material that is applied or otherwise positioned to encounter an oil product, after which the contaminated material is retrieved for proper disposal. These materials are made from natural products such as sorghum, cotton seed hulls, clays, and so on as well as synthetic materials and are deployed manually or by an air blower system. Loose sorbents are used extensively to clean up land spills but are generally not an acceptable form of response of on-water recovery as containment and retrieval of the material is problematic. One application of loose sorbents that historically has been used is the deployment of the material onto oiled surfaces such as plants and other areas where there is a high potential of secondary oiling of avian, invertebrate, and other biota. The loose material reduces the tackiness of the oil, thereby protecting species that may encounter it. Materials that cause an oil to sink, aka sinking agents, are not authorized for use in U.S. waters.
5. *Solidifiers.* Though sometimes considered within the same category as loose or other similar sorbents, these differ in that the materials are most often dry polymers that are oleophilic in nature and have physical properties similar to oil. These materials not only sorb oils but also form bonds between oil particles, resulting in masses of material that can be retrieved more easily than if contained with loose sorbent material. These products are considered a chemical response alternative and thus must be approved for use by the U.S. EPA with authorization from the applicable Regional Response Team prior to use.

E.6 WATER WASHING, FLASHING, AND DELUGE

This method is one of the most common and widely accepted forms of response to oil that has impacted man-made structures, vegetation, or other natural shorelines. Oil tends to exhibit nonpolar tendencies and thus adheres firmly to other materials. Water is polar and must exert force on an oil sufficient to break the surface tension between the oil and solid surfaces to which it may attach. Using water to remove oil that has coated such structures carries the risk of unintended or unanticipated secondary impacts. The force necessary to remove oil from a solid surface may also cause harm to biota that utilize that surface as a home or an area of forage. The water may cause erosion to the substrate, disrupting or even permanently altering the normal water flow. A consequence of water washing is that, once removed, the oil is then free to migrate further unless properly captured, potentially contaminating other areas or structures. Some common methods of water washing under different temperatures, pressures, and orientation of flow are discussed below.

1. *Low pressure washing.* This method utilizes seawater at pressures generally <10 psi to forcibly move or remove pooled and trapped oil from interstitial spaces to locations where another form of cleanup can be undertaken. Low pressures generally are less disruptive to resident and encrusting organisms and cause less erosion issues than more aggressive pressures. Low pressures do not generally remove all contamination from a surface: those areas are either allowed to be further

cleaned using manual or chemical alternatives or by natural degradation and hydrodynamic processes.

2. *High pressure washing.* This method utilizes seawater at much greater pressures, generally >100 psi. These high pressures are often sufficient to remove highly viscous or tacky oils, or oils that have weathered and adhered to surfaces. This is a highly efficient method of removing oil from a surface and relocating it to where it can be removed by some other response method. High pressures may have the detrimental consequence of removing encrusting flora and fauna or causing erosional issues. Organisms removed may take very long periods of time to recover, and areas eroded can also suffer long-lasting impacts. Another known issue is the possibility of creating oil droplets small enough to disperse into adjacent waters, causing additional impacts to aquatic organisms.
3. *Low and high temperature washing.* Though not a distinct method of washing, different seawater temperatures can be employed at both low and high pressure to aid in the removal of oils that have adhered to a solid surface. High temperature washing at water temperatures of >30°C to very hot (but below that which would be steam, discussed below) can be employed to loosen stubborn oil. Higher temperatures applied either at low or high pressure may also increase detrimental secondary impacts of the relocated oil.
4. *Steam cleaning.* This method utilizes steam that is directed onto a solid substrate to remove oil and adhered residues. Much less water is needed for this type of application, and thus less volume is generated for subsequent cleanup. This is method is highly intrusive to resident organisms; as such, it is usually reserved for areas that are being cleaned for aesthetics or when other removal responses have been attempted and found inadequate to achieve the desired level of cleaning.
5. *Deluge.* Also called flooding, this method of spill response involves the inundation of an area with copious amounts of water deployed in a sheet flow scenario meant to lift and migrate oil away from its stranded location along a shoreline. This method can be used to remove oil that has made its way deep within vegetated areas, such as mangrove roots or marsh vegetation, or deeply embedded with cobble or boulders. Care should be taken to avoid channelization of the applied waters, which may cause erosion issues.
6. *Sandblasting and dry ice blasting.* Though not often used, both sandblasting and dry ice blasting have been used to remove oil from solid substrates. Both methodologies employ high pressure air systems with a secondary material that abrades adhered oil from the surface. These methods are good for removing stubborn oils but may heavily impact the surface biota. Dry ice blasting involves the use of dry ice pellets (frozen carbon dioxide) as the abrasive. It has the benefit of not adding any additional secondary waste material that would need to be cleaned in addition to the removed oil.

E.7 VEGETATION CUTTING

This response involves the cutting and removal of all or parts of the affected vegetation that has been subjected to oiling from a release. This method increases the efficacy of oil removal from the substrate and that which has clung to the vegetation. The cutting and removal of oiled vegetation decreases the possibility of secondary impacts to wildlife using the area for foraging or shelter and aids survival of existing vegetation or regrowth of new vegetation. Vegetation cutting should only be done after consultation with resource managers familiar with this methodology as the location, season, type of oil, and species of plant and their susceptibility to the oiling must be considered to prevent longer term impacts to the area than would be seen by using other response options. The vegetation may also be considered critical habitat or be a habitat for species of concern and thus fall under the jurisdiction of state and federal agencies that must provide approval for this type of response action.

E.8 DEBRIS REMOVAL

As the name implies, this response involves the recovery of (un-oiled) debris located within the area of a potential spill impact and which may be subject to being oiled. Removal of this material must be determined to be beneficial to the overall response in some way. Debris that has been oiled becomes a hazardous material and will need to be cleaned or removed by personnel having proper training and credentials and must be transported to an appropriate waste management site. Pre-spill debris removal is a common preemptive response operation as this material can represent a significant mass given the amount of driftwood, timbers and seaweed, algae, and so forth that may be found on a shoreline and be subject to oiling. If materials can be relocated pre-spill, significant savings of response time and waste generation during spill mitigation may be realized. As pre-spill debris removal does not require personnel with hazardous waste response training, many areas have identified this effort as being amenable to volunteers during an oil spill.

Appendix F

Technical Aspects of Equations and Models for Droplet Breakup in Turbulent Flows

F.1 THEORY OF DROPLET BREAKUP IN TURBULENT FLOWS

Empirical equations for droplet breakup stem from the theoretical arguments of Hinze (1955). He developed an expression for the maximum stable droplet size D_{max} by assuming it would be small enough to lie within the inertial subrange of a turbulent flow, where the turbulent dynamics are isotropic and depend on the dissipation rate ε and kinematic viscosity ν of the turbulent, continuous phase. Because pressure fluctuations in the turbulent field generate the forces causing breakup, D_{max} also should depend on the density ρ of the continuous phase. The forces resisting breakup are represented by the interfacial tension σ between the dispersed and continuous phases and on the dynamic viscosity μ_p and density ρ_p of the dispersed phase. Hinze (1955) arranged these parameters into a non-dimensional Weber number We_p and viscosity group (see Box 5.3). For the limiting case of a dispersed phase with low viscosity relative to interfacial tension, Hinze postulated the maximum stable droplet size would be given by a constant value of the Weber number group,

$$\frac{\rho\varepsilon^{2/3}D_{max}^{5/3}}{\sigma} = C \qquad \text{(F.1)}$$

where C is an empirical constant. Hinze arrived at this form of We by recognizing that the dynamic pressure fluctuations in a turbulent flow scale as $\rho\overline{u'u'}$ and that it would be fluctuations on the scale of D_{max} that would be responsible for breakup. Here, the prime indicates velocity fluctuations of the turbulence, and the overbar is an average in time. Within the inertial subrange, Batchelor (1951) had shown for isotropic turbulence that at a scale given by D_{max}, then $\overline{u'u'} = 2.0(\varepsilon D_{max})^{2/3}$.

If we solve Equation F.1 for D_{max} and then divide by a characteristic length scale L_c describing a given flow field, we obtain the general equation

$$\frac{D_{max}}{L_c} = A'We_L^{-3/5} \qquad \text{(F.2)}$$

where A' is another empirical constant and We_L is the Weber number defined by length scale L_c, namely, $We_L = \rho\varepsilon^{2/3}L_c^{5/3}/\sigma$.

For the case where viscosity becomes an important resistive force relative to surface tension, Wang and Calabrese (1986) proposed a new viscosity number Vi based on experimental data in Calabrese et al. (1986) to replace Hinze's original viscosity group. This new, non-dimensional group is given by

$$Vi = \frac{\mu_p U}{\sigma}\left(\frac{\rho}{\rho_p}\right)^{1/2} \qquad \text{(F.3)}$$

where U is a characteristic velocity scale of the turbulence. For the case of droplet breakup within the inertial subrange of the turbulence and following the same scaling arguments as in Hinze (1955) for the Weber number, the characteristic velocity scale would be $U = (\varepsilon D_{max})^{1/3}$. Furthermore, it can be shown that any characteristic droplet size D_n can be related to D_{max} by a numerical constant (Calabrese et al., 1986; Boxall et al., 2012). Here, D_n represents the droplet size for which n percent by volume of the other droplets in the flow are smaller than D_n; similar behavior holds for number size distributions. Hence, the breakup dynamics of an immiscible droplet in a statistically stationary turbulent field is generally described by

$$\frac{D_n}{L_c} = f(We_L, Vi) \qquad \text{(F.4)}$$

Equation F.4 is the basis for all empirical equations predicting characteristic droplet sizes of dispersions in turbulent flow, and the theoretical arguments underpinning this equation describe the dynamical processes responsible for droplet breakup: turbulent pressure fluctuations on the scale of a droplet and that scale with the dissipation rate ε break up a droplet until these forces can be resisted by the droplet interfacial tension and viscosity.

Empirical equations such as these predict one characteristic droplet size of the distribution. Hinze (1955) focused on the maximum stable droplet size; other popular sizes are the volume median diameter, the size for which half of the volume is contained in smaller sizes, and the Sauter mean diameter, the size for with the surface area would represent that of the whole distribution if all droplets were of this size. To predict the whole distribution, this characteristic size should be related to an assumed probability density function. Common choices for the distribution of droplet sizes are the log-normal distribution and the Rosin–Rammler distribution, which are similar in their central tendency but differ in the shapes of their tales (Johansen et al., 2013; Wang et al., 2016, 2018; Li et al., 2017). Though some experimental studies show the width of fluid particle size distributions to be consistent across several experiments, Wang and Calabrese (1986) demonstrated that it depends on both the surface tension and viscosity of the dispersed phase through a resistance-to-breakup formulation. Falliettaz et al. (2021) also showed recently that even using the same median droplet sizes and distribution widths, the fate of oil in the water column and sea surface remained sensitive to the choice of analytical distribution when comparing Rosin–Rammler, log-normal, and one distribution from the viscous breakup model for oil droplets, VDROP. Hence, additional experiments and model calibration are required for empirical equations to predict the whole fluid particle size distribution.

F.2 APPLICATION OF TURBULENT BREAKUP FOR MULTIPHASE JETS

The equations and parameterizations in the previous section can be used to predict droplet sizes for different multiphase flows by relating the large-scale flow and turbulence production to the dissipation rate of turbulent kinetic energy at the small scales. Because turbulence is a property of the flow field, this yields a unique parameterization and different fit coefficients for different flow types. This section considers the case of breakup for turbulent jets, which resemble the flow from pipeline leaks or oil well blowouts, using both the empirical equations and population balance modeling approach to predict droplet sizes. Refer to Section 5.2.3.2 for more details, applications, and a discussion of available data for model validation.

Johansen et al. (2013) and Li et al. (2017) apply empirical equations to predict the characteristic droplet sizes for oil and gas jets. Johansen et al. (2013) handle the viscosity limitation on droplet formation using a modified Weber number that combines the normal Weber and viscosity numbers, similarly to that used by Johansen et al. (2015) for surface breaking waves. Li et al. (2016) follow the parameterization in Hinze (1955), using the Weber and Ohnesorge numbers. Johansen et al. (2013) take the turbulent eddy dissipation rate as that of the potential core in the zone-of-flow-establishment in turbulent jets, or the ZFE, of the jet release, $\varepsilon = U_0^3/D$, together with the characteristic length scale $L_C = D$ to define the Weber and viscosity numbers. U_0 is the jet exit velocity, and D is the jet orifice diameter. Li et al. (2017) take a different approach for the characteristic length scale, setting L_C equal to the smaller of the orifice diameter D or the maximum stable droplet size d_0, given by analytical stability analysis (Clift et al., 1978). Likewise, they assume $\varepsilon = U_0^3/L_C$ so that there is only one length scale used to define the Weber number. When the orifice is small, the Weber numbers in Li et al. (2017) and in Johansen et al. (2013) would be the same. As the orifice diameter exceeds the maximum stable droplet size, the Li et al. (2017) model switches to a parameterization based on d_0. Because the droplet size scales with the nozzle diameter for smaller nozzles, the Li et al. (2017) parameterization retains consistency in which the parameterization scales with droplet size despite the apparent discontinuity in switching from D to d_0 as the nozzle size increases.

Both the Johansen et al. (2013) and Li et al. (2017) equations require calibration with data. Johansen et al. (2013) take the power-law relationships in their equations from the theoretical arguments in Hinze (1955) and Wang and Calabrese (1986), only fitting a coefficient multiplying the Weber and viscosity number to data measured in a series of experiments for oil jets into water. Brandvik et al. (2018) updated the fit coefficients to their current accepted values and summarized the complete datasets used to calibrate and validate the equation fit. Li et al. (2017) take both the power-law and proportionality constants as unknown. They fit the power-law relationship to experimental data for oil slick breakup under breaking waves and argue that these scaling laws should be universal. The fit coefficient multiplying the Weber and Ohnesorge numbers for jet breakup of oil are fitted to the field measurement from the DeepSpill experiment only (Johansen et al., 2003), thus having perfect agreement with these field data. They then validate the fit to other experiments in the laboratory, including those by Brandvik et al. (2013). In this way, both models are calibrated and validated to laboratory data for oil jets into water.

Other researchers have proposed alternative empirical equations to predict droplet size. The equations used by Paris et al. (2012) and Aman et al. (2015) use correlations from Boxall et al. (2012) to predict droplet sizes from the *Deepwater Horizon* (DWH) accident. Boxall et al. (2012) evaluated the effect of viscosity on droplet breakup by using different oils in the place of water as the continuous phase and observed the breakup of water droplets in a mixing tank using a Rushton-type, six-blade turbine. Mixing tanks are often used to evaluate oil and water emulsions because their turbulence characteristics have been well characterized and parameterized in terms of the blade diameter and rotation rate of the blades. Boxall et al. (2012) found both Weber number and Reynolds number scaling, where the viscosity of the continuous phase is used to define the Reynolds number. Because relatively inviscid water was the only dispersed phase used, they did not find a dependence on the viscosity number. However, they did demonstrate that for $We^{4/5}/\mathrm{Re} > 0.1$, droplet breakup continues below the Kolmogorov scale and into the dissipation range of the turbulence. Although these physics and scaling laws likely are universal for droplet breakup in turbulent flows, the fit coefficients of Boxall et al. (2012) are only applicable to steady, constant turbulent dissipation rate, unlike that in a blowout jet, and should be used only for within-pipe flow where these turbulent conditions are present.

Zhao et al. (2014) used an experimentally derived expression for turbulent eddy dissipation rate in single-phase jets to adapt the VDROP population balance model to predict droplet evolution in subsea blowouts, yielding the jet-version of the model, VDROP-J. The original VDROP model solves for the unsteady evolution of the droplet size distribution under constant turbulent conditions of a mixing tank. The evolution of the turbulent eddy dissipation rate along the jet centerline was incorporated into the model following the classical Lagrangian integral model approach (Lee and Chu, 2003) in which a well-mixed control volume propagates along the jet centerline, advected at the mean jet velocity. Hence, the eddy dissipation rate variation with distance becomes a variation with time in the VDROP-J model. Coalescence was considered negligible, and the breakup parameter K_b, which sets the breakup efficiency, was calibrated to experiments on oil jets in water. A non-dimensional, universal calibration for K_b was not found; instead, the best-fit K_b depended on the initial dynamic momentum flux of the source. Zhao et al. (2015) apply the VDROP-J model to scenarios typical of the DWH accident. Zhao et al. (2016) adapt the model to bubble breakup, and Zhao et al. (2017) consider breakup of combined oil and gas releases. Predictions from VDROP-J for DWH were also used by Gros et al. (2017) to hindcast the fate of oil and gas released on June 8, 2010.

Appendix G

Classification of Intertidal, Subtidal, Ice, and On-Water Areas

TABLE G.1 Classification of Intertidal, Subtidal, Ice, and On-Water Areas

Coastal Habitat	Description	Predicted Oil Behavior
Intertidal		
Exposed Rocky Shores (ESI 1A)  SOURCE: NOAA (2019a).	• The intertidal zone is steep (>30° slope) and narrow with very little width. • Regular exposure to high wave energy or tidal currents. • Strong wave-reflection patterns are common. • Substrate is impermeable (usually bedrock or cement) with no potential for subsurface penetration. • Species density and diversity vary. • Barnacles, snails, mussels, polychaetes, and macroalgae can be abundant.	• Oil is held offshore by waves reflecting off the steep, hard surfaces. • Any oil that is deposited is rapidly removed from exposed faces. • The most resistant oil would remain as a patchy band at or above the high tide line. • Impacts to intertidal communities are expected to be short term. • An exception would be where heavy concentrations of a light refined product came ashore very quickly.
Exposed Rocky Banks (ESI 1A)  SOURCE: Daderot, CC0, via Wikimedia Commons.		
Exposed, Solid Man-Made Structures (ESI 1B)  SOURCE: NOAA (2019a).	• These are solid, man-made structures such as seawalls, groins, revetments, piers, and port facilities. • They are built to protect the shore from erosion by waves, boat wakes, and currents, and thus are exposed to rapid natural removal processes.	

continued

TABLE G.1 Continued

Coastal Habitat	Description	Predicted Oil Behavior
Exposed, Rocky Cliffs with Boulder Talus Base (ESI 1C)  SOURCE: NOAA (2019a).	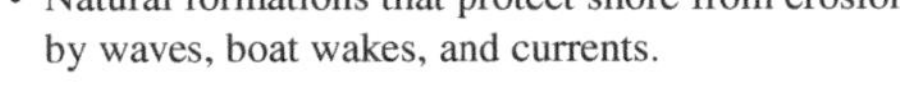 • Natural formations that protect shore from erosion by waves, boat wakes, and currents.	
Exposed, Wave-Cut Platforms (ESI 2A)  SOURCE: NOAA (2019a). Shelving Bedrock Shores (ESI 2A)  SOURCE: NOAA (2019a).	• These shores consist of a bedrock shelf or platform of variable width and very gentle slope. • The surface of the platform is irregular; tide pools are common. • Along headlands, they have only a small accumulation of sediments, mostly at the high tide line. • They often co-occur with gravel beaches; the gravel beach can be either at the upper or the lower half of the intertidal zone, depending on the nature of the bedrock outcrop. • Species density and diversity vary greatly, but barnacles, snails, mussels, and macroalgae are often abundant.	• Oil will not adhere to the wet rock surface but could penetrate crevices or sediment veneers. • Oil persistence is usually short-term, except in wave shadows or where the oil was deposited high above normal wave activity.
Fine- to Medium-Grained Sand Beach (ESI 3A) Fine-Grained  SOURCE: NOAA (2019a). Coarse-Grained  SOURCE: NOAA (2019a).	• These beaches are flat to moderately sloping and relatively hard-packed. • There can be heavy accumulations of wrack. • They can be important areas for nesting by birds and turtles. Upper beach fauna include ghost crabs and amphipods; lower beach fauna can be moderate, but highly variable.	• Light oil accumulations will be deposited as oily swashes or bands along the upper intertidal zone. • Heavy oil accumulations will cover the entire beach surface; oil will be lifted off the lower beach with the rising tide. • Maximum penetration of oil into fine- to medium-grained sand is about 10–15 cm, up to 25 cm in coarse-grained sand. Burial of oiled layers by clean sand can be rapid (within 1 day), and burial to depths as much as one meter is possible if the oil comes ashore at the beginning of a depositional period. • Organisms living in the beach sediment may be killed by smothering or lethal oil concentrations in the interstitial water. • Biological impacts include temporary declines in infauna, which can affect important shorebird foraging areas.

TABLE G.1 Continued

Coastal Habitat	Description	Predicted Oil Behavior
Scarps and Steep Slopes (Sand) (ESI 3B)  SOURCE: NOAA (2019a).		
Eroding Scarps (Unconsolidated Sediment) (ESI 3B)  SOURCE: NOAA (2019a).		
Exposed, Eroding Banks (Unconsolidated Sediment) (ESI 3B)  SOURCE: NOAA (2019a).		
Tundra Cliffs (ESI 3C)  SOURCE: NOAA (2019a).	• These are erosional features with tundra vegetation overlying peat and exposed ground ice or permafrost. • Cliff heights range from less than 1 meter to as much as 5–10 meters. • There may be a narrow beach present or just a vertical scarp. As the cliffs erode at rates of 0.5–4 meters/year, the vegetation and peat accumulate as fragmented and irregular blocks at the base of the cliff until they are reworked by waves. • The vegetation on the tundra is a living plant community that is sensitive to disturbances. • Large numbers of migratory birds can use these shorelines during the summer months.	• Oil could be stranded onshore only during the ice-free summer season. • Oil is not likely to adhere to exposed ground ice unless air temperatures are below freezing. • Oil persistence on the vegetation and peat substrates would be short in most cases, due to natural cliff erosion, provided that the oil is not stranded at the onset of freeze-up. • If the oil mixes with the peaty substrate or accumulated peat, it could create sheens until the oiled area erodes. • Biological risks would be greatest to birds feeding along oiled cliffs in summer.
Coarse-Grained Sand Beaches (4)  SOURCE: NOAA (2019a).		

continued

TABLE G.1 Continued

Coastal Habitat	Description	Predicted Oil Behavior
Sand Bars and Gently Sloping Banks (ESI 4)  SOURCE: NOAA (2019a).		
Mixed Sand and Gravel Beaches (ESI 5)  SOURCE: NOAA (2019a). Mixed Sand and Gravel Bars and Gently Sloping Banks (ESI 5)  SOURCE: NOAA (2019a).	• Because of the mixed sediment sizes on these moderately sloping beaches, there may be zones of pure sand, pebbles, or cobbles. • There can be large-scale changes in the sediment distribution patterns depending on season, because of the transport of the sand fraction offshore during storms. • Desiccation and sediment mobility on exposed beaches cause low densities of attached animals and plants. • The presence of attached algae, mussels, and barnacles indicates beaches that are relatively sheltered, with the more stable substrate supporting a richer biota.	• During small spills, oil will be deposited along and above the high tide swash. • Large spills will spread across the entire intertidal area. • Oil penetration into the beach sediments may be up to 50 cm; however, the sand fraction can be quite mobile, and oil behavior is much like on a sand beach if the sand fraction exceeds about 40 percent. • Burial of oil may be deep at and above the high tide line, where oil tends to persist, particularly where beaches are only intermittently exposed to waves. • In sheltered pockets on the beach, pavements of asphalted sediments can form if oil accumulations are not removed, because most of the oil remains on the surface.
Gravel Beaches (ESI 6A)  SOURCE: NOAA OR&R. Gravel Beaches (Granules and Pebbles) (ESI 6A)  SOURCE: NOAA (2019a).	• Gravel beaches can be very steep, with multiple wave-built berms forming the upper beach. • The degree of exposure to wave energy can be highly variable among gravel beaches. • Density of animals and plants in the upper intertidal zone is low on exposed beaches but can be high on sheltered gravel beaches and on the lower intertidal zone of all beaches.	• Stranded oil is likely to penetrate deeply into gravel beaches because of their high permeability. • Rapid burial can occur at the high tide and storm berms. • Long-term persistence will be controlled by the depth of routine reworking by the waves. • On exposed beaches, oil can be pushed over the high tide berms, pooling and persisting above the normal influence of wave washing. • Along sheltered portions of the shorelines, chronic sheening and the formation of asphalt pavements is likely where accumulations are heavy.

TABLE G.1 Continued

Coastal Habitat	Description	Predicted Oil Behavior
Gravel Bars and Gently Sloping Banks (ESI 6A)  SOURCE: NOAA (2019a).		
Riprap (ESIs 6B, 6C)  SOURCE: NOAA (2019a).  SOURCE: NOAA (2019a).	• Riprap structures are composed of cobble- to boulder-sized blocks of granite, limestone, concrete, or other materials. • Riprap structures are used as revetments and groins for shoreline protection, and as breakwaters and jetties around inlets and marinas. • Attached biota are generally sparse at the upper intertidal zone, but more common in the lower intertidal zone. • They are common in highly developed waterfront areas. • 6c applies to estuarine shorelines only in Southeast Alaska	• Deep penetration of oil between the blocks is likely, with oiling of associated debris. • Oil adheres readily to the rough surfaces of the blocks. • Uncleaned oil and debris can cause chronic leaching until the oil hardens.
Gravel Beaches (Cobbles/Boulders)—used in Alaska (ESI 6B)  SOURCE: NOAA OR&R.		
Boulder Rubble (ESI 6D)  SOURCE: NOAA (2019a).		

continued

TABLE G.1 Continued

Coastal Habitat	Description	Predicted Oil Behavior
Exposed Tidal Flats (ESI 7)  SOURCE: NOAA (2019a).  SOURCE: NOAA (2019a).	• Exposed tidal flats are broad intertidal areas composed primarily of sand and minor amounts of gravel or mud. • The presence of sand indicates that tidal currents and waves are strong enough to mobilize the sediments. • They are usually associated with another shoreline type on the landward side of the flat, though they can occur as separate shoals; they are commonly associated with tidal inlets. • The sediments are water saturated, with only the topographically higher ridges drying out during low tide. • Biological use can be very high, with large numbers of infauna, heavy use by birds for roosting and foraging, and use by foraging fish.	• Oil does not usually adhere to the surface of exposed tidal flats, but rather moves across the flat and accumulates at the high tide line. • Deposition of oil on the flat may occur on a falling tide if concentrations are heavy. • Oil does not penetrate water-saturated sediments but may penetrate coarse-grained sand and coat gravel. • Biological damage may be severe, primarily to infauna, thereby reducing food sources for birds and other predators.
Sheltered Scarps (Bedrock/Mud/Clay) (ESI 8A)  SOURCE: NOAA (2019a).  SOURCE: NOAA (2019a).  SOURCE: NOAA (2019a).	• Sheltered rocky shores are characterized by a rocky substrate that can vary widely in permeability. • Of particular concern are rocky shores that have a semi-permeable veneer of angular rubble overlying the bedrock. • Sheltered clay scarps are characterized by a steep, usually vertical scarp in hard-packed and stiff clay. Vegetation usually occurs landward of the scarp.	• Oil will adhere readily to dry, rough, rocky surfaces, particularly at the high tide line, forming a distinct oil band. • The lower intertidal zone of rocky shores is usually algae-covered and stays wet, preventing oil from adhering. • Oil will not adhere to the wet clay sediment surface but could penetrate dry sediment. • Stranded oil will persist because of the low-energy setting. • Oil will adhere readily to the rough surface, particularly along the high tide line, forming a distinct oil band. • The lower intertidal zone usually stays wet (particularly if algae covered), preventing oil from adhering to the surface.

TABLE G.1 Continued

Coastal Habitat	Description	Predicted Oil Behavior
Sheltered, Impermeable, Rocky Shores (ESI 8A)  SOURCE: NOAA (2019a).		
Sheltered, Solid Man-Made Structures (ESI 8B)  SOURCE: NOAA (2019a).	• These are structures such as seawalls, groins, revetments, piers, and port facilities, constructed of concrete, wood, or metal. • Most structures are designed to protect a single lot; thus, their composition, design, and condition are highly variable. • Often there is no exposed beach at low tide, but multiple habitats may be present. • There can be dense attachments of animal and plant life. • They are common in developed waterfront areas.	
Sheltered Riprap (ESI 8C)  SOURCE: NOAA (2019a).		
Sheltered, Rocky Rubble Shores (ESI 8D)  SOURCE: NOAA (2019a). Peat Shores (ESI 8E)  SOURCE: Gillham (1978).	• This shoreline type includes exposed peat scarps, eroded peat, and peat slurries. • Exposed peat scarps occur where the peat is frozen. • They are highly erosional (>1 meter/year), resulting from wave action, ice scour, and melting of the frozen peat. • The intertidal zone is often very complex, with slumped peat blocks and a thin (and temporary) sand layer on the peat. • Eroded peat occurs as a peat mat or veneer in a dewatered state, deposited on a sand or gravel beach; it is usually less than 20 cm thick and considered to be relatively transient. • Peat slurries (which have the appearance of coffee grounds) are up to 50 cm thick and 10 meters wide. • Peat slurries are found at the foot of eroding peat scarps and in depositional areas; they are relatively permanent features that move along the shore with the currents. • Peat shorelines comprise about 70 percent of the Beaufort Sea coast of Alaska. • The intertidal zone of this shoreline type is not particularly important as a biological habitat. • Substrate can be highly variable, from smooth, vertical bedrock to rubble slopes. • There can be dense attachments of animal and plant life.	• Oil could be stranded onshore only during the ice-free summer season. • Oil penetration and persistence are expected to be very low in frozen peat scarps. • Light oil can penetrate peat slurries, especially when the peat is dry. • Peat resists penetration by heavy oils, even when dry. • Peat slurry reacts with oil like loose granular sorbent and will partially contain and prevent the oil from spreading.

continued

TABLE G.1 Continued

Coastal Habitat	Description	Predicted Oil Behavior
 SOURCE: Chrisseufert (2008).		
Vegetated, Steeply Sloping Bluffs (ESI 8F)   SOURCE: NOAA (2019a).	• Steep shorelines with a vegetated bluff face often covered by shrubs.	• Oil adheres readily to dry, rough surfaces, forming a distinct oil band. • Stranded oil will persist at low-energy settings.
Sheltered Tidal Flats (ESI 9A)   SOURCE: NOAA (2019a).	• Sheltered tidal flats are composed primarily of mud with minor amounts of sand and shell. • They are usually present in calm-water habitats, sheltered from major wave activity, and frequently backed by marshes. • The sediments are very soft and cannot support even light foot traffic in many areas. • There can be large concentrations of bivalves, worms, and other invertebrates in the sediments. • They are heavily used by birds for feeding.	• Oil does not usually adhere to the surface of sheltered tidal flats, but rather moves across the flat and accumulates at the high tide line. • Deposition of oil on the flat may occur on a falling tide if concentrations are heavy. • Oil will not penetrate the water-saturated sediments but could penetrate burrows and desiccation cracks or other crevices in muddy sediments. • In areas of high suspended sediment concentrations, the oil and sediments could mix, resulting in the deposition of contaminated sediments on the flats. • Biological impacts may be severe.
Sheltered Sand and Mud Flats (ESI 9A)  SOURCE: NOAA (2019a).		
Vegetated Low Banks (ESI 9B)  SOURCE: NOAA (2019a).	• Vegetated low banks are composed primarily of mud with minor amounts of sand and shell. • They are usually present in calm-water habitats, sheltered from major wave activity, and frequently backed by marshes. • They are heavily used by birds for feeding.	

TABLE G.1 Continued

Coastal Habitat	Description	Predicted Oil Behavior
Hyper-Saline Tidal Flats (ESI 9C)  SOURCE: NOAA (2019a).	• Tidal flats are composed primarily of sand and shell. • They are usually present in calm-water habitats, sheltered from major wave activity, and frequently backed by marshes. • The sediments are water saturated, with only the topographically higher ridges drying out during low tide.	
Salt and Brackish Marshes (ESI 10A)  SOURCE: NOAA (2019a). Freshwater Marshes (ESI 10B)  SOURCE: NOAA (2019a). Swamps (ESI 10C)  SOURCE: NOAA (2019a). Scrub and Shrub Wetlands (ESI 10D)  SOURCE: NOAA (2019a).	• Intertidal wetlands contain emergent, herbaceous vegetation, including both tidal and muted tidal marshes. • Depending on location and interannual variations in rainfall and runoff, associated vegetation may include species tolerant or adapted to salt, brackish, or even tidal freshwater conditions. • The marsh width may vary from a narrow fringe to extensive areas. • Sediments are composed of organic muds except where sand is abundant on the margins of exposed areas. • Exposed areas are located along bays with wide fetches and along heavily trafficked waterways. • Sheltered areas are not exposed to significant wave or boat wake activity. • Abundant resident flora and fauna with numerous species and high use by birds, fish, and shellfish.	• Oil adheres readily to intertidal vegetation. • The band of coating will vary widely, depending on the water level at the time of oiling. • Large slicks will persist through multiple tidal cycles and will coat the entire stem from the high tide line to the base. • Heavy oil coating will be restricted to the outer fringe of thick vegetation, although lighter oils can penetrate deeper, to the limit of tidal influence. • Medium to heavy oils do not readily adhere to or penetrate the fine sediments but can pool on the surface or in animal burrows and root cavities. • Light oils can penetrate the top few centimeters of sediment; under some circumstances oil can penetrate burrows and cracks up to one meter.

continued

TABLE G.1 Continued

Coastal Habitat	Description	Predicted Oil Behavior
Inundated Low Lying Tundra (ESI 10E)  SOURCE: Strang (2013).	• This shoreline type occurs where very low-lying sections of the Arctic shoreline have been recently flooded by the sea, due to subsidence. • Also includes areas that are not normally in the intertidal zone but can be frequently inundated by salt water during spring tides or wind induced surges. • They have complex and convoluted shorelines comprised of tundra, vegetated flats, riverbanks, peat mats, brackish lagoons, and small streams. • These shorelines have high ice content; the surface material is mostly peat with little mineral sediments. • Where present, the vegetation is salt-tolerant and may be more adapted to drier conditions than the salt marshes. • The tundra is a living plant community and provides important feeding areas for migrating birds in the summer.	• Oil could be stranded onshore only during the ice-free summer season. • During storm surges, spilled oil could strand hundreds of meters inland. • During the summer months, the surface sediments/peat deposits are usually water-saturated, so stranded oil is likely to remain on the surface. • Physical removal rates of medium to heavy oils will be slow.
Mangroves (ESI 10F)  SOURCE: NOAA (2019a).	• The roots and trunks are intertidal, with only the lowest leaves inundated by high tide. • The width of the forest can range from one tree to many kilometers. • The substrate can be sand, mud, leaf litter, or peat, often as a veneer over bedrock. • Wrack accumulations can be very heavy. • They are highly productive, serve as nursery habitat, and support a great diversity and abundance of animal and plant species.	• Oil can wash through mangroves if oil comes ashore at high tide. • If there is a berm or shoreline present, oil tends to concentrate and penetrate into the berm sediments or accumulated wrack/litter. • Heavy and emulsified oil can be trapped in thickets of red mangrove prop roots or dense young trees. • Oil readily adheres to prop roots, tree trunks, and pneumatophores. • Re-oiling from resuspended or released oil residues may cause additional injury over time. • Oiled trees start to show evidence of effects (leaf yellowing) weeks after oiling; tree mortality may take months, especially for heavy oils.
Subtidal		
Coral Reefs  SOURCE: Wise Hok Wai Lum, CC BY-SA 4.0, via Wikimedia Commons.	• Coral reefs are structures created and maintained by the establishment and growth of populations of stony coral and coralline algae. • Coral reefs are mostly subtidal in nature, although the shallowest portions of some reefs can be exposed during very low tides. • Broad, pavement-like platforms formed by reefs when they reach sea level are a special concern. • Many coral species spawn simultaneously over a very short time period (days), a behavior that makes the entire recruitment class very vulnerable.	• Coral reefs vary widely in sensitivity to spilled oil, depending on the water depth, oil type, and duration of exposure. • There are three primary exposure pathways: direct contact with floating oil; exposure to dissolved and dispersed oil in the water column; and contamination of the substrate by oil deposited on the seafloor. • Reef-associated community of fishes, crustaceans, sea urchins, etc. can experience significant mortality.
Seagrasses  SOURCE: NOAA.	• Seagrasses are highly productive habitats that occur on intertidal flats and in shallow coastal waters worldwide from arctic to tropical climates. • Water temperature, light penetration, sediment type, salinity, and wave or current energy control seagrass distribution. • Seagrasses provide a food source for green turtles, manatees, and waterfowl, who graze on seagrasses. • Seagrasses are used by fish and shellfish as nursery areas.	• Oil will usually pass over subtidal seagrass beds, with no direct contamination. • Oil that is heavier than seawater can become trapped in the beds, coating the leaves and sediments. • Oil readily adheres to the vegetation, and the oiled blades are quickly defoliated when intertidal beds are oiled. • Floating oil stranded on adjacent beaches can pick up sediment and then get eroded and deposited in adjacent beds.

TABLE G.1 Continued

Coastal Habitat	Description	Predicted Oil Behavior
Kelp  SOURCE: NOAA.	• Kelps are very large brown algae that grow on hard subtidal substrates in cold temperate regions. • Because kelps require constant water motion to provide nutrients, they are located in relatively high-energy settings. • Kelp forests support a diverse animal community of fish, invertebrates, and marine mammals as well as important algal communities.	• Kelp has a mucous coating that prevents oil from adhering directly to the vegetation on the water surface. • Oil can be trapped in the dense surface canopy, increasing the persistence of oil within the kelp environment. • Oil persistence in kelp increases the risks of exposure to organisms concentrated in kelp forest habitats.
Soft Bottom  SOURCE: USGS.	• Soft-bottom, subtidal habitats consist of various percentages of sand, silt, and clay, occurring in sheltered bays and estuaries, and deeper offshore areas. • The presence of fine-grained sediments indicates that the substrate is not exposed to significant wave or tidal energy. • Biological resources associated with this habitat include shrimp, crabs, clams, fish, and the pelagic and benthic communities that support them (e.g., plankton, worms, amphipods, isopods).	• This habitat is not often exposed to spilled oil. • The greatest risk of exposure is from the sinking oil or the sorption of dispersed oil onto suspended sediments that are then deposited on the bottom. • Significant natural dispersion of oil and sediments into the water column occurs only during large storms and nearshore oil spills. • Shoreline cleanup can suspend oil and fine-grained sediments, causing deposition of oily sediments in nearshore habitats.
Mixed and Hard Bottom  SOURCE: NOAA.	• This habitat consists of subtidal substrates composed of rock, boulders, or cobbles, though there can be patches of sand veneer covering a hard bottom. • There may be rich, diverse communities of attached and associated algae and animals; often there is little open space. • Some of these habitats form a relief (reef or bank) several meters high that attracts a diversity of fish.	• Mixed and hard-bottom habitats are usually considered to have low risk of exposure to oil spills. • Oil in the water column seldom reaches toxic levels and benthic organisms have little exposure. • There is little risk of deposition of oil or oiled sediments in these habitats. • There could be a short-term exposure as oiled sediments are transported through the habitat into deeper areas. • Concerns about seafood contamination from dispersed oil or oiled sediments can become a significant issue. • Real, potential, or fear of contamination can close seafood harvesting activities.
Ice		
Accessible and Inaccessible Ice  SOURCE: Sandkvist (n.d.).  SOURCE: USCG photo by Petty Officer 1st Class Sara Francis, Public domain, via Wikimedia Commons.	• Ice forms on the sea surface during winter in cold climates and can persist for several months. • Most sea surface ice is floating but can be frozen to the bottom or stranded in intertidal areas during low tide. • Accessible ice can safely support the personnel and equipment suitable for response to a particular oil spill on, in, under, or adjacent to solid ice. • Inaccessible ice cannot safely support response personnel and response equipment.	• Ice along the shoreline or in the adjacent nearshore water can act as a natural barrier, reducing the amount of oil that might otherwise make contact with the shoreline substrate. • During the ice growth phase, oil in or under the ice can become encapsulated within the ice. • During a thaw, or if the surface of the ice is melting and wet, oil is unlikely to adhere to the ice surface and will tend to remain on the water surface or in leads. • In the spring, before the ice becomes inaccessible, oil in or below sea ice will often migrate through brine channels to the surface.

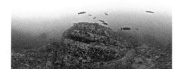

continued

TABLE G.1 Continued

Coastal Habitat	Description	Predicted Oil Behavior
On-Water		
Offshore SOURCE: NOAA OR&R.	• Offshore waters are those where the water depth is >30 feet (10 meters) with no surrounding land. • Evaluation of environmental impacts to open water habitats is focused on water column organisms and those which inhabit or use the sea surface. • Animals include marine mammals, sea turtles, pelagic birds, and many commercially and recreationally important fish and pelagic invertebrates. • Organism densities in this habitat are low on average. • Localized high densities can occur in areas such as convergence zones and upwelling areas. • Pelagic birds are at greatest risk when large numbers are concentrated for feeding, migration, overwintering, or breeding. • Biological resources in the water column are less vulnerable to spills than those at the water surface. • The sea surface microlayer is important for biochemical processes; the organisms most vulnerable to exposure are poor or passive swimmers (planktonic forms).	• Spilled oil transport is controlled primarily by wind and ocean currents than by tides and mixing with freshwater outflows. • Most of the soluble and toxic components of the spilled oil are lost through weathering within hours and days. • Dissolved or dispersed oil concentrations are likely to be greatest in the top few meters.
Bays and Estuaries SOURCE: Bruce A. Davis, U.S. Department of Homeland Security.	• Near coastal waters partially surrounded by land and more sheltered than offshore habitats. • Limited circulation and flushing, with depths frequently <30 ft. • Suspended sediment concentrations can be high. • Highly sensitive to oil spills, particularly where flushing rates are low and the probability of contact increases. • Many species spawn in these habitats during spring, and their sensitive early life stages can persist in shallow waters. • Large numbers of migratory or wintering waterfowl, wading, and diving birds are present. • Home to marine mammals and sea turtles. • Used by commercially or recreationally important finfish, shellfish, and other organisms that migrate seasonally.	• Oil can impact bottom habitats (benthic organisms) when water is shallow. • Stranded oil on nearby shorelines can become a prolonged source for oil re-released to the water column. • Tides and fresh water can substantially influence spilled oil movement.

NOTE: ESI = Environmental Sensitivity Index Code as described in NOAA, 2019a.
SOURCES: NOAA (2017). For detailed descriptions of intertidal areas, see Petersen et al. (2019).

TABLE G.2 Description of Surface Oil Distribution

T—Trace
1% cover

S—Sporadic cover
1–10% cover, seen here as brown oil bands on a white sand beachface.

P—Patchy cover
11–50% cover, seen here as black oil bands on a white sand beachface.

B—Broken cover
51–90% cover, seen here as brown oil on tan sand beach.

C—Continuous cover
91–100% cover, seen here as black oil on light sand beach.

SOURCE: Information and images are from NOAA OR&R: https://response.restoration.noaa.gov/surface-oil-distribution-percent-cover.

TABLE G.3 Surface Oiling Descriptors: Thickness

FL—Film
Transparent or iridescent sheen or oily film.

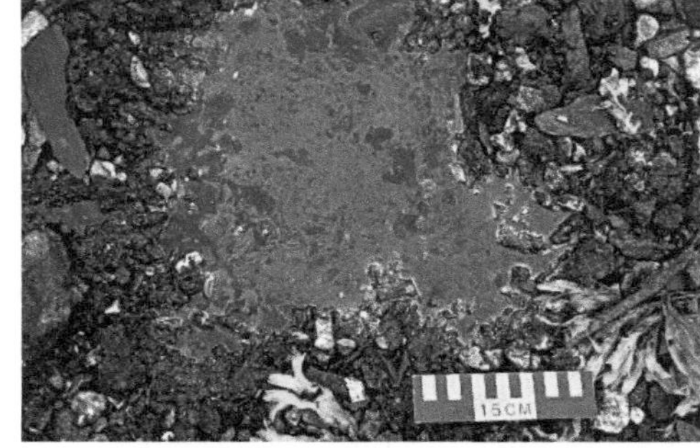

ST—Stain
Visible oil that cannot be scraped off with a fingernail.

CT—Coat
Visible coating of oil less than 0.1 cm thick; can be scraped off with a fingernail.

CV—Cover
Oil or mousse more than 0.1 cm and less than 1 cm thick.

TO: Thick Oil
Fresh oil or mousse more than 1 cm thick.

SOURCE: Information and images are from NOAA OR&R: https://response.restoration.noaa.gov/surface-oiling-descriptors-thickness.

TABLE G.4 Surface Oiling Descriptors: Type

NO—No Oil
No evidence of any type of oil.

AP—Asphalt Pavements
Cohesive, heavily oiled surface sediments.

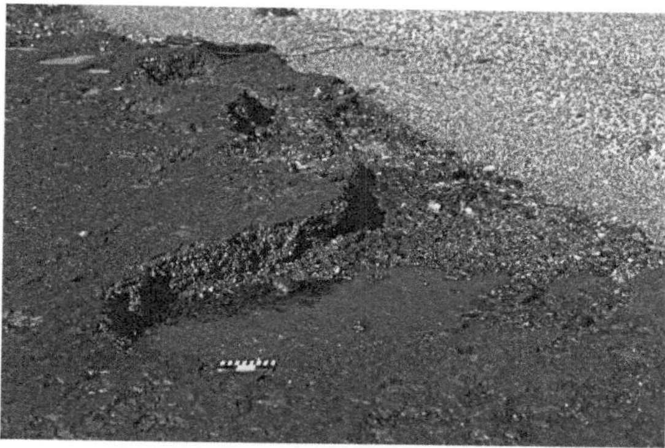

SR—Surface Oil Residue
Non-cohesive, heavily oiled surface sediments, characterized as soft incipient asphalt pavements.

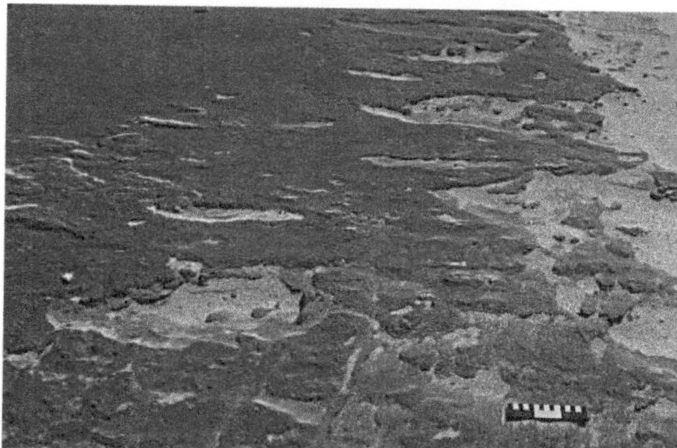

TC—Tar
Highly weathered oil of nearly solid consistency.

PT—Patties
Discrete accumulations of oil more than 10 cm in diameter.

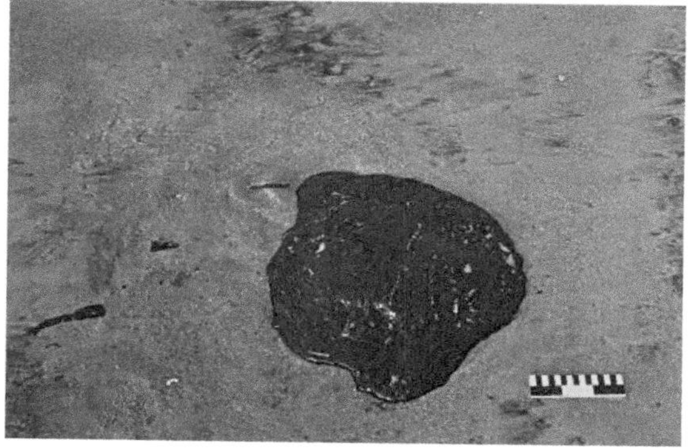

continued

TABLE G.4 Continued

NO—No Oil No evidence of any type of oil.	
TB—Tar Balls Discrete accumulations of oil less than 10 cm in diameter.	
MS—Mousse Emulsified oil. An example of mousse is seen here as orange-brown oil coating cobbles.	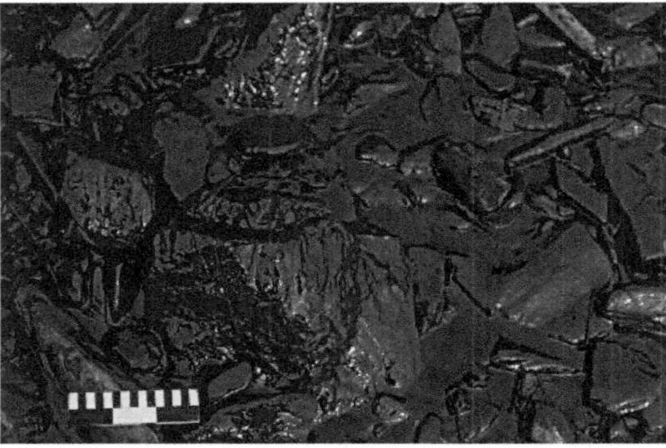
FR—Fresh Oil Unweathered, liquid oil.	

SOURCE: Information and images are from NOAA OR&R: https://response.restoration.noaa.gov/surface-oiling-descriptors-type.

TABLE G.5 Sediment Types

M—Mud
Mud can be composed of silt and/or clay.

S—Sand
Sand grains measure between 0.06 and 4 mm in diameter.

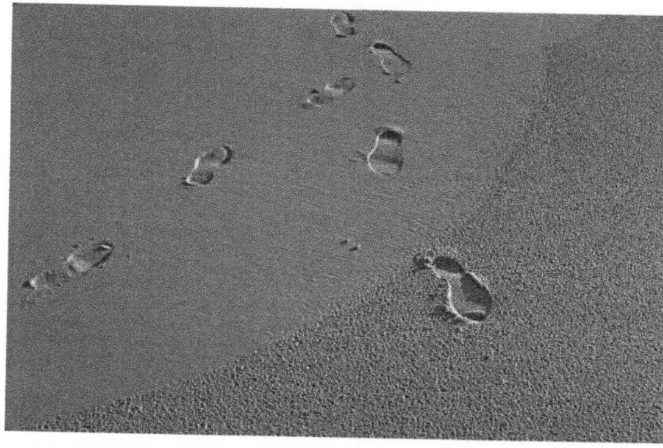

G—Granule
Granules measure between 2 and 4 mm in diameter.

P—Pebble
Pebbles measure between 4 and 64 mm in diameter.

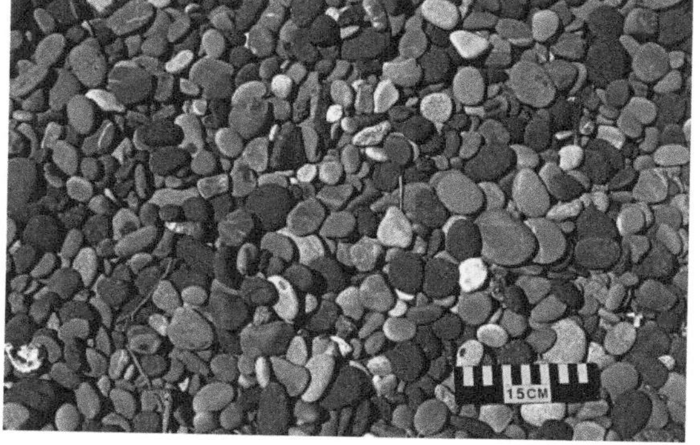

continued

TABLE G.5 Continued

C—Cobble
Cobbles measure between 64 and 256 mm in diameter.

B—Boulder
Boulders measure more than 256 mm in diameter.

R—Bedrock Outcrop

SOURCE: Information and images are from NOAA OR&R: https://response.restoration.noaa.gov/sediment-types.

BRIEF DESCRIPTION OF SHORELINES IN THE SCAT (EXCERPTS)

For detailed descriptions of intertidal areas see Petersen et al. (2019).

Rank of 1: Exposed, Impermeable Vertical Substrates. These shoreline types are exposed to large waves, which tend to keep oil offshore by reflecting waves. The substrate is impermeable so oil remains on the surface where natural processes will quickly remove any oil that does strand within a few weeks. Any stranded oil forms a band along the high tide line or splash zone.

Rank of 2: Exposed, Impermeable Substrates, Non-Vertical. These shorelines are exposed to high wave energy. They have a flatter intertidal zone, sometimes with small accumulations of sediment at the high tide line, where oil could persist for several weeks to months.

Rank of 3: Semi-Permeable Substrate, Lower Potential for Oil Penetration and Burial; infauna present but not usually abundant. These types of shorelines include exposed sand beaches on outer shores, sheltered sand beaches along bays and lagoons, and sandy scarps and banks along lake and river shores. Compact, fine-grained sand substrates minimize oil penetration, reducing the amount of oiled sediments to be removed. Fine-grained sand beaches generally accrete slowly between storms, reducing the potential for burial of oil by clean sand. On exposed beaches, oil may be buried deeply if the oil stranded right after an erosional storm or at the beginning of a seasonal accretionary period.

Rank of 4: Medium Permeability, Moderate Potential for Oil Penetration and Burial; infauna present but not usually abundant. Coarse-grained sand beaches are ranked separately and higher than fine- to medium-grained sand beaches because of the potential for higher oil penetration and burial, which can be as great as 1 m. These beaches can undergo very rapid erosional and depositional cycles, with the potential for rapid burial of oil, even after only one tidal cycle.

Rank of 5: Medium-to-High Permeability, High Potential for Oil Penetration and Burial; infauna present but not usually abundant. The gravel-sized deposits can be bedrock, shell fragments, or coral rubble. Because of higher permeability, oil can penetrate into sand and gravel beaches, making it difficult to remove contaminated sediment without causing erosion and waste disposal problems. These beaches may undergo seasonal variations in wave energy and sediment reworking, so natural removal of deeply penetrated oil may only occur during storms events. Pocket beaches with microenvironments are protected from wave energy; therefore, natural oil removal may be significantly slower.

Rank of 6: High Permeability, High Potential for Oil Penetration and Burial. Gravel beaches have the potential for very deep oil penetration and slow natural removal rates of subsurface oil. Fine-grained gravel beaches are composed primarily of pebbles and cobbles (from 4 to 256 mm), with boulders as a minor fraction. Coarse-grained gravel beaches have boulders dominating the lower intertidal zone. A boulder-and-cobble armoring of the surface of the middle to lower intertidal zone is common on these beaches. Armor may have a very important effect on oil persistence in gravel beaches. Oil beneath the armored surface can remain longer than would subsurface oil on an unarmored beach with similar grain size and wave conditions. Riprap is a man-made equivalent of this ESI class; it is usually placed at the high tide line where the highest oil concentrations are found, and the riprap boulders are sized so that they are not reworked by storm waves.

Rank of 7: High Permeability, High Potential for Oil Penetration and Burial. Exposed tidal flats commonly occur with other shoreline types, usually marshes, on the landward edge of the flat. They can occur as offshore tidal flats separate from the shoreline, particularly at tidal inlets and in tidal rivers. Oil does not readily adhere to or penetrate the compact, water-saturated sediments of exposed sand flats. Instead, the oil is pushed across the surface and accumulates at the high tide line. Even when large slicks spread over the tidal flat at low tide, the tidal currents associated with the next rising tide pick up the oil and move it alongshore. However, oil can penetrate the tops of sand bars and burrows if they dry out at low tide.

Rank of 8: Sheltered Impermeable Substrate; epibiota usually abundant. Oil can coat rough rock surfaces in sheltered settings, and can persist long-term because of the low-energy setting. While solid rock surfaces are impermeable to oil, rocky rubble slopes can trap oil beneath a veneer of coarse material. Sheltered seawalls and riprap are the man-made equivalents, with similar oil behavior and persistence patterns. In riverine settings, terrestrial vegetation along the river bluff indicates low energy and thus slow natural removal rates. Peat shorelines include peat scarps, eroded peat, and peat slurries; this shoreline type is most common along the Bering Sea.

Rank of 9: Sheltered, Flat, Semi-Permeable Substrate, Soft; infauna usually abundant. The soft substrate and limited access makes sheltered tidal flats difficult once oil reaches these habitats, natural removal rates are very slow. They can be important feeding areas for birds and rearing areas for fish, making them highly sensitive to oil-spill impacts.

Rank of 10: Vegetated Emergent Wetlands. Marshes, mangroves, and other vegetated wetlands are the most sensitive habitats because of their high biological use and value, difficulty of cleanup, and potential for long-term impacts to many organisms. They occur along the high-water line, where oil also often strands. Oil readily adheres to the vegetation. Medium to heavy oils do not readily adhere to or penetrate into the fine-grained soils but can pool on the surface or in animal burrows and root cavities, and soak into accumulated organic matter, such as wrack. Oil persistence under these conditions can be very long term, and cleanup activities can damage the vegetation or mix the oil deeper into the soils.

Appendix H

'Omics Techniques

The following text provides an overview of 'omics techniques, focusing on their application to analysis of microbial communities present in environmental samples or laboratory cultures. Note that 'omics techniques may be used differently to study individual higher organisms and their structures or organs (see Chapter 6).

Genomics is the sequencing and study of genomes (i.e., the full complement of DNA in cells, including chromosomes, plasmids, and virus genomes) including genome structure (e.g., chromosome sequence and gene order), function (e.g., potential biochemical activity) and evolution (e.g., mutability and genetic exchange). Genomes are

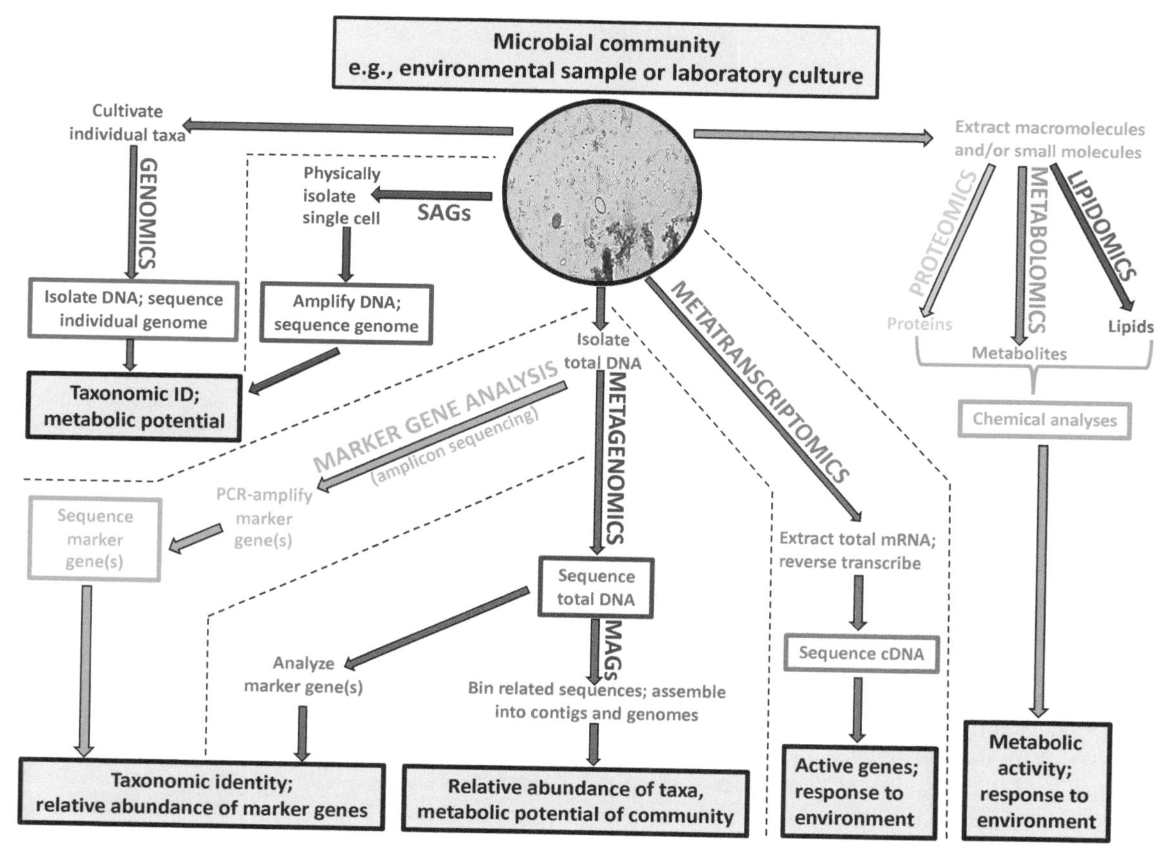

FIGURE H.1 Schematic of 'omics techniques, outputs and uses relevant to microbial communities in environmental samples or laboratory cultures.
SOURCE: Inset micrograph adapted from Reef2Reef.

obtained by sequencing the entire DNA content of a single organism or colony of identical cells then, ideally, organizing those sequences into a complete genome without sequence gaps. The genome sequence can reveal the taxonomy (classification) of an individual organism, potential enzymatic capabilities (based on gene sequences that encode enzymes or structural proteins), as well as mobile accessory genes encoded on plasmids or viral genomes (e.g., bacteriophage). The sequences encoding functional or structural products (e.g., proteins, RNA molecules used in protein synthesis and DNA replication) are commonly identified by comparison to publicly available databases using bioinformatics tools. For example, the genomes of hydrocarbon-degrading bacteria contain suites of genes encoding enzymes (catalytic proteins) that act sequentially in biochemical pathways that degrade specific hydrocarbon classes. The genomes usually contain all the genes encoding enzymes for a complete biodegradation pathway plus the regulatory genes that control when those enzymes are produced (see Transcriptomics below) as well as structural proteins for solute transport, detoxification, and sometimes motility (see Proteomics and Metabolomics below). Comparative genomics enables the discovery of novel genes in uncultivated organisms by comparison with known organisms. Pairing genomics information with classical cultivation methods and metabolomics strengthens the use of certain genomes (or selected marker genes) as proxies for biodegradation potential in a given environment.

An exciting development in genomics is the single-cell amplified genome (SAG) technique that, as the name implies, allows the genome of an isolated individual microbial cell to be sequenced without requiring cultivation of the cell. A microfluidic device is used to physically manipulate individual cells (e.g., suspended in seawater) into microscopic wells where the cell is lysed and its DNA amplified, sequenced, and assembled into contigs (contiguous DNA sequences assembled bioinformatically by overlapping multiple shorter DNA sequences) and scaffolds (long sequences comprising contigs and gaps). If sufficient sequencing "coverage" of the genome has been achieved, it is possible to assemble the contigs and scaffolds into a complete genome, but often there are gaps and the virtual organism is delineated by the inferred percentage of completeness of the genome. The process is facilitated if there is an identical or very similar genome already in a database for comparison, but it is also possible to generate a virtual genome de novo (i.e., solely from overlapping sequenced fragments of a hitherto-unknown organism). In this way, key microbes that cannot yet be cultivated in a laboratory for characterization can be sequenced and their activity inferred from their genome sequence. For obligate symbionts that cannot be cultivated alone and/or may be present in small proportions in a community, such as those implicated in anaerobic hydrocarbon biodegradation, SAG is a breakthrough technique.

Metagenomics is the study of the collective genomes comprising a mixed community of organisms, such as the microbiota in an environmental sample or a laboratory culture containing multiple species. First, the cells are lysed to release total community DNA (a step that introduces potential bias). This DNA can then be used in several ways to acquire different information: (1) Marker gene sequencing and analysis of bacterial 16S rRNA gene regions is commonly used to determine the taxonomic types and relative proportions of different microbes within a community. A specific region of this marker gene is amplified from the extracted DNA using polymerase chain reaction (PCR) then sequenced by high-throughput sequencing (aka next-generation sequencing). Various bioinformatic tools remove spurious sequences, ensure quality control, and compare the marker gene sequences to public databases such as GenBank, SILVA, RDP, or Greengenes to infer the identity of discrete groups of sequences (taxa) based on their similarity to archived sequences and their place in phylogenetic trees constructed using the same marker gene. The reference sequences may be from cultivated isolates that have been studied and characterized, or may themselves be known only as a DNA sequence (an operational taxonomic unit or amplicon sequence variant) that was also obtained by sequencing community DNA and has never been cultivated in a laboratory. In the latter case, the metabolism and behavior of the "virtual microbe" can only be inferred through its relationship to the relatively few cultivated organisms that have been studied biochemically in a lab. Although conventional (non-quantitative) rRNA gene analysis does not yield information about the absolute cell numbers of different taxa in the sample, analysis of multiple samples can reveal the taxonomic diversity of the community as well as changes in community composition that occur over time and space or in response to changing environmental conditions, such as pre- and post-exposure to oil. Quantitative PCR allows estimation of numbers of marker genes in a sample Notably, PCR amplification is known to introduce biases that can over- or under-represent particular taxa. This is one of the reasons why marker gene sequencing, which has been used extensively in the past two decades, is largely being displaced by (2) shotgun metagenome sequencing. In this technique, the entire community DNA is sequenced without PCR amplification and often without sequence assembly, and desired marker genes are sorted in silico by bioinformatic software. Selecting and analyzing taxonomic marker sequences like 16S rRNA genes will reveal the diversity of the community and the relative abundance of taxa, whereas analysis of functional marker genes (e.g., those encoding key enzymes) by using annotation software and functional gene databases can reveal the presence, diversity, and relative abundance of genes associated with potential metabolic activity if those genes are expressed (e.g., hydrocarbon biodegradation). Alternatively, using bioinformatics software the sequence information can be sorted into "bins" of related sequences that are then reconstructed into (3) metagenome-assembled genomes representing individual taxa in the community.

This generates individual genomes (complete or partial) for further study and inference. One benefit to adding assembly to the analysis is that there is context for the marker genes (e.g., associating a potential enzyme activity with a specific biochemical pathway or a particular taxon). Metagenomic analysis of samples collected at different stages of oil biodegradation can reveal which species are affiliated with biodegradation of certain oil components as well as the potential trajectory of that biodegradation activity (e.g., partial or complete oxidation of the hydrocarbons).

It is important to note that the presence of a DNA sequence in a (meta)genome, for example one encoding an enzyme, is only an indicator of the potential activity of the organism; it does not necessarily mean that the organism (or community) is actually producing that enzyme or exhibiting the biochemical function encoded by that sequence. To determine whether the genetic potential is being manifested is informed by transcriptomics, the study of the mRNA transcripts in an organism (i.e., the expression of the complement of its genes that are active under specific conditions). This is achieved by sequencing the total mRNA (or, more precisely, its complementary DNA) present in the cell at a specific sampling time. A microbe may only express (transcribe) genes for a particular biochemical pathway if the substrate for that pathway is present in the environment; in its absence, the genes are still present in the genome and would be detected in the genomic sequence as potential activity, but they are silent and will not contribute to the transcriptome. The community extension of transcriptomics is the field of metatranscriptomics, the study of the full complement of DNA transcripts in a community of organisms. This method was developed to study the aggregate gene expression of all members of the community under specific conditions. This is important because the transcriptome of a single species in isolation (e.g., in pure culture in a laboratory) likely will differ from its expression in the environment and in the presence of a community, because biochemical signals from the whole can enhance or suppress gene expression in individual community members. Metatranscriptomic studies often examine changes in transcription under two or more different conditions rather than just in a given baseline condition. That is, it is the relative abundance of transcripts, both up- and down-shifted, that provides information about how the community responds to environmental conditions such as oil incursion or oxygen depletion. Combining metatranscriptomic and metagenomic analyses yields a fuller picture of how a community behaves under baseline and perturbed conditions, something that was impossible only a decade ago.

Proteomics, lipidomics, and metabolomics study the products of gene transcription—that is, which macromolecules (proteins and lipids) and metabolites the organism(s) produce under specific conditions. Typically, these studies are conducted under two or more different growth conditions to highlight changes in cell or community response to particular stimuli (e.g., temperature, growth substrates, toxins, etc.) and, as with transcriptomics, the data are often reported as fold increases or decreases from a baseline condition. Advanced analytical chemistry techniques such as mass spectrometry and nuclear magnetic resonance are used to characterize the proteome, lipidome, and/or metabolome of organisms by comparison to public databases (e.g., UniProt, LipidMaps, and National Metabolomics Data Repository, respectively) and to infer biochemical pathways and metabolic products.

Proteomics is the study of the total complement of proteins (the proteome) produced by an organism under specific conditions. It includes protein identification, quantification, and enzymatic modification of proteins and is commonly studied using mass spectrometry of protein extracts. The proteins may be catalytic (e.g., enzymes), structural (e.g., proteins for uptake of substrates or excretion of metabolites, or for motility toward substrates like oil or away from toxins), or regulatory (controlling transcription of genes).

Lipidomics is the study of the identity, structure, and function of the complete profile of lipids produced by an organism (the lipidome), such as those comprising cell membranes (and organelles in eukaryotes) that affect surface area for substrate uptake and waste excretion, or for specialized functions like photosynthesis.

Metabolomics is the study of small molecules produced by organisms as intermediates or end products of biochemical pathways (i.e., metabolites). For example, some microbes completely oxidize hydrocarbons to CO_2 and H_2O, whereas other hydrocarbons are only partially oxidized and the pathway metabolites (e.g., acids or alcohols) are excreted to the external environment.

Metabolism, which typically can only be inferred from taxonomy, is best deduced by comparison to cultivated organisms (a highly biased set, as mentioned above) or by analyzing metatranscriptomes and/or macromolecules. In conjunction with 'omics, incubating a community with a growth substrate enriched in stable carbon or nitrogen isotopes (13C,15N; stable isotope probing) specifically labels the macromolecules and metabolites of active organism(s) directly involved in substrate assimilation so that key microbial species can be differentiated from "accessory" members of a community. Thus, 'omics can reveal significant shifts in community composition in response to environmental conditions, can predict and identify metabolic potential, and may provide clues that eventually allow cultivation of "virtual microbes" for detailed biochemical and environmental study.

Appendix I

Table of Common Hydrocarbon Degraders

TABLE I.1 Some Commonly Detected Marine Hydrocarbon-Degrading Bacteria (Aerobic Degraders Unless Otherwise Indicated)

Preferred Substrate(s)	Organism Name	Typical Marine Environment	Comments	Selected Citations
ALIPHATICS				
Methane	*Methylomonas* spp.	Natural gas seeps	Obligate methanotroph. Enriched in *Deepwater Horizon* plume during methane depletion	Dubinsky et al., 2013
n-Alkanes C_9–C_{32}; *iso*-alkanes (e.g., isoprenoids); alkyl components of alkyl-cycloalkanes and alkyl aromatics	*Alcanivorax borkumensis*	Ubiquitous in marine ecosystems (water, sediment, coastal, deep sea); rare in pristine waters; DNA detected in polar areas but only isolated from more temperate environments	Obligate hydrocarbonoclastic species; produces biosurfactants, forms emulsions; early responder; widespread; found in partnership with marine invertebrates; often early to bloom in response to oil	Yakimov et al., 1998, 2007; Dutta and Harayama, 2001; Gregson et al., 2019; Joye and Kostka 2020; Van Landuyt et al., 2020
Methane and liquid alkanes	*Candidatus* Macondimonas diazotrophica	Oil-contaminated marine sediments worldwide	Bloomed in Gulf of Mexico after *Deepwater Horizon* to 30% of total sediment microbes; N_2 fixer	Karthikeyan et al., 2019
Gaseous and liquid *n*-alkanes; cycloalkanes	Order Oceanospirillales: *Oceanospirillum* *Oceaniserpentilla* *Bermanella*	Cold marine waters	Various genera dominant during *Deepwater Horizon* spill; genes for mono-aromatic and PAH degradation detected but expressed poorly	Mason et al., 2012; Kleindienst et al., 2016; Hu et al., 2017
Alkanes	*Oleibacter* spp.	Temperate water; deep water (Mariana Trench)	Only one species has been cultivated; others detected using 'omics	Lofthus et al., 2018; Liu et al., 2019; Schreiber et al., 2021
Alkanes	*Oleiphilus messinensis*	Sponge symbiont	Obligate hydrocarbonoclastic species	Golyshin et al., 2002; Yakimov et al., 2007
Alkanes	*Oleispira antarctica*	Cold water and high latitudes; sea ice	Psychrophilic; may also degrade Corexit 9500A components	Yakimov et al., 2003; Kube et al., 2013; Boccadoro et al., 2018; Lofthus et al., 2018; McFarlin et al., 2018
>C16 *n*-alkanes; Isoprenoids (e.g., squalane)	*Alkanindiges illinoisensis*	Few reports in marine systems; Arctic beach, marine biofilms	Obligate hydrocarbonoclastic species	Røberg et al., 2011; Vergeynst et al., 2019a
n-Alkanes	*Thalassolituus oleivorans*	Marine waters and sediments	Obligate hydrocarbonoclastic species; particularly associated with cold oil biodegradation; degrades nC_{10}–nC_{32} but not pristane	Yakimov et al., 2004; Brakstad et al. 2015b; Gregson et al., 2018; Shtratnikova et al., 2018
n-Alkanes	*Halomonas neptunia* *Halomonas titanicae*	Cold marine (deep sea and Antarctic surface water)	Produces bio-emulsifier when growing on hydrocarbon	Pepi et al., 2005; Van Landuyt et al., 2020

continued

TABLE I.1 Continued

Preferred Substrate(s)	Organism Name	Typical Marine Environment	Comments	Selected Citations
n-Alkanes	*Zhongshania* spp.	Cold marine water	Early responder in cold seawater, possibly specializing in short- to medium-chain alkanes	Ribicic et al., 2018; Murphy et al., 2021
n-Alkanes	*Paraperlucidibaca*	Cold marine water	Metagenome detected in sub-Arctic marine sediments incubated with diesel or crude oil	Murphy et al., 2021
n-Alkanes	*Desulfosarcina/ Desulfococcus* clade	Marine seeps	Anaerobic degradation via sulfate reduction	Kleindienst et al., 2014
AROMATICS WITH OR WITHOUT ALIPHATICS				
Aromatics including PAHs and PACs; also ethane, propane, butane	*Cycloclasticus* spp.	Global distribution	Obligate hydrocarbonoclastic bacterial genus; associated with dilbit degradation by 'omics	Kasai et al., 2002; Yakimov et al., 2007; Brakstad et al. 2015b; Messina et al., 2016; Rubin-Blum et al., 2017; Gutierrez et al., 2018; Murphy et al., 2021; Schreiber et al., 2021
PAH and long-chain alkanes	*Marinobacter* spp.	Ubiquitous	Form biofilms and produce emulsifiers; tolerates high salt concentrations	Gauthier et al., 1992; Yakimov et al., 2007; Brakstad et al. 2015b Laio et al., 2015; Murphy et al., 2021
PAH and various alkanes	*Pseudoalteromonas*	Global distribution; versatile heterotroph	Enriched in cold North Sea water microcosms with oil	Chronopoulou et al., 2015
Short-chain alkanes (C_2–C_4), benzene, PAHs	*Colwellia* spp.	Global distribution, but certain strains are adapted to cold marine water and sea ice; also detected in deep sea sediments	Associated with marine oil snow; possibly sensitive to hydrostatic pressure; may metabolize dispersant components	Bælum et al. 2012; Redmond and Valentine, 2012; Dubinsky et al., 2013; Mason et al., 2014a,b; Brakstad et al., 2015b; Barbato and Scoma, 2020
PAH	*Dietzia* spp.	Arctic seafloor sediments; Antarctic sediments	Predominant sequence in 16S marker gene survey; degrades phenanthrene and emulsifies diesel	Dong et al., 2015; Ausuri et al., 2021

NOTE: A more comprehensive list of hydrocarbon-degrading prokaryotes, including from terrestrial and freshwater sources, has been prepared by Prince et al. (2018).

Appendix J

Committee Biographies

Kirsi K. Tikka (NAE) (*Chair*) is currently the independent non-executive director of Ardmore Shipping and Pacific Basin Shipping. She is actively involved in environmental regulatory and policy development for shipping as a member of the Royal Institute of Naval Architects' Committee for the International Maritime Organization.

Dr. Tikka has more than 30 years of shipping experience having recently retired from the American Bureau of Shipping (ABS) in July 2019. She joined ABS in 2001 and held various specialist and leadership positions, including executive vice president, global marine; Europe division president; and vice president and chief engineer, global. Her most recent ABS role was as executive vice president, senior maritime advisor, where she was responsible for aligning ABS strategic planning, client development, and product and service offerings with the industry's technical needs and requirements. She introduced and sponsored the environmental, harsh environment, and sustainability programs in ABS. She also represented ABS in the International Association of Classification Societies and was involved in the development of the Common Structural Rules for Tankers and Bulk Carriers.

From 1996 to 2001, Dr. Tikka was a professor of naval architecture at the Webb Institute in New York, where she was also awarded an honorary doctorate in 2018. In addition to teaching, she carried out research on prevention of oil pollution from shipping, tanker structural strength, and risk analysis, as well as being actively involved in the National Academies of Sciences, Engineering, and Medicine studies on double hull tankers. She served on the National Research Council's Marine Board Committee that produced the 1998 report *Double Hull Tanker Legislation: An Assessment of the Oil Pollution Act of 1990* and chaired the committee that produced the 2001 report *Environmental Performance of Tanker Designs in Collision and Grounding: Method for Comparison*. Dr. Tikka also worked for Chevron Shipping in San Francisco and Wärtsilä Shipyards in Finland. She joined Chevron at the time when the Oil Pollution Act of 1990 was being legislated, and she performed studies to evaluate its impact on tanker design and construction.

Dr. Tikka holds a doctorate in naval architecture and offshore engineering from the University of California, Berkeley (1989), and a master's degree in mechanical engineering from the University of Technology in Helsinki (1981). She is a fellow of both the Society of Naval Architects and Marine Engineers (SNAME) and the Royal Institution of Naval Architects. In 2012, she received SNAME's David W. Taylor Medal, the highest technical honor for naval architecture or marine engineering in the United States. Dr. Tikka is a foreign member of the National Academy of Engineering, and she serves on the University of California, Berkeley, Engineering Advisory Board.

Edwin "Ed" Levine (*Vice Chair*) is recently retired and is now the managing officer of Scientific Support & Coordination, LLC. Prior to retirement, he served as the regional operations supervisor—East for the National Oceanic and Atmospheric Administration's (NOAA's) Office of Response and Restoration's Emergency Response Division, managing the Scientific Support Coordinators (SSCs) from Maine to Louisiana.

From 1987 to 2015, he served as the SSC for the coastal area from Connecticut to Delaware. He has responded to several hundred incidents at the request of federal, state, and international officials, including working in Louisiana on the *Deepwater Horizon* oil spill and in Alaska on the *Exxon Valdez* oil spill. These responses ranged from crude through refined oil, to chemicals, and even floatable debris in the New York Bight. The more noteworthy incidents in this area were the Exxon Bayway pipeline failure and spill, *T/V Presidenté Rivera* and *T/V World Prodigy* tanker spills, *T/B Cibro Savannah* explosion and spill, the *C/V Santa Clara* I arsenic trioxide release, *T/B North Cape* and *Julie N* oil spills, and the B-125 barge explosion, fire and gasoline spill. Internationally, he has traveled to Uruguay, Honduras, Nicaragua, Ecuador, Mexico, Panama, China, Korea, England, Canada, Spain, and

Portugal for both planning and response activities. He worked on-scene at the oil spill from the *T/V Jessica* in the Galapagos Islands, *T/V Prestige* in Spain, and the *MT Hebei Spirit* in Korea. During the aftermath of the September 11, 2001, terrorist attacks, Mr. Levine assisted the U.S. Coast Guard (USCG) in New York City on environmental, response, and security issues. He has responded with the USCG to potential radiological and biological threats, as well as to Hurricanes Katrina, Rita, Ike, and Sandy.

The awards Mr. Levine has received include the USCG Meritorious Achievement Award (December 2019 and July 2004), USCG Certificates of Merit (May 2005, March 2004, and April 2000), the USCG Meritorious Team Commendations (November 2004 and March 2003), USCG Commander's Award for Civilian Service (August 1990), NOAA Certificate of Recognition (1995 and 1993), the U.S. Department of Transportation 9-11 Ribbon (September 2005), and the NOAA Administrators Award (2010).

As part of the contingency planning effort, Mr. Levine has helped review and comment on the Area Contingency Plans for the Coast Guard Captain of the Ports of Long Island Sound, New York City, and Philadelphia. He is also an advising member to the Regional Response Teams for Federal Regions I, II, and III.

Mr. Levine received his M.S. in marine sciences from the University of Puerto Rico, Mayagüez Campus (1981).

Akua Asa-Awuku is an associate professor at the University of Maryland, College Park, in the Department of Chemical and Biomolecular Engineering. She received her Ph.D. in chemical engineering from the Georgia Institute of Technology in 2008. She earned her M.S. in chemical engineering from the same institution in 2006 and received her B.S. in chemical engineering from the Massachusetts Institute of Technology in 2003. In 2008, Dr. Asa-Awuku served as a Camille and Henry Dreyfus Postdoctoral Fellow at the Center for Atmospheric Particle Studies and the Department of Chemical Engineering at Carnegie Mellon University.

Dr. Asa-Awuku's primary research interest is understanding and predicting aerosol sources and interactions with water. She is an expert in the physical and chemical characterization, fate, and transport of airborne nanoparticles. Her research explores the transformation of aerosol as it pertains to indoor and outdoor fuels and combustion sources. She has served as a panelist for the National Research Council's Research Associateship Program.

Cynthia Beegle-Krause is an interdisciplinary scientist committed to improving oil spill response and outcomes. She last served as a senior scientist at SINTEF Ocean, in Trondheim, Norway. She is also a full member of the International Oil Spill Control Organization. Her background is primarily in physical oceanography and biology. At SINTEF Ocean, she has been involved in a variety of projects related to oil spills from experimental field work in the Svalbard Archipelago. With funding from the Gulf of Mexico Research Initiative, she led projects developing improvements for modeling oil spills, such as calculating the dissolved oxygen consumption from the biodegradation of oil droplets and using Bayesian methods to find submerged and sunken oil. While working at a nonprofit in Seattle, Washington, she provided expert testimony for the Gitxaala Nation in Canada to the Canadian government regarding the risks of potential diluted bitumen (dilbit) spills. She was a lead oil spill response modeler in a previous position at the National Oceanic and Atmospheric Administration's (NOAA's) Office of Response and Restoration and one of the original developers on the GNOME (General NOAA Operational Modeling Environment) suite. She was also involved in *Oil in the Sea III*. Dr. Beegle-Krause holds a Ph.D. in physical oceanography from the University of Washington, an M.S. in physical oceanography from the University of Alaska Fairbanks, and a B.S. in biology from the California Institute of Technology.

Victoria Broje is an internationally recognized specialist with 20 years of experience in environmental science and emergency planning and response. She received her master's degree in offshore engineering from the Saint-Petersburg State Technical University in Russia, where she specialized in oil spill behavior under Arctic conditions. She received her doctoral degree in environmental science and management from the University of California, Santa Barbara. Her dissertation was focused on mechanical recovery of oil spills and resulted in a patented technology that later won the Wendy Schmidt Oil Cleanup X-Challenge. Since 2006 Dr. Broje has been supporting Shell businesses worldwide as a subject-matter expert for spill response technologies and environmental impacts assessments. She also leads an Environmental Unit Network providing oil spill response training to Shell staff. Dr. Broje represents Shell at the American Petroleum Institute (API), the International Association of Oil & Gas Producers, the Canadian Association of Petroleum Producers, and the International Petroleum Industry Environmental Conservation Association committees developing best practices in emergency response and environmental protection. She is a chair of the Board of the Clean Caribbean and Americas, a nonprofit organization dedicated to outreach on spill response and environmental protection topics. She also chairs the API Science and Technology Working Group for oil spill prevention and response. In 2013 Dr. Broje served on the National Academies of Sciences, Engineering, and Medicine committee that produced *A Review of Genwest's Final Report on Effective Daily Recovery Capacity (EDRC): A Letter Report*.

Steven Buschang has more than 25 years of experience working in the environmental sector along the Texas coast. Mr. Buschang began his career collecting much of the original data he still works with and oversees. The associated Minerals Management Service grant involved coordinating the accumulation of these data and the development of the first comprehensive geographic information system–Environmental Sensitivity Index biological data layer for the entire Texas coast. These data populate the state spill response atlas, commonly known as the Texas General Land Office (TGLO) Oil Spill Planning and Response Toolkit, a geospatial operational tool for oil spill response.

Mr. Buschang now works for the Bureau of Safety and Environmental Enforcement, but during his service on this committee he served on this committee he was TGLO's director of research and development and state Scientific Support Coordinator (SSC) overseeing an annual budget of more than $2 million, much of which is a dedicated funding stream for oil spill–related science research undertaken by Texas institutes of higher education.

Mr. Buschang earned a B.S. at Southwest Texas State University in marine biology and a master's degree from Texas A&M University–Corpus Christi in environmental science, where he later taught as an adjunct faculty member in the Department of Physical and Environmental Sciences, instructing both undergraduate and graduate courses in environmental regulation and environmental assessment.

Dagmar Schmidt Etkin has 45 years of experience in environmental analysis—14 years investigating issues in population biology and ecological systems and 31 years specializing in the analysis of oil spills. Since 1999, Dr. Etkin has been the president of Environmental Research Consulting (ERC), specializing in environmental risk assessment, spill response and cost analyses, and expert witness research and testimony related to oil spills. ERC's work focuses on providing regulatory agencies and industry with sound scientific data and perspectives for responsible environmental decision-making and risk assessment. Dr. Etkin has a broad range of experience related to oil spills, including environmental and socioeconomic impacts; oil behavior; oil spill costs; analysis of response strategies; and development of models of oil spill costs, environmental impacts, spill response, vessel traffic and spill risk, crude-by-rail risk, and well blowout probability.

She has been a consultant to numerous government agencies, including the U.S. Coast Guard, the National Oceanic and Atmospheric Administration, the Bureau of Safety and Environmental Enforcement, the Bureau of Ocean Energy Management, the U.S. Environmental Protection Agency, the U.S. Government Accountability Office, the Pipeline and Hazardous Materials Administration, the Maritime Administration, the Army Corps of Engineers, Environment Canada, the International Maritime Organization, the United Nations Environment Programme, the Washington State Department of Ecology, the Washington State Energy Facility Site Evaluation Council, the Minnesota Department of Commerce/Natural Resources, the California Department of Fish and Wildlife, and the Louisiana Office of Coastal Protection and Response, Plaquemines Parish (Louisiana), Mobile County (Alabama). Dr. Etkin has also been a consultant to the oil industry, including the American Petroleum Institute, ExxonMobil Upstream Research, SeaRiver Maritime, BP Shipping, Shell Oil, Woodside Energy Australia, Mitsubishi Tanker, Pipeline Research Council International, the American Salvage Association, Castrol Marine, Chevron Pipeline, Enbridge, Norbulk Shipping, Taylor Energy, and Cape Wind Associates. She has also consulted for nongovernmental organizations, including Scenic Hudson, Inc., World Wildlife Fund Canada, and the Cook Inlet Citizens Regional Advisory Council.

Dr. Etkin received her B.A. in biology from the University of Rochester (1977), her M.A. in biology from Harvard University (1980), and her Ph.D. in organismic and evolutionary biology from Harvard University (1982).

John Farrington is a dean emeritus at the Woods Hole Oceanographic Institution (WHOI), with expertise in marine chemistry and biogeochemistry. He joined WHOI in 1971 as a postdoctoral investigator. He held successive positions in the Department of Chemistry for 17 years and simultaneously served for 6 years as the director of the WHOI Coastal Research Center. He was the Michael P. Walsh Professor in the Environmental Sciences Program, at the University of Massachusetts Boston from 1988 to 1990. He returned to WHOI in 1990 as the dean of graduate studies and then the vice president for academic programs and the dean until November 2005, retiring in early 2006. He served as the interim dean of the School of Marine Science and Technology from 2009 to 2011 and as the interim provost during 2012 at the University of Massachusetts Dartmouth. His research interests include marine organic geochemistry, biogeochemistry of organic chemicals of environmental concern, and the interaction between science and policy. He has served on committees and panels for international, national, and local organizations, including the UNESCO-Intergovernmental Oceanographic Commission; the National Academies of Sciences, Engineering, and Medicine; the National Science Foundation; the Office of Naval Research; and the Commonwealth of Massachusetts. Dr. Farrington served as a member of the Gulf of Mexico Research Initiative Research Board. He has participated on seven National Academies consensus studies, chairing three, and has been a member on the National Academies' Environmental Studies Board, the Board on Environmental Studies and Toxicology, and the Marine Board. He is a Lifetime National Associate of the National Academies. Dr. Farrington holds a B.S. and an M.S. in chemistry from the University of Massachusetts Dartmouth and a Ph.D. in oceanography from the University of Rhode Island.

Julia Foght is a professor emerita at the University of Alberta, Canada, where she was a professor of petroleum microbiology in the Department of Biological Sciences from 1994 to 2014. Her expertise focuses on metagenomics of hydrocarbon-degrading microbial communities, biodegradation of petroleum hydrocarbons, fundamental studies on mechanisms of hydrocarbon transport across bacterial membranes, the use of whole-cell biocatalysts for biological upgrading of petroleum and refined products, and isolation and characterization of cold-adapted bacterial communities that live underneath glaciers. She received the Petro-Canada Young Innovators Award in 2001, a McCalla Professorship in 2011 from the University of Alberta, and the Alberta Science & Technology Foundation Award in Innovation in Oil Sands Research in 2014. She co-wrote the report *The Behaviour and Environmental Impacts of Crude Oil Released into Aqueous Environments* with the Royal Society of Canada in November 2015. Dr. Foght received her Ph.D. in environmental microbiology from the University of Alberta.

Bernard D. Goldstein (NAM) is the emeritus dean and an emeritus professor of environmental and occupational health at the University of Pittsburgh Graduate School of Public Health. He is an elected member of the National Academy of Medicine and has chaired more than a dozen National Academies of Sciences, Engineering, and Medicine committees. He has also chaired committees related to environmental health for the World Health Organization and the United Nations Environment Programme. His past experience includes service as the assistant administrator for research and development at the U.S. Environmental Protection Agency (1983–1985) and the president of the Society for Risk Analysis. His involvement in the *Deepwater Horizon* oil spill includes serving as an Advisory Board member of the National Academies' Gulf Research Program and as the original chair of the Coordinating Committee of the Gulf Research Health Outreach Program. He is also active on shale gas issues and on issues related to the science and policy interface.

Carys Mitchelmore is a professor at the University of Maryland Center for Environmental Science, Chesapeake Biological Laboratory in Solomons, Maryland. Her expertise is in environmental health and toxicology and her research emphasis is on understanding the fate and effects of chemicals and other pollutants on resident organisms. Dr. Mitchelmore's work focuses on the detection of pollutants in various environmental matrices and understanding their uptake, routes of exposure, metabolism, mechanisms of toxicity, bioaccumulation, and trophic transfer of chemical contaminants, as well as their implications to organism health, including humans. She also carries out toxicity testing and application for risk assessment, regulation, and management activities. Her investigations have focused on the chemical partitioning and fate and effects of crude oils, oil spill dispersants, organic disinfection by-products, and organic UV filters (components of sunscreens) in numerous invertebrate and vertebrate species, but especially sensitive and/or understudied species like corals and reptiles. Dr. Mitchelmore has served on two previous National Academies of Sciences, Engineering, and Medicine committees: the Committee on the Effects of Diluted Bitumen on the Environment (2016) and the Committee on Understanding Oil Spill Dispersants: Efficacy and Effects (2005), and was also a review coordinator for the recent Committee on the Use of Dispersants in Marine Oil Spill Response (2020). Dr. Mitchelmore received her Ph.D. from the University of Birmingham, United Kingdom, in 1997 for investigating the metabolism and effects of organic contaminants, including polycyclic aromatic hydrocarbons, to aquatic organisms.

Nancy Rabalais is a professor and the Shell Endowed Chair in Oceanography and Wetland Studies at the Louisiana State University College of the Coast and Environment. Dr. Rabalais's research includes the dynamics of hypoxic environments, interactions of large rivers with the coastal ocean, estuarine and coastal eutrophication, environmental effects of habitat alterations and contaminants, and the impacts of the oil through the water column from the deep benthic to the coastal and continental shelf. Dr. Rabalais is an American Association for the Advancement of Science Fellow, an Aldo Leopold Leadership Program Fellow, and a member of the National Academies of Sciences, Engineering, and Medicine. She received the 2002 Ketchum Award for coastal research from the Woods Hole Oceanographic Institution and shares the Blasker Award with R. E. Turner. She was awarded the American Society of Limnology and Oceanography Ruth Patrick Award and the National Water Research Institute Clarke Prize in summer 2008. Dr. Rabalais has served on 13 National Academies committees and served as a member and the chair of the National Academies' Ocean Studies Board (2000–2005). She received her Ph.D. in zoology from The University of Texas at Austin in 1983.

Jeffrey Short runs the consulting firm JWS Consulting in Alaska. Dr. Short began his career in oil pollution research in 1972, working for the National Oceanic and Atmospheric Administration's National Marine Fisheries Service on oil toxicity effects on Alaskan marine fauna prior to development of the Prudhoe Bay oil field and marine oil terminal in Valdez, Alaska. In investigating the *Exxon Valdez* spill, Dr. Short led numerous studies on the distribution, fate, and effects of the oil over two decades; these studies led to discovery of embryotoxic effects of oil pollution affecting fish at much lower concentrations that had been recognized previously, and quantitative assessments of lingering oil stranded on beaches and of other pollution sources in the *Exxon Valdez* spill region. He also worked on evaluating oil dispersant effectiveness under subarctic conditions and contributed to the oil budget for the *Exxon Valdez* spill, which provided a basis for evaluating the effectiveness of response measures. Dr. Short received his Ph.D. in fisheries from the University of Alaska Fairbanks in 2006, his M.S. in physical chemistry from the University of California, Santa Cruz, and his B.S. in biochemistry and philosophy from the University of California, Riverside.

Scott Socolofsky is a professor in the Zachry Department of Civil and Environmental Engineering at Texas A&M University. His research expertise is in the broad area of environmental fluid mechanics, with an emphasis on laboratory experiments and data analysis to elucidate mixing mechanisms by turbulence and coherent structures in multiphase flows. He has studied the fate and transport of oil in the offshore marine environment for more than 25 years, focusing on near field dynamics of oil spills, including the dynamics of subsea accidental oil well blowouts. He has conducted laboratory experiments of multiphase plumes, and led research cruises to study the fate of natural gas bubbles emitted a deepwater natural seeps in the Gulf of Mexico. He is the developer of the Texas A&M Oil spill/outfall Calculator (TAMOC), an open-source modeling suite for predicting the behavior and near field fate of oil and gas released from subsea spills and natural seepage. Dr. Socolofsky received his M.S. and Ph.D. in civil and environmental engineering from the Massachusetts Institute of Technology and his B.S. in civil and environmental engineering from the University of Colorado Boulder. Before joining Texas A&M University, Dr. Socolofsky worked as an engineer at Wright Water Engineering, Inc., in Denver, Colorado, and as a research associate at the University of Karlsruhe, Germany. He is currently the holder of the J. Walter "Deak" Porter '22 and James W. "Bud" Porter '51 Chair in the Zachry Department of Civil and Environmental Engineering at Texas A&M University.

Berrin Tansel is a professor in the Department of Civil and Environmental Engineering at Florida International University (FIU) in Miami, Florida. She has conducted extensive research and has published widely on oil–water emulsions remediation of contaminated media (water, sediments, and soil), sources of contaminants released to coastal waters, and the fate of petroleum hydrocarbons in the environment. Dr. Tansel's research interests include laboratory and field studies on weathering of crude and refined oils, formation and stability of oil–water emulsions, and partitioning and persistence of petroleum-based oil fractions in different phases (slick, emulsion, dispersed, sediment, and tar) and their transport and mobility. Before joining FIU, Dr. Tansel worked as a project manager at the Massachusetts Water Resources Authority on the Boston Harbor cleanup project. She has also worked at the Center for Environmental Management at Tufts University on preparation of several research reports to U.S. Congress on waste management. Dr. Tansel is the recipient of the 2009 Edmund Friedman Professional Recognition Award from the American Society of Civil Engineers (ASCE) and was named the 2007 Engineer of the Year by the ASCE Miami-Dade Branch for her commitment and impact on the vitality, perception, and future of engineering education. She is a member of the Water Environment Federation, Environmental and Water Resources Institute of ASCE and the Association of Environmental Engineering and Science Professors. She is the co-editor in chief of the *Journal of Environmental Management* and a registered professional engineer in Florida. Dr. Tansel is an elected fellow of ASCE, an elected diplomate of the American Academy of Water Resources Engineers, and an elected fellow of the Water Environment Federation. She holds an M.S. and a Ph.D. in environmental engineering from the University of Wisconsin–Madison.

Helen K. White is a professor of chemistry and environmental studies at Haverford College in Haverford, Pennsylvania. Dr. White's research examines the persistence of human-derived compounds in the marine environment, including those from oil and plastic waste. Her focus is on how the chemical structure, physical associations, and bioavailability of specific chemical compounds determine their cycling and eventual fate. She has investigated persistent oil residues for the past 20 years, including oil in the Gulf of Mexico following the *Deepwater Horizon* oil spill, and oil in Prince William Sound, Alaska. Dr. White is a recipient of the National Academies of Sciences, Engineering, and Medicine's Gulf Research Program's Early-Career Fellowship and a Henry Dreyfus Teacher-Scholar Award. She received her M.Chem. in chemistry from the University of Sussex, United Kingdom, and her Ph.D. in chemical oceanography from the Massachusetts Institute of Technology and the Woods Hole Oceanographic Institution. Following her graduate studies, Dr. White was awarded the Microbial Science Initiative Postdoctoral Fellowship at Harvard University. Her previous experience with the National Academies includes serving on the committee that produced the report *Use of Dispersants in Marine Oil Spill Response* (2020).

Michael Ziccardi is the co-director of the Karen C. Drayer Wildlife Health Center (WHC). Dr. Ziccardi has been an oil spill response veterinarian and oiled wildlife response director during more than 50 spills nationally and internationally—most notably as the marine mammal and sea turtle group supervisor for the National Oceanic and Atmospheric Administration during the *Deepwater Horizon* oil spill in 2010. He has worked as a contract veterinarian for the California Department of Fish and Wildlife, a research epidemiologist for the Lincoln Park Zoo, and as both a program coordinator and a senior wildlife veterinarian for WHC. Currently, in addition to being the co-director of WHC, he is the director for California's Oiled Wildlife Care Network as well as a Health Science Clinical Professor in the Department of Medicine and Epidemiology. Dr. Ziccardi received his D.V.M., as well as his M.S. and Ph.D. in epidemiology, from the University of California, Davis.